About Pearson

Pearson is the world's learning company, with presence across 70 countries worldwide. Our unique insights and world-class expertise comes from a long history of working closely with renowned teachers, authors and thought leaders, as a result of which, we have emerged as the preferred choice for millions of teachers and learners across the world.

We believe learning opens up opportunities, creates fulfilling careers and hence better lives. We hence collaborate with the best of minds to deliver you class-leading products, spread across the Higher Education and K12 spectrum.

Superior learning experience and improved outcomes are at the heart of everything we do. This product is the result of one such effort.

Your feedback plays a critical role in the evolution of our products and you can contact us - reachus@pearson.com. *We look forward to it.*

Basic Electrical and Electronics Engineering

Second Edition

S. K. Bhattacharya
Former Director
National Institute of Technical Teachers'
Training and Research (NITTTR)
Kolkata and Chandigarh

Pearson

Assistant Acquisitions Editor: Yogesh Dutt
Senior Production Editor: Vipin Kumar

ISBN: 978-93-325-8650-5

First Impression, 2018
Twentieth Impression, 2026
Twenty One Impression

Published by Pearson India Education Services Pvt. Ltd, CIN: U72200TN2005PTC057128.

Head Office: 1st Floor, Berger Tower, Plot No. C-001A/2, Sector 16B, Noida - 201 301, Uttar Pradesh, India.
Registered Office: Featherlite, 'The Address' 5th Floor, Survey No 203/10B, 200 Ft MMRD Road, Zamin Pallavaram, Chennai – 600 044, Tamilnadu, India
Website: in.pearson.com, Email: companysecretary.india@pearson.com

Printer: Saurabh Printers Pvt. Ltd.

Dedicated to my wife, Sumita,
without whose patience and encouragement
this work could not have been completed.

Dedicated to my wife, Sujata,

without whose patience and encouragement

this work could not have been completed.

Contents

Chapter 2 DC Networks and Network Theorems 66

Chapter 10 Synchronous Machines 525

Preface

Basic Electrical and Electronics Engineering is becoming more popular amongst the students and teachers alike.

To improve the book further, feedback were collected from various colleges. The feedback analysis has led to the inclusion of some additional topics, more solved numericals, summary in the form of points discussed in each chapter and the important formulae used chapter-wise. Further improvement of schematic diagrams, photographic views of equipments, etc., were felt necessary to be taken up in this edition.

On the basis of the above value additions, this second edition is being brought out within 5 years of its first publication.

The book includes large number of solved examples, which would help in getting an in-depth understanding of the theoretical concepts presented.

This second edition of the book has become a comprehensive, all-inclusive text book, explaining clearly the fundamental concepts and principles in a simple way. This book has been structured and pedagogically designed in such a way that students would find it very useful even for their higher classes as a ready-reference companion.

The students can test the level of their understanding of the concepts, principles and applications included in each chapter by using the material provided at the end of each chapter, like objective-type questions with answers, short- and long answer-type questions and the numerical questions with answers.

I believe that like its first edition, the second edition of the book will also be widely welcomed by students and teachers of engineering colleges.

I would like to express my sincere appreciation for all the team members of Pearson India Education Services Pvt. Ltd. who have contributed to the publication of this new edition of the book.

I would also like to assure the students that once they start using this book, they would simply enjoy learning, get motivated to know more, and as a consequence excel in their life.

S. K. Bhattacharya

Acknowledgements

List of reviewers

We would like to thank all those reviewers who took out time to go through the script and gave us their valuable feedback. Their names are listed below

Swapanjit Pattnaik
Assistant Professor
National Institute of Technology Raipur, Chattisgarh

Nutanlata Nath
Associate Professor
Birla Institute of Technology Mesra, Jharkhand

Goutam Nandi
Head of Department, Falakata Polytechnic
West Bengal

Sunil Bhooshan
Assistant Professor
Jaypee University of Information Technology
Himachal Pradesh

Dr. Bijuna Kunju K
Professor
TKM college of Engineering, Kerala

Surendra V
Assistant Professor
Kongu Engineering College, Tamil Nadu

M.A. Femina
Assistant Professor
KCG College of Technology, Tamil Nadu

Ranjani S
Assistant Professor
Anand Institute of Technology, Tamil Nadu

V S Kanchana Bhaaskaran
Professor and Dean (Academics)
VIT University, Tamil Nadu

S. Thangeval
Professor and Head
K.S.Rangasamy College of Technology, Tamil Nadu

Priyanka P Kulkarni
Associate Professor KIT's College of Engineering
Maharashtara

Nirmal Kumar Agarwal
Assistant Professor
JSS Academy of Technical Education Uttar Pradesh

V. J. Vijayalakshmi
Assistant Professor
KPR Institute of Engineering and Technology
Tamil Nadu

Kuldeep Shukla
Assistant Professor
SAL Institute of Technology and Engineering
Research Gujarat

VK Haridas
Assistant Professor
Royal college of Engineering and Technology
Thrissur, Kerala

About the Author

S. K. Bhattacharya graduated in Electrical Engineering from Jadavpur University, Calcutta, West Bengal, and obtained his masters degree from Calcutta University, specializing in electrical machines. He received his PhD degree from Birla Institute of Technology and Science (BITS), Pilani, Rajasthan.

He has acquired specialized knowledge and work experience in the areas of Educational Technology, Curriculum Development, Technical Teacher Training and Education work linkage in the Netherlands, the United Kingdom and India.

As a professor of Electrical Engineering, he has taught major subjects in Electrical Engineering to the BTech students for nearly 25 years. He preferred to teach the first-year students 'Basic Electrical and Electronics Engineering' as the subject contains lot of conceptual learning and that the first-year students need to get an experience of good teaching and learning which would motivate them to be more inquisitive and gain knowledge in the subject.

S. K. Bhattacharya is presently working as the Professor and Advisor in Sri Sukhmani Institute of Engineering & Technology, Dera Bassi, near Chandigarh, Punjab. Earlier, he worked as the Director, National Institute of Technical Teachers Training and Research (NITTTR) Chandigarh, Punjab; Director, NITTTR Kolkata, West Bengal; Director, Hindustan Institute of Technology, Greater Noida, Uttar Pradesh; Director SUS college of Engineering & Technology, Mohali, Punjab; Principal, SUS Women Engineering College, Mohali, Punjab; and Professor Emeritus, Sharda University, Greater Noida, Uttar Pradesh.

He has visited many technical institutes and industries in the United Kingdom, the Netherlands, Australia, Japan, Korea, Philippines, Malaysia, Singapore, etc., on various fellowship programs.

He has conducted a large number of Faculty Development Programs on innovative methods of teaching and learning to the delight of the participants. Towards this, he has developed teaching–learning materials in the form of educational video films, multimedia learning packages, demonstration models, experimental kits, simulations, etc. He has published a good number of popular text books for engineering students.

He has designed, developed and validated an interactive model of industry–institute interactions, involving 60 industries and four technical institutes, where all its participants (viz., students, faculty, institutions, and industries) reap optimum benefit based on their mutual interaction.

He is the Fellow of the Institution of Engineers and the Fellow of the Institution of Electronics and Telecommunication Engineers (India).

CHAPTER 1

Basic Concepts, Laws, and Principles

Topics Discussed

1. The need to study electrical and electronics engineering
2. Behaviour of materials as conductors, semiconductors, and insulators
3. Concept of current, resistance, potential, and potential difference
4. Differences between electric field and magneti.c field
5. Ohm's law
6. Effect of temperature on resistance
7. Electrostatics
8. Electromagnetism and electromagnetic induction
9. Laws of electromagnetic induction
10. Dynamically and statically induced EMF
11. Self and mutual inductance
12. Electrical circuit elements

1.1 ▶ INTRODUCTION

We see applications of electricity all around us. We observe the presence of electricity in nature. It is indeed amazing as well as interesting to know how mankind has been able to put electricity for its use. All electronic and electrical products operate on electricity. Be it your computer system, cell phones, home entertainment system, lighting, heating, and air-conditioning systems—all are examples of applications of electricity. Application of electricity is limitless and often extends beyond our imagination.

Electrical energy has been accepted as the form of energy which is clean and easy to transmit from one place to the other. All other forms of energy available in nature are, therefore, transformed into electrical energy and then transmitted to places where electricity is to be used for doing some work. Electrical engineering, therefore, has become a discipline, a branch of study which deals with generation, transmission, distribution, and utilization of electricity.

Electronics engineering is an offshoot of electrical engineering, which deals with the theory and use of electronic devices in which electrons are transported through vacuum, gas, or semiconductors. The motion of electrons in electronic devices like diodes, transistors, thyristors, etc., are controlled by electric fields. Modern computers and digital communication systems are advances of electronics. Introduction of very large scale integrated (VLSI) circuits has led to the miniaturization of all electronic systems.

Electrical and electronic engineering are, therefore, very exciting fields of study. A person who is unaware of the contribution of these fields of engineering and the basic concepts underlying the advancement, will only have to blame himself or herself for not taking any initiative in knowing the unknown.

In this chapter, we will introduce some basic concepts, laws, and principles which the students might have studied in physics. However, since these form the basis of understanding of the other chapters in this book, it will be good to study them again.

1.2 ► Atomic Structure and Electric Charge and Coulomb's Law

1.2.1 Atomic Structure

Several theories have been developed to explain the nature of electricity. The modern electron theory of matter, propounded by scientists Sir Earnest Rutherford and Niel Bohr considers every matter as electrical in nature. According to this atomic theory, every element is made up of atoms which are neutral in nature. The atom contains particles of electricity called electrons and protons. The number of electrons in an atom is equal to the number of protons.

The nucleus of an atom contains protons and neutrons. The neutrons carry no charge. The protons carry positive charge. The electrons revolve round the nucleus in elliptical orbits like the planets around the sun. The electrons carry negative charge. Since there are equal number of protons and electrons in an atom, an atom is basically neutral in nature.

1.2.2 Electric Charge

If from a body consisting of neutral atoms, some electrons are removed, there will be a deficit of electrons in the body, and the body will attain positive charge. If neutral atoms of a body are supplied some extra electrons, the body will attain negative charge. Thus, we can say that *the deficit or excess of electrons in a body is called charge.*

Charge of an electron is very small. Coulomb is the unit of charge. The charge of an electron is only 1.602×10^{-19} Coulomb (C). Thus, we can say that the number of electrons per Coulomb is the reciprocal of 1.602×10^{-19} which equals approx. 6.28×10^{18} electrons. Therefore, charge of 6.28×10^{18} electrons is equal to 1C. When we say that a body has a positive charge of 1C, it is understood that the body has a deficit of 6.28×10^{18} electrons.

Any charge is an example of static electricity because the electrons or protons are not in motion. You must have seen the effect of charged particles when you comb your hair with a plastic comb, the comb attracts some of your hair. The work of combing causes friction, producing charge of extra electrons and excess protons causing attraction.

Charge in motion is called electric current. Any charge has the potential of doing work, i.e., of moving another charge either by attraction or by repulsion. A charge is the result of separating electrons and protons. The charge of electrons or protons has potential because it likes to return back the work that was done to produce it.

1.2.3 Coulomb's Law

We have known that the deficit or excess of electrons in a body is called charge. Accordingly, a body with excess of electrons will be negatively charged while a body with deficit of electrons will be positively charged. Like charges repel and unlike charges attract each other. *Charles Coulomb*, a French Scientist, carried out a number of experiments to see the effect of placing charges near to each other. A law of electrostatics known as Coulmb's law or inverse square law was framed by him on the basis of his experiments.

According to Coulomb's law, the force of repulsion or attraction, F between two charges is expressed as

$$F = k\frac{Q_1 Q_2}{d^2} \text{ Newton}$$

where Q_1 and Q_2 are the two charges separated by a distance of d meters and k is the constant of proportionality which depends on the medium in which the charges are placed.

The constant k is expressed as

$$k = \frac{1}{4\pi\varepsilon} = \frac{1}{4\pi\varepsilon_o \varepsilon_r}$$

where ε_o is the permittivity of vacuum and ε_r is the relative permittivity.

1.3 ► Conductors, Insulators, and Semiconductors

The electrons in an atom revolve in different orbits or shells. The shells are named as K, L, M, N, etc. The number of electrons that should be in a filled inner shell is given by $2n^2$ where n is shell number 1, 2, 3, 4, etc., starting from the nearest one, i.e., first shell to the nucleus. If n = 1, the first shell will contain two electrons. If n = 2, the second shell will contain eight electrons. This way, the number of electrons in the shells are 2, 8, 18, 32, etc. The filled outermost shell should contain a maximum number of eight electrons. The outermost shell of an atom may have less than eight electrons. As for example, copper has an atomic number of 29. This means, copper atom has 29 protons and 29 electrons. The protons are concentrated in the nucleus while the electrons are distributed in the K, L, M, and N shells as 2, 8, 18, and 1 electrons, respectively. The outermost shell of a copper atom has one electron only whereas this shell could have 8 electrons.

The position occupied by an electron in an orbit signifies its energy. There exists a force of attraction between the orbiting electron and the nucleus due to the opposite charge the of electron and the proton. The electrons in the inner orbits are closely bound to the nucleus than the electrons of the outer or outermost orbit. If the electron is far away from the nucleus, the force of attraction is weak, and hence, the electrons of outermost orbit are often called free electrons. For example, a copper atom has only one atom in the last orbit which otherwise could have eight electrons.

1.3.1 Conductors

In a copper wire consisting of large number of copper atoms, the atoms are held close together. The outermost electrons of atoms in the copper wire are not sure about which atom they belong to. They can move easily from one atom to the other in a random fashion. Such electrons which can move easily from one atom to the other in a random fashion are called free electrons. It is the movement of free electrons in a material like copper that constitutes flow of current. Here, of course, the net current flow will be zero as the movement of the free electrons is in random directions. When we apply a potential, which is nothing but a force, it will direct the flow of electrons in a particular direction, i.e., from a point of higher potential towards a point of lower potential. Thus, current flow is established between two points when there exists a potential difference between the points.

When in a material the electrons can move freely from one atom to another atom, the material is called a conductor. Silver, copper, gold, and aluminium are good conductors of electricity. In general, all metals are good conductors of electricity. Although silver is the best conductor of electricity, the second best conductor, i.e., copper, is mostly used as conductor because of the cost factor. In electrical and electronic engineering fields, the purpose of using a conductor as carrier of electricity is to allow electric current to flow with the minimum of resistances, i.e., the minimum of opposition.

1.3.2 Insulators

In a material where the outermost orbit of the atoms is completely filled, the material is called an insulator. Insulators like glass, rubber, mica, plastic, paper, air, etc., do not conduct electricity very easily. In the atoms of these materials, the electrons tend to stay in their own orbits. However, insulators can store electricity and can prevent flow of current through them. Insulating materials are used as dielectric in capacitors to store electric charge, i.e., electricity.

1.3.3 Semiconductors

Carbon, silicon, and germanium having atomic numbers of 6, 14, and 32, respectively, are called semi conducting material. The number of electrons in the outermost orbit of their atoms is four instead of the maximum of eight. Thus, in the outermost orbit of a semiconductor material, there are four vacant positions for electrons. These vacant positions are called holes. In a material, the atoms are so close together that the electrons in the outermost orbit or shell behave as if they were orbiting in the outermost shells of two adjacent atoms producing a binding force between the atoms. In a semiconductor material the atoms forming a bonding, called covalent bonding, share their electrons in the outermost orbit, and thereby attain a stable state.

The condition is like an insulator having all the eight positions in the outermost orbit filled by eight electrons. However, in semiconducting materials, with increase in temperature it is possible for some of the electrons to gain sufficient energy to break the covalent bonds and become free electrons, and cause the flow of current.

1.4 ► Electric Field and Magnetic Field

When number of charges are separated, a space is created where forces are exerted on the charges. An electric field is such a space. Depending upon the polarity of the charges, the force is either attractive or repulsive. Therefore, we can say that static charges generate an electric field. An electric field influences the space surrounding it. Electric field strength is determined in terms of the force exerted on charges. A capacitor is a reservoir of charge. The two parallel plates of a capacitor, when connected to a voltage source, establishes an electric field between the plates. The positive terminal, or pole of the voltage source will draw electrons from plate 1 whereas the negative pole will push extra electrons on to plate 2. Voltage across the capacitor will rise. The capacitor gets charged equal to the voltage of the source. The capacitance of a capacitor is a measure of its ability to store charge. The capacitance of a capacitor is increased by the presence of a dielectric material between the two plates of the capacitor.

A current-carrying conductor or a coil produces magnetic field around it. The strength of the magnetic field produced depends on the magnitude of the current flowing through the conductor or the coil. There is presence of magnetic field around permanent magnets as well.

Figure 1.1 Shows an electric field in the space between a charged parallel plate capacitor and a magnetic field around a permanent magnet and an electrio magnet.

A magnet is a body which attracts iron, nickel, and cobalt. Permanent magnets retain their magnetic properties. Electromagnets are made from coils through which current is allowed to flow. Their magnetic properties will be present as long as current flows through the coil.

The space within which forces are exerted by a magnet is called a magnetic field. It is the area of influence of the magnet.

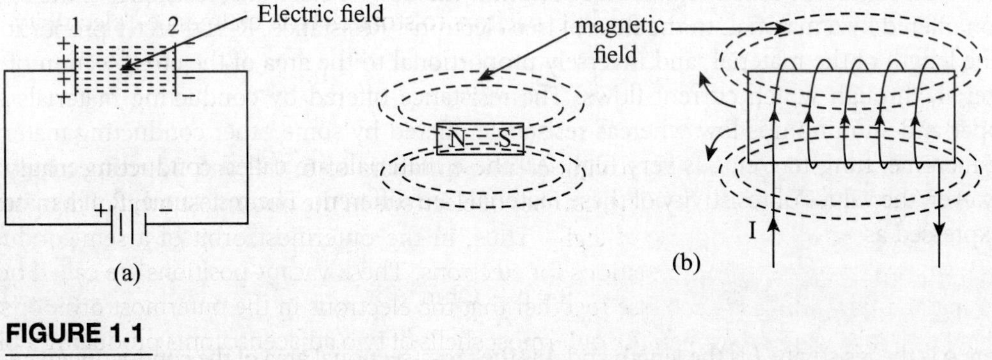

FIGURE 1.1

(a) Electric field between the plates of a capacitor. (b) Magnetic field around a permanent magnet or an electro magnet

1.5 ► Electric Current, Resistance, Potential, and Potential Difference

1.5.1 Electric Current

In any conducting material, the flow of electrons forms what is called current. Electrons have negative charge. Charge on an electron is very small. For this reason charge is expressed in terms of Coulomb. Charge of one Coulomb is equal to a charge of 6.28×10^{18} electrons. The excess or deficit of electrons in a body is called charge. Thus, electrical current is expressed as a flow of negative charge, i.e., electrons. Any substance like copper, aluminum, silver, etc., which has a large number of free electrons (i.e., loosely bound electrons in the outermost orbit of its atom) will permit the flow of electrons when electrical pressure in the form of EMF (electromotive force, i.e., voltage) is applied.

Since these materials conduct electricity, they are called conductors. They easily allow electric current to flow through them. The strength of current will depend upon the flow of charge per unit time. This is expressed as

$$\text{Current, } I = \frac{Q}{t} \text{ Coulomb per second or Ampere.} \tag{1.1}$$

where charge Q is measured in Coulomb and time, t in seconds. The unit of current, therefore, is Coulomb per second, when 1 C of charge flows in 1 s; the magnitude of current is called ampere, named after André-Marie Ampere.

Thus, 1 ampere of current is equivalent to the flow of charge of 1 Coulomb per second.

In earlier years, current was assumed to flow from positive to negative terminals. This convention is used even now although it is known that current is due to the movement of electrons from the negative to the positive terminal.

1.5.2 Resistance

Electrical resistance is the hindrance or opposition to the flow of electrons in a given material. It is measured in unit called ohm. Since current is the flow of electrons, resistance is the opposition offered by a material, to the flow of free electrons. Resistance, R, is directly proportional to the length of the material, and inversely proportional to the area of the cross section of the material, through which current flows. The resistance offered by conducting materials like copper and aluminum is low whereas resistance offered by some other conducting materials like nicrome, tungsten, etc., is very high. All these materials are called conducting materials. However, the values of resistivity of these materials are different. The resistance, R of a material is expressed as

$$R = \rho \frac{\ell}{A} \text{ Ohm} \tag{1.2}$$

where ρ is the resistivity, ℓ is the length and A is the cross-sectional area of the conducting material.

The resistivity, ρ is also called the specific resistance of the material. The most conducting material, silver has the lowest value of resistivity, i.e., 0.016×10^{-6} ohm-m. After silver, copper is most conducting. The resistivity or specific resistance of copper is somewhat more than that of silver, i.e., 0.018×10^{-6} ohm-m. That is to say, copper is less conducting than silver. We will see a little later why and how the value of resistance changes with temperature.

1.5.3 Potential and Potential Difference

EMF produces a force or pressure that causes the free electrons in a body to move in a particular direction. The unit of EMF is volt. EMF is also called electric potential. When a body is charged (i.e., either defficiency of electrons or excess of electrons is created), an amount of work is done. This work done is stored in the body in the form of potential energy. Such a charged body is capable of doing work by attracting or repelling other charges. ***The ability of a charged body to do work in attracting or repelling charges is called its potential or electrical potential***. Work done to charge a body to 1 C is the measure of its potential expressed in volts:

$$\text{Electric Potential} = \frac{\text{Work done in Joules}}{\text{Charge in Coulombs}} \text{ volt} \tag{1.3}$$

When work done is 1 joule and charge moved is 1 C, the potential is called 1 volt. If we say that a point has a potential of 6 volts, it means that 6 Joules of work has been done in moving 1 C of charge to that point. In other words, we can say that every Coulomb of charge at that point has an energy of 6 Joules.

The potential difference of two points indicates the difference of charged condition of these points. Suppose point A has a potential of 6 volts, and point B has a potential of 3 volts. When the points A and B are joined together by a conducting wire, electrons will flow from point B to point A. We say that current flows from point A towards point B. The direction of current flow is taken from higher potential to lower potential while the flow of electrons are actually in the opposite direction. The flow of current from higher potential to lower potential is similar to the flow of water from a higher level to a lower level.

1.6 ► Ohm's Law

George Simon Ohm found that the voltage, V between two terminals of a current-carrying conductor is directly proportional to the current, I flowing through it. The proportionality constant, R is the resistance of the conductor. He formulated a law, called Ohm's law relating V, I and R. According to Ohm's law

$$V = IR \qquad \text{or, } I = \frac{V}{R} \tag{1.4}$$

This relation will hold good provided the temperature and other physical conditions do not change.

Ohm's law is not applicable to non-linear devices like Zener diode, voltage regulators, etc. Ohm's law is expressed graphically on V and I-axies as a straight line passing through the origin as shown in Fig. 1.2(a).

The relationship between V and I have been shown for different values of R in Fig. 1.2(b). Here in V = RI, R indicates the slope of the line. The more the value of R is, the more will be the slope of the line as shown in Fig. 1.2(b). The slope of V–I characteristic increase with increase in value of resistance as shown in Fig. 1.2(b).

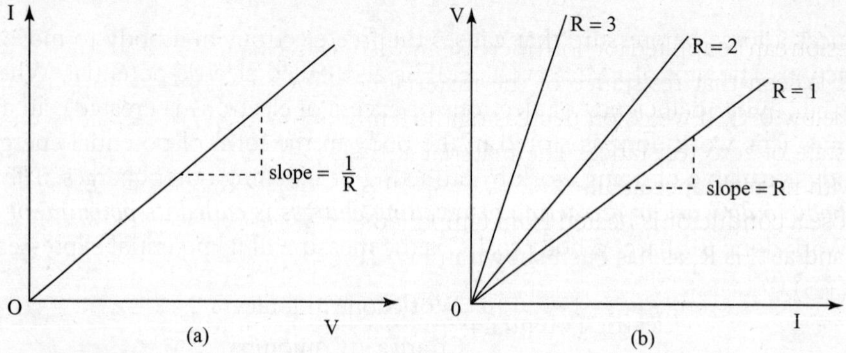

FIGURE 1.2

(a) Shows linear relationship between V and I; (b) Shows linear V–I characteristics for different values of R

1.7 ► The Effect of Temperature on Resistance

Resistance of pure metals like copper, aluminum, etc., increases with increase in temperature. The variation of resistance with change in temperature has been shown as a linear relationship in Fig. 1.3.

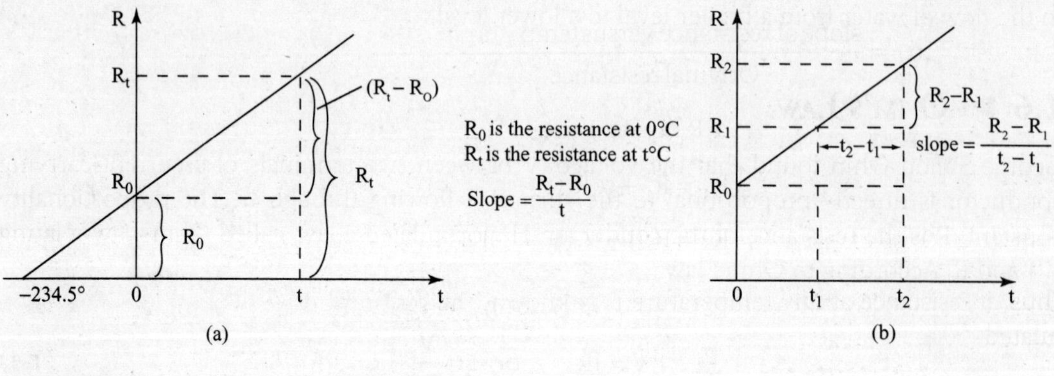

FIGURE 1.3

(a) Shows the variation of resistance with temperature; (b) resistances at two different temperatures

The change in resistance due to change in temperature is found to be directly proportional to the initial resistance, i.e., $R_t - R_0 \propto R_0$. Resistance $(R_t - R_0)$ also varies directly as the temperature rise and this change also depends upon the nature of the material. Thus, we can express the change in resistance as,

$$R_t - R_0 \propto R_0 t$$

or, $R_t - R_0 = \alpha_0 R_0 t$, where α_0 is called the temperature coefficient of resistance at 0°C.

$$\text{or, } R_t = R_0(1 + \alpha_0 t) \qquad (1.5)$$

This expression can be applied for both increase and decrease in temperature. From the graph of Fig. 1.3 (a) it is seen that resistance of the material continues to decrease with decrease in temperature below 0°C. If we go on decreasing the temperature to a very low value, the material attains a state of zero resistance. The material at that state becomes *superconducting*, i.e., conducting with no resistance at all.

Now suppose a conductor is heated from temperature t_1 to t_2. The resistance of the conductor at t_1 is R_1 and at t_2 is R_2 as has been shown in Fig. 1.3 (b).

Using eq. (1.5),

$$R_t = R_0(1 + \alpha_0 t)$$

or,
$$\alpha_0 = \frac{R_t - R_0}{R_0 t}$$

or,
$$\alpha_0 = \frac{(R_1 - R_0)/t}{R_0} = \frac{\text{slope of resistance versus temp. graph}}{\text{original resistance}} \qquad (1.6)$$

Using eq. (1.5), we can write

$$R_1 = R_0(1 + \alpha_0 t_1)$$

and
$$R_2 = R_0(1 + \alpha_0 t_2)$$

From fig 1.3 (b) using the relation in (1.6), we can write

$$\alpha_1 = \frac{\text{slope of resistance versus temp. graph}}{\text{Original resistance}} = \frac{(R_2 - R_1)/(t_2 - t_1)}{R_1}$$

or,
$$\alpha_1 R_1 (t_2 - t_1) = R_2 - R_1$$

or,
$$R_2 = R_1 + \alpha_1 R_1 (t_2 - t_1)$$

or,
$$R_2 = R_1[1 + \alpha_1(t_2 - t_1)] \qquad (1.7)$$

Thus, if resistance at any temperature t_1 is known, the resistance at t_2 temperature can be calculated.

Calculation of a at different temperatures

We have seen,

$$\alpha_0 = \frac{\text{slope of resis tance versus temp. graph}}{\text{Original resistance, } R_0}$$

If α_1 and α_2 are the temperature coefficients of resistance at t_1 and t_2 degrees, respectively, then

$$\alpha_1 = \frac{\text{slope of resistance versus temp. graph}}{R_1}$$

and
$$\alpha_2 = \frac{\text{slope of resistance versus temp. graph}}{R_2}$$

Thus, we can write,

$$\alpha_0 R_0 = \alpha_1 R_1 = \alpha_2 R_2 = \alpha_3 R_3 = \ldots \text{ and so on}$$

Therefore,

$$\alpha_1 = \frac{\alpha_0 R_0}{R_1} = \frac{\alpha_0 R_0}{R_0(1+\alpha_0 t_1)} = \frac{\alpha_0}{1+\alpha_0 t_1} \tag{1.8}$$

$$\alpha_2 = \frac{\alpha_0 R_0}{R_2}$$

$$= \frac{\alpha_0 R_0}{R_0(1+\alpha_0 t_2)} = \frac{\alpha_0}{1+\alpha_0 t_2}$$

and,
$$\alpha_2 R_2 = \alpha_1 R_1$$

or,
$$\alpha_2 = \frac{\alpha_1 R_1}{R_2} = \frac{\alpha_1 R_1}{R_1[1+\alpha_1(t_2-t_1)]}$$

or,
$$\alpha_2 = \frac{\alpha_1}{1+\alpha_1(t_2-t_1)} \tag{1.9}$$

Temperature coefficient of resistance, α at 20°C and specific resistance ρ of certain material have been shown in Table 1.1.

It is to be noted that carbon has a negative temperature coefficient of resistance. This means, the resistance of carbon decreases with increase in temperature.

TABLE 1–1 Temperature Coefficient and Specific Resistance of Different Materials

Material	Temp. coeff. of resistance a_{20}	Specific resistance r in micro–ohm
Silver	0.004	0.016
Copper	0.0039	0.018
Aluminium	0.0036	0.028
Iron	0.005	0.100
Brass	0.0015	0.070
Lead	0.0042	0.208
Tin	0.0046	0.110
Carbon	–0.00045	66.67

By this time you must be wondering as to why resistance in most materials increases with increase in temperature while resistance in some decreases with increase in temperature.

The charged particles inside a material is in the state of vibration. Temperature rise in most materials increases this vibration inside the material obstructing the flow of electrons. Obstruction to the flow of electrons is called resistance. At lower temperatures the vibration gets reduced, and hence the resistance.

1.8 ► Work, Power, and Energy

Work, power and energy are related to each other. When a force moves an object to a distance, work is done. The rate of doing work is called power. The capacity of doing work is called energy.

1.8.1 Work

When a force is applied to a body causing it to move, and if a displacement, d is caused in the direction of the force, then

$$\text{Work done} = \text{Force} \times \text{Distance} \tag{1.10}$$

or,

$$W = F \times d$$

If force is in Newtons and d is in meters, then work done is expressed in Newton–meter which is called Joules. Work will be done only when the applied force causes displacement of an object. If, for example, a person applies a force on a fixed wall, no work will be done even if the person gets exhausted.

1.8.2 Power

Power is the rate at which work is done, i.e., rate of doing work. Thus,

$$\text{Power, } P = \frac{\text{work done}}{\text{time}} = \frac{\text{Joules}}{\text{seconds}} \tag{1.11}$$

The unit of power is Joules/second which is also called Watt. When the amount of power is more, it is expressed in Kilowatt, i.e., kW.

$$1 \text{ kW} = 1 \times 10^3 \text{ W}$$

We have earlier seen in eq. (1.3), that electrical potential, V is expressed as

$$V = \frac{\text{work done}}{\text{charge}} = \frac{W}{Q}$$

or,

$$\text{Work done} = VQ = VIt \qquad \left[\because I = \frac{Q}{t}\right]$$

$$\text{Electrical Power, } P = \frac{\text{work done}}{t} = \frac{VIt}{t} = VI \text{ Watts} \tag{1.12}$$

Thus, in a circuit if I is the current flowing, and V is the applied voltage across the terminals, power, P is expressed as

$$P = VI = V\frac{V}{R} = \frac{V^2}{R}W$$

$$\text{Also, } P = VI = IR.I = I^2RW$$

Thus, electrical power can be expressed as

$$P = VI = \frac{V^2}{R} = I^2 R \text{ Watts} \quad (1.13)$$

Where V is in Volts,
I is in Amperes, and
R is in Ohms

Horse Power is also a unit of measurement of power. It was originally defined as work done at the rate of 550 ft-pound per second.
The conversion of hp into watt is done as

$$1 \text{ hp} = 746 \text{ W}$$

$$1 \text{ hp (metric)} = 135.6 \text{ W}$$

1.8.3 Energy

Energy is defined as the capacity for doing work. The total work done in an electrical circuit is called electrical energy. When a voltage, V is applied, the charge, Q will flow so that

$$\begin{aligned}\text{Electrical energy} &= V \times Q\\ &= VIt\\ &= IRIt\\ &= I^2Rt\\ &= \frac{V^2}{R}t\end{aligned}$$

$$\text{or, Electrical energy} = \text{Power} \times \text{Time} \quad (1.14)$$

If power is in kW and time is in hour, the unit of energy will be in Kilowatt hour or kWh.

1.8.4 Units of Work, Power, and Energy

In SI unit, work done is the same as that of energy.

Mechanical work or energy

When a Force, F Newton acting on a body moves it in the direction of the force by a distance d meters:

$$\text{Work done} = F \times d \qquad \text{Nm or Joules}$$

When a force F Newton is applied tangentially on a rotating body making a radius r meters, then

$$\text{Work done in 1 revolution} = F \times 2\pi r \text{ (since distance moved is } 2\pi r)$$
$$= 2\pi Fr \quad \text{Nm}$$

$$\text{Force} \times \text{Perpendicular distance} = \text{Torque, i.e., } F \times r = T$$

$$\text{Work done in 1 revolution} = 2\pi T \quad \text{Nm}$$

$\therefore$ Work done in N revolutions/second

$$= 2\pi TN \quad \text{Nm} = 2\pi NT = WT \quad \text{where } W = 2\pi N$$

If N is expressed in revolutions per minute (rpm),

$$\text{Work done} = \frac{2\pi TN}{60} \text{ Nm or Joules} \tag{1.15}$$

When a body of mass m kg is lifted to a height h meters against the gravitational force g m/sec^2, work done is converted into potential energy of the body.

$$\text{Potential energy} = \text{Weight} \times \text{Height}$$
$$= mgh \text{ Joules.} \tag{1.16}$$

Kinetic energy of a body of mass m kg moving at a speed of v meters/sec^2 $= \frac{1}{2} mv^2$ Joules. (1.17)

Electrical energy

As mentioned earlier, work done in an electrical circuit is its energy.

$$\text{Electrical energy} = \text{Applied voltage, } V \times \text{total flow of charge, } Q$$
$$= VQ = VIt$$
$$= IRIt$$
$$= I^2Rt$$
$$= \frac{V^2}{R}t$$

$$\text{Electrical power} = \frac{\text{work done}}{\text{time}} = \frac{VIt}{t} = VI$$

$$\text{Electrical energy} = \text{Electrical power} \times \text{time} \tag{1.18}$$

If electrical power is expressed is kW and time in hour, then

$$\text{Electrical energy} = \text{kWh} \tag{1.19}$$

We will now convert kWh into Calories

$$1 \text{ kWh} = 10^3 \times 60 \times 60 \text{ Watt second or Joules}$$
$$= 36 \times 10^5 \text{ Joules}$$

Since 1 Calorie = 4.2 Joules

$$1\text{ kWh} = \frac{36\times10^5}{4.2} = 860 \times 10^3 \text{ Calories}$$

or,

$$1\text{ kWh} = 860 \text{ kiloCalories} \tag{1.20}$$

Thermal energy

In SI unit* thermal energy is expressed in calories. One calorie indicates the amount of heat required to raise the temperature of 1 gm of water by 1°C. This heat is also called the specific heat. If m is the mass of the liquid, S is the specific heat, and t is the temperature rise required, then the amount of heat required, H is expressed as

$$H = \text{mst calories}$$

$$= 4.2 \times \text{mst Joules (since 1 cal = 4.2J)} \tag{1.21}$$

1 calorie = 4.2 Joules, has been established experimentally and therefore requires no deduction.

EXAMPLE 1.1

A copper wire has resistance of 0.85 ohms at 20°C. What will be its resistance at 40°C? Temperature coefficient of resistance of copper at 0°C is 0.004°C.

SOLUTION

we know,

$$\alpha_1 = \frac{\alpha_0}{1+\alpha_0 t_1}$$

Here,

$$\alpha_{20} = \frac{0.004}{1+0.004\times20} = 0.0037$$

We know,

$$R_2 = R_1[1+\alpha_1(t_2-t_1)]$$

In this case,

$$R_{40} = R_{20}[1+\alpha_{20}(40-20)]$$

$$= 0.85[1+0.0037\times20]$$

$$= 0.9129\ \Omega$$

***SI system of units**

SI stands for 'System International'. This system derives all units from seven basic units, which are: length expressed in m, mass expressed in kg, time expressed in second, electric current expressed in ampere, temperature in Kelvin, luminous intensity in candela (cd), and amount of substance in mole.

EXAMPLE 1.2

The heating element of an electric heater made of nicrome wire has value of resistivity of 1×10^{-6} Ohm-m. The diameter of the wire is 0.4 mm. What length of this nicrome wire will make a resistance of 100 Ohms?

SOLUTION

We know, $$R = \rho\frac{\ell}{a}$$

or, $$\ell = \frac{R.a}{\rho}$$

given $R = 100\ \Omega$, $\rho = 1 \times 10^{-6}$ Ohm-m, d = 0.4 mm, r = 0.2 mm.

area, $a = \pi r^2 = 3.14 \times (0.2 \times 10^{-3})^2 = 12.56 \times 10^{-8}\ m^2$

Substituting the values, length of wire, ℓ is

$$\ell = \frac{100 \times 12.56 \times 10^{-8}}{1 \times 10^{-6}} = 12.56\ m$$

EXAMPLE 1.3

What will be the value of resistance of a conductor if its length is doubled and its area of cross–section is also doubled?

SOLUTION

We know, in the first case, $$R = \rho\frac{\ell}{a}$$

In the second case, $$R = \rho\frac{2\ell}{2a} = \rho\frac{\ell}{a}$$

So, the resistance will remain the same.

EXAMPLE 1.4

An electric iron is rated 250 V, 500 W. What current will it take when connected to 250 V supply? If the iron is used for one hour daily for 30 days, what will be the monthly electricity bill at ₹ 5 per unit?

SOLUTION

We know, $$VI = W$$

or, $$I = \frac{W}{V} = \frac{500}{250} = 2\ A$$

$$\begin{aligned}\text{Energy} &= \text{Power} \times \text{Time}\\ &= 500 \times 1(\text{hr}) \times 30\ (\text{days})\\ &= 500 \times 30\ \text{Watt hr.}\\ &= \frac{500 \times 30}{1000}\ \text{Kilo Watt hr.}\\ &= 15\ \text{kWh.}\end{aligned}$$

Billing rate is ₹ 5 per kWh, i.e., unit of energy.

Bill amount per month = ₹ 5 × 15 = ₹ 75.

EXAMPLE 1.5

Two resistances, one of 30 Ω and another of unknown value are connected in parallel. The total power dissipated is 450 Watt at 100 Volts. Find the value of the unknown resistance?

SOLUTION

Let R be the value of the unknown resistance. When a resistance of 30 Ω and a resistance of R Ω are connected in parallel, the equivalent resistance is $\left(\frac{30 \times R}{30 + R}\right)\Omega$

Power dissipated, $$P = \frac{V^2}{R} = 450\ W.$$

Substituting values, $$\frac{100^2}{\dfrac{30R}{30+R}} = 450$$

or, $$\frac{100 \times 100}{450} = \frac{30R}{30+R}$$

Solving, $$R = 85.68\ \Omega$$

EXAMPLE 1.6

A copper wire of 1 cm diameter has a resistance of 0.15 Ω. It is drawn under pressure and its diameter was reduced to 50%. What is the new value of resistance of the wire?

SOLUTION

In first case, $$R_1 = \rho \frac{\ell_1}{\frac{\pi d_1^2}{4}}$$

In the second case, $$R_2 = \rho \frac{\ell_2}{\pi \frac{(0.5d_1)^2}{4}}$$

$$\frac{R_1}{R_2} = \frac{\rho \ell_1 \times 4}{\pi d_1^2} \times \frac{\pi \times 0.25 d_1^2}{\rho \ell_2 4}$$

or, $$\frac{R_1}{R_2} = \frac{0.25 \ell_1}{\ell_2} \qquad (1)$$

In both the cases, the volume of the wire will remain the same. So equating the volume, we have

$$\ell_1 a_1 = \ell_2 a_2$$

or, $$\ell_1 \frac{\pi}{4} d_1^2 = \ell_2 \frac{\pi}{4}(0.5d_1)^2$$

or, $$\frac{\ell_1}{\ell_2} = 0.25$$

From (1), $$\frac{R_1}{R_2} = 0.25 \times 0.25 = \frac{1}{16}$$

or, $$R_2 = 16\ R_1 = 16 \times 0.15 = 2.4\ \Omega$$

EXAMPLE 1.7

It is required to raise the temperature of 12 kg of water in a container from 15°C to 40°C in 30 min through an immersion rod connected to a 230 V supply mains. Assuming an efficiency of operation as 80 per cent, calculate the current drawn by the heating element (immersion rod) from the supply. Also determine the rating of the immersion rod. Specific heat of water is 4.2 kiloJoules/kg/°C.

SOLUTION

Output or Energy spent in heating the water, H is

$$H = m\,s\,(t_2 - t_1)$$

Where m is the mass of water and s is the specific heat of water.

Here, $H = 12 \times 4.2 \times 10^3 \times (40 - 15)$ Joules

$= 126 \times 10^4$ Joules

We know, efficiency $= \dfrac{\text{Output}}{\text{Input}}$

So, Energy input to immersion rod $= \dfrac{\text{Energy Spent, i.e., Output}}{\eta}$

$$= \frac{126 \times 10^4}{0.8}$$

$$= 157.5 \times 10^4 \text{ Joules}$$

The time of operation of the heater rod

$= 30$ min $= 30 \times 60$ s

$= 1800$ secs.

The power rating of the heater $= \dfrac{\text{Energy input}}{\text{Time of operation}}$ [$\because$ Energy = Power × Time]

$$= \frac{157.5 \times 10^4 \text{ Joules}}{1800 \text{ seconds}}$$

$$= 870 \text{ Joules/sec}$$

$$= 870 \text{ Watts}$$

$$= 0.87 \text{ kW}$$

Current drawn from 230 V supply

P = VI = 870 Watts.

Therefore, $I = \dfrac{870}{230} = 3.78\,\text{A}$

EXAMPLE 1.8

A motor-driven water pump lifts 64 m^3 of water per hour to an overhead tank placed at a height of 20 metres. Calculate the power of the pump motor. Assume overall efficiency of the pump as 80 per cent.

SOLUTION

Work done = mgh Joules

$$m = 64 \times 10^3 \text{ kg} \qquad (1\text{m}^3 \text{ of water weights 1000 kg})$$

$$g = 9.81 \text{ m/sec}^2$$

$$h = 20 \text{ m}$$

Substituting values

$$\text{Work done/sec} = \frac{64 \times 10^3 \times 9.81 \times 20 \text{ Joules}}{60 \times 60 \text{ seconds}}$$

$$= \frac{12.55 \times 10^6 \text{ Joules}}{60 \times 60 \text{ seconds}}$$

$$= 3.485 \times 10^3 \text{ Watts}$$

$$= 3.485 \text{ kW}$$

$$\text{Input power of the pump motor} = \frac{3.485}{0.8}$$

$$= 4.35 \text{ kW.}$$

EXAMPLE 1.9

A residential flat has the following average electrical consumptions per day:

(i) 4 tube lights of 40 watts working for 5 hours per day;
(ii) 2 filament lamps of 60 watts working for 8 hours per day;
(iii) 1 water heater rated 2 kW working for 1 hour per day;
(iv) 1 water pump of 0.5 kW rating working for 3 hours per day.

Calculate the cost of energy per month if 1 kWh of energy (i.e., 1 unit of energy) costs ₹4.50.

SOLUTION

Total kilowatt hour consumption of each load for 30 days are calculated as:

$$\text{Tube lights} = \frac{4 \times 40 \times 5 \times 30}{1000} = 24 \text{ kWh}$$

$$\text{Filament lamps} = \frac{2 \times 60 \times 8 \times 30}{1000} = 28.8 \text{ kWh}$$

$$\text{Water heaters} = 1 \times 2 \times 1 \times 30 = 60 \text{ kWh}$$

$$\text{Water pump} = 1 \times 0.5 \times 3 \times 30 = 45 \text{ kWh}$$

Total kWh consumed per month

$$= 24 \text{ kWh} + 28.8 \text{ kWh} + 60 \text{ kWh} + 45 \text{ kWh}$$
$$= 157.5 \text{ kWh}$$

One kWh of energy costs ₹4.50.
The total cost of energy per month = 157.5×4.50
= ₹708.75

EXAMPLE 1.10

An electric kettle has to raise the temperature of 2 kg of water from 30°C to 100°C in 7 minutes. The kettle is having an efficiency of 80 per cent and is supplied from a 230 V supply. What should be the resistance of its heating element?

SOLUTION

$$m = 2 \text{ kg} = 2000 \text{ gms}$$

$$t_2 - t_1 = 100 - 30 = 70°\text{C}$$

$$\text{Specific heat of water} = 1$$

$$\text{Time of heating} = 7 \text{ minutes} = \frac{7}{60} \text{ hours}$$

Output energy of the kettle = m s t

$$= 2000 \times 1 \times 70 \text{ Calories}$$
$$= 140 \text{ kilo Calories}$$
$$= \frac{140}{860} \text{ kWh}$$
$$= 0.1627 \text{ kWh. [1 kWh = 860 kCal]}$$

$$\text{Input energy} = \frac{\text{output energy}}{\text{efficiency}} = \frac{0.1627}{0.8} = 0.203 \text{ kWh}$$

$$\text{kW rating of the kettle} = \frac{0.203}{\text{time in hours}} = \frac{0.203 \times 60}{7} = 1.74 \text{ kW}$$

Supply voltage, V = 230 Volts.

Power, P = 1.74 kW = 1740 Watts.

V = 230 V

$$P = VI = V\frac{V}{R} = \frac{V^2}{R} \text{ Watts}$$

or,

$$R = \frac{V^2}{P} = \frac{230 \times 230}{1740} = 30.4 \text{ Ohms}$$

EXAMPLE 1.11

Calculate the current flowing through a 60 W lamp on a 230 V supply when just switched on at an ambient temperature of 25°C. The operating temperature of the filament material is 2000°C and its temperature coefficient of resistance is 0.005 per degree C at 0°C.

SOLUTION

We know, power,
$$W = VI = V\frac{V}{R} = \frac{V^2}{R}$$

Here,
$$W = 60 \text{ W}, V = 230 \text{ V}$$

∴
$$R = \frac{V^2}{W} = \frac{230 \times 230}{60} = 881.6 \text{ Ohms}$$

This resistance of the filament is at 2000°C. Let us call it R_{2000} = 881.6 Ohms.

At the instant of switching, resistance is at room temperature, i.e., at 25°C. Let us call it as R_{25}.

We know R_{2000}; we have to calculate R_{25} given α_0 = 0.005 ohm/°C.

We know,
$$\alpha_1 = \frac{\alpha_0}{1+\alpha_0(t_1 - t_0)}$$

∴
$$\alpha_{25} = \frac{\alpha_0}{1+\alpha_0(25-0)} = \frac{0.005}{1+0.005(25)}$$

$$= 4.44 \times 10^{-3}/°C$$

We know the relation,

$$R_2 = R_1[1+\alpha_1(t_2 - t_1)]$$

∴
$$R_{2000} = R_{25}[1+\alpha_1(2000-25)]$$

$$881.6 = R_2[1+4.44\times10^{-3}\times1975]$$

$$R_{25} = 90.25\ \Omega.$$

The current flowing through the 60 W lamp at the instant of switching will be corresponding to its resistance at 25°C.

$$\therefore \qquad I = \frac{V}{R_{25}} = \frac{230}{90.25} = 2.55 \text{ Amps}$$

EXAMPLE 1.12

A coil has a resistance of 18 Ω at 20°C and 20 Ω at 50°C. At what temperature will its resistance be 21 Ohms?

SOLUTION

$$R_{20} = 18, R_{50} = 20, R_t = 21 \text{ at what t?}$$

we know,

$$R_2 = R_1[1 + \alpha_1(t_2 - t_1)]$$

$$\therefore \qquad R_{50} = R_{20}[1 + \alpha_{20}(50 - 20)]$$

or,

$$20 = 18[1 + \alpha_{20}(30)]$$

or,

$$\alpha_{20} = 3.7 \times 10^{-3}/°C$$

We can write,

$$R_3 = R_1[1 + \alpha_1(t_3 - t_1)]$$

substituting,

$$21 = 18\ [1 + 3.7 \times 10^{-3}(t_3 - 20)]$$

or,

$$t_3 = 65°C.$$

EXAMPLE 1.13

The resistance of a wire increases from 40 Ω at 20°C to 50 Ω at 70°C. Calculate the temp. coefficient of resistance at 0°C.

SOLUTION

given

$$R_{20} = 40\Omega,\ R_{70} = 50\Omega,.\ \text{what is } \alpha_o?$$

$$R_2 = R_1[1 + \alpha_1(t_2 - t_1)]$$

$$50 = 40[1 + \alpha_1(70 - 20)]$$

$$\alpha_1 = 5\times10^{-3}/°C$$

$$\alpha_1 = \frac{\alpha_0}{1+\alpha_0 t_1}$$

or, $$5\times10^{-3} = \frac{\alpha_0}{1+20\alpha_0}$$

or, $$\alpha_0 = 5.55\times10^{-3}/°C$$

EXAMPLE 1.14

A resistance element of cross-sectional area of 10 mm² and length 10 m draws a current of 4 A at 220 V supply at 20°C. Calculate the resistivity of the material. What current will be drawn when the temperature rises to 60°C? Assume α_{20} = 0.0003/°C.

SOLUTION

$$a = 10\,\text{mm}^2$$

$$= 10\times10^{-6}\,\text{m}^2$$

$$V = IR$$

or, $$R = \frac{V}{I} = \frac{220}{4} = 55\,\Omega.$$

This resistance, we call as R_{20}

$$R = \rho\frac{\ell}{a}$$

or, $$R_{20} = \rho_{20}\frac{\ell}{a}$$

$$55 = \rho_{20}\frac{10}{10\times10^{-6}}$$

or, $$\rho_{60} = 55\times10^{-6}\,\text{ohm}-\text{m}$$

given $$\rho_{20} = 0.0003/°C$$

Since we have to calculate R_{60}, we have to calculate ρ_{60}

$$\rho_{60} = \rho_{20}[1+\alpha_{20}(60-20)]$$

$$= 55\times10^{-6}[1+0.0003\times40] = 55.66\times10^{-6}\,\Omega\text{m}$$

Now, $$R_{60} = \rho_{60}\frac{\ell}{a} = 55.66\times10^{-6}\frac{10}{10\times10^{-6}} = 55.66$$

Current, $$I = \frac{V}{R_{60}} = \frac{220}{55.66} = 3.9525\ \text{Amps}$$

1.9 ▶ ELECTROSTATICS

Electrostatics deals with statical electricity, i.e., with static charges. In this section, behaviour of static charge and the laws governing static charges are described.

We have known that the deficit or excess of electrons in a body is called charge. A body with excess of electrons is considered negatively charged while a body with deficit of electrons is taken as positively charged. Like charges repel and unlike charges attract each other.

Charles Coulomb, a French scientist, carried out a number of experiments to observe the effect of placing charges near to each other. A law of electrostatics, known as Coulomb's law or inverse square law, was framed by him on the basis of his experiments.

1.9.1 Coulombs Law

According to Coulomb's law, the force of repulsion or attraction, F between two charges can be expressed as $F = K\frac{Q_1Q_2}{d^2}$ Newton

Where Q_1 and Q_2 are the two charges separated by a distance of d meters; K is a constant of proportionality which depends on the medium in which the charges are placed.

The value of constant K as found out experimentally is given as $K = \frac{1}{4\,\pi\varepsilon} = \frac{1}{4\pi\varepsilon_0\varepsilon_r}$

The unit of Force is in newton (N); the charges Q_1 and Q_2 are expressed in coulomb; distance, d is in metres; and permittivity, ε is in farad/m. The value of $\varepsilon_o = 8.854\times10^{-12}$ F/m. Relative permittivity, ε_r for air is equal to 1.

1.9.2 Unit Charge

Coulomb's law is reproduced as

$$F = \frac{Q_1Q_2}{4\,\pi\varepsilon_o\, d^2} \text{ newton (N)}$$

Consider $Q_1 = Q_2 = 1$ Coulomb, d = 1 m and $\varepsilon_o = 8.854 \times 10^{-12}$. Then

$$F = \frac{1\times1}{4\times3.14\times8.854\times10^{-12}\times1^2}$$

or,

$$F = 8.99\times10^9 \text{ newton(N)}$$

On the basis of the above, a unit charge is defined as that charge, which when placed at a distance of 1 m from an equal and similar charge in air as the medium, will repel it with a force of 8.99×10^9 N.

1.9.3 Electric Field Intensity

Any region or space in which a charged particle experiences a force is called in electric field. The space between the two charged plates of a parallel plate capacitor is an example of an electric field.

Electric field intensity is defined as the force experienced by a unit positive charge placed at a point in an electric field .The electric field intensity is also called electric field strength. As we know, the magnitude of force on a unit charge, q placed at a distance of d metres in an electric field due to another charge Q Coulomb is given by the relation.

$$F = \frac{Qq}{4\pi\varepsilon_o d^2} \text{ newton (N)}$$

Electric field intensity, E is equal to force per unit charge. Therefore,

$$E = \frac{Q}{4\pi\varepsilon_o d^2} \text{ N/C}$$

1.9.4 Electric Field Due to a Group of Charges

The force exerted on a point charge due to number of charges is the vector sum of the individual force exerted by each charge.

Electric charge may exist in a point, in a line form, in a surface form or in a volume form.

Accordingly, they are called point charge, line charge, surface charge, and volume charge. A point charge is simply a position with no dimensions. Charges may be uniformely distributed along a line, along a surface or over the entire volume of a body.

1.9.5 Electric Flux and Electric Flux Density

The force that a point charge or test charge will experience in an electric field is represented both in magnitude and direction, by means of lines called lines of force. An electric line of force can be defined as the path along which a unit positive charge will tend to move when force to do so. The total number of lines of force emanating from a certain charge is called ***electric flux***. An electric field is represented by electric flux in the same way as a magnetic field is represented by magnetic flux. Magnetic flux are closed paths whereas electric flux are not. Electric flux lines come out from positive electric charge and terminate an equal and opposite negative charge.

We have known Electric field intensity E at a distance d from a charge Q is given as

$$E = \frac{Q}{4\pi\varepsilon_o\varepsilon_r d^2} \quad \text{Since } \varepsilon_o\varepsilon_r = \varepsilon$$

or

$$E\varepsilon = \frac{Q}{4\pi d^2} \text{ Coulomb/area of the surface.}$$

$E\varepsilon$ is called the electric flux density, D. $E\varepsilon$ has the dimension of charge/area, i.e., charge density. The product $E\varepsilon$ is called electric flux density, D

$$D = E\varepsilon = \frac{Q}{4\pi d^2}$$

1.9.6 Electric Potential and Potential Difference

Electric potential is the amount of work done in bringing a unit positive charge from infinity to a given point in an electric field.

Electric potential is work done per unit charge, i.e., W/Q Joules/Coulomb. Where W is the work done in joules in bringing a unit charge from infinity to a given point in an electric field. The potential at that point will be taken as 1 volt.

$$1 \text{ volt} = \frac{1 \text{ Joule}}{1 \text{ Coulomb}}$$

The unit of potential is Joule/Coulomb or watt/ampere. According to definition, Potential, V = Work done, W.

$$W = \int_{\infty}^{d} -\frac{Q}{4\pi\varepsilon d^2}dx = \frac{Q}{4\pi\varepsilon d}$$

Potential difference between two points in an electric field is defined as the work done in taking a unit positive charge from one point to the other against the electric field. Figure 1.4 shows a charge +Q and two points A and B at a distance of d_1, and d_2

Potential at point A, $$V_A = \frac{Q}{4\pi\varepsilon d_1}$$

Potential at point B, $$V_B = \frac{Q}{4\pi\varepsilon d_2}$$

Potential difference, $$V_A - V_B = \frac{Q}{4\pi\varepsilon}\left(\frac{1}{d_1} - \frac{1}{d_2}\right)$$

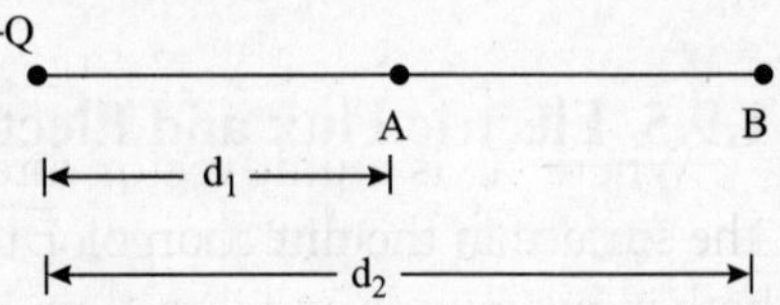

FIGURE 1.4

1.9.7 Gauss's Law

A German scientist, Karl Fredrick Gauss, developed a law which states that the total electric flux passing through a closed surface is equal to the total charge enclosed by the surface.

According to Gauss's law, the total electric flux, ψ passing through a closed surface is equal to the sum of all charges Q_1, Q_2, Q_3,……, Q_n inside the closed surface. Therefore, $\psi = \Sigma Q$ Coulomb.

Mathematically Gauss's law can be shown as

$$\psi = \int d\psi = \oint_S \bar{D} \cdot \bar{ds} = Q = \text{charge enclosed}$$

The sign $\oint_S$ indicates integration over a closed surface. The flux, d ψ passing out through the surface, ds, is

$$d\psi = |\bar{D}| \cos\theta ds = |\bar{D}| ds \cos\theta$$

$$\text{or, } d\psi = \bar{D} \cdot \bar{ds}$$

That is, $d\psi$ is the dot product of $\bar{D}$ and $\overline{ds}$.

(Remember dot product $\bar{A}$ and $\bar{B}$ is $\bar{A}\cdot\bar{B} = |A||B|\cos\theta_{AB}$)

1.9.8. Proof of Gauss's Law

Let us consider a spherical surface to prove Gauss's law. The spherical surface encloses a charge Q at the centre as shown in Fig. 1.5.

$\overline{ds}$ is a small surface area on the sphere. Its magnitude is ds and the direction normal to the surface ds is represented by $\bar{a}_n$ such that

$$\overline{ds} = ds\,\bar{a}_n$$

The electric flux density, $\bar{D}$ at a point on the surface which is at a distance, d from the point charge, Q is

$$|\bar{D}| = \frac{Q}{4\pi d^2}$$

$$D = \frac{Q}{4\pi d^2}\bar{a}_r$$

Where $\bar{a}_r$ is a unit vector directed radially outward from the point charge and normal to the surface in the direction of $\bar{D}$.

The dot product, $$\bar{D}\cdot\overline{ds} = \frac{Q}{4\pi d^2}\bar{a}_r\, ds\,\bar{a}_n$$

or, $$\bar{D}\cdot\overline{ds} = \frac{Q}{4\pi d^2}ds$$

(Since the directions of the unit vectors is the same, $\bar{a}_r\,\bar{a}_n = 1$)

Taking surface integral of both sides,

$$\oint_s \bar{D}.\overline{ds} = \oint_s \frac{Q}{4\pi d^2} = \frac{Q}{4\pi d^2}\oint_s ds$$

For a sphere, $$\oint_s ds = 4\pi d^2$$

Thus, $$\oint_s \bar{D}\cdot\overline{ds} = \frac{Q}{4\pi d^2}\times 4\pi d^2 = Q$$

So, Gauss's law, $\oint_s \bar{D}\cdot\overline{ds} = Q$ is proved.

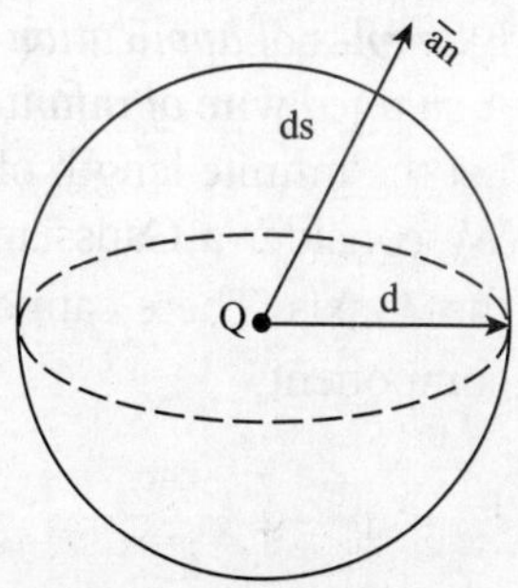

FIGURE 1.5

Charge enclosed by a spherical surface

1.9.9 Applications of Gauss's Law in Finding Electric Flux Density, $\bar{D}$, Electric Field Intensity, $\bar{E}$, and Potential, V.

The following steps can be used to obtain $\bar{D}$ and $\bar{E}$ by using Gauss's law.

i. Consider $|\bar{D}|$ and its direction
ii. Consider $|\overline{ds}|$ and its direction
iii. Take dot product, $\bar{D}\cdot\overline{ds}$
iv. Choose a Gaussian surface and integrate over the surface
v. Equate the charge enclosed with the integration obtained
vi. Evaluate for $|\bar{D}|$ and express with its direction
vii. Express $\bar{E}$ as $\bar{E} = \frac{\bar{D}}{\varepsilon_o}$.

Calculation of Potential and Potential difference: Let us consider a charge q is moved at a distance, $\overline{dl}$ in an electric field, $\bar{E}$. The force exerted, $F_e = q\bar{E}$. An applied force F_a is required to bring the charged particle into equilibrium so that $F_a = -q\bar{E}$. The differential amount of work done, $dw = F_a \, dl \cos\theta$ when $\bar{F_a}$ and $\overline{dl}$ makes an angle θ.

or,
$$dw = F_a \cdot dl = -q\bar{E}\cdot\overline{dl}$$

Work done,
$$W = \int dw = -q\int \bar{E}\cdot\overline{dl}$$

$$V = \frac{W}{q} = -\int \bar{E}\cdot\overline{dl}$$

Potential difference between two points A and B is calculated as,

$$V_{AB} = -\int_A^B \bar{E}\cdot\overline{dl} \qquad \text{(dot product)}$$

Examples of application d of Gauss's law:
A charged wire of infinite length:
Let the infinite length of wire be along the Z-axis as shown in Fig. 1.6.
We consider a Gaussian surface as a right circular cylinder of length, L and radius, r along the Z–axis. There cannot be any component of $\bar{D}$ along the Z-axis. $\bar{D}$ will have only radial component.

$$\bar{D} = |\bar{D}|\,\bar{a}_r$$

$$\overline{ds} = r\,d\phi\,dz\,\bar{a}_r$$

$$\bar{D}\cdot\overline{ds} = |\bar{D}|\,r\,d\phi dz \quad \bar{a}_r\cdot\bar{a}_r$$

$$= |\bar{D}|\,r\,rd\phi dz$$

$\overline{ds} = r\,d\phi\,dz\,\overline{a_r}$

where $\overline{a_r}$ is normal to ds.

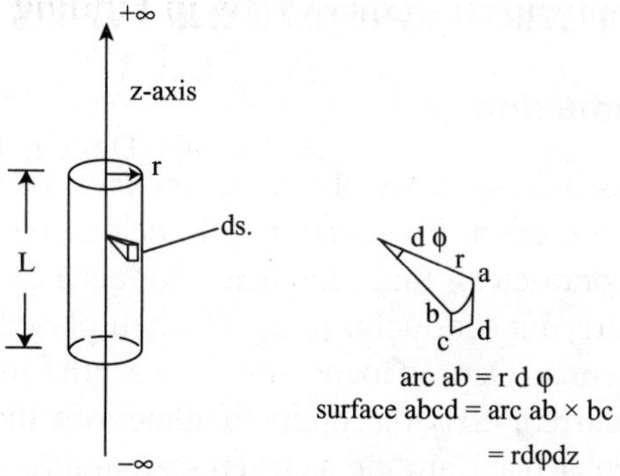

FIGURE 1.6

Determination of $\bar{D}$ and $\bar{E}$ for an infinite line charge

$$Q = \oint_S |\bar{D}|\, r\, d\phi\, dz$$

$$= \int_{z=0}^{L} \int_{\phi=0}^{2\pi} |\bar{D}|\, r\, d\phi\, dz$$

$$= r|\bar{D}|[Z]_0^L\,[\phi]_0^{2\pi}$$

or, $$Q = 2\pi r d L$$

or, $$D = \frac{Q}{2\pi r L} = \frac{Q}{L} \times \frac{1}{2\pi r} = \frac{\rho_L}{2\pi r} \left(\text{Since, } \frac{Q}{L} = \text{Charge density, } \rho_L \right)$$

$$\bar{D} = \frac{\rho_L}{2\pi r} \overline{a_r} \quad \text{Coulomb/m}$$

$$\bar{E} = \frac{\bar{D}}{\varepsilon_O} = \frac{\rho_L}{2\pi r E_O} \bar{a}_r \text{ v/m}$$

The Potential difference between any two points A and B, at a distance of a and b, respectively, from the centre of a charged wire having a charge density ρ_L Coulomb/m is calculated as

$$V_{AB} = V_A - V_B = -\int_b^a \bar{E} \cdot \overline{br} = -\int_b^a \frac{\rho_L\, dr}{2\pi r \varepsilon_0}\, \overline{a_r} \cdot \overline{a_r}$$

or $$V_{AB} = \frac{\rho_L}{2\pi \varepsilon_0} \int_b^a \frac{dr}{r} = \frac{\rho_L}{2\pi \varepsilon_0} \log \frac{b}{a}$$

Similarly, applying Gauss's law, electric flux density D, electric field intensity E, and electric potential, V can be calculated for a horizontal surface, spherical surface, etc.

1.10 ► Electromagnetism and Electromagnetic Induction

1.10.1 Introduction

Electromagnetism is the study of interaction between electric current and magnetic field, and forces produced thereof. This section will include descriptions of magnetic field around current-carrying conductors, magnetic field produced by a current-carrying coil, force produced on a current-carrying conductor or a coil when placed in a magnetic field.

A Danish scientist, Oersted in the early nineteenth century discovered that there was a magnetic field around a current-carrying conductor. Lines of force in the form of concentric circles existed on a perpendicular plane around a current-carrying conductor. This meant, magnetism could be created by electric current. It was also observed that the direction of lines of force got changed when the direction of current flowing through the conductor was changed. A few years after the discovery of Oersted, Faraday, another scientist from England discovered that a magnetic field can create an electric current in a conductor. When there is a change in flux linkage in a conductor or a coil, EMF is induced in it. This phenomenon is credited to Faraday who established famous laws of electromagnetic induction. You will observe that most of the electrical machines and devices have been developed utilizing the observations and discoveries made as mentioned above.

1.10.2 Magnetic Field Around a Current-Carrying Conductor

In Fig. 1.7 is shown a conductor carrying a current, I. Lines of force are established around the conductor on a perpendicular plane. In Fig. 1.7 (a) magnetic field around a long conductor has been shown. The lines of force are established on a perpendicular plane. In Fig. 1.7 (b) and (c), the cross-sectional views of a current-carrying conductor have been shown. The cross at the centre of the conductor indicates that current is entering the conductor which is placed perpendicular to the plane of the paper. The lines of force in the form of concentric circles

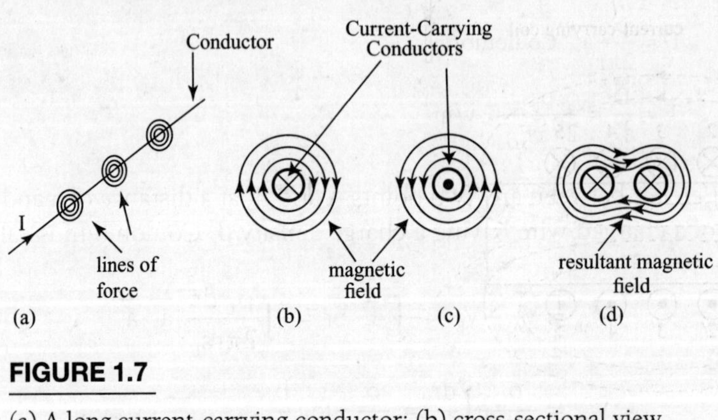

FIGURE 1.7

(a) A long current-carrying conductor; (b) cross-sectional view of a conductor with flux around it; (c) cross-sectional view of the conductor with the direction of current reversed; (d) resultant magnetic field produced by two current-carrying conductors

are on the plane of the paper. The direction of current through the conductor is reversed in Fig. 1.7 (c). The dot at the centre of the conductor indicates that the current is coming towards the observer. The direction of the lines force around the conductor also gets reversed.

The direction of flux lines around a current-carrying conductor is determined by applying the cork screw rule which is stated below.

Cork Screw Rule: Consider a right-hand screw held on one end of a current-carrying conductor and is rotated in the clockwise direction. If the advancement of the screw indicates the direction of current, the direction in which the screw is rotated will indicate the direction of the lines of force around the conductor.

In Fig. 1.7 (d) has been shown that two current-carrying conductors placed side by side produce a resultant magnetic field.

1.10.3 Magnetic Field Around a Coil

A coil is formed by winding a wire of certain cross section around a former (a hollow cylinder made of some non-magnetic material like bakelite, plastics, etc). Such a coil is often called a solenoid. When current is allowed to flow through such a coil, a magnetic field is produced by the coil. The direction of flux produced by a current-carrying coil is determined by applying the right-hand-grip rule. In Fig. 1.8 (a) is shown a current-carrying coil. If we hold the coil by our right hand in such a way that the four fingers bend towards the direction of the current flow through the coil turns, the thumb will indicate the direction of the resultant flux produced.

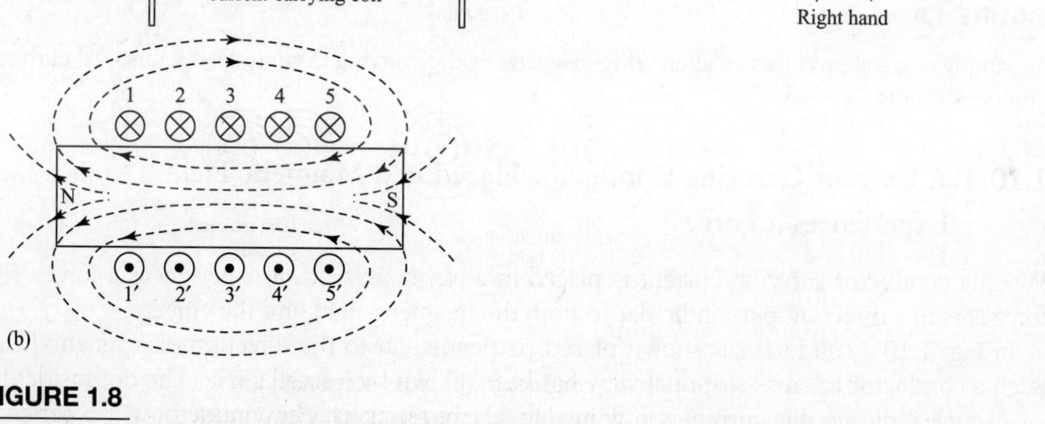

FIGURE 1.8

(a) Right-hand-grip rule applied to determine direction of flux produced by a current-carrying coil; (b) magnetic field produced by a current-carrying coil in cross-sectional view

The four fingers bend in the direction of current through the coil. The direction in which the thumb points is the direction of flux produced. In Fig. 1.8 (b), we have shown the cross-sectional view of the same coil. For the direction of current flow through the coil, cross-sections have been shown by putting cross and dot convention. The upper side of the coil turns 1, 2, 3, 4, 5 will indicate that current is entering while they will come out from the other side as shown in the bottom conductor cross-sections. By applying the cork screw rule also, we can determine the direction of the resultant magnetic field and show the positions of North and South poles formed. If the direction of current flow through the coil is reversed, the direction of the magnetic lines of force will be opposite, and hence, the positions of North and South poles will change.

If we apply some alternating voltage across the coil as shown in Fig. 1.9, the polarity of power supply will change in every half cycle of the applied voltage. If a sinusoidal ac supply is provided, both the magnitude as well as the direction of current flow will change. As a result, the magnitude of the magnetic field produced will change starting from zero value reaching its maximum value, then getting reduced again to zero, and then becoming negative. The direction of flux produced will change in every half cycle of current flow. Such a magnetic field whose magnitude as also its direction changes is called a pulsating alternating magnetic field. In case of dc supply, the magnetic field produced will be of constant magnitude and fixed polarity.

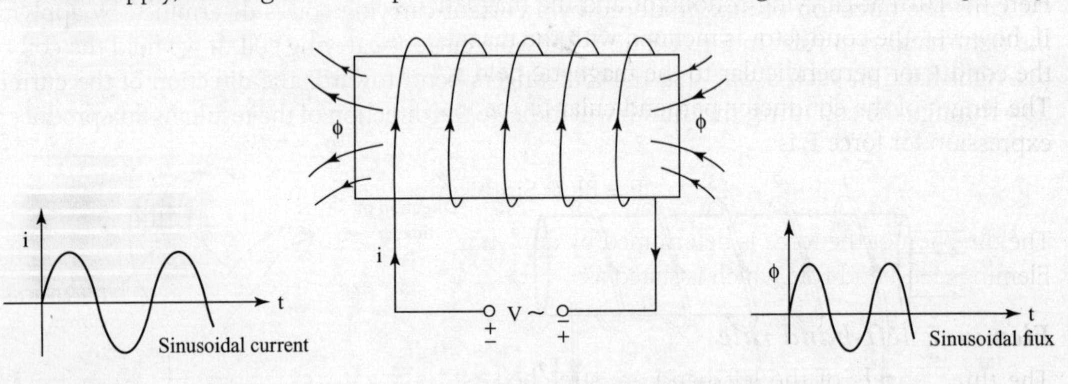

FIGURE 1.9

AC supply to a coil produces an alternating magnetic field of varying magnitude. A sinusoidal current produces a sinusoidal flux

1.10.4 A Current-Carrying Conductor Placed in a Magnetic Field Experiences a Force

When a conductor carrying current is placed in a magnetic field, it experiences a force. The force acts in a direction perpendicular to both the magnetic field and the current.

In Fig. 1.10 a conductor is shown placed perpendicular to the direction of magnetic field. Such a conductor in cross-sectional view has been shown by a small circle. The dot inside the small circle indicates that current is flowing towards the observer. The conductor will experience a force in the upward direction as has been shown. If the direction of current through the conductor is reversed, the force on the conductor will be in the downward direction.

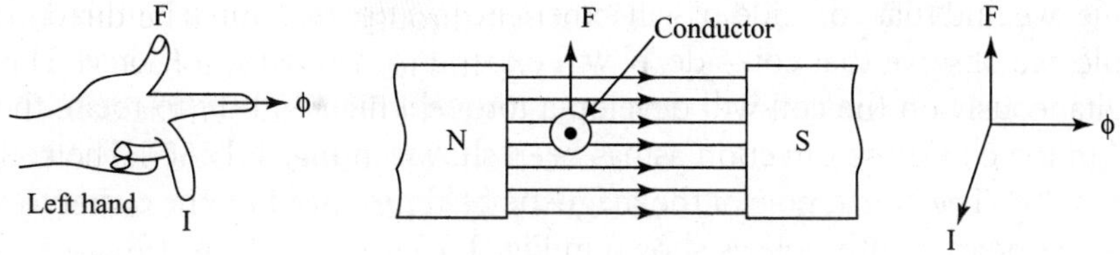

FIGURE 1.10

Force experienced by a conductor carrying current in a magnetic field

The force on the conductor will depend upon the flux, ϕ or flux density, B $\left(B = \frac{\phi}{A}\right)$ where A is the area of the magnetic poles. The force will also depend upon the effective length of the conductor in the magnetic field, i.e., ℓ and on the magnitude of current flowing, i.e., I. The force developed is expressed as

$$F = BI\ell \text{ Newtons} \tag{1.22}$$

Here the current-carrying conductor and the magnetic fields are at right angles to each other. If, however, the conductor is inclined with the magnetic field by an angle θ, then the length of the conductor perpendicular to the magnetic field is to be considered as shown in Fig. 1.11. The length of the conductor perpendicular to the magnetic field is ℓ Sinθ. Thus, the general expression for force F is

$$F = BI\ell \sin\theta \text{ Newton} \tag{1.23}$$

The direction of the force is determined by applying Fleming's left-hand rule which is stated as:

Fleming's left-hand rule

The three fingers of the left hand are stretched as shown in Fig. 1.10. If the forefinger points towards the direction of the lines of force, and the middle finger points toward the current flowing through the conductor, then the thumb will point towards the direction of force experienced by the conductor.

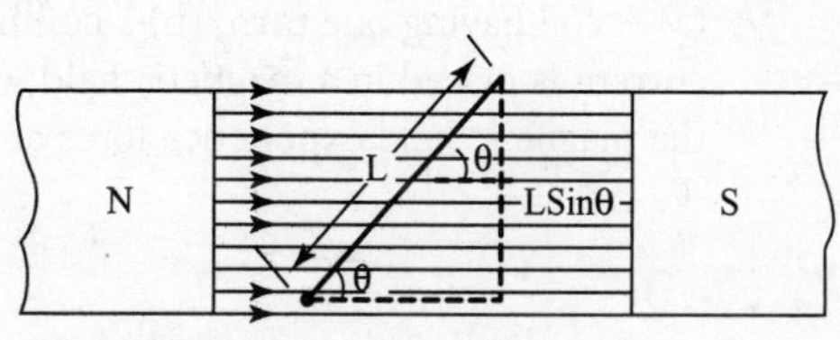

FIGURE 1.11

Force on a current-carrying conductor when the conductor is inclined with the magnetic field. ($F = BI\ell \sin\theta$ Newton)

1.10.5 A Current-Carrying Coil Placed in a Magnetic Field Experiences a Torque

Now we will consider a coil placed in a magnetic field. A coil has two coil-sides which lie in the magnetic field. These coil sides are called conductors. Thus, a coil has two conductors. If a coil has two turns, the number of conductors will be four. See Fig. 1.12 (a and b). In Fig. 1.12 (c) has been shown a single turn coil placed in a magnetic field. The direction of current through the coil has also been shown. The direction of the magnetic field is from North pole to South pole. The direction of current in coil-side 'a' is towards the observer. If we apply Fleming's left-hand rule, we find that coil-side 'a' will experience an upward' force. Similarly, by applying the same rule, we observe that coil-side 'b' will experience a downward force. The two forces acting simultaneously on the coil will develop a torque which will try to rotate the coil along an axis x-x′ in the clockwise direction as has been shown in Fig. 1.12 (c). The coil will rotate by an angle of 90°. The North pole of the magnetic field produced by the current-carrying coil will face the stationary South pole as shown in Fig. 1.13.

The two magnetic fields get aligned as shown in Fig. 1.13 (b). If it is possible to change the direction of current in the coil when it changes its position from DD′ axis to XX′ axis, the coil will continue to develop torque in the clockwise direction. We will get continuous rotation of the coil. This is the basic principle of direct-current electric motor which will be discussed in detail in a separate chapter.

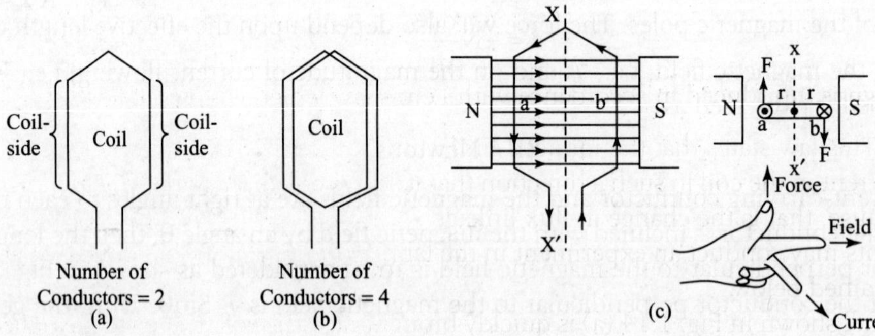

FIGURE 1.12

(a) A coil having one turn; (b) a coil having two turns; (c) a single turn coil carrying current is placed in a magnetic field; (d) the coil sides of the current-carrying coil in the magnetic field experience force and the coil experiences a torque

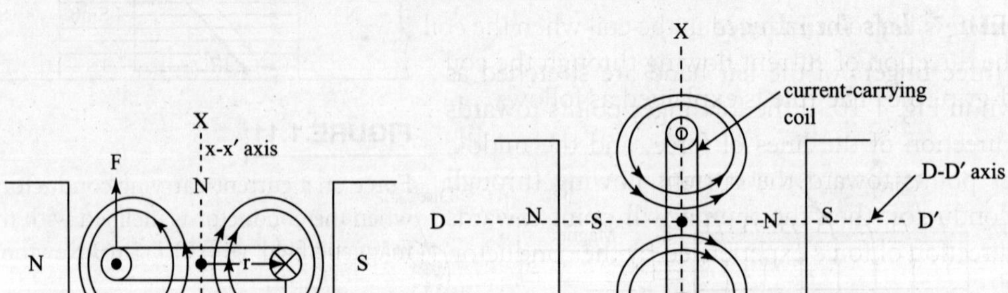

FIGURE 1.13

(a) A current-carrying coil in a magnetic field experiences a torque; (b) magnetic field produced by the current-carrying coil and the stationary magnetic field get aligned

1.11 ▶ Laws of Electromagnetic Induction

Faraday, on the basis of laboratory experiments, established that whenever there a is change in the magnetic flux linkage by a coil, EMF is induced in the coil. The magnitude of the EMF induced is proportional to the rate of change of flux linkages. Faraday's laws of electromagnetic induction are stated as:

First law: EMF is induced in a coil whenever magnetic flux linking the coil is changed.

Second law: The magnitude of the induced EMF is proportional to the rate of change of flux linkage.

The rate of change of flux linkage is expressed as $N\frac{d\phi}{dt}$ where N is the number of turns of the coil linking the flux. Thus, the induced EMF, e is expressed as

$$e = -N\frac{d\phi}{dt} \tag{1.24}$$

The minus sign is introduced in accordance with Lenz's law which is stated below.

Lenz's law: This law states that the induced EMF due to change of flux linkage by a coil will produce a current in the coil in such a direction that it will produce a magnetic field which will oppose the cause, that is the change in flux linkage.

The students may conduct an experiment in the laboratory, similar to that done by Faraday, which is explained below.

If the magnet shown in Fig. 1.14 (a) is quickly brought near the coil, there will be deflection in the galvanometer indicating EMF induced in the coil and current flow in the circuit. If the magnet is held stationary near the coil, although there is flux linking the coil, there will be no induced EMF since there is no change in the flux linkage. The induced EMF will be there only if there is increase or decrease in flux linkage by the coil. It will be observed that when the magnet is taken away quickly there will be deflection in the galvanometer. It may also be noted that EMF will also be induced in the coil when the coil is moved keeping the magnet stationary.

The direction of current flowing through the coil can be determined by applying the right-hand-grip rule. The rule is explained as follows.

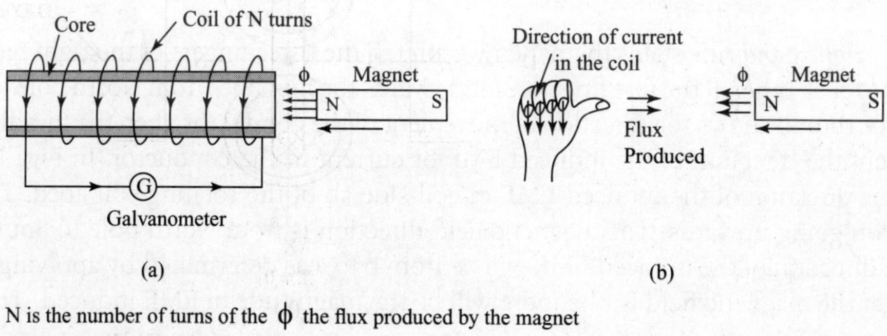

FIGURE 1.14

Faraday's experiment on electromagnetic induction. (a) A magnet is suddenly brought near a coil; (b) determination of the direction of current produced in the coil

Right-hand-grip rule

Hold the coil with your right hand with the thumb opposing the direction of movement of the magnet. The other four fingers will indicate the direction of current flow through the coil. This means that the current induced in the coil will produce flux in the direction of the thumb, thus, opposing the flux producing the induced EMF in the coil. See Fig. 1.14 (b). This is in accordance with Lenz's law.

1.12 ▶ Induced EMF in a Coil Rotating in a Magnetic Field

Now we will consider a coil rotated in a stationary magnetic field as shown in Fig. 1.15.

Here a coil, having two sides (conductors) is rotated in a uniform magnetic field as shown in Fig. 1.15. Because of the rotation of the coil in the magnetic field, flux linkage by the coil changes, i.e., the number of lines of force passing through the coil changes. Because of change of flux linkage, EMF is induced in the coil. The direction of the induced EMF in the conductors can be determined by applying ***Fleming's right-hand rule*** (FRHR).

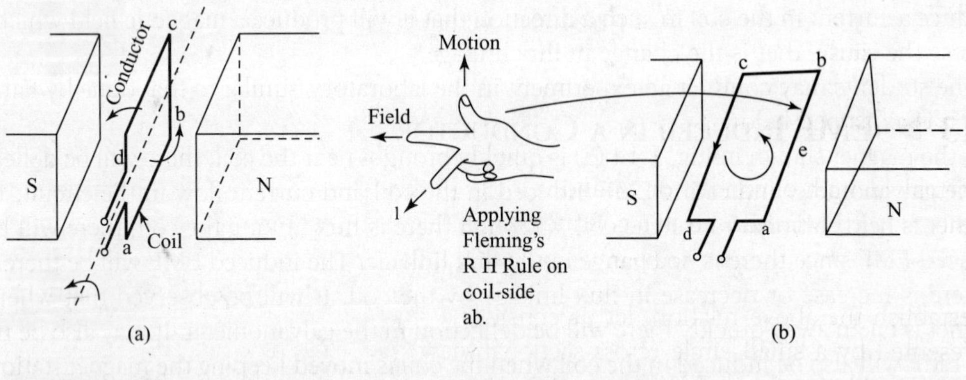

FIGURE 1.15

(a) EMF is induced in a coil when rotated in a magnetic field;
(b) determination of direction of induced EMF

Fleming's right-hand rule states that when we stretch the three fingers of the right hand perpendicular to each other, if the fore finger points towards the flux lines from North pole to South pole, and the thumb shows the direction of movement of the conductor, then the middle finger will represent the direction of the induced EMF or current in the conductor. In Fig. 1.15 (b) is shown the direction of the induced EMF in coil-side ab of the rotating coil abcd. This coil side is shown going upwards. The magnetic field direction is from North pole to South pole. Hence, the direction of the induced EMF will be from b to a as determined by applying FRHR. The stronger the magnetic field is, the more will be the magnitude of EMF induced. The more the speed of rotation of the coil is, the more will be the magnitude of the EMF induced. This is because $\frac{d\phi}{dt}$ will increase if both ϕ as well as the rate of change of linkage of ϕ are changed.

The magnitude of the EMF induced will also be directly proportional to the number of turns of the rotating coil, or the number of coils connected in series. The EMF induced can also be considered in terms of flux cut by a conductor (coil side) per second.

Here in Fig. 1.15, the number of poles is two. We can also have four poles, six poles, etc. When a conductor rotates in such magnetic field, it cuts the lines of force. The number of lines of force cut by a conductor in one revolution, when there are two poles, is 2ϕ Webers, where ϕ is the flux per pole. If there are say P number of poles, flux cut by a conductor in one revolution will be $P\phi$ Webers. If the coil makes 'n' revolutions per second, the time taken by a conductor to make one revolution will be 1/n seconds. Thus, flux cut per second will be the EMF induced, e which is

$$\text{Induced EMF, } e = \frac{\text{flux cut in 1 revolution in Wbs}}{\text{time taken in making 1 revolution in secs}}$$

or, $$e = \frac{P\phi}{1/n} \text{ Wb/sec}$$

or, $$e = P\phi n \text{ Volt}$$

$$= P\phi N/60 \quad \text{where N is in rpm.} \tag{1.25}$$

1.13 ► EMF Induced in a Conductor

In terms of length of conductor, ℓ and velocity of the conductor, v in a magnetic field of flux density, B, the EMF induced in a conductor, e is calculated as

$$e = B\,l\nu \sin\theta \text{ V} \tag{1.26}$$

To establish the above relation, let us consider a single conductor represented by a small circle (cross-sectional view) is moved in a magnetic field of strength B Wb/m^2 as shown in Fig. 1.16.

Let the conductor cut the flux at right angles by moving a distance dx meter. The area swept by the moving conductor is ℓ dx m^2. The flux density is B Wb/m^2.

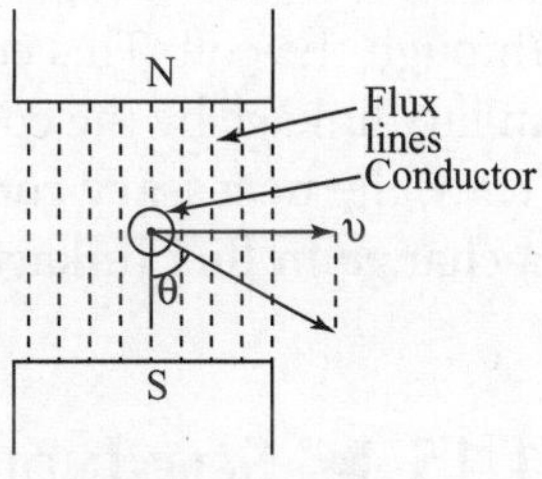

FIGURE 1.16

EMF induced in a conductor moving in a magnetic field

$$\text{Flux cut} = \text{Flux density} \times \text{area}$$

$$= B\,\ell\,dx \text{ Webers}$$

The time taken to move a distance dx m is dt seconds.

Induced emf, e = Flux cut per second

$$\text{Therefore, } e = B\,\ell \frac{dx}{dt} \text{ V}$$

Since $\frac{dx}{dt}$ is the linear velocity v of the conductor,

$$e = Bl\nu \text{ V} \tag{1.27}$$

If the conductor moves in a direction making an angle θ with the direction of magnetic field as shown in Fig. 1.16, the induced EMF will be as stated earlier in eq. 1.26.

$$e = Blv \sin\theta \text{ V} \tag{1.28}$$

1.14 ► Dynamically Induced EMF and Statically Induced EMF

When emf is induced in a coil or conductor by virtue of movement of either the conductor or the magnetic field, the emf is called dynamically induced EMF as has been explained in section 1.12.

When EMF is induced in a stationary coil by changing its flux linkage due to change in current flow through the coil, such emf is called statically induced EMF.

If a coil carries a current, flux is established around the coil. If the current is changed quickly, the flux linkage by the coil will change as shown in Fig. 1.17 (a).

In Fig. 1.17 (a), a coil of certain number of turns is wound on a former, i.e., its core. Current is supplied from a battery by closing a switch S. If the switch is continuously turned on and off, flux linkage by the coil will change. The rate of change of the flux linkage will induce EMF, e in the coil.

A similar effect will be there if an ac supply is applied across the coil as shown in Fig. 1.17 (b). The direction of current in the coil is shown for the positive half cycle of the alternating current. The direction of current will change in every half cycle, and hence, the direction of flux produced will change in every half cycle. The magnitude of current changes continuously since a sinusoidal current is flowing. This changing current will create a changing flux linkage, thereby inducing EMF in the coil in both the cases as shown in Fig. 1.17 (a) and (b). Note that in Fig. 1.17 (a), if the switch S is kept closed, a steady direct current, i.e., a constant current will flow through the coil. This constant current will produce a constant flux. There will be no change in flux linkage by the coil with respect to time, and hence, no EMF will be induced in the coil. Thus, the **necessary condition for the production of induced EMF is that there should be a change in flux linkage and not merely flux linkage by a coil.**

1.15 ► Self-Induced EMF and Mutually Induced EMF

The EMF induced in a coil due to change in flux linkage when a changing current flows through the coil is called **self-induced EMF.**

As shown in Fig. 1.17 (c), when a second coil is brought near a coil producing changing flux, EMF will be induced in the second coil due to change in current in the first coil. This is called **mutually induced EMF.** In fact, EMF will be induced in both the coils as both the coils are linking a changing flux. However, in the second coil EMF is induced due to changing flux created by coil 1. The magnitude of the induced EMF will depend upon the rate of change of flux linkage and the number of turns of the individual coils. The induced EMF in the two coils, e_1 and e_2 will be

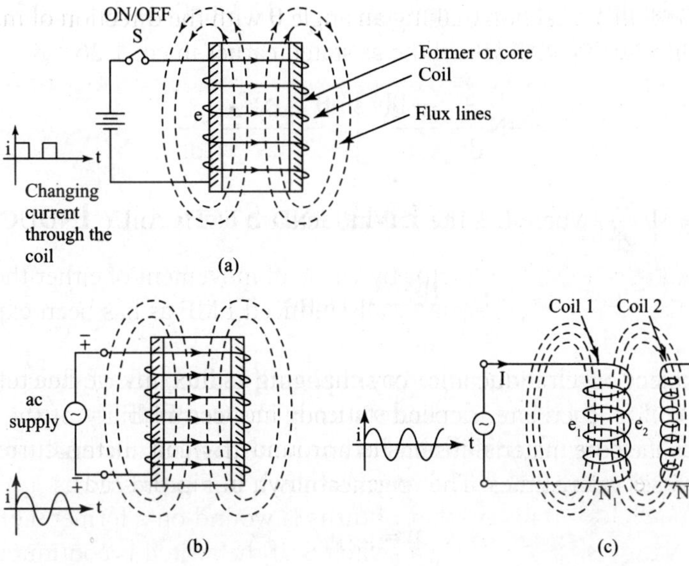

FIGURE 1.17

(a) Change in flux linkage in a coil due to switching ON and switching OFF of dc current; (b) change in flux linkage due to alternating current supply; (c) induced emf in coils 1 and 2 due to changing flux produced by alternating current flowing in coil 1

$$e_1 = -N_1 \frac{d\phi_1}{dt} \tag{1.29}$$

and

$$e_2 = -N_2 \frac{d\phi_2}{dt} \tag{1.30}$$

where N_1 and N_2 are the number of turns of coil 1 and coil 2, respectively.

You will study in a separate chapter how transformers are built utilizing the basic principle of mutually induced EMF.

1.16 ► Self-Inductance of a Coil

Consider a coil of N turns wound on a core of magnetic material. Let an alternating current i pass through the coil as shown in Fig. 1.18.

The emf induced, e will be

$e = -N\dfrac{d\phi}{dt}$ .....(i) The flux, $\phi = B \times A$

$$= \mu HA \quad [\because B = \mu H]$$

$$= \mu \frac{Ni}{\ell} A \left[\because H = \frac{Ni}{\ell}\right] \text{ or, } \phi = \frac{\mu NA}{\ell} i$$

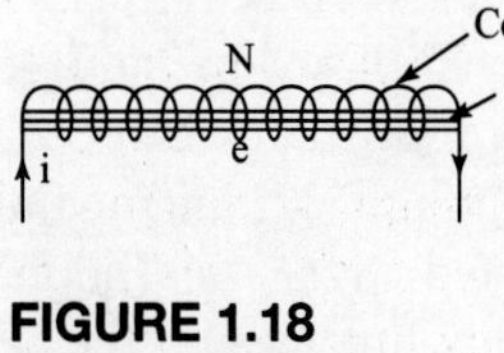

FIGURE 1.18

Inductance of a coil

where μ is the permeability of the core material; ℓ is the length of flux path; A is the area of the coil. Substituting value of ϕ in (i),

$$e = -N\frac{d}{dt}\left[\frac{\mu NA}{\ell}i\right] = -\left[\frac{\mu N^2 A}{\ell}\right]\frac{di}{dt}$$

or, $$e = -L\frac{di}{dt} \text{ where L is the self-inductance of the coil} \qquad (1.31)$$

where $$L = \frac{\mu N^2 A}{\ell} \text{ Henry} \qquad (1.32)$$

L is called the coefficient of self inductance or simply self inductance of the coil.

Inductance of a coil is, therefore, dependent upon the permeability of the core material. When we put iron as the core material instead of any non-magnetic material, or air as the core, the inductance increases many times. The (permcability), μ is expressed as

$$\mu = \mu_o \mu_r$$

where μ_r is the relative permeability and μ_o is the permeability of free space. The relative permeability of iron may be as high as 2000 times than that of air. Hence, an iron core coil may have an inductance value 2000 times more than that of an air-core one, other dimensions remaining the same. Again, inductance, L is inversely proportional to the length of the flux path and directly proportional to the area of the core material or the coil. Inductance is proportional to the square of the number of turns. To have an inductance of a large value, the number of turns should be high.

The inductance, L can be expressed in terms of the rate of change of the flux with respect to current flowing in the coil. This is shown below.

$$\text{Flux, } \phi = BA$$
$$= \mu HA$$
$$= \mu\frac{Ni}{\ell}A \cdot$$

For a small increment of di, let the increase of flux be dϕ. Therefore,

$$d\phi = \frac{\mu NA}{\ell}di$$

or, $$\frac{d\phi}{di} = \frac{\mu NA}{\ell}$$

or, $$N\frac{d\phi}{di} = \frac{\mu N^2 A}{\ell} = L$$

or, $$L = N\frac{d\phi}{di} \text{ Henry} \qquad (1.33a)$$

If ϕ and i have a linear relationship,

$$\frac{d\phi}{di} \text{ be written as } \frac{\phi}{I}$$

Therefore, $$L = N\frac{\phi}{I} \text{ Henry}$$

Again, $$L = \frac{\mu N^2 A}{\ell} = \frac{N^2}{\ell/\mu A} = \frac{N^2}{\text{Reluctance}} \tag{1.33b}$$

where $$\text{Reluctance} = \frac{\ell}{\mu A} \tag{1.34}$$

Remember that reluctance is inversely proportional to permeability. Low reluctance will give rise to a high value of inductance. That is why in order to produce high-value inductance, the number of turns should be high and the reluctance to the flux path should be low. See equation (1.33b). The core should be made of high permeability material like iron.

Again, $$L = N\frac{\phi}{I}$$

or, $$LI = N\phi$$

Considering a small increase of i producing a small increase in ϕ as $d\phi$

$$Ldi = Nd\phi$$

or, $$L\frac{di}{dt} = N\frac{d\phi}{dt} = e \tag{1.35}$$

Inductance is the property of a coil capable of inducing emf in itself due to changing current through it.

The formulae so far derived are

(1) Force on a current-carrying conductor in a magnetic field

$F = BI\ell$ Newtons.

If the conductor is inclined at an angle θ with the magnetic field,

$F = BI\ell \sin\theta$ Newtons.

(2) Induced EMF in a coil when there is change of flux linkage or change in current,

$e = N\frac{d\phi}{dt}$ V; $e = L\frac{di}{dt}$

(3) Induced EMF in a conductor rotating in a magnetic field,

$e = P\phi n$ V

where P is the number of poles, ϕ is the flux per pole and n is the revolutions per second.

(4) Induced EMF in a conductor moving in a magnetic field in a perpendicular direction,

$e = B\ell v$ V

where B is the flux density in Wb/m², ℓ is the length of the conductor in m and v is the velocity in m/sec.

If the conductor is moving at an angle of θ with the magnetic field, the induced EMF is $e = B\ell v \sin\theta$ V.

(5) Induced EMF in a coil,

$$e = -N\frac{d\phi}{dt}$$

$$e = -L\frac{di}{dt}$$

$$L = \mu\frac{N^2A}{\ell} = \frac{N^2}{\ell/\mu A} = \frac{N^2}{\text{Reluctance}}$$

$$L = N\frac{d\phi}{di} = N\frac{\phi}{I} \text{ (assuming linear magnetization)}$$

$L = \frac{N\phi}{I}$. If φ is equal to 1 Wb-turn and I is 1 ampere, then

$$L = \frac{1\text{Wb-turn}}{1\text{Ampere}} = 1\text{ Henry} \quad (1.36)$$

Thus, we can say that a coil has an inductance of 1 Henry if a current of 1 Ampere flowing through the coil produces a flux linkage of 1 Wb-turn.

1.17 ► Mutual Inductance

Consider two coils having N_1 and N_2 number of turns placed near each other as shown in Fig. 1.19. Let a changing current, i_1, flow through coil 1. The flux produced by i_1 in N_1 is ϕ_1 Since coil 2 is placed near coil 1, a part of the flux produced by coil 1 will be linked by coil 2. Let flux ϕ_2 linked by coil 2 is $\phi_2 = K_1\phi_1$ where $K_1 \le 1$.

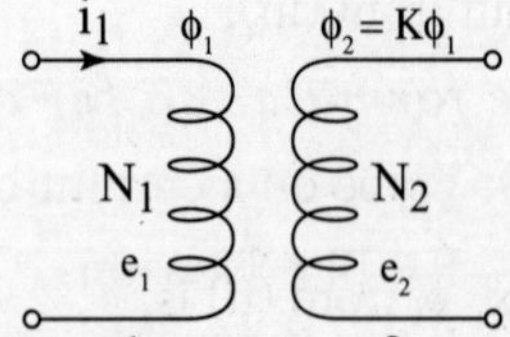

FIGURE 1.19

Mutual inductance of two coils

If magnetic coupling between the two coils is very tight, i.e., very good, the whole flux produced by coil 1 will link the coil 2, in which case the coefficient of the coupling K_1 will be 1. The induced EMF in coil 2 is e_2.

$$e_2 = N_2\frac{d\phi_2}{dt} = N_2\frac{d(K_1\phi_1)}{dt} = N_2K_1\frac{d\phi_1}{dt} \quad \text{(i)}$$

Flux, $$\phi_1 = \frac{\text{mmf}}{\text{Reluctance}} = \frac{N_1 i_1}{\ell_1/\mu A_1} = \frac{N_1\mu A_1}{\ell_1}i_1 \quad \text{(ii)}$$

From (i) and (ii),

$$e_2 = N_2K_1 \frac{d}{dt}\left(\frac{N_1\mu A_1 i_1}{\ell_1}\right) = \left(\frac{K_1N_1N_2\,\mu A_1}{\ell_1}\right)\frac{di_1}{dt}$$

or,

$$e_2 = M\frac{di_1}{dt}$$

where $M = \dfrac{K_1N_1N_2\,\mu A_1}{\ell_1}$, is called the mutual inductance of the two coils. (iii)

Similarly, if we calculate the induced EMF in coil 1, due to change in current in coil 2, we can find the induced EMF e_1 in coil 1 as

$$e_1 = \frac{K_2N_1N_2\,\mu A_2}{\ell_2}\frac{di_2}{dt}$$

where

$$M = \frac{K_2N_1N_2\,\mu A_2}{\ell_2} \tag{iv}$$

Now, multiplying the expression for M as in (iii) and (iv) above,

$$M^2 = \frac{K_1K_2N_1N_2N_1N_2\,\mu A_1\,\mu A_2}{\ell_1\ell_2}$$

or,

$$M^2 = K_1K_1\frac{\mu N_1^2A_1}{\ell_1}\frac{\mu N_2^2A_2}{\ell_2}$$

or,

$$M^2 = K_1\,K_2\,L_1\,L_2$$

Therefore,

$$M = \sqrt{K_1\,K_2}\,\sqrt{L_1\,L_2} = K\sqrt{L_1\,L_2} \tag{1.37}$$

where

$$K = \sqrt{K_1\,K_2}$$

Again from (iii),

$$M = K_1N_2\frac{N_1}{\ell_1/\mu A_1} \tag{v}$$

Flux, ϕ_1 from (ii) is,

$$\phi_1 = \frac{\text{mmf}}{\text{Reluctance}} = \frac{N_1I_1}{\ell_1/\mu A_1}]$$

or,

$$\frac{\phi_1}{I_1} = \frac{N_1}{\ell_1/\mu A_1} \tag{vi}$$

From (iii) and (vi),

$$M = \frac{K_1N_1N_2\,\mu A_1}{\ell_1} = K_1N_2\frac{N_1}{(\ell_1/\mu A_1)}$$

$$M = K_1N_2\frac{\phi_1}{I_1}$$

$$M = N_2\frac{\phi_2}{I_1}(\because \phi_2 = K_1\phi_1)$$

Thus, $$M = \frac{N_2\phi_2}{I_1} = \frac{\text{flux linkage in coil 2}}{\text{changing current in coil 2}} \tag{1.38}$$

From eq. (1.38) we can define the mutual inductance M between two coils as the flux linkage in one circuit due to change per unit of current in the other circuit.

Similarly, considering current change in the second coil

$$M = N_1 \frac{K_2\,\phi_2}{I_2}$$

or, $$M = N_1 \frac{\phi_1}{I_2} (\because \phi_1 = K_2\,\phi_2)$$

Thus, $$M = \frac{N_1\phi_1}{I_2} = \frac{\text{flux linkage in coil 1}}{\text{changing current in coil 2}} \tag{1.39}$$

1.18 ▶ Inductance of Coils Connected in Series Having a Common Core

We have two coils having self-inductance L_1 and L_2 connected in series. In Fig. 1.20 (a), they produce flux in the same direction, and in Fig. 1.20 (b), the connection is such that they produce flux in the opposite directions.

Since the two coils are connected in series, the same current flows through them.

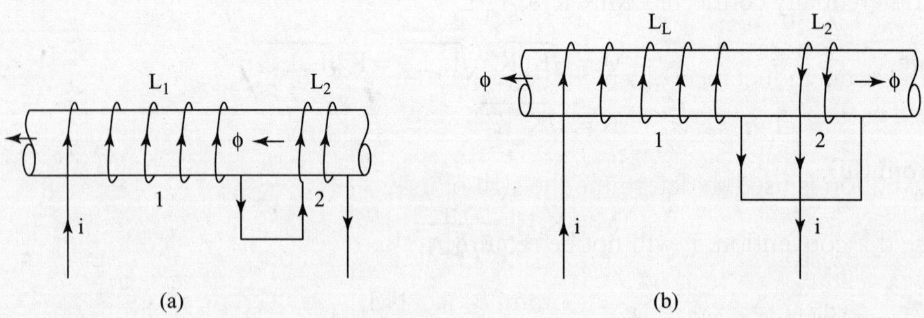

FIGURE 1.20

Coils connected in series in (a) cummulatively; in (b) differentially

If there is a change in current di amperes in time dt seconds, the EMF induced in coil 1 due to its self-inductance L_1 is

$$e_1 = -L_1 \frac{di}{dt} \tag{i}$$

Similarly, the EMF induced in coil 2 due to its self-inductance, L_2 is

$$e_2 = -L_2 \frac{di}{dt} \tag{ii}$$

Due to mutual inductance, the EMF induced in coil 1 due to change in current in coil 2 and vice versa are expressed as EMF induced in coil 1 due to change in current in coil 2 is

$$e_{12} = -M\frac{di}{dt} \qquad \text{(iii)}$$

EMF induced in coil 2 due to change in current in coil 1 is

$$e_{21} = -M\frac{di}{dt} \qquad \text{(iv)}$$

Now let the total equivalent inductance of the single circuit comprising coil 1 and coil 2 as they are connected as in Fig. 1.20 (a) be L_e

The EMF induced in the whole circuit will, therefore, be

$$e = -L_e\frac{di}{dt} \qquad \text{(v)}$$

Thus, equating the expression for e in (iv) with the total EMFs as in (i), (ii), (iii), and (iv):

$$-L_1\frac{di}{dt} - L_2\frac{di}{dt} - M\frac{di}{dt} - M\frac{di}{dt} = -L_e\frac{di}{dt}$$

Therefore,

$$L_e = L_1 + L_2 + 2M \qquad \text{(vi)}$$

When the coils are differentially connected as in Fig. 1.20 (b), the EMF induced in coil 1 due to di in time dt in coil 2, i.e., $M\frac{di}{dt}$ in opposition to the EMF induced in coil 1 due to its self-inductance. Similar is the case of the EMF induced in coil 2 due to mutual inductance. Thus, for the differentially connected coil

$$L'_e = L_1 + L_2 - 2M \qquad \text{(vii)}$$

Thus, the total inductance of an inductively coupled series-connected coil circuit can be expressed as

$$L_T = L_1 + L_2 \pm 2M \qquad (1.40)$$

Dot convention is used to determine the sign of induced voltage $M\frac{di}{dt}$.

If we use dot convention, it will not be required to know the way the coils have been actually wound.

EXAMPLE 1.15

The total inductance of two coils connected in series cumulatatively is 1.6 H and connected differentially is 0.0.4 H. The self inductance of one coil is 0.6 H. Calculate (a) the mutual inductance and (b) the coupling coefficient.

SOLUTION

We know,

$$L_T = L_1 + L_2 \pm 2M$$

Substituting the given values

$$L_1 + L_2 + 2M = 1.6 \qquad \text{(i)}$$

and

$$L_1 + L_2 - 2M = 0.4 \qquad \text{(ii)}$$

From (i) and (ii)

$$4M = 1.2$$

or, $$M = 0.3 \text{ H}$$

given, $$L_1 = 0.6$$

Therefore, $$0.6 + L_2 + 2 \times 0.3 = 1.6$$

or, $$L_2 = 1.6 - 1.2 = 0.4 \text{ H}$$

$$L_1 = 0.6 \text{ H}, L_2 = 0.4 \text{ H}, M = 0.3 \text{ H}$$

We know, $$M = K\sqrt{L_1 L_2}$$

or, $$K = \frac{M}{\sqrt{L_1 L_2}}$$

Substituting values, $$K = \frac{0.3}{\sqrt{0.6 \times 0.4}} = 0.612$$

1.19 ► Energy Stored in a Magnetic Field

Let us consider a coil as in Fig. 1.21, supplied with an alternating voltage v due to which an alternating current flows through the coil. When current increases from its zero value, the magnetic field starts increasing and reaches its maximum value when current reaches its maximum value. When current starts decreasing, the field goes on decreasing and gradually becomes zero. Then, in the negative cycle if the current flows, the field gets established in the opposite direction, which collapses when current again reaches zero. This way the field is established and then collapses in every consecutive half cycle of current flow. When the field is established, energy in the form of a magnetic field is stored and when the field collapses, the same energy is returned to the supply source. As such, no energy is consumed by the purely inductive coil. Therefore, energy stored is equal to the energy supplied.

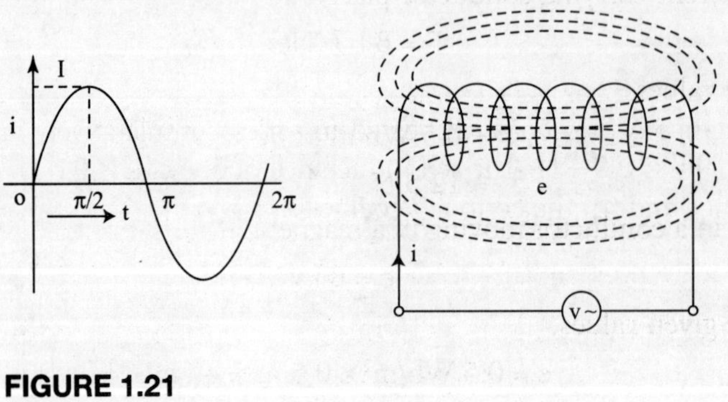

FIGURE 1.21

Magnetic field energy

Energy stored, $W = vi\,dt$

Induced EMF in the coil, $e = -L\dfrac{di}{dt}$

This induced EMF opposes the applied voltage from which it is produced. This is due to Lenz's law, so that

$$e = -v \quad \text{or,} \quad v = -e$$

Thus,

$$W = (-e)i\,dt \quad \text{and} \quad e = -L\frac{di}{dt}$$

for a current change from 0 to i

$$W = \int_O^I (-e)i\,dt$$

$$= \int_O^I L\frac{di}{dt}i\,dt$$

$$= \int_O^I Li\,di$$

$$= L\frac{I^2}{2} = \frac{1}{2}LI^2 \text{ Joules}$$

Energy stored in a coil of inductance L is

$$W = \frac{1}{2}LI^2 \text{ Joules} \qquad (1.41)$$

EXAMPLE 1.16

A conductor of length 0.5 m is placed in a magnetic field of strength 0.5 Wb/m^2. Calculate the force experienced by the conductor when a current of 50 A flows through it. If the force moves the conductor at a velocity of 20 m/sec, calculate the EMF induced in it.

SOLUTION

Force, F on a current-carrying conductor placed in a magnetic field is given as

$$F = B\,I\,\ell \text{ Newton}$$

Substituting the values,

$$F = 0.5 \text{ Wb/m}^2 \times 50 \times 0.5 \text{ m}$$

$$= 12.5 \text{ N}$$

Induced EMF, e in a conductor moving in a magnetic field is given as

$$e = B\,l\,v \text{ V}$$

Substituting the given values,

$$e = 0.5 \text{ Wb/m}^2 \times 0.5 \text{ m} \times 20 \text{ m/sec}$$

$$= 5 \text{ Wb/sec} = 5 \text{ V}$$

EXAMPLE 1.17

An iron-cored toroidal coil has 100 turns. The mean length of the flux path is 0.5 m and the cross-sectional area of the core is 10 cm^2. Calculate the inductance of the coil. Assume relative permeability of iron as 2000. Also calculate the induced emf in the coil when current of 5A is reversed in 10 ms.

SOLUTION

The expression for inductance in terms of its parameters is

$$L = \frac{\mu N^2 A}{\ell}$$

where, $\mu = \mu_o \mu_r$

Current in the coil is changed from +5 A to –5 A in 10×10^{-3} secs. Total change of current is 10 A.

Putting the given values we get,

$$L = \frac{4\pi \times 10^{-7} \times 2000 \times 100 \times 100 \times 10}{10000 \times 0.5}$$

$$= \frac{4\pi \times 10^{-4} \times 2 \times 10}{0.5}$$

$$= 16 \times 3.14 \times 10^{-3} \text{ H}$$

$$= 50.24 \times 10^{-3} \text{ H}$$

$$= 50.24 \text{ mH}$$

Induced EMF, $e = -L\dfrac{di}{dt}$

Average value of induced EMF, $E = L\dfrac{\text{change in current}}{\text{time taken}}$

$$E = L\frac{2I}{t}$$

Substituting values, $E = 50.24 \times 10^{-3} \dfrac{2 \times 5}{10 \times 10^{-3}}$

$$= 50.24 \text{ V}$$

EXAMPLE 1.18

There is mutual magnetic coupling between two coils of number of turns 500 and 2000, respectively. Only 50% of the flux produced by the coil of 500 turns is linked with the coil of 1000 turns. Calculate the mutual inductance of the two coils. Also calculate the EMF induced in the coil of 1000 turns when current changes at the rate of 10A/second in the other coil. The self-inductance of the coil of 500 turns in 200 mH.

SOLUTION

The data provided have been shown in Fig. 1.22.

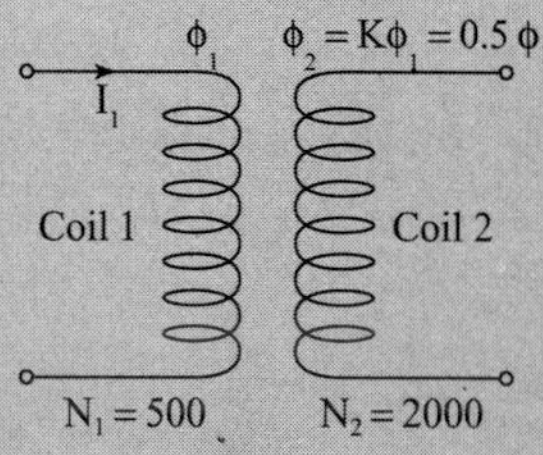

FIGURE 1.22

Mutual inductance,

$$M = \frac{\text{flux linkage in coil 2}}{\text{current change in coil 1}}$$

$$= N_2 \frac{K\phi_1}{I_1} = 2000 \times \frac{0.5 \times \phi_1}{I_1} = \frac{1000\phi_1}{I_1}$$

$$L_1 = \frac{N_1\,\phi_1}{I_1}$$

or,

$$\frac{\phi_1}{I_1} = \frac{L_1}{N_1} = \frac{200 \times 10^{-3}}{500}$$

$$M = 1000 \times \frac{\phi_1}{N_1} = \frac{1000 \times 200 \times 10^{-3}}{500}$$

or,

$$M = 400 \times 10^{-3}\ \text{H}$$

Induced EMF in the second coil, e_2 is

$$e_2 = M\frac{di}{dt}$$

$$= 400 \times 10^{-3} \times 10 = 4\ \text{V}$$

EXAMPLE 1.19

A current of 5 A flowing through a coil of 500 turns produces a flux of 1 mWb. Another coil is placed near this coil and current in this coil is suddenly reversed in 10 ms. As a result, the EMF induced in the second coil is measured as 50 V. Calculate self and mutual inductance of the coils assuming a coefficient of coupling as 60 per cent.

SOLUTION

$$e_2 = M\frac{di_1}{dt}$$

or, $$50 = M\frac{5+5}{10\times10^{-3}}$$ [+5 A current has been changed to −5 A]

or, $$M = \frac{50\times10\times10^{-3}}{10}$$

or, $M = 50 \times 10^{-3}$ H

Self-inductance of coil 1 is

$$L_1 = \frac{N_1\phi_1}{I_1} = \frac{500\times1\times10^{-3}}{5} = 10\times10^{-3}\,\text{H}$$

Using the formula, $$M = K\sqrt{L_1L_2}$$

$$50 \times 10^{-3} = 0.6\sqrt{10\times10^{-3}\times L_2}$$

or, $$L_2 = 694.4 \times 10^{-3}\ \text{H}$$

EXAMPLE 1.20

Two coils of number of turns $N_1 = 1000$ and $N_2 = 400$, respectively, are placed near each other. They are magnetically coupled in such a way that 75 per cent of the flux produced by the one of 1000 turns links the other. A current of 6 A produces a flux of 0.8 mWb in N_1 and the same amount of current produces a flux of 0.5 mWb in the coil of N_2 turns. Determine L_1, L_2, M, and K for the coils.

SOLUTION

$$L_1 = N_1\frac{\phi_1}{I_1} = 1000\times\frac{0.8\times10^{-3}}{6} = 0.133\,\text{H}$$

$$L_2 = N_2\frac{\phi_2}{I_2} = 400\times\frac{0.5\times10^{-3}}{6} = 0.033\,\text{H}$$

$$M = N_2\frac{K_1\phi_1}{I_1} = 400\times\frac{0.75\times0.8\times10^{-3}}{6}$$

$$= 0.04\ \text{H}$$

Using the relation, $$M = K\sqrt{L_1L_2}$$

substituting values,

$$0.04 = K\sqrt{0.133\times0.033}$$

or, $$K = \frac{0.04}{0.066} = 0.606$$

So, the calculated values are:
Self-inductance of coil 1 = 0.133 H
Self-inductance of coil 2 = 0.033 H
Mutual inductance of the coils = 0.04 H
Coefficient of coupling = 0.606

1.20 ▶ Electrical Circuit Elements

Resistors, inductors, and capacitors are the three basic circuit parameters or circuit components of any electrical network. Resistors can be wire-wound type or carbon-moulded type. When current flows in a resistance, heat is produced, which is dissipated. The heat is produced because friction between moving free electrons and atoms obstruct the free flow of electrons producing electric current. A resistor is an element that dissipates energy as heat when current flows through it.

Inductors are made of a coil having a number of turns. The core of the coil may be air or a magnetic material, which is placed inside the coil. When the coil is wound on an iron core, the inductor formed is called an iron-core inductor coil. Inductance of an inductor is directly proportional to the square of the number of turns of the coil used. Inductor stores energy because of current flowing through it.

A capacitor consists of two conductors or conducting plates between which a dielectric is placed. The capacitance of a capacitor is its ability to store electric charge. Different types of capacitors are available. They are named according to the dielectric placed between the conductors. Common types of capacitors are air, mica, paper, ceramic, etc.

1.20.1 Resistors

Wire-wound resistors are made of wires of constantan, manganin or nichrome wound on a ceramic tube. These resistors are available in ranges varying from a fraction of an ohm to thousands of ohms.

The power rating also varies from a fraction of a Watt to few kiloWatts. While specifying a resistance, both resistance value and power dissipating value must be mentioned. Electronic circuits require resistors of accurate values. The value of resistors used in electronic circuits is quite high, of the order of kilo ohms. Since carbon has high resistivity, carbon resistors are made with copper leads. Their power rating varies from a fraction of a Watt to several Watts. Color code is used to indicate the value of such resistors.

1.20.2 Inductors

The ability of a coil to induce EMF in itself when the current through it changes is called its inductance. The unit of inductance is Henry. 1 Henry of inductance causes 1 Volt to be induced when current changes at the rate of 1 Ampere per second. The induced emf, e is expressed as,

$$e = L\frac{di}{dt}$$

or,
$$L = \frac{e}{di/dt}$$

where L is in Henry, e is in Volt, and $\frac{di}{dt}$ is in Ampere per second.

When steady direct current flows through an inductor, no emf is induced in it as there is no change in current. The opposition offered to current flow is only the resistance of coil. Inductors are of two types viz air-core type and iron-core type. Inductors are also called chokes. Inductors are available in all current ranges. Air-core inductors are wound on bakelite or cardboard rods and are extensively used in electronic circuits in millihenry and microhenry ranges. High-value inductors are made of iron core. They are mainly used in ac power supply of frequency of 50 Hz.

The details of self and mutual inductance have been discussed earlier.

1.20.3 Capacitors

A capacitor, in its simplest form, consists of two thin parallel plates of conducting material separated by a dielectric material. A capacitor is capable of storing charge when a voltage is applied across the capacitor plates. If a voltage source, say a battery, is connected across the two plates of a parallel plate capacitor as shown in Fig. 1.23, electrons from the negative terminal of the voltage source accumulate on plate A of the capacitor. The other plate B loses electrons as they move to the positive terminal of the source of voltage. This way, the excess electrons produce negative charge on one side of the capacitor while the opposite side will have positive charge. The dielectric material placed in between the plates hold the charge because the free electrons cannot flow through an insulator (i.e., the dielectric material like air, paper, or mica). Storage of charge by a capacitor means that the charge remains in place even after the voltage source is disconnected. Capacitance of a capacitor is the ability to store charge. Charging and discharging are the two main effects of capacitors. When a voltage is applied, there is accumulation of charge in the capacitor and as a result voltage is built up across the terminals of the capacitor. This is called charging of the capacitor. The capacitor voltage becomes equal to the applied voltage when the capacitor is fully charged. The voltage across the capacitor remains even after the voltage source is disconnected. The capacitor discharges when a conducting path is provided across the plates without any applied voltage connected.

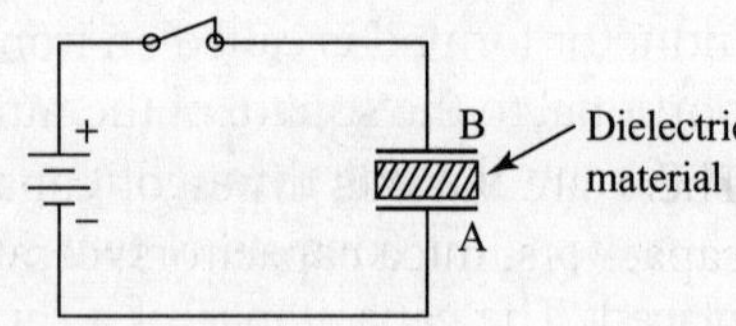

FIGURE 1.23

A capacitor stores charge in the dielectric material placed between the conducting plates

The more the charging voltage is, the more is the accumulation of charge in the capacitor. The amount of charge, Q stored in a capacitor is, therefore, proportional to the charging voltage, V. A capacitor with a large area of the parallel plates can store more charge. Capacitance of a capacitor also depend on the distance between the plates and the type of dielectric used between the plates. A large capacitor, obviously, will store more charge. Thus, we can write

$$Q = CV \text{ Coulombs}$$

where Q is the charge stored in Coulombs, V is the voltage applied across the plates, and C is the capacitance of the capacitor in Farads. The capacitance of a parallel plate capacitor is expressed as

$$C = \in \frac{A}{d} \tag{1.42}$$

where $\in$ is the absolute permittivity constant, C is the capacitance, A is the area of the plate and d is the distance between the plates.

The term absolute permittivity is expressed as

$$\in = \varepsilon_o \varepsilon_r$$

where ε_o is the permittivity constant of vacuum and ε_r is the relative permittivity of the dielectric material placed between the two plates.

The value of ε_o has been calculated experimentally as 8.85×10^{-12} Farad per meter.

Therefore, the capacitance of a parallel plate capacitor can be expressed as

$$C = \in \frac{A}{d} = \varepsilon_r \frac{A}{d} \times 8.85 \times 10^{-12} \text{ Farad}$$

There are various types of capacitors, namely air capacitors, paper capacitors, plastic film capacitors, mica capacitors, ceramic capacitors, and electrolytic capacitors.

1.21 ▶ Energy Stored in an Inductor and in a Capacitor

(a) *Energy stored in an Inductor:* When current flows through an inductor, energy is stored in it in the form of electromagnetic field. If L is the inductance of the coil in henries and I is the current flowing through the coil in amperes, then the amount of energy stored, W is given by the following expression.

$$W = \frac{1}{2} L I^2 \text{ joules}$$

Again, $$L = \frac{\mu N^2 A}{\ell}; B = \mu H; H = \frac{NI}{\ell}$$

Therefore, $$B = \frac{\mu NI}{\ell}$$

or, $$I = \frac{B\ell}{\mu N}$$

Substituting values of L and I in the expressing for W, we get

$$W = \frac{1}{2} L I^2 = \frac{1}{2} \mu \frac{N^2 A}{\ell} \times \frac{B^2 \ell^2}{\mu^2 N^2}$$

Energy stored in the inductor, $W = \dfrac{B^2 A \ell}{2\mu}$ joules.

(b) *Energy stored in a capacitor:* We have known that when a capacitor is switched on to a dc supply, the charge q can be expressed as q = Cv, where at any instant q is the change, v is the potential difference across the capacitor plates, and C is the capacitance of the capacitor.

Potential difference of v volts across the capacitor means v Joules of work has to be done in transferring 1 Coulomb of change from one plate to the other. If a small charge dq is transferred then the work done dw can be expressed as

$$dw = vdq = Cvdv$$

The total work done in raising the potential of the capacitor to the supply voltage of V volt can be expressed as

$$W = \int_0^V dW$$

$$= \int_0^V Cvdv$$

$$= C\left[\frac{\upsilon^2}{2}\right]_0^V$$

or, $$W = \frac{1}{2} CV^2$$

This work done is stored in the electrostatic field set up between the plates of the capacitor in the form of energy. Thus, the energy stored, E is expressed as

$$\text{Energy stored} = \frac{1}{2} CV^2 \text{ Joules}$$

$$= \frac{1}{2} QV \text{ Joules}$$

EXAMPLE 1.21

The current through a 100 mH inductor charges from 0 to 200 mA in 4 μs. What is the value of the induced EMF in the inductor or the choke?

SOLUTION

$$e = L\frac{di}{dt} = 100\times 10^{-3} \times \frac{200\times 10^{-3}}{4\times 10^{-6}}$$

$$= 5000 \text{ V} = 5 \text{ kV}$$

It is observed that a high voltage is induced in the choke because of very fast change of current flow through it. In a tube light circuit, a high voltage is induced in the choke by the same method and is used to ionize the gas inside the tube light, and thus start the tube light.

EXAMPLE 1.22

Self inductances of two coils are $L_1 = 2$ H and $L_2 = 8$ H. The coil L_1 produces a magnetic flux of 80 mWb of which only 60 μWb are linked with coil L_2. Calculate the mutual inductance of the two coils.

SOLUTION

The coefficient of coupling, K is given as

$$K = \frac{\text{mutual flux linkage between } L_1 \text{ and } L_2}{\text{flux produced by } L_1}$$

$$= \frac{60 \times 10^{-6}\ \text{Wb}}{80 \times 10^{-6}\ \text{Wb}} = 0.75$$

Mutual inductance M is calculated as

$$M = K\sqrt{L_1 L_2} = 0.75\sqrt{2 \times 8} = 3\ \text{H}$$

EXAMPLE 1.23

Calculate the capacitance of a capacitor made of two parallel plates of 3 m^2 having a distance between the plates of 1 cm. The dielectric is air between the plates.

SOLUTION

$$C = \varepsilon_0 \varepsilon_r \frac{A}{d} = 8.85 \times 10^{-12} \times 1 \times \frac{3}{10^{-2}} \text{F}$$

$$= 2655 \times 10^{-12}\ \text{F}$$

Note that although the area of the plates is large, the value of capacitance is very small. Instead of air as the dielectric, if we place mica or paper between the plates, capacitance will increase. If we also reduce the distance between the plates, the capacitance will increase.

EXAMPLE 1.24

A 25 microfarad capacitor is switched on to a time varying voltage source. The voltage wave is such that voltage increases at the rate of 10 V per second. Calculate the charge accumulated in the capacitor at an elapse of 1 second and the amount of energy stored in the capacitor.

SOLUTION

$$\text{Charge, } q = CV = 25\times10^{-6}\times10$$
$$= 250\times10^{-6}\text{ Coulomb}$$
$$\text{Energy stored, } W = \frac{1}{2}CV^2 = \frac{1}{2}\times25\times10^{-6}\times10^2$$
$$= 12.5\times10^{-4}\text{ Joules}$$

1.22 ► CAPACITOR IN PARALLEL AND IN SERIES

When we connect two capacitors in parallel, the plate areas are added. The total capacitance, therefore, gets added up. When capacitances C_1, C_2, C_3, etc., are connected in parallel, the total capacitance C_T becomes equal to

$$C_T = C_1 + C_2 + C_3 + \cdots$$

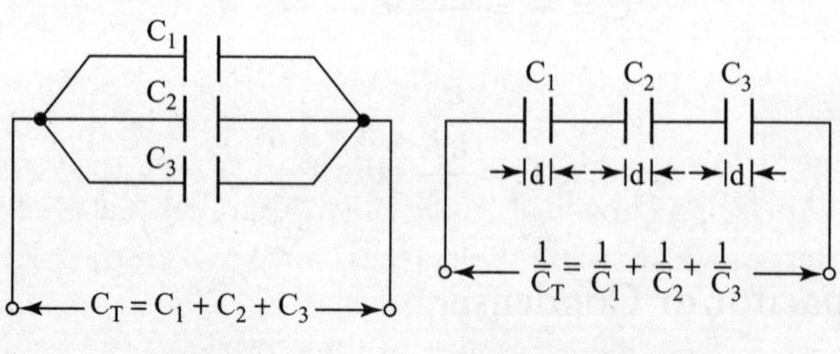

FIGURE 1.24

(a) Equivalent of capacitors connected in parallel; (b) equivalent of capacitors connected in series

This is shown in Fig. 1.24 (a).

Series connection of capacitors, as shown in Fig. 1.24 (b), is equivalent to increasing the effective distance between the plates or the thickness of the dielectric used. The combined capacitance is less than the individual value.

The value of a capacitor is always specified in either microfarad or picofarad. There are a variety of ways in which manufacturers indicate the value of a capacitor.

1.23 ► CAPACITANCE OF DIFFERENT TYPES OF CAPACITORS

The capacitance of parallel plate capacitor, spherical capacitor, isolated sphere, parallel conductors, concentric conductors, and parallel conductors can be determined using electric flux density, $\bar{D}$ and electric field intensity, $\bar{E}$ as discussed in this section.

1.23.1 Parallel Plate Capacitor

Let A be the surface area of the plates and d be the distance between them. Both the plates are separated by a dielectric material. Let Q be the charge on a plate, then charge density ρ_s is given as

$$\rho_S = \frac{Q}{A}$$

Electric flux density, $\bar{D}$ in the dielectric medium is the same as charge density ρ_s, i.e., $\bar{D} = \rho_s$.

Electric field intensity, $\bar{E} = \frac{\bar{D}}{\varepsilon} = \frac{\rho_s}{\varepsilon_o \varepsilon_r}$

$\varepsilon_o = 8.854 \times 10^{12}$ F/m and ε_r is the relative permittivity.

Potential difference between the plates is the integral of electric field $\bar{E}$ over the distance, d.

So,

$$V = \int_1^2 \bar{E} \cdot \overline{ds} = E\, d$$

Capacitance,

$$C = \frac{Q}{V} = \frac{\rho_S}{\frac{\rho_S}{\varepsilon} d} = \frac{\varepsilon A}{d}$$

$$\therefore\ C = \frac{\varepsilon A}{d} \text{ farad.}$$

1.23.2 Spherical Capacitor or Condenser

Capacitors can be made using conducting plates of different shapes. For example, a capacitor can be made using two concentric spherical shells, i.e., spheres of radius r_1 and r_2.

As shown in Fig. 1.25, the inner shell has radius r_1 and the outer shell has radius, r_2.

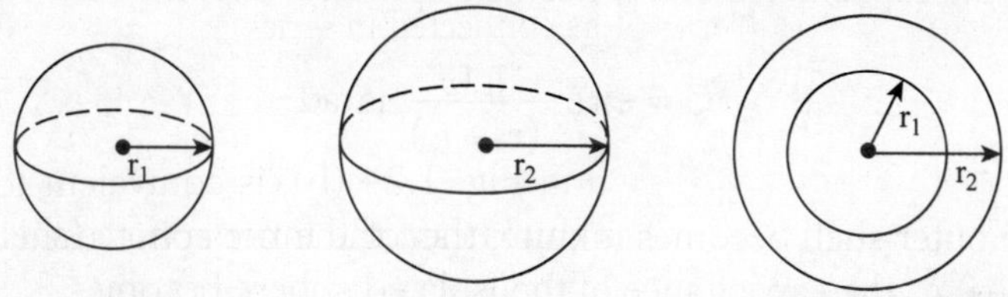

(i) inner sphere (ii) outer sphere (iii) concentric sphere

FIGURE 1.25

A spherical type capacitor

The potential difference between the surface of the two spheres is

$$V = -\int_{r_2}^{r_1} \bar{E}\, dr$$

For a spherical surface, flux density $\bar{D} = \frac{Q}{4\pi r^2}\bar{a}_r$

$$\bar{E} = \frac{\bar{D}}{\varepsilon_o} = \frac{Q}{4\pi\varepsilon_o r^2}\bar{a}_r$$

$$V = -\int_{r_2}^{r_1} \frac{Q}{4\pi\varepsilon_o}\frac{dr}{r^2} = -\frac{Q}{4\pi\varepsilon_o}\int_{r_2}^{r_1}\frac{dr}{r^2}$$

$$= -\frac{Q}{4\pi\varepsilon_o}\left[-\frac{1}{2}\right]_{r_2}^{r_1}$$

$$= \frac{Q}{4\pi\varepsilon_o}\left[\frac{1}{r_1} - \frac{1}{r_2}\right]$$

$$= \frac{Q}{4\pi\varepsilon_o}\frac{(r_2 - r_1)}{r_1 r_2}$$

$$C = \frac{Q}{V} = \frac{Q}{\frac{Q}{4\pi\varepsilon_o}\frac{(r_2 - r_1)}{r_1 r_2}} = 4\pi\varepsilon_o \frac{r_1 r_2}{(r_2 - r_1)}$$

The capacitance of a spherical condenser is

$$C = 4\pi\varepsilon_o \frac{r_1 r_2}{(r_2 - r_1)} \text{ farad}$$

1.23.3 Capacitance of an Isolated Sphere

We have seen that the capacitance of a concentric spherical capacitor is,

$$C = 4\pi\varepsilon_o \frac{r_1 r_2}{(r_2 - r_1)} \text{ farad}$$

If the radius of the outer shell becomes infinite, then the inner sphere (shell) becomes an isolated sphere. If $r_2 \to \infty$, the capacitance of the isolated sphere becomes,

$$C = 4\pi\varepsilon_o \frac{r_1}{\left(1 - \frac{r_1}{r_2}\right)}$$

$$r_2 \to \infty, \quad C = 4\pi\varepsilon_o \frac{r_1}{1 - 0}$$

or,
$$C = 4\pi\varepsilon_o r_1 \text{ farad}$$

Thus, the capacitance of an isolated sphere of radius r_1, with respect to earth as the other conducting surface is

$$C = 4\pi\,\varepsilon_o\,r_1 \text{ farad}$$

1.23.4 Capacitance of Concentric Conductors (Electric Cables)

Electric cables are used for distribution of electricity. An electric cable has a central core of conducting material such as copper or aluminium. Adequate insulation is provided around the core. A lead sheath covering is provided over the insulation to prevent entry of moisture. So, the core and the lead sheath separated by the insulating material forms a capacitor as shown in Fig. 1.26.

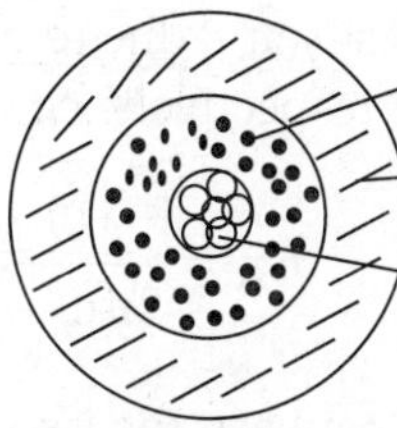

FIGURE 1.26

Cross-sectional view of an electric cable which is equivalent to two concentric conductors separated by a dielectric material

Let r be the radius of the inner conductor and R that of the sheath. The charge is uniformly distributed over the conductor surfaces. The charge density per unit length is ρ_L Coulomb/metre.

The potential difference at a point, P at a distance, x from the centre of the cable is

$V = \int E_x dx$ Where E_x is the electric field intensity at x,

$$E_x = \frac{\rho_L}{2\pi\varepsilon_o x}$$

So,

$$V = \int_r^R \frac{\rho_L\,dx}{2\pi\varepsilon_o\,x}$$

or,

$$V = \frac{\rho_L}{2\pi\varepsilon_o}\log x\Big|_r^R = \frac{\rho_L}{2\pi\varepsilon_o}\log\frac{R}{r}$$

$$C = \frac{\rho_L}{V} = \frac{\rho_L}{\dfrac{\rho_L}{2\pi\varepsilon_o}\log\dfrac{R}{r}} = \frac{2\pi\varepsilon_o}{\log\dfrac{R}{r}} \text{ F/m.}$$

So, the capacitance per unit length of an electric cable, which is in the form of concentric conductors,

$$C = \frac{2\pi\varepsilon_o}{\log\dfrac{R}{r}} \text{ F/m}$$

1.23.5 Capacitance of Parallel Conductors

Overhead electrical transmission and distribution lines form parallel conductors. We may have to calculate the capacitance between any two lines or between any line and the earth separated by air as the dielectric material in between. Fig. 1.27 shows two parallel conductors A and B separated by a distance, d. Let P be a point at a distance x from conductor A. Let the charge per unit length on the conductors be +Q and −Q, respectively.

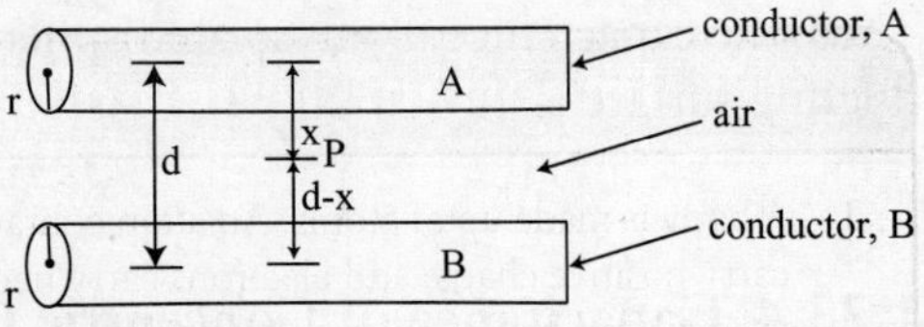

FIGURE 1.27
Capacitance between two parallel conductors

The electric field intensity at P due to charge on the two conductors will, respectively, be

$$E_{xA} = \frac{+Q}{2\pi\varepsilon_o x} \text{ and } E_{xB} = \frac{-Q}{2\pi\varepsilon_o (d-x)}$$

The potential difference between the two conductors, V is

$$V = \int_r^{d-r} E_x \, dx = \int_r^{d-r} \left[\frac{Q}{2\pi\varepsilon_o x} - \frac{(-Q)}{2\pi\varepsilon_o (d-x)} \right] dx$$

$$= \frac{Q}{2\pi\varepsilon_o} \int_r^{d-r} \left[\frac{1}{x} - \frac{1}{d-x} \right] dx$$

$$= \frac{Q}{2\pi\varepsilon_o} [\log x - \log(d-x)]_r^{d-r}$$

$$V = \frac{Q}{2\pi\varepsilon_o} [\log(d-r) - \log r - \log r + \log(d-r)]$$

$$= \frac{Q}{2\pi\varepsilon_o} \times 2 \log\left(\frac{d-r}{r}\right)$$

or,
$$V = \frac{Q}{\pi\varepsilon_o} \log\left(\frac{d-r}{r}\right)$$

If we consider $d >> r$, then $d - r \simeq d$

So,
$$V = \frac{Q}{\pi\varepsilon_o} \log \frac{d}{r}$$

Capacitance between two conducting lines of radius, r separated by a distance, d is

$$C = \frac{Q}{V} = \frac{Q}{\frac{Q}{\pi\varepsilon_o} \log \frac{d}{r}}$$

or,
$$C = \frac{\pi\varepsilon_o}{\log \frac{d}{r}} \text{ farad}$$

Important Terminologies and Points to Remember

1. A body is made up of atoms. An atom contains particles of electricity called protons and electrons. Protons carry positive charge and electrons carry negative charge. The number of protons and electrons in an atom is equal and hence, an atom is basically neutral in nature.
2. The deficit or excess of electrons in an atom is called charge. If a body consisting of neutral atoms is supplied with extra electrons, the body is said to be negatively charged. If some electrons from a body of neutral atoms are removed, the body is said to be positively charged.
3. Charge in motion is called electric current.
4. Materials are classified into conductors, insulators and semiconductors.
5. In a material when the outermost orbit of its atoms contain only one atom, they are called free electrons. Such electrons can move from one atom to the other freely. For example copper has one free electron in the outermost orbit that will constitute a current flow when a voltage is applied. Voltage is nothing but a force that will direct the movement of electrons constituting a current.
6. In a material where outermost orbit of the atoms are completely filled, the material is called an insulator. Mica, rubber, glass, plastic, etc., are insulators.
7. When the number of electrons in the outermost orbit is four instead of a maximum of eight, the material is called a semiconductor. Silicon, germanium, etc., are semiconductors.
8. A current carrying conductor or a coil produces a magnetic field around it. There exists an electric field in the space in between two charged conductors or plates.
9. Conducting materials allow flow of current through them. The magnitude of current will be the amount of charge flowing per unit of time.

   One ampere of current is equal to the flow of 1 Coulomb (6.28×10^{18}) charge per second.
10. Electrical resistance is the hindrance or opposition to the flow of electrons in the material.
11. Potential difference of two points indicates the difference in charged condition of these points.
12. Current, I flowing between two points is proportional to the potential difference, V between them, i.e., $I \propto V$ or $I = \frac{V}{R}$ where R is the resistance offered by the conductor connected between these points. This is called Ohm's law.
13. Temperature affects the resistance of a material. For most material resistance increases with temperature.
14. Unit of electrical power is Watt. One unit of power is defined as the work done per unit of time. Joules per second is Watt. Also electrical power, $P = VI = \frac{V^2}{R}$ Watts. Electrical energy is power multiplied by time.
15. A conductor carrying current and placed in a magnetic field will experience a force. A current carrying coil placed in a magnetic field will experience a torque.
16. Faraday's laws of electromagnetic induction state that (i) emf is induced in a coil when there is changing flux linkages by it; and (ii) the magnitude of induced emf is proportional to the rate of change of flux linkage.
17. When emf is induced in a coil due to movement of the coil in the magnetic field, the emf is called dynamically induced emf.
18. When a changing current flows through a coil, the coil will have an inductance, L in addition to the resistance of wire of the coil. $L = N\frac{d\phi}{di}$ where N is the number of turns of the coil and $d\phi/di$ is the rate of change of flux produced due to changing i.
19. Inductance of an individual coil is called its self inductance. Inductance between two current carrying coils is called their mutual inductance.
20. Capacitors and inductors can store energy.

Important Formulae and Equations

Equation	Remarks
1. Current, $I = \frac{\text{charge}}{\text{time}} = \frac{Q}{t}$	1. Unit of charge is Coulomb, unit of time is second, and unit of current is ampere.
2. Resistance, $R = \rho\frac{l}{a}$ Where ρ is the resistivity of the material, l is the length of the conductor and a is its cross-sectional area	2. A thiner wire will have more resistance than a thicker wire.
3. $I = \frac{V}{R}$ Where temperature and other physical conditions do not change.	3. Ohm's law.
4. $R_t = R_o(1+\alpha_o t)$ $R_2 = R_1[1+\alpha_1(t_2 - t_1)]$ $\alpha_1 = \frac{\alpha_o}{1+\alpha_o t_1}$ $\alpha_2 = \frac{\alpha_1}{1+\alpha_1(t_2 - t_1)}$ Where α is the temperature coefficient of resistance.	4. A negative temperature coefficient of resistance would mean a decrease in resistance with increase in temperature like in carbon material.
5. Power, $P = VI = \frac{V^2}{R} = I^2R$ Energy = Power × Time	5. Power is expressed in Watt and Kilo Watt(KW) Energy is expressed in KWh.
6. Force, $F = Bl\ell$ Newton	6. Force developed in a current carrying conductor in a Magnetic field in perpendicular direction.
7. Induced emf, $E = \frac{-Nd\phi}{dt}$	7. Minus sign is due to Lenz's law.
8. Dynamically induced emf, $\rho = \frac{\text{Flux cut in webers}}{\text{Time taken in seconds}}$	8. Unit of induced emf, e is in volts.
9. Inductance, $L = N\frac{d\phi}{di}$ $L = \frac{\mu N^2 A}{\rho}$ N is the number of turns of the coil, ℓ is the length of flux path and A is the area.	9. Inductance is expressed in henry μ is the permeability of the core material. If an iron core is introduced inside a coil, its inductance will increase many a times.
10. Mutual inductance, M and self inductances of coils L_1 and L_2 are related as $M = K\sqrt{L_1 L_2}$ Where K is the coefficient of coupling of the two coils.	10. The maximum value of K = 1, when two coils are very tightly magnetically coupled, i.e., one coil is placed inside the magnetic field created by the other.

Important Formulae and Equations

Equation	Remarks
11. Total inductance L_T of two coils mutually coupled is and are connected in series, $$L_T = L_1 + L_2 \pm 2\,M$$ + sign for cumulatively connected and – sign is for differentially connected.	11. Commulative connection is one when the two coils produce flux in the same direction.
12. Energy stored in a inductor, W is $W = \frac{1}{2}LI^2$ joules. Energy stored in a capacitor $W = \frac{1}{2}CV^2$ joules.	12. Capacitors and Inductors store energy.
13. Capacitance of a parallel plate capacitor, $C = \varepsilon\frac{A}{d}$ Where ε is the absolute permittivity, C is the capacitance, A is the area of the plates and d is the distance between the plates.	13. To have a large value capacitance, A should be more and d should be less and there should be a dielectric material placed between them.

REVIEW QUESTIONS

Short Answer Type Questions

1. Give an overview of the scope of electrical and electronics engineering.
2. Charge in motion is called current. Explain with the help of atomic theory.
3. Distinguish between conductors, semiconductors, and insulators.
4. Distinguish between Work, Power, and Energy.
5. Differentiate between temperature coefficient of resistance and specific resistance.
6. Distinguish between an electric field and a magnetic field.
7. Define the following terms: Volt, Ampere, Ohm.
8. Explain why two parallel current-carrying conductors attract each other when current in them flow in the same direction.
9. State Fleming's Right-Hand Rule.
10. Explain that the EMF induced in a coil depends upon the flux and the speed of rotation of the coil.
11. Distinguish between statically induced EMF and dynamically induced EMF.
12. Explain why an iron-core coil will have more inductance than an air-core coil of the same number of turns.
13. What is the meaning of coefficient of coupling between two coils? When is this value equal to unity and equal to zero?
14. What are Faraday's laws of electromagnetic induction?
15. What is the Lenz's law? Give an example.
16. What is the magnitude of force experienced by a current-carrying conductor placed in a magnetic field?
17. How do you determine the direction of force developed in a current-carrying conductor placed in a magnetic field?
18. What are the factors on which inductance of a coil depends?
19. Why does the inductance of a coil increase if the core has a magnetic material instead of air?
20. Derive the following expression for self-inductance of a coil
$$L = \frac{\mu N^2 A}{\ell}\text{Henry.}$$
21. You have to make an inductance of high value. How will you proceed?

22. What is Fleming's right-hand rule? Where is it used?
23. What rule do you apply to determine the direction of force on a current-carrying conductor placed in magnetic field?
24. What is the magnitude of force on a current-carrying conductor placed in a magnetic field?
25. Show that the energy stored in a magnetic field produced by an inductor is $\frac{B^2 A\ell}{2\mu}$ Joules.
26. Distinguish between self-inductance and mutual inductance.
27. Explain why inductance of a coil increases if an iron piece forms its core instead of air or any non-magnetic material.
28. Establish the relation, $M = K\sqrt{L_1 L_2}$ for two adjacent coils linking flux.
29. On what factors does the reluctance of a magnetic material depend?
30. What is the cork screw rule? Where do you use it?
31. Two adjacent conductors are carrying current in the opposite directions. Show that there will be force of repulsion between the conductors.
32. When capacitors are connected in parallel, their equivalent capacitance is increased. Explain why?
33. Explain why inductors and capacitors are called energy storage devices.
34. What is the meaning of relative permittivity or dielectric constant? What is it's unit?
35. Write three formulae of electrical power.
36. Prove that 1 kWh is equal to 3.6×10^6 Joules.
37. The most important property of a capacitor is its ability to block steady dc voltage while passing ac signals, explain.
38. Define the Farad unit of capacitance.
39. How is energy stored in a capacitor? On what factors does it depend?
40. What are the physical factors that affect the capacitance of a capacitor?
41. State Coulomb's law and explain the meaning of the factors that influence Coulomb's force.
42. Establish the relation between capacitance, change and potential in a capacitor.
43. What is the relationship between electric field, E and electric potential, V?
44. State and explain Gauss's law.
45. Deduce an expression for the capacitance of concentric spheres of radii a and b (b > a) using Gauss's law.
46. Calculate the inductance of a coil having 100 turns wound on a magnetic core of permeability equal to 1000, mean length of 0.25 m, and cross-sectional area of 10 cm^2.

    [**Ans** L = 50.24 mH]
47. A conductor of length 25 cm is placed in a uniform magnetic field of strength 0.5 Wb/m^2. Calculate the EMF induced in the conductor when it is moved at the rate of 10 m/sec (a) parallel to the magnetic field, (b) perpendicular to the magnetic field.

    [**Ans** (a) 0 V; (b) 1.25 V]

Multiple Choice Questions

1. The number of electrons per Coulomb is equal to
   (a) 1.602×10^{-19} (b) 6.28×10^{18}
   (c) 1.602×10^{18} (d) 6.28×10^{-19}.
2. In insulators the outermost orbit of their atoms is filled with
   (a) 4 electrons (b) 8 electrons
   (c) 1 electron (d) 18 electrons.
3. In the atoms of semiconducting materials like silicon and germanium the outermost orbit has
   (a) 1 electron (b) 2 electrons
   (c) 8 electrons (d) 4 electrons.
4. Which of the following expressions is incorrect?
   (a) Current, $I = \frac{q}{t}$ (b) Charge = current × time
   (c) $R = \rho\frac{A}{l}$ (d) Volt = joules per Coulomb.
5. Which is the following expressions does not represent power?
   (a) I^2R (b) $\frac{V^2}{R}$
   (c) VI (d) $\frac{V^2}{I}$.

6. Which of the following is not the unit of power?
   (a) Joules/second (b) Watt-hour
   (c) kW (d) Volt-ampere.
7. A conductor of length ℓ and diameter d has resistance of R ohms. If the diameter is reduced to one-third and length increased by three times, the resistance of the conductor will be
   (a) 3 R (b) 6 R
   (c) 9 R (d) 27 R.
8. Which of the following expressions is incorrect?
   (a) $e = -N\frac{d\phi}{dt}$ (b) $L = -N\frac{d\phi}{dt}$
   (c) $L = \frac{\mu N^2 A}{\ell}$ (d) $e = L\frac{di}{dt}$.
9. Which of the following expressions is incorrect?
   (a) $C = \frac{\varepsilon d}{A}$ (b) $C = \frac{Q}{V}$
   (c) $Q = \int idt$ (d) $C = \frac{\varepsilon A}{d}$.
10. Inductance of an air-core coil will increase if the core is made of
    (a) copper (b) aluminium
    (c) iron (d) porcelain.
11. Which of the following statements is not true?
    (a) Inductance of a coil will increase by four times if the number of terms eq is doubled
    (b) inductance of a coil will increase if the area of cross section of the coil, i.e., the flux path is increased
    (c) inductance of a coil will increased if the length of flux path is increased
    (d) inductance of a coil will increase if the core is made up of material having higher permeability.
12. The direction of the induced EMF in the coil sides of a coil rotating in a magnetic field can be determined by applying
    (a) Fleming's left-hand rule
    (b) Right-hand-grip rule
    (c) Fleming's-left-hand rule
    (d) Cork screw rule.
13. Which of the following is not the unit of energy?
    (a) kWh (b) Joules/second
    (c) Watt-hour (d) Joules.
14. Self-inductance of two magnetically coupled coils are 8 H and 2 H, respectively. What coefficient of coupling will make their mutual inductance equal to 4 H?
    (a) K = 0.5 (b) K = 0.25
    (c) 0.1 (d) 1.0.
15. Which of the following eq. is incorrect with respect of increase in resistance with increase in temperature of a conducting material?
    (a) $R_t = R_0(1+\alpha t)$
    (b) $R_2 = R_1[1+\alpha_1(t_2 - t_1)]$
    (c) $\alpha_1 = \frac{\alpha_0}{1+\alpha_0(t_1 - t_0)}$
    (d) $R_2 = R_1[1-\alpha_1(t_2 - t_1)]$.

Answers to Multiple Choice Questions

1. (b)	2. (d)	3. (d)	4. (c)	5. (d)	6. (b)
7. (d)	8. (d)	9. (a)	10. (c)	11. (c)	12. (c)
13. (b)	14. (d)	15. (d)			

CHAPTER 2

DC Networks and Network Theorems

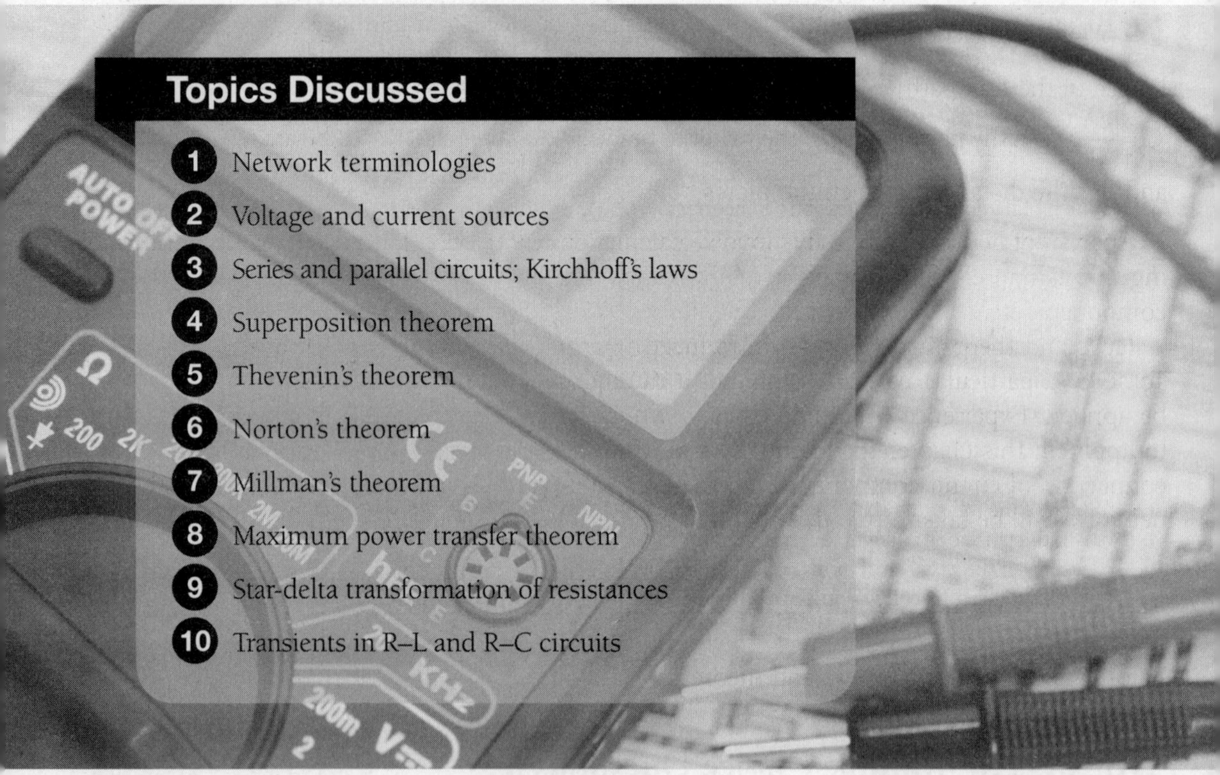

Topics Discussed

1. Network terminologies
2. Voltage and current sources
3. Series and parallel circuits; Kirchhoff's laws
4. Superposition theorem
5. Thevenin's theorem
6. Norton's theorem
7. Millman's theorem
8. Maximum power transfer theorem
9. Star-delta transformation of resistances
10. Transients in R–L and R–C circuits

2.1 ► Introduction

A network is an interconnection of elements, components, input signals, and output signals. The networks are of two types, viz active network and passive network. An active network contains one or more sources of supply whereas a passive network does not contain any source of supply, voltage or current.

There are three basic elements in a circuit or a network, viz resistor, capacitor, and inductor. Energy received by these components are either dissipated or stored in them. For example, energy dissipated in a resistor is I^2Rt where I is the current flowing, R is the resistance value, and t is the time. Energy stored in a capacitor is $\frac{1}{2}CV^2$ where C is the capacitance of the capacitor and V is the potential across it. For an inductor, the energy stored is $1/2\, L\, I^2$ where L is the inductance and I is the current flowing through it.

The formulas used to calculate the value of R, L, and C are:

$R = \rho\frac{\ell}{a}$; [ρ is the resistivity, ℓ is the length and *a* is the area of cross section of the wire]

$L = \frac{\mu N^2 A}{\ell}$; [μ is the permeability, N is the number of turns, A is the area of the coil ℓ the length of the flux path]

$C = \frac{\varepsilon A}{d}$; [ε is the permittivity of the material between the two plates, A is the area of each plate, and d is the distance between the plates.

Analysis of networks or circuits involve calculation with respect to finding out current flowing through an element, voltage across a component, power dissipated or stored in a circuit component, etc.

Laws and theorems have been introduced to make the task of network analysis simpler. To solve a particular network problem, a number of alternative methods or theorems can be applied. Experience will guide us as to which one will be the quickest or easiest method to apply. In this chapter the circuit laws and theorems, voltage sources, various methods of connection of circuit components and their transformations, etc., will be discussed. Only dc networks will be taken up in this chapter.

2.2 ► DC Network Terminologies, Voltage, and Current Sources

Before discussing various laws and theorems, certain terminologies related to dc networks are described first.

2.2.1 Network Terminologies

While discussing network theorems, laws, and electrical and electronic circuits, one often comes across the following terms.

i) **Circuit:** A conducting path through which an electric current either flows or is intended to flow is called a circuit.
ii) **Electric network:** A combination of various circuit elements, connected in any manner, is called an electric network.
iii) **Linear circuit:** The circuit whose parameters are constant, i.e., they do not change with application of voltage or current is called a linear circuit.
iv) **Non-linear circuit:** The circuit whose parameters change with the application of voltage or current is called a non-linear circuit.

v) **Circuit parameters**: The various elements of an electric circuit are called its parameters, like resistance, inductance, and capacitance.

vi) **Bilateral circuit:** A bilateral circuit is one whose properties or characteristics are the same in either direction. E.g., transmission line.

vii) **Unilateral circuit:** A unilateral circuit is one whose properties or characteristics change with the direction of its operation. E.g., diode rectifier.

viii) **Active network:** An active network is one which containts one or more sources of EMF.

ix) **Passive network:** A passive network is one which does not contain any source of EMF.

x) **Node:** A node is a junction in a circuit where two or more circuit elements are connected together.

xi) **Branch**: The part of a network which lies between two junctions is called a branch.

xii) **Loop:** A loop is a closed path in a network formed by a number of connected branches.

xiii) **Mesh:** Any path which contains no other paths within it is called a mesh. Thus, a loop contains meshes but a mesh does not contain a loop. Mesh is the smallest loop in a network.

xiv) **Lumped circuit:** The circuits in which circuit elements can be represented mutually independent and not interconnected.

For convenience, the nodes are labelled by latters.
For example in Fig. 2.1,

No. of nodes, N = 4 (i.e., A, B, C, D)
No. of branches, B = 5 (i.e., AB, BC, BD, CD, AD)
Independent meshes, M = B – N + 1
= 5 – 4 + 1 = 2 (i.e., ABDA, BCDB)

No. of loop = 3 (i.e., ABDA, BCDB and ABCDA). It is seen that a loop ABCDA encloses two meshes, i.e., mesh 1 and mesh 2.

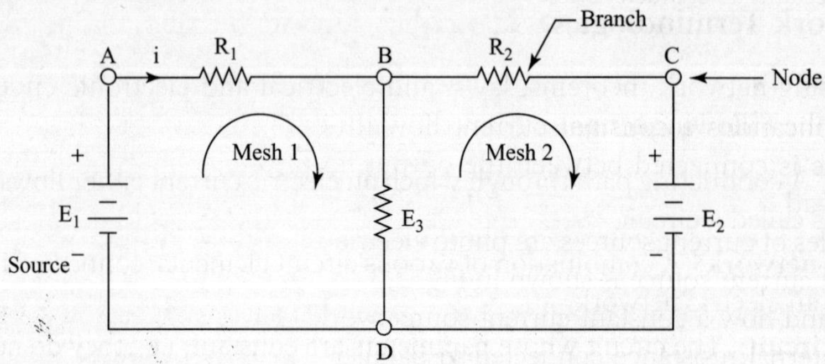

FIGURE 2.1

Different parts of an electric circuit

2.2.2 Voltage and Current Sources

A source is a device which converts mechanical, thermal, chemical or some other form of energy into electrical energy. There are two types of sources: voltage sources and current sources.

Voltage source

Voltage sources are further categorized as ideal voltage source and practical voltage source. Examples of voltage sources are batteries, dynamos, alternators, etc. Ideal voltage source is defined as the energy source which gives constant voltage across its terminals irrespective of current drawn through its terminals. The symbol of ideal voltage source is shown in Fig. 2.2 (a). In an ideal voltage source the terminal voltage is independent of the load resistance, R_L connected. Whatever is the voltage of the source, the same voltage is available across the load terminals of R_L, i.e., $V_L = V_S$ under loading condition as shown in Fig. 2.2 (b). There is no drop of voltage in the source supplying current to the load. The internal resistance of the source is therefore, zero.

In a practical voltage source, there will be a drop in voltage available across the load due to voltage drop in the resistance, R_{se} of the source itself when a load is connected as shown in Fig. 2.2 (c). If the value of source resistance R_{se} is very small, the voltage drop will be negligible.

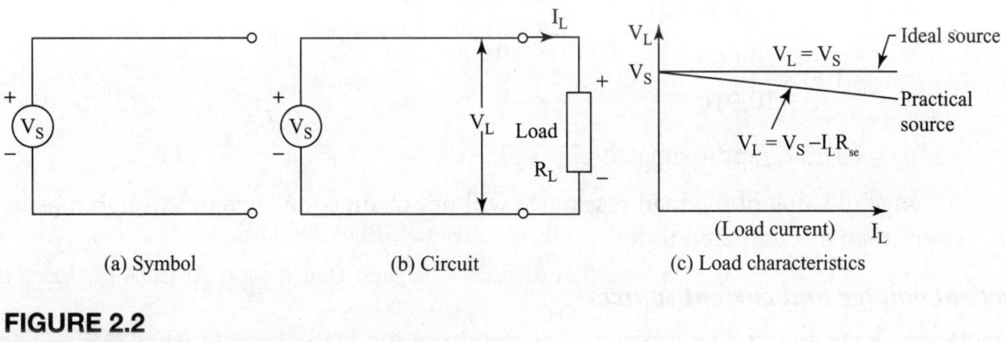

(a) Symbol (b) Circuit (c) Load characteristics

FIGURE 2.2

Voltage source and its characteristics

Current source

In certain applications a constant current flow through the circuit is required. When the load resistance is connected between the output terminals, a constant current I_L will flow through the load.

The examples of current sources are photo electric cells, collector current in transistors, etc. The symbol of current source is shown in Fig. 2.3.

To understand how a constant current source works, let us consider a current source with a very high internal resistance connected in parallel. The current source is supplying a load whose resistance is much less than the source internal resistance. As shown in Fig. 2.4. The current source of 100 A having a shunt resistance R_{sh} of 10000 ohms is supplying a load resistance R_L of 10 ohms.

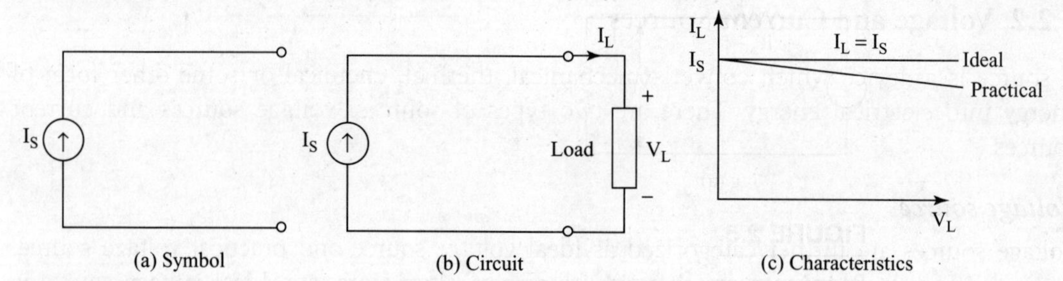

FIGURE 2.3

Current source and its characteristics

Using current division rule, we can calculate I_L as

$$I_L = I \times \frac{R_{sh}}{R_{sh} + R_L}$$

$$= 100 \times \frac{10,000}{10,000 + 10}$$

$$= 100 \times \frac{10,000}{10,010}$$

$$= 100 \text{ A (approximately)}$$

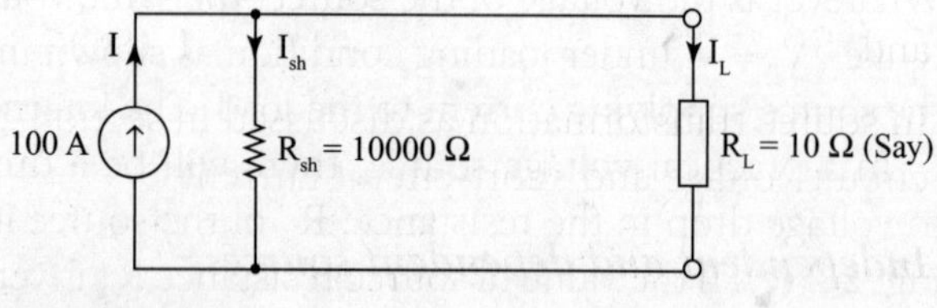

FIGURE 2.4

A current source supplying load

For any change of value of the load resistance will not change the current through it as long as R_{sh} is very high as compared to R_L.

Practical voltage and current sources

A practical voltage source like a battery has the drooping load characteristics due to some internal resistance. A voltage source has small internal resistance in series while a current source has some high-internal ***resistance*** in parallel.

For ideal voltage source $R_{se} = 0$

For ideal current source $R_{sh} = \infty$

A practical voltage source is shown as an ideal voltage source in series with a resistance. This resistance is called the internal resistance of the source as has been shown in Fig. 2.5 (a). A practical current source is shown as an ideal current source in parallel with its internal resistance as shown in Fig. 2.5 (b).

From Fig. 2.5 (a), we can write

V_L (open circuit), i.e., V_L (OC) = V_S that is, when the load R_L is removed, the circuit becomes an open circuit and the voltage across the source becomes the same as the voltage across the load terminals.

When the load is short circuited, the short-circuit current, I_L (SC) = V_S/R_{se}.

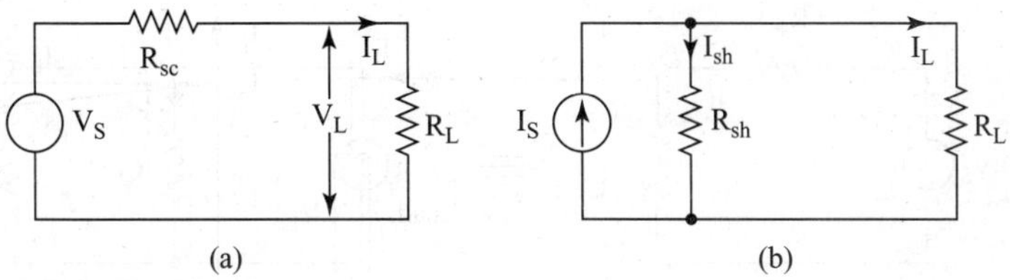

FIGURE 2.5

Representation as (a) practical voltage source (b) practical current source

In the same way, from Fig. 2.5 (b), we can write

$$V_L(OC) = I_{Sh} R_{Sh}$$

and

$$I_L\,(SC) = I_S$$

In source transformation as discussed in section 2.2.3, we shall use the equivalence of open-circuit voltage and short-circuit current.

Independent and dependent sources

The magnitude of an independent source does not depend upon the current in the circuit or voltage across any other element in the circuit. The magnitude of a dependent source gets changed due to some other current or voltage in the circuit. An independent source is represented by a circle while a dependent source is represented by a diamond-shaped symbol. Dependent voltage sources are also called controlled sources.

There are four kinds of dependent sources:

- voltage-controlled voltage source (vcvs)
- current-controlled current source (cccs)
- voltage-controlled current source (vccs)
- current-controlled voltage source (ccvs)

Dependent voltage sources find applications in electronic circuits and devices.

2.2.3 Source Transformation

A voltage source can be represented as a current source. Similarly a current source can be represented as a voltage source. This often helps the solutions of circuit problems.

Voltage source into current source and current source into voltage source

A voltage source is equivalent to a current source and a current source is equivalent to a voltage source. They should provide the same short-circuit current, and open-circuit voltage.

If voltage source is converted into current source as in Fig. 2.6, we will consider the short-circuit current equivalence. So, $\frac{V_s}{R_{se}} = I_s$ short-circuit current in the two equivalent circuits are, respectively, V_S/R_{Se} and I_Sf.

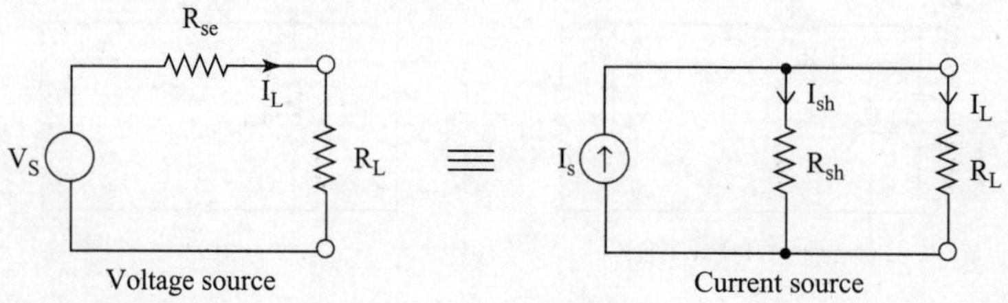

FIGURE 2.6

Voltage source converted into an equivalent current source

If a current source is converted into voltage source, as in Fig. 2.7, we consider the open-circuit voltage equivalence. So, $I_{sh}\,R_{sh} = V_s$.

A few examples will further clarify this concept.

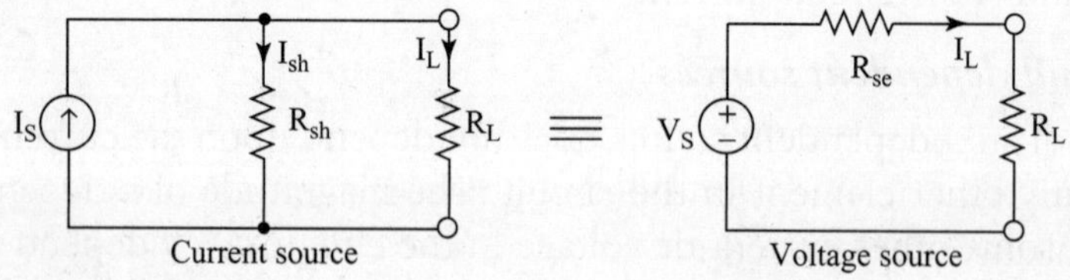

FIGURE 2.7

Equivalent voltage source

EXAMPLE 2.1

Convert a voltage source of 20 volts with internal resistance of 5 Ω into an equivalent current source.

SOLUTION

$$I_s = \frac{V_s}{R_{se}} = \frac{20}{5} = 4\text{ A}$$

The internal resistance of the current source will be the same as R_{se}

The condition for equivalence is checked from the following in the two equivalent circuits, v_{oc} should be same and I_{sc} should also be same.

In Fig. 2.8 (a), open-circuit voltage, $V_{oc} = 20$ V.

In Fig. 2.8 (b), open-circuit voltage, $V_{oc} = 4\text{ A} \times 5\ \Omega = 20$ V.

Short-circuit current, I_{sc} in Fig. 2.8 (a), is 4 A.

Short-circuit current, I_{sc} in Fig. 2.8 (b), is 4 A.

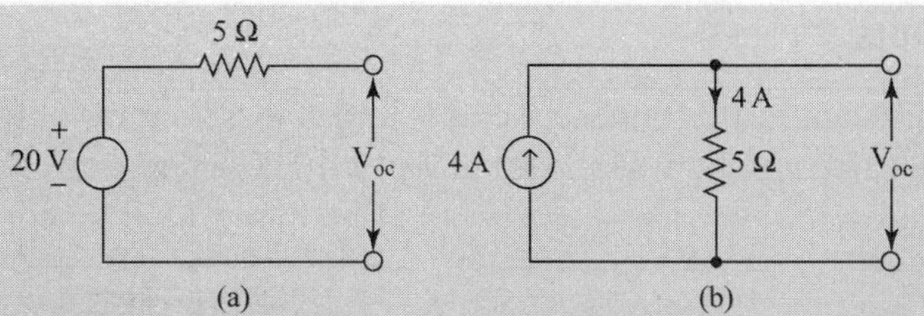

FIGURE 2.8

Conversion of a voltage source into a current source

This two circuits are equivalent because the open-circuit voltage and short-circuit current are the same in both the circuits.

EXAMPLE 2.2

Convert a current source of 100 A with internal resistance of 10 Ω into an equivalent voltage source.

SOLUTION

Here $I = 100\text{ A}, R_{sh} = 10\ \Omega$

For an equivalent voltage source

$$V = I \times R_{sh} = 100 \times 10 = 1000\text{ V}$$

$$R_{sh} = R_{se} = 10\ \Omega \text{ in series}$$

The open-circuit voltage and short-circuit current are the same in the two equivalent circuits as shown in Fig. 2.9 (a) and 2.9 (b), respectively. In Fig. 2.9 (a), V_{oc} is 100 A × 10 Ω = 100 V. In Fig. 2.9 (b), V_{oc} = 1000 V. Similarly in Fig. 2.9 (a), V_{sc} = 100 A and in Fig. 2.9 (b), V_{sc} = 100 A.

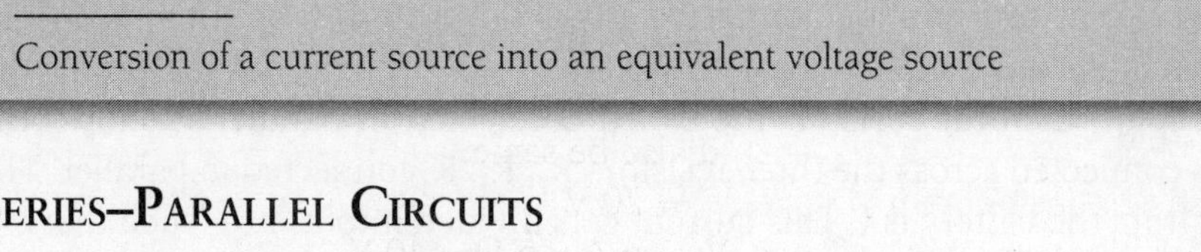

FIGURE 2.9

Conversion of a current source into an equivalent voltage source

2.3 ► SERIES–PARALLEL CIRCUITS

Resistances, capacitances, and inductances are often connected in series, in parallel, or a combination of series and parallel. We need to calculate the division of voltage and currents in such circuits.

2.3.1 Series Circuits

When a number of resistances are connected end to end across a source of supply, there will be only one path for the current to flow as shown in Fig. 2.10. The circuit is called a series circuit.

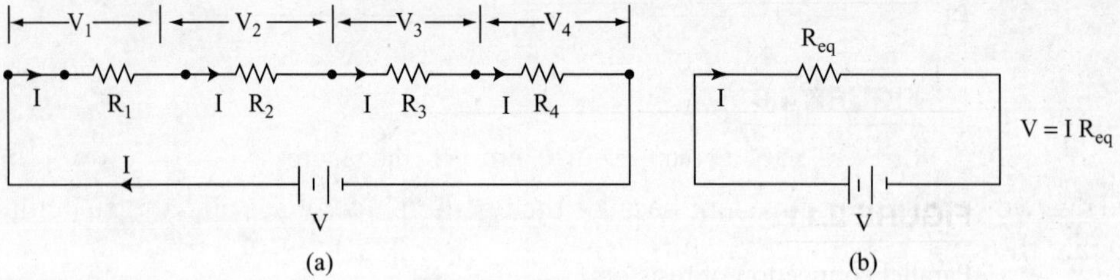

FIGURE 2.10

DC series circuit

The voltage drops across the resistances are V_1, V_2 ,V_3, and V_4, respectively. Since the same current is flowing through all the resistances, we can write

$$V_1 = IR_1,\ V_2 = IR_2,\ V_3 = IR_3,\ \text{and}\ V_4 = IR_4$$

Again, the total voltage, V applied is equal to the sum of the voltage drops across the resistances, Thus, we can write

$$V = V_1 + V_2 + V_3 + V_4$$

To find the value of equivalent resistance of a number of resistances connected in series, we equate the voltage, V of the two equivalent in units as shown in Fig. 2.10 (a) and Fig. 2.10 (b) as

or, $$I\,R_{eq} = IR_1 + IR_2 + IR_3 + IR_4$$

or, $$R_{eq} = R_1 + R_2 + R_3 + R_4$$

Assuming R_{eq} as equal to R, $R = R_1 + R_2 + R_3 + R_4$. (2.1)

Thus, when resistances are connected in series, the total equivalent resistance appearing across the supply can be taken as equal to the sum of the individual resistances.

2.3.2 Parallel Circuits

When a number of resistors are connected in such a way that both the ends of individual resistors are connected together and two terminals are brought out for connection to other parts of a circuit, then the resistors are called connected in parallel as shown in Fig. 2.11. Voltage V is connected across the three resistors R_1, R_2, R_3 connected in parallel. The total current drawn from the battery is I. This current gets divided into I_1, I_2, I_3 such that $I = I_1 + I_2 + I_3$. As voltage V is appearing across each of these three resistors, applying Ohm's law we write

$$I = I_1 + I_2 + I_3 = \frac{V}{R_1} + \frac{V}{R_2} + \frac{V}{R_3} \qquad \text{(i)}$$

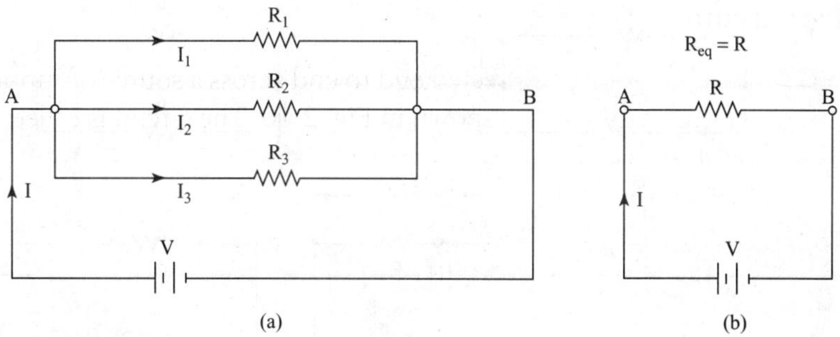

FIGURE 2.11

Parallel connection of resistors

Let the equivalent resistance of the three resistors connected in parallel across terminals A and B be R as shown in Fig. 2.11 (b). Then,

$$I = \frac{V}{R} \tag{ii}$$

From (i) and (ii),

$$\frac{V}{R} = \frac{V}{R_1} + \frac{V}{R_2} + \frac{V}{R_3}$$

or,

$$\frac{1}{R} = \frac{1}{R_1} + \frac{1}{R_2} + \frac{1}{R_3}$$

In general, if there are n resistors connected in parallel, the equivalent resistance R is expressed as

$$\frac{1}{R} = \frac{1}{R_1} + \frac{1}{R_2} + \frac{1}{R_3} + \cdots + \frac{1}{R_n} \tag{2.2}$$

2.3.3 Series–Parallel Circuits

Figure 2.12 shows a number of resistors connected in series–parallel combinations. Here, two parallel branches and one resistance, all connected in series have been shown. To determine the equivalent resistance across the end terminals of the entire circuit, we first calculate the equivalent resistance of parallel branches and then put them in series along with any individual resistance already connected in series.

$$\frac{1}{R_{AB}} = \frac{1}{R_1} + \frac{1}{R_2} = \frac{R_2 + R_1}{R_1 R_2}$$

or,

$$R_{AB} = \frac{R_1 R_2}{R_1 + R_2}$$

and

$$\frac{1}{R_{CD}} = \frac{1}{R_4} + \frac{1}{R_5}$$

$$R_{CD} = \frac{R_4 R_5}{R_4 + R_5}$$

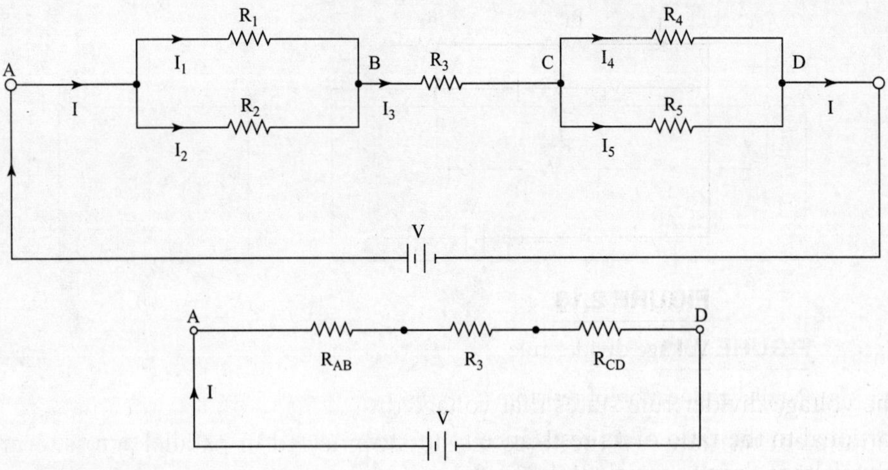

FIGURE 2.12

DC series–parallel circuit

Total resistance, R = Series combination of $R_{AB} + R_{BC} + R_{CD}$

$$R = \frac{R_1 R_2}{R_1 + R_2} + R_3 + \frac{R_4 R_5}{R_4 + R_5}$$

In any electrical circuit we will find a number of such resistances connected in series–parallel combinations.

2.4 ► Voltage and Current Divider Rules

2.4.1 Voltage Divider Rule

For easy calculation of voltage drop across resistors in a series circuit, a voltage divider rule is used which is illustrated in Fig. 2.13.

$$I = \frac{V}{R_1 + R_2 + R_3}$$

$$V_1 = IR_1 = \frac{V}{R_1 + R_2 + R_3} \times R_1 = \frac{V}{R_T} \times R_1$$

where $R_T = R(\text{Total}) = R_1 + R_2 + R_3$

Similarly,

$$V_2 = IR_2 = \frac{V}{R_T} \times R_2$$

and

$$V_3 = IR_3 = \frac{V}{R_T} \times R_3$$

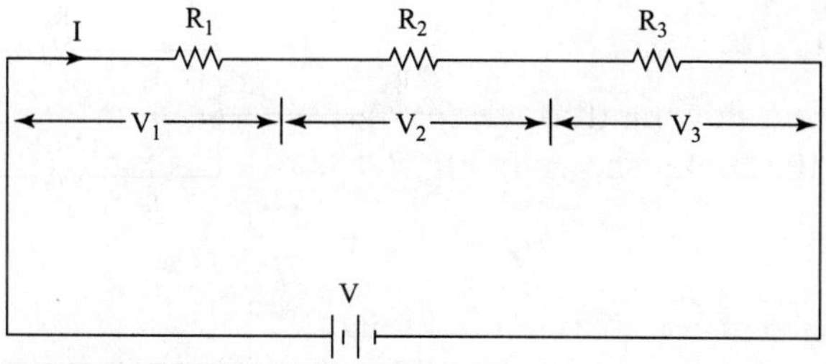

FIGURE 2.13

Voltage divider rule

Thus, the voltage divider rule states that voltage drop across any resistor in a series circuit is proportional to the ratio of its resistance to the total resistance of the series circuit. Alternately, the voltage drop is equal to circuit current multiplied by the resistance.

2.4.2 Current Divider Rule

Current divider rule is used in parallel circuits to find the branch currents if the total current is known. To illustrate, this rule is applied to two parallel branches as in Fig. 2.14.

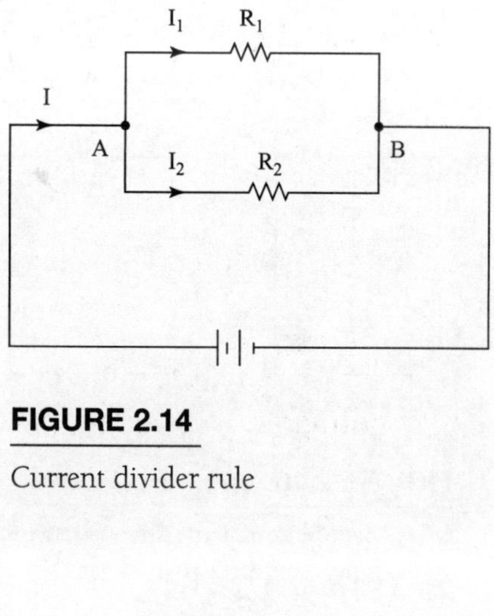

FIGURE 2.14

Current divider rule

$$V_{AB} = I_1R_1 = I_2R_2$$

and $$I = I_1 + I_2$$

$$I_1R_1 = (I - I_1)R_2$$

or, $$I_1(R_1 + R_2) = IR_2$$

or, $$I_1 = I\frac{R_2}{R_1 + R_2} \quad \text{(i)}$$

And, $$I_2 = I - I_1 = I - I\frac{R_2}{R_1 + R_2}$$

$$= I\left[1 - \frac{R_2}{R_1 + R_2}\right]$$

or, $$I_2 = I\frac{R_1}{R_1 + R_2} \quad \text{(ii)}$$

Thus, in a parallel circuit of two resistances, current through one branch is equal to line current multiplied by the ratio of resistance of the other branch divided by the total resistance as have been shown in expressions (i) and (ii).

EXAMPLE 2.3

Calculate the current flowing through the various resistances in the circuit shown in Fig. 2.15.

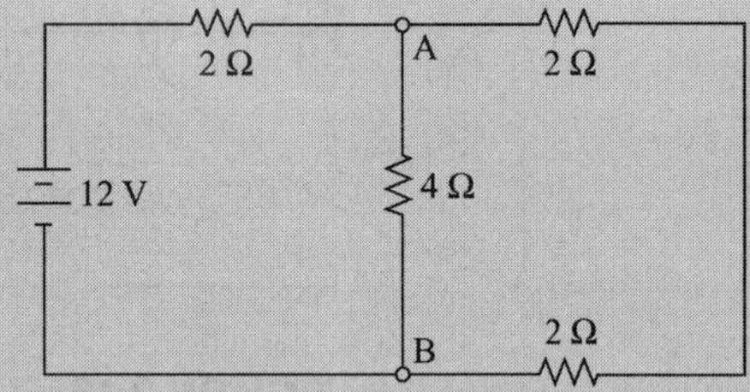

FIGURE 2.15

SOLUTION

The circuit is reduced to a simple circuit through the following step: across terminals A and B, the 4 Ω resistor is connected in parallel with two 2 Ω resistor in series. Thus, we have two 4 Ω resistors connected in parallel across terminal AB as has been shown in Fig. 2.16 (b). In Fig. 2.16 (c) is shown the equivalent to two 4 Ω resistances in parallel. In Fig. 2.16 (d) is shown the total resistance 4 Ω connected across the 12 V supply. The current is, therefore, 3 A. This 3 A will get divided equally in the two parallel branches as can be seen from Fig. 2.16 (b) and (a).

In Fig. 2.16 (d), current is $\dfrac{12\text{ V}}{4\,\Omega} = 3\text{ A}.$

In Fig. 2.16 (c) current is 3 A as it is a series circuit.

In Fig. 2.16 (b) current 3 A gets divided equally.

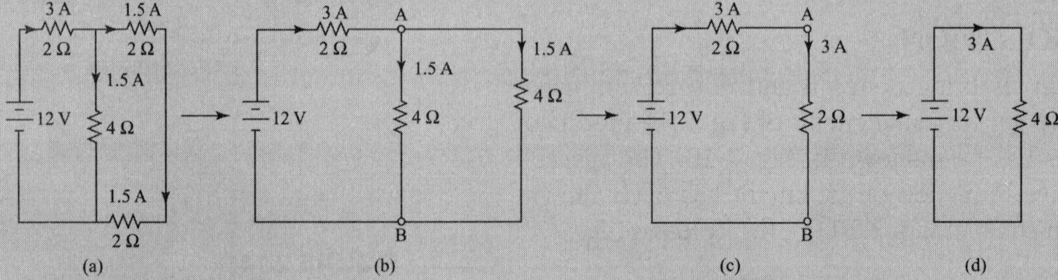

FIGURE 2.16

EXAMPLE 2.4

Calculate the current supplied by the battery in the network shown in Fig. 2.17.

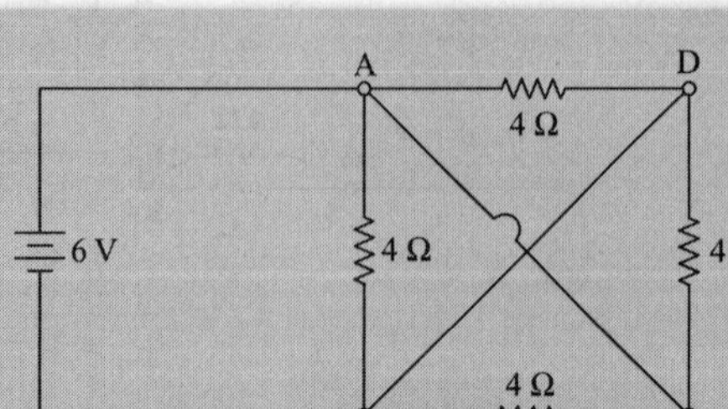

FIGURE 2.17

SOLUTION

Points A and C are joined together. Similarly points B and D are joined together. Thus, we can first bring point A and C together and the circuit will look like as has been shown in Fig. 2.18.

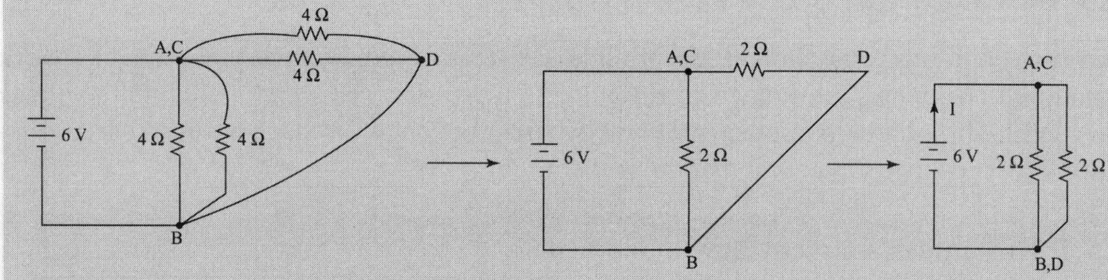

FIGURE 2.18

The circuit is further simplified by paralleling the two 4 Ω resistances between A and B and another two 4 Ω resistances between A and D. Now by bringing D and B together and paralleling the two 2 Ω resistors, the current I is calculate as 6 A.

EXAMPLE 2.5

Calculate the resistance between the terminals P and Q of the network shown in Fig. 2.19.

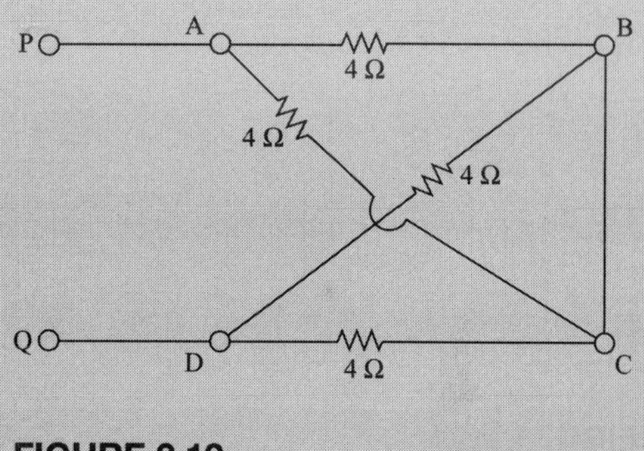

FIGURE 2.19

SOLUTION

Let us bring points B and C together. Then we get the same circuit of Fig. 2.19 modified as has been shown in Fig. 2.20. The successive reduction of the circuit has been shown in steps in Fig. 2.20 (a), (b), (c), and (d).

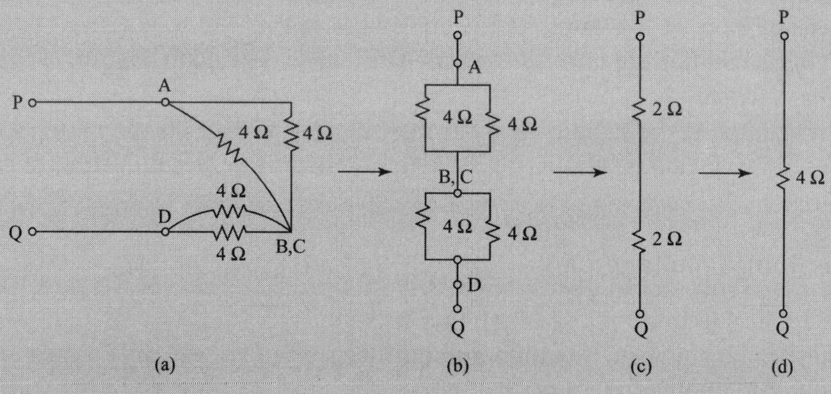

FIGURE 2.20

2.5 ▶ KIRCHHOFF'S LAWS

Two laws given by Gustav Robert Kirchhoff (1824–1887) are very useful in writing network equations. These laws are known as Kirchhoff's current law (KCL) and Kirchhoff's voltage law (KVL). These laws do not depend upon, whether the circuit is made of resistance, inductance or capacitance, or a combination of them.

2.5.1 Kirchhoff's Current Law

This law is applied at any node of an electric network. ***Kirchhoff's current law states that the algebraic sum of currents meeting at a junction or a node in a circuit is zero.*** KCL can be expressed mathematically as

$$\sum_{j=1}^{n} I_j = 0$$

where n is the number of branches meeting at a node and I_j represents the current in the jth branch as has been shown in Fig. 2.21.

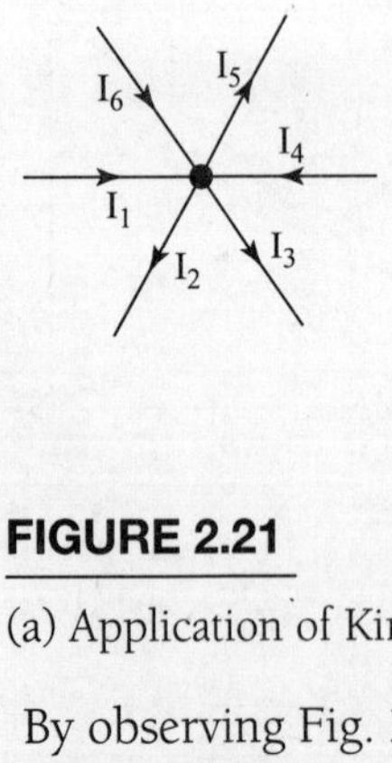

$$I_1 + I_6 + I_4 - I_5 - I_3 - I_2 = 0$$

or

$$I_1 + I_6 + I_4 = I_5 + I_3 + I_2$$

(a)

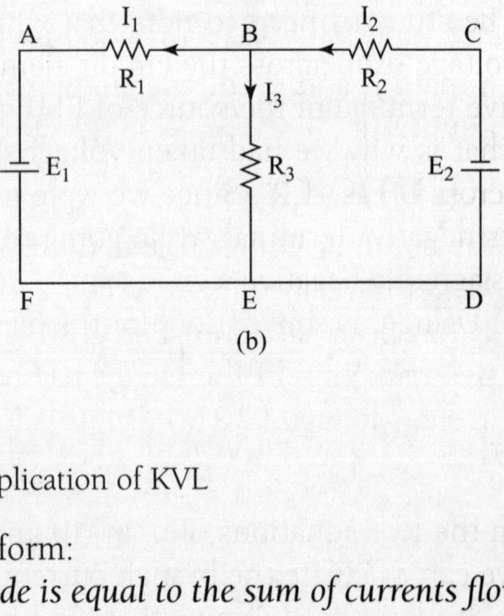

(b)

FIGURE 2.21

(a) Application of Kirchhoff's current law; (b) circuit for application of KVL

By observing Fig. 2.21, we can state KCL in another form:

The sum of current flowing towards a junction or a node is equal to the sum of currents flowing out of the junction.

The current entering the junction has been taken as positive while the currents leaving the junction have been taken as negative. That is to say there is no accumulation of current in a junction.

2.5.2 Kirchhoff's Voltage Law

This law is applicable to any closed loop in a circuit.

KVL states that at any instant of time the algebraic sum of voltages in a closed loop is zero.

In applying KVL in a loop or a mesh a proper sign must be assigned to the voltage drop in a branch and the source of voltage present in a mesh. For this, a positive sign may be assigned to the rise in voltage and a negative sign may be assigned to the fall or drop in voltage.

KVL can be expressed mathematically as

$$\sum_{j=1}^{n} V_j = 0$$

where V_j represents the voltages of all the branches in a mesh or a loop, i.e., in the jth element around the closed loop having n elements.

Let us apply KCL and KVL in a circuit shown in Fig. 2.22 (b). The current flowing through the branches have been shown.

Applying KCL at node B, we can write

$$I_1 + I_3 = I_2 \quad \text{(i)}$$

Now, let us apply KVL in mesh ABEFA and mesh CBEDC, respectively.

For the mesh ABEFA, starting from point A, the sum of voltage drops and voltage rise are equated to zero as

$$+ I_1 R_1 - I_3 R_3 - E_1 = 0$$

or,

$$I_1 R_1 - (I_2 - I_1) R_3 - E_1 = 0 \quad \text{(ii)}$$

The students need to note that while we move in the direction of the flow of current, the voltage drop across the circuit element is taken as negative. While we move from the negative terminal of the source of EMF to the positive terminal, the voltage is taken as positive. That is why we had taken voltage drop across the branch AB in Fig. 2.22 (b) as $+I_1R_1$ and across BE as $-I_3R_3$. Since we were moving from the positive terminal of the battery towards its negative terminal while going round the mesh we had considered it as voltage drop and assigned a negative sign.

Using this convention, for the mesh CBEDC, applying KVL we can write

$$-I_2 R_2 - I_3 R_3 - E_2 = 0$$

or,

$$-I_2 R_2 - (I_2 - I_1) R_3 - E_2 = 0 \quad [\because I_3 = (I_2 - I_1)] \quad \text{(iii)}$$

In the two equations, i.e., in (ii) and (iii), if the values of R_1, R_2, R_3, E_1, and E_2, are known, we can calculate the branch currents by solving these equations.

Students need to note that Kirchhoff's laws are applicable to both dc and ac circuits.

Let us apply KVL in a circuit consisting of a resistance, an inductance, and a capacitance connected across a voltage source as has been shown in Fig. 2.22.

We will equate the voltage rise with the voltage drops.

The voltage equation is $e = Ri + L\frac{di}{dt} + \frac{1}{C}\int idt$ (iv)

While solving network problems using Kirchhoff's laws we frame a number of simultaneous equations. These equations are solved to determine the currents in various branches in a circuit. We will discuss solving of simultaneous equations by the method of determinants or Cramer's Rule.

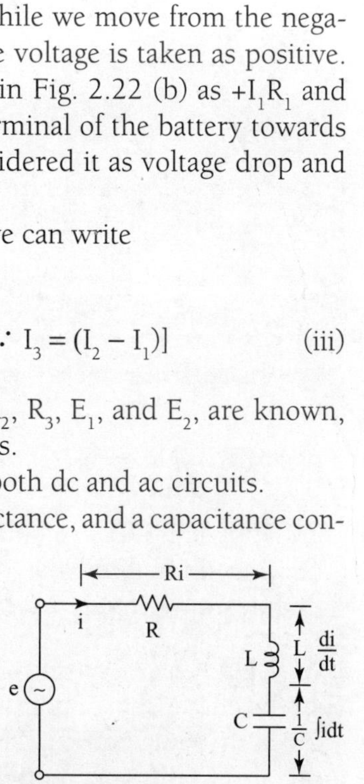

FIGURE 2.22

Application of KVL

EXAMPLE 2.6

By applying Kirchhoff's current law, determine the current flowing through all the branches in the circuit shown in Fig. 2.23.

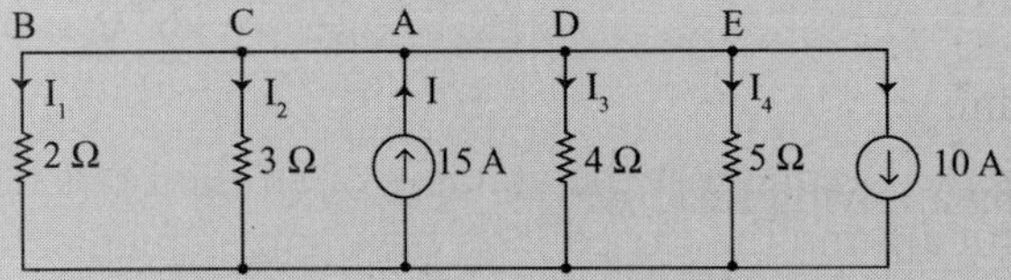

FIGURE 2.23

SOLUTION

By considering the node A, and by applying KCL, we take incoming currents as equal to outgoing currents so that

$$15\ A = I_1 + I_2 + I_3 + I_4 + 10\ A$$

or, $$I_1 + I_2 + I_3 + I_4 = 5\ A. \quad \text{(i)}$$

By assuming the potential at A as V, we note that potential at A is the same as at B, C, D, and E. So, we can find currents through the resistors by applying Ohm's law as

$$I_1 = \frac{V}{2};\ I_2 = \frac{V}{3};\ I_3 = \frac{V}{4};\ I_4 = \frac{V}{5}$$

Substituting these values of currents in (i),

$$\frac{V}{2} + \frac{V}{3} + \frac{V}{4} + \frac{V}{5} = 5$$

or, $$V(0.5 + 0.333 + 0.25 + 0.2) = 5$$

or, $$V = \frac{5}{1.283} = 3.9\ V.$$

so, $I_1 = \frac{3.9}{2} = 1.95\ A;\ I_2 = \frac{3.9}{3} = 1.3\ A;\ I_3 = \frac{3.9}{4} = 0.97\ A$ and $I_4 = \frac{3.9}{5} = 0.78\ A$

EXAMPLE 2.7

By applying KCL, calculate the current flowing through all the branches in the circuit shown in Fig. 2.24.

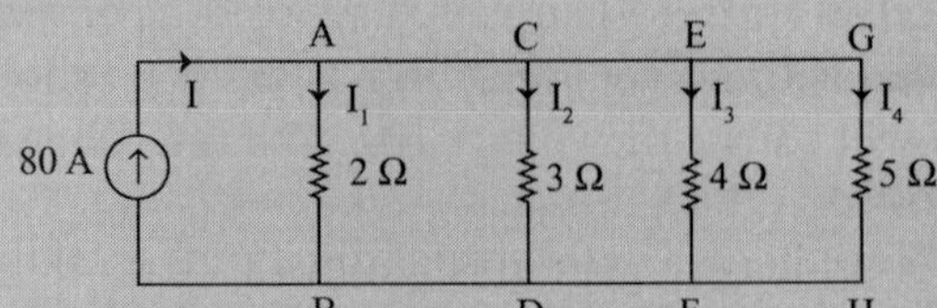

FIGURE 2.24

SOLUTION

In the circuit shown points A, C, E, and G are at the same potential. Points B, D, F, H are also at the same potential. Therefore, potential difference between points A and B is the same as between C and D; E and F; and G and H.

Therefore, $I_1 = \frac{V_{AB}}{2}; I_2 = \frac{V_{CD}}{3}; I_3 = \frac{V_{EF}}{4};$ and $I_4 = \frac{V_{GH}}{5}$

and $V_{AB} = V_{CD} = V_{EF} = V_{GH} = V$ (say)

By applying KCL, we get

$$I = I_1 + I_2 + I_3 + I_4$$

On substituting the values, we get

$$80 = \frac{V}{2} + \frac{V}{3} + \frac{V}{4} + \frac{V}{5} = V\left(\frac{1}{2} + \frac{1}{3} + \frac{1}{4} + \frac{1}{5}\right)$$

or, $80 = V(0.5 + 0.33 + 0.25 + 0.2)$

$$80 = 1.283\ V$$

or, $V = \frac{80}{1.283} = 62.35$ volts.

Now, $I_1 = \frac{62.35}{2} = 31.17\,A$

$$I_2 = \frac{62.35}{3} = 20.77\,A$$

$$I_3 = \frac{62.35}{4} = 15.58\,A$$

$$I_4 = \frac{62.35}{5} = 12.48\,A$$

EXAMPLE 2.8

Calculate the voltage, V_s across the open switch in the circuit shown in Fig. 2.25.

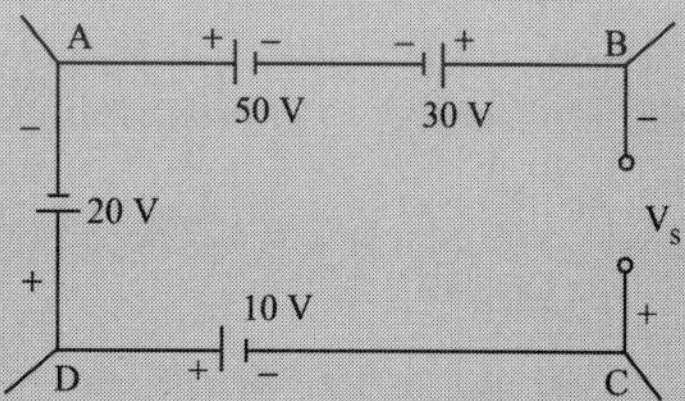

FIGURE 2.25

SOLUTION

By applying KVL, starting from point A in the clockwise direction and considering the polarity of the source, we get the following:

$$-50 + 30 + V_s + 10 - 20 = 0$$

or, $V_s = 30$ V.

EXAMPLE 2.9

Calculate the voltage, V across the resistor R_1 in the circuit shown in Fig. 2.26.

SOLUTION

By applying KVL in the loop ABCDEFA,

$$-16 \times 3 - 2 \times 4 + 40 - V = 0$$

or, $V = -16$ V.

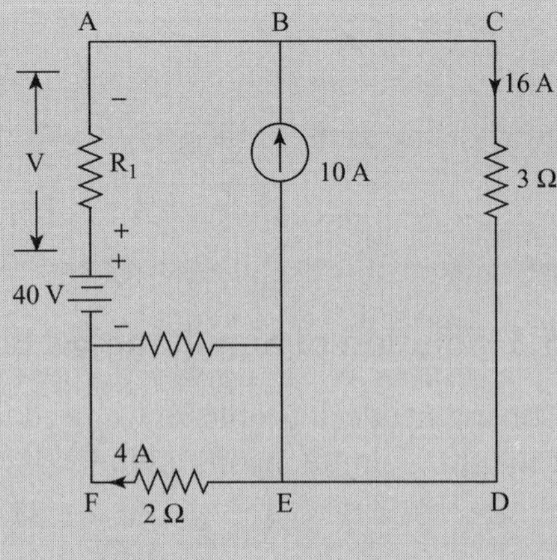

FIGURE 2.26

EXAMPLE 2.10

Find the voltage difference between points C and E in the circuit shown in Fig. 2.27.

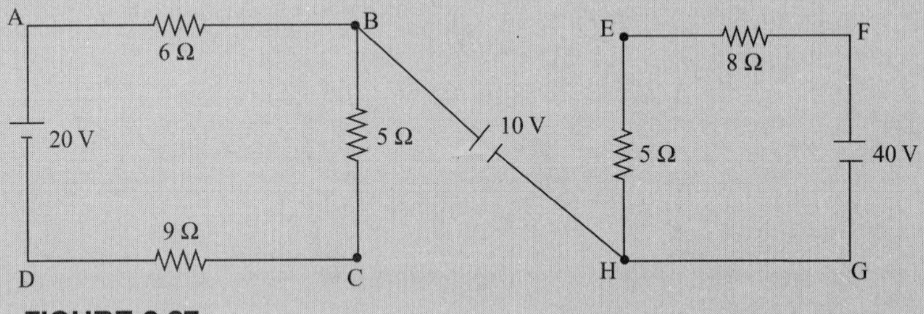

FIGURE 2.27

SOLUTION

The given circuit can be redrawn as shown in Fig. 2.28. Let I_1 and I_2 be the loop currents as shown. By applying KVL in loop CDABC, we get

$$-9\, I_1 + 20 - 6\, I_1 - 5\, I_1 = 0$$

or, $I_1 = 1$ A

By applying KVL in loop FEHGF, we get

$$-8\, I_2 - 5\, I_2 + 40 = 0$$

or, $I_2 = \dfrac{40}{13} = 3.07$ A

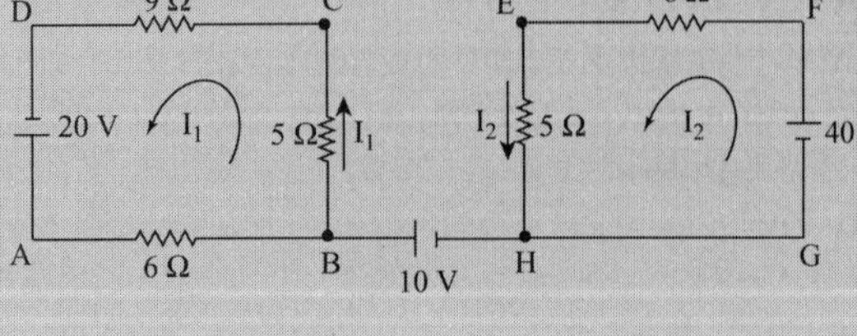

FIGURE 2.28

Now, by considering voltage drop and voltage rise between points C and E we calculate V_{CE} as

$$\begin{aligned} V_{CE} &= +5\,I_1 - 10 + 5\,I_2 \\ &= 5 \times 1 - 10 + 5 \times 3.07 \\ &= 5 - 10 + 15.35 \\ &= 10.35 \text{ V} \end{aligned}$$

2.5.3 Solution of Simultaneous Equations Using Cramer's Rule

In solving network problems we need to solve a number of simultaneous equations. Let the three simultaneous equations written for a network problem be of the form

$$\begin{aligned} a_1x + b_1y + c_1z &= d_1 \\ a_2x + b_2y + c_2z &= d_2 \\ a_3x + b_3y + c_3z &= d_3 \end{aligned}$$

Where x, y, z are the three variables.
We can determine the values of x, y, and z using Cramer's rule as

$$x = \frac{\Delta_a}{\Delta} \quad \text{Where, } \Delta = \begin{vmatrix} a_1 & b_1 & c_1 \\ a_2 & b_2 & c_2 \\ a_3 & b_3 & c_3 \end{vmatrix}$$

and

$$\Delta_a = \begin{vmatrix} d_1 & b_1 & c_1 \\ d_2 & b_2 & c_2 \\ d_3 & b_3 & c_3 \end{vmatrix}.$$

Similarly,

$$y = \frac{\Delta_b}{\Delta} \quad \text{Where, } \Delta_b = \begin{vmatrix} a_1 & d_1 & c_1 \\ a_2 & d_2 & c_2 \\ a_3 & d_3 & c_3 \end{vmatrix}$$

and

$$z = \frac{\Delta_c}{\Delta}$$

and

$$\Delta_c = \begin{vmatrix} a_1 & b_1 & d_1 \\ a_2 & b_2 & d_2 \\ a_3 & b_3 & d_3 \end{vmatrix}$$

The determinants expressed as Δ, Δ_a, Δ_b, and Δ_c are to be evaluated so as to find x, y, and z.

2.5.4 Method of Evaluating Determinant

i) Let us write two simultaneous equations as

$$a_1x + b_1y = m$$
$$a_2x + b_2y = n$$

Where x and y are the variables.
The common determinant Δ is evaluated as

$$\Delta = \begin{vmatrix} a_1 & b_1 \\ a_2 & b_2 \end{vmatrix} = a_1b_2 - a_2b_1$$

The determinant for x, i.e., Δ_a is evaluated as

$$\Delta_a = \begin{vmatrix} m & b_1 \\ n & b_2 \end{vmatrix} = mb_2 - nb_1$$

The determinant for y, i.e., Δ_b is evaluated as

$$\Delta_b = \begin{vmatrix} a_1 & m \\ a_2 & n \end{vmatrix} = a_1n - a_2m$$

$$x = \frac{\Delta_a}{\Delta} \text{ and } y = \frac{\Delta_b}{\Delta}$$

ii) If there are more than two simultaneous equations, say three, the method of evaluating the determinants would be as illustrated below. Let the common determinant be

$$\Delta = \begin{vmatrix} 5 & 100 & 10 \\ 7 & -50 & -2 \\ 3 & -50 & -3 \end{vmatrix}$$

The procedure followed is like this. Select any row and the first column. Multiply each element in the row or the column by its minor and by + sign or – sign and then add the product. The multiplication by + sign or – sign is decided by a factor $(-1)^{j+k}$ where the minor of the element is appearing in row j and column k. Therefore, the determinant Δ given above is calculated as

$$\Delta = (5)\times\begin{vmatrix} -50 & -2 \\ -50 & -3 \end{vmatrix} - (100)\times\begin{vmatrix} 7 & -2 \\ 3 & -3 \end{vmatrix} + (10)\begin{vmatrix} 7 & -50 \\ 3 & -50 \end{vmatrix}$$
$$= 5(150 - 100) - 100(-21 + 6) + 10(-350 + 150)$$
$$= 5\times 50 - 100(-15) + 10(-200)$$
$$= 250 + 1500 - 2000$$
$$= -250$$

It can be observed that multiplication of the minor is done alternately as positive and negative starting from the first row or the first column. Now we will take up a few numerical problems to calculate branch currents in electric networks.

EXAMPLE 2.11

Use KCL and KVL to calculate the branch currents in the circuit shown in Fig. 2.29.

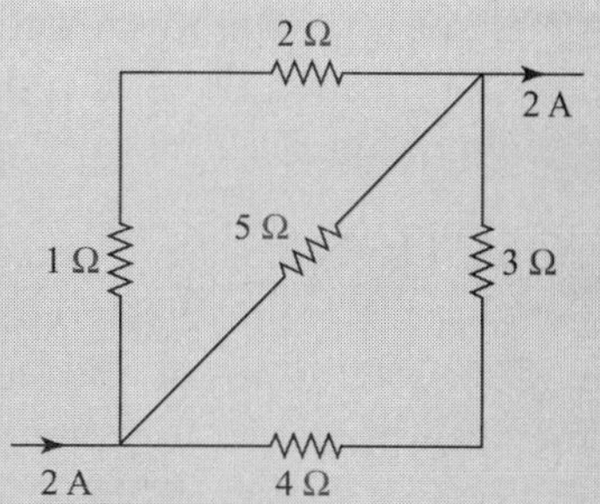

FIGURE 2.29

SOLUTION

We first indicate the branch currents applying KCL as shown in Fig. 2.30.

We have assumed I_1 and I_2 flowing respectively through branches AD and AC. Since 2 A is entering the node A, $2 - (I_1 + I_2)$ must be coming out through the branch AB. Similarly, applying KCL at node C we see that the sum of currents entering the node is equal to the current coming out of the node. Thus, current distribution in the various branches is perfectly done.

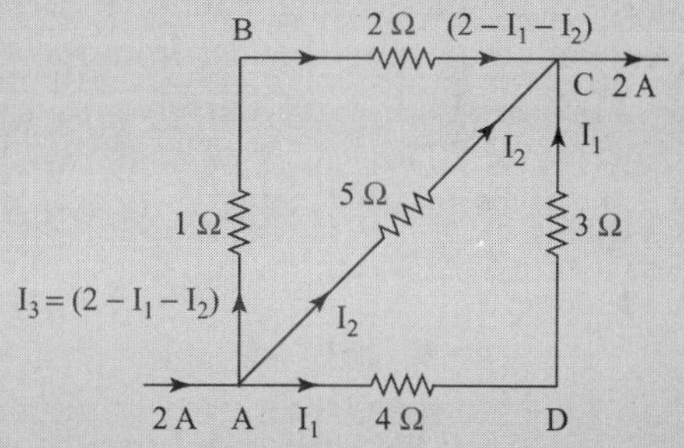

FIGURE 2.30

Now, we will apply KVL to the loop ABCA and loop ACDA.

From loop ABCA we can write the voltage equation as

$$-1(2 - I_1 - I_2) - 2(2 - I_1 - I_2) + 5\,I_2 = 0$$

or, $$3I_1 + 8I_2 = 6 \quad \text{(i)}$$

From loop ACDA we can write

$$-5I_2 + 3I_1 + 4I_1 = 0$$

or, $$7I_1 - 5I_2 = 0 \quad \text{(ii)}$$

To solve eqs. (i) and (ii), multiply eq. (i) by 7 and (ii) by 3 and subtract as

$$21I_1 + 56I_2 = 42 \quad \text{(i)}$$

$$21I_1 - 15I_2 = 0 \quad \text{(ii)}$$

From which

$$I_2 = \frac{42}{71}\,\text{A}$$

$$I_1 = \frac{30}{71}\,\text{A}$$

and

$$I_3 = 2 - (I_1 + I_2) = 2 - \left(\frac{30}{71} + \frac{42}{71}\right)$$

$$= \frac{70}{71}\,\text{A}$$

sum of $$I_1 + I_2 + I_3 = 2\ \text{A}$$

EXAMPLE 2.12

Calculate applying Kirchhoff's laws the current flowing through the 8 Ω resistor in the circuit shown in Fig. 2.31.

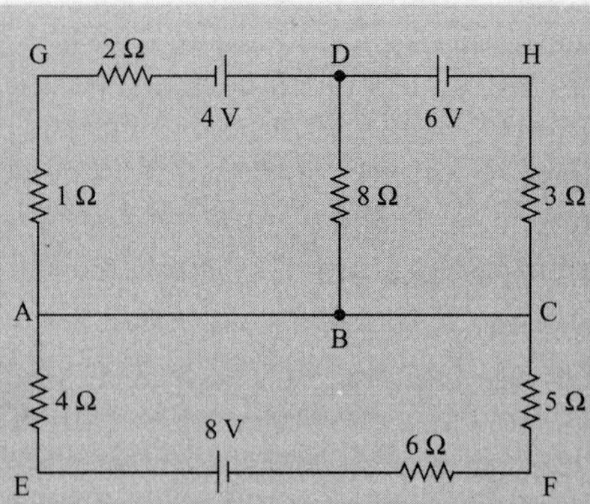

FIGURE 2.31

SOLUTION

By observing the given circuit, we see that nodes A, B, C are at the same potential and they can be joined together so that the circuit will be like shown in Fig. 2.32.

In the loop EA FE, current I_1 will flow. No current from this loop will flow to the other two loops. Current flowing from E to A is to be the same as the current flowing from A to F.

The distribution of currents in loop GDAG and HDAH have been shown. By applying KVL in these loops, we write:

for loop GDAG

$$-2I_2 + 4 - 8(I_2 + I_3) - I_2 = 0$$

or, $$11I_2 + 8I_3 = 4 \qquad \text{(i)}$$

for loop HDAH

$$6 - 8\,(I_2 + I_3) - 3\,I_3 = 0 \qquad \text{(ii)}$$

or, $$8I_2 + 11I_3 = 6$$

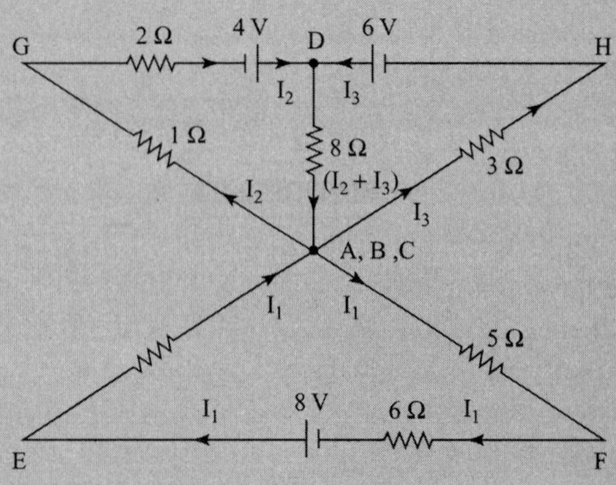

FIGURE 2.32

Solving eqs. (i) and (ii)

$$I_3 = 0.6\,A \text{ and } I_2 = -0.07\,A$$

Current through the 8 Ω resistor = $I_2 + I_3$

$= -0.07 + 0.6$

$= 0.53\,A$

So far, we have assumed branch currents in a network and applying KVL written the voltage equations. From the loop equations, we have calculated the branch currents. Two other methods, namely Maxwell's mesh current method and Node voltage method, are described in the following sections.

2.6 ► Maxwell's Mesh Current Method

A mesh is a smallest loop in a network. KVL is applied to each mesh in terms of mesh currents instead of branch currents. As a convention, mesh currents are assumed to be flowing in the clockwise direction without branching out at the junctions. Applying KVL, the voltage equations are framed. By knowing the mesh currents, the branch currents can be determined. The procedure followed is explained through an example. Let us calculate the current flowing through the branches in the circuit given in Fig. 2.33.

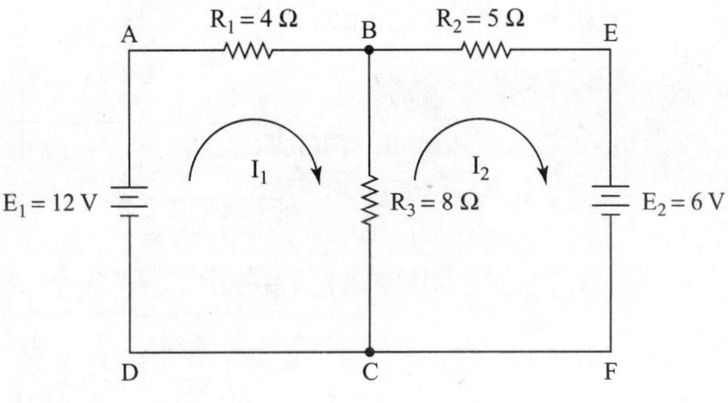

FIGURE 2.33

We have assumed loop currents I_1 and I_2 flowing in the clockwise direction as shown.

It may be noted that current flowing through the resistor R_3 is the algebraic sum of the two currents I_1 and I_2. Here I_1 is flowing in the downward direction while I_2 is flowing in the upward direction.

We will now write the voltage equations for the two loops applying KVL and then solve the equations. If the value of any mesh currents is calculated as negative, we will take the direction of that mesh current opposite to the assumed clockwise direction.

For loop DABCD, the voltage equation is

$$12 - 4\,I_1 - 8(I_1 - I_2) = 0$$

or, $$12 - 12\,I_1 + 8\,I_2 = 0$$

or, $$3\,I_1 - 2I_2 = 3 \qquad \text{(i)}$$

For loop BEFCB, the voltage equation is

$$-5\,I_2 - 6 - (I_2 - I_1)\,8 = 0$$

or, $$8\,I_1 - 13\,I_2 = 6 \qquad \text{(ii)}$$

solving eqs. (i) and (ii), we get

$$I_1 = 1.17\text{ A},\ I_2 = 0.26\text{ A}$$

and current flowing through R_3 is $(I_1 - I_2) = 0.91$ A

I_1 is flowing through R_1, I_2 is flowing through R_2 and $(I_1 - I_2)$ is flowing through R_3.

EXAMPLE 2.13

Using the mesh current method, calculate the current flowing through the resistors in the circuit shown in Fig. 2.34.

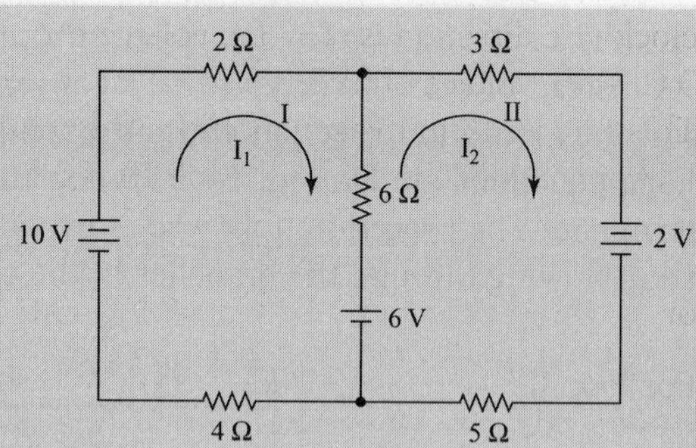

FIGURE 2.34

SOLUTION

Applying KVL in mesh I,

$$10 - 2\,I_1 - 6\,(I_1 - I_2) - 6 - 4\,I_1 = 0$$

or, $$6\,I_1 + 3\,I_2 = 2 \qquad \text{(i)}$$

Applying KVL in mesh II,

$$-3I_2 - 2 - 5I_2 + 6 - 6\,(I_2 - I_1) = 0$$

or, $$-6I_1 + 14\,I_2 = 4 \qquad \text{(ii)}$$

Adding eqs. (i) and (ii)

$$11\,I_2 = 6$$

or, $$I_2 = \frac{6}{11}\,\text{A}$$

and $$I_1 = \frac{2+3I_2}{6} = \frac{2+3\times 6/11}{6} = \frac{20}{33}\,\text{A}$$

Current through the 6 Ω resistor is $(I_1 - I_2)$ which is equal to $\left(\frac{20}{33} - \frac{6}{11}\right) = \frac{2}{33}\,\text{A}$.

EXAMPLE 2.14

A network with three meshes has been shown in Fig. 2.35. Applying Maxwell's mesh current method determine the value of the unknown voltage, V for which the mesh current, I_1 will be zero.

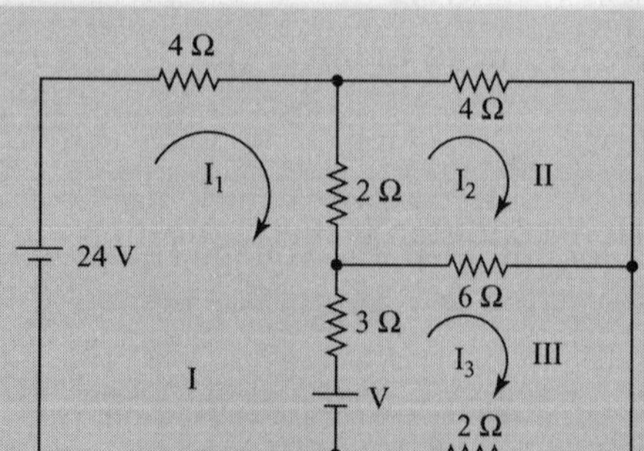

FIGURE 2.35

SOLUTION

In applying KVL, we follow sign convention as:

While completing a loop in the assumed direction of current flow, voltage drops are taken as –ve and voltage rise while going from –ve terminal of the source to the positive terminal as +ve.

In this case, in loop I, voltage drop 4 I_1 is –ve. In the resistor 2 Ω, current flow in the clockwise direction is $(I_1 - I_2)$, voltage drop is $-2(I_1 - I_2)$; voltage drop in 3 Ω resistor is $3(I_1 - I_3)$, voltage of battery V is taken as –ve and voltage of 24 V battery is taken as +ve. This way we form the equations for the other loops.

Applying KVL in loop I, II, and III respectively, we get

$$-4I_1 - 2(I_1 - I_2) - 3(I_1 - I_3) - V + 24 = 0$$

or, $$9I_1 - 2I_2 - 3I_3 = 24 - V \qquad \text{(i)}$$

For loop II, $$-4I_2 - 6(I_2 - I_3) - 2(I_2 - I_1) = 0$$

or, $$I_1 - 6I_2 + 3I_3 = 0 \qquad \text{(ii)}$$

For loop III, $$-6(I_3 - I_2) - 2I_3 + V - 3(I_3 - I_1) = 0$$

or, $$3I_1 + 6I_2 - 11I_3 = -V \qquad \text{(iii)}$$

The three equations are:

$$9I_1 - 2I_2 - 3I_3 = 24 - V \qquad \text{(i)}$$

$$I_1 - 6I_2 + 3I_3 = 0 \qquad \text{(ii)}$$

$$3I_1 + 6I_2 - 11I_3 = -V \qquad \text{(iii)}$$

From (i), (ii), and (iii) the determinants Δ and Δ_a or Δ_1 are

$$\Delta = \begin{vmatrix} 9 & -2 & -3 \\ 1 & -6 & 3 \\ 3 & 6 & -11 \end{vmatrix}$$

$$\Delta_1 = \begin{vmatrix} 24 - V & -2 & -3 \\ 0 & -6 & 3 \\ -v & 6 & -11 \end{vmatrix}$$

According to Cramer's rule,

$$I_1 = \frac{\Delta_1}{\Delta},\ I_2 = \frac{\Delta_2}{\Delta}\ I_3 = \frac{\Delta_3}{\Delta}$$

Here condition is that I_1 must be zero.

$$I_1 = \frac{\Delta_1}{\Delta} = 0. \text{ So, } \Delta_1 \text{ must be zero}$$

Thus, we equate Δ_1 to zero.

$$\begin{vmatrix} 24 - V & -2 & -3 \\ 0 & -6 & 3 \\ -v & 6 & -11 \end{vmatrix} = 0$$

or, $$(24-v)\begin{vmatrix}-6 & 3\\ 6 & -11\end{vmatrix}-(-2)\begin{vmatrix}0 & 3\\ -V & -11\end{vmatrix}-3\begin{vmatrix}0 & -6\\ -V & 6\end{vmatrix}=0$$

$$(24 - V)\,[66 - 18] + 2\,[0 - (-3V)] - 3[0 - 6V] = 0$$

or, $$(24 - V)\,48 + 6V + 18V = 0$$

or, $$24 \times 48 - 48V + 24V = 0$$

or, $$24V = 24 \times 48$$

or, $$V = 48 \text{ Volts}$$

2.7 ► Nodal Voltage Method (Nodal Analysis)

In the nodal analysis method a reference node in the network is chosen. Then the unknown voltages at the other nodes are determined with respect to the reference node. After determining the node voltages, currents in all branches are calculated. This method of circuit analysis is suitable where a network has a number of loops, for which a large number of simultaneous equations are to be solved which is some what a lengthy process. The procedure for the node voltage method is explained through an example.

EXAMPLE 2.15

For the circuit shown in Fig. 2.36 determine the voltages at nodes B and C and calculate the current through the 8 Ω resistor.

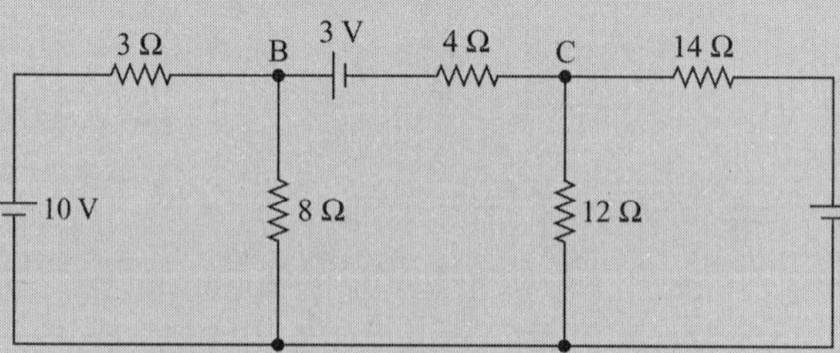

FIGURE 2.36

SOLUTION

We will take one reference node at zero potential. Generally the node at which maximum branches are meeting is taken as the reference node. Let G is the reference node as shown in Fig. 2.37. The reference node will be called ground node or zero potential node.

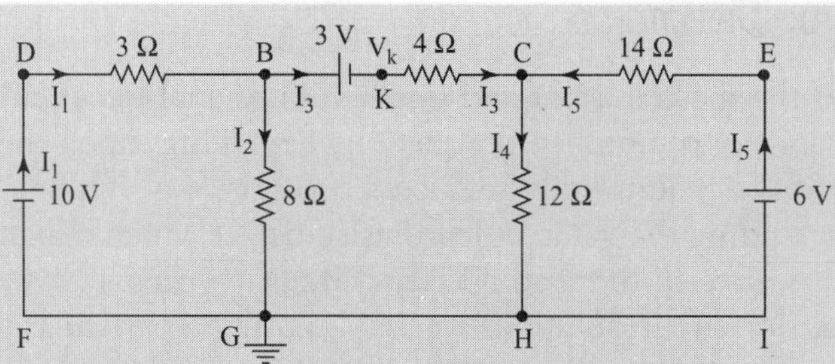

FIGURE 2.37

Points F, G, R, H, I are at zero reference potential. Let us now assign potential at all nodes with respect to the reference node. Let V_D, V_B, V_C, V_E are the potentials at points D, B, C, and E, respectively. Let us also assume unknown currents I_1, I_2, I_3, I_4, and I_5 flowing through the branches.

Applying Ohm's law currents I_1, I_2, I_3, I_4, and I_5 are expressed as

$$\frac{V_D - V_B}{3} = I_1; \ \frac{V_B}{8} = I_2; \quad I_4 = \frac{V_c}{12}; \quad I_5 = \frac{V_c - V_E}{14}; \quad V_E = 6; \quad V_D = 10.$$

or, $$\frac{10 - V_B}{3} = I_1$$

To find I_3, we assume potential at point k as V_K. We can write, $V_K + 3 = V_B$

and $$I_3 = \frac{V_K - V_C}{4} = \frac{V_B - 3 - V_C}{4}$$

Applying KCL at node B,

$$I_1 = I_2 + I_3$$

or, $$\frac{10 - V_B}{3} = \frac{V_B}{8} + \frac{V_B - V_C - 3}{4}$$

or, $$17\ V_B - 6\ V_C = 98 \tag{i}$$

Applying KCL at node C,

$$I_4 = I_3 + I_5$$

or, $$\frac{V_C}{12} = \frac{6 - V_C}{14} + \frac{V_B - V_C - 3}{4}$$

or, $$21\ V_B - 34\ V_C = 27 \tag{ii}$$

Solving eqs. (i) and (ii), we get

$$V_B = 7.01\ \text{V};\ V_C = 3.537\ \text{V}$$

and current in 8 Ω resistor, $$I_2 = \frac{V_B}{8} = \frac{7.01}{8} = 0.88\ \text{A}$$

More problems using this method have been solved separately, at a later stage.

2.8 ► NETWORK THEOREMS

We described earlier the mesh current and nodal voltage analysis of circuit problems. The procedure involves solving of number of equations depending upon the complexity of the network. Many networks require only restricted analysis, e.g., finding current through a particular resistor or finding the value of load resistance at which maximum power will be transferred from the source to the load. Certain circuit theorems have been developed to solve such problems. For circuit solutions we will be using a particular theorem or method depending upon which method will require less time in calculations. The circuit theorems being discussed in this chapter are as follows:

1. Superposition theorem
2. Thevenin's theorem
3. Norton's theorem
4. Millman's theorem
5. Reciprocity theorem
6. Maximum power transfer theorem
7. Tellegen's theorem

In addition, circuit simplification using the star-delta transformation method has also been discussed with plenty of examples.

2.8.1 Superposition Theorem

An electrical circuit may contain more than one source of supply. The sources of supply may be a voltage source or a current source. In solving of circuit problems having multiple sources of supply, the effect of each source is calculated separately and the combined effect of all the sources are taken into consideration. While considering the effect of any one source, the other voltage sources are kept short circuited and current sources are kept open circuited. This is the essence of the superposition theorem.

The superposition theorem states that in a linear network containing more than one source, the current flowing in any branch is the algebraic sum of currents that would have been produced by each source taken separately, with all the other sources replaced by their respective internal resistances. In case the internal resistance of a source is not provided, the voltage sources will be short circuited and current sources will be open circuited.

The procedure for solving circuit problems using the above stated superposition theorem is illustrated through a few examples.

EXAMPLE 2.16

Using the superposition theorem, find the value of current, I_{BD} in the circuit shown in Fig. 2.38.

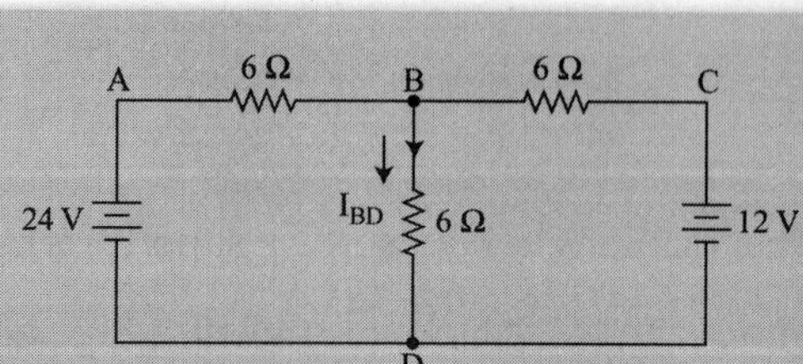

FIGURE 2.38

SOLUTION

We shall consider each source separately and calculate the current flowing through the branch BD. First the 24 V source is taken by short circuiting the 12 V source as shown in Fig. 2.39.

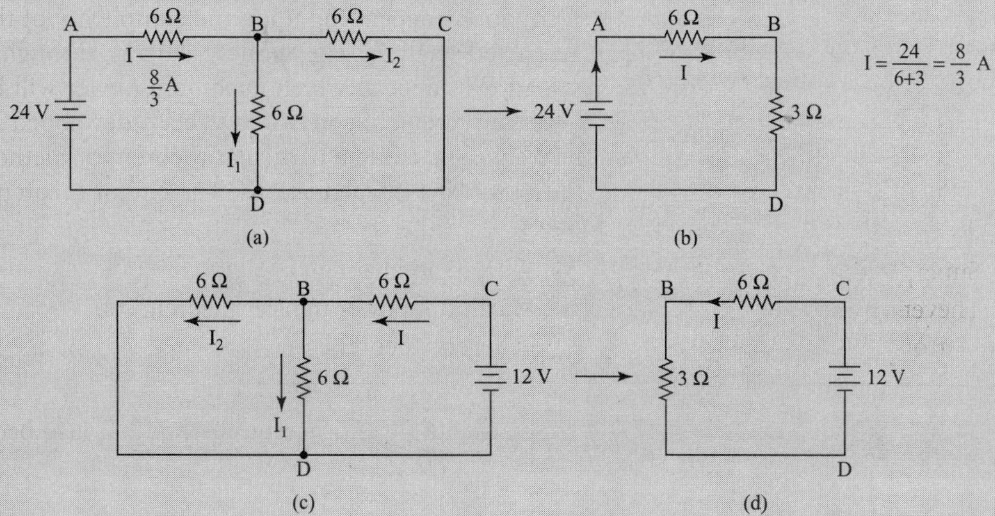

FIGURE 2.39

Between terminals B and D, two 6 **Ω** resistors are connected in parallel. The equivalent resistance is 3 **Ω.**

The current flowing from the battery is calculated $I = \frac{8}{3}A$ as shown in Fig. 2.39 (b). This $I = \frac{8}{3}A$ in Fig. 2.39 (a) gets divided into two parts as I_1 and I_2. Current through the resistor across BD is I_1. To find I_1, we can use the current division rule as

$$I_1 = I\frac{R_2}{R_1 + R_2} = \frac{8}{3} \times \frac{6}{6+6} = \frac{4}{3}A$$

Now, consider the 12 V source and short circuit the 24 V source as shown in Fig. 2.39 (c). The current supplied by the 12 V source is calculated as

$$I = \frac{12}{6+3}$$

$$= \frac{4}{3}A$$

The total current I due to 12 V supply has been calculated as $\frac{4}{3}A$. This current gets divided into I_1 and I_2 as has been shown in Fig. 2.39 (a). Current I_1 is calculated using the current division rule as

$$I_1 = I \times \frac{R_2}{R_1 + R_2} = \frac{4}{3} \times \frac{6}{6+6} = \frac{2}{3} A$$

To determine the current flowing through the resistor across BD, the combined effect of the two voltage sources will be taken. Therefore,

$$I_{BD} = I_1 \text{ due to 24 V source} + I_1 \text{ due to the 12 V source}$$

$$= \frac{4}{3} + \frac{2}{3} = 2\,A, \text{ flowing from node B towards node D}$$

EXAMPLE 2.17

For the circuit shown in Fig. 2.40, calculate the current, I using the superposition theorem.

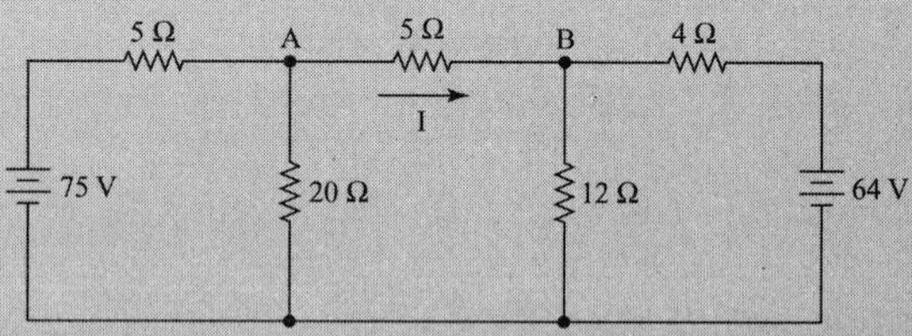

FIGURE 2.40

SOLUTION

We will consider the 75 V source first and short circuit the 64 V source. The current supplied by the 75 V source will be calculated. From the total current, current flowing through the resistor across terminals A and B will be calculated. The steps are illustrated in Fig. 2.41. When 64 V is short circuited, the 12 Ω and 4 Ω resistors get connected in parallel.

From Fig. 2.41 (e), the battery current calculated has been 7 A and the current through the 5 Ω resistor across terminals A and B is calculated using the current divider rule in Fig. 2.41 (b) as

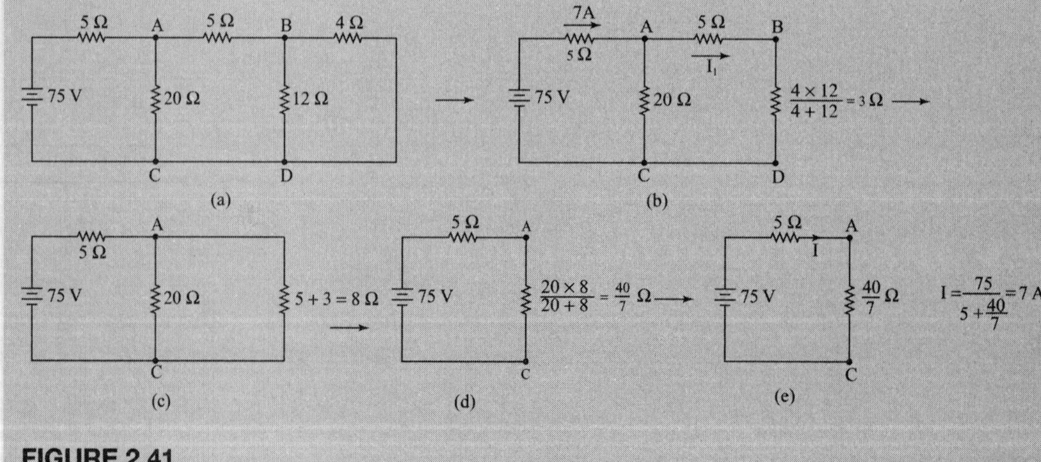

FIGURE 2.41

$$I_1 = 7 \times \frac{20}{20+5+3} = 5\ \text{A}$$

This 5 A through the resistor is due to the voltage source of 75 V. Now, we will calculate the current through the same resistor due to the other voltage source. We will short circuit the 75 V source and proceed as follows.

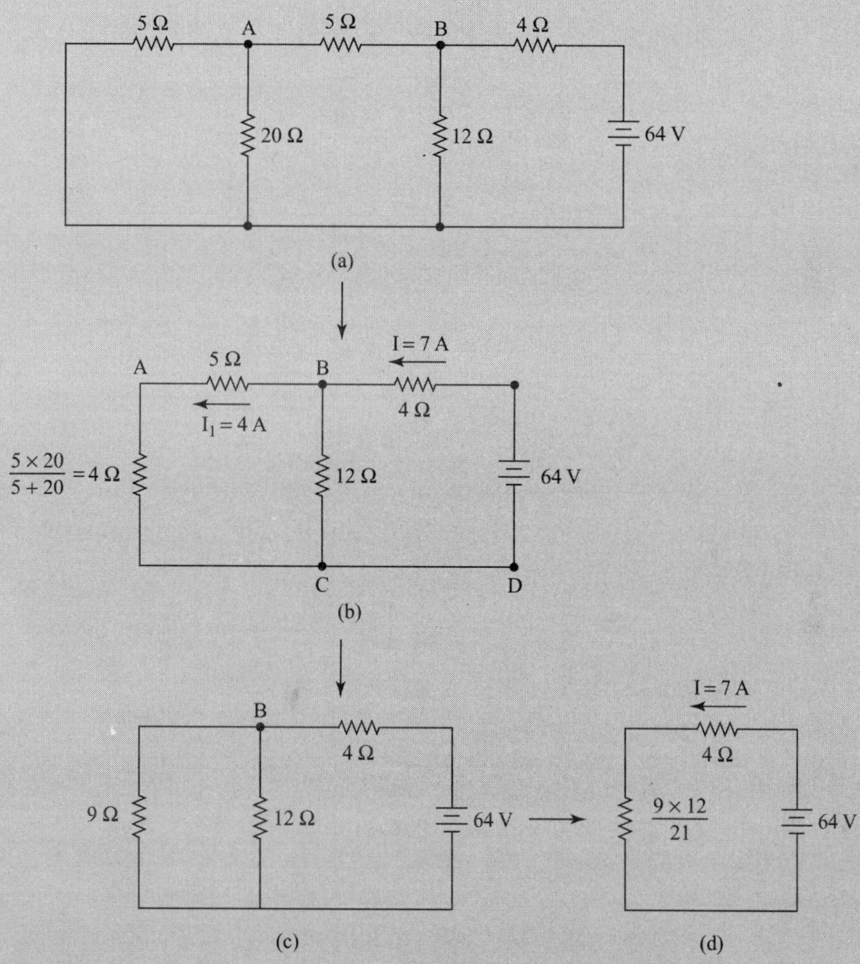

FIGURE 2.42

Current,

$$I = \frac{64}{4+108/21} = \frac{64}{84+108/21} = \frac{64 \times 21}{192} = 7\ \text{A}$$

Current I_1 through parallel circuit BAC in Fig. 2.42 (b), is calculated as

$$I_1 = I \times \frac{12}{9+12} = 7 \times \frac{12}{9+12} = 4 \text{ A}$$

Through the resistor across terminals AB, current of 5 A flows from A to B due to voltage source of 75 V and a current of 4 A flows from B to A due to voltage source of 64 V. The net current I_{AB} is equal to I_{AB} = 5 − 4 = 1 A when the effect of both the voltage sources are superimposed.

EXAMPLE 2.18

Determine current through a 8 Ω resistor in the network shown in Fig. 2.43 using the superposition theorem.

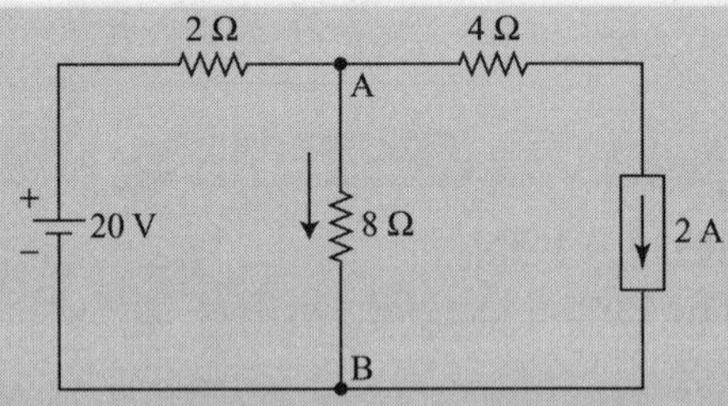

FIGURE 2.43

SOLUTION

Step 1: First the effect of current source will be considered. The voltage source is replaced by a short circuit. Using the current division rule, we determine current I_1 flowing from B to A as

$$I_1 = 2 \times \frac{2}{2+8}$$

$$= 0.4 \text{ A}$$

Current I_1 is flowing from B towards A as in Fig. 2.44 (a)

Step 2: Now if only the voltage source is considered, the current source has to be kept open circuited as shown in Fig. 2.44 (b). The current flowing through the 8 Ω resistor from A to B is determined as

$$I_2 = \frac{20}{2+8} = 2 \text{ A}$$

Step 3: Hence, total current in the 8 Ω resistor from A to B is resistor,

$$I = -I_1 + I_2$$

$$= -0.4 + 2 = 1.6 \text{ A}$$

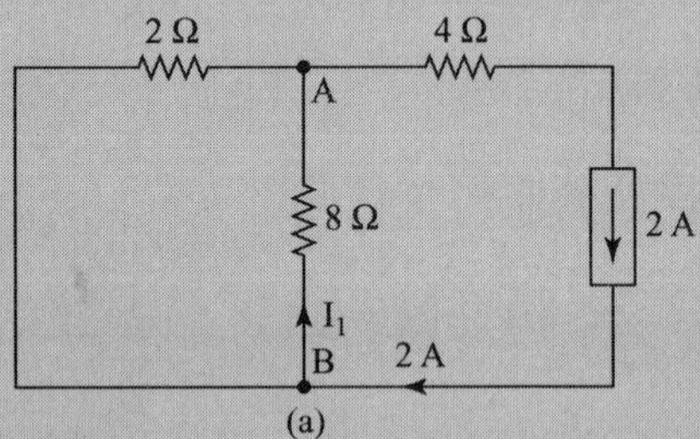

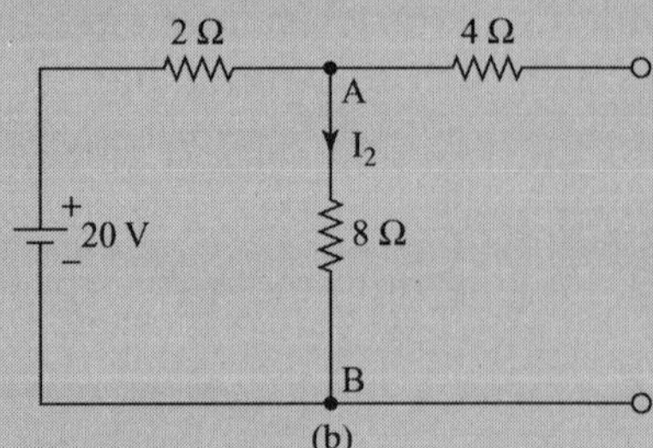

FIGURE 2.44

EXAMPLE 2.19

Two batteries are connected in parallel, each represented by an emf along with its internal resistance. A load resistance of 6 Ω is connected across the batteries. Calculate the current through each battery and through the load.

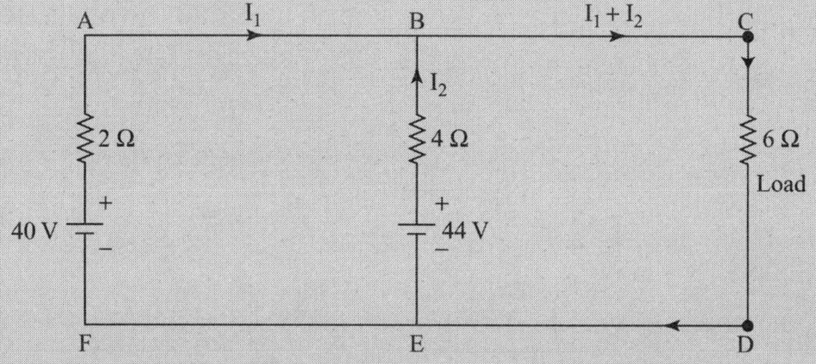

FIGURE 2.45

SOLUTION

The circuit diagram of two batteries supplying a common load has been shown in Fig. 2.45.

Apply KVL to mesh ABEFA and BCDEB after arbitrarily showing the branch current directions. Current through the batteries are I_1 and I_2 and through the load is $I_1 + I_2$, respectively, as has been shown.

From mesh ABEFA,

$$40 - 2I_1 + 4I_2 - 44 = 0$$

$$-2I_1 + 4I_2 = 4$$

or,

$$2I_1 - 4I_2 = -4 \quad \text{(i)}$$

From mesh BCDEB

$$-6(I_1 + I_2) + 44 - 4I_2 = 0$$

$$-6I_1 - 10I_2 = -44 \text{ or, } 6I_1 + 10I_2 = 44 \quad \text{(ii)}$$

Students are once again reminded that while going through the loop or the mesh, the following sign convention has been followed.

1. Moving from negative terminal towards positive terminal of a battery is considered positive voltage, i.e., as voltage rise
2. Voltage drop in the resistor is taken as negative when moving in the direction of current flow.

Solving eq. (i) and (ii) we get,

$$\cancel{6I_1} - 12\,I_2 = -12 \text{ (multiplying eqn (i) by 3)}$$

$$\cancel{6I_1} + 10\,I_2 = 44$$

$$-22\,I_2 = -56$$

$$I_2 = \frac{28}{11} A$$

Substituting value of I_2 in eqn. (i), $$I_1 = \frac{34}{11} A$$

Total current through the load

$$= I_1 + I_2 = \frac{34}{11} + \frac{28}{11} = \frac{62}{11} = 5.636 \text{ A}.$$

EXAMPLE 2.20

Calculate the current through the galvanometer having a resistance of 4 Ω in the bridge circuit shown in Fig. 2.46 (a)

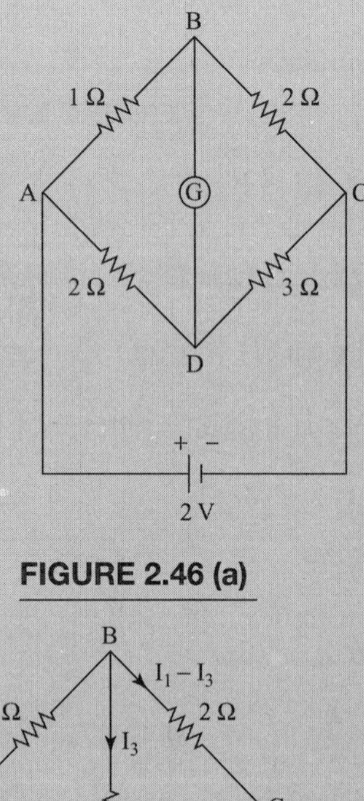

FIGURE 2.46 (a)

SOLUTION

Step 1:– We apply mesh analysis in ABDA, BCDB, ABCA as in Fig. 2.46 (b).

In mesh ABDA

$$-I_1 - 4I_3 + 2I_2 = 0 \quad \text{(i)}$$

In mesh BCDB

$$-2(I_1 - I_3) + 3(I_2 + I_3) + 4I_3 = 0$$

or, $$-2I_1 + 2I_3 + 3I_2 + 3I_3 + 4I_3 = 0$$

or, $$-2I_1 + 3I_2 + 9I_3 = 0 \quad \text{(ii)}$$

In mesh ABCA

$$-I_1 - 2(I_1 - I_3) + 2 = 0$$

or, $$-3I_1 + 2I_3 = -2 \quad \text{(iii)}$$

The three equations are written again as

$$-I_1 + 2\,I_2 - 4\,I_3 = 0 \quad \text{(i)}$$

$$-2I_1 + 3\,I_2 + 9I_3 = 0 \quad \text{(ii)}$$

$$-3I_1 \qquad + 2I_3 = -2 \quad \text{(iii)}$$

$$\begin{bmatrix} -1 & 2 & -4 \\ -2 & 3 & 9 \\ -3 & 0 & 2 \end{bmatrix} \begin{bmatrix} I_1 \\ I_2 \\ I_3 \end{bmatrix} = \begin{bmatrix} 0 \\ 0 \\ -2 \end{bmatrix}$$

$$\Delta = \begin{bmatrix} -1 & 2 & -4 \\ -2 & 3 & 9 \\ -3 & 0 & 2 \end{bmatrix}$$

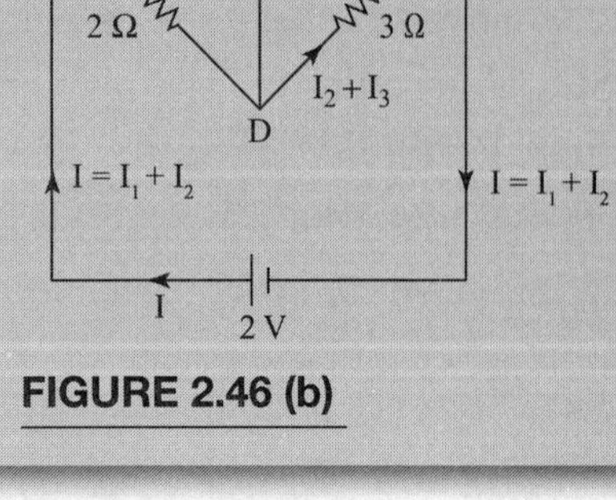

FIGURE 2.46 (b)

$$= -1[6-0] - 2[-4+27] - 4[0+9]$$
$$= -88$$
$$\Delta_1 = \begin{bmatrix} 0 & 2 & -4 \\ 0 & 3 & 9 \\ -2 & 0 & 2 \end{bmatrix}$$
$$= 0\,[6-0] - 2\,[0+18] - 4\,[0+6]$$
$$= -60$$
$$I_1 = \frac{\Delta_1}{\Delta} = \frac{\not{-}60}{\not{-}88} = \frac{30}{44}\text{ A}$$

Similarly,

$$I_2 = \frac{\Delta_2}{\Delta} = \frac{17}{44}\text{ A}$$

Similarly,

$$I_3 = \frac{\Delta_3}{\Delta} = \frac{1}{44}\text{ A}$$

Current through galvanomeneter

$$= I_3 = \frac{1}{44}\text{ A}$$

EXAMPLE 2.21

Applying KCL, determine current I_S in the electric circuit to make V_0 = 16 V in the network shown in Fig. 2.47.

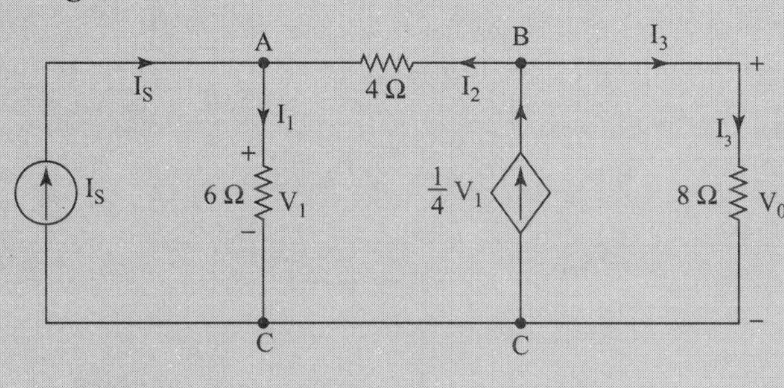

FIGURE 2.47

SOLUTION

Applying Kirchhoff's current law at nodes A and B we have

$$I_1 = I_2 + I_S \quad \text{(i)}$$

and

$$I_2 + I_3 = \frac{V_1}{4} \quad \text{(ii)}$$

also voltage of node B = V_0 =16 V (as required).

Voltage across AC plus voltage across AB = voltage at node B.

$$V_1 + 4I_2 = 16 \text{ V} \qquad \text{(iii)}$$

$$I_1 = \frac{V_1}{6} \qquad \text{(iv)}$$

Solving eq. (i), (ii), (iii), and (iv) we have

$$V_1 = 12 \text{ V} \qquad I_1 = \frac{12}{6} = 2 \text{ A} \qquad 4\,I_2 = 16 - 12 = I_2 = 1 \text{ A}$$

Therefore, $\quad I_S = I_1 - I_2 = 1 \text{ A} \qquad I_3 = \frac{V_1}{4} - I_2 = 3 - 1 = 2 \text{ A}.$

EXAMPLE 2.22

Two batteries A and B are connected in parellel to a load of 10 Ω. Battery A has an EMF of 12 V and an internal resistance of 2 Ω and Battery B has an EMF of 10 V and internal resistance of 1 Ω. Using nodal analysis to determine the current supplied by each battery and the load current.

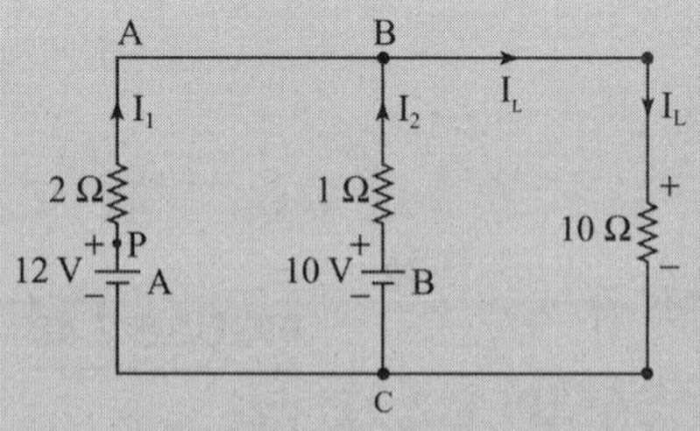

FIGURE 2.48

SOLUTION

The circuit diagram of two batteries supplying a load has been shown in Fig. 2.48. Taking node C as a reference node and potential of nodes A and B be V_A and V_B. Current distribution, I_1, I_2 and I_L has been shown. The potential of point P with respect to the reference point C is 12 V. So,

$$I_1 = \frac{12 - V_A}{2}$$

$$I_2 = \frac{10 - V_B}{1}$$

$$I_L = \frac{V_B}{10}$$

at node B, using KCL,

$$I_L = I_1 + I_2$$

$$\frac{V_B}{10} = \frac{12 - V_A}{2} + \frac{10 - V_B}{1}$$

$$V_A = V_B = 10 \text{ V}$$

Current supplied by battery, A $\qquad I_1 = \frac{12 - V_A}{2} = 1 \text{ A}$

Current supplied by battery, B $\quad I_2 = \frac{10-10}{1} = 0$

Load current $\quad I_L = \frac{V_B}{10} = 1\text{ A}$

EXAMPLE 2.23

For the circuit shown in Fig. 2.49, find voltages of nodes B and C and determine current through the 8 Ω resistor.

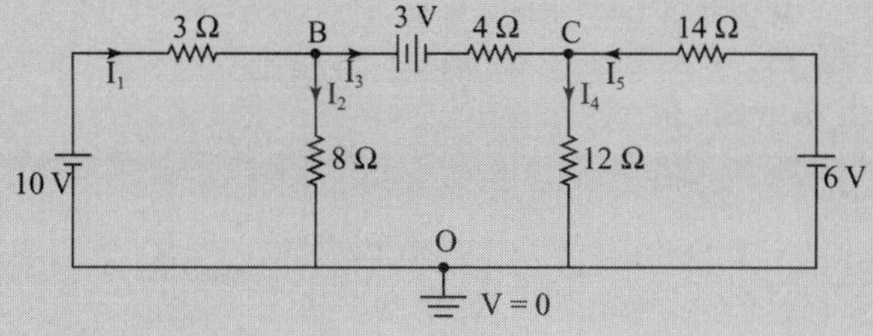

FIGURE 2.49

Let the reference point be at 0 which is taken at zero potential. By applying KCL at node B, we get

$$I_1 = I_2 + I_3$$

or,

$$\frac{10 - V_B}{3} = \frac{V_B}{8} + \frac{V_B - 3 - V_C}{4}$$

$$17\,V_B - 6\,V_C = 98 \qquad \text{(i)}$$

Applying KCl at node C, we get

$$I_4 = I_3 + I_5$$

$$\frac{V_C}{12} = \frac{V_B - V_C - 3}{4} + \frac{6 - V_C}{14}$$

$$21\,V_B - 34\,V_C = 27 \qquad \text{(ii)}$$

Solving eq. (i) and (ii), we get

Voltage of node B is calculated as, $\quad V_B = 7.0133\text{ V}$

Voltage of node C is calculated as, $\quad V_C = 3.5376\text{ V}$

Current in 8 Ω Resistor $\quad = I_2 = \frac{V_B}{8} = \frac{7.0133}{8}$

$$= 0.87666\text{ A}$$

EXAMPLE 2.24

Two batteries of EMF 2.05 V and 2.15 V having internal resistances of 0.05 Ω and 0.04 Ω, respectively, are connected togather in parallel to supply a load resistance of 1 Ω as has been shown in Fig. 2.50. Calculate using the superposition theorem, current supplied by each battery and also the load current.

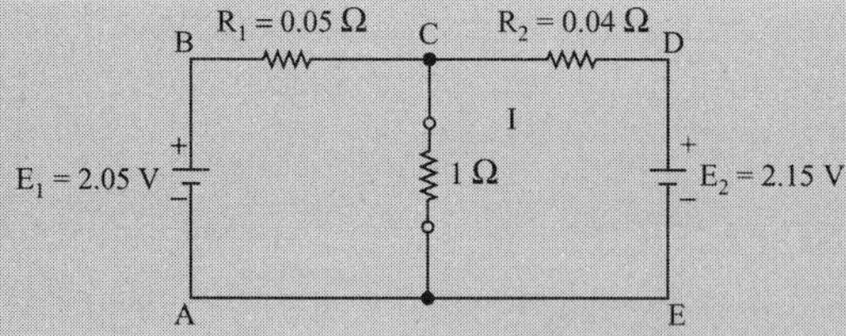

FIGURE 2.50

SOLUTION

First we will consider the effect of voltage source E_1 by short circuiting E_2. The currents will be calculated by considering the series parallel connections of resistances as in Fig. 2.51. Thus,

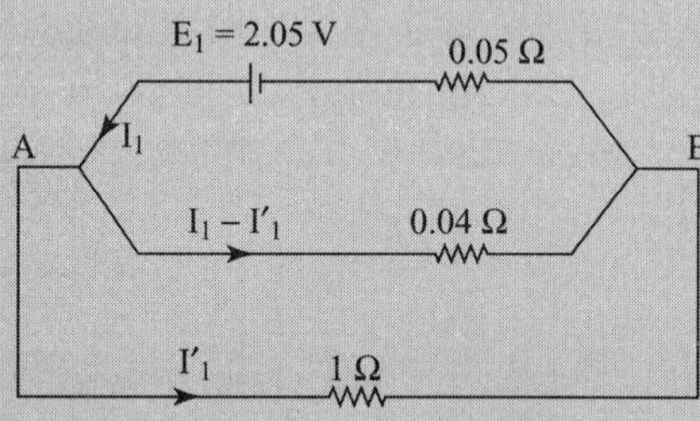

FIGURE 2.51

$$I_1 = \frac{2.05}{0.05 + 0.04 \times 1/1.04} = \frac{2.05}{0.085} = 23.2 \text{ A}$$

Using current divider rule,

$$I_1' = I_1 \times \frac{0.04}{0.04 + 1} = 23.2 \times \frac{0.04}{1.04} = 0.892 \text{ A from A to B.}$$

Now, we will consider the effect of voltage source E_2 by short circuiting E_1. The circuit will be, as shown in Fig. 2.52.

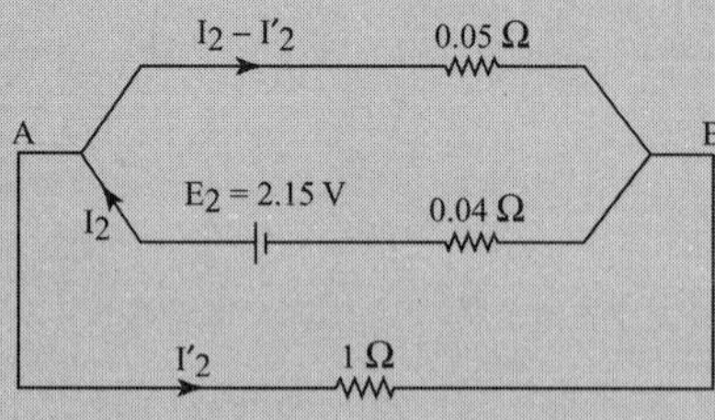

FIGURE 2.52

$$I_2 = \frac{2.15}{0.04 + 0.05 \times 1/1.05}$$

$$= 24.54 \text{ A}$$

current through the 1 Ω resistor,

$$I_2' = I_2 \times \frac{0.05}{0.05 + 1}$$

$$= 24.54 \times \frac{0.05}{1.05}$$

$$= 1.169 \text{ A} \quad \text{from A to B}$$

By superimposing the effect of two voltage sources, the current through the 1 Ω resistor, I is calculated as

$$I = I_1' + I_2' = 0.892 + 1.169$$

or, $$I = 2.061 \text{ A}$$

EXAMPLE 2.25

Determine the current through the 8 Ω resistor in the network shown in Fig. 2.53. Use the superposition theorem.

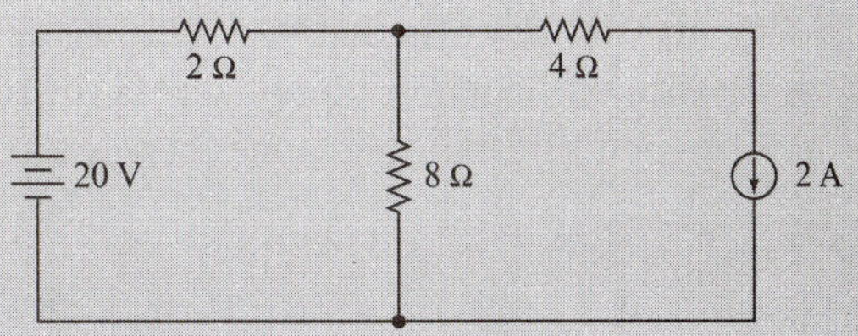

FIGURE 2.53

SOLUTION

First remove one of the sources, say the voltage source, short circuit the terminals as in Fig. 2.54, and calculate the current flow through the 8 Ω resistor.

Current of 2 A will be divided into two parts, one part going through the 8 Ω resistor from B to A and the other part going through the 2 Ω resistor. Current I_1' going through the 8 Ω resistor is calculated as

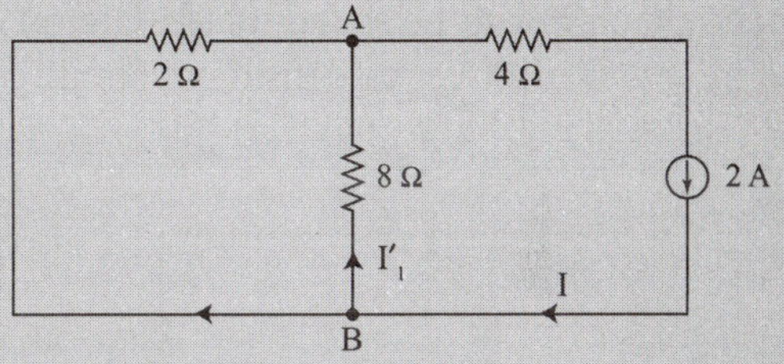

FIGURE 2.54

$$I_1' = I \times \frac{2}{2+8} = 2 \times \frac{2}{10} = 0.4 \text{ A from B to A}$$

Now, we will consider the voltage source, keeping the current source open circuited as shown in Fig. 2.55 and find the current through the 8 Ω resistor. Using Ohm's law, the current through the 8 Ω resistor is calculated as

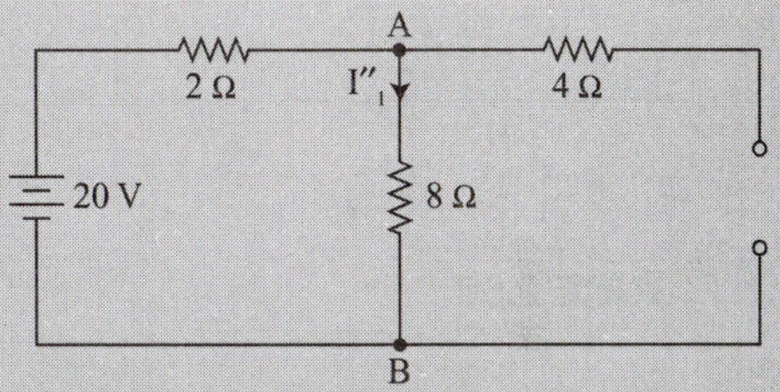

FIGURE 2.55

$$I_1'' = \frac{20}{2+8} = 2 \text{ A from A to B}$$

The combined effect of the two sources when superimposed is

$$I = -I_1' + I_1'' = -0.4 + 2 = 1.6 \text{ A from A to B}$$

2.8.2 Thevenin's Theorem

Application of this theorem often comes useful when we want to determine the current flowing through any branch or component of a network. We can conveniently determine the current through any component when it is required that the component has to be replaced. The use of kirchchoffs laws to calculate the branch current for the changed value of a resistor becomes time consuming as we have to repeat the calculations.

Here, the whole circuit across the terminals of the resistor, through which current flow is to be calculated, is converted into a voltage source with an internal resistance. The voltage is the open-circuit voltage of the network across the terminals and internal resistance is the equivalent resistance of the whole circuit across the open-circuited terminals by replacing the voltage sources by their internal resistances. The current through the resistor R, is

$$I = \frac{V_{OC}}{R_{eq} + R} \tag{2.3}$$

Where V_{OC} is the open-circuit voltage across the terminals of resistor R (when R is removed from the circuit); R_{eq} is the equivalent circuit resistance across the terminals of R.

The theorem will be stated a little later after a specific problem is solved applying the procedure mentioned. Let us consider a circuit as in Fig. 2.56.

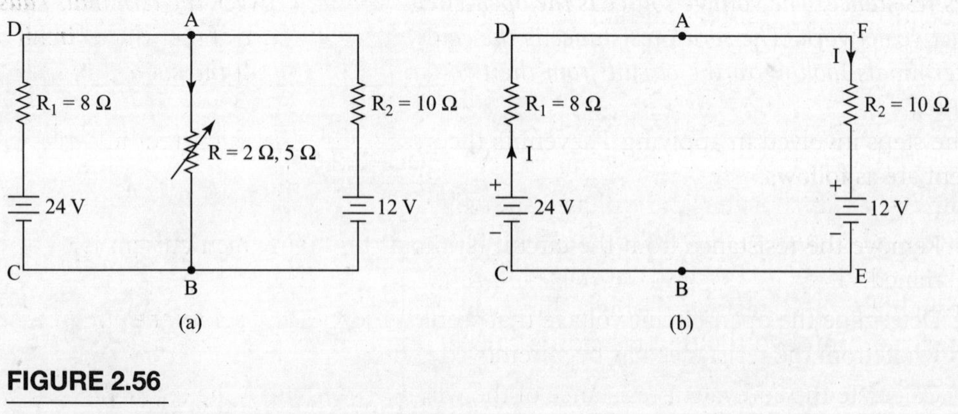

FIGURE 2.56

To calculate the current through the variable resistor R, the first step would be to take away the resistor R and calculate the V_{OC} across terminals A and B. Applying KVL in the circuit of Fig. 2.56 (b)

$$+24 - 8\,I - 10\,I - 12 = 0$$

or,
$$I = 0.67 \text{ A}$$

Approaching from B to A through the path CD,

$$V_{OC} = 24 - 8 \times 0.67 = 18.64 \text{ V}$$

The same result will be arrived at when we approach from A to B through the path EF.

Now we calculate R_{eq} across terminals A and B by short circuiting the voltage sources or by replacing them by their internal resistances. Across terminals AB we find two resistances of 8 Ω and 10 Ω connected

in parallel. R_{eq} across terminals A and B $= \dfrac{8 \times 10}{8 + 10} = 4.44\ \Omega$

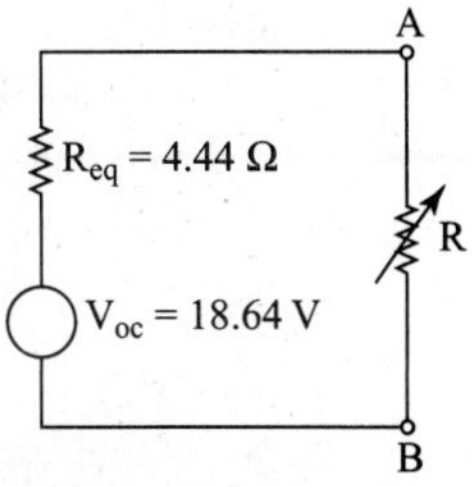

FIGURE 2.57

Thus, Thevenin's equivalent circuit is represented as in Fig. 2.57 Current through R with its variable values i.e., for R = 2 Ω and R = 5 Ω are calculated as

$$I = \frac{V_{OC}}{R_{eq} + R} = \frac{18.64}{4.44 + 2} = 2.89 \text{ A}$$

and

$$I = \frac{V_{OC}}{R_{eq} + R} = \frac{18.64}{4.44 + 5} = 1.97 \text{ A}$$

Now, we are in a position to state Thevenin's Theorem. *The Thevenin's theorem is stated as:*

Any two terminals of an electrical network consisting of active and passive elements (i.e., voltage sources and resistors) can be replaced by an equivalent voltage source and an equivalent series resistance. The voltage source is the open-circuit voltage between the terminals caused by the active network. The series resistance is the equivalent resistance of the whole circuit across the terminals looking to the circuit from the two terminals with all the sources of EMF short circuited.

The steps involved in applying Thevenin's theorem in calculating current in a circuit component are as follows:

1. Remove the resistance from the circuit terminals through which current is to be determined.
2. Determine the open-circuit voltage that would be appearing across the circuit terminals wherefrom the resistance has been removed. This is called V_{OC}.
3. Calculate the equivalent resistance of the whole circuit across the terminals after replacing the sources of EMFs by their internal resistances (or by simply short–circuiting them if internal resistance is not provided or not known) and by keeping the current sources open–circuited (i.e., considering having infinite resistance).
4. Draw the Thevenins equivalent circuit with V_{OC} as the voltage source, R_{eq} as the internal resistance of the voltage source, and R is the load resistance connected across the voltage source.
5. Calculate the current through R using the relation.

$$I = \frac{V_{oc}}{R_{eq} + R}$$

EXAMPLE 2.26

Using Thevenin's theorem, calculate the range of current flowing through the resistance R of Fig. 2.58, when its value is varied from 6 Ω to 36 Ω.

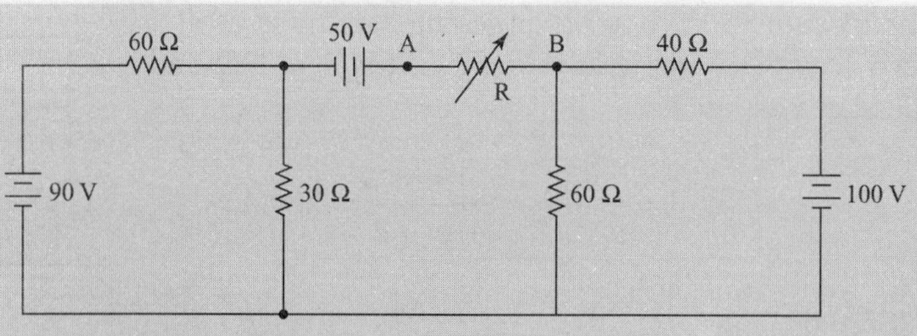

FIGURE 2.58

SOLUTION

The open-circuit voltage, V_{OC} across the terminals A and B by removing the resistance R as shown in Fig. 2.59, is calculated as

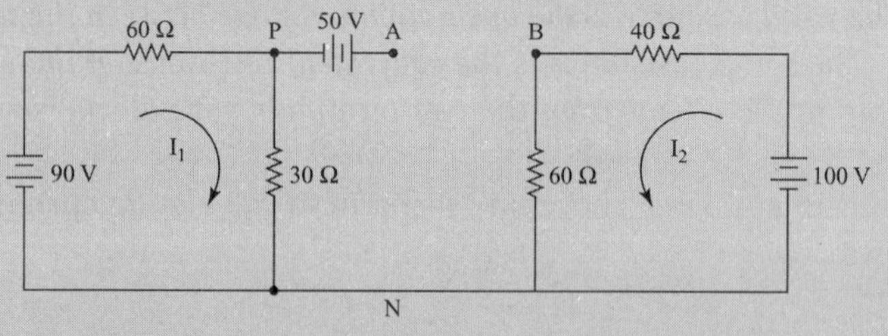

FIGURE 2.59

$$I_1 = \frac{90}{60+30} = 1\text{ A},\quad I_2 = \frac{100}{60+40} = 1\text{ A}$$

Voltage drop across the 30 Ω resistor = 30 × 1 = 30 V. The potential of point P with respect to N is +30 V. The potential of point A with respect of point P is +50 volts. Therefore, the to potential of point A with respect to point N is +30 +50 = +80 V.

Voltage drop across the 60 Ω resistor in the right-hand-side loop is calculated as

$$I_2 = \frac{100}{40+60} = 1\text{ A}$$

$$V_{NB} = 60 \times 1 = 60\text{ V}$$

Potential of point B with respect to point N is + 60 V and that of A w.r.t. N = $30I_1$ + 50 = +80 V.

Now we observe that the potential of point A with respect to point N is +80 V and the potential of point B with respect to point N is +60 V. Therefore, the potential of point A with respect to point B, i.e., V_{OC} becomes +20 V, point A being at higher a potential than point B.

To calculate R_{eq}, we redraw the circuit by short circuiting the sources of EMFs as in Fig. 2.60.

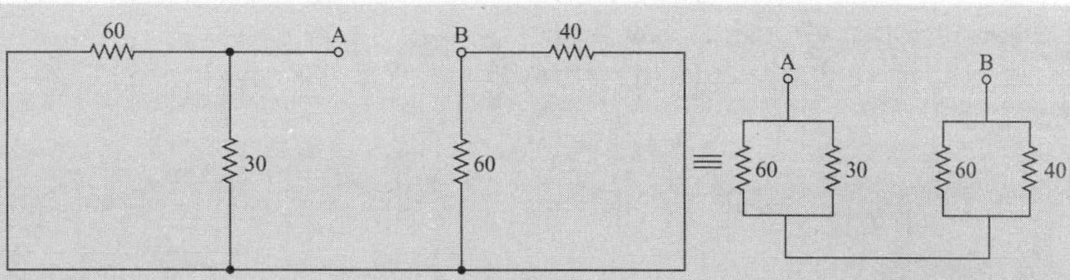

FIGURE 2.60

$$R_{eq} = \frac{60 \times 30}{60+30} + \frac{60 \times 40}{60+40} = 44\ \Omega$$

Thevenin's equivalent circuit is drawn as in Fig. 2.61

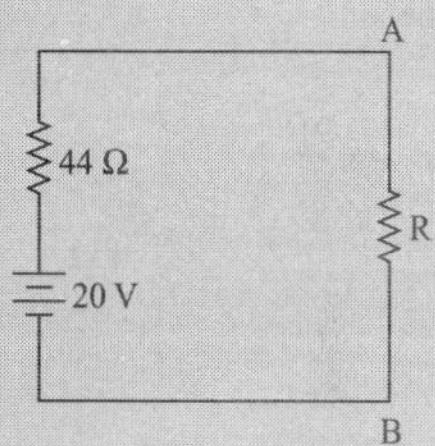

FIGURE 2.61

$$I = \frac{20}{44+6} \text{ when } R = 6\ \Omega$$

$$= 0.4\text{ A}$$

$$I = \frac{20}{44+36} \text{ when } R = 36$$

$$= 0.25\text{ A}$$

The value of current through the resistor R will vary from 0.4 A to 0.25 A when its value is changed from 6 Ω to 36 Ω keeping the other circuit conditions unchanged.

EXAMPLE 2.27

Using Thevenin's theorem, calculate the current flowing through the load resistance R_L connected across the terminals A and B as shown in Fig. 2.62.

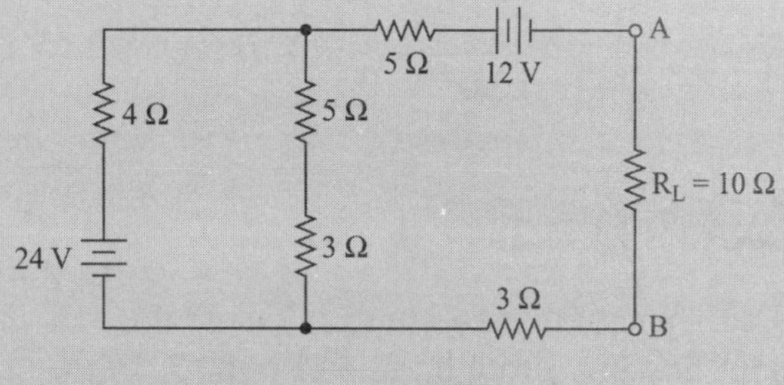

FIGURE 2.62

SOLUTION

First we remove the load resistance R_L from the circuit and calculate the open-circuit voltage, V_{OC} across its terminals A and B as in Fig. 2.63,

Let current I be known through the loop as shown.

$$I = \frac{24}{4+8} = 2\text{ A}$$

voltage drop across PQ is equal to IR = 2 × 8 = 16 V. Terminal P is at higher potential than terminal Q. No current is flowing through the 3 Ω and 5 Ω resistors in the circuit because these are open circuited. The potential of A with respect to B is calculated starting from point B as

$$V_{AB} = V_{BQ} + V_{QP} + V_{PS} - V_{SA}$$
$$= 0 + 16\text{ V} + 0 - 12\text{ V}$$
$$= +4\text{ V}$$
$$= V_{OC}$$

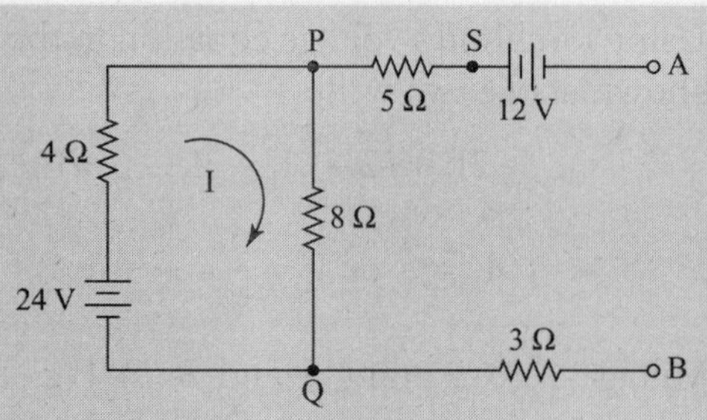

FIGURE 2.63

For calculating R_{eq}, we short circuit the sources of EMFs as shown in Fig. 2.64.

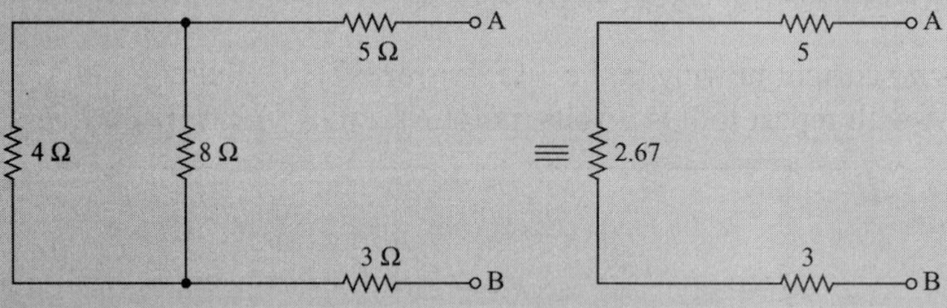

FIGURE 2.64

$$R_{eq} = R_{AB} = 10.67\ \Omega$$

Thevenin's equivalent circuit and current through the load resistor is calculated from Fig. 2.65, as

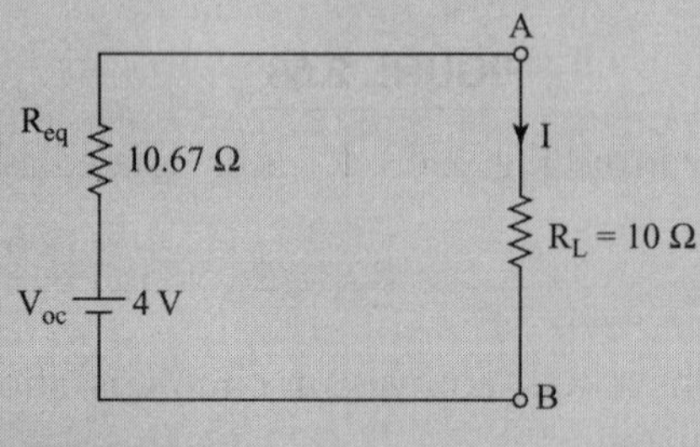

FIGURE 2.65

$$I = \frac{V_{OC}}{R_{eq} + R_L}$$
$$= \frac{4}{10.67 + 10} = 0.193\text{ A}$$

EXAMPLE 2.28

Use Thevenin's theorem to calculate the current flowing through the 5 Ω resistor in the circuit shown in Fig. 2.66.

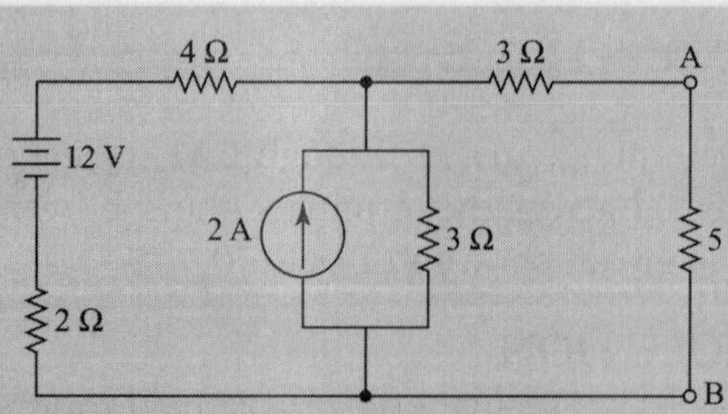

FIGURE 2.66

SOLUTION

We shall convert the current source into its equivalent voltage source and also take away the 5 Ω resistor from the terminals A and B. We will calculate the V_{oc} across terminals AB as in Fig. 2.67.

Using kirchhoff's voltage equation in the loop as shown above, we write

$$-2I_1 + 12 - 4I_1 - 3I_1 - 6 = 0$$

$$9I_1 = 6 \text{ or } I_1 = \frac{6}{9} = \frac{2}{3} \text{ Amps}$$

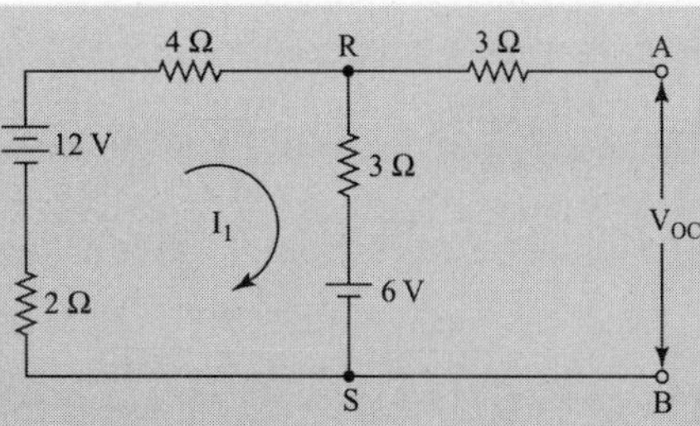

FIGURE 2.67

Voltage across point S and R in Fig. 2.67, is calculated by moving upwords from point S to point R.

Considering the voltage rise we get, $+6\text{ V} + 3 \times \frac{2}{3}\text{V} = +8$ volts

Since there is no current flowing in the 3 Ω resistor between point R and A, the potential across point A with repeat to B is 8 Volts, point A being at higher potential than point B.

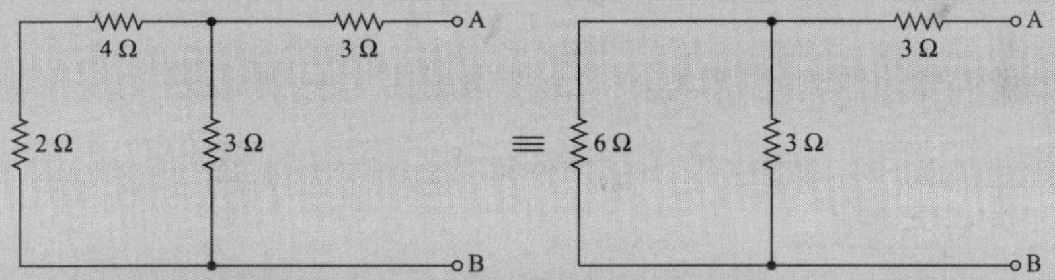

FIGURE 2.68

From Fig. 2.68, R_{eq} is calculated as

$$R_{AB} = R_{eq} = 3 + \frac{3 \times 6}{9} = 5\ \Omega$$

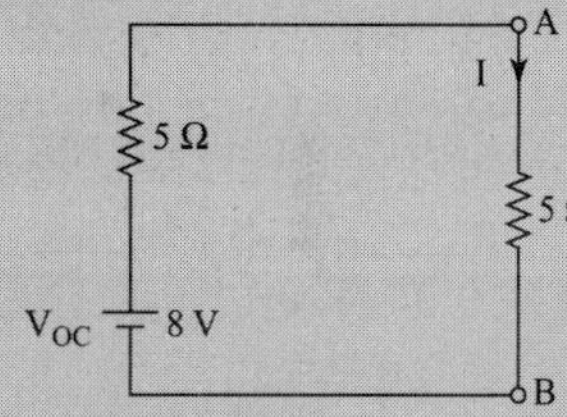

FIGURE 2.69

Thevenins equivalent circuit is shown in Fig. 2.69. The current, I is

Calantated as, $$I = \frac{V_{OC}}{R_{eq} + R} = \frac{8}{5+5} = 0.8 \text{ Amps}$$

EXAMPLE 2.29

Determine current through 6 Ω resistance connected across the terminals A and B in the electric circuit shown in Fig. 2.70.

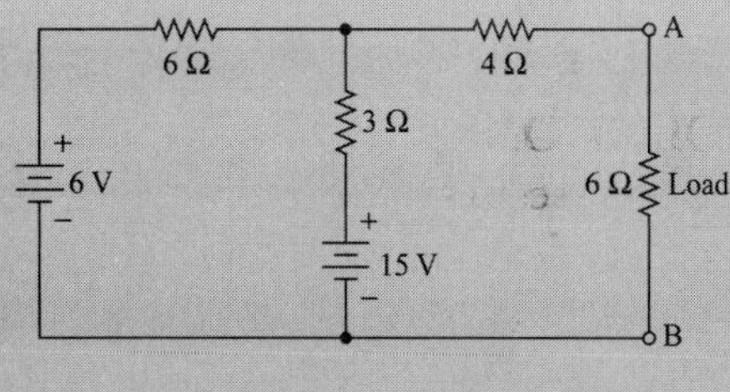

FIGURE 2.70

SOLUTION

Step 1:– Remove load resistance through which current is required to be calculated. As shown in Fig. 2.71.

Applying KVL in the loop CDEFC,

$$15 - 3I - 6I - 6 = 0$$

or, $9I = 9$

or, $I = 1$

voltage across CD

$$= 15 - 3I$$
$$= 15 - 3 \times 1$$
$$= 12 \text{ V}$$

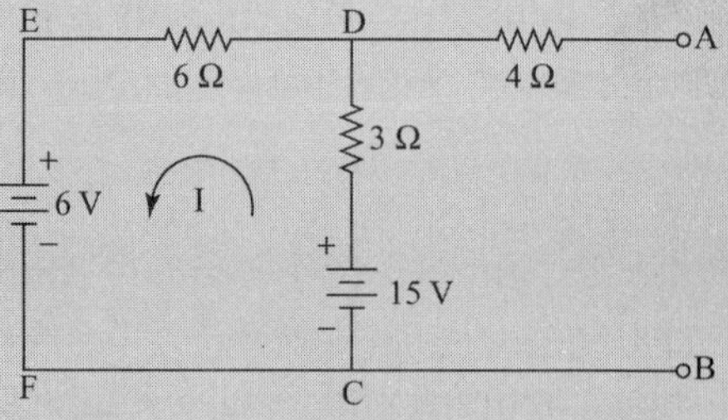

FIGURE 2.71

Since no current is flowing through the 4 Ω resistor, $V_{CD} = V_{BA} = V_{OC}$. Point A is at a higher potential than point B. Equivalent resistance across terminals A and B is,

$$R_{th} = R_{eq} = 4 + \frac{3 \times 6}{3 + 6} = 6\ \Omega$$

Current through the load, $$I = \frac{V_{oc}}{R_{eq} + R} = \frac{12}{6 + 6} = 1\text{ A}$$

EXAMPLE 2.30

The resistance of the various arms of a Wheatstone bridge are shown in Fig. 2.72. The battery has an EMF of 2 V and negligible internal resistance. Using Thevenin's theorem, determine the value and direction of the current in the galvanometer circuit BD.

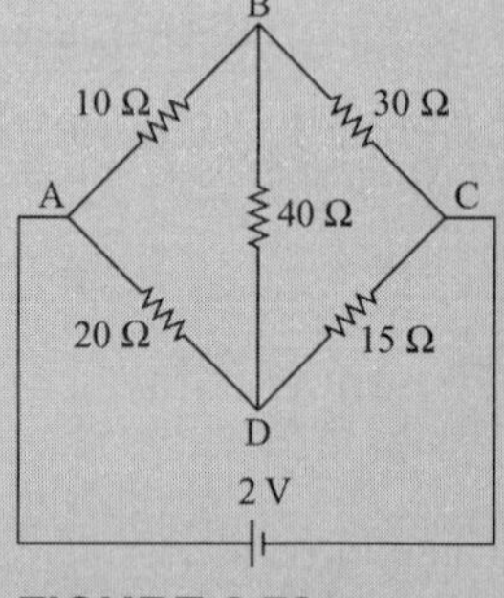

FIGURE 2.72

SOLUTION

Remove the load resistance of 40 Ω between terminals B and D as shown in Fig. 2.73. Calculate V_{BD} as V_{OC}.

$$I_1 = \frac{2}{10 + 30} = \frac{1}{20}\text{A}$$

$$I_2 = \frac{2}{30 + 15} = \frac{2}{45}\text{A}$$

$$V_{BC} = 30I_1 = \frac{30 \times V}{30 + 10} \therefore V_{BC} = \frac{30 \times 2}{30 + 10} = 1.5\text{ V}$$

$$V_{DC} = 15I_2 = \frac{15 \times V}{20 + 15} \therefore V_{DC} = \frac{15 \times 2}{20 + 15} = \frac{30}{35} \text{ or } \frac{6}{7}\text{ V}$$

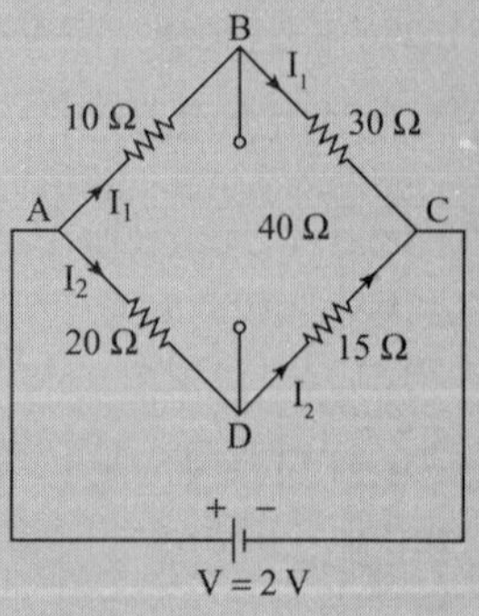

FIGURE 2.73

Potential difference between terminals B and D = $V_{OC} = -15I_2 + 30I_1 = -15\times\frac{2}{45} + 30\times\frac{1}{20}$

$$= -\frac{2}{3} + \frac{3}{2} = \frac{5}{6} = 0.833\ \text{V}$$

The equivalent resistance, R_{eq} which is also called R_{th} is calculated by short circuiting the voltage source and rearranging the circuit components as shown in Fig. 2.74.

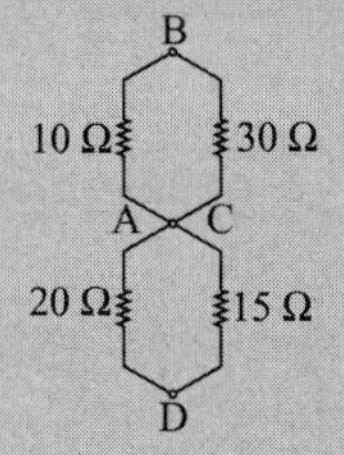

FIGURE 2.74

$$R_{th} = R_{eq} = \frac{10\times 30}{10+30} + \frac{20\times 15}{20+15} = \frac{225}{14} = 16\ \Omega$$

$$I_L = \frac{Vth}{R_{th}+R_L} = \frac{0.833}{16+40}\text{A} = \frac{0.833\times 1000}{56}\text{mA}$$
$$= 14.9\ \text{mA}$$

Since point B is at a higher potential than point D, I_L will flow from B to D.

EXAMPLE 2.31

In the circuit shown in Fig. 2.75, find the branch current I2 which will flow through RL, when RL is taken as 5 Ω, 15 Ω, and 50 Ω, respectively.

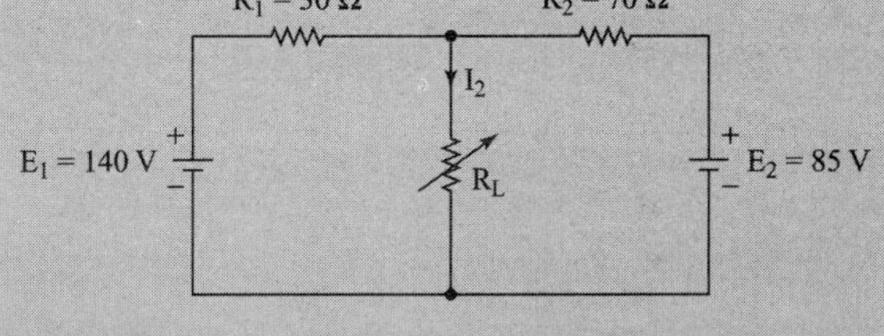

FIGURE 2.75

SOLUTION

Step 1:– Remove load resistance and draw the circuit as shown in Fig. 2.76. Calculate the current flowing in the circuit applying KVL in the loop ABCDEA as,

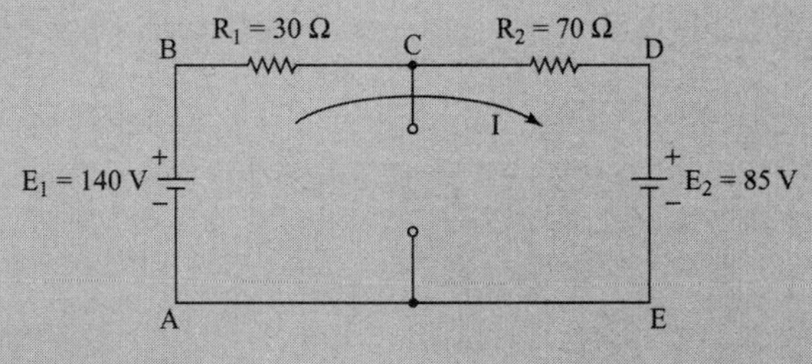

FIGURE 2.76

$$R_{eq} = \frac{30 \times 70}{30 + 70} = 21\ \Omega$$

$$140 - 30I - 70I - 85 = 0$$

$$+100I = +55$$

$$I = \frac{55}{100} = 0.55\ A$$

$$V_{OC} = V_{th} = E_1 - IR_1$$

$$= 140 - 0.55 \times 30 = 123.5\ V$$

for, $R_L = 5\ \Omega$, $$I = \frac{V_{OC}}{R_{eq} + R_L} = \frac{123.5}{21 + 5} = 4.75\ A$$

for $R_L = 15\ \Omega$, $$I = \frac{123.5}{21 + 15} = 3.43\ A$$

for $R_L = 50\ \Omega$, $$I = \frac{123.5}{21 + 50} = 1.74\ A$$

EXAMPLE 2.32

Calculate current through a 1,000 Ω resistor connected between terminals A and B in the circuit shown in Fig. 2.77 (a).

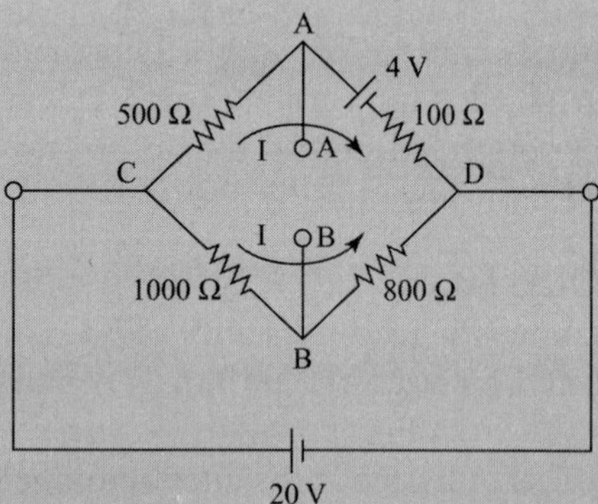

FIGURE 2.77 (a)

SOLUTION

Current through branch CBD $= \frac{20}{1000 + 800}$

$V_{CB} = -\frac{20 \times 1000}{1000 + 800} = -11.11\ V$ (minus sign is for voltage drop)

Applying KVL in the loop starting from battery to CAD and back to battery,

$$20 - 500\,I - 4 - 100\,I = 0$$

or, $$I = \frac{16}{600}\ A$$

The equivalent resistance across terminals A and B is calculated by shorting the battery terminals as shown in Fig. 2.77(b).

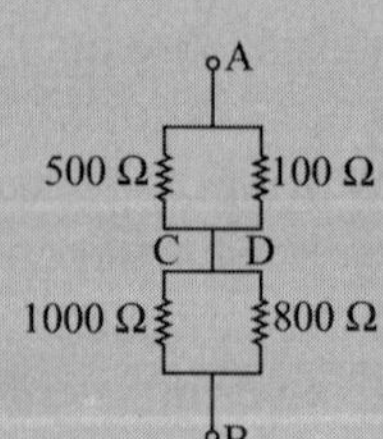

FIGURE 2.77 (b)

$$V_{CA} = -\frac{16}{600} \times 500 - \frac{80}{60} = -13.33\ V$$

$$V_{CB} = -11.11\ V \text{ and}$$

$$V_{CA} = -13.33\ V$$

Point B is at higher potential than point A.

$$V_{BA} = 2.22 \text{ V}$$

i.e.,

$$V_{OC} = 2.22 \text{ V}$$

R_{eq} across terminals A and B is calculated considering 500 Ω and 100 Ω resistance in parallel plus 1000 Ω and 800 Ω resistances in parallel as shown in Fig. 2.77 (b).

$$R_{eq} = \frac{500 \times 100}{500 + 100} + \frac{1000 \times 800}{1000 + 800} = \frac{9500}{18} = 52.8\ \Omega$$

Current

$$I = \frac{Voc}{R_{eq} + R_L} = \frac{2.22}{52.8 + 1000} = \frac{2.22}{1052.8} \text{A}$$

$$= \frac{2.22 \times 1000}{1052.8} \text{mA} = 2.11 \text{ mA}$$

EXAMPLE 2.33

Calculate using Thevenin's theorem the current flowing through the 5 Ω resistor connected across the terminals A and B as shown in Fig. 2.78.

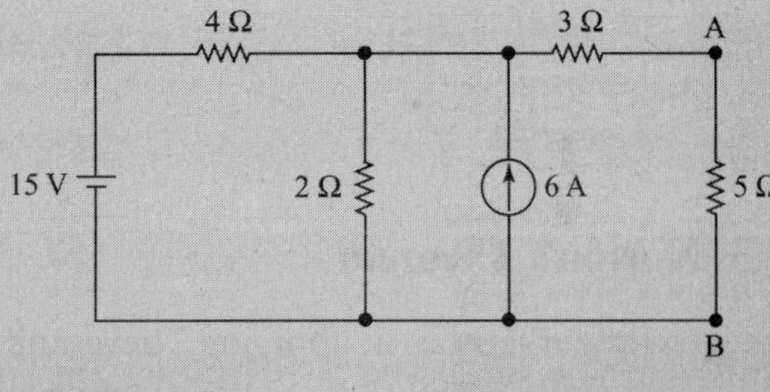

FIGURE 2.78

SOLUTION

Remove the load resistance of 5 Ω connected between the terminals A and B as shown in Fig. 2.79.

The equivalent resistance across terminals A and B is calculated by short circuiting the voltage source and open circuiting the current source. The equivalent resistance will be equal to 3 Ω plus the parallel combination of 2 Ω and 3 Ω resistors. This comes to equal to 4.33 Ω.

Now we will convert the current source into an equivalent voltage source.

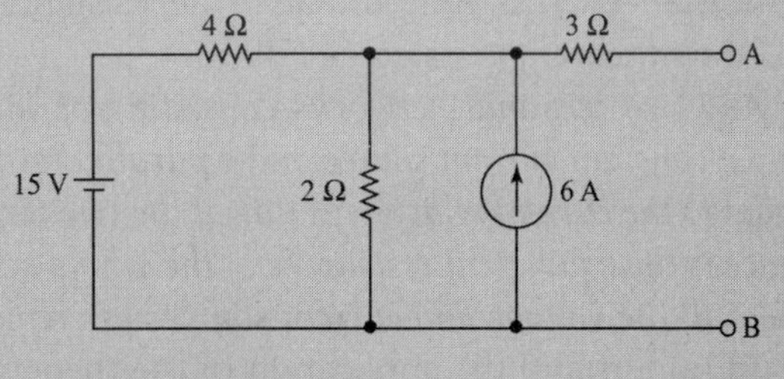

FIGURE 2.79

Converting current source of 6 A connected across the 2 Ω resistor, the equivalent voltage source is represented as shown in Fig. 2.80.

Now, we will calculate the open-circuit voltage V_{oc} across terminals A and B. Applying KVL in the loop,

$$15 - 4I - 2I - 12 = 0$$

or, $$-6I = -3$$

or, $$I = 0.5 \text{ A}$$

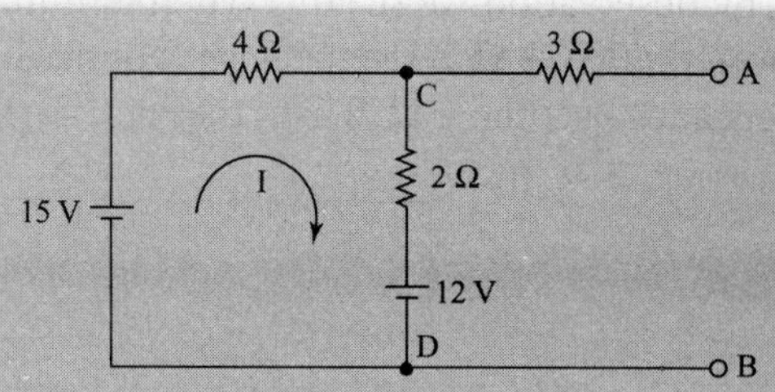

FIGURE 2.80

Since no current flows through the 3 Ω resistor, $V_{AB} = V_{OC} = V_{CD}$.

$$V_{AB} = 12 + 2I$$
$$= 12 + 2 \times 0.5$$
$$= 13 \text{ V}$$

Point C is at a higher potential than point D. Hence, point A is at a higher potential than point B. Current through the load resistor will flow from A to B.

$$I = \frac{V_{AB}}{R_{eq} + R_L}$$

Current through the load resistor $$= \frac{13}{4.33 + 5}$$
$$= 1.39 \text{ A}$$

2.8.3 Norton's Theorem

We have seen earlier that in applying Thevenin's theorem, a network is converted across two terminals of a resistance into a voltage source and an equivalent series resistance.

In applying Norton's theorem, a network is converted into a constant current source and a parallel resistance across the terminals of the resistance through which current has to be calculated. The Norton's theorem is stated as follows:

Any two terminal networks consisting of voltage sources and resistances can be converted into a constant current source and a parallel resistance. The magnitude of the constant current is equal to the current which will flow if the two terminals are short circuited and the parallel resistance is the equivalent resistance of the whole network viewed from the open-circuited terminals after all the voltage and current sources are replaced by their internal resistances.

To understand the application of the theorem, let us consider a simple circuit as shown is Fig. 2.81.

Applying Norton's theorem, let us calculate the current that would flow through the load resistance, $R_L = 5\ \Omega$ as in Fig. 2.81. The first step is to remove the load resistance and then short circuit the terminals AB as shown in Fig. 2.82, and calculate I_{SC}. This is done as follows.

By observation we see that when terminals A and B are shorted, terminals C and D also get shorted. Two 4 Ω resistances get connected in parallel across terminals EF, their equivalent resistance become 2 Ω. Thus, the total current, I supplied by the battery becomes

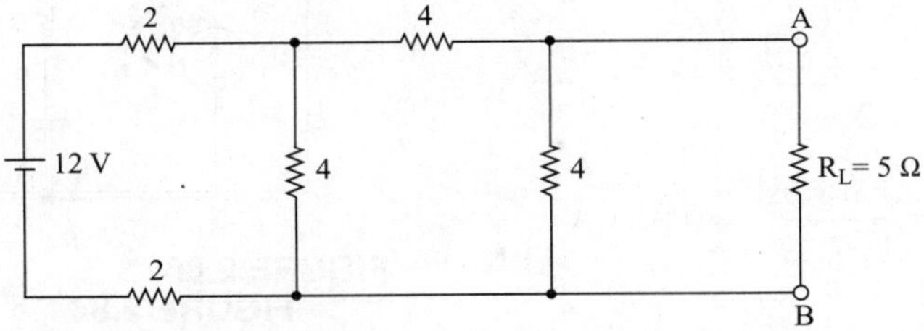

FIGURE 2.81

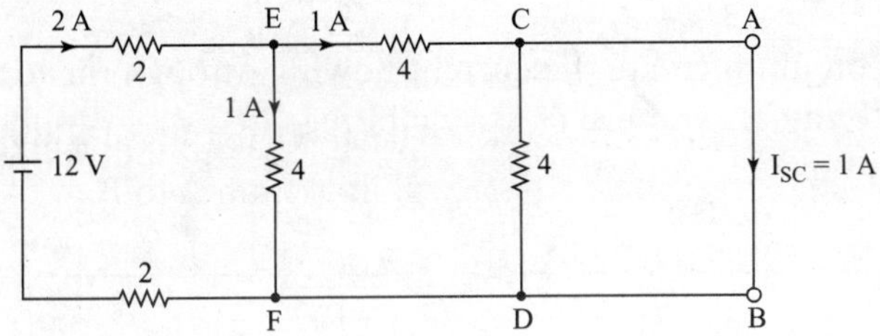

FIGURE 2.82

$$I = \frac{12}{2+2+2} = 2 \text{ A}$$

At point E, this I = 2 A gets divided equally; 1 A going in branch EF and 1 A to branch EC (CD being shorted). At point C, 1 A current will flow through the short-circuited path provided between terminals A and B. Therefore, I_{SC} = 1 A.

Now, the resistance of the network viewed from the terminals AB when the battery is short circuited is

$$R_{AB} = \frac{6 \times 4}{10} = 2.4\ \Omega \qquad \text{(see also Fig. 2.83)}$$

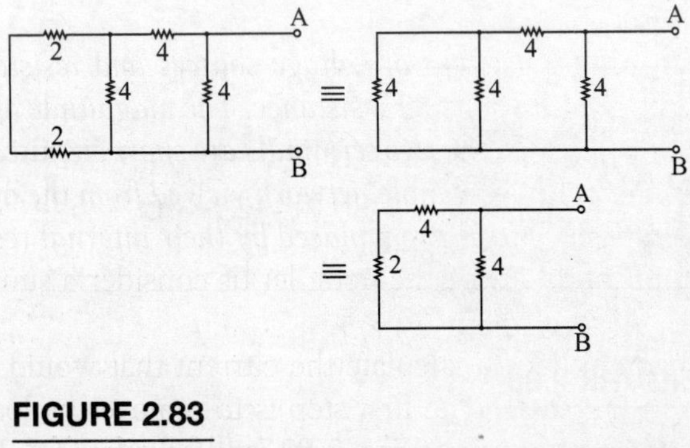

FIGURE 2.83

Norton's equivalent circuit is shown in Fig. 2.84. Now, using the current divider rule, the current through the load resistance is calculated as,

$$I_L = I\frac{2.4}{2.4+5}$$

$$= \frac{1.0\times 2.4}{7.4} = 0.324\ \text{A}$$

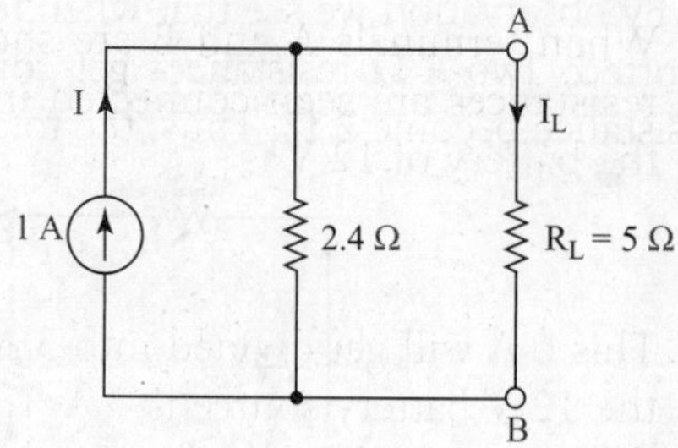

FIGURE 2.84

EXAMPLE 2.34

Using Norton's theorem, calculate the current flowing through the load resistance connected across the terminals A and B as shown in Fig. 2.85. Also apply Thevenin's theorem to calculate the same.

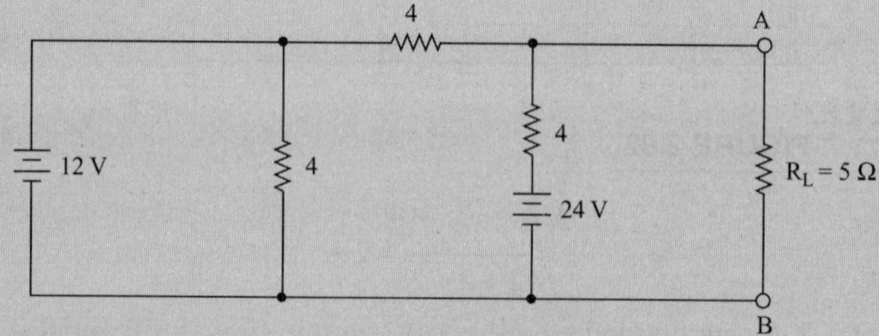

FIGURE 2.85

SOLUTION

Apply Norton's theroem

The first step is to remove R_L and short circuit the terminals A and B, and calculate I_{SC} due to the two voltage sources as in Fig. 2.86.

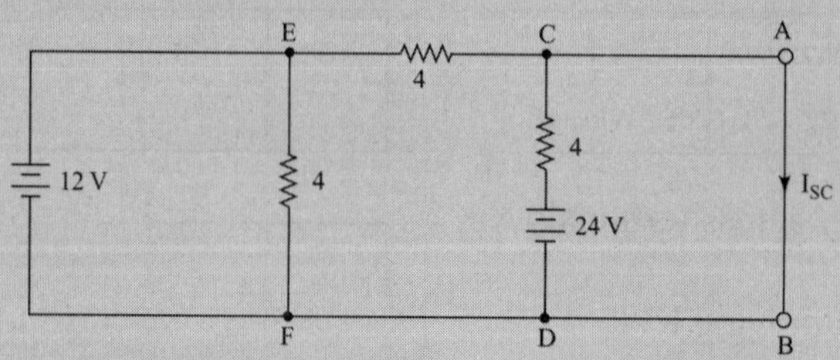

FIGURE 2.86

When terminals A and B are shorted, terminals C and D are also shorted. Two 4 Ω resistances are seen connected in parallel across terminals E and F. The current from the battery of 12 V is

$$I = \frac{12}{2} = 6 \text{ A}$$

This 6 A will get divided into 3 A each at branch EF and E C A B D F. Thus, I_{SC} due to the 12 V battery source is 3 A. I_{SC} due to the 24 V battery is calculated by considering the loop DC ABD which is 24/4 = 6 A. Thus, total I_{SC} due to both the sources of EMF is 3 + 6 = 9 Amps. The equivalent resistance of the network across terminals A and B is calculated as (after short circuiting the sources of EMFs) as in Fig. 2.87.

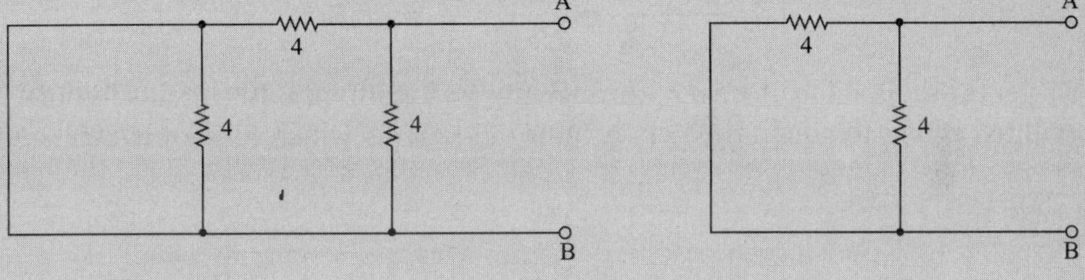

FIGURE 2.87

$$R_{AB} = R_{eq} = \frac{4 \times 4}{4+4} = 2\ \Omega$$

Norton's equivalent circuit is formed by considering the short-circuit current as the magnitude of the current source and the equivalent resistance as the parallel resistance representing the internal resistance as shown in Fig. 2.88.

Using current divider rule, $I_L = 9 \times \frac{2}{2+5} = \frac{18}{7} = 2.57$ A

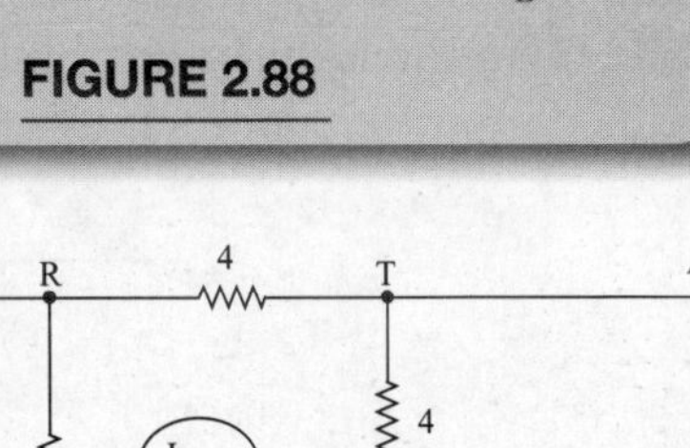

FIGURE 2.88

Apply thevenin's theorem

By shorting the voltage sources, we see $R_{eq} = R_{AB} = R_{Th} = 2\ \Omega$

Let us calculate V_{OC} across terminals A and B, in Fig. 2.89.

From loop PQRS, applying KVL,

$$+12 - (I_1 + I_2)\,4 = 0$$

or, $$(I_1 + I_2) = \frac{12}{4} = 3 \text{ A}$$

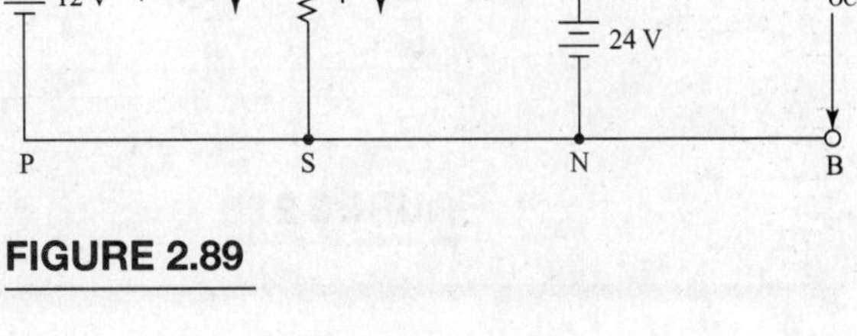

FIGURE 2.89

Similarly from loop N T R S,

$$+24 - 4I_2 - 4I_2 - 4(I_1 + I_2) = 0$$

or,

$$8\,I_2 = 24 - 4(I_1 + I_2)$$

Substituting the value of $(I_1 + I_2)$

$$I_2 = \frac{24 - 4\times 3}{8} = \frac{12}{8} = \frac{3}{2}\text{ A}$$

$$V_{OC} = V_{NT} = 24 - 4I_2 = 24 - 4\times\frac{3}{2} = 18\text{ V}$$

According to Thevenin's,

$$I_L = \frac{V_{OC}}{R_{Th} + R_L} = \frac{18}{2+5} = \frac{18}{7} = 2.57\text{ A}$$

Thus, we get the same value of current through the load resistance. In circuit solutions, experience will tell you as to which theorem is more suitable for which kind of circuit solution.

EXAMPLE 2.35

Apply Norton's theorem to determine the current flowing through the resistance of 6 Ω connected across the terminals. A and B as shown in Fig. 2.90. Also calculate the potential of point A. What will be the current through the 6 Ω resistor across AK. Solve this problem using Thevenin's theorem also.

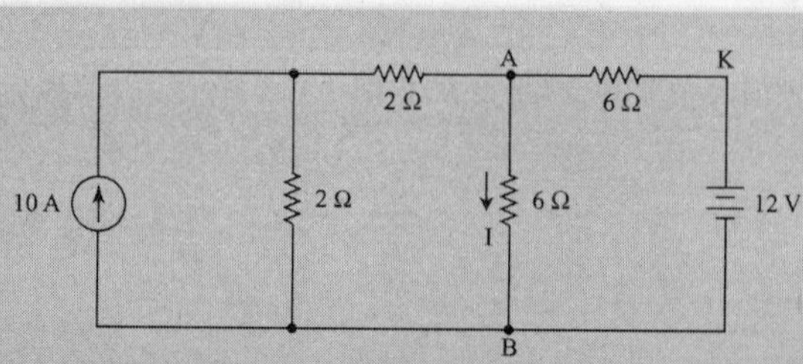

FIGURE 2.90

SOLUTION

After removing the 6 Ω resistor across terminals A and B, we short circuit the terminals and determine I_{sc} due to the current source and also due to the voltage source and then add them to find their combined effect. Terminals A and B have been shown short circuited separately for each of the sources as shown in Fig. 2.91.

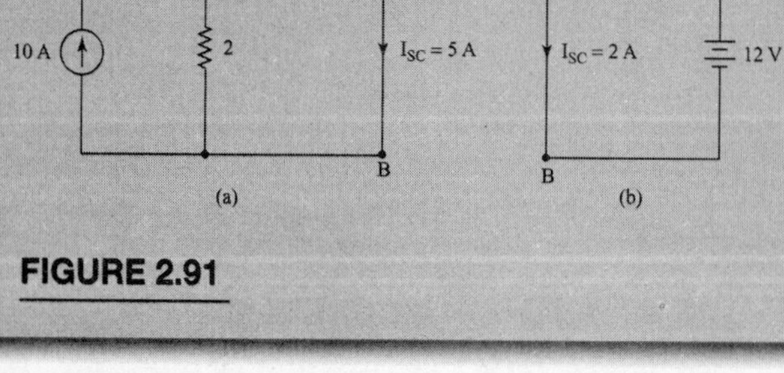

FIGURE 2.91

Total current due to two sources, $I_{sc} = 5 + 2 = 7\text{ A}$

Equivalent resistance across A and B in Fig. 2.92 is determined by open circuiting the current source and short circuiting the voltage source.

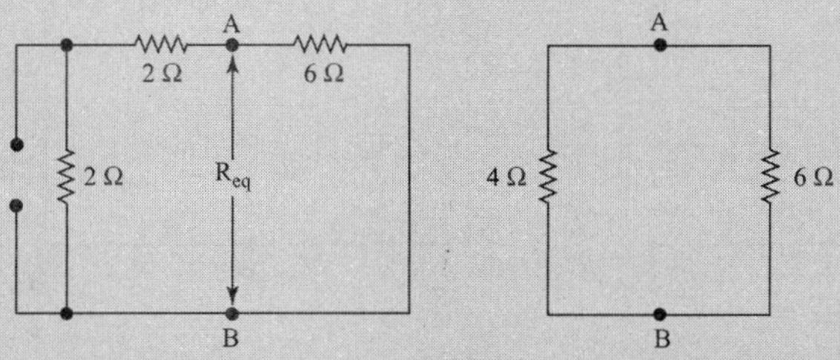

FIGURE 2.92

$$R_{eq} = R_{AB} = \frac{6 \times 4}{6 + 4} = 2.4\ \Omega$$

Norton's equivalent circuit connected across terminals A and B is shown in Fig. 2.93. Using the current divider rule current, I is calculated as

$$I_L = I \times \frac{R_{eq}}{R_{eq} + R_L} = 7 \times \frac{2.4}{2.4 + 6} = 2\text{ A}$$

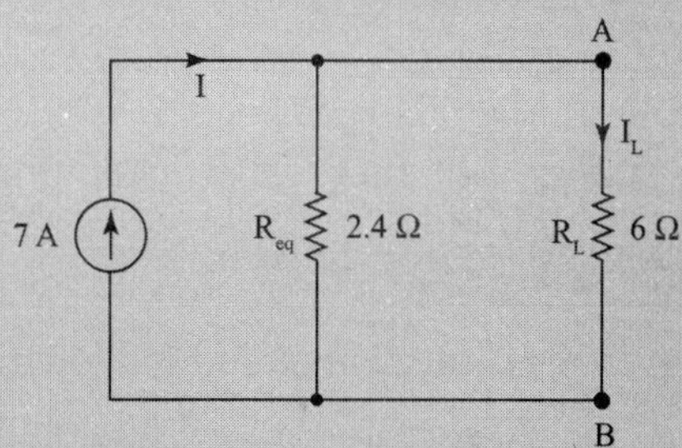

FIGURE 2.93

The potential of A with respect to B = 6 × 2 = 12 V

Since potential across A and B is 12 V and the potential of point K with respect to point B is again 12 V, no current will flow through the resistor connected between terminals A and K.

Now, let us apply thevenin's theorem

We convert the current source of 10 A and a parallel resistance of 2 **Ω**, into its equivalent voltage source and the network is redrawn as shown in Fig. 2.94.

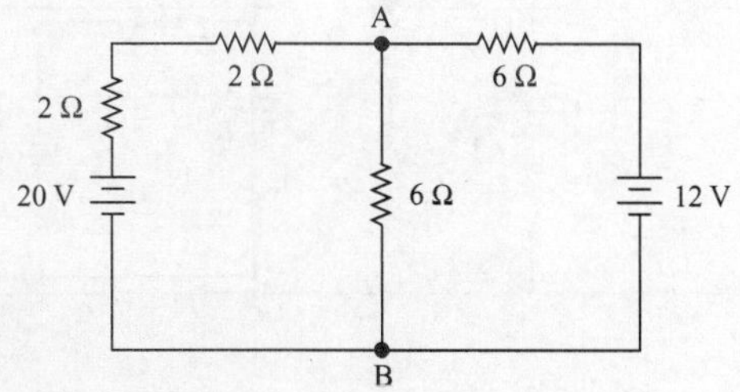

FIGURE 2.94

Remove the 6 Ω resistor across AB and determine V_{OC}, as in Fig. 2.95.

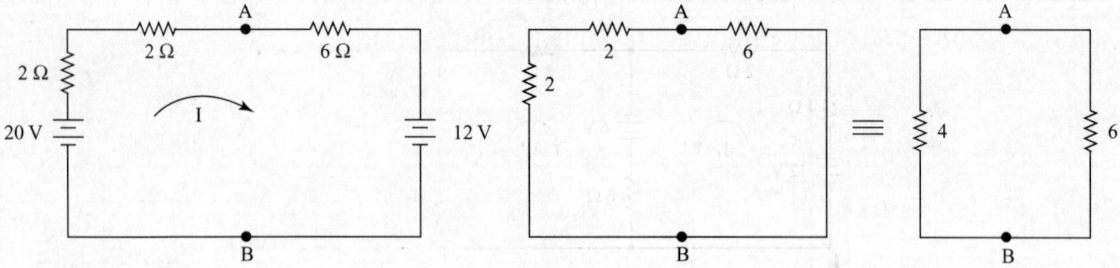

FIGURE 2.95

We write,

$$+20\text{ V} - 2I - 2I - 6I - 12\text{ V} = 0$$

or,

$$10\,I = 8\text{ V}$$

$$I = 0.8\text{ A}$$

$$V_{OC} = +12 + 6 \times 0.8 = 16.8\text{ V}$$

R_{eq} across AB after short circuiting the source of EMFs,

$$R_{eq} = R_{AB} = \frac{6 \times 4}{6+4} = 2.4\ \Omega$$

Thevenin's equivalent circuit is shown in Fig. 2.96.

$$I = \frac{V_{OC}}{R_{Th} + R} = \frac{16.8}{2.4+6} = 2\text{ A}$$

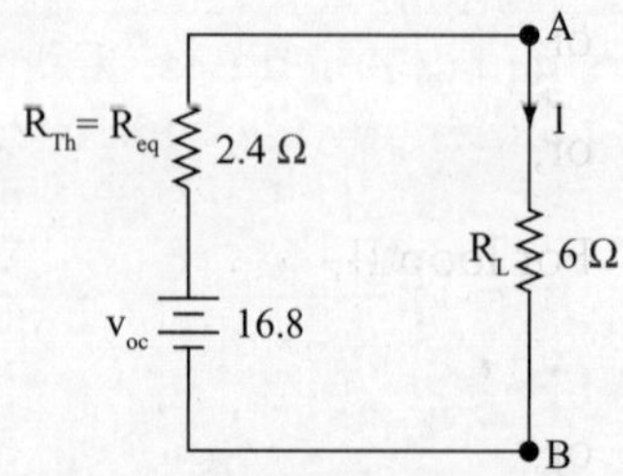

FIGURE 2.96

EXAMPLE 2.36

By applying Thevenin's as well as Norton's theorems, show that current flowing through the 16 Ω resistance in the network shown in Fig. 2.97 is 0.5 A.

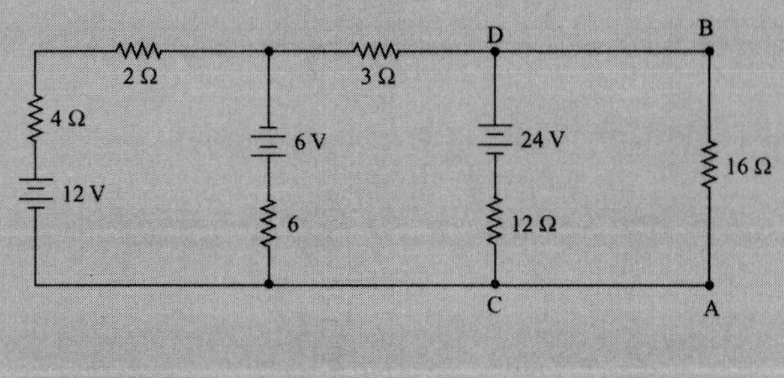

FIGURE 2.97

Apply thevenin's theorem

Remove the 16 Ω resistor and calculate V_{OC} as shown in Fig. 2.98.

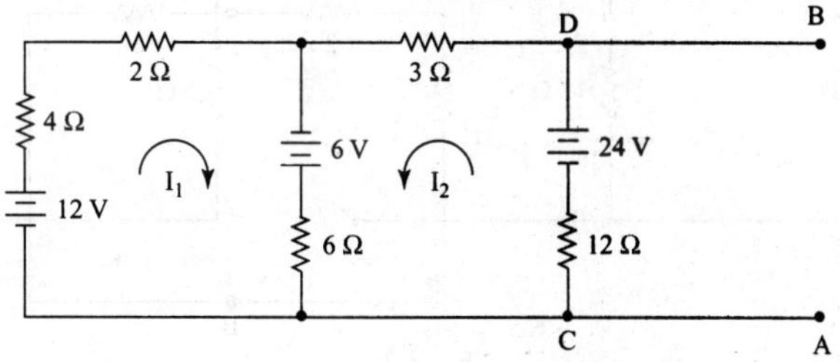

FIGURE 2.98

Kirchhoff's voltage equations are

For loop I,

$$12 - 6I_1 + 6 - 6(I_1 + I_2) = 0$$

or, $$6(2I_1 + I_2) = 18$$

or, $$2I_1 + I_2 = 3 \qquad \text{(i)}$$

For loop II,

$$24 - 3I_2 + 6 - (I_1 + I_2)6 - 12I_2 = 0$$

or, $$21I_2 + 6\,I_1 = 30$$

or, $$7I_2 + 2\,I_1 = 10 \qquad \text{(ii)}$$

$$2I_1 + 7I_2 = 10 \qquad \text{(iii)}$$

$$2I_1 + I_2 = 3 \qquad \text{(i)}$$

Subtracting,

$$6I_2 = 7$$

or, $$I_2 = \frac{7}{6}\text{ A}$$

From (i)

$$2I_1 = 3 - I_2 = 3 - \frac{7}{6} = \frac{11}{6}$$

$$I_1 = \frac{11}{12}\text{ A}$$

$$V_{CD} = V_{OC} = -12\,I_2 + 24$$

$$= -12 \times \frac{7}{6} + 24 = -14 + 24 = 10\text{ V}$$

R_{th} is calculated as in Fig. 2.99.

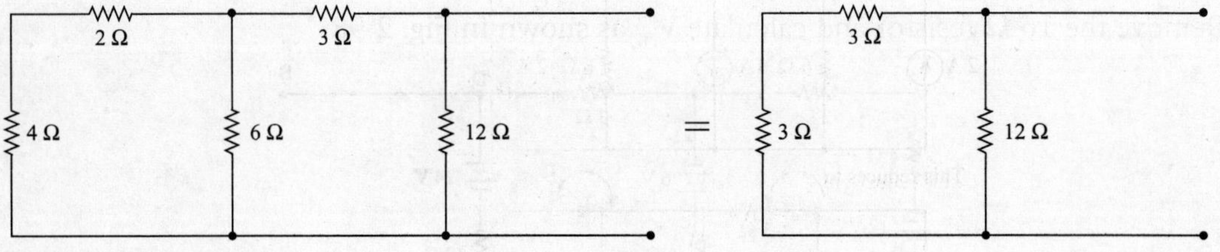

FIGURE 2.99

$$R_{th} = \frac{6 \times 12}{6+12} = 4\,\Omega$$

Current through the 16 Ω resistor is

$$I = \frac{V_{OC}}{R_{th}+16} = \frac{10}{4+16} = \frac{10}{20} = 0.5\text{ A}$$

Now, we will apply Norton's theorem to calculate the current in Fig. 2.100.

First, we will convert the voltage sources into equivalent current sources as in Fig. 2.101. The circuit becomes the one shown in Fig. 2.102.

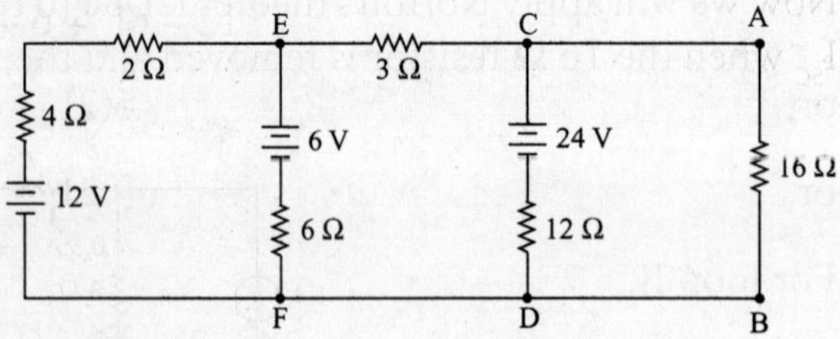

FIGURE 2.100

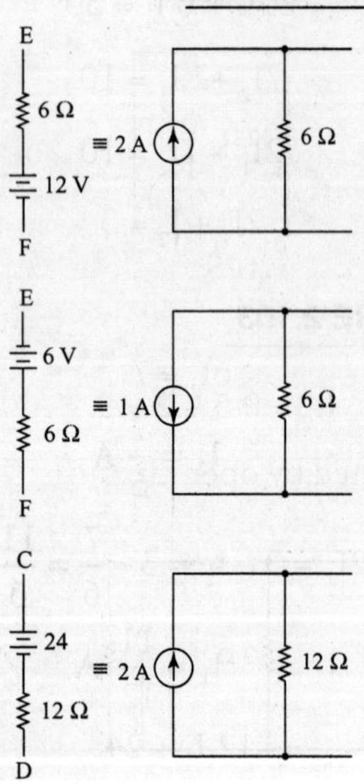

FIGURE 2.101

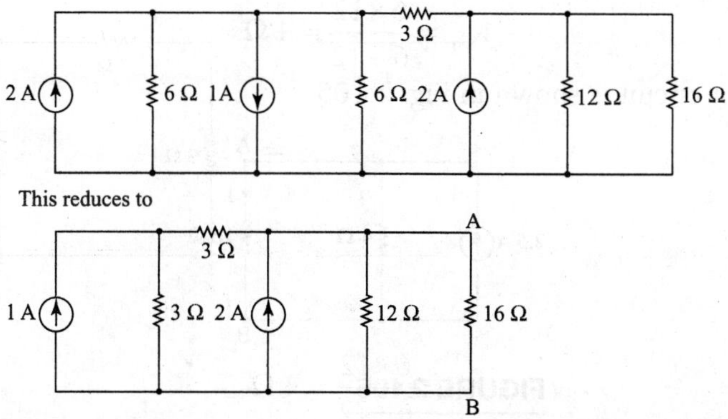

FIGURE 2.102

Now we will apply Norton's theorem. Due to the two current sources let us calculate the total I_{SC} when the 16 Ω resistor is removed and the terminals A and B are shorted as in Fig. 2.103.

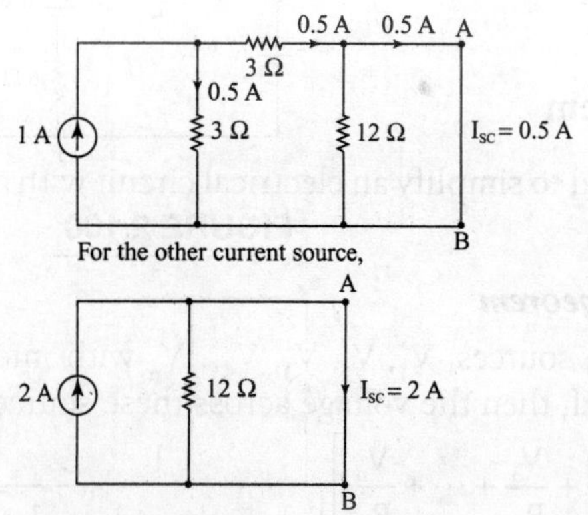

FIGURE 2.103

Total $$I_{SC} = 0.5 + 2 = 2.5 \text{ A}$$

The equivalent resistance is calculated by open circuiting the current sources as in Fig. 2.104.

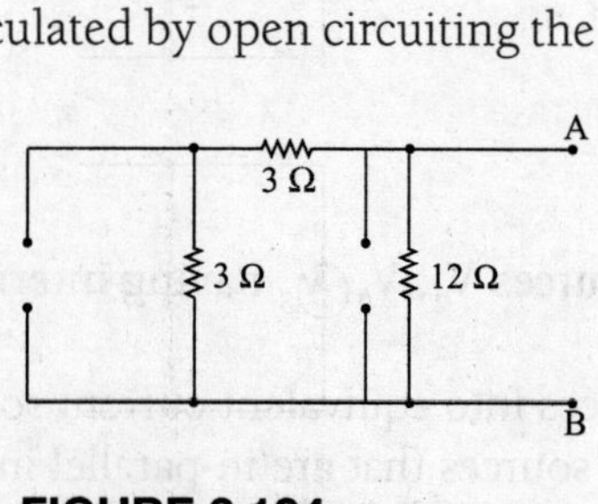

FIGURE 2.104

$$R_{eq} = \frac{6 \times 12}{6+12} = 4\ \Omega$$

Norton's equivalent circuit is shown in Fig. 2.105.

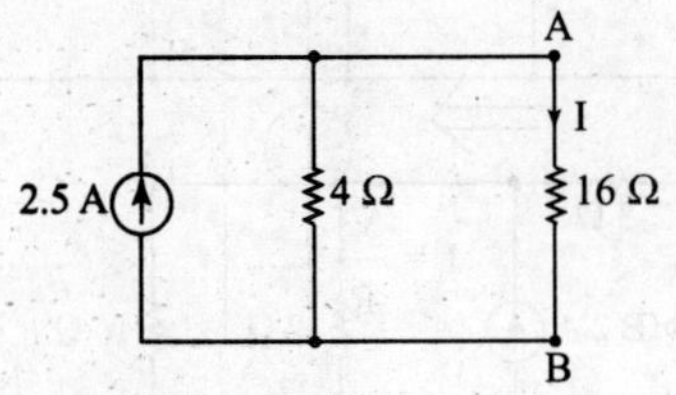

FIGURE 2.105

$$I = \frac{2.5 \times 4}{4+16}$$

or, $$I = \frac{2.5}{5} = 0.5\ \text{A}$$

Thus, the current through the 16 **Ω** resistor calculated using Thevenin's theorem and Norton's theorem is the same.

2.8.4 Millman's Theorem

Millman's theorem is applied to simplify an electrical circuit with a number of voltage sources connected in parallel.

According to millman's theorem

When a number of voltage sources, V_1, V_2, V_3, $V_4 \ldots V_n$ with internal resistances R_1, R_2, R_3, R_4, are connected in parallel, then the voltage across these sources, V can be calculated as

$$V = \left[\frac{V_1}{R_1} + \frac{V_2}{R_2} + \frac{V_3}{R_3} + \cdots + \frac{V_n}{R_n}\right] \frac{1}{\left[\frac{1}{R_1} + \frac{1}{R_2} + \frac{1}{R_3} + \frac{1}{R_4}\right]}$$

$$= \left[V_1G_1 + V_2G_2 + V_3G_3 + \cdots + V_nG_n\right] \left[\frac{1}{G_1 + G_2 + G_3 + \cdots + G_n}\right]$$

Where $$G = \frac{1}{R}$$

Proof of millman's theorem

Let us consider three voltage sources V_1, V_2, V_3 having internal resistance R_1, R_2, R_3 respectively, as shown in Fig. 2.106(a).

We convert these voltage sources into equivalent current sources as shown in Fig 2.106(b).

We then combine the current sources that are in parallel into a single source. We also take the equivalent of the three resistance that are in parallel. We get an equivalent circuit as shown in Fig. 2.106 (c). From Fig. 2.106(c), we find the voltage, V across terminals AB which is the resultant of the three voltage sources across their common terminals.

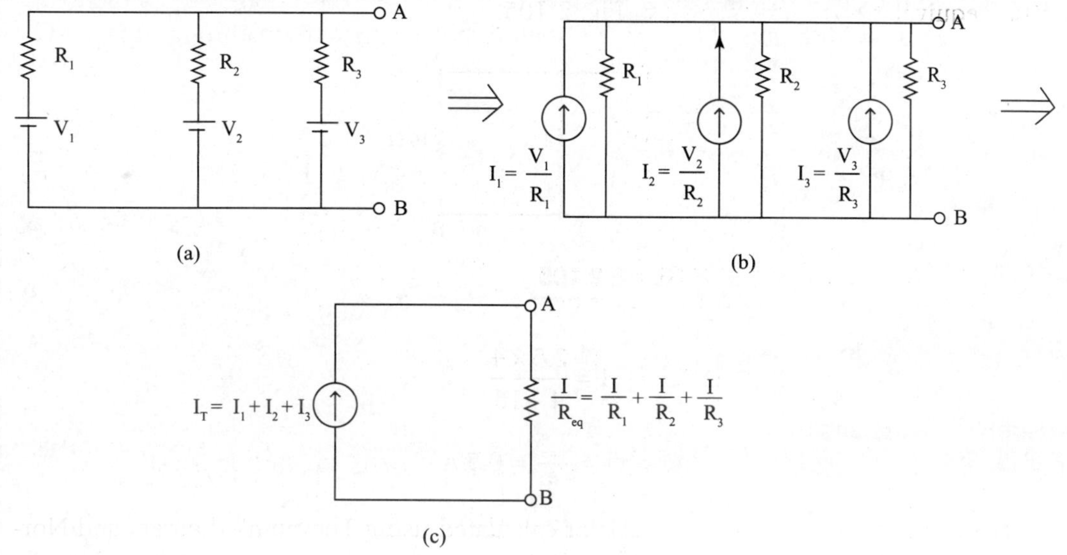

FIGURE 2.106

$$V_{AB} = I_T \times R_{eq} = (I_1 + I_2 + I_3)\, R_{eq}$$

Hence,

$$V_{AB} = \left[\frac{V_1}{R_1} + \frac{V_2}{R_2} + \frac{V_3}{R_3}\right]\left[\frac{1}{\left(\frac{1}{R_1} + \frac{1}{R_2} + \frac{1}{R_3}\right)}\right]$$

EXAMPLE 2.37

Find the current flowing through the load resistance of 10 ohms in the circuit shown in Fig. 2.107, using Millman's theorem.

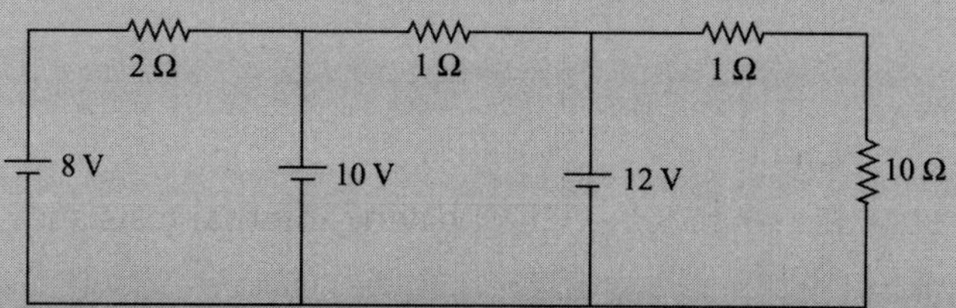

FIGURE 2.107

SOLUTION

We redraw the circuit as in Fig. 2.108 and then calculate the voltage across the load resistance using Millman's theorem. We then calculate the current through the 10 Ω load resistance.

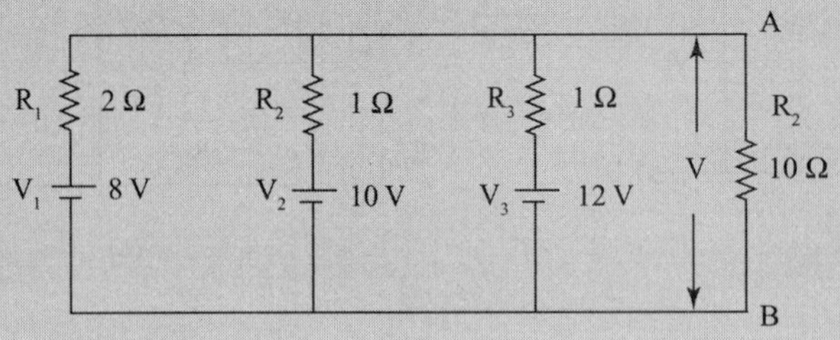

FIGURE 2.108

Applying Millman's theorem,

$$V = \left[\frac{V_1}{R_1} + \frac{V_2}{R_2} + \frac{V_3}{R_3}\right] \times \left[\frac{1}{\frac{1}{R_1} + \frac{1}{R_2} + \frac{1}{R_3}}\right]$$

Substituting values,

$$V = \left[\frac{8}{2} + \frac{10}{1} + \frac{12}{1}\right]\left[\frac{1}{\frac{1}{2} + \frac{1}{1} + \frac{1}{1}}\right]$$

$$= 26 \times \frac{1}{5/2}$$

$$= 26 \times \frac{2}{5}$$

$$= 10.4 \text{ V.}$$

Current through the 10 Ω resistor is

$$I = \frac{V}{R_L} = \frac{10.4}{10} = 1.04 \text{ A.}$$

EXAMPLE 2.38

The voltage sources of 12 V, 13 V, and 14 V having internal resistances of 1 Ω, 2 Ω, and 3 Ω, respectively, are connected in parallel. Prove that the voltage available across their common terminals is 12.63 V.

SOLUTION

The circuit is shown in Fig. 2.109 below.

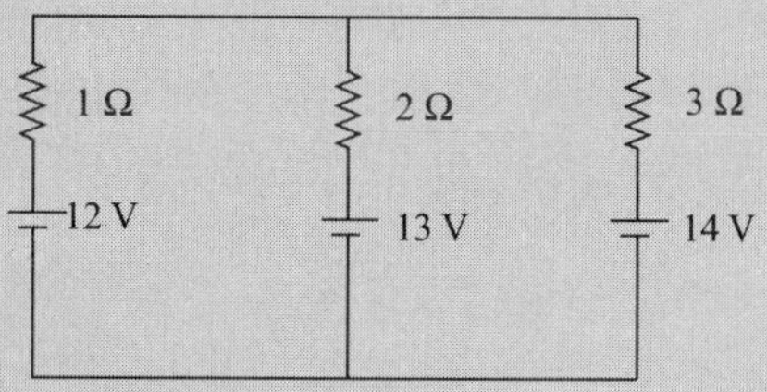

FIGURE 2.109

We will convert the voltage sources into equivalent current sources as shown in Fig. 2.110.

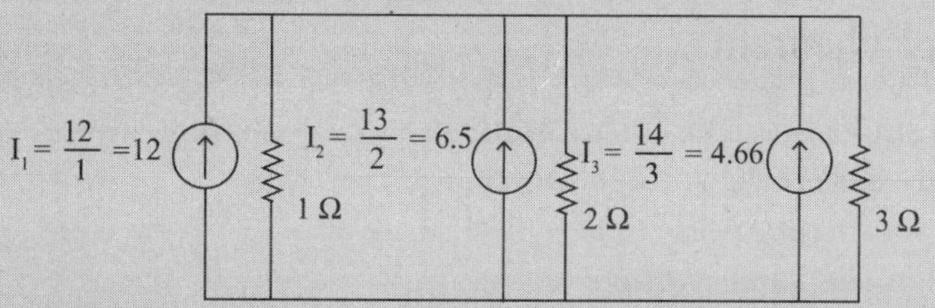

FIGURE 2.110

Now, the current sources are added and equivalent resistance of the three parallel resistance taken as shown in Fig. 2.111.

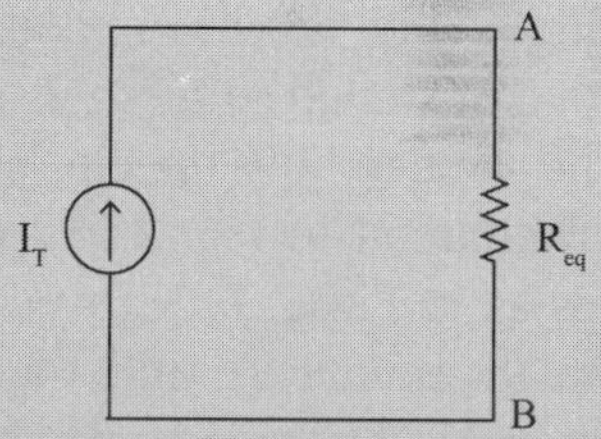

FIGURE 2.111

$$I_T = I_1 + I_2 + I_3$$

$$= 12 + 6.5 + 4.66$$

$$= 23.16 \text{ A}$$

$$\frac{1}{R_{eq}} = \frac{1}{R_1} + \frac{1}{R_2} + \frac{1}{R_3}$$

$$= \frac{1}{1} + \frac{1}{2} + \frac{1}{3}$$

$$= \frac{6+3+2}{6}$$

$$= \frac{11}{6}$$

$$R_{eq} = \frac{6}{11}\Omega$$

$$V_{AB} = I_T \times R_{eq}$$

$$= 23.16 \times \frac{6}{11}$$

$$= \frac{138.96}{11}$$

$$= 12.63 \text{ V}$$

2.8.5. Reciprocity Theorem

Reciprocity theorem states that in a linear bilateral single source network, if the voltage source, V connected in one mesh produces a current, I in another mesh, then the same voltage source connected to that mesh will produce the same current, I in the first mesh.

This is illustrated through the circuit shown in Fig. 2.112 (a). In Fig. 2.112 (b) the voltage source in shifted to mesh 2. The current flow, I is the same in both the circuits as shown.

The circuit in Fig. 2.112 (a) is simplified in steps as in Fig. 2.113, to find the total current from the source, V and then the current I in branch AB. The same current will flow through branch CD when the voltage source is shifted as in Fig. 2.112 (b)

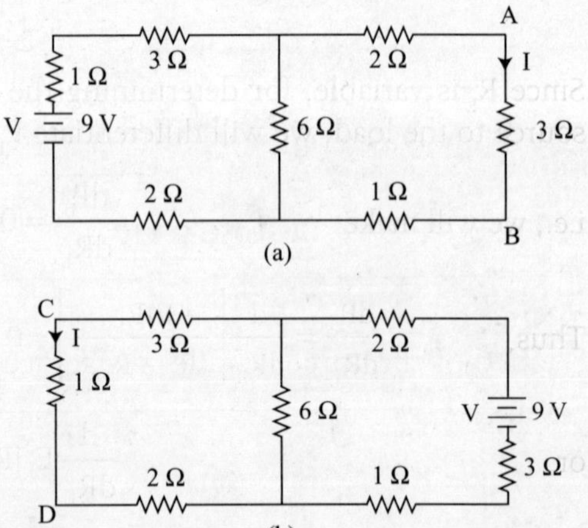

FIGURE 2.112

Reciprocity theorem illustrated

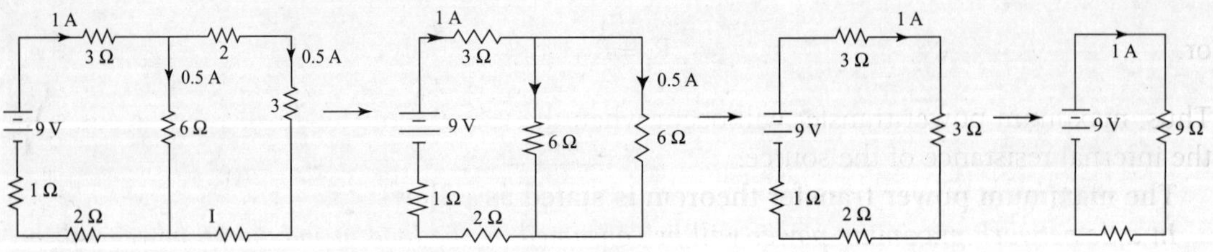

FIGURE 2.113

2.8.6 Maximum Power Transfer Theorem

Power is supplied from a source to a load. Figure 2.114 shows a generator with internal resistance, R_i supplying power to a load resistance, R_L. Maximum power transfer theorem tells us at what load (i.e., at what value of R_L) maximum power will be transferred from the source to the load.

Let us arrive at the condition of maximum power transfer through the following calculations and thereafter state the related theorem.

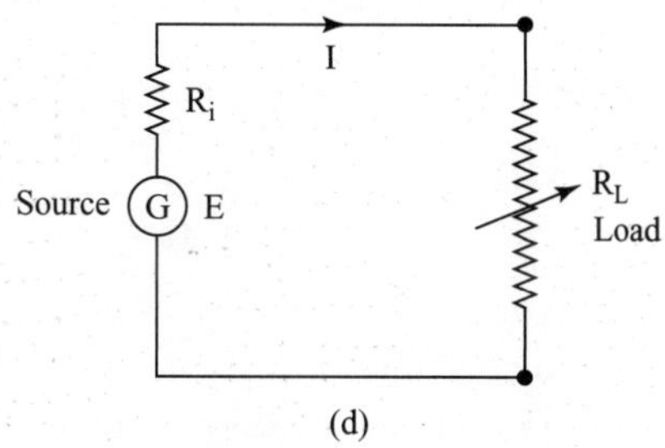

FIGURE 2.114

Current in the circuit, $$I = \frac{E}{R_i + R_L}$$

Let power consumed by, i.e., power delivered to the load be P_L. Since power = I^2R,

$$P_L = I^2 R_L = \left(\frac{E}{R_i + R_L}\right)^2 R_L$$

Since R_L is variable, for determining the condition for maximum power transfer from the source to the load, we will differentiate P_L with respect to R_L.

i.e., we will make $$\frac{dP_L}{dR_L} = 0$$

Thus, $$\frac{dP_L}{dR_L} = \frac{d}{dR_L}\left[\frac{E^2 R_L}{(R_L + R_i)^2}\right] = 0 \quad \text{or,} \quad \frac{d}{dR_L} E^2 \left[\frac{R_L}{(R_L + R_i)^2}\right] = 0$$

or, $$\frac{d}{dR_L} E^2 [R_L (R_L + R_i)^{-2}] = 0$$

or, $$E^2 \left[\frac{1}{(R_L + R_i)^2} - \frac{2R_L}{(R_L + R_i)^3}\right] = 0$$

or, $$2R_L (R_L + R_i)^2 = (R_L + R_i)^3$$

or, $$2R_L = R_L + R_i$$

or, $$R_L = R_i$$

Thus, maximum power transfer will occur when the value of the load resistance is equal to the internal resistance of the source.

The maximum power transfer theorem is stated as follows:

In a dc network maximum power will be consumed by the load or maximum power will be transferred from the source to the load when the load resistance becomes equal to the internal resistance of the network as viewed from the load terminals.

The value of maximum power when $R_L = R_i$ is calculated as

$$P_L(\max) = \frac{E^2 R_L}{(R_L + R_L)^2} \quad (\text{since } R_i = R_L)$$

$$= \frac{E^2}{4R_L} = \frac{E^2}{4R_i} \qquad (2.4)$$

When a complex dc network is to be analysed for maximum power transfer, the circuit can first be converted into a voltage source with one internal resistance by applying Thevenin's theorem.

Efficiency of a circuit is the ratio of power supplied to the load to the power supplied by the source.

Accordingly, for maximum power transfer, the

$$\text{Power transfer efficiency,} \qquad \eta = \frac{\text{Power supplied to the load}}{\text{Power supplied by the source}}$$

$$= \frac{\text{output}}{\text{output} + \text{losses}}$$

$$= \frac{I^2 R_L}{I^2 R_L + I^2 R_i}$$

We know, condition for maximum power transfer; $\quad R_i = R_L$

$$\therefore \% \, \eta = \frac{I^2 R_L \times 100}{I^2 R_L + I^2 R_L} = 50$$

So, the power transfer efficiency is 50 percent only, which is not desirable.

Maximum power transfer is not tried by the electricity departments because transmission efficiency becomes low. Half the power generated is wasted as loss. Equal amount of power is lost in the load and also as loss in the transmission system including the losses in the generator.

However, maximum power transfer is tried in the area of electronic circuits. In electronics, we often deal small power. For example, maximum power transfer is desirable from the output amplifier to the speaker in an audio sound system and the circuit design is made for maximum power transfer.

Another example is a TV antenna. A TV antenna receives power from radio waves. The power collected by the antenna is very small. The receiver circuit should be designed to receive maximum power from the antenna.

2.8.7 Tellegen's Theorem

Tellegen's theorem states that the algebraic sum of powers in all the branches in a network at any instant of time is zero. This theorem is valid for any network that may be linear or non-linear, active or passive with the condition that all branch currents and voltages in the network must satisfy Kirchhoff's laws.

According to the Tellegen's theorem, the rate of supply of energy by the active elements of a network equals the rate of energy dissipated or stored by the passive elements of the network.

EXAMPLE 2.39

A 12 V battery is supplying power to a resistive load R_L through a network as shown in Fig. 2.115. Calculate at what value of R_L power transferred to the load will be maximum and what would be the value of that maximum power.

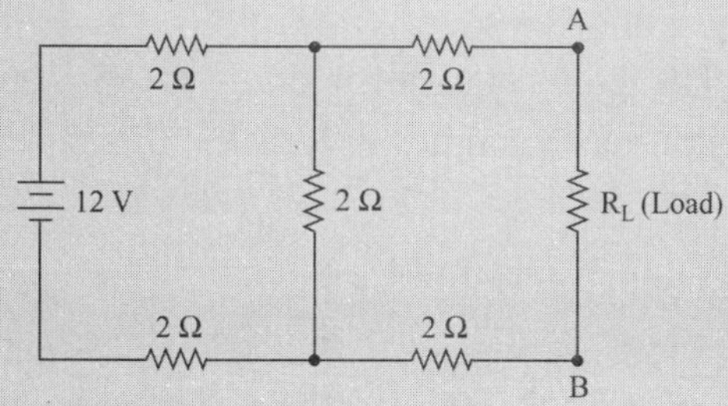

FIGURE 2.115

SOLUTION

Let us convert this circuit into a Thevenin's equivalent circuit through the following steps:

Open circuit voltage V_{AB} is calculated as in Fig. 2.116.

$$I = \frac{12}{2+2+2} = \frac{12}{6} = 2 \text{ A}$$

Voltage drop across terminals C and D = 2 × 2

= 4 V

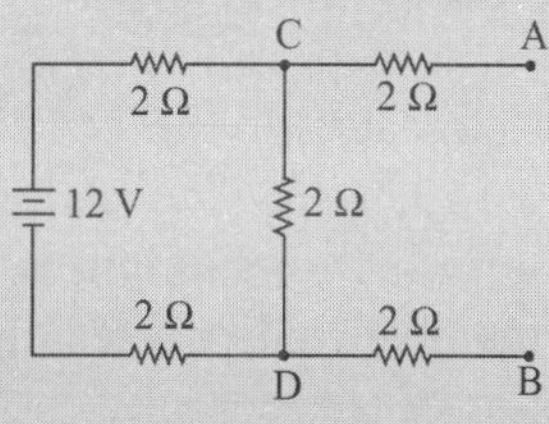

FIGURE 2.116

Voltage across terminals C and D is the same as the voltage across terminals A and B since no current is flowing beyond terminals C and D when the load resistance has been removed as has been shown in Fig. 2.117.

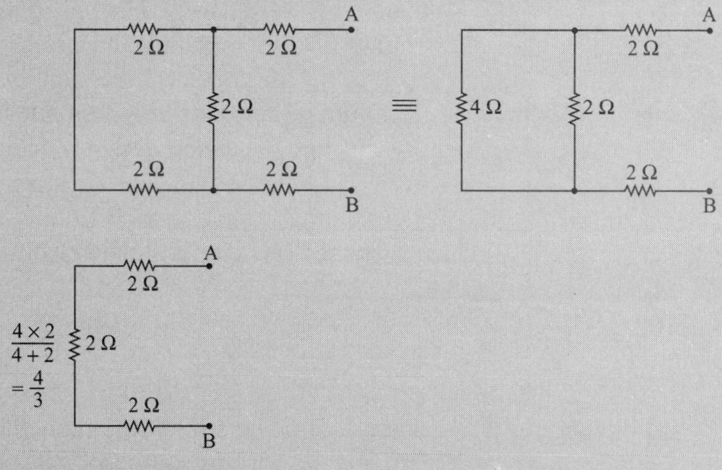

FIGURE 2.117

The equivalent resistance of the circuit as viewed from the load end, after short circuiting the voltage source is calculated as,

$$R_{eq} = 2 + 2 + \frac{4}{3} = \frac{16}{3}$$
$$= 5.33\ \Omega$$

Thevenin's equivalent circuit is represented as in Fig. 2.118.

For maximum power transfer, $R_L = R_i$

Therefore, $R_L = 5.33\ \Omega$

Value of maximum power

$$P_L(\max) = \frac{E^2}{4R_L} = \frac{4^2}{4 \times 5.33} = \frac{4}{5.33} \text{ Watts}$$
$$= 0.75 \text{ Watts}$$

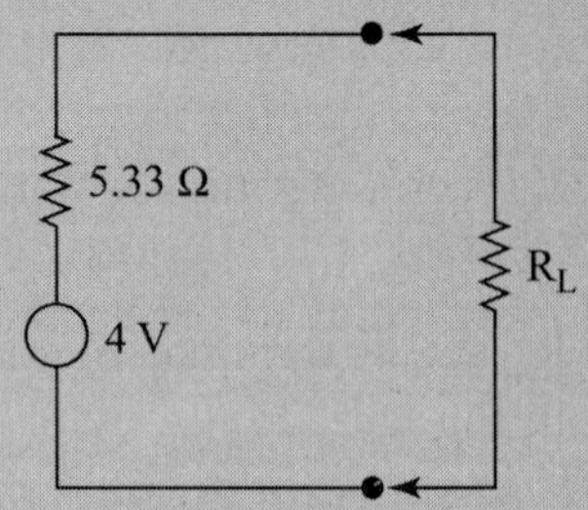

FIGURE 2.118

EXAMPLE 2.40

Calculate for the circuit shown in Fig. 2.119, the value of load resistance, R_L for which maximum power will flow to the load. Also calculate the maximum power transfer efficiency, i.e., the power transmission efficiency when maximum power is transferred.

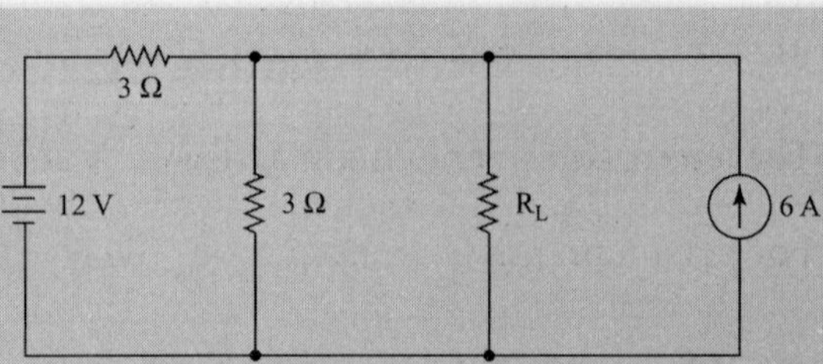

FIGURE 2.119

SOLUTION

By shifting the load resistance, R_L to the left of the circuit becomes as shown in Fig. 2.120 (a).

By converting the current source into equivalent voltage source the circuit becomes as shown in Fig. 2.120 (b).

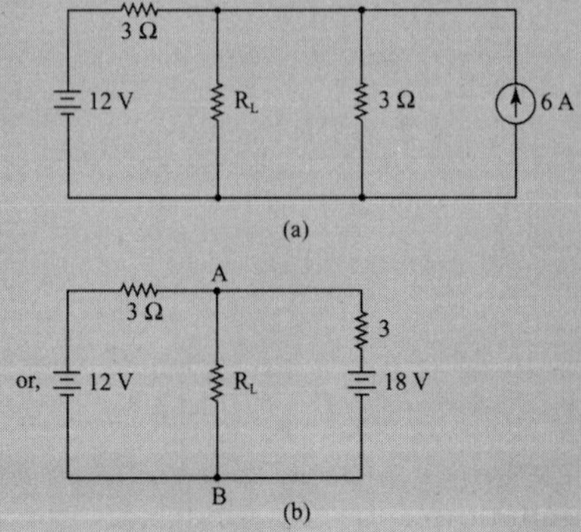

FIGURE 2.120

Open-circuit voltage across terminals A and B is calculated from Fig. 2.121.

Applying KVL,

$$+12 - 3\,I - 3\,I - 18\ V = 0$$

$$-6V = 6\,I$$

$$I = -1\ A$$

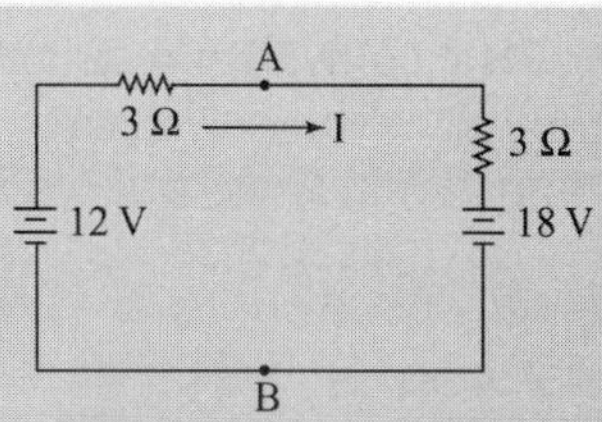

FIGURE 2.121

V_{OC} across terminals A and B = 18 V − 3 × 1

= 15 V

R_{eq} across terminals A and B by short-circuiting voltage sources is, $R_{eq} = \frac{3}{2}\ \Omega$ as shown in Fig. 2.122.

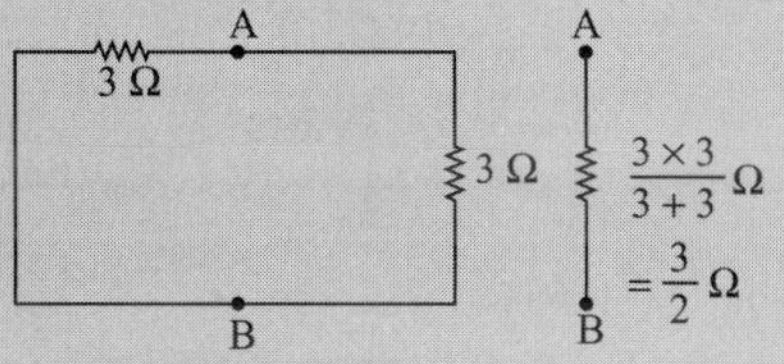

FIGURE 2.122

Thevenin's equivalent circuit across R_L is shown in Fig. 2.123.

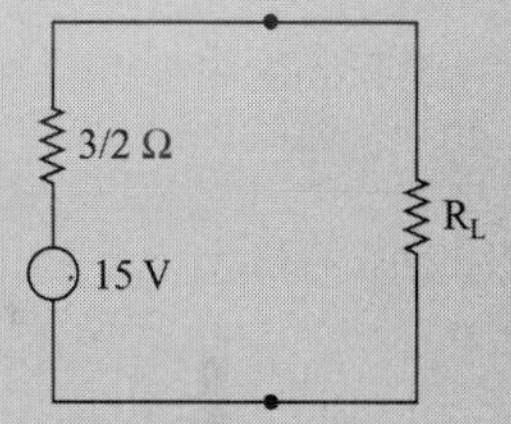

FIGURE 2.123

For maximum power transfer, $R_L = \frac{3}{2}\Omega = 1.5\ \Omega$

Current, $I = \frac{15}{1.5 + 1.5}$ $[\because R_L = R_i]$

$= 5\ A$

Maximum power $= I^2R_L = 5^2 \times 1.5 = 37.5$ Watts

Power transfer efficiency, $\eta = \frac{\text{Power supplied to the load}}{\text{Total power supplied by the source}}$

$$= \frac{I^2R_L}{I^2R_i + I^2R_L} = \frac{R_L}{R_i + R_L}$$

$$= \frac{1}{1 + (R_i/R_L)}$$

when $R_i = R_L$,

$$\%\eta = \frac{1}{1+1} \times 100 = 50 \text{ per cent}$$

This shows that for maximum power transfer conditions, the power transfer efficiency is 50 per cent only.

The maximum power transferred from the source to the load and also the power transfer efficiency are important in practical applications. If we change R_L, the value of power transferred and the transmission efficiency will change.

EXAMPLE 2.41

Calculate the value of load resistance, R_L for which maximum power will be transferred from the source to the load in the circuit shown in Fig. 2.124.

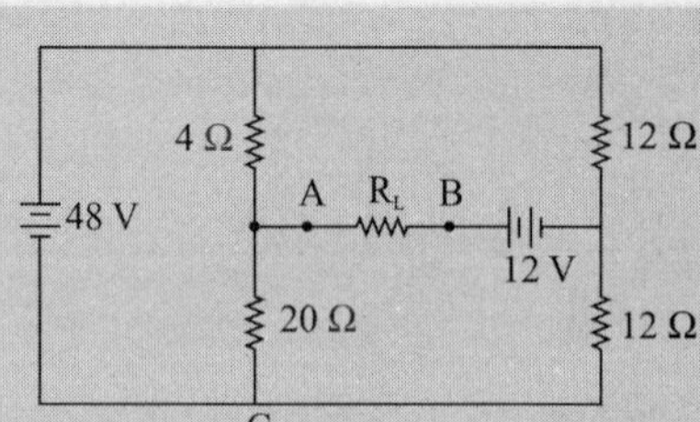

FIGURE 2.124

SOLUTION

We will apply Thevenin's theorem to reduce the circuit into a single voltage source connected across the load terminals A and B.

After removing R_L from terminals A and B we will calculate the open-circuit voltage across these terminals as Voltage drops across the 4 Ω resistor is 8 V, 20 Ω resistor is 40 V, across the 12 Ω resistors is 24 V each as have been shown in Fig. 2.125. We have to determine the voltage across terminals A and B. As in Fig. 2.125.

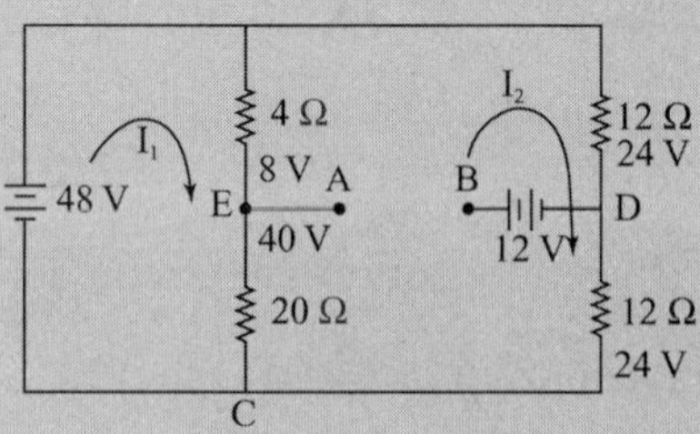

FIGURE 2.125

$$I_1 = \frac{48}{4+20} = 2 \text{ A and } I_2 = \frac{48}{12+12} = 2 \text{ A}$$

Potential difference between point C and E is 40 V. Point E is at a higher potential than C.

Potential difference between point C and D is 24 V. Point D is at a higher potential than C.

Potential of B with respect to C is 24 V + 12 V = 36 V.

Since the potential of point E or A is 40 V with respect to point C and the potential of point B with respect to point C is 36 V, point A is at a higher potential than point B. The potential difference between points A and B is 4 V. This is called V_{OC}.

Now let us calculate the equivalent resistance of the circuit shown in Fig. 2.126. Across terminals A and B after short circuiting the sources of EMFs.

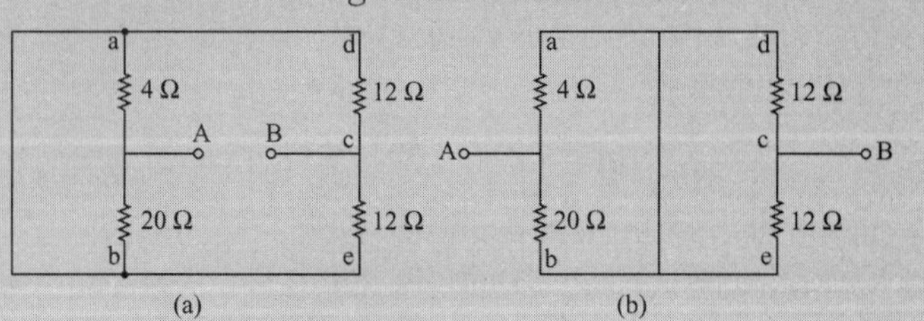

FIGURE 2.126

Rearranging, we draw the circuit as in Fig. 2.127.

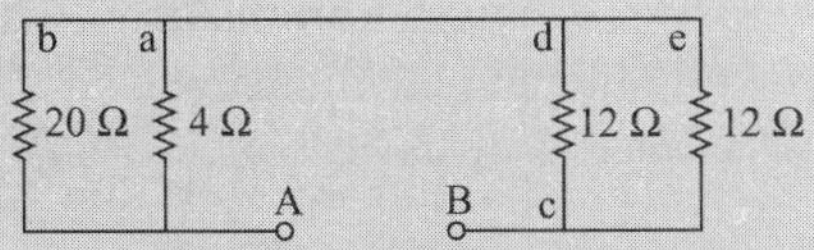

FIGURE 2.127

Note that ABDE are all connected together.

$$R_{thevenin} = \frac{4\times20}{4+20} + \frac{12\times12}{12+12}$$

$$= 3.33 + 6 = 9.33\ \Omega$$

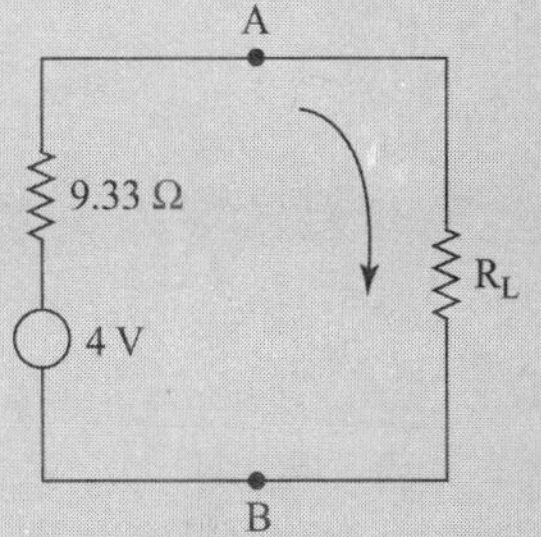

FIGURE 2.128

Thevenin's equivalent circuit is, shown in Fig. 2.128,

For maximum power transfer, $R_L = R_i$

∴ $R_L = 9.33\ \Omega$

Current, $I = \dfrac{4}{9.33+9.33}$

$= 0.214$ A

Maximum power transferred

$P_L = I^2R_L$

$= (0.214)^2 \times 9.33$ Watts

$= 0.42$ Watts

P_L (max) can also be calculated using the relation, $P_L(\text{max}) = \dfrac{E^2}{4R_L} = \dfrac{4^2}{4\times9.33} = 0.42$ Watts

EXAMPLE 2.42

Calculate the value of load resistance R_L in the circuit of Fig. 2.129, for which maximum power transfer will occur from source to load. Also calculate the value of maximum power and power transfer efficiency.

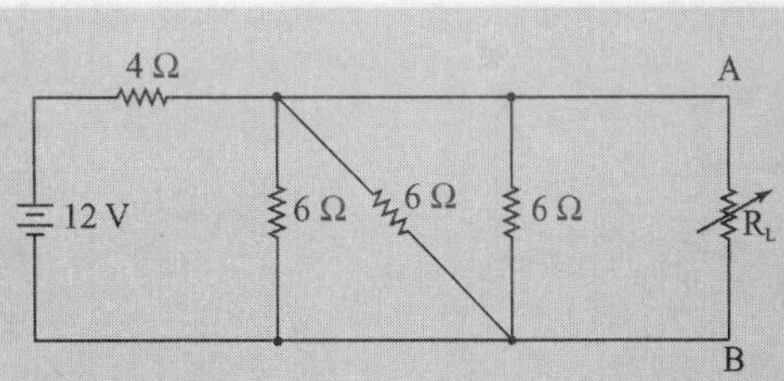

FIGURE 2.129

SOLUTION

We will apply the procedure for determining Thevenin's equivalent network. Open-circuit voltage across terminals A and B in the circiut of Fig. 2.130, is first calculated by removing R_L as

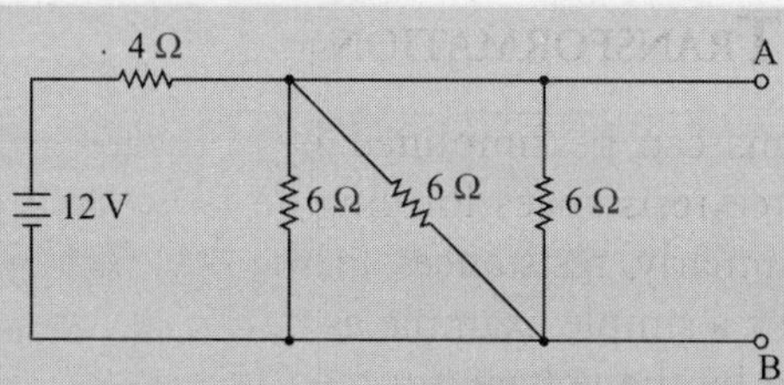

FIGURE 2.130

Since all the three 6 Ω resistors are in parallel, the above circuit gets reduced to the one shown in Fig. 2.131

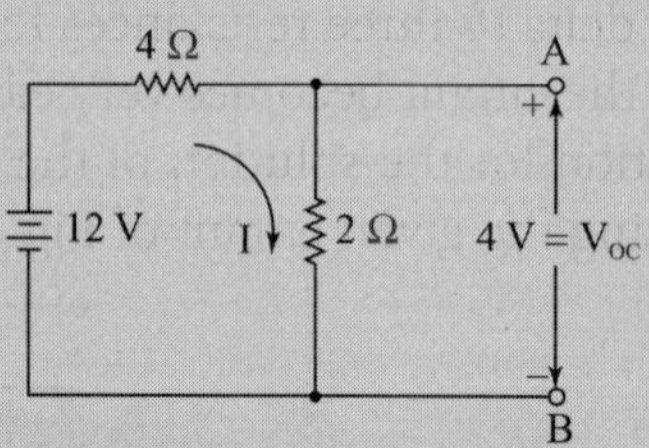

FIGURE 2.131

$$I = \frac{12}{4+2} = 2\text{ A}$$

Voltage across AB, i.e., $V_{OC} = IR = 2 \times 2 = 4\text{ V}$

Equivalent resistance R_{Th} is calculated as shown in Fig. 2.132.

$$R_{Th} = \frac{4 \times 2}{4+2} = \frac{4}{3}\ \Omega$$

Thus, Thevenin's equivalent circuit is shown in Fig. 2.133.

For maximum power transfer, $R_L = R_i$ as shown in Fig. 2.134.

$$\therefore \qquad R_L = \frac{4}{3}\Omega$$

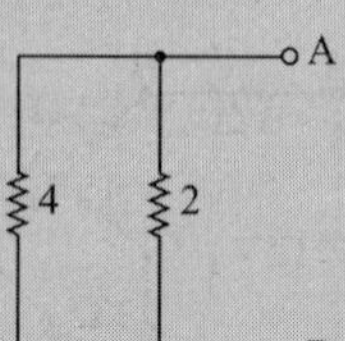

FIGURE 2.132

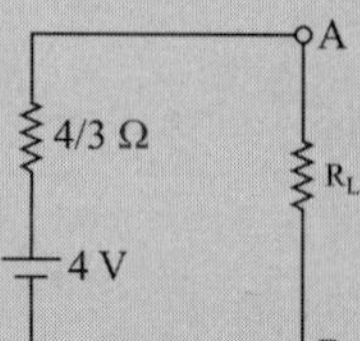

FIGURE 2.133

Current through the circuit, $I = \dfrac{4}{4/3 + 4/3} = 1.5\text{ A}$

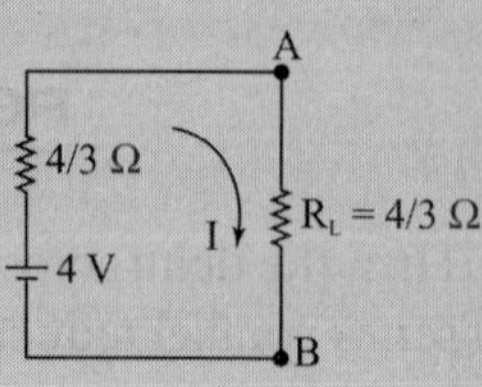

FIGURE 2.134

Power transferred or consumed = I^2R_L

$$= (1.5)^2 \times \frac{4}{3}$$

$$= 2.25 \times \frac{4}{3}\text{ Watts}$$

$$= 3\text{ Watts}$$

Transfer Efficiency, $\eta = \dfrac{R_L}{R_L + R_i} = \dfrac{R_i}{2R_i} = \dfrac{1}{2}$

Percent Efficiency = 50 per cent.

2.9 ▶ Star–Delta Transformation

Electrical network problems can be simplified by converting three resistances forming a delta to corresponding three resistances forming an equivalent star between the three terminals of the network. Similarly, resistances in the star formation can be converted into equivalent delta. Let us take a simple example as in Fig. 2.135 (a). Suppose we want to calculate the current supplied by the voltage source to the network. As such we have to write the equations for the three loops using Kirchhoff's laws and solve these equations to find the total current supplied by the battery. However, simply by transformation of three resistances in delta to three resistances forming an equivalent star, the circuit is simplified and solution of the circuit becomes very easy. This process of transformation of a delta to star and simplication for the solution of the problem is illustrated in Fig. 2.135. Fig. 2.135(b) is redrawn using the given circuit of Fig. 2.135(a).

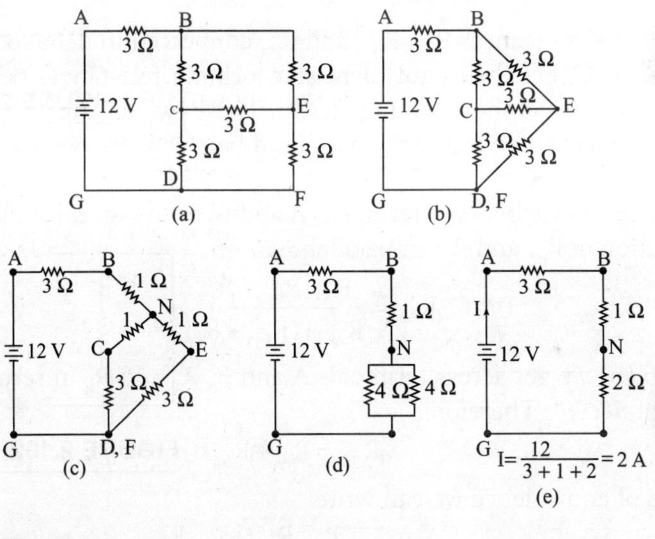

FIGURE 2.135

Then the delta across terminals BCE is changed to equivalent star as in Fig. 2.135(c). Then by series parallel reduction the circuit gets reduced to a simple series circuit as in Fig. 2.135 (e)

It can be seen from Fig. 2.135 (a) and (b) that between terminals B, C, and E three resistances of 3 Ω each are forming a delta. This delta is converted into an equivalent star between the same terminals replacing the 3 Ω resistances by equivalent 1 Ω resistances as shown in Fig. 2.135 (c). The equivalence of delta into star will be discussed a little later. By just one transformation of delta-forming resistances into star-forming resistances, the circuit is simplified and the total resistance across the battery terminals AG becomes 6 Ω, and hence, the battery current is calculated as shown in Fig. 2.135 (d) and (e). This shows that star to delta and delta to star transformation of resistances is advantageous in solving electrical circuit problems.

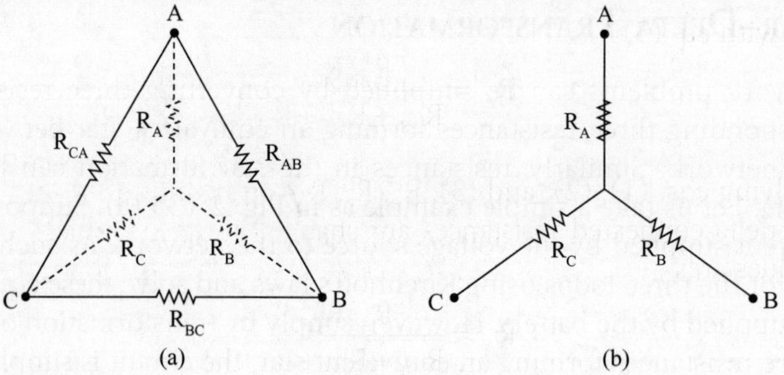

FIGURE 2.136

2.9.1 Transforming Relations for Delta to Star

Let us consider three resistances R_{AB}, R_{BC}, and R_{CA} connected in delta formation between the terminals A, B, and C. Let their equivalent star-forming resistances between the same terminals be R_A, R_B, and R_C as shown in Fig. 2.136. These two arrangements of resistances can be said to be equivalent if the resistance measured between any two terminals is the same in both the arrangements.

If we measure resistance between terminals A and B, from Fig. 2.136 (a) we will get R_{AB} and a series combination of R_{BC} and R_{CA} in parallel, i.e.,

$$R_{AB} = \frac{R_{AB}(R_{BC} + R_{CA})}{R_{AB} + (R_{BC} + R_{CA})}$$

From Fig. 2.136 (b) we get across terminals A and B, R_A and R_B in series, terminal C being open and not connected. Therefore,

$$R_{AB} = R_A + R_B$$

For the purpose of equivalence we can write

$$R_A + R_B = \frac{R_{AB}\,R_{BC} + R_{AB}R_{CA}}{R_{AB} + R_{BC} + R_{CA}} \tag{1}$$

In the same way between terminals B and C, the equivalence can be expressed as

$$R_B + R_C = \frac{R_{BC}\,(R_{CA} + R_{AB})}{R_{BC} + (R_{CA} + R_{AB})} \tag{2}$$

Between terminals C and A, the equivalence can be expressed as

$$R_A + R_C = \frac{R_{CA}\,(R_{BC} + R_{AB})}{R_{CA} + (R_{BC} + R_{AB})} \tag{3}$$

Subtracting eq. (2) from eq. (1)

$$R_A - R_C = \frac{R_{AB}\,R_{CA} - R_{BC}R_{CA}}{R_{AB} + R_{BC} + R_{CA}} \tag{4}$$

Adding eq. (4) with eq. (3)

$$R_A = \frac{R_{AB} R_{CA}}{R_{AB} + R_{BC} + R_{CA}}$$

This way by solving eqs. (1), (2), and (3), R_A, R_B, R_C can be found.

Thus, when delta-connected resistances are changed to an equivalent star-forming resistances, their values are:

$$R_A = \frac{R_{AB} R_{CA}}{R_{AB} + R_{BC} + R_{CA}} \quad (5)$$

$$R_B = \frac{R_{AB} R_{BC}}{R_{AB} + R_{BC} + R_{CA}} \quad (6)$$

$$R_C = \frac{R_{BC} R_{CA}}{R_{AB} + R_{BC} + R_{CA}} \quad (7)$$

To remember this expressions of R_A, R_B, and R_C, let us look at Fig. 2.136 (a) again. The star equivalence of delta-forming resistances can be shown through dotted lines. The value of R_A is equal to the product of the two resistances of delta touching point A, i.e., R_{AB} and R_{CA} divided by the sum of all the delta-forming resistances, i.e., sum of R_{AB}, R_{BC} and R_{CA}. Similarly the values of R_B and R_C can be remembered.

2.9.2 Transforming Relations for Star to Delta

Now let us consider star to delta transformation as shown in Fig. 2.137. The basic equations guiding the conversion from delta to star remains the same in this case also. As such we can use eqs. (1), (2), and (3). Solving eqs. (1), (2), and (3) we got eqs. (5), (6), and (7).

Multiplying eq. (5) by eq. (6),

$$R_A R_B = \frac{R_{AB}^2 R_{BC} R_{CA}}{(R_{AB} + R_{BC} + R_{CA})^2} \quad (8)$$

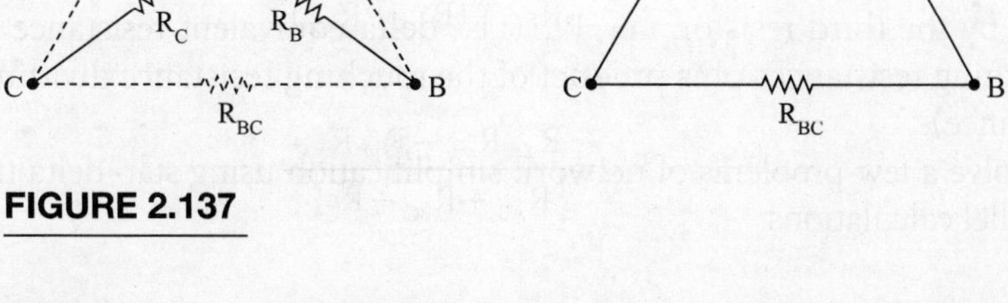

FIGURE 2.137

Multiplying eq. (5) by eq. (7), $$R_A R_C = \frac{R_{CA}^2 R_{AB} R_{BC}}{(R_{AB} + R_{BC} + R_{CA})^2} \qquad (9)$$

Multiplying eq. (6) by eq. (7), $$R_B R_C = \frac{R_{BC}^2 R_{AB} R_{CA}}{(R_{AB} + R_{BC} + R_{CA})^2} \qquad (10)$$

Now, adding eqs. (8), (9), and (10) $$R_A R_B + R_B R_C + R_C R_A = \frac{R_{AB} R_{BC} R_{CA} (R_{AB} + R_{BC} + R_{CA})}{(R_{AB} + R_{BC} + R_{CA})^2}$$

or, $$R_A R_B + R_B R_C + R_C R_A = \frac{R_{AB} R_{BC} R_{CA}}{R_{AB} + R_{BC} + R_{CA}} = R_{AB}\left[\frac{R_{BC} R_{CA}}{R_{AB} + R_{BC} + R_{CA}}\right] \qquad (11)$$

From eqn (7) we have, $$R_C = \frac{R_{BC} R_{CA}}{R_{AB} + R_{BC} + R_{CA}}$$

Substituting the value of R_C in eqn (11), $R_A R_B + R_B R_C + R_C R_A = R_{AB} R_C$

Dividing both sides by R_C and rearranging, we have $R_{AB} = R_A + R_B + \dfrac{R_A R_B}{R_C}$

Similarly R_{23} and R_{31} can be calculated.

Thus, from eqs. (5), (6), and (7), R_{AB}, R_{BC}, and R_{CA} in terms of R_A, R_B, and R_C across terminals A, B, and C are calculated.

Thus, when-star-connected resistances are changed to equivalent delta-forming resistances, the values are

$$R_{AB} = R_A + R_B + \frac{R_A R_B}{R_C} \qquad (12)$$

$$R_{BC} = R_B + R_C + \frac{R_B R_C}{R_A} \qquad (13)$$

$$R_{CA} = R_C + R_A + \frac{R_C R_A}{R_B} \qquad (14)$$

Remembering of these expressions is easy as R_{AB} is the sum of R_A and R_B plus product of R_A and R_B divided by the third resistor, i.e., R_C. (i.e., delta-equivalent resistance of one side is the sum of touching resistances plus product of the touching resistance dividing by the non-touching resistance).

Now let us solve a few problems of network simplification using star–delta transformation and series–parallel calculations.

EXAMPLE 2.43

Calculate the equivalent resistance of the network shown in Fig. 2.138, across terminals P and Q.

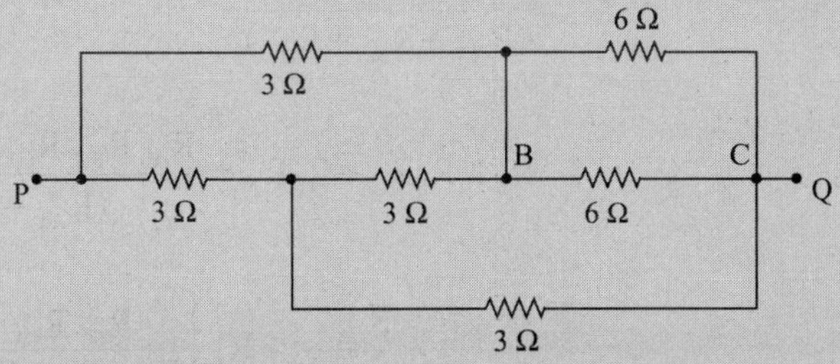

FIGURE 2.138

SOLUTION

Two 6 Ω resistors are in parallel. Their equivalent resistance is 3 Ω. So the circuit is redrawn as shown in Fig. 2.139. By pulling point Q downwords, this circuit of Fig. 2.139 is drawn as in Fig. 2.140 (a).

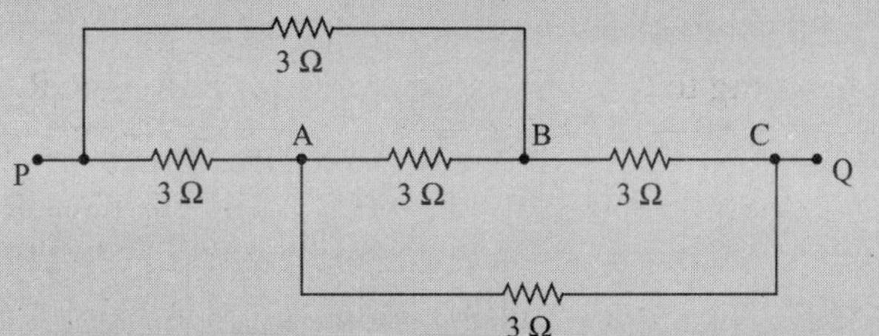

FIGURE 2.139

Between the terminals A, B, and C, the three resistances of 3 Ω each are connected in delta. This delta is now transformed into an equivalent star with the values calculated using the transformation relationship. The equivalent resistance in star are calculated as R_A, R_B, and R_C as shown in Fig. 2.140 (b).

$$R_A = \frac{R_{AB}\,R_{AC}}{R_{AB}+R_{BC}+R_{CA}}$$

$$= \frac{3\times 3}{3+3+3} = 1\,\Omega$$

$$R_B = \frac{R_{AB}\,R_{BC}}{R_{AB}+R_{BC}+R_{CA}}$$

$$= \frac{3\times 3}{3+3+3} = 1\,\Omega$$

$$R_C = \frac{R_{BC}\,R_{CA}}{R_{AB}+R_{BC}+R_{CA}}$$

$$= \frac{3\times 3}{3+3+3} = 1\,\Omega$$

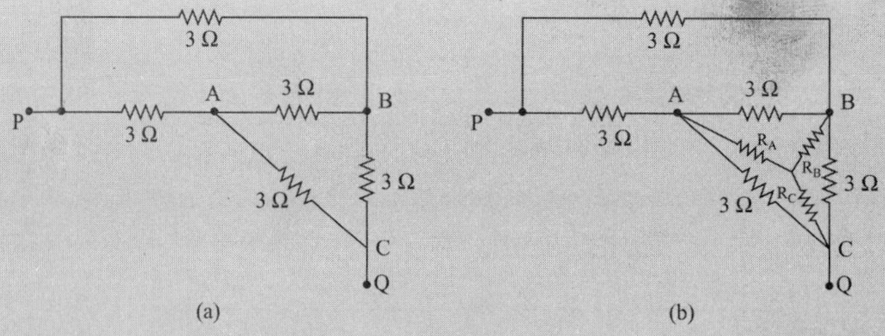

FIGURE 2.140

The circuit is drawn with the equivalent star as shown in Fig. 2.141.

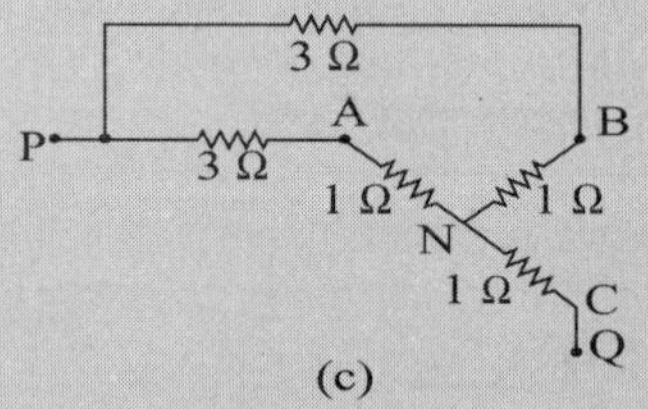

FIGURE 2.141

By considering series and parallel connection of resistances, the circuit is further simplified as in Fig. 142.

$$R_{PQ} = 3\ \Omega$$

Students should remember that the terminals between which the equivalent resistance has to be calculated have to be kept intact in the transformation process.

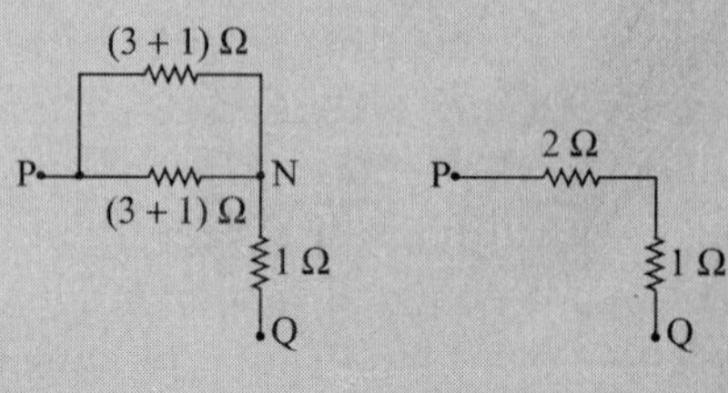

FIGURE 2.142

EXAMPLES 2.44

Calculate the current, I supplied by the battery in the circuit shown in Fig. 2.143.

SOLUTION

Instead of applying Kirchhoff's laws and writing the loop equations, we will convert the delta between ABC or BCD of Fig. 2.143, into equivalent star and then make simplifications to calculate I as has been shown in Fig. 2.144.

$$R_A = \frac{3\times3}{3+3+3} = 1\ \Omega$$

$$R_B = \frac{3\times3}{3+3+3} = 1\ \Omega$$

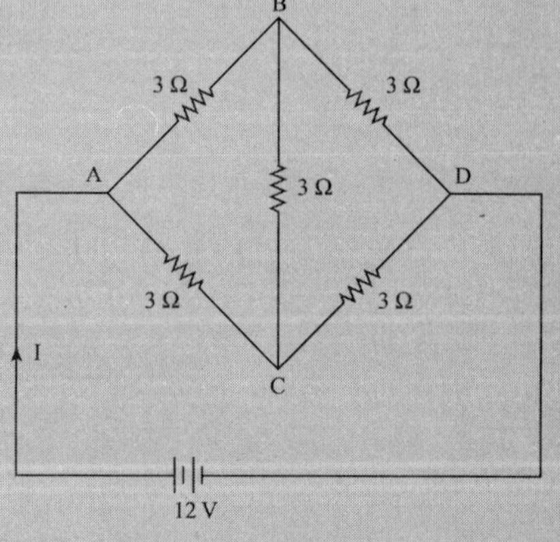

FIGURE 2.143

$$R_C = \frac{3 \times 3}{3+3+3} = 1\ \Omega$$

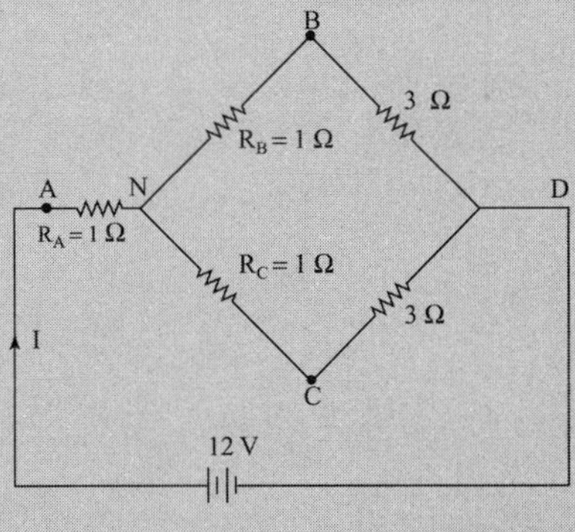

FIGURE 2.144

The circiut of Fig. 2.144 is further reduced as shown in Fig. 2.145.

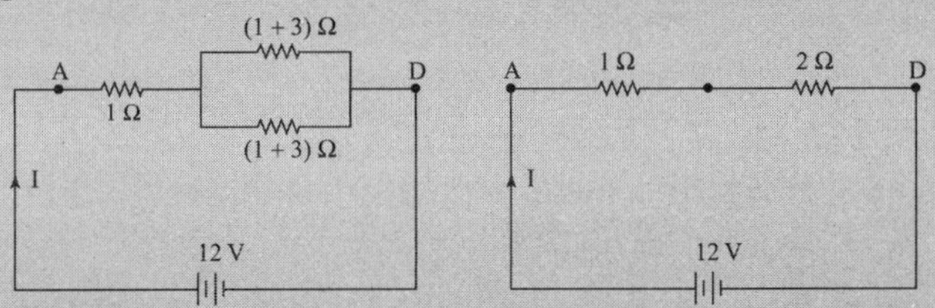

FIGURE 2.145

$$I = \frac{12}{2+1} = 4\ A$$

EXAMPLE 2.45

Six resistances each of value R Ω are connected as shown in Fig. 2.146. Calculate the equivalent resistance across the terminals B and C

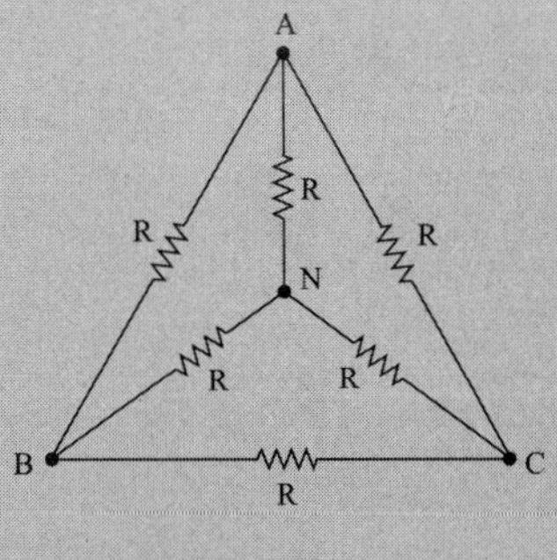

FIGURE 2.146

SOLUTION

We transform the three resistances forming a star point at N into an equivalent delta as shown in Fig. 2.147.

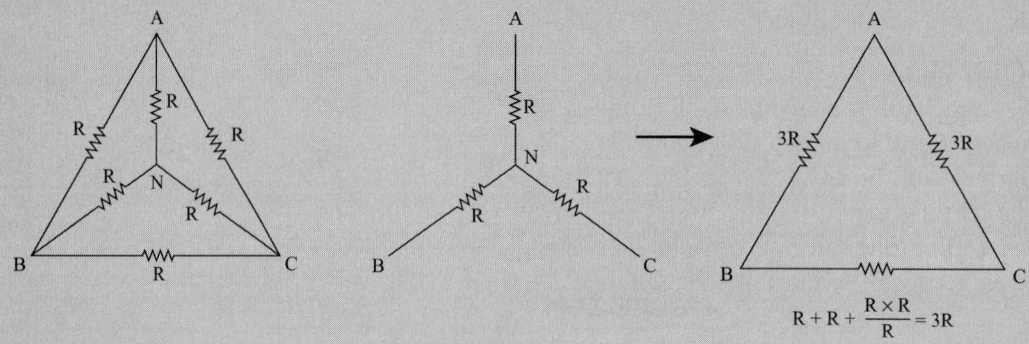

FIGURE 2.147

The circuit is redrawn and reduced subsequently to find the resistance between the terminals B and C as shown in Fig. 2.148.

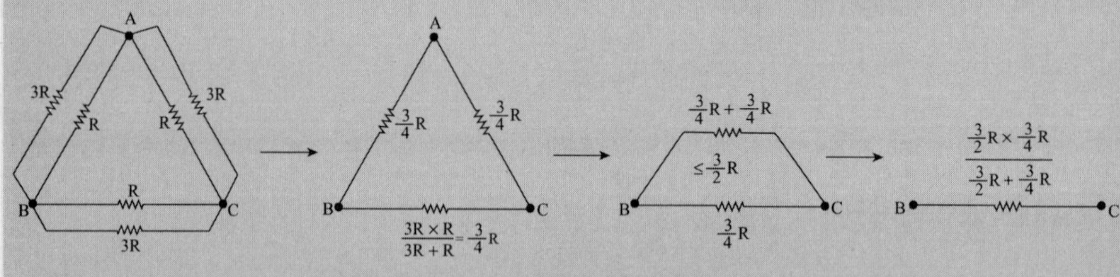

FIGURE 2.148

The equivalent resistance of the network between the terminals B and C is

$$\frac{\frac{3}{2}R\times\frac{3}{4}R}{\frac{3}{2}R+\frac{3}{4}R}=\frac{\frac{9}{8}R^2}{\frac{6R+3R}{4}}=\frac{9}{8}R^2\times\frac{4}{9R}=\frac{R}{2}\Omega.$$

EXAMPLES 2.46

Using star–delta transformations, calculate the equivalent resistance of the network between terminals A and B. All resistances in Fig. 2.149 are in Ohms.

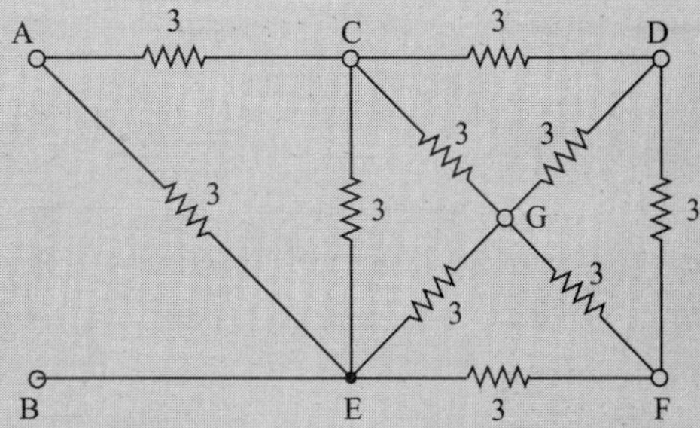

FIGURE 2.149

SOLUTION

We transform the two delta forming resistances across terminals CDG and EFG. The network will be as shown in Fig. 2.150 (a). The circuit is then simplified in steps as shown in 2.150(b). The value of equivalent resistance across terminals A and B is 1.81 Ohms.

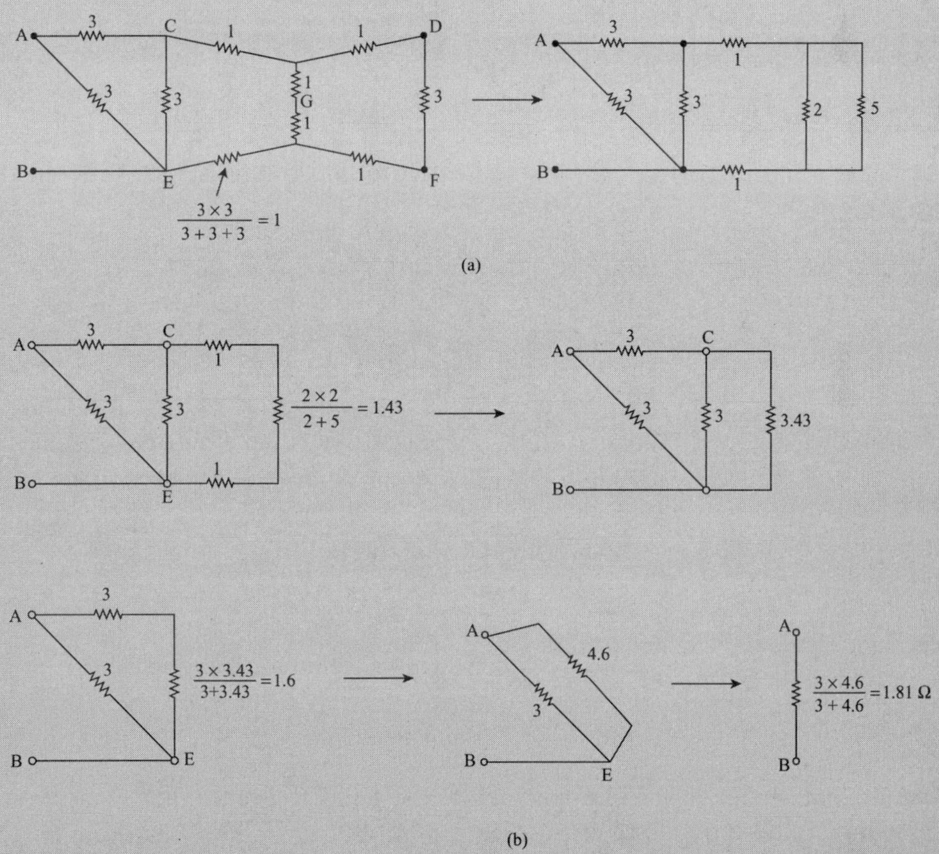

FIGURE 2.150

EXAMPLE 2.47

Find the voltage drop across the 10 Ω resistor in the network shown in Fig. 2.151.

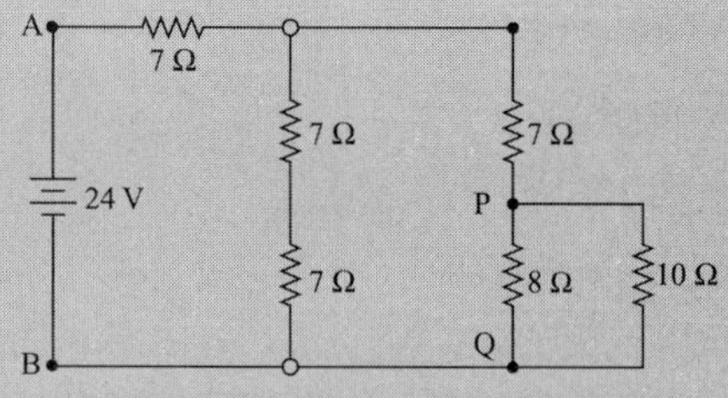

FIGURE 2.151

SOLUTION

Firstly we calculate the total current supplied by the battery by determining the equivalent resistance of the circuit across terminals AB. The total resistance of the circuit is calculated by successively reducing the circuit as shown in Fig. 2.152.

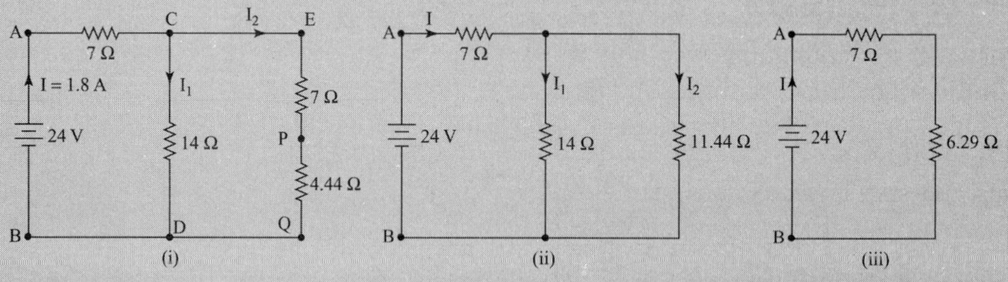

FIGURE 2.152

The total current supplied by the battery is

$$I = \frac{24}{13.29} = 1.8 \text{ A}$$

Working backwards, the branch currents are calculated. If total current supplied as calculated is 1.8 A, then current I_2 in Fig. 2.152 (ii) is calculated using the current division rule as

$$I_2 = I\frac{14}{14 + 11.44} = 1.8\frac{14}{25.44} = 0.99 \text{ A}$$

As the same current I_2 is flowing from P to Q, the voltage drop across PQ which is the same as voltage drop across the 10 Ω resistor is calculated as

$$V_{PQ} = V_{10\,\Omega} = I_2 R = 0.99 \times 4.44 = 4.39 \text{ V}$$

The battery voltage of 24 V is dropped across the series resistance of 7 Ω, across the combination of 7 Ω and 4.44 Ω resistors as

$$7 \times I + 7 \times I_2 + 4.44 \times I_2 = 7 \times 1.8 + 11.44 \times 0.99 = 24 \text{ V}$$

This problem can be solved in another way like, total current, I = 1.8 A.

Drop across AC = 7 × 1.8 = 12.6 V.

Voltage across CD = 24 − 12.6 = 11.4 V

Voltage across EQ = Voltage across CD = 11.4 V

Voltage across $$PQ = \frac{V_{EQ} \times 4.44}{7+4.44} = \frac{11.4 \times 4.44}{7+4.44} = 4.39\ V$$

EXAMPLE 2.48

Calculate the equivalent resistance between the terminals A and B of the network shown in Fig. 2.153.

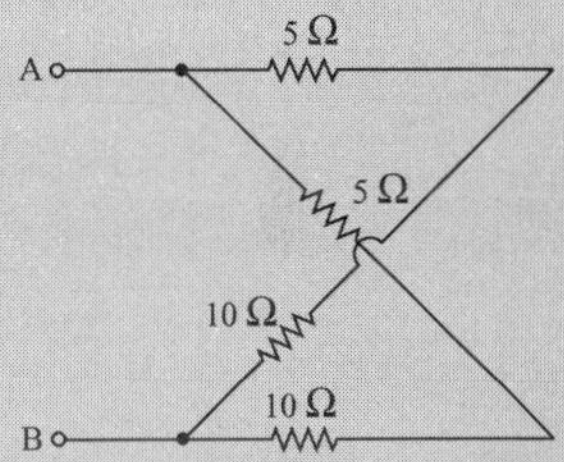

FIGURE 2.153

SOLUTION

This is a simple series and parallel connection of resistors. The circuit is redrawn as in Fig. 2.154.

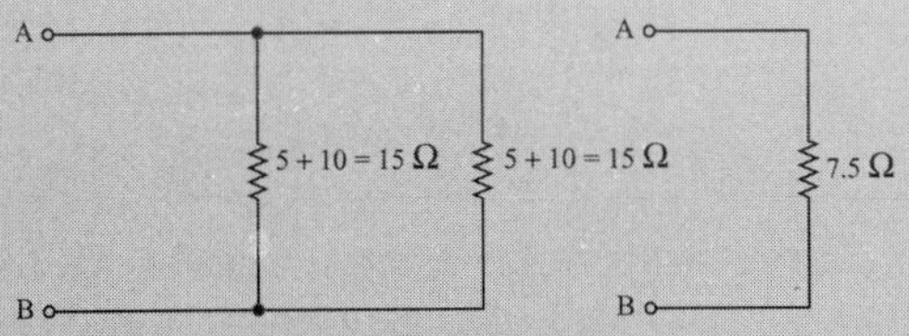

FIGURE 2.154

The equivalent resistance is calculated as 7.5 Ω.

EXAMPLE 2.49

Calculate the total current supplied by the battery in the network shown in Fig. 2.155. All resistances shown are in Ohms.

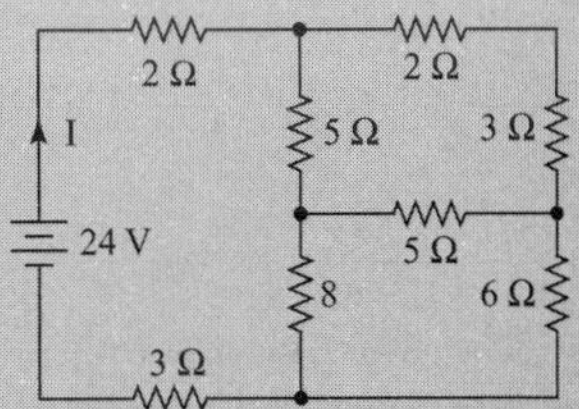

FIGURE 2.155

SOLUTION

The circuit is redrawn as shown in Fig. 2.156 after adding two series resistors. The three resistors of 5 Ω each forming a delta across terminals A, B, and C is converted into equivalent star across three terminals and then the circuit is further simplified through series and parallel operations as shown in Fig. 2.157.

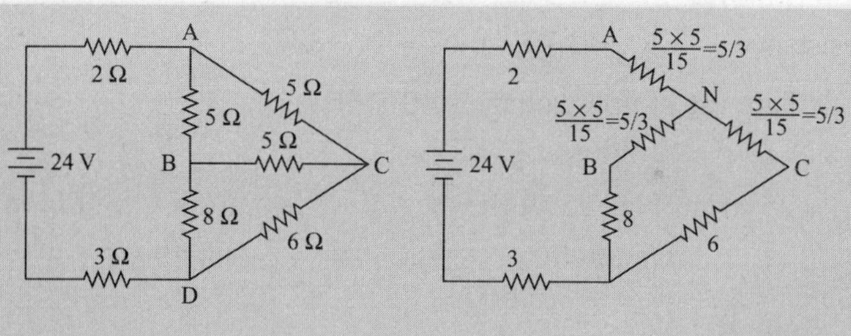

FIGURE 2.156

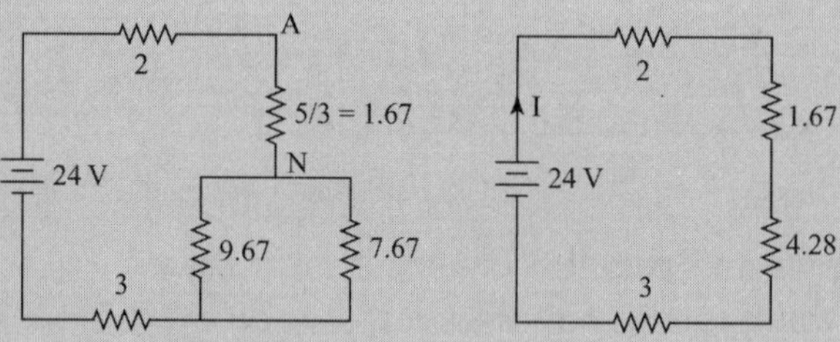

FIGURE 2.157

$$I = \frac{24\text{ V}}{2+3+1.67+4.28}$$

$$= 2.2\text{ A}$$

EXAMPLE 2.50

Four resistances are connected as shown in Fig. 2.158. Calculate the equivalent resistance across terminals A and B. What voltage is required to be applied across terminals AB so that potential drop across the terminals A and P is 25 V?

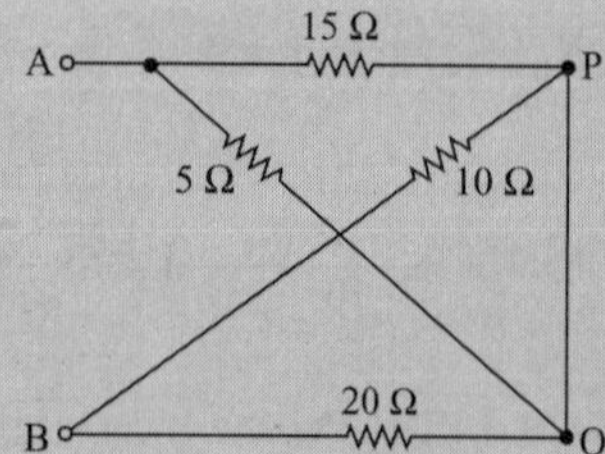

FIGURE 2.158

SOLUTION

By examining the given network, we see that terminals P and Q are at the same potential. The circuit is then redrawn as shown by connecting the 10 Ω resistor between B and Q. Similarly a 5 Ω resistor is shown connected between points A and P. The circuit is then simplified in steps as shown in Fig. 2.159.

Equivalent resistance $R_{AB} = 3.75 + 6.67 = 10.42\ \Omega$.

Now let us calculate the supply voltage so that P.D across AP is 25 V.

Assuming $V_{AP} = 25$, from Fig. 2.159 (ii),

$$I = \frac{V_{AP}}{R_{AP}} = \frac{25}{3.75} = 6.66\ \text{A}$$

$$V_{BQ} = R_{BQ} \times I = 6.67 \times 6.66 = 44.46\ \text{V}$$

Therefore
$$V = V_{AP} + V_{BQ}$$
$$= 25\ \text{V} + 44.46\ \text{V}$$
$$= 69.46\ \text{V}$$

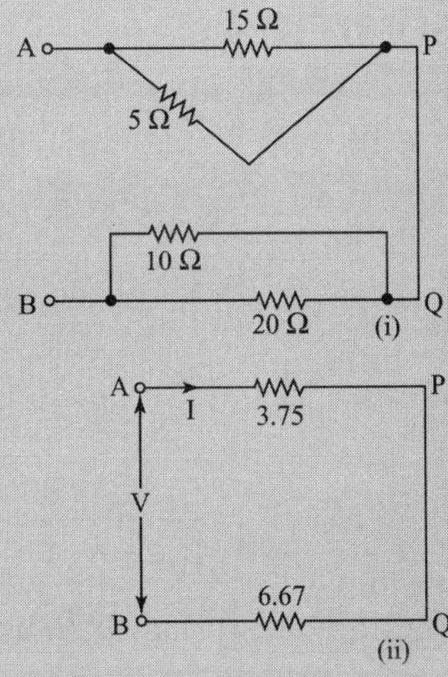

FIGURE 2.159

EXAMPLE 2.51

Three Resistances r, 2r, 3r are connected in delta. Determine the resistance of an equivalent star connection.

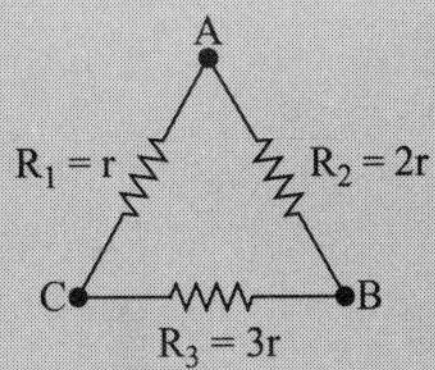

FIGURE 2.160

SOLUTION

The circuit is drawn as in Fig. 2.160. The equivalent star resistances have been calculated and shown in Fig. 2.161.

$$R_A = \frac{R_1 \times R_2}{R_1 + R_2 + R_3} = \frac{2r^2}{6r} = \frac{r}{3}$$

$$R_B = \frac{R_2 \times R_3}{R_1 + R_2 + R_3} = \frac{6r^2}{6r} = r$$

$$R_C = \frac{R_3 \times R_1}{R_1 + R_2 + R_3} = \frac{3r \times r}{6r} = \frac{r}{2}$$

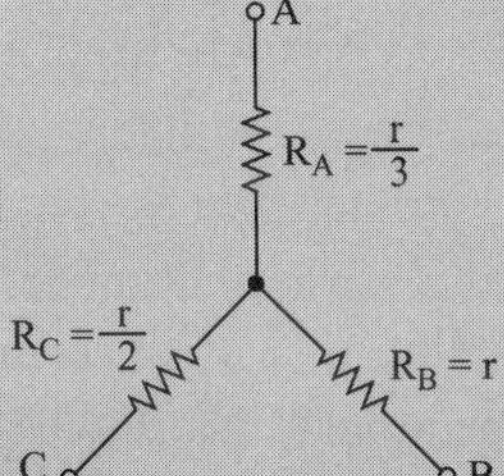

FIGURE 2.161

EXAMPLE 2.52

Find the resistance between terminal XY of the bridge circuit shown in Fig. 2.162, by using delta–star conversion.

SOLUTION

Let us change the resistances forming a delta across terminals A, B, and C into equivalent star.

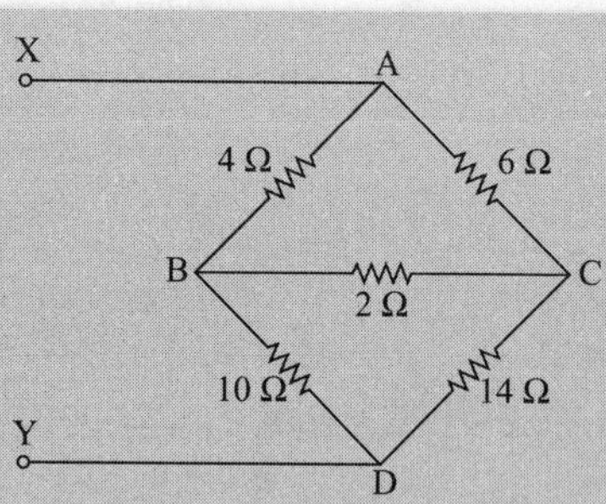

FIGURE 2.162

$$R_A = \frac{R_{AB} \times R_{AC}}{R_{AB} + R_{BC} + R_{AC}} = \frac{4 \times 6}{4+6+2} = \frac{24}{12} = 2\,\Omega$$

$$R_B = \frac{R_{AB} \times R_{BC}}{R_{AB} + R_{BC} + R_{AC}} = \frac{4 \times 2}{12} = \frac{2}{3}\,\Omega$$

$$R_C = \frac{R_{BC} \times R_{AC}}{R_{AB} + R_{BC} + R_{AC}} = \frac{2 \times 6}{12} = 1\,\Omega$$

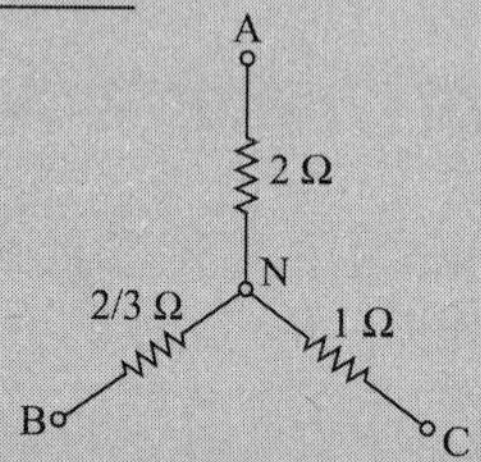

FIGURE 2.163

The equivalent star–forming resistances are shown in Fig. 2.163.

By replacing the delta resistance into equivalent star resistance, the circuit is drawn as in Fig. 2.164.

The resistances of the two parallel paths between N and D are $1 + 14 = 15\,\Omega$ and $\frac{2}{3} + 10 = \frac{32}{3}\,\Omega$, respectively.

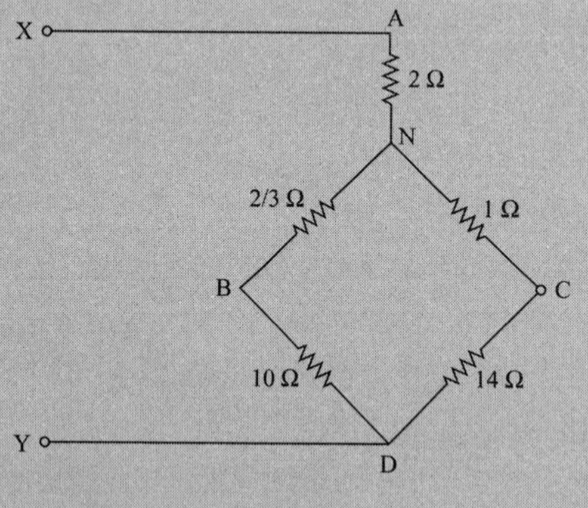

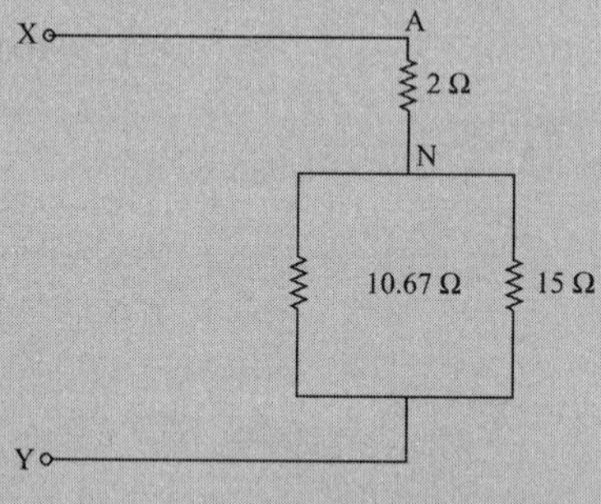

FIGURE 2.164

Total Resistance Network Terminal X and Y

$$= 2 + \frac{15 \times 10.67}{15 + 10.67} = 8.23\,\Omega$$

EXAMPLE 2.53

Find the resistance between terminals A and B in the electric circuit of Fig. 2.165 using Δ–Y transformation.

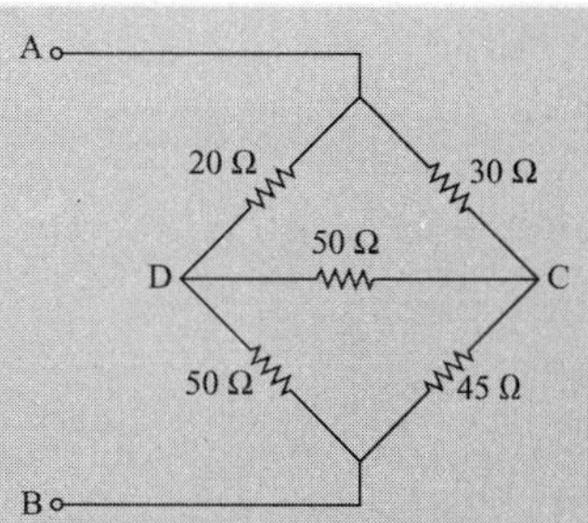

FIGURE 2.165

SOLUTION

We convert the delta forming resistances between the terminals A, B, and C into an equivalent star. The resistances between the terminals A, B, and C and the star point N are R_{AN}, R_{CN}, and R_{DN}. These are calculated as

$$R_{AN} = \frac{20 \times 30}{100} = 6\ \Omega$$

$$R_{CN} = \frac{30 \times 50}{100} = 15\ \Omega$$

$$R_{DN} = \frac{20 \times 50}{100} = 10\ \Omega$$

After transformation of the delta into star, the circuit becomes as shown in Fig. 2.166.

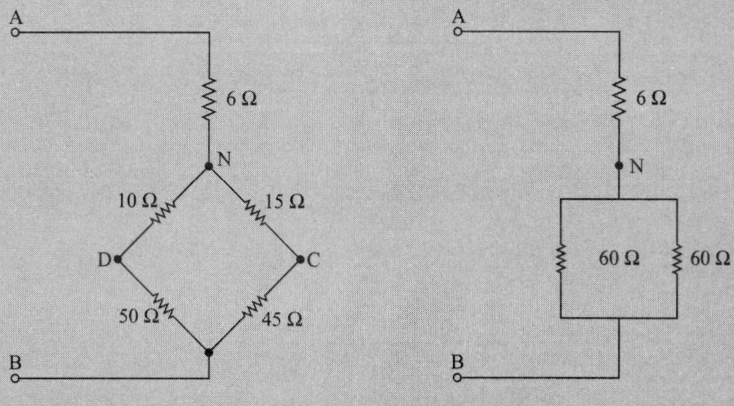

FIGURE 2.166

Total resistance,
$$R_{AB} = 6 + \frac{60 \times 60}{120}$$
$$= 36\ \Omega$$

EXAMPLE 2.54

For the circuit shown in Fig. 2.167, calculate the current flowing through the 5 Ω resistor by using the nodal method.

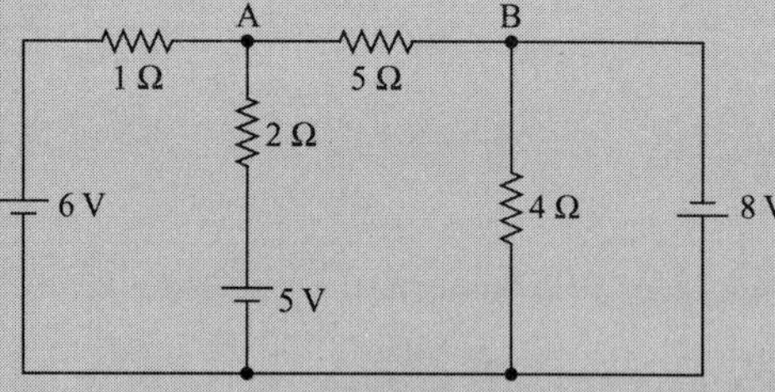

FIGURE 2.167

SOLUTION

Let V_A and V_B be the potentials at node A and node B, respectively. Let the reference node be at C. Let us assume current directions at node A as shown in Fig. 2.168.

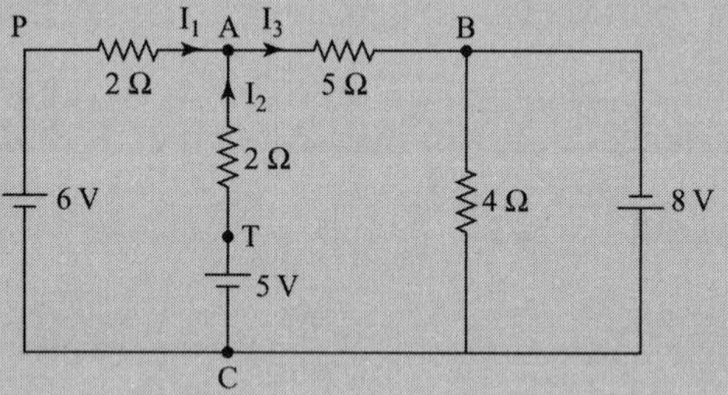

FIGURE 2.168

We will have incoming currents as equal to outgoing current, i.e.,

$$I_1 + I_2 = I_3 \tag{i}$$

Current, $$I_1 = (V_P - V_A)/R = \frac{6 - V_A}{2} \tag{ii}$$

Current, $$I_2 = \frac{V_T - V_A}{R} = \frac{5 - V_A}{2} \quad [\because \text{ potential of point T is +5 V}] \tag{iii}$$

Current, $$I_3 = \frac{V_A - V_B}{R} = \frac{V_A - (-8)}{5} \quad [\because \text{ potential of point B is −8 V}] \tag{iv}$$

Therefore, from (i), (ii), (iii) and (iv),

$$\frac{6 - V_A}{2} + \frac{5 - V_A}{2} - \frac{V_A + 8}{5} = 0$$

or, $$\frac{5(6 - V_A) + 5(5 - V_A) - 2(V_A + 8)}{10} = 0$$

or, $$-12\,V_A + 39 = 0$$

or, $$V_A = \frac{39}{12} = 3.25\text{ V}$$

Current through the 5 Ω resistor is I_3.

$$I_3 = \frac{V_A - V_B}{5} = \frac{3.25-(-8)}{5}$$

$$= \frac{3.25+8}{5} = 2.25 \text{ A}$$

EXAMPLE 2.55

Use nodal analysis to determine the current flowing through the various branches in the circuit shown in Fig. 2.169. All resistances shown are in Ohms.

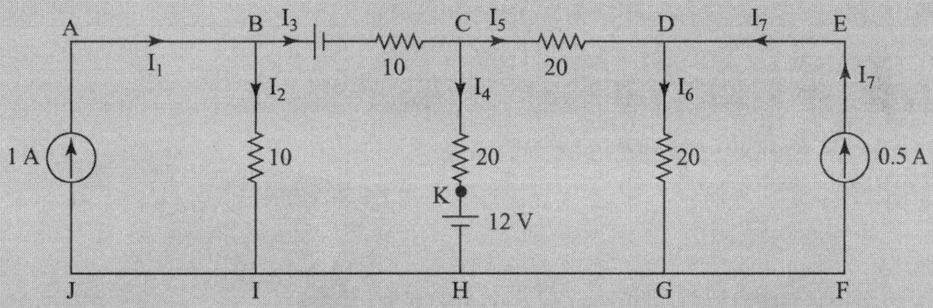

FIGURE 2.169

SOLUTION

We have shown the current directions in the various branches of the circuit and will apply KCL at node B, node C, and node D, respectively. Let V_B, V_C, V_D be the voltages at these nodes.

We have

$$I_1 = I_2 + I_3$$

$$I_1 = 1 = \frac{V_B}{10} + \frac{V_B - V_C}{10}$$

or,
$$2V_B - V_C - 10 = 0 \tag{i}$$

Again,

$$I_3 = I_4 + I_5 \text{ at node C}$$

Putting values

$$\frac{V_B - V_C}{10} = \frac{V_C - 12}{20} + \frac{V_C - V_D}{20}$$

or,
$$2V_B - 4V_C + V_D + 12 = 0 \tag{ii}$$

Again

$$I_5 + I_7 = I_6$$

$$\frac{V_C - V_D}{20} + 0.5 = \frac{V_D}{20}$$

or, $$V_C - V_D + 10 = V_D$$

or, $$V_C - 2V_D + 10 = 0 \qquad \text{(iii)}$$

Solving Eqs. (i), (ii) and (iii), we get

$$V_B = 10.4\text{ V}, \quad V_C = 10.8\text{ V}, \quad V_D = 10.4\text{ V}$$

$$I_2 = \frac{V_B}{10} = \frac{10.4}{10} = 1.04\text{ A}$$

$$I_3 = \frac{V_B - V_C}{10} = \frac{10.4 - 10.8}{10} = -\frac{0.4}{10} = -0.04\text{ A}$$

(direction of I_2 is opposite to that shown)

$$I_5 - \frac{V_C - V_D}{20} - \frac{10.8 - 10.4}{20} - \frac{0.4}{20} - 0.02\text{ A}$$

$$I_6 = \frac{V_D}{20} = \frac{10.4}{20} = 0.52\text{ A}$$

Again $$I_6 = I_7 + I_5 = 0.5 + 0.02 = 0.52\text{ A}$$

again $$I_2 = I_1 + I_3 = 1.0 + 0.04 = 1.04\text{ A}$$

$$I_4 = \frac{V_c - 12}{20} = \frac{10.8 - 12}{20} = -0.06$$

Potential of point k is +12 V and potential of point C is 10.8 V. The directions of currents at node C are shown in Fig. 2.170 below.

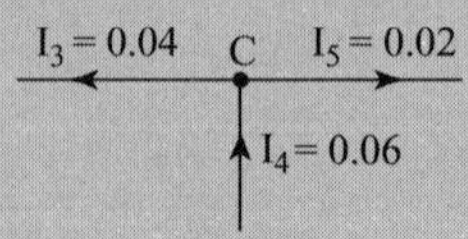

FIGURE 2.170

EXAMPLE 2.56

Using nodal analysis, calculate the current flowing through all the branches in the network shown in Fig. 2.171. Cross-check the results by any other method.

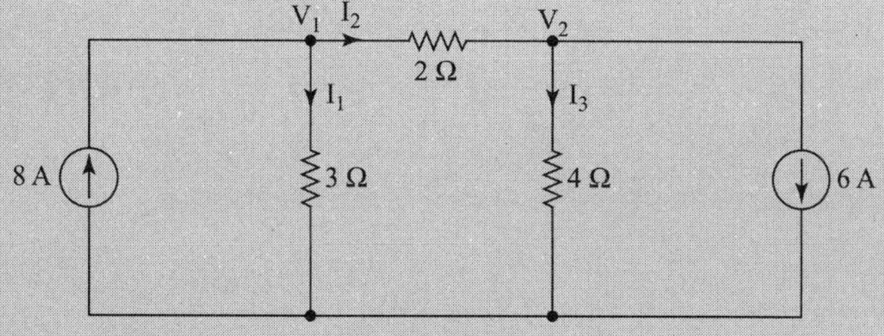

FIGURE 2.171

SOLUTION

Applying KCL, we can write

$$8 = I_1 + I_2$$

or, $$8 = \frac{V_1}{3} + \frac{V_1 - V_2}{2}$$

or, $$5V_1 - 3V_2 = 48 \qquad \text{(i)}$$

again $$I_2 = I_3 + 6$$

or, $$\frac{V_1 - V_2}{2} = \frac{V_2}{4} + 6$$

or, $$2V_1 - 3V_2 = 24 \qquad \text{(ii)}$$

from (ii)

$$3V_2 = 2V_1 - 24$$

$$V_2 = \frac{2V_1 - 24}{3}$$

Substituting V_2 in (i)

$$5V_1 - 3\left(\frac{2V_1 - 24}{3}\right) = 48$$

or, $$V_1 = 8\text{ V}$$

Putting value of V_1 in (i)

$$5V_1 - 3V_2 = 48$$

or, $$5 \times 8 - 3V_2 = 48$$

$$3V_2 = 40 - 48 = -8$$

$$V_2 = -\frac{8}{3}\text{ V}$$

$$I_2 = \frac{V_1 - V_2}{2} = \frac{8 - (-8/3)}{2} = 5.33\text{ A}$$

$$I_1 = \frac{V_1}{3} = \frac{8}{3} = 2.66 \text{ A}$$

$$I_3 = \frac{V_2}{4} = -\frac{8}{3\times 4} = -\frac{2}{3} = -0.67 \text{ A}$$

To cross-check

$$I_2 = I_3 + 6 = -0.67 + 6 = 5.33 \text{ A}$$

Alternate method

Now, we will apply superposition theorem. We will consider one source at a time and then superimpose them to get the combined effect.

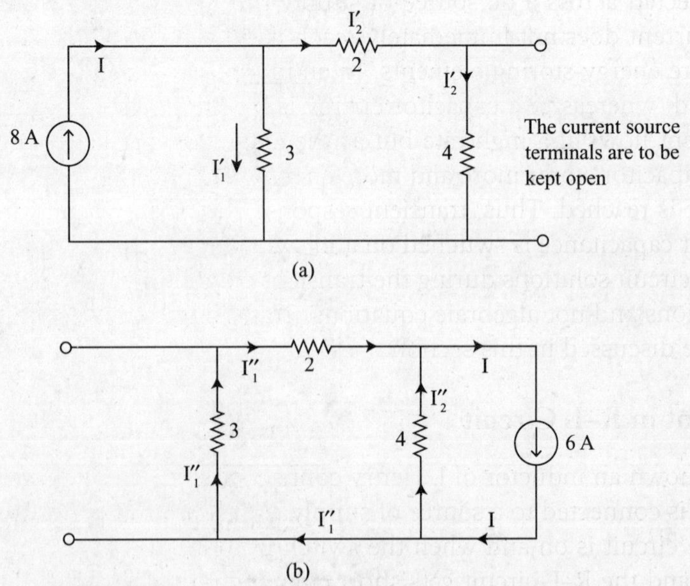

FIGURE 2.172

Using current divided rule in Fig. 2.172 (a),

$$I_1' = I\times\frac{2+4}{3+2+4} = 8\times\frac{6}{9} = 5.33\text{A}.$$

$$I_2' = I - I_1' = 8 - 5.33 = 2.67\text{ A}.$$

Using current divider rule in Fig. 2.172(b),

$$I_1'' = I\times\frac{4}{4+3+2} = 6\times\frac{4}{9} = 2.67 \text{ A}$$

$$I_2'' = I - I_1'' = 6 - 2.67 = 3.33 \text{ A}.$$

Current through the 3 Ω resistor = $I_1' - I_1'' = 5.33 - 2.67 = 2.66$ A
Current through the 2 Ω resistor = $I_2' - I_1'' = 2.67 + 2.67 = 5.34$ A
Current through the 4 Ω resistor = $I_2'' - I_2' = 3.33 - 2.67 = 0.66$ A
The results are the same by both the methods.

2.10 ► DC TRANSIENTS

2.10.1 Introduction

When a circuit containing inductance, capacitance, and resistance is switched on to a dc supply, the time taken for the current to attain a steady-state condition is called its transient response or transient time. Let a circuit contain a resistance and an inductance or a capacitance connected across a dc source of supply through a switch. When the switch is turned on, the current does not immediately reach its final value. Both the inductance and the capacitance are energy-storing elements. In an inductance, energy is stored in the form of a magnetic field, whereas in a capacitor energy is stored in the form of an electric field. Initially, the current flows at a high rate but as the energy-storing elements, i.e., either the inductor or the capacitor stores more and more energy, the rate of flow of current decreases, and a steady state is reached. Thus, transient response is studied when a circuit containing inductance and/or capacitance is switched on and switched off. Unlike a purely resistive circuit, R–L or R–C circuit solutions during the transient condition will involve the solution of differential equations and not algebraic equations. Transient response of an R–L circuit and an R–C circuit are discussed in this section.

2.10.2 Transient in R–L Circuit

In Fig. 2.173 is shown an inductor of L Henry connected in series with a resistance of R Ω. The combination is connected to a source of supply, V. When a two-way switch S connected to terminal 1, the circuit is on and when the switch is connected to terminal 2, the supply voltage is cut off and the R–L circuit gets short circuited. Let at time t = 0, the switch S be connected to terminal 1. The supply voltage V will be the sum of the voltage drop across the resistance and the voltage developed across the inductor. So we can write

$$V = Ri + L\frac{di}{dt} \tag{i}$$

or,

$$\frac{V}{R} = i + \frac{L}{R}\frac{di}{dt}$$

or,

$$\frac{V}{R} - i = \frac{L}{R}\frac{di}{dt}$$

or,

$$\frac{Rdt}{L} = \frac{di}{\left(\frac{V}{R} - i\right)}$$

or, $$\frac{R}{L}dt = \frac{di}{(V/R)-i}$$

Integrating,

$$\frac{R}{L}t = \log\left(\frac{V}{R}-i\right)+K$$

at $$t=0,\ i=0$$

therefore, $$K = -\log\frac{V}{R}$$

Substituting

$$\frac{R}{L}t = \log\frac{(V/R)-i}{(V/R)}$$

or, $$\frac{V}{R}-i = \frac{V}{R}e^{\frac{-R}{L}t}$$

or, $$i = \frac{V}{R}\left(1-e^{\frac{-R}{L}t}\right)$$

or, $$i = \frac{V}{R}\left(1-e^{\frac{-t}{\tau}}\right) \tag{2.5}$$

where $\tau = \frac{L}{R}$, called the time constant of the circuit.

At time $t=\infty$, the current is the steady-state current. Then $i=\frac{V}{R}=I_0$.

The eq. (2.5) can then be expressed as $i = I_0(1-e^{(-t/\tau)})$.

This current has two components, i.e.,

$$i = I_0 \text{ and } I_0e^{-t/\tau}$$

I_0 is the steady-state current and $I_0e^{-t/\tau}$ is the transient component of the current which goes on decreasing exponentially with the passage of time. The rise in current in the R–L circuit when the switch is closed has been shown in Fig. 2.173. The rise in current in the circuit is initially rapid but gradually the rise becomes slower and finally comes to a steady-state value. Although theoretically speaking, the current would reach its steady-state value after infinite time, but practically this time is too small a time—a fraction of a second only. The time taken by the current to reach 63.2 per cent of its final value is called the *time constant of the circuit* where, $\tau = \frac{L}{R}$.

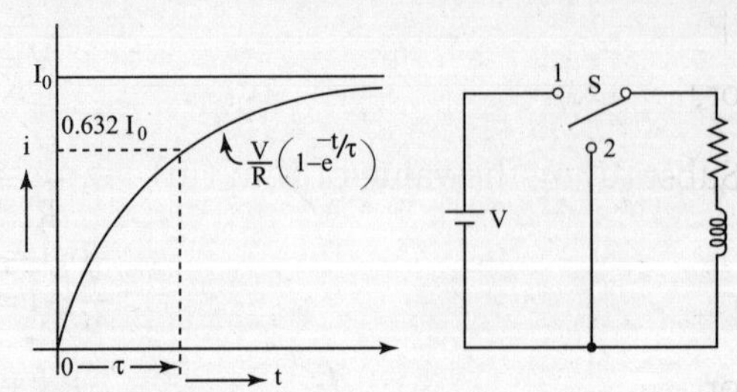

FIGURE 2.173

Rise in current in an R–L circuit

If we put $t=\tau=\frac{L}{R}$, the value of i will become 0.632 of I_0 as

$$i=I_0\left(1-e^{\frac{-t}{\tau}}\right)$$

Put $$t=\tau$$

then $$i=I_0\left(1-e^{-1}\right)$$

$$=I_0\left(1-0.368\right)$$
$$=0.632\,I_0$$

Thus, $i=63.2$ per cent of I_0. This has been shown in Fig. 2.173.

Now let us consider what happens when the switch S in Fig. 2.173 is changed to position 2 as shown. We are of course assuming that the switch was in position 1 and current had attained its steady-state value of I_0 A. From position 1, the switch is moved to position 2. Applying KVL we can write the voltage equation as

$$L\frac{di}{dt}+Ri=0$$

or, $$\frac{di}{i}=-\frac{R}{L}dt$$

Integrating with respect to time t

$$\log i=-\frac{R}{L}t+K$$

The value of K is determined by applying the initial conditions. Here the initial condition is that at t = 0 $i=I_0=\frac{V}{R}$.

Therefore, $$\log i=-\frac{R}{L}t+K \qquad \text{(iii)}$$

or, $$\log I_0=K$$

or, $$K=\log\frac{V}{R}$$

Substituting the value of K in (iii)

$$\log i=-\frac{R}{L}t+\log\frac{V}{R}$$

or, $$i=\frac{V}{R}e^{\frac{-R}{L}t}=I_0e^{\frac{-R}{L}t}=I_0e^{\frac{-t}{\tau}}$$

where, $$\tau=\frac{L}{R}$$

The decay of current in the circuit has been shown in Fig. 2.174.

Like rise in current, the decay in current in an R–L circuit depends on the time constant $\tau = \frac{L}{R}$ seconds.

Time constant for decaying current is obtained by putting $t = \tau$ the expression for i as

$$i = I_0 e^{\frac{-t}{\tau}} = I_0 e^{-1} = 0.368 I_0$$

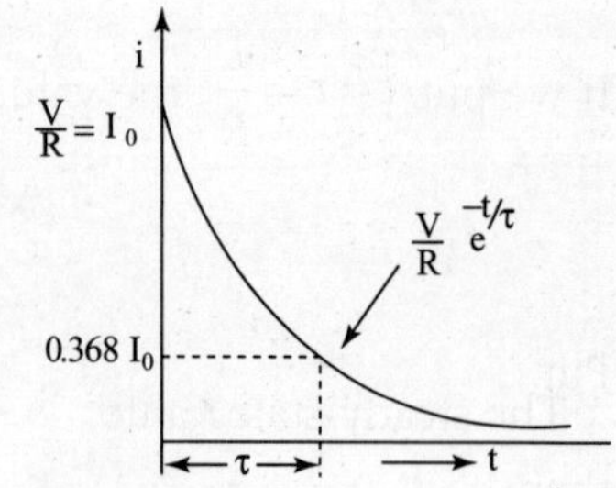

FIGURE 2.174

Decay in current in on R–L circuit

As time is increased, the current goes on reducing. As for example the magnitude of decaying current at time $t = 5\tau$ will be $i = I_0 e^{\frac{-t}{\tau}} = I_0 e^{\frac{-5\tau}{\tau}} = I_0 e^{-5} = 0.0067 I_0$.

This shows that as time increases, the current practically becomes very very small to ultimately becoming zero.

EXAMPLE 2.57

A coil of resistance 5 Ω and inductance 0.1 H is switched on to a 230 V supply. Calculate the rate of rise of current at t = 0 and at t = 2τ where τ = L/R. What is the steady-state value of the circuit current?

SOLUTION

For the rise in current

$$i = \frac{V}{R}\left(1 - e^{\frac{-R}{L}t}\right)$$

Substituting values, $$i = \frac{230}{5}\left(1 - e^{\frac{-5}{0.1}t}\right)$$

or, $$i = 46\left(1 - e^{-50t}\right)$$

The rate of change of current, $$\frac{di}{dt} = 46 \times 50 e^{-50t}$$

$$= 2300 e^{-50t}$$

at $$t = 0$$

$$\frac{di}{dt} = 2300 \text{ A/sec}$$

at t equal to two times the time constant, $\tau = 2\frac{L}{R}$

$$t = 2\frac{L}{R} = 2\frac{0.1}{5} = 0.04 \text{ sec}$$

$$\frac{di}{dt} = 2300e^{-50\times0.04} = 2300e^{-2} \text{ A/sec}$$

The steady-state value, $$I_0 = \frac{V}{R} = \frac{230}{5} = 46 \text{ A}$$

The larger the time constant is, the more is the time taken by the current to rise or decay in a L–R circuit during transients. Current almost decays to zero at five times the time constant. Beyond this time the magnitude of current is less than one per cent of its steady-state value. When the circuit is switched on, energy is stored in the inductor in the form of a magnetic field and during switching off, the stored energy gets dissipated.

EXAMPLE 2.58

A coil having an inductance of 1.4 H and a resistance of 1 Ω is connected to a 12 V dc source through a switch. What will be the value of current after 400 m sec of switching on the supply? How much time will it take for the steady-state current to drop to half its value after the switch is turned on?

SOLUTION

We have,

$$i = I_0\left(1 - e^{\frac{-t}{\tau}}\right)$$

$$I_0 = \frac{V}{R} = \frac{12}{1} = 12 \text{ A}$$

$$\tau = \frac{L}{R} = \frac{1.4}{1} = 1.4$$

$$t = 400 \text{ ms} = 0.4 \text{ sec}$$

Therefore,

$$i = 12\left(1 - e^{\frac{-0.4}{1.4}}\right)$$

$$= 12\left(1 - e^{-.285}\right)$$

$$= 3 \text{ A}$$

For the second part of the problem we will use the expression for decaying current.

$$i = I_0 e^{-t/\tau}$$

We have to calculate t for i to become $I_0/2$

Therefore, $$6 = 12e^{\frac{-t}{1.4}}$$

or, $$0.5 = e^{-t/1.4}$$

or, $$-\frac{t}{1.4} = \log 0.5$$

Therefore $$t = 0.75 \text{ seconds}$$

2.10.3 Transient in R–C Circuit

In Fig. 2.175 is shown a capacitor, C and a resistor, R connected in series across a dc voltage source, V through a two-way switch S. When the switch S is connected to position 1, the circuit is switched on. Current i will start flowing. Let the instantaneous voltage across the capacitor be v_c, and charge be q. Applying KVL, we can write

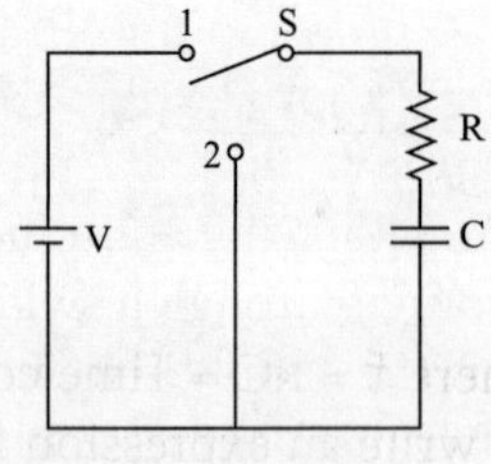

FIGURE 2.175

RC circuit supplied from a dc source through a two-way switch

$$V = Ri + v_c \qquad \text{(i)}$$

and $$C = \frac{q}{v_c}$$

or, $$q = Cv_c$$

and $$\int i\,dt = Cv_c \qquad \left[\because\; i = \frac{dq}{dt} \text{ or, } q = \int i dt\right]$$

or, $$i = C\frac{dv_c}{dt}$$

From (i)

$$V = Ri + v_c$$

$$= RC\frac{dv_c}{dt} + v_c$$

or, $$V - v_c = RC\frac{dv_c}{dt}$$

or, $$\frac{dt}{RC} = \frac{dv_c}{V - v_c}$$

Integrating with respect to t

$$\frac{t}{RC} = l_n(V - v_c) + K \tag{ii}$$

To find the value of K, we put the initial conditions. At time t = 0, the voltage across the capacitor $v_c = 0$.

Putting this value, we find the integration constant

$$K = l_n V$$

The eq. (ii) is written as

$$\frac{t}{RC} = -l_n(V - v_c) + l_n V$$

or,
$$\frac{V - v_c}{V} = e^{\frac{-t}{RC}}$$

or,
$$V - v_c = Ve^{\frac{-t}{RC}}$$

or,
$$v_c = V\left(1 - e^{\frac{-t}{RC}}\right) \tag{2.6}$$
$$= V\left(1 - e^{-t/\tau}\right)$$

where τ = RC = Time constant.

To write an expression for current i, we will use eq. (i) as

$$V = Ri + v_c$$

or,
$$i = \frac{V}{R} - \frac{v_c}{R}$$
$$= \frac{V}{R} - \frac{V\left(1 - e^{-t/\tau}\right)}{R}$$
$$= \frac{V}{R} - \frac{V}{R} + \frac{V}{R}e^{-t/\tau}$$

or,
$$i = \frac{V}{R}e^{-t/\tau} = I_0 e^{-t/\tau} \tag{2.7}$$

Current I_0 is the initial charging current when the switch is just turned on. Because at that instant voltage across the capacitor, $v_c = 0$.

Therefore,
$$i = \frac{V}{R} = I_0$$

This is the initial circuit current, I_0 when the voltage across the capacitor is zero. As the capacitor gets charged, current flowing through the circuit which is, $i = \frac{V - v_c}{R}$ goes on decreasing from its initial value I_0.

Thus, when the circuit is switched on, while current in the circuit goes on decreasing, the voltage across the capacitor goes on increasing as shown in Fig. 2.176.

The time constant τ is defined as the time taken in seconds for the voltage across the capacitor, v_c to attain its final value, V if the rate of rise of voltage were the same as its initial value as indicated by the dotted line OC. The initial rate of rise of voltage across the capacitor is

$$\frac{BC}{OB} = \frac{V}{RC} = \frac{V}{\tau}$$

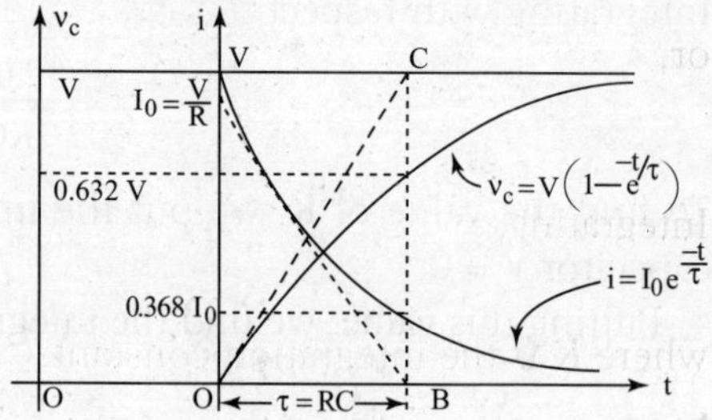

FIGURE 2.176

Transient current, i and capacitor voltage vc in a RC circuit

Now let us find out the value of the voltage across the capacitor at $t = \tau = RC$. Using the voltage equation,

$$v_c = V\left(1 - e^{\frac{-t}{\tau}}\right)$$

$$= V\left(1 - e^{-1}\right) \qquad \text{as } t = \tau$$

$$= V(1 - 0.368)$$

$$= 0.632 \text{ V}$$

Thus, *time constant can also be defined as the time taken in seconds for the voltage across the capacitor to reach 63.2 per cent of its final value as has been shown in the Fig.* 2.176.

The above explanation relates to charging of the capacitor upto a voltage V when the switch S is put in position 1. Now we will study the discharging of the capacitor when the switch is put in position 2. The capacitor voltage now at time $t = 0$ is equal to the supply voltage V. When the switch has been put in position 2, the supply voltage is cut off and the RC circuit is short circuited. The current flowing will be due to the voltage build up across the capacitor. The current $i = -\frac{v_c}{R}$ will flow in the opposite direction to the direction when the capacitor was being charged. That is why we have put a negative sign for i.

Again, $$i = \frac{dq}{dt} = \frac{d}{dt} C v_c = C \frac{dv_c}{dt}$$

thus, $$C \frac{dv_c}{dt} = -\frac{v_c}{R} \qquad \left[\because i = -\frac{v_c}{R}\right]$$

or, $$\frac{dv_c}{dt} = \frac{v_c}{RC}$$

or, $$-\frac{dv_c}{v_c} = \frac{dt}{RC}$$

Integrating, $$-l_n v_c = \frac{t}{RC} + K \quad \text{(iii)}$$

where K is the integration constant.

So, to find the value of K we will put the initial conditions, i.e., when t = 0, v_c = V

therefore, $$-l_n V = 0 + K$$

or, $$K = -l_n V$$

The eq. (v) is written as

$$-l_n v_c = \frac{t}{RC} - l_n V$$

or, $$\ln v_c - \ln V = \frac{t}{RC}$$

or, $$-v_c = Ve^{\frac{-t}{RC}} \quad \text{(iv)}$$

and $$i = -\frac{v_c}{R} = -\frac{V}{R}e^{\frac{-t}{RC}} = -I_0 e^{\frac{-t}{RC}} \quad (2.8)$$

The time constant, τ of the circuit is the time in seconds taken for the voltage across the charged capacitor to become zero if the initial rate of decay is maintained as shown by the dotted line AB. OB is the time constant, τ as has been shown.

If we calculate the value of v_c at $t = \tau$, we get

$$v_c = Ve^{\frac{-t}{\tau}} = Ve^{-1} \quad \text{for } t = \tau \quad (2.9)$$
$$= 0.368\,V$$

and $$i = I_0 e^{\frac{-t}{\tau}} = I_0 e^{-1} = 0.368\,I_0 \quad (2.10)$$

Thus, for a time equal to time constant τ, both capacitor voltage v_c and the circuit current i are reduced to 36.8 per cent of their initial values with initial rate of decay remaining unchanged as has been shown in Fig. 2.177.

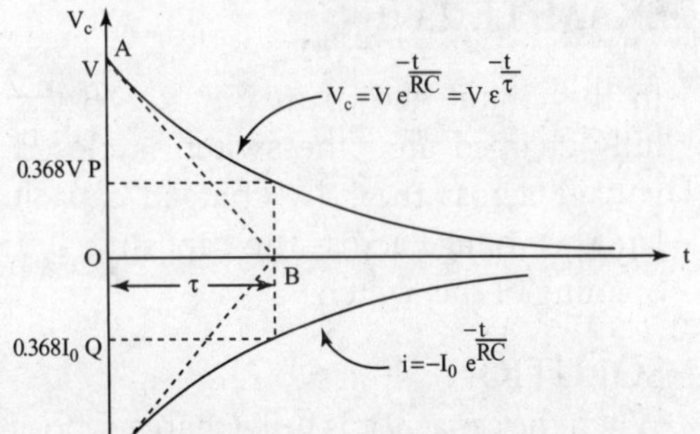

FIGURE 2.177

Transients in RC circuit, decay of voltage across the charged capacitor and the circuit current

EXAMPLE 2.59

A capacitor of value 1 μF and a series resistor of 5.45 MΩ are switched on across a 220 V dc supply. Calculate the time by which the capacitor will be charged to 60 per cent of the supply voltage.

SOLUTION

$$R=5.45\times10^{6}\,\Omega,\quad C=1\times10^{-6}\,F$$

Time constant,

$$\tau=RC=5.45\times10^{6}\times1\times10^{-6}\,\text{sec}$$

$$=5.45\text{ seconds}$$

$$v_c=V\left(1-e^{\frac{-t}{\tau}}\right)$$

$$0.6\times230=230\left(1-e^{\frac{-t}{5.45}}\right)$$

$$0.6=1-e^{\frac{-t}{5.45}}$$

or,

$$e^{-t/5.45}=1-0.6=0.4$$

or,

$$e^{t/5.45}=\frac{1}{0.4}=2.5$$

or,

$$\frac{t}{5.45}=\ln 2.5=0.915$$

or,

$$t=0.915\times5.45=4.98\text{ seconds.}$$

EXAMPLE 2.60

In the circuit shown in Fig. 2.178 the capacitor gets fully charged after the switch is closed. Calculate the voltage across the fully charged capacitor. Also calculate the voltage across the capacitor 0.05 seconds after opening of the switch.

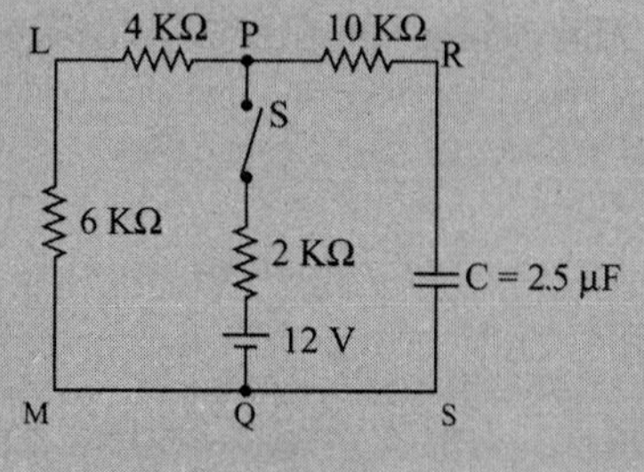

FIGURE 2.178

SOLUTION

When the capacitor is fully charged, no current will flow through the capacitor. The capacitor will be charged to a voltage equal to V_{PQ}. However, current, I flowing through the loop PLMQP will be

$$I=\frac{12}{(2+4+6)\times10^{3}}=1\text{ mA}$$

Voltage across terminals P and Q is equal to the battery voltage of 12 V minus the voltage drop in 2k Ω resistor due of current flow of 1 mA.

$$\therefore \quad V_{PQ} = 12 - 1 \times 10^{-3} \times 2 \times 10^{3} = 10 \text{ V}$$

$$\text{or} \quad V_{PQ} = v_c = 10 \text{ V}$$

Now, when the switch is opened, at $t = 0$, $v_c = 10$ V. The capacitor will be getting discharged through the resistors 10 kΩ, 4 kΩ, and 6 kΩ in the loop RPLMQSR. The time constant of the circuit, $\tau = RC$.

$$\tau = RC = (10+4+6)\times 10^3 \times 2.5\times 10^{-6} = 20\times 10^3 \times 2.5\times 10^{-6}$$
$$= 0.05 \text{ seconds}$$

Let the capacitor voltage at $t = 0.05$ sec be v_c' and discharging current at $t = 0.05$ be i'.

$$v_c' = v_c e^{-t/\tau} = 10e^{\frac{-0.05}{0.05}} = 10\times e^{-1} = 10\times 0.368 = 3.68 V$$

Important Terminologies and Points to Remember

1. Active network is one which contain one or more sources of emf.
2. A passive network does not contain any source of emf.
3. A linear circuit is one whose parameters are of constant value.
4. A non-linear circuit is one whose parameters change with operating conditions.
5. An Ideal voltage source has zero internal impedance, i.e., zero series resistance, i.e., R_{se} = u
6. An Ideal current source has infinite shunt resistance, i.e., $R_{se}h = \infty$
7. A voltage source can be converted into an equivalent current source and vice-versa. The open-circuit voltage and short-circuit current in the two equivalent circuits must be equal.

   Example:

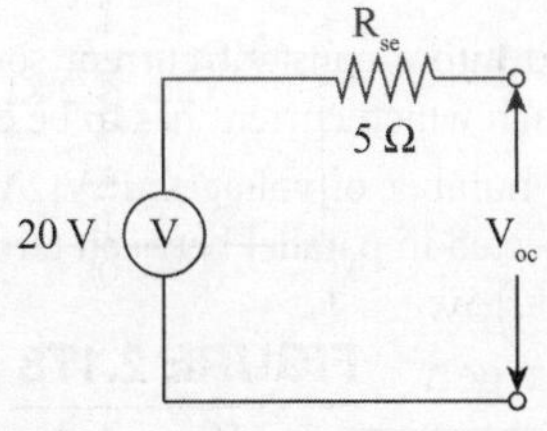

Voltage source

$\Rightarrow$

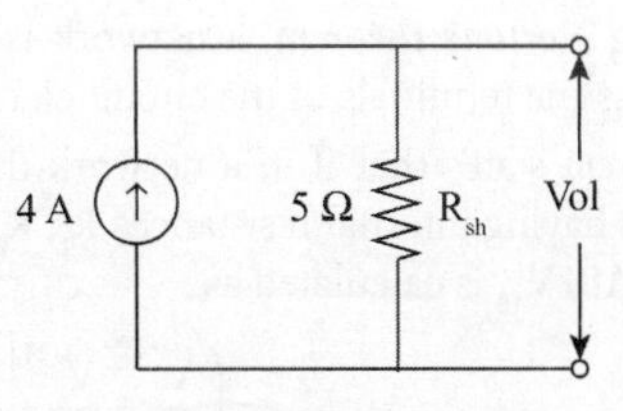

Equivalent current source

$$V_{OC} = 20 \text{ V} \qquad V_{OC} = 5\times 4 = 20 \text{ V}$$

$$I_{sh} = \frac{20 \text{ V}}{5\,\Omega} = 4 \text{ A} \qquad I_{sh} = 4 \text{ A}$$

8. *Voltage divider rule*: In a series circuit containing a number of resistances, the voltage drop across any resistor say $V_x = IR_x$ where I is the circuit
9. *Current divider rule:* In a parallel circuit of two resistances R_1 and R_2, the branch currents I_1 and I_2 in terms of the total current I, are calculated as

$$I_1 = I\frac{R_2}{R_1 + R_2} \text{ and } I_2 = I\frac{R_1}{R_1 + R_2}$$

10. *Kirchhoffs current law (KCL):* It states that algebraic sum of current in a junction or node in a circuit is zero, or the incoming currents is equal to outgoing currents.
11. *Kirchhoffs voltage law (KVL):* It states that algebraic sum of all the voltages in a closed loop is zero. In other words, the total voltage rise is equal to the total voltage drops in a closed loop.

    Remember the sign conversion: Positive sign is to be assigned to the rise in voltage and negative sign to drop in voltage present in a loop. While writing the voltage equation for a loop, moving from –ve terminal towards +ve terminal of the voltage source, the voltage be taken as positive. While moving from +ve terminal towards –ve terminal, voltage be taken as negative. While moving in the direction of current the voltage drop in a resistor, i.e., IR be taken as negative which moving opposite to the direction of current the voltage drop be taken as positive.
12. If a network contains both voltage source and current source, the current source be first converted to an equivalent voltage source before applying KCL and KVL.
13. When a network has large number of loops, the number of equations framed will be large and time taken to solve the equations will be more. In such cases, nodal voltage method of analysis will be more suitable.
14. In nodal analysis, a reference node called ground node or zero potential node is taken and potential at all other nodes is assigned to calculate the branch currents.
15. ***Superposition theorem*** is applied to a network where there is more than one source of supply. The current flowing in any branch is calculated by taking the effect of each source separately and then adding them for the combined effect.
16. ***Thevenin's theorem*** is applied to a circuit to calculate the current flowing through an element between any two terminals. Between these terminals, the whole circuit is converted into a voltage source with an internal resistance. The voltage, V_{oc} is the open-circuit voltage of the network across the terminals and the internal resistance, R_{eq} is the equivalent resistance of the whole circuit across the open-circuited terminals by replacing the voltage sources by their internal resistances and by open circuiting the current sources.

    Current flowing through the resistance R between the terminals is calculated by

$$I = \frac{V_{oc}}{R_{eq} + R}$$

17. While applying ***Norton's theorem***, a network is converted into a constant current source and a parallel resistance across the terminals of the circuit element through which current has to be calculated.
18. ***Millman's theorem*** states that if in a network there are a number of voltages. If V_1, V_2, V_3 are the three voltage sources having internal resistances R_1, R_2, R_3 connected in parallel between terminals A and B, the voltage across AB, V_{AB} is calculated as

$$V_{AB} = \left(\frac{V_1}{R_1} + \frac{V_2}{R_2} + \frac{V_3}{R_3}\right) \times \frac{1}{\frac{1}{R_1} + \frac{1}{R_2} + \frac{1}{R_3}}$$

19. According to ***Reciprocity theorem***, in a linear bilateral network, if a voltage source, V in branch A produces a current in branch B having a resistance R, a same amount of current I will flow through R in branch CD when V and R are interchanged between AB and CD.

20. By applying Maximum Power Transfer theorem, we find out at what value of load resistance, maximum power coil be transferred from source to the load.

According to this theorem, maximum power will be transferred when load resistance will be equal to the resistance of the source.

21. In solving complex network problem, we convert star connected resistances into equivalent delta connected resistances and vice versa. The circuit gets simplified this way.

22. When a circuit containing resistance, inductance and capacitance are switched on to dc supply, the time taken by the current to attain a steady state value is called its transient response time.

Unlike a purely resistive circuit, solutions of RL and RC circuit will involve solving differential equations and not algebraic equations.

Important Formulae and Equations

Equation	Remarks
1. $C = \frac{\varepsilon A}{d}$	1. Capacitance of parallel path capacitor
2. $L = \frac{\mu N^2 A}{\ell}$	2. Factors on which inductance of a coil depends
3. $E = \frac{1}{2} CV^2$	3. Energy stored in a capacitor
4. $R_{eq} = R_1 + R_2 + R_3 + \cdots$.	4. Equivalant resistance for series connected resistances
5. $\frac{1}{R_{eq}} = \frac{1}{R_1} + \frac{1}{R_2} + \frac{1}{R_3} + \cdots$	5. Equivalant resistance for parallel connected resistances
6. $V_1 = \frac{V}{R_{eq}} R_1$	6. Voltage drop in a resistor for series connected resistance across A voltage, V. (voltage divider rule)
7. $I_1 = I\frac{R_2}{R_1 + R_2}$, $I_2 = I\frac{R_1}{R_1 + R_2}$	7. Current divider rule for two resistances connected in parallel
8. $\sum_{j=1}^{\eta} I_j = 0, \sum_{j=1}^{\eta} V_j = 0$	8. Kirchhoffs current law (KCL) and Kirchhoffs voltage law) (KVL)
9. $\rho = Ri + L\frac{di}{dt} + \frac{1}{c}\int idt$	9. Voltage equation of a RLC series circuit
10. $I = \frac{V_{oc}}{R_{Th} + R}$	10. Current, I through any branch in a circuit is calculated using Thevenin's theorem.
11. $P(max) = \frac{E^2 R_L}{(R_L + R_i)^2} = \frac{E^2}{4R_L} = \frac{E^2}{4R_1}$	11. Following maximum power transfer theorem where $R_i = R_L$
12. $R_{12} = R_1 + R_2 + \frac{R_1 R_2}{R_3}$ $R_{23} = R_2 + R_3 + \frac{R_2 R_3}{R_1}$ $R_{31} = R_3 + R_1 + \frac{R_3 R_1}{R_2}$	12. Three resistances R_1, R_2, R_3 connected in star formation. Their equivalent resistance in delta formation between terminals 12, 23, and 31.

Important Formulae and Equations

Equation	Remarks
13. $R_1 = \frac{R_{12}R_{31}}{R_{12}+R_{23}+R_{31}}$ $R_2 = \frac{R_{12}R_{23}}{R_{12}+R_{23}+R_{31}}$ $R_3 = \frac{R_{23}R_{31}}{R_{12}+R_{23}+R_{31}}$	13. Three resistances R_{12}, R_{23}, R_{31} are connected in delta formation. Their equivalent resistances in star formation between star point and between terminals 1, 2, 3, respectively
14. $i = \frac{V}{R}(1 - e^{\frac{-t}{\tau}})$ where $\tau = \frac{L}{R}$ (time constant) at $t \to \infty, i = \frac{V}{R} = I_o$ So, $i = I_o(1 - e^{-t/\tau})$	14. Rise in current in an RL series circuit and steady state current
15. $i = I_o e^{-t/\tau}$	15. Decay in current in an RL series circuit
16. $i = \frac{V}{R}e^{-t/\tau} = I_o e^{-t/\tau}$ where $\tau = RC$ (time constant) voltage across capacitor, $V_C = V(I - e^{-t/\tau})$	16. Decay in transient current in an RC series circuit and the voltage developed across the capacitor.

REVIEW QUESTIONS

Short Answer Type Questions

1. Define Ohm's law and state if there are any conditions.
2. Explain the concept of voltage and current source transformation with an example.
3. Give the concept of current, voltage, and resistance.
4. State the factors on which resistance of a wire depends. What is meant by resistivity of a conducting material?
5. Explain why silver is more conducting than copper.
6. Draw the V–I characteristics of a variable resistor whose value of resistance has been fixed at 5 Ω, 8 Ω, and 10 Ω.
7. Explain the effect of change of temperature on the resistance of most of the conducting materials.
8. What is meant by superconducting materials?
9. Prove that $R_t = R_0(1 + \alpha_0 t)$ for a conducting material where R_t is the resistance at t°C, R_0 is the resistance at 0°C, α_0 is the temperature coefficient of resistance at 0°C and t is the rise in temperature.
10. If α_1 and α_2 are the temperature coefficients of resistance at t_1 and t_2 degrees, respectively, then prove that

$$\alpha_2 = \frac{\alpha_1}{1 + \alpha_1(t_2 - t_1)}$$

11. Explain why resistance of most of the conducting materials increase with temperature.

12. From the definition of power as rate of doing work, show that power in an electric circuit

$$P = VI \text{ Watts}$$

13. Distinguish between work, power, and energy.
14. Establish the relationship between resistance, current, voltage, power, and energy in an electric circuit.
15. Show that 1 kWh is equal to 860 kilo Calories.
16. What is meant by a constant voltage source and a constant current source?
17. State with example the current divider rule and the voltage divider rule as applicable to parallel circuits and series circuits, respectively.
18. State Kirchhoff's current law and Kirchhoff's voltage law.
19. What is Cramer's Rule?
20. State and explain Thevenin's theorem.
21. With a simple example show how, by applying Thevenin's theorem, current flowing through any branch of an electrical network can be calculated.
22. Write the steps of application of Thevenin's theorem.
23. State and explain Norton's theorem.
24. Distinguish between Thevenin's theorem and Norton's theorem.
25. What is maximum power transfer theorem? Prove the theorem.
26. Write the relationship of star–delta transformation of three resistors.
27. Distinguish between an ideal voltage source and a practical voltage source.
28. Write the conversion formula for delta to star conversion of three resistors.
29. What is the relationship between power, torque, and speed?
30. What are the limitations of Ohm's law? Is Ohm's law applicable in both dc and ac circuit?
31. State two fundamental laws of circuit analysis.
32. What is meant by time constant of an R–L circuit?
33. Derive equations that relate the resistance of a material at two different temperatures.
34. If n number of resistances each of value R are connected in parallel, then what will be the value of their equivalent resistance?
35. Derive the formula used in calculating the temperature coefficient of resistance at any temperature from its given value at any particular temperature.

Numerical Problems

36. A long copper wire has a resistance of 25 Ω at 40°C. Its resistance becomes 45 Ω when the temperature is 100°C. Calculate the value of its resistance at 0°C.

[Ans 11.7 Ω]

37. At what value of load resistance R_L, maximum power will be transferred in the circuit shown in Fig. 2.179.

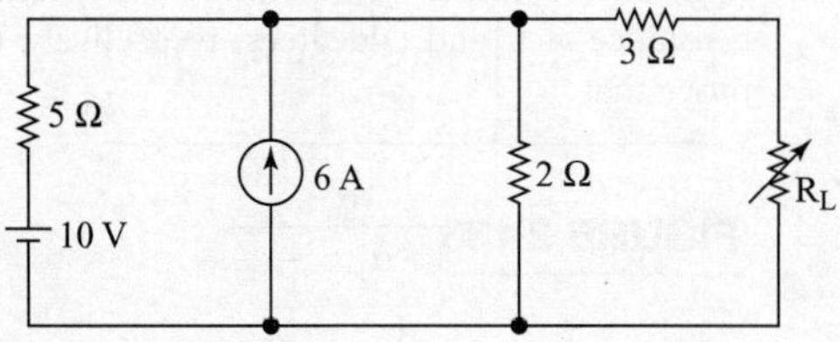

FIGURE 2.179

38. Calculate the value of R_L for which maximum power will be transferred from the source to the load in the network shown in Fig. 2.180. Also calculate the value of maximum power transferred.

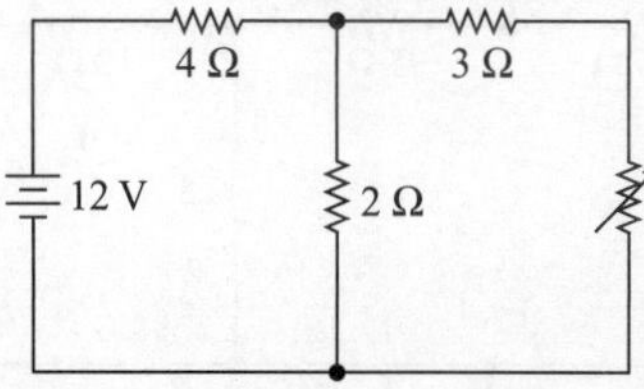

FIGURE 2.180

[Ans R_L = 7.33 Ω, P_{max} = 0.545 W]

39. Apply Norton's theorem to calculate the current through the 5 Ω resistor in the circuit shown in

Fig. 2.181. Also verify by applying Thevenin's theorem.

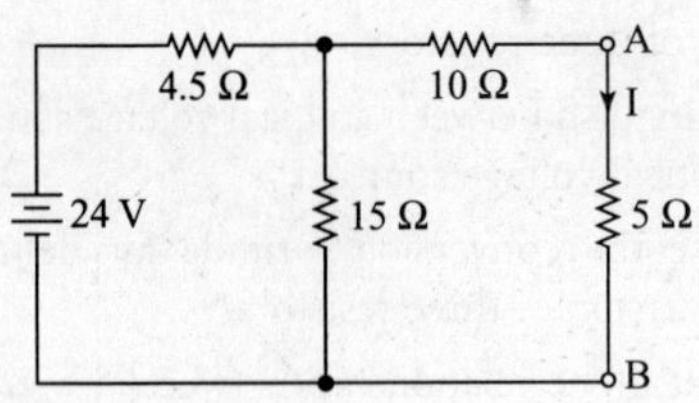

FIGURE 2.181

[**Ans** I = 1 A]

40. Apply Thevenin's theorem to calculate the current flowing through the 30 Ω resistor connected across terminals AB in the network shown in Fig. 2.182.

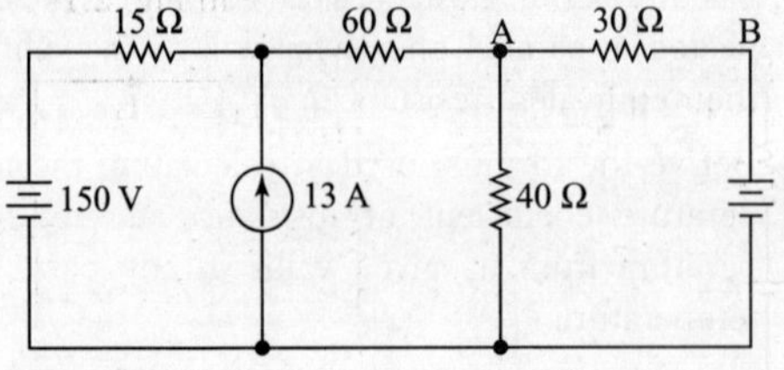

FIGURE 2.182

[**Ans** I_{AB} = 1.25 A]

41. By using the superposition theorem, calculate the current flowing through the 10 Ω resistor in the network shown in Fig. 2.183.

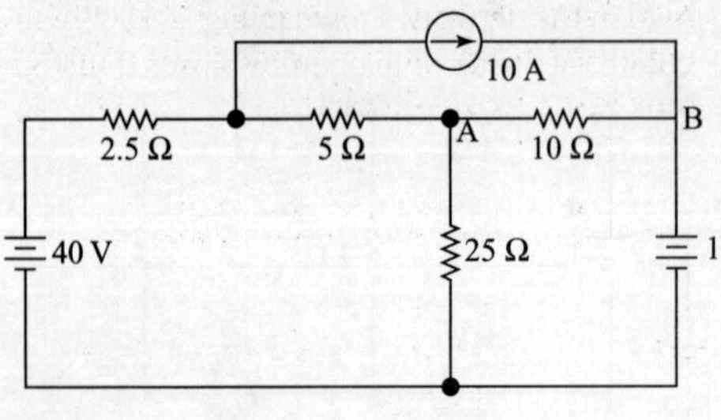

FIGURE 2.183

[**Ans** 0.054 A]

42. By applying Kirchhoff's laws or otherwise, calculate the current flowing through the 6 Ω resistor in the network shown in Fig. 2.184.

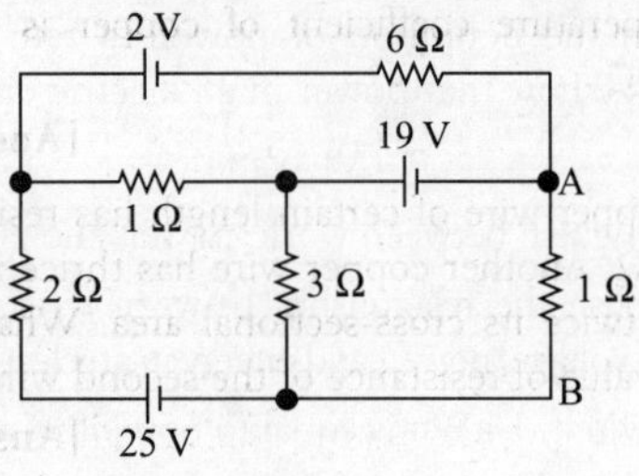

FIGURE 2.184

[**Ans** 2.95 A]

43. Three resistances of 25 Ω, 50 Ω, and 100 Ω are connected in parallel. If the total current drawn in 32 A, calculate the current drawn by each resistor.

[**Ans** 18.284 A, 9.144 A, 4.572 A]

44. Determine the current drawn from the battery in the circuit shown in Fig 2.185, using Kirchhoff's laws.

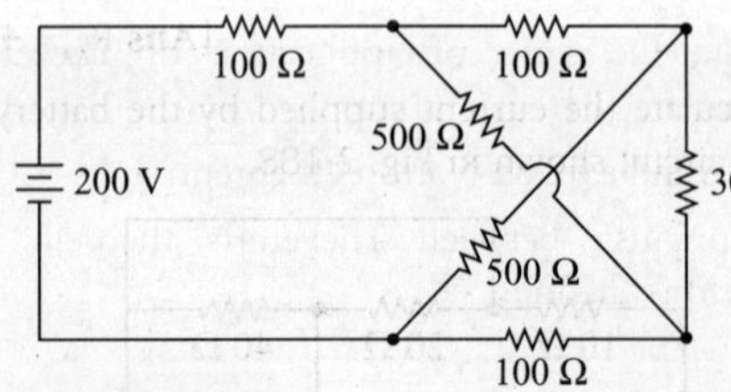

FIGURE 2.185

[**Ans** 0.6 A]

45. Use star–delta conversion of resistors to determine the current delivered by the battery in the network shown in Fig. 2.186.

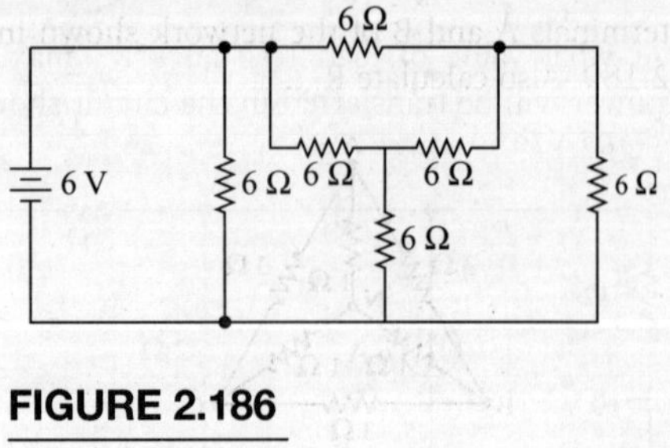

FIGURE 2.186

[**Ans** 2 A]

46. A winding wire made of copper has a resistance of 80 Ω at 15°C. Calculate its resistance at 50°C.

Temperature coefficient of copper is 0.004/°C at 0°C.

[Ans 90.6 Ω**]**

47. A copper wire of certain length has resistance of 4.5 Ω. Another copper wire has thrice its length and twice its cross-sectional area. What will be the value of resistance of the second wire?

[Ans 6.75 Ω**]**

48. Calculate the equivalent resistance of the network shown in Fig. 2.187 across the terminals A and B.

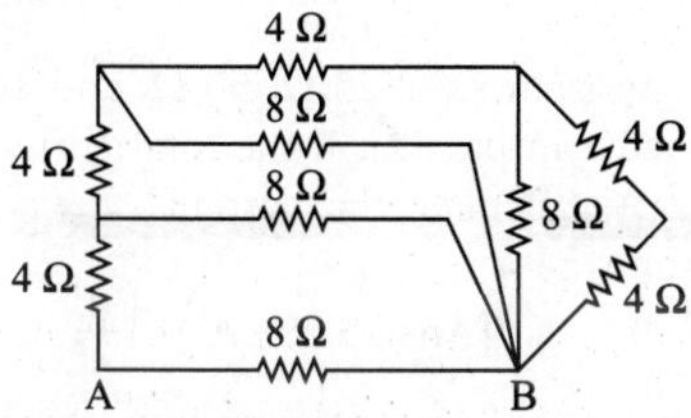

FIGURE 2.187

[Ans $R_{AB} = 4\ \Omega$**]**

49. Calculate the current supplied by the battery in the circuit shown in Fig. 2.188.

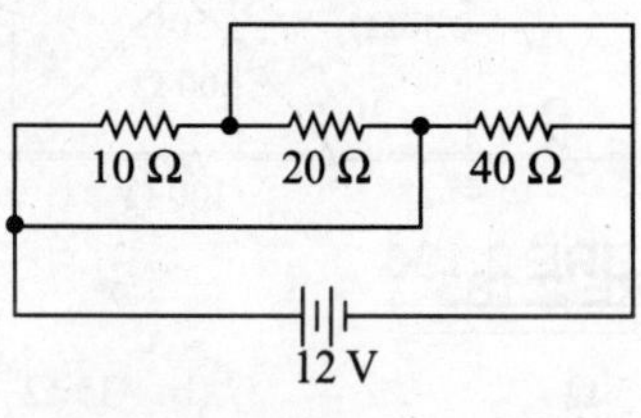

FIGURE 2.188

[Ans 2.099 A**]**

50. Calculate the equivalent resistance between the terminals A and B of the network shown in Fig. 2.189. Also calculate R_{AN}.

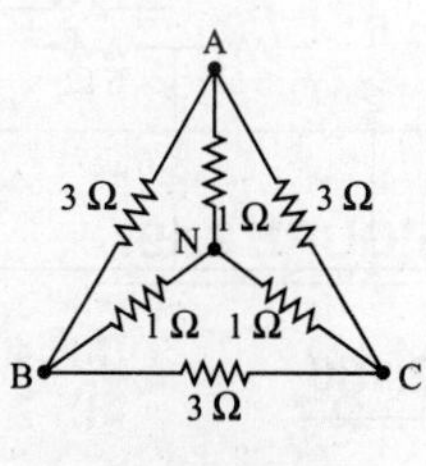

FIGURE 2.189

[Ans $R_{AB} = 1.0\ \Omega$ $R_{AN} = 0.664\ \Omega$**]**

51. Calculate the current supplied by the battery in the circuit shown in Fig. 2.190. All resistances are in Ohms.

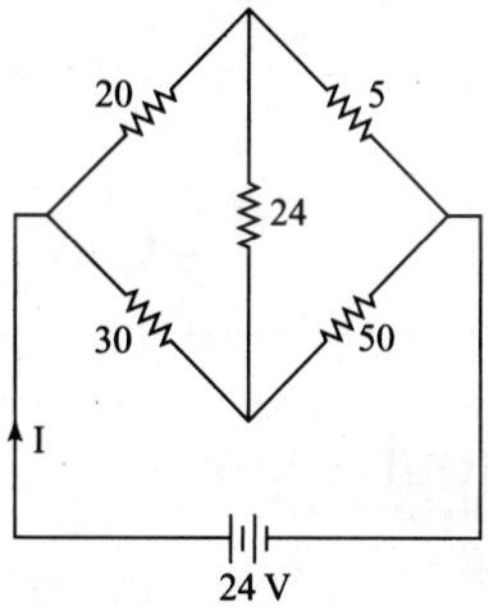

FIGURE 2.190

[Ans I = 1.35 A**]**

52. Using modal voltage analysis, calculate the current flowing through the resistor connected across the terminals A and B as shown in Fig. 2.191.

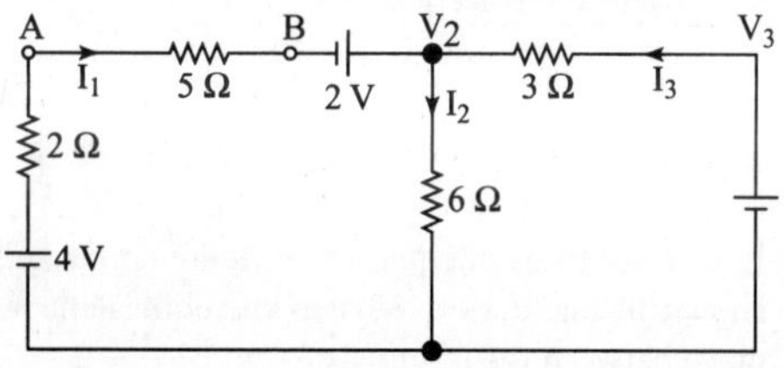

FIGURE 2.191

[Ans $I_1 = 0.371$ A**]**

53. Calculate the current flowing through the 2 Ω resistor connected across terminals A and B in the network shown in Fig. 2.192 by (i) applying Kirchhoff's laws; (ii) applying Thevenin's theorem; (iii) nodal voltage analysis. Compare the time taken by you in each case.

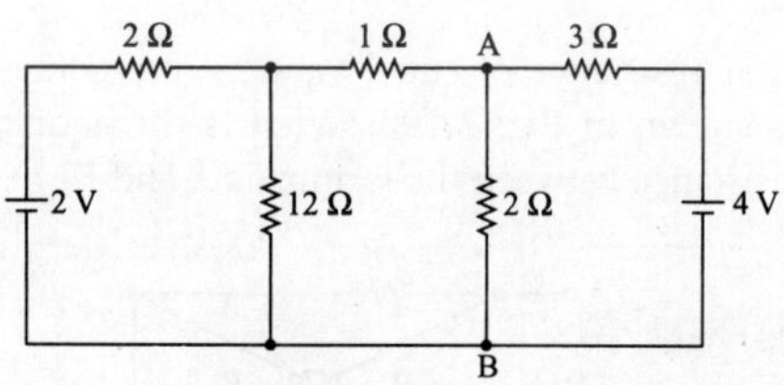

FIGURE 2.192

[Ans I = 0.817 A Applying Kirchhoff's laws takes the maximum time**]**

54. Calculate the current flowing through the 5 Ω resistor as shown in the network shown in Fig. 2.193.

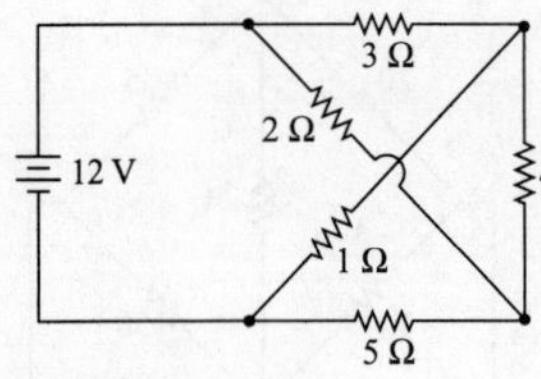

FIGURE 2.193

[**Ans** 0.663 A]

Multiple Choice Questions

1. Three resistances of equal value, R are connected such that they form a triangle having terminals A, B, and C. The equivalent value of the resistances across terminal A and B is equal to
(a) R/3 (b) 3/2 R
(c) $\frac{2}{3}$ R (d) 3 R.

2. Four resistances of equal value, R are connected as shown in Fig. 2.194. What is the equivalent resistance between the terminals A and B?

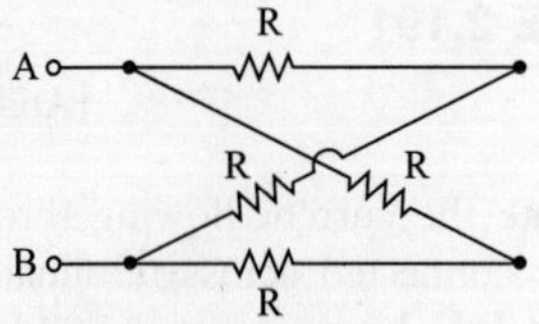

FIGURE 2.194

(a) R/4 (b) R/2
(c) 4 R (d) R.

3. Four resistance of equal value, R are connected as shown in Fig. 2.195. What is the equivalent resistance between the terminals A and B?

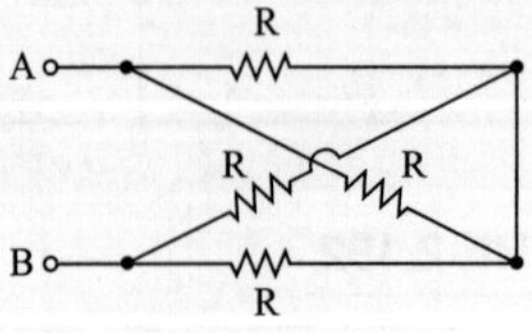

FIGURE 2.195

(a) 0 (b) 4 R
(c) $\frac{R}{4}$ (d) R.

4. What will be the equivalent resistance of the circuit shown in Fig. 2.196, between the terminals A and B?

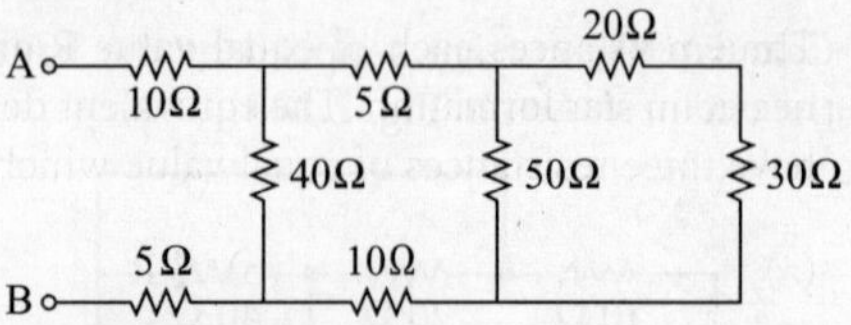

FIGURE 2.196

(a) 45 Ω (b) 35 Ω
(c) 170 Ω (d) 80 Ω.

5. What are the values of R_1 and R_2 in the circuit shown in Fig 2.197?

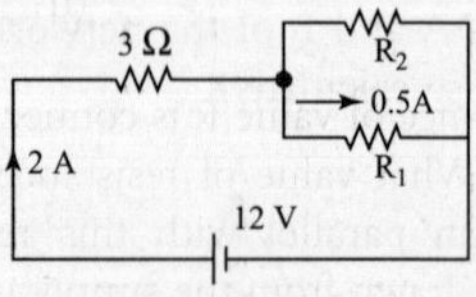

FIGURE 2.197

(a) 12 Ω and 4 Ω (b) 4 Ω and 12 Ω
(c) 6 Ω and 2 Ω (d) 2 Ω and 6 Ω.

6. The voltage applied across a 230 V, 60 W lamp is reduced to 115 V. What will be power consumed at this reduced voltage?

(a) 60 W (b) 30 W
(c) 120 W (d) 15 W.

7. Two resistances of equal value R and a wire of negligible resistances are connected in star formation as shown in Fig. 2.198. What will be the resistance between the terminals A and B when terminal C touches terminal A?

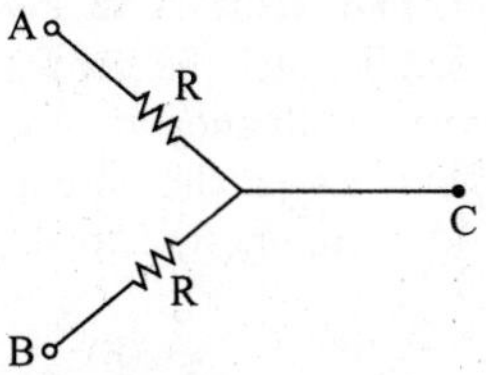

FIGURE 2.198

(a) 2 R (b) $\frac{R}{2}$
(c) R (d) 0.

8. Ampere second can be the unit of
(a) charge (b) voltage
(c) power (d) resistivity.

9. Three resistances each of equal value R are connected in star formation. The equivalent delta will have three resistances of equal value which is
(a) $\frac{R}{3}$ (b) 3 R
(c) $\frac{2}{3}$R (d) $\frac{R}{2}$.

10. Three resistance each of equal value R are connected in delta formation. The equivalent star will have three resistances of equal value which is
(a) $\frac{R}{3}$ (b) 3 R
(c) $\frac{2}{3}$ R (d) $\frac{R}{2}$.

11. A resistance of value R is connected across a voltage, V. What value of resistance should be connected in parallel with this resistance so that current drawn from the supply is doubled?
(a) R (b) 2R
(c) $\frac{R}{2}$ (d) $\frac{R}{4}$.

12. A current of 10 A gets divided into two parallel paths having resistance of 2 Ω and 3 Ω, respectively. The current through the 2 Ω and 3 Ω resistance will be, respectively
(a) 4 A and 6 A (b) 6 A and 4 A
(c) 5 A and 5 A (d) 2 A and 8 A.

13. Eight, Ω resistances are connected in parallel across terminals A and B. What is the equivalent resistance across AB?
(a) 64 Ω (b) 32 Ω
(c) 1 Ω (d) 4 Ω.

14. Two unequal resistances when connected in series gives a value of 10 Ω and when connected in parallel gives a value of 2.4 Ω. The value of the resistances are
(a) 4 Ω and 6 Ω (b) 1 Ω and 9 Ω
(c) 8 Ω and 2 Ω (d) 7 Ω and 3 Ω.

15. A wire of a particular length and cross-sectional area, a_1 is elongated to twice its length and the cross-sectional area gets reduced to a_2. Its resistance will increase by a factor of
(a) $2\left(\frac{a_1}{a_2}\right)$ (b) $2\left(\frac{a_2}{a_1}\right)$
(c) $\frac{a_1}{a_2}$ (d) $\frac{a_2}{a_1}$.

16. In the circuit shown in Fig. 2.199, what are the values of currents I_1 and I_2?

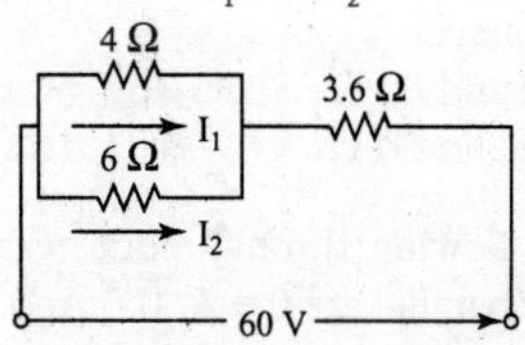

FIGURE 2.199

(a) 4 A and 6 A (b) 6 A and 4 A
(c) 10 A and 5 A (d) 2 A and 3 A.

17. Specific resistance of a conductor depends on
(a) length of the conductor
(b) area of cross sections of the conductor
(c) resistance of the conductor
(d) the nature of the material of the conductor.

18. In the circuit shown in Fig. 2.200, what is the value of R when the power dissipated in the 5 Ω resistor is 45 W?

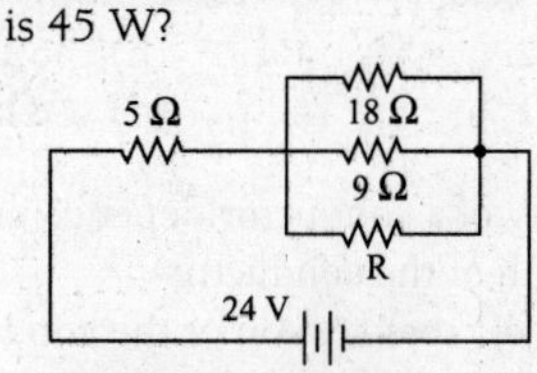

FIGURE 2.200

(a) 9 Ω (b) 3 Ω
(c) 6 Ω (d) 18 Ω.

19. The equivalent resistance of the circuit of Fig. 2.201 between the terminal A and B is

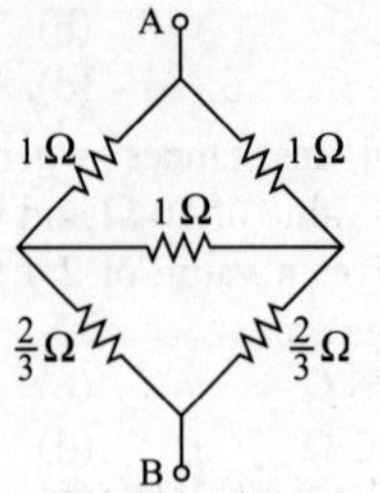

FIGURE 2.201

(a) 1.66 Ω (b) 0.833 Ω
(c) 2.66 Ω (d) 1.33 Ω.

20. The seem of two resistances connected in series across a supply voltage is 100 Ω. What are the values of the individual resistance if voltage drop across one of the resistors is 40 per cent of the supply voltage.
(a) 20 Ω and 80 Ω (b) 50 Ω and 50 Ω
(c) 40 Ω and 60 Ω (d) 10 Ω and 90 Ω.

21. Currents flowing through four resistances connected in parallel are 0.4 A, 0.3 A, 0.2 A, and 0.1 A, respectively. The equivalent resistance of the parallel circuit is 12 Ω. The value of resistances are
(a) 30 Ω, 40 Ω, 60 Ω, 120 Ω
(b) 15 Ω, 30 Ω, 60 Ω, 120 Ω
(c) 15 Ω, 20 Ω, 25 Ω, 30 Ω
(d) 4 Ω, 3 Ω, 2 Ω, 1 Ω.

22. The resistance of a wire is 6 Ω. The wire is drawn such that its length increases three times. The resistance of the elongated wire will be (note volume remaining same, if length is increased, area of cross section will decrease proportionately)
(a) 18 Ω (b) 36 Ω
(c) 54 Ω (d) 6 Ω.

23. Resistivity of a conductor depends upon
(a) length of the conductor
(b) area of cross-section of the conductor
(c) type or nature of the material
(d) all the factors as in (a), (b), and (c).

24. A conductor of length l and diameter d has a resistance of R Ohms. The diameter of the conductor is halved and its length is doubled. What will be the value of resistance of the conductor?
(a) 2R Ω (b) 4R Ω
(c) 8R Ω (d) $\frac{R}{4}$ Ω.

25. Two resistances of value 12 Ω and 8 Ω are connected in parallel and the combination is connected in series with another resistance of value 5.2 Ω. This series parallel circuit is connected across a 100 V supply. The total current drawn will be
(a) 10 A (b) 3.96 A
(c) 20 A (d) 5 A.

26. With the increase in temperature
(a) the resistance of metal will increase and that of insulator will decrease
(b) the resistance of metal will decrease and that of insulator will increase
(c) resistance of both metal and insulator will increase
(d) resistance of both metal and insulator will decrease.

27. Two resistances when connected in parallel across a 6 V battery draws a total current of 6 A. When one of the resistances is disconnected, the current drawn becomes 3 A. The resistances are of values
(a) 1 Ω and 1 Ω (b) 2 Ω and 2 Ω
(c) 6 Ω and 6 Ω (d) 3 Ω and 2 Ω.

28. Two resistances, 40 Ω and 10 Ω are connected in parallel and the combination is connected in series with a 2 Ω resistor. When the whole network is connected across a 100 V supply, the current drawn by the network will be
(a) 10 A (b) 1.92 A
(c) 20 A (d) 2.2 A.

29. Three resistance of 20 Ω each are connected in star. The resistance of each branch of the equivalent delta will be equal to
(a) 40 Ω (b) 60 Ω
(c) 400 Ω (d) 6.67 Ω.

30. Three resistances of 10 Ω each are connected in delta. The value of each of resistance of the equivalent star will be equal to
(a) 6.67 Ω (b) 3.33 Ω
(c) 30 Ω (d) 10 Ω.

31. The equivalent resistance of the network across terminals A and B, as shown in Fig. 2.202, will be

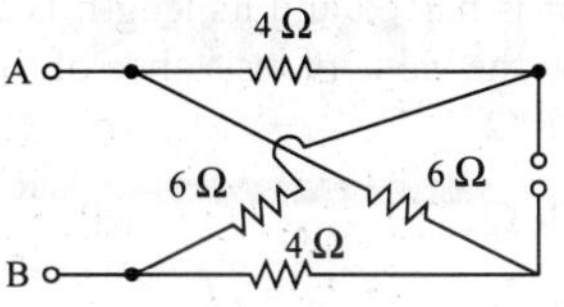

FIGURE 2.202

(a) 10 Ω (b) 20 Ω
(c) 2.5 Ω (d) 5 Ω.

32. In the circuit shown in Fig. 2.203, the voltage across terminals A and B is

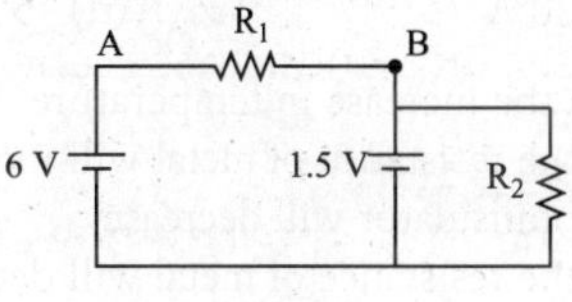

FIGURE 2.203

(a) 4.5 V (b) 6 V
(c) 1.5 V (d) 7.5 V.

33. The voltage across the resistances in the circuit shown in Fig. 2.204 is

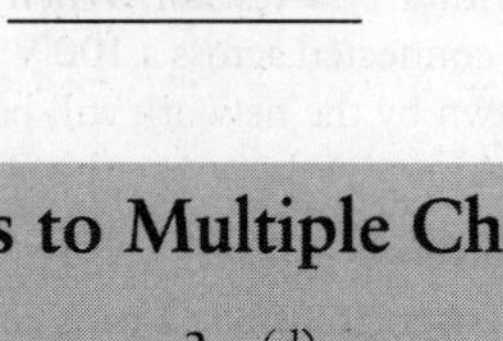

FIGURE 2.204

(a) 24 V (b) 12 V
(c) 48 V (d) 4.8 V.

34. The resistance between the terminals A and B of the circuit shown in Fig. 2.205 is

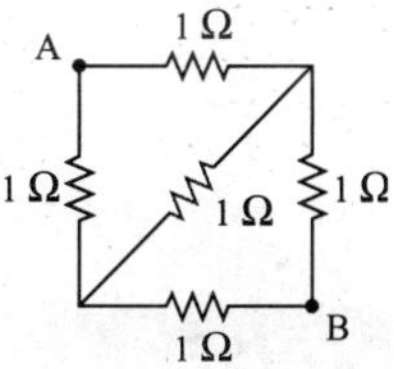

FIGURE 2.205

(a) 1 Ω
(b) 2 Ω
(c) 5 Ω
(d) 1.5 Ω.

35. The resistance between the terminals A and B of the circuit shown in Fig. 2.206 is

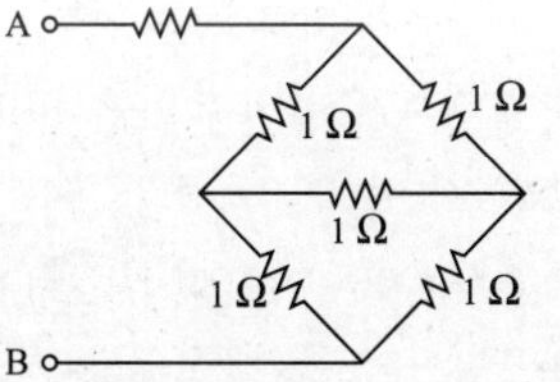

FIGURE 2.206

(a) 1 Ω (b) 2 Ω
(c) 6 Ω (d) 3 Ω.

Answers to Multiple Choice Questions

1. (c)	2. (d)	3. (d)	4. (b)	5. (a)	6. (d)
7. (c)	8. (a)	9. (b)	10. (a)	11. (a)	12. (b)
13. (c)	14. (a)	15. (a)	16. (b)	17. (d)	18. (c)
19. (b)	20. (c)	21. (a)	22. (c)	23. (c)	24. (c)
25. (a)	26. (a)	27. (b)	28. (a)	29. (b)	30. (b)
31. (d)	32. (d)	33. (a)	34. (a)	35. (b)	

CHAPTER 3

AC Fundamentals and Single-phase Circuits

Topics Discussed

1. Concept of dc and ac
2. Concept of frequency
3. Time period
4. Instantaneous value
5. Average value and maximum value of an alternating quantity
6. Sinusoidal and non-sinusoidal wave forms
7. Concept of root mean square value
8. Concept of phase and phase difference
9. Single-phase ac circuits
10. Series–parallel circuit containing resistance inductance and capacitance
11. Concept of apparent power
12. Real power and reactive power
13. Resonance in ac circuits
14. Solution of ac circuit problems

3.1 ► AC Fundamentals

Now days electricity is generated in the form of ac (alternating current). The generated electricity is transmitted, distributed, and mostly utilized in the form of ac. In this chapter, the fundamental concepts of ac and ac circuits have been discussed.

3.1.1 Introduction

We have known that current drawn from a battery is unidirectional. The polarities of the battery are marked +ve and –ve. When a particular load, say a lamp (represented by its resistance) is connected across the two terminals of the battery, current flows through the lamp in a particular direction. The magnitude of current as well as its direction remains constant with respect to time as long as the battery voltage remains constant. Such a current is known as steady or constant-value direct current.

When the direction of current through a circuit changes periodically, such a current is called alternating current. The polarities of the ac supply source changes alternately and causes alternating current to flow through the load connected across the terminals. Fig. 3.1 (a) shows dc flowing through the load when connected across the battery terminals. Since the magnitude is assumed constant, it is represented through a graph as shown in Fig. 3.1 (b). Fig. 3.1 (c) shows the nature of the current when the switch is turned '*ON*' and '*OFF*' at regular constant intervals. By using a reverse switching arrangement, as shown in Fig. 3.1 (d), we can have an ac flowing through the load. Here, through a reversing switch, supply terminal 1 is connected to load terminal 2′ and supply terminal 2 is connected to load terminal 1′. It can be noted that if the period of switching *ON* in both the directions in kept constant, the load current will be alternating in nature and its wave shape will be square type as shown in Fig. 3.1 (e).

To generate ac from an available dc source, we need an automatic switching arrangement. This is achieved by using electronic circuitry as in the case of inverters used as emergency lighting arrangement.

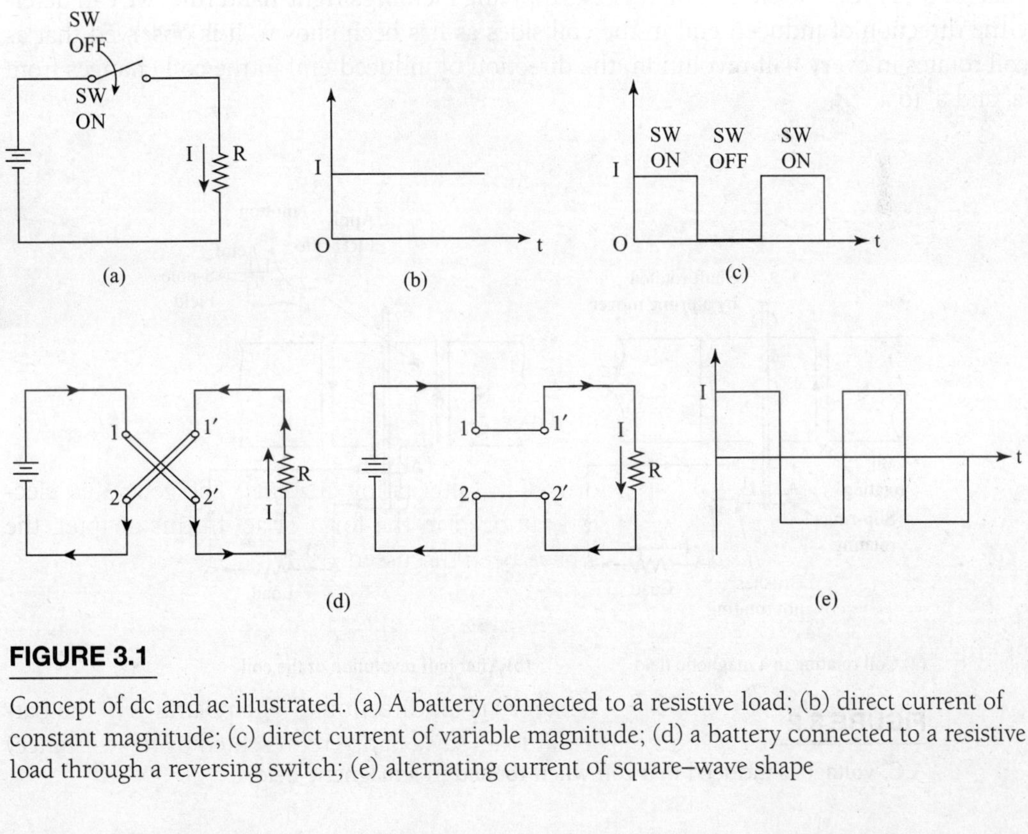

FIGURE 3.1

Concept of dc and ac illustrated. (a) A battery connected to a resistive load; (b) direct current of constant magnitude; (c) direct current of variable magnitude; (d) a battery connected to a resistive load through a reversing switch; (e) alternating current of square–wave shape

However, alternating current on a large scale is made available by using ac generators installed in power houses. AC generators are driven by turbines (gas, steam, water). Turbines are used to create a relative motion between a set of magnets and a set of coils. The rate of change of magnetic flux linkages or the rate of cutting of flux by the conductors of the coils causes EMF to be induced in the coil windings. The relative motion between the coils and the magnets producing a magnetic field can be created by making one system rotating with respect to the other. For example, we may have a stationary magnetic field system and inside the magnetic field we can place the coils which will be rotated by a prime mover (i.e., a turbine).

Alternately, the coils could be kept stationary and a set of magnets could be made rotating, thus causing EMF to be induced in the coils. We shall see the nature of EMF induced when we create a relative motion between a set of coils and a magnetic field. For simplicity we will consider only one coil rotating in a magnetic field created by a North and a South Pole.

3.1.2 Generation of Alternating Voltage in an Elementary Generator

We have known that when a coil rotates in a magnetic field, emf is induced in it. In Fig. 3.2, it is shown that a coil a – a′ rotating in a magnetic field that exists between two poles N and S. The position of the coil sides a and a′ are as shown in Fig. 3.2 (a) and (b), respectively, after each half of every revolution the coil makes. Applying Fleming's right-hand rule, we can determine the direction of induced emf in the coil sides as has been shown. It is observed that as the coil rotates in every half revolution, the direction of induced emf in the coil changes from a to a′ and a′ to a.

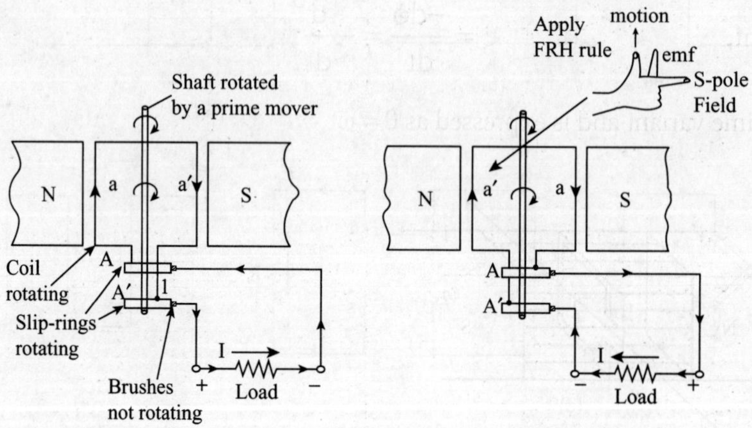

(a) Coil rotating in a magnetic field (b) After half revolution of the coil

FIGURE 3.2

A.C. voltage is induced in a coil when rotated in a magnetic field

If we now take out connections from the coil sides to an external load resistance, R through brush and slip-ring arrangement, current through the load will change every half cycle of rotation.

Slip-rings are solid circular conducting rings mounted on the same shaft as that of the coil. Two ends of the coil a and b are connected solidly with the two slip-rings A and B, respectively. Two brushes made of carbon are placed on the two slip-rings. Brushes do not rotate. The coil and the slip-rings rotate as the shaft is rotated. Brush and slip-ring arrangement helps us connect an external load resistance with the rotating coil. As the coil rotates an alternating current flows through the load resistance. The direction of current changes every half cycle of coil rotation.

To determine the wave shape of the induced emf, we will examine the rate of change of flux linkage by the coil as it rotates. Figure 3.3 shows that a coil a-b is being rotated in the magnetic field between two poles. The coil is shown perpendicular to the direction of the flux lines.

Flux linkage by a coil means the amount of flux passing through it. For the position of coil shown in Fig. 3.3(a), flux linkage is maximum as the whole of the flux passes through the coil when the coil rotates by an angle θ, the flux linkage becomes less. When $\theta = 90°$, the flux linkage becomes zero.

Thus, we can express flux linkage, ϕ as

$$\phi = \phi_m \quad \text{when} \quad \theta = 0$$

$$\phi = 0 \quad \text{when} \quad \theta = 90°$$

The expression for the flux, ϕ will be $\phi = \phi_m \cos\theta$. This satisfies the above conditions. Induced emf is the rate of change of flux linkage and not the flux linkage.

$$\text{Induced emf,} \qquad e = \frac{-d\phi}{dt} = -\frac{d}{dt}\phi_m \cos\theta.$$

Angle θ is time variant and is expressed as $\theta = \omega t$, where ω is the angular velocity of the coil.

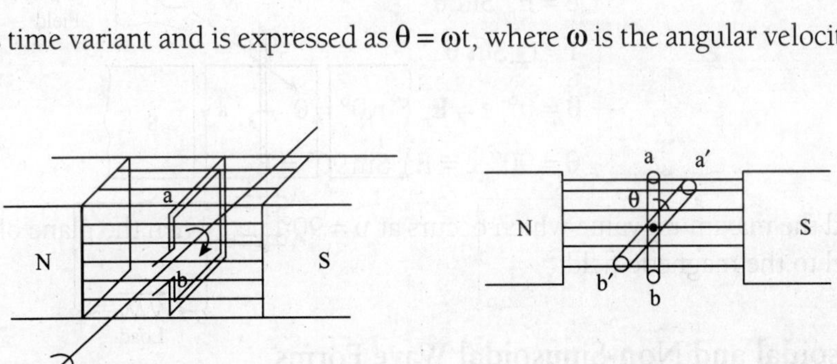

FIGURE 3.3

A coil a-b is shown rotated in an uniform magnetic field. The change of flux linkage changes as the coil position gets changed

Therefore, induced emf,
$$e = -\frac{d}{dt}\phi_m \cos\omega t$$
$$= (\omega\phi_m)\sin\omega t$$
or,
$$e = E_m \sin\omega t.$$

The emf induced in a coil rotating in an uniform magnetic field produce an alternating emf whose waveshape is sinusoidal in nature. Note that when the coil is perpendicular to the flux lines, flux linkage is maximum but rate of change of flux linkage is minimum, and hence, induced emf is minimum.

3.1.3 Concept of Frequency, Cycle, Time Period, Instantaneous Value, Average Value, and Maximum Value

One positive value and the subsequent one negative value of an alternating quantity constitute a **cycle**. The time taken for the generation of one cycle of EMF or flow of current caused due to such an EMF is called the **time period,** T. The total number of cycles of EMF or current produced per second is called the **frequency, f**. The relationship between time period, T and frequency, f can be found out as follows:

In T seconds the cycle produced is 1
In 1 second the cycle produced is 1/T
Since f is the cycle produced per second,

$$f = 1/T \tag{3.1}$$

The value of an alternating quantity at any instant of time is called its **instantaneous value**. Such values are expressed in small lettering like e, i, etc. For sinusoidal waves, we may write

$$e = E_m \text{ Sin } \theta$$
and
$$i = I_m \text{ Sin } \theta$$
at
$$\theta = 0°,\ e = E_m \text{ Sin} 0° = 0$$
at
$$\theta = 90°,\ e = E_m \text{ Sin} 90° = E_m$$

E_m is called the maximum value which occurs at $\upsilon = 90°$, i.e., when the plane of the rotating coil is parallel to the magnetic field.

3.1.4 Sinusoidal and Non-Sinusoidal Wave Forms

We have seen earlier that when a coil rotates in a uniform magnetic field the EMF induced in the coil is sinusoidal in nature. The wave shape of an alternating voltage or current produced in an ac generator having uniform flux distribution is also sinusoidal in nature. However, an alternating voltage or current available may be non-sinusoidal also. Any non-sinusoidal wave can be seen as consisting of a number of sinusoidal waves of different frequencies. Such component sine waves of a non-sinusoidal wave are called harmonic waves. In Fig. 3.4 (a) have been shown non-sinusoidal waves. In Fig. 3.4 (b) have been shown non-sinusoidal waves and their corresponding component sine waves (shown using dotted lines). The component sine wave which has the same frequency as the original wave is called fundamental wave and the sine waves of higher frequencies are called harmonic waves or simply harmonics.

In Fig. 3.4 (a) have been shown a trapezoidal and a triangular-type non-sinusoidal wave. In Fig. 3.4 b (i) has been shown a non-sinusoidal wave which is the sum of two component sine waves of different frequencies. One has the same frequency as the non-sinusoidal wave. This is called the fundamental. The other harmonic wave has twice the frequency as the fundamental. This is called the second harmonic. The non-sinusoidal wave shown in 3.4 b (ii) and (iii) are composed of a fundamental wave and a third harmonic. A third harmonic wave has three times the frequency as the fundamental wave. The number of harmonics present in an alternating non-sinusoidal quantity will depend upon the complexity of the wave shape. A symmetrical wave is the one whose positive half is identical to its negative half. Whether a wave is symmetrical or not can be tested

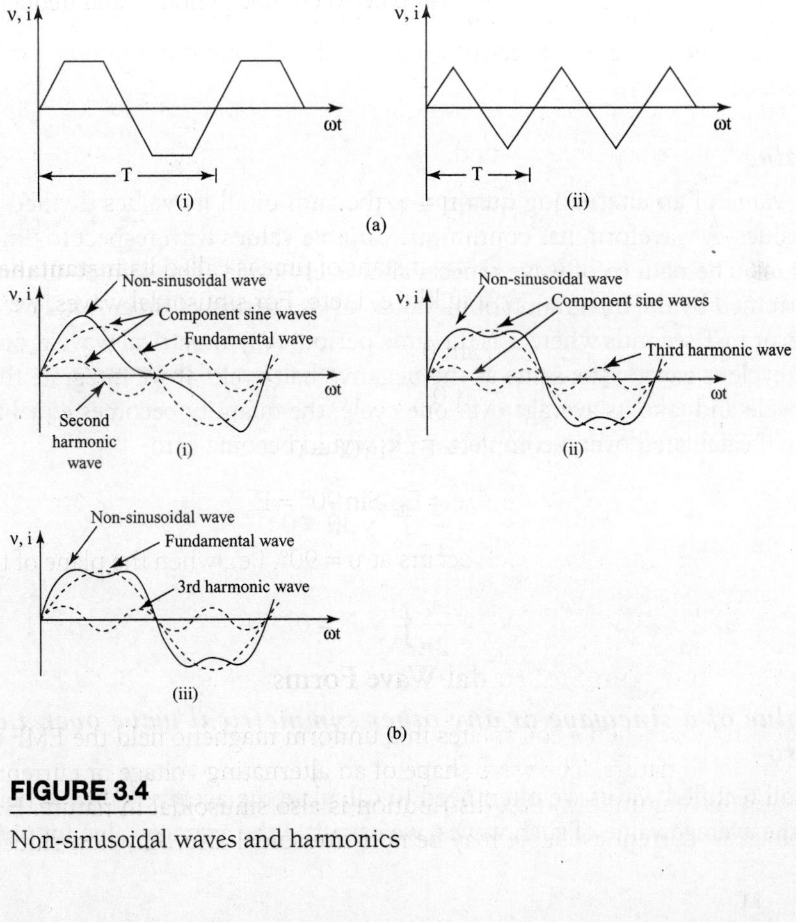

FIGURE 3.4

Non-sinusoidal waves and harmonics

by lifting the negative half and shifting it to the positive half axis and placing it just over the positive half. If both the half waves match each other, the wave shape is symmetrical. When generators are built physically symmetrical, the EMF wave shape induced in the coils in such machines are symmetrical in nature. A symmetrical wave will contain fundamental and odd harmonics only. The presence of even harmonics, i.e., 2nd harmonic, 4th harmonic, etc., will be there in non-symmetrical, non-sinusoidal ac waves. Harmonics in voltage and current is undesirable. Reduction of harmonics is essential because they cause increase heating of equipment, torque pulsation in electric motors, increased losses in electrical system, etc.

3.1.5 Concept of Average Value and Root Mean Square (RMS) Value of an Alternating Quantity

For a symmetrical alternating voltage or current wave, the positive half is identical to the negative half, and hence, the average value of the quantity for a complete cycle is zero. In earlier days the usefulness of such ac was questioned and only dc was considered effective. However, it was observed that when ac is allowed to pass through a resistance element, heat is produced. The question that arose was that if the average value of an alternating quantity is zero, why then was it producing heat. The concept of effective value was then brought in from the point of view of heat equivalence.

Average value

The average value of an alternating quantity is the sum of all its values divided by the total number of values. A waveform has continuous variable values with respect to time, t or angle θ where $\theta = \omega t$. The pattern of wave repeats after every cycle. The sum of all the values in a cycle is determined by the integration of its values over a period of time. A full cycle is formed in 2π radians or in T seconds where T is the time period. A symmetrical wave is one where the positive half cycle is exactly the same as the negative half cycle. If we integrate the values for a complete cycle and take its average over one cycle, the quantity becomes equal to zero. The average value if calculated over a complete cycle would become zero.

$$V_{av} = \frac{1}{2\pi}\int_0^{2\pi} v\, d\theta = 0$$

or,

$$V_{av} = \frac{1}{2\pi}\int_0^{T} v\, dt = 0$$

Average value of a sinewave or any other symmetrical wave over a complete cycle is zero

For half or full rectified waves we often need to calculate the average value. When we intend to calculate the average value of such waves, we calculate the average value for one-half cycle.

The average value is calculated for half cycle as

$$V_{av} = \frac{1}{\pi}\int_0^{\pi} v\, d\theta$$

or,
$$V_{av} = \frac{2}{T}\int_0^{T/2} v\, dt$$

Effective or RMS value

The effective or RMS value of an alternating quantity is determined by considering equivalent heating effect.

The effective or rms value of an alternating quantity (say current) is that the value of dc current which when flowing through a given circuit element (say a resistance element) for a given time will produce the same amount of heat as produced by the alternating current when flowing through the same circuit element for the same time.

Let I be the effective or rms value of the ac flowing through a resistance element R for a time t, then the amount of heat produced, H is expressed as

$$H \propto I^2 Rt = KI^2 Rt \text{ Joules} \tag{3.2}$$

Now, let the alternating current i, be passed through the same resistance R for the same time t, as shown in Fig. 3.5. Current i has been shown divided into n intervals and the

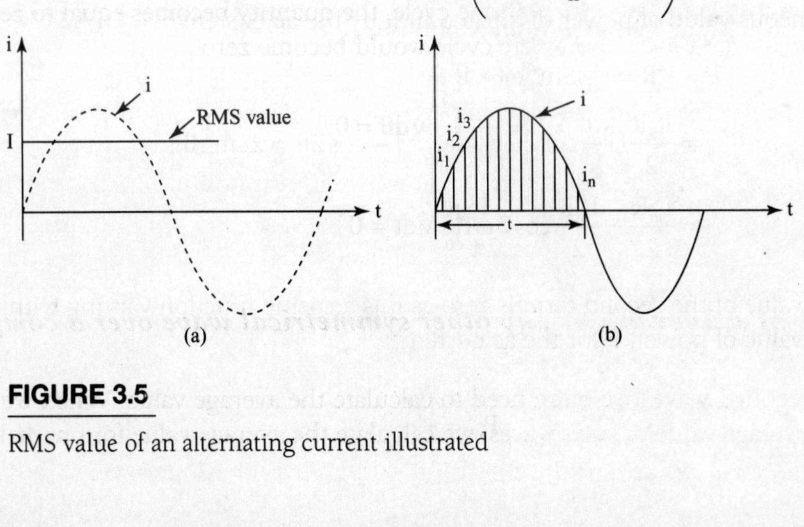

FIGURE 3.5
RMS value of an alternating current illustrated

magnitudes are i_1, i_2, i_3, etc. Heat produced in t seconds by the ac is equal to the sum of heat produced in n intervals of time during the time t. This can be expressed as

$$H \propto i_1^2 R\, t/n + i_2^2 R\, t/n + i_3^2 R\, t/n + \cdots + i_n^2 R\, t/n$$

$$\propto \frac{(i_1^2 + i_2^2 + i_3^2 + \cdots + i_n^2)}{n} Rt$$

$$H = K \frac{(i_1^2 + i_2^2 + i_3^2 + \cdots + i_n^2)}{n} Rt \text{ Joules} \tag{3.3}$$

Equating expressions (3.2) and (3.3),

$$I^2 Rt = \frac{(i_1^2 + i_2^2 + i_3^2 + \cdots + i_n^2)}{n} Rt$$

or,

$$I = \sqrt{\frac{i_1^2 + i_2^2 + i_3^2 + \cdots + i_n^2}{n}}$$

Thus, the effective value of an alternating current is equal to the square root of the mean of the squares of instantaneous values of the alternating current. Alternately, this can be read as *square mean root* value or *root mean square* value, i.e., RMS value. While expressing alternating quantities we always use RMS values and write in capital letters as E, I, V, etc. To further clarify the concept of the effective or RMS value, let us find out the value of dc current, I which gives the same amount of heating as that of ac when it flows through a resistance of value, say, R. The alternating quantity is represented as, say, $i = I_m \sin \omega t$.

The power dissipated in R by the dc current,

$$P_{av} = I^2 R \text{ watts} \tag{i}$$

The instantaneous value of power dissipated in R by the ac current,

$$P = i^2 R = I_m^2 \sin^2 \omega t \times R$$

$$= \frac{I_m^2 R}{2}(1 - \cos 2\omega t) \quad \left[\because 1 - \cos 2\theta = 2\sin^2\theta\right]$$

$$= \frac{I_m^2 R}{2} - \frac{I_m^2 R}{2}\cos 2\omega t$$

The average value of the second term is zero as it is a cosine function varying with time. The average value of power, P for the ac current,

$$P_{av} = \frac{I_m^2 R}{2} \tag{ii}$$

Equating the power dissipated due to dc current, I and the ac. current, i we can get the effective value as

$$I^2R = \frac{I_m^2 R}{2}$$

or,

$$I = \frac{I_m}{\sqrt{2}} = 0.707\ I_m$$

The effective or RMS value of an alternating current for a full cycle is calculated as

$$I = \sqrt{\frac{1}{T}\int_0^T i^2 dt} \quad \text{or} \quad I = \sqrt{\frac{1}{2\pi}\int_0^{2\pi} i^2 dt}$$

The RMS values of an alternating quantity of any other type of wave shapes can be calculated using analytical methods.

3.1.6 Analytical Method of Calculation of RMS Value, Average Value, and Form Factor

Suppose we have a sinusoidal alternating current, we have to first square it, then take its mean over one cycle or half cycle, and then take the square root (note that RMS value is calculated by making reverse operation i.e., first square, then take mean and then take square root).

Square of the current $i = I_m \sin\theta$ is $I_m^{\ 2} \sin^2\theta$

Its mean over one cycle is calculated by integrating it from 0 to 2π and dividing by the time period of 2π as follows:

Mean of square $= 1/2\pi\int_0^{2\pi} I_m^2 \sin^2\theta\, d\theta$

RMS value, $I = \sqrt{1/2\pi\int_0^{2\pi} I_m^2 \sin^2\theta d\theta}$

To calculate, let us put $\sin^2\theta = \dfrac{1-\cos 2\theta}{2}$

then

$$I = \sqrt{\frac{I_m^2}{4\pi}\int_0^{2\pi}(1-\cos 2\theta)d\theta}$$

$$= \sqrt{\frac{I_m^2}{4\pi}\left[\theta - \frac{\sin 2\theta}{2}\right]_0^{2\pi}}$$

$$= \sqrt{\frac{I_m^2}{4\pi}\times 2\pi}$$

$$= \sqrt{\frac{I_m^2}{2}} = \frac{I_m}{\sqrt{2}}$$

Therefore, $$I = \frac{I_m}{\sqrt{2}} = 0.707 I_m \tag{3.4}$$

i.e., $$\text{RMS value} = \frac{\text{Maximum value}}{\sqrt{2}} \quad \text{(for a sinusoidal wave)}$$

If we calculate the RMS value for half cycle, it can be seen that we will get the same value by calculating as

$$I = \sqrt{\frac{1}{\pi}\int_0^{\pi} I_m^2 \sin^2\theta d\theta} = \frac{I_m}{\sqrt{2}} = 0.707 I_m$$

Average value

Average value of a sinusoidally varying quantity over one cycle is zero because for the first half cycle current flows in the positive direction and for the second half cycle same current flows in the negative direction, i.e., in the opposite direction.

Average value has to be calculated by considering one-half cycle as

$$I_{av} = \frac{1}{\pi}\int_0^{\pi} I_m \sin\theta d\theta$$

$$= \frac{I_m}{\pi}\int_0^{\pi} I_m \sin\theta d\theta$$

$$= \frac{I_m}{\pi}[-\cos\theta]_0^{\pi}$$

$$= -\frac{I_m}{\pi}[-1-1]$$

or, $$I_{av} = \frac{2I_m}{\pi} = 0.637 I_m$$

Similarly average value for sinusoidal voltage, $V_{av} = \dfrac{2V_m}{\pi} = 0.637 V_m$

Therefore, average value of a sinusoidal voltage or current $= \dfrac{2I_m}{\pi}$ or $= \dfrac{2V_m}{\pi}$ (3.5)

Form factor

As the name suggests, form factor is an indicator of the shape or the form of the ac wave. It is the ratio of the RMS value to the average value of an alternating quantity. For a sinusoidal varying quantity, the form factor K_f is calculated as

$$K_f = \frac{\text{RMS value}}{\text{Average value}} = \frac{0.707 I_m}{0.637 I_m} = 1.11 \text{ (for a sine wave)} \tag{3.6}$$

Significance of form factor

The sharper the wave shape, the more will be the value of the form factor. For example, for a triangular wave, form factor will be more than 1.11 and for a rectangular wave form factor will be less than 1.11 (in fact, its value will be 1).

The peak or crest factor is the ratio of peak or maximum value to its rms value.

For a sinsusoidal wave, peak factor $= \dfrac{I_m}{I_m/\sqrt{2}} = 1.414$

If the peak factor is known, the maximum value can be calcuated from a given rms value.

It is obvious that by knowing the value of the form factor, the RMS value can be calculated if the average or mean value is known.

3.1.7 RMS and Average Values of Half-Wave-Rectified Alternating Quantity

A half-rectified sine wave is shown in Fig. 3.6. A half-wave-rectified quantity, whether voltage or current will have its one half cycle blocked by using a diode rectifier, as shown. Since the diode allows current to flow in one direction, current through the load resistance will flow, in one direction only. During the negative half cycle of the input voltage the diode will block current flow, and hence, no voltage will be appearing across the load during all negative half cycles. For half-wave-rectified current or voltage, we have to consider the current or voltage which is available for the positive half cycles and average it for the complete cycle. For a complete cycle, i.e., from 0 to 2π, current flows only from 0 to π. To calculate the RMS value, we have to square the current, take its sum from 0 to π and then take the average for the whole cycle, i.e., from 0 to 2π. Thus, the RMS value for a half-wave-rectified current, is calculated as

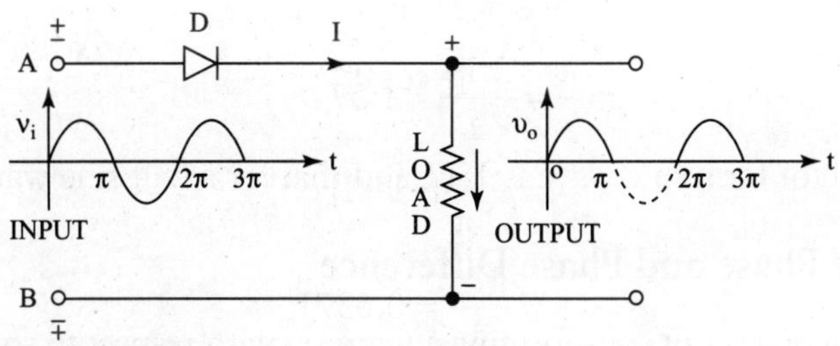

FIGURE 3.6

Half-rectified sine wave

$$I = \sqrt{\frac{1}{2\pi}\int_0^{\pi} I_m^2 \sin^2\theta\, d\theta} = \sqrt{\frac{I_m^2}{4\pi}\int_0^{\pi}(I - \cos 2\theta)\, d\theta}$$

$$[\text{since} 2\sin^2\theta = 1 - \cos 2\theta]$$

$$= \sqrt{\frac{I_m^2}{4\pi}\left[\theta - \frac{\sin 2\theta}{2}\right]_0^{\pi}}$$

$$= \sqrt{\frac{I_m^2}{4\pi} \times \pi} = \frac{I_m}{2}$$

Thus, the RMS value of a full sine wave is $\dfrac{I_m}{\sqrt{2}}$ and for a half wave, $\dfrac{I_m}{2}$.

Average value of current for half sine wave is

$$I_{av} = \frac{1}{2\pi}\int_0^{\pi} i d\theta = \frac{1}{2\pi}\int_0^{\pi} I_m \sin\theta d\theta$$

$$= \frac{I_m}{2\pi}[-\cos\theta]_0^{\pi}$$

$$= \frac{I_m}{2\pi}\times 2 = \frac{I_m}{\pi}$$

Note that for a complete sine wave, the average value was calculated as

$$I_{av} = \frac{1}{\pi}\int_0^{\pi} i d\theta = \frac{2I_m}{\pi}$$

and for a half-rectified sine wave, the average value has been calculated as

$$I_{av} = \frac{1}{2\pi}\int_0^{\pi} i d\theta = \frac{I_m}{\pi}. \tag{3.7}$$

Obviously, we note that for a half-rectified wave, the average value is half of that of a full sine wave.

Form factor for a half sine wave quantity is

$$K_f = \frac{\text{RMS value}}{\text{Average value}} = \frac{I_m}{2}\times\frac{\pi}{I_m}$$

$$= \frac{\pi}{2} = \frac{3.14}{2} = 1.57$$

Note that form factor for a sine wave is 1.11 and that for a half sine wave it is 1.57.

3.1.8 Concept of Phase and Phase Difference

The position of a coil or a set of coils forming a winding with respect to some axis of reference is called its *phase*. If three coils are placed at different angles with respect to the reference axis, there exists a *space phase difference* between these three coils AA′, BB′, and CC′,. When EMFs will be induced in these coils due to the cutting of the magnetic flux or due to change in flux linkages, the EMFs will have similar ***time phase difference*** between them as shown in Fig. 3.7.

A magnet has been shown rotating in the anticlockwise direction. Maximum flux will be cut by the coil AA′ at time, t = 0. Hence, maximum voltage will be in the coil AA′ at time, t = 0 as has been shown as v_A in Fig. 3.7(b) Maximum flux will be cut by the coil BB′ after an elapse of angle 30°, i.e., by the time the rotating magnet rotates by an angle of 30°. Similarly, maximum flux will be cut by the coil CC′ after an elapse of time represented by 60°.

The voltage waves in coil AA′, BB′, and CC′ will, therefore, have a time phase difference of 30°. (30° corresponds to the time taken by the rotating magnet to rotate by 30°). Since voltage v_A is appearing earlier than v_B, v_A is said to be leading voltage v_B. Voltage induced in the three coils AA′, BB′, and CC′ will have a time phase difference of 30°.

Such phase difference may exist between the voltage and current in an electrical circuit. If current in a circuit changes in accordance with the voltage, i.e., when the voltage is at its maximum value, the current is also at its maximum value, and when the voltage starts increasing

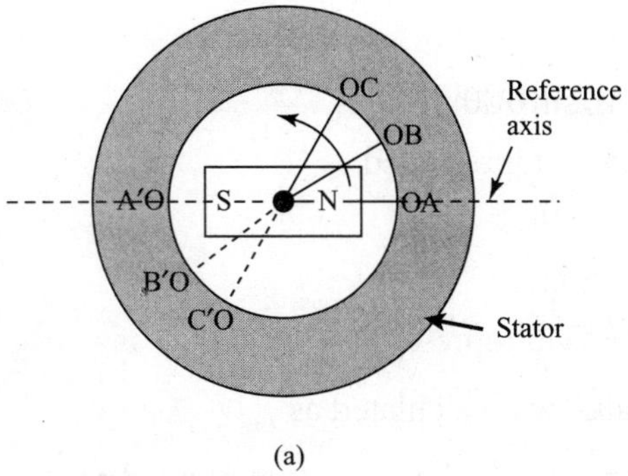

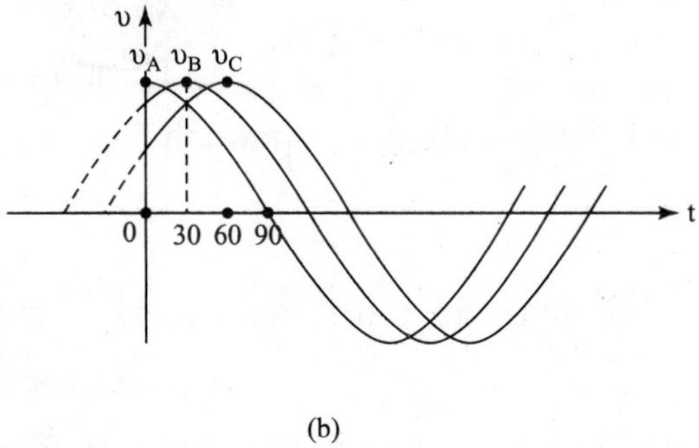

FIGURE 3.7

(a) Three coils AA′, BB′, CC′ are placed in stator at a space phase difference of certain angle.
(b) EMF induced in these coils will have a time phase difference

in the positive direction from its zero value, the current also starts increasing in the positive direction from its zero value; then, the voltage and current are said to be in phase as shown in Fig. 3.8 (a). Note that the magnitudes of voltage and current may be different. In Fig. 3.8 (b) is shown current, i lagging the voltage by 90°, i.e., by an angle π/2. The expressions for voltage and current as shown in Fig. 3.8 (a) can be written as

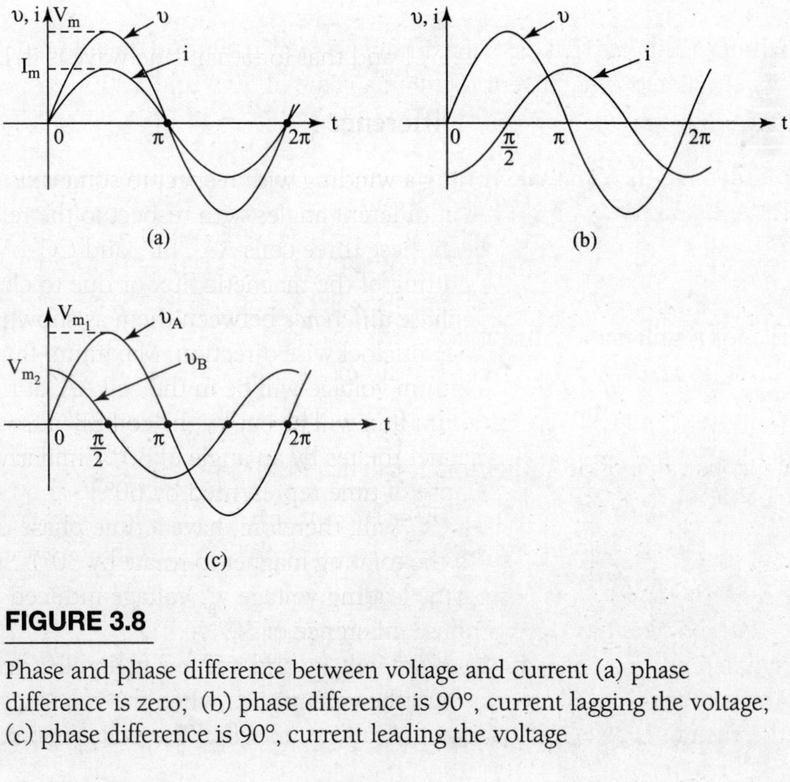

FIGURE 3.8

Phase and phase difference between voltage and current (a) phase difference is zero; (b) phase difference is 90°, current lagging the voltage; (c) phase difference is 90°, current leading the voltage

$$\upsilon = V_m \sin \omega t$$

$$i = I_m \sin(\omega t + 0) = I_m \sin \omega t$$

The voltage and current shown in Fig. 3.8 (b) can be represented as

$$v = V_m \sin \omega t$$

$$i = I_m \sin(\omega t - \pi/2)$$

If current is leading the voltage by $\pi/2$ degrees, we will represent the current, i as

$$i = I_m \sin(\omega t + \pi/2)$$

If two voltages v_A and v_B are represented as in Fig. 3.8 (c), they can be expressed as

$$v_A = V_{m1} \sin \omega t$$

$$v_B = V_{m_2} \sin(\omega t + \pi/2)$$

This is because voltage v_B is leading the voltage v_A. The maximum value of v_B is appearing $\pi/2$ degrees before the maximum value of v_A appears. However, if v_B is taken as the reference voltage we can express v_B and v_A as

$$v_B = V_m \sin \omega t$$

$$v_A = V_m \sin(\omega t - \pi/2)$$

EXAMPLE 3.1

A sinusoidal alternating current has a maximum value of 100 A and the frequency is 50 Hz. How long will it take for the current to attain a value of 50 A and 100 A, respectively?

SOLUTION

We have to calculate the time taken for the current to reach its instantaneous values of 50A and 100A, respectively.

Given, $I_m = 100 \text{ A} \quad \text{and} \quad f = 50 \text{ Hz}$

$$w = 2\pi f = 2 \times 3.14 \times 50 = 314$$

We can write for a sinusoidal current,

$$i = I_m \sin \omega t$$

So, here, $i = 100 \sin 314t$

For instantaneous value of 50A, the time is calculated as

$$50 = 100 \sin 314t$$

or, $\sin 314t = 0.5 = \sin 30°$

$$314t = 30$$

$$t_1 = \frac{30}{314} \sec = \frac{30 \times 1000}{314} \text{ms} = 95.5 \text{ ms}.$$

Time period, $T = \frac{1}{f} = \frac{1}{50} = 0.02 \text{ s} = 20 \text{ ms}.$

Current will reach its maximum value of 100 A in $\frac{T}{4}$ seconds.

$$\therefore \; t_2 = \frac{T}{4} = \frac{20}{4} = 5 \text{ ms}.$$

EXAMPLE 3.2

A 50 Hz sinusoidal voltage has a maximum value of 325 V. Calculate its instantaneous value after 2 ms from its zero value. What is its rms value? At what time the voltage will reach its maximum value?

SOLUTION

$$w = 2\pi f = 2 \times 180° \times 50$$

The equation for the sinusoidal voltage is

$$v = V_m \sin\omega t$$

or, $$v = 325 \sin 2\pi ft$$

The instantaneous value after 2 ms is calculated as

$$v = 325 \sin 2 \times 180° \times 50 \times 2 \times 10^{-3}$$

or, $$v = 325 \sin 36°$$
$$= 325 \times 0.6 = 195 \text{ V}.$$

$$\text{Rms Value} = \frac{V_m}{\sqrt{2}} = \frac{325}{\sqrt{2}} = 230 \text{ V}.$$

It will reach its maximum value at time $t = \frac{T}{4}$

Here, $T = \frac{1}{f} = \frac{1}{50} \text{sec} = \frac{1000}{50} = 20 \text{ ms. so, } t = \frac{20}{4} = 5 \text{ ms}.$

EXAMPLE 3.3

A 50 Hertz alternating current has its wave shape such that the value of current increases in steps of 10 V after every 20 degrees starting from 0 to a maximum value of 40 V and then getting reduced to zero in same steps completing the positive half cycle. The negative half cycle is symmetrical with the positive half cycle. Draw the wave shape, calculate its average and rms values.

SOLUTION

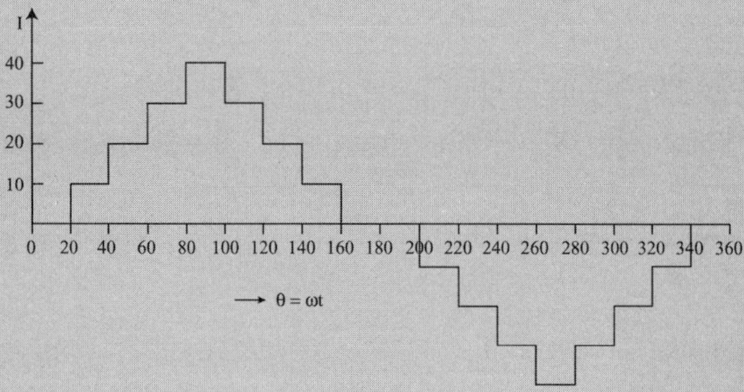

FIGURE 3.9

$$I_{av} = \frac{0+10+20+30+40+30+20+10+0}{9} = \frac{160}{9} = 17.77 \text{ A.}$$

$$I_{rms} = \sqrt{\frac{0^2+10^2+20^2+30^2+40^2+30^2+20^2+10^2+0^2}{9}}$$

$$= \sqrt{\frac{0+100+400+900+1600+900+400+100+0}{9}}$$

$$= \sqrt{\frac{4400}{9}}$$

$$= \sqrt{488.8}$$

$$= 22.1 \text{ A.}$$

EXAMPLE 3.4

An alternating voltage of 100 sin 314 t is applied to a half-wave diode rectifier which is in series with a resistance of 20 Ω. What is the RMS value of the current drawn from the supply source?

SOLUTION

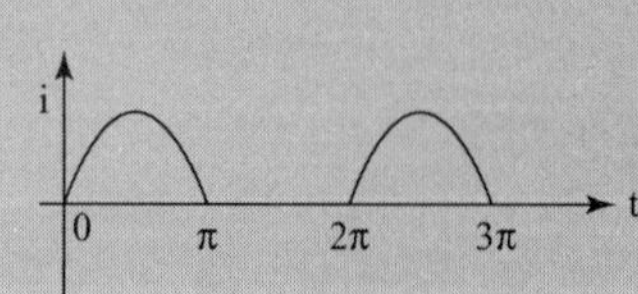

FIGURE 3.10

Circuit diagram of example 3.4

We have, $v = V_m \sin \omega t = 100 \sin 314\, t$

$$V_m = 100\text{ V}$$

For a full sine wave, RMS value

$$V_{rms} = \frac{V_m}{\sqrt{2}}$$

For half-rectified wave

$$V_{rms} = \frac{V_m}{2}$$

Therefore,

$$I_{rms} = \frac{V_{rms}}{R} = \frac{V_m}{2 \times 20} = \frac{100}{2 \times 20} = 2.5\text{ A}$$

EXAMPLE 3.5

An alternating sinusoidal voltage of $v = 150 \sin 100\pi t$ is applied to a circuit which offers a resistance of 50 Ω to the current in one direction and prevents the flow of any current in the opposite direction. Calculate the RMS and average values of the current and the form factor. What is the frequency of the supply?

SOLUTION

$$v = 150 \sin 100\pi t$$

standard form, $v = V_m \sin \omega t$

The maximum value of voltage, $V_m = 150$

$$\omega = 100\pi = 2\pi f$$

frequency, $f = 50\text{ Hz}$

The circuit in the question is a half-wave-rectified one.

For half sine wave, the RMS value

$$V_{rms} = \frac{V_m}{2} \text{ and } V_{av} = \frac{V_m}{\pi}$$

Therefore,

$$I_{rms} = \frac{V_{rms}}{R} = \frac{V_m}{2R} = \frac{150}{2 \times 50} = 1.5\text{ A}$$

$$I_{av} = \frac{V_{av}}{R} = \frac{V_m}{\pi R} = \frac{150}{3.14 \times 50} = 0.95\text{ A}$$

Form factor,

$$K_f = \frac{I_{rms}}{I_{av}} = \frac{1.5}{0.95} = 1.57$$

EXAMPLE 3.6

Calculate the RMS value, average value and form factor of a half-rectified square voltage shown in Fig. 3.11.

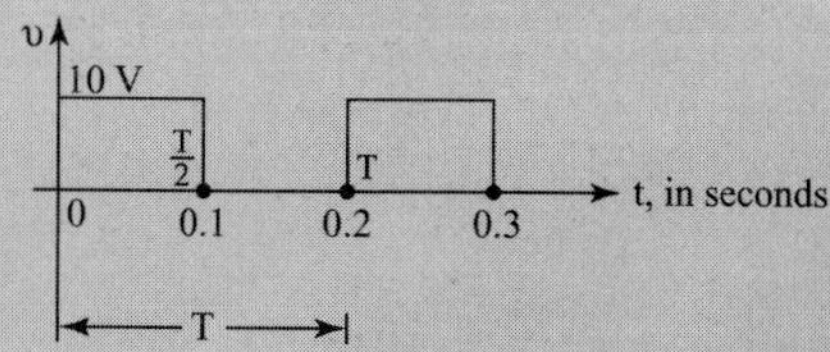

FIGURE 3.11

Circuit diagram of example 3.6

SOLUTION

For half-rectified wave, $V_{av} = \frac{1}{T}\int_0^{T/2} v\,dt$

Here, $T = 0.2,\ v = 10\text{ V}$

Therefore,

$$V_{av} = \frac{1}{0.2}\int_0^{0.1} 10\,dt = \frac{1}{0.2}[10t]_0^{0.1} = \frac{1}{0.2}\times 10\times 0.1 = 5\text{ V}$$

$$V_{rms} = \sqrt{\frac{1}{T}\int_0^{T/2} v^2 dt} = \sqrt{\frac{1}{0.2}\int_0^{0.1} 10^2 dt}$$

$$= \sqrt{\frac{1}{0.2}\left[[10]^2 t\right]_0^{0.1}} = \sqrt{\frac{1}{0.2}[100\times 0.1]}$$

$$= \sqrt{\frac{10}{0.2}} = \sqrt{50} = 7.09\text{ V}$$

Form factor, $K_f = \frac{V_{rms}}{V_{av}} = \frac{7.09}{5} = 1.4$

EXAMPLE 3.7

Calculate the RMS value and average value of the elevated saw-tooth-type current wave shown in Fig. 3.12.

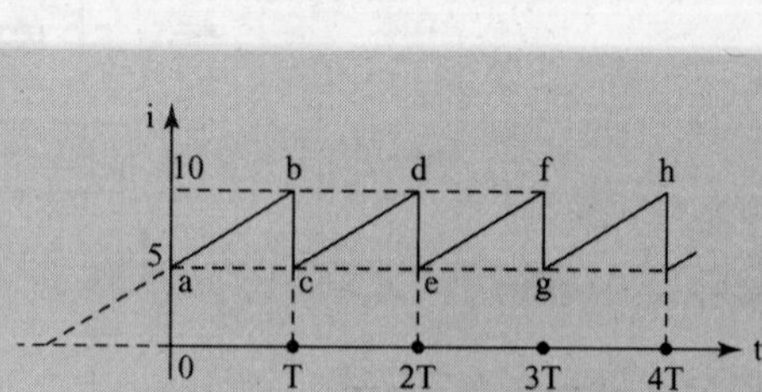

FIGURE 3.12

Saw-tooth wave of example 3.7

SOLUTION

It can be seen from the wave shape that 0abcT makes one cycle. The same pattern is being repeated for each time period of T. The equation for the line ab is of the form y = mx + c. Here the slope m is bc/ac, i.e., equal to 5/T. The value of c is 5 and y is represented by i and x by t.

Therefore, the equation of the line ab is

$$i = \frac{5t}{T} + 5$$

The value of average current is calculated as

$$I_{av} = \frac{1}{T}\int_0^T i\,dt = \frac{1}{T}\int_0^T \left[\frac{5t}{T}+5\right]dt$$

$$= \frac{1}{T}\left[\frac{5t^2}{2T}+5t\right]_0^T$$

$$= \frac{1}{T}\left[\frac{5T^2}{2T}+5T\right] = \frac{1}{T}\times 7.5T = 7.5\text{ A}$$

(In this case by actual observation of the wave shape as shown in Fig. 3.11, the average value can also be determined)

The RMS value is calculated as

$$I_{rms} = \sqrt{\frac{1}{T}\int_0^T i^2\,dt}$$

$$= \sqrt{\frac{1}{T}\int_0^T \left(\frac{5t}{T}+5\right)^2 dt}$$

$$= \sqrt{\frac{1}{T}\int_0^T \left(\frac{25t^2}{T^2}+5^2+2\times\frac{5t}{T}\times 5\right)dt}$$

$$= \sqrt{\frac{1}{T}\left[\frac{25t^3}{3T^2}+25T+\frac{50t^2}{2T}\right]_0^T}$$

$$= \sqrt{\frac{1}{T}\left[\frac{25T}{3}+25T+25T\right]}$$

$$= \sqrt{\frac{175}{3}} = 7.68\text{ A}$$

EXAMPLE 3.8

Find the average value, RMS value and form factor of the saw-tooth current wave shown in Fig. 3.13.

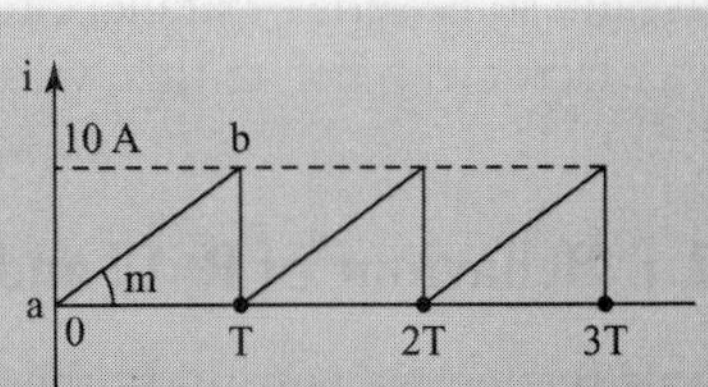

FIGURE 3.13

Diagram for example 3.5

SOLUTION

The equation of the line ab is of the form

$$y = mx$$

Here, $y = i$, $m = 10/T$, and $x = t$

Therefore, we can write

$$i = \frac{10t}{T}$$

RMS value of i,

$$I = \sqrt{\frac{1}{T}\int_0^T \left(\frac{10t}{T}\right)^2 dt}$$

$$= \sqrt{\frac{1}{T}\int_0^T \frac{100t^2}{T^2} dt}$$

$$= \sqrt{\frac{1}{T}\left[\frac{100}{T^2}\frac{t^3}{3}\right]_0^T}$$

$$= \sqrt{100/3} = 5.78 \text{ A}$$

Average value of a right-angle triangle is half of its height, i.e., equal to

$$I_{av} = 10/2 = 5 \text{ A}$$

Form factor

$$= \frac{\text{RMS value}}{\text{Average value}} = \frac{5.78}{5} = 1.15$$

The students are to note that the form factor of a saw-tooth wave has been calculated as 1.15 whereas for a sine wave the value was 1.11. Since a saw-tooth wave is stiffer than a sine wave, its form factor is higher than that of a sine wave.

3.2 ► Single-Phase AC Circuits

A resistance, an inductance, and a capacitance are the basic elements of an ac circuit. These elements are connected in series and parallel combinations to form an actual circuit. Circuits may include any two or three elements. For example, we may have one resistance and one inductance connected in series across an ac supply source. We may have one resistance connected in series with one inductance and one capacitance in parallel. Accordingly circuits are named as L-R circuits, L-R-C circuits, etc. We will take up few series circuits, few parallel circuits, and some series–parallel circuits and calculate the main current, branch currents, power, power factor, etc. Before this, we will discuss the behaviour of R, L, and C in ac circuits

3.2.1 Behaviour of R, L, and C AC Circuits

In this section we will study the relationship of applied voltage and current in an ac circuit involving only a resistance, an inductance, and a capacitance. When a resistance is connected across an ac supply, we call it a purely resistive circuit. Similarly, an inductance coil connected across an ac supply is called a purely inductive circuit and a capacitance connected across an ac supply is called a purely capacitive circuit. We shall study the phase relationship between the applied voltage and current flowing in each case under steady-state condition.

AC applied across a pure resistor

When we say a pure resistance, we assume that the resistance wire does not have any inductance or capacitance. Fig. 3.14 shows a pure resistance connected across an ac supply. The voltage and current wave forms as well as the phasor diagram showing the positions of voltage and current have been shown. The instantaneous value of

Voltage, v of the source is $v = V_m \text{ Sin } \omega t$

Where, V_m is the maximum value of the voltage in Volts; $\omega = 2\pi f$ rad/sec; and f is the frequency of supply voltage in cycles per second.

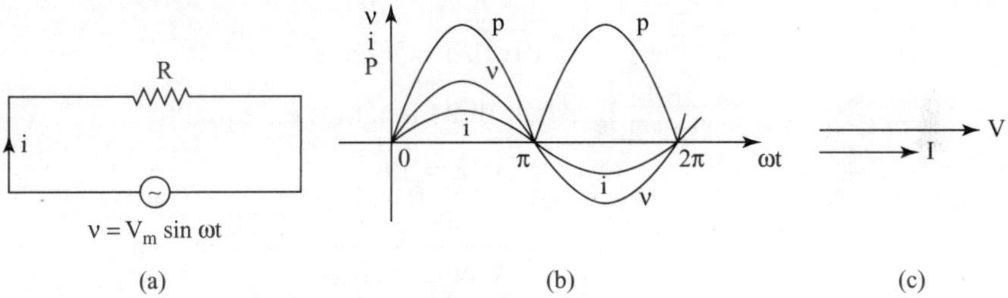

FIGURE 3.14

(a) Resistive circuit with a sinusoidal voltage source; (b) voltage and current wave shapes; (c) phasor diagram

The current flowing though the circuit will be $i = \dfrac{v}{R} = \dfrac{V_m \sin \omega t}{R}$

$$\text{At,} \quad \omega t = \frac{\pi}{2}, \; i = \frac{V_m}{R} = I_m$$

The maximum value of i is I_m

Therefore,

$$I_m = \frac{V_m}{R}$$

Thus, i can be written as

$$i = \frac{V_m}{R} \sin \omega t$$

or,

$$i = I_m \sin \omega t$$

with $v = V_m \sin \omega t$, we have current flowing $i = I_m \sin \omega t$.

The steady-state response of the circuit is also sinusoidal of the same frequency of the voltage applied. As shown in Fig. 3.14 (b), both voltage and current wave shapes are sinusoidal and their frequency is also the same. Since current is proportional to the voltage all the time, the two wave forms are in phase with each other.

The phasor diagram is drawn with the RMS values of the time-varying quantities. As shown in Fig. 3.14 (c), V and I are the RMS values of voltage and current. They have been shown in phase. For the sake of clarity only, the two phasors have been shown with a gap between them.

In a purely resistive circuit, current and voltage are in phase. Power is the product of voltage and current. The product, P = vi has been calculated for all instants of time and has been shown in Fig. 3.14 (b). Power in a resistive circuit is taken as the average power which is

$$P = \frac{1}{2\pi}\int_0^{2\pi} V_m \sin\theta \, I_m \sin\theta \, d\theta \qquad [\theta = \omega t]$$

$$= \frac{1}{2\pi} V_m I_m \int_0^{2\pi} \sin^2 \theta d\theta$$

$$= \frac{V_m I_m}{2\pi \times 2} \int_0^{2\pi} 2\sin^2 \theta d\theta$$

$$= \frac{V_m I_m}{4\pi} \int_0^{2\pi} (1 - \cos 2\theta) d\theta$$

Power in a resistive circuit, $P = \frac{V_m I_m}{2} = \frac{V_m}{\sqrt{2}} \frac{I_m}{\sqrt{2}} = VI$ (3.8)

Power factor is the cosine of the phase angle between voltage and current.

In a resistive circuit the phase difference between voltage and current is zero, i.e., they are in phase. So the phase angle $\theta = 0$.

Power factor, $P\,f = \cos\theta = \cos 0° = 1$

AC applied across a pure inductor

A pure inductor means that the resistance of the inductor coil is assumed to be zero. The coil has only inductance, L. Such an inductor is connected across a sinusoidally varying voltage, $v = V_m \sin \omega t$ as has been shown in Fig. 3.15 (a).

As a result of application of voltage, v an alternating current, i will flow through the circuit. This alternating current will produce an alternating magnetic field around the inductor. This alternating or changing field flux will induce an EMF in the coil:

$$e = L\frac{di}{dt}$$

This EMF will oppose the voltage applied (remember Lenz's law). Therefore, we can write

$$v = e = L\frac{di}{dt}$$

or, $L\,di = v\,dt = V_m \sin \omega t\, dt$

or, $di = \frac{V_m}{L} \sin \omega t\, dt$

Integrating

$$i = \frac{V_m}{L}\int \sin \omega t \, dt$$

$$= \frac{V_m}{\omega L}(-\cos \omega t)$$

or,
$$i = \frac{V_m}{\omega L}\sin\left(\omega t - \frac{\pi}{2}\right)$$

or,
$$i = I_m \sin\left(\omega t - \frac{\pi}{2}\right)$$

where
$$I_m = \frac{V_m}{\omega L}$$ ωL is called the inductive reactance x_L

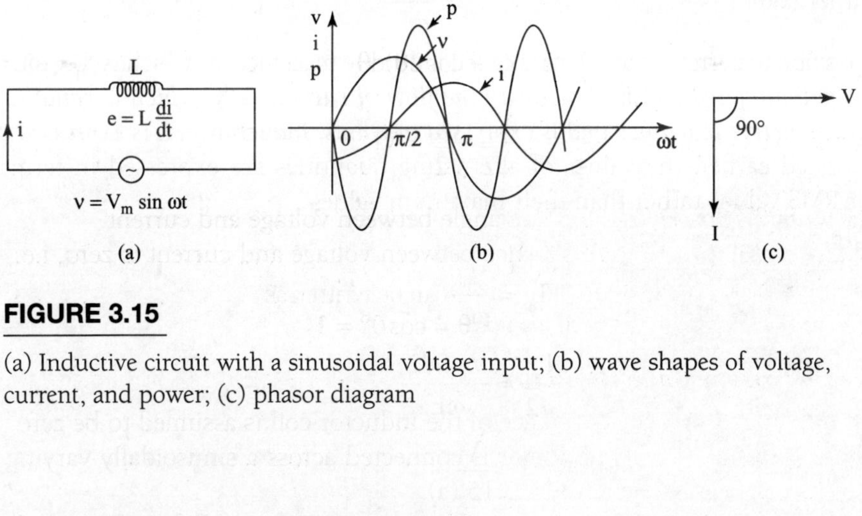

FIGURE 3.15

(a) Inductive circuit with a sinusoidal voltage input; (b) wave shapes of voltage, current, and power; (c) phasor diagram

Thus, we observe that in a purely inductive circuit

$$v = V_m \sin\omega t$$

and
$$i = I_m \sin\left(\omega t - \frac{\pi}{2}\right)$$

$$= I_m \sin(\omega t - 90°)$$

The current, i is also sinusoidal but lagging behind, v by 90°. The voltage and current wave shapes have been shown in Fig. 3.15 (b). The instantaneous power, p is the product of v and i. The wave shape of instantaneous power has also been shown in the figure. The phasor diagram of RMS values of v and i has been shown in Fig. 3.15 (c). In a purely inductive circuit current, I lags the voltage, V by $\frac{\pi}{2}$ degrees, i.e., 90°.

Power factor, $\cos\phi = \cos 90° = 0$

Average power
$$P = \frac{1}{2\pi}\int_0^{2\pi} V_m \operatorname{Sin}\omega t\, I_m \operatorname{Sin}(\omega t - \pi/2)\,d\omega t$$
$$= \frac{V_m I_m}{2\pi}\int_0^{2\pi} \operatorname{Sin}\omega t \operatorname{Sin}(\omega t - \pi/2)\,d\omega t$$
$$= \frac{V_m I_m}{4\pi}\int_0^{2\pi} 2\operatorname{Sin}\omega t \operatorname{Sin}(\omega t - \pi/2)\,d\omega t$$
$$= \frac{V_m I_m}{4\pi}\int_0^{2\pi} (0 - \operatorname{Sin}2\omega t)\,d\omega t$$
$$= 0$$

Average power in a purely inductive circuit, $P = 0$ (3.9)

Hence, the average power absorbed by a pure inductor is zero.

We had earlier taken, $I_m = \dfrac{V_m}{\omega L}$

The opposition to current is ωL. This is called inductive reactance, X_L which is $X_L = \omega L = 2\pi f L\ \Omega$. *The opposition offered by an inductor to the flow of current is X_L which is equal to ωL This is called the inductive reactance and is expressed in Ohms. Inductance, L is expressed in Henry.*

As mentioned earlier, the values of alternating quantities are expressed in terms of their effective or RMS values rather than their maximum values.

Therefore,

$$I_m = \frac{V_m}{\omega L} \text{ can be written as}$$
$$\frac{I_m}{\sqrt{2}} = \frac{V_m/\sqrt{2}}{\omega L}$$

or,
$$I = \frac{V}{X_L}$$

or,
$$V = IX_L$$

If V is taken as the reference axis, we can represent V as a phasor and represent as $V\angle 0°$.

Since current, I is lagging voltage, V by 90°, we represent the current as $I\angle -90°$ or $-jI$ for a purely inductive circuit. Again, if I is taken as the reference axis, then I and V can be represented as $I\angle 0°$ and $V\angle +90°$ or $+jV$, respectively, as shown in Fig. 3.16.

Note that j is an operator which indicates rotation of a phasor by 90° in the anti clockwise direction from the reference axis.

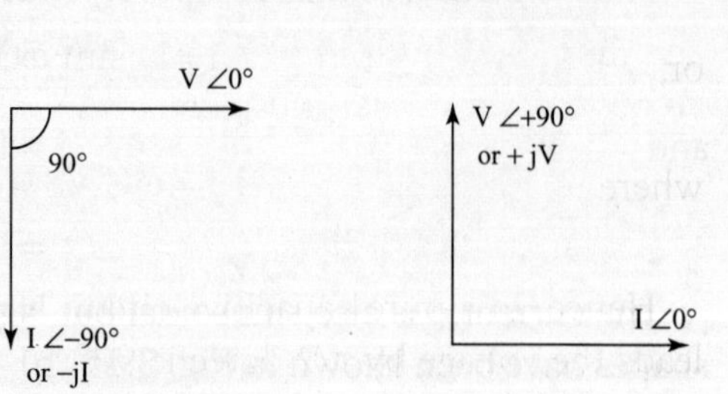

FIGURE 3.16

Phasor diagram of V and I in a purely inductive circuit

Power absorbed by a pure inducter is zero

Let us examine why the power absorbed by a pure inductive circuit is zero. We refer back to Fig. 3.15 (b) where it is seen that for one half cycle power is negative and for the next half cycle power is positive. The average value for a complete cycle, the power consumed is zero. Positive power indicates that power is drawn by the circuit from the supply source. When current rises in the circuit, energy is required to establish a magnetic field around the inductor coil. This energy is supplied by the source and is stored in the magnetic field. As the current starts reducing, the magnetic field collapses and the energy is returned to the supply source. Thus, in one half cycle power is drawn by the inductor and in the next half cycle power is returned to the source. This way the net power absorbed by the inductor becomes zero. The power which is being circulated from the source to the inductor and back to the source is called reactive power which will also be discussed in a separate section.

AC applied across a pure capacitor

A sinusoidal voltage source has been shown connected across a pure capacitor in Fig. 3.17 (a). When current starts flowing, the capacitor starts getting charged. The charge, q of the capacitor in terms of capacitance of the capacitor, C and supply voltage, v is expressed as

$$q = Cv$$

Current, i is the rate of flow of charge. Therefore,

$$i = \frac{dq}{dt}$$

$$= C\frac{dv}{dt}$$

$$= C\frac{d}{dt} V_m \sin \omega t$$

$$= \omega C V_m \cos \omega t$$

or,

$$i = \omega C V_m \sin\left(\omega t + \frac{\pi}{2}\right)$$

or,

$$i = I_m \sin\left(\omega t + \frac{\pi}{2}\right)$$

where

$$I_m = \omega C V_m = \frac{V_m}{1/\omega C} = \frac{V_m}{X_c} \text{ Where } X_c = \frac{1}{\omega C}$$

Hence, in a pure capacitive circuit, $v = V_m \sin \omega t$ and current $i = I_m \sin (\omega t + \frac{\pi}{2})$. Current leads the voltage by $\pi/2$ dgree ie 90°:

$X_C = \frac{1}{\omega C}$ *is called the capacitive reactance of the capacitor*. To express in terms of RMS values,

We take

$$\frac{I_m}{\sqrt{2}} = I \text{ and } \frac{V_m}{\sqrt{2}} = V$$

Therefore, $I_m = \frac{V_m}{X_C}$ can be written as

$$I = \frac{V}{X_C}$$

X_c is the opposition offered by the capacitor to the flow of current and is called capacitive reactance.

Like an inductor, in a capacitor also the average power absorbed for a complete cycle is zero. When voltage is applied, the capacitor starts getting charged, energy gets stored in the capacitor in the form of electro-static field. When the applied voltage starts falling from its maximum value, the energy starts getting returned to the supply. This way, the power is absorbed from and then returned to the supply source The net power absorbed by a pure capacitor is zero. Since current leads the voltage by 90°, the power factor of the circuit is

$$\text{P.f} = \cos\phi = \cos 90° = 0$$

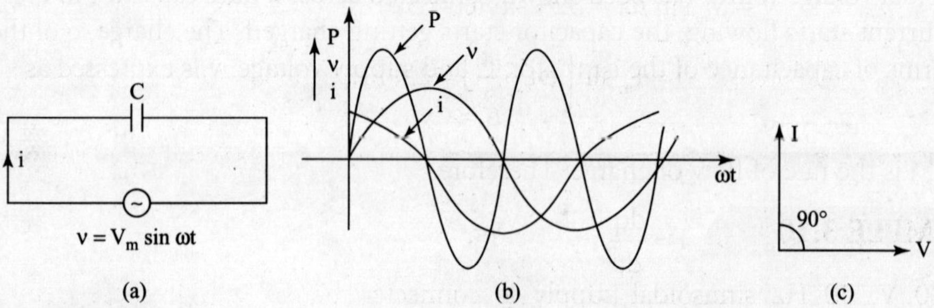

FIGURE 3.17

(a) Pure capacitive circuit; (b) wave shapes of voltage, current, and power; (c) phasor diagram

The average or net power in a pure capacitor circuit can be calculated as

$$P = \frac{1}{2\pi}\int_0^{2\pi} p\, d\omega t = \frac{1}{2\pi}\int_0^{2\pi} vi\, d\omega t$$

$$= \frac{1}{2\pi}\int_0^{2\pi} V_m \sin\omega t\, I_m \sin(\omega t + \pi/2)\, d\omega t$$

$$= \frac{V_m I_m}{4\pi}\int_0^{2\pi} 2\sin\omega t \sin(\omega t + \pi/2)\, d\omega t$$

or, $$P = \frac{V_m I_m}{4\pi}\int_0^{2\pi} \sin 2\omega t\, d\omega t$$

$$= 0$$

Average power in a purely capacitive circuit, $P = 0$ (3.10)

Hence, it is proved that the average power obsorbed by a pure capacitor is zero.

EXAMPLE 3.9

An inductor of 0.5 H is connected across a 230 V, 50 Hz supply. Write the equations for instantaneous values of voltage and current.

SOLUTION

$$V = 230\text{ V},\ V_m = \sqrt{2}\,V = 1.414 \times 230 = 324\text{ V}$$

$$X_L = \omega L = 2\pi fL = 2 \times 3.14 \times 50 \times 0.5\ \Omega = 157\ \Omega$$

$$I = \frac{V}{X_L} = \frac{230}{X_L} = \frac{230}{157} = 1.46\text{ A}$$

$$I_m = \sqrt{2}\,I = 1.414 \times 1.46 = 2.06\text{ A}$$

The equations are,

$$V = V_m \sin \omega t = 324 \sin \omega t = 324 \sin 2\pi ft = 324 \sin 314t$$

and

$$i = I_m \sin\left(\omega t - \frac{\pi}{2}\right) = 2.06 \sin\left(314t - 90^\circ\right)$$

EXAMPLE 3.10

A 230 V, 50 Hz sinusoidal supply is connected across a (i) resistance of 25 Ω; (ii) inductance of 0.5 H; (iii) capacitance of 100 μF. Write the expressions for instantaneous current in each case.

SOLUTION

given

$$V = 230\text{ V}$$

$$V_m = \sqrt{2}\,V = 1.414 \times 230 = 324.3\text{ V}$$

$$\omega = 2\pi f = 2 \times 3.14 \times 50 = 314\text{ rad/sec}$$

Voltage equation is

$$v = V_m \sin \omega t$$

or,

$$v = 324.3 \sin 314t$$

Inductive reactance,

$$X_L = \omega L = 314 \times 0.5 = 157\ \Omega$$

Capacitive reactance,

$$X_C = \frac{1}{\omega C} = \frac{1}{314 \times 100 \times 10^{-6}}$$

$$= \frac{10^{-6}}{314 \times 100} = 32.2\ \Omega$$

When the voltage is applied across a 25 Ω resistor, the current will be

$$i = \frac{V_m}{R}\sin\omega t = \frac{324.3}{25}\sin 314t$$

or,

$$i = 12.97\sin 314t\,A$$

Current through the inductor is

$$i = \frac{V_m}{X_L}\sin\left(\omega t - \frac{\pi}{2}\right)$$

$$= \frac{324.3}{157}\sin\left(314t - \frac{\pi}{2}\right)$$

or,

$$i = 2.06\sin(314t - 90°)\,A$$

Current through the capacitor is

$$i = \frac{V_m}{X_c}\sin\left(\omega t + \frac{\pi}{2}\right)$$

$$= \frac{324.3}{32.2}\sin\ (314t + 90°)$$

or,

$$i = 10.07\sin\ (314t + 90°)\,A$$

EXAMPLE 3.11

An alternating voltage of RMS value 100 V, 50 Hz is applied separately across a resistance of 10 Ω, an inductor of 100 mH, and a capacitor of 100 μF. Calculate the current flow in each case. Also draw and explain the phasor diagrams.

SOLUTION

$$R = 10\ \Omega$$

$$X_L = \omega L = 2\pi fL = 2 \times 3.14 \times 50 \times 100 \times 10^{-3}\ \Omega$$

$$= 31.4\ \Omega$$

$$X_C = \frac{1}{\omega C} = \frac{1}{2\pi fC} = \frac{1}{2 \times 3.14 \times 50 \times 100 \times 10^{-6}}$$

$$= \frac{10^6}{314 \times 100} = 31.8\ \Omega$$

Current through R,

$$= \frac{100}{10} = 10\ A$$

Current through L,

$$= \frac{100}{X_L} = \frac{100}{31.4} = 3.18\ A$$

Current through C, $= \frac{100}{X_C} = \frac{100}{31.8} = 3.1$ A

We know that in a resistive circuit current is in phase with the applied voltage; in a purely inductive circuit current lags the voltage by 90°; and in a purely capacitive circuit current leads the voltage by 90°. The phasor diagrams have been shown in Fig. 3.18.

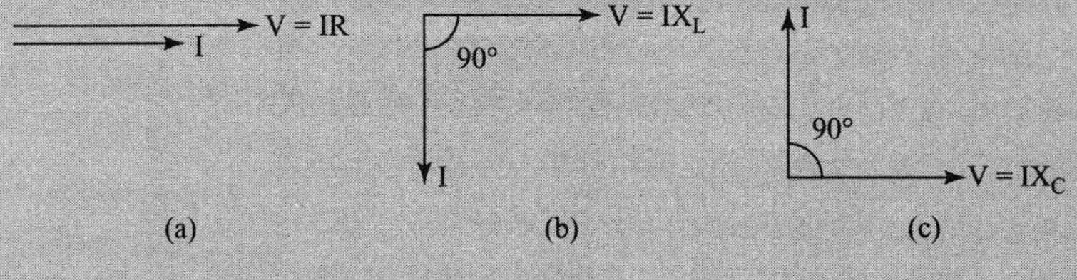

FIGURE 3.18

Phasor diagrams (a) resistive circuit; (b) purely inductive circuit; (c) purely capacitive circuit

3.2.2 R–L Series Circuit

Let us consider a resistance element and an inductor connected in series as shown in Fig. 3.19. A voltage, V of frequency, f is applied across the whole circuit. The voltage drop across the resistance is V_R and across the inductor is V_L. Current flowing through the circuit is I.

$$V_R = IR,\ V_L = IX_L \text{ where } X_L = \omega L = 2\pi fL$$

We have to add V_R and V_L to get V. But these are to be added vectorially as they are all not in phase, i.e., these vectors are not along the same direction. To draw the current and voltage phasor, we take the current I as the reference phasor as shown in Fig. 3.19 (b), since current I is common to V_R and V_L, i.e., since the same current is flowing through both resistance and inductance. We have, therefore, chosen I as the reference phasor. Voltage drop across the resistance and the current flowing through it are in phase. This is because, as we have seen earlier that in a resistive circuit,

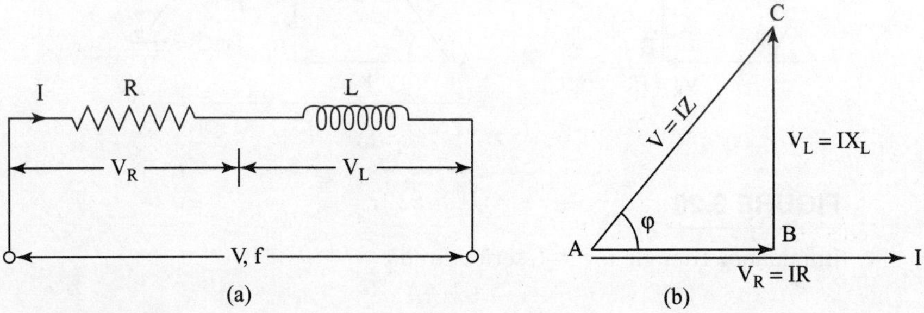

FIGURE 3.19

(a) R–L series circuit; (b) phasor diagram

voltage and current are in phase. The current flowing through an inductor lags the voltage across it by 90°. That is to say, voltage drop across L, i.e.,V_L will lead the current by 90°. Again $V_L = IX_L$ and $X_L = \omega L$. The vector sum of V_R and V_L is equal to V. The angle between V and I is called the power factor angle ϕ. Power factor is $\cos\phi$. Considering the triangle ABC we can express

$$V^2 = V_R^2 + V_L^2$$

or,

$$V = \sqrt{V_R^2 + V_L^2} = \sqrt{(IR)^2 + (IX_L)^2}$$

$$= I\sqrt{R^2 + X_L^2}$$

or,

$$I = \frac{V}{\sqrt{R^2 + X_L^2}} = \frac{V}{Z} \text{ or, } V = IZ$$

where

$$Z = \sqrt{R^2 + X_L^2}$$

Z is called the impedance of the total circuit. Triangle ABC in Fig. 3.19 (b) is also called the impendance triangle which is redrawn as in Fig. 3.20. From the impedance triangle

$$Z = \sqrt{R^2 + X_L^2}$$

Also,

$$Z = R + jX_L$$

Where j indicates rotation by 90° in anti-clockwise direction.

$$\cos\phi = \frac{R}{Z}$$

or,

$$Z\cos\phi = R$$

and

$$Z\sin\phi = X_L$$

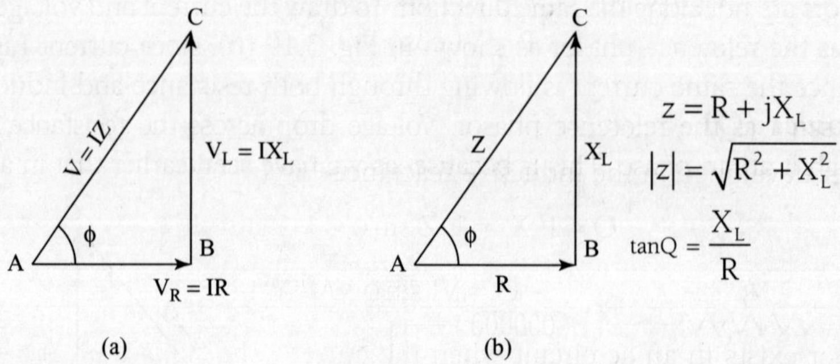

FIGURE 3.20

Impedance triangle for R–L series circuit

Fig. 3.20 (a) is the same as Fig. 3.20 (b). In Fig. 3.20 (b) the current, I has been kept aside which is common to all the sides. Impedance Z can be represented as the vector sum of R and X_L. since IX_L is leading I by 90° and R is in phase with I, we can write

$$Z = R + jX_L, \text{ and } \cos\phi = \frac{R}{Z}, \tan\phi = \frac{V_L}{V_R} = \frac{IX_L}{IR} = \frac{X_L}{R}$$

$$\phi = \tan^{-1}\frac{X_L}{R}, \cos\phi = \frac{R}{Z}$$

Power is, $$P = VI\cos\phi$$

or, $$\text{Power} = \text{Volt–Ampere} \times \text{Power factor}$$

3.2.3 Apparent Power, Real Power, and Reactive Power

We know that reactive loads such as inductors and capacitors dissipate zero power, yet they draw current and they drop voltage. This phantom power (i.e., not real power) is called reactive power. The actual power dissipated or used in a circuit is called real or true power or active power. The combination of reactive power and true power is called apparent power.

Apparent power (S)

It is defined as the product of the RMS value of voltage (V) and current (I). It is denoted by S

Apparent power, $$S = VI = \text{Voltage} \times \text{Current}$$

Unit of apparent power is VA or kVA. (3.11)

Real or true power or active power (P)

It is the power which is actually dissipated in the circuit resistance. (watt-full power)

$$\text{Active power, } P = \text{Apparent power} \times \text{power factor} \quad (3.12)$$

or, $$P = VI\cos\phi \text{ Watts or kW}$$

Reactive power (Q)

It is the power developed in the inductive reactance of the circuit. (watt-less power)

$$Q = I^2 X_L = I^2 Z\sin\phi = I\,(ZI)\sin\phi$$

or, $$Q = VI\sin\phi \text{ VAR} \quad (3.13)$$

Reactive power exists in an ac circuit when the current and voltage are not in phase. These three powers are shown in the power triangle of Fig. 3.22 (b) from where it can be seen that

$$S^2 = P^2 + Q^2$$

$$S = \sqrt{P^2 + Q^2}$$

$$\text{kVA} = \sqrt{(\text{kW})^2 + (\text{kVAR})^2} \quad (3.14)$$

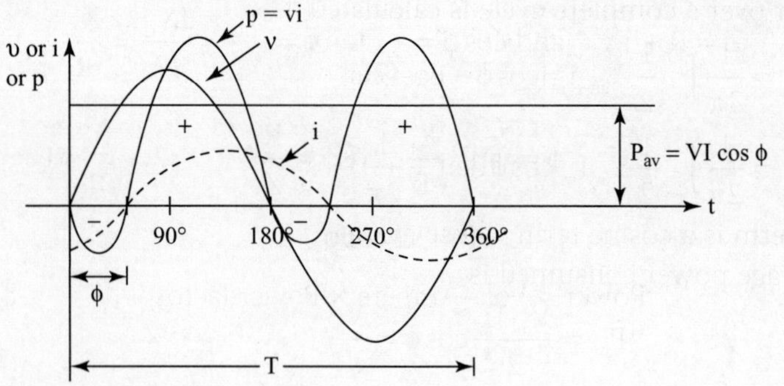

FIGURE 3.21

Wave forms of voltage, current, and power in an R–L series circuit

3.2.4 Power in an R-L Series Circuit

Let us now develop a general expression for power in an R-L series circuit by considering the instantaneous values of voltage and current. A sinusoidal voltage ν is expressed as

$$\nu = V_m \sin \omega t$$

In a R-L circuit where current is lagging the voltage by an angle ϕ, current i is expressed as

$$i = I_m \sin(\omega t - \phi)$$

Sinusoidal waveforms of voltage and current are shown in Fig. 3.21. It is seen that the current wave is lagging the voltage wave by an angle, ϕ which is the power factor angle.

Fig. 3.21 clearly shows that current in an R–L circuit lags voltage by an angle ϕ, which is called the power factor angle.

The expression for the voltage and current in series R–L circuit is,

$$\nu = V_m \sin \omega t$$

$$i = I_m \sin(\omega t - \phi), \text{ as i lags v}$$

The power is product of instantaneous values of voltage and current,

$$p = \nu \times i$$

$$= V_m \sin \omega t \times I_m \sin(\omega t - \phi)$$

$$= \frac{1}{2} V_m I_m [2 \sin \omega t \sin(\omega t - \phi)]$$

$$= \frac{1}{2} V_m I_m [\cos\phi - \cos(2\omega t - \phi)]$$

$$= \frac{1}{2} V_m I_m \cos\phi - \frac{1}{2} V_m I_m \cos(2\omega t - \phi)$$

The average power over a complete cycle is calculated as

$$P_{av} = \frac{1}{2\pi}\int_0^{2\pi}\frac{1}{2}V_m I_m[\cos\phi - \cos(2\omega t - \phi)]d\omega t$$

$$= \frac{1}{2\pi}\int_0^{2\pi}\frac{1}{2}V_m I_m \cos\phi\, d\omega t - \frac{1}{2\pi}\int_0^{2\pi}\frac{1}{2}V_m I_m \cos(2\omega t - \phi)d\omega t$$

Now, the second term is a cosine term whose average value over a complete cycle is zero.

Hence, the average power consumed is

$$P_{av} = \frac{V_m I_m}{2}\cos\phi = \frac{V_m}{\sqrt{2}} \times \frac{Im}{\sqrt{2}}\cos\phi$$

$$P_{av} = V_{rms} \times I_{rms} \cos\phi = VI\cos\phi \text{ W}$$

Power factor

It may be defined as the cosine of the phase angle between the voltage and current; $\cos\phi$ is known as power factor. Power factor can also be expressed as the ratio, R/Z = resistance/impedance = $\cos\phi$

In Fig. 3.22, the power triangle diagram has been developed from the simple voltage–current relationship in an R–L series circuit. First we have shown I laggingV by the power factor angle ϕ. The in-phase component of I is I $\cos\phi$ and quadrature component is I $\sin\phi$ as have been shown in Fig. 3.22 (a).

Multiplying all the sides of the triangle ABC by KV (kilo-volt), we can draw the power triangle as in Fig. 3.22 (b)

$$\text{kVA}\cos\phi = \text{kW}$$

$$\text{kVA}\sin\phi = \text{kVAR}$$

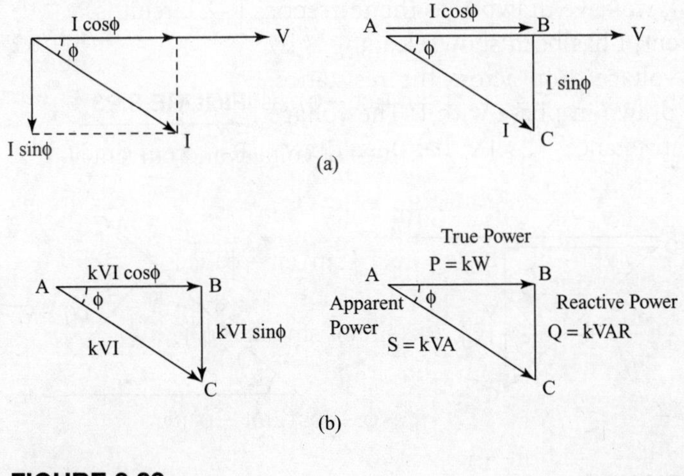

FIGURE 3.22

Power triangle diagram

In the power triangle diagram, if ϕ is taken as zero, i.e., if the circuit is resistive, reactive power, Q becomes zero. If the circuit is having pure inductance or capacitance, $\phi = 90$, active power, P becomes zero. Reactive power will be present whenever there is inductance or capacitance in the circuit. Inductors and capacitors are energy-storing and energy-releasing devices in the form of magnetic and electric fields, respectively, and are of importance in the field of electrical engineering.

3.2.5 R—C Series Circuit

Consider a circuit consisting of a pure resistance R and connected in series with a pure capacitor C across an ac supply of frequency f as shown in Fig. 3.23.

When the circuit draws a current I, then there are two voltage drops.

(i) Drop across pure resistance $V_R = I \times R$

(ii) Drop across pure capacitance $V_C = I \times X_C$

where $X_C = \dfrac{1}{2\pi fC}$ and I, V_R, V_C are the RMS values.

The phasor diagram for such a circuit can be drawn by taking the current as a reference phasor represented by OA as shown in Fig. 3.24. The voltage drop V_R across the resistance is in phase with current and is represented by OB. The voltage drop across the capacitance V_C lags the current by 90° and is represented by BC. The phasor OC is the phasor sum of two voltages V_R and V_C. Hence, the OC represents the applied voltages. Thus, in a capacitive circuit, current leads the voltages by an angle ϕ. The same phasor diagram can be drawn by taking voltage, V as the reference vector as shown in Fig. 3.24 (b).

In Fig. 3.24 (b), we have drawn V as the reference vector. Then current, I has been shown leading V by an angle ϕ. The voltage drop across the resistance, $V_R = IR$ has been drawn in phase with I. The voltage drop across the capacitance $V_C = IX_C$ has been drawn

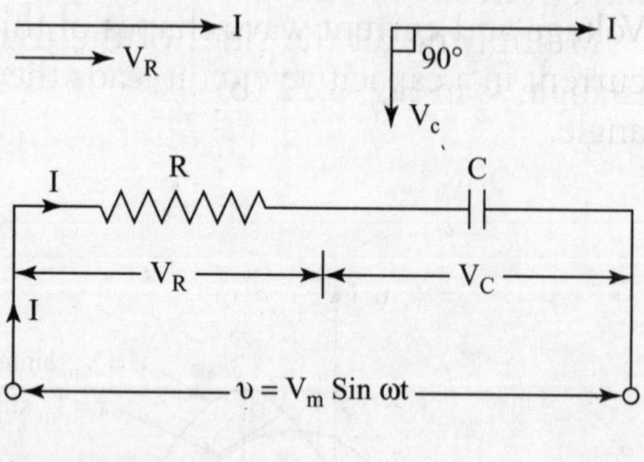

FIGURE 3.23

R–C series circuit

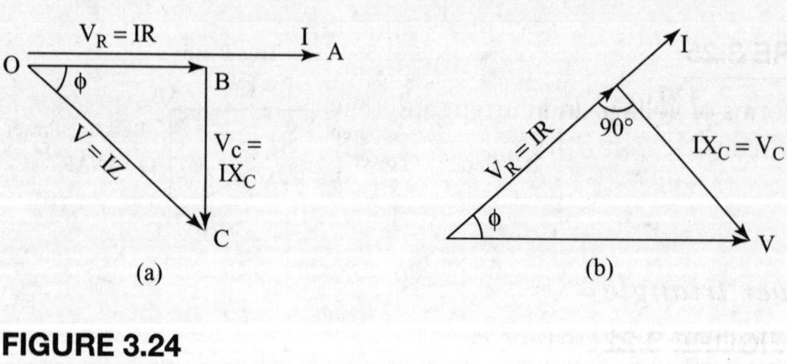

FIGURE 3.24

Phasor diagrams of R–C series circuit

lagging I by 90° (V_C lagging I is the same as I leading V_C). The length of V_R and V_C are such that they make an angle of 90°.

In an R–C series circuit, I leads V by an angle ϕ or supply voltage V lags current I by an angle ϕ as shown in the phasor diagram in Fig. 3.24 (b).

$$\tan\phi = \frac{V_C}{V_R} = \frac{IX_C}{IR} = \frac{X_C}{R}$$

$$\phi = \tan^{-1}\frac{X_C}{R}$$

Applied voltage, $$V = \sqrt{V_R^2 + V_C^2}$$

$$= \sqrt{(IR)^2 + (IX_C)^2}$$

$$= I\sqrt{R^2 + X_C^2}$$

$$V = IZ$$

where $$Z = \sqrt{R^2 + X_C^2} = \text{impedance of the circuit}$$

Voltage and current wave shapes of this circuit are shown in Fig. 3.25, which shows that the current in a capacitive circuit leads the voltage by an angle ϕ, which is called the power factor angle.

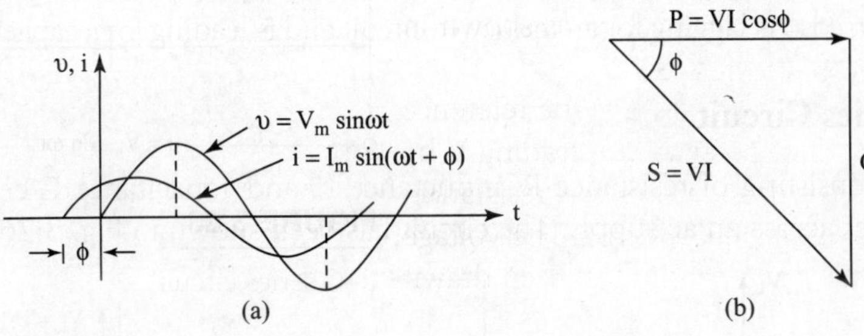

FIGURE 3.25

Wave forms of voltage and current and their phase relationship in an R–C series circuit

Power and power triangle

The expression for voltage and current is

$$v = V_m \text{ Sin } \omega t$$

$$i = I_m \text{ Sin } (\omega t + \phi) \text{ as I leads V}$$

Power is the product of voltage and current. The instantaneous power is

$$\begin{aligned} P &= v \times i \\ &= V_m \operatorname{Sin} \omega t \times I_m \operatorname{Sin}(\omega t + \phi) \\ &= \frac{1}{2} V_m I_m [2 \sin \omega t \sin(\omega t + \phi)] \\ &= \frac{1}{2} V_m I_m [\cos(-\phi) - \cos(2\omega t + \phi)] \\ &= \frac{1}{2} V_m I_m \cos\phi - \frac{V_m I_m}{2} \cos(2\omega t + \phi) \end{aligned}$$

$$\text{as } \cos(-\phi) = \cos\phi$$

The second term is a cosine term whose average value over a complete cycle is zero. Hence, average power consumed by the circuit is

$$P_{av} = \frac{V_m I_m}{2} \cos\phi = \frac{V_m}{\sqrt{2}} \frac{I_m}{\sqrt{2}} \cos\phi$$

$$P_{av} = V_{rms} I_{rms} \cos\phi = VI \cos\phi \text{ W}$$

The power triangle has been shown in Fig. 3.25 (b).

Thus, various powers are

Apparent power S = VI Volt Amperes or VA

Active power P = VI cosϕ W

Reactive power Q = VI sinϕ VAR

where cosϕ = Power factor of the circuit.

Note: Power factor, cosϕ is lagging for an inductive circuit and is leading for a capacitive circuit.

3.2.6 R–L–C Series Circuit

Consider a circuit consisting of resistance R, inductance L, and capacitance C connected in series with each other across an ac supply. The circuit has been shown in Fig. 3.26.

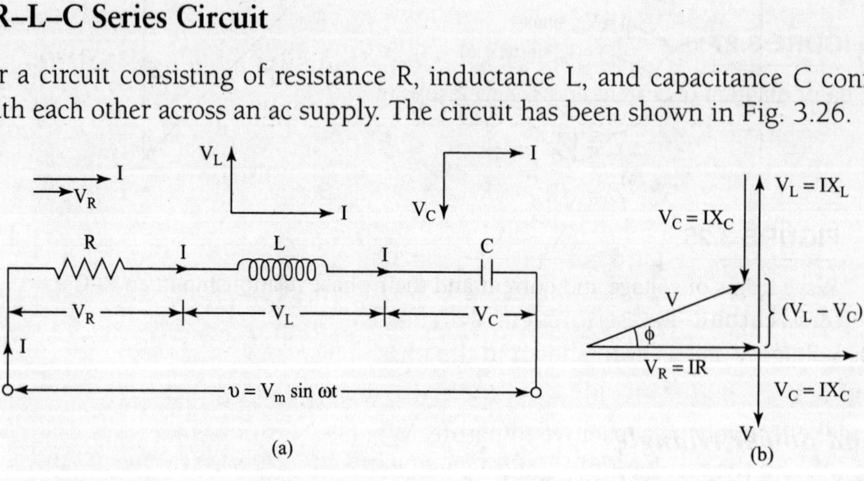

FIGURE 3.26

(a) R–L–C series circuit; (b) phasor diagram

The circuit draws a current I. Due to flow of current I, there are voltage drops across R, L, and C which are given by

(i) drop across resistance R is $V_R = IR$
(ii) drop across inductance L is $V_L = IX_L$
(iii) drop across capacitance C is $V_C = IX_C$

where I, V_R, V_L, and V_C are the RMS values.

The phasor diagram depends on the magnitude of V_L and V_C, which obviously depends upon X_L and X_C. Let us consider the different cases.

(a) When $\mathbf{X_L > X_C}$, i.e., when inductive reactance is more than the capacitive reactance. The circuit will effectively be inductive in nature. When $X_L > X_C$, obviously, IX_L, i.e., V_L is greater than IX_C, i.e., V_C. So the resultant of V_L and V_C will be $V_L - V_C$ so that V is the phasor sum of V_R and $(V_L - V_C)$. The phasor sum of V_R and $(V_L - V_C)$ gives the resultant supply voltage V. This is shown in Fig. 3.26 (b) and again redrawn as in Fig. 3.27.

Applied voltage is
$$OB = \sqrt{OA^2 + AB^2}$$
$$V = \sqrt{V_R^2 + (V_L - V_C)^2}$$
$$= \sqrt{(IR)^2 + (IX_L - IX_C)^2}$$
$$= I\sqrt{R^2 + (X_L - X_C)^2}$$

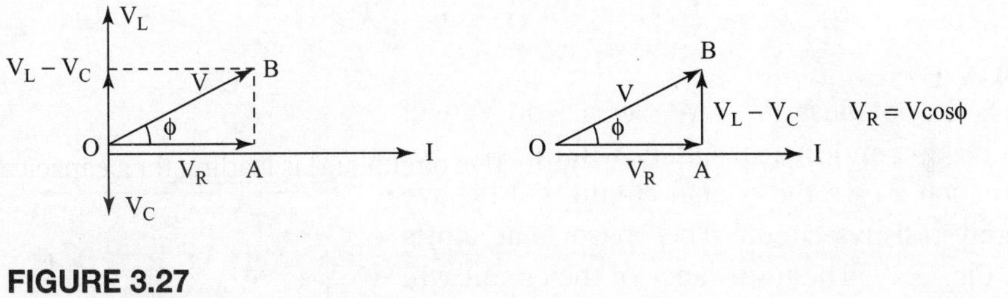

FIGURE 3.27

Phasor diagram of current and voltage drops in an R–L–C circuit where $X_L > X_C$

or,
$$V = IZ$$
where
$$Z = \sqrt{R^2 + (X_L - X_C)^2}$$
$$\tan\phi = \frac{(X_L - X_C)}{R}, \quad \phi = \tan^{-1}\frac{(X_L - X_C)}{R}$$

Note when $X_L > X_C$, the R–L–C series circuit will effectively be an inductive circuit where current I will lag the voltage V as has been shown in the phasor diagram of Fig. 3.27.

(b) When $X_L < X_C$

The circuit will effectively be capacitive in nature. When $X_L < X_C$, obviously, IX_L, i.e., V_L is less than IX_C, i.e., V_C. So the resultant of V_L and V_C will be directed towards V_C. Current I will lead $(V_C - V_L)$. The phasor sum of V_R and $(V_C - V_L)$ gives the resultant supply voltage V. This is shown in Fig. 3.28.

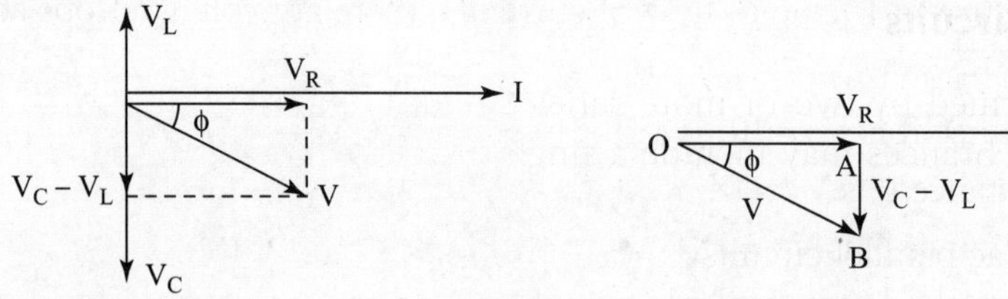

FIGURE 3.28

Phasor diagram of an R–L–C series circuit when $X_L < X_C$

Applied voltage represented by OB $= \sqrt{OA^2 + AB^2}$

$$V = \sqrt{V_R^2 + (V_C - V_L)^2}$$

$$= \sqrt{(IR)^2 + (IX_C - IX_L)^2}$$

$$= I\sqrt{R^2 + (X_C - X_L)^2}$$

or, $$V = IZ$$

where $$Z = \sqrt{R^2 + (X_C - X_L)^2}$$

Phase angle, $$\phi = \tan^{-1}\left(\frac{X_C - X_L}{R}\right)$$

(c) When $X_L = X_C$

When $X_L = X_C$, obviously, $V_L = V_C$. So, V_L and V_C will cancel each other and their resultant will be zero. So, $V_R = V$. In such a case the overall circuit will behave like a purely resistive circuit. The phasor diagram is shown in Fig. 3.29. The impedance of the circuit will be minimum, i.e., equal to R.

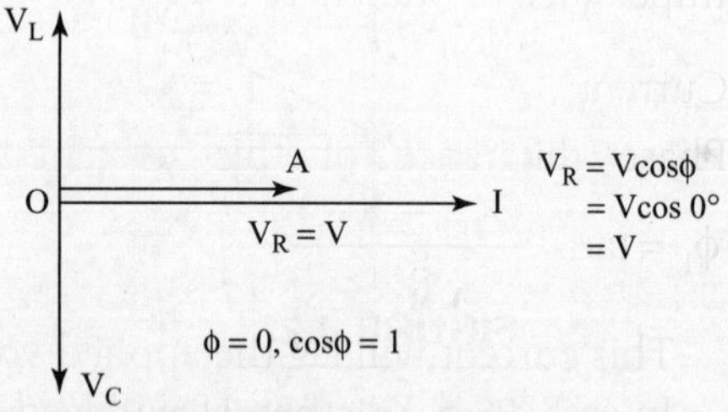

FIGURE 3.29

Phasor diagram of an R–L–C series circuit when $X_L = X_C$

Power and power triangle

The average power consumed by the circuit is

P_{av} = Average power consumed by R + Average power consumed by L + Average power consumed by C.

But pure L and C never consume any power.

Therefore, P_{av} = Power taken by R = $I^2R = IV_R$

But $V_R = V\cos\phi$ in all the cases.

Therefore, $P = VI\cos\phi$ W.

Thus, for any condition, that is when $X_L > X_C$ or $X_L < X_C$ or $X_L = X_C$, power can be expressed as P = Voltage × Component of I in phase with V = VI cos ϕ

Note that when $X_L = X_C$, the component of I in phase with V is I only because $I\cos\phi = I$ (as $\cos\phi = 1$).

3.2.7 AC Parallel Circuits

Parallel circuits are formed by two or more series circuits connected to a common source of supply. The parallel brances may include a single element or a combination of elements in series.

Methods for solving ac parallel circuits:

The following three methods are available for solving ac parallel circuits:

1. phasor or vector method
2. admittance method
3. using vector algebra (symbolic method or j-operator method)

These methods are explained with examples as follows.

1. *Phasor or vector method*

A parallel circuit consisting of three branches has been shown in Fig. 3.30. Branch 1 consists of R_1, L_1, and C_1 in series. Branch 2 is resistive and capacitive and branch 3 is resistive and inductive. Let the current be I_1, I_2, and I_3 in the branch 1, 2, and 3, respectively. The total current drawn by the circuit is the phasor sum of I_1, I_2, and I_3.

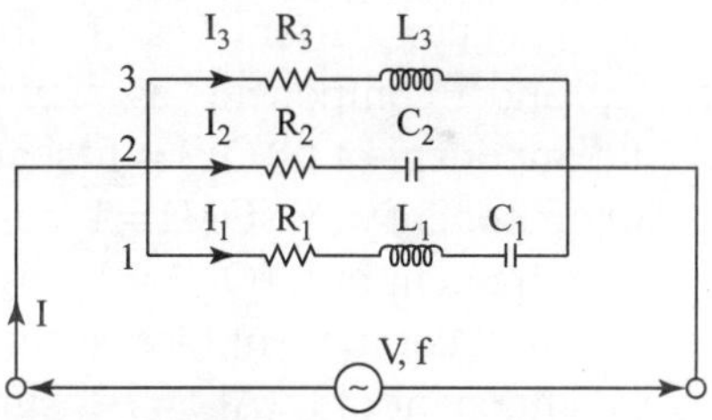

FIGURE 3.30

AC parallel circuit

Branch 1

Impedance of branch 1, $= \sqrt{(R_1)^2 + (X_{L1} - X_{C1})^2} = Z_1$

Current $I_1 = V/Z_1$

Phase difference of this current with respect to the applied voltage is given by $\phi_1 = \tan^{-1}\frac{(X_{L1} - X_{C1})}{R_1}$

This current will lag the applied voltage by an angle ϕ_1, if $X_{L1} > X_{C1}$.

In case $X_{C1} > X_{L1}$, then I_1, will lead V.

Branch 2

Capacitive branch (I_2 leads V)

Impedance of branch 2, $Z_2 = \sqrt{(R_2)^2 + (X_{C2})^2}$

Current $I_2 = V/Z_2$

The branch current I_2 leads the applied voltage V, by an angle ϕ_2, given by

$$\phi_2 = \tan^{-1}\frac{(X_{C2})}{R_2}$$

Branch 3

Inductive branch 3, $Z_3 = \sqrt{(R_3)^2 + (X_{L3})^2}$

Current $I_3 = V/Z_3$

This current will lag the applied voltage by an angle ϕ_3,

$$\phi_3 = \tan^{-1}\frac{X_{L3}}{R_3}$$

Choose a current scale and draw to the scale the current vectors with the voltage as the reference axis. Add vectorially any two currents, say I_1 and I_2. The vector sum of I_1 and I_2 is OE as shown in Fig. 3.31 (a). Add vectorially OE with the other branch current, i.e., with I_3 to get the sum of the three currents as OF. Convert this length OF to amperes using the current scale choosen earlier.

An alternate method is to show the three currents with the voltage as the horizontal reference axis as shown in Fig. 3.31 (b). Calculate the sum of the horizonal components and vertical components of the currents and then determine the resultant.

The branch currents with their phase angles with respect to V has been shown (not to the scale) separately in Fig. 3.31 (b).

The resultant current I can be found out by resolving the branch currents I_1, I_2, and I_3 into their X and Y components as shown in Fig. 3.31 (b).

X component of I_1 (OL) = $I_1 \cos\phi_1$
X component of I_2 (OM) = $I_2 \cos\phi_2$
X component of I_3 (ON) = $I_3 \cos\phi_3$
Sum of X component (active component) of branch currents = $I_1 \cos\phi_1 + I_2 \cos\phi_2 + I_3 \cos\phi_3$
Y component of I_1 (AL) = $-I_1 \sin\phi_1$
Y component of I_2 (BM) = $+I_2 \sin\phi_2$
Y component of I_3 (ON) = $-I_3 \sin\phi_3$

Sum of Y component (reactive component) of branch currents = $-I_1 \sin\phi_1 + I_2 \sin\phi_2 - I_3 \sin\phi_3$
Active component of resultant current I = $I \cos\phi$
Reactive component of resultant current I = $I \sin\phi$

Active and reactive components of resultant current must be equal to the sum of active and reactive components of branch currents.

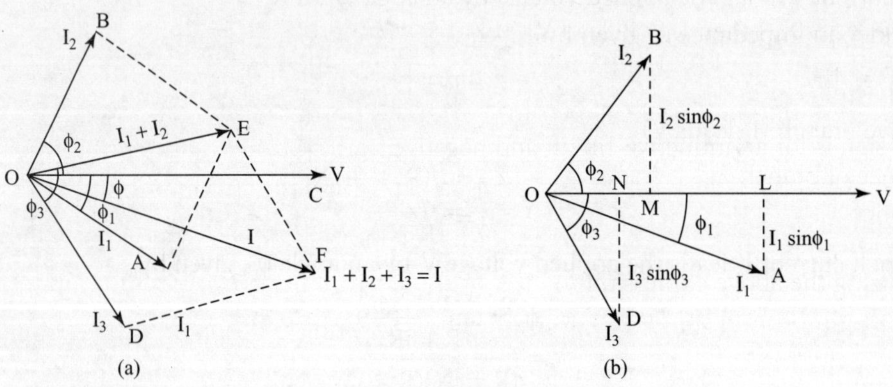

FIGURE 3.31

Phasor diagrams of parallel circuit shown in Fig. 3.30

∴ $$I\cos\phi = I_1\cos\phi_1 + I_2\cos\phi_2 + I_3\cos\phi_3 \quad \text{(i)}$$

$$I\sin\phi = -I_1\sin\phi_1 + I_2\sin\phi_2 - I_3\sin\phi_3 \quad \text{(ii)}$$

Resultant current $$I = \sqrt{(I\cos\phi)^2 + (I\sin\phi)^2}$$

$$= \sqrt{(I_1\cos\phi_1 + I_2\cos\phi_2 + I_3\cos\phi_3)^2 + (-I_1\sin\phi_1 + I_2\sin\phi_2 - I_3\sin\phi_3)^2}$$

$$\tan\phi = \frac{I\sin\phi}{I\cos\phi}$$

$$\phi = \tan^{-1}\frac{(-I_1\sin\phi_1 + I_2\sin\phi_2 - I_3\sin\phi_3)}{(I_1\cos\phi_1 + I_2\cos\phi_2 + I_3\cos\phi_3)}$$

Resultant current lags the applied voltage if ϕ is –ve, and leads the voltage in case ϕ is +ve.

Power factor of the circuit as a whole is

From (i), $$\cos\phi = \frac{I_1\cos\phi_1 + I_2\cos\phi_2 + I_3\cos\phi_3}{I}$$

$$= \frac{\text{sum of active components of branch currents}}{\text{resultant current}}$$

2. *Admittance method*

Concept of Admittance Method: Admittance is defined as the reciprocal of the impedance. It is denoted by Y and is measured in unit mho or siemens.

Components of admittance:

If the circuit contains R and L, $Z = R + jX_L$;

If the circuit contains R and C, $Z = R - jX_C$.

Considering X_L and X_C as X we can write $Z = R \pm jX$.

Consider an impedance as given by

$$Z = R \pm jX$$

Positive sign is for an inductive circuit and negative sign is for a capacitive circuit.

Admittance $$Y = \frac{1}{Z} = \frac{1}{R \pm jX}$$

Rationalizing the above expression,

$$Y = \frac{R \mp jX}{(R \pm jX)(R \mp jX)}$$

$$= \frac{R \mp jX}{R^2 + X^2} \quad [\because \ j^2 = -1]$$

$$= \frac{R}{R^2 + X^2} \mp j\frac{X}{R^2 + X^2}$$

$$= \frac{R}{Z^2} \mp j\frac{X}{Z^2} \quad \left[\because \; Z = \sqrt{R^2 + X^2}\right]$$

$$Y = G \mp jB$$

Where G = Conductance $= \frac{R}{Z^2}$ mho

and B = Susceptance $= \frac{X}{Z^2}$ mho

B is negative if the circuit is inductive and B is positive if the circuit is capacitive. The impedance triangle and admittance triangle for the circuit have been shown in Fig. 3.32.

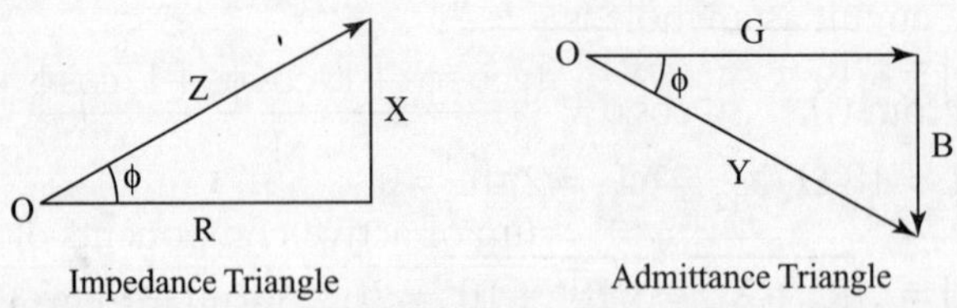

FIGURE 3.32

Impedance and admittance triangles

Application of admittance method

Consider a parallel circuit consisting of two branches 1 and 2. Branch 1 has R_1 and L_1 series while Branch 2 has R_2 and C_1 series, respectively. The voltage applied to the circuit is V Volts as shown in Fig. 3.33.

Total conductance is found by adding the conductances of two branches. Similarly, the total susceptance is found by algebraically adding the individual susceptance of different branches.

$$\frac{1}{Z} = \frac{1}{Z_1} + \frac{1}{Z_2} \qquad I_1 = \frac{V}{Z_1} = VY_1$$

$$Y = Y_1 + Y_2 \qquad I_2 = \frac{V}{Z_2} = VY_2$$

Total conductance $G = G_1 + G_2$

Total susceptance $B = (-B_1) + B_2$

Total current $I = VY$

Power factor $\cos\phi = G/Y$

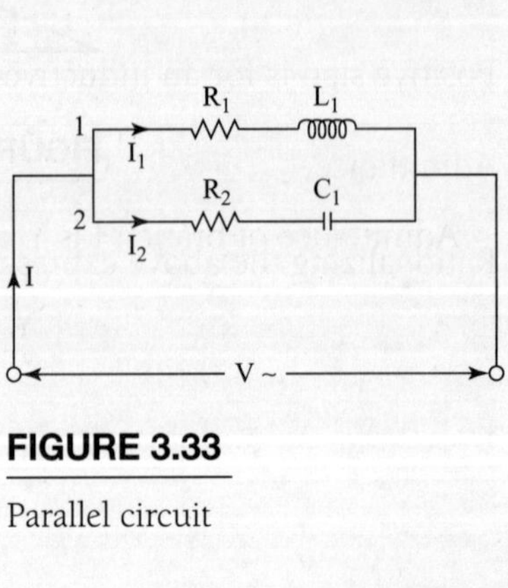

FIGURE 3.33

Parallel circuit

It is quite clear that this method requires calculations which are time consuming. To illustrate this method, we will take one example.

EXAMPLE 3.12

Two impedences Z_1 and Z_2 are connected in parallel across a 230 V, 50 Hz supply. The impelance, Z_1 consists of a resistance of 14 Ω and an inductance of 16 mH. The impedance, Z_2 consists of a resistance of 18 Ω and an inductance of 32 mH. Calculate the branch currents, line current, and total power factor. Draw the phasor diagram showing the voltage and currents.

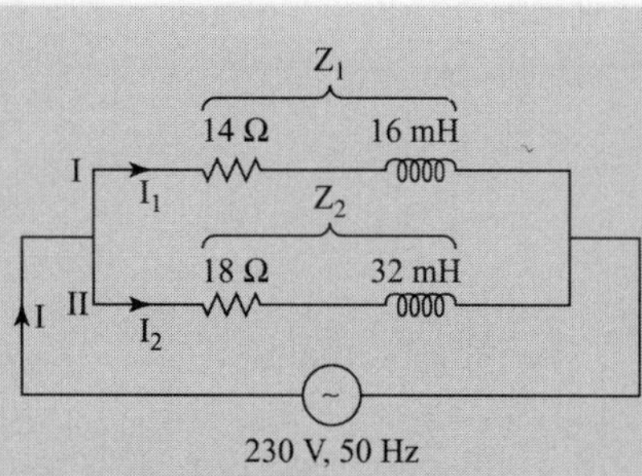

FIGURE 3.34

SOLUTION

Let $R_1 = 14\ \Omega,\ X_L = \omega L_1 = 2\pi f L_1 = 2 \times 3.14 \times 50 \times 16 \times 10-^{3} = 5\ \Omega$

$$|Z_1| = \sqrt{R_1^2 + X_{L_1}^2} = \sqrt{14^2 + 5^2} = 14.9\ \Omega$$

$$R_2 = 18\ \Omega,\ X_{L2} = \omega L_2 = 2\pi f L_2 = 2 \times 3.14 \times 50 \times 32 \times 10^{-3} = 10\ \Omega$$

$$|Z_2| = \sqrt{R_2^2 + X_{L_2}^2} = \sqrt{18^2 + 10^2} = 20.6\ \Omega$$

The phase angles of Z_1 and Z_2 are calculated from the impedance triangles as

$$\phi_1 = \tan^{-1}\frac{X_{L1}}{R_1} = \tan^{-1}\frac{5}{14} = 19.6°$$

$$\phi_2 = \tan^{-1}\frac{X_{L2}}{R_2} = \tan^{-1}\frac{10}{18} = 29°$$

Thus, $Z_1 = 14.9\angle 19.6°$ and $Z_2 = 20.6\angle 29°$

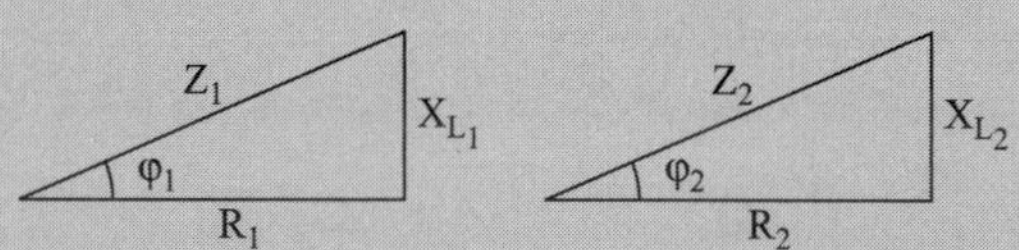

FIGURE 3.35 (a)

Admittance of branch I is Y_1 and admittance of branch II is Y_2

$$Y_1 = \frac{1}{Z_1} = \frac{1}{14.9\angle 19.6°} = 0.067\angle -19.6°$$

$$Y_2 = \frac{1}{Z_2} = \frac{1}{20.6\angle 29°} = 0.0485\angle -29°$$

Taking voltage, V as the reference axis,

$$I_1 = \frac{V}{Z_1} = VY_1 = 230\angle 0 \times 0.067\angle -19.6° = 15.41\angle -19.6° \text{ A}$$

$$I_2 = \frac{V}{Z_2} = VY_2 = 230\angle 0° \times 0.0485\angle -29° = 11.15\angle -29° \text{A}$$

The phasor diagram showing V, I_1, I_2 has been shown in Fig. 3.35 (b). The sum of I_1 and I_2 gives total current, I. The cos of angle between V and I gives the value of total power factor

Taking the cosine and sine components of the branch currents and the line current

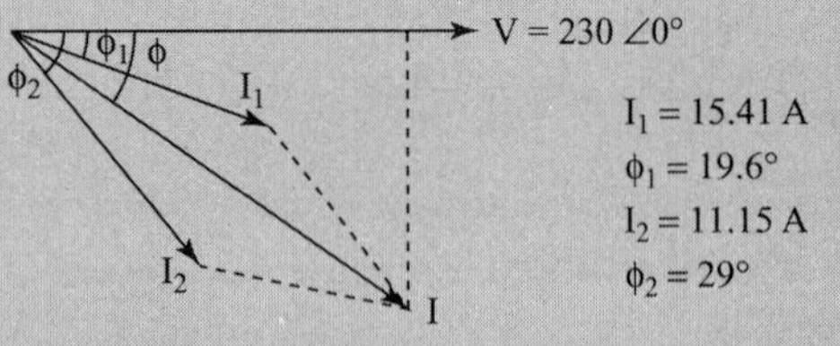

FIGURE 3.35 (b)

$$I \cos\phi = I_1 \cos\phi_1 + I_2 \cos\phi_2$$

and $$I \sin\phi = I_1 \sin\phi + I_2 \sin\phi_2$$

Substituting values

$$I \cos\phi = 15.41 \times 0.942 + 11.15 \times 0.335 = 18.24$$

$$I \sin\phi = 15.41 \times 0.325 + 11.15 \times 0.485 = 10.4$$

$\tan\phi = \frac{I\sin\phi}{I\cos\phi} = \frac{10.4}{18.24} = 0.57$ $\phi = \text{ten}^{-1}0.57 = 30°$ Power factor $= \cos\phi = \cos 30° = 0.866$ lagging

$$I = \sqrt{(I \sin\phi)^2 + (I \cos\phi)^2} = \sqrt{(10.4)^2 + (18.24)^2} = 21 \text{ A}$$

$$I = 21\angle -30° \text{ A}$$

Current, I can also be calculated as

$$I = VY$$

Where $$Y = Y_1 + Y_2$$

EXAMPLE 3.13

In Fig. 3.36 is shown a parallel circuit, an inductance L and a parallel R connected across 200 V, 50 Hz ac supply. Calculate:

(a) the current drawn from the supply; (b) apparent power; (c) real power; (d) reactive power.

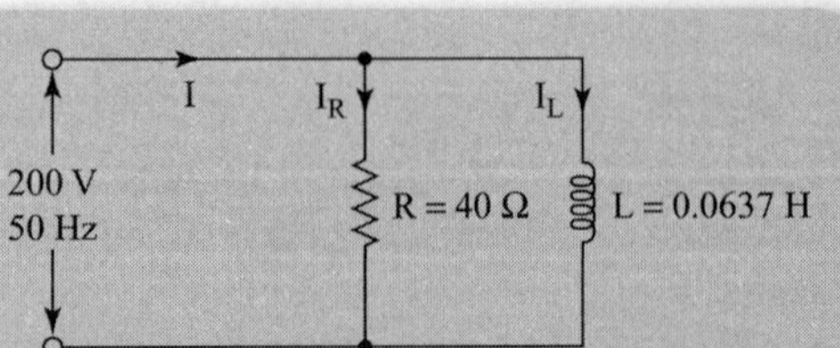

FIGURE 3.36

SOLUTION

Resistance of resistive branch, $R = 40\ \Omega$

Inductive reactance of inductive branch.

$$X_L = 2\pi fL$$
$$= 2\pi \times 50 \times 0.0637 = 20\ \Omega$$

Current drawn by resistive branch, $I_R = \dfrac{V}{R} = \dfrac{200}{40} = 5\ \text{A}$

Current drawn by inductive branch, $I_L = \dfrac{V}{X_L} = \dfrac{200}{20} = 10\ \text{A}$

(i) Current drawn from the supply (see Fig. 3.37),

$$I = \sqrt{I_R^2 + I_L^2}$$
$$= \sqrt{5^2 + 10^2} = 11.18\ \text{A}$$

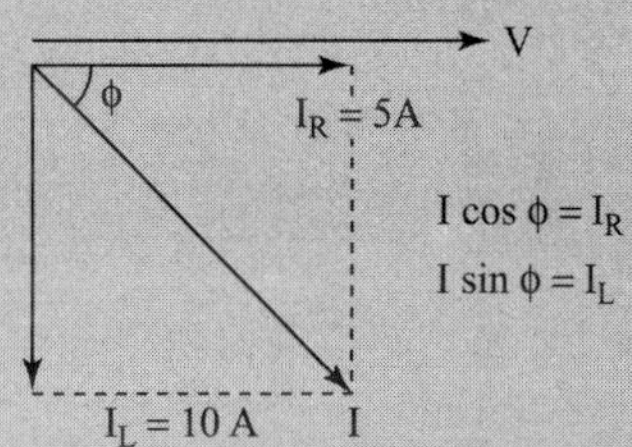

FIGURE 3.37

(ii) Apparent power, $S = V \times I = 200 \times 11.18 = 2.236$ kVA

(iii) Real power, $P = V\,I\,\text{Cos}\phi = V\,I_R = 200 \times 5 = 1.0$ kW

(iv) Reactive power, $Q = V\,I\,\text{Sin}\phi = V \times I_L = 200 \times 10 = 2.0$ kVAR

EXAMPLE 3.14

The parallel circuit shown in the Fig. 3.38 is connected across a single-phase 100 V, 50 Hz ac supply. Calculate:

(i) the branch currents
(ii) the total current
(iii) the supply power factor
(iv) the active and reactive power supplied by the source.

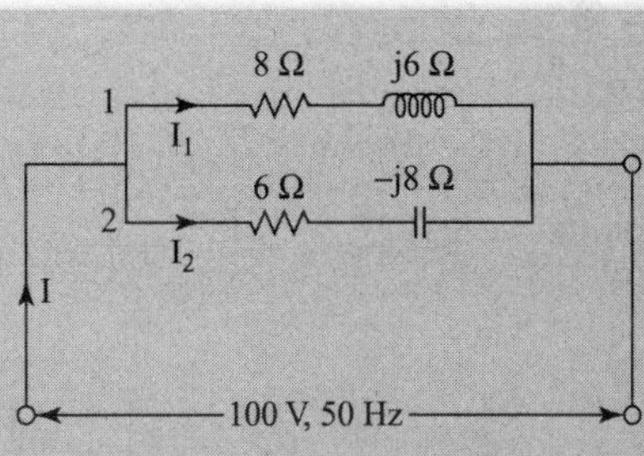

FIGURE 3.38

SOLUTION

It is assumed that the students are aware of the method of representation of a complex number in the forms of a + ib or a + jb. However, this has been explained in the next section.

$$Z_1 = R + jX_L = 8 + j6 = \sqrt{8^2 + 6^2}\ \angle \tan^{-1}\frac{6}{8} = 10\angle 40°$$

$$Z_2 = R - jX_C = 6 - j8 = \sqrt{6^2 + 8^2}\angle -\tan^{-1}\frac{8}{6} = 10\angle -48^\circ$$

$$I_L = \frac{V}{Z_1} = \frac{100\angle 0}{10\angle 40^\circ} = 10\angle -40^\circ A$$

$$I_2 = \frac{V}{Z_2} = \frac{100\angle 0}{10\angle -48^\circ} = 10\angle 48^\circ A$$

$$I = I_1 + I_2 = 10\angle -40^\circ + 10\angle 48^\circ$$

$$I = 10\cos 40^\circ - j10\sin 40^\circ + 10\cos 48^\circ + j10\sin 48^\circ$$

$$= (10\cos 40^\circ + 10\cos 48^\circ) + j(10\sin 48^\circ - 10\sin 40^\circ)$$

$$= 10\times 0.766 + 10\times 0.669 + j(10\times 0.743 - 10\times 0.642)$$

$$= 7.66 + 6.69 + j(7.43 - 6.48)$$

$$= 14.35 + j\,0.95$$

$$= \sqrt{(14.35)^2 + (0.95)^2}$$

$$= 14.45 \text{ A}$$

Power factor angle, $\phi = \tan^{-1}\frac{0.95}{14.35} = 4^\circ$

Power factor, $\cos\phi = 0.99$

Active power

$$= V I \cos\phi$$

$$= 100 \times 14.45 \times 0.99$$

$$= 1330 \text{ W}$$

Reactive power

$$= V I \sin\phi$$

$$= 100 \times 14.45 \times 0.069$$

$$= 99.7 \text{ VAR}$$

3. Use of phasor algebra

Alternating quantities like voltage, current, etc., can be represented either in the polar form or in the rectangular form on real and imaginary axis. In Fig. 3.39 is shown a voltage, V represented in the complex plane.

The voltage, V can be represented as $V\angle\phi$. This is called the polar form of representation. Voltage V can also be represented as $V = a + jb = V\cos\phi + jV\sin\phi$. This is called the rectangular form of representation using a j operator.

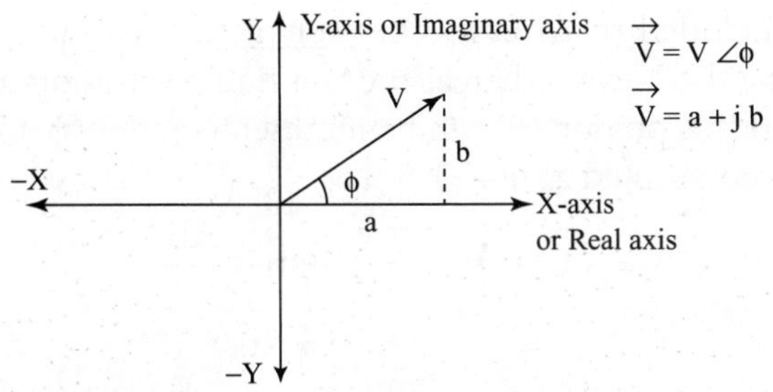

FIGURE 3.39

Representation of a phasor

Significance of operator j

The operator j used in the above expression indicates a real operation. This operation when applied to a phasor, indicates the rotation of that phasor in the counter-clockwise direction through 90° without changing its magnitude. As such it has been referred to as an operator. For example, let a phasor A drawn from O to A be in phase with the X-axis as has been shown in Fig. 3.40 (a). This phasor when represented by jA shows that the phasor A has been rotated in the anticlockwise direction by an angle 90° and as such its position now is along the Y-axis. If the operator j is again applied to phasor jA, it turns in the counter-clockwise direction through another 90°, thus giving a phasor j^2A which is equal and opposite to the phasor A, i.e., equal to –A. See Fig. 3.40 (a).

Thus, j^2 can be seen as equal to –1. Therefore, the value of j becomes equal to $\sqrt{-1}$.
Hence,

$$j = +\sqrt{-1},\ 90^\circ \text{ CCW rotation from OX-axis}$$

$$j^2 = j \times j = (\sqrt{-1})^2 = -1, 180^\circ \text{ CCW rotation from OX-axis}$$

$$j^3 = (\sqrt{-1})^3 = -\sqrt{-1}, 270^\circ \text{ CCW rotation from OX-axis}$$

and

$$j^4 = (\sqrt{-1})^4 = (-1)^2 = 1, 360^\circ \text{ CCW rotation from OX-axis}$$

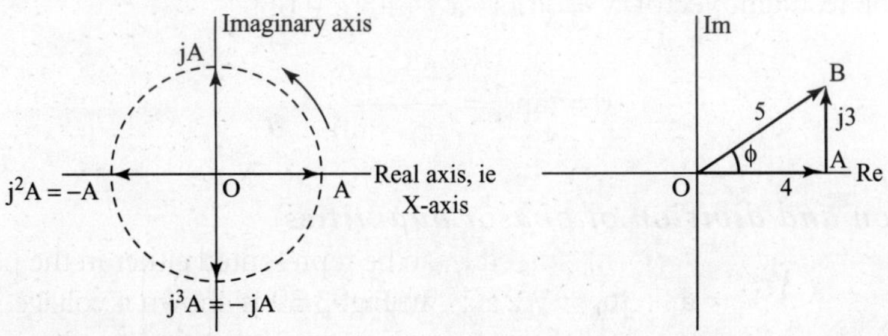

FIGURE 3.40

Use of operator j to represent a phasor

From above, it is concluded that j is an operator rather than a real number. However, it represents a phasor along the Y-axis, whereas the real number is represented along the X-axis.

As shown in Fig. 3.40 (b), phasor OB can be represented as $5\angle\phi$ in the polar form. In the rectangular form OB is represented as 4 + j3

$$\overrightarrow{OB} = \left|\sqrt{OA^2 + AB^2}\right| \angle \tan^{-1}\frac{AB}{OA}$$

$$= \sqrt{4^2 + 3^2}\angle\tan^{-1}\frac{3}{4}$$

$$= 5\angle 37° = 5\cos 37° + j\,5\sin 37°$$

$$= 5\times 0.8 + j5\times 0.6 = 4 + j3$$

Addition and subtraction of phasor quantities

Refer to Fig. 3.41.

Let $V_1 = a_1 + jb_1$ and $V_2 = a_2 + jb_2$

Addition: $V = V_1 + V_2$

$$= (a_1 + jb_1) + (a_2 + jb_2)$$

$$= (a_1 + a_1) + j(b_1 + b_2)$$

Magnitude of the resultant vector $V = \sqrt{(a_1 + a_2)^2 + (b_1 + b_2)^2}$

Phase angle, $\theta = \tan^{-1} = \dfrac{(b_1 + b_2)}{(a_1 + a_2)}$

Subtraction: $V = V_1 - V_2$

$$= (a_1 + jb_1) - (a_2 - jb_2)$$

$$= (a_1 - a_2) + j(b_1 - b_2)$$

Magnitude of the resultant vector $V = \sqrt{(a_1 - a_2)^2 + (b_1 - b_2)^2}$

Phase angle, $\theta = \tan^{-1} = \dfrac{(b_1 - b_2)}{(a_1 - a_2)}$

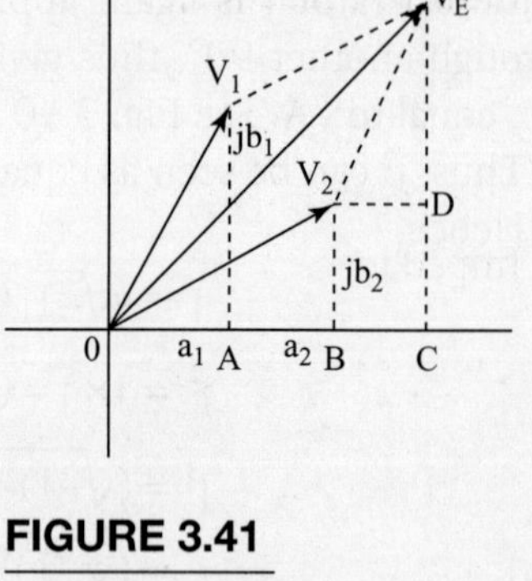

FIGURE 3.41

Addition of phasor quantities

Multiplication and division of phasor quantities

Let $V_1 = a_1 + jb_1 = V_1\angle\theta_1;$ where $\theta_1 = \tan^{-1}\left(\dfrac{b_1}{a_1}\right)$

$$V_2 = a_2 + jb_2 = V_2\angle\theta: \text{ where } \theta_2 = \tan^{-1}\left(\frac{b_2}{a_2}\right)$$

Multiplication

$$V_1 = V_1 \times V_2 = V_1 \angle\theta_1 \times V_2 \angle\theta_2 \quad \text{in the polar form}$$
$$= V_1 V_2 \angle(\theta_1 + \theta_2), \quad \text{angles are added algebraically}$$

Division

$$\frac{V_1}{V_2} = \frac{a_1 + jb_1}{a_1 + jb_2} = \frac{V_1\angle\theta_1}{V_2\angle\theta_2} = \frac{V_1}{V_2}\angle\theta_1 - \theta_2$$

angles are subtracted algebraically

EXAMPLE 3.15

A coil having a resistance of 5 Ω and inductance of 30 mH in series are connected across a 230 V, 50 Hz supply. Calculate current, power factor, and power consumed.

SOLUTION

$$R = 5\,\Omega$$
$$L = 30 \times 10^{-3}\,\text{H}$$

Inductive reactance, $X_L = \omega L = 2\pi f L$

$$X_L = 2 \times 3.14 \times 50 \times 30 \times 10^{-3}\,\Omega$$
$$= 9.42\,\Omega$$

Impedance, $Z = R + jX_L = 5 + j9.42$

$$= \sqrt{5^2 + (9.42)^2}\angle\tan^{-1}\frac{9.42}{5}$$
$$= 10.66\angle\tan^{-1}1.884$$
$$= 10.66\angle 62°\,\Omega$$

Current, $I = \frac{V}{Z} = \frac{230\angle 0}{10.66\angle 62°}$

$$= 21.57\angle -62°\,\text{A}$$

Magnitude of $I = 21.57$ A

Current I is lagging the voltage, V by 62°. Power factor = cosϕ

Here, p.f. = cos 62° = 0.47 lagging. The phasor diagram along with its circuit has been shown in Fig. 3.42.

Power consumed $= V I \cos\phi$

$$= 230 \times 21.57 \times 0.47$$
$$= 2331.7\,\text{W}$$

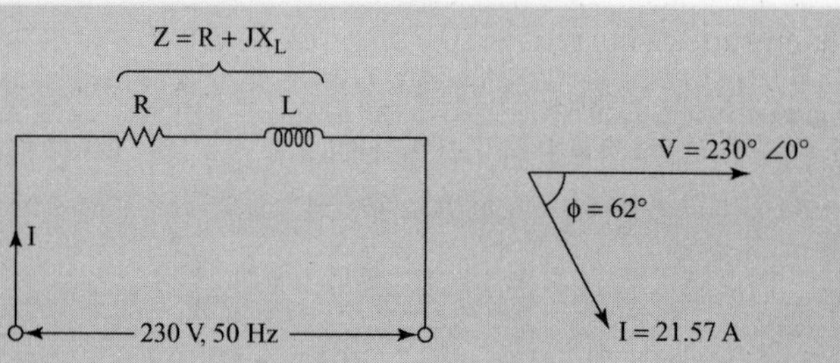

FIGURE 3.42

EXAMPLE 3.16

For the R–L–C series circuit shown in Fig. 3.43 (a), calculate current, power factor, and power consumed.

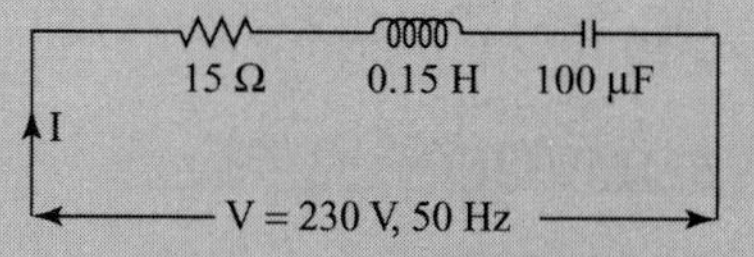

FIGURE 3.43 (a)

SOLUTION

Inductive reactance,

$$X_L = 2\pi fL = 2 \times 3.14 \times 50 \times 0.15$$
$$= 47.1\ \Omega$$

Capacitive reactance,

$$X_C = \frac{1}{2\pi fC} = \frac{1}{2 \times 3.14 \times 50 \times 100 \times 10^{-6}}$$
$$= 31.84\ \Omega$$

Here, X_L is greater than X_C. Therefore, the circuit reactance will be $X_L - X_C$.

Impedance,

$$Z = R + j(X_L - X_C)$$
$$= 15 + j(47.1 - 31.84)$$
$$= 15 + j15.26\ \Omega$$

Current,

$$I = \frac{V}{Z} = \frac{230\angle 0}{15 + j15.26} = \frac{230\angle 0°}{\sqrt{(15)^2 + (15.26)^2}\angle \tan^{-1} 15.26/15}$$
$$= \frac{230\angle 0°}{21.39\angle \tan^{-1} 1.01}$$
$$= \frac{230\angle 0°}{21.39\angle 45.3°}$$
$$= 10.75\angle -45.3°\text{A}$$

This shows that magnitude of current is 10.75 A and the current lags the voltage by 45.3°.

Power factor, $\cos\phi = \cos 45.3° = 0.703$ lagging

Power consumed,
$$P = V I \cos\phi$$
$$= 230 \times 10.75 \times 0.703 \text{ W}$$
$$= 1738.16 \text{ W}$$

EXAMPLE 3.17

Two coils having impedance Z_1 and Z_2 are connected in series across a 230 V, 50 Hz power supply as shown in Fig. 3.43 (b).

The voltage drop across Z_1 is equal to 120/30° V. Calculate the value of Z_2.

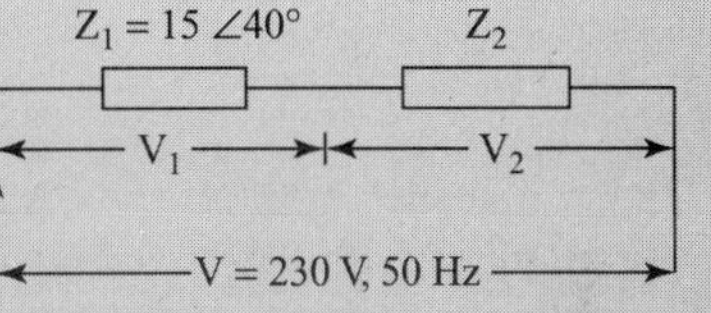

FIGURE 3.43 (b)

SOLUTION

We have, $V = V_1 + V_2$

or,
$$V_2 = V - V_1 = 230\angle 0 - 120\angle 30°$$
$$= 230(\cos 0° + j\sin 0°) - 120(\cos 30° + j\sin 30°)$$
$$= 230 - 120 \times 0.866 - j\,120 \times 0.5$$
$$= 126.1 - j\,60 = \sqrt{126.1^2 + 60^2}\,\angle -\tan^{-1}\frac{60}{126.1}$$
$$= 139.6\angle -25.4°$$

Since this is a series circuit, the current flowing through the circuit is the same.

The circuit current can be calculated by using any of the following relations

$$I = \frac{V_1}{Z_1} \text{ or } I = \frac{V_2}{Z_2} \text{ or } I = \frac{V}{Z_1 + Z_2}$$

current,
$$I = \frac{V_1}{Z_1} = \frac{120\angle 30}{15\angle 40} = 8\angle -10°\text{A}$$

since the same current will flow through Z_2,

$$Z_2 = \frac{V_2}{I} = \frac{139.6\angle -25.4°}{8\angle -10} = 17.45\angle -15.4°$$
$$= 17.45(\cos 15.4° - j\sin 15.4°)$$
$$= 17.45 \times 0.964 - j17.45 \times 0.2656$$
$$= 16.82 - j4.6\ \Omega$$

Note that the impedance of an R–L circuit is R + j X_L and impedance of an R–C circuit is R – jX_C. Since, Z = 16.82 – j4.63, it must be an R–C circuit.

Thus, impedance coil Z_2 is written as

$$Z_2 = R - jX_c$$

where, $R = 16.82\ \Omega$

and $X_C = 4.63\ \Omega$

$$X_C = \frac{1}{2\pi fC} = \frac{1}{314 \times C}$$

Substituting

$$X_C = 4.63 = \frac{1}{314 \times C}$$

or,

$$C = \frac{1}{314 \times 4.63} F$$

$$= \frac{10^6}{314 \times 4.63} \mu F$$

$$= 687.8 \mu F$$

EXAMPLE 3.18

A 50 Hz alternating voltage, V = (160 + j170)V is connected across an R-L series circuit.

A current of I = (12 – j5) A flows through the circuit. Calculate impedance, power factor, and power consumed. Draw the phasor diagram.

SOLUTION

Voltage,

$$V = 160 + j170 = \sqrt{(160) + (170)^2}\ \angle \tan^{-1}\frac{170}{160} = 233\angle 46.8°$$

Current,

$$I = 12 - j5 = \sqrt{(12)^2 + (5)^2}\ \angle -\tan^{-1}\frac{5}{12} = 13\angle -22.6°$$

Impedance,

$$Z = \frac{V}{I} = \frac{233\ \angle 46.8°}{13\ \angle -22.60°} = 17.92\ \angle 69.4°$$

$$= 17.92 \cos 69.4° + j\ 17.92 \sin 69.4°$$

$$= 17.92 \times 0.35 + j\ 17.92 \times 0.93$$

$$= 6.27 + j\ 16.66$$

$$= R + j\ X_L$$

The R–L series circuit consists of a resistance of 6.27 Ω and an inductive reactance of 16.66 Ω. The phasor diagram is shown in Fig. 3.44. The horizontal axis has been assumed as reference axis. The voltage of $233\angle 46.8°$ has been shown leading the reference axis by 46.8°. Current of $13\ \angle -22.6°$ has been drawn lagging the reference axis by 22.6°. The phase angle between voltage, V and current, I becomes equal to (46.8° + 22.6°) or 69.4°.

The circuit power factor, $\cos\phi = \cos.\ 69.4° = 0.35$ lagging

Power consumed, $P = VI\cos\phi = 233 \times 13 \times 0.35 = 1060.1$ W

Impedance $= 17.92\ \angle 69.4° = 6.27 + j\ 16.66 = R + j\ \omega L$

$$X_L = w\,L = 16.66$$

or, $$L = \frac{16.66}{w} = \frac{16.66}{2\pi f} = \frac{16.66}{2\times 3.14\times 50} = \frac{16.66\times 1000}{314}\ \text{mH} = 53\ \text{mH}$$

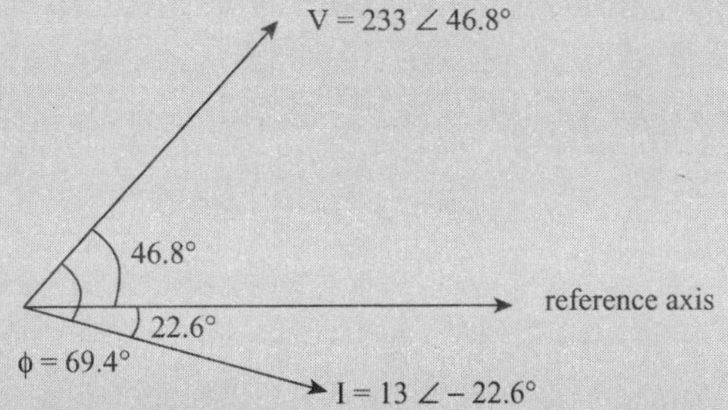

FIGURE 3.44

EXAMPLE 3.19

A sinusoidal voltage of v = 325 sin 314t when applied across an R-L series circuit causes a current of $i = 14.14 \sin(314t - 60°)$ flowing through the circuit. Calculate the value of L and R of the circuit. Also calculate power consumed.

SOLUTION

given $v = 325 \sin 314t$

comparing $v = V_m \sin \omega t$

$$V_m = 325\ \text{V}$$

RMS value, $$V = \frac{V_m}{\sqrt{2}} = \frac{325}{1.414} = 230\ \text{V}$$

$$\omega = 314$$

or, $2\pi f = 314$

$$f = \frac{314}{2\pi} = 50 \text{ Hz}$$

given $i = 14.14 \sin(314t - 60°)$

comparing $i = I_m \sin(\omega t - \phi)$

$$I_m = 14.14$$

RMS value, $I = \frac{I_m}{\sqrt{2}} = \frac{14.14}{1.414} = 10 \text{ A}$

Power factor angle, $\phi = 60°$

Power factor, $\cos\phi = \cos 60° = 0.5$ lagging

Power, $P = VI\cos\phi = 230 \times 10 \times 0.5$

$$= 1150 \text{ W}$$

Impedance $Z = \frac{V}{I} = \frac{230\angle 0}{10\angle -60°} = 23\angle 60°\ \Omega$

In complex form,

$$Z = 23(\cos 60° + j\sin 60°)$$
$$= 23 \times 0.5 + j23 \times 0.866$$
$$= 11.5 + j22.99$$
$$= R + jX_L$$

Thus, resistance of the circuit, $R = 11.5\ \Omega$

Inductive reactance, $X_L = 22.99\ \Omega$

or, $\omega L = 22.99$

$$L = \frac{22.99}{\omega} = \frac{22.99}{2\pi f} = \frac{22.99}{2 \times 3.14 \times 50}$$
$$= \frac{22.99}{314} \text{H} = \frac{22.99 \times 10^3}{314} \text{mH}$$
$$= 73.21 \text{ mH}$$

EXAMPLE 3.20

A variable resistance R and an inductance L of value 100 mH in series are connected across at 50 Hz supply. Calculate at what value of R the voltage across the inductor will be half the supply voltage.

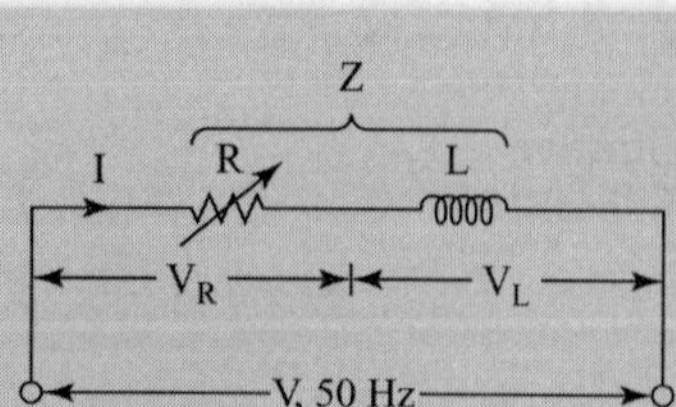

FIGURE 3.45

SOLUTION

Given, $L = 100 \text{ mH}$

$$X_L = \omega L = 2\pi f L$$

$$= 2 \times 3.14 \times 50 \times 100 \times 10^{-3}$$
$$= 31.4\ \Omega$$

We have to find R for which

$$V_L = \frac{1}{2}V$$

$$V_L = IX_L \text{ and } V = IZ = I\sqrt{R^2 + X_L^2}$$

Equating,
$$V_L = \frac{1}{2}V$$

or,
$$IX_L = \frac{1}{2}I\sqrt{R^2 + X_L^2}$$

or,
$$\sqrt{R^2 + X_L^2} = 2\,X_L = 2 \times 31.4 = 62.8\ \Omega$$

Squaring,

$$R^2 + X_L^2 = (62.8)^2$$
$$R^2 = (62.8)^2 - (31.4)^2$$
$$R^2 = 3943.8 - 985.9$$
$$R = 54.4\ \Omega$$

EXAMPLE 3.21

A voltage of v = 100 sin (314t + 0) is applied across a resistance and inductance in series. A current of 10 sin (314t 2 π/6) flows through the circuit. Calculate the value of R and L of the circuit. Also calculate power and power factor.

SOLUTION

$$v = 100 \sin 314t$$
$$= V_m \sin \omega t$$
$$V_m = 100\text{ V}$$

$$V(\text{RMS value}) = \frac{V_m}{\sqrt{2}} = \frac{100}{1.414} = 70.7\text{ V}$$

$$\omega = 314 \text{ or, } 2\pi f = 314,\ f = \frac{314}{2\pi} = 50\text{ Hz}$$

Similarly,
$$i = 10 \sin (\omega t - 30°)$$
$$= I_m \sin (\omega t - 30°)$$

$$I_m = 10,\ I = \frac{I_m}{\sqrt{2}} = \frac{10}{1.414} = 7.07\text{ A}$$

Current, I is lagging V by 30°.

Power factor,
$$\cos\phi = \cos 30°$$
$$= 0.866 \text{ lagging}$$

Impedance, $$Z = \frac{V}{I} = \frac{70.7\angle 0}{7.07\angle -30^\circ} = 10\angle 30^\circ$$

In rectangular form,

$$Z = 10(\cos 30^\circ + j\sin 30^\circ)$$
$$= 10 \times 0.866 + j\,10 \times 0.5$$
$$= 8.66 + j5.0$$
$$= R + jX_L$$
$$R = 8.66\ \Omega$$

and $$X_L = 5.0\ \Omega$$

Again $$X_L = \omega L = 314\ L$$

$$L = \frac{X_L}{314} = \frac{5}{314}\,H = \frac{5000}{314}\,mH$$
$$= 15.92\ mH$$

EXAMPLE 3.22

The expression of applied voltage and current flowing through an ac series R-L circuit are

$$v = 200\sin\left(314t + \frac{\pi}{3}\right) \text{ and } i = 20\sin\left(314t + \frac{\pi}{6}\right)$$

Calculate for the circuit (i) power factor; (ii) average power; (iii) impedance; (iv) R and L

SOLUTION

We will compare the voltage, v and current, i with the standard form

$$v = V_m \sin \omega t \text{ and}$$
$$I = I_m \sin \omega t$$

From the data provided

$$V_m = 200,\ I_m = 20,\ \omega = 314$$

RMS values, $$V = \frac{V_m}{\sqrt{2}} = \frac{200}{1.414} = 141.4\ V$$

$$I = \frac{I_m}{\sqrt{2}} = \frac{20}{1.414} = 14.14\ A$$

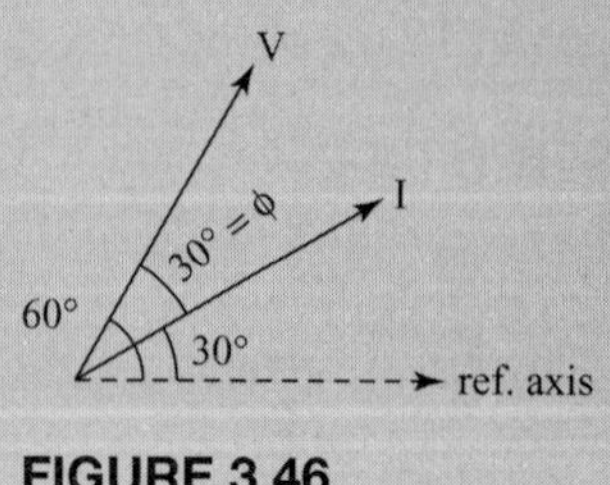

FIGURE 3.46

We have represented in Fig. 3.46 the voltage and current with respect to a common reference axis. The voltage, V is leading the reference axis by $\pi/3^\circ$, i.e., 60°, while current I is leading the reference axis by 30°. The phase angle between V and I is 30°. The current in the circuit lags the voltage by 30°.

Power factor, $\cos\phi = \cos 30° = 0.866$ lagging

Average power, $P = V I \cos\phi = 141.4 \times 14.14 \times 0.866$

$= 1732$ W

The impedance of the circuit, $Z = \sqrt{R^2 + X_L^2}$

Again, $$Z = \frac{V}{I} = \frac{141.4}{14.14} = 10\ \Omega$$

In the polar form,

$$Z = \frac{V \angle 60°}{I \angle 30°} = \frac{141.4 \angle 60°}{14.14 \angle 30°}$$

$$= 10\angle 30°\ \Omega.$$

Expressing in the rectangular form,

$$Z = Z\cos\phi + j\,Z\sin\phi$$

$$= 10\cos 30° + j\,Z\sin 30°$$

$$= 10 \times 0.866 + j10 \times 0.5$$

$$= 8.66 + j\,5$$

$$= R + j\,X_L$$

$$R = 8.66\ \Omega$$

$$X_L = \omega L$$

$$L = \frac{X_L}{\omega} = \frac{5}{314}\text{H} = \frac{5000}{314} = 15.92\ \text{mH}$$

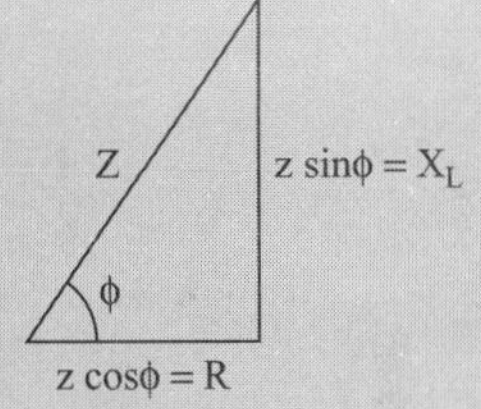

FIGURE 3.47

EXAMPLE 3.23

In an R-L-C series circuit the voltage drops across the resistor, inductor, and capacitor are 20 V, 60 V, and 30 V, respectively. Calculate the magnitude of the applied voltage and the power factor of the circuit.

SOLUTION

In Fig. 3.48 the voltage drops across the circuit components and their phase relationship have been drawn. Since it is a series circuit, it is always convenient to take current as the reference axis. The voltage drop across the resistor and the current are in phase. Voltage across the inductor will lead the current, and voltage across the capacitor will lag the current. The circuit diagram and the phasor diagram have been shown.

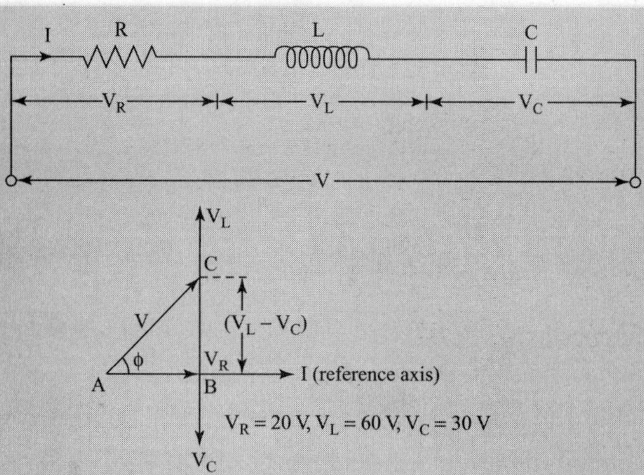

FIGURE 3.48

From triangle ABC

$$AC^2 = AB^2 + BC^2$$

or,
$$V^2 = (V_R)^2 + (V_L - V_C)^2$$
$$= (20)^2 + (60 - 30)^2$$
$$= 1300$$

or,
$$V = 36 \text{ V}$$

Power factor
$$= \cos\phi = \frac{AB}{AC} = \frac{V_R}{V} = \frac{20}{36} = 0.55$$

Power factor angle,
$$\phi = 56.6°$$

Since current I is lagging V by an angle $\phi = 56.6°$, the power factor is taken as lagging power factor.

EXAMPLE 3.24

In the circuit shown in Fig. 3.49, calculate the value of R and C.

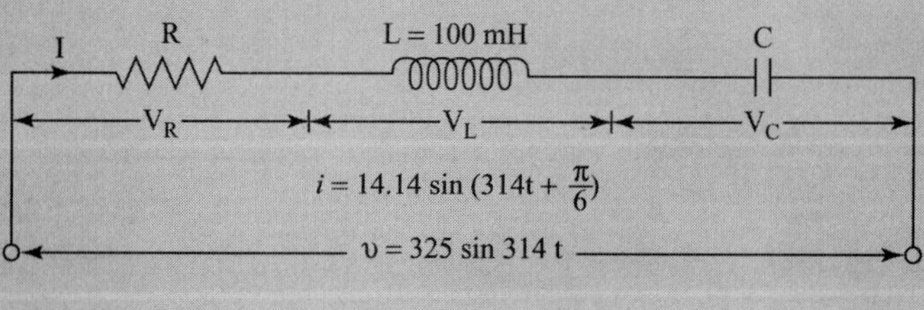

FIGURE 3.49

SOLUTION

$$v = 325 \sin 314t, \omega = 314$$

$$V_m = 325, V = \frac{V_m}{\sqrt{2}} = \frac{325}{1.414} = 230\,V$$

$$I = \frac{I_m}{\sqrt{2}} = \frac{14.14}{1.414} = 10\,A$$

Current I is leading V by $\frac{\pi}{6}$ degrees.

Taking V as the reference axis,

$$Z = \frac{V}{I} = \frac{230\angle 0}{10\angle +30^\circ} = 23\angle -30^\circ$$

In an R-L-C series circuit, if current I is leading the voltage V, we have to consider the circuit as leading p.f circuit. This means the capacitive reactance is more than the inductive reactance (i.e., the circuit is effectively an R–C circuit. We will draw the phasor diagram by taking current on the reference axis. Here we see that V is lagging I by the power factor angle. That is, I is leading V by an angle ϕ. The phasor diagram taking I as the reference axis has been shown in Fig. 3.50.

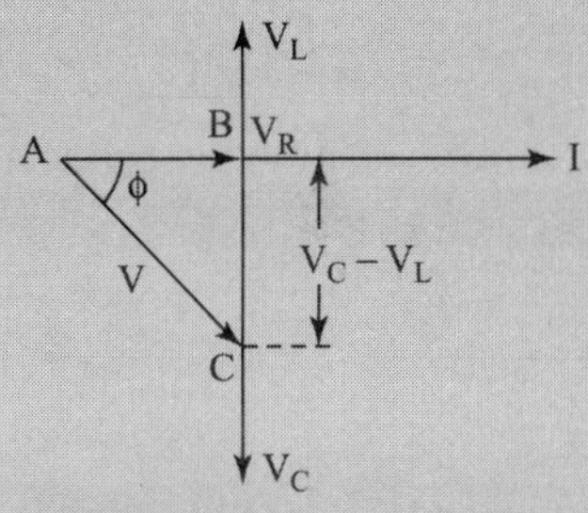

FIGURE 3.50

If we take V as the reference axis,

$$Z = \frac{V\angle 0}{I\angle +30^\circ} = \frac{230\angle 0}{10\angle +30} = 23\angle -30^\circ$$

$$Z = 23 \cos 30^\circ - j\,23 \sin 30^\circ$$

$$= 23 \times 0.866 - j\,23 \times 0.5$$

$$= 19.9 - j\,11.5$$

$$Z = R - j\,(X_C - X_L)$$

$$R = 19.9\,\Omega$$

$$X_C - X_L = 11.5\,\Omega$$

$$X_L = \omega L = 314 \times 100 \times 10^{-3} = 31.4\,\Omega$$

$$X_C - 31.4 = 11.5$$

$$X_C = 42.9\,\Omega$$

$$X_C = 42.9 = \frac{1}{\omega C} = \frac{1}{314C}$$

or,

$$C = \frac{1}{314 \times 42.9}\,F = 72.23\,\mu F$$

Power factor = $\cos \phi = \cos 30° = 0.866$ leading.

In Fig. 3.51, AB = IR = V_R has been drawn in the direction of current I. $I(X_c - X_L)$ is effectively a voltage drop which is capacitive in nature. I will lead $I(X_c - X_L)$, or we can say that $I(X_c - X_L)$ will lag I by 90°. BC has been shown lagging AB by 90°. The sum of AB and BC is AC which the total voltage, V and V = IZ. By taking away I from all the sides of the triangle ABC, the impedance triangle has be drawn.

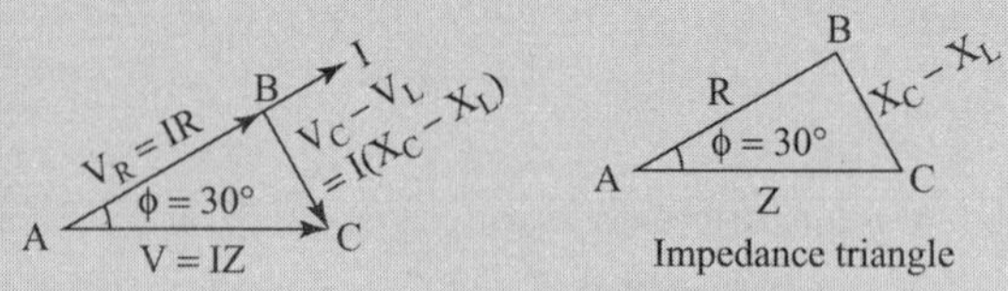

$$\cos\phi = \frac{IR}{IZ} = \frac{R}{Z}$$

FIGURE 3.51

EXAMPLE 3.25

A resistance of 15 Ω and an inductance of 100 mH are connected in parallel across at 230 V, 50 Hz supply. Calculate the branch currents, line current, and power factor. Also calculate the power consumed in the circuit. .

SOLUTION

The circuit diagram and the phasor diagram have been shown in Fig. 3.52. We note that in a parallel circuit the voltage applied across the branches is the same. The current in the resistive branch is in phase with the voltage while current in the inductive branch lags the voltage by 90°. The phasor sum of the branch currents gives us the total line current. Since in a parallel circuit voltage, V is common to the parallel branches, we generally take V as the reference axis while drawing the phasor diagram. Current through the resistive branch, I_R has been drawn in phase with V. Current through the inductive branch, I_L is lagging V by 90°. The sum of I_R and I_L gives I as has been shown in Fig. 3.52.

Power factor, $\qquad \cos\phi = \dfrac{I_R}{I}$

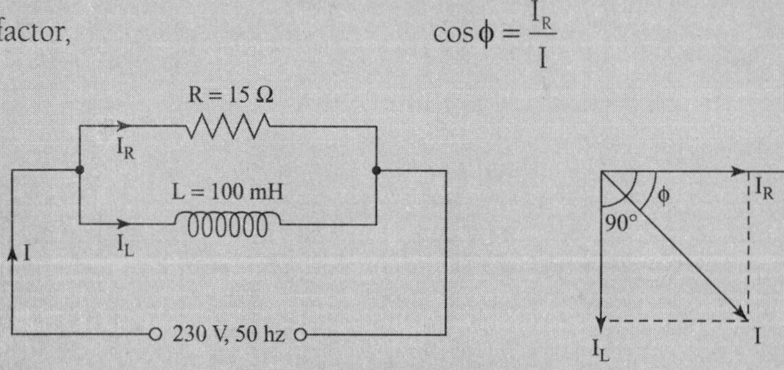

$$\cos\phi = \frac{I_R}{I}$$

FIGURE 3.52

Now using the given values, calculations are made as follows.

Inductive reactance, $X_L = \omega L = 2\pi f L$

$= 2 \times 3.14 \times 50 \times 100 \times 10^{-3}\,\Omega$

$= 31.4\,\Omega$

$$I_R = \frac{V}{R} = \frac{230}{15} = 15.33 \text{ A}$$

$$I_L = \frac{V}{X_L} = \frac{230}{31.4} = 7.32 \text{ A}$$

$$I = \sqrt{I_R^2 + I_L^2} = \sqrt{(15.33)^2 + (7.32)^2} = \sqrt{288.89} = 17 \text{ A}$$

$$\text{Power factor} = \cos\phi = \frac{15.33}{17} = 0.9 \text{ lagging}$$

Power factor angle, $\phi = \cos^{-1} 0.9 = 25°$

Since the line current I is lagging the voltage V by 25°, the power factor is mentioned as lagging. The students should note that while mentioning power factor, it is essential to indicate whether the same is lagging or leading.

Power consumed, $P = VI\cos\phi = 230 \times 17 \times 0.9 = 3519$ W

EXAMPLE 3.26

For the circuit shown in Fig. 3.53 calculate the total current drawn from the supply. Also calculate the power and power factor of the circuit.

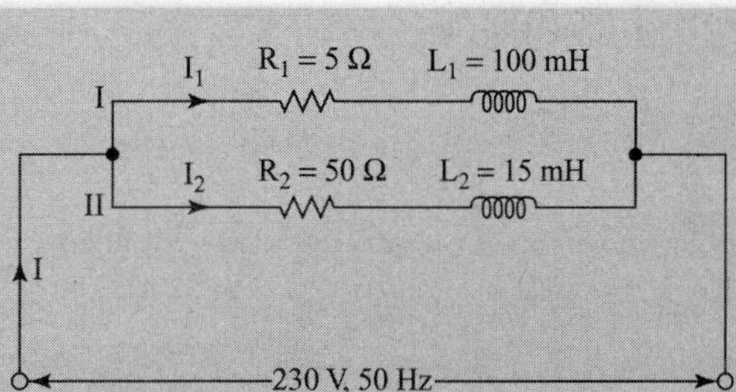

FIGURE 3.53

SOLUTION

For branch I, the impedance Z_1 is calculated as

$$Z_1 = R_1 + jX_{L_1} = 5 + j\omega L_1 = 5 + j\,2\pi \times 50 \times 100 \times 10^{-3}$$

$$= 5 + j\,31.4$$

$$= \sqrt{5^2 + 31.4^2} \angle \tan^{-1}\frac{31.4}{5}$$

$$= 31.8\angle 81°\,\Omega$$

Similarly for branch II,

$$Z_2 = R_2 + jX_{L_2} = 50 + j\,2\pi \times 50 \times 15 \times 10^{-3}$$

$$= 50 + j4.71$$

$$= \sqrt{(50)^2 + (4.71)^2} \angle \tan^{-1} \frac{4.71}{50}$$

$$= 50.22\angle 5.4° \ \Omega$$

Current, $$I_1 = \frac{V}{Z_1} = \frac{230\angle 0°}{31.8\angle 81°} = 7.23\angle -81° \text{ A}$$

Current, $$I_2 = \frac{V}{Z_2} = \frac{230\angle 0}{50.22\angle 5.4°} = 4.58\angle -5.4° \text{ A}$$

Total Current, $$I = I_1 + I_2 = 7.23\angle -81° + 4.58\angle -5.4°$$

$$= 7.23\cos 81° - j7.23\sin 81° + 4.58\cos 5.4° - j4.58\sin 5.4°$$

$$= 7.23 \times 0.156 - j7.23 \times 0.987 + 4.58 \times 0.995 - j4.58 \times 0.09$$

$$= 1.127 - j7.136 + 4.557 - j0.414$$

$$= 5.68 - j7.55 = \sqrt{(5.68)^2 + (7.55)^2} \angle -\tan^{-1} \frac{7.55}{5.68}$$

Line current, $$I = 9.44\angle -53° \text{ A}$$

The phasor diagram representing the branch currents and the line current with respect to the supply voltage has been shown in Fig. 3.54. The line current lags the applied voltage by an angle, $\phi = 53°$.

Thus, $$\text{power factor} = \cos\phi = \cos 53° = 0.6 \text{ lagging}$$

It may be noted that the branch II is more resistive and less inductive than branch I. That is why current I_1 is more lagging than current I_2.

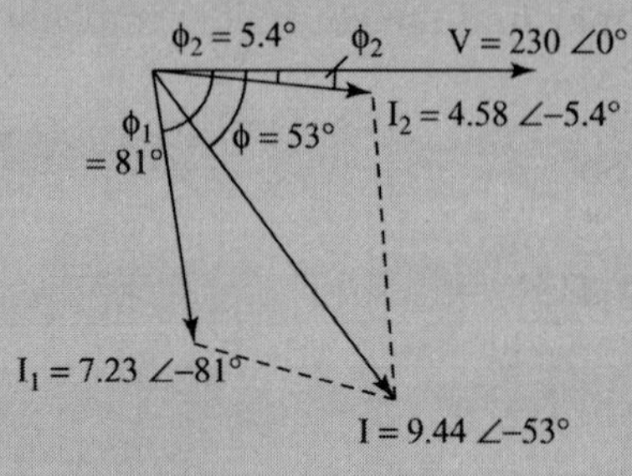

FIGURE 3.54

Power, $$P = VI\cos\phi$$

$$= 230 \times 9.44 \times \cos 53°$$

$$= 230 \times 9.44 \times 0.6$$

$$= 1302.7 \text{ W}$$

EXAMPLE 3.27

Two impedances $Z_1 = 10 + j12$ and $Z_2 = 12 - j10$ are connected in parallel across a 230 V, 50 Hz supply. Calculate the current, power factor, and power consumed.

SOLUTION

The two impedances are of the form, $Z_1 = R_1 + jX_L$ and $Z_2 = R_2 - jX_C$

Z_1 is composed of a resistor and an inductor while Z_2 is composed of a resistor and a capacitor as shown in Fig. 3.55.

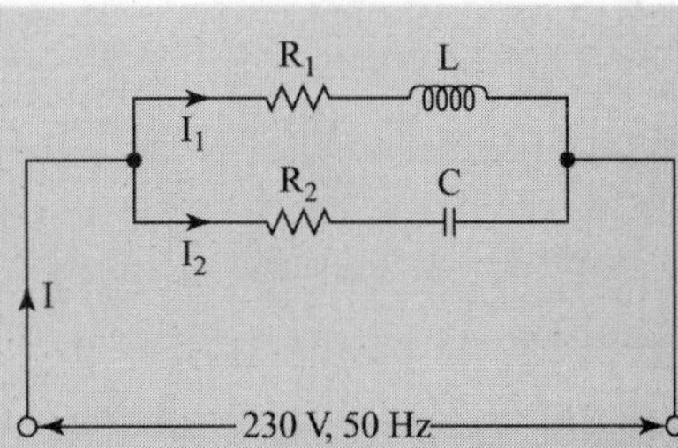

FIGURE 3.55

$$\begin{aligned} Z_1 &= 10 + j12 \\ &= \sqrt{(10)^2 + (12)^2}\angle \tan^{-1} 12/10 \\ &= 15.62\angle \tan^{-1} 1.2 \\ &= 15.62\angle 50^\circ\ \Omega \end{aligned}$$

$$\begin{aligned} Z_2 &= 12 - j10 \\ &= \sqrt{(12)^2 + (10)^2}\angle -\tan^{-1} 10/12 \\ &= 15.62\angle -40^\circ \Omega \end{aligned}$$

We may calculate, $I_1 = \frac{V}{Z_1}$ and $I_2 = \frac{V}{Z_2}$ and then add I_1 and I_2 to get I. Alternately, we may find the equivalent impedance of the circuit, Z and then find, $I = \frac{V}{Z}$

So,
$$Z = \frac{Z_1 Z_2}{Z_1 + Z_2} = \frac{15.62\angle 50^\circ \times 15.62\angle -40^\circ}{15.62\angle 50^\circ + 15.62\angle -40^\circ}$$

$$= \frac{243.98\angle 10^\circ}{15.62\cos 50^\circ + j15.62\sin 50^\circ + 15.62\cos 40^\circ - j15.62\sin 40^\circ}$$

$$= \frac{243.98\angle 10^\circ}{15.62 \times 0.64 + j15.62 \times 0.76 + 15.62 \times 0.76 - j15.62 \times 0.64}$$

or,
$$Z = \frac{243.98\angle 10^\circ}{21.86 - j1.88} = \frac{243.98\angle 10^\circ}{21.94\angle -\tan^{-1} 1.88/21.86}$$

$$= \frac{243.98\angle 10}{21.94\angle -5^\circ} = 11.12\angle 15^\circ\ \Omega$$

Total line current,
$$I = \frac{V}{Z} = \frac{230\angle 0^\circ}{11.12\angle 15^\circ} = 20.68\angle -15^\circ \text{A}$$

Current, I lags voltage, V by 15°

Power factor = $\cos\phi = \cos 15^\circ = 0.96$ lagging

Magnitude of current, I = 20.68 A

Supply voltage, V = 230 V

Power consumed, $P = VI \cos \phi$

$$= 230 \times 20.68 \times 0.96$$
$$= 4566 \text{ W}$$
$$= 4.566 \text{ kW}$$

EXAMPLE 3.28

For the circuit shown in Fig. 3.56 calculate the total current, power, and power factor of the whole circuit. Also calculate the reactive power and apparent power of the circuit. Draw the phasor diagram.

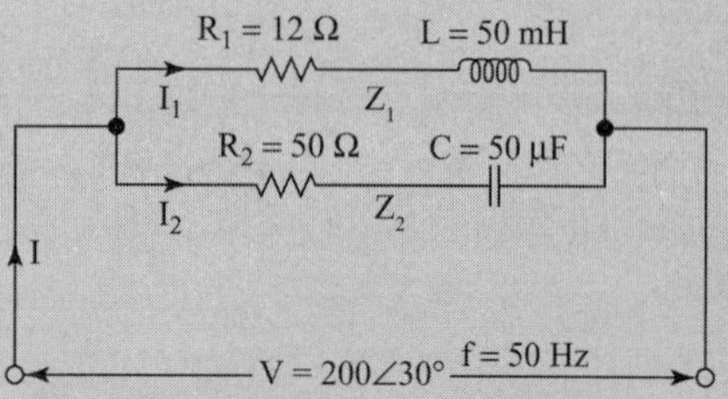

FIGURE 3.56

SOLUTION

$$\omega = 2\pi f = 6.28 \times 50 = 314$$

$$X_L = \omega L = 314 \times 50 \times 10^{-3} = 15.57\ \Omega$$

$$X_C = \frac{1}{\omega C} = \frac{1}{314 \times 5 \times 10^{-6}} = \frac{10^6}{314 \times 50} = 63.7\ \Omega$$

$$Z_1 = R_1 + jX_L = 12 + j15.57 = 19.6\angle 52.2^\circ$$

$$Z_2 = R_2 - jX_C = 50 - j63.7 = 80.9\angle -52^\circ$$

$$I_1 = \frac{V}{Z_1} = \frac{200\angle 30^\circ}{19.6\angle 52.2^\circ} = 10.2\angle -22.2^\circ$$

$$I_2 = \frac{V}{Z_2} = \frac{200\angle 30^\circ}{80.9\angle -52^\circ} = 2.47\angle 82^\circ$$

$$I = I_1 + I_2 = 10.2\angle -22.2^\circ + 2.47\angle 82^\circ$$
$$= 10.2 \cos 22.2^\circ - j10.2 \sin 22.2^\circ + 2.47 \cos 82^\circ + j2.47 \sin 82^\circ$$
$$= 10.2 \times 0.92 - j10.2 \times 0.37 + 2.47 \times 0.14 + j2.47 \times 0.99$$
$$= 9.72 - j1.33$$
$$= 9.8\angle -7.8^\circ$$

Total Current, $I = 9.8$ A

The voltage V is making an angle of +30° with the reference axis as shown in Fig. 3.57. Current I_2 is making 82° with the reference axis; current I_1 is making –22.2° with the reference axis. The resultant of I_1 and I_2 is I. Current, I is making an angle of –7.8° with the reference axis. The phase difference between V and I is 37.8° as has been shown in Fig. 3.57. Therefore, power factor, $\cos \phi = \cos 37.8^\circ = 0.79$ lagging

Power, $P = VI\cos\phi$

$= 200 \times 9.8 \times 0.79$

$= 1548.4\ W = 1.5484\ kW$

Reactive Power $Q = VI\sin\phi$

$= 200 \times 9.8 \times \sin 37.8°$

$= 200 \times 9.8 \times 0.61$

$= 1195.6\ VARs = 1.1956\ kVARs$

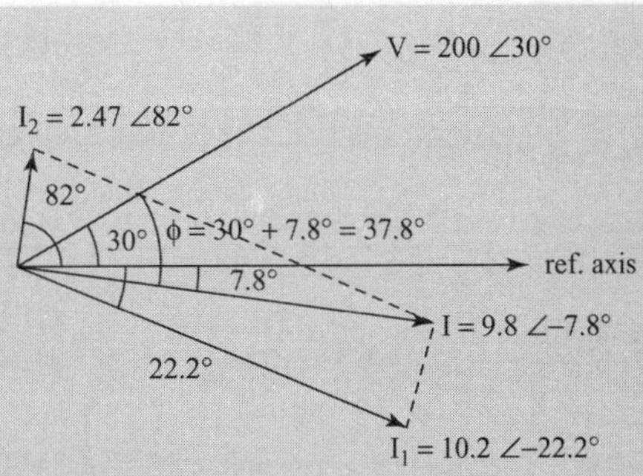

FIGURE 3.57

Apparent Power $S = VI$

$= 200 \times 9.8$

$= 1960\ VA$

$= 1.96\ kVA$

To check,

$$(kVA)^2 = (kW)^2 + (kVAR)^2$$

$$kVA = \sqrt{(kW)^2 + (kVAR)^2}$$

$$= \sqrt{(1.5484)^2 + (1.1956)^2} = 1.96$$

EXAMPLE 3.29

Three impedances, Z_1, Z_2, Z_3 are connected in parallel across a 230 V, 50 Hz supply. The values are given $Z_1 = 12\angle 30°$; $Z_2 = 8\angle -30°$; $Z_3 = 10\angle 60°$. Calculate the total admittance, equivalent impedance, total current, power factor, and power consumed by the whole circuit.

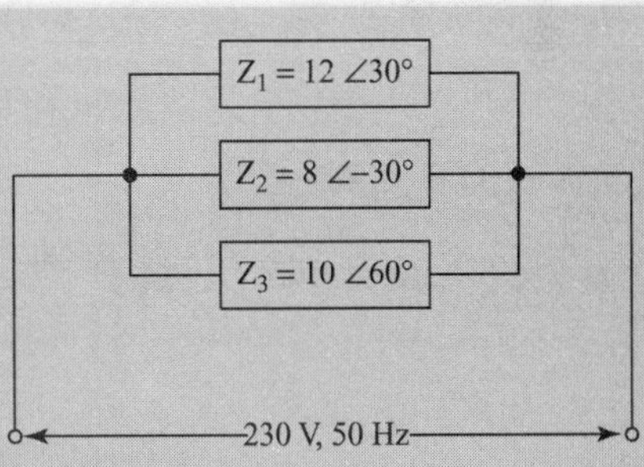

FIGURE 3.58

SOLUTION

Admittance

$$Y_1 = \frac{1}{Z_1} = \frac{1}{12\angle 30°}$$

$$Y_2 = \frac{1}{Z_2} = \frac{1}{8\angle -30°}$$

$$Y_3 = \frac{1}{Z_3} = \frac{1}{10\angle 60°}$$

Total admittance, $Y = Y_1 + Y_2 + Y_3$

Substituting,

$$Y = \frac{1}{12\angle 30^\circ} + \frac{1}{8\angle -30^\circ} + \frac{1}{10\angle 60^\circ}$$

$$= 0.08\angle -30^\circ + 0.125\angle 30^\circ + 0.1\angle -60^\circ$$

$$= 0.08(\cos 30^\circ - j\sin 30^\circ) + 0.125(\cos 30^\circ + j\sin 30^\circ) + 0.1(\cos 60^\circ - j\sin 60^\circ)$$

$$= (0.227 - j\,0.064)\,\text{mho}$$

$$= 0.235\angle -14^\circ\ \text{mho}$$

Impedance, $Z = \dfrac{1}{Y} = \dfrac{1}{0.235\angle -14^\circ} = 4.25\angle 14^\circ\ \Omega$

Total current, $I = \dfrac{V}{Z} = VY = 230\angle 0^\circ \times 0.235\angle -14^\circ$

$$= 54.05\angle -14^\circ\,\text{A}$$

Power factor, $\cos\phi = \cos 14^\circ = 0.97$ lagging

$$P = VI\cos\phi = 230 \times 54.05 \times 0.97 = 12058\ \text{W} = 12.058\ \text{kW}$$

EXAMPLE 3.30

For the circuit shown in Fig. 3.59 calculate the current in each branch and total current by the admittance method. Also calculate power and power factor of the total circuit.

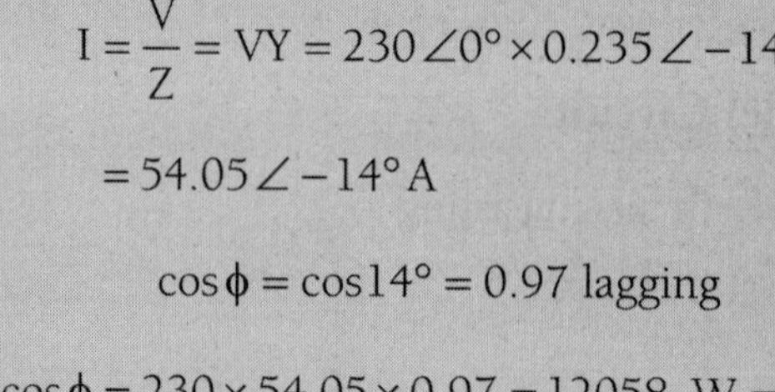

FIGURE 3.59

SOLUTION

$$Y_1 = \frac{1}{Z_1} = \frac{1}{12 + j12} = \frac{1}{16.96\angle 45^\circ} = 0.0589\angle -45^\circ$$

$$Y_2 = \frac{1}{Z_2} = \frac{1}{8 + j16} = \frac{1}{17.88\angle 64^\circ} = 0.0559\angle -64^\circ$$

$$I_1 = VY_1 = 230 \times 0.0589\angle -45^\circ = 13.54\angle -45^\circ \text{ A}$$

$$I_2 = VY_2 = 230 \times 0.0559\angle -64^\circ = 12.85\angle -64^\circ \text{ A}$$

$$I = I_1 + I_2 = 13.54\angle -45^\circ + 12.85\angle -64^\circ$$

or, $$I = 15 - j21 = 25.9\angle -54^\circ \text{ A}$$

Power factor = cos ϕ = cos 54° = 0.58 lagging

Power $= VI\cos\phi = 230 \times 25.9 \times 0.58$

$= 3455 \text{ W} = 3.455 \text{ kW}$

3.2.8 AC Series—Parallel Circuits

Consider the series–parallel circuit consisting of three branches A, B, and C as shown. In Fig. 3.60. How such circuit problems can be solved is shown below.

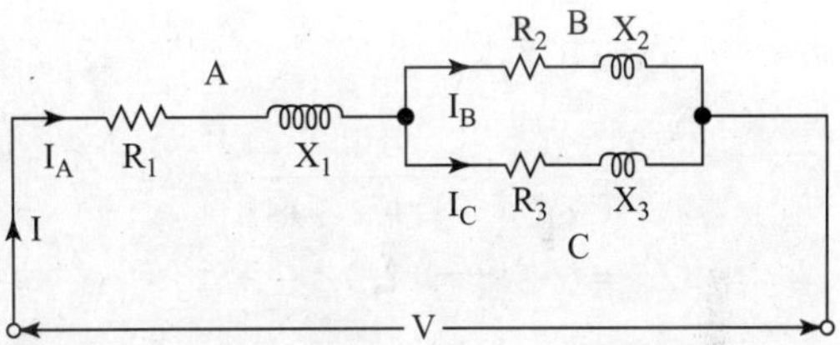

FIGURE 3.60

Impedance of branch A, $Z_A = R_1 + jX_1$

Impedance of branch B, $Z_B = R_2 + jX_2$

Impedance of branch C, $Z_C = R_3 + jX_3$

Total impedance of the circuit Z is

$$Z = Z_A + \frac{Z_B Z_C}{Z_B + Z_C}$$

Total current, $$I = \frac{V}{Z} = I_A$$

Current, $$I_B = I\frac{Z_C}{Z_B + Z_C} \quad \text{(applying current divider rule)}$$

Current, $$I_C = \frac{Z_B}{Z_B + Z_C}$$

By applying the admittance method, we can also solve the problem as

$$Y_A = \frac{1}{Z_A}; Y_B = \frac{1}{Z_B}; Y_C = \frac{1}{Z_C}$$

Total admittance of the parallel branches B and C

$$Y_{BC} = Y_B + Y_C$$

Impedance, $$Z_{BC} = \frac{1}{Y_{BC}}$$

Total impedance, $$Z = Z_A + Z_{BC}$$

Total current, $$I = \frac{V}{Z} = \frac{V}{Z_A + Z_{BC}}$$

EXAMPLE 3.31

Determine the total current drawn from the supply by the series–parallel circuit shown in Fig. 3.61. Also calculate the power factor of the circuit.

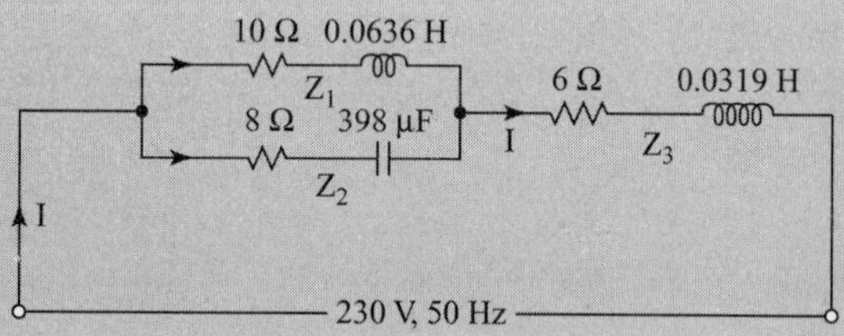

FIGURE 3.61

$$\omega = 2\pi f = 2\times 3.14\times 50 = 314$$

$$Z_1 = 10 + j314\times 0.0638 = 10 + j20$$

$$= 22.36\angle 64^\circ$$

$$Z_2 = 8 - jX_C$$

$$X_C = \frac{1}{\omega C} = \frac{10^6}{314\times 398} = 8\ \Omega$$

$$Z_2 = 8 - j8 = 11.3\angle -45^\circ$$

$$Z_3 = 6 + j314\times 0.0319 = 6 + j10 = 11.66\angle 59^\circ$$

Equivalent impedance, $$Z = \frac{Z_1 Z_2}{Z_1 + Z_2} + Z_3$$

$$= \frac{22.36\angle 64^\circ \times 11.3\angle -45^\circ}{10 + j20 + 8 - j8} + 11.66\angle 59^\circ$$

$$= \frac{252.7\angle 19^\circ}{18 + j12} + 11.66\angle 59^\circ$$

$$= \frac{252.7\angle 19^\circ}{21.63\angle 34^\circ} + 11.66\angle 59^\circ$$

$$= 11.68\angle -15^\circ + 11.66\angle 59^\circ$$

$$= 11.21 - j3 + 6 + j10$$

$$= 17.21 + j7$$

$$= 18.58\angle 22^\circ\Omega$$

Current, $I = \frac{V}{Z} = \frac{230\angle 0°}{18.58\angle 22°}$

$= 12.37\angle -22° A$

Total current, $I = 12.37$ A

Power factor, $\cos\phi = \cos 22° = 0.92$ lagging

EXAMPLE 3.32

What should be the value of R for which a current of 25 A will flow through it in the circuit shown in Fig. 3.62.

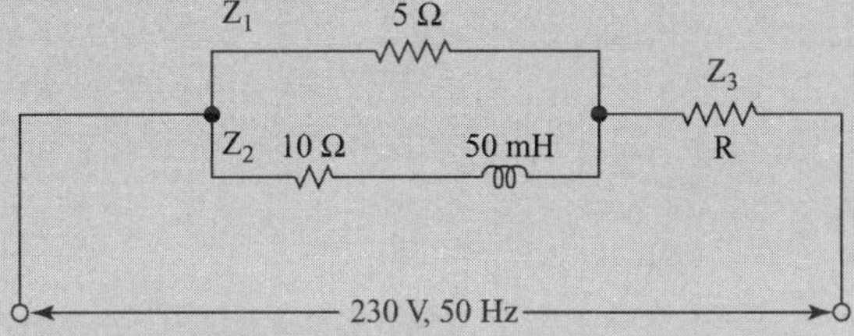

FIGURE 3.62

SOLUTION

$$Z_1 = 5\ \Omega, \omega = 2\pi f = 2\pi \times 50 = 314$$

$$Z_2 = 10 + j314 \times 50 \times 10^{-3}$$

$$= 10 + j15.7$$

$$= 18.6\angle 57.5°\ \Omega$$

$$Z_3 = R\ \Omega.$$

Total impedance, $Z = \frac{Z_1 Z_2}{Z_1 + Z_2} + Z_3$

$$= \frac{5 \times 18.6\angle 57.5°}{5 + 10 + j51.7} + R$$

$$= \frac{93.0\angle 57.5°}{15 + j15.7} + R$$

$$= \frac{93.0\angle 57.5°}{21.7\angle 46.5°} + R$$

or, $Z = 4.28\angle 11° + R$

Again, $Z = \frac{V}{I} = \frac{230}{25} = 9.2$

Equating the above two expressions for Z,

$$4.28\angle 11^\circ + R = 9.2$$

or,

$$R = 9.2 - 4.28\angle 11^\circ$$
$$= 9.2 - 4.28(\cos 11^\circ + j\sin 11^\circ)$$
$$= 9.2 - 4.19 + j\,0.8$$
$$= 5.01 + j\,0.8$$

Considering the real parts, $R = 5.01\ \Omega$

EXAMPLE 3.33

In the series–parallel circuit shown in Fig. 3.63, the parallel branches A and B are in series with branch C. The impedances are $Z_A = (4 + j3)\ \Omega$, $Z_B = (10 - j7)\ \Omega$ and $Z_C = (6 + j5)\ \Omega$. If the voltage applied to the circuit is 200 V at 50 Hz, calculate (a) current I_A, I_B, and I_C; (b) the power factor for the whole circuit. Draw also the phasor diagram.

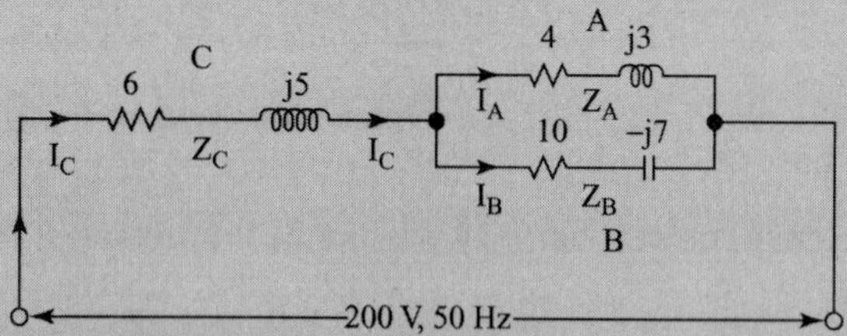

FIGURE 3.63

SOLUTION

$$Z_A = (4 + j3) = 5\angle 36.9^\circ\ \Omega$$
$$Z_B = (10 - j7) = 12.2\angle -35^\circ\ \Omega$$
$$Z_C = (6 + j5) = 7.8\angle 39.8^\circ\ \Omega$$
$$Z_A + Z_B = 4 + j3 + 10 - j7$$
$$= 14 - j4$$
$$= 14.56\angle -16^\circ$$

$$Z_{AB} = \frac{Z_A Z_B}{Z_A + Z_B}$$
$$= \frac{5\angle 36.9^\circ \times 12.2\angle -35^\circ}{4 + j3 + 10 - j7}$$
$$= \frac{5\angle 36.9^\circ \times 12.2\angle -35^\circ}{14 - j4} = \frac{61\angle 1.9^\circ}{14.56\angle -16^\circ}$$
$$= 4.19\angle 17.9^\circ$$
$$= 4.19(\cos 17.9^\circ + j\sin 17.9^\circ)$$

$$= 4 + j1.3$$

$$Z = Z_C + Z_{AB} = 6 + j5 + 4 + j1.3$$

$$= 10 + j6.3 = 11.8\angle 32.2°$$

Let

$$V = 200\angle 0°$$

$$I_C = \frac{V}{Z} = \frac{200\angle 0°}{11.8\angle 32.2°} = 16.35\angle -32.2°\text{A}$$

Using current divider rule,

$$I_A = I_C \frac{Z_B}{Z_A + Z_B} = 16.35\angle -32.2° \times \frac{12.2\angle -35°}{14.56\angle -16°} = 13.7\angle -51.2\text{ A}$$

$$I_B = I_C \frac{Z_A}{Z_A + Z_B} = 16.35\angle -32.2° \times \frac{5\angle 36.9°}{14.56\angle -16°} = 5.7\angle 20.7°\text{A}$$

Phase angle between applied voltage V and line current I is –32.2°

Hence, power factor of the whole circuit = cos ϕ = cos 32.2°
= 0.846 lagging

Voltage drop across series branch C, $V_C = I_C Z_C$

$$= 16.3\angle -32.2° \times 7.8\angle 39.8°$$

$$= 127.53\angle 7.6°\text{V}$$

Voltage drop across parallel branches, $V_A = V_B = I_C \cdot Z_{AB}$

$$= 16.35\angle -32.2° \times 4.19\angle 17.9$$

$$= 68.5\angle -14.3°\text{V}$$

Note that voltage across the parallel branches is also equal to $I_A Z_A$ or $I_B Z_B$.
The complete phasor diagram is shown in Fig. 3.64.

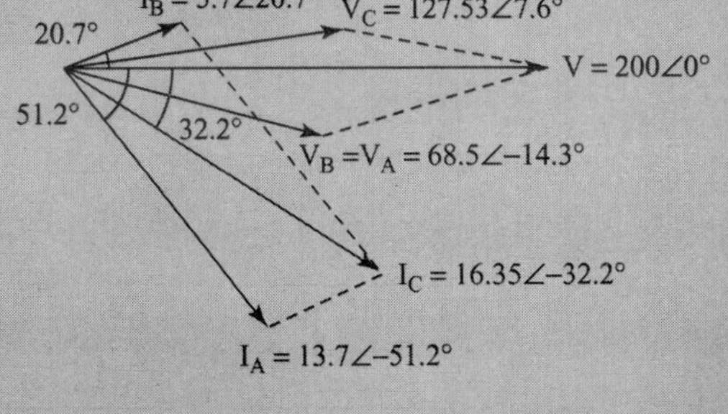

FIGURE 3.64

Phasor diagram representing voltages and currents in the circuit of Fig. 3.56

EXAMPLE 3.34

In the circuit shown in Fig. 3.65, determine the voltage at 50 Hz to be applied across terminals AB in order that a total current of 10 A flows through the circuit.

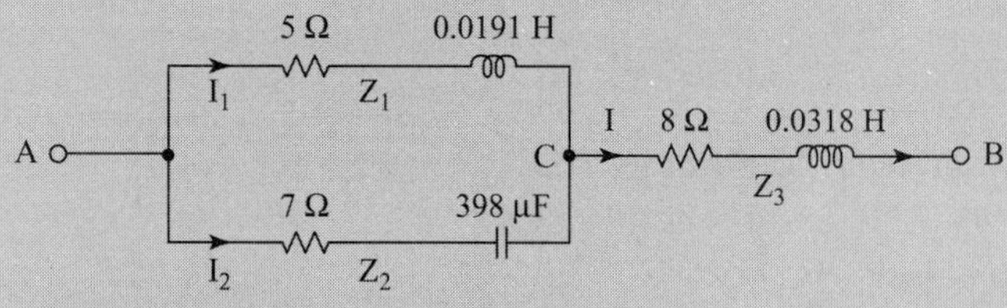

FIGURE 3.65

SOLUTION

$$Z_1 = 5 + j2\pi \times 50 \times 0.0191 = 5 + j6 = 7.81\angle 50.2°\ \Omega$$

$$Z_2 = 7 - j\frac{1}{2\pi \times 50 \times 398 \times 10^{-6}} = (7 - j8)\Omega = 10.63\angle -48.8°\ \Omega$$

$$Z_3 = 8 + j2\pi \times 50 \times 0.0318 = (8 + j10)\Omega = 12.8\angle 51.34°\ \Omega$$

Total impedance of the circuit, $Z = \dfrac{Z_1 Z_2}{Z_1 + Z_2} + Z_3.$

Substituting values,

$$Z = \frac{7.81\angle 50.2° \times 10.63\angle -48.8°}{5 + j6 + 7 - j8} + 8 + j10$$

$$= \frac{83.82\angle 3.4°}{12 - j2} + 8 + j10$$

$$= \frac{83.82\angle 3.4°}{12.16\angle -9.5°} + 8 + j10$$

$$= 6.81\angle 12.9° + 8 + j10$$

$$= 6.81 \times \cos 12.9° + j\,6.81\ \text{Sin}\ 12.9° + 8 + j10$$

$$= 6.81 \times 0.9748 + j\,6.81 \times 0.2232 + 8 + j10$$

$$= 6.64 + j\,1.52 + 8 + j10$$

$$= 14.64 + j11.52 = \sqrt{(14.64)^2 + (11.52)^2}\angle \tan^{-1}\frac{11.52}{14.64}$$

$$= 18.65\ \angle 38.2°$$

$$V_{AB} = I \times Z = 10\ \angle 0° \times 18.65\ \angle 38.2°$$

or, $$V_{AB} = 186.5\ \angle 38.2°\ \text{Volts.}$$

V_{AB} is leading I by 38.2°. Alternately, we can say that I is lagging V_{AB} by 38.2°.

Power factor, $\cos\phi = \cos 38.2° = 0.78$ lagging.

EXAMPLE 3.35

In a series–parallel circuit shown in Fig. 3.66, the parallel branches A, B, and C are in series with branch D. Calculate:

(i) the impedance of the overall circuit,

(ii) current taken by the circuit, and

(iii) power consumed by each branch and the total power consumed.

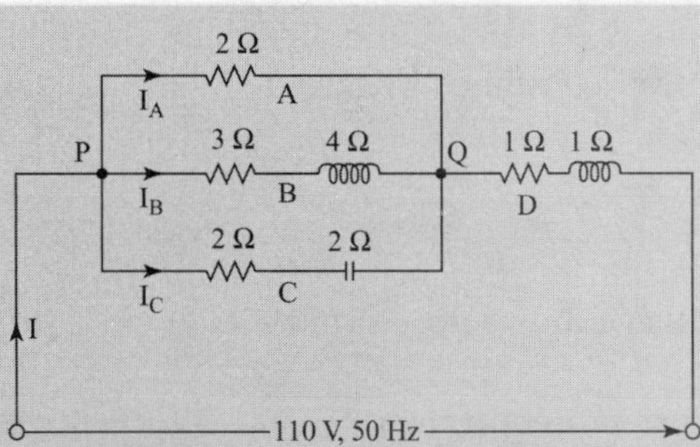

FIGURE 3.66

SOLUTION

(i) Impedance of branch A,

$$Z_A = 2 + j0 = 2\ \Omega$$

Impedance of branch B,

$$Z_B = 3 + j4$$

$$Z_{AB} = \frac{Z_A Z_B}{Z_A + Z_B} = \frac{2(3+j4)}{2+3+j4} = \frac{6+j8}{5+j4}$$

$$= \frac{6+j8}{5+j4} \times \frac{5-j4}{5-j4} = \frac{62-j16}{5^2+4^2} = 1.51 + j0.39$$

Impedance of branch C, $Z_C = (2 - j2)\ \Omega$

Equivalent impedance of the parallel circuit,

$$Z_P = \frac{Z_{AB} Z_C}{Z_{AB} + Z_C}$$

$$= \frac{(1.51+j0.39)(2-j2)}{(1.51+j0.39)+(2-j2)}$$

$$= 1.136 - j0.118 = 1.142\ \Omega$$

Impedance of branch D, $Z_D = 1 + j1$

Total impedance of the overall circuit $Z, = Z_P + Z_D = 1.136 - j\,0.118 + 1 + j\,1 = 2.136 + j\,0.882$

$$= 2.311\angle 22.46°\ \Omega$$

(ii) $V = 110\text{ V}$

$Z = 2.311\,\Omega$

$$I = \frac{110}{2.311} = 47.6\text{ A}$$

(iii) $I_D = 47.6\text{ A}$

$R_D = 1\,\Omega$

Power consumed by branch $D = I_D^2 R_D$

$$= (47.6)^2 \times 1$$

$$= 2265.8\text{ W}$$

Voltage drop across terminals $PQ = IZ_P$

Current in branch A, $I_A = \dfrac{IZ_P}{Z_A} = \dfrac{47.6 \times 1.142}{2} = 27.18\text{ A}$

$R_A = 2\,\Omega$

Power consumed by branch A, $= I_A^2 R_A = (27.18)^2 \times 2 = 1477.5\text{ W}$

Current in branch B, $I_B = \dfrac{IZ_P}{Z_B} = \dfrac{47.6 \times 1.142}{5} = 10.87\text{ A}$

$R_B = 3\,\Omega$

Power consumed by branch B, $= I_B^2 R_B = (10.87)^2 \times 3 = 354.5\text{ W}$

Current in branch C, $I_C = \dfrac{IZ_P}{Z_C} = \dfrac{47.6 \times 1.142}{2.83} = 19.21\text{ A}$

$R_C = 2\,\Omega$

Power consumed by branch C, $= I_C^2 R_C = (19.21)^2 \times 2 = 738\text{ W}$

Total power consumed by circuit $= 1477.5 + 354.5 + 738 + 2265.8$

$$= 4835.8\text{ W}$$

3.3 ► Resonance in AC Circuits

In ac circuits resonance occurs when two independent energy storing devices are capable of interchanging energy from one another. For example, inductance and capacitance are the two energy-storing devices or elements of an ac circuit, which may create a condition of resonance.

Resonance occurs in other systems also, like in a mechanical system, where mass and spring are the two energy-storing elements and they may create a condition of resonance. Mass stores energy when in motion $\left(=\frac{1}{2}mv^2\right)$ and a spring stores energy when it is elongated or compressed.

An electric circuit generally consists of circuit elements like resistance, inductance, and capacitance. The voltage and frequency are generally constant at the supply terminal. However, in electronic communication circuits, the supply voltage may have variable frequency. When frequency is variable, the inductive and capacitive reactance of the circuit elements will change $\left(X_L = 2\pi fL; X_C = \frac{1}{2\pi fC}\right)$ The current in the circuit will depend upon the values of X_L and X_C and that of R. A condition may occur at a particular frequency that the impedance offered to the flow of current is maximum or minimum. The circuit elements, namely the inductance element and the capacitance element are often connected in series or in parallel. It will be interesting and also useful to study the effect of varying input frequency on the circuit condition when these elements are connected in series or in parallel.

3.3.1 Resonance in RLC Series Circuit

Let us consider a series circuit consisting of a resistor, an inductor, and a capacitor as shown in Fig. 3.67 (a). The supply voltage is constant but its frequency is variable.

The impedance of the circuit, $Z = R + jX_L - jX_C = R + j\omega L - j\frac{1}{\omega C}$

The current flowing through the circuit is

$$I = \frac{V}{R + j\left(\omega L - \frac{1}{\omega C}\right)} = \frac{V}{R + j(X_L - X_C)}$$

$$= \frac{V}{R + j\left(2\pi fL - \frac{1}{2\pi fC}\right)}$$

As the frequency is changing, both X_L and X_C will change. Inductive reactance X_L will increase as the frequency, f is increasing while the capacitive reactance, X_C will decrease with increasing frequency. The value of R is independent of frequency. The variation of R, X_L, and X_C with variation of frequency, f has been shown in Fig. 3.67 (b). It may be noted that inductive reactance is jX_L and capacitive reactance is $-jX_c$, i.e., vectorially they should be shown in opposite directions. However, in Fig. 3.67 (b) we have shown their magnitudes only. At a frequency f_0, it is seen that the magnitude of X_L is equal to X_C as the two curves cut at point P. Since X_L and X_C are vectorially jX_L and $-jX_c$, the two reactances will cancel each other when frequency is f_0. At f_0 the impedance of the series R–L–C circuit is equal to R which is the minimum value of Z. The total impedance graph of Z shows that at $f = f_0$, $Z = R$, i.e., at f_0 the circuit offers minimum impedance, and hence, maximum current will flow through the circuit.

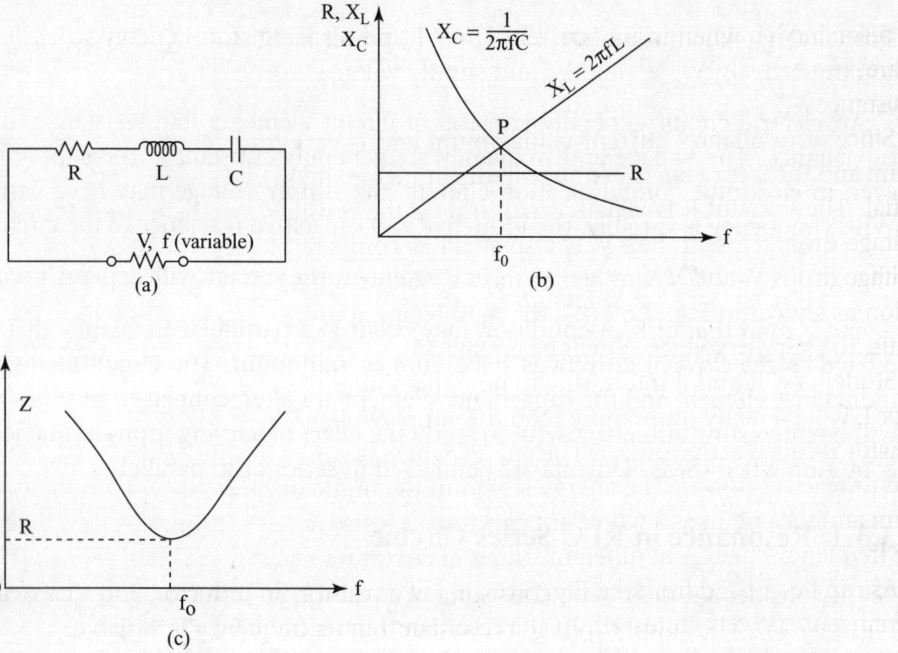

FIGURE 3.67

Resonance in R–L–C series circuit: (a) circuit diagram; (b) variation of R, X_L, X_C with frequency; (c) variation of impedance

At minimum value of Z, the current in the circuit will be maximum as I = V/R. This condition of the circuit when X_L equals X_C, Z = R, current is maximum and is called the resonant condition and the frequency, f_0 at which resonance occurs is called the resonant frequency. At resonance, since X_L equals X_C, we can write

$$2\pi f_0 L = \frac{1}{2\pi f_0 C}$$

or, Resonant frequency,

$$f_0 = \frac{1}{2\pi\sqrt{LC}}. \tag{3.15}$$

Alternately

$$\omega_0 L = \frac{1}{\omega_0 C}$$

or,

$$\omega_0^2 = \frac{1}{LC}$$

or,

$$\omega_0 = \frac{1}{\sqrt{LC}}$$

At resonance, frequency is f_0, current $I_0 = V/R$, power factor is unity, voltage drops across R, L, C are, respectively, V_R, V_L, and V_C and supply voltage V is equal to the voltage drop across the resistance V_R.

Since at resonance, current is maximum and is very high, power dissipation I_0^2R is maximum and the rate of energy storage in the inductor and the capacitor is maximum and they are equal. The value of R is usually small (this is the resistance of the inductive coil), and hence, voltage drop across it, i.e., V_R is also small as compared to the voltage drops across L and C. Voltage drops V_L and V_C are higher than V_R. However, as $V_L = V_C$ and they are in phase opposition as shown in Fig. 3.69 (b), the net voltage across L and C in series, V_X is equal to zero. Thus, the supply voltage, V will be equal to V_R.

Students will find it interesting to note that under the resonant condition the voltage across C or L will be many times more than the supply voltage. The power which is dissipated in the resistor is called active power. The energy which is stored in the inductor and the capacitor are due to reactive power. The energy stored in the inductor and the capacitor oscillates between them and the circuit as a whole appears to be a resistive only. The variation of circuit current as the frequency changes at different values of circuit resistance have been shown in Fig. 3.68 (a).

As can be noticed from the Fig. 3.68, at lower values of R, i.e., when $R = R_1$ the sharpness of the current curve is increased. At the resonant frequency when current is at its maximum, the reactive power which oscillates between the inductor and the capacitor is much higher than the resistive power.

Quality factor

The ratio of the reactive power to the resistive power is called the quality factor. It can also be defined as the ratio of energy stored in the circuit to the energy lost or consumed in the circuit. Quality factor is also defined as the ratio of voltage drop appearing across the inductor or the capacitor to the supply voltage.

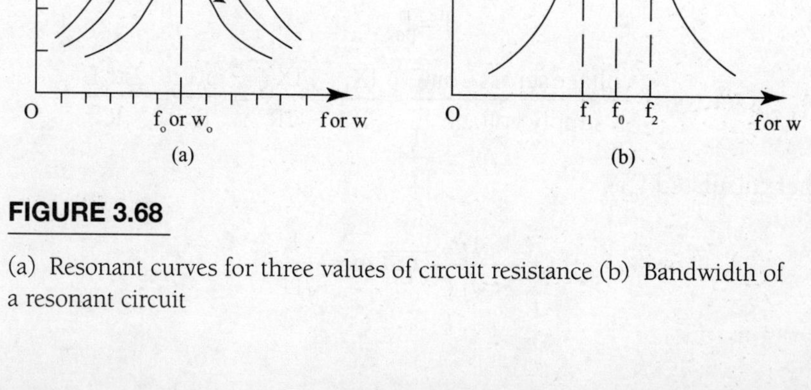

FIGURE 3.68

(a) Resonant curves for three values of circuit resistance (b) Bandwidth of a resonant circuit

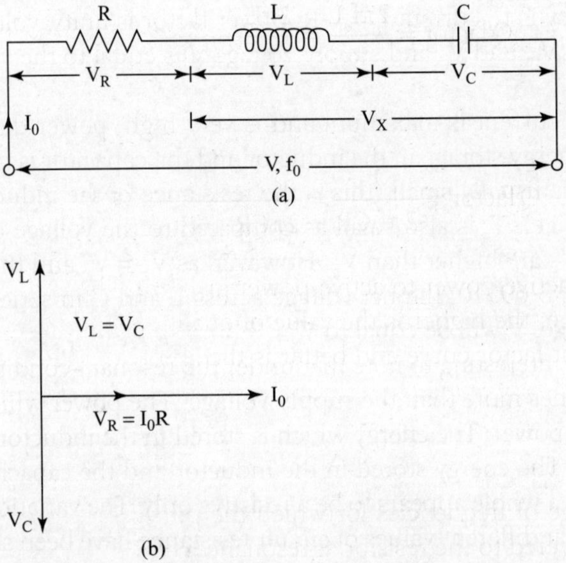

FIGURE 3.69

Resonance condition in R–L–C series circuit: (a) circuit diagram; (b) phasor diagram

The Q–factor of a resonant circuit is a measure of 'goodness' or quality of a resonant circuit. A higher value for this figure of merit corresponds to a more narrow bandwidth which is desirable in many applications. Resonant circuits are used to respond selectively to signals of a given frequency while discriminating against signals of different frequencies. If the response of the circuit is more narrowly peaked around the chosen frequency, we say that the circuit has higher "selectivity". A quality factor, Q is a measure of selectivity. A circuit having a high Q is more narrowly selective. For a series resonant circuit, lower the value of resistance R, higher will be value of Q for given values of L and C as shown in Fig. 3.69 (a).

We now calculate the value of Q factor as:

$$\text{Q factor} = \frac{\text{reactive power}}{\text{resistive power}} = \frac{I^2X_L}{I^2R} = \frac{I^2X_C}{I^2R} = \frac{\omega_0 L}{R} = \frac{1}{\omega_0 CR}$$

when $\qquad f = f_0$

or, $$\text{Q factor} = \frac{\text{voltage across L or C}}{\text{supply voltage}} = \frac{IX_L}{IR} = \frac{IX_C}{IR} = \frac{\omega_0 L}{R} = \frac{2\pi f_0 L}{R}$$

We had earlier calculated f_0 as

$$f_0 = \frac{1}{2\pi\sqrt{LC}} \text{ and } \omega = \frac{1}{\sqrt{LC}}$$

Thus,
$$Q\,\text{factor} = \frac{2\pi f_0 L}{R} = \frac{2\pi L}{R}\frac{1}{2\pi\sqrt{LC}} = \frac{1}{R}\sqrt{\frac{L}{C}}$$

In terms of X_C

$$Q\,\text{factor} = \frac{1}{2\pi f_0 CR} = \frac{1}{2\pi CR \cdot 1/2\pi\sqrt{LC}} = \frac{1}{R}\sqrt{\frac{L}{C}} \tag{3.16}$$

Higher the ratio of reactive power to active power or higher the ratio of voltage across L and C to the supply voltage, the higher is the value of quality factor. Higher quality factor means sharper is the resonant factor curve and better is the ability of the network to accept current or power signals.

Bandwidth

Bandwidth is the range of frequencies for which the power delivered to the resistor is equal to half the power delivered to the resistor at resonance. As can be seen from Fig. 3.69 (b), the range of frequencies is $(f_2 - f_1)$ and the corresponding current is $\frac{1}{\sqrt{2}}I_0$, i.e., 0.707 I_0

Power delivered at resonance = $I_0^2 R$

Half the power delivered at resonance $= \frac{1}{2}I_0^2 R$

$$\frac{1}{2}I_0^2 R = I_{BW}{}^2 R$$

or,
$$I_{BW} = I_0/\sqrt{2} = 0.707 I_0 \tag{3.17}$$

Therefore, the range of frequencies within which current does not drop below 0.707 times the maximum value of current, i.e., I_0 is called the bandwidth. See Fig. 3.68 (b). The frequencies f_1 and f_2 are often called the lower and upper cut-off frequencies. Bandwidth, BW = $f_2 - f_1$.

From Fig. 3.68 (b) it is seen that at both f_1 and f_2, the power delivered to R is equal to half the power delivered to R at resonance.

Using equation (3.17), we can write

$$I_{BW} = 0.707\, I_0 = 0.707\frac{V}{R}$$

or,
$$\frac{V}{\sqrt{R^2 + (X_L - X_C)^2}} = \frac{V}{\sqrt{2}R}$$

or,
$$R^2 + (X_L - X_C)^2 = 2R^2$$

or,
$$(X_L - X_C)^2 = R^2$$

or,
$$\omega L - \frac{1}{\omega C} = \pm R$$

$$\frac{\omega^2 LC - 1}{\omega C} = \pm R$$

or,
$$\omega^2 \pm \frac{R}{L}\omega - \frac{1}{LC} = 0$$

From the above, the two values of frequencies, i.e., ω_1 and ω_2 are

$$\omega_2 = \sqrt{\frac{R^2}{4L^2} + \frac{1}{LC}} - \frac{R}{2L} \quad \text{and} \quad \omega_2 = \sqrt{\frac{R^2}{4L^2} + \frac{1}{LC}} - \frac{R}{2L}$$

Normally, $\frac{R^2}{4L^2}$ is very small as compared to $\frac{1}{LC}$ in a resonant circuit.

Therefore,

$$\omega_1 = \frac{1}{\sqrt{LC}} - \frac{R}{2L} = \omega_0 - \frac{R}{2L}$$

$$\omega_2 = \frac{1}{\sqrt{LC} + 2L} \frac{R}{} = \omega_0 + \frac{R}{2L}$$

$$2\pi f_1 = 2\pi f_0 - \frac{R}{2L}$$

or,
$$f_1 = f_0 - \frac{R}{4\pi L} \tag{3.18 a}$$

Similarly,
$$f_2 = f_0 + \frac{R}{4\pi L} \tag{3.18 b}$$

$$\omega_2 - \omega_1 = \omega_{BW}$$

$$f_2 - f_1 = \frac{R}{2\pi L} = f_{BW} \tag{3.18 c}$$

f_2 and f_1 are the higher and lower bandwidth frequencies. Let us now calculate the value of $f_0 / (f_2 - f_1)$

$$\frac{f_0}{f_2 - f_1} = \frac{f_0}{R/2\pi L} = 2\pi f_0 \frac{L}{R} = \frac{\omega_0 L}{R}$$

Quality factor Q as we have seen earlier is

$$Q = \frac{\text{voltage across L}}{\text{supply voltage}} = \frac{I_0 V_L}{I_0 R}$$

$$= \frac{I_0 \omega_0 L}{I_0 R} = \frac{\omega_0 L}{R}$$

Thus,

$$Q = \frac{f_0}{f_2 - f_1} = \frac{f_0}{BW} \tag{3.19}$$

or,

$$BW = \frac{f_0}{Q}$$

From eq. (3.19), it is seen that if the quality factor is high, bandwidth will be narrow. The circuit will, therefore, allow only a narrow band of signal frequencies which are close to the resonant frequency. Such a circuit is, therefore, highly selective in allowing signals to pass through. High-quality factor resonant circuits are also called '*tuned circuits*', which will be studied in detail in electronic circuit design.

We have seen earlier that power dissipated at cutoff frequencies is half the power dissipated at resonant frequency. Hence, the cutoff frequencies f_1 and f_2 are also called half-power frequencies.

EXAMPLE 3.36

An R–L–C series circuit has R = 10 Ω, L = 0.1 H, and C = 8 μF.

Calculate the following:

(a) resonant frequency;
(b) Q-factor of the circuit at resonance;
(c) half-power frequencies and bandwidth.

SOLUTION

(a) We know that the condition for the series resonance is

$$X_L = X_C$$

i.e.,

$$2\pi f_0 L = \frac{1}{2\pi f_0 C}$$

or,

$$f_0 = \frac{1}{2\pi\sqrt{LC}}$$

Substituting values

$$f_0 = \frac{1}{2\pi\sqrt{0.1 \times 8 \times 10^{-6}}} = 178\,\text{Hz}$$

(b) $\text{Q-factor} = \frac{\text{voltage across L}}{\text{supply voltage}} = \frac{I_0 X_L}{I_0 R} = \frac{2\pi f_0 L}{R} = \frac{2 \times 3.14 \times 178 \times 0.1}{10}$

$$= 11.2$$

Alternately Q factor $= \frac{1}{R}\sqrt{\frac{L}{C}}$

Substituting values

$$\text{Q-factor} = \frac{1}{10}\sqrt{\frac{.01}{8 \times 10^{-6}}} = 11.2$$

(c) Half-power frequencies are the frequencies corresponding to current which is 0.707 of the resonant current. They are f_1 and f_2, i.e., the lower, and upper frequencies forming the bandwidth.

From equation (3.19 a), $f_1 = f_0 - \frac{R}{4\pi L}$

and from equation(3.19 b), $f_2 = f_0 + \frac{R}{4\pi L}$

Substituting $f_1 = 178 - \frac{10}{12.56 \times 0.1} = 169 \text{ Hz}$

$$f_2 = 178 + \frac{10}{12.56 \times 0.1} = 187 \text{ Hz}$$

Bandwidth, $BW = f_2 - f_1 = 18 \text{ Hz}$

EXAMPLE 3.37

A circuit of R = 4 Ω, L = 0.5 H, and a variable capacitance C in series is connected across a 100 V, 50 Hz supply. Calculate:

(a) the value of capacitance for which resonance will occur;
(b) the voltage across the capacitor at resonance and the Q-factor of the circuit

SOLUTION

(a) Applying condition for series resonance,

$$X_C = X_L$$

or, $\frac{1}{2\pi f_0 C} = 2\pi f_0 L = 6.2 \times 50 \times 0.5 = 157\ \Omega$

or, $$C = \frac{1}{6.28 \times 50 \times 157} F = \frac{10^6}{314 \times 157} \mu F = 20.3 \mu F$$

(b) Resonant Current, $I_0 = \frac{V}{R} = \frac{100}{4} = 25 \text{ A}$

Voltage across the capacitor,

$$V_C = I_0 X_C = \frac{I_0}{2\pi f_0 C}$$

Substituting values $$V_C = \frac{25}{2 \times 3.14 \times 50 \times 20.3 \times 10^{-6}}$$

$$= \frac{25 \times 10^3}{3.14 \times 2.03}$$

$$= 3925 \text{ V}$$

The students should note that a very high voltage of 3925 V is appearing across the capacitor while the supply voltage is only 100 V.

$$\text{The Q-factor of the circuit} = \frac{V_C}{V} = \frac{3925}{100}$$

$$= 39.25$$

Thus, the voltage multiplication across the capacitor as also across the inductor is 39.25 times at resonance.

EXAMPLE 3.38

A resistor, a variable iron-core inductor, and a capacitor are connected across a 230 V, 50 Hz supply. By varying the position of the iron core inside the inductor coil, its inductance is changed. Maximum current of 1.5 A was obtained in the circuit by changing the inductance of the coil. At that time the voltage across the capacitor was measured as 600 V. Calculate the values of circuit parameters.

SOLUTION

We know that the maximum current flows at resonance when $X_L = X_C$ and $Z = R$

Maximum current, $$I_0 = I_m = \frac{V}{Z} = \frac{V}{R} [\because Z = R]$$

Therefore, $$R = \frac{V}{I_m} = \frac{230}{1.5} = 153.3\ \Omega$$

At resonance,
$$V_L = V_C = 600\,V$$
$$V_L = I_m X_L = 600\,V$$
$$X_L = \frac{600}{1.5} = 400\ \Omega$$
or,
$$2\pi f L = 400$$
$$L = \frac{400}{6.28 \times 50} = 1.27\,H$$
$$X_C = X_L = 400$$
or,
$$\frac{1}{2\pi f C} = 400$$
or,
$$C = \frac{1}{314 \times 400} = 7.96 \times 10^{-6}\,F$$

EXAMPLE 3.39

An inductor, a variable capacitor, and a resistor are connected in series across a constant voltage, 100 Hz power supply. When the capacitor value is fixed at 100 μF, the current reaches its maximum value. Current gets reduced to half its maximum value when the capacitor value is 200 μF. Calculate the values of circuit parameters and the Q-factor of the circuit.

SOLUTION

Let resonant frequency be f_0.

At resonance,
$$X_L = X_C$$
$$2\pi f_0 L = \frac{1}{2\pi f_0 C}$$
or,
$$LC = \frac{1}{(2\pi f_0)^2}$$
or,
$$L = \frac{1}{(2\pi f_0)^2 C}$$

Substituting values,
$$L = \frac{1}{(6.28 \times 100)^2 \times 100 \times 10^{-6}}$$
$$= 25.3 \times 10^{-3}\,H$$

Since $X_L = X_C$, the value of impedance is equation to R

Maximum value of current, $$I_m = I_0 = \frac{V}{R} A$$

At a frequency of 100 Hz, $C = 200 \times 10^{-6}$ F, current is reduced to half,

Impedance, $$Z = \sqrt{R^2 + (X_L - X_C)^2}$$

Current, $$I = \frac{V}{Z} = \frac{V}{\sqrt{R^2 + (X_L - X_C)^2}}$$

According to the problem

$$I = \frac{I_m}{2}$$

or, $$\frac{V}{\sqrt{R^2 + (X_L - X_C)^2}} = \frac{I_m}{2} = \frac{V}{2R}$$

or, $$\sqrt{R^2 + (X_L - X_C)^2} = 2R$$

or, $$(X_L - X_C)^2 = 3R^2$$

or, $$\sqrt{3}R = X_L - X_C$$

$$X_L = 2\pi f_0 L = 628 \times 25.3 \times 10^{-3} = 15.88\ \Omega$$

$$X_C = \frac{1}{2\pi f_0 C} = \frac{1}{6.28 \times 200 \times 10^{-6}} = 7.96\ \Omega$$

$$R = \frac{X_L - X_C}{\sqrt{3}} = \frac{15.88 - 7.96}{1.732} = 4.57\ \Omega$$

EXAMPLE 3.40

An inductive coil of resistance 10 Ω and inductance 20 mH are connected in series with a capacitor of 10 μF. Calculate the frequency at which the circuit will resonate. If a voltage of 50 V at resonant frequency was applied across the circuit, calculate the voltage across the circuit components and the Q-factor.

SOLUTION

$$R = 10\ \Omega,\ L = 20 \times 10^{-3}\ H,\ C = 10 \times 10^{-6} F$$

at resonance, $$X_L = X_C$$

from which we get resonance frequency

$$f_0 = \frac{1}{2\pi\sqrt{LC}} = \frac{1}{6.28\sqrt{20\times10^{-3}\times10\times10^{-6}}}$$
$$= 356 \text{ Hz}$$

At resonance, impedance of the circuit is equal to R. Therefore, the maximum current that will flow is equal to

$$I_0 = \frac{V}{R} = \frac{50}{10} = 5 \text{ A}$$

To calculate the voltage drop across the circuit components, we calculate X_L and X_C at the resonance frequency first.

$$X_L = 2\pi f_0 L = 2\times3.14\times356\times20\times10^{-3}\ \Omega$$
$$= 44.7\ \text{V}$$

voltage drop across L, $\quad V_L = I_0 X_L = 5\times44.7 = 223.5\ \text{V}$

voltage drop across R, $\quad V_R = I_0 R = 5\times10 = 50\ \text{V}$

Note that the voltage drop across R is the same as the supply voltage, i.e., 50 V. Voltage drop across the capacitor should be the same as the voltage drop across inductive reactance X_L. Let us calculate V_C.

$$V_C = I_0 X_C = \frac{I_0}{2\pi f_0 c}$$

Substituting values

$$V_C = \frac{5}{6.28\times356\times10\times10^{-6}}$$
$$= \frac{5\times10^3}{6.28\times3.56} = 223.5\ \text{V}$$

Thus, $\quad V_L = V_C = 223.5\ \text{V}$

Q-factor $\quad = \frac{V_L}{V} = \frac{223.5}{50} = 4.47$

EXAMPLE 3.41

A coil of inductance 1 mH and resistance 50 Ω connected in series with a capacitor is fed from a constant voltage, variable frequency supply source. If the maximum current of 5 A flows at a frequency of 50 Hz, calculate the value of C and the applied voltage.

SOLUTION

Resonant frequency, $f_0 = \frac{1}{2\pi\sqrt{LC}}, f_0 = 50 \text{ Hz (given)}$

or, $$LC = \frac{1}{(2\pi f_0)^2}$$

or, $$C = \frac{1}{(2\pi f_0)^2 L} = \frac{1}{(2\times 3.14\times 50)^2 \times 1\times 10^{-3}}$$

$$= 0.0101 \text{ F}$$

At resonance, $X_L = X_C,\ Z = R$

Voltage drop across R = Supply voltage = $I_m R = I_0 R$

Thus, the applied voltage $= 5\times 50 = 250$ V

EXAMPLE 3.42

An inductive coil has a resistance of 2.5 Ω and an inductive reactance of 25 Ω. This coil is connected in series with a variable capacitance and a voltage of 200 V at 50 Hz is applied across the series circuit. Calculate the value of C at which the current in the circuit will be maximum. Also calculate the power factor, impedance, and current in the circuit under that condition.

SOLUTION

When current is maximum in an R, L, C series circuit, the circuit is under the resonance condition. At resonance, $X_L = X_C$ and $Z = R$.

Here

$$X_L = 25\ \Omega \text{(given)}, f_0 = 50 \text{ Hz}$$

$$X_C = \frac{1}{2\pi f_0 C} = X_L = 25\ \Omega$$

or, $$C = \frac{1}{2\pi f_0 \times 25} = \frac{1}{6.28\times 50\times 25} = 127.4\times 10^{-6} \text{F}$$

At resonance, $X_L = X_C$, $Z = R$. The circuit behaves like a resistive circuit. Therefore, the power factor = 1. Impedance, Z = R, and current is maximum.

$$I_m = \frac{V}{R} = \frac{200}{2.5} = 80 \text{ A}$$

3.3.2 Resonance in AC Parallel Circuits

Let us consider an inductive coil and a capacitor in parallel connected across a constant voltage and variable frequency supply source as shown in Fig. 3.70 (a). Practically, both the capacitor and the inductor will have some losses which should be represented by a small resistance in series. Here we assume the capacitor as a loss-less one while the inductor coil has some resistance which has been shown separately. The phasor diagram of voltage and current components have been shown in Fig. 3.70 (b). The line current I is equal to the in-phase component of I_L with the voltage V, i.e., $I = I_L \cos\phi_L$. At resonance, the current through the capacitor I_C is balanced by $I_L \sin\phi_L$ as shown. Thus, the reactive component of line current which is the phasor sum of I_C and $I_L \sin\phi_L$ is zero. In parallel resonance circuit, resonance is defined as the frequency at which current I_c is balanced by reactive component of inductive current I_L. The condition for resonance is

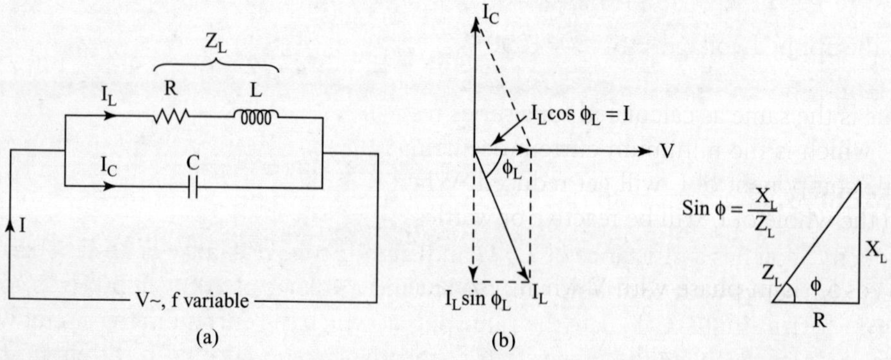

FIGURE 3.70

$$I_L \sin\phi_L = I_C$$

or,

$$\frac{V}{Z_L}\frac{X_L}{Z_L} = \frac{V}{X_C}$$

The condition of resonance is $X_L X_C = Z_L^2$

To calculate resonance frequency, f_0 we take, $X_L X_C = Z_L^2 = R^2 + X_L^2$

or,

$$2\pi f_0 L \frac{1}{2\pi f_0 C} = R^2 + X_L^2$$

or,

$$\frac{L}{C} = R^2 + (2\pi f_0 L)^2$$

or,

$$(2\pi f_0 L)^2 = \frac{L}{C} - R^2$$

or,
$$(2\pi f_0)^2 = \frac{1}{LC} - \frac{R^2}{L^2}$$

or,
$$2\pi f_0 = \sqrt{\frac{1}{LC} - \sqrt{\frac{R^2}{L^2}}}$$

or,
$$f_0 = \frac{1}{2\pi}\sqrt{\frac{1}{LC} - \frac{R^2}{L^2}} \tag{3.20}$$

This is the resonance frequency of a parallel L and C circuit.

If we consider the value of R as negligible, then the resonance frequency is

$$f_0 = \frac{1}{2\pi\sqrt{LC}}$$

This value is the same as calculated for a series resonance circuit. The line current I is equal to $I_L \cos\phi_L$ which is the minimum current occurring at resonance. If the value of R is reduced, the cosine component of I_L will get reduced. When R is made equal to zero, $I_L \cos\phi_L$ will be zero and the whole of I_L will be reactive or wattless component and will be equal and opposite to I_C.

$I = I_L \cos\phi_L$ is in phase with V when resonance occurs. The circuit impedance, Z_0 is calculated as

$$Z_0 = \frac{V}{I} = \frac{V}{I_L \cos\phi_L} = \frac{V}{V/Z_L \times R/Z_L} = \frac{Z_L^2}{R}$$

From the condition of resonance

$$X_L X_C = Z_L^2$$

or,
$$Z_L^2 = X_L X_C = 2\pi f L \times \frac{1}{2\pi f C} = \frac{L}{C}$$

Thus,
$$Z_0 = \frac{Z_L^2}{R} = \frac{L/C}{R}$$

or,
$$Z_0 = \frac{L}{CR} \tag{3.21}$$

Z_0 is known as the equivalent impedance or '*dynamic impedance*' of the parallel resonant circuit. It can be noticed that $I = I_L \cos\phi_L$ is in phase with the supply voltage. This shows that the circuit behaves like a resistive circuit only since the reactive component currents cancel each other. The impedance Z_0 is therefore resistive only. *Since current is minimum, impedance of the circuit, $Z_0 = L/CR$ is maximum under the resonant condition. Since current at resonance is minimum, a parallel resonant circuit is often referred to as a 'rejector circuit' meaning that a parallel resonant circuit tends to reject current at resonant frequency.*

It may be noted that the current drawn from the supply at resonance, i.e., $I = I_L \cos\phi_L$ is minimum. The current circulating through the capacitor and the inductor, i.e., I_C which is equal to $I_L \sin\phi_L$ is very high ($I_C >> I$ or $I_L \cos\phi_L$ Since I_C is many times more than I, we can say that parallel resonance is a case of current resonance. Here we recall that series resonance is a case of voltage resonance as voltage across the capacitor or the inductor is many times higher than the supply voltage. In parallel resonance current through the capacitor or the inductor is many times more than the line current. Current flowing within the circuit is much more than the current supplied. In series resonance there is voltage magnification while in parallel resonance there is current magnification.

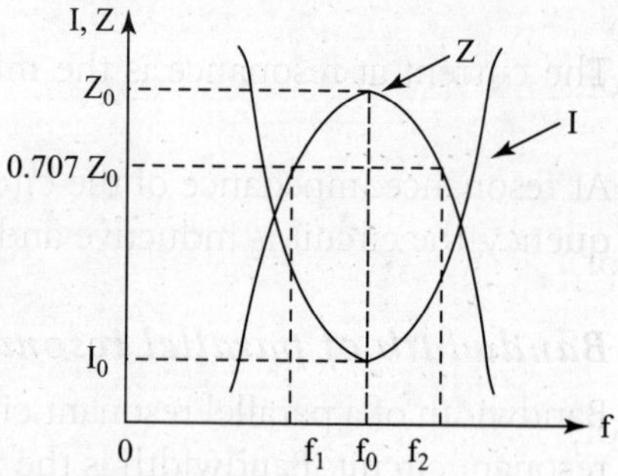

FIGURE 3.71
Variation of impedance and current in a parallel resonant circuit

Q-factor of parallel circuit

The ratio of circulating current between the two parallel branches, i.e., the capacitor and the inductor to the circuit line current is called the Q-factor or current magnification factor of the parallel circuit. Thus,

$$\text{Q-factor} = \frac{I_C}{I}$$

$$I_C = \frac{V}{X_C} = \omega CV$$

$$I = \frac{V}{Z_0} = \frac{V}{L/CR} = \frac{VCR}{L}$$

Now,

$$\text{Q-factor} = \frac{I_C}{I} = \frac{\omega CV}{VCR/L} = \frac{\omega L}{R} = \frac{2\pi f_0 L}{R} \tag{3.22}$$

Q - factor of a parallel resonant circuit, $\text{Q-factor} = \dfrac{2\pi f_o L}{R}$ (3.23)

and

$$f_0 = \frac{1}{2\pi\sqrt{LC}} \text{ if R is neglected.}$$

So,

$$\text{Q-factor} = \frac{2\pi L}{R} \times \frac{1}{2\pi\sqrt{LC}}$$

or,

$$\text{Q-factor} = \frac{1}{R}\sqrt{\frac{L}{C}}$$

The effect of variation of frequency on circuit impedance and current has been shown in Fig. 3.71. At resonant frequency f_0, the impedance is maximum, i.e., equal to $Z_0 = \dfrac{L}{CR}$.

The current at resonance is the minimum and is equal to I_0 where $I_0 = \frac{V}{Z_0} = \frac{V}{L/CR} = \frac{VCR}{L}$.

At resonance impedance of the circuit is maximum, current is minimum. Below resonant frequency, the circuit is inductive and above resonant frequency, the circuit is capacitive.

Bandwidth of parallel resonant circuit

Bandwidth of a parallel resonant circuit is determined the same way as in the case of the series resonant circuit. Bandwidth is the range of frequencies $(f_2 - f_1)$ where the power dissipated is half of the power dissipated at the resonant frequency.

The critical parameters of series and parallel resonant circuits have been compared and shown in Table 3.1

We have seen that resonance in ac series and parallel circuits can take place at a particular frequency when a constant voltage, variable frequency supply source is applied across the circuit. The frequency at which resonance occurs are

$$f_0 = \frac{1}{2\pi\sqrt{LC}} \text{ for series circuit}$$

and

$$f_0 = \frac{1}{2\pi}\sqrt{\frac{1}{LC} - \frac{R^2}{L^2}} \text{ for parallel circuit}$$

If we neglect the small value of R, f_0 for series and parallel resonance is the same.

TABLE 3–1 Comparison Between Series and Parallel Resonance

Parameters	Series circuit	Parallel circuit
Current at resonance	Maximum, $I_0 = V/R$ (Acceptor type)	Minimum, $I_0 = \frac{V}{L/CR}$ (Rejector type)
Impedance at resonance	Minimum, $Z_0 = R$	Maximum, $Z_0 = L/CR$
Power factor at resonance	Unity, circuit is resistive	Unity, circuit is resistive
Resonant frequency	$f_0 = \frac{1}{2\pi\sqrt{LC}}$	$f_0 = \frac{1}{2\pi}\sqrt{1/LC - R^2/L^2}$
Magnification element	Voltage	Current $= \frac{1}{2\pi\sqrt{LC}}$ if $R = 0$
Magnification factor or Q-factor	Q-factor $\frac{1}{R}\sqrt{\frac{L}{C}}$	Q-factor $= 2\pi f_0 \frac{L}{R}$

If we have a constant frequency supply source, a resonance condition can also be achieved if we change the value of L or C creating a condition when

$$2\pi f L = \frac{1}{2\pi f C}$$

A resonance condition is created in tuning circuits of radio receiver sets by adjusting the values of circuit parameters.

EXAMPLE 3.43

An inductive coil has a resistance of 10 Ω and inductance of 100 mH. This coil is connected in parallel with a capacitor of 20 μF. A variable frequency power at 100 V is applied across this parallel circuit. Calculate the frequency at which the circuit will reasonate. Also calculate the Q-factor, dynamic impedance of the circuit, and resonant current.

SOLUTION

Resonant frequency, $f_0 = \frac{1}{2\pi}\sqrt{\frac{1}{LC} - \frac{R^2}{L^2}}$

Substituting the given values,

$$f_0 = \frac{1}{2\pi}\sqrt{\frac{1}{100\times10^{-3}\times20\times10^{-6}} - \frac{(10)^2}{(100\times10^{-3})^2}}$$

$$= 112.48 \text{ Hz}$$

$$\text{Q-factor} = \frac{2\pi f_0 L}{R} = \frac{6.28\times112.48\times100\times10^{-3}}{10} = 7.06$$

Dynamic impedance, $Z_0 = \frac{L}{CR} = \frac{100\times10^{-3}}{20\times10^{-6}\times10} = 500\,\Omega$

Current at resonance, $I_0 = \frac{V}{Z_0} = \frac{100}{500} = 0.2\text{ A } (I_{minimum})$

EXAMPLE 3.44

An inductive coil of resistance 5 Ω and inductive reactance 10 Ω is connected across a voltage of 230 V at 50 Hz. Calculate the value of the capacitor which when connected in parallel with the coil will bring down the magnitude of the circuit current to a minimum. Draw the phasor diagram.

SOLUTION

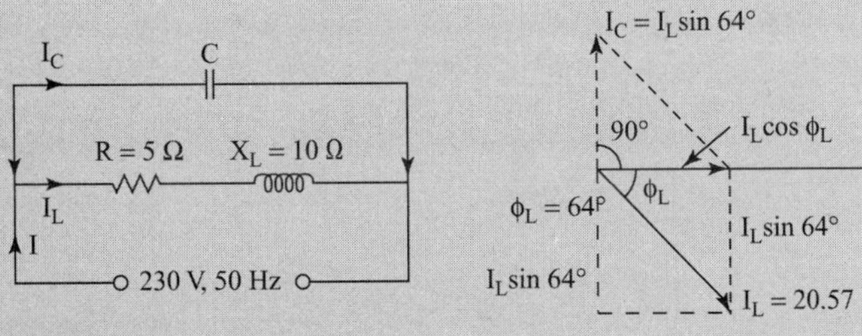

FIGURE 3.72

Before a capacitor is connected, current flowing through the inductor, I_L is

$$I_L = \frac{V}{Z} = \frac{230}{5 + j10} = \frac{230}{11.18\angle 64°} = 20.57\angle -64°\ \text{A}$$

$$\cos\phi_L = \cos 64° = 0.438$$

$$\sin\phi_L = \sin 64° = 0.835$$

If a capacitor is now connected in parallel, it must draw a current I_C which will lead V by 90°. The magnitude of I_C must be equal to $I_L \sin\phi_L$ so that these two currents cancel each other. In such a case, the resultant current, I is the in-phase current, i.e., $I_L \cos\phi_L$.

$$I_C = I_L \sin\phi_L = 20.57 \times 0.895 = 18.4\ \text{A}$$

$$I_C = \frac{V}{X_C} \text{ or } X_C = \frac{V}{I_C} = \frac{230}{18.4} = 12.5\ \Omega$$

$$X_C = \frac{1}{\omega C} = 12.5$$

or,

$$C = \frac{1}{2\pi f \times 12.5} = \frac{1}{6.28 \times 50 \times 12.5}\text{F} = \frac{10^6}{314 \times 12.5}\mu\text{F}$$

$$= 254.7\ \mu\text{F}$$

Magnitude of the in-phase current, i.e., the current which is in phase with the voltage, $I = I_L \cos\phi$, is

$$I = I_L \cos\phi_L = 20.57 \times 0.438 = 9\ \text{A}$$

This is the minimum current drawn by the circuit and is called the resonant current, I_0.

EXAMPLE 3.45

An inductor having a resistance of 4 Ω and an inductance of 20 mH are connected across a 230 V, 50 Hz supply. What value of capacitance should be connected in parallel to the inductor to produce a resonance condition? What will be the value of the resonant current?

SOLUTION

$$Z_L = R + j\omega L,\ \omega = 2\pi f = 2\times 3.14\times 50 = 314$$

$$Z_L = 4 + j314\times 20\times 10^{-3} = 4 + j6.28 = 7.44\angle 57.5°\ \Omega$$

$$I_L = \frac{V}{Z_L} = \frac{230\angle 0}{7.44\angle 57.5°} = 30.9\angle -57.5°\ \text{A}$$

For resonance, the current drawn by the capacitor in parallel must be equal to $I_L \sin\phi_L$.

$$I_C = I_L \sin\phi_L = 30.9\times \sin 57.5° = 30.9\times 0.843 = 26\ \text{A}$$

$$I_C = \frac{V}{X_C}\ \text{or,}\ X_C = \frac{V}{I_C} = \frac{230}{26} = 8.84\ \Omega.$$

$$X_C = \frac{1}{\omega C} = \frac{1}{314C} = 8.84\ \Omega$$

or,

$$C = \frac{1}{314\times 8.84}\text{F} = \frac{10^6}{314\times 8.84}\mu\text{F}$$

$$= 360.2\mu\text{F}$$

Resonant current for parallel resonance is the minimum current which is the in-phase component, i.e., $I_L \cos\phi_L$.

$$I_0 = I_L \cos\phi_L = 30.9\times \cos 57.5° = 30.9\times 0.537$$

$$= 16.6\ \text{A}$$

The phasor diagram representing the resonant condition has been shown in Fig. 3.73.

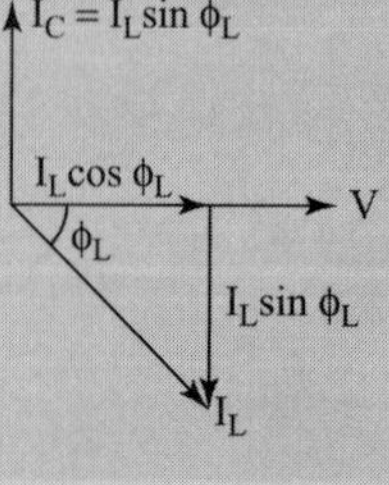

FIGURE 3.73

EXAMPLE 3.46

Calculate the value of R_1 in the circuit given in Fig. 3.74 such that the circuit will resonate.

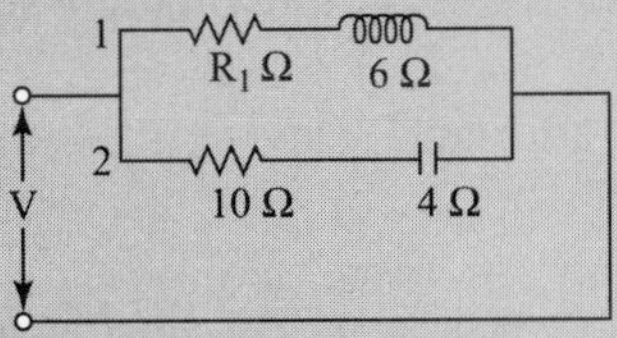

FIGURE 3.74

SOLUTION

We know that at resonance the impedance of the circuit will be resistive only. We will calculate the value of impedance in a complex form and equate its imaginary part to zero to determine the value of R_1.

Here, $Z = R_1 + j6$ and $Z_2 = 10 - j4$

$$Z = \frac{Z_1 Z_2}{Z_1 + Z_2} = \frac{(R_1 + j6)(10 - j4)}{(R_1 + j6) + (10 - j4)} = \frac{(R_1 + j6)(10 - j4)}{(R_1 + 10) + j(6 - 4)}$$

$$= \frac{(10R_1 + 24) + j\,(60 - 4R_1)}{[(R_1 + 10) + j2]\,[(R_1 + 10) - j2]}$$

$$= \frac{[(10R_1 + 24) + j\,(60 - 4R_1)]\,[(R_1 + 10) - j2]}{[(R_1 + 10) + j2]\,[(R_1 + 10) - j2]}$$

$$= \frac{[(10R_1 + 24)(R_1 + 10) + 2(60 - 4R_1)] + j(60 - 4R_1)(R_1 + 10) - j2(10R_1 + 24)}{(R_1 + 10)^2 + 2)}$$

During resonance, the imaginary part of Z will be zero.
Therfore,

$$j\,(60 - 4R_1)\,(R_1 + 10) - j2\,(10R_1 + 24) = 0$$

or, $$60R_1 - 4R_1^{\,2} + 600 - 40R_1 - 20R_1 - 48 = 0$$

or, $$4R_1^{\,2} = 552$$

or, $$R_1^{\,2} = 138$$

or, $$R_1 = 11.74\ \Omega$$

3.3.3 Applications of Resonant circuits

An electric circuit which has either a very low impedance or a very high impedance at a certain frequency is called a resonant circuit.

With very low value of impedance, the circuit will allow the highest amount of current flow at the resonant frequency.

Resonant circuits are used in radio and television tuners to pick up broadcast signals of specific frequencies.

A resonant or tuned circuit combines an inductor and a capacitor to make a circuit, that is, responsive to a particular frequency. Depending on the configuration (i.e., series or parallel), the circuit will have high or low impedance at the resonant frequency and operate as band pass or band-stop filter, or an oscillator.

Application of Series Resonance: A series LC circuit as shown in Fig. 3.75, is used in a radio receiver as intermediate frequency trap. If a signal of intermediate frequency is picked up at the aerial, it may severely interfere with the desired or wanted signal. Intermediate frequency lies outside the reception band of the radio. An IF trap is included in the form of an LC circuit which has resonance frequency equal to the intermediate frequency. L the winding of the aerial coil and a capacitor in series form the LC circuit. The LC series circuit at resonant frequency, i.e., at the intermediate frequency, gives almost zero impedance to the IF signal, and hence, such signals at IF are shunted to the ground and cannot appear at the input of the amplifier.

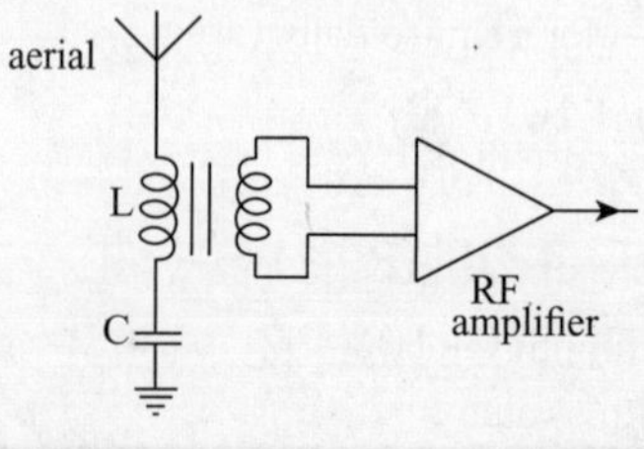

FIGURE 3.75

A series LC resonant circuit used in a radio receiver

Application of Parallel Resonance: A parallel resonant circuit offers a very high impedance to signal flow at resonant frequency. At all other frequencies, the impedance will be low. Since at resonant frequency, the impedance of the parallel LC network, as shown in Figure 3.76, will have high impedance, even a small signal current will produce a significant voltage across the parallel network. At all other frequencies the impedance will be low thereby creating a smaller voltage drop.

At resonant frequency, the high output voltage can be utilized for rejecting signals of other frequencies. This way, we can select the desired radio signal from the many that are caught by the antenna.

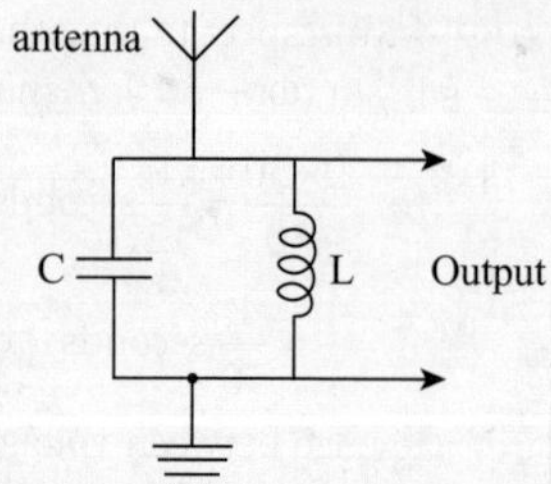

FIGURE 3.76

Tuned circuit, an example of parallel resonant circuit

Important Terminologies and Points to Remember

1. The value of an alternating quantity like voltage or current at any instant of time is called its instantaneous value.
2. A sinusoidally varying alternating quantity like voltage is expressed as $v = V_m \sin \omega t$
   Where v expressed by small letter is the instantaneous value, V_m is the maximum value and $\omega = 2\pi f$ where f is the frequency.
3. Number of cycles per second made by an alternating quantity is called its frequency. The time taken to complete one cycle is called the time period, T.
4. When f cycle is made in one second, then the time taken for making one cycle is $\frac{1}{f}$ seconds. Thus, $T = \frac{1}{f}$.
5. When a voltage is expressed as say v = 100 sin 314 t volts, its maximum value is 100 V, and frequency is 50 Hz since $\omega = 2\pi f = 314$, $f = \frac{314}{2\pi} = 50$.
6. Average value of an alternating quantity for a complete cycle is zero. Its average value is to be calculated for one-half cycle.
7. Rms value is the Root Mean square value. To calculate the rms value, we go by reverse way, i.e., we square the current, take its mean and then take the root. For example, rms values I of $i = I_m \sin \omega t$ is calculated as

$$I = \sqrt{\frac{1}{2\pi}\int_0^{2\pi} I_m^2 \sin^2 \omega t} = \frac{I_m}{\sqrt{2}}$$

   So $\text{Rms value} = \frac{\text{Maximum value}}{\sqrt{2}}$ for a sinusoidally varying current.
8. Rms value of sinusoidally varying current is $I_m/\sqrt{2}$, i.e., 0.707 I_m and average value is $\frac{2I_m}{\pi}$, i.e., 0.637 I_m.
   The ratio of rms value to average value is called the Form factor. For sine wave $FF = \frac{0.707 I_m}{0.637 I_m} = 1.11$.
9. Peak factor or crest factor $= \frac{\text{Maximum value}}{\text{Rms value}}$.
   For sine wave $P.F. = \frac{I_m}{I_m/\sqrt{2}} = \sqrt{3} = 1.414$.
10. Rms value is an indicator of the sharpness of the alternating quantity.
11. Phase difference between two alternating quantities indicates the amount of lag or lead by one quantity with respect to the other. For example, in a circuit if the applied voltage is $v = V_m \sin \omega t$ and current is $i = I_m \sin(\omega t - 30°)$, then the current is seen lagging the voltage by 30°. This 30° when expressed in time will be $\frac{T \times 30}{360} = \frac{T}{12}$ seconds. Where T is the time taken to complete one cycle. If frequency f = 50 Hz, then $T = \frac{1}{f} = \frac{1}{50}$ seconds or $T = \frac{1000}{50}$ ms = 20 ms. For 30°, time is $\frac{20}{12} = 1.66$ miliseconds. So time-phase difference between voltage and current is 30° or 1.66 ms.
12. In a purely resistive circuit, voltage and current are in phase; in a purely inductive circuit, current lags the voltage by 90° and in a purely capacitive circuit, current leads the voltage by 90°.
13. In an ac circuit average power, $P = VI \cos\phi$ where V and I are the rms values of voltage and current and $\cos\phi$ is the power factor. ϕ is the angle between, i.e., the phase difference between voltage and current in the circuit.

For a resistive circuit, P.f. $\cos\phi = \cos 0° = 1$

For an inductive circuit, $\cos\phi = \cos 90° = 0$

For a capacitive circuit, $\cos\phi = \cos 90° = 0$

The maximum value of P.f. is 1 and the minimum value of P.f is 0.

Depending upon the phase relationship between voltage and current, power factor may be lagging or leading.

14. To determine the power factor, we consider the phase relationship of voltage and current with respect to a reference axis. Suppose the voltage is represented as $v=100 \sin wt+30°$ and current is represented as $i = 10 \sin \omega t - 30°$. Then the phase difference between v and i is 60° and power factor, $\cos\phi = \cos 60° = 0.5$ lagging because current i is lagging the voltage by 60°.

15. In an RL circuit, the opposition offered to current flow, i.e., the impedance, Z is expressed as

$$Z = R + j\omega L \text{ where } \omega = 2\pi f.$$
$$= R + jX_L$$
$$= \sqrt{R^2 + X^2_L} \angle \tan^{-1}\frac{X_L}{R}$$

Power factor, $\cos\phi = \dfrac{R}{Z}$

Power, $P = VI \cos\phi$.

$$I = \frac{V}{Z}$$

16. The impedance of an R-C series circuit is

$$Z = R - jx_c \quad \text{where } X_C = \frac{1}{\omega c} = \frac{1}{2\pi fc}$$

$$|Z| = \sqrt{R^2 + X^2_C}$$

$$\phi = \tan^{-1}\frac{X_C}{R}$$

$$Z = \sqrt{R^2 + X^2_C} \angle -\tan^{-1}\frac{X_C}{R}$$

17. Apparent power, S = VI VA
Active power, $P = VI \cos\phi$ W
Reactive power, $Q = VI \sin\phi$ VAR

18. Reciprocal of Impedance Z is called Admittance λ

$$\lambda = \frac{1}{z} = \frac{1}{R \pm jX} = \frac{R}{2^2} \mp \frac{X}{2^2}$$

or, $\lambda = G \pm B$

Where G is called conductance and B is called susceptance.

19. All complex quantities like impedance, current, etc., are expressed either in a + jb form or $A\angle\theta$ form. Conversion of quantities from one form to the other is required in solving ac circuit problems.

20. Phasor diagram is the representation of all the time varying quantities of a circuit with respect to a reference axis.

21. Resonance is a condition in an RLC series circuit when current is the maximum, i.e., the effective impedance of the circuit is the minimum for a particular frequency.

22. Power dissipated in the resistor in an RLC series circuit is the active power. Energy stored in the capacitor and the inductor are due to reactive power. At particular frequency, F_0 when resonance occurs, energy

stored in the inductor and the capacitor oscillates between them and the circuit as a whole appears as a resistive circuit only. The current in the circuit reaches its maximum value. Such resonant circuits, find practical use in electronic devices and circuits to allow a signal of particular frequency to pass through.

23. Quality factor and Bandwidth are two quality parameters of a resonant circuit. Quality factor is the ratio of reactive power to the resistive power in the circuit at resonant frequency f_0 where $f_0 = \frac{1}{2\pi\sqrt{LC}}$ condition for resonance is $X_L = X_C$ and $Z = R$. Q-factor is more when active power is less, that is, value of R in the RLC series circuit is low. Then we get a very sharp I versus f characteristic as shown in figure. Such a circuit becomes more suitable as a fine tuning circuit.

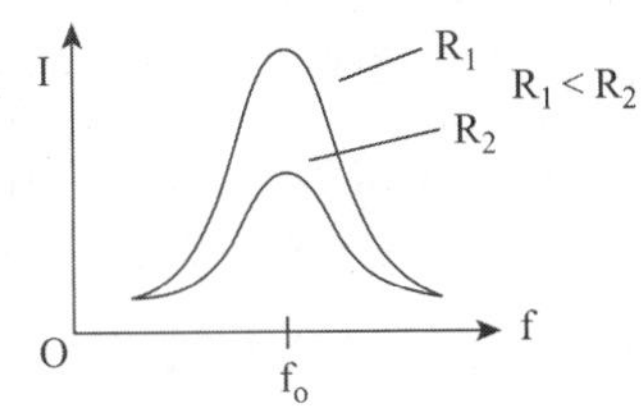

24. Bandwidth is a range of frequencies within which the current (signal) strength does not drop below 70.7 percent of its maximum value. At these frequency limits, half power is delivered to the circuit resistance.

25. An inductor and a capacitor connected in parallel across a variable frequency supply source forms a parallel resonance circuit when the capacitor current equals the inductor current. At resonance frequency these two currents, i.e., I_C and I_L are equal and: opposite to each other, the net current drawn from the supply source will be zero. If the resistance of the inductor is considered, then I_C will be equal to $I_L \sin\phi_L$.

Resonance frequency, $f_0 = \frac{1}{2\pi}\sqrt{\frac{1}{LC} - \frac{R^2}{L^2}}$ where R is the resistance of the inductor coil. If R is neglected, then $f_0 = \frac{1}{2\pi\sqrt{LC}}$.

A parallel resonant circuit virtually does not allow the signal to path high because the circuit impedance is infinite. Such a resonant circuit is referred to as a rejector circuit.

Important Formulae and Equations

Equation	Remarks
1. $I = \sqrt{\frac{1}{2\pi}\int_0^{2\pi} i^2 d\theta}$ If $i = I_m \sin\theta$, Rms value I is $I = \sqrt{\frac{1}{2\pi}\int_0^{2\pi} i_m^2 \sin^2\theta d\theta} = \frac{I_m}{\sqrt{2}}$	1. Rms value of current is calculated by squaring the current, taking the average and taking the square root.
2. $I_{av} = \frac{1}{\pi}\int_0^{\pi} i_m \sin\theta d\theta$ $I_{av} = \frac{2I_m}{\pi}$	2. Average value is calculated over one-half of a cycle. Average value, if calculated for one complete cycle, will be zero.
3. $\text{Form factor} = \frac{\text{Rms value}}{\text{Average value}}$ $= \frac{I_m/\sqrt{2}}{2I_m/\pi} = 1.11$	3. Form factor of a sinusoidal quantity, i.e., current or voltage is 1.11.

Important Formulae and Equations

Equation	Remarks
4. Form factor for a half sine wave $= \dfrac{\text{Rms value}}{\text{Average value}} = \dfrac{I_m}{2} \times \dfrac{\pi}{I_m} = 1.57$ Form factor for a rectangular wave = 1.0	4. Form factor indicates the shapness of a wave. For example, F.F. for rectangular wave is 1.0, for a sine wake is 1.11, for a saw-tooth wave is 1.15, and for a half-wave rectified sine wave is 1.5%.
5. Peak factor or crest factor $= \dfrac{\text{Maximum value}}{\text{Rms value}}$	5. Peak factor for a sine wave is 1.414.
6. In an R-L series circuit, if $v = V_n \sin wt$ and $i = I_m \sin(\omega t - \theta)$, their phase difference is θ degree lagging, i.e., current lags the voltage by a time equivalent to θ degree.	6. Phase difference between two time varying quantities indicates the time lag or time lead between them, expressed in degrees.
7. Power in an ac circuit = VI cos ϕ where cos ϕ is called the power factor. P.f = 1 (for a resistive circuit) = 0 (for a pure inductive or capacitive circuit)	7. In a resistive circuit, voltage and current are in phase, so that $\cos\theta = \cos 0° = 1$. For inductive or capacitive circuit, $\phi = 90°$ so that $\cos 90° = 0$
8. $X_L = \omega L = 2\pi f L$ $X_C = \dfrac{1}{\omega C} = \dfrac{1}{2\pi f C}$	8. Opposition offered by an inductor to current flow is called the inductive reactance, X_L opposition offered by a capacitor to the current flow is called capacitive reactance, X_C.
9. Average power in a pure inductive or capacitive circuit, P=VI cos ϕ =VI cos 90°=0.	9. In a pure inductor or a pure capacitor, the average power absorbed is zero.
10. Impedance of an R-L series circuit: $Z = \sqrt{R^2 + X_L^2}$ $Z = R + jX_L$ $X_L = \omega L = 2\pi f L$ $\cos\phi = P.f = \dfrac{R}{Z}$ 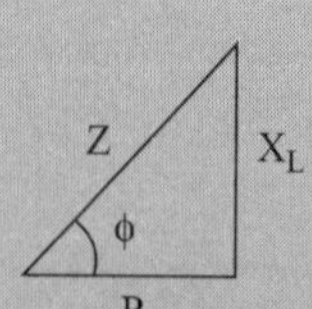 Impedence Triangle $Z\cos\phi = R$ $Z\sin\phi = X_L$ $\tan\phi = \dfrac{X_L}{R}$ or $\phi = \tan^{-1}\dfrac{X_L}{R}$ Power $P = VI\cos\phi$ Power = Volt − Ampere × Power factor	10. The combined effect of opposition offered to flow of current by a R–L series circuit is the impedance, Z. From the impedance triangle, the power factor, and power factor angle, ϕ can be calculated.

Important Formulae and Equations

Equation | **Remarks**

Equation

11. In an ac circuit,
Apparent power, S = VI
Expressed in VA or KVA.
i.e., volt ampere or kilo volt ampere.
Active power, P = Apparent power × power factor
$= VI\cos\phi$
Expressed in watts or kilo watts.
Reactive power, $Q = VI\sin\phi$

$$S^2 = P^2 + Q^2$$

$$S = \sqrt{P^2 + Q^2}$$

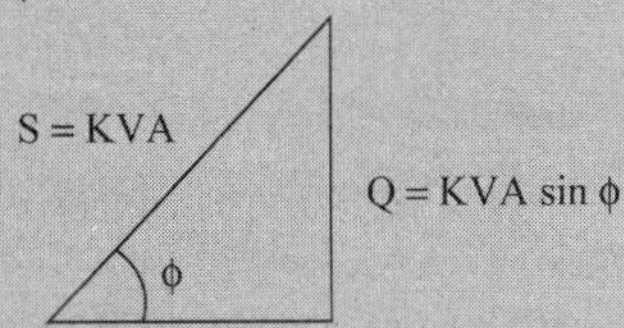

$KVA\cos\phi = KW.$

$KVA\sin\phi = KVAR.$

$P = KW = KVA\cos\phi$
Power Triangele.

Remarks

11. In a circuit containing resistance, inductance and capacitance, the resistance element contributes to active power while the inductance and capacitance elements contribute to reactive power. For example, in a R–L circuit if R is reduced to zero, the active power is reduced to zero because the power factor angle is 90° and power factor $\cos\phi = \cos 90° = 0$, so that active power $P = VI\cos\phi = 0$ and reactive power $Q = VI\sin\phi = VI\sin 90° = VI$.

Equation

12. For a RLC series circuit, impedance Z is calculated as:

$$Z = R + j(X_L - X_C)$$

$$|Z| = \sqrt{R^2 + (X_L - X_C)^2}$$

when $X_L > X_C$ and, $\phi = \tan^{-1}\dfrac{(X_L - X_C)}{R}$

when $X_L = X_C$, $Z = R$

when $X_C > X_L$

$$Z = R + j(X_C - X_L)$$

$$|Z| = \sqrt{R^2 + (X_C - X_L)^2}$$

and $\phi = \tan^{-1}\dfrac{(X_C - X_L)}{R}$

Power, $P = VI\cos\phi$

Remarks

12. In a RLC series circuit when X_L equals X_C, the circuit becomes purely resistive one, i.e., Z becomes equal to R. The circuit offers minimum impedance, and hence, current becomes maximum. The unit of X_L and X_C in ohms while inductance L is in henry and capacitance C is in farad.

Equation

13. Impedance of RL circuit, and RC circuit respective are:
$Z = R + jX_L$ and $Z = R - jX_C$

Where $X_L = \omega L = 2\pi fL$ and $X_C = \dfrac{1}{\omega C} = \dfrac{1}{2\pi fC}$

Admittance, $\lambda = \dfrac{1}{Z} = \dfrac{1}{R \pm jX}$

Considering X_L and X_C as X.

Remarks

13. Admittance method is used for solving parallel ac circuits conveniently.

Important Formulae and Equations

Equation	Remarks
Simplifying, $Y = \frac{R}{Z^2} \mp j\frac{X}{Z^2}$ $= G \mp jB.$ Where G = Emductance $= \frac{R}{Z^2}$ mho and B = Susceptance $= \frac{X}{Z^2}$ mho. B is –ve if the circuit is inductive and B is +ve if the circuit is capacitive.	
14. When one RL series circuit and one RC series circuit are connected in parallel, their total impedance is calculated as $Z_1 = R + jX_L$ $Z_2 = R - jX_C$ $Z = \frac{Z_1 - Z_2}{Z_1 + Z_2}$ Branch currents, $I_1 = \frac{V}{Z_1}$ and $I_2 = \frac{V}{Z_2}$ Line current, $I = \frac{V}{Z}$.	14. While adding, subtracting, multiplying and dividing impedances, it will be required to convert them from complex form of a + jb to polar form as $a + jb = \sqrt{a^2 + b^2} \angle \tan^{\frac{-b}{a}}$ From polar form of $A\angle\theta$, the complex form is written as $A\angle\theta = A\cos\theta + jA\sin\theta$.
15. For series resonance circuit, X_L equals X_C and $Z_0 = R$ The resonance frequency, f_0 is given as $f_0 = \frac{1}{2\pi\sqrt{LC}}$ $\omega_0 = \frac{1}{\sqrt{LC}}$ and Q-factor $= \frac{1}{R}\sqrt{\frac{L}{C}}$ resonance current, $I_0 = \frac{V}{R}$ which is the maximum current Bandwidth $= f_2 - f_1 = \frac{R}{2\pi L}$ Q factor × Bandwidth $= \frac{1}{R}\sqrt{\frac{L}{C}} \times \frac{R}{2\pi L} = \frac{1}{2\pi\sqrt{LC}} = f_0$	15. From the relation, f_0 = Q-factor × Bandwidth it is seen that for a particular f_0 if Q-factor is high, bandwidth will be low. Low bandwidth means the range of frequency will be very close to resonance frequency and the circuit will be very selective in allowing signals to pass through. High quality factor resonant circuits are therefore used as tuning circuits in many electronic devices.

Important Formulae and Equations

Equation	Remarks
16. For parallel resonance, resonance frequency f_0 is given as $f_0 = \frac{1}{2\pi}\sqrt{\frac{1}{LC} - \frac{R^2}{L^2}}$ and circuit impedance $Z_0 = \frac{L}{CR}$ $Q-\text{factor} = \frac{1}{R}\sqrt{\frac{L}{C}}$	16. In parallel resonance, the impedance Z_0 is the maximum and current drawn by the circuit is the minimum. Since current is minimum at resonance, such a circuit is called "rejector circuit."

REVIEW QUESTIONS

Short Answer Type Questions

1. Explain frequency, time period, instantaneous value, maximum value, and average value for a sinusoidal voltage.
2. What do you understand by harmonic waves of a non-sinusoidal wave?
3. Why do we use RMS value instead of average value for an alternating quantity?
4. Show that for a sinusoidal voltage, RMS value is 0.707 times its maximum value.
5. What is the value of form factor for a sine wave? What is the significance of the value of form factor for an alternating quantity?
6. The form factors for different kinds of voltage wave shapes have been calculated as 1.0, 1.11, and 1.15. Is it possible to predict the type of the voltage wave shapes?
7. What is inductive reactance, capacitive reactance, and impedance of an L–R–C circuit?
8. What is meant by power factor of an ac circuit? What is its minimum value and its maximum value?
9. Prove that average power in an ac circuit is $VI\cos\phi$, where V is the RMS value of voltage, I is the RMS value of current, and $\cos\phi$ the power factor.
10. What is the significance of very low (poor) power factor of a circuit?
11. A resistance R, an inductance L, and a capacitance C are connected in series across an alternating voltage, V. A current I flows through the circuit. Draw a phasor diagram showing the voltage drops across the circuit parameters with respect to V and I.
12. A resistance of 10 Ω and an inductive reactance of 10 Ω are connected in series. Calculate the value of impedance and draw the impedance triangle.
13. Show that current in a pure inductive circuit lags the voltage by 90°.
14. What is the power factor of a purely resistive circuit, purely inductive circuit, and purely capacitive circuit?
15. State the condition for maximum current in an L–R–C series circuit.
16. State the condition for series resonance in an L–R–C circuit.
17. State the condition for parallel resonance. How do we calculate the value of capacitor to be shunted to create a resonant condition?
18. What is the value of resonant frequency in the case of series resonance and in the case of parallel resonance?

19. What is meant by Q-factor of a series resonant circuit? What does Q-factor signify?
20. What do you mean by bandwidth in a series circuit?
21. A resonant circuit with high Q-factor is also called a tuned circuits. Explain why.
22. Write the expression for resonant frequency, Q-factor, and dynamic impedance for a parallel resonant circuit.
23. Explain what is meant by phase and phase difference of alternating quantities.
24. A sinusoidal current is expressed as i = 100 sin 314t. What is the maximum value, RMS value, frequency, and time period of the alternating current?
25. A sinusoidal voltage, $v = 300\sin(314t + 30°)$ when connected across an ac series circuit produces a current, $i = 20\sin(314t - 30°)$. What is the power factor of the circuit? Draw the phasor diagram.
26. Define apparent power, active power, and reactive power of an ac circuit.
27. Define the terms impedance, inductive reactance, capacitive reactance, admittance, active power, reactive power, and power factor for an ac circuit.
28. Two $Z_1 = 10\angle 30°$ and $Z_2 = 20\angle 60°$ are connected in series. What is the value of equivalent impedance?
29. Two impedances $Z_1 = 10\angle 30$ and $Z_2 = 20\angle 30$ are connected in parallel. What is the equivalent impedance?
30. An impedance of 10 + j10 is connected across a voltage of $230\angle 60°$ V. What is the magnitude of current and the value of power factor?
31. Two impedances $Z_1 = 10\angle 30°$ and $Z_2 = 10\angle 60°$ are connected in series. Calculate the equivalent impedance.
32. What is resonant frequency? Why is series resonance called voltage resonance?

Numerical Problems

33. Calculate the RMS value of an alternating current $i = 20(1 + \sin\theta)$.

[**Ans** 24.5 A]

34. Calculate the RMS value of a half-wave-rectified voltage of maximum value of 100 V.

[**Ans** 50 V]

35. An alternating voltage is expressed as v =141.1 sin 314t. What is the RMS value, time period, and frequency?

[**Ans** 100 V, 20 msec, 50 Hz]

36. An alternating current of frequency 50 Hz has its maximum value of 5 A and lagging the voltage by 30°. Write the equation for the current.

[**Ans** i = 5 sin (314t – 30°)]

37. An alternating voltage is expressed as, v = 100 sin 314t. Determine the time taken for the voltage to reach half its maximum value, time counted from t = 0. At what time will voltage reach its maximum value?

[**Ans** t = 1.66 msec; t = 5 msec]

38. Determine the average value of the voltage wave form shown in Fig. 3.77.

[**Ans** 7.5 V]

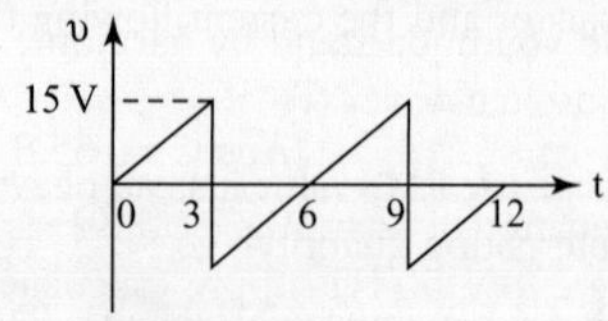

FIGURE 3.77

39. An alternating voltage is defined as:

$v = 100\sin\theta$ V $0 < \theta < \pi$

$v = 0$V $\pi < \theta < 2\pi$

What is the RMS value of this voltage? [**Ans** 50 V]

40. Find the RMS value of the sinusoidal voltage waveform shown in Fig. 3.78.

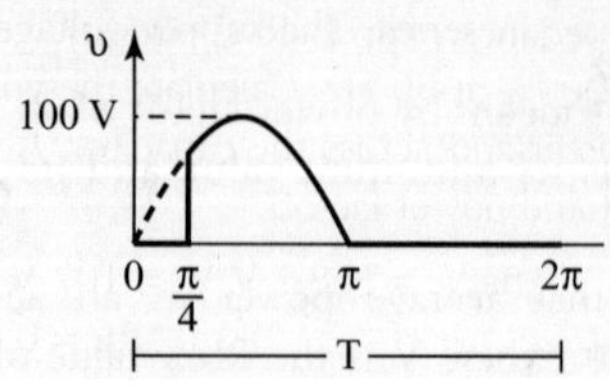

FIGURE 3.78

[**Ans** 47.6 V]

41. Find the RMS value of the voltage wave shown in Fig. 3.79.

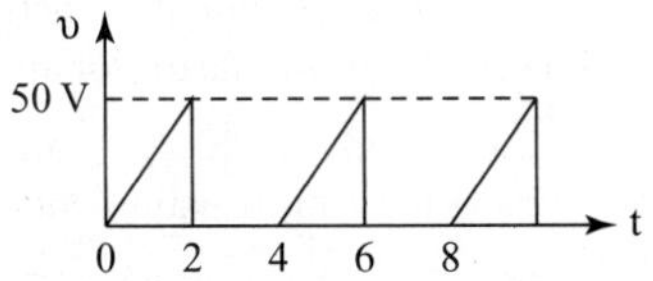

FIGURE 3.79

[**Ans** 20.4 V]

42. A resistance of 50 Ω, an inductance of 0.1 H, and a capacitance of 50 μf are connected in series across a 230 V, 50 Hz supply. Calculate (i) the value of impedance; (ii) current flowing; (iii) power factor; (iv) power consumed.

[**Ans** Z = 59.5 Ω, I = 3.86 A, P.f. = 0.84 leading; P = 746.7 W]

43. In an R–L–C series circuit, the voltage across R is 160 V, across L is 240 V, and power consumed is 1000 W when a voltage of 200 V at 50 Hz is applied across the circuit. Calculate the value of the capacitor and the current flowing through the circuit.

[**Ans** C = 165.8 μF, 6.25 A]

44. A impedance of $Z = 50\angle-60°\Omega$ is connected across a 230 V, 50 Hz supply. Calculate the values of circuit elements, current, and power factor.

[**Ans** R = 25, C = 73.54 μF, I = 4.6, P.f. = 0.5 leading]

45. A coil of resistance 5 Ω and inductance 20 mH is connected across a voltage of v = 230 sin 314t. Write an expression for the current flowing through the circuit.

[**Ans** i = 28.75 sin (314t – 51°]

46. A resistance of 50 Ω and a capacitor of 100 μf are connected in series. The supply voltage to the circuit is 200 V at 50 Hz. Calculate the voltage across the resistor and across the capacitor. Also calculate current and power factor.

[**Ans** V_R = 168.7 V, V_C = 107 V, I = 3.37 A, P.f. = 0.84 leading]

47. In an R–L–C series circuit a maximum current of 0.5 A is obtained by varying the value of inductance L. The supply voltage is fixed at 230 V, 50 Hz. When maximum current flows through the circuit, the voltage measured across the capacitor is 350 V. What are the values of the circuit parameters?

[**Ans** R = 460 A, L = 2.229 H, C = 4.549 μF]

48. A 200 V, a variable frequency supply is connected across an L–R–C series circuit with R = 10 Ω, L = 10 mH, and C = 1 μF. Calculate the frequency at which reasonance will occur. Also calculate the Q-factor and bandwidth.

[**Ans** f = 1591.5 Hz, Q-factor = 10, Bandwidth = 159 H_z]

49. A coil of R = 10 Ω, L = 0.023 H connected in parallel with another coil of R = 5 Ω, L = 0.035 H. The combination is connected across at 200 V, 50 Hz supply. Calculate the current drawn from the supply and the power factor.

[**Ans** I = 31.4 A, P.f. = 0.63 lagging]

50. A coil of resistance 20 Ω and inductance of 300 mH is connected in parallel with a capacitance of 200 μF. The combination is connected across 200 V, variable frequency power supply. At what frequency will the parallel circuit resonate and what would be the current at resonance?

[**Ans** 20.5 Hz, 2.66 A]

51. Calculate the value of R in the circuit shown in figure below such that the circuit will resonate.

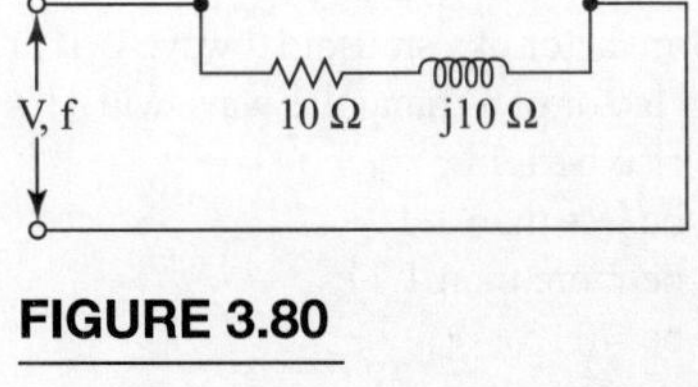

FIGURE 3.80

[**Ans** R = 6 Ω]

52. Calculate the half power frequencies of a series resonance circuit in which the resonant frequency is 150 KHz and bandwidth 75 KHz.

[**Ans** f_1 =117 KHz, f_2=.19 KHz]

Multiple Choice Questions

1. The voltage and current in an ac circuit is represented by $\upsilon = V_m \sin(wt + 30°)$ and $i = I_m \sin(wt - 45°)$. The power factor angle of the circuit is
(a) 15° (b) 75°
(c) 45° (d) 30°.

2. A current is represented by $i = 100 \sin(314t - 30°)$ A. The RMS value of the current and the frequency are, respectively,
(a) 100 A and 314 Hz
(b) 100 A and 50 Hz
(c) 70.7 A and 314 Hz
(d) 70.7 A and 50 Hz.

3. A current of 10 A is flowing through a circuit. The power factor is 0.5 lagging. The instantaneous value of the current can be written as
(a) $i = 10 \sin 60°$ A
(b) $i = 10 \sin(\omega t - 30°)$A
(c) $i = 14.14 \sin(\omega t - 60°)$
(d) $i = 14.14 \sin(\omega t + 60°)$.

4. In a purely inductive circuit
(a) current lags the voltage by 90°
(b) current leads the voltage by 90°
(c) voltage lags the current by 90°
(d) current lags the voltage by 180°.

5. Form factor of an ac wave indicates
(a) low sharp or steep the wave shape is
(b) low flat the wave shape is
(c) low symmetrical the wave shape is
(d) the degree of its conformily to sinusoidal form.

6. Power consumed by a pure inductor is
(a) infinite (b) very high
(c) zero (d) very small.

7. If form factor of a sinusoidal wave is 1.11, then the form factor of a triangular wave will
(a) also be 1.11
(b) be less than 1.11
(c) be more than 1.11
(d) be = 1.

8. A voltage of $\upsilon = 100 \sin(314t - 30°)$ is connected across a 10 Ω resistor. The power dissipated in the circuit will be
(a) 100000 W
(b) 1000 W
(c) 500 W
(d) 250 W.

9. The average value of a sinusoidal current is
(a) $\frac{2I_m}{\pi}$ (b) $\frac{I_m}{\pi}$
(c) $\frac{I_m}{2\pi}$ (d) $\frac{I_m^2}{2\pi}$.

10. Form factor of an alternating wave form is the ratio of
(a) RMS value and average value
(b) average value and RMS value
(c) maximum value and average value
(d) maximum value and RMS value.

11. A capacitance of C Farad is connected to a 230 V, 50 Hz supply. The value of capacitive reactance is
(a) 314 C Ω (b) $\frac{1}{314C}$ Ω
(c) 628 C Ω (d) $\frac{1}{628C}$ Ω.

12. The form factor of a square wave is
(a) 1.11 (b) 1.0
(c) 0 (d) 1.414.

13. Two sinusoidal waves are represented as $v_1 = 100 \sin(wt + 30°)$ and v_2 200 $\sin(wt - 60°)$. The phasor relationship between the voltages can be expressed as
(a) v_1 lags v_2 by 90°
(b) v_2 lags v_1 by 90°
(c) v_1 leads v_2 by 30°
(d) v_2 lags v_1 by 60°.

14. Inductive reactance of a coil of 0.1 H at 50 Hz is
(a) 31.4 Ω (b) 62.8 Ω
(c) 314 Ω (d) 5 Ω.

15. The power factor of a purely resistive circuit is
(a) 1.0 (b) 0
(c) 0.1 (d) 0.5.

16. A sinusoidal voltage is represented as $v = 141.4 \sin\left(628t - \frac{\pi}{3}\right)$. The RMS value, frequency and phase angle are, respectively
(a) 141.4, 628, 60°
(b) 100, 100, –60°
(c) 141.4, 50, 60°
(d) 141.4, 100, 60°.

(On single-phase ac circuits)

17. In an R–L series circuit, the power factor of the circuit is increased if
   (a) X_L, the inductive reactance is increased
   (b) X_L, the inductive reactance is decreased
   (c) R, resistance is decreased
   (d) the supply frequency is increased.

18. The power factor of an R–L circuit can be expressed as
   (a) $\cos\phi = \frac{R}{Z}$ (b) $\cos\phi = \frac{X_L}{Z}$
   (c) $\cos\phi = \frac{R}{X_L}$ (d) $\cos\phi = \frac{X_L}{R}$.

19. An R–L series circuit consists of R = 3 Ω and X_L = 4 Ω. The impedance of the circuit is
   (a) Z = 7 Ω (b) Z = 1 Ω
   (c) Z = 5 Ω (d) Z = $\sqrt{7}$ Ω.

20. The impedance of a circuit is $\angle 30°$. The value of resistance and inductive reactance of the circuit, respectively, are
   (a) 10 Ω and 17.32 Ω
   (b) 17.32 Ω and 10 Ω
   (c) 10 Ω and 10 Ω
   (d) 10 Ω and 8.66 Ω.

21. The impedance q an R–L series circuit is 628 $\angle 30°$, when the supply frequency is 50 Hz. The value of inductance, L is
   (a) 314 H (b) 1 H
   (c) 2 H (d) 628 H.

22. An impedance of z = 6 + j8 Ω is connected across a 200 V, 50 Hz supply. The power factor of the circuit is
   (a) 0.6 lagging (b) 0.6 leading
   (c) 0.75 lagging (d) 0.8 lagging.

23. The current flowing through the circuit of question 6 is
   (a) 10 A (b) 20 A
   (c) 14–28 A (d) 2 A.

24. An R–L series circuit has an impedance of 10 + j10 Ω. The power factor angle of the circuit is
   (a) 30° lagging (b) 30° leading
   (c) 45° leading (d) 45° lagging.

25. An R–C series circuit has a resistance of 10 Ω and a capacitive reactance of 10 Ω. What will be the phase difference between the voltage and current in the circuit?
   (a) Current will lead the voltage by 90°
   (b) Current will lag the voltage by 90°
   (c) Current will lead the voltage by 45°
   (d) Current will lag the voltage by 45°.

26. The impedance of an R–L series circuit is (50 + j100) Ω at 50 Hz. When the supply frequency is 100 Hz, the value of impedance will be
   (a) (50 + j 1000) Ω (b) (50 + j 200) Ω
   (c) (100 + j 100) Ω (d) (100 + j 200) Ω.

27. A voltage of υ = 10 sin (314t + 15°) is applied across an R–L–C series circuit, where R = 5 Ω, X_L = 15 Ω, and X_C = 10 Ω. The current flowing the circuit will be
   (a) 0.33 A (b) 1 A
   (c) 1.414 A (d) 2 A.

28. The resonant frequency in R–L–C series circuit is
   (a) $f_0 = \frac{2\pi}{\sqrt{LC}}$ (b) $f_0 = \frac{\sqrt{LC}}{2\pi}$
   (c) $f_0 = \frac{1}{2\pi\sqrt{LC}}$ (d) $f_0 = \frac{1}{2\pi}\sqrt{\frac{L}{C}}$.

29. A series RLC circuit has R = 50 Ω, L = 50 μH, C = 2 μF. The Q-factor of the circuit is
   (a) 0.1 (b) 1
   (c) 10 (d) 2.

30. When a parallel circuit is in resonance, which of the following of the circuit is maximum?
   (a) Current (b) Impedance
   (c) Admittance (d) Power factor.

31. In an R–L–C series circuit for a frequency less than the resonant frequency,
   (a) $X_L > X_C$ (b) $X_C > X_L$
   (c) $X_C = X_L$ (d) $X_C \alpha \frac{1}{X_L}$.

32. In an R–L–C series circuit, if the frequency is made more than the resonant frequency the circuit will effectively be
   (a) inductive (b) capacitive
   (c) resistive (d) oscillatory.

33. Two impedances, Z_1 = 4 + j4 Ω and Z_2 = 4 – j4 Ω are connected in parallel. Their equivalent impedance is
   (a) 8 + j8 Ω (b) 4 + j0 Ω
   (c) 8 – j8 Ω (d) 8 + j0 Ω.

34. When an inductance, L and a resistance, R are connected in parallel across an ac supply, the current drawn by the two parallel branches will be out of phase by

(a) 0° (b) 90°
(c) 180° (d) 45°.

35. When an inductance, L and a capacitance, C are connected in parallel across an ac supply, the current drawn by the two parallel branches will be out of phase by
(a) 0° (b) 90°
(c) 180° (d) 45°.

36. In an R–L circuit, $X_L = R$. The power factor angle q the circuit is
(a) 30° (b) 45°
(c) 60° (d) 0°.

37. In a series resonant circuit, a change in supply voltage will cause a change in
(a) the current drawn
(b) the Q-factor q the circuit
(c) the bandwidth of the circuit
(d) the resonant frequency.

38. Which of the following conditions is true for both series and parallel resonance?
(a) impedance is minimum
(b) power factor is unity
(c) power factor is zero
(d) power is low.

39. The bandwidth of a series R–L–C circuit is
(a) $\frac{C}{2\pi L}$ (b) $\frac{R}{2\pi L}$
(c) $\frac{C}{2\pi R}$ (d) $\frac{L}{2\pi R}$.

40. The product of voltage and current in an ac circuit is called
(a) active power (b) apparent power
(c) average power (d) reactive power.

41. In a series resonance circuit
(a) L = C (b) L = R
(c) $X_L = X_C$ (d) R = L = C.

Answers to Multiple Choice Questions

1. (b)	2. (d)	3. (c)	4. (a)	5. (a)	6. (c)
7. (c)	8. (c)	9. (a)	10. (a)	11. (b)	12. (b)
13. (b)	14. (a)	15. (a)	16. (b)		

(On single-phase ac circuits)

17. (b)	18. (a)	19. (c)	20. (b)	21. (b)	22. (a)
23. (b)	24. (d)	25. (c)	26. (b)	27. (b)	28. (c)
29. (a)	30. (b)	31. (b)	32. (a)	33. (b)	34. (b)
35. (c)	36. (b)	37. (a)	38. (b)	39. (b)	40. (b)
41. (c)					

CHAPTER 4

Three-phase System

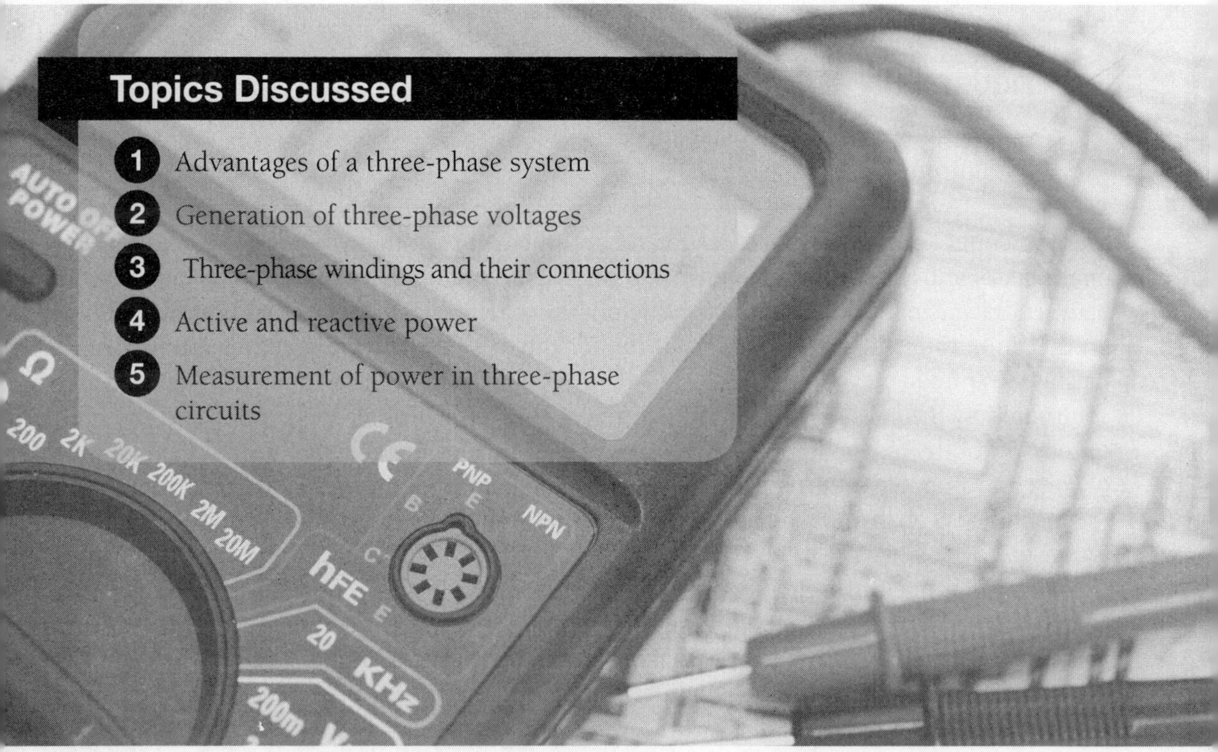

Topics Discussed

1. Advantages of a three-phase system
2. Generation of three-phase voltages
3. Three-phase windings and their connections
4. Active and reactive power
5. Measurement of power in three-phase circuits

4.1 ► Introduction

Generation, Transmission, and Distribution of electricity is done by three-phase electrical networks consisting of generators, transformers, transmission and distribution lines forming a system, called the power system. In a three-phase system we have three independent voltages induced in the three windings of the generator. To understand the difference between a single-phase voltage and a three-phase voltage, let us consider how these voltages are generated in ac generators. We have known that EMF is induced in the conductors of a coil if they cut lines of force. In Fig. 4.1 we have placed one coil in slots of a hollow cylindrical stator core. A two-pole magnet is rotated at a particular speed by some means. The flux lines will cut the conductors and EMF will be induced in the coil. Since the flux under North and South poles will cut the conductors alternately, an alternating single-phase voltage will be induced in the coil.

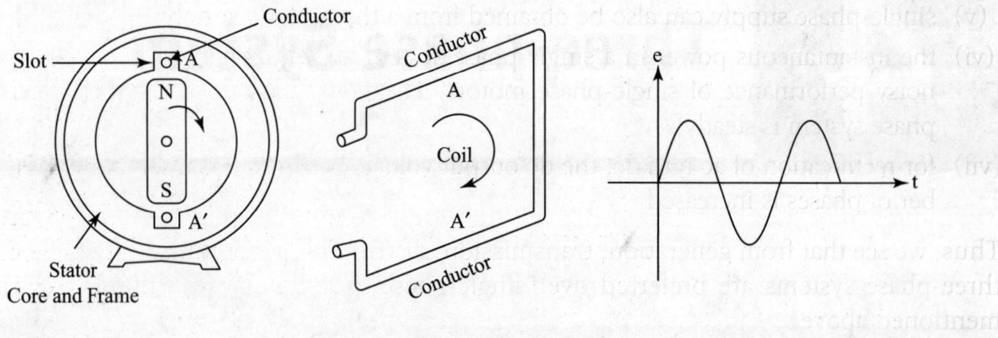

FIGURE 4.1

Generation of single-phase voltage

Here, we have seen that the coil is stationary and the magnetic field is being rotated. However, we could have had the field stationary and the coil rotating by placing the field magnets on the stator and the coil on a cylindrical rotor. What is required is to produce a relative motion between the magnetic field and the conductor. The magnitude of the induced EMF will depend on the number of coils, the strength of the magnetic field, and the speed of rotation of the magnet. The frequency of the induced EMF will depend upon the number of magnetic poles confronted by the coils per revolution. Normally, the number of magnetic poles and the speed of rotation of the magnetic poles by a drive, usually a turbine, is kept constant. The number of coils and number of turns used in each coil are kept as per design and are constant once the machine is constructed. That is why the magnitude of the induced EMF and its frequency is constant. Thus, we get a single-phase voltage from the single-phase winding which can be used to supply an electric circuit comprising resistance, inductance, and capacitance elements.

In a three-phase system we will have three-phase voltages induced in the three-phase windings of the generator. In all generating stations three-phase generators are installed. A three-phase winding consists of three separate windings placed physically at an angle of 120°. A single-phase winding consist of a single winding.

4.2 ► Advantages of Three-Phase Systems

A three-phase system of power generation, transmission, distribution and utilization has a number of advantages over a single-phase system. Some of these are mentioned as follows:

(i) the output of a three-phase machine generating electricity is more than the output of a single-phase machine of the same size;

(ii) the most commonly used three-phase induction motors are self starting. For single-phase motors, a separate starting winding is required;

(iii) electrical power transmission from the generating station to the places of use is done by transmission lines. Three-phase power transmission is more economical than single-phase power transmission;

(iv) the power factor of three-phase systems is better than that of the single-phase systems;

(v) single-phase supply can also be obtained from a three-phase supply;

(vi) the instantaneous power in a single-phase system is fluctuating with time giving rise to noisy performance of single-phase motors. The power output of a symmetrical three-phase system is steady;

(vii) for rectification of ac into dc, the dc output voltage becomes less fluctuating if the number of phases is increased.

Thus, we see that from generation, transmission, distribution, and utilization points of view, three-phase systems are preferred over single-phase systems due to a number of reasons mentioned above.

4.3 ► Generation of Three-Phase Voltages

Due to a number of practical considerations, generators are built to generate three-phase voltages. Commercial generators are built to generate three-phase voltages. In three-phase generators, three separate windings are made. Windings are made of coils. These windings are placed in stator slots at an angle of 120° apart as shown in Fig. 4.2. RR′ is one phase winding.

YY′ is the second phase winding and BB′ is the third phase winding. The three-phase windings are placed at an angular distance of 120°. For simplicity only one coil per phase has been shown. In practice a number of coils connected in series makes one phase winding.

When the magnetic poles are rotated by a prime mover (say a turbine), the magnetic flux of North and South poles will cut the windings in sequence. For clockwise rotation, flux will be cut by coil RR′ first, then by coil YY′, and then by coil BB′. Therefore, EMF will be induced in these coils in sequence. There will be a time phase difference between the EMFs induced in these coils (windings). The time phase difference will be 120°. In terms of time, the phase difference will be the time taken by the magnetic poles to rotate by 120°, i.e., one-third of a revolution. Thus, across the three-phase windings we will get three voltages which are equal in magnitude and frequency but having a time phase difference of 120° between them, as shown in Fig. 4.3.

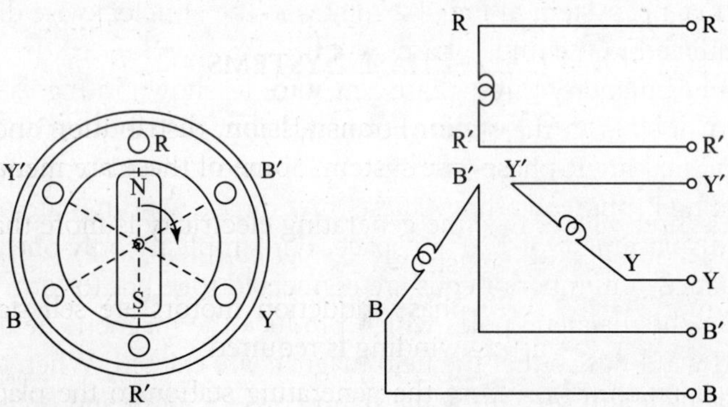

FIGURE 4.2

Generation of three-phase voltages

The equation of voltages are

$$e_R = E_m \sin \omega t \quad (4.1)$$

$$e_Y = E_m \sin (\omega t - 120°) \quad (4.2)$$

$$e_B = E_m \sin (\omega t - 240°) \quad (4.3)$$

Resultant EMF = $E_m \sin \omega t + E_m \sin (\omega t - 120°) + E_m \sin (\omega t - 240°) = 0$

Since the three voltages are equal in magnitude but displaced in time phase by 120°, their phasor sum is zero as shown in Fig. 4.3.

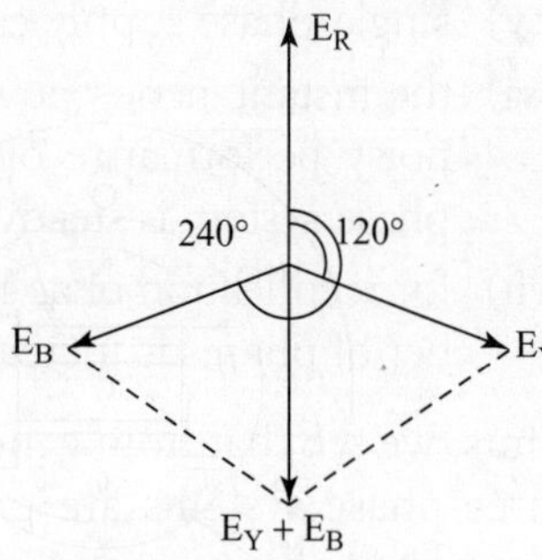

FIGURE 4.3

Three-phase voltages displaced in time phase by 120°

Three-phase supply is required for large-capacity electrical loads. These loads could be three-phase motors used in industrial, commercial, agricultural, and other sectors. For example, the water pump used for irrigation purpose is invariably a three-phase-motor-driven pump requiring a three-phase supply. So, like three-phase supply, we will have three-phase loads. Three-phase supply will be supplying electrical power to three-phase loads. A number of terms are used in connection with three-phase supply and three-phase loads. These are described as follows. Further the three-phase windings can be connected together in the form of star or delta. The voltage between the two-phase windings and current flowing through the phase windings, and the supply line will be different in star and delta connections. These will be studied in detail.

By now we must have realized that by phase, we mean a winding. A phase difference between two windings is the physical angular displacement between them. In a three-phase winding, the phase difference between the windings is 120°. Phase sequence is the order in which maximum voltage is induced in the windings. For example, if the magnetic field cuts the conductors of the phase RR′ first and then cuts the conductors of phase YY′, and lastly cuts the conductors of phase BB′, then EMF will be induced in all the phases of equal magnitude but their maximum value will appear in a sequence RYB. Then we call the phase sequence of EMF as RYB. If the magnet system of Fig. 4.2 rotates in the anticlockwise direction, the phase sequence of EMF induced in the three phases will be RBY.

Let us consider an elementary three-phase generator as shown in Fig. 4.4 (a). Three-phase windings are placed in slots in the stator. For simplicity, only one coil per phase has been used. R-R′ is one coil making R-phase winding. Y-Y′ is another coil forming Y-phase winding. B-B′ is one coil forming B-phase winding. These three-phase windings are placed in the stator slots at an angular displacement of 120° in space. For simplicity only one coil per phase has been shown. In practice, a number of coils are connected together to form each phase winding. The rotor carries the magnetic poles which produce the magnetic field. Direct current is supplied to the field windings so that the field magnets are excited. When the rotor is rotated by a turbine, the magnetic flux are cut by the stator coils RR′, YY′, and BB′ in a sequence. The emfs induced in these coils are sinusoidal in nature because of the nature of flux distribution. The voltage induced in the three-phase windings will be identical in nature but they will

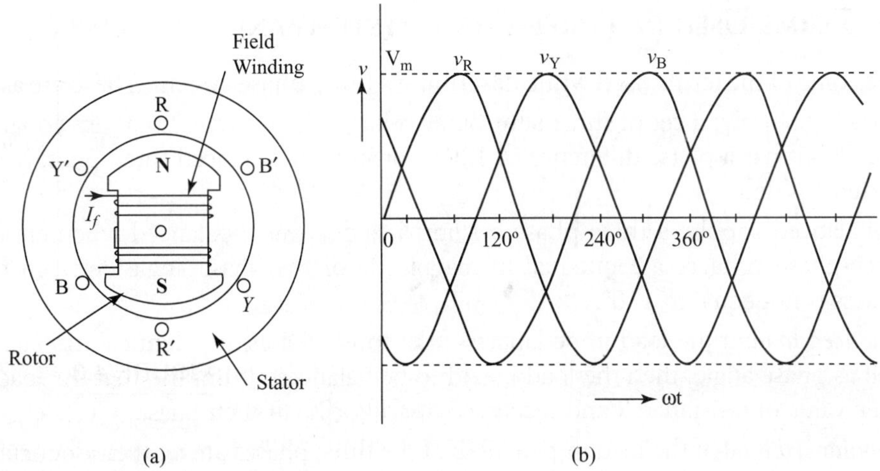

FIGURE 4.4

(a) Three-phase two-pole generator; (b) three-phase voltages induced in the windings

be displaced in time-phase by 120° as has been shown. The order in which the phase voltages attain their maximum value or peak value, V_m is called the phase sequence. If the rotor rotates in the clockwise direction, voltages in the phases will be induced in the sequence, V_R, V_Y, V_B. If the rotor poles rotate in the opposite direction the phase sequence of the induced voltage will change from RYB to RBY.

The phase voltages should supply power to electrical loads for which the two end terminals of each phase are to be connected to loads as shown in Fig. 4.5. Six wires are required to be taken out from the generator to the load which is not followed in practice. Instead of taking out six wires from the generator and connecting them to the loads separately as shown in Fig. 4.5, the *three-phase windings are connected either in star or in delta so that only three wires are to be taken out from the generator to the load. The loads are also connected either in star or in delta.* In the case of the star connection a fourth wire may be taken out from the neutral point.

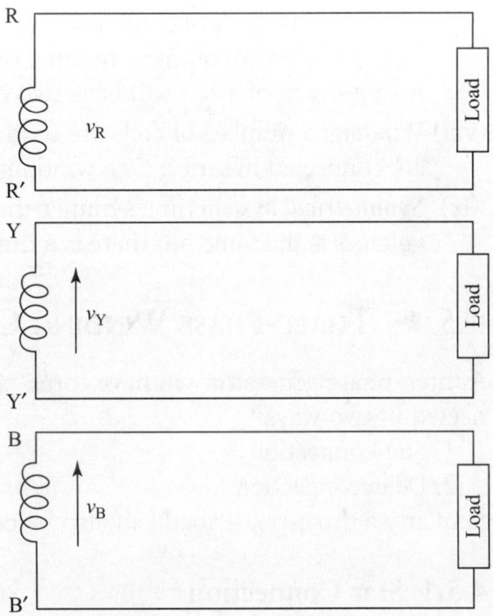

FIGURE 4.5

The three-phase windings are shown connected to the load independently through six wires which is not followed in practice

4.4 ► Terms Used in Three-Phase Systems and Circuits

These are some of the terms used while describing a three-phase system. These are as follows:

(i) Balanced supply: a set of three sinusoidal voltages (or currents) that are equal in magnitude but has a phase difference of 120°, constitute a balanced three-phase voltage (or current) system.

(ii) Unbalanced supply: a three-phase system is said to be unbalanced when either of the three-phase voltages are unequal in magnitude or the phase angle between the three phases is not equal to 120°.

(iii) *Balanced load:* if the load impedances of the three phases are identical in magnitude as well as phase angle, then the load is said to be balanced. It implies that the load has the same value of resistance R and reactance X_L and/or X_C in each phase.

(iv) *Unbalanced load:* if the load impedances of the three phases are neither identical in magnitude nor in phase angle then the load is said to be unbalanced.

(v) *Single phasing:* when one phase of the three-phase supply is not available then the condition is called single phasing.

(vi) *Phase sequence:* the order in which the maximum value of voltages of each phase appear is called the phase sequence. It can be RYB or RBY.

(vii) *Coil:* a coil is made of conducting wire, say copper, having an insulation cover. A coil can be of a single turn or many number of turns. Normally a coil will have a number of turns. A single turn of a coil will have two conductors on its two sides called coil sides

(viii) *Winding:* a number of coils are used to make one winding. Normally the winding coils are connected in series. One winding forms one phase.

(ix) *Symmetrical system:* in a symmetrical three-phase system the magnitude of three-phase voltages is the same but there is a time phase difference of 120° between the voltages.

4.5 ► Three-Phase Winding Connections

A three-phase generator will have three-phase windings. These phase windings can be connected in two ways:

1. Star connection
2. Delta connection

Similarly a three-phase load can also be connected either in star or in delta.

4.5.1 Star Connection

The star connection of windings is formed by connecting the starting or finishing ends of all the three windings together. A fourth conductor which is taken out of the star point is called the neutral point. The remaining three ends are brought out for connection to load. These ends are generally referred to as R–Y–B, to which load is to be connected. The star connection is shown in Fig. 4.6 (a). This is a three-phase, four-wire star-connected system. If no neutral conductor is taken out from the system it gives rise to a three-phase, three-wire star-connected system as shown in Fig. 4.6(b).

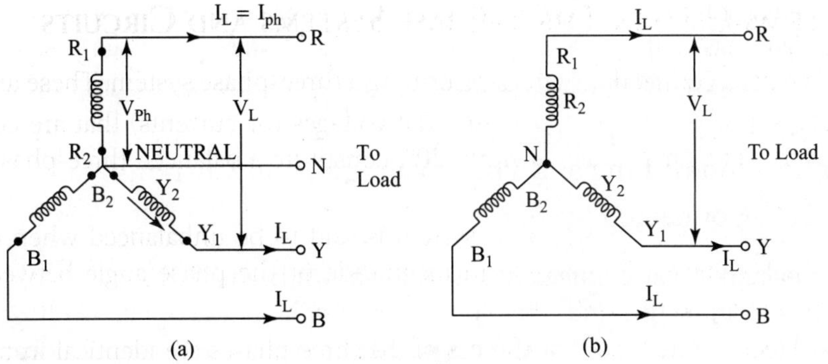

FIGURE 4.6

Star connection of phase windings: (a) three-phase four-wire system; (b) three-phase three-wire system

The current flowing through each line conductor is called line current, I_L. In the star connection the line current is also the phase current. Similarly, voltage across each phase is called phase voltage, V_{Ph}. Voltage across any two line conductors is called line voltage, V_L. When a balanced three-phase load is connected across the supply terminals R,Y,B, currents will flow through the circuit. The sum of these currents, i.e., I_R, I_Y, and I_B will be zero. The neutral wire connected between the supply neutral point and the load neutral point will carry no current for a balanced system.

4.5.2 Delta Connection

The delta connection is formed by connecting the end of one winding to the starting end of the other and connections are continued to form a closed loop. In this case, the current flowing

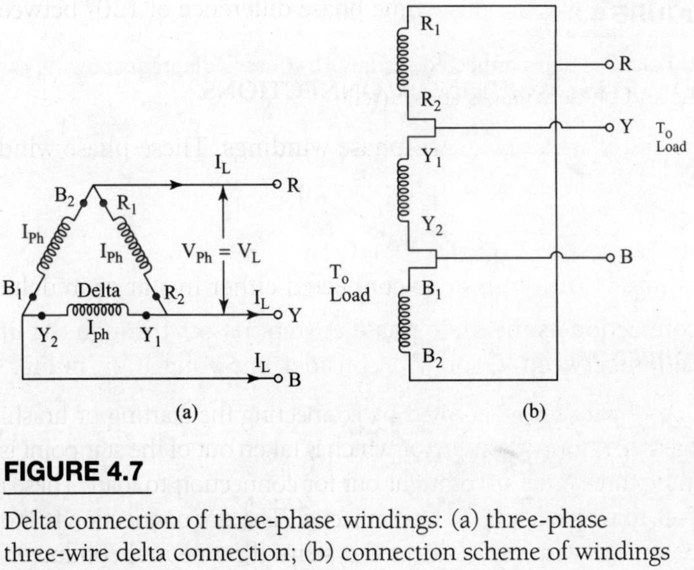

FIGURE 4.7

Delta connection of three-phase windings: (a) three-phase three-wire delta connection; (b) connection scheme of windings forming a delta

through each line conductor is called line current I_L and the current flowing through each phase winding is called phase current, I_{Ph}. However, we find that the phase voltage is the same as the line voltage in a delta connection. The delta connection of windings has been shown in Fig. 4.7.

4.5.3 Relationship of Line and Phase Voltages, and Currents in a Star-Connected System

Consider the balanced star-connected system as shown in Fig. 4.8.

Suppose load is inductive and, therefore, current will lag the applied voltage by angle ϕ. Consider a balanced system so that the magnitude of current and voltage of each phase will be the same.

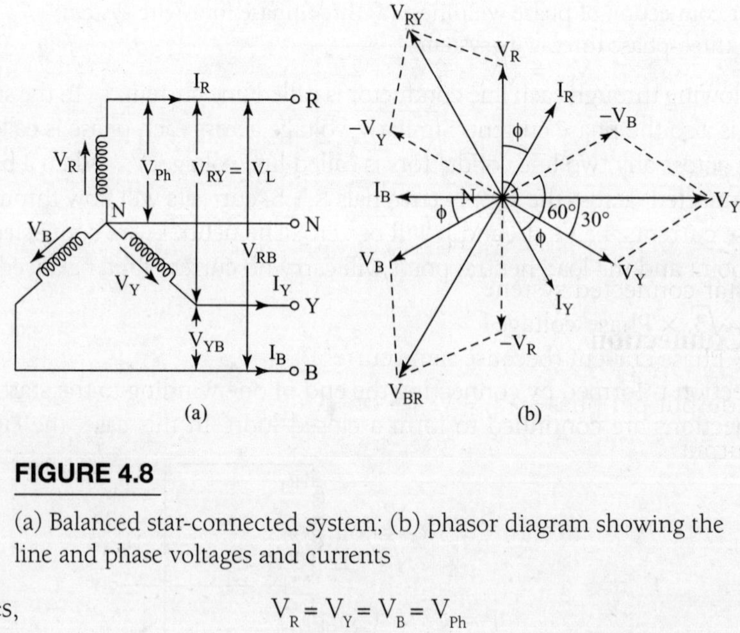

FIGURE 4.8

(a) Balanced star-connected system; (b) phasor diagram showing the line and phase voltages and currents

Phase voltages, $V_R = V_Y = V_B = V_{Ph}$

Line currents, $I_R = I_Y = I_B = I_L$

Line voltages, $V_L = V_{RY} = V_{YB} = V_{BR}$

Phase currents, $I_{Ph} = I_R = I_Y = I_B$

$I_L = I_{PH}$ for star connection as the same phase current passes through the lines to the load.

To derive the relation between V_L and V_{Ph}, consider line voltage V_{RY} in Fig. 4.8 (b),

$$V_{RY} = V_{RN} + V_{NY}$$

$$V_{RY} = V_{RN} + (-V_{YN})$$

Similarly $$V_{YB} = V_{YN} + (-V_{BN})$$

and $$V_{BR} = V_{BN} + (-V_{RN})$$

The procedure for drawing the phasor diagram of Fig. 4.8 (b) is as follows.

Draw three phasors V_R, V_Y, and V_B representing the phase voltages. These voltages are of equal magnitude but displaced by 120°. The line voltage phasors, V_{RY}, V_{YB}, V_{BR} are drawn by vectorially adding the phase voltages. For example, to draw line voltage V_{RY} we have to add the phase voltages as

$$V_{RY} = V_{RN} + V_{NY} = V_{RN} + (-V_{YN})$$

The phasor V_{YN} is obtained by reversing V_{NY}. V_{RY} is obtained by vectorially adding V_{RN} and V_{YN} as has been shown in Fig. 4.8 (b). Similarly the other line voltages have been drawn. The phase currents I_R, I_B, and I_Y have been shown lagging the phase voltages by the power factor angle ϕ.

From the phasor diagram shown in Fig. 4.8 (b), the phase angle between phasors V_R and $(-V_Y)$ is 60°.

$$\therefore \quad V_{RY} = \sqrt{V_R^2 + V_Y^2 + 2V_R V_Y \cos 60°}$$

$$V_{RY} = V_L = \sqrt{V_{Ph}^2 + V_{Ph}^2 + 2V_{Ph} V_{Ph} \times \frac{1}{2}}$$

$$V_L = \sqrt{3}\, V_{Ph} \tag{4.4}$$

Thus, for the star-connected system

Line Voltage = $\sqrt{3}$ × Phase voltage

Line Current = Phase current (because same current flows from the phase winding to the line)

Power: Power output per phase $= V_{Ph} I_{Ph} \cos\phi$

Total power output $= 3\, V_{Ph} I_{Ph} \cos\phi$

$$= 3 \times \frac{V_L}{\sqrt{3}} \times I_L \cos\phi$$

$$P = \sqrt{3}\, V_L I_L \cos\phi \tag{4.5}$$

Power = $\sqrt{3}$ × line voltage × line current × power factor

4.5.4 Relationship of Line and Phase Voltages and Currents in a Delta-Connected System

Consider the balanced delta-connected system as shown in Fig. 4.9.

In a delta-connected system the voltage aeros the winding, i.e., the phases is the same as that across the line terminals. However, the current through the phases is not the same as through the supply lines.

Therefore, in the case of the delta-connected circuit, phase voltage is equal to the line voltage, but line current is not equal to phase current.

Line voltage	$V_L = V_{RY} = V_{YB} = V_{BR}$
Line current	$I_L = I_R = I_Y = I_B$
Phase voltage	$V_{Ph} = V_{RY} = V_{YB} = V_{BR}$
Phase current	$I_{Ph} = I_{RY} = I_{YB} = I_{BR}$

$\therefore V_{Ph} = V_L$ for delta-connected load. (4.6)

In Fig. 4.9 (a) is shown a three-phase delta-connected supply system connected to a three-phase delta-connected load. The line currents are I_R, I_Y, and I_B, respectively. The phase currents are I_{RY}, I_{YB}, and I_{BR}. The phasor diagram in Fig. 4.9 (b) has been developed by first showing the three-phase voltages V_{YB}, V_{BR}, and V_{RY} of equal magnitude but displaced by 120° from each other. Then the phase currents I_{YB}, I_{BR}, and I_{RY} have been shown lagging respective phase voltages by power factor angle ϕ. The line currents are drawn by applying KCL at the nodes R, Y, and B and adding the phasors, as has been shown.

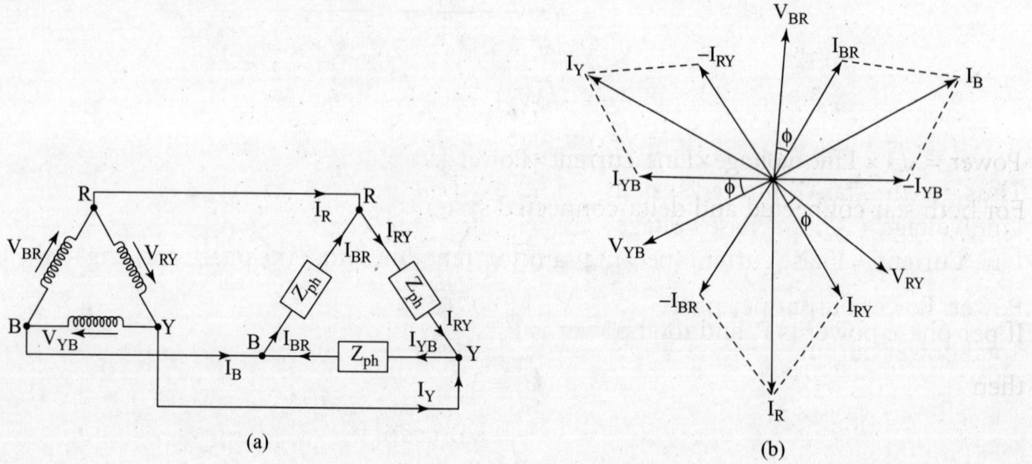

FIGURE 4.9

Delta-connected system: (a) a three-phase delta-connected load supplied from a delta-connected supply source; (b) phasor diagram of voltages and currents

To derive the relation between I_L and I_{Ph}, apply KCL at node R, as shown in Fig. 4.9 (b)

$$I_R + I_{BR} = I_{RY}$$

$$\therefore \quad I_R = I_{RY} - I_{BR}$$

Similarly at node Y and B, we can write

$$I_Y = I_{YB} - I_{RY}$$

$$I_B = I_{BR} - I_{YB}$$

Since phase angle between phase currents I_{RY} and $-I_{BR}$ is 60°

$$\therefore \quad I_R = \sqrt{I_{RY}^2 + I_{BR}^2 + 2I_{RY}\, I_{BR} \cos 60°}$$

$$I_R = I_L = \sqrt{I_{Ph}^2 + I_{Ph}^2 + 2I_{Ph}\, I_{Ph} \times \frac{1}{2}}$$

$$I_L = \sqrt{3} \times I_{ph}$$

Thus, for a three-phase delta-connected system,

Line current = $\sqrt{3}$ × Phase current

Line voltage = Phase voltage

Power: Power output per phase, $P_h = V_{Ph}\, I_{Ph} \cos \phi$

Total power output, $P_T = 3V_{Ph}\, I_{Ph} \cos \phi$

$$= 3 \times V_L \times \frac{I_L}{\sqrt{3}} \cos \phi$$

$$= \sqrt{3} V_L\, I_L \cos \phi$$

Power = $\sqrt{3}$ × Line voltage × Line current × Power factor

For both star-connected and delta-connected systems, the total power P is

$$P = \sqrt{3}\, V_L\, I_L \cos \phi \tag{4.7}$$

If per phase power is P_h and total power is P_T,

then

$$P_T = 3\, P_h \tag{4.8}$$

4.6 ▶ Active and Reactive Power

Electrical circuits consists of resistance, inductance and capacitance elements. In alternating current, circuits power transferred from the source to the load through any phase varies periodically. In energy storage elements like inductors and capacitors periodic reversal of the direction of energy flow occurs. The net transfer of energy in one direction is known as active power or real power. The portion of power due to stored energy, which gets returned to the source in each cycle is called the reactive power. If the load is purely reactive, i.e., if the load consists of inductance and capacitance, the current lags and leads the voltage by 90 degrees. The product of voltage and current is positive for half the cycle and negative for next half cycle in every cycle. This indicates that whatever energy flows into the circuit is immediately returned back to the source. There is no net energy flow to the load.

Practical electrical loads have resistance, inductance and capacitance. Therefore, both active and reactive power will flow to real loads.

Apparent power in the product of rms values of voltage and current.

Voltage, V_{ph} on the reference axis, current I_{ph} lagging V by an angle ϕ is shown in Fig. 4.10.

The in-phase component of I_{Ph} along V has been shown in Fig. 4.10 as I_{Ph} cos ϕ and the perpendicular component as I_{Ph} sin ϕ. If we multiply all the sides of the triangle ABC by V_{Ph}, the triangle becomes a power triangle where AB = V_{Ph} I_{Ph} cos ϕ is called the *active power*, BC = V_{Ph} I_{Ph} sin ϕ is called the *reactive power*, and V_{Ph} I_{Ph} is called the *apparent power*.

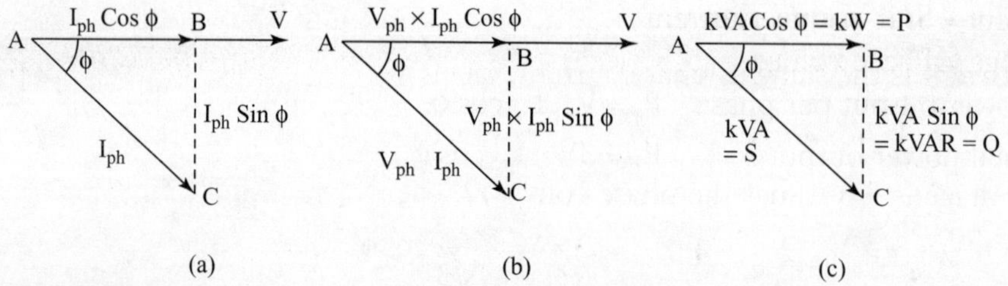

FIGURE 4.10

Relationship between active power, reactive power, and apparent power

As shown in Fig. 4.10(c),

Apparent power = kVA = S (4.9)

Apparent power × cos ϕ = Active power = kW = P (4.10)

Apparent power × sin ϕ = Reactive power = kVAR = Q (4.11)

Multiplying all the sides of the power triangle by 10^3, i.e., expressing the power in terms of 'kilo', the power triangle of Fig. 4.10 (b) is redrawn as has been shown in Fig. 4.10 (c), kVA is kilo Volt Ampere.

$$\text{kVA}\cos\phi = \text{kW (kilo Watt)} \tag{4.12}$$

$$\text{kVA}\sin\phi = \text{kVAR (kilo Var)} \tag{4.13}$$

From Figure 4.10 (c), it is observed that AB represents the *Active power,* BC represents the *Reactive power* and AC represents the Apparent power. Angle ϕ is the power factor angle, while cos ϕ is the power factor. If power factor is low, i.e., when angle ϕ is more due to the presence of reactive elements (L or C) in the load circuit, for the same active power, apparent power will increase. More apparent power means more line current and more load on the lines supplying power and more I^2R loss in the conductors.

Electrical engineers therefore has to take into consideration the apparent power while designing electrical systems rather than the active power.

4.7 ► Comparison Between Star Connection and Delta Connection

As mentioned earlier, the three windings of a generator can be connected either in star or in delta. Same is the case with transformers. Three-phase electrical loads and the windings of three-phase motors can also be connected in star or in delta.

The relationship between voltages, currents and their phase relationship along with some other related factors have been compared and are presented in a tabular form.

Star connection	Delta connection
1. Line current is the same as phase current, i.e., $I_L = I_{Ph}$	1. Line current is $\sqrt{3}$ times the phase current, i.e., $I_L = \sqrt{3}\, I_{Ph}$
2. Line voltage is $\sqrt{3}$ times the phase voltage, i.e., $V_L = \sqrt{3}\, V_{Ph}$	2. Line voltage is the same as phase voltage, i.e., $V_L = V_{ph}$
3. Total power = $\sqrt{3}\, V_L I_L \cos\phi$	3. Total power = $\sqrt{3}\, V_L I_L \cos\phi$
4. Per phase power = $V_{Ph} I_{Ph} \cos\phi$	4. Per phase power = $V_{Ph} I_{Ph} \cos\phi$
5. Three-phase three-wire and three-phase four-wire systems are possible	5. Three-phase three-wire system is possible
6. Line voltages lead the respective phase voltages by 30°	6. Line currents lag the respective phase currents by 30°

EXAMPLE 4.1

A 400 V, three-phase, 50 Hz power supply is applied across the three terminals of a delta-connected three-phase load. The resistance and reactance of each phase is 6 Ω and 8 Ω, respectively. Calculate the line current, phase current, active power, reactive power, and apparent power of the circuit.

SOLUTION

The load is delta connected. Hence,

$$V_{Ph} = V_L = 400 \text{ V}$$

$$Z_{Ph} = R + jX = 6 + j8 = \sqrt{6^2 + 8^2}\,\angle \tan^{-1}\frac{8}{6} = 10\angle 53° \ \Omega$$

$$I_{Ph} = \frac{V_{Ph}}{Z_{Ph}} = \frac{400\angle 0°}{10\angle 53°} = 40\angle -53° \text{ A}$$

Power factor, $\cos\phi = \cos 53° = 0.6$ lagging

$$I_L = \sqrt{3} I_{Ph} = 1.732 \times 40 = 69.28 \text{ A}$$

Power factor, $\cos\phi = \frac{R}{Z} = \frac{6}{10} = 0.6$ lagging

$$\sin\phi = 0.8$$

Active Power $= \sqrt{3} V_L I_L \cos\phi = 1.732 \times 400 \times 69.28 \times 0.6$

$$= 28798 \text{ W} = 28.798 \text{ kW}$$

$$\simeq 28.8 \text{ kW}$$

Reactive Power $= \sqrt{3} V_L I_L \sin\phi = 1.732 \times 400 \times 69.28 \times 0.8$

$$= 38397 \text{ VAR} = 38.397 \text{ kVAR}$$

$$\simeq 38.4 \text{ kVAR}$$

Apparent Power $= 3 V_{Ph} I_{Ph} = \sqrt{3} V_L I_L$

$$= 1.732 \times 400 \times 69.28$$

$$= 47997 \text{ VA} = 47.997 \text{ kVA}$$

$$\simeq 48 \text{ kVA}$$

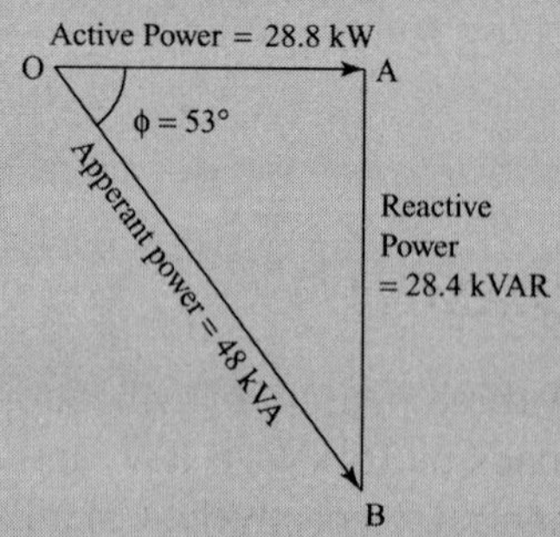

FIGURE 4.11

The power triangle is shown in Fig. 4.11.

Apparent Power in $\text{kVA} = \sqrt{(\text{Active Power})^2 + (\text{Reactive Power})^2}$

$$= \sqrt{(28.8)^2 + (38.4)^2}$$

$$= \sqrt{829.44 + 1474.56}$$

$$= \sqrt{2304} = 48$$

EXAMPLE 4.2

A balanced star-connected load of (8 + j6) Ω per phase is connected to a balanced three-phase, 400 V supply. Find the line current, power factor, power, and total volt-amperes.

SOLUTION

Phase voltage, $V_{ph} = \frac{\text{Line Voltage, } V_L}{\sqrt{3}} = \frac{400}{\sqrt{3}} = 231 \text{ V}$

Impedence per phase, $Z_{Ph} = \sqrt{R^2 + X_L^2} = \sqrt{8^2 + 6^2} = 10\ \Omega$

Phase current, $I_{Ph} = \dfrac{V_P}{Z_P} = \dfrac{231}{10} = 23.1\ A$

Line current $I_L = I_{Ph} = 23.1\ A$

Power factor $\cos\phi = \dfrac{R}{Z} = \dfrac{8}{10} = 0.8(\text{lagging})$

Total power, $P = \sqrt{3}\, V_L\, I_L \cos\phi$

$= \sqrt{3} \times 400 \times 23.1 \times 0.8$

$= 12{,}800\ W$

Total volt amperes $= \sqrt{3}\, V_L\, I_L$

$= \sqrt{3} \times 400 \times 23.1 = 16{,}000\ VA$

EXAMPLE 4.3

A three-phase four-wire supply system has a line voltage of 400 V. Three non-inductive loads of 16 kW, 8 kW, and 12 kW are connected between R, Y, and B phases and the neutral, respectively. Calculate the current flowing through the neutral wire.

SOLUTION

Loads connected between the different phases and the neutral are of 16 kW, 8 kW, and 12 kW, respectively.

The current through the neutral wire line is the phasor sum of all the line currents. We will first calculate the line currents and then add them vectorially.

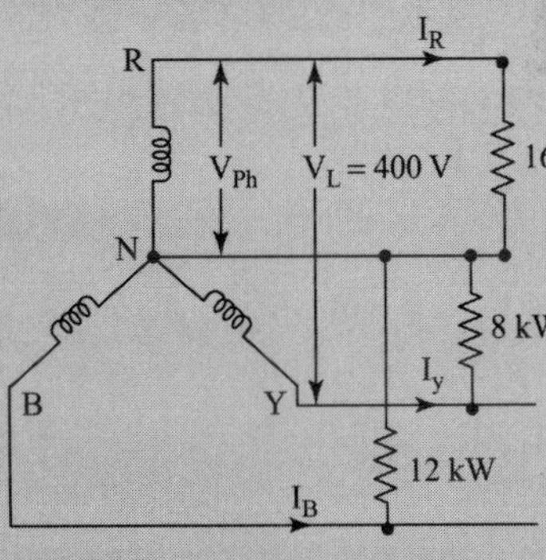

FIGURE 4.12

For star connection, $V_{Ph} = \dfrac{V_L}{\sqrt{3}} = \dfrac{400}{VB} = 231\text{V}$

The three-phase voltages are equal but have a phase difference of 120° between them.

$V_R = 231\angle 0°, V_Y = 231\angle -120°, \text{and } V_B = 231\angle -240°$

$I_R = \dfrac{10 \times 1000}{V_R} = \dfrac{16 \times 1000}{231\angle 0} = 69.3\angle 0°$

$$I_Y = \frac{8\times 1000}{V_Y} = \frac{8\times 1000}{231\angle -120°} = 34.6\angle 120°$$

$$I_B = \frac{12\times 1000}{V_B} = \frac{12\times 1000}{231\angle -240°} = 52\angle 240°$$

Current through the neutral wire, I_N is

$$I_N = I_R + I_Y + I_B$$
$$= 69.3\angle 0° + 34.6\angle 120° + 52\angle 240°$$
$$= 69.3(\cos 0° + j\sin 0°) + 34.6(\cos 120° + j\sin 120°)$$
$$+52(\cos 240° + j\sin 240°)$$
$$= 69.3(1 + j0) + 34.6(-0.5 + j0.866)$$
$$+52(-0.5 - j0.866)$$
$$I_N = 69.3 - 17.3 + j30 - 26 - j45$$
$$= 26 - j15$$
$$I_N = \sqrt{(26)^2 + (15)^2}\angle -\tan^{-1}\frac{15}{26}$$
$$= 30\angle -30° \text{ A}$$

Note: If the load connected to the three phase are equal, then current through the neutral wire, I_N will be zero.

EXAMPLE 4.4

A three-phase load has a resistance and reactance of 6 Ω each for all the three phases. The load is connected in star. A 400 V, 50 Hz, three-phase supply is connected across the load. Calculate phase voltage, phase current, power factor, power consumed per phase, and the total power consumed by the load.

SOLUTION

$$V_{Ph} = \frac{V_L}{\sqrt{3}} = \frac{400}{\sqrt{3}} = 231 \text{ V}$$

$$Z_{Ph} = 6 + j\,6\ \Omega = 8.48\ \angle 45°\ \Omega$$

$$I_{Ph} = \frac{V_{Ph}}{Z_{Ph}} = \frac{231\angle 0°}{8.48\angle 45°} = 27.2\angle -45° \text{ A}$$

$$I_{Ph} = I_L = 27.2 \text{ A}$$

Angle between V_{Ph} and I_{Ph} is 45°.

Power factor = cos 45° = 0.7 lagging

Power absorbed by each phase of the load = $V_{Ph}\, I_{Ph} \cos \phi$

= 231 × 27.2 × 0.7

= 4398 W = 4.398 kW

(Total power consumed = 3 × 4398 = 13194 W = 13.194 kW

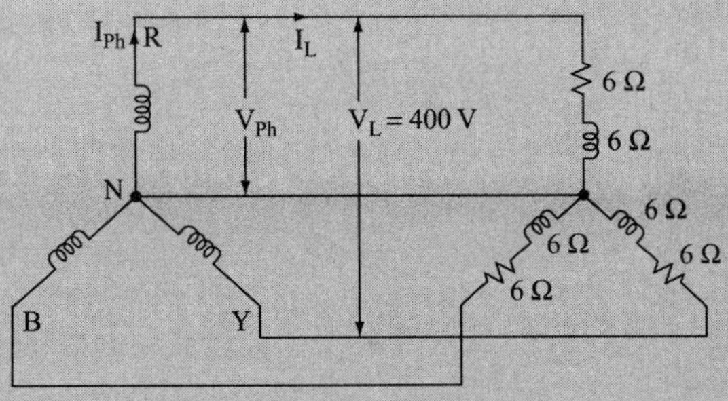

FIGURE 4.13

EXAMPLE 4.5

A 400 V, 50 Hz, three-phase supply is provided to a three-phase star-connected load. Each phase of the load absorbs a power of 2000 W. The load power factor is 0.8 lagging. Calculate the total power supplied to the load and the line current.

SOLUTION

Power consumed by each phase = 2000 W

Power consumed by all the three phases = 3 × 2000 W

= 6000 W

Total power supplied = total power consumed.

$$V_L = 400 \text{ V}, \; V_{ph} = \frac{V_L}{\sqrt{3}} = \frac{400}{\sqrt{3}} = 231 \text{ V}$$

Power consumed per phase $= V_{Ph}\, I_{Ph} \cos \phi$

$$V_{Ph}\, I_{Ph} \cos \phi = 2000$$

$$I_L = I_{Ph} = \frac{2000}{231 \times 0.8} = 10.82 \text{ A}$$

EXAMPLE 4.6

A 400 V, 50 Hz, three-phase supply is provided to a three-phase delta-connected load. The resistance and inductance of each phase of the load is 8 Ω and 0.04 H, respectively. Calculate the phase current and the line current drawn by the load. Also calculate the total power consumed.

SOLUTION

The impedance of load per phase,

$$Z_{Ph} = R + j\,\omega L = 8 + j\,2\pi \times 50 \times 0.04$$
$$= 8 + j\,12.56\ \Omega.$$

Since the load is delta connected, $V_{Ph} = V_L = 400$ V

$$I_{ph} = \frac{V_{Ph}}{Z_{Ph}} = \frac{400}{8 + j\,12.56} = \frac{400\angle 0}{14.89\angle \tan^{-1}(12.56/8)} = 26.86\ \angle -58°;$$

$$I_L = \sqrt{3}\ I_{ph} = 1.73 \times 26.86 = 36\ \text{A}$$

Total power consumed $= 3\ V_{Ph}\ I_{Ph} \cos\phi$

$$= 3 \times 400 \times 26.86 \cos 58°\ \text{W}$$

$$= 3 \times 400 \times 26.86 \times 0.52 = 1736\ \text{W}$$

EXAMPLE 4.7

A three-phase star-connected load consumes a total of 12 kW at a power factor of 0.8 lagging when connected to a 400 V, three-phase, 50 Hz power supply. Calculate the resistance and inductance of load per phase.

SOLUTION

Total power consumed = 12 kW

Per phase power consumed = 4 kW

So, $V_{Ph}\ I_{Ph} \cos\phi = 4000$ W

$$I_{ph} = \frac{4000}{V_{ph} \cos\phi}$$

or,

$$I_{Ph} = \frac{4000}{(400/\sqrt{3}) \times 0.8} = 21.6\ \text{A}$$

$$I_{Ph} = \frac{V_{Ph}}{Z_{Ph}}$$

$$Z_{Ph} = \frac{V_{Ph}}{I_{Ph}} = \frac{231}{21.6} = 10.7\ \Omega.\ \text{Power factor,}\ \cos\phi = 0.8,\ \sin\phi = 0.6$$

$$R = Z_{Ph} \cos\phi = 10.7 \times 0.8 = 8.56\ \Omega$$

$$X = Z_{Ph} \sin\phi = 10.7 \times 0.6 = 6.42\ \Omega$$

The impedance triangle is shown in Figure 4.14.

$$X = 6.42$$

$$2\pi fL = 6.42$$

$$L = \frac{6.42}{2 \times 3.14 \times 50} = 20.4 \times 10^{-3}\ \text{H}$$

$$= 20.4\ \text{mH}$$

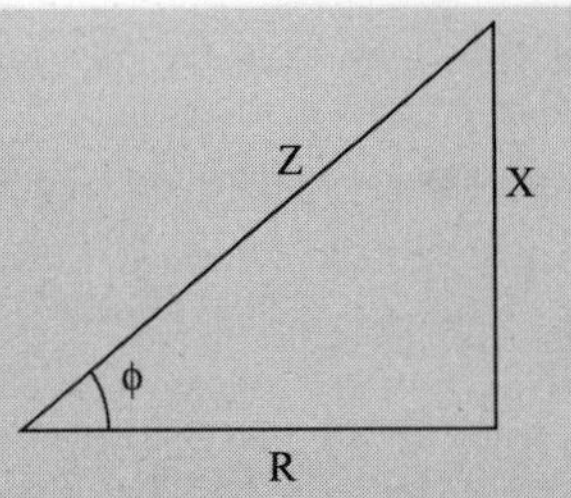

FIGURE 4.14

$\text{Cos}\,\phi = \frac{R}{Z}$

$\therefore R = z\cos\phi$

$\text{Sin}\,\phi = \frac{x}{z}$

$\therefore x = z\sin\phi$

EXAMPLE 4.8

A balanced three-phase star-connected load of 8 + j6 Ω per phase is supplied by a 400 V, 50 Hz supply. Calculate the line current, power factor, active, and reactive power.

SOLUTION

$$V_L = 400\ \text{V, For star connection, } V_{Ph} = \frac{V_L}{\sqrt{3}} = \frac{400}{\sqrt{3}} = 231$$

$$Z_{Ph} = 8 + j6 = \sqrt{8^2 + 6^2}\ \angle \tan^{-1}\frac{6}{8} = 10\angle 37°$$

$$I_{Ph} = \frac{V_{Ph}}{Z_{Ph}} = \frac{231}{10\angle 37°} = 23.1\angle -37°$$

For star connection,

$$I_{Ph} = I_L = 23.1\ \text{A}$$

Angle of lag of I_{Ph} with V_{Ph} is 37°

Power factor $= \cos\phi = \cos 37° = 0.8$ lagging

Active power $= \sqrt{3}\, V_L I_L \cos\phi$

$= 1.732 \times 400 \times 23.1 \times 0.8$

$= 12802$ W $= 12.802$ kW

Reactive power $= \sqrt{3}\, V_L I_L \sin\phi$

$= 1.732 \times 400 \times 23.1 \times 0.6$

$= 9602$ VAR $= 9.602$ kVAR

EXAMPLE 4.9

A delta-connected three-phase motor load is supplied from a 400 V, three-phase, 50 Hz supply system. The line current drawn is 21 A. The input power is 11 kW. What will be the line current and power factor when the motor windings are star connected?

SOLUTION

Line voltage, $V_L = 400$ V, $V_{Ph} = V_L$ for delta connection

Line current, $I_L = \sqrt{3} I_{Ph} = 21$ A

Impedance of each winding, $Z_{Ph} = \dfrac{V_{Ph}}{I_{Ph}} = \dfrac{400}{I_L/\sqrt{3}} = \dfrac{400 \times \sqrt{3}}{21} = 33\ \Omega$

$$P = \sqrt{3} V_L I_L \cos\phi = \sqrt{3} \times 400 \times 21 \times \cos\phi$$

or, $\cos\phi = \dfrac{P}{\sqrt{3} \times 400 \times 21} = \dfrac{11 \times 1000}{\sqrt{3} \times 400 \times 21} = 0.756$

When the motor windings are star connected

$$V_{Ph} = \frac{V_L}{\sqrt{3}} = \frac{400}{1.732} = 231 \text{ V}$$

Z_{Ph} will remain the same as the same windings are connected in star

$$I_{Ph} = \frac{V_{Ph}}{Z_{Ph}} = \frac{231}{33} = 7 \text{ A}$$

In star connection, the line current is the same as phase current, hence, $I_L = I_{Ph} = 7$ A.

Power factor, depends on the circuit parameters

$$\cos\phi = \frac{R}{Z}$$

Since both R and Z remain unchanged, the power factor will remain the same at 0.756. Students may note that line current in a star connection is one-third of the line current in a delta connection.

EXAMPLE 4.10

A balanced star-connected load of 4 + j6 Ω per phase is connected across a 400 V, three-phase, 50 Hz supply. Calculate line current, phase current, line voltage, phase voltage, power factor, total power, and reactive power.

SOLUTION

$$Z_{Ph} = 4 + j6 = \sqrt{4^2 + 6^2}\angle\tan^{-1}\frac{6}{4} = 7.21\angle 56°$$

For star connection,

$$V_{Ph} = \frac{V_L}{\sqrt{3}} = \frac{400}{\sqrt{3}} = 231 \text{ V}$$

$$I_{Ph} = \frac{V_{Ph}}{Z_{Ph}} = \frac{231\angle 0}{7.21\angle 56°} = 32\angle{-56°}$$

Power factor,

$$\cos\phi = \cos 56° = 0.56 \text{ lagging}$$

$$I_{Ph} = 32 \text{ A}$$

$$I_{Ph} = I_L = 32 \text{ A}$$

$$V_L = 400 \text{ V}$$

$$V_{Ph} = 231 \text{ V}$$

Total Power

$$= \sqrt{3}V_L I_L \cos\phi$$

$$= 3\ V_{Ph} I_{Ph} \cos\phi$$

$$= 3 \times 231 \times 32 \times 0.56$$

$$= 12418 \text{ W}$$

$$= 12.418 \text{ kW}$$

Total reactive power

$$= \sqrt{3}V_L I_L \sin\phi$$

$$= 3\ V_{Ph} I_{Ph} \sin\phi$$

$$= 3 \times 231 \times 32 \times 0.83$$

$$= 18406 \text{ VAR}$$

$$= 18.406 \text{ kVAR}$$

4.8 ► Measurement of Power in Three-Phase Circuits

We have known that in dc circuits power is calculated as the product of voltage and current, i.e., power, P = VI. DC power can be measured using a voltmeter and an ammeter. In ac circuits power, P = VI cos ϕ. In three-phase ac circuits, total power is three times the power per phase. Wattmeter is an instrument used for measurement of power in ac circuits. Wattmeters are available as single-phase wattmeters and three-phase wattmeters. Single-phase wattmeters can also be used to measure three-phase power. In case of star-connected balanced load with neutral connection, only one single-phase wattmeter can be used to measure the three-phase power. The three-phase power is three times the single-phase power. For unbalanced three-phase loads, i.e., if the currents in the three phases are not the same, two wattmeters are to be used to measure the three-phase power. These methods are described in the following sections.

4.8.1 One-Wattmeter Method

In this method, only one single-phase wattmeter can be used to measure the total three-phase power. In this method, the current coil (CC) of the wattmeter is connected in series with any phase and the pressure coil (PC) is connected between that phase and the neutral as shown in Fig. 4.15. One-wattmeter method has a demerit that even a slight degree of unbalance in the load produces a large error in the measurement. In this method one wattmeter will measure only the power of one phase. Hence, total power is taken as three times the wattmeter reading.

$$\therefore \qquad \text{Total Power} = 3 \times V_{ph}\, I_{ph} \cos\phi$$

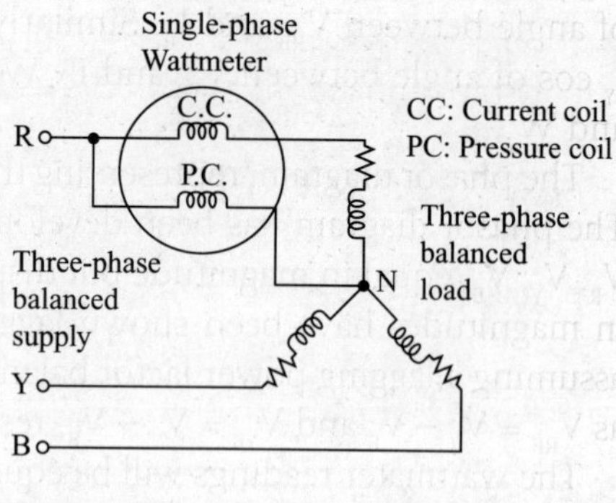

FIGURE 4.15

One-wattmeter method of measuring power of a star-connected balanced three-phase load

4.8.2 Two-Wattmeter Method

This method requires only two wattmeters to measure three-phase power for balanced as well as unbalanced loads. In this method cc of the two wattmeters are connected in two phases and their pressure coils togather are connected to the remaining third phase as has been shown in Fig. 4.16.

This method of measurement is useful for balanced and unbalanced loads.

Let us consider the measurement of three-phase power of a star-connected load using two single-phase wattmeters as has been shown in Fig. 4.17(a). We will calculate the power measured by the two wattmeters separately. Let W_1 and W_2, respectively, be the two wattmeter

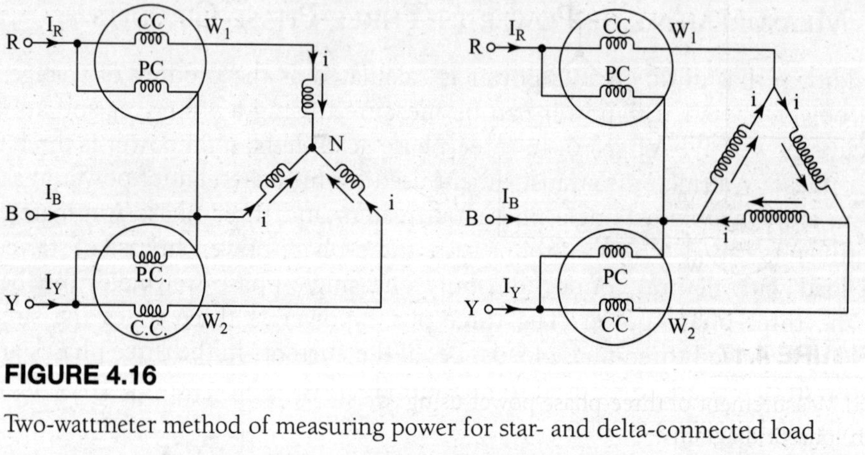

FIGURE 4.16

Two-wattmeter method of measuring power for star- and delta-connected load readings. Current flowing through the current coil of wattmeter W_1 is I_R. The voltage appearing across its pressure coil is V_{RB}. The wattmeter reading will be equal to, $W_1 = V_{RB} I_R$. cos of angle between V_{RB} and I_R. Similarly, the wattmeter reading W_2 will be equal to, $W_2 = V_{YB}$ I_Y cos of angle between V_{YB} and I_Y. We will now draw the phasor diagram, and calculate W_1 and W_2.

The phasor diagram, representing the voltages and currents, has been shown in Fig. 4.17 (b). The phasor diagram has been developed by first considering the three-phase supply voltages V_R, V_Y, V_B, equal in magnitude but displaced by 120°. The phase or line currents I_R, I_Y, I_B equal in magnitudes have been shown lagging the respective phase voltages by an equal angle ϕ, assuming a lagging power factor balanced load. The line voltages V_{RB} and V_{YB} have been drawn as $V_{RB} = V_R - V_B$ and $V_{YB} = V_Y - V_B$, respectively.

The wattmater readings will be equal to VI cos ϕ in each case where V is the voltage applied across the pressure coils, I is the current flowing through the current coils, and ϕ is the angle between V and I.

From the phasor diagram as shown in Fig. 4.17 (b), the wattmeters readings will be.

$$W_1 = V_{RB} I_L \cos(30-\phi) = \sqrt{3} V_{ph} I_{ph} \cos(30-\phi) = V_L I_L \cos(30-\phi) \quad (4.14)$$

And

$$W_2 = V_{YB} I_Y \cos(30+\phi) = \sqrt{3} V_{ph} I_{ph} \cos(30+\phi) = V_L I_L \cos(30+\phi) \quad (4.15)$$

We know that the total power in a three-phase circuit is $3V_{ph} I_{ph} \cos\phi$ or equal to $\sqrt{3} V_L I_L \cos\phi$.

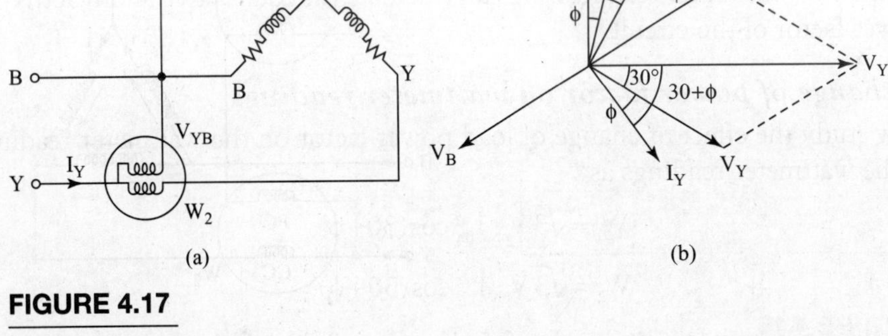

FIGURE 4.17

(a) Measurement of three-phase power using two single-phase wattmetters;
(b) phasor diagram

In star connection, phase current I_{ph} is the same as line current, I_L.

Let us add the two wattmeter readings, i.e., W_1 and W_2.

$$W_1 + W_2 = \sqrt{3}\,V_{ph}\,I_{ph}\cos(30-\phi) + \sqrt{3}\,V_{ph}\,I_{ph}\cos(30+\phi)$$
$$= \sqrt{3}\,V_{ph}\,I_{ph}\,[\cos(30-\phi) + \cos(30+\phi)]$$
$$= \sqrt{3}\,V_{ph}\,I_{ph}\,2\cos 30^\circ \cos\phi$$
$$= \sqrt{3}\,V_{ph}\,I_{ph}\,2\frac{\sqrt{3}}{2}\cos\phi$$
$$= 3V_{ph}\,I_{ph}\cos\phi$$

or, $$W_1 + W_2 = \sqrt{3}\,V_L\,I_L\cos\phi \text{ (Active power)} \qquad (4.16)$$

Thus, it is proved that the sum of the two wattmeter readings is equal to the three-phase power.

Now, let us see what we get when the two wattmeter readings are subtracted from each other

$$W_1 - W_2 = \sqrt{3}\,V_{ph}\,I_{ph}\,[\cos(30^\circ - \varphi) - \cos(30^\circ + \varphi)]$$
$$= \sqrt{3}\,V_{ph}\,I_{ph}\,2\sin\varphi\,\sin 30^\circ$$

or, $$\sqrt{3}(W_1 - W_2) = 3\,V_{ph}\,I_{ph}\sin\varphi$$

or, $$\sqrt{3}(W_1 - W_2) = \sqrt{3}\,V_L\,I_L\sin\varphi \text{ (Reactive power)} \qquad (4.17)$$

Dividing eq. (4.17) by eq. (4.16)

$$\frac{\sqrt{3}(W_1 - W_2)}{W_1 + W_2} = \frac{\sqrt{3}\,V_L\,I_L\sin\phi}{\sqrt{3}\,V_L\,I_L\cos\phi} = \tan\phi$$

or,

$$\varphi = \tan^{-1}\frac{\sqrt{3}(W_1 - W_2)}{W_1 + W_2}$$

Power factor, $$\cos\varphi = \cos\tan^{-1}\frac{\sqrt{3}(W_1 - W_2)}{W_1 + W_2} \qquad (4.18)$$

Thus, from the two wattmeter readings, we can calculate the total active and reactive powers and the power factor of the circuit.

Effect of change of power factor on wattmeter readings

We will now study the effect of change of load power factor on the wattmeter readings. Let us rewrite the wattmeter readings as

$$W_1 = \sqrt{3}\,V_{ph}\,I_{ph}\cos(30-\phi)$$
$$W_2 = \sqrt{3}\,V_{ph}\,I_{ph}\cos(30+\phi)$$

We will consider a power factor of unity, 0.5, less than 0.5, and 0 and study the effect on the wattmeter readings.

(i) *At unity power factor, i.e.*, when $\cos\phi=1$, i.e., $\phi=0$

$$W_1 = \sqrt{3}\,V_{Ph} I_{Ph} \cos(30-0°) = \sqrt{3} V_{Ph} I_{Ph} \cos 30°$$
$$W_2 = \sqrt{3} V_{Ph} I_{Ph} \cos(30+0°) = \sqrt{3}\,V_{Ph} I_{Ph} \cos 30°.$$

Thus, at Power factor = 1, both the wattmeter readings will be positive and of equal value.

(ii) *At 0.5 power factor*, i.e., $\cos\phi = 0.5$, i.e., $\phi = 60°$.

$$W_1 = \sqrt{3}\,V_{ph}\,I_{ph} \cos(-30°)$$
$$= \sqrt{3}\,V_{ph}\,I_{ph} \cos 30°$$
$$W_2 = \sqrt{3}\,V_{ph}\,I_{ph} \cos(30° + 60°) = 0$$

Thus, at power factor equal to 0.5, one of the wattmeters will give zero reading.

(iii) *When the power factor is less than 0.5*, i.e., when $\phi > 60°$. Let us observe the wattmeter readings.

$$W_1 = \sqrt{3}\,V_{ph}\,I_{ph} \cos(30-\phi)$$
$$W_2 = \sqrt{3}\,V_{ph}\,I_{ph} \cos(30+\phi)$$

When $\phi > 60$, W_1 will give positive readings but W_2 will give a negative reading.

Thus, for power factor less than 0.5, i.e., for $\phi > 60°$, one of the wattmeters will give a negative reading.

(iv) *When load is purely inductive or capacitive*, so power factor will be zero, i.e., $\phi = 90°$

$$W_1 = V_L I_L \cos(30° - 90°) = V_L I_L \cos 60°$$
$$W_2 = V_L I_L \cos(30° + 90°) = -V_L I_L \sin 30°$$

Both the wattmeters show equal but opposite readings. Hence, the total power consumed will, as the sum of W_1 and W_2, be zero.

4.8.3 Three-Wattmeter Method

In this method, three wattmeters are used to measure three-phase power. Three wattmeters are connected in each phase as has been shown in Fig. 4.18. Their pressure coils are connected between each phase and the neutral. This method is valid for the measurement of three-phase power for balanced and unbalanced loads. The main drawback of this method is the requirement of three wattmeters. Availability of the neutral wire is also essential.

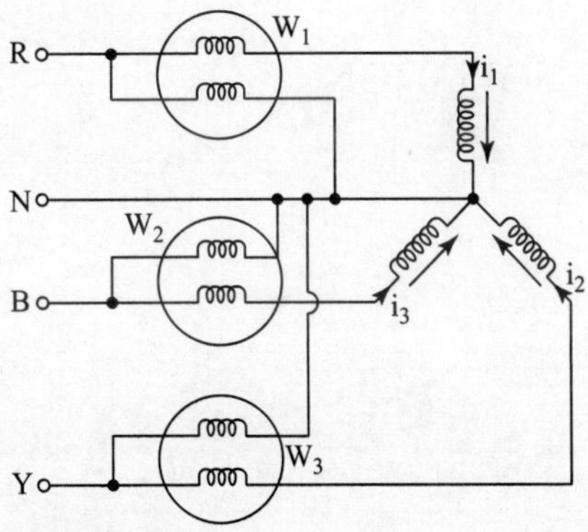

FIGURE 4.18

Measurement of three-phase balanced or unbalanced power using three single-phase wattmeters

EXAMPLE 4.11

In the two-wattmeter method of power measurement for a three-phase load, the readings of the wattmeter are 1000 W and 550 W. What is the power factor of the load?

SOLUTION

$$W_1 = 1000 \text{ W}$$

$$W_2 = 550 \text{ W}$$

Power factor of load,

$$\cos\phi = \cos\tan^{-1}\sqrt{3}\frac{W_1 - W_2}{W_1 + W_2} \quad \text{(see eqn 4.18)}$$

$$= \cos\tan^{-1}\sqrt{3}\frac{1000 - 550}{1000 + 550}$$

$$= \cos 26.695^\circ$$

$$= 0.893 \text{ lagging}$$ (lagging p.f. load was assumed for eqn 4.18)

EXAMPLE 4.12

In the measurement of three-phase power by the two-wattmeter method, for a certain load, one of the wattmeters reads 20 kW and the other 5 kW after the current coil connection of one of the wattmeters reading 5 kW has been reversed. Calculate the power and power factor of the load.

SOLUTION

$$W_1 = 20 \text{ kW}$$

$$W_2 = -5 \text{ kW}$$ (If connection is reversed, use –ve sign)

$$P = W_1 + W_2 = 20 - 5 = 15 \text{ kW}$$

Power factor of the load

$$= \cos\tan^{-1}\frac{W_1 - W_2}{W_1 + W_2}\sqrt{3}$$

$$= \cos\tan^{-1}\frac{20 - (-5)}{20 + (-5)}\sqrt{3}$$

$$= 0.327 \text{ lagging}$$

EXAMPLE 4.13

Draw the connection diagram for measurement of power in a three-phase star-connected load using the two-wattmeter method. In one such a measurement the load connected was 30 kW at 0.7 pf lagging. Find the reading of each wattmeter.

SOLUTION

The connection diagram for measurement of power in a three-phase Y-connected load using the two-wattmeter method has been shown in Fig. 4.17. We know that the reading of the two wattmeters will be $\sqrt{3}\ V_{ph}\ I_{ph}\ \cos(30 - \phi)$ and $\sqrt{3}\ V_{ph}\ I_{ph}\ \cos(30 + \phi)$, respectively. For star connection, $\sqrt{3}\ V_{ph} = V_L$ and $I_{ph} = I_L$.

The total load, $P = 30$ kW

Power factor, $\cos\phi = 0.7$ lagging

Phase angle, $\phi = \cos^{-1}(0.7) = 45.57°$ lagging

So,

$$V_L I_L = \frac{P\,\text{in}\,\text{kW} \times 1000}{\sqrt{3}\cos\phi}$$

$$= \frac{30 \times 1000}{\sqrt{3} \times 0.7} = 24743.6\,\text{VA}$$

Reading of wattmeter,

$$W_1 = V_L I_L \cos(30 - \phi)$$
$$= 24743.6 \cos(30 - 45.57°)$$
$$= 23.835\ \text{kW}$$

Reading of wattmeter,

$$W_2 = V_L I_L \cos(30 + \phi)$$
$$= 24743.6 \cos(30 + 45.57°)$$
$$= 6.165\ \text{kW}$$

To check, total power,

$$P = W_1 + W_2 = 23.835 + 6.165$$
$$= 30\ \text{kW}$$

EXAMPLE 4.14

A three-phase balanced load connected across a 3ϕ, 400 V ac supply draws a line current of 10 A. Two wattmeters are used to measure input power. The ratio of two wattmeter readings is 2:1. Find the readings of the two wattmeters.

SOLUTION

Let the ratio of wattmeter readings be X, i.e., $\dfrac{W_2}{W_1} = X$

and $$\tan\phi = \sqrt{3}\left(\frac{W_1 - W_2}{W_1 + W_2}\right)$$

Now we will divide both numerator and denominator by W_1. Then by putting $\frac{W_2}{W_1}$ as X,

we get, $$\tan\phi = \sqrt{3}\left(\frac{1 - W_2/W_1}{1 + W_2/W_1}\right) = \sqrt{3}\left(\frac{1-X}{1-X}\right)$$

and power factor, $$\cos\phi = \frac{1}{\sec\phi} = \frac{1}{\sqrt{1+\tan^2\phi}}$$

$$= \frac{1}{\sqrt{1+3[(1-X)/(1+X)]^2}}$$

Substituting $$\frac{W_2}{W_1} = \frac{1}{2} = 0.5$$

$$\cos\phi = \frac{1}{\sqrt{1+3(1-0.5/1+0.5)}} = 0.866$$

or, $$\phi = \cos^{-1}(0.866)$$
$$= 30°$$

Wattmeter reading, $W_1 = V_L I_L \cos(30° - 30°) = 400 \times 10 \times \cos 0° = 4000$ W

Wattmeter reading, $W_2 = V_L I_L \cos(30° + 30°) = 400 \times 10 \times \cos 60° = 1000$ W

EXAMPLE 4.15

Three equal impedances, each consisting of R and L in series are connected in star and are supplied from a 400 V, 50 Hz, three-phase, three-wire balanced supply system. The power input to the load is measured by the two-wattmeter method and the two wattmeters read 3 kW and 1 kW, respectively. Determine the values of R and L connected in each phase.

SOLUTION

Reading of wattmeter 1, $W_1 = 3$ kW

Reading of wattmeter 2, $W_2 = 1$ kW

Total power $P = W_1 + W_2 = 3 + 1 = 4$ kW

Power factor of the circuit, $$\cos\phi = \cos\tan^{-1}\frac{w_1 - w_2}{w_1 + w_2}\sqrt{3}$$

$$= \cos\tan^{-1}\frac{3-1}{3+1}\sqrt{3}$$

$$= \cos 40.89$$

$$= 0.7559 \text{ lagging}$$

Line current, $$I_L = \frac{P}{\sqrt{3}\, V_L \cos\phi}$$

$$= \frac{4 \times 1000}{\sqrt{3} \times 400 \times 0.7559} = 7.64 \text{ A} = \text{Phase current, } I_{Ph}$$

Impedance of the circuit per phase $$Z = \frac{V_{Ph}}{I_{Ph}} = \frac{400\sqrt{3}}{7.64}$$

$$= 30.237\ \Omega$$

$\therefore$ $$R = Z\cos\phi = 30.237 \times 0.7559 = 22.856\ \Omega$$

Reactance per phase, $$X_L = \sqrt{Z^2 - R^2}$$

$$= \sqrt{(30.237)^2 - (22.856)^2}$$

$$= 19.796\ \Omega$$

Inductance per phase, $$L = \frac{X_L}{2\pi f}$$

$$= \frac{19.796}{2\pi \times 50}$$

$$= 0.063 \text{ H}$$

$$= 63 \text{ mH}$$

EXAMPLE 4.16

The power input to a three-phase motor is measured by two single-phase wattmeters. The total input power has been measured as equal to 15 kW and the power factor calculated as 0.5. What have been the readings of the two wattmeters?

SOLUTION

Total power = $W_1 + W_2 = 15$ kW

We have to calculate W_1 and W_2

When we measure three-phase power by the two-wattmeter method, the readings of the two wattmeters are

$$W_1 = V_L I_L \cos(30 - \phi)$$

and $$W_2 = V_L I_L \cos(30 + \phi)$$

$$\cos\phi = 0.5,\ \phi = 60°$$

$$\sqrt{3}\, V_L I_L \cos\phi = 15 \text{ kW}$$

$$V_L I_L = \frac{15}{\sqrt{3} \times 0.5} = 17.3 \text{ kVA}$$

$$W_1 = V_L I_L \cos(30 - \phi)$$

$$= 17.3 \cos(30 - 60°)$$

$$= 17.3 \times 0.866 = 15 \text{ kW}$$

$$W_2 = V_L I_L \cos(30 + \phi)$$

$$= 17.3 \cos 90°$$

$$= 17.3 \times 0$$

$$= 0$$

Total power $= W_1 + W_2 = 15 + 0 = 15$ kW

Thus, at a load power factor of 0.5, one of the wattmeters has given zero reading. This has been explained earlier under the effect of change of power factor on wattmeter readings.

Important Terminologies and Points to Remember

1. Three-phase voltage are induced in three-phase windings of the generator. These three voltages will have a time-phase difference of 120 degrees. The physically displaced positions of the three windings produces a time-phase difference in the emf induced in them.
2. A three-phase system has a number of advantages over single-phase system like economy in power generation and transmission, improve power factor, steady power output, etc.
3. A balanced three-phase power supply means that the magnitude of voltage in the three phases are equal and they are displaced in time by 120°. The balanced voltages of the three phases R, Y, B are

   $V_R = V_m \sin \omega t$, $V_Y = V_m \sin(\omega t - 120°)$.

   and $V_Y = V_m \sin(\omega t - 240°)$.
4. A balanced load is constituted by three load impedances of equal magnitude as well as phase angle.
5. Phase sequence is the order in which maximum value of the three phases appear, i.e., as $V_R V_Y V_B$ or $V_R V_B V_Y$.
6. Three-phase windings can be connected either in delta or in star formation. Three terminals are called the phase terminals. A fourth conductor is taken out from the neutral point, in the case of star connection.
7. In star connection, the voltage between any two lines is $\sqrt{3}$ times the voltage of each phase, i.e., $V_L = \sqrt{3}\, V_{ph}$. The phase current, I_{ph} is the same as the line current, I_L.
8. In delta connection, phase voltage is equal to the line voltage, i.e., $V_{ph} = V_L$. Line current, I_L is $\sqrt{3}$ times the phase current I_{ph}, i.e., $I_L = \sqrt{3}\, I_{ph}$.

9. Total power in both star connection and delta connection is $\sqrt{3}\ V_L I_L \cos\phi$ or $3\ V_{ph} I_{ph} \cos\phi$. Where $\cos\phi$ is the power factor. Per phase power is $V_{ph}\ I_{ph}\ \cos\phi$.
10. The product of voltage and current, i.e., VA or volt-ampere or kVA (kilo volt ampere) is called the apparent power. KVA $\cos\phi$ is called the active power and kVA $\sin\phi$ is called the reactive power. Reactive power is due the presence of inductance and/or capacitance in the circuit.
11. One single-phase wattmeter can be used to measure power in a single-phase as well as three-phase star connected balanced circuit. For three-phase power measurement, we multiply the wattmeter reading by 3.
12. For measurement of three-phase power for balanced as well as unbalanced circuit either in star connected or in delta connected, two single-phase wattmeters can be used.

Important Formulae and Equations

Equation	Remarks
1. Power output per phase $= V_{ph}\ I_{ph}\ \cos\phi$	1. Unit is watt or kilowatt.
2. Total power in a 3-phase system $= \sqrt{3}\,V_L I_L \cos\phi$.	2. Valid for both star and delta connected system.
3. Apparent power = kVA Active power = kVA $\cos\phi$ Reactive power = kVA $\sin\phi$	3. For a resistive circuit, reactive power is zero, and hence, apparent-power is the same as active power.
4. Apparent power, $\text{kVA}=\sqrt{(\text{Active Power})^2+(\text{Reactive Power})^2}$	4. Use power triangle as 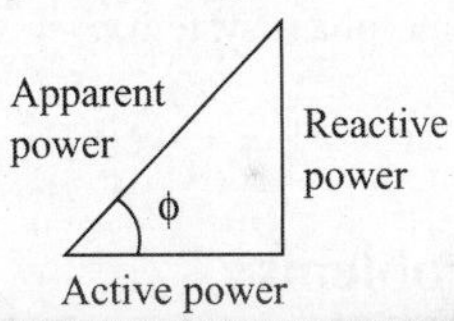
5. In two-wattmeter method of measurement of three-phase power, (i) $W_1=\sqrt{3}\,V_{ph}\,I_{ph}\cos(30-\phi)$ and $W_2=\sqrt{3}\,V_{ph}\,I_{ph}\cos(30+\phi)$ (ii) $W_1+W_2=\sqrt{3}\,V_L\,I_L\cos\phi$ (iii) $\sqrt{3}(W_1-W_2)=\sqrt{3}\,V_L\,I_L\sin\phi$ (iv) $\tan\phi=\dfrac{\sqrt{3}(W_1-W_2)}{W_1+W_2}$ (v) $\cos\phi=\cos\tan^{-1}\dfrac{\sqrt{3}(W_1-W_2)}{(W_1+W_2)}$ (vi) $W_1=W_2$ when power factor, $\cos\phi=1$, i.e., $\phi=0$. (vii) When $\cos\phi=0.5$, i.e., $\phi=60°$, one of the Wattmeters will give zero reading. (viii) When $\cos\phi<0.5$, i.e., $\phi>60°$, W_1 will give positive reading and W_2 will give negative reading.	5. Product of ammeter and voltmeter reading gives an account of the apparent power only.

REVIEW QUESTIONS

Short Answer Type Questions

1. What is the difference between a single-phase winding and a three-phase winding?
2. Draw wave shapes of a three-phase supply.
3. What is the difference between a balanced load and an unbalanced load?
4. Show that the phasor sum of the three-phase balanced voltages is zero.
5. What do you mean by phase sequence of three-phase voltages?
6. Distinguish between star connection and delta connection of three-phase windings.
7. Derive the relationship between line current, line voltage, phase current, and phase voltage in case of star and delta connection of three-phase windings.
8. Prove that the power in a three-phase circuit is equal to $\sqrt{3}\,V_L I_L \cos\phi$.
9. Distinguish between active power and reactive power in a three-phase system.
10. What is the significance of low-power factor of any load on the system?
11. Draw the circuit diagram for measurement of three-phase power with two single-phase wattmeters.
12. At what value of load power factor the reading of one of the wattmeters, in the two-wattmeter method of measurement of three-phase power, will be zero?
13. What are the advantages of the three-phase system over the single-phase system?
14. Write the relationship between phase voltage and current in a delta-connected load.
15. Draw the connection diagram for three-phase resistive–inductive balanced load connected across a three-phase supply. Also draw the phasor diagram showing voltages and currents.

Numerical Problems

16. Three coils having same resistance and inductance are connected in star. A three-phase 400 V supply is connected across the three coils. The power consumed by each coil is 800 W and the load power factor is 0.8 lagging. What is the total power consumed by the coils. If now the coils are connected in delta across the supply what would be the total power consumed? Also calculate the line current when delta connected.

    [**Ans** 2400 W, 7200 W, 13 A]
17. A balanced three-phase star-connected load supplied from a 400 V, 50 Hz, three-phase supply system. The current drawn by each phase is 20 $\angle -60°$ A. Calculate the line current, phase voltage, and total power consumed.

    [**Ans** 20 A, 230.94 V, 6928 W]
18. A delta-connected load has a resistance of 15 Ω and inductance of 0.03 H per phase. The supply voltage is 400 V, 50 Hz. Calculate line current, phase current, phase voltage, and total power consumed.

    [**Ans** 39.1 A, 22.5 A, 400 V, 22.94 kW]
19. Three identical coils of resistance 20 Ω and inductance 500 mH are connected first in star and then in delta across a 400 V, 50 Hz power supply. Calculate phase current, line current, phase voltage, and power consumed per phase.

    [**Ans** 1.46 A, 1.46 A, 230.94 V, 42.6 W; 2.53 A, 4.38 A, 400 V, 127.8 W]
20. Calculate the phase current and line current of a delta-connected load drawing 75 kW at 0.8 power factor from a 440 V, three-phase supply.

    [**Ans** 71 A, 122.97 A]
21. The power consumed by a three-phase balanced load has been measured by two single-phase wattmeters. The readings of the two wattmeters are 8.2 kW and

7.5 kW. Calculate the total power consumed and the load power factor.

[**Ans** 15.7 kW, 0.997 lagging]

22. In the measurement of three-phase power by two single-phase wattmeters, it has been observed that the ratio of the two wattmeter readings is 3:1. What is the power factor of the load?

[**Ans** 0.75]

23. Three identical coils are connected in star across a three-phase 415 V, 50 Hz supply. The total power drawn is 3 kW at a power factor of 0.3. Calculate the resistance and inductance of each coil.

[**Ans** 5.16 Ω, 52.3 mH]

24. Two single-phase wattmeters are used to measure three-phase power. The readings of the two wattmeters are 2000 W and 400 W, respectively. Calculate the power factor of the circuit. What would be the power factor if the reading of the second wattmeter is negative?

[**Ans** 0.65, 0.36]

25. Three identical coils each having a resistance of 10 Ω and inductive reactance of 10 Ω are first connected in star and then connected in delta across a 400 V, 50 Hz power supply. Calculate in each case the line current and the readings of two wattmeters connected for the measurement of power.

[**Ans** 16.33 A, 49 A, 6309 W and 1690 W, 18931 W and 5072 W]

Multiple Choice Questions

1. Three-phase system is used
(a) For transmission of electrical power
(b) For generation of electrical power
(c) For distribution of electrical power
(d) For generation, transmission, and distribution of electrical power.

2. In a three-phase system the phase sequence indicates.
(a) The amplitude of voltages
(b) The order in which the voltages obtain their maximum values
(c) The phase difference between the three voltages
(d) The frequency in which the phase voltages are changing.

3. In a star-connected system the relationship between the phase and line quantities are
(a) $V_{ph} = V_L$ (b) $V_{ph} = \sqrt{3}\, V_L$
(c) $3I_{ph} = I_L$ (d) $I_{ph} = I_L$.

4. In a delta-connected system the relationship between the phase and line quanties are
(a) $\sqrt{3}\, V_{ph} = V_L$ (b) $V_{ph} = \sqrt{3}\, V_L$
(c) $I_{ph} = I_L$ (d) $I_L = \sqrt{3}\, I_{ph}$.

5. Line currents drawn by a three-phase star-connected bnalanced load is 12 A when connected to a balanced three-phase four-wire system. The neutral current will be
(a) 36 A (b) 4 A
(c) 0 A (d) 3 A.

6. Power in a balanced three-phase system circuit is
(a) $3\, V_p I_p \cos\phi$ (b) $3\, V_L I_L \cos\phi$
(c) $\sqrt{3}\, V_p I_p \cos\phi$ (d) $V_p I_p \cos\phi$.

7. Reactive power of a three-phase circuit is
(a) $\sqrt{3}\, V_p I_p \sin\phi$ (b) $\sqrt{3}\, V_L I_L \sin\phi$
(c) $\sqrt{3}\, V_L I_L \cos\phi$ (d) $\sqrt{3}\, V_L I_L$.

8. power in a single-phase ac circuit can be expressed as
(a) $\sqrt{3}\, V_p I_p \cos\phi$ (b) $3\, V_p I_p \cos\phi$
(c) $V_p I_p \cos\phi$ (d) $V_p I_p \cos\phi$.

9. One single-phase wattmeter can be used to measure power in a three-phase circuit when
(a) The load is balanced
(b) The load is delta connected and is balanced
(c) The load is balanced, star connected, and the neutral wire is available
(d) The load is balanced and star connected.

10. A balanced three-phase sinusoidal power supply means
(a) Three sinusoidal voltages of the same frequency and maximum value displaced in 120° time phase
(b) Three sinusoidal voltages of any frequency but having the same maximum value
(c) Three sinusoidal voltages of some frequency and maximum value with no time phase displacement between them
(d) Three sinusoidal voltages of any value but having a time phase displacement of 120° between them.

11. An unbalanced three-phase supply system will have
    (a) Three unequal voltages
    (b) Three voltages having unequal time phase displacement between them
    (c) Three voltages of unequal magnitude and angular displacement among them
    (d) All of the above.

12. In the two-wattmeter method of measuring three-phase power, the reading of the two wattmeters will be equal when the power factor of the circuit is
    (a) 0 (b) 1
    (c) 0.5 (d) 0.866.

13. In the two-wattmeter method of measuring three-phase power, the reading of one of the wattmeters can be negative when the power factor angle is
    (a) More than 60° (b) Less than 60°
    (c) More than 30° (d) Less than 30°.

14. Four equal resistance of 100 Ω each connected in delta is supplied from 400 V three-phase star-connected supply, the line current drawn will be
    (a) 12 A (b) 4 A
    (c) 6.928 A (d) 13.856 A.

Answers to Multiple Choice Questions

1. (d)	2. (b)	3. (d)	4. (d)	5. (e)	6. (a)
7. (b)	8. (c)	9. (c)	10. (a)	11. (d)	12. (b)
13. (a)	14. (c)				

CHAPTER

5 Electromagnetism and Magnetic Circuits

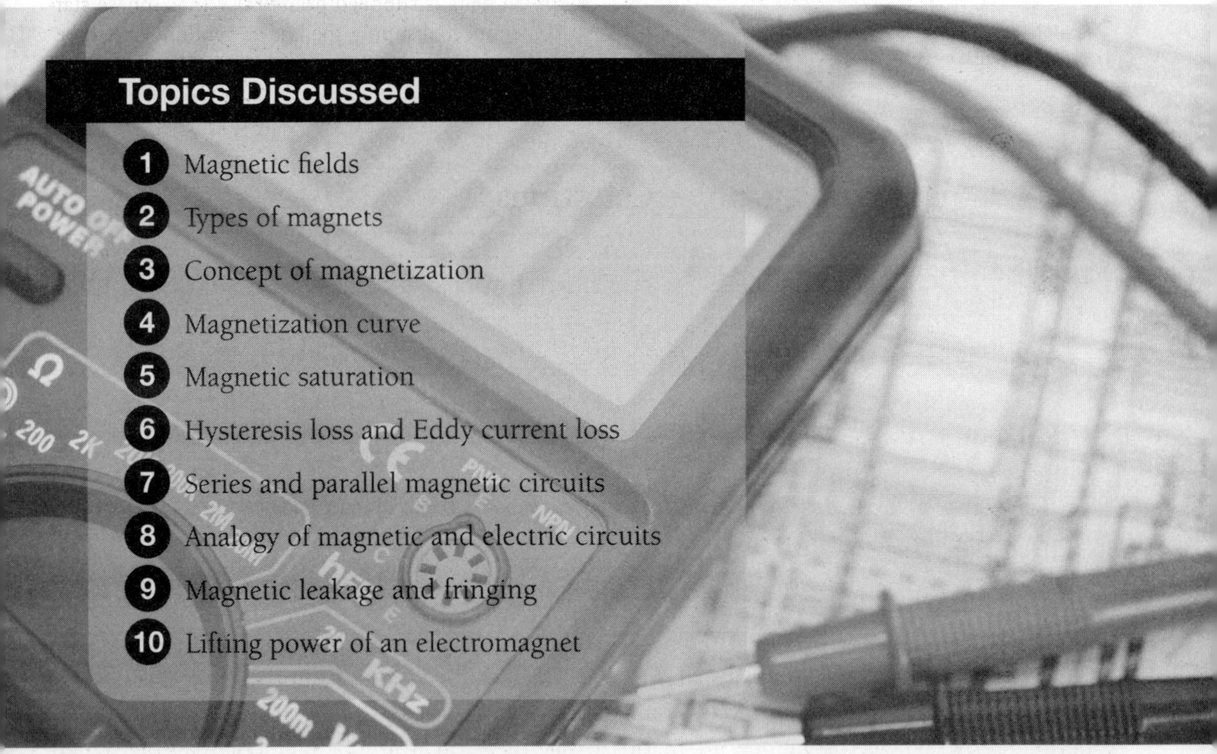

Topics Discussed

1. Magnetic fields
2. Types of magnets
3. Concept of magnetization
4. Magnetization curve
5. Magnetic saturation
6. Hysteresis loss and Eddy current loss
7. Series and parallel magnetic circuits
8. Analogy of magnetic and electric circuits
9. Magnetic leakage and fringing
10. Lifting power of an electromagnet

5.1 ► Magnets and Magnetic Fields

Magnetism is a force of attraction or repulsion caused by a magnet on magnetic materials. Magnetism plays an important role in the field of electrical engineering. Construction of almost all electrical gadgets, equipment, and machines are done using the properties of magnetism, like in transformers, electrical rotating machines, i.e., generators and motors, relays, cutouts, electrical bells, etc.

The word "magnetism" originated from the city of Magnesia (now called Manisa in Turkey) where iron ores were discovered which had the property of adhering to each other in lumps.

Magnets show the property of magnetism. Magnets are of two types, viz permanent magnets and electromagnets. Magnets attract all ferromagnetic material which contain iron, nickel, and cobalt.

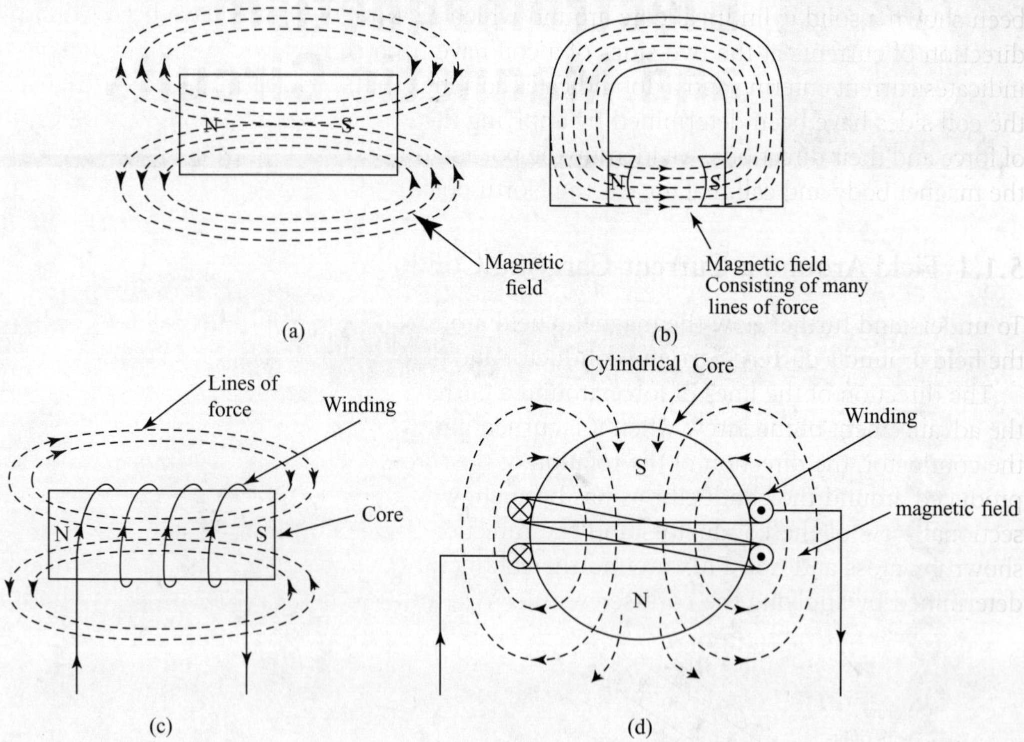

FIGURE 5.1

(a) and (b) permanent magnets; (c) and (d) electromagnets

Permanent magnets are made of material like alnico (alloy of aluminium, nickel, and cobalt) in which the magnetism once created is retained for a very long time, i.e., the magnetic property is permanently set. Electromagnets are made by placing a coil around a magnetic material which forms the core. They demonstrate magnetic properties as long as current flows through the coil.

Magnetic field is the area around a magnet in which there is influence of the magnet. This can be tested by bringing a magnetic needle near a magnet and observing the deflection of the needle. The magnetic field around a magnet is shown through lines of force. There are very large number of lines of force around a magnet. Lines of force are closed curves. The lines of force come out of the magnet body from the North pole and enter the South pole and close their path through the magnet body. The pattern of lines of force is the same for a permanent magnet and for an electromagnet. Fig. 5.1 shows magnets and magnetic fields of permanent magnets and electromagnets of different shapes.

Fig. 5.1 (a) shows a bar-type permanent magnet, while Fig. 5.1 (b) shows a horse-shoe-type permanent magnet. Fig. 5.1 (c) and (d) are electromagnets. It should be noticed that the North pole is the one wherefrom the magnetic lines of force come out of the magnet body and the South pole is the one where the lines of force enter the magnet body. In Fig. 5.1 (d) has been shown a solid cylindrical core around which a coil of two turns have been wound. The direction of currents at the two sides of a coil have been shown by crosses and dots. A cross indicates current entering and a dot indicates current coming out. The direction of flux around the coil sides have been determined by applying the cork screw rule. After showing the lines of force and their directions, we identify the portion of the core from where the flux lines leave the magnet body and call that area as the North pole.

5.1.1 Field Around a Current-Carrying Conductor

To understand further how the magnetic field around a coil is established, we will first draw the field around a current-carrying conductor and then show the magnetic field around a coil.

The direction of the lines of force around a current-carrying conductor has been shown. If the advancement of the screw when it is turned indicates the direction of the current through the conductor, the direction of the rotation of the screw will indicate the direction of the flux produced around the conductor as has been shown in Fig. 5.2 (a). In Fig. 5.2 (b) the cross-sectional view of the conductors and the direction of the current through them have been shown by cross and dot. Flux around the conductor have been shown and their direction is determined by applying the cork-screw rule. When two conductors or coil sides appear side

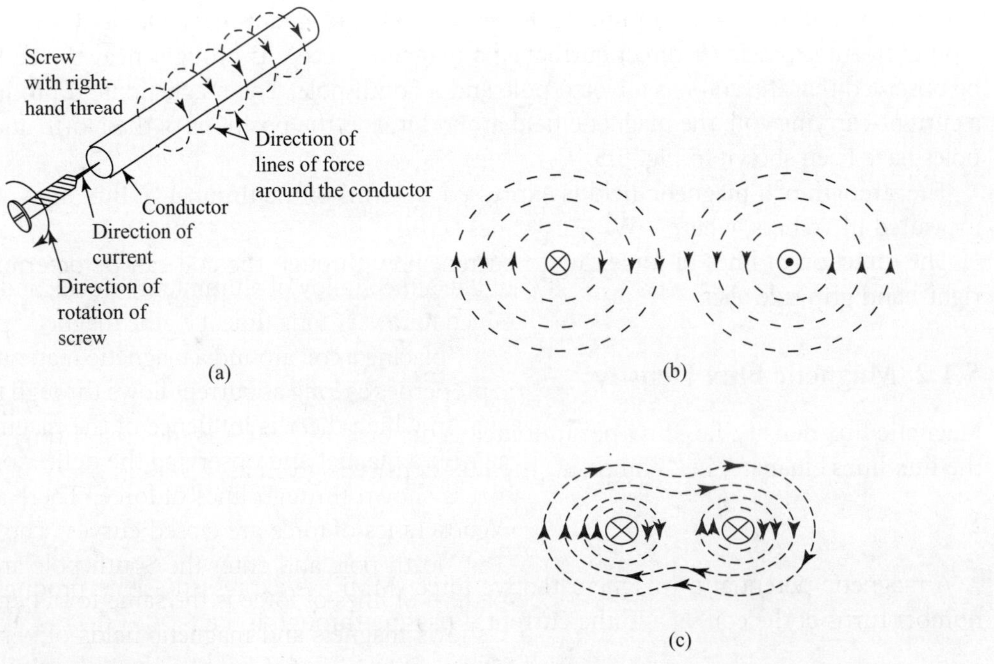

FIGURE 5.2

(a) Magnetic field around a current-carrying conductor; (b) two conductors placed side by side carrying current in opposite directions; (c) two conductors placed side by side carrying current in the same direction

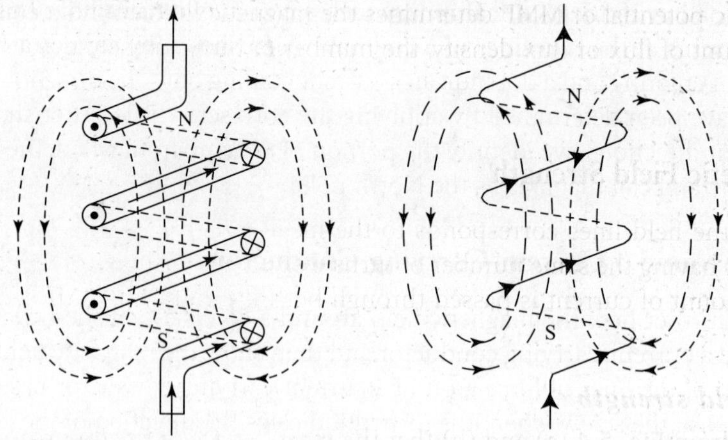

FIGURE 5.3

Magnetic field around a current-carrying coil

by side carrying current in the same direction, a resultant magnetic field gets established as has been shown in Fig. 5.2 (c).

When a conductor is wound in the form of a coil, a resultant magnetic field is established around it. After passing a direct current, if a magnetic needle is brought near the coil, it will be observed that the coil has a North pole and a South pole. The magnetic field produced by a current-carrying coil, the magnetic field around it, and the positions of the North and South poles have been shown in Fig. 5.3.

The strength of a magnetic field is expressed in terms of the number of flux lines, ϕ and is measured in Webers, where 1 Wb = 10^8 lines.

The direction of lines of force due to current flow through the coil can be determined by right-hand grip rule also.

5.1.2 Magnetic Flux Density

Magnetic flux density, i.e., flux per unit area is denoted by B. If A is the area through which the flux lines emanate, i.e., come out, the flux density is given as

$$B = \frac{\phi}{A} \text{ Wb/m}^2 \text{ or Tesla} \tag{5.1}$$

A magnetic potential or magneto motive force (MMF) is expressed as the product of the number turns of the coil, N and the current, I passing through it, i.e.,

$$\text{MMF} = N\,I \tag{5.2}$$

The unit of I is Amperes, and number of turns has no unit. Thus, logically the unit of MMF should be Amperes only. However, to differentiate the unit of MMF from the unit of current, the unit of MMF is expressed as Ampere Turns (AT).

The magnetic potential or MMF determines the magnetic flux around a coil. To produce a particular amount of flux or flux density, the number of turns or the current flowing, can be varied.

5.1.3 Magnetic Field Strength

The length of the field lines corresponds to the mean length of a coil. Fig. 5.4 shows two electromagnets having the same number of turns wound on two cores of different diameters. If the same amount of current is passed through both the coils, the MMF, i.e., N × I will be the same.

Magnetic field strength

It can be seen from Fig. 5.4 (a) and (b) that the mean length of the flux path is different. In the case of Fig. 5.4 (b) the length is higher. The MMF which is a product of N and I is the same.

For the same MMF, the magnetic flux, ϕ or the magnetic flux density, B will be dependent upon the length of the flux lines. The longer is the length, l the weaker will be the flux density. Magnetic field strength, H is expressed as

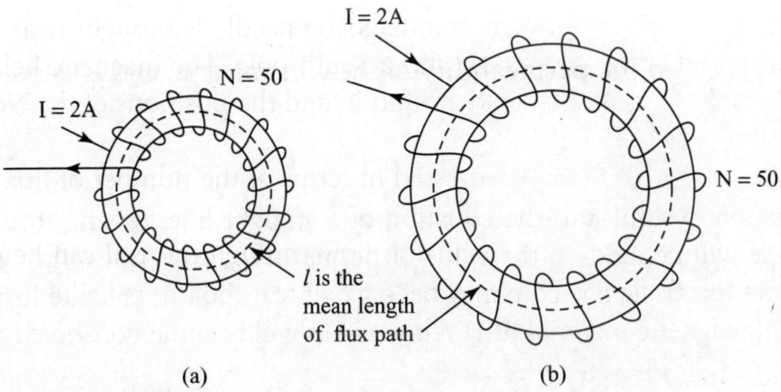

FIGURE 5.4

Mean length of the flux path in toroidal cores. The mean length of the flux path is more in (b) as compared to that in (a)

$$H = \frac{NI}{l} \text{AT/m} \tag{5.3}$$

The longer is the flux line length, the weaker will be the magnetic field strength for the same amount of MMF. Since the magnetic field strength, H depends on the number of turns of the magnetizing coil, the current flowing through the coil, and the flux line length, there would exist a relationship between the magnetic field strength H and the magnetic flux density B, when a magnetic material is magnetized by applying MMF. If we plot the flux density, B against the magnetizing force applied, H, we will get a linear relationship as shown in Fig. 5.5 so that

$$B \propto H$$

or,

$$B = K H$$

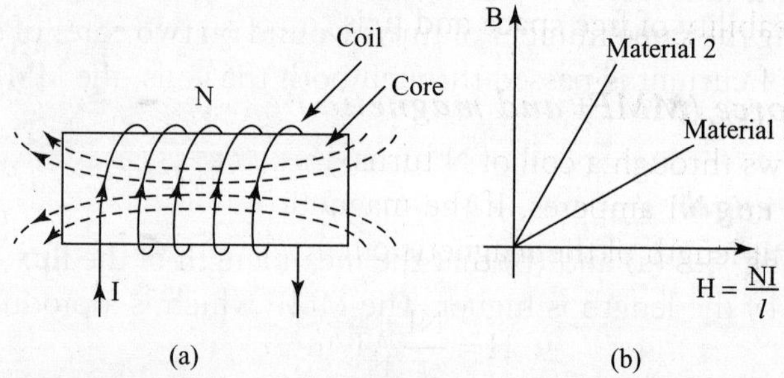

FIGURE 5.5

(a) Magnetizing force applied to a magnetic material;
(b) incomplete B–H curve of two magnetic materials

This constant K is called the **permeability** of the core material and is denoted by μ. Thus, we can write

$$B = \mu H \tag{5.4}$$

Since this equation is similar to the equation of a straight line passing through the origin (y = mx), the slope will depend on the value of permeability, μ. Permeability is the magnetic property of the core material. The core may be, e.g., air, cardboard, bakelite former, or iron. If the core is made of magnetic material, B-H relationship will become non-linear at certain stage of magnetization.

5.1.4 Permeability

Permeability is the measure of how well the material conducts magnetism or allows the establishment of magnetic field through it. As shown in Fig. 5.5 (b), material 2 is more permeable than material 1.

The graphic representation of the relationship between B and H is called the magnetization curve. In the case of air core coil, the magnetization curve is a straight line with a small slope with the H-axis showing that a large value of H is required to produce a small amount of flux or flux density. If iron is used as the core material the slope will greatly increase indicating that a small magnetizing force produces a large amount of flux and a high value of flux density.

5.1.5 Relative Permeability

The permeability of a magnetic material is compared with that of free space and is called relative permeability. The relative permeability indicates how many times the material is more permeable than air. Permeability, μ is expressed as

$$\mu = \mu_o \mu_r \tag{5.5}$$

where μ_o is the permeability of free space and μ_r is the relative permeability.

Magneto motive force (MMF) and magnetic field strength, H

When a current, I flows through a coil of N turns, the MMF, is the total current linked with the magnetic circuit, i.e., NI amperes. If the magnetic circuit is of uniform cross-sectional area, the MMF per unit length of the magnetic circuit is called magnetic field strength, H.

Thus, $$H = \frac{NI}{l}\ \text{AT/m}$$

Relation between flux density, B and field intensity, H

The ratio of B and H is called the permeability of free space and is represented by μ_0

$$\mu_0 = \frac{B}{H}\ \text{henry per meter}$$
$$= 4\pi \times 10^{-7}\ \text{H/m}$$

This value of μ_0 is for free space or air or for any non-magnetic material like paper, wood, oil, etc.

The relation, $B = \mu_0 H$ is true when the medium is free space or air. When the medium through which flux or flux density is established is changed, the value of flux, ϕ or flux density, B increases. It has been observed that when the core of a current-carrying coil is made of iron instead of being an air-core one, the value of the flux density produced increases many times. The ratio of the flux density produced with iron core to the flux density produced with air core by the same magnetic field strength is called the relative permeability, μ_r.

For air, $\mu_r = 1$ and for iron or for some alloys of iron, μ_r can be very high. Relative permeability, as mentioned earlier, indicates how many times the material is more permeable than air (permeability is similar to conductivity in an electrical circuit. Permeability is the ability of the material to allow the establishment of flux through it)

For a non-magnetic material or air or vacuum as the medium, we write

$$B = \mu_0 H \tag{5.6}$$

While for a material of the medium having a relative permeability of μ_r we write,

$$B = \mu_0\mu_r H = \mu H \tag{5.7}$$

where $$\mu = \mu_0\mu_r = 4\pi \times 10^{-7} \times \mu_r$$

5.2 ► Magnetic Field Due to Current-Carrying Conductor-Laws of Electromagnetism

When a conductor carries current, a magnetic field is established around it in a perpendicular plane in the form of concentric circles. The relationship between the current, I and the magnetic field intensity, H is obtained by using Amperes circuital law.

The magnetic field strength at a point due to an incremental length dl of current-carrying conductors is determined by using Biot–Savart law. These two laws are explained below.

5.2.1 Ampere's Circuital Law

This law states that the line integral of magnetic field intensity, H around a closed path is equal to the current enclosed by the path. Consider a current-carrying conductor, I producing a magnetic field an a perpendicular plane as shown in Fig. 5.6. The magnetic field intensity, H at a distance, r from the current-carrying conductor is expressed as the line integral of H multiplied by dl as equal to the total enclosed current. That is

$$\oint H \cdot dl = I \qquad (5.8)$$

or,

$$H(2\pi r) = I$$

or,

$$H = \frac{I}{2\pi r}$$

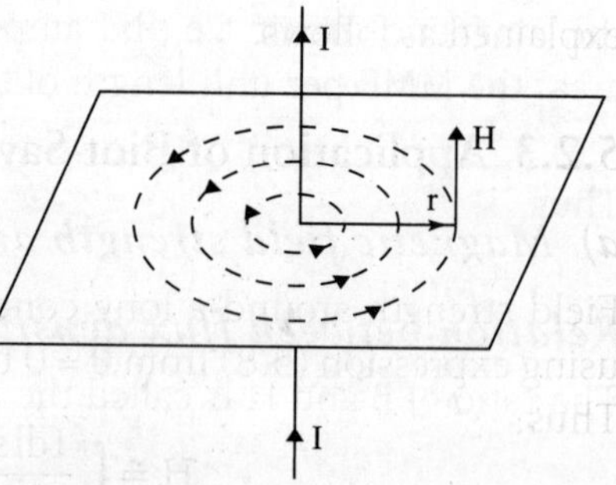

FIGURE 5.6

Magnetic field around a current-carrying conductor

If there are N number of current-carrying conductors enclosed then, H is expressed as

$$H = \frac{NI}{2\pi r}$$

where $2\pi r$ is the length of the flux path. This expression for H is the same as in eq. (5.3).

5.2.2 Biot-Savart Law

This law is also used to determine the magnetic field intensity, H around a current-carrying conductor. According to Biot-Savart law, the magnetic field intensity at a point P due to current flowing through an extremely small element of length dl is

(i) directly proportional to the current, I

(ii) directly proportional to the length of the element, dl

(iii) directly proportional to the sine of angle θ where θ is the angle between the direction of the current and the line joining the element dl with the point P as shown in Fig. 5.7.

(iv) inversely proportional to the square of the distance r of the point from the element of length dl.

The law can be expressed mathematically as

$$dH \propto \frac{Idl}{r^2}\sin\theta$$

or,
$$dH = \frac{Idl\sin\theta}{4\pi r^2} \quad (5.9)$$

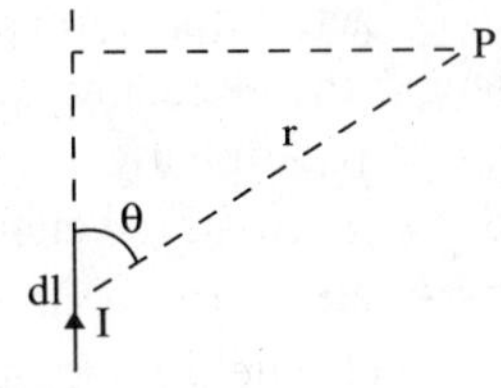

FIGURE 5.7

Biot-Savart law

where $1/4\pi$ is the constant of proportionality.

By applying Biot-Savart law, we can determine the magnetic field strength around current-carrying conductors arranged in different manner. Some of these are explained as follows.

5.2.3 Application of Biot-Savart Law

a) Magnetic field strength around a long straight current-carrying conductor

Field strength around a long conductor will be calculated by using expression (5.8) from $\theta = 0$ to $\theta = \pi$ as shown in Fig. 5.8. Thus,

$$H = \int_0^{\pi} \frac{Idl\sin\theta}{4\pi r^2}$$

From Fig (5.8),

$$dl\sin\theta = r\,d\theta$$

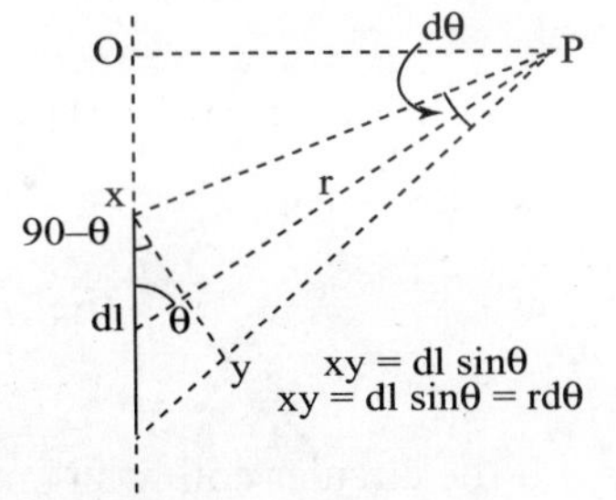

FIGURE 5.8

Substituting

$$H = \int_0^{\pi} \frac{Ir\sin\theta\,d\theta}{4\pi r^2}$$

$$= \int_0^{\pi} \frac{I\sin\theta\,d\theta}{4\pi r}$$

$$= \frac{I}{4\pi r}\int_0^{\pi} \sin\theta\,d\theta$$

$$= \frac{I}{4\pi r}[-\cos\theta]_0^{\pi}$$

or,
$$H = \frac{I}{2\pi r}\ \text{AT/m}$$

Flux density in a medium of air is

$$B = \mu_0 H = \frac{\mu_0 I}{2\pi r}\ \text{Wb/m}^2$$

b) Field strength around a circular loop

A circular coil of radius r metres carrying a current of I amperes has been shown in Fig. 5.9. The field strength at a point P which is situated at a distance of d metres from the centre of the coil is to be determined. Let dl_1 and dl_2 be the two elements of length diametrically opposite to each other on the circular path. Applying Biot-Savart law, the field strength at P due to current, the elements of length dl_1 and dl_2, respectively, are

$$dH_1 = \frac{I dl_1}{4\pi x^2} \text{ acting along PQ}$$

and $$dH_2 = \frac{I dl_2}{4\pi x^2} \text{ acting along PR}$$

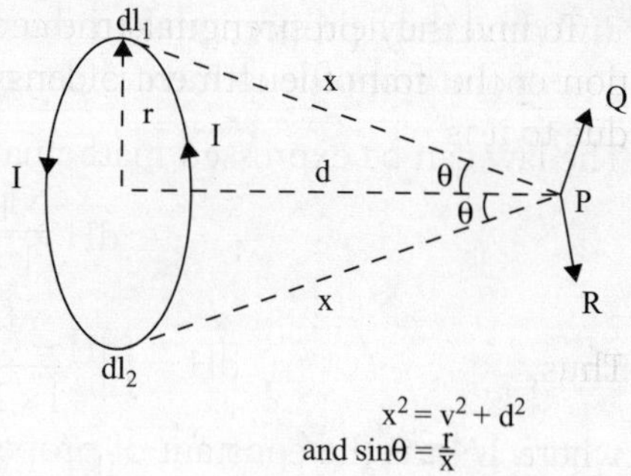

FIGURE 5.9

The resultant of the two vectors PQ and PR were give the net magnetizing field intensity

$$dH = \frac{2 I dl_1}{4\pi x^2}\sin\phi = \frac{I dl_1 \sin\phi}{2\pi x^2}$$

To determine the total H, we have to integrate the above expression as

$$H = \int_0^{\pi r}\frac{I dl_1 \sin\phi}{2\pi x^2} = \int_0^{\pi r}\frac{I\, r}{2\pi x^2 \times x}dl_1 = \frac{Ir \times \pi r}{2\pi x^3}$$

$$H = \frac{IR^2}{2(r^2+d^2)^{3/2}} \quad \left[\begin{array}{l}\because x^2 = r^2 + d^2 \\ \text{or } x = (r^2+d^2)^{3/2}\end{array}\right]$$

If the circular coil has N number of turns, then

$$H = \frac{NIr^2}{2(r^2+d^2)^{3/2}} \text{ AT/m} \tag{5.10}$$

For determining the field strength at the centre of the current-carrying coil, we put d = 0. Thus, the magnetic field strength, the H_c at the centre of the coil is

$$H_c = \frac{NIr^2}{2r^3} \text{ AT/m}$$

$$H_c = \frac{NI}{2r} \text{ AT/m} \tag{5.11}$$

c) Field strength inside a solenoid carrying current

Fig. 5.10 (a) shows a solenoid with N number of turns and having a length *l*. The cross-sectional view has been shown in Fig. 5.10 (b).

We had calculated the field intensity at the centre of a circular coil as

$$H = \frac{Ir^2}{2(r^2+x^2)^{3/2}}$$

To find the field strength at the centre of the solenoid, a section of the coil of length dx is considered. The field strength due to it is

$$\frac{NI}{l}dx$$

Thus,
$$dH = \int_{-l/2}^{+l/2} \frac{NI\,dx\,r^2}{l \times 2(r^2 + x^2)^{3/2}}$$

or,
$$H = \frac{NIr^2}{2l} \times \frac{l}{r^2(r^2 + l^2/4)^{1/2}}$$

or,
$$H = \frac{NI}{(4r^2 + l^2)^{1/2}} \text{ AT/m}$$

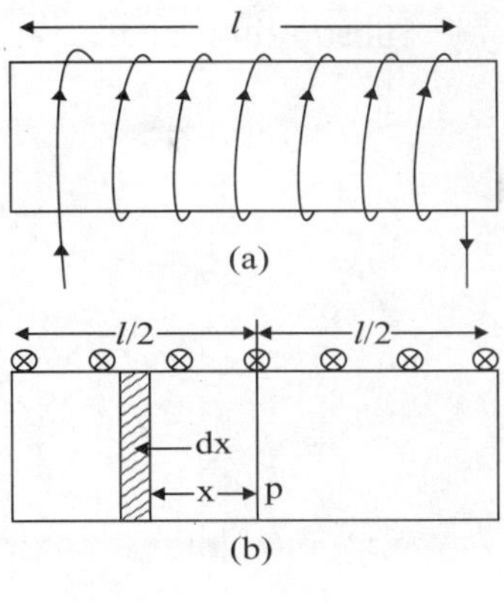

FIGURE 5.10

If the length of the solenoid is large as compared to its radius, r then, the flux density at the centre of the solenoid is

$$H = \frac{NI}{l} \text{ AT/m}$$

EXAMPLE 5.1

A single turn coil of radius 10 cm is carrying a current of 100 A. Calculate (i) the flux density at the centre of the coil; (ii) the flux density in the perpendicular plane at a distance of 5 cm from the coil.

SOLUTION

Earlier we had calculated the magnetic field strength at the centre of a current-carrying coil of single turn as

$$H_c = \frac{I}{2r} \text{ AT/m}$$

With air as the medium, the flux density at the centre

$$B_c = \frac{\mu_0 I}{2r}$$

Substituting values

$$B_c = \frac{4\pi \times 10^{-7} \times 100}{2 \times 10 \times 10^{-2}}$$
$$= 628 \times 10^{-6} \text{ Wb/m}^2 \text{ or Tesla}$$

Field intensity at P at a perpendicular distance of from the coil of radius r is

$$B = \frac{\mu_0 Ir^2}{2(r^2 + d^2)^{3/2}}$$

Substituting value

$$B = \frac{4\pi \times 10^{-7} \times 100 \times (0.1)^2}{2[0.1^2 + 0.05^2]^{3/2}}$$

or,

$$B = \frac{6.28 \times 10^{-7}}{(0.0125)^{3/2}} \text{Wb/m}^2$$

$$= \frac{62.8 \times 10^{-6}}{(0.0125)^{-3/2}} \text{Wb/m}^2$$

$$= 268 \times 10^{-6} \text{ Wb/m}^2$$

5.3 ► Magnetization Curve of a Magnetic Material

Electromagnets are produced by winding a coil around a piece of magnetic material, say iron, and passing current through the coil. When current is increased gradually, the flux will increase. The rate of increase of the flux produced with increase in current will slow down after a sufficient increase of current as shown in Fig. 5.11. The core is said to be saturated when further increase of current through the coil does not cause further increase of flux. The flux per unit area is expressed as flux density B.

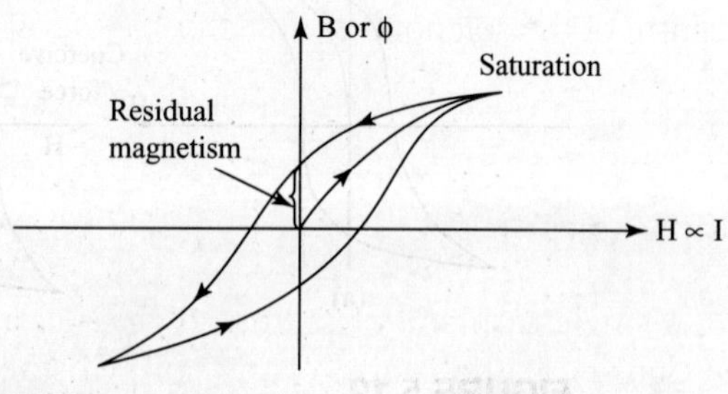

FIGURE 5.11

B–H characteristic of a magnetic material

Flux density,

$$B = \frac{\text{Flux}}{\text{Area}} = \frac{\phi}{A}$$

where A is the cross-sectional area of the core. The magnetizing force is expressed in terms of ampere-turns per unit length. That is, magnetizing force H is given as

$$H = \frac{NI}{l}$$

Where l is the length of the flux path.

5.4 ► Hysteresis Loss and Eddy Current Loss in Magnetic Materials

Magnetization of the magnetic material in opposite directions due to application of alternate magnetizing force, involves certain amount of work done. The work done is represented by the area of the hysteresis loop. The hysteresis loop area depends upon the nature of the magnetic material. In selecting the material for the core of any electrical machine and equipment, a study of the hysteresis loop is made. For example, if we want that high flux should be produced by applying a low-magnetizing force, and the loop area also be small then we must use a material whose hysteresis loop area should be as shown in Fig. 5.12 (a). A certain percentage of silicon when added to steel provides this kind of B–H characteristic. Silicon steel is used as the core material of transformers, as will be studied in a separate chapter. In Fig. 5.12 (b) is shown the B–H characteristic of the material used for making permanent magnets like alnico (alloy of aluminium, nickel, and cobalt). Here the residual magnetism is large and the negative magnetizing force required to bring down the residual magnetism to zero (which is called the coercive force) is also large.

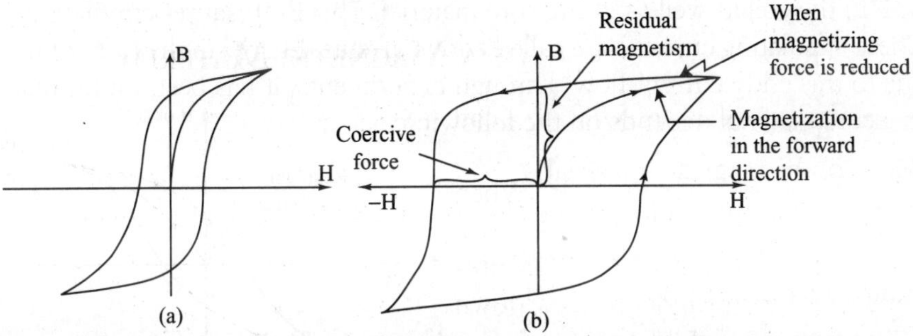

FIGURE 5.12

(a) Hysteresis loop for silicon steel; (b) hysteresis loop for alnico

5.4.1 Hysteresis Loss

The energy spent in magnetization of the core in opposite directions in every half cycle of current flow appears as heat in the magnetic material. An emperical formula to calculate hysteresis loss in a magnetic material has been developed on the basis of experiments.

After magnetization if the magnetizing force is reduced, the curve traces a somewhat different path as shown in Fig 5.12(b). The negative current provides magnetization in the opposite direction as shown. When the magnetizing force is reduced to zero, there is some magnetism left in the magnetic material, which is known as residual magnetism. In electrical machines, this residual magnetism plays an important role, as will be studied in chapters on electrical machines.

When, instead of direct current through the coil, an alternating current flows, the core will get magnetized in the opposite directions alternately. The work done in this process will cause power loss, which is called hysteresis loss. In most electrical machines and equipment attempt is made to reduce hysteresis loss to increase the efficiency of machines and equipment. The magnetization characteristic due to alternating current passing through the coil is represented by a loop, called hysteresis loop as shown in Fig. 5.11 and Fig. 5.12.

The expression for hysteresis loss, W_h is given as

$$W_h = K_h B_m^{1.6} f W. \tag{5.12}$$

Where, K_h is a constant which depends upon the material and the range of flux density, and the volume of the core material, f is the frequency of alternation of the current passing through the magnetizing coil B_m is the maximum value of flux density in the core in Wb/m^2.

The power of B_m is generally 1.6. However, depending on the quality of the material, the power of B_m may vary from 1.5 to 2.0.

5.4.2 Eddy Current Loss

When a magnetic material like iron is magnetized by passing alternating current through a coil wound around it, an alternating magnetic field is produced. This alternating magnetic field induces EMF in the coil as well as in the core material. This EMF causes circulating currents in the core. These circulating currents are called eddy currents. Energy in the form of heat is lost in the core due to this eddy current flow. Through experiments, it has been found that eddy current loss in a core material depends on the following:

(i) thickness of the magnetic material, t;

(ii) frequency of current producing the alternating magnetic field, i.e., the frequency of current flowing through the magnetizing coil, f;

(iii) maximum flux density, B_m

$$\text{Eddy current loss, } W_e = K_e B_m^2 f^2 t^2 W \tag{5.13}$$

Where k_e is the eddy current coefficient which depends upon the type of the magnetic material. Thickness of lamination varies from 0.5 to 5 mm for electrical machines and from 0.01 to 0.5 mm for electronic devices operating at higher frequencies.

To reduce eddy current loss in a magnetic material, the thickness is reduced. Laminated sheets are used to build a core instead of using one piece solid core. ***The sum of hysteresis loss and eddy current loss is called core loss or iron loss.***

5.5 ► Magnetic Circuits

All electrical machines and equipment are made of magnetic material as their core. A winding which will carry current is placed around the core. The core and the current-carrying coil around the core forms an electro magnet. Thus, we can say that an electro magnet is made using a piece of magnetic material as the core and around which a current-carrying coil is placed. Let us consider a bar-type piece of magnetic material, i.e., a piece of iron being made in to an electro magnet. A coil has been wound around it. Current is being allowed to flow through the coil from a source of supply, say a battery, as shown in Fig. 5.13 (b). The magnetizing force applied is the product of the number of turns of the coil (N) and the current flowing through the coil (I). Let us examine how the magnetizing force, NI magnetizes the magnetic material.

It is known that a magnetic material is composed of tiny magnets called magnetic dipoles oriented in a random fashion in all directions as shown in Fig. 5.13 (a). The magnetizing force

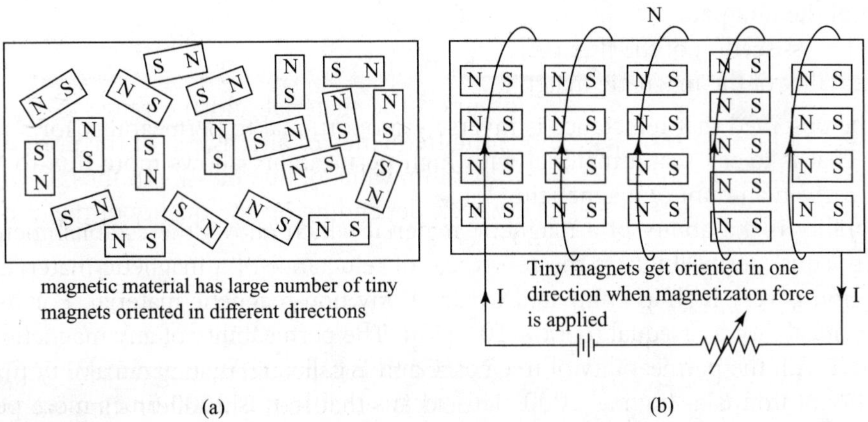

FIGURE 5.13

(a) Orientation of magnetic dipoles before the application of magnetizing force; (b) on application of the magnetizing force, NI, the magnetic dipoles get oriented in one direction

orients these tiny magnets in the direction of magnetization. When the tiny magnets get oriented in a particular direction, the material becomes a strong magnet as the magnetism of all the tiny magnets get summed up. In Fig. 5.13, the tiny magnets forming the magnetic material have been shown very much enlarged only to help understanding and bring clarity. In fact, their number is huge and they are very very tiny and cannot be observed through naked eyes.

Thus, when magnetized, one side of the bar magnet becomes a strong North pole and the other side becomes a strong South pole. The strength of this electro magnet produced by the magnetizing force is directly proportional to the magnetizing force and inversely proportional to the reluctance of the flux path. Reluctance is the opposition offered to the establishment of flux.

The amount of flux produced by the magnet indicates the strength of the magnet. The more the magnetizing force (MMF), more is the flux produced. The more the opposition to flux path (i.e., reluctance or magnetic resistance) less is the flux produced. This relationship is expressed as

$$\text{Flux} = \frac{\text{MMF}}{\text{Reluctance}}$$

or,
$$\phi = \frac{NI}{S} \tag{5.14}$$

Reluctance is the opposition offered by the material in the flux path to the establishment of the flux. Reluctance in a magnetic circuit is similar to the resistance in an electric circuit.

We have known that resistance, $R = \rho \frac{l}{A}$

Similarly, reluctance,
$$S = K\frac{l}{A} = \frac{l}{\mu A} \tag{5.15}$$

Where $K = \frac{1}{\mu}$

l = length of the flux path
A = area of cross section of the flux path
μ = permeability of the magnetic material.

It can be observed that reluctance is inversely proportional to permeability for a particular material. That is to say that a material with high permeability allows more flux to be established for a given amount of magnetizing force.

Permeability is the ability of a magnetic material which allows the establishment of flux through it. Thus, permeability is the reciprocal of reluctance of a magnetic material. Permeability of iron is very high as compared to air or any non-magnetic material. For free space, i.e., air, permeability μ_o is equal to $4\pi \times 10^{-7}$ H/m. The permeability of any magnetic material is compared with the permeability of free space and is called relative permeability μ_r. Relative permeability of iron is as high as 2000. This means that iron is 2000 times more permeable than air. For the same amount of ampere turns, an iron-core coil will produce about 2000 times more flux than an air-core coil as shown in Fig. 5.14.

For the same amount of ampere turns, the flux produced by an iron-core coil is much more than that produced by an air-core one. Actually, the amount of flux produced in an iron-core coil is much more than what has been shown.

Now, if we make a cut in the magnetic material to create an air gap as shown in Fig. 5.14 (c), the flux produced for the same ampere turns will be somewhat less as in the case of Fig. 5.14 (a), because the total reluctance of the flux path is now increased. The flux has to cross the air gap whose reluctance is very high as compared to iron. The flux produced will be calculated as

$$\text{Flux, } \phi = \frac{\text{MMF}}{\text{Reluctance of iron path + Reluctance of air gap}}$$

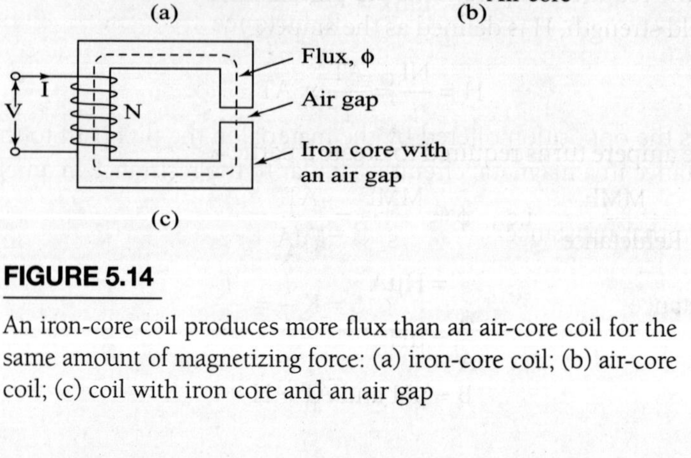

FIGURE 5.14

An iron-core coil produces more flux than an air-core coil for the same amount of magnetizing force: (a) iron-core coil; (b) air-core coil; (c) coil with iron core and an air gap

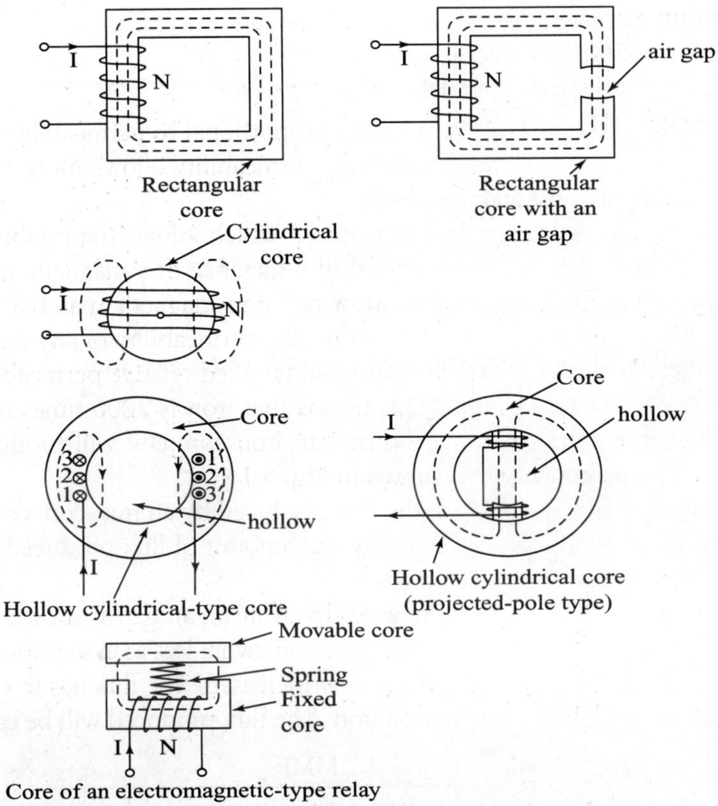

FIGURE 5.15

Magnetic circuits of different types and shapes used in making electrical machines and devices

The magnetic circuits of electrical machines, transformers, electromagnetic relays, and other electrical equipment are of different shapes and sizes as shown in Fig. 5.15. The current-carrying coil providing the required ampere turns are placed at various convenient locations as shown.

Magnetic field strength, H is defined as the ampere turn per unit length, i.e., as AT/m. Thus,

$$H = \frac{NI}{l} = \frac{AT}{l} \text{ or } AT = H \times l.$$

To calculate the ampere turns required to create a particular amount of flux, we use the relation

$$\text{Flux} = \frac{\text{MMF}}{\text{Reluctance}} \quad \text{i.e.,} \quad \phi = \frac{\text{MMF}}{s} = \frac{AT}{l/\mu A} = \frac{AT}{l}\mu A$$

$$= H\mu A$$

or,

$$\frac{\phi}{A} = \mu H$$

or,

$$B = \mu H \text{ and } AT = H \times l$$

The various quantities associated with magnetic circuits are stated as follows.

Magneto motive force (MMF) = Ampere turns = N I

Magnetic field strength (H) = AT/m= $\frac{NI}{l}$

Flux density (B) = Flux per unit area = $\frac{\phi}{A}$

Permeability $\mu = \mu_o \mu_r$

Permeability of free space (air), $\mu_o = 4\pi \times 10^{-7}$

Relative permeability (μ_r) = The number of times the material is more permeable than air.

Flux density, $B = \mu H$

Flux, $\phi = BA = \mu HA = \mu_o \mu_r \frac{NI}{l} A$

$$= \frac{NI}{l/\mu_o \mu_r A} = \frac{NI}{l/\mu A} = \frac{NI}{S}$$

∴ Flux, $Q = \frac{MMF}{Reluctance}$

Reluctance, $S = \frac{l}{\mu A} = \frac{l}{\mu_o \mu_r A}$

5.6 ► Comparison Between Magnetic and Electric Circuits

Now, we will consider two simple circuits, namely an electric circuit and a magnetic circuit, and establish their similarity as shown in Fig. 5.16.

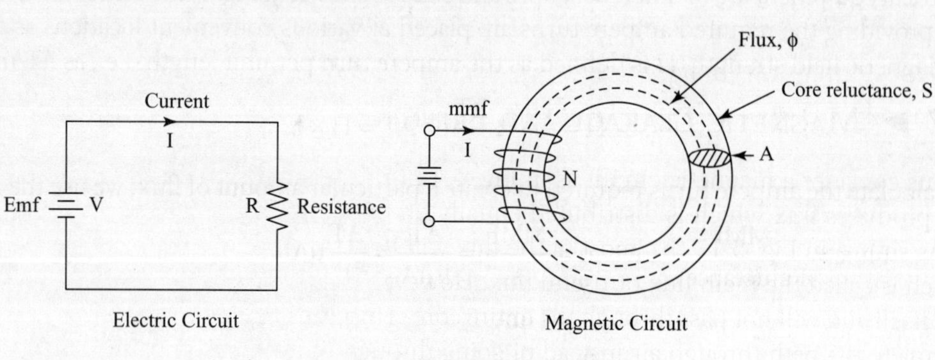

FIGURE 5.16

Comparison between an electric circuit and a magnetic circuit

The comparison has been shown in a tabular form in Table 5.1.

There are, however, few points of dissimilarities between the magnetic and electric circuits. For example, flux can pass through air although the reluctance is high whereas current will flow through air only when the air gets ionized; there will be some residual magnetism left in iron when the magnetizing force is removed whereas no current is left in the circuit when the source of EMF is removed; flux does not actually flow in a magnetic circuit (magnetic field is established) whereas current flows in an electric circuit.

Calculation of current or ampere turns required to create a magnetic field of a particular strength will be necessary while designing a magnetic circuit for any electrical equipment. Accordingly, a few solved numerical examples have been included in this chapter.

TABLE 5-1 Comparison Between an Electric Circuit and a Magnetic Circuit

Electric circuit	Magnetic circuit
Current, I	Flux, ϕ
EMF, V	MMF, N I
Resistance, R	Reluctance, S
$I = \frac{EMF}{Resistance} = \frac{V}{R}$	$\phi = \frac{MMF}{Reluctance} = \frac{NI}{S}$
$R = \rho \frac{l}{a}$	$S = \frac{l}{\mu A}$
Conductance $= \frac{1}{Resistance}$	Permeance $= \frac{1}{Reluctance}$
Electric field intensity $= \frac{V}{l}$	Magnetic field intensity, $H = \frac{AT}{l}$ or, $H = \frac{NI}{l}$

5.7 ► Magnetic Leakage and Fringing

Let us consider a magnetic circuit as in Fig. 5.17 (a). The current passing through the winding produces flux which is distributed equally on both sides of the coil as shown.

As shown in Fig. 5.17 (b), most of the flux will flow or pass through the magnetic material which is called the main flux or useful flux. However, a certain percentage of flux will link the coil itself and will not pass through the entire core. This flux is called leakage flux. Leakage flux completes its path through air instead of going through the entire iron path.

In a magnetic circuit with some air gap, when flux has to pass through the air gap, there is tendency of the magnetic flux to spread out at the two edges or sides. This effect is called fringing.

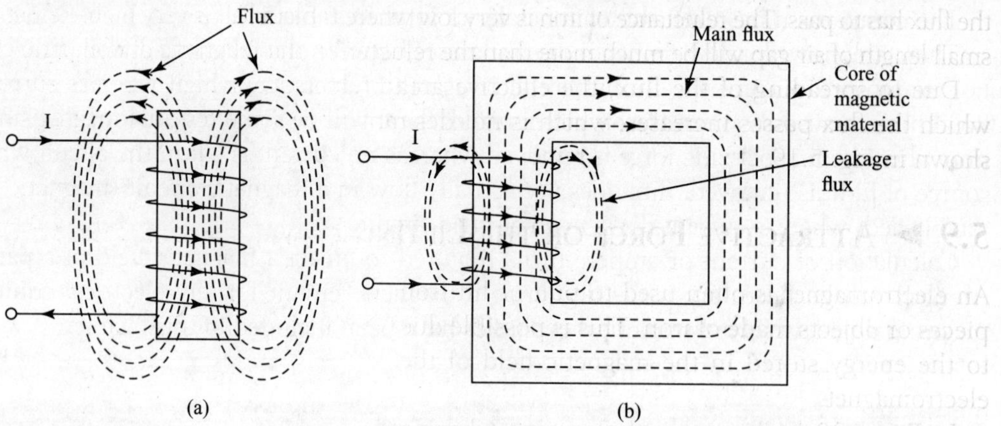

FIGURE 5.17

(a) Uniform distribution of flux around a current-carrying coil; (b) main flux and leakage flux

5.8 ► Series and Parallel Magnetic Circuits

Let us consider two magnetic circuits as shown in Fig. 5.18 (a) and (b) and their equivalent electrical circuits. The equivalence of MMF NI, of the magnetic circuit is EMF E or V of the electric circuit. Magnetic reluctance of the iron path and that of air gap are represented by their equivalent resistances R_i and R_g in the electric circuit. Flux in the magnetic circuit is represented by current in the electric circuit. The magnetizing force, $\frac{NI}{l}$ produces the flux in the core. The opposition to the flux path is provided by iron and the air in the air gap through which

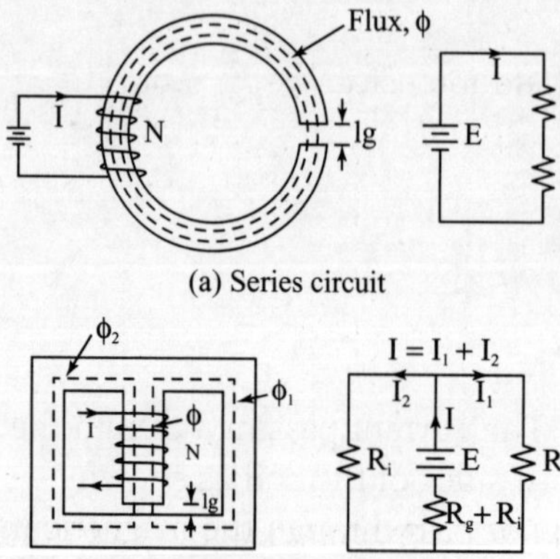

FIGURE 5.18

Series and parallel magnetic circuits with equivalent electric circuits

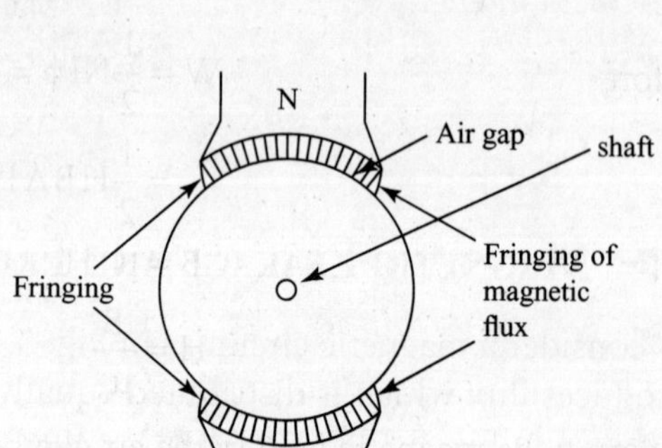

FIGURE 5.19

Fringing of flux at the corners of the North pole and South pole of an electrical machine

the flux has to pass. The reluctance of iron is very low whereas that of air is very high. Reluctance of small length of air gap will be much more than the reluctance of total length of iron path.

Due to spreading of the flux, the effective area of cross section of the air gap through which the flux passes increases, which is not desirable. The fringing effect of flux has been shown in Fig. 5.19.

5.9 ▶ ATTRACTIVE FORCE OR THE LIFTING POWER OF ELECTROMAGNETS

An electromagnet is often used to pull or lift iron pieces or objects made of iron. This is possible due to the energy stored in the magnetic field of the electromagnet.

In Fig. 5.20 is shown an electromagnet having N number of turns and the coil carrying a current, I. The cross-sectional area of the core is A. An iron piece is separated from the electromagnet by a distance, l. We will calculate the energy stored in the magnetic field and the lifting power of the electromagnet.

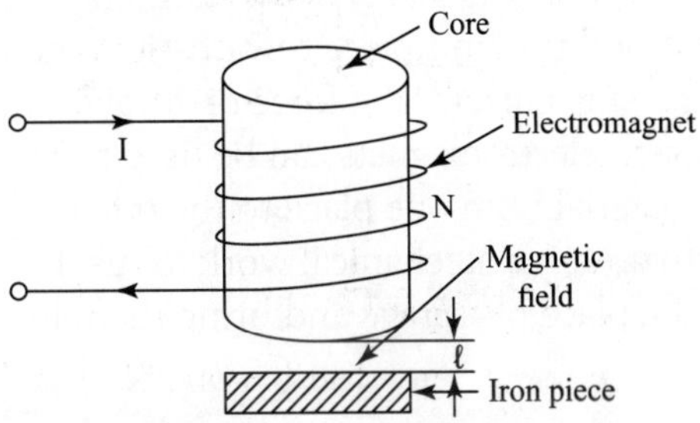

FIGURE 5.20

Lifting power of an electromagnet

Energy stored in a magnetic field is given by

$$W = \frac{1}{2}LI^2 \text{ joules}$$

Putting $L = N\frac{\phi}{I}$ we get

$$W = \frac{1}{2}N\phi I \text{ joules}$$

Again $H = \frac{NI}{l}$ and $\phi = B \times A$

Therefore,
$$W = \frac{1}{2}NI\phi = \frac{1}{2}H \times l \times B \times A$$

$$= \frac{1}{2}HBAl \text{ joules}$$

Again, $B = \mu_o H$,

or, $H = \frac{B}{\mu_o}$

Considering the magnetic field in the air gap between the magnet and the piece of iron,

Energy stored,
$$W = \frac{1}{2}\frac{B}{\mu_o}BAl \text{ joules}$$

$$= \frac{B^2}{2\mu_o}(A \times l) \text{ joules} \tag{5.16}$$

(A × l) is the volume of the air gap between the electromagnet and the iron piece.

Force, $$F = \text{Work done per unit length} = \frac{W}{l}$$

or, $$F = \frac{B^2 A}{2\mu_o} \text{ Newton} \quad (5.17)$$

The stored energy of the magnetic field between the electromagnet and the piece of iron is able to pull the iron piece towards it. When the exciting coil of the electromagnet is energized by passing current, the force of attraction will pull the iron near to the face of the electromagnet. Such electromagnets can be used to pull or lift large amount of magnetic material and shift the material from one place to the other. Thus, the lifting power of the electromagnet can be used to do some mechanical work for us. For example, electromagnets are used to lift iron ores from the place of storage and bring them for processing when required.

EXAMPLE 5.2

A circular iron ring of mean diameter 25 cm and cross-sectional area 9 cm² is wound with a coil of 100 turns and carries a current of 1.5 A. The relative permeability of iron is 2000. Calculate the amount of flux produced in the ring.

SOLUTION

Mean length of flux path, $l = \pi D = 3.14 \times 25 \text{ cm} = \frac{3.14 \times 25}{100} \text{ m} = 0.785 \text{ m}$

Flux, $$\phi = \frac{\text{MMF}}{\text{Reluctance}} = \frac{NI}{l/\mu_o \mu_r A} = \frac{100 \times 1.5}{0.785/4\pi \times 10^{-7} \times 2000 \times 9 \times 10^{-4}}$$

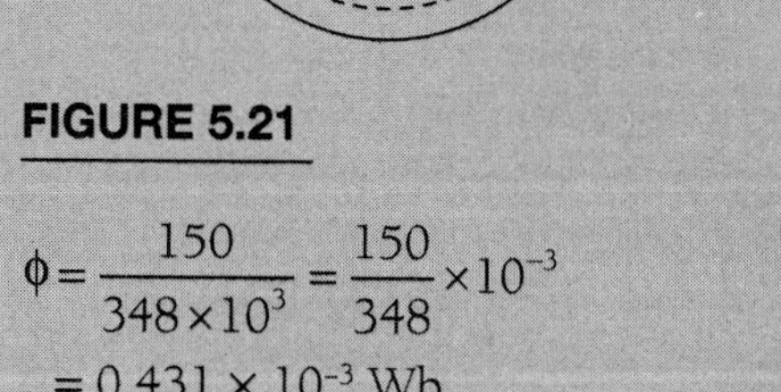

FIGURE 5.21

or, $$\phi = \frac{150}{348 \times 10^3} = \frac{150}{348} \times 10^{-3}$$

$= 0.431 \times 10^{-3}$ Wb

$= 0.431$ mWb

EXAMPLE 5.3

A rectangular shape iron core has an air gap of 0.01 cm. The mean length of the flux path through iron is 39.99 cm. The relative permeability of iron is 2000. The coil has 1000 turns. The cross-sectional area of the core is 9 cm². Calculate the current required to produce a flux of 1 mWb in the core.

SOLUTION

Total reluctance of the flux path = Reluctance of iron path + Reluctance of air gap,

i.e., $$S = S_i + S_g$$

$$S = \frac{l_i}{\mu_o \mu_r A} + \frac{l_g}{\mu_o A}$$

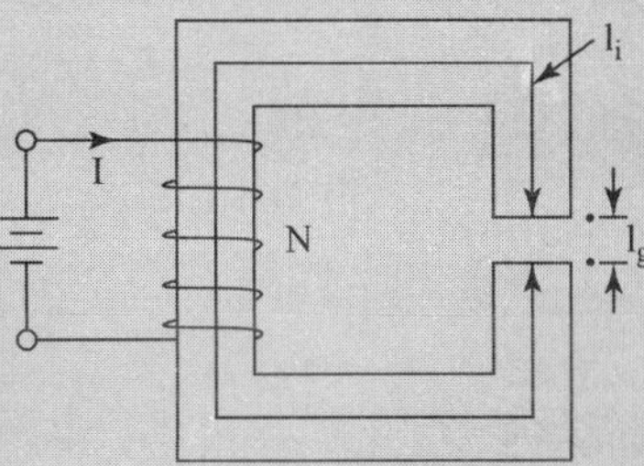

FIGURE 5.22

Note: the iron path permeability is μ which is equal to $\mu_0\mu_r$ whereas for the air gap the permeability is μ_0 only.

Substituting the given values,

$$S = \frac{39.09 \times 10^{-2}}{4\pi \times 10^{-7} \times 2000 \times 9 \times 10^{-4}} + \frac{0.01 \times 10^{-2}}{4\pi \times 10^{-7} \times 9 \times 10^{-4}}$$

$$= \frac{10^6}{4\pi}\left[\frac{39.09}{18} + \frac{100}{10}\right] = \frac{295.45 \times 10^5}{36\pi}$$

Flux, $$\phi = \frac{NI}{S} = \frac{1000\,I}{S}$$

$\therefore$ $$I = \frac{\phi \times S}{1000} = \frac{1 \times 10^{-3} \times 295.45 \times 10^5}{36\pi \times 1000} = \frac{29.545}{36\pi} = 0.26\text{ A.}$$

EXAMPLE 5.4

A magnetic circuit is having its winding on its central limb. The cross-sectional area of the central limb is 10 cm² whereas the cross-sectional area of the outer limbs is 5 cm². The effective length of the central limb is 16 cm and that of the outer limbs is 25 cm. Calculate the current required to flow through the winding which has 1000 turns to produce a flux of 1.2 mWb in the central limb. Assume that for a flux density of 1.2 Wb/m², the magnetizing force required is 750 AT/m. Draw the equivalent electric circuit.

SOLUTION

The details of the magnetic circuit are shown in Fig. 5.23.

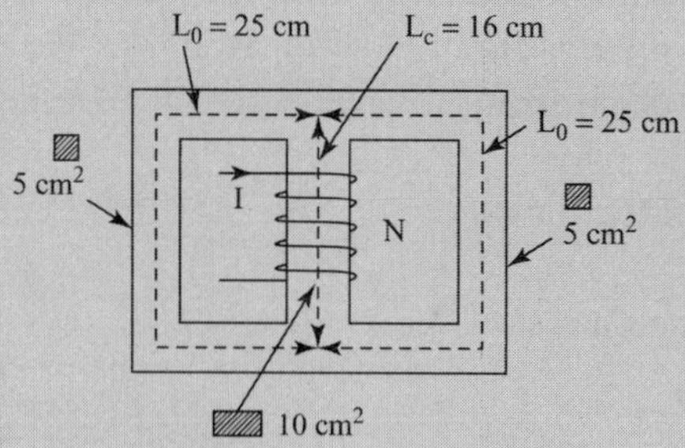

FIGURE 5.23

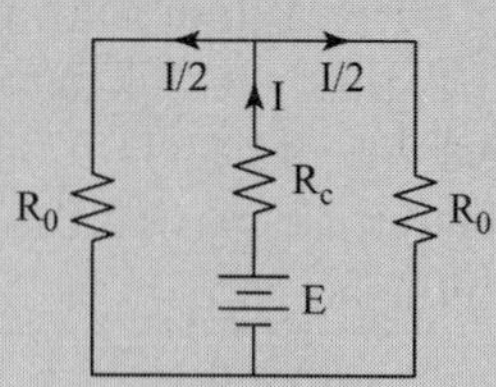

FIGURE 5.24

The equivalent electric circuit is drawn as shown in Fig. 5.24.

This is an example of a parallel circuit. As current in an electric circuit gets divided into two parallel branches, the flux produced in the central limb will get divided into the two outer limbs.

We will calculate the MMF required for the central limb as also for any of the outer limbs which will maintain the desired flux in the core. For a flux density of 1.2 Wb/m^2, the value of H has been given. Let us calculate the flux density in the central limb first.

Flux density in the central limb, $B_c = \dfrac{\phi_c}{A_c} = \dfrac{1.2\times10^{-3}}{10\times10^{-4}} = 1.2 \text{ Wb/m}^2$

The Flux density in the outer limb will be the same as that in the central limb since half the flux is available in each of the outer limbs and their cross-sectional area is half of that of the central limb.

$$B_o = \frac{\phi_o}{A_o} = \frac{0.6\times10^{-3}}{5\times10^{-4}} = 1.2\,\text{Wb/m}^2$$

The corresponding H, i.e., AT/m for flux density of 1.2 Wb/m^2 has been given as 750.

The total MMF required = MMF required for the central limb + MMF required for one outer limb (and not for both the limbs).

Since $\quad H = \dfrac{\text{MMF}}{l}, \ \text{MMF} = H\times l.$

Total MMF (NI) required $\quad = \dfrac{750\times16}{100} + \dfrac{750\times25}{100} = 307.5$ [considering length in m] AT

$$AT = 307.5 = NI$$

$$N = 1000$$

$$I = \frac{307.5}{1000} = 0.3075\,\text{A}.$$

EXAMPLE 5.5

An iron ring of mean length of an iron path of 100 cm and having a uniform cross-sectional area of 10 cm^2 is wound with two magnetizing coils as shown. The direction of current flowing through the two coils are such that they produce flux in the opposite directions. The permeability of iron is 2000. There is a cut in the ring creating an air gap of 1 mm. Calculate the flux available in the air gap.

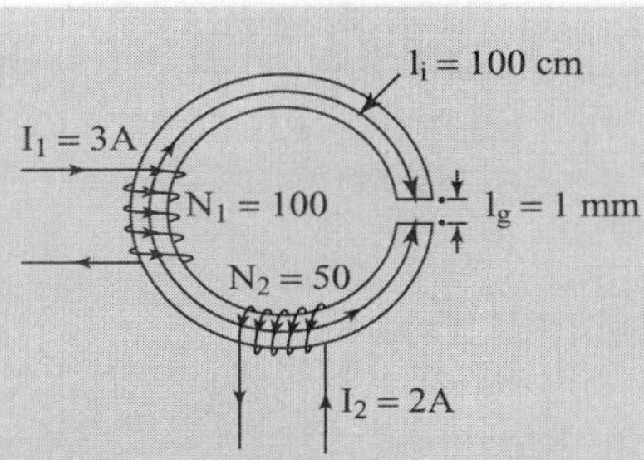

FIGURE 5.25

SOLUTION

The net MMF will be the resultant effect of MMF of the two coils in producing the flux in the core. As in Fig. 5.25, the flux produced by the MMF of the two coils are in opposite directions. Thus, the resultant MMF will be the difference of these two MMFs.

$$\text{Resultant MMF} = N_1 I_1 - N_2 I_2 = 100 \times 3 - 50 \times 2 = 200$$

Total reluctance, $S = S_i + S_g$ (i.e., reluctance of iron path and reluctance of air gap)

$$= \frac{l_i}{\mu_o \mu_r A_i} + \frac{l_g}{\mu_o A_g}$$ [we have considered the reluctance of $\mu_o \mu_r$ for iron and μ_o for air]

Since $A_i = A_g$, i.e., the cross-sectional area of the iron path is the same as that of the air gap,

$$S = \frac{1}{\mu_o A}\left[\frac{l_i}{\mu_r} + l_g\right]$$

Substituting values

$$S = \frac{1}{4\pi \times 10^{-7} \times 10 \times 10^{-4}}\left[\frac{50 \times 10^{-2}}{2000} + 1 \times 10^{-3}\right]$$

$$S = \frac{10^{10}}{4\pi}[25 \times 10^{-5} + 1 \times 10^{-3}]$$

$$= \frac{10^{10}}{4\pi} 10^{-3}[25 \times 10^{-2} + 1]$$

$$= \frac{10^{7}}{4\pi}[1.25] = 10^{6}$$

Flux,

$$\phi = \frac{\text{MMF}}{S} = \frac{200}{10^{6}} = 2 \times 10^{-4} = 0.2 \text{ mWb.}$$

EXAMPLE 5.6

Calculate the flux produced in the air gap in the magnetic circuit shown in Fig. 5.26, which is excited by the MMF of two windings. The mean length of the flux path is 40 cm. The permeability of iron is 2000. The uniform core cross-sectional area is 10 cm^2. Air gap length in 1 mm.

SOLUTION

Applying Fleming's thumb rule we can see that the ampere turns of coil 2 produce flux in the opposite direction as the flux produced by the ampere turns of coil 1.

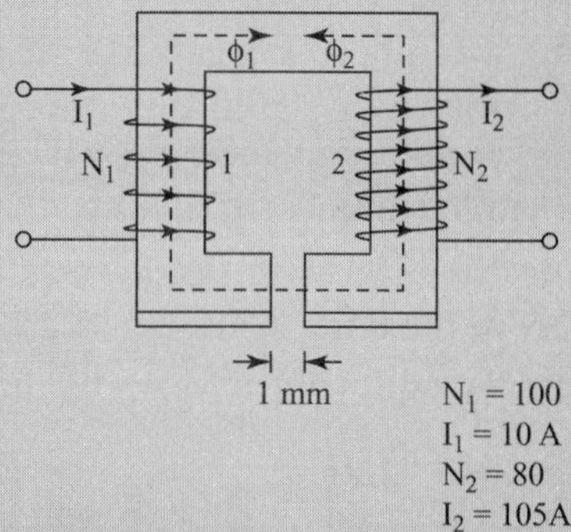

FIGURE 5.26

Total MMF $= N_1 I_1 - N_2 I_2$

$= 100 \times 10 - 80 \times 1.5$

$= 880$ AT

Total reluctance = Reluctance of iron + Reluctance of air gap

$$= \frac{l_i}{\mu_o \mu_r A} + \frac{l_g}{\mu_o A} = \frac{l}{\mu_o A}\left[\frac{l_i}{\mu_r} + l_g\right]$$

Substituting values

$$\text{Total Reluctance} = \frac{1}{4\pi \times 10^{-7} \times 10 \times 10^{-4}}\left[\frac{(40\times10^{-2} - 1\times10^{-3})}{2000} + 1\times10^{-3}\right]$$

$$= \frac{10^{10}}{4\pi}\left[\frac{399\times10^{-3}}{2000} + 1\times10^{-3}\right]$$

$$= \frac{10^{7}}{4\pi}\left[\frac{399}{2000} + 1\right]$$

$= 0.955 \times 10^6$ AT/Wb.

$$\text{Core Flux} = \frac{\text{MMF}}{\text{Reluctance}} = \frac{880}{0.955\times10^{6}} = 921.5\times10^{-6}$$

$= 0.9215 \times 10^{-3}$ Wb

The air gap flux is the same as the core flux calculated above, as the whole of the core flux crosses the air gap and there is no fringing.

EXAMPLE 5.7

A parallel magnetic circuit with 2000 turns on its central limb has been shown in Fig. 5.27. The air gaps are 2 mm each. The mean diameter of the circular magnetic path is 20 cm.

The cross-sectional area of the central limb is 10 cm^2 while the cross section of the outer limbs is 5 cm^2. A few readings from the magnetization curve are given below.

B in Wb/m^2	1.0	1.1	1.2	1.3	1.4
H in AT/m	550	650	750	820	870

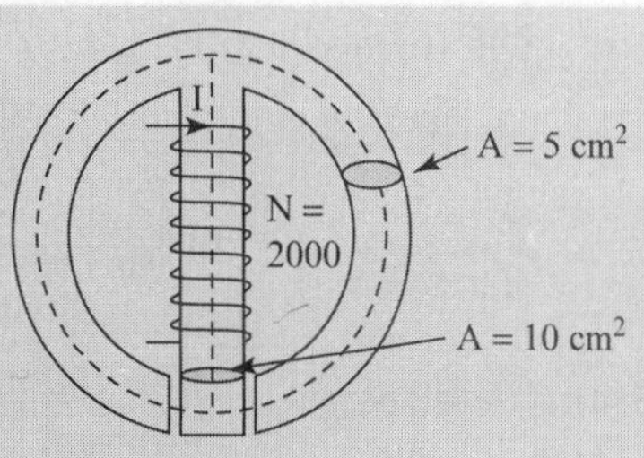

FIGURE 5.27

Calculate the current, I which must flow through the coil so as to produce a flux of 1.1 mWb in the central limb.

SOLUTION

The flux from the central limb will get equally divided in the two outer limbs. So that flux in the outer limbs will be 0.55 mWbs.

The flux density will be the same throughout as flux density, B is

for central limb $$B = \frac{\phi}{A} = \frac{1.1 \times 10^{-3}}{10 \times 10^{-4}} = 1.1\,\text{Wb/m}^2\text{, or Tesla}$$

for outer limbs $$B = \frac{\phi}{A} = \frac{0.55 \times 10^{-3}}{5 \times 10^{-4}} = 1.1\,\text{Wb/m}^2\text{, or Tesla}$$

The magnetic field strength, H corresponding to B = 1.1 Wb/m^2 is 650 AT/m.

We have to calculate the AT required to be provided for one of the parallel paths, i.e., for the central limb, one outer limb, and one air gap.

length of central limb, l_c = diameter = 20 cm

length of the outer limbs, (including air gap) $l_o = \frac{\pi d}{2} = \frac{3.14 \times 20}{2} = 31.4\,\text{cm}$

length of air gap, l_g = 2 mm = 0.2 cm

$$H = \frac{NI}{l},\ H = 1.1\ \text{Wb/m}^2$$

(a) NI for central limb $$= H \times l_c = 650 \times \frac{20}{100} = 130$$

(b) NI for outer limb $$= H \times (l_o - l_g) = 650\frac{(31.4 - 0.2)}{100}$$

$$= 650 \times \frac{31.2}{100} = 202.8$$

NI for air gap $= H_g \times l_g$

we have to calculate H_g for air

$$B = \mu_o H_g \text{ as } \mu_r = 1$$

$$H_g = \frac{B}{\mu_o} = \frac{1.1}{4\pi \times 10^{-7}} = 8.758 \times 10^5 \text{ AT/m}$$

(c) NI for air gap $= H_g \times l_g = 8.758 \times 10^5 \times 2 \times 10^{-3} = 1751.6$

Total AT required $= a + b + c$

$= 130 + 202.8 + 1751.6$

$= 2084.6$

Total number of turns of the exciting coil placed on the central limb, N = 2000.

Current, $$I = \frac{\text{Total AT}}{N}$$

$$= \frac{2084.6}{2000} = 1.04 \text{ A}$$

EXAMPLE 5.8

An iron ring is made up of two different materials A and B having a relative permeability of 1000 and 1500, respectively. The length L_A and L_B of the two materials used are 75 cm and 25 cm, respectively. The air gap length is 2 mm. The cross-sectional area of the core is 10 cm^2. The magnetizing coil has 1000 turns and a current of 5 A is allowed to flow through it. Calculate the flux produced in the air gap.

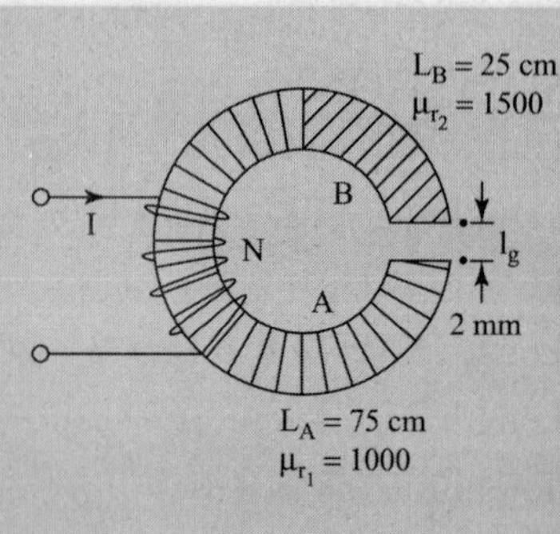

FIGURE 5.28

SOLUTION

We know, $$\phi = \frac{\text{MMF}}{\text{Total reluctance}}$$

Total reluctance, S = Reluctance of part A + Reluctance of part B + Reluctance of air gap

$$S = \frac{L_A}{\mu_o \mu_{r_1} A} + \frac{L_B}{\mu_o \mu_{r_1} A} + \frac{l_g}{\mu_o A}$$

$$= \frac{0.75}{4\pi \times 10^{-7} \times 1000 \times 10 \times 10^{-4}} + \frac{0.25}{4\pi \times 10^{-7} \times 1500 \times 10 \times 10^{-4}}$$

$$+ \frac{2 \times 10^{-3}}{4\pi \times 10^{-7} \times 10 \times 10^{-4}}$$

$$= 10^4 \, [59.7133 + 13.2696 + 1592.3566]$$

$$= 10^4 \times 1665.3395$$

$$= 16.653395 \times 10^6 \text{ AT/Wb}$$

Flux, $$\phi = \frac{\text{mmf}}{\text{Reluctance}} = \frac{NI}{S} = \frac{1000 \times 5}{16.653395 \times 10^6}$$

$$= 0.3 \times 10^{-3} \text{ Wb}$$

$$= 0.3 \text{ mWb}$$

EXAMPLE 5.9

For the core shown in Fig. 5.29, it is required to produce a flux of 2 mWb in the limb CD. The entire core has a rectangular cross section of 2cm × 2cm. The magnetizing coil has 800 turns. The relative permeability of the material is 1200. Calculate the amount of magnetizing current required.

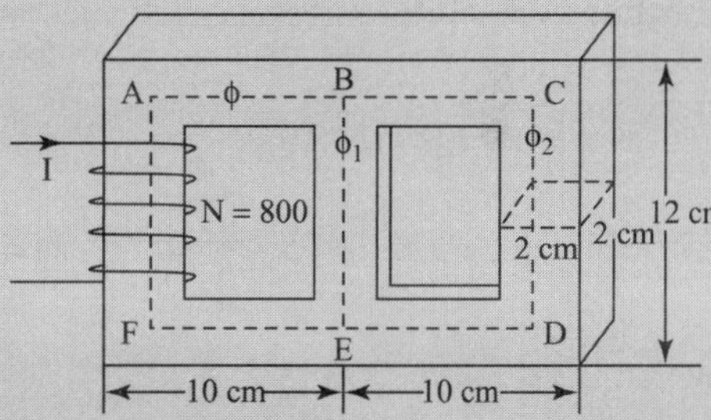

FIGURE 5.29

SOLUTION

Length CD = BE = AF = 10 cm

Length BC = ED = AB = EF = 8 cm

Length BCDE = 8 + 10 + 8 = 26 cm

Length BAFE = 8 + 10 + 8 = 26 cm

Length BE = 10cm; $\mu_r = 1200$

Total flux, $\phi = \phi_1 + \phi_2$

N = 800, $\phi_2 = 2 \times 10^{-3}$ Wb, current, I = ?

Let us draw the equivalent electrical circuit of the given magnetic circuit. The equivalent electric circuit will be as shown in Fig. 5.30.

The voltage drop across CD is $I_2 R_2$.

The voltage drop across BE is equal to the voltage drop cross CD.

Therefore,

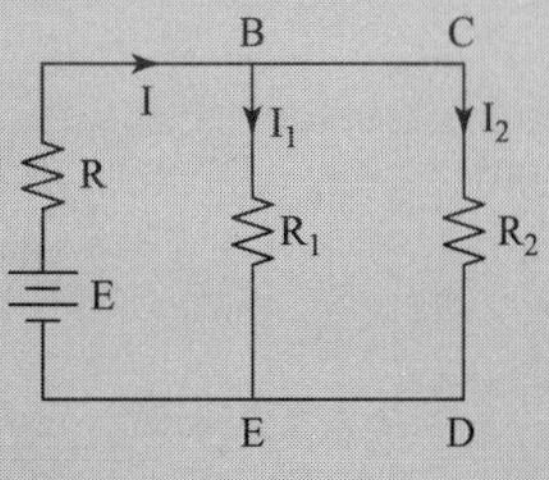

FIGURE 5.30

$$I_1 R_1 = I_2 R_2$$

or, $$I_1 = I_2 \frac{R_2}{R_1}$$

For the magnetic circuit, from the analogy of the above equivalent electric circuit, we can write

$$\phi_1 = \phi_2 \frac{S_2}{S_1}$$

S_2 is the reluctance of path BCDE

S_1 is the reluctance of path BE

$$S_2 = \frac{l}{\mu_o \mu_r A} = \frac{26 \times 10^{-2}}{4\pi \times 10^{-7} \times 1200 \times 4 \times 10^{-4}}$$

$$S_1 = \frac{10 \times 10^{-2}}{\mu_o \mu_r A} = \frac{10 \times 10^{-2}}{4\pi \times 10^{-7} \times 1200 \times 4 \times 10^{-4}}$$

$$\phi_1 = \phi_2 \frac{S_2}{S_1} = 2 \times 10^{-3} \frac{26}{10} = 5.2 \times 10^{-3} \text{ Wb}$$

$$\phi = \phi_1 + \phi_2 = 2 \times 10^{-3} + 5.2 \times 10^{-3} = 7.2 \times 10^{-3} \text{ Wb}$$

AT required for portion BAFE (=26 cm) = $\phi \times S_3$

$$= \frac{7.2 \times 10^{-3} \times 26 \times 10^{-2}}{4\pi \times 10^{-7} \times 1200 \times 4 \times 10^{-4}} = 3105$$

AT required for portion BE $= \phi_1 \times S_1$

$$= \frac{5.2 \times 10^{-3} \times 10 \times 10^{-2}}{4\pi \times 10^{-7} \times 1200 \times 4 \times 10^{-4}}$$

$$= 862$$

In the electric circuit, we see that by applying KCL

$$E - IR - I_1 R_1 = 0$$

or,

$$E = IR + I_1 R_1$$

Similarly, for the magnetic circuit

Total AT = AT required for portion BAFF + AT required for the portion BE

$$= 3105 + 862$$

$$= 3967.$$

The number of turns of the exciting coil is 800.

$$AT = NI = 3967$$

$$I = \frac{3967}{N} = \frac{3967}{800} = 4.95 \text{A}$$

EXAMPLE 5.10

An electromagnetic-type relay shown in Fig. 5.31 has the following particulars. The mean length of the flux path through the iron is 20 cm and the length of air gaps is 1 mm each. The exciting coil has 8000 turns and carries a current of 50 mA when excited. The cross-sectional area of the core is 0.5 cm². The permeability of iron is 500. Calculate the flux density and the magnetic pull produced in the armature (i.e., on the moving part).

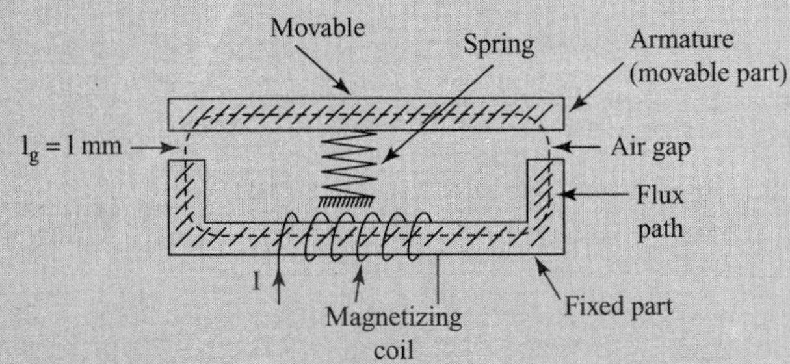

FIGURE 5.31

SOLUTION

Total reluctance of the flux path = Reluctance of the path through the iron + Reluctance of the two air gaps. The length of each of the air gaps, l_g = 1 mm. We have to take into account $2l_g$ as the total air gap to the flux path.

$$\text{Total reluctance} = \frac{l_i}{\mu_o \mu_r A} + \frac{2l_g}{\mu_o A}$$

$$= \frac{20 \times 10^{-2}}{4\pi \times 10^{-7} \times 500 \times 0.5 \times 10^{-4}} + \frac{2 \times 1 \times 10^{-3}}{4\pi \times 10^{-7} \times 0.5 \times 10^{-4}}$$

$$= 3.82 \times 10^7 \text{ AT/Wb}$$

$$\text{Flux,} \quad \phi = \frac{\text{MMF}}{\text{Reluctance}} = \frac{\text{NI}}{\text{Reluctance}} = \frac{8000 \times 50 \times 10^{-3}}{3.82 \times 10^7} = 104.7 \text{ mWb}$$

$$\text{Flux, density,} \quad B = \frac{\phi}{A} = \frac{104.7 \times 10^{-3}}{0.5 \times 10^{-4}} = 0.2094 \text{ Wb/m}^2$$

From equation 5.17, the force or the pull on the armature by each pole,

$$F = \frac{B^2 A}{2\mu_o} = \frac{(0.2094)^2 \times 0.5 \times 10^{-4}}{2 \times 4\pi \times 10^{-7}} = 4.16 \text{ N Newton}$$

EXAMPLE 5.11

An overhead electrical power transmission line carries a current of 100A. The direction of current flow through the line is from the West to the East direction. What is the direction of the magnetic field produced and what is its value 1 m below the overhead line?

SOLUTION

We know, $H = \frac{NI}{l}$ and $B = \mu H$

Here $\mu = \mu_o$ (air)

$N = 1$ (straight conductor)

$l = 2\pi r$

The magnitude of the magnetic field, B is expressed as

$$B = \frac{\mu_0 I}{2\pi r}$$

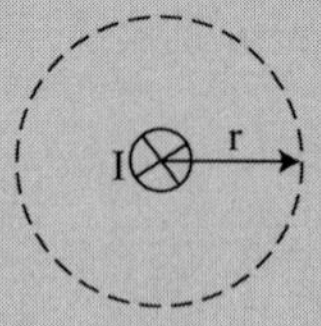

FIGURE 5.32

Substituting values of μ_o, I, and r as $4\pi \times 10^{-27}$, 100 A and 1 m, respectively,

$$B = \frac{4\pi \times 10^{-7} \times 100}{2\pi \times 1} = 20 \times 10^{-6}\ \text{Wb/m}^2 = 20 \times 10^{-6}\ \text{Tesla}$$

The field produced is at right angles to the length of the current-carrying conductor as has been shown in Fig. 5.32.

EXAMPLE 5.12

A rectangular coil ABCD of size 20 cm × 10 cm is placed 1 cm away from a long current-carrying conductor with its longer sides parallel to the conductor. A current of 100 A is flowing through the long conductor while the coil is carrying a current of 10 A. Calculate the resultant force developed on the coil.

SOLUTION

Force will be developed between the longer sides of the coil and the conductor as they are parallel to each other. There will be force of attraction with one coil side and force of repulsion with the other. The difference between the two forces will be the resultant force acting on the coil. There will be no force developed on the sides which are perpendicular to the conductor as shown in Fig. 5.33.

Force experienced per unit length by a current-carrying conductor due to flux density created by the other current-carrying conductor is

$$F_1 = BI_2 = \frac{\mu_0 I_1 \times I_2}{2\pi r_1}\ \text{N}$$

Force of attraction between the conductor and side AB

$$F_1 = \frac{\mu_0 I_1, I_2 l}{2\pi r_1}$$

Force of repulsion between the conductor and side CD

$$F_2 = \frac{\mu_0 I_1 I_2 l}{2\pi r_2}$$

$$\text{The net force} = F_1 - F_2 = \frac{\mu_0 I_1 I_2 l}{2\pi}\left[\frac{1}{r_1} - \frac{1}{r_2}\right]$$

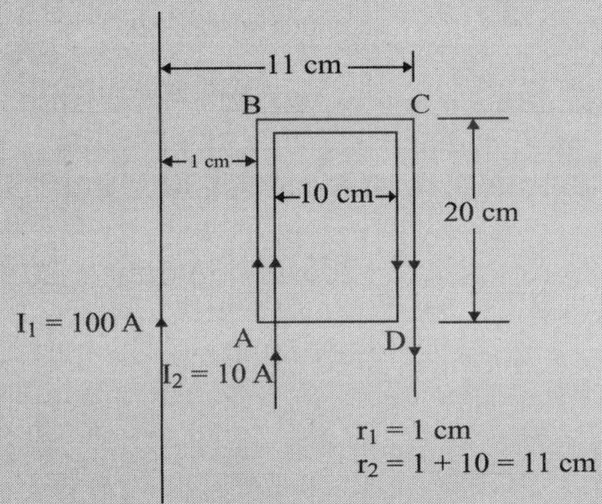

FIGURE 5.33

Substituting the values

$$F_1 - F_2 = \frac{4\pi \times 10^{-7} \times 100 \times 10 \times 0.2}{2\pi}\left[\frac{1}{0.01} - \frac{1}{0.11}\right]$$

$$= 3.63 \times 10^{-3} \text{ N}$$

EXAMPLE 5.13

A horse-shoe-type iron-core electromagnet is wound with 500 turns and is required to lift heavy iron bars of 200 kg each time. The area of cross section of each of the poles of the horse-shoe magnet is 0.01 m^2. The mean length of the flux path through the electromagnet is 0.5 m. Calculate the value of the exciting current through the coil. The relative permeability of the flux path is 1000.

SOLUTION

A weight of 200 kg has to be lifted by both the poles of the electromagnet. Thus, each pole will have to lift a load of 100 kg. The force of attraction which is also called the lifting power of the electromagnet is

$$F = \frac{B^2 A}{2\mu_0}$$

or,

$$B = \sqrt{\frac{F \times 2\mu_0}{A}}$$

Putting the values,

$$F = 100 \text{ kg} = 100 \times 9.8 \text{ N}$$

$$B = \sqrt{\frac{100 \times 9.8 \times 2 \times 4\pi \times 10^{-7}}{0.01}}$$

$$= 0.5 \text{ Wb/m}^2$$

Again

$$B = \mu H$$

or,

$$H = \frac{B}{\mu} = \frac{B}{\mu_0 \mu_r} = \frac{0.5}{4\pi \times 10^{-7} \times 1000} = 400$$

We know

$$H \times l = AT = NI$$

or,

$$I = \frac{H \times l}{N} = \frac{400 \times 0.5}{500} = 0.4 \text{ A}$$

5.10 ▶ ELECTROMAGNETIC INDUCTION

This topic has already been discussed in chapter 1. However, because of the importance of this topic in electrical engineering, the salient points are briefly described in this section.

Whenever a conductor cuts magnetic lines of force or the conductor is cut by the magnetic lines of force, emf is induced in the conductor.

The magnitude of induced emf is directly proportional to the rate of change of magnetic flux cut.

These two basic laws were discovered, on the basis of his laboratory experiments, by Faraday.

The change of flux linkage by a coil or the rate of cutting of flux by a conductor can be done in two ways, viz., (1) by a motion of the coil or the conductor in the magnetic field. This is called *dynamically induced emf*. Electrical generators are built on the basis of this principle. In certain generators, coils are kept stationary while the magnetic poles producing magnetic flux are made rotating. (2) by passing a changing current through a coil thereby changing the flux linking to coil. If another coil is placed near this coil, emf is induced in both the coils. This emf is called *statically induced emf*. Statically induced emf may have self induced emf and mutually induced emf as its components. Transformers are built using this principle.

Important Terminologies and Points to Remember

1. A current flowing through a coil which is wound around a piece of magnetic material, makes it an electromagnet. Flux is produced around the coil creating a magnetic field. The strength of the magnetic field will depend upon the product of number of turns, N of the coil and the current, I flowing through it. The strength will also depend on the resistance (reluctance) offered by the path through which magnetic lines of force (flux) pass.
2. Magnetic flux density, B is the flux per unit area, i.e., $B = \frac{\phi}{A}$ where ϕ is the total flux and A is the area through which this flux pass.
3. Magnetising force, H is the ampere-turn per meter, i.e., $H = \frac{NI}{l}$ A_T/m. where l is the length of the flux line.
4. Permeability, μ is the measure of how will a magnetic material conducts magnetism or allows establishment of flux through it.
5. Relative permeability, μ_r is the permeability of a magnetic material as compared to that of air or free space.
6. The permeability of free space μ_0 is equal to $4\pi \times 10^{-7}$ H/m and $\mu = \mu_0 \mu_r$.
7. Hysteresis loss in a magnetic material is due to the magnetization of the magnetic material in opposite directions in every half cycle of the current flowing through the magnetising coil wound around the material.
8. Eddy current loss takes place in the magnetic core material due to the emf induced in the material when subjected to a changing magnetic field.
9. MMF is the product of number of turns, N of the magnetising coil and the current, I flowing through it.
10. Flux, MMF and Reluctance are related as

$$\text{Flux} = \frac{\text{MMF}}{\text{Reluctance}}$$

11. Reluctance, S is expressed as, $S = \frac{\ell}{\mu A}$ where ℓ is the length of flux path, A is the cross-sectional area through which flux passes and μ is the permeability of the material. Hence, reluctance is inversely proportional to permeability.
12. Current in an electric circuit is analogous to flux in a magnetic circuit. Similarly emf is analogous to mmf in a magnetic circuit.

Important Formulae and Equations

Formula/Equation	Remarks
1. $B = \mu H, B = \frac{\phi}{A}$ $\mu = \mu_0 \mu_r$ $H = \frac{NI}{l}$ $\mu_0 = 4\pi \times 10^{-7}$ H/m	1. Flux, Flux density, Magnetising force, permeability are important considerations in the design of electrical machines.

Important Formulae and Equations

Formula/Equation	Remarks
2. $\phi = \dfrac{\text{mmf}}{\text{reluctance}}$	2. Analogous to Ohm's law for an electrical circuit.
3. $W_h = K_h f B_m^{1.6} v$ watts Where v is the volume of the material. $W_e = k_t B_m^2 f^2 t^2 v$ watts	3. Hysteresis loss and Eddy current loss togather is called Iron loss.
4. Reluctance, $S = \dfrac{l}{\mu A}$	4. Reluctance is analogous to resistance in electrical circuit. Reluctance is the inverse of permeability.
5. Lifting power of a magnet, F is given as $F = \dfrac{B^2 A}{2\mu_0}$ Newton.	5. Electromagnets are used to lift material in industry.
6. Reluctance of the magnetic material with an air-gap of length l_g is equal to $S = S_i + S_g = \dfrac{l_i}{\mu_0 \mu_r A} + \dfrac{l_g}{\mu_0 A}$.	6. Useful in solving problems related to magnetic circuits.

REVIEW QUESTIONS

Short Answer Type Questions

1. Explain why the core of an electromagnet is made of magnetic material like iron.
2. Distinguish between an electromagnet and a permanent magnet.
3. What is meant by magnetization of a magnetic material?
4. Explain the following terms: MMF, reluctance, flux density, permeability, relative permeability, magnetic field intensity.
5. What is meant by magnetic saturation of a magnetic material?
6. Differentiate between electric resistance and magnetic reluctance.
7. Compare a magnetic circuit with an equivalent electric circuit.
8. What is meant by Hysteresis loss and Eddy current loss? On what factors does hysteresis loss depend?
9. On what factors does attractive power of an electromagnet depend?
10. Explain briefly magnetic leakage and fringing.
11. Derive the expression for lifting power of an electromagnet.
12. Explain the process of dynamically induced emf and statically induced emf.

Numerical Problems

12. An iron ring of 19.1 cm mean diameter has a cross-sectional area of 8 cm². The ring has an air gap of 5 mm. The winding on the ring has 500 turns and carries a current of 5 A. Calculate the flux produced in the air gap. The relative permeability of iron is 750.

[**Ans** Air gap flux = 0.434 mWb]

13. The air gap of a magnetic circuit is 2 mm long and 25 cm^2 in cross section. Calculate the reluctance of the air gap. How much ampere turns will be required to produce a flux of 1.2 mWb in the air gap?

[**Ans** 0.636×10^6 AT/Wb; 764]

14. An iron ring of mean length 50 cm has an air gap of 1 mm. The ring is provided with a winding of 200 turns through which a current of 1 A is allowed to flow. Find the flux density across the air gap. Assume the relative permeability of iron as 300.

[**Ans** B = 0.094 Wb/m^2]

15. A magnetic core has a cross-sectional area of 16 cm^2. The air gap length is 2 mm. Length of the iron path is calculated as 73.8 cm. The exciting coil has 2000 turns. Calculate the current which is required to flow through the winding to create an air gap flux of 4 mWb. Assume relative permeability of the core material as 2000.

[**Ans** I = 2.356 A]

16. The armature and the field magnets of an electrical machine has been shown in Fig. 5.34. The air gap between the poles and the armature has been kept as 10 mm. The pole area is 0.1 m^2 and the flux per pole is 0.15 Wb. Calculate the mechanical force exerted by each pole on the armature. Also calculate the energy stored in the air gaps.

[**Ans** F = 89523 N; W = 1655.3 J]

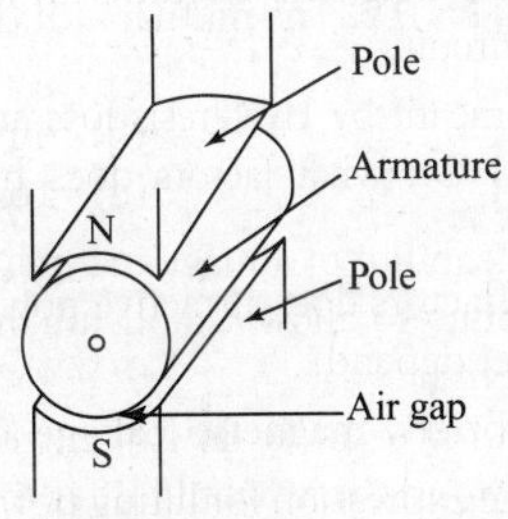

FIGURE 5.34

17. A circular iron ring having a rectangular cross section is wound with a coil of 500 turns. When a current of 3 A flows through the winding, a flux density of 1.2 Wb/m^2 is produced in the 1 mm air gap. The inner diameter of the ring is 20 cm and the outer diameter is 25 cm. The thickness of the ring is 2 cm. Calculate the magnetic field intensity in the material and in the air gap. Also calculate the relative permeability of the magnetic material, i.e., iron.

[**Ans** 771 AT/m; 9.55×10^5 AT/m; 1238]

18. A steel ring has a mean diameter of 159.23 mm and cross-sectional area of 3 cm^2. The ring has an air gap of 1 mm. Determine the current required in the exciting coil having 250 turns to produce a flux of 0.2 mWb in the air gap. Take the permeability of iron to be 1200.

[**Ans** I = 3.3 A]

19. An iron ring of mean diameter of 10 cm and cross-sectional area of 8 cm^2 is wound with a wire having 300 turns. The permeability of iron is 500. What current should be passed through the winding wire so that a flux density of 1.2 Wb/m^2 is produced in the core?

[**Ans** I = 2 A]

20. Fig. 5.35 shows a magnetic circuit. Calculate the current required to be passed through the central limb winding so as to produce a flux of 1.6 mWb in this limb. Length of iron in the central limb is 15 cm. Cross-sectional area of the central limb is 8 cm^2 and that of the outer limbs is 4 cm^2.

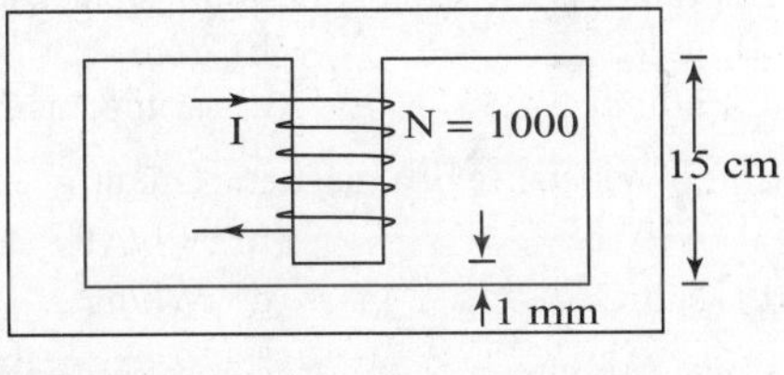

FIGURE 5.35

The mean length of iron of the outer limbs is 32 cm each. Given that for iron for a flux density of 2.0 Wb/m2 the value of H is 800 AT/m.

[**Ans** I = 1.905 Ampere]

Multiple Choice Questions

1. A magnetic circuit and an electric circuit can be compared as
(a) flux is analogous to current
(b) reluctance is analogous to resistance
(c) MMF is analogous to EMF
(d) all these as in (a), (b), and (c).

2. The relationship of flux, reluctance and MMF is
   (a) flux $\frac{\text{MMF}}{\text{Reluctance}}$ (b) flux = MMF × reluctance
   (c) flux $\frac{\text{Reluctance}}{\text{MMF}}$ (d) flux $\frac{\text{EMF}}{\text{Reluctance}}$.
3. Relative permeability of air is
   (a) equal to 0 (b) equal to 1
   (c) equal to ∞ (d) around 2000.
4. A magnetic circuit is said to be saturated when an increase in the field intensity results in
   (a) decrease in flux density
   (b) proportional increase in flux density
   (c) very marginal increase in flux density
   (d) sudden increase in flux density.
5. Which of the following is applicable for magnetic circuits?
   (a) Thevenin's Theorem
   (b) Maximum Power Transfer Theorem
   (c) Norton's Theorem
   (d) Kirchhoff's Laws.
6. Material used for making a permanent magnet should have
   (a) large hysteresis loop area
   (b) small hysteresis loop area
   (c) low-coercieve force
   (d) low-saturation flux density.
7. The value of permeability of iron may be taken as
   (a) unity (b) zero
   (c) infinity (d) as high as 2000.
8. Unit of reluctance of a magnetic circuit is
   (a) AT/Wb (b) Wb/AT
   (c) Wb-Cm (d) Wb/m^2.
9. A magnetic circuit of uniform cross-sectional area of length 50 cm is wound uniformly by 250 turns of a coil and carries a current of 4 A. The magnetic field strength or the magnetizing force produced is
   (a) 1000 AT/m (b) 2000 AT/m
   (c) 1000 AT/cm (d) 2000 AT/cm.
10. Which of the flowing relations is False.
   (a) $L = N\frac{d\phi}{dt}$ (b) $B = \mu H$
   (c) $H = \frac{NI}{l}$ (d) $\mu = \mu_o \mu_r$.
11. Which of the flowing similarities in electrical and magnetic circuits is False
   (a) Conductance and Permeance
   (b) Current and Flux
   (c) Conductance and Reluctance
   (d) Resistance and Reluctance
12. Hysteresis loss and Eddy current loss in a magnetic material can be expressed, respectively, as
   (a) $W_h = k_h B_m^{1.6} f$ and $W_e = k_e B_m^2 f^2$
   (b) $W_h = k_h B_m^{1.6} f^2$ and $W_e = k_e B_m^2 f$
   (c) $W_h = k_h B_m^{1.6} f^2$ and $W_e = k_e B_m^2 f^2$
   (d) $W_h = k_h B_m f^2$ and $W_e = k_e B_m^{1.6} f^2$
13. Self inductance of an air-core coil can be increased by introducing
   (a) a copper rod inside the core
   (b) a wooden rod inside the core
   (c) an iron rod inside the core
   (d) none of these.
14. The coefficient of coupling of two coils of 4 mH and 16 mH is 0.5. The mutual inductance between them is
   (a) 2 mH (b) 4 mH
   (c) 8 mH (d) 16 mH.
15. The permeability of a magnetic material means
   (a) its ability to allow establishment of lines of force in it
   (b) its ability to allow flow of current through it
   (c) the opposition it would offer to establishment of flux in it
   (d) its ability to provide high reluctance path to the magnetic flux.

Answers to Multiple Choice Questions

1. (d)	2. (a)	3. (b)	4. (c)	5. (d)	6. (a)
7. (d)	8. (a)	9. (b)	10. (a)	11. (c)	12. (a)
13. (c)	14. (b)	15. (a)			

CHAPTER

6 Transformers

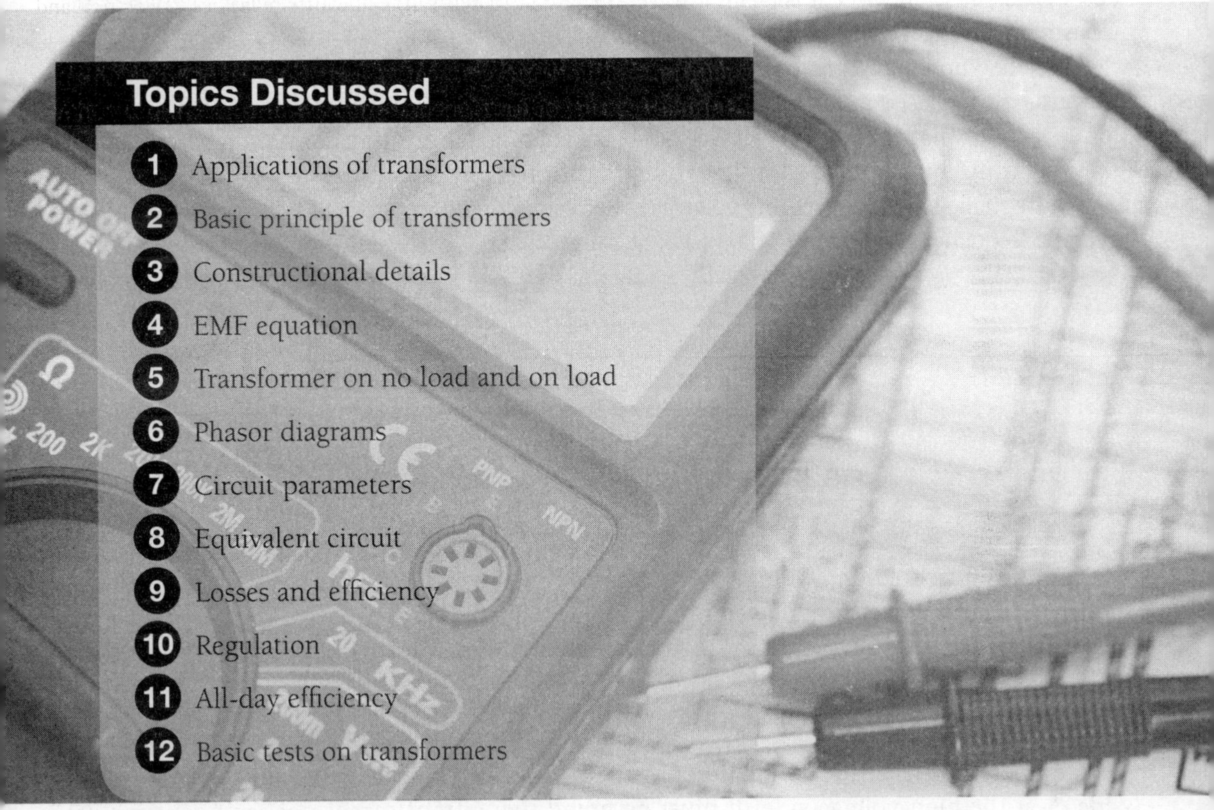

Topics Discussed

1. Applications of transformers
2. Basic principle of transformers
3. Constructional details
4. EMF equation
5. Transformer on no load and on load
6. Phasor diagrams
7. Circuit parameters
8. Equivalent circuit
9. Losses and efficiency
10. Regulation
11. All-day efficiency
12. Basic tests on transformers

6.1 ▶ Introduction

Electricity is generated in power houses. Power houses are situated at distant places wherefrom the generated power is to be taken to industries, commercial centres, residential colonies, and villages. The generators used in power houses generate electricity at a particular voltage. Normally the generation voltage is 11,000 V or 11 kV. Electricity is transmitted from the place of generation to the place of its use through electrical transmission lines which are taken overhead using transmission towers. The voltage level of the power to be transmitted is raised to higher values, say from 11 kV to 220 kV to reduce the cost of transmission. High-voltage transmission reduces the size of transmission line conductors thereby reducing the weight of the conducting material used. Since power is the product of voltage and current, for the same power if voltage is increased, the magnitude of current will decrease. For transmitting at this reduced current, the size of the line conductors will reduce, and hence, the cost gets reduced. The power house wherefrom electricity is transmitted through transmission lines is called the 'sending end' whereas the other end of the transmission lines where the electricity is received for use is called the 'receiving end'. At the sending end the voltage level is increased using step-up transformers before connecting to the transmission lines, while at the receiving end the voltage level is lowered before the distribution of electricity for use. The voltage level is raised for reducing the cost of transmission while it is again lowered before supplying to consumers, for safety reasons. Consumers use electricity at 230 V or at 400 V.

A device known as transformer which either raises or lowers the voltage level of electrical power is always used at both the ends of the transmission lines. When voltage is raised from lower level to higher level, the device used is called a step-up transformer. When voltage is lowered from a high level to a low level, the transformer used is called a step-down transformer.

The frequency of the alternating voltage on both sides of the transformer will not change. Whatever the frequency of the input voltage is, the same will be the frequency of the output voltage.

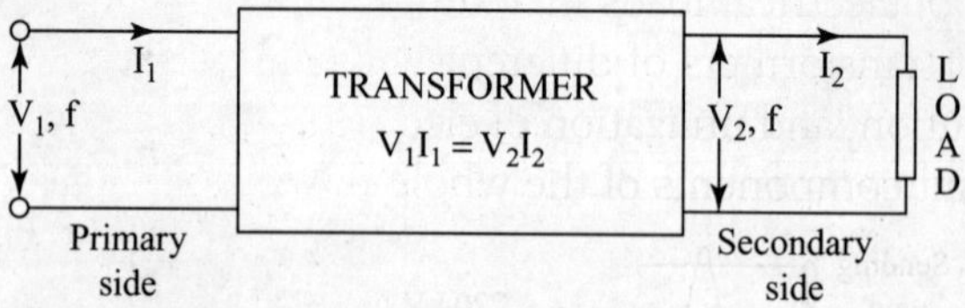

FIGURE 6.1

Block diagramatic representation of a transformer

With this introduction, we may define a transformer as a device which raises or lowers the voltage level of any electrical power input without changing the frequency. The block diagrammatic representation of a transformer has been shown in Fig. 6.1.

6.2 ► Applications of Transformers

Volt–ampere rating of the transformer is the same whether calculated on the low-voltage side or at the high-voltage side. It must be noted that a transformer does not generate any electricity. It only transforms and transfers electrical power from one circuit to the other at different voltage levels. Depending upon the requirement, transformers are made for various voltage and current ratings. Transformers used to raise the voltage level at the sending end of the transmission lines, and to lower the voltage level at the receiving end are called *power transformers*. These are very big transformers rated at, say, 11 kV/220 kV and several MVAs (mega volt ampers) as their power ratings.

Smaller transformers are used in lowering the voltage level for the purpose of distribution of electricity to consumers. The transformer which feeds electricity to your house will have specifications like 11 kV/400 V, 50 Hz, 500 kVA. Here, the voltage is being reduced to 400 V or 230 V at the user end for safe supply to residences. These are called *distribution transformers*. Supply of power at high voltage to residences may lead to chances of fatal accidents and other problems. Equipment to be used also have to be manufactured for higher voltage ratings.

Insulation of wires used in house wiring will have to be sustaining for higher voltages. Therefore, electricity is supplied at residences at 230 V for single-phase appliances and at 400 V for three-phase equipment. Very small transformers are used in many electrical and electronic equipment, and gadgets to lower the voltage level from 230 V on one side to, say, 6 V or 3 V, on the other side. For example, if you are to construct a battery eliminator for your transistor radio, or your tape recorder, you need to get 6 V dc supply from the available 230 V ac supply. A transformer is required to step down the voltage and then a diode rectifier and filter are required to get the steady 6 V dc output. Fig. 6.2 shows the use of transformers of different voltage ratings.

Electricity is generated in the power house when a turbine (T) rotates a generator (G). The generation voltage is 11 kV which is stepped up to 220 kV by a step-up transformer. The output of the transformer is connected to the high-voltage transmission line. At the receiving end of the transmission line, a step-down transformer is used to bring the voltage level again to a lower level. The power received is further stepped-down to lower voltage by use of distribution transformers and is connected to the load of the consumer, i.e., to industry, commercial buildings, and residential houses. Fig. 6.2 (b) shows the use of distribution transformers for supply to various kinds of electrical loads including low-voltage dc supply after rectification. Thus, we have seen that transformers of different voltage levels and power capacity are used for transmission, distribution, and utilization of electricity. Transformers are, therefore, seen as one of the most important components of the whole power system network.

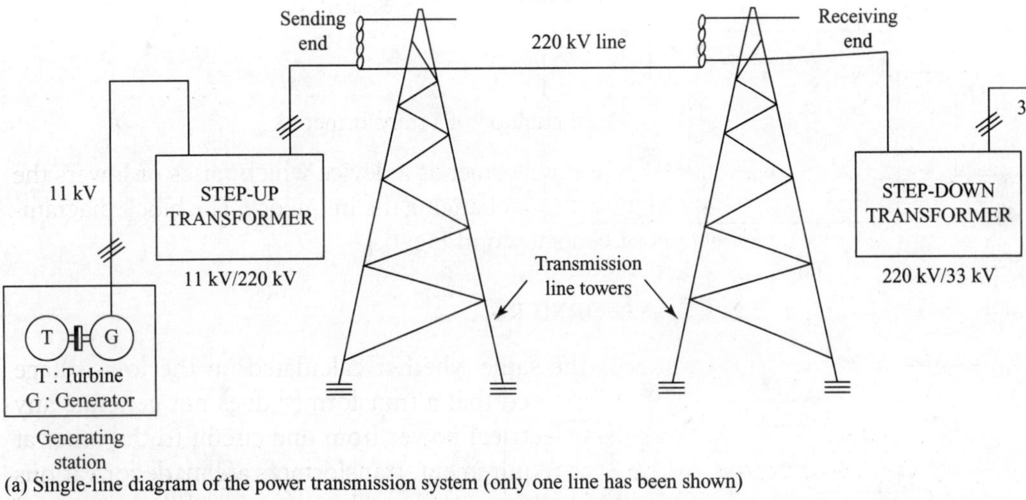

(a) Single-line diagram of the power transmission system (only one line has been shown)

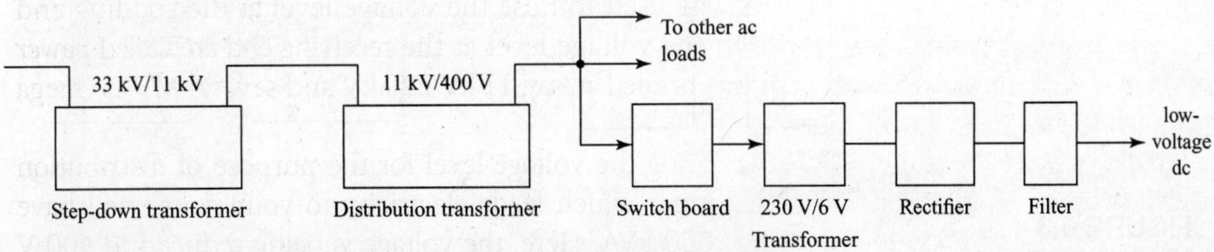

(b) Block-diagramatic representation of the distribution of electricity for end use

FIGURE 6.2

Illustrates the use of transformers in stepping-up and stepping-down operations

6.3 ► Basic Principle and Constructional Details

Transformers work on the principle of electromagnetic induction. To understand the principle of working of a transformer, one small experiment can be performed as explained below. We will need one battery, one single-way and one two-way switch, two coils, and a PMMC-type dc low-range center zero voltmeter. The connections are as shown in Fig. 6.3. In Fig. 6.3 (a), when the switch S is turned on and off quickly, the voltmeter neddle will get deflected. In Fig. 6.3 (b), a two-way switch is used. When the switch is quickly turned on and off, the current through coil 1 will flow in opposite directions every time the switch is operated. There will be voltage induced in the second coil as would be indicated by the deflection of the needle of the voltmeter in opposite directions.

Operation of the two-way switch in opposite directions changes the polarity of supply voltage to coil 1. The frequency of change of polarity will depend upon how quickly the switch is repeatedly operated. This is equivalent to connecting an ac supply of certain frequency to coil 1, which has been shown in Fig. 6.3 (c).

Voltage is induced in the second coil due to changes of current flowing in the first coil. When current flows through the first coil, a flux is produced around the coil. If current is changing, the flux produced will also change. If the second coil is placed near the first coil,

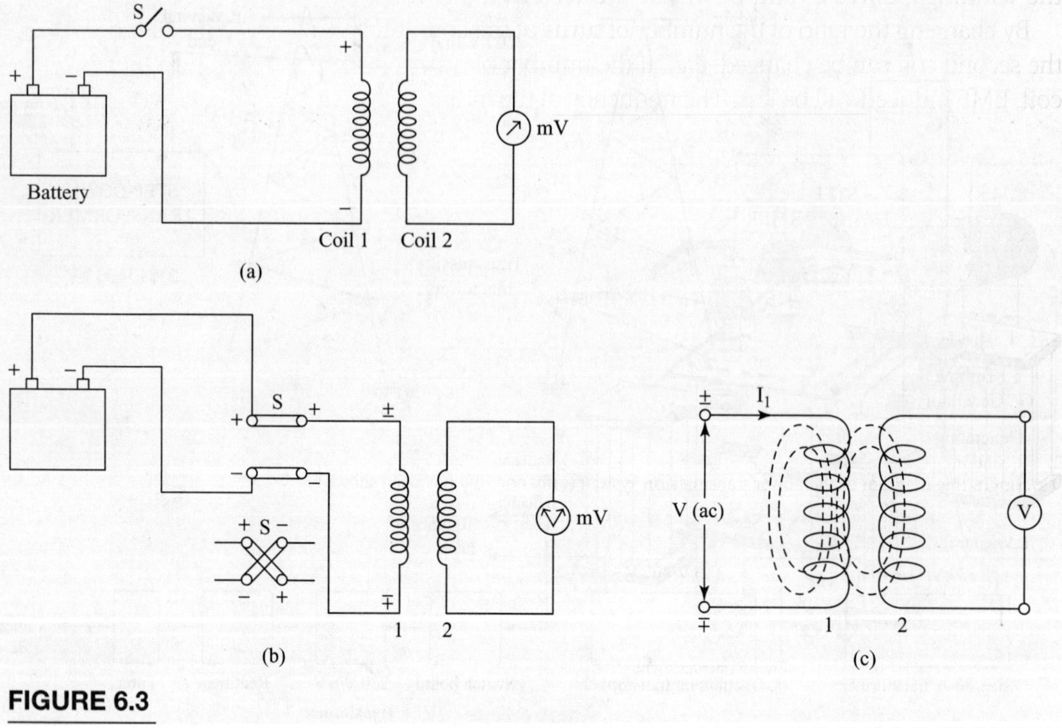

FIGURE 6.3

Principle of electromagnetic induction: (a) changing current in coil 1 produces EMF in coil 2; (b) changing current flowing in reverse direction in coil 1 produces alternating voltage in the second coil; (c) alternating voltage applied to coil 1 induces alternating EMF in the second coil

there will be a changing linkage of the flux by the coils. This will induce EMF in both the coils. The magnitude of EMF induced will depend upon the rate of change of the flux linkage and the number of turns of the coil.

6.3.1 Basic Principle

The basic principle of the transformer is that EMF is induced in a coil due to the rate of change of flux linkage by it as has been shown in Fig. 6.3 (c).

In a transformer, two coils, which are also called windings wound on an iron core, are used. Coil 1 is called the primary winding and coil 2 is called the secondary winding. [See Fig. 6.4 (b)]. These windings are made on an iron core which is made of a magnetic material. The magnetic material permits easy establishment of the flux through the core, and hence through the windings. Since it is necessary to produce more flux by using small current, the reluctance of the flux path must be low. Iron has high permeability, i.e., low reluctance. Use of iron as the core material for the windings improves the magnetic coupling between the windings, which is essential for the transfer of power from one circuit to the other efficiently.

Primary winding has N_1 number of turns and the secondary, winding has N_2 number of turns. When an alternating voltage, V_1 is applied across the primary winding, a current I_0 will flow through it. This current will produce alternating flux which will link (pass through) both the windings. EMFs E_1 and E_2 will be induced in the two windings, respectively.

By changing the ratio of the number of turns of the coils, the magnitude of the induced EMF in the second coil can be changed, e.g., if the number of turns of the second coil is less than the first coil, EMF induced will be less. The frequency of the induced EMF in the two coils will be the same

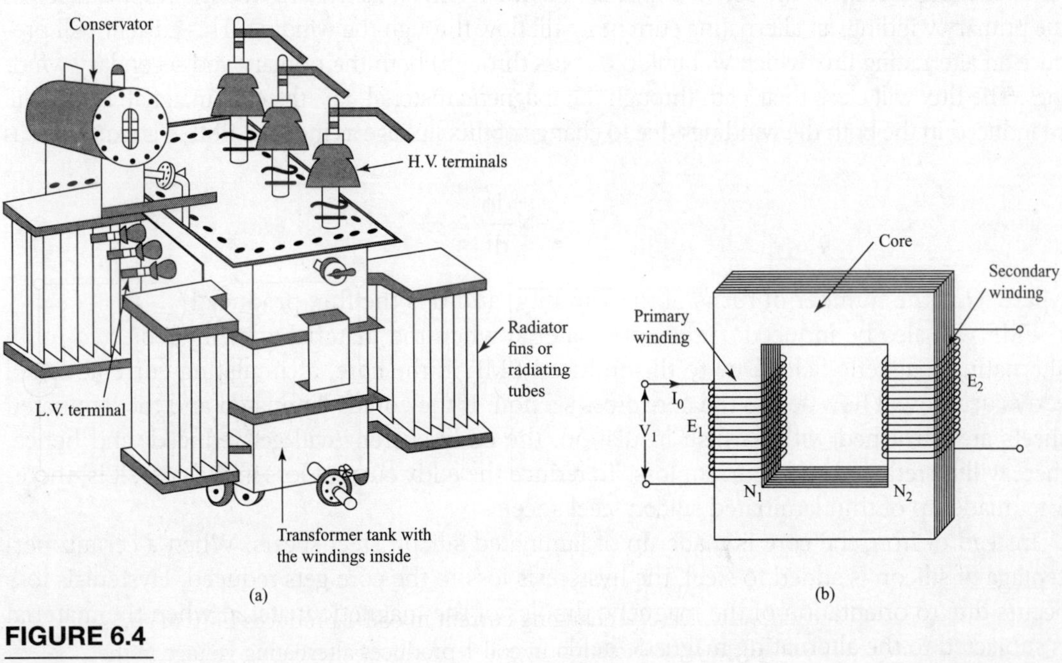

FIGURE 6.4

Constructional details of a transformer. (a) Outside view of a transformer; (b) view of core and windings

as that of the frequency of power supply to the first coil. Now, it should be possible to connect an electrical load across the second coil, and power will be delivered to the load. Thus, when electrical supply is connected to the primary circuit or the winding, power gets transferred to the second circuit via the magnetic circuit. This device, called transformer, is based on the same principle of magnetic coupling of two coils. The constructional details of a transformer is described as follows.

6.3.2 Constructional Details

Fig. 6.4 (a) shows the outside view of a transformer. It may be noted that the transformer is placed inside an iron tank filled with oil. The tank has some radiating tubes and fins so that oil inside the tank gets circulated and the heat from the transformer is radiated to the atmosphere. The transformer consists of a core made up of a magnetic material around which two coils are placed. One coil is connected to supply voltage as shown in Fig. 6.4 (b). This coil is called the primary winding. The other coil is called the secondary winding. Supply is taken from the secondary winding by connecting any electrical load like fan, tube light, electrical motors, etc.

The transformer core is made up of thin steel sheets called laminations. The laminated silicon steel sheets are cut into proper size from a big sheet and are placed one above the other to form the core of required width and cross section. The laminated sheets are tightly fastened to form the core. If the laminations are not tightly fastened, they will vibrate in the magnetic field and give rise to humming noise. This magnetic vibration of laminations is known as magnetostriction which is not desirable. The core is made up of a magnetic material using thin laminated sheets instead of a solid one. This is done to reduce power loss in the core due to circulating current flowing in the core and producing undesirable heating of the core as well as of the windings which are wound on the core. The reason for circulating current in the core is explained below. When an alternating voltage is applied across the primary winding, an alternating current I_0 will flow through the winding. This current will produce an alternating flux which will link (i.e., pass through) both the primary and secondary windings. The flux will close their path through the magnetic material, i.e., the core material. EMFs will be induced in the both the windings due to change of flux linkage as induced EMF, e is expressed as

$$e = -N\frac{d\phi}{dt} \tag{6.1}$$

where, N is the number of turns of the winding, and ϕ is the flux produced.

EMF will also be induced in the core material when the material is being subjected to an alternating magnetic field. Due to this induced EMF in the core, a circulating current, called eddy current will flow across the core cross section. If the core is laminated and the laminated sheets are varnished with varnish insulation, the eddy current will get reduced, and hence, there will be reduced eddy current loss. To reduce the eddy current loss in the core, it is, therefore, made up of thin laminated silicon steel sheets.

Instead of iron, the core is made up of laminated silicon steel sheets. When a certain percentage of silicon is added to steel, the hysteresis loss in the core gets reduced. Hysteresis loss occurs due to orientation of the magnetic dipoles of the magnetic material when the material is subjected to the alternating magnetic field.

The windings of the transformer are made up of insulated copper wires. The cross-sectional area of the winding wires will depend upon the requirement of current-carrying capacity, and the number of turns are calculated according to the voltage ratio of primary and secondary windings. The core with and the winding assembly are placed inside a tank filled with transformer oil for the purpose of providing insulation to the windings and also for cooling purpose. Transformer oil used is mineral oil having high-dielectric strength. The tank is provided with radiating tubes so that heated oil gets circulated through the tubes and heat produced in the transformer is radiated to the atmosphere through the oil circulating from the tank through the radiating tubes.

Heat is produced in the transformer due to I^2R loss in the windings and hysteresis and eddy current loss in the core. The I^2R loss in the windings will depend upon the magnitude of current flow through the windings when the transformer is supplying some electrical load. The core loss which is the sum of hysteresis loss and eddy current loss remains constant at any load. Even when the transformer output circuit, i.e., the secondary winding, is not connected to any load, there will be core loss once an alternating voltage is applied to its primary winding. As long as the primary voltage is kept constant, the core loss will remain constant. That is why the core loss is called constant loss. The I^2R loss which is also called copper loss is a variable loss, as it varies with the magnitude of load current.

These losses will overheat the transformer unless the heat generated is radiated out to the atmosphere. If a transformer gets too overheated, its insulation strength will reduce and ultimately there may be short circuit inside the transformer damaging it completely.

Transformers are manufactured as single-phase transformers and as three-phase transformers. In three-phase transformers three separate windings are made for both primary and secondary sides. The windings are connected either in star or in delta. Terminal connections are brought out through low-voltage (L.V.) terminals and high-voltage (H.V.) terminals. A conservator tank fitted with a breather is placed above the tank. The conservator is connected to the transformer tank with a pipe and carries transformer oil. The empty space above the level of oil in the conservator is provided to allow the expansion of oil in the tank due to heating and for the removal of gas formed.

6.4 ► Core-Type and Shell-Type Transformers

There are two types of core constructions, namely core type and shell type. In core-type construction, as shown in Fig. 6.5(a), the primary and secondary windings are placed around the limbs of the transformer core. The windings are made in cylindrical form and are placed around the core limbs. First, the low-voltage winding is placed around the limbs. Over the low-voltage windings are placed the high-voltage windings. The high-voltage winding is placed somewhat away from the core so as to reduce the insulation problem. The windings are insulated from the core through insulating cylindrical disc made of insulating material. The windings are made up of insulated copper wires in two sections or parts and are connected after they are placed in positions to form primary and secondary windings, respectively.

In a shell-type construction, the windings are placed in the central limb as shown in Fig. 6.5(b). The windings are wound in the form of a number of circular discs, and are placed one above the other. The extreme two discs on the central limb are low-voltage winding sections. These, l.υ sections and h.υ. sections are then connected to form low-voltage and

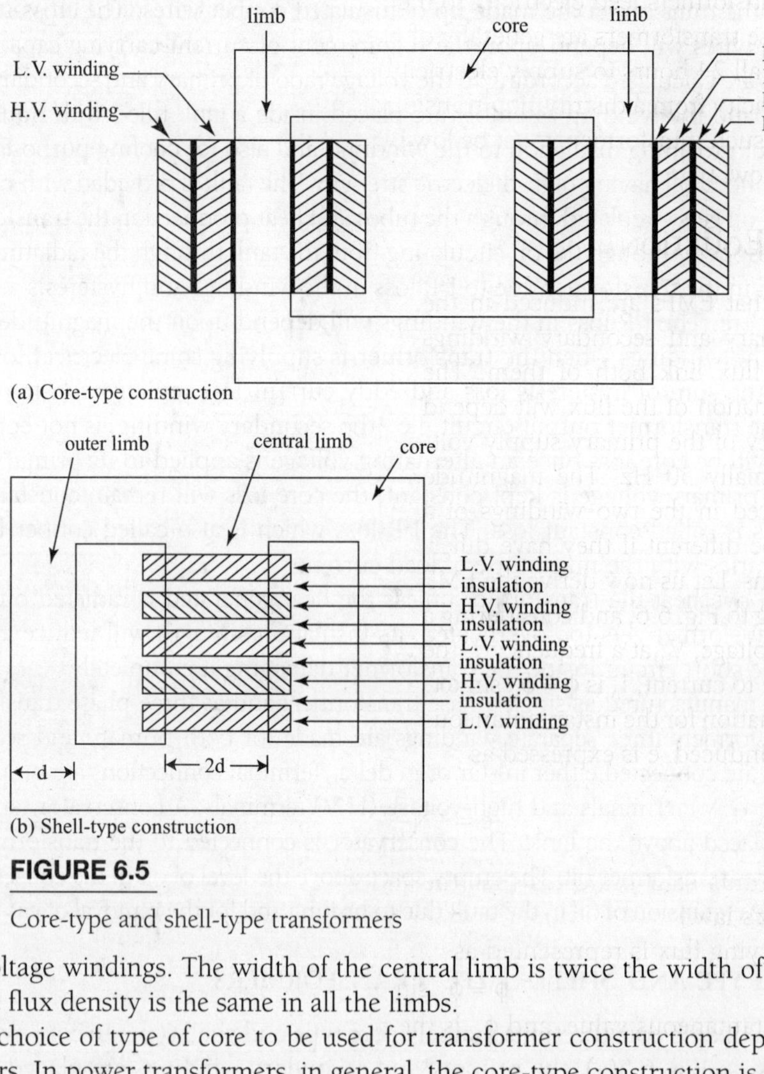

FIGURE 6.5

Core-type and shell-type transformers

high-voltage windings. The width of the central limb is twice the width of the outer limbs so that the flux density is the same in all the limbs.

The choice of type of core to be used for transformer construction depends on a number of factors. In power transformers, in general, the core-type construction is preferred while for distribution transformers, the shell-type construction is preferred. The leakage flux and leakage reactance of a transformer depends upon the magnetic coupling between the primary and secondary windings. Voltage regulation and short-circuit impedance depend upon the leakage reactance of a transformer. In the shell-type construction the magnetic coupling is better than in the core-type construction.

6.4.1 Power Transformers and Distribution Transformers

Power transformers are connected at the two ends of the transmission lines to step up or to step down the voltage. A number of such transformers are connected in parallel depending upon the amount of power to be transmitted. They are rated for high voltages, e.g., 11 kV/220 kV, 100 MVA. The size of such transformers is very large. They are installed outdoors in a substation.

Distribution transformers feed electricity to the consumers. They are rated for voltages like 11 kV/400 V. These transformers are generally of pole-mounted type and always remain energized being ready all 24 hours to supply electricity to the consumers. Even if there is no consumption of electricity from a distribution transformer, it has to remain energized all the time. The core losses of such transformers must be low by design. Otherwise, their all-day operating efficiency will be low.

6.5 ► EMF EQUATION

We have known that EMFs are induced in the transformers primary and secondary windings when alternating flux link both of them. The frequency of alternation of the flux will depend upon the frequency of the primary supply voltage which is normally 50 Hz. The magnitude of the EMF induced in the two windings of a transformer will be different if they have different number of turns. Let us now derive the EMF equation. Referring to Fig. 6.6, and considering a sinusoidal input voltage, V_1 at a frequency, f, the flux produced due to current, I_0 is $\phi = \phi_m \sin \omega t$.

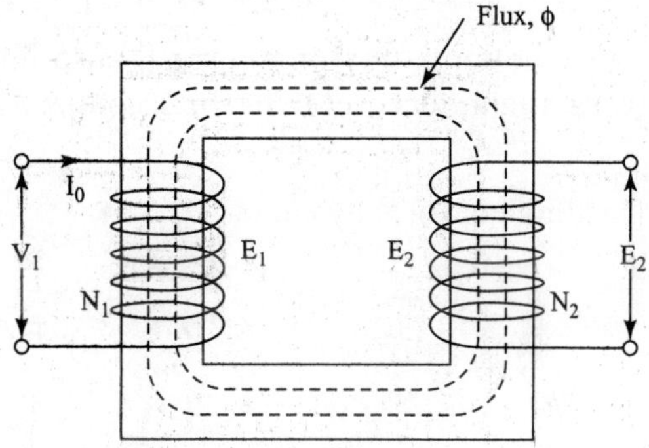

FIGURE 6.6

Two windings of a transformer wound on a common core

The general equation for the instantantaneous value of the EMF induced, e is expressed as

$$e = -N\frac{d\phi}{dt}$$

[Note that the minus sign indicates that the induced EMF opposes the supply voltage according to Lenz's law.]

A sinusoidally varying flux is represented as

$$\phi = \phi_m \sin \omega t$$

where ϕ is the instantaneous value, and ϕ_m is the maximum value

Considering $\phi = \phi_m \sin \omega t$,

$$e = -N\frac{d(\phi_m \sin \omega t)}{dt}$$

we get

$$e = -N \phi_m \omega \cos \omega t$$

or,
$$e = -N \phi_m \omega[-\sin (\omega t - \pi/2)]$$

or,
$$e = N \phi_m \omega \sin (\omega t - 90°)$$

This equation of e, is of the form

$$e = E_m \sin (\omega t - 90°)$$

where
$$E_m = N \phi_m \omega$$

or,
$$E_m = N \phi_m 2\pi f$$

We know for a sinusoidal voltage, the RMS value is $\frac{1}{\sqrt{2}}$ times its maximum value. If, E is the RMS value of the induced EMF, then

$$E = \frac{E_m}{\sqrt{2}} = \frac{N\phi_m 2\pi f}{\sqrt{2}} = \frac{N\phi_m 2 \times 3.14 f}{1.414}$$

or, $$E = 4.44\, \phi_m\, f\, N\ \text{V} \tag{6.2}$$

Primary winding has N_1 turns. The induced EMF in the primary winding E_1 is

$$E_1 = 4.44\, \phi_m\, f\, N_1\ \text{V} \tag{6.3}$$

Since the same flux producing E_1 also links the secondary winding having N_2 number of turns, the induced EMF in the secondary winding, E_2 is

$$E_2 = 4.44\, \phi_m\, f\, N_2\ \text{V} \tag{6.4}$$

Dividing eq. (6.3) by eq. (6.4),

$$\frac{E_1}{E_2} = \frac{N_1}{N_2}$$

or, $$\frac{E_2}{E_1} = \frac{N_2}{N_1} = K$$

K is called the ratio of voltage transformation. For a step-up transformer, $N_2 > N_1$, and hence, the value of K is more than 1. For a step-down transformer, N_2 is less than N_1, and hence, $K < 1$.

6.6 ▶ Transformer on No-Load

Transformer on no-load means that the secondary winding is open and no electrical load (like a motor, a heater, a fan, an air-conditioner, etc.) is connected across its terminals for supply of electrical power. Since the transformer on no-load is not doing any useful work except that it remains energized and is ready to supply electricity when required, its output on no-load is considered to be zero, On the input side, the transformer will draw some small amount of current, I_0. Simplified representation of a transformer on no-load has been shown in Fig. 6.7 (a). Since there is no load connected across the secondary winding, the circuit is open, and hence, no current will flow through the secondary winding as has been shown. The primary supply voltage is V_1 and the current flowing through this winding is I_0. What will be the phasor relationship between V_1 and I_0? If the winding is a purely inductive one, I_0 will lag the voltage V_1 by 90°. However, there will be hysteresis loss and eddy current loss in the core. Thus, I_0 should have a component I_C in phase with V_1. The core loss is equal to $V_1 I_C$ Watts. Therefore, I_0 will lag V_1 by an angle somewhat less than 90°. If V_1 is shown vertical, I_0 will lag V_1 by an angle ϕ_0 of say 85° as has been shown in Fig. 6.7 (b).

The induced EMFs E_1 and E_2 which are due to a time-varying flux ϕ will lag ϕ by 90°. This can be observed from the EMF equation where it was shown that when $\phi = \phi_m \sin \omega t$, the induced EMF $e = E_m \sin (\omega t - \pi/2)$. Again, according to Lenz's law, the induced EMF E_1 must oppose the cause from which it is due, i.e., V_1. The magnitude of E_1 will be somewhat

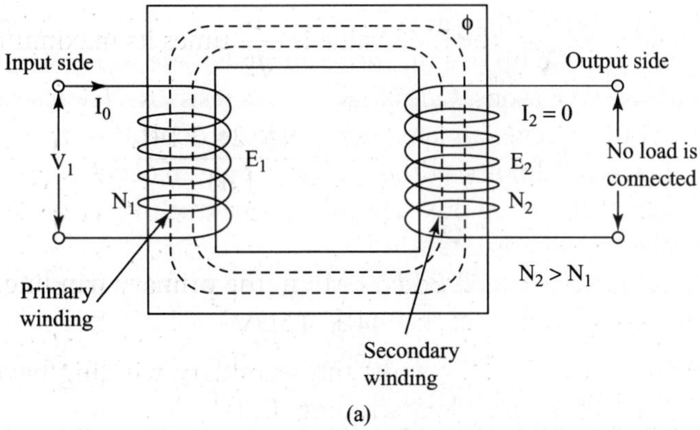

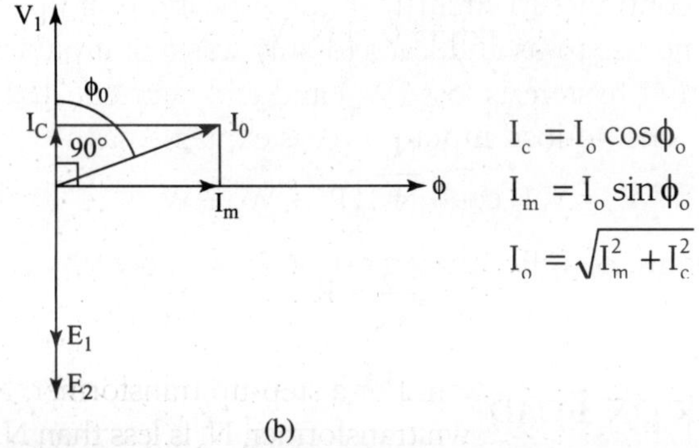

FIGURE 6.7

Transformer on no-load: (a) core and the windings; (b) no-load phasor diagram

less than V_1. This can be observed from Fig. 6.7 (a) where it is seen that current I_0 flows from a higher potential V_1 to a comparatively lower potential E_1. Thus, E_1 opposes V_1, E_1 is created by ϕ and lags ϕ by 90°. The phasor relationship of V_1, I_0, ϕ, E_1, E_2 has been shown in Fig. 6.7 (b). It has also been shown that I_0 can be resolved into two components I_m and I_c, where I_m is in phase with ϕ and is responsible for producing ϕ. This I_m is also called the magnetizing current because this current magnetizes the core, i.e., produces the required flux in the core. I_m is equal to $I_0 \sin \phi_0$. I_c is equal to $I_0 \cos \phi_0$. Therefore, $I_0 = \sqrt{I_m^2 + I_c^2}$. The no-load power input W_0 is equal to $V_1 I_0 \cos \phi_0$ which equals $V_1 I_c$. The induced EMF in the secondary winding, i.e., E_2 has been shown lagging flux ϕ by 90°. It has been assumed that $N_2 > N_1$, and hence, $E_2 > E_1$.

Here, $\cos \phi_0$ is the no-load power factor which is very low (may be 0.1 or so).

The no-load power input is wasted as a loss as there is no output.

$$\text{Input} = \text{output} + \text{losses}$$

If output is zero, input equals losses. Let us see what are the no-load losses. Since current I_0 is flowing through the primary winding which has a resistance of, say R_1, there will be some amount of copper loss in the winding as $I_0^2 R_1$ Watts. However, since I_0 is small, $I_0^2 R_1$ will also be small. Since the core is made up of a magnetic material there will be loss in the core para.

The core loss is due to two reasons. One is called hysteresis loss. *Hysteresis loss* is caused due to the magnetization of the magnetic material in alternate directions in every half cycle of the supply voltage. The magnetic dipoles of the magnetic core material align themselves in alternate directions producing alternating flux. The work done due to this is equivalent to the input energy spent and is called hysteresis loss.

The other loss component is called Eddy current loss which is due to eddy current. Large number of small eddy currents flow in the magnetic core material due to the EMF induced in the core, which is subjected to alternating magnetic field similar to the two windings. EMFs get induced in the core material for the same reason as for the coils. This EMF induced in the core creates current which continues to circulate in the core and heat up the core unnecessarily. This eddy current is of no use to us and leads to only waste of input energy. This is called eddy current loss. The sum of hysteresis loss (W_h) and eddy current loss (W_e) is called core loss (W_c) or iron loss (W_i). The no-load input power is expressed as

$$W_0 = V_1 I_0 \cos\phi_0 = I_0^2 R_1 + W_h + W_e$$

If we neglect the small amount of $I_0^2 R_1$, then, no-load input power

$$W_0 = W_h + W_e = W_c, \text{ i.e., equal to core loss}$$

6.7 ► Transformer on Load

When some electrical load is connected across the secondary terminals, power will be supplied to the load from the primary supply via the magnetic circuit. A current of I_2 will flow in the secondary circuit as shown in Fig. 6.8(a). The voltage available across the load, V_2 will be somewhat less than E_2.

When the transformer is loaded, the secondary current I_2 will create flux in the core in the opposite direction to that of the original core flux ϕ which was produced on no-load. Thus, the resultant flux will get reduced momentarily. This will reduce the induced EMF E_1 and E_2. As E_1 is reduced, the difference between V_1 and E_1 will increase and due to this more current of amount I_1' will flow from the supply mains through the primary winding. This current will produce an opposing flux to that produced by I_2 such that $I_2 N_2 = I_1' N_1$. Then, the two fluxes will balance each other, and hence, the original flux ϕ will remain unchanged in the core. Irrespective of the magnitude of the load current, the net core flux remains practically constant at all load conditions. Using ampere-turns balance equation,

$$I_2 N_2 = I_1' N_1$$

or,

$$I_1' = \left(\frac{N_2}{N_1}\right) I_2 = K I_2$$

It may be noted that under load condition the primary current I_1 is equal to the sum of no-load current I_0 and I_1', which is K times I_2. I_1' is the additional current drawn by the primary winding due to the loading of the transformer.

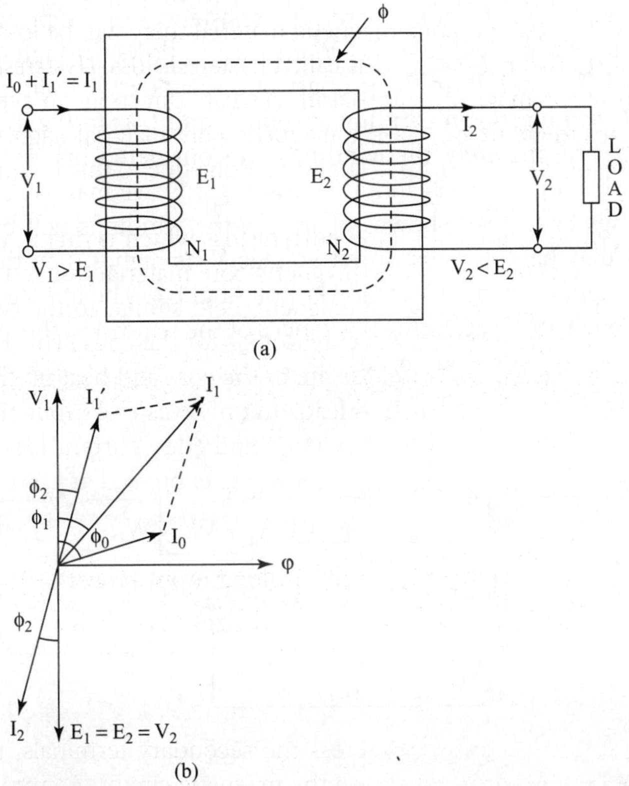

FIGURE 6.8

(a) Transformer on load; (b) phasor diagram of a transformer on resistive–inductive load without considering the winding resistances and leakage reactances

Thus, $$I_1 = I_0 + I_1' = I_0 + KI_2$$

The phasor diagram relating all the parameters under loading condition neglecting the voltage drop due to winding resistances and leakage reactances has been shown in Fig. 6.8 (b). The phasor diagram is for some resistive–inductive load when the load power factor angle is ϕ. That is why I_2 has been shown lagging the load voltage V_2 by an angle ϕ_2. I_1' is the additional primary current drawn from the supply source to counter balance the magnetizing effect of $I_2 (I_2 N_2 = I_1' N_1)$. To sum up, the transformer draws a current I_o on no load. When a load current I_2 flows under loading condition, an additional current I'_1 is drawn from the supply through the primary circuit.

6.8 ► Transformer Circuit Parameters and Equivalent Circuit

Earlier, while explaining transformer on load, we had neglected the winding resistances and leakage reactances of the transformer. Since the windings are made of copper wire of certain cross-sectional area, they will have some resistance. Thinner the wire, higher will be the resistance, for a particular length of the wire. The resistances of the primary and secondary windings are called R_1 and R_2, respectively. When current flows through the windings, there will be voltage drop $I_1 R_1$ and $I_2 R_2$ in the primary and secondary windings, respectively. The resistances R_1 and R_2 can be shown carrying current I_1 and I_2 as in Fig. 6.9. When the resistances have been shown separately, the windings are considered made up of some number of turns but without having any resistances. In other words, we may say that when the resistances of the windings have been shown separately, the windings will be assumed as having no resistance. It may be noted that input voltage V_1 is higher than the induced EMF E_1 (current flows from higher potential V_1 to lower potential E_1). The induced EMF E_2 is greater than load terminal voltage V_2. Neglecting reactances of the windings, the voltage equation are

$$V_1 - I_1 R_1 = E_1$$

and

$$E_2 - I_2 R_2 = V_2$$

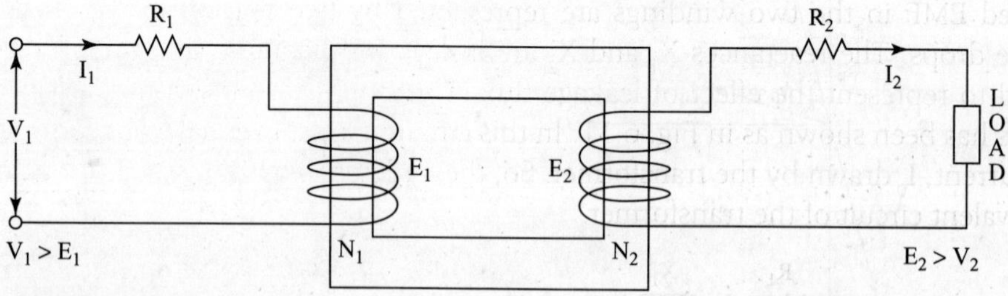

FIGURE 6.9

Transformer on load

The power loss in the primary and secondary windings are, respectively, $I_1^2 R_1$ and $I_2^2 R_2$. These are also called copper losses. Note that the above two voltage equations have been written considering only the resistances of the windings. Voltage drop due to reactances of the windings has been neglected.

In addition to the resistance of the windings, the windings will have leakage reactances due to the leakage flux in the core. The concept of leakage reactance due to leakage flux is explained below. Fig. 6.10 shows a transformer on load. As shown in the figure, the main

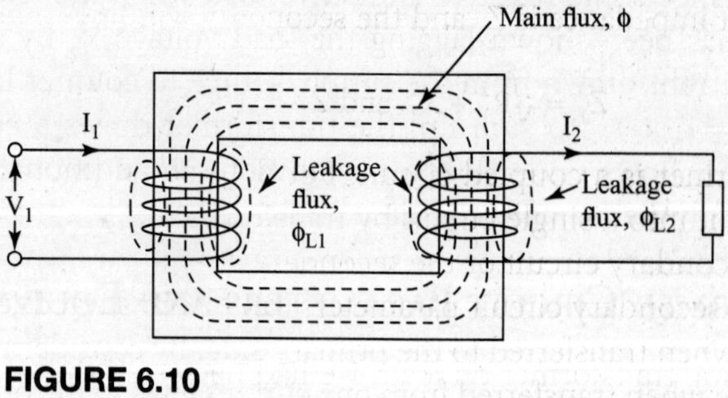

FIGURE 6.10

Leakage flux and leakage reactance of a transformer

flux ϕ is common to both the windings. This flux links both the primary and secondary windings. In addition, some more flux, called leakage flux, ϕ_L is created due to current I_1 and I_2 flowing through the two windings. These fluxes will also create some induced EMFs in the two windings. Since ϕ_{L1} is proportional to I_1, ϕ_{L1} will be in phase with I_1. Similarly, ϕ_{L2} is proportional to I_2 and will be in phase with I_2. ϕ_{L1} will induce an EMF E_{L1}, and ϕ_{L2} will induce an EMF E_{L2} in the windings, respectively.

E_{L1} will lag I_1 by 90° and E_{L2} will lag I_2 by 90°. These induced voltages are balanced by the reactance voltage drop in the two windings, respectively. The primary and secondary winding leakage reactances are called, respectively, X_1 and X_2 so that I_1X_1 is considered voltage drop in the primary winding due to leakage reactance X_1 and I_2X_2 is considered voltage drop in secondary winding due to leakage reactance X_2. It is to be noted that EMFs E_1 and E_2 are induced in primary and secondary windings due to main flux ϕ whereas E_{L1} and E_{L2} are induced in these windings due to their leakage fluxes ϕ_{L1} and ϕ_{L2}. The effect of leakage flux and the resulting induced EMF in the two windings are represented by two reactances X_1 and X_2 creating voltage drops. The reactances X_1 and X_2 are in fact two fictitious (imaginary) reactances considered to represent the effect of leakage flux. The complete transformer circuit with its parameters has been shown as in Fig. 6.11. In this circuit diagram we have not considered the no-load current, I_0 drawn by the transformer. So, the circuit shown in Fig. 6.11 is an approximate equivalent circuit of the transformer.

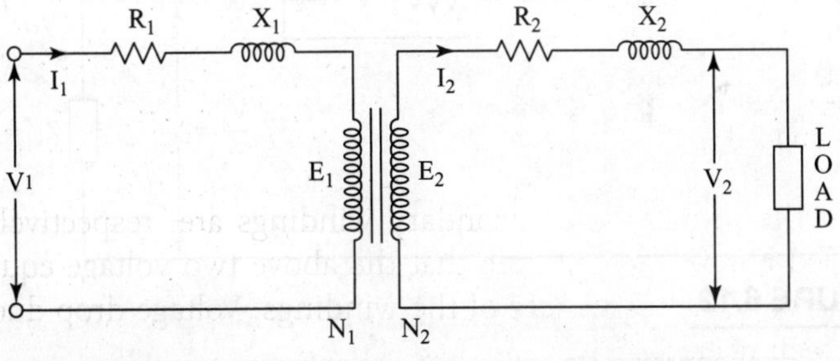

FIGURE 6.11

Transformer approximate equivalent circuit neglecting the small amount of no-load current

The primary circuit impedance is Z_1 and the secondary winding impedance is Z_2.

$$Z_1 = \sqrt{R_1^2 + X_1^2} \text{ and } Z_2 = \sqrt{R_2^2 + X_2^2}$$

Note that the transformer is a coupled circuit. For the sake of simplicity in calculation, we might like to convert it into a single circuit by transferring the circuit parameters of the primary circuit to the secondary circuit or the secondary circuit parameters to the primary side.

Let us see how the secondary circuit parameters can be transferred to the primary side. Let R_2' be the value of R_2 when transferred to the primary side. By considering the same amount of losses in the resistance when transferred from one current level to the other, we can equate the copper losses as where R_2' is the value of R_2 when transferred to primary side.

$$I_2^2 R_2 = I_1^2 R_2'$$

or,
$$R_2' = R_2\left(\frac{I_2}{I_1}\right)^2 = R_2\left(\frac{N_1}{N_2}\right)^2 = R_2 / \left(\frac{N_2}{N_1}\right)^2$$

∴
$$R_2' = \frac{R_2}{K^2}, \quad \text{where} \quad K = \frac{N_2}{N_1}$$

Reactance $X = \omega L$ and $L = \frac{\mu N^2 A}{l}$

Therefore, $X \propto N^2$

Let X_2, when transferred from the secondary circuit to the primary circuit, be X_2'

$$X_2' = X_2\left(\frac{N_1}{N_2}\right)^2 = \frac{X_2}{K^2} \qquad \because \frac{N_2}{N_1} = K.$$

The secondary voltage, V_2 and load impedance, Z_L have also been transferred to the primary side as V_2' and Z_L'. By transferring the circuit parameter on the primary side, the approximate equivalent circuit of the transformer can be represented as shown in Fig. 6.12.

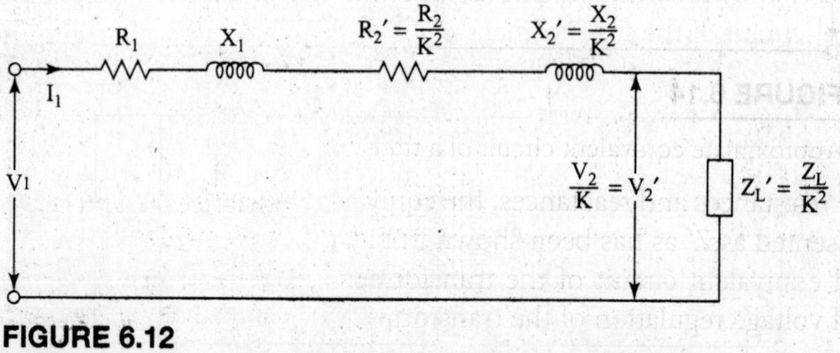

FIGURE 6.12

Transformer approximate equivalent circuit having transferred the secondary circuit parameters to the primary side

This circuit can further be simplified by adding the resistances and the reactances for the sake of calculations.

Now we will consider the no-load current of the transformer along with the load currents I_2 and I_1 to draw the complete equivalent circuit.

It may be noted that I_1 is the sum of I_1' and I_0. I_0 has two components, I_m and I_c. I_m lags V_1 by 90° whereas I_c is in phase with V_1. I_m can be shown as a current flowing through an inductive reactance called the magnetizing reactance X_m whereas I_c can be shown as a current flowing through a resistance R_c as shown. The sum of I_m and I_c is I_0. Sum of I_0 and I_1' is I_1. The complete equivalent circuit representing all the parameters has been shown in Fig. 6.13.

The circuit can be simplified by neglecting I_0 which is about three to five per cent of the rated current of the transformer. So by removing the parallel branch and adding the resistances and reactances we draw the approximate equivalent circuit as shown in Fig. 6.14. Thus, the circuit becomes the same as was drawn in Fig. 6.12 earlier.

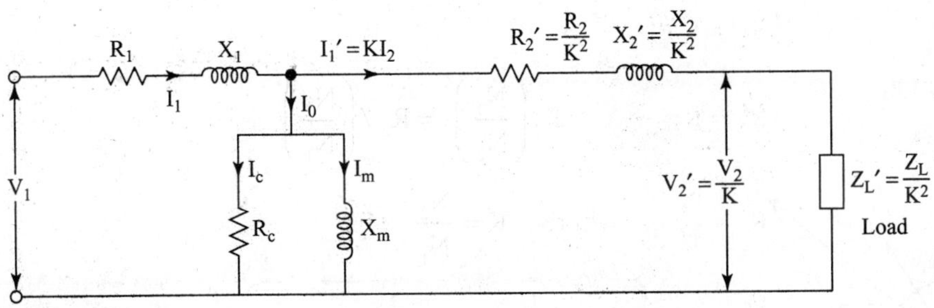

FIGURE 6.13

Exact equivalent circuit of a transformer considering the no load component current

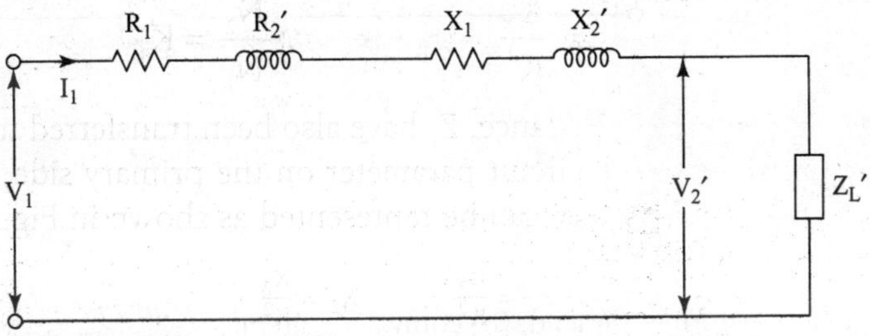

FIGURE 6.14

Approximate equivalent circuit of a transformer

Combining the resistances and reactances, the equivalent resistances R_e' and equivalent reactance X_e' are represented as Z_e' as has been shown in Fig. 6.15.

This simplified equivalent circuit of the transformer can be used to calculate the performance in terms of voltage regulation of the transformer under various loading conditions.

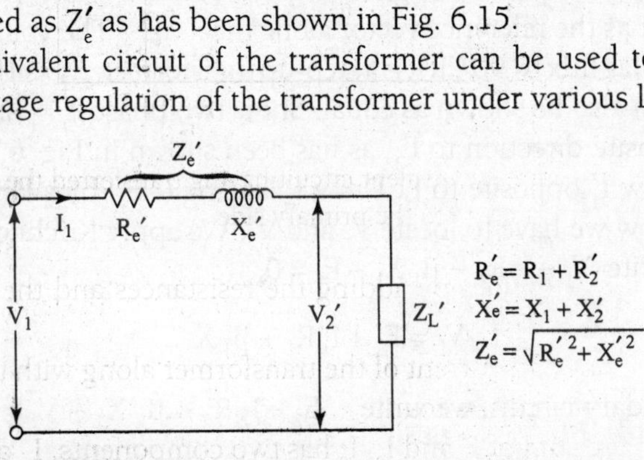

FIGURE 6.15

Approximate equivalent circuit of a transformer with secondary parameters referred to the primary side

6.9 ▶ Phasor Diagram of a Transformer

The exact equivalent circuit of the transformer and the complete phasor diagram have been shown in Fig. 6.16.

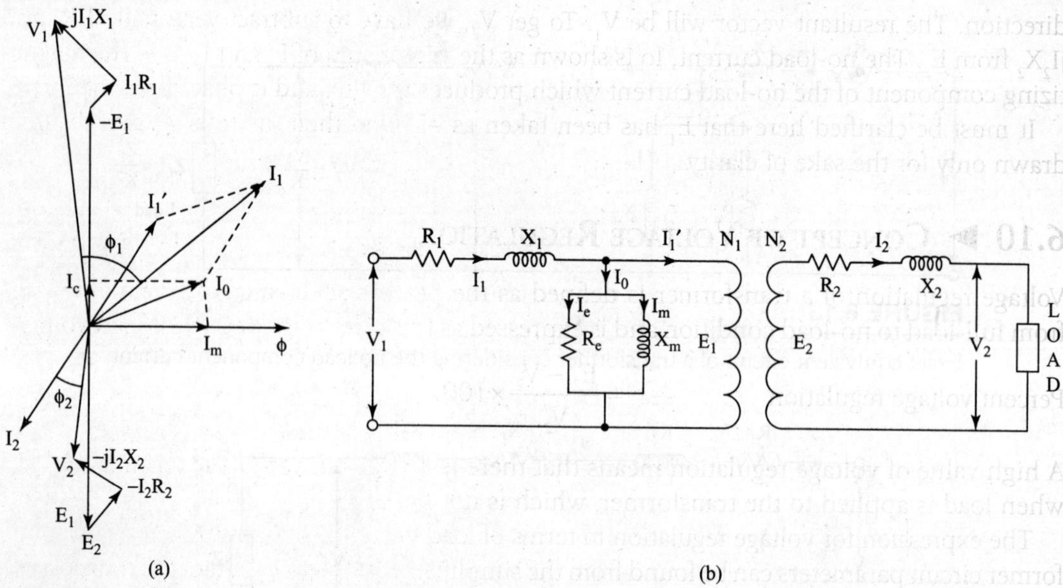

FIGURE 6.16

(a) Phasor diagram of a transformer on load; (b) equivalent circuit

The complete phasor diagram of a transformer at a lagging Pf load has been shown in Fig. 6.16 (a).

Procedure for drawing the phasor diagram is given below.

Draw the flux vector as the reference vector along the x-axis. The voltage induced in the two windings, E_1 ans E_2 will lag flux by 90°. If it is a step-up transformer, E_2 will be bigger in length than E_1. Otherwise, they can also be shown as equal. Show two phasors E_1 and E_2 lagging flux ϕ by 90 °. Draw $-E_1$ in opposite direction to E_1, as has been shown in Fig. 6.16. Draw I_2 lagging E_2 by a certain angle. Draw I_1' opposite to I_1. Draw I_o lagging $(-E_1)$ by a large angle. Add I_1' and I_o vectorially to get I_1. Now we have to locate V_1 and V_2. We apply Kirchhoff's law in the primary winding circuit and write $V_1 - I_1 R_1 - jI_1 X_1 - E_1 = 0$

or, $$V_1 = E_1 + I_1 R_1 + jI_1 X_1 \quad \text{(i)}$$

Similarly, for the secondary circuit, we write $E_2 - I_2 R_2 - jI_2 X_2 - V_2 = 0$

or, $$V_2 = E_2 - I_2 I_2 - jI_2 X_2. \quad \text{(ii)}$$

Using equation (i), develop the phasor diagram step by step.

We know, E_1 will be opposing V_1. So, it will be drawn on the opposite side of E_1. Further to get V_1 we have to add I_1R_1 and j I_1X_1 with E_1. For this purpose, we will show E_1 in the reversed direction as $-E_1$ and then do the phasor addition.

Add $I_1 R_1$ with E_1 (E_1 drawn as $-E_1$) vectorially. The direction of I_1R_1 is in the direction of I_1. Then add j I_1X_1. This vector will make 90 o with the direction of I_1 in the anticlockwise direction. The resultant vector will be V_1. To get V_2, we have to subtract vectorially I_2R_2 and jI_2X_2 from E_2. The no-load current, Io is shown as the vector sum of I_m and I_c. I_m is the magnetizing component of the no-load current which produces the flux and is phase with the flux ϕ.

It must be clarified here that E_1 has been taken as $-E_1$ and then the total phasor diagram drawn only for the sake of clarity.

6.10 ▶ Concept of Voltage Regulation

Voltage regulation of a transformer is defined as the percentage change in terminal voltage from full-load to no-load condition and is expressed as the percentage of the full-load voltage.

$$\text{Percent voltage regulation} = \frac{E_2 - V_2}{V_2} \times 100.$$

A high value of voltage regulation means that there is a large change in the terminal voltage when load is applied to the transformer, which is not desirable.

The expression for voltage regulation in terms of load current, load power factor, and transformer circuit parameters can be found from the simplified equivalent circuit of the transformer. For convenience, we will consider the approximate equivalent circuit of the transformer with primary circuit parameters referred to the secondary side as shown in Fig. 6.17. The phasor diagram relating the various quantities has also been shown. R_e'' and X_e'' are the equivalent resistance and reactance, respectively, of the transformer referred to the secondary side.

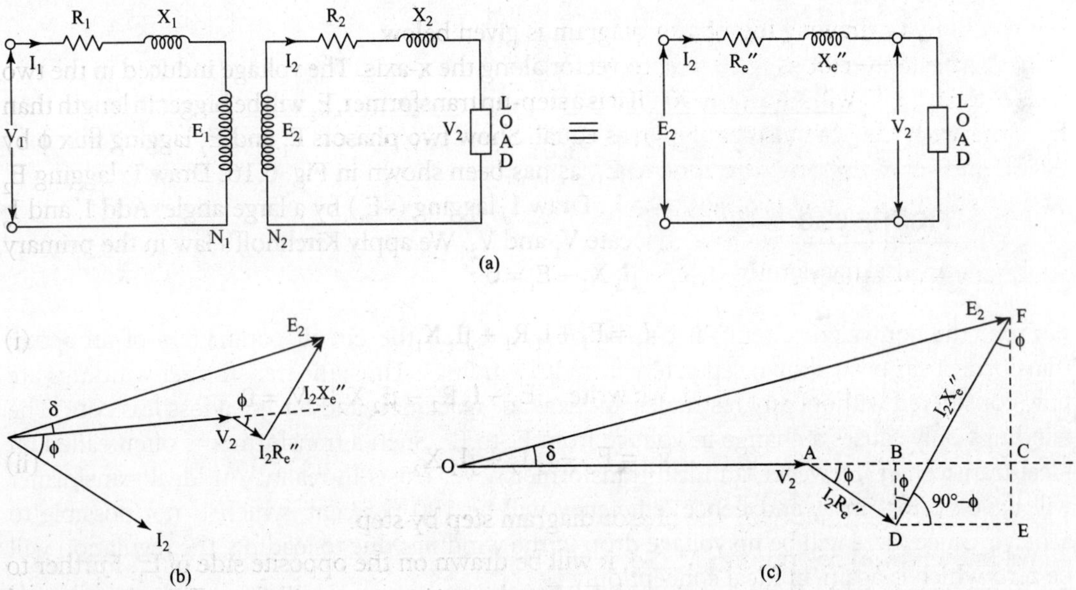

FIGURE 6.17

(a) Simplified equivalent circuit of a transformer; (b) phasor diagram; (c) use of the phasor diagram for calculation of the voltage drop

The equation relating E_2 and V_2 is

$$E_2 = V_2 + I_2R_e'' + jI_2X_e''$$

The phasor diagram shown in Fig. 6.17 (b) has been drawn using the above equation. I_2 has been shown lagging the voltage V_2 by the power factor angle ϕ. We add I_2R_e'' with V_2, I_2R_e'' being taken in the same direction as I_2. Then jI_2X_e'' is added which is at 90° anticlockwise with I_2. The phasor sum of V_2, I_2R_e'', jI_2X_e'' gives E_2. The approximate value of the voltage drop, i.e., the difference between E_2 and V_2 can be calculated. The phasor diagram shown in Fig. 6.17 (c) will be used to develop an expression for voltage regulation as has been done in section 6.16. Expression for voltage regulation of a transformer in terms of load current and load power factor has been developed.

6.11 ► Concept of an Ideal Transformer

The no-load current I_o is about three to five per cent of the rated current, i.e., I_1 or I_2. The voltage drop due to I_o on $(R_1 + jX_1)$ will therefore be small. If we neglect this small effect of I_o, then the parallel branch of R_c and X_m can be shifted towards the left as shown in Fig. 6.18.

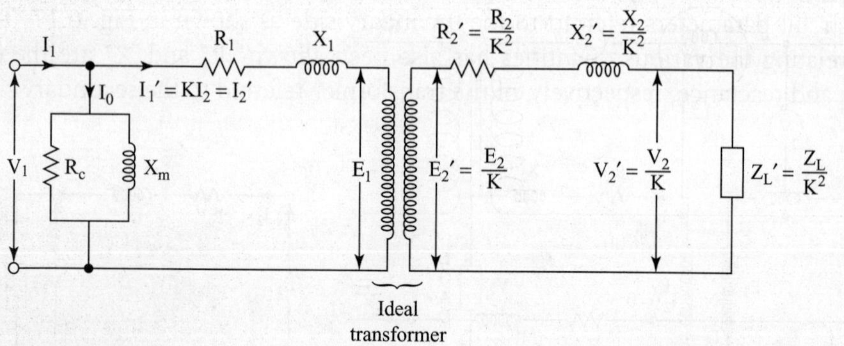

FIGURE 6.18

An ideal transformer

From the equivalent circuit, it can be observed that the circuit parameters of an actual transformer has been shown separately from its windings. Thus, the transformer windings are now considered without any resistance and leakage reactance, and having a loss less core. The windings only causes a change in voltage from E_1 to E_2. Such a transformer is often called an ideal transformer. However, an ideal transformer never exists in reality. An ideal transformer will have no loss in it, and hence, efficiency will be 100 per cent, which is not possible to achieve. Since there will be no voltage drop in the windings due to loading, the regulation will be zero which is again an ideal concept only.

An ideal transformer is one which has no core loss, no winding losses, no resistance of its windings, no winding leakage reactances, and no voltage drop in the windings. The efficiency will be 100 per cent and voltage regulation will be zero. Such ideal conditions are not possible to achieve.

6.12 ▶ Transformer Tests

Transformers can be tested for voltage regulation and efficiency by actually winding the transformers. The performance of a transformer in terms of its voltage regulation and efficiency under various loading conditions can also be calculated using the approximate equivalent circuit explained earlier. However, the circuit parameters like R_e and X_e, R_c and X_m have to be known. These parameters and the losses in the transformer can be determined by performing two tests, viz the open-circuit test or the no-load test, and the short-circuit test. These tests are explained as follows.

6.12.1 Open-Circuit Test or No-Load Test

In this test the transformer primary winding is supplied with its rated voltage at rated frequency keeping the secondary winding unconnected to the load, i.e., with no-load on the secondary. Normally, the supply is given to the low-voltage winding. The high-voltage winding is kept open. Three measuring instruments, viz a wattmeter, a voltmeter, and an ammeter are connected to the low voltage side as shown in Fig. 6.19 (a).

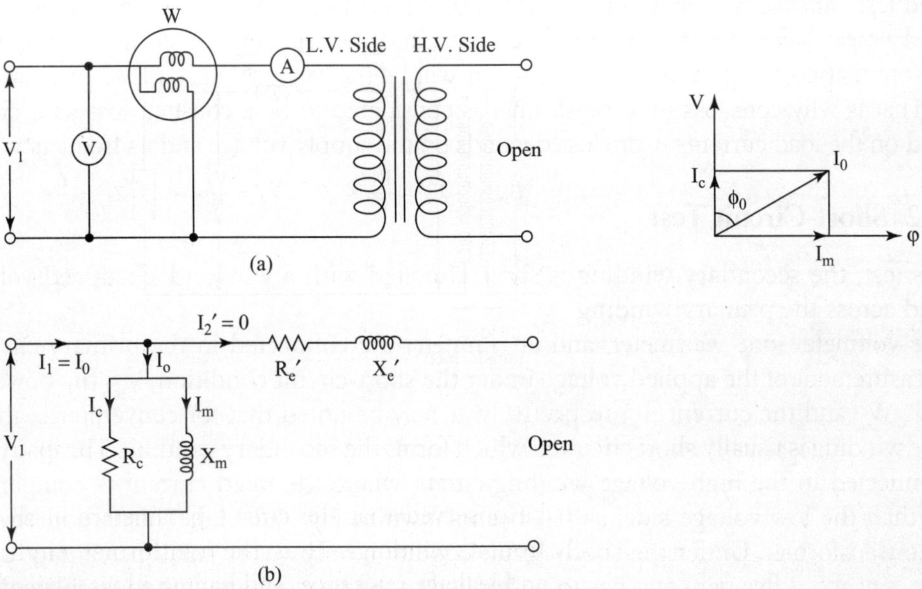

FIGURE 6.19

(a) Transformer on no load; (b) equivalent circuit on no load

The equivalent circuit of the transformer has also been shown under no-load condition in Fig. 6.19 (b). The wattmeter connected on the L.V. side will record the input power, W_o to the transformer. The supply voltage is measured by the voltmeter and the no-load line current is measured by the ammeter reading, I_o. The input power, W_o is

$$W_o = V_1 I_0 \cos\phi_0 \text{ W}$$

$$\cos\phi_0 = \frac{W_O}{V_1 I_0}$$

$$I_c = I_0 \cos\phi_0$$

$$I_m = I_0 \sin\phi_0$$

From the equivalent circuit on no-load [see Fig. 6.19 (b)]

$$R_c = \frac{V_1}{I_c}\ \Omega \text{ and } X_m = \frac{V_1}{I_m}\ \Omega$$

Since on no-load the output is zero, the input power is utilized in supplying the no-load losses. At no-load there will be no current in the secondary winding, and hence, no copper loss will take place in that winding. The primary winding current on no-load is small. There will be losses in the iron core which will have two components, viz hysteresis loss and eddy current loss. Thus, the wattmeter reading on no-load can be approximately equated to core loss only. From the no-load test data we, will be able to know the core loss of the transformer, no-load current, the no-load power factor, the magnetizing reactance, and the resistance R_c corresponding to core loss. Note that core loss calculated on no-load will be the same as on full-load or at any other load. That is why core loss of a transformer is considered to be a constant loss as it does not depend on the load currents. Core loss depends on the supply voltage and its frequency.

6.12.2 Short-Circuit Test

In this test, the secondary winding is short circuited with a wire and a reduced voltage is applied across the primary winding.

One voltmeter, one wattmeter, and an ammeter are connected in the primary circuit for the measurement of the applied voltage under the short-circuit condition, V_{sc}, the power consumed, W_{sc} and the current, I_{sc}, respectively. It may be noted that for convenience, the low-voltage winding is usually short circuited which forms the secondary winding. The instruments are connected in the high-voltage winding circuit where the rated current is comparatively lower than the low-voltage side, as has been shown in Fig. 6.20 (a). This is done by using an auto transformer. Under the short-circuit condition only a very small percentage of rated voltage, say about five per cent, has to be applied to the primary winding to circulate the full-load current through the windings. Current in the primary winding (H.V. winding) will also be lower than that of the low-voltage winding. Thus, by conducting the short-circuit test from the high-voltage winding side with the low-voltage winding short-circuited, we can have an accurate measurement.

It may also be noted that in the short-circuit test we create a condition when a full-load rated current will flow through both the primary and secondary windings. Therefore, the copper losses in the two windings will be equal to the amount of copper loss that would otherwise

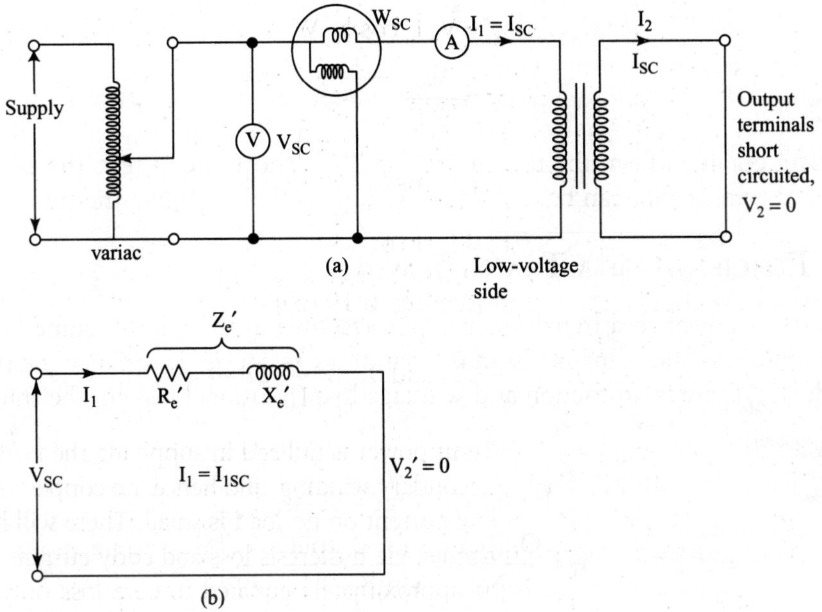

FIGURE 6.20

Short-circuit test on a transformer: (a) circuit diagram; (b) equivalent circuit under the short-circuit condition

occur when the transformer is actually supplying full-load at the rated voltage. Creating this type of a loading condition of a transformer is called phantom loading or fictitious loading. The actual circuit diagram for the short-circuit test and the equivalent circuit of the transformer under the short-circuit condition have been shown in Fig. 6.20 (a) and (b), respectively.

From the equivalent circuit shown, it may be observed that the core loss component of the equivalent circuit as was shown on the no-load test has been neglected here. This is because the voltage applied in the short-circuit test is not the full-load rated voltage but a small fraction of it. Since core loss is proportional to the applied voltage, for a small voltage applied under short circuit, the core loss component has been neglected. Thus, we can equate the wattmeter reading to copper losses in the two windings. The short-circuit test readings are taken by adjusting the voltage applied to the primary through a variac, i.e., an auto transformer so that the rated current flows (to be noted from the transformer name plate) through the windings. From the readings of the three instruments, the following calculations are made.

W_{sc} = Copper losses in the two windings having the rated current flowing through them.

$$W_{sc} = I_{1(sc)}^2 R_e'$$

or,

$$R_e' = \frac{W_{sc}}{I_{1(sc)}^2}$$

$$Z_e' = \frac{V_{sc}}{I_{1(sc)}} \Omega$$

$$Z_e'^2 = R_e'^2 + X_e'^2$$

or,
$$X_e' = \sqrt{Z_e'^2 - R_e'^2}.$$

From the data obtained from the no-load test and the short-circuit test, the efficiency and regulation of a transformer can be calculated without actually loading the transformer.

6.13 ► Efficiency of a Transformer

The whole input power to a transformer is not available at the output, some is lost in the iron core as core loss and some is lost in the windings as copper loss. Since the transformer is a static device, there is no friction and windage loss (rotational loss) in the transformer.

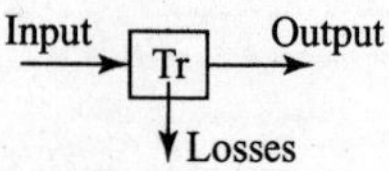

$$\text{Input} = \text{Output} + \text{Losses}$$

Efficiency,
$$\eta = \frac{\text{Output}}{\text{Input}} = \frac{\text{Output}}{\text{Output} + \text{Losses}}$$

$$= \frac{\text{Output power}}{\text{Output power} + W_i + W_{cu}}$$

or, Efficiency,
$$\eta = \frac{V_2 I_2 \cos\phi_2}{V_2 I_2 \cos\phi_2 + W_i + W_{cu}} \tag{6.5}$$

where W_i = iron loss or core loss
and W_{cu} = copper loss in primary and secondary windings
V_2 = output voltage
I_2 = load current
$\cos\phi_2$ = load power factor.

If the losses in a transformer at a particular load is known, its efficiency at that load can be calculated.

The load at which efficiency will be maximum, that is the condition for maximum efficiency, can be determined by differentiating the expression for efficiency with respect to the load current, I_2.

6.14 ► Condition for Maximum Efficiency

The expression for η after dividing the numerator and denominator of the right-hand side of the expression for η by I_2 can is written as

$$\eta = \frac{V_2 \cos\phi_2}{V_2 \cos\phi_2 + (W_i/I_2) + (I_2^2 R_e''/I_2)}$$

where R_e'' is the equivalent resistance of the transformer windings referred to the secondary side. Efficiency η will be maximum if the denominator is minimum. We can, therefore, minimize the denominator by differentiating it with respect to the load current, I_2 and equating to zero as

$$\frac{d}{dI_2}\left[V_2\cos\phi_2 + \frac{W_i}{I_2} + I_2R_e''\right] = 0$$

or,

$$-\frac{W_i}{I_2^2} + .R_e'' = 0$$

or,

$$I_2^2 R_e'' = W_i \tag{6.6}$$

When copper loss at a particular load equals the core loss, efficiency will be maximum at that load. The condition for maximum efficiency of a transformer is

Copper loss = Iron loss or core loss.

The value of the load current at maximum efficiency is determined as

$$I_2^2 = \frac{W_i}{R_e''}$$

or,

$$I_2 = \sqrt{\frac{W_i}{R_e''}} \tag{6.7}$$

If we want to know at what percentage of full-load, the efficiency will be maximum, (ie if we would like to know if efficiency is maximum at 75% of the full load or at 80% of the full load or at 100% of the full load) we can determine as follows.

Determination of load at which efficiency will be maximum

Let us assume that x is the fraction of the full load at which the efficiency is maximum. The core loss or the iron loss W_i will remain constant at all loads. The copper loss will vary as the square of the load current.

This means if at full load copper loss is $= W_{cu}$

at half of full load copper loss $= \frac{1}{4}W_{cu}$

at one-third of full load copper loss $= \frac{1}{9}W_{cu}$.

Thus, at x load, copper loss $= x^2 W_{cu}$

satisfying the condition for maximum efficiency

$$W_i = x^2 W_{cu}$$

or, Fraction of full-load at which efficiency will be maximum,

$$x = \sqrt{\frac{W_i}{W_{cu}}} \tag{6.8}$$

where W_{cu} is the full-load copper loss and x is the fraction of the full-load at which efficiency will be maximum.

If we want to know the kVA of the transformer at maximum efficiency, we would determine it as follows:

$$\text{kVA at } \eta_{max} = x \times \text{Full-load kVA}$$

$$\text{Therefore, kVA at maximum efficiency} = \sqrt{\frac{W_i}{W_{cu}}} \times \text{Full-load kVA.}$$

Efficiency of a transformer is often expressed in terms of energy output in 24 hours in a day to the energy input. Such a calculated efficiency is known as all-day efficiency which is explained as follows.

6.15 ► All-Day Efficiency

A transformer when connected to the load has to remain energized all the time ready to supply the load connected to it. Even when all the loads are switched off, i.e., no one is utilizing any electricity, the transformer has to remain on. Thus, irrespective of the load on the transformer, the core loss will occur for all the 24 hours of the day. However, the copper loss will depend on the magnitude of the load current, and will eventually vary from time to time. All-day efficiency is calculated by considering the energy output (power multiplied by time, i.e., energy) in 24 hours to the energy input in 24 hours as

$$\text{All-day Efficiency} = \frac{\text{Output energy in 24 hours.}}{\text{Output energy in 24 hours} + \text{Iron loss in 24 hours} + \text{copper loss in 24 hours}}$$

Both commercial efficiency which is the ratio of the output power to the input power and the all-day efficiency as stated above are calculated for distribution transformers. Distribution transformers are connected to the load all the time.

All-day efficiency of such transformers which are always connected to the load at the output side is somewhat less than their commercial efficiency, which is calculated on the basis of output power and the corresponding input power. All day efficiency provides feedback to the electricity distribution authority regarding the level of utilization of the transformer.

6.16 ► Calculation of Regulation of a Transformer

When a transformer is not supplying any load, the voltage across the output terminals is the same as that induced in the secondary winding, i.e., E_2. Now, when the transformer is connected to the load, the voltage available across the output terminals, V_2 becomes somewhat less than E_2.

The reduction in the output voltage from no-load to load is due to the voltage drop in the winding resistance and leakage reactance. The students are to refer to the phasor diagram as shown in Fig. 6.17 (c) for determination of voltage regulation which has been redrawn here.

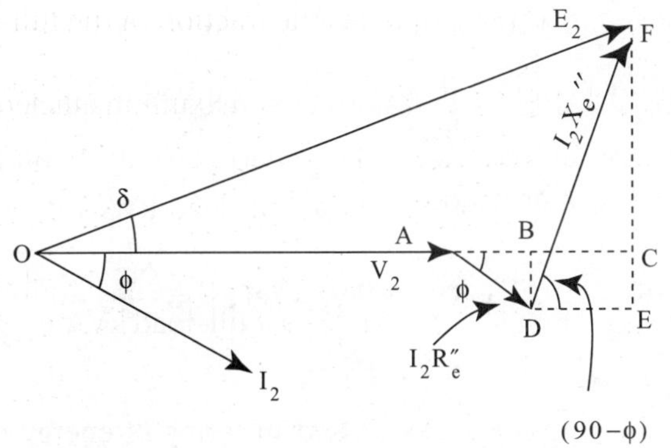

FIGURE 6.21

Phasor diagram of a transformer as in Fig. 6.17 (c)

In the phasor diagram shown in Fig. 6.21, we will consider E_2, i.e., length OF as equal to length OC as the angle δ is actually very small. This approximation is made to simplify the determination of an expression for voltage regulation.

Thus,

$$E_2 = OF = OC = OA + AB + BC = OA + AB + DE$$
$$= V_2 + I_2 R_e'' \cos\phi + DF \cos(90 - \phi)$$
$$= V_2 + I_2 R_e'' \cos\phi + I_2 X_e'' \sin\phi$$

$$\text{Percentage regulation} = \frac{E_2 - V_2}{V_2} \times 100$$

$$= \frac{I_2 R_e'' \cos\phi + I_2 X_e'' \sin\phi}{V_2} \times 100$$

If the power factor is leading, then we will have

$$\text{Percentage regulation} = \frac{I_2 R_e'' \cos\phi - I_2 X_e'' \sin\phi}{V_2} \times 100. \quad (6.9)$$

In this expression R_e'' and X_e'' are the resistance and reactance of the transformer windings refcrred to the secondary side. Regulation can also be calculated by transferring the secondary circuit parameters to the primary side.

6.17 ► Factors Affecting Losses in a Transformer

Since a transformer is a static device, there is no rotational part in it, and hence, there is no rotational or frictional losses. Due to current flow through the windings, there will be $I^2 R$ loss in both the primary and secondary windings. Thus,

$$\text{Copper loss} = I_1^2 R_1 + I_2^2 R_2$$

Copper loss is proportional to the square of the current. With the secondary circuit resistance referred to the primary side, the total effective resistance, $R'_e = R_1 + \frac{R_2}{K^2}$

Similarly, the primary circuit resistance when referred to the secondary side, the total effective resistance of the transformer windings

$$R''_e = R_2 + K^2 R_1 \quad \text{where } K = \frac{N_2}{N_1}$$

The copper loss $= I_2^2 R''_e = I_1^2 R'_e$

When the load current is changed, say from full load, I_2 to half load $\frac{I_2}{2}$, the copper loss becomes one-fourth of its value at full load.

The losses that take place in the iron core is called iron loss or core loss. Iron loss consists of two parts, viz Hysteresis loss and eddy-current loss. These are explained in detail as follows.

(a) Hysteresis loss

When alternating voltage is applied to the primary winding of the transformer, the core gets magnetized. The magnetization of the core takes place in alternate directions every half cycle of the frequency of the supply voltage. Magnetization in alternate directions basically means that the magnetic dipoles of the magnetic material changes their orientation in opposite directions every half cycle. This gives rise to loss of energy which is expressed as

$$W_h = k_h B_m^{1.6} f \nu W \tag{6.10}$$

where W_h is the hysteresis loss in Watts
B_m is the maximum value of flux density in the core in Wb/m^2
f is the supply frequency
ν is the volume of the iron core in m^3
K_h is the steinmetz constant.

To keep the hysteresis loss low, the material with a lower value of steinmetz constant, such as silicon steel, is chosen as the core material.

(b) Eddy current loss

When the core is subjected to an alternate magnetic field, EMF is induced in the core material also. This EMF causes circulating current in the core, and thereby producing loss-resulting generation of heat. If the core gets heated up, it produces an undesirable effect on the insulation material used in the windings. Eddy current loss is expressed as

$$W_e = K_e B_m^2 f^2 t^2 W \tag{6.11}$$

where W_e is the eddy current loss in Watts
B_m is the maximum value of flux density

f is the supply frequency
t is the thickness of the core material.
K_e is a constant.

Eddy current loss is minimized by using a thin laminated sheet steel as the core material instead of a solid core. The laminated steel sheets are assembled together and are insulated from each other using insulating varnish. This creates an obstruction to the flow of eddy current, and hence reduces the eddy current loss.

The sum of hysteresis loss and eddy current loss is called core loss or iron loss.

$$W_i = W_h + W_c$$

$$= k_h B_m^{1.6} f \nu + k_c B_m^{2} f^2 t^2$$

or,

$$W_i = k_h B_m^{1.6} \nu f + k_c B_m^{2} t^2 f^2$$

or,

$$W_i = Af + Bf^2$$

If W_i is calculated at different frequencies, keeping the ratio of input voltage V to the frequency f constant so that B_m remain constant, the hysteresis and eddy current losses can be separated out from the iron loss.

6.18 ► Solved Numerical Problems

EXAMPLE 6.1

A transformer has 1000 turns on its primary and 500 turns on the secondary. When a voltage, V of frequency f is connected across the primary winding a maximum flux of 2×10^{-3} Wb is produced in the core which links both the windings. Calculate the value of the EMF induced in the two windings.

SOLUTION

Let E_1 and E_2 be the EMFs induced in primary and secondary windings, respectively. Here, $N_1 = 1000$ and $N_2 = 500$.

$$E_1 = 4.44\phi_m f N_1$$

$$= 4.44 \times 2 \times 10^{-3} \times 50 \times 1000.$$

$$= 444 \text{ V}$$

and

$$E_2 = 4.44\phi_m f N_2$$

$$= 4.44 \times 2 \times 10^{-3} \times 50 \times 500$$

$$= 222 \text{ V}$$

We observe that, $\frac{E_1}{E_2} = \frac{444}{222} = 2$

and $\frac{N_1}{N_2} = \frac{1000}{500} = 2$

Therefore $\frac{E_1}{E_2} = \frac{N_1}{N_2}$

EXAMPLE 6.2

A transformer has 900 turns on its primary winding and 300 turns on its secondary. A voltage of 230 V at 50 Hz is connected across its primary winding. The cross-sectional area of the core is 64 cm². Calculate the magnitude of the induced EMF in the secondary winding. Also calculate the value of maximum flux density in the core.

SOLUTION

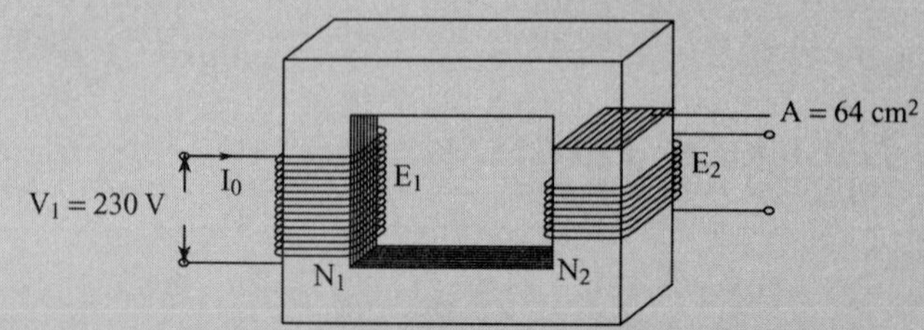

FIGURE 6.22

Given V_1 = 230 V.

The induced EMF in the primary windings is E_1. E_1 is slightly less than V_1 because there will be some voltage drop in the winding.

$V_1 > E_1$, $V_1 - E_1$ = Voltage drop in the primary winding due to current, I_0 flowing through it.

The no-load current, I_0 is very small as compared to the current that would flow when some electrical load is connected across the secondary winding. Here, the transformer is on no-load, i.e., no load has been connected to its secondary winding.

If we neglect the no-load voltage drop in the winding, we can write $V_1 = E_1$.

$$E_1 = 4.44\ \phi_m\ f\ N_1$$

and $$E_2 = 4.44\ \phi_m\ f\ N_2$$

therefore, $$\frac{E_1}{E_2} = \frac{N_1}{N_2}$$

or, $$E_2 = E_1\left(\frac{N_2}{N_1}\right) = 230\left(\frac{300}{900}\right) = 76.7\text{ V}$$

Again, $$E_1 = 4.44\,\phi_m\, f\, N_1$$
$$= 4.44\, B_m\, A\, f\, N_1$$

when B_m is the maximum flux density and A is the cross-sectional area of the core.

Substituting values
$$230 = 4.44 B_m \times \frac{64}{10^4} \times 50 \times 900$$

or, $$B_m = \frac{230 \times 10^4}{4.44 \times 64 \times 45000} = \frac{2300}{4.44 \times 64 \times 45}$$

or, $$B_m = 0.18\text{ Wb/m}^2$$

EXAMPLE 6.3

A 110 V/220 V transformer is supplied with 110 V, 50 Hz supply to its low-voltage side. It is desired to have maximum value of core flux as 4.2 mWbs. Calculate the required number of turns in its primary winding.

SOLUTION

$V_1 = 110$ V. Neglecting the winding voltage drop under no-load condition, $V_1 = E_1 = 110$ V.

$$E_1 = 4.44\,\phi_m\, f\, N_1$$

Substituting values, $$110 = 4.44 \times 4.2 \times 10^{-3} \times 50 \times N_1$$

or, $$N_1 = \frac{110 \times 10^3}{4.44 \times 4.2 \times 50} = 119\text{ turns.}$$

EXAMPLE 6.4

A 100 kVA, 1100/220 V, 50 Hz transformer has 100 turns on its secondary winding. Calculate the number of turns of the primary winding; the currents that would flow in both the windings when fully loaded, and the maximum value of flux in the core.

SOLUTION

Given,
$$N_2 = 100$$
$$V_1 \approx E_1 \text{ and } V_2 \approx E_2$$
$$\frac{E_1}{E_2} = \frac{N_1}{N_2}$$
$$N_1 = N_2\left(\frac{E_1}{E_2}\right) = 100\left(\frac{1100}{220}\right) = 500 \text{ turns}$$

Rating of the transformer is = 100 kVA = 100 × 10^3 VA.

Primary current,
$$I_1 = \frac{VA}{V_1} = \frac{100 \times 10^3}{1100} = 90.1 \text{ A.}$$

Since volt-ampere rating is the same for the transformer on both the sides,

scondary current,
$$I_2 = \frac{VA}{V_2} = \frac{100 \times 10^3}{220} = 450.5 \text{ A}$$

For calculating ϕ_m, we will use the EMF equation.
$$E_1 = 4.44\, \phi_m\, f\, N_1$$

Substituting values
$$1100 = 4.44\, \phi_m \times 50 \times 500$$

or,
$$\phi_m = 9.9 \times 10^{-3} \text{ Wb}$$

EXAMPLE 6.5

The maximum flux density in the core of a 1100/220 V, 50 Hz, 100 kVA transformer is 3.5 Wb/m^2. Calculate the area of cross section of the core and the number of turns of the primary and secondary windings if the EMF per turn is 5.5 V.

SOLUTION

$$V_1 = E_1 = 1100 \text{ V}$$

EMF per turn × No. of turns = total induced EMF

or,
$$5.5 \times N_1 = E_1$$

or,
$$N_1 = \frac{1100}{5.5} = 200 \text{turns.}$$
$$\frac{E_1}{E_2} = \frac{V_1}{V_2} = \frac{N_1}{N_2}$$
$$N_2 = N_1\left(\frac{E_2}{E_1}\right) = 200\left(\frac{220}{1100}\right) = 50 \text{turns}$$

Again, $E_1 = 4.44\,\phi_m\, f\, N_1$

or, $E_1 = 4.44\, B_m\, A\, f\, N_1$

$$A = \frac{E_1}{4.44 B_m f N_1} = \frac{1100}{4.44 \times 3.5 \times 50 \times 200}$$

$$= 70 \times 10^{-4}\ m^2$$

$$= 70\ cm^2$$

EXAMPLE 6.6

The no-load input power to a transformer is 100 W. The no-load current is 3 A when the primary applied voltage is 230 V at 50 Hz. The resistance of the primary winding in 0.5 Ω. Calculate the value of iron loss and no-load power factor.

SOLUTION

At no-load, the input power is lost as a small amount of $I^2 R$ loss in the winding and as core loss.

In an ac circuit, power = VI cos ϕ

Let No-load input power be W_0

$$W_0 = V_1 I_0 \cos \phi_0$$

$$\cos\phi_0 = \frac{W_0}{V_1 I_0}$$

$$\cos\phi_0 = \frac{100}{230 \times 3} = 0.15 \text{ lagging.}$$

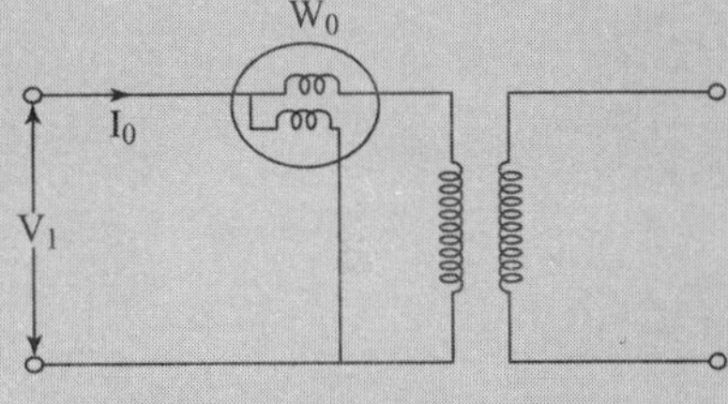

FIGURE 6.23

$\cos \phi_0$ is the no-load power factor of the transformer. At no-load, primary winding copper loss is equal to

$$I_0^2 R_1 = 3^2 \times 0.5 = 4.5\ W$$

The wattmeter reading W_0 indicates the power loss in the core as also in the primary winding.

Core loss, or iron loss, $W_i = W_0 - I_0^2 R_1$

$$= 100 - 4.5$$

$$= 95.5\ W$$

EXAMPLE 6.7

A 100 kVA, 2400/240 V, 50 Hz transformer has a no-load current of 0.64 A and a core loss of 700 W, when its high-voltage side is energized at rated voltage and frequency. Calculate the components of the no-load current and no-load branch parameters of the equivalent circuit.

SOLUTION

Neglecting the small amount of copper loss at no-load,

$$V_1 I_0 \cos\phi_0 = \text{Iron loss}$$

$$I_0 \cos\phi_0 = \frac{\text{Iron loss}}{V_1} = \frac{700}{2400} = 0.29 \text{ A}$$

$$I_c = I_0 \cos\phi_0 = 0.29 \text{ A}$$

$$\cos\phi_0 = \frac{0.29}{I_0} = \frac{0.29}{0.64} = 0.45$$

$$\sin\phi_0 = 0.89$$

$$I_m = I_0 \sin\phi_0 = 0.64 \times 0.89 = 0.57 \text{ A.}$$

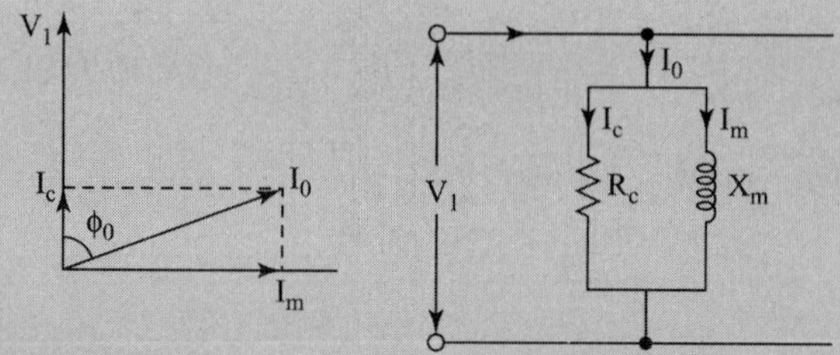

FIGURE 6.24

$$X_m = \frac{V_1}{I_m} = \frac{2400}{0.57}$$

$$= 4.2 \times 10^3 \,\Omega.$$

$$R_c = \frac{V_1}{I_c} = \frac{2400}{0.29} = 8.27 \times 10^3 \,\Omega$$

EXAMPLE 6.8

A 400/200 V, 50 Hz transformer draws a no-load current of 6 A at 0.2 power factor lagging. The transformer supplies a current of 100 A at 200 V to the load. The load power factor is 0.8 lagging. What is the magnitude of current drawn by the transformer from the supply mains?

SOLUTION

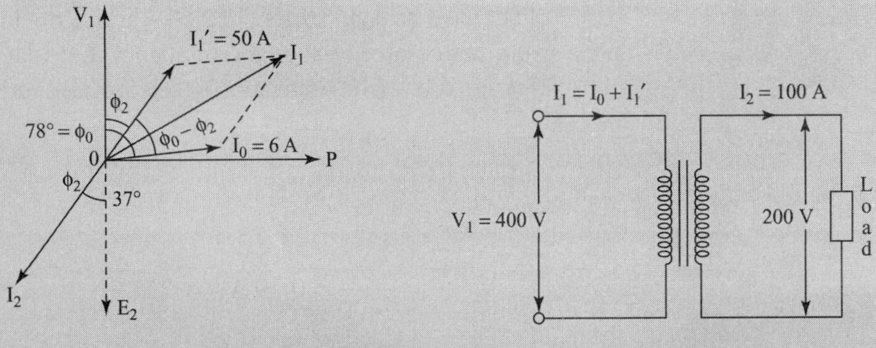

FIGURE 6.25

The circuit diagram and the phasor diagram showing the currents with reference to supply voltage, V_1 have been shown in Fig. 6.25. I_o is the no-load current making an angle of lag ϕ_0 with V_1 where $\cos \phi_o = 0.2$ or $\phi_o = 78°$ (lagging). The load power factor, $\cos \phi_2 = 0.8$ or, $\phi_2 = 37°$. I_2 is the load current. I'_1 is the additional current drawn by the primary to balance the load current I_2 such that

$$I_1' N_1 = I_2 N_2$$

substituting values

$$I_1' = 100\left(\frac{N_2}{N_1}\right) = 100\left(\frac{V_2}{V_1}\right) = 100\left(\frac{200}{400}\right)$$

$$= 50 \text{ A.}$$

As shown in Fig. 6.25, it is observed that the phasor sum of I_o and I_1' is the primary current when the transformer is loaded. The angle between I_o and I_1' is $\phi_o - \phi_2$, i.e., $(78° - 37°) = 41°$.

Using law of parallelogram,

$$I_1^2 = (I_1')^2 + (I_o)^2 + 2I_1' I_o \cos(\phi_o - \phi_2)$$

substituting values

$$I_1^2 = (50)^2 + (6)^2 + 2 \times 50 \times 6 \cos 41°.$$

or, $$I_1^2 = 2986$$

or, $$I_1 = 54.6 \text{ A}$$

It is also possible to find out I_1 as the resultant of I_1' and I_0 by resolving the in-phase and perpendicular components of I_1 with V_1 and then equating them with the corresponding components of I_1' and I_0 as

$$I_1 \cos\phi_1 = I_1' \cos \phi_2 + I_0 \cos \phi_0$$

and
$$I_1 \sin\phi_1 = I_1' \sin \phi_2 + I_0 \sin \phi_0$$

Putting the values of I_1', I_0, $\cos \phi_2$, $\sin \phi_2$, $\cos \phi_0$, $\sin \phi_0$ then by squaring and adding both sides of these equations we can find I_1.

To find the primary circuit power factor, $\cos\phi_1$ we will divide $I_1 \sin\phi_1$ by $I_1 \cos\phi_1$ to find $\tan\phi_1$ first. Then we find ϕ_1 and $\cos\phi_1$.

EXAMPLE 6.9

A 400/200 V, 50 Hz, 10 kVA transformer has primary and secondary winding resistances of 2.5 Ω and 0.5 Ω and winding leakage reactances of 5 Ω and 1 Ω, respectively. Calculate the equivalent resistance and reactance of the transformer referred to the secondary side. What amount of power will be lost in the windings?

SOLUTION

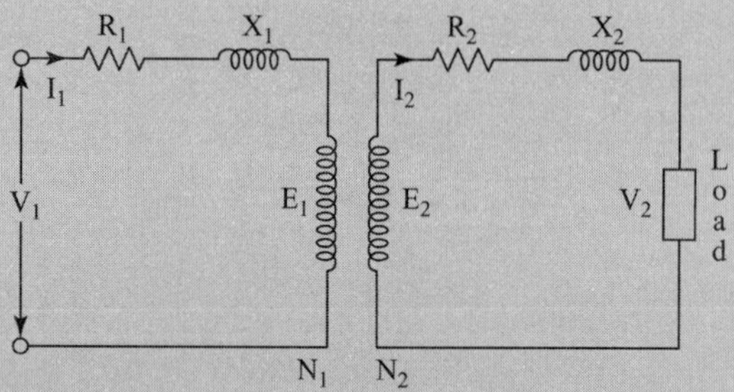

FIGURE 6.26 (a)

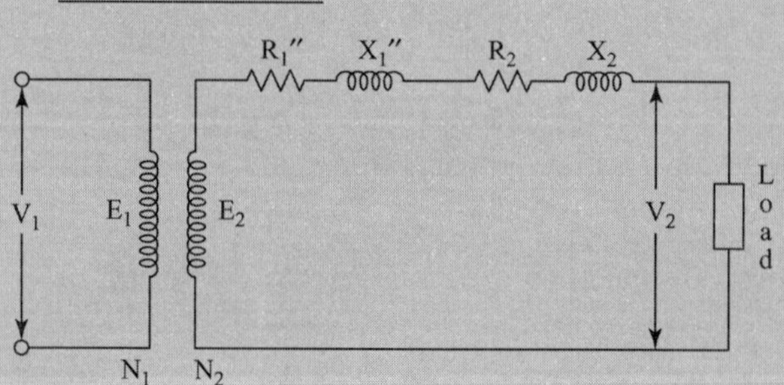

FIGURE 6.26 (b)

Given, $R_1 = 2.5\ \Omega, X_1 = 5\ \Omega, R_2 = 0.5\ \Omega, X_2 = 1.0\ \Omega.$

Equating the I²R losses, $\frac{E_2}{E_1} = \frac{N_2}{N_1} = \frac{V_2}{V_1} = \frac{200}{400} = \frac{1}{2} = K$. (assuming $V_1 = E_1$ and $V_2 = E_2$)

$$I_1^2 R_1 = I_2^2 R_1'' \text{ or}, R_1'' = R_1 \left(\frac{I_1}{I_2}\right)^2 = R_1 \left(\frac{N_2}{N_1}\right)^2 = K^2 R_1$$

thus,

$$R_1'' = \left(\frac{1}{2}\right)^2 R_1 = \frac{2.5}{4} = 0.625\ \Omega$$

Similarly,

$$X_1'' = X_1 K^2 = \frac{5}{4} = 1.25\ \Omega.$$

$$R_e'' = R_1'' + R_2 = 0.625 + 0.5.$$

$$= 1.125\ \Omega.$$

$$X_e'' = X_1'' + X_2 = 1.25 + 1.0$$

$$= 2.25\ \Omega.$$

Given,

$$\text{kVA} = 10$$

$$\text{VA} = 10 \times 1000$$

$$V_2 I_2 = 10 \times 1000$$

$$I_2 = \frac{10 \times 1000}{V_2} = \frac{10 \times 1000}{200} = 50\,\text{A}.$$

This current I_2 is passing through the equivalent resistance, $R_e'' = 1.125\ \Omega$.

Total copper loss = $I_2^2 R_e'' = (50)^2 \times 1.125 = 2.8\,\text{kW}$.

EXAMPLE 6.10

A 25 kVA, 2000/200 V transformer has constant loss, i.e., iron loss of 350 W and full-load copper loss called the variable loss of 400 W. Calculate the efficiency of the transformer at full load and at half load 0.8 power factor lagging.

SOLUTION

Output in kVA = 25, output in kW = kVA cosϕ = 25 × 0.8 = 20 kW

Efficiency,

$$\eta = \frac{\text{Output}}{\text{Input}} \times 100 = \frac{\text{Output} \times 100}{\text{Output} + \text{Losses}}$$

Output in Watts $= 20 \times 1000$ W

Core loss or iron loss $= 350$ W

Full-load copper loss $= 400$ W

$$\eta = \frac{\text{Output} \times 100}{\text{Output} + \text{Core loss} + \text{Copper loss}}$$

$$= \frac{20 \times 1000 \times 100}{20 \times 1000 + 350 + 400} = 96.4 \text{ per cent}$$

At half load

Output $= 10 \times 10^3$

Core loss $= 350$ W (remains constant at all loads)

Copper loss $= \frac{400}{4} = 100$ W (variable loss, varies as square of the load)

$$\eta = \frac{10 \times 1000 \times 100}{10 \times 1000 + 350 + 100} = 95.7 \text{ per cent.}$$

EXAMPLE 6.11

A 5 kVA, 1000/200 V, 50 Hz single-phase transformer has the following no-load test, i.e., the open-circuit test and the short-circuit test data.

No-load test conducted at the low-voltage side:

$$W_0 = 90 \text{ W}, \quad I_0 = 1.2 \text{ A}, \quad V = 200 \text{ V}$$

The short-circuit test conducted at the high-voltage side:

$$W_{sc} = 110 \text{ W}, \quad I_{sc} = 5 \text{ A}, \quad V_{sc} = 50 \text{ V}$$

Calculate the efficiency of the transformer at full-load 0.8 p.f. lagging. What will be the equivalent resistance of the transformer windings referred to high-voltage side?

SOLUTION

Power consumed on no-load test can be taken approximately equal to the core loss, and power loss on the short-circuit test when the rated current flows through the windings can be taken as equal to full-load copper loss.

$I_{sc} = 5$ A; and full-load current at the high-voltage side $= \dfrac{\text{VA rating}}{\text{voltage}}$

$$= \frac{5 \times 1000}{1000} = 5\text{ A}$$

Thus, we see that the short-circuit test was conducted under the full-load condition.

Therefore, copper loss $= W_{SC} = 110\text{ W} = \frac{110}{1000}\text{ kW}$

and iron loss $= W_0 = 90\text{ W} = \frac{90}{1000}\text{ kW}$

Efficiency

$$\eta = \frac{\text{kVA}\cos\phi \times 100}{\text{kVA}\cos\phi + W_{sc} + W_0}$$

$$= \frac{5 \times 0.8 \times 100}{5 \times 0.8 + (110/1000) + (90/1000)} = 95.24\text{ per cent}$$

Full load copper loss $= I_1^2 R_e' = 110\text{ W}$ or $R_e' = \frac{110}{5^2} = 4.4\ \Omega.$

EXAMPLE 6.12

A 20 kVA, 1000/200 V, 50 Hz has core loss and copper loss as 400 W and 600 W, respectively, under the full-load condition. Calculate the efficiency at full load 0.8 lagging power factor. At what percentage of full load will the efficiency be maximum and what is the value of maximum efficiency?

SOLUTION

Full-load efficiency at 0.8 lagging p.f.,

$$\eta = \frac{\text{kW output} \times 100}{\text{kW output} + \text{losses}}$$

$$= \frac{\text{kVA}\cos\phi \times 100}{\text{kVA}\cos\phi + W_c + W_{cu}}$$

Substituting values,

$$\eta = \frac{20 \times 0.8 \times 100}{20 \times 0.8 + 0.4 + 0.6}$$

$$= \frac{16 \times 100}{17} = 94\text{ per cent}$$

Now we have to determine at what load the efficiency will be maximum.

Let x be the fraction of full load at which the efficiency will be maximum. Since copper loss is proportional to the square of the load, and at maximum efficiency core loss equals copper loss

$$x^2 \times W_{cu} = W_c$$

$$x = \sqrt{\frac{W_c}{W_{cu}}}$$

Substituting values

$$x = \sqrt{\frac{400}{600}} = 0.82$$

Therefore, efficiency will be maximum when the load is 0.82 of 20 kVA, i.e., 0.82 × 20 = 16.4 kVA. At this load, core loss equals copper loss; and maximum efficiency is

$$\eta_m = \frac{16.4 \times 0.8 \times 100}{16.4 \times 0.8 + (400/1000) + (400/1000)} = 95 \text{ per cent}$$

EXAMPLE 6.13

Efficiency of 400/200 V, 200 kVA transformer is 98.5 per cent at full load at 0.8 lagging power factor. At half load, 0.8 power factor lagging the efficiency is 97.5 per cent. Calculate the values of core loss and full-load copper loss.

SOLUTION

Let W_c be the core loss and W_{cu} be the full-load copper loss.

$$\eta_{F.L} = 0.985 = \frac{200 \times 0.8}{200 \times 0.8 + W_c + W_{cu}}$$

or,

$$0.985 = \frac{160}{160 + W_c + W_{cu}}$$

or,

$$\frac{160}{0.985} = 160 + W_c + W_{cu}$$

or,

$$W_c + W_{cu} = 2.43 \text{ kW} = 2430 \text{ W} \qquad \text{(i)}$$

At half load, W_c will remain the same but W_{cu} will be one-fourth its full-load value. Efficiency at half load is

$$\eta_{H.L} = \frac{100 \times 0.8}{100 \times 0.8 + W_c + (W_{cu}/4)} = \frac{80}{80 + W_c + (W_{cu}/4)}$$

or,

$$0.975 = \frac{320}{320 + 4W_c + W_{cu}}$$

or, $$\frac{320}{0.975} = 320 + 4W_c + W_{cu}$$

or, $$4\,W_c + W_{cu} = 8.2 \text{ kW} = 8200 \text{ W} \quad \text{(ii)}$$

From (i) and (ii), $$W_c = 1923 \text{ W}$$

and $$W_{cu} = 507 \text{ W}$$

EXAMPLE 6.14

The equivalent circuit parameters of a 300 kVA, 2200/200 V, 50 Hz single-phase transformer are: primary winding resistance, $R_1 = 0.1\ \Omega$; secondary winding resistance, $R_2 = 0.01\ \Omega$; primary leakage reactance, $X_1 = 0.4\ \Omega$; secondary leakage reactance, $X_2 = 0.03\ \Omega$; resistance representing core loss, $R_c = 6 \times 10^3\ \Omega$, magnetizing reactance $X_m = 2 \times 10^3\ \Omega$. Calculate the voltage regulation and efficiency of the transformer at full load at 0.8 power factor lagging.

SOLUTION

With the given data, the equivalent circuit is drawn as in Fig 6.27.

We will consider $E_1 = 2200$ V and $E_2 = 220$ V

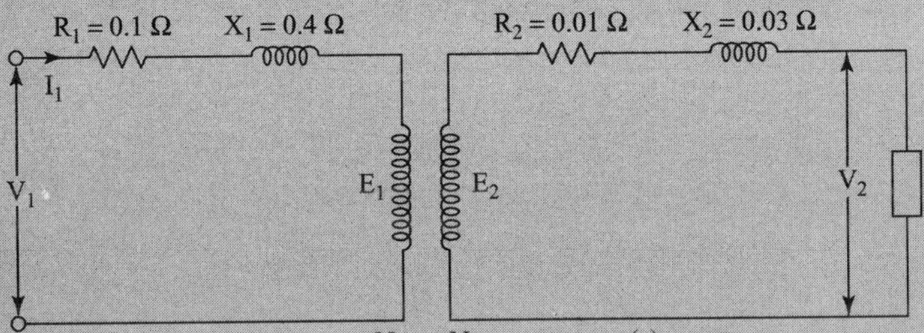

FIGURE 6.27

Transformation ratio, $$K = \frac{N_2}{N_1} = \frac{E_2}{E_1} = \frac{220}{2200} = \frac{1}{10} = 0.1$$

By transferring the secondary quantities to the primary side, we will calculate the equivalent resistance and equivalent reactance of the transformer as

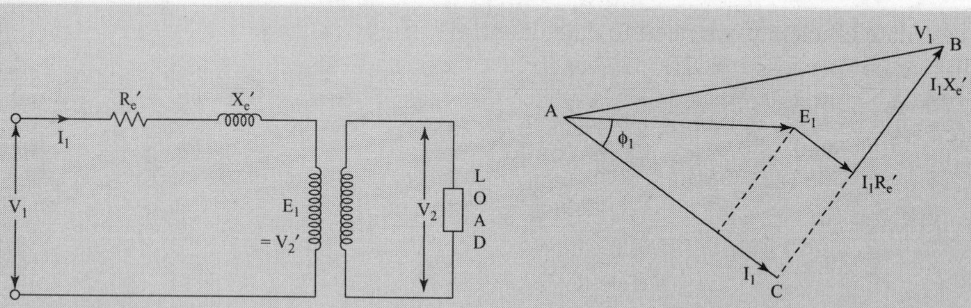

FIGURE 6.28

$$R'_e = R_1 + \frac{R_2}{K^2} = 0.1 + \frac{0.01}{(0.1)^2} = 1.1\ \Omega$$

$$X'_e = X_1 + \frac{X_2}{K^2} = 0.4 + \frac{0.04}{(0.1)^2} = 1.4\ \Omega$$

kVA rating = 300

VA rating = 300 × 1000

$$I_1 = \frac{VA}{E_1} = \frac{300 \times 1000}{2200} = 136\,A.$$

$$p.f = \cos\phi = 0.8;\ \sin\phi = 0.6$$

The equivalent circuit with the secondary quantities referred to the primary side and the phasor diagram have been shown in Fig. 6.28. R'_e and X'_e are the equivalent resistance and reactance with the secondary quantities referred to the primary side.

In the phasor diagram, we have assumed angle between E_1 and I_1 as ϕ_1 because the angle between V_1 and E_1 is practically very small.

From triangle ABC,
$$V_1^2 = AC^2 + CB^2$$
$$= (E_1\cos\phi_1 + I_1R'_e)^2 + (E_1\sin\phi_1 + I_1X'_e)^2$$
$$= (2200 \times 0.8 + 136 \times 1.1)^2 + (2200 \times 0.6 + 136 \times 1.4)^2$$
$$= (1909)^2 + (1510.4)^2$$

Voltage regulation
$$V_1 = 2400\,V$$
$$= \frac{V_1 - E_1}{V_1} \times 100 = \frac{(2400 - 2200)}{2400} \times 100$$
$$= 8.3 \text{ per cent.}$$

To calculate efficiency, we need to calculate the copper loss and core loss.
Full-load copper loss,

$$W_{cu} = I_1^2 R'_e = (136)^2 \times 1.1 = 20.345 \text{ kW}.$$

Core loss

$$= V_1 I_c = V_1 \frac{V_1}{R_c} = \frac{V_1^2}{R_c} = \frac{(2400)^2}{6\times 10^3} = 960 \text{ W}$$

$$= 0.96 \text{ kW}.$$

Efficiency,

$$\eta = \frac{\text{Output} \times 100}{\text{Output} + W_{cu} + W_c}$$

$$= \frac{300 \times 0.8 \times 100}{300 \times 0.8 + 20.345 + 0.96}$$

$$= \frac{240 \times 100}{261.3} = 91.84 \text{ per cent.}$$

EXAMPLE 6.15

A 10 kVA 440/220 V, 50 Hz single-phase transformer gave the following test results when both the following tests were conducted on the high-voltage side:

Open-circuit test : 440 V, 1.0 A, 100 W
Short-circuit test : 20 V, 22.7 A, 130 W.

Using the test data, calculate the efficiency and voltage regulation at 0.8 power factor lagging.

SOLUTION

Full-load current on the high-voltage side $= \dfrac{10 \times 1000}{440} = 22.7$ A

This shows that the short-circuit test has been conducted on full load. The wattmeter reading, therefore, represents the full-load copper loss.

$$W_{cu} = 130 \text{ W and from oc test data,}$$

$$W_c = 100 \text{ W}$$

Full-load efficiency is calculated as

$$\eta = \frac{\text{Output}}{\text{Output} + W_{cu} + W_c} \times 100$$

$$= \frac{10 \times 0.8}{10 \times 0.8 + 0.13 + 0.1} \times 100$$

$$= 97.2 \text{ per cent}$$

Calculation of voltage regulation

Current, $$I_1 = \frac{\text{kVA}}{V_1} = \frac{10 \times 1000}{440} = 22.7 \text{ A.}$$

From the short-circuit test data, wattmeter reading can be taken as equal to copper losses in the windings.

$$W_{SC} = 130 = I_1^2 R'_e = (22.7)^2 R'_e$$

Therefore, $$R'_e = \frac{130}{(22.7)^2} = 0.25 \ \Omega.$$

$$Z'_e = \frac{V_{sc}}{I_{sc}} = \frac{20}{22.7} = 0.88 \ \Omega.$$

$$Z_e^2 = R_e^{2\prime} + X_e^{2\prime}$$

$$X'_e = \sqrt{(0.88)^2 - (0.25)^2} = 0.84 \ \Omega.$$

$$\cos\phi = 0.8, \sin\phi = 0.6$$

Regulation $$= \frac{(I_1 R'_e \cos\phi + I_1 X'_e \sin\phi)}{V_1} \times 100$$

$$= \frac{(22.7 \times 0.25 \times 0.8 + 22.7 \times 0.84 \times 0.6)}{440} \times 100$$

$$= \frac{(4.54 + 11.44)}{440} \times 100$$

$$= 3.63 \text{ per cent}$$

EXAMPLE 6.16

A 440/220 V single-phase transformer has percentage resistance drop and reactance drop of 1.2 per cent and 6 per cent, respectively. Calculate the voltage regulation of the transformer at 0.8 power factor lagging.

SOLUTION

Percentage voltage regulation $= \dfrac{(I_2 R_e'' \cos\phi + I_2 X_e'' \sin\phi)}{V_2} \times 100$

$$= \frac{I_2 R_e''}{V_2}\cos\phi \times 100 + \frac{I_2 X_e''}{V_2}\sin\phi \times 100$$

Given, $\dfrac{I_2 R_e''}{V_2} \times 100 = 1.2,\ \dfrac{I_2 X_e''}{V_2} \times 100 = 6.0,\ \cos\phi = 0.8$

$\sin\phi = 0.6.$

Substituting values regulation $= \dfrac{I_2 R_e''}{V_2}\cos\phi + \dfrac{I_2 X_e''}{V_2}\sin\phi$

$= 1.2 \times .0.8 + 6.0 \times 0.6$

$= 4.56$ per cent

EXAMPLE 6.17

le 6.17 A 230/115 V, 5 kVA transformer has circuit parameters as $R_1 = 0.2\ \Omega$, $X_1 = 0.8\ \Omega$, $R_2 = 0.1\ \Omega$, $X_2 = 0.2\ \Omega$. Calculate the regulation of the transformer at 0.8 power factor lagging. At what value of power factor will the regulation be zero? Can the value of regulation be negative for any power factor load?

SOLUTION

The transformation ratio, $K = \dfrac{N_2}{N_1} = \dfrac{E_2}{E_1} = \dfrac{115}{230} = 0.5$

Referring all the circuit parameters to the secondary side,

$$R_e'' = R_2 + R_1''$$
$$= R_2 + K^2 R_1$$
$$= 0.1 + (.5)^2 \times 0.2$$
$$= 0.15\ \Omega$$

$$X_e'' = X_2 + X_1'' = X_2 + K^2 X_1 = 0.2 + (0.25)^2 \times 0.8 = 0.4\ \Omega.$$

$$\cos\phi = 0.8, \sin\phi = 0.6, I_2 = \frac{5 \times 1000}{115} = 43.8\,A$$

$$\text{Regulation} = \frac{(I_2 R_e'' \cos\phi + I_2 X_e'' \sin\phi)}{V_2} \times 100$$

$$= \frac{(43.8 \times 0.15 \times 0.8 + 43.8 \times 0.4 \times 0.6) \times 100}{115}$$

= 13.7 per cent

We know regulation can be negative only when the power factor is leading and when the expression for regulation is

$$\text{Regulation} = \frac{(I_2 R_e'' \cos\phi - I_2 X_e'' \sin\phi) \times 100}{V_2}$$

Regulation will be zero when the numerator of the above expression will be zero. That is

$$I_2 R_e'' \cos\phi = I_2 X_e'' \sin\phi$$

or,

$$\tan\phi = \frac{R_e''}{X_e''} = \frac{0.15}{0.4} = 0.375, \quad \text{or} \quad \phi = 20.6°$$

The power factor at which regulation will be zero is cos 20.6° = 0.94 leading

Regulation will be negative if the power factor angle is more than 20.6° leading. To verify, let the angle of lead of load current be higher than 20.6°. If we assume this angle as 37°, the power factor is 0.8 leading. The regulation at 0.8 leading is calculated as

$$\text{Regulation} = \frac{(I_2 R_e'' \cos\phi - I_2 X_e'' \sin\phi)}{V_2} \times 100$$

$$= \frac{(43.8 \times 0.15 \times 0.8 - 43.8 \times 0.4 \times 0.6) \times 100}{115}$$

$$= \frac{(5.256 - 10.512)}{115} \times 100$$

= –4.57 per cent

EXAMPLE 6.18

Calculate the all-day efficiency of a 25 kVA distribution transformer whose loading pattern is as follows:

15 kW at 0.8 power factor for 6 hours
12 kW at 0.7 power factor for 6 hours
10 kW at 0.9 power factor for 8 hours
Negligible load for 4 hours.

The core loss is 500 W and full-load copper loss is 800 W.

SOLUTION

Output energy in kWh of the transformer in 24 hours is calculated as

$$15 \text{ kW} \times 6 + 12 \text{ kW} \times 6 + 10 \text{ kW} \times 8 + 0 \times 4$$
$$= 242 \text{ kWh} = \text{output energy}$$

We will calculate the core loss and copper loss for 24 hours at different loading conditions. Core loss remains constant at all loads. Therefore,

$$\text{Core loss for 24 hours} = 0.5 \times 24 = 12 \text{ kWh}$$
$$= \text{Energy lost in the core}$$

Copper loss varies as the square of the load. The loads on transformer have to be calculated in terms of kVA.

$$15 \text{ kW at } 0.8 \text{ p.f.} = \frac{15 \text{ kW}}{0.8} = 18.75 \text{ kVA}$$

$$12 \text{ kW at } 0.7 \text{ p.f.} = \frac{12 \text{ kW}}{0.7} = 17.14 \text{ kVA}$$

$$10 \text{ kW at } 0.9 \text{ p.f.} = \frac{10 \text{ kw}}{0.9} = 11.11 \text{ kVA}$$

at 25 kVA load copper loss $= 0.8$ kW.

at 18.75 kVA load copper loss $= 0.8 \times \left(\frac{18.75}{25}\right)^2 = 450 \text{ W} = 0.45 \text{ kW}$

at 17.14 kVA load copper loss $= 0.8 \times \left(\frac{17.14}{25}\right)^2 = 376 \text{ W} = 0.376 \text{ kW}$

at 11.11 kVA load copper loss $= 0.8 \times \left(\frac{11.11}{25}\right)^2 = 157 \text{ W}$

$= 0.157$ kW

All day efficiency

$$= \frac{\text{Energy output in 24 hours}}{\text{Energy output in 24 hours} + \text{Energy loss in core in 24 hours} + \text{Energy loss in copper for 24 hours}}$$

$$= \frac{242}{242 + 12 + (0.45 \times 6 + 0.376 \times 6 + 0.157 \times 10 + 0 \times 4)}$$

$$= \frac{242}{242 + 12 + 6.47} = \frac{242}{260.47} = 0.929$$

Thus, $\eta_{\text{all-day}} = 0.929 = 92.9$ per cent

EXAMPLE 6.19

No-load test conducted on a transformer gave the flowing readings

	Supply Voltage	Frequency	Wattmeter reading
(i)	250 V	50 Hz	55 W
(ii)	200 V	40 Hz	40 W

Calculate the hysteresis loss and eddy current loss at 250 V, 50 Hz.

SOLUTION

To maximum constant flux density, B_m, we must have $\frac{V_1}{f_1}$ equal to $\frac{V_2}{f_2}$

In this case, $$\frac{V_1}{f_1} = \frac{250}{50} = 5$$

and $$\frac{V_2}{f_2} = \frac{200}{40} = 5$$

Hence, the core flux density B_m is constant. We know, iron loss, W_i is the sum of hysteresis loss W_h and eddy current loss, We

So, $$W_i = W_h + W_c$$
$$= k_h B_m^{1.6} f + k_c B_m^2 t^2 f^2$$

or, $$W_i = Af + Bf^2$$ Where A and B are constants.

Substituting the given values,

$$55 = A \times 50 + B \times 50^2 \quad \text{(i)}$$

and $$40 = A \times 40 + B \times 40^2 \quad \text{(ii)}$$

From (i) and (ii), $A = 0.6$ and $B = 0.01$.

At 250 V, 50 Hz, $$W_h = Af = 0.6 \times 50 = 30 \text{ Watts}$$

$$W_c = Bf^2 = 0.01 \times 2500 = 25 \text{ Watts}$$

$$W_i = W_h + W_c = 30 + 25 = 55 \text{ Watts}$$

Important Terminologies and Points to Remember

1. Transformers are used to step-up or step-down the voltage level so as to transfer power from one circuit to the other without changing the frequency. Transformers do not generate electricity.
2. The volt-ampere rating of a transformer on both primary and secondary side is the same. So, if voltage level is changed, current level will also change. If voltage is stepped up, current will be low.
3. Transformers are manufactured for a few VA rating to several MVA (Mega Volt Ampere) ratings. You may find a tiny transformer used in a battery charger and a very large transformer in an electrical sub-station.
4. A transformer basically has a core made up of magnetic material on which two coils, called primary and secondary are wound it the two limbs of the core.
5. When an alternating voltage is applied to primary winding of a transformer, the current flowing produces an alternating flux in the core. This alternating flux links with the primary and secondary windings and emf are induced in these two windings. High voltage will be induced in the coil having more number of turns.
6. The core of a transformer is made of laminated steel sheets to reduce eddy current loss in the core.
7. Hysteresis loss in the core material gets reduced if silicon steel is used. Hysteresis loss depends on the area of the hysteresis loop.
8. Transformers are made as core type and shell type.
9. Power transformers are connected to the two ends of transmission lines to step-up or step-down the voltage level. Distribution transformers feed electricity to the consumers.
10. Most of the no-load current is used in producing the flux, i.e., in magnetizing the core.
11. The no-load current drawn by a transformer is only 3 to 5 percent of its full-load current.
12. On no-load, the input to the motor is equal to core loss (Hysteresis loss and Eddy current loss) plus a small amount I^2R loss in the primary winding. The output at no load is zero.
13. In a transformer, the flux which links, i.e., passes through the primary and secondary winding is called the useful flux. The small amount of flux which only links any one of the windings is called leakage flux. Leakage flux causes leakage reactance of the windings. The core flux or the mutual flux causes magnetizing reactance.
14. Equivalent circuit is drawn and then simplified so as to enable calculation of voltage regulation of a transformer as different loads and load power factor without actually loading the transformer.
15. An ideal transformer is only a theoretical concept which has no losses and no voltage drop in the windings. For such an ideal transformer, efficiency is 100 percent and voltage regulation is zero.
16. Short-circuit test is conducted to find the full-load copper loss in a transformer without actually loading the transformer. A reduced voltage is applied so as to circulate rated current through the transformer windings.
17. Efficiency and voltage regulation of a transformer is calculated from the data of no-load test and short-circuit test.
18. Efficiency of a transformer is maximum at a load at which core loss equals copper loss.
19. All day efficiency of distribution transformers is calculated on the basis of energy output for 24 hours to the energy input in 24 hours. All day efficiency is somewhat less than commercial efficiency which is calculated on the basis of output power to the corresponding input power.

Important Formulae and Equations

Equation	Remarks
1. $E = 4.44\,\phi_m f N$ volts where N is the number of turns of the winding.	1. Emf equation for both primary and secondary windings
2. $\eta = \dfrac{V_2 I_2 \cos\phi_2}{V_2 I_2 \cos\phi_2 + W_{iron} + W_{copper}}$	2. Iron loss is constant but copper loss varies as the square of the load current.
3. $X = \sqrt{\dfrac{W_i}{W_{cu}}}$ Where X is the fraction of full-load at which efficiency will be maximum.	3. Condition for maximum efficiency is that core loss becomes equal to copper loss at that load.
4. Voltage Regulation $= \dfrac{I_2 R''_e \cos\phi - I_2 X''_e \sin\phi}{V_2} \times 100$ Where R''_e and X''_e are the total resistance and reactance of the transformer by transferring the secondary quantities to the primary side.	
5. All day Efficiency $= \dfrac{\text{Energy output in 24 hours}}{\text{Energy output in 24 hours + Energy loss in core for 24 hours + Energy loss in winding for 24 hours}}$	

REVIEW QUESTIONS

Short Answer Type Questions

1. Explain with examples why transformers are required in transmission and distribution of electrical power.
2. Distinguish between a step-up transformer and a step-down transformer.
3. What is the expression for voltage per turn of a transformer?
4. Derive the EMF equation of a transformer.
5. What are the losses in a transformer and how can these be kept low?
6. What is eddy current loss and how can this loss be reduced?
7. Why do we use laminated sheets to build the core of a transformer instead of using a solid core?
8. Distinguish between core-type and shell-type construction of the transformer core.
9. Explain why the frequency of output voltage is the same as input voltage in a transformer.
10. Distinguish between magnetizing reactance and leakage reactance of a transformer.
11. Draw the equivalent circuit of a transformer under the no-load condition.
12. Explain the concept of an ideal transformer.

13. What is meant by voltage regulation of a transformer? Is it desirable to have a high-voltage regulation of a transformer? Justify your answer.

14. How can we calculate the efficiency of a transformer by knowing its losses?

15. Draw the no-load phasor diagram of a transformer. What are the two components of a no-load current ?

16. Draw the full-load phasor diagram of a transformer neglecting the voltage drop in the windings.

17. Derive the condition for maximum efficiency of a transformer.

18. What is all-day efficiency of a transformer? What is its significance?

19. How can you determine the efficiency of a transformer indirectly, i.e., without actually loading the transformer?

20. What is the expression for voltage regulation of a transformer in terms of its equivalent resistance, equivalent reactance, power factor, and the output voltage?

21. How can you determine the efficiency of a transformer of a given rating at any load if the values of full-load losses are known?

22. Explain how the short-circuit test on a transformer is to be conducted. What information do you get from the short-circuit test data?

23. Draw and explain the exact equivalent circuit of a transformer.

24. Explain how in a transformer, the primary current increases as the secondary current, i.e., load current increases.

25. Why is the core of a transformer made of magnetic material?

26. What is the difference between a practical transformer and an ideal transformer?

27. Why do we consider core loss as a constant loss and copper loss as a variable loss?

28. Distinguish between a power transformer and a distribution transformer.

29. Is efficiency of a transformer same at a particular load but at different power factors?

30. Is efficiency of a transformer at a particular load same at 0.8 power factor lagging and 0.8 power factor lagging?

31. What may be the main reason for a constant humming noise in a transformer when it is supplying some load?

32. Why is the efficiency of a transformer higher than any rotating electrical machine of similar ratings?

33. How does regulation of a transformer get affected by load power factor?

34. Write short notes on the following: (i) magnetizing reactance; (ii) leakage reactance; (iii) eddy current loss; (iv) hysteresis loss; (v) all-day efficiency.

Numerical Problems

35. A 40 kVA 3200/400 V, single phase, 50 Hz transformer has 112 turns on the secondary winding. Calculate the number of turns on the primary winding. What is the secondary current at full load? What should be the cross-sectional area of the core for a core flux density of 1.2 Wb/m^2?

[**Ans** 896, 100 A, 01362 m^2]

36. A 400 kVA transformer has a full-load core loss of 800 W and copper loss of 2500 W. What will be the values of these losses at $\frac{1}{2}$ load?

[**Ans** 800 W, 625 W]

37. A single-phase transformer is required to step down the voltage from 1100 V to 400 V at 50 Hz. The core has a cross-sectional area of 25 cm^2 and the maximum flux density is 5Wb/m^2. Determine the number of turns of the primary and secondary windings.

[**Ans** 396, 144]

38. A single phase 40 kVA transformer has primary and secondary voltages of 6600 V and 230 V, respectively. The number of turns of the secondary winding is 30. Calculate the number of turns of the primary winding. Also calculate the primary and secondary winding currents.

[**Ans** 860, 6.06 A, 173.9 A]

39. A transformer on no load takes 4.5 A at a power factor of 0.25 lagging when connected to a 230 V, 50 Hz supply. The number of turns of the primary winding is 250. Calculate (a) the magnetizing current, (b) the core loss, and (c) the maximum value of flux in the core.

[**Ans** $I_m = 4.35$, $P_c = 259$ W, $\phi_m = 4.14 \times 10^{-3}$ Wb]

40. A 660 V/ 220 V single-phase transformer takes a no-load current of 2A at a power factor of 0.225 lagging. The transformer supplies a load of 30 A at a power factor of 0.9 lagging. Calculate the current drawn by the primary from the mains and primary power factor. Resistance and reactance of the windings may be neglected.

[**Ans** $I_1 = 11.38$ A, $\cos\phi_1 = 0.829$ lag]

41. A 100 kVA transformer has 400 turns on the primary and 80 turns on the secondary. The primary and secondary resistances are 0.3 Ω and 0.01 Ω, respectively, and the corresponding leakage reactances are 1.1 Ω and 0.035 Ω, respectively. Calculate the equivalent impedance referred to the primary side.

[**Ans** $Z'_e = 2.05$ Ω]

42. A 660 V/220 V single-phase transformer takes a no-load current of 2 A at a power factor of 0.255 lagging. The transformer supplies a load of 30 A at a power factor of 0.9 lagging. Calculate the current drawn by the primary from the mains and primary power factor. Neglect winding resistances and reactances.

[**Ans** $I_1 = 11.4$ A, $\cos\phi_1 = 0.83$ lagging]

43. The primary and secondary windings of a 500 kVA transformer have $R_1 = 0.4$ Ω and $R_2 = 0.001$ Ω, respectively. The primary and secondary voltages are 6600 V and 400 V, respectively. The iron loss is 3 kW. Calculate the efficiency on full load at 0.8 power factor lagging.

[**Ans** 98.3 per cent]

44. A 5 kVA 200/400 V, 50 Hz single-phase transformer gave the following test data:

(i) L.V. side open-circuit test –220 V, 0.7 A, 60 W

(ii) H.V. side short-circuit test –22 V, 16 A, 120 W

Calculate the regulation of the transformer under the full-load condition.

[**Ans** 3 per cent]

45. The no-load current of a transformer is 15 A at a power factor of 0.2 lagging when connected to a 460 V, 50 Hz supply. If the primary winding has 550 turns, calculate (i) magnetizing component of the no-load current, (ii) the iron loss, and (iii) maximum value of flux in the core.

[**Ans** $I_m = 14.67$ A, 780 W, 2.129 mwb]

46. A single-phase, 100 kVA distribution transformer is loaded as mentioned during 24 hours:

4 hours : no load

8 hours : 50 per cent load at power factor = 1

6 hours : 75 per cent load at power factor = 0.9

6 hours : full-load at 0.9 power factor

The full load copper loss and core loss is 5 kW and 2 kW, respectively. Calculate the all-day efficiency of the transformer.

[**Ans** 92.5 per cent]

47. A 12 kVA, 200/400 V, 50 Hz single-phase transformer gave the following readings on the open-circuit test and the short-circuit test:

open-circuit test : 200 V, 1.3 A, 120 W

short-circuit test conducted on the H.V. side: 22 V, 30 A, 200 W

Calculate the equivalent circuit parameters as referred to the low-voltage side. Also calculate the magnetizing component of the no-load circuit.

[**Ans** $R_c = 333$ Ω, $X_m = 174$ Ω, $R'_e = 0.055$ Ω, $X'_e = 0.175$ Ω, $I_m = 1.15$A]

Multiple Choice Questions

1. A transformer having number of turns in the primary and secondary winding of 1000 and 500, respectively, is supplied with 230 V at 50 Hz. The induced EMF in the secondary winding will be

(a) 460 V at 50 Hz (b) 115 V at 25 Hz
(c) 115 V at 50 Hz (d) 500 V at 50 Hz.

2. The core of the transformers is made of laminated steel sheets so as to

(a) Reduce hysteresis loss
(b) Reduce eddy current loss
(c) Increase output voltage
(d) Reduce both hysteresis loss and eddy current loss.

3. The EMF induced in the windings of a transformer
(a) Lags the core flux by 90°
(b) Leads the core flux by 90°
(c) Is in phase with the core flux
(d) Is in opposition to the core flux.

4. To reduce the core losses in a transformer
(a) The core is made of silicon steel laminations
(b) The core is fastened very tight so that the core flux do not fly away
(c) The core is made of solid steel
(d) The core is made of copper laminations.

5. The no-load current of a 10 kVA, 230 V/115 V transformer wire will be about
(a) 5 per cent of its rated current
(b) 20 per cent of its rated current
(c) 30 per cent of its rated current
(d) 0.1 per cent of its rated current.

6. The no-load current of a 15 kVA, 230/1100 V single-phase transformer will be about
(a) 15.33 A (b) 3 A
(c) 12 A (d) 73.3 A.

7. Efficiency of a transformer is higher tham that of a motor or a generator of similar rating because
(a) There is no hysteresis and eddy current loss in a motor or a generator
(b) Designs of transformers are superior than motors and generators
(c) Transformer is a static device and there is no rotational loss in it
(d) Transformers are connected to high-voltage transmission lines whereas motors and generators are connected to low-voltage supply lines.

8. Which of the following losses in a transformer vary with load
(a) Hysteresis loss
(b) Eddy current loss
(c) Copper losses in the windings
(d) Iron loss.

9. Open-circuit test and short-circuit test on a transformer are performed to determine, respectively, the following losses
(a) Copper loss and core loss
(b) Core loss and copper loss
(c) Eddy current loss and hysteresis loss
(d) Hysteresis loss and eddy current loss.

10. Maximum efficiency of a transformer is obtained at a load at which its
(a) Core loss becomes the minimum
(b) Copper loss becomes the minimum
(c) Copper loss equals core loss
(d) Core loss becomes negligible.

11. A transformer has 350 primary turns and 1050 secondary turns. The primary winding is connected across a 230 V, 50 Hz supply. The induced EMF in the secondary will be
(a) 690 V, 50 Hz (b) 690 V, 100 Hz
(c) 690 V, 150 Hz (d) 115 V, 50 Hz.

12. The rating of a transformer is expressed in
(a) kW (b) kVA
(c) kWh (d) $\cos\phi$.

13. The ampere-turns balance equation for a transformer on load can be expressed as
(a) $I_1N_1 = I_2N_2$ (b) $I_1N_1 >> I_2N_2$
(c) $I_1N_1 = I_2N_2$ (d) $I_1N_1 < I_2N_2$.

14. Which of the following characteristic assumptions for an ideal transformer are true?
(a) Coupling coefficient between the windings is unity
(b) There are absolutely no core and copper losses
(c) The core is made up material having infinite permeability
(d) All the characteristics as in (a), (b), and (c).

15. The EMF equation for a transformer is
(a) $E = 4.44\ \phi\ f\ N^2$ (b) $E = 4.44\ \phi\ fN$
(c) $E = 4.44\ \phi\ f^2N$ (d) $E = 4.44\ \phi_m\ f N$.

16. Which of the following statements is true for a transformer?
(a) A transformer is an energy conversion device
(b) A transformer changes the voltage and frequency from one level to the other
(c) A transformer is a static variable frequency device
(d) A transformer transforms or changes voltage level of energy utilization keeping the frequency constant.

17. Which of the following effects will the secondary load current of a transformer will have on the main flux created by the magnetizing current?

(a) Magnetization (b) Demagnetization
(c) Polarization (d) No effect.

18. On which of the following does the voltage regulation of a transformer depend?
(a) Load power factor
(b) Magnitude of load
(c) Winding resistance and reactance
(d) all factor as in (a), (b), and (c).

19. The no-load current of a transformer is
(a) The algebraic sum of I_m and I_c
(b) Phasor sum of I_m and I_c
(c) Algebraic sum of I_o and I_1^1
(d) Phasor sum of I_o and I_1^1.

20. The no-load current of a transformer as compared to its full-load current can be expressed as
(a) 0 to 2 per cent
(b) 2 to 5 per cent
(c) 10 to 20 per cent
(d) 20 to 30 per cent.

21. Power factor of a transformer of no load is low due to
(a) Large component of magnitizing current, which lags the voltage by 90°
(b) Large component of loss component of current
(c) Secondary ampere-turns interfering with the primary ampere-turns
(d) The fact that the primary and secondary windings are not firmly coupled.

22. The core loss and copper loss of a transformer on full load are 400 W and 600 W, respectively. Their values at one-third full load will be
(a) 133.3 W and 200 W
(b) 400 W and 66.66 W
(c) 133.3 W and 600 W
(d) 400 W and 200 W.

23. The full-load core loss and copper loss of a transformer are 400 W and 600 W, respectively. At approximately what percentage of full-load will the efficiency be maximum?
(a) 81 per cent (b) 91 per cent
(c) 95 per cent (d) 99 per cent.

24. The full load output of a transformer at unity power factor is 800 W. Its output at half load 0.8 power factor will be
(a) 400 W (b) 320 W
(c) 160 W (d) 640 W.

25. Large capacity transformers are placed in tanks filled with transformer oil. Which of the following are not valid for the above?
(a) Oil provides insulation between the two windings.
(b) Oil cools the transformer.
(c) Oil prevents magnetizing flux from getting reduced with time.
(d) Temperature rise of the winding is kept under control.

26. The full-load copper loss of a transformer is 1200 W. At half load the copper loss will be
(a) 600 W (b) 1200 W
(c) 300 W (d) 900 W.

27. The full-load core loss of a transformer is 1200 W. At half load the core loss will be
(a) 600 W (b) 1200 W
(c) 300 W (d) 900 W.

28. The no-load current of a certain transformer is 2 A. Its magnetizing component may be
(a) 1.8 A (b) 0.2 A
(c) 0.4 A (d) 0.02 A.

29. When the primary and secondary windings of a transformer are perfectly magnetically coupled
(a) The leakage reactance will be high and voltage regulation will be high (i.e., poor)
(b) The leakage reactance will be low and voltage regulation will be low (i.e., good)
(c) The leakage reactance will be low and voltage regulation will be high (i.e., poor)
(d) The leakage reactance will be high and voltage regulation will be low (i.e., good).

Answers to Multiple Choice Questions

1. (c)	2. (b)	3. (a)	4. (a)	5. (a)	6. (b)
7. (c)	8. (c)	9. (b)	10. (c)	11. (a)	12. (b)
13. (c)	14. (d)	15. (d)	16. (d)	17. (b)	18. (d)
19. (b)	20. (b)	21. (a)	22. (b)	23. (a)	24. (b)
25. (c)	26. (c)	27. (b)	28. (a)	29. (b)	

CHAPTER 7 DC Machines

Topics Discussed

1. Working of a dc machine as a generator and as a motor
2. Basic principle of dc machines
3. Constructional details
4. Need of brush and commutator
5. Types of windings
6. EMF equation
7. Types of dc machines and their characteristics
8. Starting of a dc motor
9. Speed control
10. Losses and efficiency

7.1 ▶ Introduction and Principle of Working

DC machines work either as a dc generator or as a dc motor. In a dc generator, a set of conductors or coils placed on a rotating body, called armature, are rotated continuously inside a magnetic field with the help of a prime mover (prime mover is another machine, may be a diesel engine or a turbine, which rotates the armature). The magnetic field is created by passing dc current through the windings of a set of field magnets. When conductors pass under alternate North and South poles, alternating EMF is induced in the armature winding. The ac generated gets converted into dc when the voltage is collected from the rotating armature through the brush and commutator arrangement. The brush and commutator arrangement, therefore, works like a full-wave rectifier which converts generated ac into dc for the output circuit. A dc

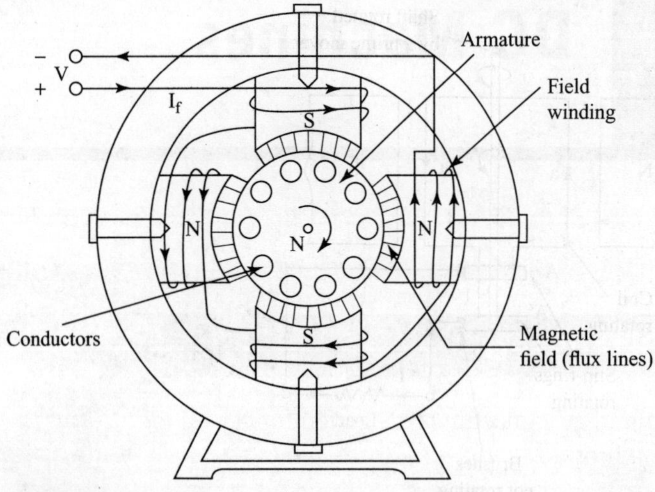

FIGURE 7.1

Cross-sectional view of a four-pole dc machine

generator, therefore, converts mechanical energy supplied through the prime mover to electrical energy to be supplied from the generator armature to an electrical load. We, of course, will need to create a magnetic field by a field system as shown in Fig. 7.1. It can be noticed from the figure that the magnetic poles are electromagnets fixed on a hollow cylindrical frame. Depending on the direction of the winding and the direction of current flow through the field windings, alternate North and South poles are formed, creating a magnetic field inside which the cylindrical rotor called armature is placed and is rotated.

The brush and commutator arrangement has not been shown in the figure. The armature conductors have been shown rotated in the magnetic field by a prime mover at N rpm (revolutions per minute).

All the field–pole windings have been shown connected in series. The directions of current flow in the pole windings are such that alternate North and South poles have been formed. The field system have been energized by passing current, I_f through their windings from a separate dc source. To understand the working of a dc generator, we will consider, for simplicity, a two-pole construction having only one coil on the armature. The two terminals of the armature coil are to be brought out for connection to the load circuit. For this, two arrangements are possible, i.e., (i) through brush and slip-ring arrangement, and (ii) through brush and commutator arrangement. We will consider these two arrangements side by side and see the nature of output voltage and load current.

7.1.1 Nature of Load Current When Output Is Taken Out Through Brush and Slip-Ring Arrangement

In brush and slip-ring arrangement the coil terminals are connected to two conducting slip rings which are fitted on one side of the rotating shaft. The carbon brushes are placed on the slip rings and they are fixed. Connections from the brushes are taken to the load. When the

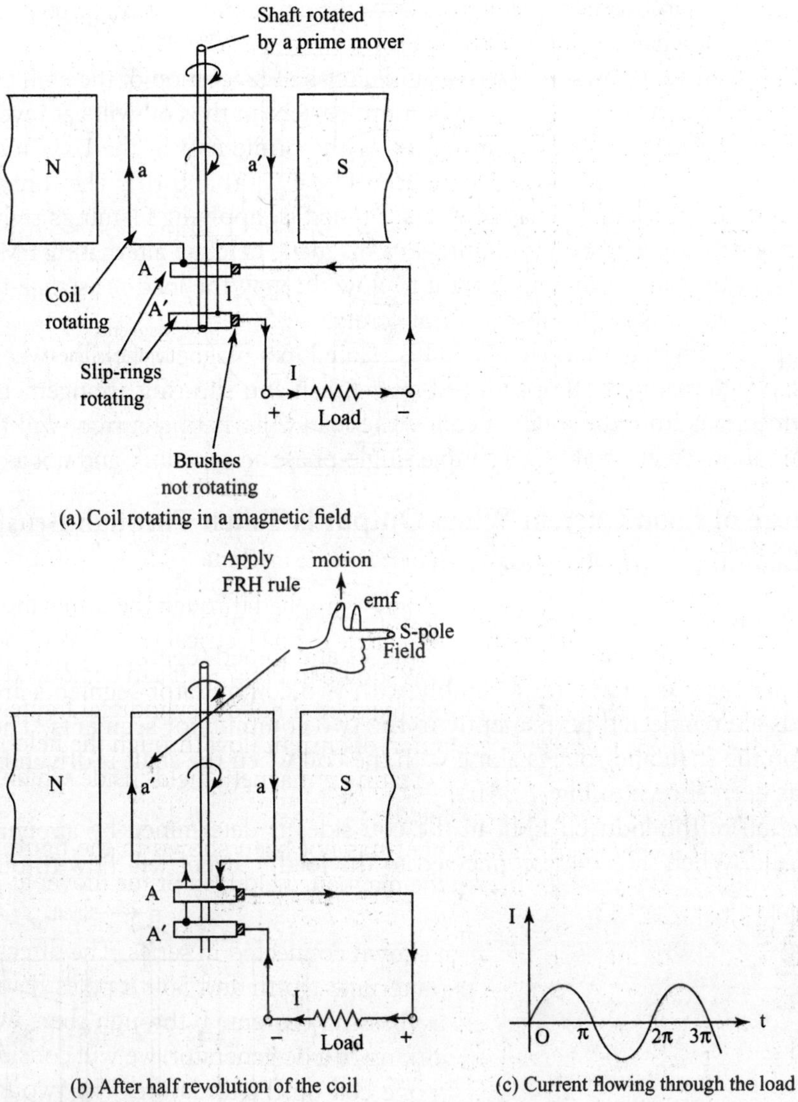

FIGURE 7.2

AC generated in the coil causes ac to flow through the load when connection is through brush and slip-ring arrangement

shaft is rotated by a prime mover, the coil rotates. The slip rings are nothing but extensions of the coil-end connections. The slip rings also rotate as the coil rotates. The brushes sit on the slip rings and make sliping contact with the coil and are able to connect the EMF generated to the load for supply of current.

In Fig. 7.2 (a), the coilside a of coil a–a′ is under North pole and the coilside a′ is under South pole. After half revolution the positions of the coilsides change as shown in Fig. 7.2 (b). The direction of EMF induced in the coilsides have been shown. The EMF induced will cause current to flow through the load resistance as shown. It is noted that the direction of current through the load resistance has changed after half revolution of the coil.

In Fig. 7.2 (a), current flows from a to a′ and after half revolution of the coil, current in the coil flows from a′ to a. After every half revolution, current in the coil will get reversed.

When the armature is rotated continuously by the prime mover, the EMF induced in the coil is alternating in nature as can be seen from Fig. 7.2 (a) and (b). The direction of EMF induced in the coil sides a and a′ has been determined by applying Fleming's right-hand rule. We can conclude that when a coil is rotating in a magnetic field, an alternating EMF is induced in it which will cause an alternating current to flow through the load resistance if connection is made through the brush and slip-ring arrangement.

To Sum up, we can state that when a coil is rotated in a magnetic field between North and South poles, an alternating EMF is induced in it. Brush and slip-ring arrangement is made to take out connections from the rotating coil to the load. Alternating current will flow through the load. This arrangement makes it a simple single-phase ac generator and not a dc generator.

7.1.2 Nature of Load Current When Output Is Taken Through Brush and Commutator Arrangement

Now we will consider a brush and commutator arrangement of connecting the rotating coil to the load. When a slip ring is cut into two pieces and joined together by putting insulation in the joint, a simple commutator assembly with two commutator segments are produced. The coil ends are connected permanently to the two commutator segments. The commutator is fixed on the shaft and rotates along with the coil when the shaft is driven by the prime mover as has been shown in Fig. 7.3 (b).

The direction of the induced EMF in the coil sides is determined by applying Fleming's right-hand rule. When the coil is connected to the load, current will flow through the load

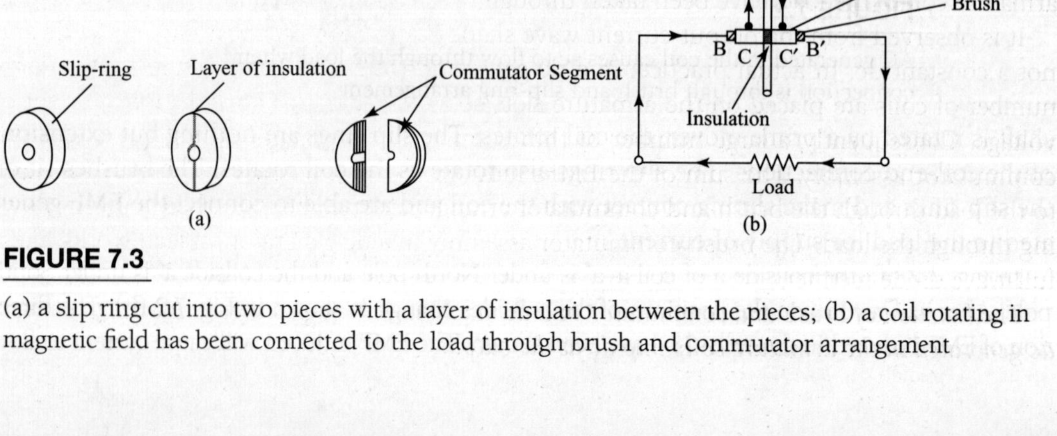

FIGURE 7.3

(a) a slip ring cut into two pieces with a layer of insulation between the pieces; (b) a coil rotating in magnetic field has been connected to the load through brush and commutator arrangement

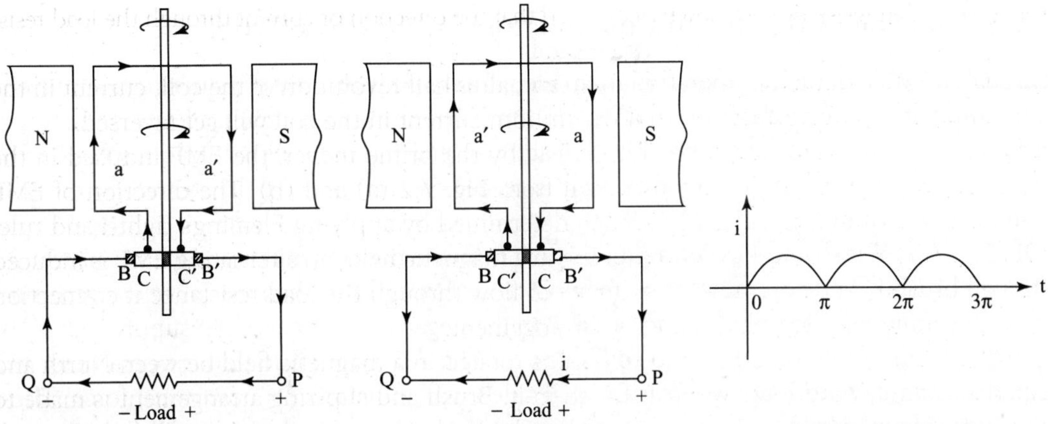

(a) A coil is shown rotated in a magnetic field by a prime mover. The direction of current flowing through the load is from P to Q

(b) After half revolution the position of coil sides along with commutator segments change. The position of brushes are fixed.Load current flows from P to Q

(c) Output wave shape of the load current

FIGURE 7.4

(a) A coil is shown rotated in a magnetic field; (b) after half revolution the position of the coil side along with commutator segments change, the position of brushes remain unchanged; (c) output dc wave shape of the load current

as has been shown. Current from coil side a′ flows to the load through commutator segment C′ and brush B′. From the load, the current returns to the coil through brush B, commutator segment C and then to the coil side a. No current can flow from commutator segment C to C′ or from C′ to C as there exits a layer of insulation between them. With such an arrangement when the coil rotates in the magnetic field, a unidirectional current will flow through the load as has been shown in Fig. 7.4 (a) and (b). In Fig. 7.4 (a), current in the coil flows from coil side a to coil side a′. After half revolution the direction of current is changed from a′ to a. In every half revolution of the coil, this change in direction of current will take place. The current through the load resistance, however, will be unidirectional because the connections from the armature coil to the load have been taken through brush and commutator.

It is observed from the output current wave shape that we are getting a fluctuating dc and not a constant dc. In actual practice, in a dc generator, instead of using a single coil a large number of coils are placed on the armature slots so as to generate a considerable amount of voltage. Consequently, a large number of thin commutator segments are used to make the commutator assembly. The sum of the EMFs induced in the armature coils when connected to the load through the brush and commutator arrangement will be a steady dc current flowing through the load. The brush commutator assembly in a dc generator, in fact, works like a full-wave diode rectifier.

The function of brush and commutator in a dc machine working as a generator is to convert ac generated in the armature coils into dc at the output.

7.1.3 Function of Brush and Commutators in Motoring Action

We shall now examine the operation of an elementary dc motor by considering a single coil (for simplicity) on its armature. In a dc machine, operating as a motor, electrical energy is converted into mechanical energy. Fig. 7.5 shows the armature coil fed from a source of dc supply. The armature having the coil on it is placed in a magnetic field created by the field system as shown in Fig. 7.5 (a).

In Fig. 7.5 (b) is shown how current will flow from the positive polarity of the supply source through brush B′ and commutator segment C′ to coil side a′ and then to coil side a, returning through commutator segment C and brush B to the negative terminal of the supply source. By applying Fleming's left-hand rule to coil-side a′ in Fig. 7.5 (b) we find that the conductor will experience an upward force whereas the coil side a will experience a downward force. These two forces would create a torque to rotate the armature in the anticlockwise direction.

After every half revolution, i.e., for every rotation of 180° mechanical, coil side a′ along with the connected commutator segment C′ will change positions with the coil side a and the connected commutator segment C. After half revolution it is seen that the direction of current in the coil has reversed. Earlier as in Fig. 7.5 (b), current was flowing from a′ to a and after half revolution current is flowing from a to a′ as shown in Fig. 7.5 (c).

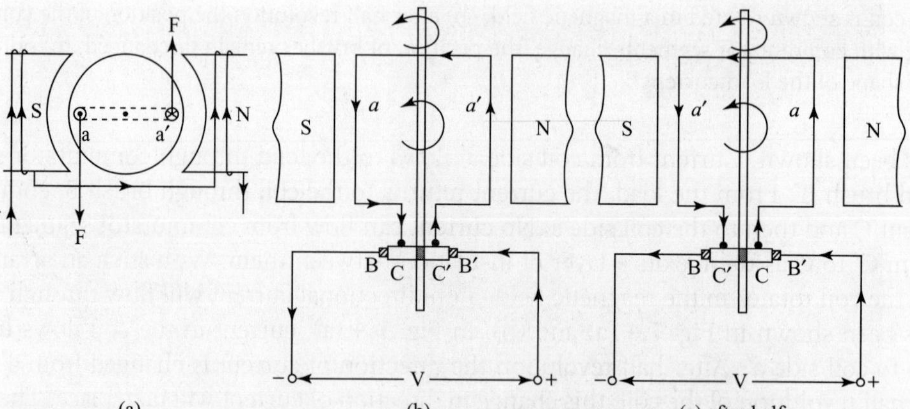

FIGURE 7.5

Illustrates motoring operation. Supply is through brush and commutator arrangement for achieving continuous rotation

Supply polarities remaining fixed, it is seen that current in the armature coil is alternating its direction. However, the direction of the rotation of the coil, as obtained, is unidirectional, i.e., in this case in the anti clockwise direction. The students are advised to check the nature of torque developed in the armature if the supply is given from a dc source but through the brush and slip-ring arrangement. It will be seen that the torque developed will be alternating in every half cycle thereby net torque being zero.

Thus, it can be concluded that *the function of brush and commutator in a dc motor is to produce unidirectional torque, i.e., to cause rotation of the armature of the motor in a particular direction.*

We shall now study the constructional details of a dc machine.

7.2 ► Constructional Details

A photographic view of a dc machine is shown in Fig. 7.6 (a).

A dc machine consists of a field system which produces the magnetic field, the armature which carries the armature conductors placed in slots, the brush and commutator arrangement, the shaft, and the bearings.

These are explained in brief in the following sections.

FIGURE 7.6 (a)

Photographic view of a dc motor-generator set connected to panel board

7.2.1 The Field System

The purpose of the field system is to produce a magnetic field inside which a set of conductors will be rotating. The field system consist of a set of electromagnets fixed on the inside periphery of a hollow cylindrical structure called the yoke as shown in Fig. 7.6 (b). The field poles have field windings wound on a laminated iron core. The number of poles of a dc machine may be two or multiples of two. A dc current supplied from a dc source magnetizes the field system. Alternate North and South poles are formed on the basis of the direction of the current flowing through the field windings. Small poles, called interpoles are often fixed between two main poles, particularly in case of large dc machines.

The field windings are made of thin insulated copper wire of a large number of turns. The resistance of field winding is fairly high of the order of 100 Ω or so. The side view of such a field system has been shown in Fig. 7.6 (b).

7.2.2 The Armature

The armature of a dc machine is built by using circular laminated steel sheets to form a solid cylindrical structure with a shaft passing through its centre.

A simplified cross-sectional view of a dc machine with the armature placed inside the field system has been shown in Fig. 7.7.

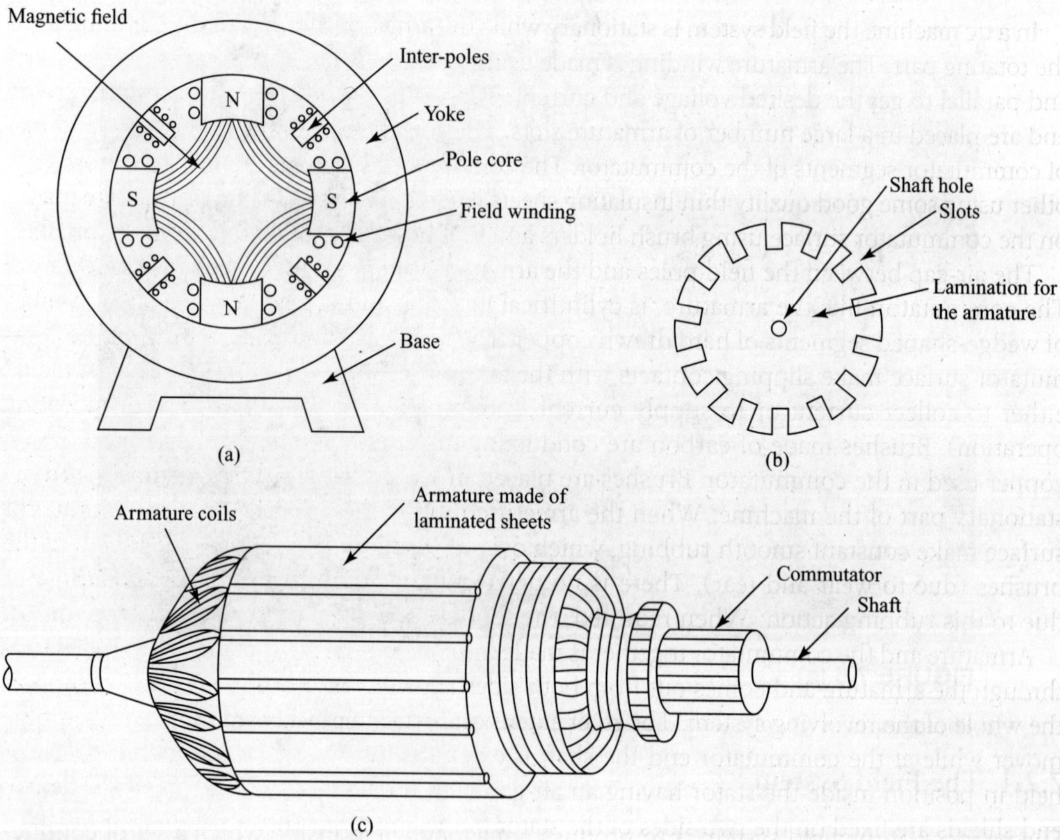

FIGURE 7.6 (b)

(b) Parts of a dc machine: (a) the field system creating a magnetic field when current will flow through the field windings; (b) laminated sheets used to make the cylindrical armature; (c) the armature made of laminated sheets and the commutator assembly

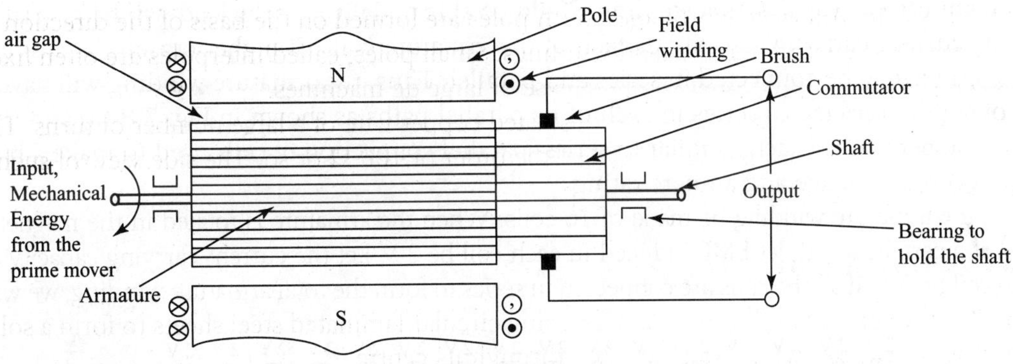

FIGURE 7.7

Simplified diagram of a dc machine working as a generator

In a dc machine the field system is stationary while the armature along with the commutator is the rotating part. The armature winding is made using a large number of coils connected in series and parallel to get the desired voltage and current. The coils are made of insulated copper wires and are placed in a large number of armature slots. The coil ends are connected to large number of commutator segments of the commutator. The commutator segments are insulated from each other using some good quality thin insulating sheets like mica sheets. Carbon brushes are placed on the commutator surface using brush holders and terminals are brought out from the brushes.

The air-gap between the field poles and the armature is kept small, of the order of few mm. The commutator, like the armature, is cylindrical in shape and is made up of a large number of wedge-shaped segments of hard drawn copper. The fixed carbon brushes sitting on the commutator surface make slipping contacts with the armature coils via the commutator segments either to collect current or to supply current from an external source (in case of motoring operation). Brushes made of carbon are conducting material but softer than the hard drawn copper used in the commutator. Brushes are placed in brush holders which are fixed with the stationary part of the machine. When the armature rotates, the brushes and the commutator surface make constant smooth rubbing, which over a period of time reduces the length of the brushes (due to wear and tear). There is no deterioration of the surface of the commutator due to this rubbing action. When required, the set of brushes can be replaced by new ones.

Armature and the commutator together is made into one unit. A shaft made of mild steel runs through the armature and comes out from both sides. Two sets of bearings are used to support the whole of the revolving system. The shaft extension on one end is used to connect the prime mover while at the commutator end the shaft is extended for use of the bearing. The shaft is held in position inside the stator having an air-gap with the help of two end shields. The two end shields are fixed on the two sides of the stator frame and holds the armature in position.

7.2.3 Armature Winding

When the armature coils rotate in the magnetic field, EMF is induced in each coil. If all the coils are connected in series, the total EMF available will be the sum of all the EMFs in all the coils. But the current that this winding would be able to deliver to the load will be governed by the current-carrying capacity of each of the armature coils. If higher current to desired, the coils have to be connected in series and parallel. Thus, the armature winding will have a set of coils connected in series in each of its parallel paths as shown in Fig. 7.8 (a) and (b). The arrangement is exactly similar to series–parallel connection of cells used to make a battery of certain voltage and ampere rating.

Let the armature winding be made of 16 coils. When the armature is rotated in the magnetic field at a certain speed, let EMF induced in each coil be 2 V. Let the current-carrying capacity of each coil be 3 A. If all the coils are connected in series to form the total armature winding, we will

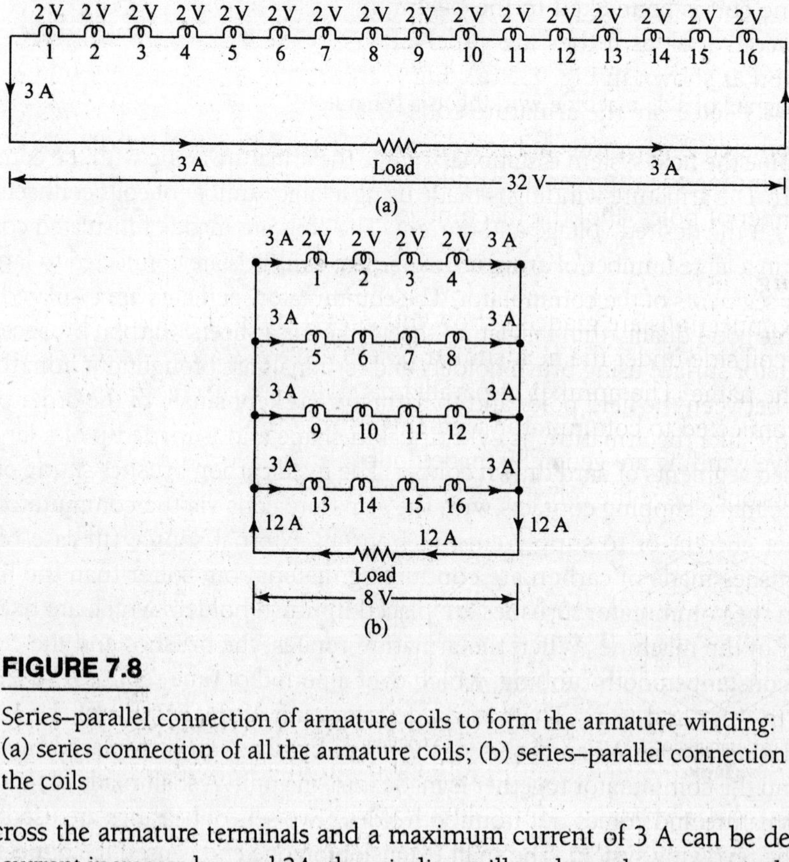

FIGURE 7.8

Series–parallel connection of armature coils to form the armature winding: (a) series connection of all the armature coils; (b) series–parallel connection of the coils

get 32 V across the armature terminals and a maximum current of 3 A can be delivered to the load. If the current increases beyond 3A, the winding will get heated up excessively, which is not desirable. If four coils are connected in series, there will be four series circuits which will induce an EMF of 8 V in each circuit. If the four series circuits are connected in parallel to supply current to the load, a maximum of 12 A can be supplied to the load although the rated current of 3A will flow through each coil as shown in Fig. 7.8 (a) and (b). Here the parallel paths of the armature winding is four. However, depending on the voltage and current requirements, the number of parallel paths could be different, say 2, 4, 6, etc. The output power delivered to the load will be the same. In first case, P = 32V × 3A = 96 W and is the second case, p = 8 × 12 = 96 W.

7.2.4 Types of Armature Winding

All the coils placed in armature slots are connected together in a particular manner to form the armature winding. Two basic types of winding connections are made. They are (i) lap winding and (ii) wave winding. In all cases, the coil ends are connected to commutator segments. The commutator segments are, in fact, extension of coil end connections. Brushes placed on commutator touching the commutator segments make sliping contact with the coils. The two types of armature windings are explained below.

(a) Lap winding

Here, the end of one coil is connected to the beginning of the next coil. If connections are made this way the coils look as if they are super imposed on each other and then given a push in one direction as shown in Fig. 7.9 (a). Fig. 7.9 (b) shows a wave winding.

1 – 1′, 2 – 2′, 3 – 3′, etc., are the armature coils. In lap winding coil side 1′ is connected with coil side 2, coil side 2′ is connected to coil side 3 and this way the whole winding is completed. In lap winding the number of parallel paths, A formed in the armature winding is equal to the number of poles, P of the machine is A = P for lap winding.

(b) Wave winding

Here the winding connections are made as show in Fig. 7.9 (b). The end of one coil is connected to the next coil side under the next similar poles. The winding so formed looks like a wave, and hence the name. The number of parallel paths formed is always equal to 2. Note that the coils are connected to commutator segments in both types of windings.

Both lap and wave winding are generally made double layer winding placing the coil sides inside armature slots one above the other.

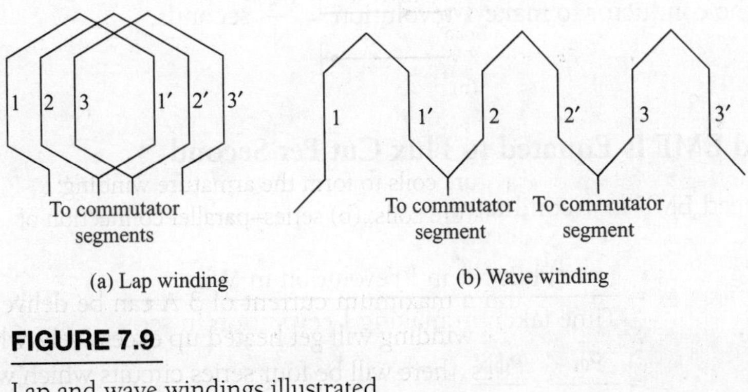

FIGURE 7.9

Lap and wave windings illustrated

7.3 ► EMF EQUATION OF A DC MACHINE

A large number of conductors placed in armature slots when rotated in a magnetic field produced induced EMF.

The total induced EMF in a dc machine, as can be visualized should depend on the speed at which the conductors are rotated in the magnetic field, the total number of conductors connected in series, and the total magnetic field flux being cut by the conductors. The exact equation in now being developed as under.

Fig. 7.10 shows a single conductor rotated by a prime mover in a magnetic field.
Let us assume the following:
No. of poles = P.
Flux per pole = ϕ Wb.
Speed of the driving prime mover = N rpm.
Actual number of armature conductors = Z.
Number of parallel paths of the armature winding = A.
Induced EMF in the conductors will be due to relative velocity between the conductor and the flux produced by the field poles.

When the conductor in Fig. 7.10 makes 1 revolution, the flux cut by the conductor = Pϕ Wbs.

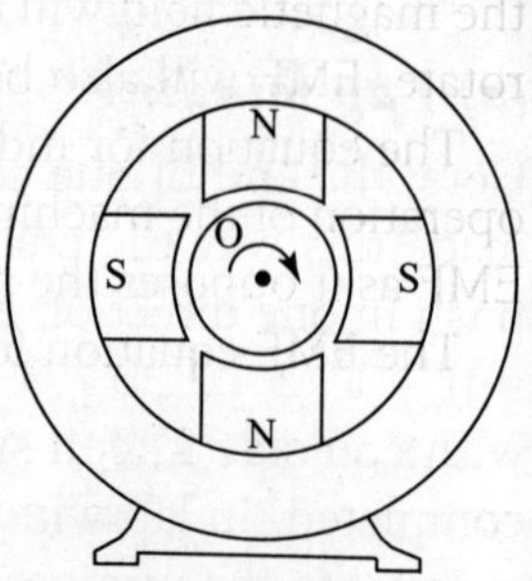

FIGURE 7.10
Development of equation for EMF induced

(The students will appreciate that if there were two poles and flux per pole was ϕ Wbs, the conductor would cut 2ϕ flux, for a four-pole system, the flux cut per revolution would be 4 ϕ Wbs, and so on). Thus, if ϕ is the flux per pole, and P is the number of poles, flux cut by a conductor in 1 revolution will be Pϕ Wbs.

The conductor is rotating at a speed of N rpm or $\dfrac{N}{60}$ rps (revolutions per second)

Time taken by the conductor to make 1 revolution = $\dfrac{60}{N}$ seconds.

7.3.1 Induced EMF Is Equated to Flux Cut Per Second

Therefore, induced EMF in 1 conductor

$$= \frac{\text{Flux cut in 1 revolution in Wbs}}{\text{Time taken in making 1 revolution in secs}}$$

$$= \frac{P\phi}{60/N} = \frac{P\phi N}{60} \text{ V}$$

Z is the total number of armature conductors and they are connected in A number of parallel paths. The total number of conductors per parallel path would be equal to $\dfrac{Z}{A}$.

The induced EMF available across the output terminals will be equal to the induced EMF per parallel path.

Thus, the total induced EMF, E is

$$E = \frac{P\phi N}{60}\left(\frac{Z}{A}\right) V$$

or,

$$E = \frac{\phi ZNP}{60A} V \tag{7.1}$$

When the dc machine is also working as a motor, the current-carrying conductors placed in the magnetic field will develop torque and rotate in a particular direction. When they would rotate, EMF will also be induced in them.

The equation for induced EMF will be the same both in generating and motoring mode of operation of the machine. The induced EMF in the armature of a dc motor is often called back EMF as it opposes the applied voltage.

The EMF equation for a dc machine can also be expressed as

$$E = \left(\frac{ZP}{60\ A}\right)\phi N \quad V$$

or,

$$E = k\,\phi\,N \quad V$$

Where $K = \frac{ZP}{60A}$ a machine constant.

For a particular dc machine induced EMF will depend on magnetic field strength as indicated by flux per pole, ϕ and the speed of rotation N of conductors in the magnetic field.

7.4 ► Types of DC Machines

In the dc generator shown in Fig. 7.11, the armature has to be rotated by some prime mover and the field windings have to be excited by giving a dc supply to them. We, therefore, would need a separate dc source of supply for the field winding. Such a generator where the field winding is supplied from a separate dc source for its excitation is called a ***separately excited dc generator***.

EMF E will be induced in the armature when it is rotated by a prime mover and the field windings are excited from a separate dc source.

Now, let us consider what would happen when the armature is rotated by a prime mover without exciting the field windings, i.e., when field winding current, $I_f = 0$.

Even when the field windings are not excited, there is some residual magnetizm in the field poles due to their earlier excitation.

A feeble magnetic field will, therefore, exist due to which a very small amount of voltage will be induced in the armature winding when rotated. Let us assume that this induced EMF is 5 V. If the field winding having a resistance of say 100 Ω is connected across the armature, a small amount of current, 5 V/100 Ω = .05 A will flow through the field windings, which will produce some more flux, and as a consequence more EMF will be induced in the armature. This way voltage will be built up across the armature terminals when the field windings are connected in parallel with the armature as shown in Fig. 7.12 (a). Such a generator is called a ***shunt generator*** or a ***self-excited generator***.

Fig. 7.12 (b) shows the connection diagram of field and armature windings for a *series generator*. The field current I_f is equal to I_a, i.e., a very high current will now flow through the field windings. For a series generator, therefore, field windings are made of thick wires of a few turns to provide the required ampere turns needed for production of magnetic field of a particular strength.

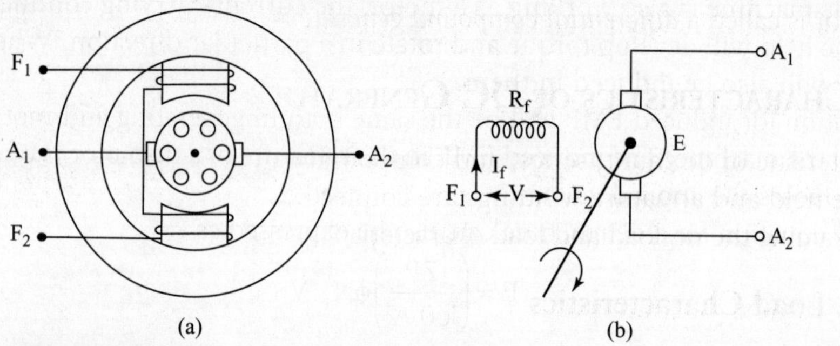

FIGURE 7.11

(a) A dc generator with field and armature winding terminals brought out; (b) the field windings of the generator are excited from a dc supply source

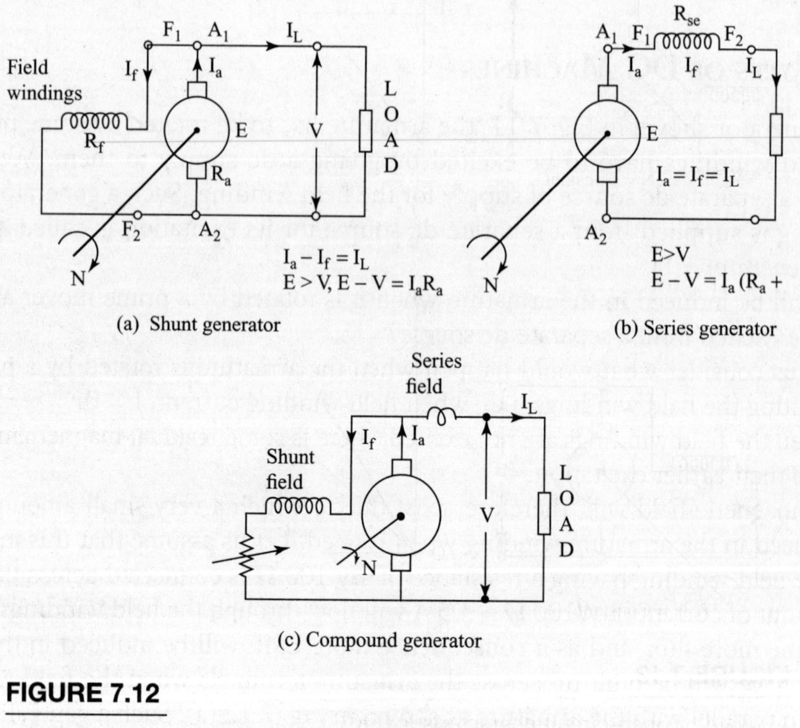

FIGURE 7.12

(a) Shunt generator; (b) series generator; (c) compound generator

A *compound generator* will have both shunt field winding and series field winding. Both the field windings are wound around the pole core. Shunt field winding is connected in parallel with the armature while the series field winding is connected in series with the armature. The resultant field produced will be equal to the sum of the field produced by the two field ampere turns. Such a generator is called a *cumulative compound generator.*

When the flux produced by the series field opposes the flux produced by the shunt field, the generator is called a *differential compound generator.*

7.5 ► Characteristics of DC Generators

The characteristic of dc shunt, series, and compound generators will be different because of the way the field and armature windings are connected.

Let us examine the no-load and load characteristics of dc generators.

7.5.1 No-Load Characteristics

Fig. 7.13 (a) shows a separately excited dc generator. The generator armature is rotated at a constant speed by a prime mover. When no field current is there, a small amount of EMF will be induced due to residual magnetism. By increasing the field current, I_f gradually, we will get the no-load or open-circuit characteristic as shown in Fig. 7.13 (a).

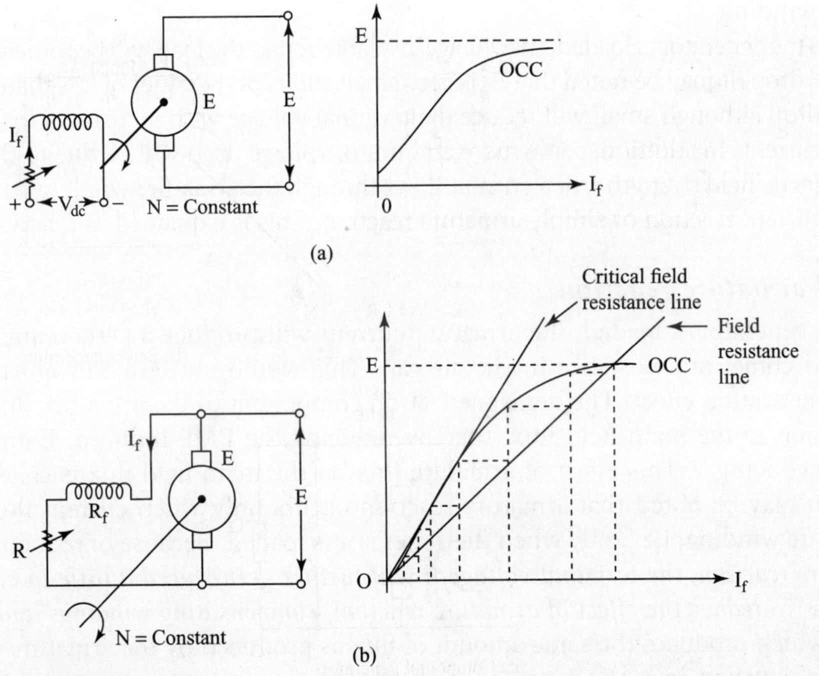

FIGURE 7.13

No-load or open-circuit characteristic (OCC) of dc generators: (a) separately excited generator; (b) shunt generator

In the case of a shunt generator, the voltage gets built up due to residual magnetism as has been shown in Fig. 7.13 (b). The induced voltage will be the value at which the OCC and field resistance line cross each other. By adjusting the value of the field-circuit resistance, i.e., by adding an extra resistance in the field circuit, the value of E can be adjusted. The speed of the prime mover is assumed constant. A self-excited dc shunt generator will fail to build up its voltage if there is no residual magnetism in the field poles and if the value of the field resistance in higher than the critical field resistance. The value of the critical field resistance can be determined by drawing a line tangent to the OCC and finding its slope. The students should notice that the OCC is initially linear, but later becomes somewhat horizontal. This shows that with increase of field current, I_f the induced EMF increases linearly but later the core saturates. Further increase of I_f does not give rise to much increase in induced EMF.

7.5.2 Load Characteristics

The load characteristic is drawn between the voltage available across the output terminals, V when the generator is loaded, against the load current I_L. The equation relating E, V, and I_a for a dc generator is given as

$$V = E - I_a R_a \tag{7.2}$$

Where R_a is the armature winding resistance and I_a is the current flowing through the armature winding.

Thus, as the generator is loaded, the voltage available across the load will be somewhat reduced due to $I_a R_a$ drop. It may be noted that R_a is very small and is of the order of less than 1 Ω. Therefore, $I_a R_a$ drop although small will reduce the terminal voltage with increase of load current or armature current. In addition, some more amount of voltage drop will be due to the reduction of the magnetic field strength when current flows through the armature winding which is called armature current reaction or simply armature reaction. This is explained as follows.

Effect of armature reaction

When the generator is loaded, the armature current will produce a certain amount of flux that would come into existance in the air gap. This would create a demagnetization and cross-magnetization effect. The demagnetization component of the armature flux will work in opposition to the main field flux, thereby reducing the EMF induced, E and as a consequence reducing V. This effect of armature flux on the main field flux is called *armature reaction*. It may be noted that armature reaction occurs only when current flows through the armature winding, i.e., only when the generator is loaded. Because of reduction of E due to *armature reaction, the terminal voltage E will further get reduced a little in case of shunt* generators. To ***reduce the effect of armature reaction, compensating windings and inter poles are used,*** which produces the same amount of flux as produced by the armature current but in opposite direction so as to eliminate the effect of armature flux on the main field flux. The load characteristics of dc generators have been shown in Fig. 7.14. The voltage drop from E to V is very small as compared to E. The shunt generators can be considered as constant voltage output generators for practical purposes.

In case of series generators, the field current is the same as the load current. Initially there is some induced EMF due to residual magnetizm of the field poles. As the generator is loaded, current through the field increases and E and V increase and later saturation effect takes place. Further loading increases the armature reaction effect and voltage starts falling.

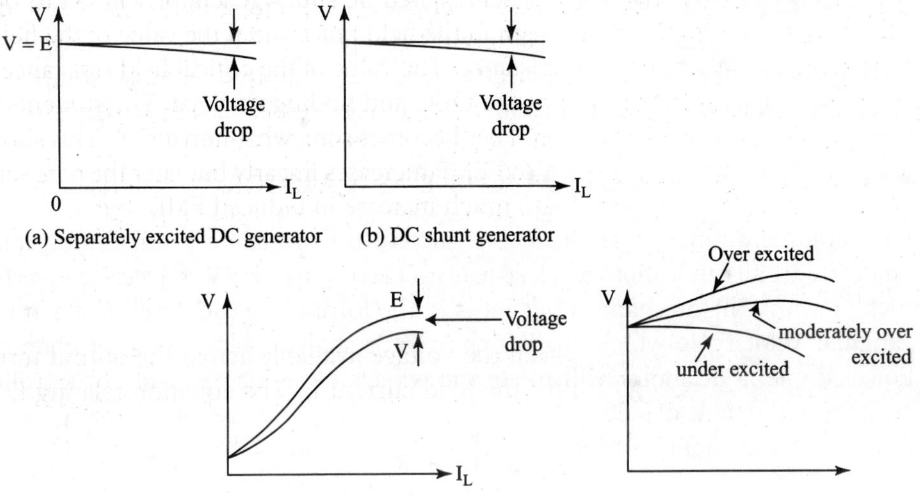

FIGURE 7.14

Load characteristics of dc generators: (a) separately excited dc generator; (b) dc shunt generator; (c) dc series generator (d) dc compound generator

7.6 ► Applications of DC Generators

Applications of different types of dc generators are mentioned below.

1. *Separately excited dc generators*: In separately excited dc generators, by controlling the field current, the output voltage can be varied. Such a generator can be used as a supply source to dc motor whose speed is to be controlled. Such a generator can be used in the laboratories for testing of equipment and instrument or for performing experiments where a variable dc supply is required.
2. *DC shunt generators*: The load characteristic of a dc shunt generator shows that there is small amount of voltage drop when the generator is loaded. For practical purpose, a shunt generator may be considered as a constant output voltage generator. Shunt generators are used in applications like battery charging, electroplating, and to supply dc voltage to the field winding of alternators. Small dc shunt generators are used as exciters for alternators (ac generators).
3. *DC Series generators*: We have seen that terminal voltage of a dc series generator increases with increase of load. Dc series generators can be used to boost up voltage of an existing system where voltage drop takes place due to loading. Such a series generators is called a voltage booster. Series generators are also used as welding generator and in arc lamps.

4. *DC Compound generators*: Compound generators have both shunt field and series field windings made on their poles. The flux produced by the series field either aids the shunt field or opposes the shunt field. According, compound generators are called cummulative compound generator or differential compound generators. In cummulative compound generators, the terminal voltage increases with increase in load similar to dc series generators. Hence, cummulative compound generators can be used as voltage booster in existing system like in feeders of electric supply system. In differential compound generator, the voltage drop due to loading high and are suitable for dc welding machines.

7.7 ► Operation of a DC Machine as a Motor

Like dc generators, dc motors are also constructed to work as dc shunt, series, and compound motors. In all such motors, electrical power input, i.e., V × I gets converted into mechanical output. The mechanical output is in the form of torque developed T multiplied by the angular velocity, ω which enables the motor shaft to carry some mechanical load on it. For example, a dc motor will rotate the wheels of an electric train or a trolley bus. The constructional details of a dc motor is similar to that of a dc generator except for certain minor changes in the cooling system.

7.7.1 Working Principle of a DC Motor

A dc motor works on the basic principle that when a current-carrying conductor is placed in a magnetic field it experiences a force. In a dc motor, the armature carries a number of conductors placed in slots and the armature is placed inside the magnetic field created by field magnets as shown in Fig. 7.15. The field magnets are excited by field current passing through them. The field windings may either be excited from a separate dc source or could be excited from the source of supply provided to the armature. The armature is fed from a dc supply source of V Volts as shown. The direction of current flowing through the winding have been shown by cross ⊗ and dot ⊙ A cross indicates current flowing in the direction perpendicular to the plane of the paper downwards and a dot indicates that current is coming towords the observer looking into the paper.

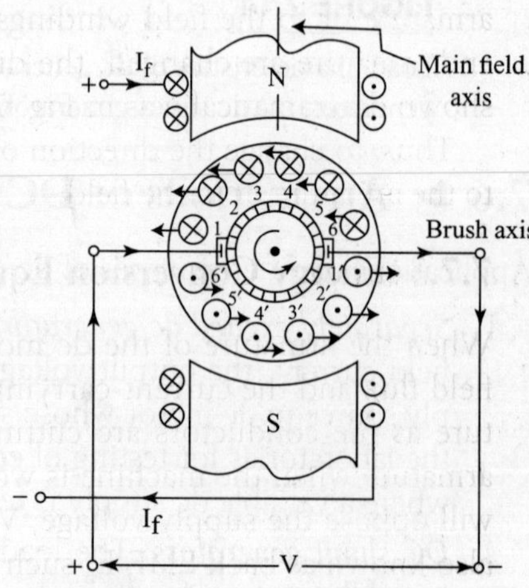

FIGURE 7.15 Working principle of a dc motor illustrated

There are six coils 1–1′, 2–2′, 3–3′, 4–4′, 5–5′, and 6–6′ placed in armature slots. Current in coil sides 1, 2, 3, etc., are shown by cross and current at the other coil sides, i.e., 1′, 2′, 3′, etc., are shown by dots. The flow of current through the coils has been shown diagrammatically as in Fig. 7.16. By applying Fleming's left-hand rule, it is seen that force developed on the upper half of the armature conductors in Fig. 7.15 is from right to left and for the lower half of the armature conductors it is from left to right. These forces lead to development of torque which causes rotation of the armature in the anticlockwise direction as shown in Fig. 7.15. It may be noted carefully from the figure that as the armature rotates in the anticlockwise direction, current in the conductors of a coil passing under the brushes would change. This will ensure that at any point of time any conductor on the upper half of the armature will have cross current and on the lower half will have dot current. There will be continuous rotation of the armature in one direction. ***The reversal of current in the coils passing under the brushes is called commutation.***

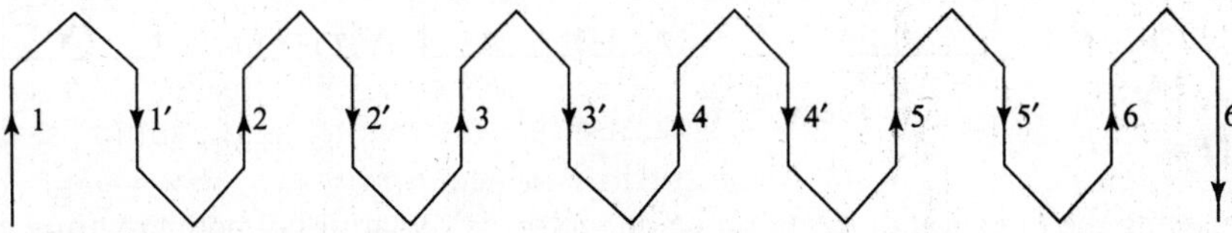

FIGURE 7.16

Coils 1–1′, 2–2′, 3–3′, 4–4′, 5–5′, and 6–6′ of the armature are shown connected in series

7.7.2 Changing the Direction of Rotation

The direction of rotation of the armature will change if the polarities of supply to either the armature or to the field windings are reversed. If the polarities of supply to both the field and armature are changed, the direction of rotation will remain unchanged. This has been shown diagramatically as in Fig. 7.17.

Thus, to change the direction of rotation, we have either to change the polarities of supply to the armature or to the field.

7.7.3 Energy Conversion Equation

When the armature of the dc motor starts rotating because of the interaction between the field flux and the current-carrying armature conductors, EMF will be induced in the armature as the conductors are cutting the field flux. This EMF and the EMF induced in the armature when the machine is working as a generator is the same. However, now this EMF will oppose the supply voltage, V. Since the EMF induced opposes the supply voltage, it is also known as back EMF, E_b such that

$$V - E_b = I_a R_a \tag{i}$$

or,

$$E_b = V - I_a R_a \tag{ii}$$

or,

$$I_a = \frac{V - E_b}{R_a} \tag{7.3}$$

and from (i)

$$VI_a = E_b I_a + I_a^2 R_a \tag{iii}$$

From equation (iii) we can write for a dc motor,

(a) Electrical Input power = Electrical equivalent of mechanical power developed + Armature copper loss

(b) Electrical equivalent of mechanical power developed – Rotationed losses = Mechanical power output

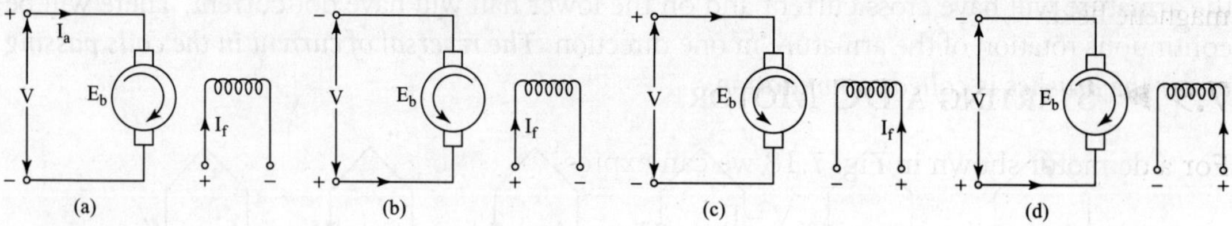

FIGURE 7.17

Method of changing the direction of rotation of a dc motor: (a) polarities of both armature and field positive, leading to clockwise rotation; (b) polarities of armature changed, leading to anticlockwise rotation; (c) polarities of supply to field changed, leading to anticlockwise rotation; (d) polarities of both armature and field reversed, no change in direction of rotation

The power flow in a dc motor is shown below.

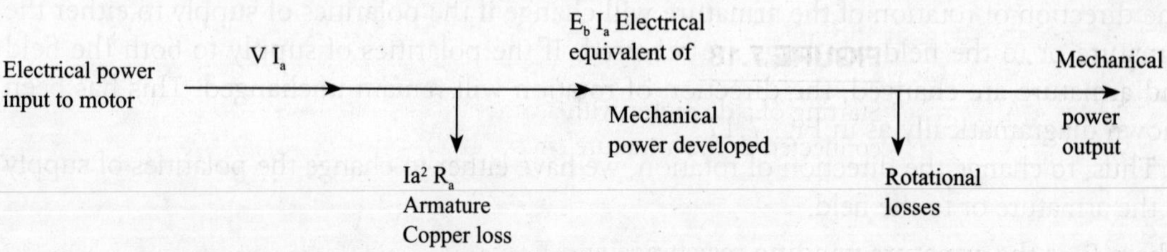

7.8 ► Torque Equation

Torque developed T, angular velocity ω, and mechanical power developed, P are related as

$$P = T\omega$$

where T is in Nm and ω is in rad/sec.

In a dc motor, electrical power converted into mechanical power can be expressed as

$$E_b I_a = T\omega = T\frac{2\pi N}{60}$$

or,

$$T = \frac{60}{2\pi N} E_b I_a$$

$$= \frac{60}{2\pi N}\frac{\phi ZNP}{60A} I_a = \left(\frac{ZP}{2\pi A}\right)\phi I_a$$

Torque, $$T = K_t \phi I_a \tag{7.4}$$

where $$K_t = \frac{ZP}{2\pi A}$$

Thus, we can say that torque developed is proportional to the magnetic field strength or the magnetic flux ϕ and the magnitude of current I_a flowing through the conductors placed in the magnetic field.

7.9 ► Starting a DC Motor

For a dc motor shown in Fig. 7.18 we can express armature current as (see eq. 7.3)

$$I_a = \frac{V - E_b}{R_a} \quad \text{and} \quad E_b = E = \frac{\phi ZNP}{60A}$$

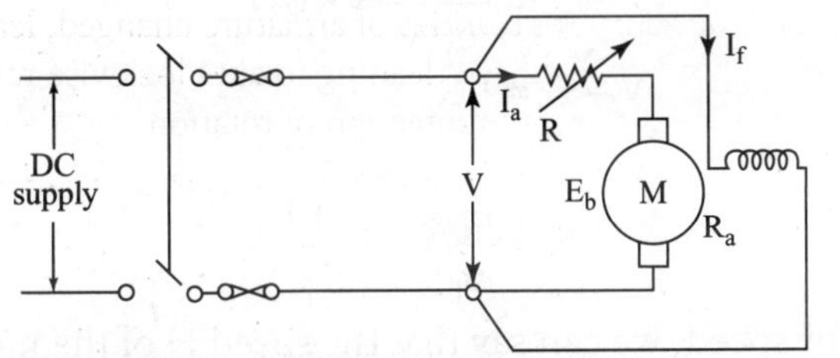

FIGURE 7.18

Starting of a dc motor with a variable resistance connected in the armature circuit

where R_a is the armature winding resistance and E_b is the back EMF or induced EMF in the armature.

At the moment of start, the speed N of the motor is zero. If N = 0, $E_b = 0$.

Thus, $$I_a = \frac{V}{R_a}$$

The value of armature resistance for a dc motor is very small. Let us assume that V = 220 V and $R_a = 0.5\ \Omega$. Then at start,

$$I_a = \frac{220}{0.5} = 440\,\text{A}$$

This is a huge current to be allowed to flow through the armature in a small dc motor. To restrict this high amount of current to flow through the armature, a variable resistance can be connected in series with the armature so that eq. (7.3) gets modified as

$$I_a = \frac{V - E_b}{R_a + R} \tag{7.5}$$

Once the motor starts rotating, back EMF E_b starts increasing and the numerator of the expression for I_a as in eq. (7.5) gets reduced. As the numerator goes on reducing, the denominator of expression (7.5) can be gradually reduced by reducing the value of variable resistance, R. This resistance is completely cut out of the circuit, once the motor picks up sufficient speed.

This variable resistance connected in the armature circuit is called a starter. ***Thus, a starter is a variable resistance connected in series with the armature circuit to limit the initial current drawn by the motor.*** Once the motor picks up speed, back EMF E_b comes into full existence, and automatically the armature current gets reduced even when the extra resistance is cut out. Removing the extra variable resistance from the circuit when it is not required is essential to avoid unnecessary wastage of energy as I^2R loss.

7.10 ► Speed Control of DC Motors

The basic equations for a dc motor are

$$E_b = \frac{\phi ZNP}{60A} = K\phi N \tag{7.6}$$

$$V - E_b = I_a R_a \tag{7.7}$$

from (7.6) and (7.7),

$$N = \frac{E_b}{K\phi} = \frac{V - I_a R_a}{K\phi} \tag{7.8}$$

From this expression for speed, we can say that the speed N of the motor can be changed by any one or a combination of the following three methods.

7.10.1 Voltage Control Method

By changing the supply voltage V, the speed can be changed. As supply voltage can only be reduced, speed N can also be reduced from its rated value by this method.

7.10.2 Field Control Method

By varying the flux ϕ, speed can be changed. The field current produces the flux ϕ, and hence, this method is called the field control method. By putting a variable resistance in the field circuit, field current can be reduced, and hence, flux produced can be reduced. When flux is reduced, speed is increased. Thus, by the field control method the speed of the motor can only be increased. See eq. (7.8).

By a combination of voltage control and field control methods, the speed of the motor can be increased and also decreased above and below its normal speed.

7.10.3 Armature Control Method

By putting an extra variable resistance in the armature circuit, the speed can be reduced as

$$N = \frac{V - I_a(R_a + R^{\nearrow})}{K\phi} \quad \text{where } R^{\nearrow} \text{ is a variable resistance}$$

It can be seen here that the speed can only be reduced because the numerator will get reduced.

7.11 ► Starter for a DC Motor

We have mentioned earlier that a starter is a variable resistance connected in series with the armature circuit during starting to reduce the starting current. This resistance is gradually cut out as the motor starts running. Two types of starters, namely a three-point starter and a four-point starter are described below.

7.11.1 Three-Point Starter

A three-point starter circuit is described as follows. To start the motor, the starter arm, as shown in Fig. 7.19 (a) is moved in the clockwise direction. The arm will touch the point 1 of the starting resistance R. The whole of the resistance will appear in the armature circuit. The field winding will also get full supply through the coil of the NVR. The motor will develop torque and start rotating with full starting resistance in the armature circuit. The resistance will be cut in succession by moving the starter arm in the clockwise direction and will be brought

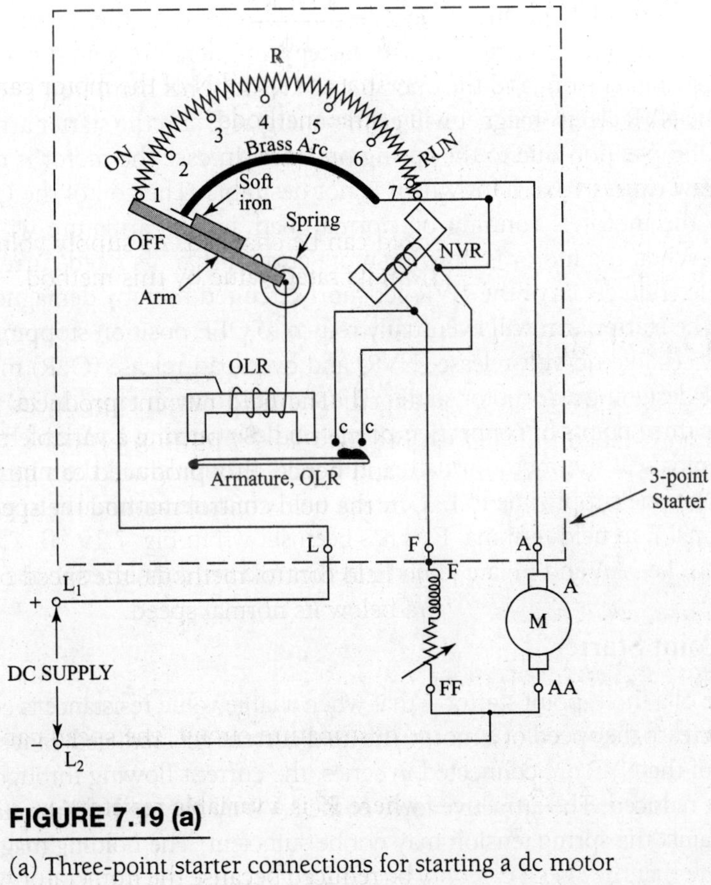

FIGURE 7.19 (a)

(a) Three-point starter connections for starting a dc motor

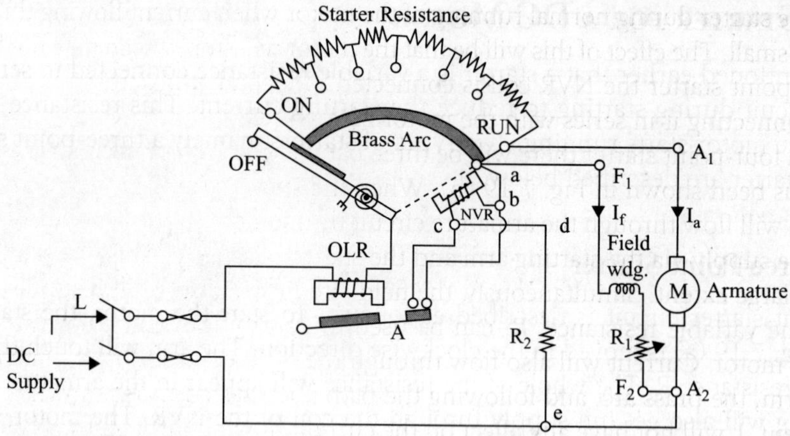

FIGURE 7.19 (b)

(b) A four-point starter for a dc motor

to RUN position. In the RUN position, the soft iron piece fixed on the starter arm will face the NVR magnet piece and remain attracted. The starter arm, therefore, will stay in the RUN position against the spring tension, and the operator can remove his hand from the arm. In case of supply failure, the NVR electromagnet will get de-energized, and the starter arm will automatically return to OFF position due to the spring pressure. In case the motor is over loaded, the armature will draw excessive current which is not desirable. The coil of the OLR will remain energized when the motor is running on normal load and its armature will remain in the position shown. When the motor is drawing more current than its rated current the armature gets lifted and the contacts cc of the NVR get short-circuited thereby demagnetizing the NVR electromagnet. The starter arm will eventually return to OFF position stopping the motor.

Protecting devices like no-volt-release (NVR) and over-load-release (OLR) mechanisms have been added while designing a dc motor starter. The starter shown in Fig. 7.19 (a) is called a three-point starter. The three points or connection points are designated as L, A, and F. The connection of the starter terminals to the motor armature and field terminals and the supply terminals are as follows. Connect one supply line to L. Connect one armature terminal A to point A and connect starter terminal F to field terminal F as has been shown in Fig. 7.19 (a). The other ends of armature and field, i.e., AA and FF are joined together and are connected to the supply line L_2.

7.11.2 Four-Point Starter

The disadvantage of a three-point starter is that when a large value resistance is connected in the field circuit to increase the speed of a motor, the field current gets reduced. Since the field winding and the coil of the NVR are connected in series, the current flowing through the coil of the NVR will also get reduced. The attractive force of the NVR magnet to hold the starter arm in the RUN position against the spring tension may not be sufficient. The holding magnet may release the arm of the starter during normal running of the motor when current flowing through its coil becomes too small. The effect of this will be that the motor will stop, which may not be desirable.

In a four-point starter the NVR coil is connected independently across the supply voltage instead of connecting it in series with the motor field winding.

Thus, in a four-point starter there will be three parallel circuits connected across the supply voltage as has been shown in Fig. 7.19 (b). When the starter arm is brought to the ON position, current will flow through the armature circuit through the starter resistance. Current will flow from the supply via the starting arm and the starter resistance. This will limit the starting current to a large extent. Simultaneously, the field circuit will also get full supply through the brass arc. The variable resistance, R_1 can be used to control the field current to control the speed of the motor. Current will also flow through the NVR coil as supply will come through the starter arm, the brass arc, and following the path abcde as has been shown. Thus, change in field current, I_f will not have any effect on the current flowing through the NVR coil circuit. When the starter arm is brought to the RUN position, the armature will attain full speed and remain connected to the supply via the starter arm. The field circuit and the NVR circuit will get full voltage independently. The NVR will keep the starter arm in the RUN position even after the hand is released. The spring tension cannot bring back the arm to OFF position because of the attractive force of the NVR. In case of overload, the NVR terminals will be short circuited due to the attractive force of the OLR and the starter arm will get released

7.12 ► Types and Characteristics of DC Motors

DC machines are available as shunt, series, and compound machines. In motoring mode of operation, they are called dc shunt motors, dc series motors, and dc compound motors. The relationship between three variables, namely torque, speed, and load (load current) are studied to find the suitability of each type of motor for different applications. For example, if a mechanical load has to be rotated at a constant speed, we would need a motor as a drive whose speed will remain constant at all loads, i.e., there should not be any variation of its speed from no-load condition upto full-load condition. Again, if a set of dc motors are to drive an electric train, the starting torque developed by the motors should be very high. The motors have to develop sufficient torque so as to start the train from rest condition with a large number of passengers and other loads inside the train.

The characteristics of all types of motors are drawn as follows.

7.12.1 Characteristics of DC Shunt Motors

The basic equations of a dc motor are

$$E_b = K\phi N$$

$$E_b = V - I_a R_a$$

$$T = K\phi I_a$$

$$N = \frac{V - I_a R_a}{K\phi}$$

For a shunt motor flux produced by the field current is directly proportional to the field current. Hence,

$$\phi \propto I_f$$

From Fig. 7.20 (a), $$I_f = \frac{V}{R_f}$$

As can be seen from Fig. 7.20 (a), when V and R_f are constant, I_f will be constant.

Since I_f is constant, ϕ is constant. Then

$$T = K\phi I_a = K_1 I_a \quad \text{where } K\phi = K_1$$

$$\therefore \quad T \propto I_a$$

Torque T versus current I_a characteristic has been shown in Fig. 7.20 (bi).

$$N = \frac{V - I_a R_a}{K_2 I_f} \quad \text{where } \phi = \frac{K_1}{K} = K_2$$

Since $I_a R_a$ is very small as compared to V, change of $(V - I_a R_a)$ with change of I_a will be very small.

Thus, N will remain constant as I_a, which is proportional to load, increases as has been shown in Fig. 7.20 (bi).

By knowing the variation of T and N against the load current, I_a, the relationship of T versus N can be drawn as shown in Fig. 7.20 (bii).

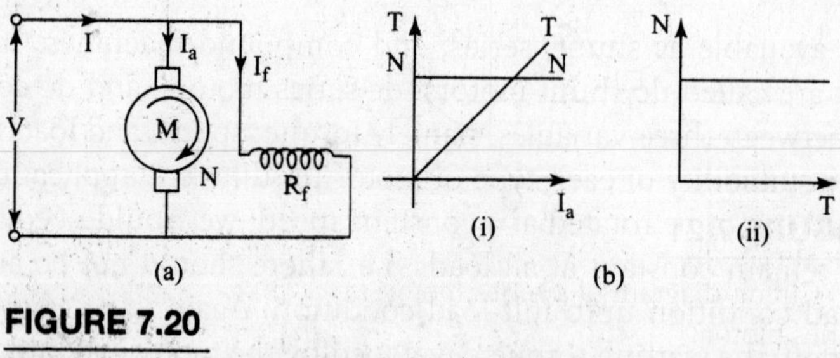

FIGURE 7.20

(a) DC shunt motor; (b) characteristics of a dc shunt motor

It can be seen that dc shunt motors are approximately constant speed motors and can be used in applications like lathe machines, drilling machines, milling machines, in printing press, paper mills, etc.

7.12.2 Characteristics of DC Series Motors

In a dc series motor the field winding is connected in series with the armature so that same current flows through the field and armature windings. The flux produced, ϕ is proportional to the field current which is equal to I_a. The relevant equations are written as

$$N = \frac{V - I_a R_a}{K\phi}$$

$$\phi \propto I_f \quad \text{and} \quad I_f = I_a$$

Therefore,

$$\phi \propto I_a$$
$$T \propto \phi I_a$$
$$T \propto I_a^2$$

Since $I_a R_a$ is very small as compared to V, $V - I_a R_a$ = constant, and $\phi \propto I_a$

Therefore, from
$$N = \frac{V - I_a R_a}{K\phi}$$

we can write
$$N \alpha \frac{1}{I_a}$$

i.e.,
$$N \times I_a = \text{constant}$$

Thus, the relations of N versus I_a and T versus I_a are drawn as shown in Fig. 7.21 (b). The relation between N and T is also drawn. From N versus T characteristics it is seen that at n = o, i.e., at starting T is very high. That is, a series motor develops a very high torque at starting.

Therefore, a series motor is suitable for application as a drive motor in electric trains, cranes, hoists, trolley bus, etc., where the drive motor should develop very high starting torque.

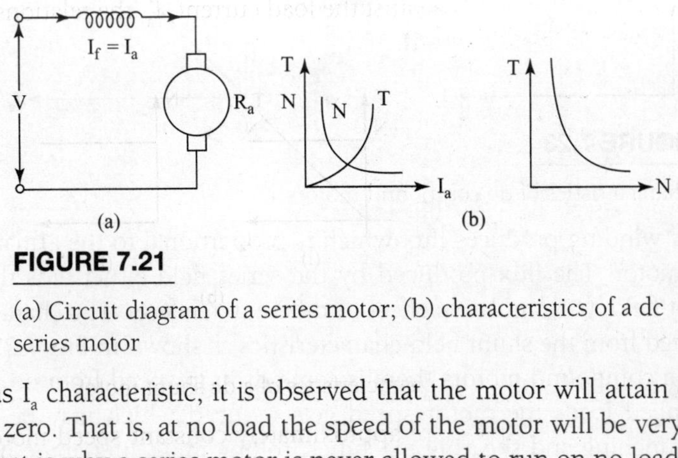

FIGURE 7.21

(a) Circuit diagram of a series motor; (b) characteristics of a dc series motor

From N versus I_a characteristic, it is observed that the motor will attain dangerously high speed when I_a is zero. That is, at no load the speed of the motor will be very high which may be dangerous. That is why a series motor is never allowed to run on no load. A load is always connected to its shaft before starting.

7.12.3 Characteristics of DC Compound Motors

In a compound motor two separate field windings are wound around each pole. One is shunt field winding and the other is series field winding. The shunt field winding is connected in parallel with the armature while the series field winding is connected in series with the armature as shown in Fig. 7.22. The flux produced due to the shunt field current remains constant but the flux produced by the series field current increases with the load current. The characteristic curves of a compound motor will be in between those of shunt and series motors.

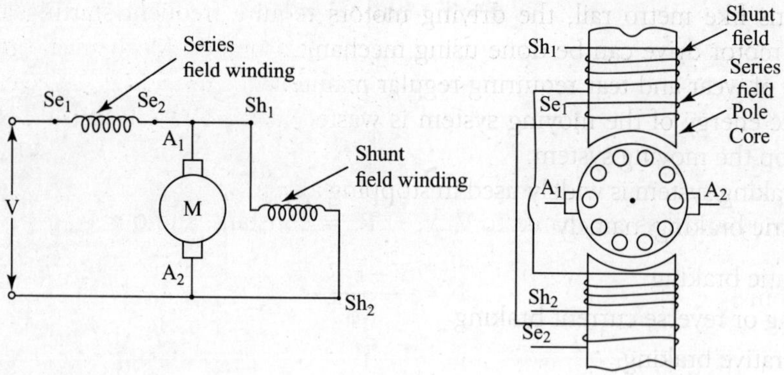

FIGURE 7.22

Shows the connections of shunt and series field windings of a compound motor

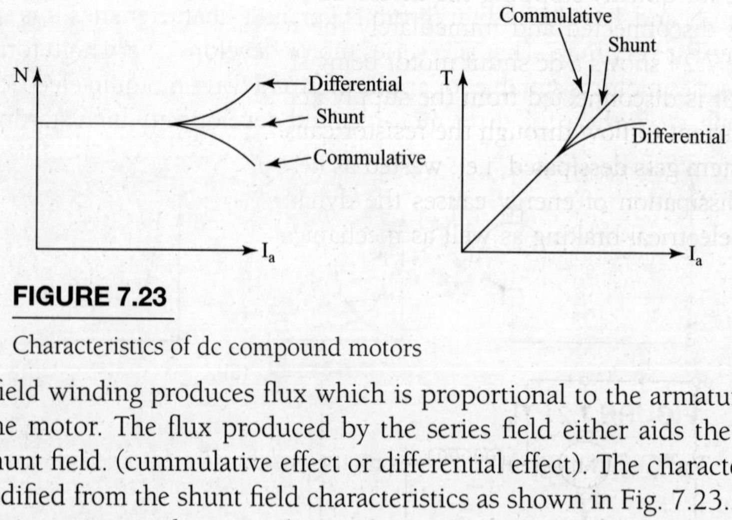

FIGURE 7.23

Characteristics of dc compound motors

The series field winding produces flux which is proportional to the armature current, i.e., the load on the motor. The flux produced by the series field either aids the shunt field or opposes the shunt field. (cummulative effect or differential effect). The characteristics relating T, N, I_a get modified from the shunt field characteristics as shown in Fig. 7.23.

In cummulative compound motors there is some drop in speed from no load to full load. For suddenly applied loads the motor speed gets reduced which may be advantageous in application like punching and shearing machines, rolling mills, lifts, mine hoists, etc. In differential compound motors, the resultant flux gets reduced as load increases, and hence, the speed increases. This is seen from the expression for speed, N which is

$$N = \frac{V - I_a(R_a + R_{sc})}{K\phi}$$

From the above expression it can be seen that if flux ϕ is reduced, speed N will increase.

7.13 ► Electric Braking of DC Motors

A dc motor may be used to drive an electric train or a street car (also called tram car). Braking of motors is required to quickly and accurately stop the motor with its load. Also in applications like metro rail, the driving motors require frequent starting and stopping. Braking of a motor drive can be done using mechanical brakes. Mechanical brakes have the disadvantage of wear and tear requiring regular maintenance including replacement of brake shoes. Kinetic energy of the moving system is wasted in the brakes as heat when a brake is applied to stop the moving system.

Electric braking system is widely used in stopping a dc motor driven system. There are three types of electric braking, namely

(a) Rheostatic braking
(b) Plugging or reverse current braking
(c) Regenerative braking.

7.13.1 Rheostatic Braking

In this method, for quickly stopping the motor driving a heavy mechanical load, the supply to the motor is disconnected and immediately the motor terminals are connected across a rheostat. Figure 7.24 shows a dc shunt motor being stopped quickly using rheostatic braking. When the motor is disconnected from the supply and is connected across rheostat, the back emf causes a current to flow through the resistor causing $I^2 R$ loss. The mechanical energy of the rotating system gets dessipated, i.e., wasted as heat energy in the resistance.

The quick dissipation of energy causes the dynamic system to stop quickly. As a safety measure, both electrical braking as well as mechanical braking system are made available to the operator.

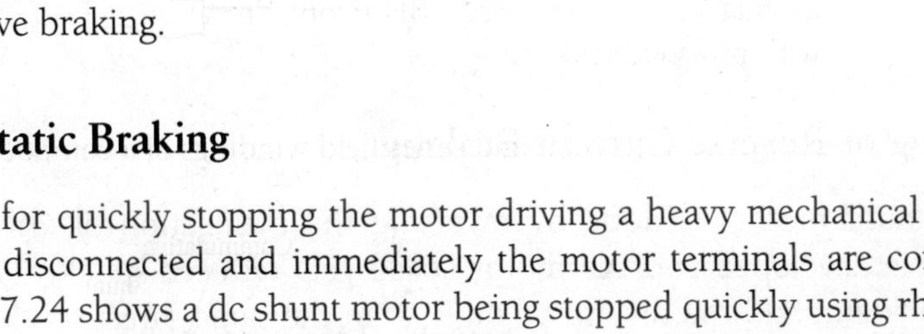

FIGURE 7.24

(a) A dc motor when in running condition driving a mechanical load; (b) Rheostatic braking is applied to the motor where a reverse current causes a reverse torque developed to stop the drive

In Rheostatic braking for series motor, the motor is first disconnected from the supply, the field connection is reversed and then the motor is connected across a rheostat as shown in Fig. 7.25. By reversing the field connection, the direction of current flowing through the field winding is kept unchanged. If this is not done, reverse current through the field winding during braking operation will destroy the field flux and no braking torque will be developed.

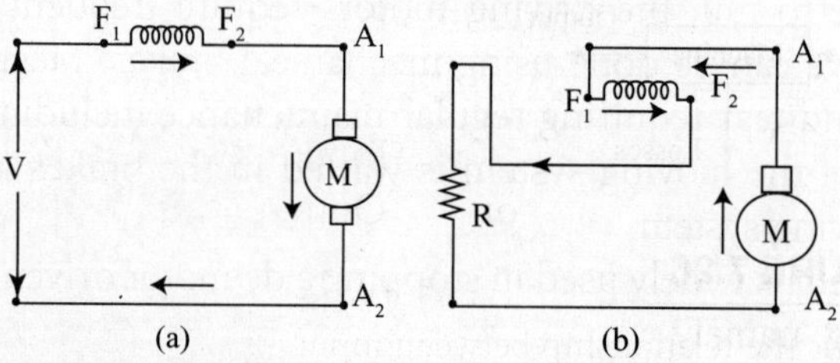

FIGURE 7.25

(a) A dc series motor when in running condition driving a mechanical load; (b) braking operation with reversed field connection.

7.13.2 Plugging or Reverse Current Braking

Plugging of dc shunt motors is done by reversing the connection of armature terminals. A reverse torque is developed and the motor speed gets reduced quickly. During plugging operation, the back emf and the reverse voltage act in the same direction thereby increasing the current in the armature circuit beyond the rated of value. To limit that reverse current, a resistance is connected in the armature circuit during plugging period.

Plugging operation is not only done for stopping the motor quickly but also is done to reverse the direction of rotation. Forward and reverse operation of the motor frequently is required on applications like rolling mills, overhead crane operation, etc.

7.13.3 Regenerative Braking

In both rheostatic braking and plugging operation, the kinetic energy of the driving system is wasted. In regenerative braking, energy of the rotating system is converted into electrical energy and returned back to the supply source. The necessary condition for regenerative braking is that the back emf should be higher than the supply voltage so that current will flow from the motor to the supply system.

7.14 ► LOSSES AND EFFICIENCY

The efficiency of a dc machine, like any other machine is the ratio of output power to the input power. The efficiency can never be 100 per cent because output is never equal to input. Some energy is lost in the machine during conversion of energy from mechanical to electrical or vice-versa. To achieve higher efficiency, the designer of the machine tries to keep the losses as low as possible.

7.14.1 Losses in a DC Machine

In a dc machine, like any other machines, the whole of input energy does not get converted into output energy. A portion of the input energy gets lost in the machine as shown in Fig. 7.26.

The various losses that take place in a dc machine are described as follows.

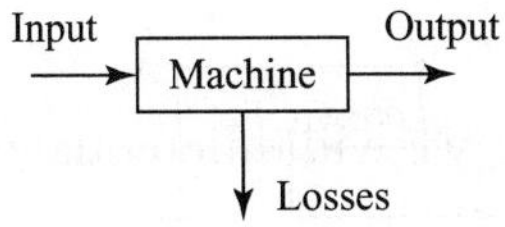

$$\text{Input} = \text{Output} + \text{Losses}$$

or

$$\text{Output} = \text{Input} - \text{Losses}$$

FIGURE 7.26

Shows the relationship between input, output, and losses

(a) I^2 R loss in the armature winding

Due to current flow in the armature winding a good amount of power gets lost as $I_a^2 R_a$, where I_a is the armature current and R_a is the resistance of the armature circuit. As load on the machine changes, I_a also changes. Hence, $I_a^2 R_a$ is called the variable loss as this loss varies with the variation of load on the motor.

(b) Core loss or iron loss in the armature

Iron loss or core loss consists of hysteresis loss and eddy current loss. The core is made up of magnetic material and is subjected to variations in magnetic flux. When the armature rotates, it comes under North and South poles alternately. *Hysteresis loss* occurs due to the alternate magnetization of the magnetic material. *Hysteresis loss depends upon the flux density, the frequency of variation of flux, and the volume of the core material.*

Eddy current loss is due to the presence of circulating current in the core material. When the armature rotates in the magnetic field, EMF is induced in the armature core also. This EMF causes a circulating current i_c in the core which is wasted as $i_c^2 r_c$ and produces heat. To reduce eddy current loss in the core, the core is made up of varnished, laminated steel sheets instead of a solid core. This causes increase of resistance, r_c through which the eddy current flows. *Eddy current loss depends upon flux density, frequency of alternation of flux, thickness of laminations used, and the volume of the core material.*

(c) Loss in the field windings

Losses take place in the field windings due to flow of current. This loss is equal to (VI_f) W where V is the applied voltage and I_f is the field current.

(d) Friction and windage losses

Due to rotation of the armature, air-friction loss which is also called windage loss, takes place. Frictional loss occurs due to brush and commutator rubbing and loss due to bearing friction.

7.14.2 Efficiency of DC Machine

The efficiency of a dc machine is expressed as

$$\eta = \frac{\text{Output}}{\text{Input}} = \frac{\text{Output}}{\text{Output} + \text{Losses}}$$

$$= \frac{VI_a}{VI_a + I_a^2 R_a + VI_f + C}$$

Where C is the sum of iron, friction, and windage losses.

7.14.3 Condition for Maximum Efficiency

To determine the condition for maximum efficiency, we will differentiate the expression for η w.r.t. I_a and equate to zero as

$$\frac{d}{dI_a}\left[\frac{V}{V+I_aR_a+\frac{(VI_f+C)}{I_a}}\right]=0$$

From which,

$$I_a^2 R_a = (VI_f + C)$$

that is, *variable loss = constant loss*.

Thus, the efficiency of a dc machine will be maximum at a load at which the variable loss becomes equal to the constant loss of the machine.

7.14.4 Testing of DC Machines: Determination of Efficiency

Efficiency of a dc machine can be determined by directly loading the machine. The output is measured and input is recorded. The ratio of output power to input power will give the value of efficiency. This method of determining efficiency is called direct loading method. The output and input are to be expressed in the same unit.

Efficiency of large machines are calculated by indirect method, i.e., by measuring the losses. Indirect method is preferred because for large machines, loading of the machine may be difficult in the laboratory. Further energy will be wasted during experimentation. A popular method, known as *Swinberne's method* of determining efficiency is described as follows.

Swinberne's method

In this method the dc machine is run as a motor. The applied voltage and the speed is adjusted to their rated values as shown in Fig. 7.27. There is no load connected to the motor shaft.

When the motor is running on no load, the input power is wasted as losses. The losses at no load are (i) iron loss; (ii) friction and windage loss, and (iii) $I_{ao}^2 R_a$ loss.

The armature current, I_{ao} at no load is small, and hence, $I_{ao}^2 R_a$ will be very small. However, this value can be calculated. Iron loss, and friction and windage loss depend upon supply voltage and motor speed, respectively. The supply voltage is kept constant and speed of the motor is approximately constant at all loads. These losses are called constant losses as they remain constant at all loads. Thus, we can calculate the constant losses by subtracting $I_{ao}^2 R_a$ from the no-load input to the motor. By knowing the constant losses, efficiency of the machine can be calculated at any load current.

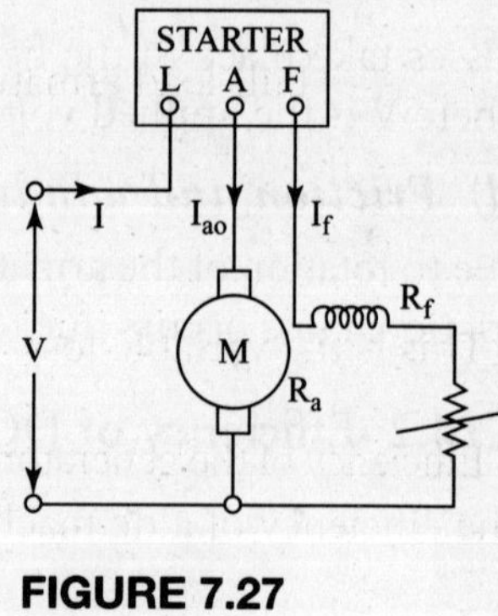

FIGURE 7.27

Swineberne's test for determining efficiency

EXAMPLE 7.1

A 220 V, 50 kW dc shunt generator was run as a motor on no load at rated speed. The current drawn from the line was 8 A and the shunt field current was 2 A. The armature resistance of the machine is 0.1 Ω. Calculate the efficiency of the generator at full load.

SOLUTION

Input power at no load = VI

$= 220 \times 8$

$= 1760$ W

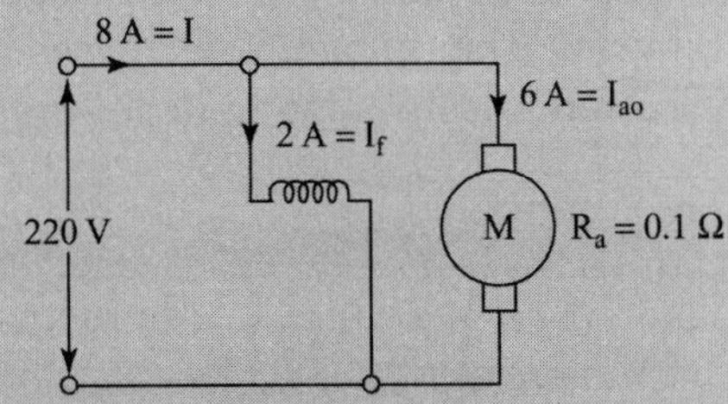

FIGURE 7.28

Iron, friction and windage, and field copper losses
= No-load input – $I_{ao}^2 R_a$

$= 1760 - 6^2 \times 0.1$

$= 1760 - 3.6$

$= 1756.4$ W

These are constant losses.

Note that $I_{ao}^2 R_a$ loss in the armature at no load is very small.

The generator is rated at 50 kW and 220 V.

The generator output current at full load.

$$= \frac{50 \times 1000}{220} = 227.2\,\text{A}$$

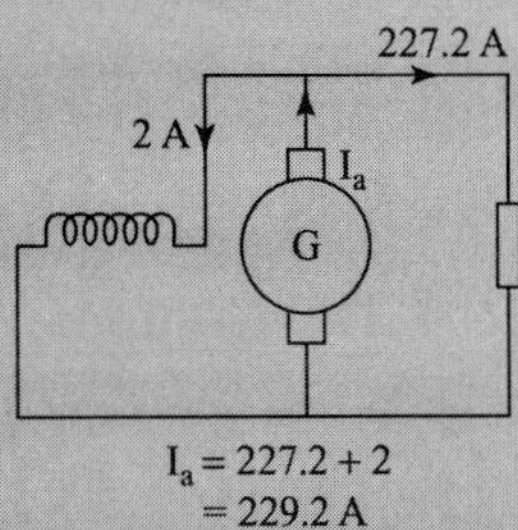

FIGURE 7.29

Full-load armature Copper loss = $I_a^2 R_a$

$= (229.2)^2 \times 0.1$

$= 5253$ W

This is the variable loss.

Efficiency of the generator in percentage = Output/(Output + Constant losses + Variable loss)

$$= \frac{50 \times 1000 \times 100}{50 \times 1000 + 1756.4 + 5253}$$

= 89 per cent

7.15 ► Applications of DC Machines

7.15.1 DC Generators

In earlier days dc generators were used to generate electricity and the power was supplied to consumers through dc distribution networks. At present, use of dc generators for generation and distribution of electricity is rare. All commercial generators are ac generators which are also called alternators. Generation and transmission of alternating current has a number of advantages. The use of dc generators is confined to supplying excitation current to ac generators and to convert ac to dc for industrial applications.

7.15.2 DC Motors

DC motors are available as dc shunt motors, dc series motors, and dc compound motors. DC shunt motors are more or less constant-speed motors. They may be used in driving a line shaft where speed has to be kept fairly constant between no load and full load. In situations where variable load has to be driven but at constant speed, such as driving a lathe, the speed change can be obtained using a shunt field regulator.

7.15.3 DC Series Motors

In applications where high starting torque is required, such as in driving hoists, cranes, electric trains, etc., series motors are used. Series motors are also used where the motor can be permanently coupled to the load, such as fans, where the torque requirement increases with speed.

Series motors attain very high speed at light load. That is why series motors should never be run on no load.

7.15.4 DC Compound Motors

DC compound motors are used in applications where large starting torque is required but there is a chance for the load to fall to a very low value. In such applications dc series motors cannot be used.

7.16 Solved Numerical Problems

EXAMPLE 7.2

A four-pole dc generator having wave-wound armature winding has 51 slots, each slot containing 20 conductors. Calculate the voltage generated in the armature when driven at 1500 rpm. Assume flux per pole to be 0.5 mWb.

SOLUTION

$P = 4$, $A = 2$ (because the armature winding is wave wound)

$$N = 1500 \text{ rpm}$$

Total number of armature conductors, $Z = 20 \times 51$

$$= 1020$$

Equation for induced EMF, $E = \frac{\phi ZNP}{60A} V$

Substituting values

$$E = \frac{0.5 \times 10^{-3} \times 1020 \times 1500 \times 4}{60 \times 2}$$

$$= 255 \text{ V}$$

EXAMPLE 7.3

Example 7.3 A six-pole, lap-connected dc generator has a total of 650 conductors. The flux per pole is 0·05 Wb. Calculate the speed at which the armature is to be driven to generate an EMF of 220 V.

SOLUTION

P = 6, A = P = 6 (because the armature winding is lap connected)

$$\phi = 0.05 \text{ Wb}; E = 220 \text{ V}$$

$$Z = 650, N = ?$$

$$E = \frac{\phi ZNP}{60A}$$

Substituting values

$$220 = \frac{0.05 \times 650 \times N \times 6}{60 \times 6}$$

or,

$$N = \frac{220 \times 60}{650 \times 0.05} = 406 \text{rpm}$$

EXAMPLE 7.4

A four-pole 220 V dc shunt generator supplies a load of 3 kW at 220 V. The resistance of the armature winding is 0.1 Ω and that of the field winding is 110 Ω. Calculate the total armature current, the current flowing through armature conductors, and the EMF induced. Assume that the armature winding is wave wound.

SOLUTION

Output power = 3 kW = 3000 W

Power = output voltage, V × Output current, I_L

$$I_L = \frac{3000}{220} = 13.6\,A$$

From the figure it can be seen that

$$I_f = \frac{V}{R_f} = \frac{220}{110} = 2\,A$$

$$I_a = I_L + I_f$$

$$= 13.6 + 2 = 15.6\,A$$

The armature winding is wave wound. The number of parallel paths is 2. That is, all the armature conductors are connected in such a way that half the armature current flows through each path. Thus, current flowing though each armature conductor will be $I_a/2$, i.e., $\frac{15.6}{2} = 7.8$ Amps. E is the EMF induced in the armature. A voltage drop of I_aR_a takes place in the armature winding when it is supplying current. The remaining voltage, V is available across the load terminals.

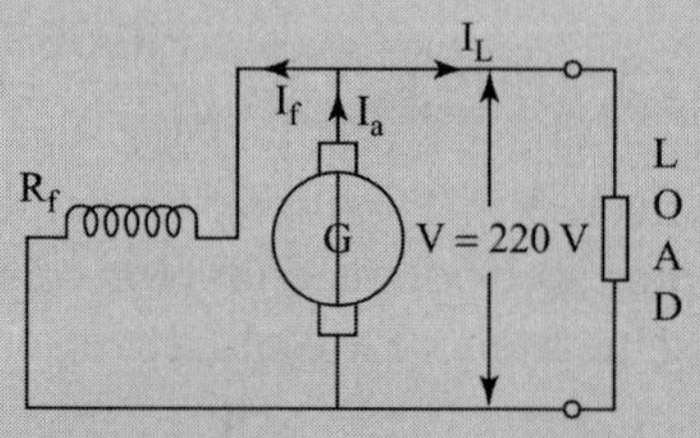

FIGURE 7.30

Thus,

$$E - I_aR_a = V$$

or,

$$E = V + I_aR_a = 220 + 15.6 \times 0.1 = 218.44\,V$$

EXAMPLE 7.5

A four-pole, 12 kW, 240 V dc generator has its armature coils wave connected. If the same machine is lap connected, all other things remaining constant, calculate the voltage, current, and power rating of the generator.

SOLUTION

In a wave winding all the armature coils are arranged in two parallel paths. The current-carrying capacity of each conductor, therefore, will be half of the total armature current.

$$I \times V = P$$

$$I = \frac{12 \times 1000}{240} = 50\,A$$

Current per path

$$= \frac{50}{2} = 25\,A$$

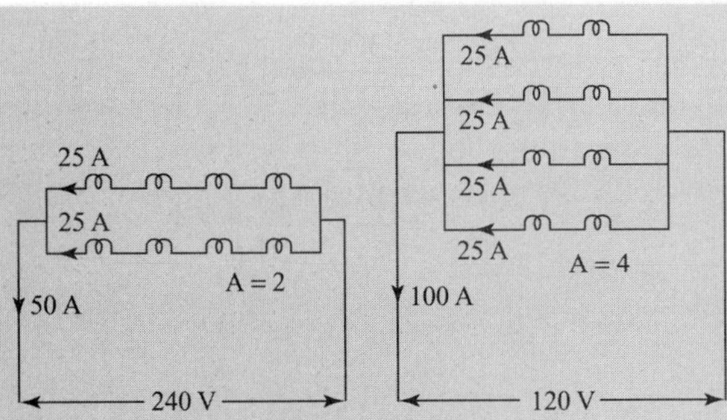

FIGURE 7.31

When there will be lap connection of windings, armature coils will be connected in P number of parallel paths. That is, in this case there will be four parallel paths. If each conductor or coil carries 25 A, the total output current will be 100 A. The number of coils in each path will be reduced to half, and hence, the induced EMF per parallel path will be $\frac{240}{2} = 120$ V. The output power = 120 × 100 W $= \frac{120 \times 100}{1000}$ kW = 12 kW.

Thus, we observe that output power remains the same but the voltage and current ratings change.

EXAMPLE 7.6

A dc shunt generator delivers 12 kW at 240 V while running at 1500 rpm. Calculate the speed of the machine when running as a shunt motor and taking 12 kW at 240 V. The armature resistance is 0.1 Ω and field resistance is 80 Ω.

SOLUTION

As a generator,

$$I_f = \frac{V}{R_f} = \frac{240}{80} = 3\text{ A}$$

$$I_L = \frac{12 \times 1000}{240} = 50\text{ A}$$

$$I_a = I_L + I_f = 50 + 3 = 53\text{ A}$$

$$E_g = V + I_a R_a = 240 + 53 \times 0.1$$
$$= 245.3\text{ V}$$

As a motor,

$$I_L = I_a + I_f$$
$$I_a = I_L - I_f$$

$$I_L = \frac{12 \times 1000}{240} = 50\text{ A}, \quad I_f = \frac{V}{R_f} = \frac{240}{80} = 3\text{ A}$$

$$I_a = 50 - 3 = 47\text{ A}$$

$$E_m = V - I_a R_a = 240 - 47 \times 0.1$$
$$= 235.3\text{ V}$$

Let the speed of the machine as generator be N_1 and as motor be N_2

$$E_g = \frac{\phi Z N_1 P}{60A} \text{ and } E_m = \frac{\phi Z N_2 P}{60A}$$

or, $$\frac{Eg}{Em} = \frac{N1}{N2}$$

or, $$N_2 = N_1 \frac{Em}{Eg} = 1500 \times \frac{235.3}{245.3} = 1439\text{ rpm.}$$

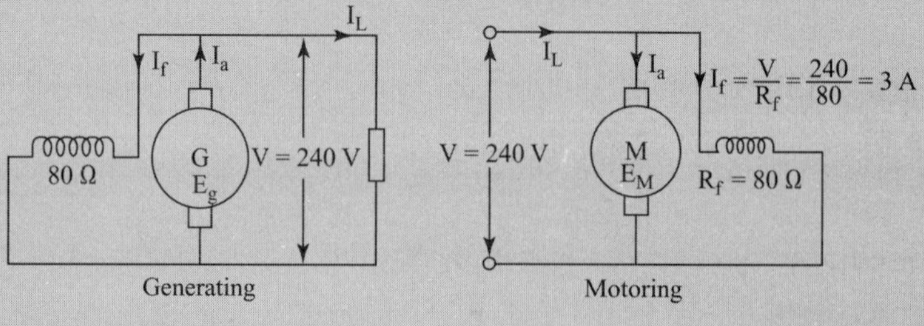

FIGURE 7.32

EXAMPLE 7.7

A four-pole 220 V dc series motor has 240 slots in the armature and each slot has six conductors. The armature winding is wave connected. The flux per pole is 1.75×10^{-2} Wb when the motor takes 80 A. The field resistance is 0.05 Ω and the armature resistance is 0.1 Ω. The iron and friction losses 440 W. Calculate the speed of the motor. Also calculate the output horse power.

SOLUTION

In a series motor the armature winding and the field winding are connected in series across the supply voltage. Thus, the line current, field current, and the armature current are the same,

i.e., $$I_a = I_f = I_L = 80\text{ A}$$

The total member of armature conductors, $Z = 240 \times 6$

$$= 1440$$

Armature is wave connected, and hence, A = 2

No. of poles = 4

From Fig 7.33, $V - E = I_a (R_a + R_{se})$

or, $E = V - I_a (R_a + R_{se})$

$= 220 - 80 (0.1 + 0.05)$

$= 208$ V

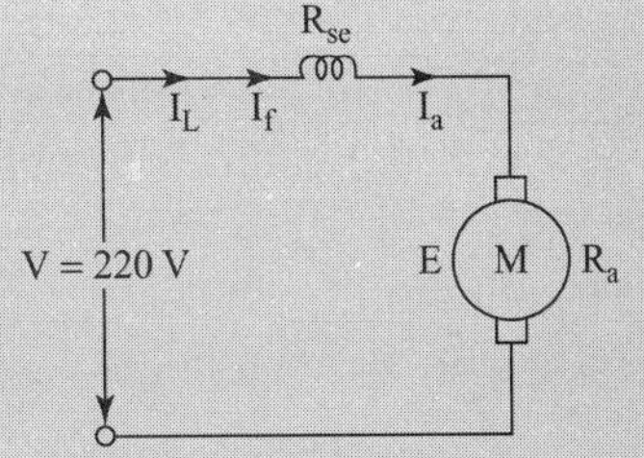

FIGURE 7.33

Again, $$E = \frac{\phi ZNP}{60A}$$

substituting values $$208 = \frac{1.75 \times 10^{-2} \times 1440 \times N \times 4}{60 \times 2}$$

or, $$N = \frac{208 \times 60 \times 2 \times 10^2}{1.75 \times 1440 \times 4} = 248 \text{ rpm}$$

Power developed by the armature = $E \times I_a$

$$= \frac{248 \times 80}{1000} = 19.84 \text{ kW}$$

Power output = Power developed – Iron and Frictional losses

= 19.84 – 0.44

= 19.4 kW

If we want to convert in horse power, we use the relation 1 kW = 0.735 hp.

Thus, power output = 19.4 × 0.735 = 14.26 hp.

EXAMPLE 7.8

A 220 V dc shunt motor takes 5 A at no load. The armature resistance is 0.2 Ω and field resistance is 110 Ω. Calculate the efficiency of the motor when it takes 40 A on full load.

SOLUTION

From Fig. 7.34, $$I_f = \frac{V}{R_f} = \frac{220}{110} = 2 \text{ A}$$

$$I_L = I_a + I_f$$

or, $$I_a = I_L - I_f$$

At no-load, $I_L = 5$ A

therefore, $I_a = 5 - 2 = 3$ A

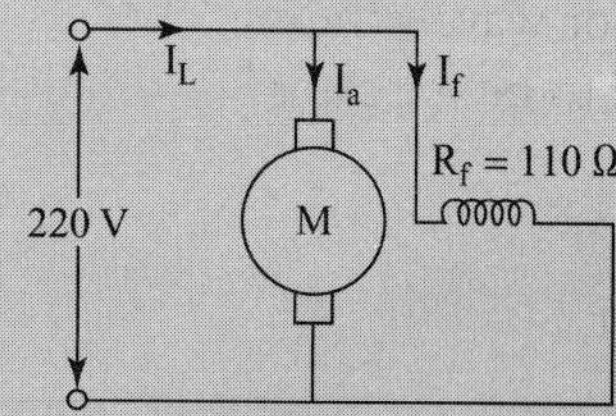

FIGURE 7.34

At no-load, when the motor output is zero, the input $= V\,I_L$ (No-load) $= 220 \times 5 = 1100$ W.

The whole of input is lost as $I_a^2 R_a$ loss + $I_f^2 R_f$ loss + Iron loss + Friction and Windage loss.

$$I_a^2 R_a = 5^2 \times 0.2 = 5 \text{ W}$$

$$I_f^2 R_f = 2^2 \times 110 = 440 \text{ W}$$

Iron, friction, and windage losses $= 1100 - 5 - 440 = 655$ W

These losses are constant losses and are same at any load. This means, on full load these losses will remain at 655 W.

At full-load,

$$I_L = 40 \text{ A}$$

$$I_a = I_L - I_f = 40 - 2 = 38 \text{ A}$$

$$I_a^2 R_a = (38)^2 \times 0.2 = 289 \text{ W}$$

$$I_f^2 R_f = 2^2 \times 110 = 440 \text{ W}$$

Iron, friction, and windages losses $= 655$ W

Total losses $= 289 + 440 + 655 = 1384$ W

Efficiency $= \dfrac{\text{Output}}{\text{Input}} = \dfrac{\text{Input} - \text{losses}}{\text{Input}}$

$$= \frac{(220 \times 40 - 1384) \times 100}{220 \times 40} = 84.3 \text{ per cent}$$

EXAMPLE 7.9

A four-pole dc generator has 1000 conductors. The flux per pole is 25 mWb. Calculate the EMF induced when the armature is lap connected and run at 1500 rpm. At what speed the generator must be driven to produce the same EMF with the armature winding wave connected?

SOLUTION

Case I

$$P = 4,\ Z = 1000,\ \phi = 25 \times 10^{-3} \text{ Wb}$$

$$A = 4,\ N = 1500 \text{ rpm}$$

Induced EMF, $E = \dfrac{\phi Z N p}{60A} = \dfrac{25 \times 10^{-3} \times 1000 \times 1500 \times 4}{60 \times 4}$

For wave winding, $A = 2 = 625$ V

E to be the same,

$$625 = \frac{\phi ZNP}{60A} = \frac{25\times10^{-3}\times1000\times N\times4}{60\times2}$$

or, $N = 750$ rpm

EXAMPLE 7.10

Calculate the output power of a 12-pole separately excited having 1200 lap-connected conductors each carrying a current if 15 A. The armature is being driven at 300 rpm. The flux per pole is 60 mWb. Resistance if armature circuit is 0.1 **Ω**.

SOLUTION

Current flowing in each parallel path is the same as current flowing through each conductor in the path. Since there are 12 parallel paths in the armature (since A = P = 12), the total armature current is 15 × 12 = 180 A.

Induced EMF,
$$E = \frac{\phi ZNP}{60A} = \frac{60\times10^{-3}\times1200\times300\times12}{60\times12}$$
$$= 360 \text{ V}$$
$$V = E - I_a R_a$$
$$= 360 - 180\times0.1$$
$$= 342 \text{ V}$$

Power output
$$= VI = \frac{342\times180}{1000}\text{kW} = 61.56 \text{ kW}$$

EXAMPLE 7.11

A four-pole, 500 V, wave-wound dc shunt motor has 900 conductors on its armature. Calculate the speed of the motor if its armature current is 80 A, the flux per pole is 21 mWb and armature resistance is 0.1 **Ω**.

SOLUTION

The back EMF induced in the armature of the motor is E_b.

$$E_b = \frac{\phi ZNP}{60A} = \frac{21\times10^{-3}\times900\times N\times4}{60\times2} \quad \text{(i)}$$

For motor

$$V - E_b = I_a R_a$$

or,

$$E_b = V - I_a R_a$$
$$= 500 - 80 \times 0.1$$
$$= 492 \text{ V} \quad \text{(ii)}$$

Equating (i) and (ii)

$$\frac{21\times10^{-3}\times900\times N\times4}{60\times2} = 492$$

or,

$$N = \frac{60\times2\times492}{21\times10^{-3}\times900\times4} = 780 \text{ rpm}$$

EXAMPLE 7.12

A dc machine induces an EMF of 240 V at 1500 rpm. Find the developed torque for an armature current of 25 A.

SOLUTION

Power developed,

$$P = E \times I_a$$
$$= 240 \times 25$$
$$= 6000 \text{ W}$$

Again, $P = T \times \omega$ where ω is the angular velocity in radians per second.

$$\omega = \frac{2\pi N}{60} \text{ where N is in rpm}$$

Therefore,

$$T = \frac{P}{2\pi N/60} = \frac{60}{2\pi}\frac{P}{N} = 9.55\times\frac{6000}{1500} = 38.2 \text{ Nm}$$

EXAMPLE 7.13

A dc shunt machine connected to 220 V supply has armature resistance of 0.1 Ω and field resistance of 110 Ω. Find the ratio of the speed of the machine working as a generator to the speed of the machine when working as a motor when the line current is 100 A in both the cases.

SOLUTION

$$I_{sh} = \frac{V}{R_f} = \frac{220}{100} = 2\text{ A}$$

As a generator, the machine will supply 100 A to the supply mains.

$$I_a = I_f + I_L = 2 + 100 = 102\text{ A}$$

Output voltage = 220 V.

$$E_g = V + I_a R_a = 220 + 102 \times 0.1 = 230.2\text{ V}$$

As a motor, the machine will draw 100 A from the supply out of which 2 A will go to the field circuit.

Therefore, $I_a = 100 - 2 = 98\text{ A}$

$$E_m = V - I_a R_a = 220 - 98 \times 0.1 = 210.2\text{ V}$$

$$\frac{\text{Speed of gen}, N_g}{\text{Speed of motor}, N_m} = \frac{E_g}{E_m}$$

or,

$$\frac{N_g}{N_m} = \frac{230.2}{210.2} = 1.095$$

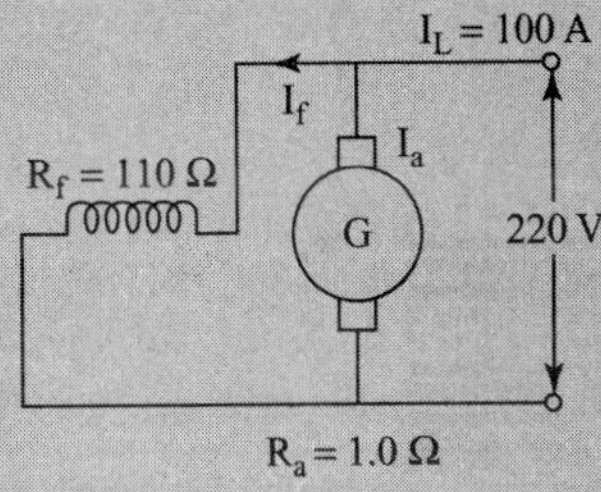

FIGURE 7.35

Important Terminologies and Points to Remember

1. A dc machine has stationary poles and rotating armature. Emf induced in the coils rotating in the magnetic field is alternating in nature. Brush and commutation arrangement helps in changing ac into dc at the output.
2. The function of brush and commutator in a dc generator is to convert ac to dc. In motoring operation, brush and commutator arrangement helps in producing unidirectional torque.
3. Armature coils are connected in series and parallel so as to get the required magnitude of voltage and current.
4. There are two basic types of armature windings, namly Lap Winding and Wave Winding.
5. Induced emf in a dc machine is calculated on the basis of amount of flux cut per second.
6. DC machines are classified as shunt, series, and compound.
7. Residual magnetism plays a very important function by voltage build up in a dc shunt generator.
8. In a shunt generator, the field circuit resistance should be less than critical field resistance. Critical field resistance is that value at which field resistance line becomes tangent to the open-circuit characteristic.
9. Induced emf in a dc machine is directly proportional to the magnetic field flux, ϕ and the speed, N at which the armature winding is rotated, i.e., $E \propto \phi N$.
10. Torque, T developed in a dc motor is proportional to the product of flux, ϕ and current flowing through the armature conductors, I_a, i.e., $T \propto \phi I_a$.

11. The basic voltage equations for dc machines are:
(i) $E = V + I_a R_a$.......for generator
(ii) $E = V - I_a R_a$.......for motor.
12. A dc motor at starting draws a very heavy current because for a motor, $I_a = \frac{V-E}{R_a}$ and $E \propto \phi N$. At starting $N = 0$, and hence, $E = 0$ and hence, $I_a = \frac{V}{R_a}$. Armature resistance R_a is very small, and hence, I_a is very high at starting time.
13. A starter is a variable resistance connected in series with the armature of a dc motor so as to limit the current during starting period.
14. Characteristics of dc motors for shunt, series and compound types are drawn between any two variables of speed, load and torque.
15. For a series motor, line current, field current and armature current are the same. So torque T is $T \propto \phi I_a$, $\phi \propto I_a$, so $T \propto I_a^2$. I_a is proportional to load. Again $N = \frac{V - I_a R_\phi}{K\phi}$. So, $N \propto \frac{1}{I_a}$. In a series motor starting torque is very high because $T \propto I^2_a$ or $T \propto \frac{1}{N^2}$. Again on no-load (i.e., I_a being very small), speed is very high.
On the basis of the above, it is concluded that a dc series motor should not be started without connecting any load on its shaft. Also dc series motors are suitable for applications where a very high starting torque is required.
16. Efficiency of a dc machine can be determined by Swinburne's test or Hopkinson's test.
17. Speed of dc motors can be controlled by any of the following methods: (i) Voltage control method, (ii) Field control method, (iii) Armature circuit resistance control method.
18. Starter is a variable resistance connected in the armature circuit to control the initial high-current flowing through the armature. Three-point starter and four-point starter are available for starting a dc motor. Four point starter is superior than three-point starter because in four-point starter, supply to the No-Volt. Release (NVR) is made directly and not in series with the field circuit. Otherwise, if field circuit current is reduced, the NVR becomes weak and releases the handle unwantedly.

Important Formulae and Equations

Equation	Remarks
1. $E = \frac{\phi ZNP}{60\ A}$ where Z is the total number of armature conductors, A is the number of parallel paths in the armature winding, P is the number of poles, N is the revolutions per minute at which the armature is rotated and ϕ is the flux per pole.	1. Emf equation for a dc machine. In case of operation as a motor the induced emf, E is called back emf, E_b because it acts in opposite direction to the applied voltage, V.
2. $E = V + I_a R_a$ (i) $E = V - I_a R_a$ (ii)	2. Voltage equations for (i) generator and (ii) for motor.
3. $T = \left(\frac{ZP}{2\pi A}\right)\phi I_a$ or, $T = K_t \phi I_a$	3. Torque developed is directly proportional to strength of the magnetic field and the amount of current flowing through the coils.

Important Formulae and Equations

Equation	Remarks
4. $I_a = \dfrac{V - E_b}{R_a + R}$	4. I_a is the starting current of a dc motor when E_b=0. E_b starts growing as the motor starts picking up speed.
5. $N = \dfrac{V - I_a R_a}{K\phi}$ for series motor, $\phi \propto I_a$ for series motor, $\phi \propto I_f$	5. $I_a R_a$ drop is very small as compared to V.

REVIEW QUESTIONS

Short Answer Type Questions

1. Draw a neat sketch of a dc machine and name the component parts.
2. What is the function of the following parts of a dc machine: (i) field poles; (ii) armature; (iii) brush and commutator; (iv) shaft?
3. Explain the function of brush and commutator in a dc machine for generating action.
4. With a simple example show how lap winding and wave windings are made.
5. Deduce the EMF equation for a dc machine.
6. Describe various methods of speed control of dc motors.
7. Explain why dc motors should require starters.
8. Draw the connection diagram of a dc motor starter.
9. Draw characteristics of dc series motors and mention applications.
10. State the various losses that occur in a dc machine.
11. What is hysteresis loss and eddy current loss?
12. Derive the torque equation for a dc motor.
13. Draw the connection diagrams for dc shunt, series, and compound motors.
14. Explain the working principle of a dc generator.
15. How can you determine the efficiency of a dc machine without actually loading the machine?
16. Why do we use laminated sheets for the armature and the field cores?
17. How can you change the direction of rotation of a dc motor?
18. Why is it advisable not to start a dc series motor without having any load on it?
19. Why do we connect the coils of the armature in series parallel?
20. What are the various losses in a dc machine. Which losses are called constant losses and why?
21. How can you determine the efficiency of a dc machine without actually loading the machine?

Numerical Problems

22. The wave-connected armature of a two-pole 200 V generator has 400 conductors and runs at 300 rpm. Calculate the useful flux per pole.

    [Ans ϕ = 10 mWb]

23. The induced EMF in a dc machine while running at 500 rpm is 180 V. Calculate the induced EMF when the machine is running at 600 rpm. Assume constant flux.

    [Ans E = 216 V]

24. A 250 V shunt motor draws 5 A while running on no load at 1000 rpm. Calculate the speed of the motor when it is loaded and draws a current of

50 A. The armature circuit resistance is 0.2 Ω and field circuit resistance in 250 Ω.

[**Ans** N = 964 rpm]

25. A dc shunt machine has armature resistance of 0.5 Ω and field resistance of 750 Ω. When seen as a motor on no load at 500 V, the line current drawn is 3 A. Calculate the efficiency of the machine when it operates as a generator with an output of 2 kW at 500 V.

[**Ans** η = 89.6 per cent]

26. A four-pole wave-connected dc armature has 50 slots with 10 conductors per slot. The armature is rotated at 1000 rpm of the useful flux per pole is 30 mwb, calculate the amount of EMF induced.

[**Ans** 500 V]

27. A six-pole armature has 410 wave-connected conductors. The flux per pole is 0.02 wb. Calculate the speed at which the armature must be rotated so as to generate 400 V.

[**Ans** 975 rpm]

28. A 200 V dc shunt motor having an armature resistance of 0.2 Ω and field resistance of 100 Ω draws a line current of 50 A at full-load at a speed of 1500 rpm. What will be its speed at half load?

[**Ans** 1539 rpm]

29. A 500 V shunt motor takes a current of 5 A on no-load. Calculate the efficiency of the motor when it takes 100 A. Take R_a = 0.5 Ω and R_f = 250 Ω.

[**Ans** 85.4 per cent]

30. A shunt motor takes 125 A at 400 V at 1000 rpm at a particular load. If the total torque remains unchanged, Calculate the speed and armature current when the magnetic field is reduced 80 per cent of its original value. Take R_a = 0.25 Ω.

[**Ans** 1224 rpm, 156.25 A]

31. A dc shunt generator has a field resistance of 60 Ω and armature resistance of 0.03 Ω. As a generator, the machine delivers 40 kW at 240 V when driven at a speed of 450 rpm. Calculate the speed of the machine when running as a motor taking 40 kW of power input at 240 V.

[**Ans** N = 424 rpm]

Multiple Choice Questions

1. The expression for EMF induced in a dc machine is

   (a) $E = \dfrac{\phi ZNP}{60A}$ (b) $E = \dfrac{\phi ZNP}{60}$

   (c) $E = \dfrac{\phi ZNP}{60N}$ (d) $E = \dfrac{ZNPA}{60\phi}$.

2. Which of the following statements is not true for a dc machine?
   (a) EMF induced is directly proportional to air-gap flux
   (b) EMF induced is directly proportional to number of armature conductors
   (c) EMF induced in inversely proportional to number of parallel paths of the armature conductors
   (d) EMF induced is inversely proportional to the number of poles.

3. The poles and armature of a dc machine is made of laminated steel sheets to
   (a) reduce hysteresis loss
   (b) reduced eddy current loss
   (c) reduce I^2R loss
   (d) reduce humming noise of the core.

4. The EMF induced in the armature of a dc generator is alternating in nature but in the output circuit dc is made available by
   (a) Brush and slip-ring arrangement
   (b) Brush and commutator arrangement
   (c) diode rectifiers
   (d) converter circuit.

5. The windings of a dc machine are either
   (a) lap or wave connected
   (b) lap or spirally connected
   (c) wave or spirally connected
   (d) made of concentric coils or of short-pitched coils.

6. Critical field resistance of a dc generator is that value of the resistance at which

(a) the field resistance line always lies below the OCC
(b) the field resistance line is tangent to the OCC
(c) the field resistance line crosses the OCC atleast at two points
(d) the field resistance line does not touch the OCC at all.

7. The EMF induced in a four-pole dc generator having 1000 conductors, lap-connected windings, flux per pole of 10 mwb and rotated at 600 rpm is
(a) 1000 V (b) 500 V
(c) 250 V (d) 100 V.

8. The brush and commutator arrangement in a dc motor is used to achieve
(a) unidirectional current in the armature
(b) unidirectional torque to achieve continuous rotation of the armature
(c) change in the direction of rotation of the armature
(d) high-starting torque.

9. The number of parallel paths of the armature winding of an eight-pole, 250 V, wave wound dc machine having 1500 armature conductors is
(a) 4 (b) 2
(c) 6 (d) 8.

10. Which of the following statements is not true for the EMF induced in a dc machine?
(a) EMF induced is directly proportional to speed of the armature
(b) EMF induced is inversely proportional to flux per pole
(c) EMF induced is directly proportional to number of armature conductors
(d) EMF is directly proportional to the flux per pole.

11. The direction of rotation of a dc motor can be changed
(a) by reversing the polarities of the supply
(b) by reversing either the polarities of the supply to armature or to the poles
(c) by reversing only the polarities of the supply to the armature
(d) by reversing only the polarities of the supply to the field poles.

12. The speed of a dc motor is
(a) directly proportional to back EMF and inversely proportional to flux
(b) inversely proportional to back EMF and directly proportional to flux
(c) directly proportional to both back EMF and flux
(d) inversely proportional to both back EMF and flux.

13. The nature of EMF induced in the armature coils in a dc machine is
(a) dc (b) ac
(c) pulsating dc (d) variable dc.

14. A dc machine is connected to 220 V supply mains. Its armature resistance is 0.2 Ω. What should the magnitude of EMF generated so that it may feed 100 A to the supply?
(a) 200 V (b) 220 V
(c) 240 V (d) 260 V.

15. Residual magnetizm of the field poles is necessary for the voltage built up in
(a) dc shunt motor
(b) dc shunt generator
(c) dc series motor
(d) dc separately excited generator.

16. A dc motor when connected directly to the supply would draw a very heavy current because
(a) the back EMF at starting is zero
(b) the back EMF at starting is maximum
(c) the back EMF is opposing the supply voltage
(d) torque required at starting is high.

17. The speed of a dc shunt motor can be reduced by
(a) decreasing the field current
(b) increasing the supply voltage to the motor
(c) decreasing the supply voltage to the motor
(d) by decreasing the supply voltage to the motor and simultaneously decreasing the field current.

18. The relationship between torque, T and armature current, I_a for a dc series motor is
(a) $T \propto I_a$ (b) $T \propto \frac{1}{I_a}$
(c) $T \propto I_a^2$ (d) $T \propto \frac{1}{I_a^2}$.

19. The relationship between torque, T and armature current, I_a for a dc shunt motor is

(a) $T \propto I_a$ (b) $T \propto \frac{1}{I_a}$

(c) $T \propto I_a^2$ (d) $T \propto \frac{1}{I_a^2}$.

20. Efficiency of a dc machine is less than that of an equivalent transformer because

(a) there is friction and windage losses in dc machines
(b) core losses are more in dc machines than in transformers
(c) copper losses are more in dc machines than in transformers
(d) for all the reasons mentioned in (a), (b) and (c).

Answers to Multiple Choice Questions

1. (a)	2. (d)	3. (b)	4. (b)	5. (a)	6. (b)
7. (d)	8. (b)	9. (b)	10. (b)	11. (b)	12. (a)
13. (b)	14. (c)	15. (b)	16. (a)	17. (c)	18. (c)
19. (a)	20. (a)				

CHAPTER 8

Three-phase Induction Motors

Topics Discussed

1. Basic principle of a three-phase induction motor
2. Constructional details
3. Rotating magnetic field
4. Synchronous speed
5. Slip speed
6. Rotor-equivalent circuit
7. Rotor current
8. Rotor frequency
9. Power-flow diagram
10. Torque
11. Torque-slip characteristics
12. Effect of change of rotor circuit resistance on torque-slip characteristics
13. Starting torque
14. Condition for maximum torque
15. Starting
16. Speed control
17. Losses
18. Efficiency
19. Applications

8.1 ► Introduction

Three-phase induction motors are used in many industrial applications such as a drive motor. These motors are very rugged, and hence, there is virtually no maintenance required. Only three-phase supply is required for the stator. No supply is to be provided to the rotor. The rotor gets energized due to electromagnetic induction.

As the name suggests, a three-phase induction motor will have three windings placed in stator slots 120° apart connected either in star or in delta formation. Three-phase supply is provided to these three windings. Due to electromagnetic induction, EMF will be induced into the rotor winding, and if the rotor winding is closed, current will flow through the rotor winding. The interaction between the field produced, due to current flow in the stator windings, when fed from a three-phase supply, and the current-carrying rotor conductors will produce a torque which will rotate the rotor. This is the basic principle of an induction motor. We will now discuss the constructional details and the principle of working of a three-phase induction motor in detail.

8.2 ► Constructional Details

The main parts of any rotating electrical machine, are the stator and the rotor. The stator is a hollow cylindrical structure while the rotor is a solid cylindrical body which is placed inside the stator and is supported at the two ends by two end shields. A small air gap is maintained between the stator and the rotor so that the rotor can rotate freely. The rotor shaft is held at the two ends by two bearings so that the frictional loss is minimum.

Fig. 8.1 (a) shows the stator and the rotor with two end shields from two sides to be brought nearer after placing the rotor inside the stator. When the end shields are fitted to the stator from two sides with the rotor shaft passing through the bearings, the rotor will rest on the bearings and the rotor will remain separated from the stator by a small air gap. The three-phase windings are made on the stator. The windings, made of a number of coils, are placed in slots in the stator as shown in Fig. 8.1(b). Three-phase winding consists of three identical windings

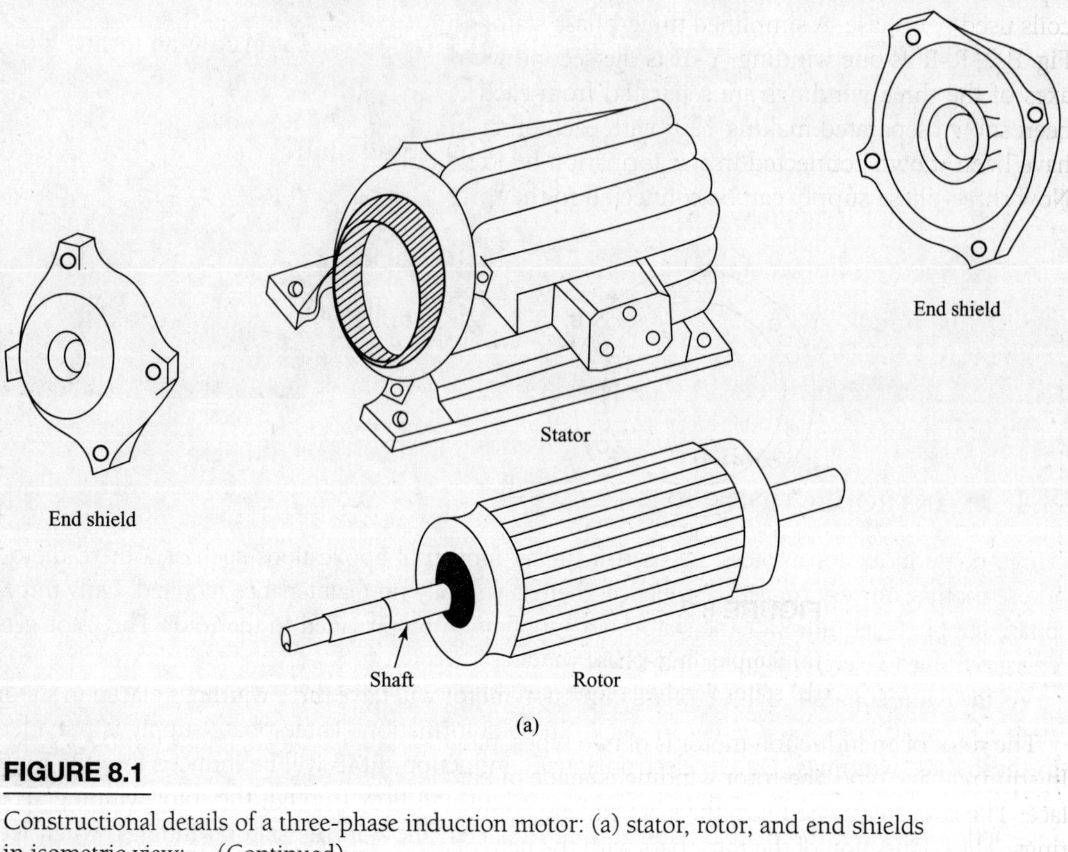

FIGURE 8.1

Constructional details of a three-phase induction motor: (a) stator, rotor, and end shields in isometric view; (Continued)

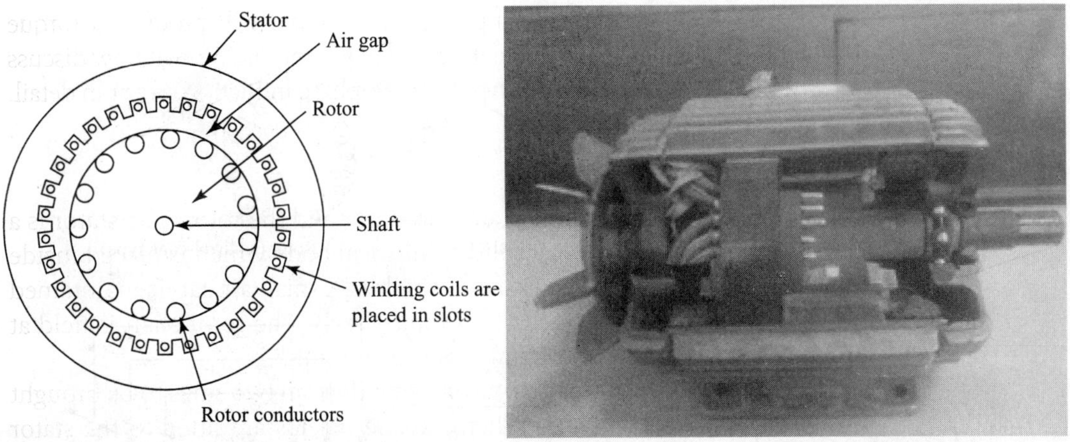

(b) (c)

FIGURE 8.1

(b) cross-sectional view of stator and rotor (c) cut model of an induction motor

separated from each other by 120° in space. In actual practice, there will be large number of coils used per phase. A simplified three-phase stator winding is shown in Fig. 8.2. As shown is Fig. 8.2, R–R′ is one winding, Y–Y′ is the second winding, and B–B′ is the third winding. The axes of the three windings are separated from each other by 120°. The three windings have been shown separated making 120° with each other in Fig. 8.2 (b). The three-phase windings have been shown connected in star formation by joining the end terminals R′, Y′, B′ together. Now three-phase supply can be connected to the three open terminals R, Y, B.

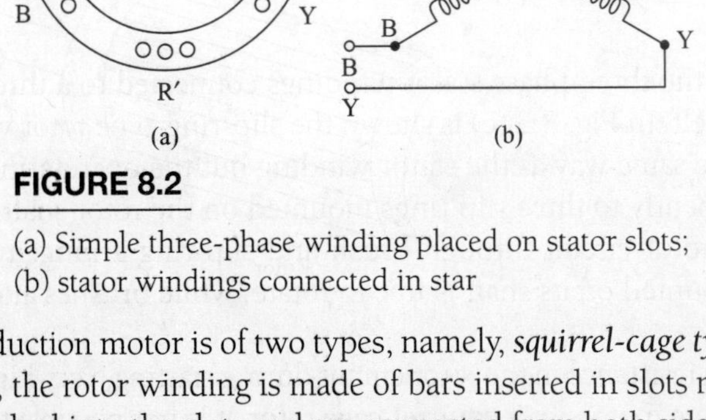

(a) (b)

FIGURE 8.2

(a) Simple three-phase winding placed on stator slots;
(b) stator windings connected in star

The rotor of an induction motor is of two types, namely, *squirrel-cage type* and *slip-ring type*. In squirrel-cage type, the rotor winding is made of bars inserted in slots made on the rotor surface. The bars are pushed into the slots and are connected from both sides through conducting rings. The connection of the rotor bars with the help of *end rings* has been shown in Fig. 8.3 (a).

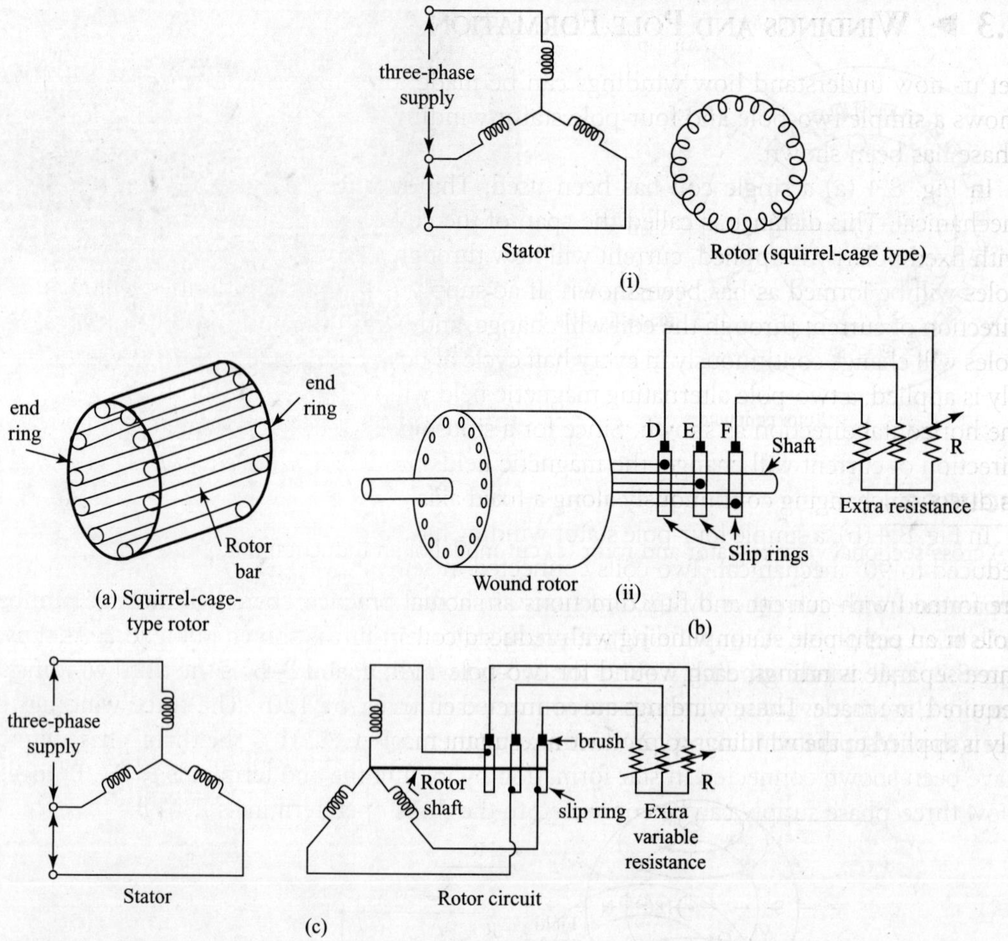

FIGURE 8.3

(a) Squirrel-cage-type rotor; (b) stator and rotor circuits (i) squirrel-cage type, (ii) slip-ring type; (c) slip-ring-type induction motor-stator and rotor circuits

Fig. 8.3 (b) shows the three-phase stator windings connected to a three-phase supply with the rotor closed on itself. In Fig. 8.3 (c) is shown the slip-ring-type rotor where the rotor winding is also made in the same way as the stator winding but the open terminals of the windings are connected permanently to three slip rings mounted on the rotor shaft. Extra resistance can be connected in the rotor circuit through brush and slip-ring arrangement. The rotor along with the slip rings mounted on its shaft is free to rotate, while brushes and the extra resistance are stationary.

By connecting extra resistance in the rotor circuit during starting, very high-starting torque can be developed in slip-ring motors. For squirrel-cage rotor, it is not possible to add any extra resistance in the rotor as the circuit is closed by itself permanently and no terminals are brought out.

8.3 ► Windings and Pole Formation

Let us now understand how windings can be made for different number of poles. Fig. 8.4 shows a simple two-pole and four-pole stator winding. For simplicity, winding for only one phase has been shown.

In Fig. 8.4 (a) a single coil has been used. The distance between the coil sides is 180° mechanical. This distance is called the span of the coil or simply coil span. When dc supply with fixed polarity is applied, current will flow through the coil and a magnetic field with two poles will be formed as has been shown. If ac supply is given, the polarities will change, the direction of current through the coil will change, and hence, the positions of North and South poles will change continuously in every half cycle of power supply. That is to say, when ac supply is applied, a two-pole alternating magnetic field will be produced whose axis will lie along the horizontal direction as shown. Since for a sinusoidal ac supply both magnitude as well as direction of current will change, the magnetic fields produced will have its magnitude as well as direction changing continuously along a fixed axis.

In Fig. 8.4 (b), a simple four-pole stator winding has been shown. Here the coil span has been reduced to 90° mechanical. Two coils connected in series has formed the winding. Four poles are formed with current and flux directions as shown. You can now easily draw a simple six-pole or an eight-pole stator winding with reduced coil spans. In a three-phase induction motor, three separate windings, each wound for two-pole, four-pole, or any even number of poles as required, are made. These windings are connected either in star or in delta and three-phase supply is applied to the windings to produce a resultant magnetic field. The resultant magnetic field

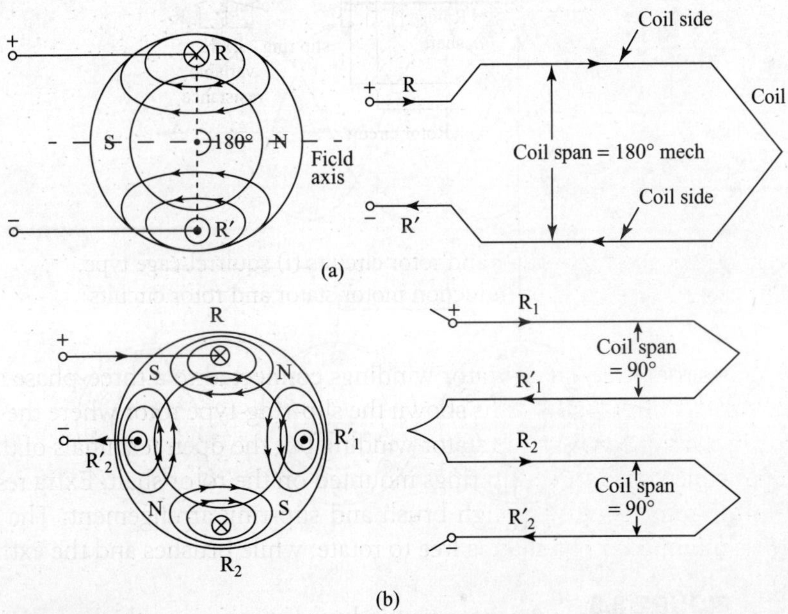

FIGURE 8.4

Simplified two-pole and four-pole stator winding

is the sum total of the magnetic fields produced by the three winding ampere turns. We will soon see that the resultant magnetic field, when a three-phase supply is applied to a three-phase stator winding, is a rotating magnetic field. A rotating magnetic field is one whose axis goes on rotating continuously when supply is given. That is, the angular position of North and South poles goes on shifting, as time passes, at a fast speed. The speed N_s depends upon frequency of power supply f, and the number of poles P, for which the winding has been made as

$$N_s = \frac{120f}{P} \tag{8.1}$$

If

$$f = 50 \text{ Hz}, P = 2,$$

$$N_s = \frac{120 \times 50}{2} = 3000 \text{ rpm (revolutions per minute)}$$

$$= 50 \text{ rps (revolutions per second)}$$

8.4 ► Production of Rotating Magnetic Field

Now let us actually see how the magnetic field rotates when a three-phase supply is connected to a three-phase stator winding. For the sake of understanding, we will consider only three consecutive instants of time of the three-phase supply voltage, show the direction of current flowing through each of the stator windings, and then draw the resultant magnetic field produced.

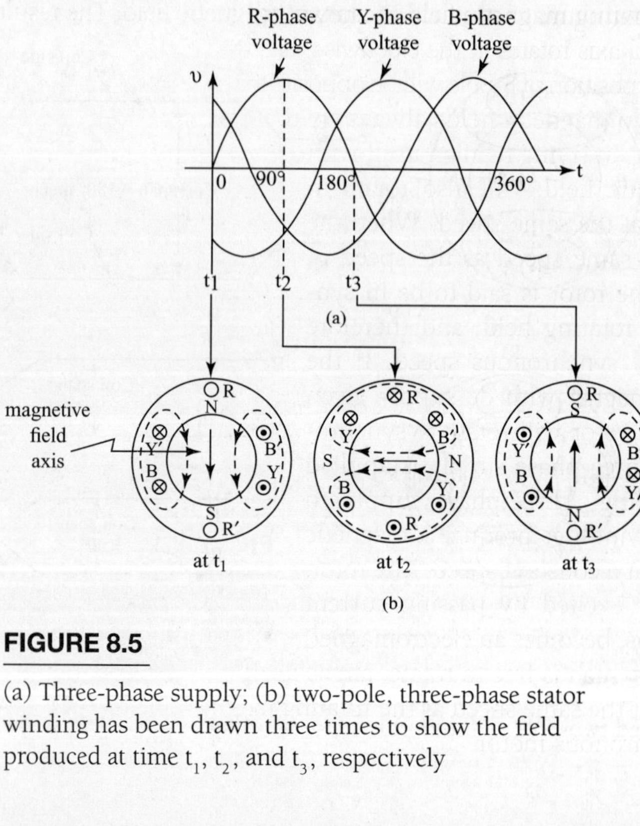

FIGURE 8.5

(a) Three-phase supply; (b) two-pole, three-phase stator winding has been drawn three times to show the field produced at time t_1, t_2, and t_3, respectively

In a three-phase supply three separate sinusoidal voltages having a displacement of 120° with respect to time is available. We will represent a three-phase supply and assume that these are connected to a two-pole three-phase stator winding as shown in Fig. 8.5 (b).

When three-phase supply is connected to R, Y, and B terminals of the stator which are connected in star, current flowing through the phases will be as follows:

at time t_1 R-phase: zero; y-phase: –ve; B-phase: +ve
at time t_2 R-phase: +ve; y-phase: –ve; B-phase: –ve
at time t_3 R-phase: zero; y-phase: +ve; B-phase: –ve

Positive current in a phase, say R-phase, will be shown as entering through conductor R and leaving through conductor R′, and for negative current direction will be just reverse. Positive current will be represented by a cross and negative current will be represented by a dot. Accordingly, all the current directions have been shown and the resultant field drawn. It is observed that for half cycle of current flow, i.e., in time t_1 to t_3, the magnetic-field axis has rotated by 180°. For one complete cycle of current flow, the resultant magnetic field produced will rotate by 360°, i.e., one revolution. As current continues to flow through the three windings, the magnetic-field axis will go on rotating at a speed given by

$$N_s = \frac{120f}{P}$$

If frequency is high, N_s will be high and if number of poles, P is high, N_s will be low.

Now let us consider an electromagnet, or a permanent magnet, which is free to rotate, placed inside the rotating magnetic field as shown in Fig. 8.6.

The magnetic-field axis rotates in the clockwise direction and shifts its position from 1 to 2, 3, 4. . . , and so on. The position of S-pole will be opposite to the position of N-pole, as the poles rotate.

We know that two magnetic fields always try to align with each other.

The rotor magnet, in trying to align itself with the rotating magnetic field, will also rotate in the same direction at the same speed. When the rotor rotates at the same speed as the speed of the rotating field, the rotor is said to be in synchronism with the rotating field, and therefore this speed is called synchronous speed. If the rotor is an electromagnet with dc supply given to its windings, the rotor will rotate at synchronous speed when three-phase supply is applied to the stator winding. Three-phase supply to three-phase stator windings produce a magnetic field which rotates at a constant speed. The rotor, when energized or excited by passing current through its windings, becomes an electromagnet. The rotor, magnet, which is free to rotate, aligns itself with the rotating magnetic field. It will continue to rotate at the same speed as the rotating magnetic field. Such a motor will be called a three-phase synchronous motor.

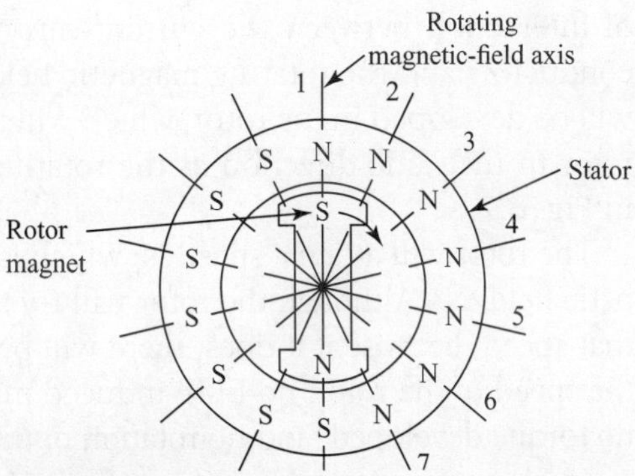

FIGURE 8.6

Two magnetic fields always try to align with each other

In a three-phase induction motor, however, only three-phase supply is applied across the stator windings. No supply is provided to the rotor winding. The rotor is made a closed winding either directly as in the case of squirrel-cage winding; or, through extra resistance inserted in the rotor circuit as in the case of slip-ring motors. The rotor gets excited due to electromagnetic induction. We will now study the principle of working of a three-phase induction motor.

8.5 ► Principle of Working of an Induction Motor

In a three-phase induction motor, the stator is wound with a three-phase winding for P number of poles. The poles for which the winding is made could be 2, 4, 6, 8, … etc. The rotor which is placed inside the stator is either squirrel-cage type or slip-ring type. In both cases, current flowing through the three-phase stator winding produces a rotating magnetic field which will be rotating at a speed, N_s where

$$N_S = \frac{120f}{P}$$

For a 50 Hz supply and P = 2, N_s is 3000 rpm. The rotating field will be rotating continuously at a very high speed. The rotating flux will cut the stationary rotor windings at that speed. Due to this cutting of flux, EMF will be induced in the rotor winding. As the rotor circuit has been made to be a closed winding, current will flow through the rotor-winding conductors. Thus, the rotor circuit gets excited due to electromagnetic induction effect. Because of interaction between the current-carrying rotor conductors and the rotating magnetic field, torque will be developed in the rotor, which will rotate the rotor in the same direction as the rotating magnetic field at a speed N_r as has been shown in Fig. 8.7.

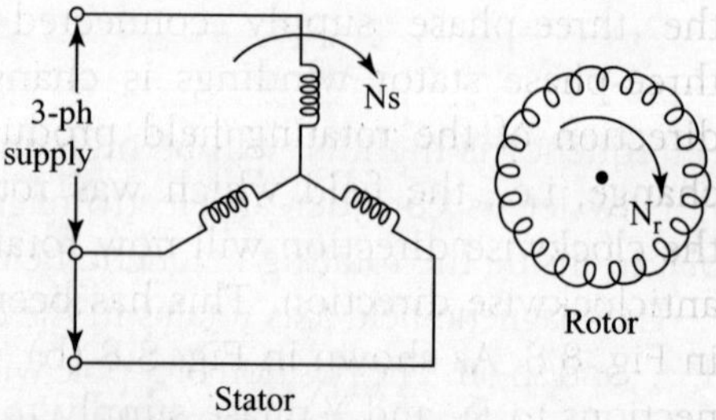

FIGURE 8.7

Stator and rotor windings of a three-phase induction motor

The rotor will attain a speed N_r which is somewhat less than the speed of the rotating magnetic field, N_s. Although the rotor will try to attain a speed of N_s, it will never be able to attain that speed, because if it does, there will be no relative velocity between the rotating field and the speed of the rotor, no EMF induced in the rotor, no current flow in the rotor conductors, no torque developed, and no rotation of the rotor. That is why an induction motor cannot run at synchronous speed, N_s as it is to be excited by electromagnetic induction, which is possible only if there exists a relative velocity between the rotating magnetic field and the rotor.

The difference between the speed of the rotating magnetic field, N_s and the rotor speed N_r is the slip S. Slip is usually expressed as the percentage of N_s, thus,

slip, $$S = \frac{N_s - N_r}{N_s}$$

or, $SN_s = N_s - N_r$ or, $N_r = (1 - S)\,N_s$ (8.2)

Percentage Slip $= \dfrac{N_s - N_r}{N_s} \times 100$

Slip of a three-phase induction motor is generally 3 to 4 per cent. For example, when a 400 V, 3-ph, 50 Hz supply is connected to a four-pole three-phase induction motor, the speed of the rotating field will be 1500 rpm. The rotor will rotate at a speed less than 1500 rpm, may be, say 1440 rpm. In such a case slip S is

$$S = \frac{(1500 - 1440) \times 100}{1500}$$
$$= 4 \text{ per cent}$$

Changing the direction of rotation

It has been observed that if the sequence of the three-phase supply connected to the three-phase stator windings is changed, the direction of the rotating field produced will change, i.e., the field which was rotating in the clockwise direction will now rotate in the anticlockwise direction. This has been shown in Fig. 8.8. As shown in Fig. 8.8 (b), the connections to R- and Y-phase supply to the stator windings have been interchanged but the connection to phase B remains same as in Fig. 8.8 (a). The direction of rotation of the rotating magnetic field will reverse, and hence, the motor will rotate in the opposite direction. Thus, to change the direction of rotation of a three-phase induction motor we have to simply interchange the connections of supply of any two phases.

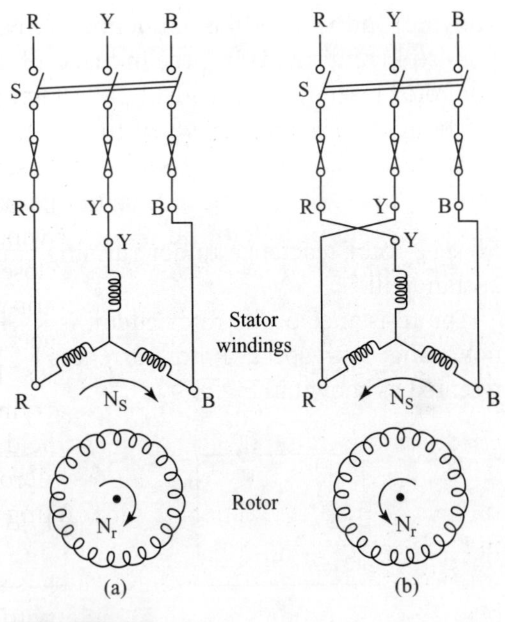

FIGURE 8.8

Method of changing the direction of rotation of a three-phase induction motor

8.6 ► ROTOR-INDUCED EMF, ROTOR FREQUENCY, ROTOR CURRENT

The frequency of supply to the stator is f Hz. The EMF induced in the rotor when the rotor is rotating at a speed N_r is due to the difference of speed of the rotating field and the rotor speed, i.e., $N_s - N_r$.

When the rotor is at standstill, i.e., not yet rotating, the difference between speed of rotation of the stator field and the rotor is $N_s - 0 = N_s$. The EMF induced in the rotor will be maximum and will have a frequency f. When the rotor attains a speed of N_r, the frequency of the induced EMF will get reduced. Let this rotor frequency be called, f_r. When the relative speed between the rotating magnetic field and the rotor is N_s, the frequency of the induced EMF in the rotor is f. When the relative speed is $(N_s - N_r)$, the frequency is f_r. The relationship between f and f_r is found as follows:

when relative speed is N_s, frequency is f.

When relative speed is $(N_s - N_r)$, the frequency, f_r is $\dfrac{f(N_s - N_r)}{N_s} = \dfrac{N_s - N_r}{N_s} \times f$

$$= Sf$$

therefore,

$$f_r = Sf \tag{8.3}$$

i.e., rotor frequency = Slip × Stator frequency

The rotor-induced EMF is a function of the frequency. If the induced EMF in the rotor under standstill condition is E_{20}, the induced EMF, E_2 in the rotor when it is rotating is, $E_2 = S\,E_{20}$. The rotor reactance at standstill, $X_{20} = 2\pi fL$.

The rotor reactance X_2 when the rotor is rotating will be corresponding to the rotor frequency f_r. Thus,

$$X_2 = 2\pi f_r L = 2\pi SfL = S2\pi fL = SX_{20}$$

That is, rotor reactance under running condition is equal to slip times the rotor reactance at standstill.

The resistance of the rotor circuit is R_2. R_2 is independent of rotor speed. Resistance does not change with speed. Rotor current is I_2. The rotor circuit when the rotor is rotating is represented as shown in Fig. 8.9 (a).

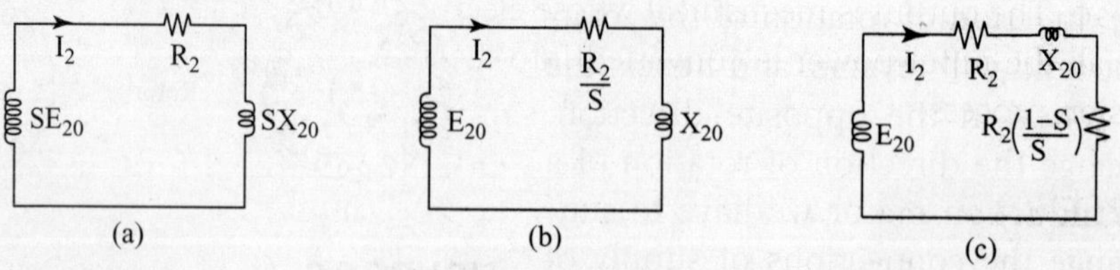

FIGURE 8.9

Rotor circuit when the rotor rotates at a slip, S

The EMF induced in the rotor when it is rotating is SE_{20}, the rotor-circuit resistance is R_2, the rotor-circuit reactance is SX_{20}, and the rotor current is I_2.

From Fig. 8.9 (a) the current in the rotor circuit is

$$I_2 = \frac{S\,E_{20}}{\sqrt{R_2^2 + S^2X_{20}^2}} \tag{8.4}$$

Dividing both numerator and denominator by S,

$$I_2 = \frac{E_{20}}{\sqrt{(R_2/S)^2 + X_{20}^2}}$$

The equivalent circuit representation of the rotor for the above equation has been shown in Fig. 8.9 (b). This is the condition of the circuit when the rotor is at standstill with E_{20} as the EMF induced, I_2 as current flowing, X_{20} as the rotor-circuit reactance, and R_2/S the rotor-circuit resistance. But we know that the rotor-circuit resistance is R_2 and not R_2/S.

To get R_2 as the resistance, we write $\frac{R_2}{S}$ as

$$\frac{R_2}{S} = R_2 + \frac{R_2}{S} - R_2 \text{ (By adding and subtracting } R_2)$$

$$= R_2 + R_2\left(\frac{1}{S} - 1\right)$$

$$= R_2 + R_2\left(\frac{1-S}{S}\right)$$

This has been shown in Fig. 8.9 (c).

In $\frac{R_2}{S}$ we see two resistances, viz rotor-circuit resistance R_2 and another resistance $R_2\left(\frac{1-S}{S}\right)$. The question that arises is what is the significance of resistance $R_2\left(\frac{1-S}{S}\right)$. The power lost in this resistance will be equal to $I_2^2 R_2\left(\frac{1-S}{S}\right)$. This power is the electrical equivalent of the mechanical load on the motor shaft.

This power is, therefore, called fictitious (not real but equivalent to) electrical load representing the mechanical power output of the motor. In an induction motor the input power is $\sqrt{3}\, V_L I_L \cos\phi$. The output is mechanical whose electrical-equivalent power has been found as above. Not all the input power is converted into mechanical power: some power gets lost in the conversion process.

EXAMPLE 8.1

A four-pole, three-phase induction when supplied with 400 V, 50 Hz supply rotates at a slip of 4 per cent. What is the speed of the motor?

SOLUTION

$$N_s = \frac{120f}{p} = \frac{120 \times 50}{4} = 1500 \text{ rpm}$$

$$S = \frac{N_s - N_r}{N_s} \quad \text{Substituting values, } 0.04 = \frac{1500 - N_r}{1500}$$

or,

$$N_r = 1500 - 60 = 1440 \text{ rpm}$$

8.7 ► Losses in Induction Motors

The input power to an induction motor is taken from the supply mains. For a three-phase induction motor the input power is $3\,V_{Ph}\,I_{ph}\cos\phi$, or $\sqrt{3}\,V_L I_L \cos\phi$, where V_{ph}, I_{ph} are the phase values and V_L, I_L are the line values. This electrical power input is converted into mechanical power output at the shaft. Power is transferred from the stator to the rotor shaft via the air gap. The whole of the input does not get converted into mechanical output. Some power gets lost in the stator windings as I^2R loss, in the stator core as core loss. Similarly, power is also lost in the rotor as I^2R loss and as core loss. When the rotor rotates against the wind (air) friction, power is lost as windage loss and bearing-friction loss and brush-friction loss (in case of slip-ring motors only). Iron loss is the sum of hysteresis loss and eddy curent loss. Hysteresis and eddy current losses depend upon the frequency and flux density. The frequency of the induced EMF and current in the rotor is very small ($f_r = sf$). Therefore, rotor core loss is considered negligible. Stator core loss is constant at all loads since the supply voltage and frequency are normally constant. I^2R losses in the stator and rotor windings are variable, i.e., they vary with change of load (i.e., load current).

Thus, the various losses are

(a) stator copper loss;
(b) rotor copper loss;
(c) iron loss in stator;
(d) iron loss in the rotor (very small);
(e) air-friction loss due to rotation of the rotor;
(f) bearing-friction loss;
(g) brush- and slip-ring-friction loss (in case of slip-ring motors only).

8.8 ► Power Flow Diagram

The flow of power in an induction motor from stator to rotor is depicted in the form of a diagram as in Fig. 8.10. In the figure we have considered rotor core loss as negligible. Power transferred from stator to rotor is through the rotating magnetic field. The rotating magnetic field is rotating at synchronous speed, N_s. Therefore, the power transferred is $\frac{2\pi TN_s}{60}$ W. The same power is transferred to the rotor through the magnetic field and there is no loss in the air gap in this transfer. Therefore, stator output is taken as equal to rotor input. When the rotor rotates at a speed N_r, the power developed by the rotor is $=\frac{2\pi TN_r}{60}$ W. The difference between rotor input and rotor power developed is expressed as (see Fig. 8.10),

$$\frac{2\pi TN_s}{60} - \frac{2\pi TN_r}{60} = \text{Rotor } I^2R\text{-loss (neglecting rotor core loss)}$$

Mechanical power output is somewhat less than the power developed. This is because of friction and windage losses.

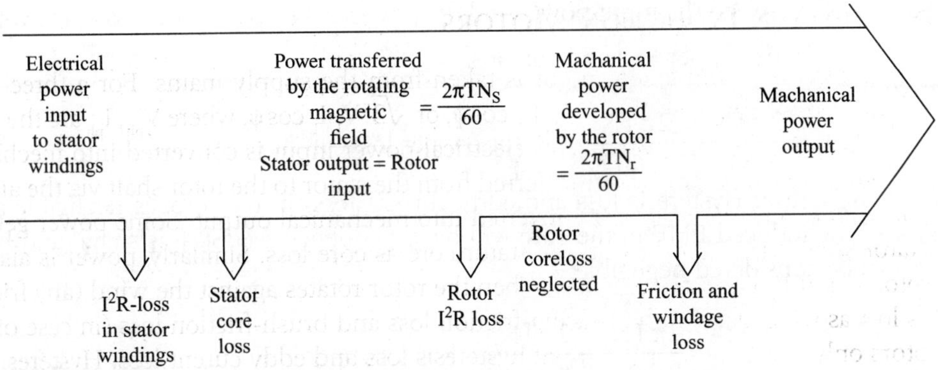

FIGURE 8.10

Power flow diagram of an induction motor

8.9 ▶ Torque Equation

When three-phase supply is applied to the three-phase stator windings of an induction motor, as shown in Fig. 8.11, a rotating magnetic field is produced which rotates at synchronous speed, N_s where

$$N_s = \frac{120f}{P}$$

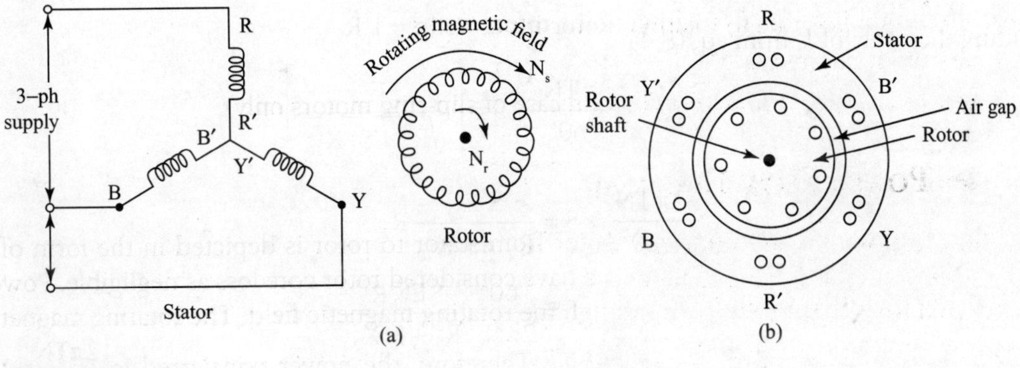

FIGURE 8.11

(a) Stator and rotor windings of a three-phase induction motor;
(b) simplified corss-sectional view

Through this rotating magnetic field, power is transferred from stator to the rotor via the air gap. The power transferred from stator to rotor through the magnetic field in equal to $\frac{2\pi TN_s}{60}$. This is the input power to the rotor since there in no power loss in the air gap. Torque will be developed in the rotor which will cause the rotor to rotate at a speed N_r. The power developed by the rotor is $\frac{2\pi TN_r}{60}$.

The difference between the input power and the power developed is the loss in the rotor. In the rotor there will be copper loss and core loss.

Therefore,

$$\frac{2\pi TN_s}{60} - \frac{2\pi TN_r}{60} = \text{Rotor copper loss + Rotor core loss}$$

The core loss (sum of Hysteresis loss and eddy current loss) in the rotor is negligible as the frequency of the induced EMF in the core will have frequency, $f_r = Sf$. f_r is small, and hence, core loss can be considered negligible.

Thus,

$$\frac{2\pi TN_s}{60} - \frac{2\pi TN_r}{60} = \text{Rotor copper loss}$$

Dividing both sides by rotor input

$$\frac{(2\pi TN_s/60) - (2\pi TN_r/60)}{(2\pi TN_s/60)} = \frac{\text{Rotor copper loss}}{\text{Rotor input}}$$

$$\frac{N_s - N_r}{N_s} = \frac{\text{Rotor copper loss}}{\text{Rotor input}}$$

$\therefore$

$$\text{Rotor copper loss} = \text{Slip} \times \text{Rotor input} \tag{8.5}$$

The above is an important relationship which can also be used to arrive at the expression for torque as has been shown below.

$$\text{Rotor input} \times \text{slip} = I_2^2 R_2$$

Putting the value of I_2 from eq. (8.4)

$$\frac{2\pi TN_s}{60} \times S = \left[\frac{SE_{20}}{\sqrt{R_2^2 + S^2 X_{20}^2}}\right]^2 R_2$$

$$\frac{2\pi TN_s}{60} \times S = \frac{S^2 E_{20}^2}{R_2^2 + S^2 X_{20}^2} R_2$$

or,

$$T = \frac{60}{2\pi N_s} \frac{SE_{20}^2 R_2}{R_2^2 + S^2 X_{20}^2}$$

If the supply voltage, V is constant, then the induced EMF E_{20} in the rotor at standstill will be constant (like in a transformer)

Therefore,

$$T = K\frac{SR_2}{R_2^2 + S^2 X_{20}^2}, \quad \text{where} \quad K = \frac{60 \times E_{20}^2}{2\pi N_s} \tag{8.6}$$

The value of S varies from 0 to 1. Slip is 0 if the rotor is able to rotate at synchronous speed, N_s. At the moment of start, rotor speed N_r is zero, and hence, slip is 1 or 100 per cent. Rotor resistance R_2 is much smaller than the standstill rotor reactance, X_{20}.

Using the expression for torque, we can draw the torque versus slip (or speed) characteristic of the three-phase induction motor.

8.10 ► Starting Torque

At start, rotor speed $N_r = 0$, therefore, $S = \dfrac{N_s - N_r}{N_s} = \dfrac{N_s}{N_s} = 1$

From the torque equation with S = 1

starting torque,
$$T_s = \frac{KR_2}{R_2^2 + X_{20}^2} \tag{8.7}$$

The value of starting torque will depend upon the rotor circuit parameters, i.e., R_2 and X_{20}. The value of X_{20} is generally higher than R_2. Let us assume $R_2 = 1\ \Omega$ and $X_{20} = 8\ \Omega$.

Let us, for the sake of interest, study the value of T_s if R_2 is increased (which can be done by including an extra resistance in the rotor circuit while starting the motor).

With $R_2 = 1$ and $X_{20} = 8$, and using K to be having a value, say 100

$$T_s = \frac{100 \times 1}{1 + 8^2} = \frac{100}{65} = 1.54 \text{ Nm}$$

Now let us make $R_2 = 2\ \Omega$, Then T_s is

$$T_s = \frac{100 \times 2}{2^2 + 8^2} = \frac{200}{68} = 2.95 \text{ Nm}$$

We further increase R_2 and make $R_2 = 4\ \Omega$

then,
$$T_s = \frac{100 \times 4}{4^2 + 8^2} = \frac{400}{80} = 5 \text{ Nm}$$

Thus, we see that if rotor-circuit resistance is increased, we can achieve higher starting torque developed by the motor. High-starting torque of the drive motor is a requirement in many applications.

8.11 ► Condition for Maximum Torque

Condition for maximum torque can be found out by maximizing the expression for torque, i.e., by differentiating T with respect to slip, S and equating to zero. Let us consider the torque equation which is

$$T = K\frac{SR_2}{R_2^2 + S^2 X_{20}^2}$$

or,
$$T = K\frac{R_2}{(R_2^2/S) + SX_{20}^2}$$

To maximize T, we have to minimize the denominator we differentiate the denominator with respect to S and equate to zero, i.e.,

$$\frac{dT}{dS}\left(\frac{R_2^2}{S} + SX_{20}^2\right) = 0$$

or, $$\frac{dT}{dS}\left(R_2^2 S^{-1} + S X_{20}^2\right) = 0$$

or, $$-R_2^2 S^{-2} + X_{20}^2 = 0$$

or, $$-\frac{R_2^2}{S^2} + X_{20}^2 = 0$$

or, $$R_2^2 = S^2 X_{20}^2$$

or, $$R_2 = S X_{20}, \quad \text{or} \quad S = \frac{R_2}{X_{20}}$$

Thus, the condition for maximum torque is established. Maximum torque in an induction motor will occur at a slip at which,

$$S = \frac{R_2}{X_{20}} \tag{8.8}$$

If we take $R_2 = 1\ \Omega$ and $X_{20} = 8\ \Omega$, maximum torque will occur when the slip is $\frac{1}{8}$, i.e., 0.125.

If $R_2 = 2\ \Omega$ and $X_{20} = 8\ \Omega$, maximum torque will occur when the slip is $\frac{1}{4}$, i.e., 0.25.

When value of R_2 equals the value of X_{20}, i.e., when $R_2 = 8\ \Omega$, and $X_{20} = 8\ \Omega$, maximum torque will occur at $S = 1$, i.e., at starting.

EXAMPLE 8.2

A four-pole, three-phase induction motor is supplied with 400 V, 50 Hz supply. The rotor-circuit resistance is 2 Ω and standstill rotor-circuit reactance is 8 Ω. Calculate the speed at which maximum torque will be developed.

SOLUTION

Given, f = 50 Hz, P = 4. $R_2 = 2\ \Omega$, $X_{20} = 8\ \Omega$

$$N_s = \frac{120f}{P} = \frac{120 \times 50}{4} = 1500 \text{ rpm}$$

Condition for maximum torque is given by

$$R_2 = S X_{20}$$

substituting values $$S = \frac{R_2}{X_{20}} = \frac{2}{8} = 0.25$$

at this slip, rotor speed, $$N_r = (1 - S)N_s$$
$$= (1 - 0.25)1500 = 1125 \text{ rpm}$$

The value of maximum torque

We will put the condition for maximum torque, i.e., $R_2 = S X_{20}$ in the torque equation as

$$T = K\frac{SR_2}{R_2^2 + S^2 X_{20}^2}$$

putting $R_2 = SX_{20}$, Maximum torque T_m is

$$T_m = \frac{K.S.SX_{20}}{S^2 X_{20}^2 + S^2 X_{20}^2}$$

or,

$$T_m = \frac{KS^2 X_{20}}{2S^2 X_{20}^2}$$

or,

$$T_m = \frac{K}{2X_{20}} \tag{8.9}$$

Thus, we see that the value of maximum torque, T_m is independent of rotor resistance R_2 but the slip at which maximum torque is developed changes with value of R_2.

8.12 ► Torque–Slip Characteristic

The variation of torque with slip or speed can be studied using the torque equation. The shape of the torque–slip characteristic is predicted as follows:

Let us write the torque equation derived earlier. Referring to (8.7), we have

$$T = \frac{KSR_2}{R_2^2 + S^2 X_{20}^2}$$

Let us assume $R_2 = 1\ \Omega$ and $X_{20} = 8\ \Omega$.

The value of S changes from 0 to 1. When the value of S is very small, say 0.01, 0.02, 0.03, 0.04, etc., the value of $S^2 X_{20}^2$ will be small as compared to R_2^2. Thus, for small values of S, we can write

$$T = \frac{K}{R_2}S$$

or,

$$T \propto S$$

and for larger values of S, say 0.2, 0.3, 0.4, etc., R_2^2 is smaller than $S^2 X_{20}^2$, and hence, we can write

$$T = \frac{KSR_2}{S^2 X_{20}^2} = \frac{KR_2}{X_{20}^2}\,\frac{1}{S}$$

or,

$$T \propto \frac{1}{S}$$

Thus, we see that for lower values of slip, torque is directly proportional to slip, and for higher values of slip, torque is inversely proportional to slip. We further calculate the values of torque at S = 0 and S = 1. Also, we calculate the value of maximum torque and the slip at which maximum torque occurs. These values will help us draw the torque–slip characteristic.

At $$S = 0, T = 0 \text{ Nm}$$

With assumed value of $$K = 100,\ T_m = \frac{K}{2X_{20}} = \frac{100}{2 \times 8} = 6.25 \text{ Nm}$$

The value of slip at which T_m occurs is

$$S = \frac{R_2}{X_{20}} = \frac{1}{8} = 0.125$$

and the value of starting torque at S = 1 is

$$T_s = \frac{KR_2}{R_2^2 + X_{20}^2} = \frac{100 \times 1}{1^2 + 8^2} = \frac{100}{65} = 1.54 \text{ Nm}$$

We will now draw the torque–slip characteristic using the above values.

Since just at start, speed is zero, and slip is 1, we can represent S = 1 as $N_r = 0$. Again when S = 0, N_r must be equal to N_s since

$$S = \frac{N_s - N_r}{N_s}$$

Thus, the torque–slip characteristic can also be shown as torque–speed characteristic as has been shown in Fig. 8.12.

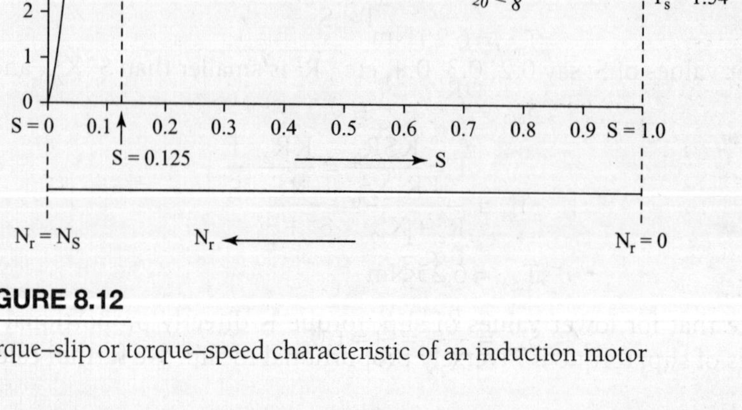

FIGURE 8.12

Torque–slip or torque–speed characteristic of an induction motor

8.13 ► Variation of Torque–Slip Characteristic with Change in Rotor–Circuit Resistance

We will now study the effect of variation of rotor-circuit resistance on the torque-slip characteristics of an induction motor. Variation of rotor circuit resistance is possible in slip-ring type motors only.

We have known that starting torque, maximum torque, and the slip, S at which maximum torque occurs changes with change of rotor-circuit resistance, the reactance remaining unchanged. The basic shape of the torque–slip charactesic however remains the same. We will draw four torque–slip characteristic of a three-phase induction motor for

(i) $R_2 = 1\ \Omega, X_{20} = 8\ \Omega$ (iii) $R_2 = 4\ \Omega, X_{20} = 8\ \Omega$

(ii) $R_2 = 2\ \Omega, X_{20} = 8\ \Omega$ (iv) $R_2 = 8\ \Omega, X_{20} = 8\ \Omega$

Firstly, we calculate T_s, T_m, S at which T_m occurs.

(i)
$$T_{s1} = \frac{KR_2}{R_2^2 + X_{20}^2} = \frac{100 \times 1}{1^2 + 8^2} = 1.54 \quad (\text{assuming } K = 100)\ \text{Nm}$$

$$T_{m1} = \frac{K}{2X_{20}} = \frac{100}{2 \times 8} = 6.25\ \text{Nm}$$

$$S_{m1} = \frac{R_2}{X_{20}} = \frac{1}{8} = 0.125$$

(ii)
$$T_{s2} = \frac{KR_2}{R_2^2 + X_{20}^2} = \frac{100 \times 2}{2^2 + 8^2} = \frac{200}{68} = 2.95\ \text{Nm}$$

$$T_{m2} = \frac{K}{2 \times 20} = \frac{100}{2 \times 8} = 6.25\ \text{Nm}$$

$$S_{m2} = \frac{R_2}{X_{20}} = \frac{2}{8} = 0.25$$

(iii)
$$T_{s3} = \frac{KR_2}{R_2^2 + X_{20}^2} = \frac{100 \times 6}{4^2 + 8^2} = 5.0\ \text{Nm}$$

$$T_{m3} = 6.25\ \text{Nm}$$

$$S_{m3} = \frac{R_2}{X_{20}} = \frac{4}{8} = 0.5$$

(iv)
$$T_{s4} = \frac{KR_2}{R_2^2 + X_{20}^2} = \frac{100 \times 8}{8^2 + 8^2} = 6.25\ \text{Nm}$$

$$T_{m4} = 6.25\ \text{Nm}$$

$$S_{m4} = \frac{R_2}{X_{20}} = \frac{8}{8} = 1.0$$

To draw the torque–slip characteristics, we will take the value of T_{s1}, T_{m1}, and S_{m1} in each case. For example, with $R_2 = 1\ \Omega$, $X_{20} = 8\ \Omega$, $T_{s1} = 1.54$. $S_{m1} = 0.125$ and $T_{m1} = 6.25$ and with T = 0 at S = 0, we can draw the T–S characteristic with its slope on T α S for lower values of slip and $T\alpha\frac{1}{S}$ for higher values of slip.

The calculated values of starting torque, maximum torque, the slip at which maximum torque occurs for different values of rotor circuit resistance, with an assumed value of constant k, are shown below.

Rotor circuit resistance R_2	Stand still value of rotor circuit reactance X_{20} (constant)	Starting torque, T_{st}	Maximum torque, T_m (remain constant)	Slip at maximum torque S_m
1.0 Ω	8.0 Ω	1.54 Nm	6.25 Nm	0.125
2.0 Ω	8.0 Ω	2.95 Nm	6.25 Nm	0.25
4.0 Ω	8.0 Ω	5.0 Nm	6.25 Nm	0.50
8.0 Ω	8.0 Ω	6.25 Nm	6.25 Nm	1.0
12.0 Ω	8.0 Ω	5.77 Nm	6.25 Nm	1.5
16.0 Ω	8.0 Ω	5.00 Nm	6.25 Nm	2.0

We can now draw the four characteristics as shown in Fig. 8.13.
By observing the characteristics, we can conclude that

(i) increasing rotor-circuit resistance increases the starting torque;
(ii) maximum torque remains the same for all values of rotor-circuit resistance;

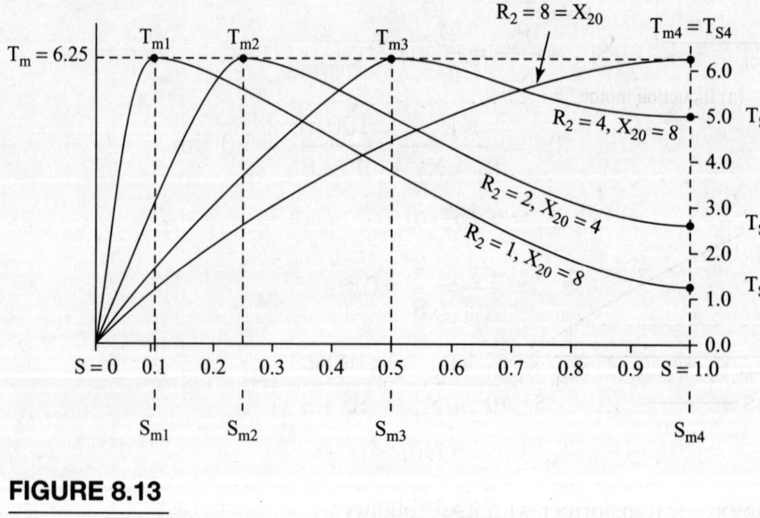

FIGURE 8.13

Torque–slip characteristics with increasing rotor-circuit resistance

(iii) starting torque becomes equal to maximum torque when rotor-circuit resistance is made equal to rotor-circuit reactance;

(iv) slip at which maximum torque is developed changes with change of rotor-circuit resistance;

(v) increasing rotor-circuit resistance beyond the value of rotor-circuit reactance will reduce the starting torque (this can be verified by calculating T_s at $R_2 = 12\ \Omega$, $16\ \Omega$ and $X_{20} = 8\ \Omega$ as shown in the table.

8.14 ► STARTING OF INDUCTION MOTORS

There is similarity between a transformer and an induction motor. An induction motor is like a short-circuited transformer. This has been shown in Fig. 8.14. The only difference is that the secondary, i.e., the rotor in an induction motor is cylindrical in shape and is free to rotate while a transformer is totally a static device. The secondary winding, i.e., the rotor winding of an induction motor is closed, i.e., shorted. The similarity between an induction motor and a short-circuited transformer has been shown in Fig. 8.14 (a and b).

With the secondary winding of a transformer short circuited, if full voltage is applied across the primary windings, very high current will flow through the windings. As in the case of transformers, when full voltage applied across stator terminals of an induction motor and the rotor is stationary, very high current will flow through the rotor and stator windings. If this

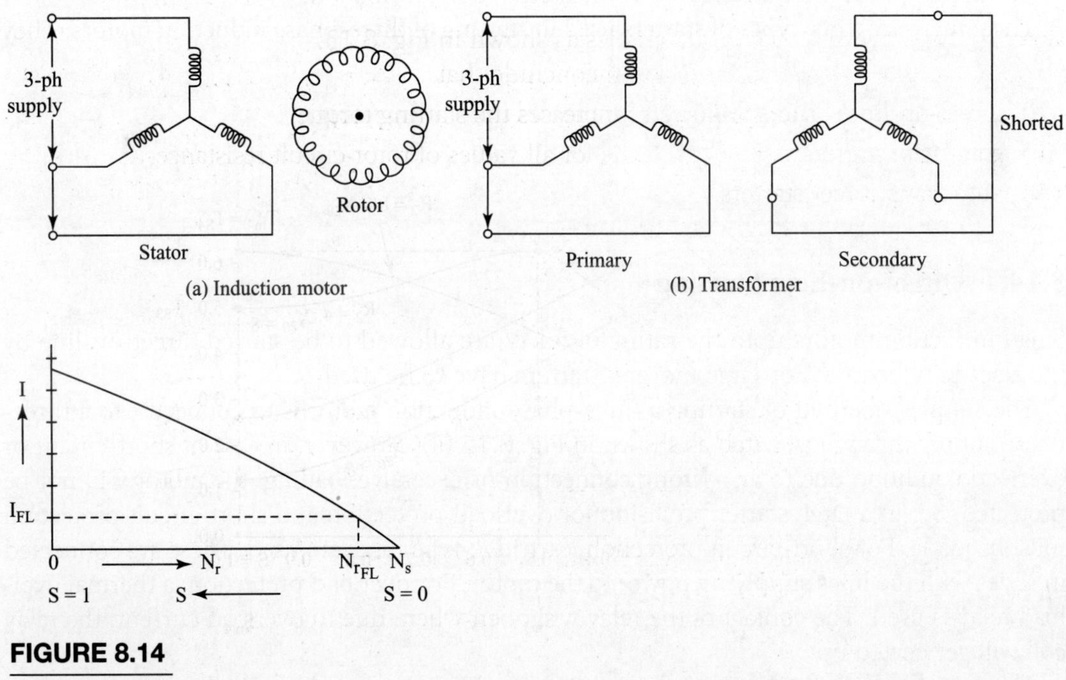

FIGURE 8.14

(a) Induction motor; (b) transformer with its secondary windings shorted;
(c) high current drawn by an induction motor if started with full voltage

high current is allowed to flow for a considerable time the motor windings will be burnt out. However, as the motor picks up speed, the EMF induced in it will be SE_{20} so that

$$I_2 = \frac{SE_{20}}{\sqrt{R_2^2 + S^2 X_{20}^2}}$$

As speed increases, i.e., slip decreases, I_2 will go on decreasing; accordingly the current drawn from the supply will also be gradually reducing as has been shown in Fig. 8.14 (c). An induction motor at starting, therefore, will draw very high current if started with full voltage applied across its stator terminals. The starting current may be as high as six-times its full-load current. Thus, it becomes necessary to limit the starting current of an induction motor.

Starting of three-phase induction motors by applying full voltage directly to the stator windings is restricted to small motors upto 5 kW rating. If large motors with large inertia load on shaft are started this way, heavy current will be drawn (usually six to eight times the rated current) by the motors. This will not only be harmful to the motors but will also create heavy voltage drop in the electrical distribution lines, which will disturb the working of other electrical gadgets and machines connected to the line. Higher the rating of the motor, higher will be the disturbance of the line voltage. Starting of motors upto 5 kW rating may be done by applying full voltage. This is called direct-on-line or DOL starting of motors. Reduced voltage starting should be done for larger motors by using a three-phase auto-transformer or by connecting the stator windings of the motor first in star formation and giving the supply. As the motor picks up sufficient speed, the windings are connected in delta formation.

Thus, there are three types of starters used in starting of three-phase induction motors. They are

(i) direct-on-line starters; (allowed for motors upto 5 kw rating);
(ii) star–delta starters;
(iii) auto-transformer starters.

8.14.1 Direct-on-Line Starting

Small induction motors upto the rating of 5 kW are allowed to be started direct-on-line by the electricity boards. For large motors, starters have to be used.

The simplest method of starting a three-phase induction motor is to connect it to a three-phase supply through a switch as shown in Fig. 8.15 (a). However, in case of short-circuit or overload condition due to any wrong connection or excessive loading, the motor will not be protected. So, in a DOL starter provision for overload protection and short-circuit protection must be made. For short-circuit protection, cartridge-type fuses are used. These are connected in series with the lines supplying power to the motor. For overload protection, a thermal overload relay is used. The contact of the relay will open when, due to overload current, the relay coil will get heated up.

The switching of supply to the motor is done through a contactor and not by a manual switch. Push-button switches are provided for starting and stopping the motor. When the

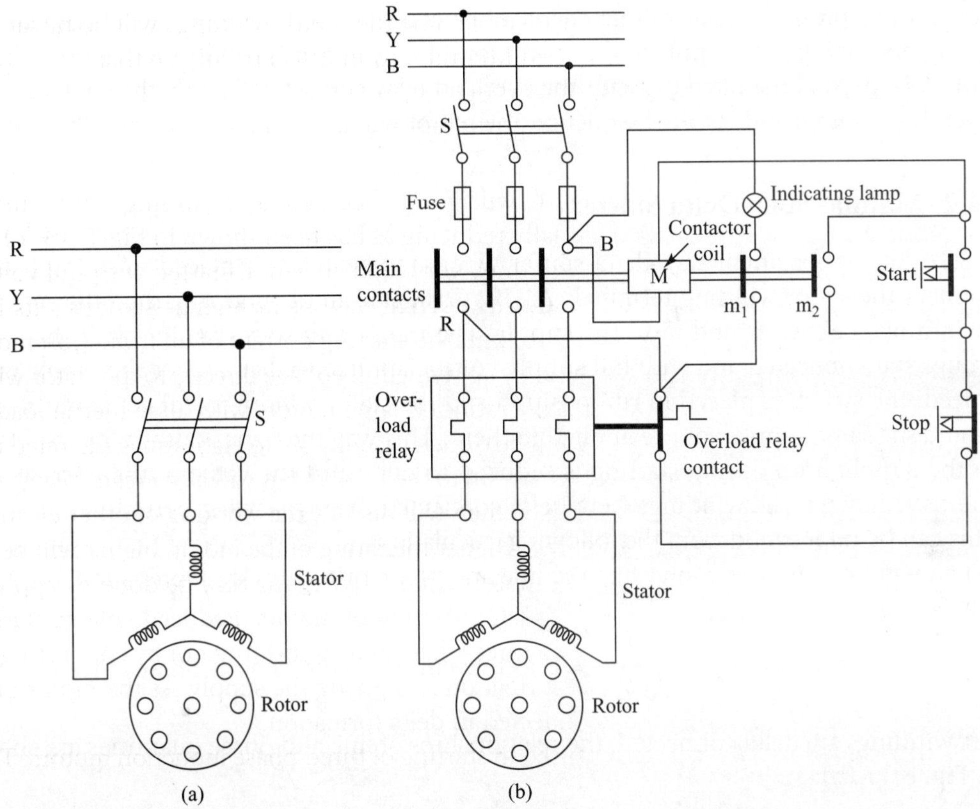

FIGURE 8.15

DOL starting of three-phase induction motors: (a) direct connections to supply through a switch but without any overload or short-circuit protection; (b) push-button-operated DOL starter with overload and short-circuit protection

START push button is pressed, the contactor coil gets energized, its main contacts close, and the motor starts running. For stopping the motor the STOP push button is to be pressed due to which the contactor coil will not receive supply and will be de-energized. The contacts of the contactor will open, thereby stopping the motor. In case of over load also the contactor will get de energized due to opening of the overload relay contact. The working of the DOL starter is further explained using Fig. 8.15 (b) as follows.

When the start push button is pressed the contactor coil, M gets energized as the coil gets connected between R-phase and B-phase. The main contacts of the contactor as well as the auxiliary contacts m_1 and m_2 will close. The motor will get full supply and start rotating. Since the auxiliary contact m_2, is closed, pressure on the start push button can be released and the contactor will continue to remain energized. Indicating lamp will get supply through contact m_1, and will glow, showing that the motor is running. To switch off the motor the stop push button is to be pressed which will de-energize the contactor coil thereby opening its contacts and disconnecting power supply to the motor. In case the motor is overloaded, current drawn will increase beyond the rated capacity, the overload relay contact will open, the contactor coil will get de-energised and, as a consequence, the motor will get disconnected from the supply.

8.14.2 Manual Star–Delta Starter

Fig. 8.16 Shows a manual star–delta starter where the switch S is placed in start position to connect the stator winding terminals A_2, B_2, C_2 in star and connecting the other ends of the windings, i.e., A_1, B_1 and C_1 to the supply. The motor will start running with the stator windings star connected and with full supply voltage applied to them. Once the motor picks up speed, the switch is placed in run position and the stator windings will get connected in delta and the same supply voltage applied to them. This way the current drawn by the motor from the supply lines during starting is reduced to one-third the value of the current that would have flown if the windings were delta connected during starting period.

This can be understood from the following calculations.

When windings are star connected, the line and phase quantities are represented as shown in Fig. 8.17 (a).

Here, from Fig. 8.17 (a), $I'_{Ph} = \dfrac{V}{\sqrt{3}\, Z_{ph}} = I'_L$ as line current is the same as phase current.

When windings are delta connected, the relationships of line and phase quantities are shown as in Fig. 8.17 (b).

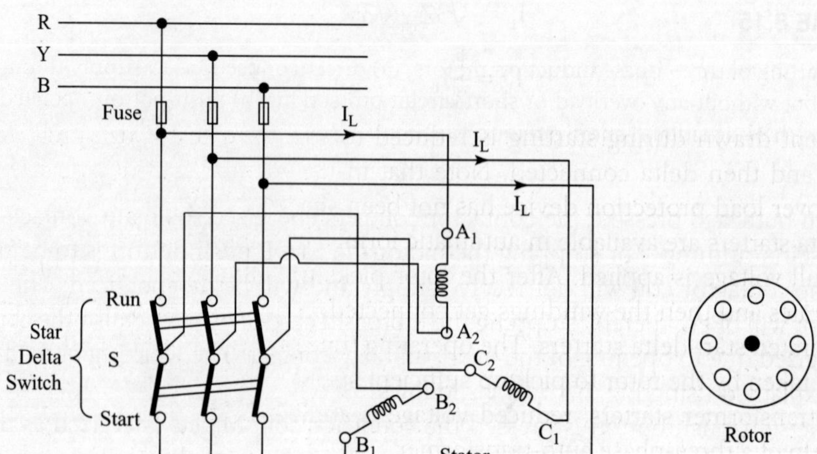

FIGURE 8.16

Manual star–delta starter

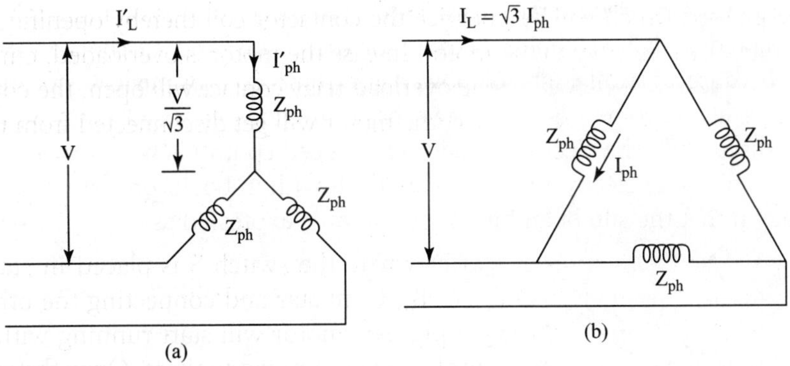

FIGURE 8.17
(a) Stator windings star connected during starting;
(b) stator windings delta connected in running condition

Here, $$I_{Ph} = \frac{V}{Z_{Ph}}$$

$$I_L = \sqrt{3}\, I_{Ph} = \frac{\sqrt{3}V}{Z_{Ph}}$$

If we take the ratio of line current drawn when the windings are star connected to the line current drawn when they are delta connected, we get

$$\frac{I_L'}{I_L} = \frac{V \times Z_{Ph}}{\sqrt{3}Z_{Ph}\,\sqrt{3}V} = \frac{1}{3}$$

or, $$I_L' = \frac{1}{3} I_L$$

Thus, current drawn during starting is reduced to one-third if the windings are first star connected and then delta connected. Note that in the manual star–delta starter shown in Fig. 8.16, over load protection device has not been shown.

Star–delta starters are available in automatic form. The windings are first connected in star and then full voltage is applied. After the rotor picks up sufficient speed, a time-delay relay (TDR) operates and then the windings get connected in delta. Such starters are called push-button-operated star–delta starters. The operating time of the TDR can be adjusted according to the time taken by the rotor to pick up sufficient speed.

In auto-transformer starters, reduced voltage is applied to the stator windings at starting with the help of a three-phase auto-transformer.

8.15 ▶ Speed Control of Induction Motors

Different types of dc and ac electric motors are manufactured for use as drives in various kinds of industrial applications. The selection criterion for use are their ruggedness, cost, ease of use, starting torque, supply requirement, and control of speed. Induction motors have become universally accepted as a first choice because of their satisfying all the above-mentioned requirements. Because of availability of power electronic control devices, speed control of induction motors has become easy now. The basic methods of speed control of induction motors will be discussed. The electronic method of speed control is beyond the scope of this book. The details of use of electronic control are dealt with in books on power electronics.

We have known that the slip of an induction motor is expressed as

$$S = \frac{N_s - N_r}{N_s}$$

or,
$$SN_s = N_s - N_r$$

or,
$$N_r = N_s(1 - S)$$

or,
$$N_r = \frac{120f}{P}(1 - S)$$

where N_r is the speed of the motor, f is the frequency of supply, and P is the number of poles of the motor.

This equation indicates that the speed N_r of the motor can be changed by any of the following methods:

(a) by changing the supply frequency f;
(b) by changing the number of poles, p for which the stator windings are wound;
(c) by changing the slip;
(d) in addition, speed of slip-ring induction motors can be changed by changing the rotor-circuit resistance.

These methods are described in brief as follows:

8.15.1 Control of Speed by Changing Supply Frequency

By changing the supply frequency, the speed of the motor can be increased or decreased smoothly. However, the supply frequency available from the electricity supply authority is at 50 Hz which is fixed. Frequency conversion equipment is, therefore, needed for speed control of motors. Variable frequency supply can be obtained from a separate motor generator set, rotary convertors, or solid-state electronic devices. The frequency changing device should change frequency and applied voltage simultaneously as a direct ratio. If frequency is increased, the supply voltage must also be increased and if frequency is decreased, supply voltage must also be decreased proportionally. This will keep the torque developed constant and the operating efficiency high.

8.15.2 Control of Speed by Pole Changing

Three techniques, viz (i) use of separate stator windings wound for two different number of poles; (ii) use of consequent pole technique, and (iii) use of pole-amplitude modulation can be used.

Instead of one winding for, say eight poles *on the stator we may use two separate windings* insulated from each other, say one for eight poles and the other for 10 poles. The synchronous speed corresponding to P = 8 and P = 10 will be 750 rpm and 600 rpm, respectively. The rotor speed corresponding to these synchronous speeds will be somewhat less than these synchronous speeds. Thus, we will get two rotor speeds by having a switching arrangement of power supply to the two stator windings as per our speed requirement. It may be noted that the rotor poles are formed by induction effect, and hence will be the same as the stator number of poles. If the stator number of poles are increased, the rotor poles will also increase. The number of poles of the stator and rotor must be the same.

In the *consequent pole technique*, terminals are brought out from the stator winding and by proper switching arrangement the number of poles formed by the stator current is changed. The technique is illustrated in Fig. 8.18. For, simplicity, only two coils have been shown forming the stator winding. Through switching arrangement, supply to the windings are changed as shown in Fig. 8.18 (a) and (b).

The direction of flux around the current-carrying conductors have been shown. The effective number of poles when supply is given at terminal P and taken out from terminal Q is 4. When supply is given at R and taken out from P and Q, the number of poles formed is 2. Thus, we will get two speeds corresponding to P = 4 and P = 2 by changing the power supply points.

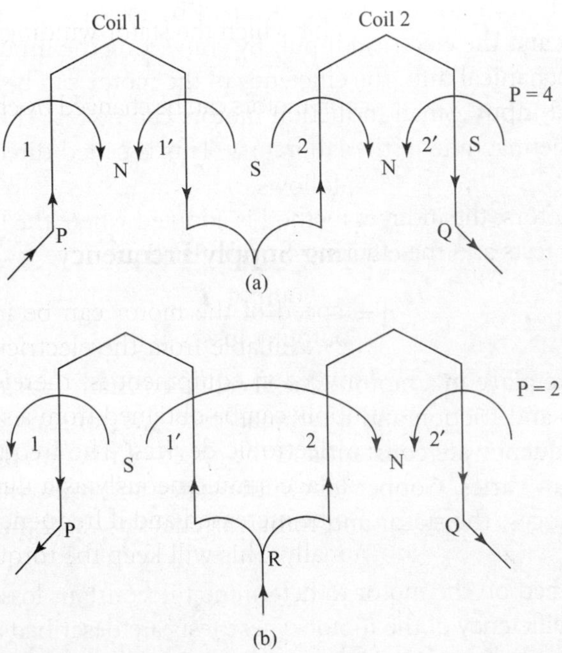

FIGURE 8.18

Speed control by using the consequent pole technique: (a) four-pole formation; (b) two-pole formation

Pole-amplitude modulation technique involves reversing the connection of one-half of the windings. It is possible to obtain a different ratio of pole formation, and hence of the rotor speed.

8.15.3 Speed Control by Changing the Slip Using Voltage Injection Method

Slip of the motor, and hence, the rotor speed can be changed by introducing, i.e., injecting a voltage of slip frequency directly into the rotor circuit. Depending upon the phase of the injected EMF with respect to the rotor EMF, the slip, and hence, the speed will change.

8.15.4 Speed Control by Changing Rotor-Circuit Resistance

In slip-ring induction motors, it is possible to add extra resistance in the rotor circuit. To obtain high-starting torque and also to control speed when the motor is running, the extra resistance connected in the rotor circuit is changed. Higher the rotor-circuit resistance, lower will be the speed obtained. However, starting torque will increase with increase of rotor-circuit resistance.

8.16 ▶ Testing of Induction Motor for Determination of Efficiency

Efficiency of an induction motor can be determined by loading the motor and measuring the mechanical output and the electrical input. By converting the input and output in either electrical unit or in mechanical unit, the efficiency of the motor can be determined by taking the ratio of output and input. Small induction motors can be tested by this method where it is possible to load the machine in the laboratory. This is called direct loading method for determining efficiency.

In testing of large motors, the indirect method is adopted where the losses of the motor are determined from some tests and the efficiency is calculated as

$$\eta = \frac{\text{output}}{\text{output} + \text{losses}}$$

Two types of losses take place in a motor. One is called constant loss and the other is called variable loss. Iron loss and friction and rotational losses are constant at all loads as long as supply voltage and frequency are constant and the speed of rotation does not change much as the load on the motor varies. Copper loss in the stator and rotor varies with load. As the load on the motor changes, the stator and rotor currents vary and the copper losses in the windings also vary.

Two tests are performed on the motor to determine the constant losses and variable losses so as to determine the efficiency of the motor. These tests are described as follows.

8.16.1 No-Load Test

The motor is run on no load with full voltage applied across its stator terminals as shown in Fig. 8.19. Two single-phase wattmeters are connected to measure the three-phase power input. Since with no load connected on the motor shaft, the input power is completely wasted as loss, the sum of the wattmeter reading can be considered equal to the various losses. Therefore,

$$W_1 + W_2 = I^2 R \text{ loss in the windings at no load}$$
$$+ \text{Iron loss in the core}$$
$$+ \text{Friction and windage loss due to rotation of the rotor}$$

By subtracting the no-load I^2R loss from the input power, the constant losses can be calculated. Note that $I^2 R$ loss in the rotor at no load is very small. Thus, we will consider only the $I^2 R$ loss in the stator windings.

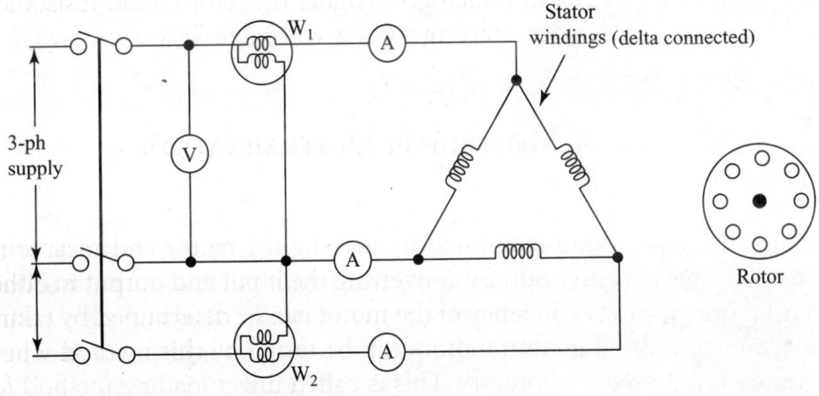

(a) No-load test on an induction motor

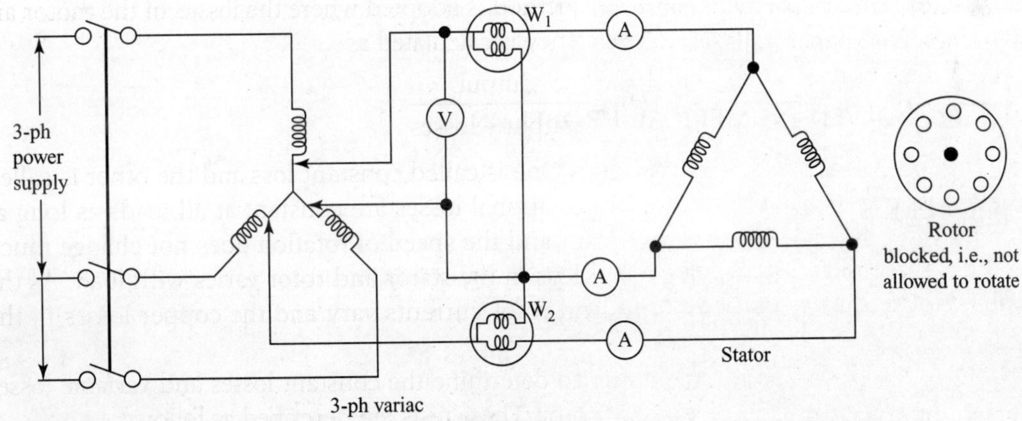

(b) Blocked-rotor test on an induction motor

FIGURE 8.19

Circuits for no-load and blocked-rotor tests

8.16.2 Blocked-Rotor Test

The blocked-rotor test is performed by blocking the rotor so that it is not allowed to rotate. Low voltage is applied to the stator through a three-phase auto-transformer (Variac). The Output voltage of the variac is adjusted so that full-load-rated current flows through the windings. The core loss, which is proportional to this low-input voltage, is small. There is no friction and windage loss as the rotor is not allowed to rotate. The sum of the wattmeter readings can approximately be taken as equal to full-load I^2R loss in the windings.

Thus, we observe that the no-load test and the blocked-rotor test together provide us the account for losses that would take place when the motor is fully loaded with full voltage applied across its terminals. By knowing the losses, we will be able to calculate the efficiency of the motor.

In these two tests, we have created a loading condition that would happen when the motor is actually loaded. That is why this method of finding out the efficiency without actually applying load on the motor shaft is called an indirect method.

8.17 ▶ Applications of Induction Motors

Around 90 per cent of the electrical motors used in industry and domestic appliances are either three-phase induction motors or single-phase induction motors. This is because induction motors are rugged in construction requiring hardly any maintenance, that they are comparatively cheap, and require supply only to the stator. No supply is required to be given to the rotor. The rotor gets excited by virtue of electromagnetic induction. Further, there is no requirement of brush, slip rings, or commutator. However, slip-ring-type induction motors where extra resistance is added to the rotor circuit are used in applications where high-starting torque is required.

Three-phase induction motors are used as drive motors in pumps, lifts, cranes, hoists, lifts, compressors, large capacity exhaust fans, driving lathe machines, crushers, in oil extracting mills, textile and paper mills, etc.

8.18 ▶ Solved Numerical Problems

EXAMPLE 8.3

A 3 hp, three-phase, four-pole, 400 V, 50 Hz induction motor runs at 1440 rpm. What will be the frequency of the rotor-induced EMF?

SOLUTION

Synchronous speed, $N_s = \dfrac{120f}{p} = \dfrac{120 \times 50}{4} = 1500 \text{ rpm}$

Rotor speed, $N_r = 1440$ rpm

Slip, $S = \frac{N_s - N_r}{N_s} = \frac{1500 - 1400}{1500} = 0.04$

Stator supply frequency, $f = 50$ Hz

Rotor induced EMF frequency, $f_r = Sf$

$= 0.04 \times 50$

$= 2$ Hz

EXAMPLE 8.4

The frequency of the rotor-induced EMF of 400 V, three-phase, 50 Hz, six-pole induction motor is 2 Hz. Calculate the speed of the motor.

SOLUTION

$$N_s = \frac{120f}{P} = \frac{120 \times 50}{6} = 1000 \text{ rpm}$$

Rotor frequency, $f_r = Sf$

or, $2 = S \times 50$

or, $S = 0.04$

Again, $S = \frac{N_s - N_r}{N_s}$

Substituting values $0.04 = \frac{1000 - N_r}{1000}$

or, $N_r = 960$ rpm

EXAMPLE 8.5

A slip-ring-type three-phase induction motor rotates at a speed of 1440 rpm when a 400 V, 50 Hz is applied across the stator terminals. What will be the frequency of the rotor-induced EMF?

SOLUTION

We know the synchronous speed of the rotating magnetic field produced is expressed as

$$N_s = \frac{120f}{P}$$

Here, f = 50 Hz but number of poles of the stator winding has not been mentioned. The number of poles can be 2, 4, 6, 8, etc.

for $$P = 2, \quad N_s = \frac{120 \times 50}{2} = 3000 \text{ rpm}$$

for $$P = 4, \quad N_s = \frac{120 \times 50}{4} = 1500 \text{ rpm}$$

for $$P = 6, \quad N_s = \frac{120 \times 50}{6} = 1000 \text{ rpm}$$

for $$P = 8, \quad N_s = \frac{120 \times 50}{8} = 750 \text{ rpm}$$

We know that an induction motor runs at a speed slightly less than the synchronous speed. Here, N_r = 1440 rpm (given). Synchronous speed corresponding to this rotor speed must, therefore, be 1500 rpm Thus, N_r = 1440 rpm and N_s = 1500 rpm.

Slip, $$S = \frac{N_s - N_r}{N_s} = \frac{1500 - 1440}{1500} = 0.04$$

Rotor frequency, $$f_r = S \times f = 0.04 \times 50 = 2\,\text{Hz}$$

EXAMPLE 8.6

A six-pole, three-phase synchronous generator driven at 1000 rpm supplies power to an induction motor which runs at a speed of 1440 rpm on full load. Calculate the percentage slip of the motor and the number of poles of the motor.

SOLUTION

The synchronous generator is rotated at a synchronous speed of 1000 rpm by a prime mover. The synchronous speed N_s is expressed as

$$N_s = \frac{120f}{P} \quad \text{where f is the frequency of the generated EMF}$$

or, $$f = \frac{N_s \times P}{120}$$

$$= \frac{1000 \times 6}{120} = 50 \text{ Hz}$$

This synchronous generator now supplies power to the induction motor at a frequency of 50 Hz. The synchronous speed of the rotating magnetic field produced in the motor is

$$N_s = \frac{120 \times f}{p} = \frac{120 \times 50}{2} = 3000 \text{ rpm (for P = 2)}$$

and $$N_s = \frac{120 \times 50}{4} = 1500 \text{ rpm (for P = 4)}$$

The motor speed is 1440 rpm, which should be slightly less than the synchronous speed. Logically, the number of poles of the motor must be 4.

Percentageslip, $$S = \frac{N_s - N_r}{N_s} \times 100 = \frac{(1500 - 1440) \times 100}{1500} = 4 \text{ per cent}$$

EXAMPLE 8.7

A 400 V, 50 Hz, three-phase induction motor is rotating at 1440 rpm on full load. Calculate the following for the motor:

Number of poles; full-load slip; frequency of rotor-induced EMF; speed of the rotor magnetic field with respect to the rotor.

Also show that both the stator field and the rotor field are stationary with respect to each other.

SOLUTION

$$f = 50 \text{ Hz}$$

$$P = 2, 4, 6, \text{ etc.}$$

$$N_s = \frac{120f}{p} = \frac{120 \times 50}{2} = 3000 \text{ rpm (for P = 2)}$$

$$= \frac{120 \times 50}{4} = 1500 \text{ rpm (for P = 4)}$$

Rotor speed is somewhat less than the synchronous speed N_s. Logically, here N_s can only be 1500 rpm, when N_s = 1500 rpm, P = 4.

Full load slip, $$S = \frac{N_s - N_r}{N_s} = \frac{1500 - 1440}{1500} = 0.04$$

Frequency of rotor induced EMF $f_r = S \times f$

$$= 0.04 \times 50$$

$$= 2 \text{ Hz}$$

Speed of rotor field with respect to rotor, N is

$$N = \frac{120 \times f_r}{P} = \frac{120 \times 2}{4} = 60 \text{ rpm}$$

The rotor rotates at a speed of 1440 rpm. This means the speed of the rotor with respect to stator, which is stationary, is 1440 rpm.

The speed of the rotor field with respect to the rotor is 60 rpm.

Therefore, the speed of the rotor field with respect to the stator is 1440 + 60 = 1500 rpm. And, the speed of the rotating magnetic field produced by the stator rotates at synchronous speed, N_s with respect to the stator. In this case the speed of rotating field produced by the stator is 1500 rpm. Thus, we see that both the magnetic fields of the stator and rotor are stationary with respect to each other, which is, of course, the essential condition for production of torque.

EXAMPLE 8.8

A three-phase, four-pole, 50 Hz induction motor rotates at a full-load speed of 1470 rpm. The EMF measured between the slip-ring terminals when the rotor is not rotating is 200 V. The rotor windings are star connected and has resistance and stand-still reactance per phase of 0.1 Ω and 1.0 Ω, respectively. Calculate the rotor current on full load.

SOLUTION

$$N_r = 1470 \text{ rpm}$$

$$N_s = \frac{120f}{P} = \frac{120 \times 50}{4} = 1500 \text{ rpm}$$

$$S = \frac{N_s - N_r}{N_s} = \frac{1500 - 1470}{1500} = 0.02$$

Rotor-induced EMF between the slip rings at standstill, E_{20} = 200 V.

As the rotor windings are star connected,

$$E_{20} \text{ per phase} = \frac{200}{\sqrt{3}} = 115.4 \text{ V}$$

When the rotor is rotating at a speed of 1470 rpm the rotor-induced EMF per phase, E_2 is

$$E_2 = S\,E_{20} = 0.02 \times 115.4 = 2.3 \text{ V}$$

Rotor current when the rotor is rotating at 1470 rpm,

$$I_2 = \frac{SE_{20}}{Z_2} = \frac{SE_{20}}{\sqrt{R_2^2 + (SX_{20})^2}}$$

Substituting values

$$I_2 = \frac{2.3}{\sqrt{(0.1)^2 + (0.02 \times 1)^2}} = \frac{2.3}{0.102} = 22.5 \text{ A}$$

EXAMPLE 8.9

A 15 hp, three-phase, four-pole, 50 Hz induction motor has full-load speed of 1455 rpm. The friction and windage loss of the motor at this speed is 600 W. Calculate the rotor copper loss.

SOLUTION

$$N_s = \frac{120f}{P} = \frac{120 \times 50}{4} = 1500 \text{ rpm}$$

Slip, $$S = \frac{N_s - N_r}{N_s} = \frac{1500 - 1455}{1500} = 0.03$$

Motor output available at the shaft = 15 hp

= 15 × 735.5 W

= 11032 W

Now, let us look at the power flow diagram

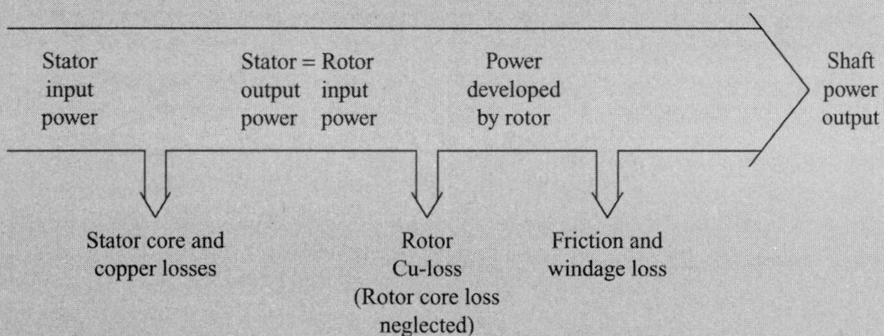

Power developed by rotor = Shaft power output + Friction and Windage losses

= 11032 + 600

= 11632 W

We know the relation

Rotor copper loss = S × Rotor input (See eq. 8.3)

(Slip × Rotor input) is lost as rotor copper loss

The remaining power, i.e., (1–S) Rotor input, is developed as the rotor power.

Therefore,

Power developed by rotor = (1–S) Rotor input

$$= \frac{(1-S) \text{ Rotor copper loss}}{S}$$

Thus, $$\text{Rotor copper loss} = \frac{S}{(1-S)} \times \text{Rotor power developed}$$

$$= \frac{0.3 \times 11632}{(1-0.03)} \text{ W}$$

$$= 360 \text{ W}$$

EXAMPLE 8.10

A 10 hp, four-pole, 50 Hz, three-phase induction motor has friction and windage loss of 3 per cent of output. Calculate at full load the rotor copper loss, rotor input for a full-load slip of 4 per cent. If at this load stator loss is 6 per cent of the input power, calculate the efficiency.

SOLUTION

Output power $= 10 \text{ hp} = 10 \times 735.5 \text{ W}$

$= 7355 \text{ W}$

and slip, $S = 0.04$

Friction and windage loss $= 0.03 \times 7355 \text{ W}$

$= 220.6 \text{ W}$

Power developed $= 7355 - 220.6 = 7134.4 \text{ W}$

Rotor copper loss $= S \times \text{Rotor input}$, (i)

Power developed $= (1–S) \text{ Rotor input}$ (ii)

From (i) and (ii),

$$\text{Power developed} = (1 - S)\,\frac{\text{Rotor copper loss}}{S}$$

or,

$$\text{Rotor copper loss} = \frac{S}{(1-S)}\text{Power developed}$$

Substituting values,

$$\text{Rotor copper loss} = \frac{0.04}{0.96} \times 7134.4 \text{ W}$$

$$= 297.3 \text{ W}$$

Rotor input–Rotor copper loss = Power developed

Therefore, Rotor input = Power developed + Rotor copper loss

$= 7134.4 + 297.3$

$= 7431.7 \text{ W}$

Let input power be = X Watts

Out of this, 0.06 X W is wasted as stator losses. Stator output = Rotor input = 0.94 X W. Equating with actual values, 0.94 X = 7431.7 W

or, X = 7906 W

$$\text{Percentage Efficiency} = \frac{\text{Output}}{\text{Input}} \times 100 = \frac{7355 \times 100}{7906} = 93 \text{ per cent}$$

EXAMPLE 8.11

A four-pole 50 Hz, three-phase induction motor has rotor resistance of 0.5 Ω phase. The maximum torque occurs at a speed or 1200 rpm. Calculate the ratio of starting torque to maximum torque.

SOLUTION

The synchronous speed, $N_s = \frac{120f}{P} = \frac{120 \times 50}{4} = 1500 \text{ rpm}$

Rotor speed, $N_r = 1200 \text{ rpm}$

Slip, $S = \frac{N_s - N_r}{N_s} = \frac{1500 - 1200}{1500} = 0.2$

At this slip, torque is maximum. Condition for maximum torque is $R_2 = SX_{20}$.

Thus, $X_{20} = \frac{R_2}{S} = \frac{0.5}{0.2} = 2.5\,\Omega$

For constant supply voltage, the expression for torque, T is given as

$$T = \frac{KSR_2}{R_2^2 + S^2X_{20}^2}$$

Value of maximimum torque T_m is

$$T_m = \frac{K}{2X_{20}}$$

At starting, S = 1, Starting torque T_{st} is

$$T_{st} = \frac{KR_2}{R_2^2 + X_{20}^2}$$

Substituting values, $\frac{T_{st}}{T_m} = \frac{KR_2}{R_2^2 + X_{20}^2} \times \frac{2X_{20}}{K} = \frac{2X_{20}R_2}{R_2^2 + X_{20}^2}$

$$\frac{T_{st}}{T_m} = \frac{2 \times 2.5 \times 0.5}{(0.5)^2 + (2.5)^2} = 0.98$$

That is, the starting torque is 98 per cent of the maximum torque.

EXAMPLE 8.12

No-load test and blocked-rotor test were performed on a 10 hp, four-pole, 400 V, 50 Hz, three-phase induction motor to determine its efficiency. The test data are given as follows:

no-load test: V = 400 V, I = 6 A, P = 300 W

blocked-rotor test: V = 40 V, I = 24 A, P = 700 W

Calculate the efficiency of the motor on full load.

SOLUTION

Losses of 700 W under blocked-rotor test is considered equal to I^2R losses in the two windings. If R_e is the equivalent resistance of the two windings referred to stator side, then total copper loss

$$3\,I^2\,R'_e = 700\text{ W}$$

$$R'_e = \frac{700}{3\times24\times24} = 0.4\ \Omega$$

At no load, I^2 R loss in the windings

$$= 3\,I_0^2\,R'_e = 3\times6^2\times0.4$$

$$= 43.2\text{ W}$$

R'_e is the equivalent resintance of the windings, the rotor circuit resintance referred to the stator side.

No-load power input = 300 W

Core loss + Frictional and Windage loss = 300 – 43.2 = 256.8 W

Output = 10 hp = 10 × 735.5 = 7355 W

Efficiency, $$\eta = \frac{\text{Output}}{\text{Output} + \text{Losses}} = \frac{10\times735.5\times100}{10\times735.5+700+256.8}$$

= 88.5 per cent

EXAMPLE 8.13

A three-phase, 20-pole slip-ring induction motor runs at 291 rpm when connected to a 50 Hz supply. Calculate slip for full-load torque if the total rotor-circuit resistance is doubled. Assume $R_2 >> SX_2$.

SOLUTION

$$N_s = \frac{120f}{P} = \frac{120 \times 50}{20} = 300 \text{ rpm}$$

slip,

$$S = \frac{N_s - N_r}{N_s} \times 100 = \frac{(300-291)}{300} = 0.03 = 3 \text{ per cent}$$

Full-load torque equation is,

$$T = \frac{KSE_{20}^2 R_2}{R_2^2 + S^2 X_{20}^2}$$

If $R^2 >> S^2 X_{20}^2$

$$T = \frac{KSE_{20}^2 R_2}{R_2^2} = \frac{KE_{20}^2 S}{R_2} \quad \text{or,} \quad R_2 = \frac{KE_{20}^2}{T} S$$

For a given torque, T and rotor-induced EMF at standstill, i.e., E_{20},

$$R_2 \propto S$$

If R_2 is doubled, S will be doubled. The slip at doubled R_2 will be 6 per cent.

Important Terminologies and Points to Remember

1. In a three-phase induction motor, a three-phase winding is made and placed in stator slots. The rotor is either squirrel cage type or slip-ring type. The rotor circuit is to be kept closed.
2. A three-phase supply is applied to the stator windings which are either connected in star or delta formation. A rotating magnetic field rotating at synchronous speed is produced. Emf is induced in the rotor due to electromagnetic induction. The rotor circuit being closed, current flows with the rotor conductors and a torque is developed. The rotor rotates at a speed N_r which is somewhat less than the synchronous speed, N_S of the rotating magnetic field.
3. An induction motor cannot run at synchronous speed. Because at synchronous speed, there will be no relative velocity between the rotating field and the rotor of the motor. Hence, there will be no induced emf in the rotor and no rotor current and no torque developed.
4. The difference between the synchronous speed, N_S and the rotor speed, N_r is the slip and is expressed as $S = \frac{N_s - N_r}{N_s} \times 100$. Slip is around 3 to 5 percent.
5. Frequency, f_r of the induced emf in the rotor at starting is the same as stator frequency, f. But when the rotor is rotating at a speed, N_r the rotor frequency f_r is $f_r = S \times f$ and rotor induced emf, E_2 is given as $E_2 = SE_{20}$ where E_{20} is the induced emf in the rotor at stand-still condition. Similarly, rotor reactance, $X_2 = SX_{20}$.
6. The direction of rotation of the motor can be reversed by interchanging the supply to any of the two phases of the stator winding.
7. By increasing the rotor circuit resistance, the starting torque developed can be increased.

8. Condition for maximum torque developed is $R_Z = SX_{20}$ and $T_m = \frac{K}{2X_{20}}$.

Slip at which torque will be maximum is given by the relation, $S_m = \frac{R_2}{X_{20}}$. If $R_2 = X_{20}$, i.e., if rotor circuit resistance is made equal to the rotor reactance at stand still, maximum torque will be developed when S_m = 1, i.e., when $N_r = 0$ $\left(\because S = \frac{N_S - N_r}{N_S}\right)$. Slip is equal to 1 at starting when $N_r = 0$.

9. An induction motor before it starts rotating is like a short-circuited transformer. If full rated voltage is applied to start the motor, which is called DOL (Direct-on-line) starting, a heavy current will flow through the stator and rotor windings.
10. DOL starting is restricted to only small motors upto 5 kW rating. Large motors must be started with the help of either star-delta starters or auto-transformer starters to reduce the starting current when a reduced voltage gets applied across the windings.
11. Starters for large capacity induction motors are available in push-button operated type. Start-delta starters are available as manual and automatic type.

    An automatic starter is made using contactors, push-buttons, TDR (time-delay relay), and thermal overload relay.
12. Speed of an induction motor can be controlled by changing the frequency of power supply, by changing the poles, i.e., by having two separate windings on the stator wound for different number of poles, by changing the slip and by changing the rotor circuit resistance. These days speed control of induction motors is done conveniently with the help of thyristors. By controlling the time of triggering the thyristor gate, the output voltage and frequency can be changed and the motor speed can be controlled.
13. Efficiency of an induction motor can be determined by measuring its losses as efficiency, η is expressed as:

$$\eta = \frac{\text{Output}}{\text{Input}} = \frac{\text{Output}}{\text{Input} + \text{losses}}$$

The losses can be determined by conducting two tests, namely No-load Test and Blocked rotor Test.

Important Formulae and Equations

Equation	Remarks
1. $N_S = \frac{120f}{P}$	1. N_S is called synchronous speed. If f = 50 Hz, N_S is 3000 rmp, 1500 rmp, 1000 rmp,..., for P = 2, 4, 6, etc.
2. $S = \frac{N_S - N_r}{N_S} \times 100$	2. Slip is usually expressed in percentage.
3. $f_r = Sf$ $E_2 = SE_{20}$ $X_2 = SX_{20}$	3. E_{20}, X_{20} are function of f. If f is replaced by rotor frequency f_r, then E_2 and X_2 under running condition will be corresponding to f_r = Sf.

Important Formulae and Equations

Equation	Remarks
$I_2 = \frac{E_2}{Z_2} = \frac{SE_{20}}{\sqrt{R_2^2 + S^2X_{20}^2}}$ $\cos\phi_2 = \frac{R_2}{\sqrt{R_2^2 + S^2X_{20}^2}}$	
4. $T = \frac{60}{2\pi N_S} - \frac{SE_{20}^2R_2}{R_2^2 + S^2X_{20}^2}$ or, $T = K\frac{SE_{20}^2R_2}{R_2^2 + S^2X_{20}^2}$	4. E_{20} is the induced emf in the rotor at stand still condition. E_{20} is directly proportional to the applied voltage, V. So, $T \propto V^2$.
5. Rotor copper loss = S × Rotor Input.	5. Rotor input is equal to stator output which is $\frac{2\pi TN_S}{60}$.
6. Condition for maximum torque is: $R_Z = SX_{20}$ and maximum torque, $T_m = \frac{K}{ZX_{20}}$	6. Maximum torque is independent of rotor circuit resistance. It remains constant but the value of slip S at which it occurs changes with change of R_2.
7. $S = \frac{N_S - N_r}{N_S}$ or, $N_r = N_S(1-S)$ or, $N_r = \frac{120f(1-S)}{P}$	7. Speed of motor, N_r can be changed by changing frequency, f or slip S, or number of poles, P.

REVIEW QUESTIONS

Short Answer Type Questions

1. Explain the principle of working of a three-phase induction motor.
2. Explain why a three-phase induction motor can not run at synchronous speed.
3. What is meant by slip of an induction motor? What is the value of slip at starting and at synchronous speed?
4. How can you change the direction of rotation of a three-phase induction motor? What is meant by phase sequence of power supply?
5. Explain the purpose of making two types of rotor construction for three-phase induction motors.
6. Write an expression for torque developed by an induction motor and write the meaning of each term.
7. Draw the torque-speed characteristic of a three-phase induction motor and explain its shape.
8. Show that stator magnetic field and the field produced by the rotor mmf are stationary with respect to each other.
9. What is the condition for maximum torque developed at starting?
10. Show how a rotating magnetic field is produced when a three-phase supply is connected across a three-phase winding.

11. What is the expression for maximum torque developed in a three-phase induction motor?
12. Draw the power-flow diagram in an induction motor.
13. What are the various losses in an induction motor? State the factors on which they depend.
14. Explain how efficiency of an induction motor can be determined by performing no-load test and blocked-rotor test.
15. Explain why a starter is required to start a large three-phase induction motor.
16. Draw a push-button-operated direct-on-line starter for an induction motor.
17. Draw the connection diagram for a manual star–delta starter.
18. Show the effect of variation of rotor-circuit resistance on the torque-speed characteristic of an induction motor.
19. What is the limit of increasing the rotor-circuit resistance for achieving a high-starting torque? Explain your answer.
20. If the applied voltage is reduced to half, what will be the reduction in torque developed?
21. Prove that in an induction motor the rotor copper loss is slip times the rotor input.
22. Show how the maximum torque developed in an induction motor is independent of rotor-circuit resistance.
23. How can you determine experimentally, the full-load copper loss of an induction motor?
24. Establish the relation rotor frequency, $f_r = S \times f$.
25. Mention various applications of three-phase induction motors.
26. What are the advantages and disadvantages of squirrel-cage induction motors over slip-ring induction motors?
27. Explain why the rotor-circuit reactance of an induction motor varies with speed?
28. Establish the relation, $X_2 = S X_{20}$ where S is the slip, X_2 is the rotor-circuit reactance under running condition, and X_{20} is the rotor-circuit reactance when the rotor is at rest.
29. Establish the similarity between a transformer and an induction motor.
30. At what slip will torque developed by an induction motor be maximum when rotor resistance equals half the rotor reactance at standstill.
31. In an induction motor, slip is always positive, why?
32. Explain why an induction motor draws heavy current at starting when started on full voltage.

Numerical Problems

33. A four-pole, three-phase, 50 Hz induction motor rotates at a speed of 1440 rpm. Calculate its slip in percentage. Also calculate the frequency of the induced EMF in the rotor circuit.

    [**Ans** 4 per cent; 2 Hz]

34. A six-pole, three-phase, 400 V, 50 Hz induction motor is running at a speed of 940 rpm. Calculate its slip.

    [**Ans** 6 per cent]

35. A four-pole, three-phase, 400 V, 50 Hz induction motor develops an induced EMF in the rotor of 2 Hz. What is the speed of the motor?

    [**Ans** 1440 rpm]

36. A four-pole, three-phase induction motor in connected to a 50 Hz supply. Calculate synchronous speed; the rotor speed when slip is 4 per cent, and the rotor frequency when the rotor is running at 1425 rpm.

    [**Ans** 1500 rpm, 1440 rpm, 2.5 Hz]

37. The input power to the rotor of a 400 V, 50 Hz, six-pole, three-phase induction motor is 75 kW. The frequency of the rotor induced EMF is 2 Hz. Calculate slip; rotor speed, and power developed by the rotor; rotor $I^2 R$ loss.

    [**Ans** 0.04, 960 rpm, 72 kW; 1 kW]

38. A 400 V, 50 Hz, six-pole, three-phase induction motor running at 975 rpm draws 40 kW from the mains. The stator loss is 1kW. The friction and windage loss is 2 kW. Calculate (i) slip; (ii) $I^2 R$ loss in the rotor; (iii) the shaft output in hp, and (iv) the efficiency of the motor.

    [**Ans** 2.5 per cent; 975 W; 49 hp; 90 per cent]

Multiple Choice Questions

1. For production of a rotating magnetic field using stationary windings we must connect
   (a) a single-phase supply to a single-phase winding
   (b) a three-phase supply to a three-phase winding
   (c) a single-phase supply to a two-phase winding
   (d) either a single-phase winding or a two-phase winding to a single-phase supply.
2. When a three-phase 50 Hz, 400 V supply is applied across a four-pole, three-phase winding, a rotating magnetic field is produced which is rotating at
   (a) 200 rpm (b) 1600 rpm
   (c) 1500 rpm (d) 3000 rpm.
3. A four-pole, three-phase induction motor is rotating at 1440 rpm when a 400 V, 50 Hz supply is applied to its stator terminals. The slip of the motor expressed in percentage is
   (a) 4 per cent (b) 2 per cent
   (c) 5 per cent (d) 6 per cent.
4. The synchronous speed N_s, frequency f, and number of poles, P are related as
   (a) $N_s = \frac{120P}{f}$ (b) $N_s = \frac{120f}{P}$
   (c) $N_s = \frac{fP}{120}$ (d) $\frac{Pf}{60}$.
5. When an induction motor is yet to start, i.e., at standstill, its slip is
   (a) 0 (b) 1
   (c) infinity (d) near to 0.
6. The speed of a three-phase induction motor when supplied with three-phase, 400 V, 50 Hz supply is 1440 rpm. The number of poles for which the windings are made must be
   (a) 2 (b) 4
   (c) 6 (d) 8.
7. A three-phase induction motor develops a torque of 500 Nm at normal supply voltage. If the supply voltage is reduced to half, what will be the torque developed?
   (a) 250 Nm
   (b) 125 Nm
   (c) will remain constant at 500 Nm
   (d) 62.5 Nm.
8. The slip of an induction motor can be expressed as
   (a) $S = \frac{Nr - Ns}{Nr}$ (b) $S = \frac{Ns - Nr}{Nr}$
   (c) $\frac{Ns - Nr}{Ns}$ (d) $\frac{Nr}{Ns - Nr}$.
9. A four-pole, 50 Hz, 400 V, three-phase induction motor is running at 1440 rpm. The frequency of rotor-induced EMF is
   (a) 2 Hz (b) 4 Hz
   (c) 1 Hz (d) 50 Hz.
10. A four-pole, three-phase induction motor when fed from a 400 V, 50 Hz supply, runs at 1440 rpm. The frequency of EMF induced in the rotor is
    (a) 2 Hz (b) 45 Hz
    (c) 50 Hz (d) 3 Hz.
11. If an induction motor by some means is rotated at synchronous speed, then
    (a) the EMF induced in the rotor will be maximum
    (b) the EMF induced in the rotor will be zero
    (c) the torque developed by the rotor will be maximum
    (d) the frequency of induced EMF will be slip times the supply frequency.
12. Increase in the rotor-circuit resistance of an induction motor will
    (a) increase the starting torque
    (b) decrease the starting torque
    (c) increase the maximum torque developed
    (d) decrease the maximum torque developed.
13. Torque developed by a three-phase induction motor depends on
    (a) the supply voltage
    (b) the rotor-circuit resistance
    (c) the slip of the rotor
    (d) all the above.
14. A large capacity three-phase induction motor is started using a star–delta starter instead of starting direct-on-line. The starting current
    (a) is increased three times
    (b) remains constant
    (c) is reduced to one-third its value
    (d) is reduced to half its value.

15. With rotor-circuit resistance, R_2 of 2 Ω, the maximum torque of an induction motor is developed at 10 percent slip. When the maximum torque is developed at 20 percent slip, the value of R_2 will be
(a) 1 Ω (b) 2 Ω
(c) 0.5 Ω (d) 4 Ω.

16. The speed of a three-phase slip-ring-type induction motor can be controlled by
(a) changing the frequency of supply voltage to the motor
(b) changing the rotor-circuit resistance
(c) changing the number of poles
(d) all the methods as in (a), (b) and (c).

17. The four no-load speeds of three-phase induction motors operating on 400 V, 50 Hz supply are 576 rpm, 720 rpm, 1440 rpm, and 2880 rpm. The number of poles of the motor, respectively, are
(a) 8, 6, 4, 2 (b) 10, 8, 4, 2
(c) 12, 8, 4, 2 (d) 16, 8, 4, 2.

18. The power factor of a three-phase induction motor on no load is found to be 0.15 lagging. When the motor is fully loaded, its power factor would be around
(a) 0.05 lagging (b) 0.15 lagging
(c) 0.85 lagging (d) 0.85 leading.

19. The three supply terminals R, Y, B when connected, respectively, to the three stator terminals a, b, c of a three-phase induction motor, the motor rotates in the clockwise direction. The motor will rotate in the reverse direction if
(a) R is connected to c, Y is connected to a, and B is connected to b
(b) R is connected to b, Y is connected to a, connection of B to c remains unchanged
(c) the supply voltage is reduced to half its value
(d) the frequency of supply is gradually reduced.

20. A three-phase induction motor has standstill rotor-circuit resistance, X_{20} of 8 Ω and resistance of 2 Ω. The maximum torque will be developed at starting if the circuit parameters are
(a) $R_2 = 2\ \Omega$, $X_{20} = 4\ \Omega$
(b) $R_2 = 4\ \Omega$, $X_{20} = 8\ \Omega$
(c) $R_2 = 8\ \Omega$, $X_{20} = 8\ \Omega$
(d) $R_2 = 16\ \Omega$, $X_{20} = 8\ \Omega$.

Answers to Multiple Choice Questions

1. (b)	2. (c)	3. (a)	4. (b)	5. (b)	6. (b)
7. (b)	8. (c)	9. (a)	10. (d)	11. (b)	12. (a)
13. (d)	14. (c)	15. (d)	16. (d)	17. (b)	18. (c)
19. (b)	20. (c)				

CHAPTER 9

Single-phase Motors

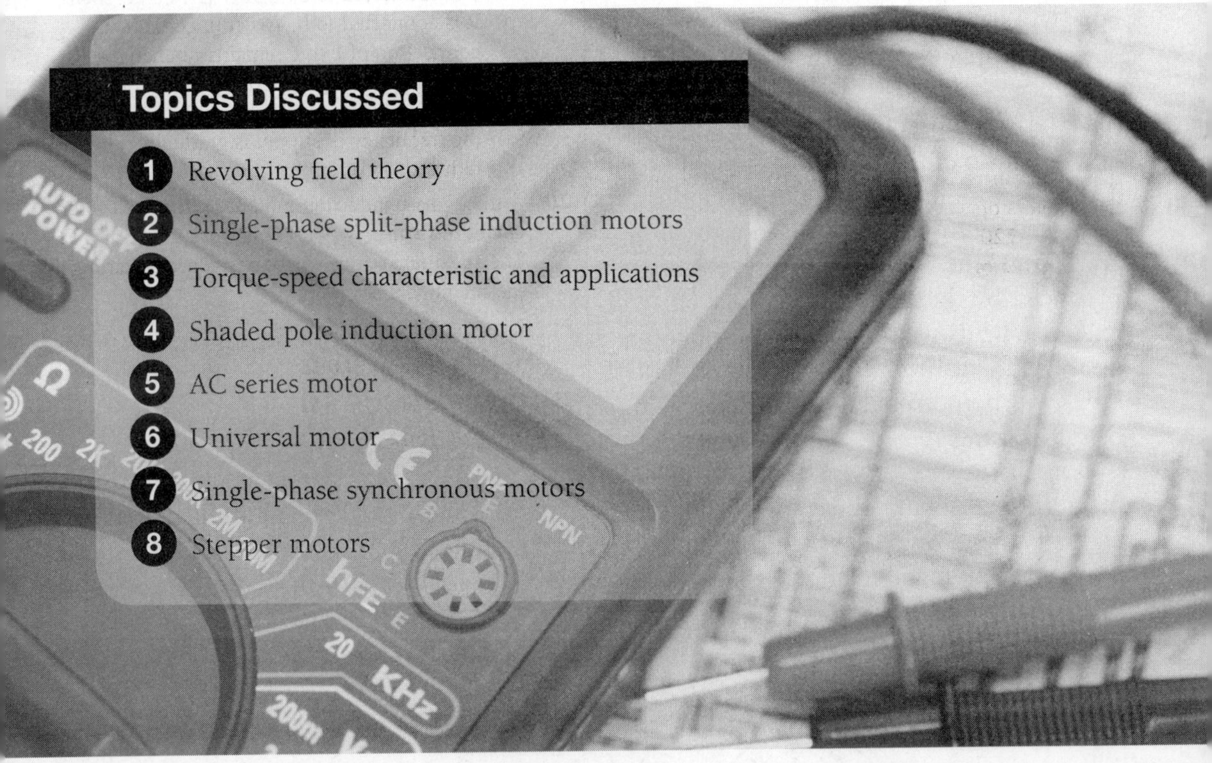

Topics Discussed

1. Revolving field theory
2. Single-phase split-phase induction motors
3. Torque-speed characteristic and applications
4. Shaded pole induction motor
5. AC series motor
6. Universal motor
7. Single-phase synchronous motors
8. Stepper motors

9.1 ► Introduction to Single-Phase Induction Motors

Single-phase induction motors are widely used in electrical appliances and gadgets like ceiling fans, exhaust fans, refrigerators, washing machines, etc. They require only single-phase supply to the stator. The rotor is squirrel-cage type and does not require any supply from a separate source.

9.2 ► Constructional Details

The constructional details of single-phase induction motors are more or less the same as that of three-phase induction motors except that the stator will have a single-phase winding and the rotor is of squirrel-cage type. Slip-ring-type rotor construction is not done for single-phase motors. Single-phase induction motors as such do not develop starting torque unless some mechanism for starting the motor is provided. An auxiliary winding is provided in the stator for developing starting torque. Let us now examine as to why a single-phase supply given to a single-phase stator winding of the motor does not lead to the development of any torque. Fig. 9.1 shows a single-phase induction motor in cross-sectional view and a single-phase supply connected to its stator terminals.

The stator winding shown has been made with only three coils. In actual practice more coils will be used. The rotor has a squirrel cage winding. When supply from a single-phase source is applied, current will flow through the stator winding for the instantaneous polarity of voltage shown. The North and South poles formed in the stator along with the magnetic field axis have been shown. Since the supply voltage is varying sinusoidally the magnitude and direction of the flux produced will change. EMF will be induced in the rotor winding. The rotor winding being a closed winding current will flow through it. The direction of flux produced will be such that the rotor flux will oppose the stator flux (according to Lenz's law). Thus, the two magnetic fields, i.e., one produced by the stator current and the other produced by the rotor-induced current will be aligned to each other. The axis of the two magnetic fields will be along the horizontal axis. See Fig. 9.1 (a). Since there is no angle of non-alignment between the two magnetic fields, no torque will be developed, and hence, there will be no rotation of the rotor. The single-phase induction motor as such will not be self-starting.

It has however been observed that if the rotor is given some initial torque in any direction it picks up speed in that direction and continues to rotate. Thus, a single-phase induction motor without any starting mechanism will not develop any torque at starting but will pick up speed if it is given an initial rotation.

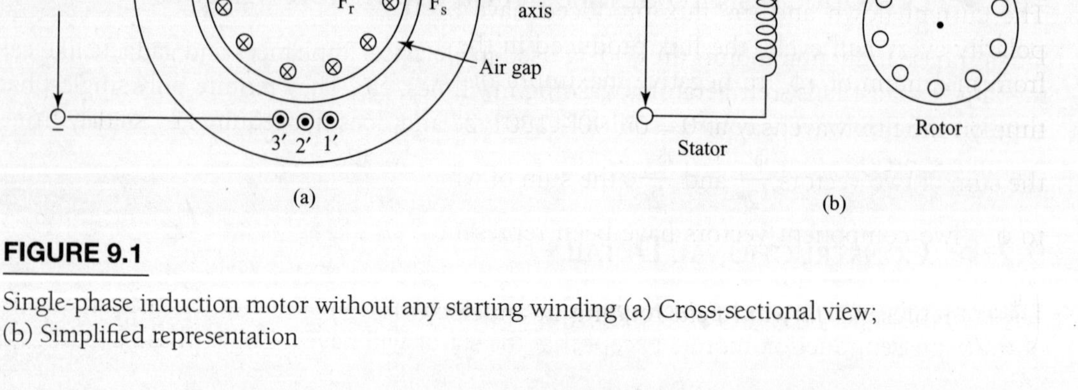

FIGURE 9.1

Single-phase induction motor without any starting winding (a) Cross-sectional view; (b) Simplified representation

To explain this, two theories were developed. These theories are

(i) double revolving field theory;
(ii) cross-field theory.

We will explain one of these theories to show why a single-phase induction motor does not develop any starting torque but requires an initial torque to be provided for it to continue to rotate.

9.3 ▶ Double Revolving Field Theory and Principle of Working of Single-Phase Induction Motors

According to double revolving field theory an alternating magnetic field can be considered equivalent to two revolving fields of constant magnitude (half the magnitude of the alternating field) rotating in opposite directions at synchronous speed.

Thus, an alternating field can be seen as a resultant of two revolving magnetic fields which will have an effect on the rotor which is placed inside that alternating field. The component rotating fields will develop torque on the rotor due to the induction effect and the rotor will rotate on the basis of the resultant torque developed due to these two revolving fields.

Thus, a single-phase induction motor having a single-phase supply on its stator winding can be visualized as two three-phase induction motors trying to rotate the rotor in opposite directions. The rotor, however, rotates in one direction due to the effect of resultant of the two torques developed in opposite directions.

The double or two revolving field theory is explained with the help of vector representations as in Fig. 9.2. The alternating magnetic field is represented by a vector, ϕ, which varies from its positive maximum value to negative maximum value, in every half cycle of current flow through the stator winding. This maximum value of ϕ as ϕ_m has been shown by two vectors $\frac{\phi_m}{2}$ and $\frac{\phi_m}{2}$; the sum of these two vectors will always be equal to ϕ_m. As the flux produced varies sinusoidally, it will be observed that this magnetic flux at every instant of time is the vector sum of two rotating magnetic fields.

In Fig. 9.2 is shown the stator winding connected to a single-phase supply voltage V. The current flown and the flux produced have been shown. As the current changes its polarity every half cycle, the flux produced in the air gap will change sinusoidally varying from maximum of $+\phi_m$ to negative maximum of $-\phi_m$. Let us consider different instants of time on the flux wave, say at $\theta = 0°, 90°, 180°, 270°$, and $360°$. We have considered ϕ_m as the sum of two vectors $\frac{\phi_m}{2}$ and $\frac{\phi_m}{2}$, the sum of which at every instant of time will be equal to ϕ_m. Two component vectors have been represented in the figure as $\left(\frac{\phi_m}{2}\right)_f$ and $\left(\frac{\phi_m}{2}\right)_b$. $\left(\frac{\phi_m}{2}\right)_f$ represents *forward field* rotating in anticlockwise direction and $\left(\frac{\phi_m}{2}\right)_b$ represents

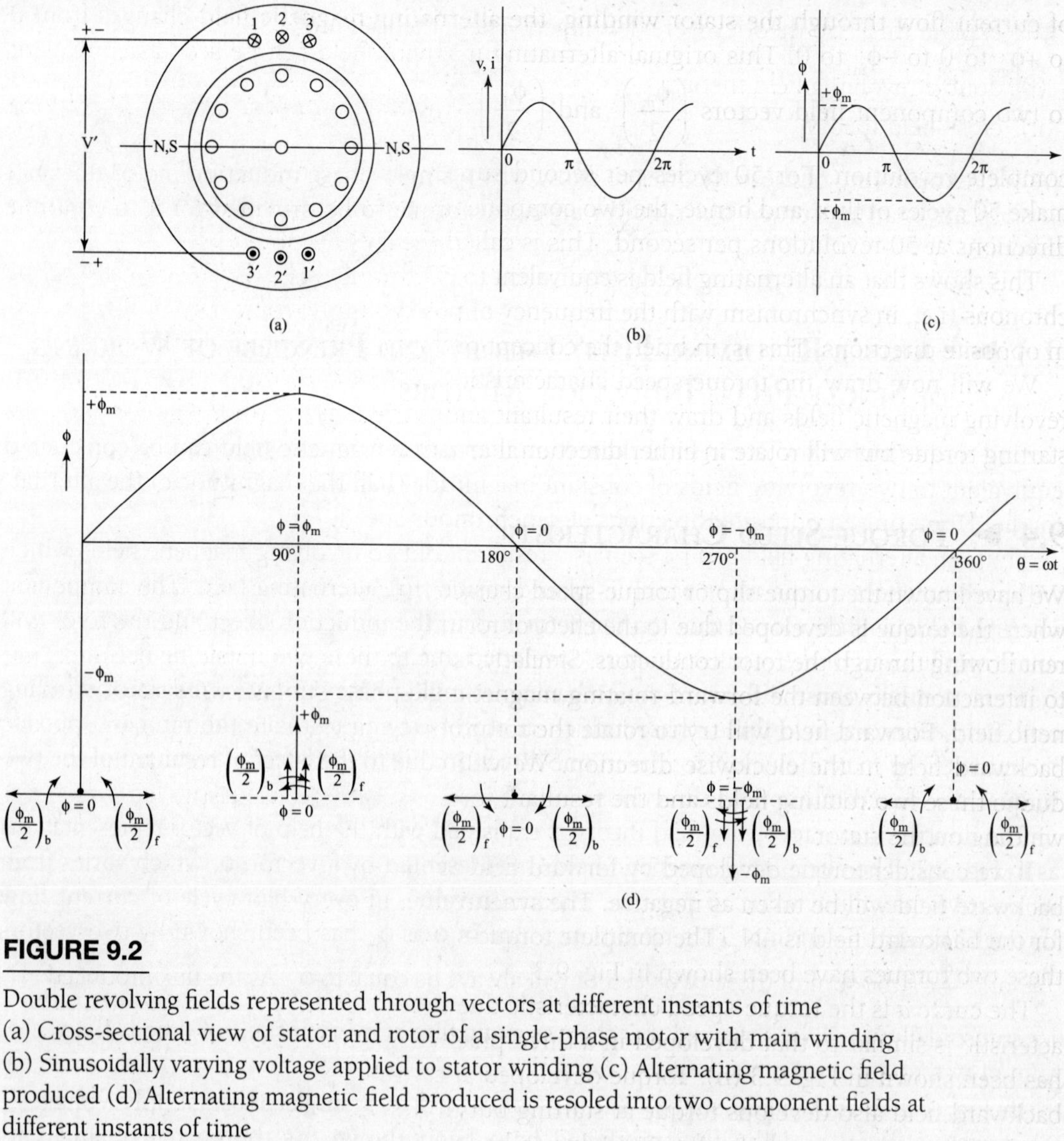

FIGURE 9.2

Double revolving fields represented through vectors at different instants of time
(a) Cross-sectional view of stator and rotor of a single-phase motor with main winding
(b) Sinusoidally varying voltage applied to stator winding (c) Alternating magnetic field produced (d) Alternating magnetic field produced is resoled into two component fields at different instants of time

backward field rotating in the clockwise direction at the same speed. The speed of rotation of the two fields is the same and is synchronous with the alternating magnetic field. At $\theta = 0$, the two fields are shown opposite to each other so that the resultant flux $\phi = 0$. At $\theta = 90°$, the two flux vectors have rotated by 90° in opposite directions and the resultant flux is $+\phi_m$. At $\theta = 180°$, the two component flux vectors have rotated by another 90° and the resultant flux $\phi = 0$. At $\theta = 270°$, the vectors have rotated in opposite directions by another 90° and the sum of the two vectors is $-\phi_m$. At $\theta = 360°$, the two vectors have again rotated by another 90° and their sum is now equal to zero. Thus, for one complete cycle of current flow through the stator winding, the alternating magnetic field changes from 0 to $+\phi_m$ to 0 to $-\phi_m$ to 0. This original alternating magnetic field can be seen as equivalent to two component field vectors $\left(\frac{\phi_m}{2}\right)_f$ and $\left(\frac{\phi_m}{2}\right)_b$ rotating in opposite directions by one complete revolution. For 50 cycles per second supply, the alternating flux produced will make 50 cycles of flux, and hence, the two component magnetic fields will rotate in opposite directions at 50 revolutions per second. This is called the synchronous speed.

This shows that an alternating field is equivalent to two revolving fields which rotate at synchronous (i.e., in synchronism with the frequency of power supply or the current flow) speed in opposite directions. This is, in brief, the concept of double recovering field theory.

We will now draw the torque-speed characteristics of the motor due to the effect of two revolving magnetic fields and draw their resultant and prove that the rotor will not have any starting torque but will rotate in either direction if an initial torque is provided.

9.4 ▶ Torque-Speed Characteristic

We have known the torque-slip or torque-speed characteristic of a three-phase induction motor where the torque is developed due to the effect of rotating magnetic field and the induced current flowing through the rotor conductors. Similarly, torque will be developed on the rotor due to interaction between the forward rotating magnetic field and the backward rotating magnetic field. Forward field will try to rotate the rotor in the anticlockwise direction while the backward field in the clockwise direction. We will draw the torque-speed characteristics due to these two rotating fields and the resultant torque-speed characteristics with only one winding on the stator.

If we consider torque developed by forward field as positive, the torque developed by the backward field will be taken as negative. The synchronous speed for forward field is $+N_S$ and for the backward field is $-N_S$. The complete torque-speed characteristics and the resultant of these two torques have been shown in Fig. 9.3.

The curve ***a*** is the torque-speed characteristic due to forward field. The shape of the characteristic is similar to that developed in a three-phase induction motor, which for reference has been shown in Fig. 9.3 (b). Torque developed at starting, i.e., at zero speed, is $+T_S$. The backward field also develops torque at starting but in the reverse direction which is shown as $-T_S$. The torque-speed characteristic due to backward field has been shown by curve *b*. It must be noted that while forward field is trying to rotate the rotor in one direction, the backward field is trying to rotate in the opposite direction. Thus, while the effect of one field is motoring action, the effect of the other field on the rotor will be braking action. This has been shown by drawing the complete torque-speed characteristic due to both forward field and backward field.

From the resultant torque-speed characteristic it is observed that the effective starting torque, T_s, is zero. However, the motor will pick up speed in whichever direction a small torque is provided by some means. The torque-speed characteristic of a single-phase induction motor for one direction of rotation has been redrawn as in Fig. 9.4.

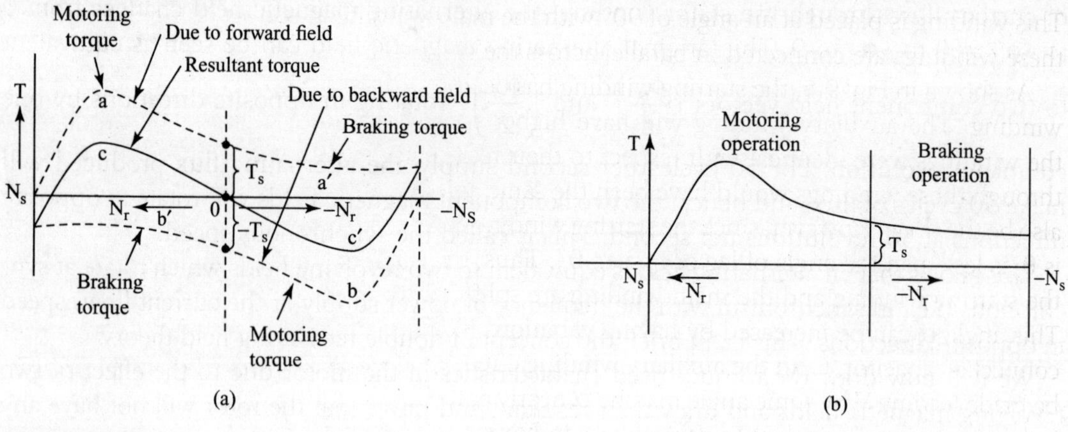

FIGURE 9.3

(a) Torque-speed characteristic of a single-phase induction motor without having any starting winding; (b) complete torque-speed characteristic of a three-phase induction motor

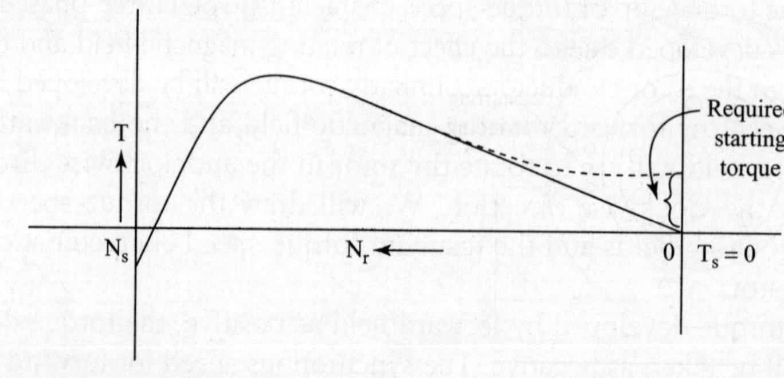

FIGURE 9.4

Torque-speed characteristic with no starting torque

Such a motor has to be provided with some starting torque, otherwise the rotor will not rotate.

Various methods have been developed to make a single-phase induction motor self-starting. The names of the motors have been given according to the starting methods employed. We shall describe one of the popular methods, called the *split-phase method* and the other used in very small motors which is known as the *shaded pole method*.

9.5 ► Split-Phase Induction Motors

Single-phase induction motors are made self-starting by using an additional winding in the stator.

Thus, in addition to the main single-phase winding in the stator a separate winding, called the auxiliary winding is provided. This auxiliary winding is also called the starting winding.

This winding is placed at an angle of 90° with the main winding as shown in Fig. 9.5 (a). Both these windings are connected in parallel across the single-phase supply.

As shown in Fig. 9.5, the starting winding has been wound with thinner wires than the main winding. The auxiliary winding will have higher resistance than the main winding. If both the windings were identical with respect to their resistance and reactance, the current flowing through these windings would have been the same and the angle of lag with the voltage would also be the same. However, since the starting winding is more resistive, the angle of lag of I_a which is θ_a is less than the angle of lag of I_m, i.e., θ_m. Thus, the two currents I_a and I_m flowing through the starting winding and the main winding are split by an angle α which is equal to $(\theta_m - \theta_a)$. This angle α can be increased by having variations in L and R ratio of the two windings. If we connect a capacitor, C, in the auxiliary winding circuit as has been shown in Fig. 9.5 (d), I_a can be made leading V by some angle making α nearly equal to 90°. This will make the single-phase induction motor equivalent to a two-phase induction motor but fed from a single-phase supply

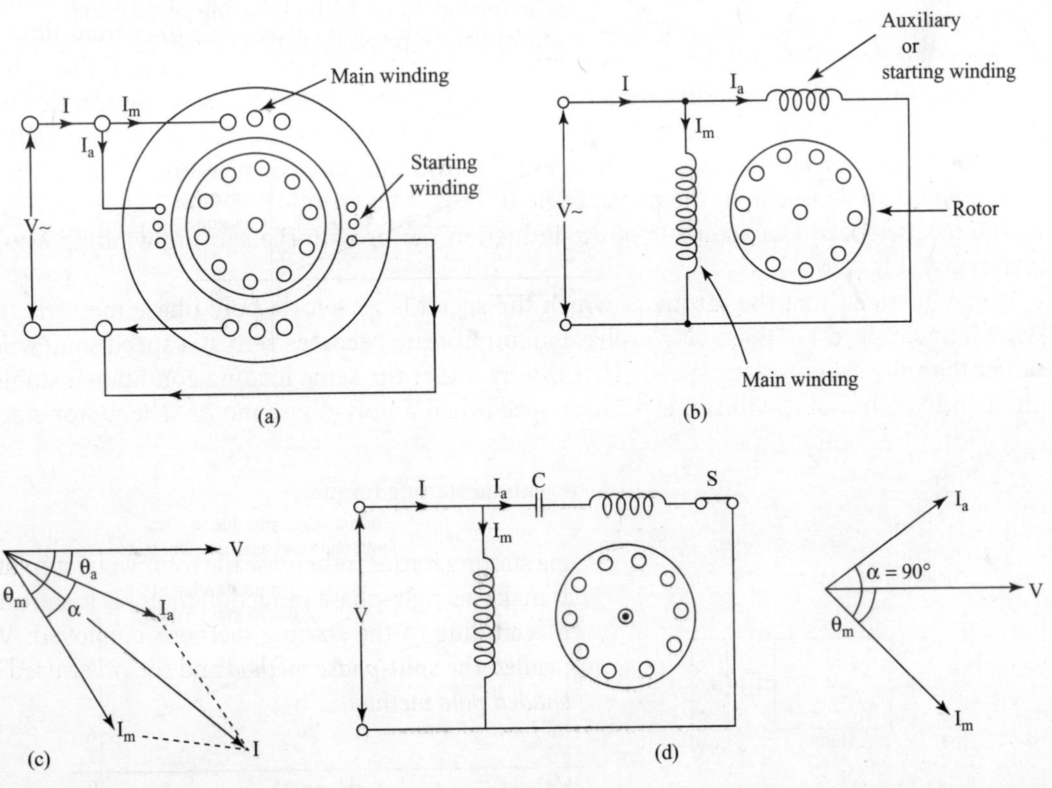

FIGURE 9.5

Split-phase single-phase induction motor: (a) single-phase induction motor with main winding and starting winding; (b) connection diagram; (c) phasor diagram showing splitting of two currents, I_m and I_a; (d) split phasing with a capacitor in the starting winding circuit to increase the angle of phase splitting (in a two-phase motor there will be two-phase windings in the stator and the windings are identical. A two-phase supply is connected to the two windings. The current flowing through the windings will have a phase difference of 90°).

When a poly-phase supply is given to a poly-phase winding, a rotating magnetic field is produced. This we have seen in the case of three-phase induction motors. For a two-phase motor also, a rotating magnetic field will be produced. A single-phase induction motor with an auxiliary winding is similar to a two-phase motor. The current flowing through the windings will have a phase difference of 90° or somewhat less than 90°.

Creating a phase split in the currents flowing through the two windings will help produce a rotating magnetic field effect on the rotor. The rotor will develop a starting torque and start rotating. The direction of rotation will depend on the way the connections of windings are made across the supply. Thus, use of auxiliary winding with or without a capacitor makes the induction motor self-starting. If the phase-split angle is more the magnitude of torque developed will be more (torque is proportional to sin α). Once the motor picks up speed we may disconnect the starting winding from the supply through a centrifugal switch or a relay. The motor will continue to develop torque due to current flow in the main winding. When the motor is stopped, the switch should close again so that while restarting, the auxiliary winding gets connected to help develop the starting torque. For improved power factor during running condition, however, the auxiliary winding can be kept connected for all the time. In that case the resultant current of I_a and I_m, i.e., I_I will have a phase angle less than the phase angle θ_m.

The torque-speed characteristic of the induction motor with the starting winding in use is shown in Fig. 9.6.

It may be noed that the torque at which the speed is zero for a poly-phase motor is the synchronous speed N_S. For a single-phase motor, torque becomes zero at a speed somewhat earlier than the synchronous speed. That is why under the same loading condition a single-phase induction motor will run at a lower speed than a poly-phase motor. The motor starts

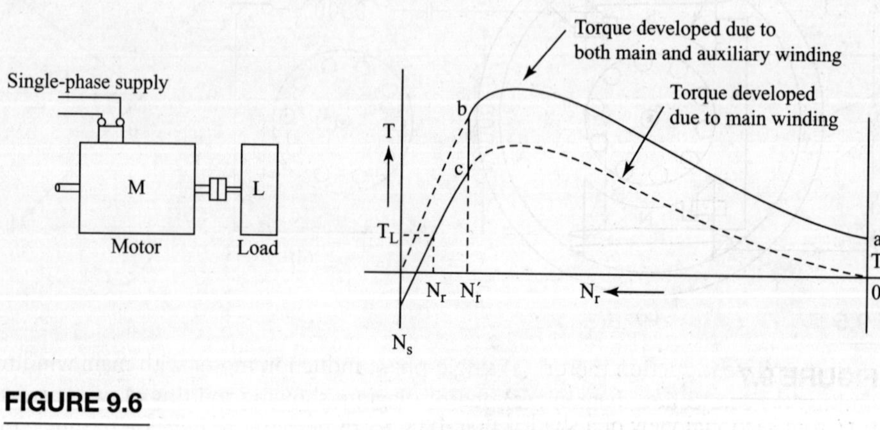

FIGURE 9.6

Torque-speed characteristic of a single-phase induction motor with its auxiliary winding connected to the supply. The starting torque developed, T_S, is shown as Oa. The motor starts rotating with the mechanical load connected to its shaft. When the rotor attains a speed of say N_r', the centrifugal switch disconnects the auxiliary winding and the motor continues to drive the load and attains a speed at which motor torque equals the load torque requirement, T_L. Note that N_r' is the speed at which the auxiliary winding is disconnected automatically and the motor continues to work with the main winding only.

9.6 ► Shaded Pole Induction Motor

Shaded pole-type single-phase induction motors are provided with shading rings on their poles which are the projected type of poles. The stator of such motors have projected poles like dc machines as shown in Fig. 9.7. The rotor is a squirrel cage type similar to that of split-phase-type motors. The poles are excited by giving single-phase ac supply. Single-turn thick coil in the form of a ring, called the shading ring is fitted on each side of every pole as shown. The portion of the poles where the shading ring is fitted is called the shaded portion, while the other portion is called the unshaded portion. When a single-phase supply is connected across the field windings, an alternating current will flow and produce an alternating flux. An EMF will be induced in the rotor conductors due to transformer action, in the same way as in the case of split-phase-type induction motors. Since the rotor conductors are connected together, current will flow through them. If the rotor is given an initial rotation, it will pick up speed like any other single-phase motors.

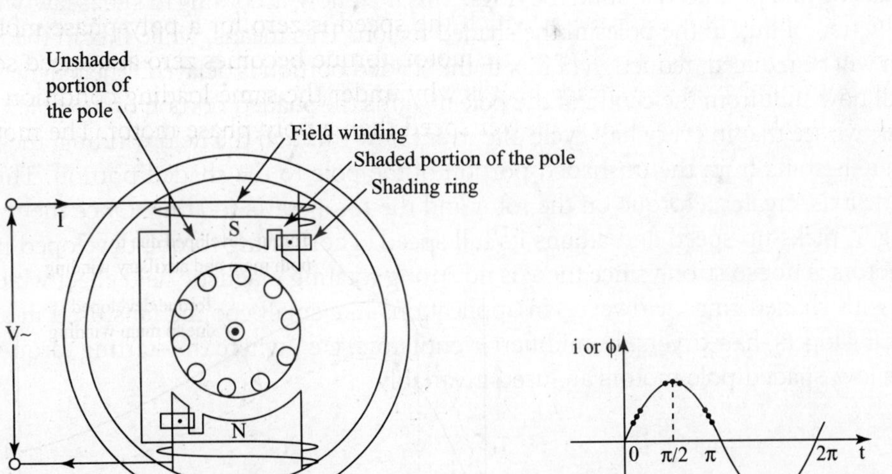

FIGURE 9.7

(a) Cross-sectional view of a shaded pole-type single-phase induction motor;
(b) sinusoidal flux produced by the stator current

Here, in shaded pole motors, the starting torque is produced due to the presence of shading rings. How the shading rings help produce a rotating magnetic field effect resulting a small starting torque to start the motor is explained as follows.

Let us assume that the current through the field winding is increasing from zero value towards its maximum value sinusoidally during the first quarter of the cycle. The flux produced will also be rising as shown in Fig. 9.7 (b). This change of flux with respect to time will induce EMF in the shading ring. Current will flow through the shading ring. This current flow through the shading ring will produce a flux around the ring. This flux, by Lenz's law will oppose the main field flux produced by the rising alternating current flowing through the field winding. The opposition of shading ring flux on the main field flux will cause reduction of flux in the shaded region. As a result there will be more flux in the unshaded region than in the shaded region. The magnetic neutral axis therefore will lie towards the unshaded region of the pole.

When current through the field winding reaches its maximum value, the rate of change of the current, and hence, the rate of change of the flux produced will be nearly zero. There will be no induced EMF in the shading ring, and hence, the shading ring will have no effect on the flux distribution in the main pole. The magnetic neutral axis will lie at the centre of the pole, i.e., at the geometrical neutral axis of the poles. Thus, by the time the current through the field coils has reached its maximum, the magnetic neutral axis has shifted from unshaded side to the centre of the poles.

Now when the current starts falling, the flux in the poles will also be collapsing, i.e., go on reducing. This changing flux will produce EMF in the shaded rings which will induce EMF in the rings causing flux produced around the rings. This flux, now, according to Lenz's law, will oppose the reduction of flux in the poles in the shaded region. This means, while flux in the unshaded portion will be reduced, reduction of flux in the shaded portion is delayed. The magnetic neutral axis will now shift from the centre of the pole towards the shaded portion of the pole.

Thus, we see that in every half cycle of current flow through the field winding, the magnetic neutral axis shifts from the unshaded portion of the pole to the shaded portion. This shift of magnetic axis, creates a torque on the rotor and the rotor starts rotating. Once the rotor starts rotating, it picks up speed and attains its full speed. The starting torque developed in shaded pole motors is not so strong since there is no strong rotating magnetic field effect which is produced with shaded rings. However, in applications like small cooling fans used in almost all electrical gadgets, hair dryers, humidifiers, ventilators, etc., where the starting torque requirement is low, shaded pole motors are used invariably.

9.7 ► Single-Phase AC Series Motors

We have known that in a dc series motor if we change the supply polarities of either the field winding or the armature winding, the direction of rotation changes. If we reverse the polarities of both the field winding and the armature winding to the power supply, the direction of rotation of a dc series motor remains unchanged. From this, it can be said that a dc series

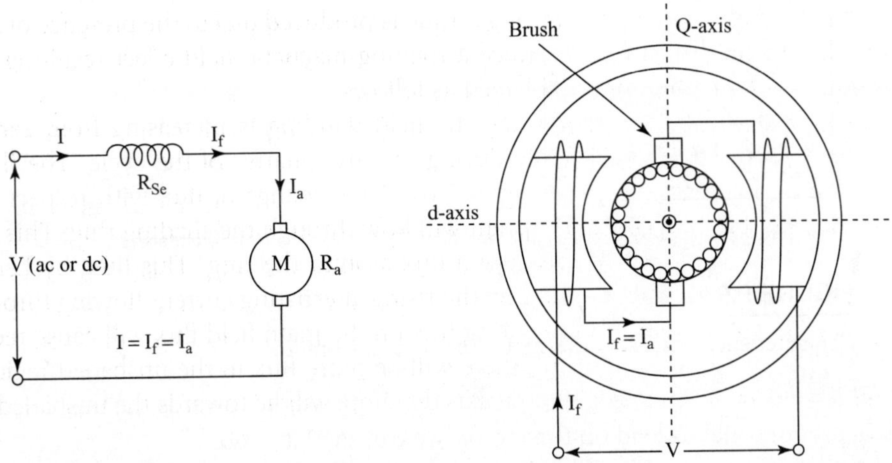

FIGURE 9.8
AC series motor or universal motor

motor should also work on ac supply as well. A series motor which will work on both dc supply and single-phase ac supply is called an universal motor. Universal motors in fractional kilowatt ratings are used in many domestic electrical appliances like food mixtures, vacuum cleaners, portable drills, etc. These small motors are usually light in weight and operate at very high speeds varying from 1,500 rpm to 10,000 rpm.

It can be noticed from Fig. 9.8 that in a series motor, the line current, the field current, and the armature current is the same. The current flowing through the field windings produces a flux ϕ_d along the pole axis, i.e., the direct axis or simply the d-axis. The current flowing through the armature I_a will also produce a flux ϕ_q in the quadrature axis (Q-axis), i.e., along the brush axis.

The torque developed is expressed as

$$T = K_t \phi_d I_a \tag{i}$$

Since $\phi_d \propto I_f$ and $I_f = I_a$, $T \propto I_a^2$ and the speed of the motor N is expressed as

$$N = \frac{V - I_a(R_a + R_{se})}{K\phi_d} \tag{ii}$$

Considering $I_a(R_a + R_{se})$ very small as compared to V,

we can write $$N \propto \frac{V}{\phi_d}$$

or, $$N \propto \frac{1}{I_a} \quad \text{since} \quad \phi_d \propto I_f \propto I_a \tag{iii}$$

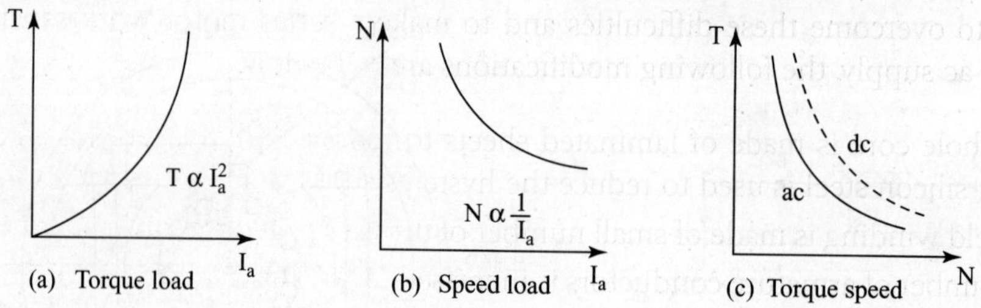

FIGURE 9.9

Characteristics of an ac series motor or a universal motor

Since I_a is proportional to load on the motor, we can say $I_a \alpha$ load.

Using the relation in (i) and (iii) above, we can draw the characteristics such as torque versus load, speed versus load, and torque versus speed as shown in Fig. 9.9.

$T \propto I_a^2$ is the equation of a parabola of the form $y = x^2$. $N \propto \frac{1}{I_a}$ is the equation of a rectangular hyperbola of the form $y \propto \frac{1}{x}$ or $xy = C$. These are shown in Figs. 9.9 (a) and (b), respectively. From the relationship of T versus I_a and N versus I_a, the relationship between T versus N can be developed as has been shown in Fig. 9.9 (c).

A series motor, designed only for dc operation, will have the following drawbacks when it is used for ac supply.

(i) There will be hysteresis loss and eddy current loss in the magnetic circuit. As the total losses will increase, the efficiency of operation on ac will be low.

(ii) On ac operation, there will be reaction of the field winding and armature winding. So, the power factor will be low on ac operation. Low-power factor will cause increase in magnitude of the current drawn.

(iii) There will be sparking at the brushes.

9.8 ► Operation of a Series Motor on DC and AC (Universal Motors)

The speed of a series motor on ac operation is somewhat lower than that for dc operation due to the effect of magnetic saturation, i.e., ϕ_d on ac operation will be less than ϕ_d on dc operation. Hence, the torque developed will be somewhat lower in ac operation as shown in Fig. 9.9 (c). As observed from the characteristic at Fig. 9.9 (b) at no load, the series motor will attain very high speed which may be dangerous. From Fig. 9.9 (c), it is observed that the motor develops high torque at low speed and low torque at high speed.

For satisfactory operation of the dc series motor on both dc and ac supply, certain modifications are to be made.

In order to overcome these difficulties and to make a series motor work satisfactorily on both dc and ac supply, the following modifications are to be done:

(a) The whole core is made of laminated sheets to reduce eddy current loss. For laminate sheets, silicon steel is used to reduce the hysteresis loss.
(b) The field winding is made of small number of turns to reduce the reactance voltage drop.
(c) The number of armature conductors is increased to develop required torque even though flux is low.
(d) Series motors are provided with a compensating winding which is connected in series with the armature. The compensating winding is placed on the stator at 90° with main field axis. The compensating winding mmf will neutralize the effect of armature reaction thereby reducing the sparking of the brushes.

The armature reactance voltage will also get reduced thereby making commutation process more or less sparkless.

AC series motors, also called universal motors, because of their high speeds are used in appliances such as blenders, vacuum cleaners, and hair dryers where high speed and light weight is the requirement. They are also used in drills and circular saws.

9.9 ▶ Single-Phase Synchronous Motors

These are very small motors suitable for use in clocks, timers, etc. They are available as reluctance motors and hysteresis motors. These two types of motors are described in brief as follows.

9.9.1 Reluctance Motors

Reluctance motors are single-phase motors where the stator construction is similar to that of an induction motor. That is, the stator has one main winding and one auxiliary winding. Both the windings are connected in parallel. The rotor construction is somewhat different than a single-phase induction motor. Some of the tooths of the rotor are removed so as to make the air gap between the stator and rotor non-uniform. This way the reluctance of the motor across the air gap becomes variable. The squirrel cage bars and the end rings of the rotor remain the same.

When single-phase supply is applied across the stator winding, the rotor starts rotating as an induction motor. At about 70% of the synchronous speed, the starting winding is cut off automatically. However, the rotor continues to speed up and attain synchronous speed due to reluctance torque developed. The rotor aligns itself with the synchronously rotating field and runs at synchronous speed.

In Fig. 9.10 (a) is shown the constructional details of a reluctance motor, where mm′ is the main winding while aa′ is the auxiliary winding or the starting winding. These two windings are wound at right angles to each other on the stator, exactly similar to a single-phase induction motor.

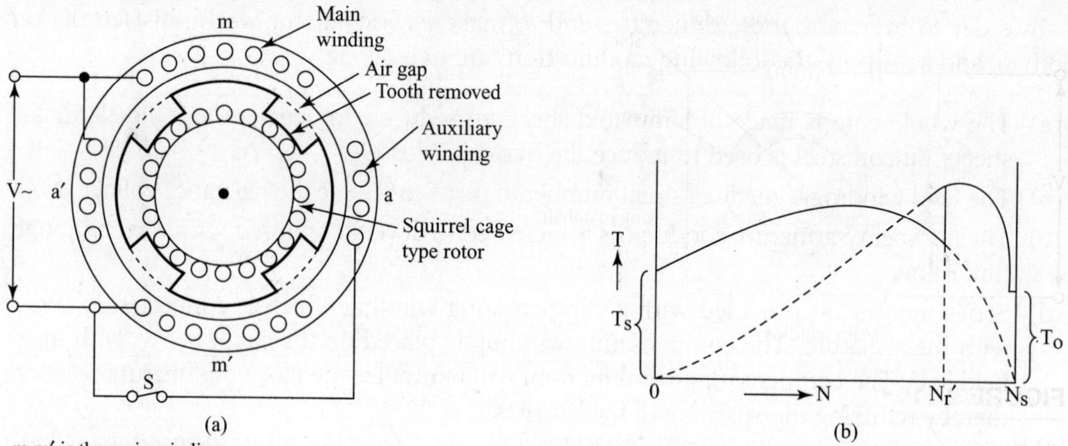

m m′ is the main winding
a a′ is the auxiliary winding
S is the centrifugal switch

FIGURE 9.10

(a) Constructional details of a reluctance motor; (b) torque-speed characteristics

In Fig. 9.10 (b), T_o is the operating torque of the motor at the synchronous speed. At a speed of N_r', the centrifugal switch S is opened. The motor will continue to develop torque and run on its main winding.

Reluctance motors are available from fractional horse power (few watts) to several KW range. Small size reluctance motors are used for instrumentation applications as hard disc drive motor, control rod drive in nuclear reactors.

9.9.2 Hysteresis Motors

Hysteresis motors are single-phase small size synchronous motors.

The stator windings are similar to the stator windings of single-phase induction motors. In the auxiliary winding a permanent value capacitor is connected. Like the main winding the auxiliary winding is always connected to the supply. When the stator windings are connected to a single-phase supply, a rotating field is produced which is rotating at synchronous speed. There is no winding provided on the rotor. The rotor is simply made of aluminium or other non-magnetic material having a ring of a special magnetic material such as cobalt or chromium mounted on it.

The rotating field produced by the stator will induce eddy currents in the rotor. The rotor will get magnetized. But the magnetization of the rotor will lag the inducing revolving field by some angle due to the hysteresis effect. The rotating magnetic field will pull the rotor along with it and the rotor will rotate at synchronous speed. A constant torque will be developed upto the synchronous speed as shown in Fig. 9.11 (b). The performance of a single-phase hysteresis motor is silent (no noise) because there is no slot on the rotor and the rotor surface is smooth.

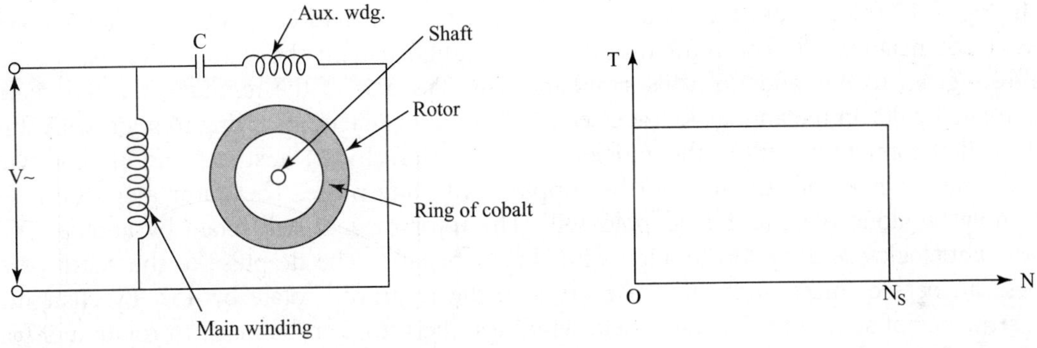

FIGURE 9.11

(a) Hysteresis motors; (b) torque-speed characteristic

9.10 ▶ Stepper Motors

Stepper motors are also called the step motors. They rotate in steps by a certain angle depending upon the design. The rotor of such motors may be made of a set of permanent magnets or with a soft magnetic material with salient poles. The stator will have a set of poles with winding as shown in Fig. 9.12. The stator poles are excited by a sequence of dc pulses. The poles get magnetized one after the other in a clockwise or anticlockwise direction. Torque is developed on the rotor as the rotor magnets try to align with the stator poles.

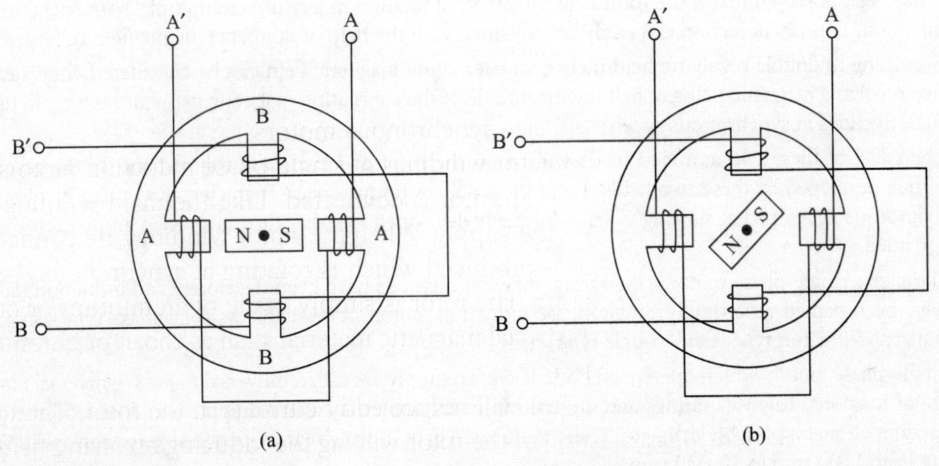

FIGURE 9.12

Simple illustration of a stepper motor: (a) dc pulse given to stator winding AA′; (b) dc pulse given to both the stator windings, i.e., to AA′ and BB′

In Fig. 9.12 (a) suppose a dc pulse is given to the stator field coils AA′. The stator poles AA will be magnetized. The rotor magnet will get aligned with the stator poles. Next a dc pulse is given to AA′ and BB′ coils simultaneously. The axis of the resultant magnetic field will rotate by 90° in the anticlockwise direction. The rotor magnet is trying to align with this field will also rotate by 45° in the anticlockwise direction. In the next step the BB′ coil will be energized while coil AA′ will not be supplied with any pulse. The stator magnetic field will now be along the stator field poles BB′. The rotor magnet will rotate by another 90° in the anticlockwise direction to align with the stator field. The dc pulse to the stator field poles can be sequenced such that in every step the rotor will rotate by 45°. By changing the sequence of supply to the stator field windings, the rotor can be made to rotate in steps in the clockwise direction. The step by which the rotor will rotate can be chosen by a proper design, i.e., by choosing the proper number of stator and rotor poles.

Stepper motors can be rotated to a specific angle in discrete steps, and hence, such motors are used for read/write head positioning in computer floppy diskette drives. Stepper motors are also used in computer printers, optical scanners, and digital photocopiers to move the optical scanning element. The quartz analogue watches contain the smallest stepping motors.

Important Terminologies and Points to Remember

1. When a single-phase supply is connected across a single-phase stator winding, an alternating magnetic field is produced. A squirrel cage rotor placed inside the stator does not rotate. This is because the axes of the stator magnetic field and that of the induced rotor magnetic field are aligned with each other.
2. It has been observed that if the rotor is given an initial rotation in any direction, it picks up speed in that direction. This phenomenon can easily be explained with the help of double revolving field theory.
3. According to double revolving field theory, an alternating magnetic field can be considered equivalent to two revolving magnetic fields of half the magnitude of the alternating magnetic field but rotating in opposite directions at synchronous speed.
4. The effect of these two assumed fields will be on the rotor and the rotor will respond to the resultant torque developed by these two fields. From the resultant torque-speed characteristics, it can be observed that torque developed at starting is zero. Torque is developed in either direction when an initial torque is provided.
5. To make a single-phase motor self-starting, different methods have been developed and induction motors have been named accordingly. Single-phase induction motor are available as split-phase type, capacitor start-capacitor-run type, shaded-pole type, etc.
6. Single-phase motors which operate on both dc and ac supply are called universal motors. Universal motors are of fractional Kilowatt ratings and are used in many domestic electrical appliances like food mixures, vacuum cleaners, portable drill, etc. These motors are light in weight and operate at very high speed varying from 1500 rmp to 10,000 rmp.
7. Reluctance motors and hysteresis motors are very small single-phase synchronous motors.
8. Stepper motors or step motors rotate in steps by a certain angle. Stepper motor can be rotated by specific angle in steps. They are used in floppy disc driver, computer printers, photo copiers, etc.

REVIEW QUESTIONS

Short Answer Type Questions

1. Explain the constructional details and principle of working of a split-phase-type single-phase induction motor.
2. Explain double revolving field theory and show that a single-phase induction motor without the auxiliary winding will not develop any starting torque.
3. Explain the need for connecting a capacitor in the auxiliary winding of a single-phase induction motor.
4. Explain how an alternating magnetic field can be considered equivalent to two revolving fields.
5. Draw and explain the complete torque-speed characteristics of a single-phase induction motor.
6. Explain 'the speed of a single-phase induction motor is somewhat less than an equivalent three-phase motor'.
7. Show the constructional details and explain the principle of working of a shaded pole induction motor. How do you determine the direction of rotation?
8. Explain the working principle of an universal motor, draw the torque-speed characteristic, and mention its applications.
9. Explain the principle of working of a one type of single-phase synchronous motor.
10. What is a reluctance motor? How does it attain synchronous speed? Draw and explain its torque-speed characteristic.
11. Explain the construction and working principle of a hysteresis motor.
12. How do we make a single-phase induction motor self-starting?
13. Explain the working principle of a stepper motor. Mention two applications of such a motor.

Multiple Choice Questions

1. A split-phase single-phase induction motor has
   (a) one stator winding
   (b) two stator windings placed at an angle of 90°
   (c) wound type rotor
   (d) two stator windings connected in series.
2. In a resistance split-phase induction motor, phase difference between the currents flowing through the two windings of the stator is created by
   (a) giving two-phase supply to the two windings
   (b) creating a space-phase difference between the two windings, i.e., by placing the two windings at right angles
   (c) connecting the two stator windings in series opposition across a single-phase supply
   (d) having different ratios of resistance to inductive reactance for the two windings and connected across a single-phase supply.
3. When a single-phase sinusoidal ac supply is connected to a single-phase stator winding, the magnetic field produced is
   (a) pulsating in nature
   (b) rotating in nature
   (c) constant in magnitude but rotating at synchronous speed
   (d) constant in magnitude but changing in direction.
4. In a split-phase capacitor-start induction motor a time-phase difference between the currents flowing through the two windings of the stator is produced by
   (a) placing the two windings at an angle of 90° in the stator slots
   (b) applying two-phase supply across the two windings
   (c) introducing capacitive reactance in the auxiliary winding circuit
   (d) connecting the two windings in series opposition across a single-phase supply.
5. The direction of rotation of a split-phase-type single-phase induction motor can be reversed by

(a) reversing the connections of either the main winding or the auxiliary winding terminals
(b) reversing the supply terminal connections
(c) reversing the connections of main winding only
(d) reversing the connections of auxiliary winding only.

6. A dc series motor when connected across an ac supply will
(a) develop torque in the same direction
(b) draw dangerously high current
(c) develop a pulsating torque
(d) not develop any torque at all.

7. A dc series motor will work satisfactory on ac supplying provided
(a) the yoke and the poles are completely laminated
(b) only the poles are laminated
(c) the air gap is reduced
(d) compensating poles are introduced.

8. The ceiling fan in your home has a
(a) shaded pole-type motor
(b) dc series motor
(c) universal motor
(d) capacitor-start motor.

9. In a capacitor start induction motor, the capacitor is connected
(a) in series with the main winding
(b) in series with the auxiliary winding
(c) across the supply terminals
(d) in parallel with the auxiliary winding.

10. The rotor of a stepper motor
(a) has no winding
(b) has no commutator
(c) has no slip rings
(d) all these as in (a), (b), and (c).

11. According to double revolving field theory, an alternating field can be considered equivalent to
(a) two revolving fields of constant magnitude rotating at synchronous speed in the same direction
(b) two revolving fields of constant magnitude rotating at synchronous speed but in opposite directions
(c) two revolving fields of variable magnitude rotating at synchronous speed but in opposite directions
(d) two revolving fields of variable magnitude rotating at synchronous speed in the same direction.

12. A universal motor is
(a) a series motor designed to operate on ac
(b) a series motor designed to operate on both ac and dc
(c) a series motor designed to operate on dc
(d) a dc shunt motor modified to work on both dc and ac.

13. A dc series motor has
(a) very high-starting torque
(b) very low-starting torque
(c) constant torque developed at all speeds
(d) constant speed-load characteristic.

14. A fractional kW ac series motor has
(a) very high speed and high-starting torque
(b) constant speed and high-starting torque
(c) very high speed and low-starting torque
(d) maximum torque developed on full-load condition.

Answers to Multiple Choice Questions

1. (b)	2. (d)	3. (a)	4. (c)	5. (a)	6. (a)
7. (a)	8. (d)	9. (b)	10. (d)	11. (b)	12. (b)
13. (a)	14. (a)				

CHAPTER 10

Synchronous Machines

Topics Discussed

1. Constructional details of synchronous machines
2. Armature winding
3. Induced EMF
4. Distribution factor and pitch factor
5. Open-circuit and short-circuit tests
6. Synchronous impedance
7. Armature reaction
8. Voltage regulation
9. Parallel operation of synchronous generators
10. Synchronous motors
11. Effect of change of excitation of synchronous motors
12. Applications

10.1 ► INTRODUCTION

Electricity is generated in power houses. The source of energy which is converted into electrical energy could be potential energy of water, energy of high-pressure steam, or gas. In an hydroelectric power house, e.g., Bhakra dam in Punjab or Idukki dam in Kerala, water head has been created by constructing a high-rise dam over the river. The potential energy of water is utilized in running a number of water turbines located at the base of the dam. Large capacity ac generators are coupled with these turbines. When the turbines rotate, electricity is generated in the ac generators which is brought out through wire connections, stepped up to a higher voltage and transmitted through long transmission lines to be taken to places where electricity is required. The ac generators, also called the alternators, used for generation of electricity on a large scale are invariably three-phase ac generators. The generation of voltage is based on the basic principle that when there is relative motion between a conductor and a magnetic field, EMF is induced in the conductor. This is called the generating action. The same machine will work as a motor when electrical energy is the input and mechanical energy is the output. A motor works on the basic principle that when a current carrying conductor is placed in a magnetic field, it experiences a force. Thus, the electro-mechanical energy conversion that takes place in an electrical machine is a reversible process. That is the same machine works as a generator when mechanical energy is converted into electrical energy and as a motor when electrical energy is converted into mechanical output.

10.2 ► Constructional Details of Synchronous Machines

Synchronous machines may be a synchronous generator or a synchronous motor. Synchronous generators used for generation of electricity are three-phase generators.

In a three phase synchronous generator, a set of coils are placed in slots inside a hollow cylindrical stator. The coils are wound for different number of poles. Magnetic poles are formed on the rotor and are rotated by a prime mover, i.e., a turbine. The rotating poles produce a flux which cuts the stator conductors. Because of the cutting of flux by the coil sides, i.e., conductors, EMF is induced in them.

The poles forming the rotor are rotated at a constant speed, called the synchronous speed so that EMF of constant frequency is generated. Normally the electricity generated is for 50 cycles per second. The relationship between the rotor speed, i.e., synchronous speed, N_S the number of poles, P and the frequency of induced EMF, f is given by

$$N_S = \frac{120f}{P}$$

If the poles for which the machine is made is 2, and the frequency of generated EMF to be 50 Hz, then the turbine speed must be

$$N_S = \frac{120 \times 50}{2} = 3000 \text{ rpm}$$

It can been calculated that for P = 4, N_s = 1500 rpm; for P = 6, N_s = 1000 rpm and so on.

The construction of the rotor forming the magnetic poles which are rotated by the turbine are of two types. One type is of projected pole-type rotor construction where number of poles are made by passing field current through the windings of the pole cores. Where a large number of poles are required to be formed, this type of projected (or salient) poles are made.

In hydroelectric generating stations, the speed of the turbines is comparatively lower than the steam turbines used in thermal power stations. For example, the turbine speed in Bhakra Hydroelectric power generating station is only 167 rpm. To generate electricity at 50 Hz, the number of magnetic poles required on the rotor is as many as 36. Projected type of poles are used in the rotors when a large number of poles are to be fixed on the rotor. Such rotors are

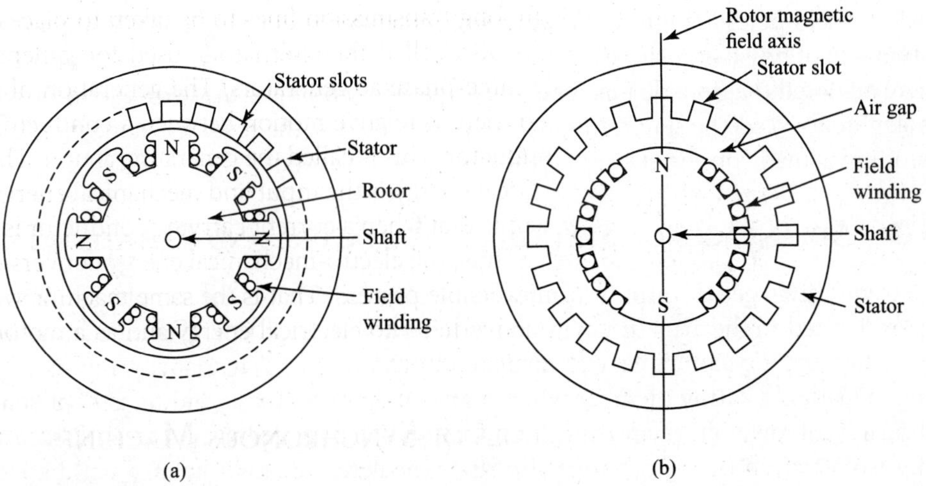

FIGURE 10.1

(a) Salient or projected-pole-type synchronous machine;
(b) cylindrical-type synchronous machine

called salient-type rotors. When P = 36 and the frequency of the induced EMF is to be 50 Hz, then the turbine speed, N_s is

$$N_S = \frac{120 \times 50}{36} = 167 \text{ rpm}$$

In thermal power stations the turbine speed is usually maintained at 3000 rpm so that the number of rotor poles is only 2.

$$N_S = \frac{120 \times 50}{2} = 3000 \text{ rpm}$$

High-speed rotors are made cylindrical type, or non-projected type and are also called non-salient type. Thus, two types of rotor construction are made, one is the salient-type rotor and the other is the non-salient or cylindrical-type rotors. The stator construction is the same in both the cases. Three-phase windings, displaced at 120° apart are made on the stator slots. Cross-sectional view of the two types of synchronous machines are shown in Fig. 10.1. The stator windings that are made on stator slots have not been shown in the figures.

Direct current supply is provided to the field windings so that the poles are magnetized. Current to the rotating field windings is supplied through brush and slip-ring arrangement. The field winding current, I_f produces the flux. The rotor carrying the field poles with their windings carrying current is rotated with the help of a prime mover, i.e., a turbine. Since the rotor is rotating, current from fixed supply terminals is to be provided to the rotor field windings through the brush and slip-ring arrangement.

A set of two brushes and slip rings are required to supply dc current to the field windings. Three-phase armature windings are made on the stator slots, and connections from these windings to the load can be taken directly. Thus, we have a three-phase winding placed on the stator slots. Field poles are formed on the rotor which are excited by supplying direct current using two sets of brush and slip rings.

It may be noted that in a dc machine, the armature winding is made on the rotor while the field poles are fixed on the stator. In synchronous machines, the reverse is done, i.e., the armature winding is made on the stator and the field poles are rotated.

10.3 ► Advantages of Stationary Armature and Rotating Field

The field windings get dc supply from a low-voltage dc source of supply, say 250 V. The voltage generated in the armature winding is normally at 11,000 V. If the armature winding is kept stationary, it becomes easy to insulate the conductors. That is why, low-voltage field winding is made a rotating member while high-voltage armature winding is kept stationary. Two slip rings of low voltage and current rating will be required in this case. If the armature winding is placed on the rotor, three slip rings insulated for high voltage will be required. The rotor with field poles and windings will have less weight and inertia as compared to armature winding with its iron core on the rotor. Further cooling of the armature windings carrying high currents can easily be done when they are stationary.

10.4 ► Use of Laminated Sheets for the Stator and the Rotor

The stator is made up of thin laminated silicon steel sheets with varnish insulation. These laminated sheets are placed one above the other and are pressed together and held tightly. Loosely held laminations would cause magnetic vibration resulting in humming noise. A large number of slots are made on the inner side of the laminations by punching. After putting an insulated paper or some other insulating sheets inside the slots, windings are placed inside the slots. The windings are held tightly inside the slots. The slots could be open type or semi-closed type. Similarly, the rotor is also made of laminated steel sheets.

10.5 ► Armature Windings

Insulated copper wires are used to form coils which are placed inside the slots made on the stator. The windings are made for a different number of poles depending upon the design. The number of poles for which the stator winding is made and the number of rotor poles are the same. For the generation of three-phase voltages three separate windings are made on the stator and are joined together. The three windings are displaced in space by 120°. For simplicity in Fig. 10.2 (a) we have shown only one coil per phase. The three-phase windings, i.e., R–R′, Y–Y′, B–B′ are shown placed at an angle of 120°. The winding has been made for two poles as in Fig. 10.2 (a) and for four poles as in Fig. 10.2 (b).

The three windings are connected in star by joining R′ Y′ B′ forming the star points and the terminals R, Y, B are brought out for external connections.

The electrical diagram for the stator windings for two-pole and four-pole formation and that of the field winding have been shown in Fig. 10.3 (a) and (b), respectively. The field winding is provided with current I_f from a dc voltage source through brush and slip-ring

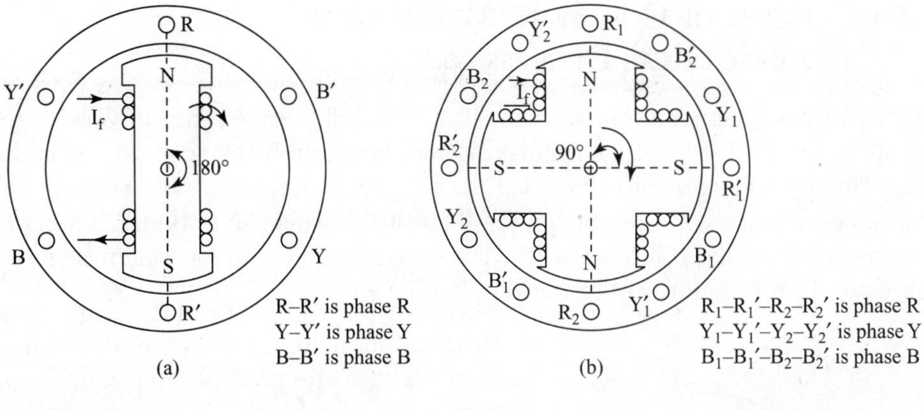

FIGURE 10.2

(a) Three-phase two-pole stator winding; (b) three-phase four-pole winding

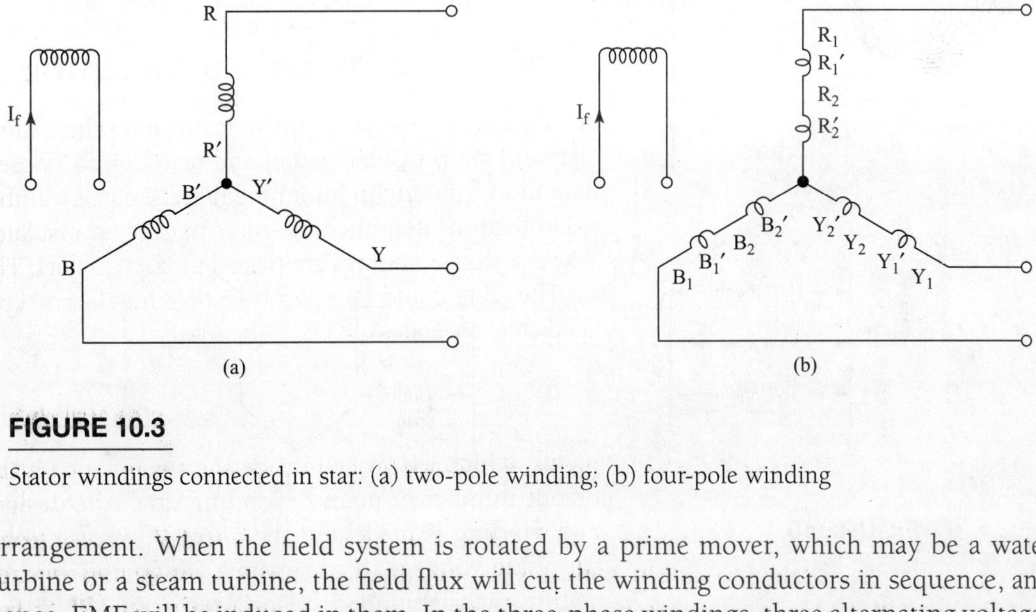

FIGURE 10.3

Stator windings connected in star: (a) two-pole winding; (b) four-pole winding

arrangement. When the field system is rotated by a prime mover, which may be a water turbine or a steam turbine, the field flux will cut the winding conductors in sequence, and hence, EMF will be induced in them. In the three-phase windings, three alternating voltages will be available. There will be a time phase difference between the voltages induced in the three phases as they are physically displaced at an angle of 120 electrical degrees.

Time phase difference between the voltages induced in the three-phase windings is created because of the time difference in which the revolving field pole flux cut the stator phase windings. Phase sequence is the sequence in which the voltages are induced in the three phases R, Y and B. If the field flux rotating in clockwise direction cuts the phase winding in the sequence R, Y and B, the phase sequency will be RYB. If the field poles are rotated in anticlockwise direction the phase sequence of voltage induced will be RBY.

10.6 ► Concept of Coil Span, Mechanical, and Electrical Degrees

The angular distance between the two coil sides of a coil is called the coil span. From Fig. 10.2 (a), it can be observed that the angular distance between the two coil sides of coil R–R′ is 180°. The winding is for two poles. From Fig. 10.2 (b) which has a four-pole winding, the angular distance between the coil sides of coil R_1–R_1' is 90°. If we make a winding for eight poles, the coil span will be reduced further to 45°. The change of coil span for windings of a different number of poles has been further illustrated in Fig. 10.4.

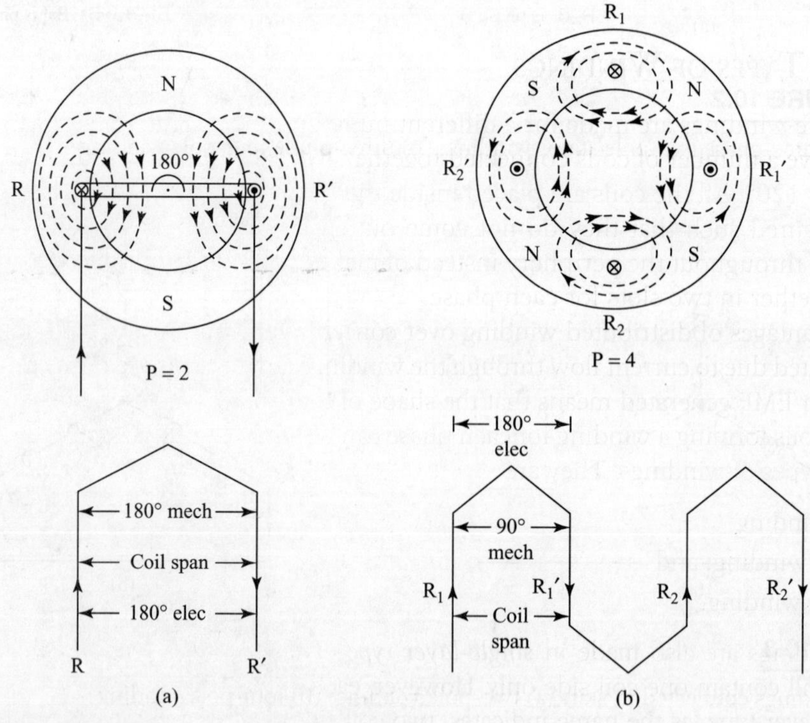

FIGURE 10.4

(a) Two-pole stator winding; (b) four-pole stator winding

The direction of current flowing through the coils and the flux produced have been shown. It is observed that when the coil span is 180° mechanical, as in Fig. 10.4 (a), two poles are formed. In Fig. 10.4 (b), the coil span has been reduced to 90° mechanical. Two coils have been used to complete the winding distributed throughout the stator. The directions of the flux produced show that four poles are formed in the stator. This shows how by using coils of different spans, a winding can be made for a different number of poles. The students are advised to draw a simple six-pole stator winding by using three coils connected together. The coil span here should be 60° mechanical.

The distance between the two coil sides of a coil is always expressed as 180° electrical irrespective of the number of poles for which the winding is made. For a two-pole winding, the coil span is 180° electrical which is also equal to 180° mechanical. For a four-pole winding, the coil span is again 180° electrical which is equal to 90° mechanical. If the winding is made for six-poles, the coil span will be counted as 180° electrical but will be equal to 60° mechanical.

In general, the relationship between electrical degrees and mechanical degrees is expressed as

$$1^\circ \text{ Mechanical} = \frac{P^\circ}{2} \text{ Electrical}$$

When P = 2, 1° mechanical is equal to 1° electrical or 180° mechanical is equal to 180° electrical.

10.7 ▶ Types of Windings

Three-phase windings are made for a different number of poles. Each phase winding generally will have a number of coils connected together. The three-phase windings are displaced in space by 120°. All the coils are placed inside the slots made in the stator inside periphery and are secured such that they do not come out easily. The whole winding is distributed uniformely throughout the periphery instead of making big slots and placing a large number of coils together in two slots for each phase.

The advantages of distributed winding over concentrated winding are better dissipation of heat generated due to current flow through the windings and better wave form of the generated EMF (better EMF generated means that the shape of the voltage wave should be sinusoidal).

All the coils forming a winding for each phase can be connected in a number of ways. These are called types of windings. They are

(i) lap winding;
(ii) wave winding; and
(iii) spiral winding.

The windings are also made in ***single-layer type*** or ***double-layer type***. In single-layer type each slot will contain one coil side only. However, each coil will have a large number of turns. In double-layer type, as the name indicates, two coil sides will occupy one slot in the whole of the armature winding. The coils used for winding may be of ***full-pitch type*** or ***short-pitch type***. In a full-pitch coil, the distance between the two coil sides is 180° electrical. The coil span of short-pitch coils is reduced by a certain angle. Windings made with short-pitch coils is called ***fractional-pitch winding***. By use of fractional-pitch winding, any specific harmonic present in the generated EMF can be eliminated so that we get a sinusoidal EMF. Fig. 10.5 shows a single-layer distributed stator winding where the windings can be connected in the lap, wave, or spiral form. The connections for only one phase has been shown. The number of coils used per phase is three only. Full-pitch coils have been used in the winding shown in Fig. 10.5 (a).

Use of short-pitch coil improves the wave shape of the induced EMF making it more like a sine-wave. However, the EMFs induced in the two coil sides when added vectorially in a short-pitch coil will be less than that of a full-pitch coil.

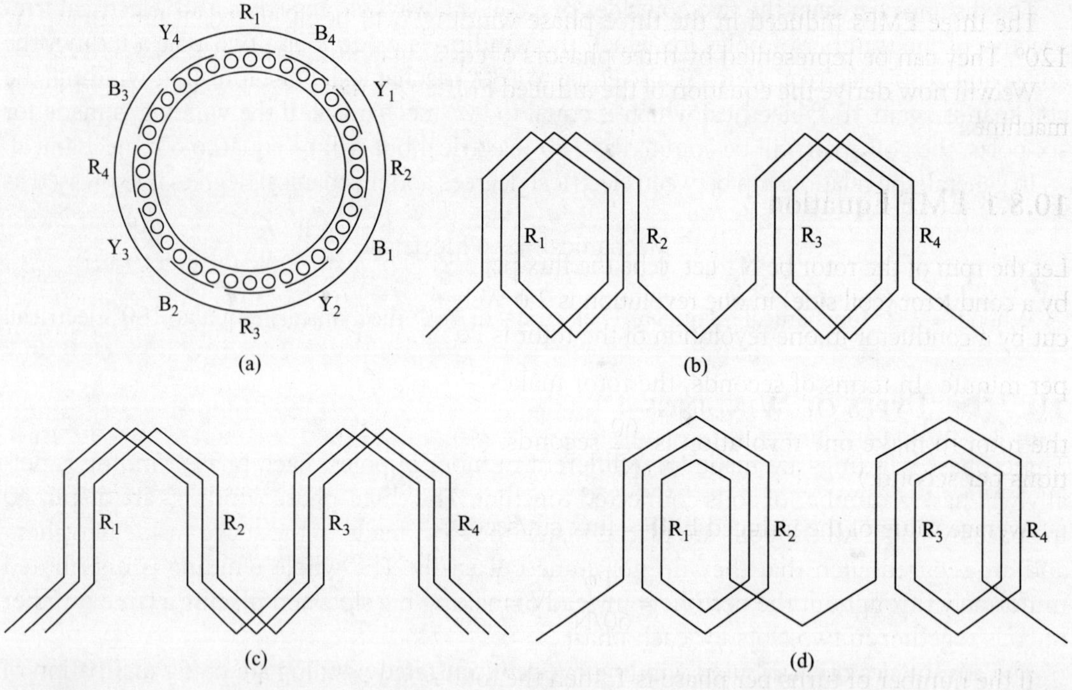

FIGURE 10.5

(a) Single layer stator winding; (b) lap-type winding; (c) wave-type winding; (d) spiral-type winding

10.8 ► Induced EMF in a Synchronous Machine

In synchronous machines, the armature winding is made on the stator. The rotor consists of magnetic poles excited by dc field current. The rotor poles are rotated by a prime mover, may be a steam turbine or a water turbine, as the case may be. The poles when rotating, as shown in Fig. 10.6, will induce EMF in the armature coils because the magnetic lines will cut the coil sides. The EMF induced in phase R_1–R_2, Y_1–Y_2, and B_1–B_2 will be identical but will have a time phase difference of 120°. Phase difference of 120° degrees corresponds to the time taken by the rotor to rotate by 120 electrical degrees. The generated voltages in the R, Y, and B phases can be expressed as

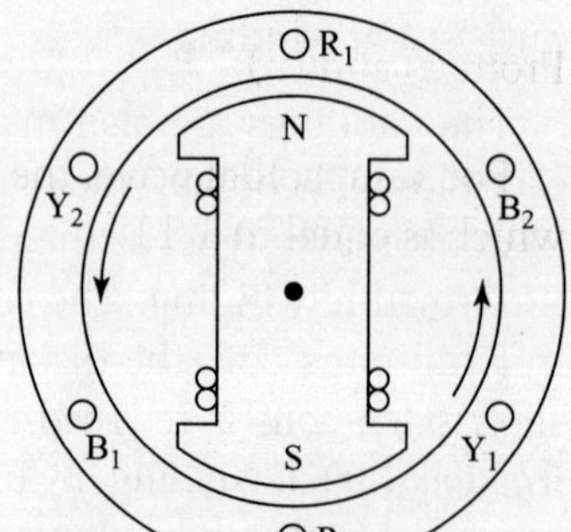

FIGURE 10.6

EMF is induced in the stator winding due to rotation of poles

$$e_R = E_m \sin \omega t$$

$$e_Y = E_m \sin(\omega t - 120°)$$

$$e_B = E_m \sin(\omega t - 240°)$$

The three EMFs induced in the three-phase windings will be displaced in time phase by 120°. They can be represented by three phasors of equal magnitude but displaced by 120°.

We will now derive the equation of the induced EMF in each of the phases of a synchronous machine.

10.8.1 EMF Equation

Let the rpm of the rotor be N_s. Let ϕ be the flux per pole. For a two-pole machine, the flux cut by a conductor (coil side) in one revolution is 2 ϕ Webers. If P is the number of poles, then flux cut by a conductor in one revolution of the rotor is Pϕ webers. The rotor makes N_s revolutions per minute. In terms of seconds, the rotor makes $\frac{N_s}{60}$ revolutions. Thus, the time taken by the rotor to make one revolution is $\frac{60}{N_s}$ seconds. (Since N is the rpm, or $\frac{N_s}{60}$ is the revolutions per second.)

Average value of the induced EMF = flux cut/second

$$= \frac{P\phi}{60/N_s}\ V = \frac{P\phi N_s}{60}\ V$$

If the number of turns per phase is T, then the total number of conductors Z will be 2T.

Hence, Average EMF induced in each phase $E_{av} = \frac{P\phi N_s 2T}{60}\ V$ (i)

If, f is the frequency of the induced EMF, the relationship between, f, P, and N_S is given by

$$N_s = \frac{120f}{P} \quad \text{(ii)}$$

From (i) and (ii)

$$E_{av} = \frac{PN_s}{120} \times 4\phi T = 4\phi f T\ V$$

For a sinusoidal wave, the ratio of the rms value to the average value is called the form factor which is equal to 1.11.

$$\frac{E_{rms}}{E_{av}} = 1.11$$

$$E_{rms} = 1.11\,E_{av} = 1.11 \times 4\ \phi f T\ V$$

If we write E_{rms} = E, the EMF eq. is

$$E = 4.44\ \phi f T\ V \quad (10.1)$$

It is interesting to note that this EMF equation is the same as that developed for transformers where the EMF in the primary and secondary windings were derived, respectively, as $E_1 = 4.44\ \phi_m f N_1$ and $E_2 = 4.44\ \phi_m f N_2$.

In the case of the synchronous machine the EMF induced is called the dynamically induced EMF while in the case of the transformer the EMF induced is called statically induced EMF.

In synchronous machines, EMF is induced due to the relative motion the between the rotor flux and the stator conductors. In the case of transformers, EMF is induced in the winding due to the linkage of the time-varying flux with stationary coils.

The EMF equation derived as above is to be multiplied by two factors, namely the ***distribution factor***, K_d and the ***pitch factor***, K_p. Because of the distribution of the coils in the armature, the EMFs induced in the individual coils cannot be added arithmetically. They have to be added vectorially. The vector sum of voltages is less than the algebraic sum of the voltages in the coils. Hence, the ratio is less than 1. The value of K_d is somewhat less than 1.

If the whole winding is concentrated in two slots with all the coil sides placed in one slot, then the value of K_d will be 1. That is, there would be no reduction of the total EMF induced due to the use of number of coils to form the winding on the stator.

The pitch factor K_p is due to the use of short-pitch coils. The vector sum of the voltages induced in the two sides of a coil is not equal to the their algebraic sum. With K_d and K_p into consideration, the equation for the induced EMF is

$$E = 4.44\,\phi f\,T K_d K_p \text{ V} \tag{10.2}$$

The factor K_d is used in the EMF equation due to the use of winding being distributed rather than concentrated in two slots. The factor K_p is due to the use of short-pitch coils rather than full-pitch coils.

The values of K_d and K_p are derived as follows.

10.8.2 Distribution Factor

Distribution factor is defined as the ratio of the EMF induced in the distributed winding in a phase to the EMF induced in a concentrated winding. In Fig. 10.7, the stator with a number of slots have been shown. The conductors are placed in the slots. The EMFs induced in the conductors are e_1, e_2, e_3, etc. These EMFs are to be added vectorially as shown in Fig. 10.7.

Let α be the angle between two slots.

$$CA = 2CM = 2\,OC\cos\left(90 - \frac{3\alpha}{2}\right)$$

$$CB = 2CP = 2OC\cos(90 - \alpha/2)$$

$$CB = BD = DA = 2OC\sin\alpha/2$$

and

$$CA = 2\text{ OC }\sin\frac{3\alpha}{2}$$

Distribution factor,

$$K_d = \frac{\text{Vector sum of the voltages}}{\text{algebraic sum of the voltages}} = \frac{CA}{CA + BD + DA} = \frac{CA}{3CB}$$

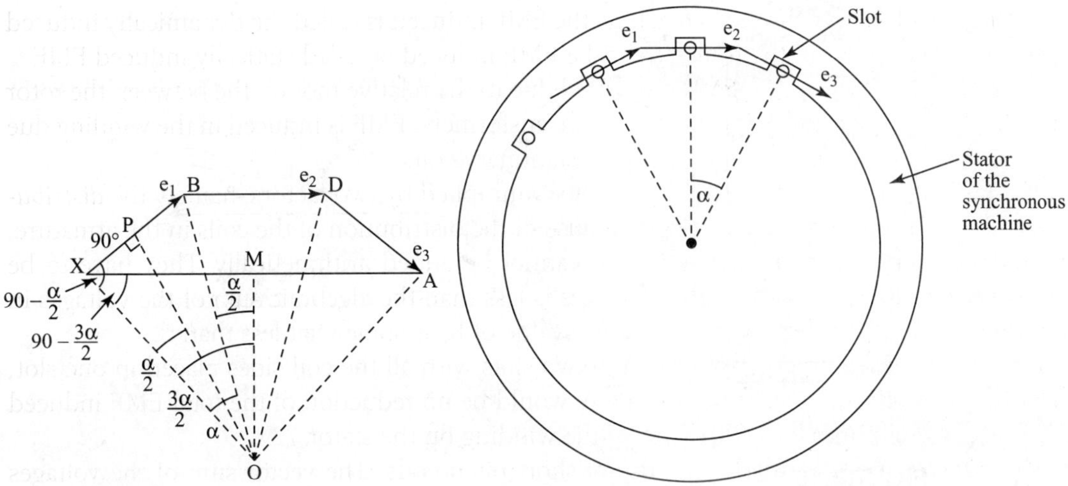

FIGURE 10.7

Distribution factor of the EMF equation

or, $$K_d = \frac{2OC \sin 3\alpha/2}{3 \times 2OC \sin \alpha/2}$$

or, $$K_d = \frac{\sin 3\alpha/2}{3 \sin \alpha/2}$$

Here, we have considered three slots per pole per phase. If m is the number of slots per pole per phase, then

Distribution factor, $$K_d = \frac{\sin m\alpha/2}{m \sin \alpha/2} \tag{10.3}$$

10.8.3 Pitch Factor

The pitch factor is due to the use of short-pitch coils as has been shown in Fig. 10.8. If the winding is made with full-pitch coils then, pitch factor K_p is equal to 1. The pitch factor is defined as the ratio of the EMF induced in a short-pitch coil to the EMF induced in a full-pitch coil. Let β be the angle through which the coil is made less than the full pitch. The pitch factor, K_p is the ratio of the vector sum of the EMFs induced in the coil sides to the algebraic sum of the EMFs.

$$K_p = \frac{AC}{AB + BC} \text{ [from Fig. 10.8 (b)]}$$

or, $$K_p = \frac{2AB \cos \beta/2}{2AB} \text{ since AB = BC}$$

Pitch factor, $$K_p = \cos \beta/2 \tag{10.4}$$

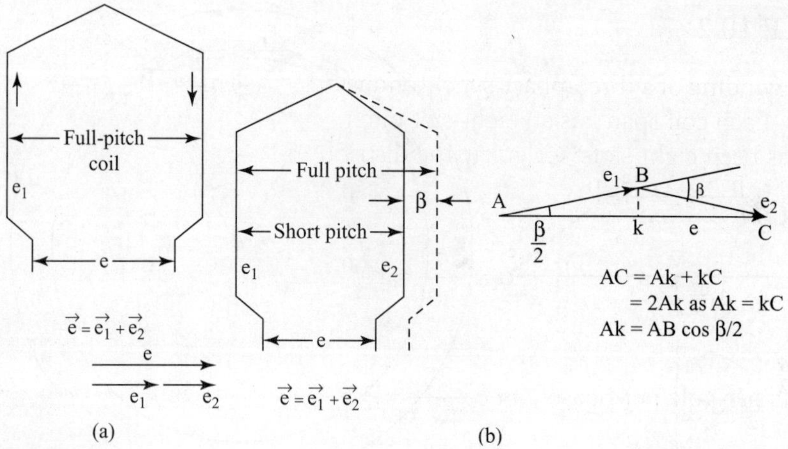

FIGURE 10.8

Pitch factor due to use of short-pitch coil

The value of the pitch factor is somewhat less than unity. For example, if the coil is short pitched by an angle say 30°, then $\beta = 30°$:

$$K_p = \cos\frac{30}{2} = \cos 15° = 0.96$$

EXAMPLE 10.1

Calculate the distribution factor for a single-layer 36-slot two-pole three-phase stator winding of a synchronous machine.

SOLUTION

Slots are made distributed throughout the whole of stator periphery. The angle between two slots, i.e., slot angle α is calculated as

$$\alpha = \frac{360°}{\text{No. of slots}} = \frac{360°}{36} = 10°$$

Number of slots per pole per phase, i.e., m is calculated as

$$m = \frac{\text{No. of slots}}{\text{No. of poles} \times \text{No. of phases}}$$

$$= \frac{36}{2 \times 3} = 6$$

$$K_d = \frac{\sin m\alpha/2}{m \sin \alpha/2} = \frac{\sin(6 \times 10)/2}{6 \sin 10/2} = \frac{\sin 30°}{6 \sin 5°} = \frac{0.5}{6 \times 0.087}$$

or, $$K_d = 0.958$$

EXAMPLE 10.2

The stator winding of a three-phase synchronous machine has been wound for four-poles in 36 slots. Each coil span has an eight-slot pitch, i.e., the distance between the coil sides of a coil has been eight slots. Calculate the distribution factor and the pitch factor.

SOLUTION

$$\alpha = \frac{360}{36} = 10°\text{ mechanical}$$

No. of slots per pole per phase, $m = \frac{36}{4 \times 3} = 3$

1° mechanical $= \frac{P°}{2}$ electrical. We have to convert all mechanical degrees into electrical degrees.

Since, P = 4,

1° mech = 2° electrical

10° mech = 20° electrical

$\alpha = 20°$, m = 3

$$K_d = \frac{\sin m\alpha/2}{m \sin \alpha/2} = \frac{\sin(3 \times 20)/2}{3 \sin 20/2} = \frac{\sin 30°}{3 \sin 10°} = \frac{0.5}{3 \times 0.1736}$$

or, $K_d = 0.96$

For calculation of pitch factor,

No. of slots per pole $= \frac{36}{4} = 9$

This means the coil sides of a full-pitch coil will be at a distance of nine slots. Then the coil sides will lie under opposite poles occupying similar locations under each pole. Since the pole pitch is eight, the slot pitching has been done for one slot angle, i.e., by 20° electrical.

Thus, $\beta = 20°$

$$K_p = \cos\frac{\beta}{2} = \cos 10° = 0.98.$$

EXAMPLE 10.3

The induced EMF in a synchronous machine is 11,000 V with a distributed fractional pitch winding. If a concentrated full-pitch winding were made, what would have been the induced EMFs. Assume the distribution factor and the pitch factor as 0.96 and 0.98, respectively.

SOLUTION

If concentrated winding and full-pitch coils are used, the EMF induced will be

$$E = 4.44\,\phi f\,T\ \text{V}$$

But with distributed winding and use of short-pitch coils

$$E = 4.44\phi f\,T\,K_d\,K_p\ \text{V}$$

with the given

$$11{,}000 = 4.44\phi f\,T \times 0.96 \times 0.98$$

$$\therefore \qquad E = 4.44\phi f\,T = \frac{11{,}000}{0.96 \times 0.98} = 11{,}692\ \text{V}$$

This shows that due to the distribution of winding throughout the stator periphery and use of short-pitch coils, the EMF induced has been reduced in this case by 692 V. However, the use of distributed winding and short-pitch coils improves the heat dissipation and wave shape of the voltage generated, respectively.

EXAMPLE 10.4

A three-phase 36-pole synchronous generator is rotated by a water turbine at 167 rpm. The stator has 324 slots and each slot has 10 conductors. The flux per pole is 20 mWb. Calculate the EMF induced per phase if full-pitch coils are used for the winding.

SOLUTION

Synchronous speed,

$$N_s = \frac{120f}{P}$$

$$167 = \frac{120 \times f}{36}$$

$$f = 50\ \text{Hz}$$

No. of slots per pole per phase, m is calculated as

$$m = \frac{\text{No. of slots}}{\text{No. of poles} \times \text{No. of phases}} = \frac{324}{36 \times 3} = 3$$

$$\text{Slot angle} = \frac{360^\circ}{324} \text{ mechanical}$$

$$1^\circ \text{ mechanical} = \frac{P^\circ}{2} \text{ Electrical}$$

$$\text{Here } 1^\circ \text{ mechanical} = \frac{36}{2} = 18^\circ \text{ Electrical}$$

Slot angle, α, in electrical degrees is

$$\alpha = \frac{360 \times 18}{324} = 20^\circ$$

Distribution factor, K_d, is calculated as

$$K_d = \frac{\sin m\alpha/2}{m \sin \alpha/2} = \frac{\sin(3 \times 20)/2}{3 \sin 20/2} = \frac{\sin 30^\circ}{3 \sin 10^\circ}$$

or,

$$K_d = \frac{0.5}{3 \times 0.1736} = 0.96$$

$$\text{Total no. of conductors per phase} = \frac{\text{No. of slots} \times \text{No. of conductors in each slot}}{\text{No. of phases}}$$

Number of T will half the number of conductors

or,

$$T = \frac{324 \times 10}{3} = 1080$$

$K_p = 1$ because full-pitch coil has been used.
EMF induced per phase is calculated as

$$E = 4.44 \phi f T K_p K_d \text{ V}$$

Substituting all values

$$E = 4.44 \times 20 \times 10^{-3} \times 50 \times 540 \times 1 \times 0.96 \text{ V}$$

or,

$$E = 2301 \text{ V}$$

10.9 ▶ Open-Circuit or No-Load Characteristic

EMF induced in a synchronous generator is expressed as, $E = 4.44\,\phi f T$ volts. Flux, ϕ is produced by the field current, I_f. Frequency of induced EMF, f depends upon the speed of rotation of the field poles, N_s and the number of poles, P Freqnency, $f = \frac{N_s \times P}{120}$.

Thus, for a particular generator of P poles and being driven at a constant speed, N_S, $E \propto I_f$.

If the field current is increased, the EMF induced will increase till the magnetic circuit gets saturated when further increase of I_f will not cause increase in E. This relationship of I_f versus E is called the open-circuit or no-load characteristic. How this no-load characteristic is determined is explained here.

When a synchronous machine is driven by a prime mover at synchronous speed, N_s, it will generate an induced EMF if its field winding is excited. The field winding is excited by supplying a dc voltage through the brush and slip-ring arrangement. When the field system is rotated at a constant speed N_s and the field current, I_f is gradually increased, keeping the output terminals open as shown in Fig. 10.9, the induced EMF will go on increasing but will have a saturation effect as shown in Fig. 10.9 (b).

The open-circuit characteristic or OCC as shown in Fig. 10.9 is the relationship between the field current and the induced EMF when the rotor is rotated at a constant speed. Since no load is connected across the stator output terminals, the OCC is also called the no-load characteristic.

If the rotor is driven of any other constant speed, similar open-circuit characteristics can be drawn. However, OCC is normally drawn at rated synchronous speed.

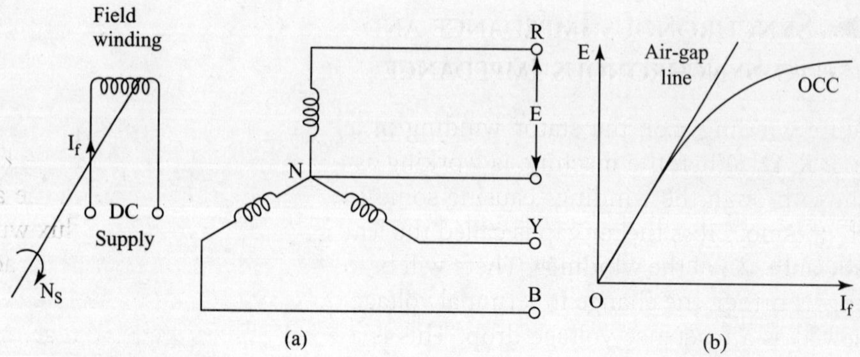

FIGURE 10.9

Open-circuit characteristic: (a) field system is rotated by prime mover; (b) field current versus EMF-induced characteristic

10.10 ▶ Synchronous Generator on Load

We first run the synchronous generator on no-load at rated speed. We then connect the electrical load across its terminals.

When a synchronous generator is loaded, current will flow through its winding as well as through the load. Since three-phase currents will flow through the three-phase windings, these currents will develop a resultant rotating magnetic field. This rotating field is due to the currents flowing through the armature windings, i.e., the stator windings. This stator field will rotate at synchronous speed, N_s. The field magnets are also rotating at speed, N_s. Thus, these two fields, i.e., the rotor field and the field produced by the stator which is also called the armature will rotate at the same speed, i.e., these two fields are stationary with respect to each other. The armature field flux and the main field flux produced by the field windings will rotate at the same speed, called the synchronous speed. The air-gap flux will be the resultant of these two fluxes. The effect of the armature flux on the main field flux is called ***armature reaction***. Depending on the power factor of the load, the armature flux will oppose, aid, or distort the main field flux.

If the load is purely inductive, the armature flux will be opposing the main field flux. If the load is purely capacitive, the armature flux will aid the main field flux.

For resistive loads, the armature flux will distort the distribution of the main field flux.

The armature reaction will, therefore, have an effect on the magnitude of the induced EMF. The more is the load current, the more will be the effect of armature reaction. At no load, there is no effect of armature reaction.

When the synchronous generator is loaded, there will be a voltage drop in the windings as well as an armature reaction effect. At unity power factor load, the voltage drop due to loading will be less than at lagging power factor load. For capacitive load, since the armature flux will aid the main field flux, the air-gap flux will increase, and hence, the EMF induced will go on increasing as the capacitive loading increases.

10.11 ▶ Synchronous Impedance and Voltage Drop Due to Synchronous Impedance

The armature winding, i.e., the stator winding of a synchronous machine has a winding resistance of R_a Ω. When the machine is working as a generator supplying some load, current will flow through the windings causing some I_aR_a voltage drop. Some of the armature flux which does not cross the air gap is called the leakage flux. This leakage flux will lead to leakage reactance, X_l, of the windings. There will be voltage drop due to leakage reactance of the windings. Further, the change in terminal voltage due to the armature reaction effect can also be viewed as a reactance voltage drop. This is a fictitious reactance voltage drop. This reactance due to the armature flux is called X_a. The reactance due to the armature leakage flux is called X_l. The voltage drop due to resistance, R_a is in phase with the armature current, I_a. The reactance voltage drops are in quadrature with the armature current, such that

$$E = V + I_aR_a + j\,I_a\,(X_l + X_a)$$

where, E is the induced EMF per phase at no-load, I_a is the armature current flowing through each phase, R_a is the armature resistance per phase, X_l is the leakage reactance of the armature winding due to the leakage flux, and X_a is a fictitious (not real) armature reactance which replaces the effect of armature reaction. Again

$$\begin{aligned} E &= V + I_a R_a + jI_a (X_l + X_a) \\ &= V + I_a R_a + jI_a X_s \\ &= V + I_a (R_a + jX_s) \\ &= V + I_a Z_s \end{aligned}$$

where $X_s = X_a + X_l$ is called the ***synchronous reactance*** and $Z_s = R_a + jX_s$ is called the ***synchronous impedance***.

The vector sum of R_a and X_s is called the synchronous impedance. The effect of armature voltage drop due to armature resistance and synchronous reactance, i.e., synchronous impedance at different power factor load, has been shown in Fig. 10.10. It is interesting to note that at leading power factor load the terminal voltage of the synchronous generator increases with increase in load.

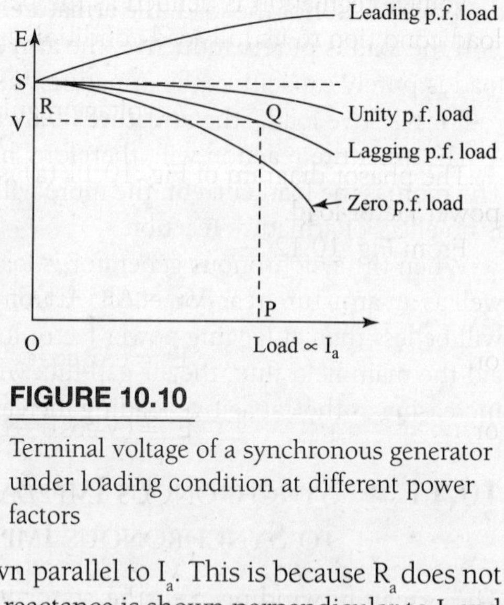

FIGURE 10.10

Terminal voltage of a synchronous generator under loading condition at different power factors

Let OS represent the voltage induced at no load, i.e., E. When a load current I_a equivalent to OP flows at a lagging power factor load, the terminal voltage available across the load terminals gets reduced and will be equal to OR which we call as V. The phasor diagram representing E, V, I_a, and voltage drop due to $I_a Z_s$ has been shown in Fig. 10.11 (a). In Fig. 10.11 (b), the phasor diagram has been drawn for leading power factor load.

In the phasor diagram in Fig. 10.11 (a), I_a is shown lagging V by the power factor angle ϕ. Voltage drop in the armature winding resistance R_a is equal to $I_a R_a$. This voltage drop of $I_a R_a$ has been shown parallel to I_a. This is because R_a does not have any direction. Voltage drop across an inductive reactance is shown perpendicular to I_a and

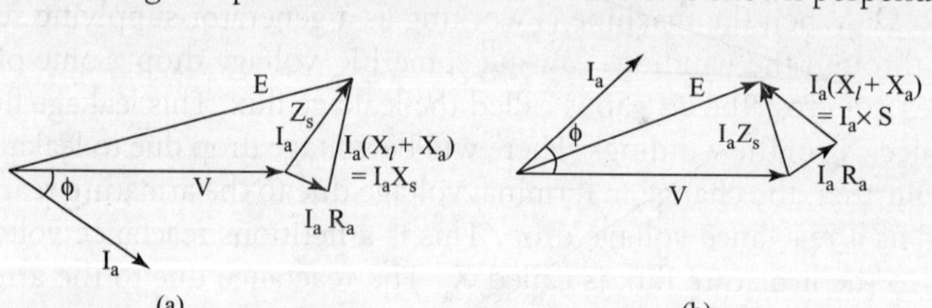

FIGURE 10.11

(a) Phasor diagram at lagging power factor load; (b) phasor diagram at leading power factor load

leading I_a. Therefore, drops I_aX_l and I_aX_a have been shown perpendicular to I_a. The sum of $I_aR_a + j I_a (X_l + X_a)$ is I_aZ_s, where $Z_s = R_a + j (X_l + X_a) = R_a + j X_s$. The sum of V and I_aZ_s is equal to E.

In the same way the phasor diagram for leading Power factor load current I_a has been drawn as in Fig. 10.11 (b).

It is interesting to note that for leading power factor load, the terminal voltage V is more than the induced EMF at no load.

10.12 ▶ Voltage Regulation of a Synchronous Generator

Synchronous generators supply power to various loads at a particular terminal voltage. The generator has to maintain its terminal voltage within a specified limit. If there is very large change in the terminal voltage when load on the generator changes, it will give rise to difficulties in operations of various electrical machines and gadgets connected to the system. Voltage regulation tells us about the health of the machine in terms of its voltage stability.

Voltage regulation is defined as the percentage change of the terminal voltage from its no-load condition to its full-load condition as a percentage of full-load voltage. Thus,

$$\text{voltage regulation} = \frac{(E - V)}{V} \times 100$$

The phasor diagram of Fig. 10.11 (a) is redrawn to calculate voltage regulation at a lagging power factor load.

From Fig. 10.12

$$AG^2 = AK^2 + KG^2 = (AD + DK)^2 + (KC + CG)^2$$

or,
$$E^2 = (AB\cos\phi + I_aR_a)^2 + (AB\sin\phi + I_aX_s)^2$$

or,
$$E^2 = (V\cos\phi + I_aR_a)^2 + (V\sin\phi + I_aX_s)^2$$

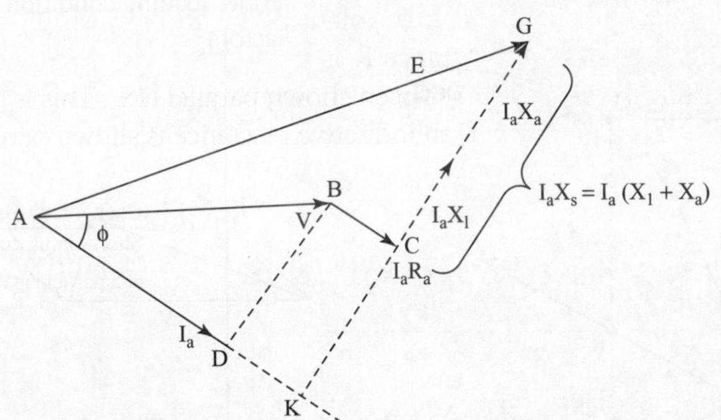

FIGURE 10.12

Phasor diagram of a synchronous generator at lagging power factor load

or, $$E = \sqrt{(V\cos\phi + I_aR_a)^2 + (V\sin\phi + I_aX_s)^2} \tag{10.5}$$

For leading power factor load, it will be seen that

$$E = \sqrt{(V\cos\phi + I_aR_a)^2 + (V\sin\phi - I_aX_s)^2} \tag{10.6}$$

To calculate voltage regulation, we have to calculate E using the above relation. While V, ϕ, R_a, I_a will be known, the value of synchronous reactance has to be determined. The vector sum of armature resistance and synchronous reactance will give the value of synchronous impedance. Voltage regulation depends on the load and load power factor.

10.13 ► Determination of Voltage Regulation by the Synchronous Impedance Method

The direct method of determining voltage regulation is to load the generator and measure its full-load terminal voltage. Then remove the load and measure its no-load voltage, without changing the excitation. Then, calculate voltage regulation using the relation, Voltage Regulation = $\frac{E - V}{V}$.

However, for large machines, it may not be possible to find such loads in the testing laboratory. Further, there will be huge power loss due to loading during the testing time. For these reasons an indirect method of testing is always used.

In the synchronous impedance method, we perform no-load test, i.e., the open-circuit test and short-circuit test on the generator.

For the open-circuit test, the generator is driven at its rated speed by the prime mover. It's field current I_f is gradually increased using a field regulator. The values of I_f and corresponding induced EMF, E are recorded and plotted. This will give us the open-circuit characteristic, i.e., OCC.

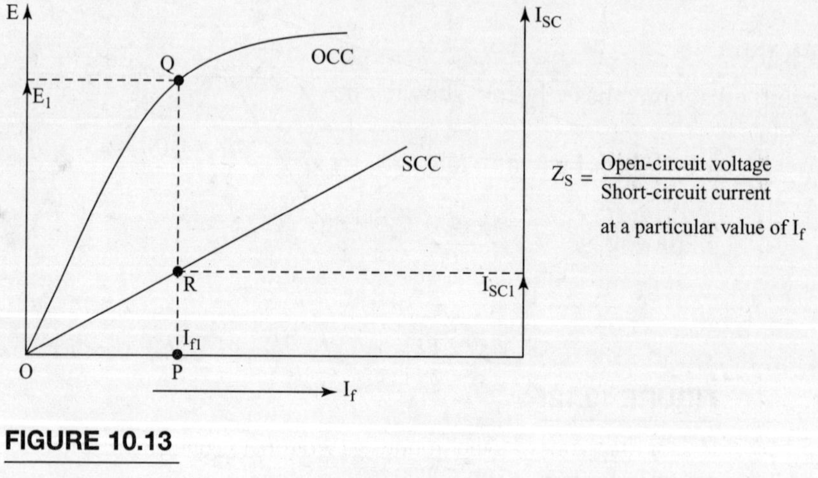

FIGURE 10.13

Open-circuit and short-circuit characteristics of a synchronous generator

The short-circuit test is conducted by running the generator at rated speed, keeping the output terminals short circuited through an ammeter. Reduced excitation current I_f is allowed to flow through the field winding so that the short-circuit current, I_{sc} does not exceed the rated current of the generator. The short-circuit characteristic, SCC is plotted as I_{sc} as a function of I_f by taking a few readings. Both OCC and SCC are drawn as a function of I_f as shown in Fig. 10.13.

At a particular value of field current, say I_f = OP, the open-circuit voltage is PQ Volts. When the output terminals are kept short circuited under the short-circuit test, the voltage E will cause a short-circuit current I_{sc1} to flow. The EMF E_1 on open circuit is regarded as being responsible for circulating a short-circuit current of I_{sc1} through the synchronous impedance Z_s. Thus, Z_s can be calculated as

$$Z_s = \frac{\text{OC voltage}}{\text{SC current}} \text{ at a particular } I_f$$

$$= \frac{E_1}{I_{sc1}} \text{ at } I_f = I_{f1}$$

The per-phase armature winding resistance, R_a can be measured by the ammeter voltmeter method. When the synchronous generator is at rest, from Z_s the value of X_s can be calculated as $X_s = \sqrt{Z_s^2 - R_a^2}$.

Voltage regulation for a particular load current and power factor can then be calculated by calculating the value of E using equation (10.6).

EXAMPLE 10.5

A 3 MVA, 6600 V, three-phase, star-connected synchronous generator has a resistance of 0.2 Ω and synchronous reactance of 3.5 Ω per phase. Calculate the regulation at rated output at 0.8 power factor lagging. The speed and excitation remain constant.

SOLUTION

The circuit diagram has been shown in Fig. 10.14 (a).

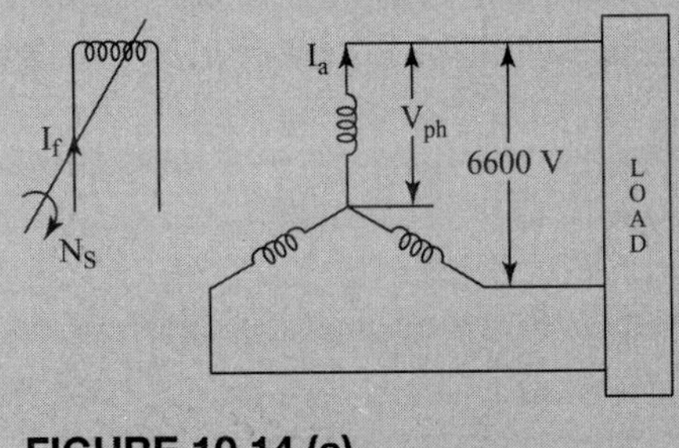

FIGURE 10.14 (a)

Given line voltage = 6600 V

$$\text{phase voltage } V_{Ph} = \frac{6600}{\sqrt{3}}$$

So, $V = V_{ph} = 3810$ V

$R_a = 0.2\ \Omega$

$X_s = 3.5\ \Omega$, $\cos\phi = 0.8$ lagging, $\sin\phi = 0.6$

Total MVA = 3

Total VA = 3×10^6

This VA is for the three phases. VA per phase will be one-third of the total VA

VA per phase = 1×10^6

$$\text{Current per phase} \quad I_a = \frac{1\times10^6}{V_{ph}} = \frac{10,00,000}{3,810} = 262.5 \text{ A}$$

$$\text{Induced EMF,} \quad E = \sqrt{(V\cos\phi + I_aR_a)^2 + (V\sin\phi + I_aX_s)^2}$$

$$= \sqrt{(3810\times0.8 + 262.5\times.2)^2 + (3810\times0.6 + 262.5\times3.5)^2}$$

$$= 4049.4 \text{ V}$$

$$\text{Percentage regulation} = \frac{(E-V)}{V}\times100 = \frac{(4049.4-3810)}{3810}\times100 = 6.28 \text{ per cent}$$

EXAMPLE 10.6

A 1500 kVA, 3300V, 50Hz, three-phase, star-connected synchronous generator has an armature resistance of 0.2 Ω per phase. A field current of 50 A produces a short-circuit current of 262 A and an open-circuit EMF of 1200 V between the lines. Calculate voltage regulation of the generator on full load at 0.8 power factor lagging and at 0.8 power factor leading.

SOLUTION

Total kVA = 1500

$$\text{kVA per phase} = \frac{1500}{3} = 500$$

$$\text{Per-phase voltage} = \frac{V_L}{\sqrt{3}} = \frac{3300}{\sqrt{3}} \text{ V} = 1905 \text{ V}$$

$$\text{Per-phase current,} \quad I_a = \frac{500\times1000}{3300/\sqrt{3}} = 262 \text{ A}$$

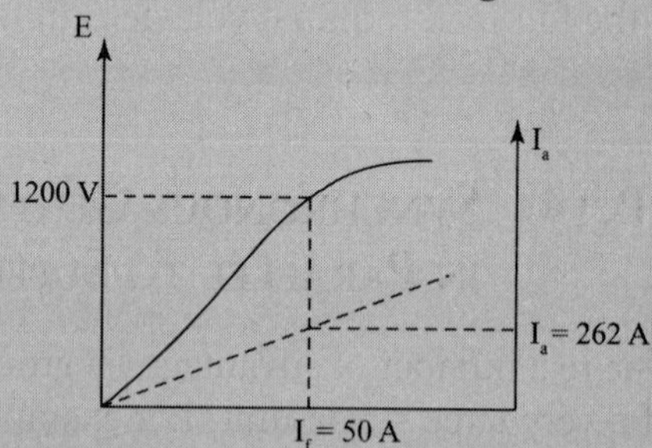

FIGURE 10.14 (b)

Given that at a field current of 50 A, short-circuit I_a is 262 A and the open-circuit line voltage is 1200 V as has been shown in Fig. 10.14 (b).

$$\text{Synchronous impedance/phase } Z_s = \frac{\text{open-circuit voltage per phase}}{\text{short-circuit current}} = \frac{1200}{\sqrt{3}\times262} = 2.64\ \Omega$$

$$R_a \text{ per phase} = 0.2\ \Omega$$

$$\cos\phi = 0.8,\ \phi = 37°,\ \sin\varphi = 0.6$$

At lagging power factor load

$$E_1 = \sqrt{(V\cos\phi + I_aR_a)^2 + (V\sin\phi + I_aX_s)^2}$$

and at leading power factor load

$$E_2 = \sqrt{(V\cos\phi + I_aR_a)^2 + (V\sin\phi - I_aX_s)^2}$$

$$E_1 = \sqrt{(1905\times 0.8 + 262\times 0.2)^2 + (1905\times 0.6 + 262\times 2.64)^2}$$
$$= 2418\text{ V}$$

Percentage regulation at full load 0.8 power factor lagging

$$= \frac{(2418-1905)\times 100}{1905} = 26.9 \text{ per cent}$$

$$E_2 = \sqrt{(1905\times 0.8 + 262\times 0.2)^2 + (1905\times 0.6 - 262\times 2.64)^2}$$
$$= 1640\text{ V}$$

Percentage regulation at full load 0.8 power factor leading

$$\frac{E_2 - V}{1905}\times 100 = \frac{(1640-1905)}{1905}\times 100 = -13.9 \text{ per cent.}$$

This shows that regulation is negative at 0.8 leading power factor load. This is because the full-load terminal voltage is more than the no-load voltage.

10.14 Synchronous Generators Connected in Parallel to Supply a Common Load

Due to a number of advantages, a group of alternators are installed in the power house instead of a very large single unit. For example, instead of installing a 1000 MVA alternator, we may decide to instal five 200 MVA alternators and connect them in parallel to supply common load. There are a few advantages of parallel connection and operation of number of alternators.

10.14.1 Advantages of Parallel Operation

(i) If instead of one very large alternator, a number of smaller units are installed, it is possible to switch off any alternator for repair and maintenance without disrupting the power supply completely.

(ii) Additional sets can be added depending upon the need.

(iii) It may not be possible to build generators for very high capacity, i.e., the capacity of a power plant. For example, a thermal power plant near Talwandi Sabo in Panjab is being set up to generate 1980 MW. A single alternator of such a high capacity may be physically difficult to construct and transport to the site.

(iv) Alternators connected in parallel can be operated near to full load rather than running a big alternator on low load when the demand for electricity changes.

10.14.2 Parallel Connection of Alternators

At the power-generating station, a number of alternators are connected in parallel on a common busbar. The load is supplied from the busbar as shown in Fig. 10.15. The procedure of connection of alternators to the busbar is called synchronization. Although load has been shown connected with the busbar near to the generators, in practise the generated power is sent to the places of its use through high-voltage transmission lines not shown in the figure. Synchronization refers to inter-connection of alternators with a busbar in which a large number of alternators have already been connected. Such busbars are called *infinite busbars*. These days all the generating stations as well as all the loads are being interconnected through a huge network of transmission and distribution system. This is called the National Grid system. Synchronization of an alternator is the connection of an incoming alternator to the infinite busbar. A few conditions must be satisfied before a new machine is connected to the system.

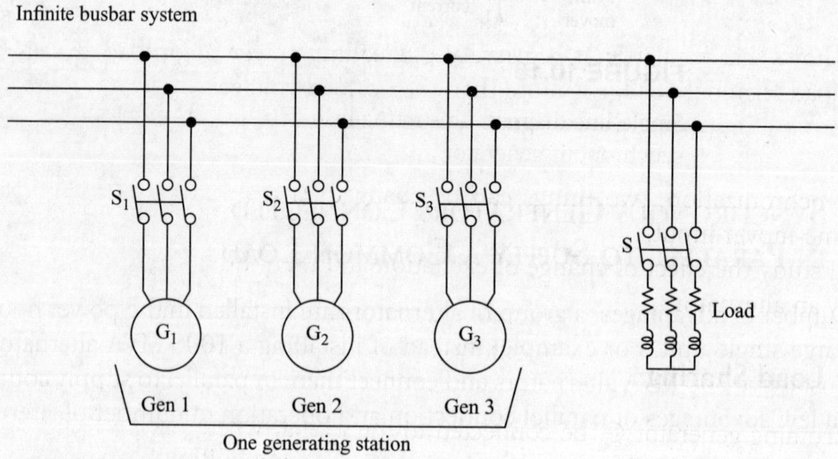

FIGURE 10.15

Parallel connection of alternators (synchronous generators)

10.14.3 Conditions for Parallel Connection and Synchronization

For satisfactory parallel connection of a synchronous generator to the busbar, the following three conditions must be met:

(i) The generated voltage of the incoming machine should be equal to the busbar voltage.

(ii) The frequency of generated voltage of the incoming generator should be equal to the frequency of the busbar voltage.

(iii) The phase sequence of the voltages of the incoming generator should be the same as the phase sequence of busbar voltages.

It may be noted that the kVA or MVA rating of the alternators connected in parallel need not be the same.

Synchronization of alternators is done using a synchroscope or synchronizing lamps to make sure that the conditions of parallel operation are met.

Once an alternator is synchronized, it gets connected to the busbar. Now it has to share a portion of the common load. This is called *load sharing*. If load sharing is not done, the generator will simply remain connected to the busbar; this condition is called floating of the generator with the busbar.

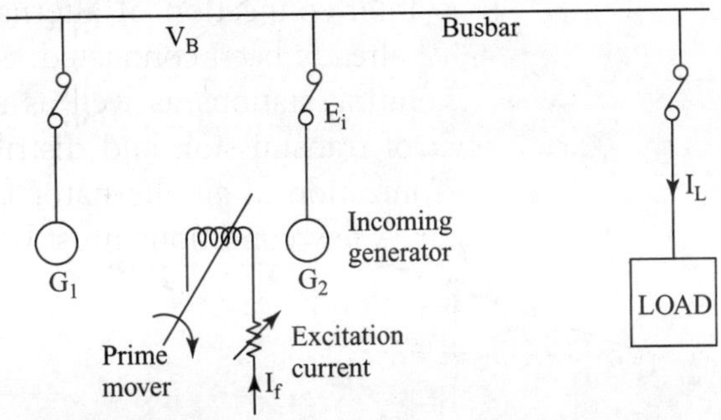

FIGURE 10.16

Single-line diagram illustrating load sharing by a synchronous generator

After synchronization, two things can be changed: (i) the excitation of the generator or (ii) its prime-mover input.

We will study the effect of change of excitation and that of the prime-mover input on load sharing by an alternator.

10.14.4 Load Sharing

Let the incoming generator G_2 be connected to the busbar where one generator G_1 is already connected. See Fig. 10.16. After synchronization, G_2 will simply float on the busbar. It will neither draw any current nor supply any current. For load sharing by the incoming alternator two things can be done. We can change the field current (excitation) or we can change the input to the prime mover driving the generator.

(a) Effect of change of excitation

Let V_B represents the busbar voltage and E_1' represents the induced EMF of the incoming generator, G_2. Since the incoming generator is connected in parallel, the two voltages V_B and E_i' are equal and opposite to each other as shown in Fig. 10.17(a).

Let us increase the excitation current so that the induced EMF is increased from E_i' to E_i'' as has been shown in Fig. 10.17 (a). A resultant voltage equal to $E_i'' - V_B = E_R$ will appear which will circulate current I_c between the incoming alternator and the existing one. I_c will lag E_R because the reactance of the alternators is very high as compared to their resistances. The effect of I_c will be to reduce the reactive component of the load current so that the load current will change from I_L to I_L'. Thus, it is seen that change of excitation current of the incoming generator will only cause reactive load sharing but not active load sharing. Change of excitation of the incoming machine will only change the reactive power delivered by the existing machines.

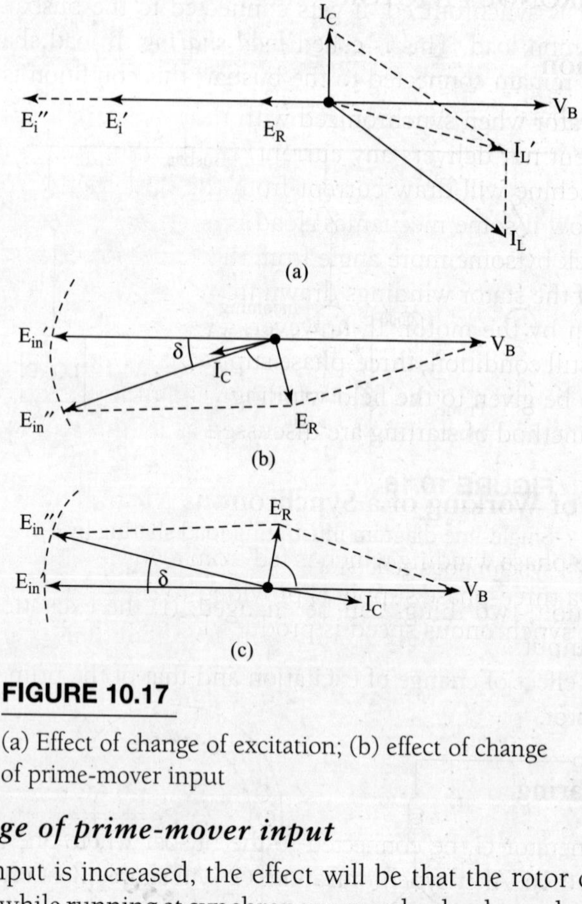

FIGURE 10.17

(a) Effect of change of excitation; (b) effect of change of prime-mover input

(b) Effect of change of prime-mover input

If the prime-mover input is increased, the effect will be that the rotor of the generator will advance by an angle δ while running at synchronous speed as has been shown in Fig. 10.17 (b):

The induced EMF phasor E_{in} has moved to an advanced position.

Note that phasors are rotating phasors; their relative positions have been shown.

The resultant voltage E_R will circulate a current I_c which will lag E_R by about 90°. It is observed that I_c has a strong in-phase component with E_{in}' so that the machine will be working as a generator supplying load. It will be possible to reduce the prime-mover input to the existing machine. If the prime-mover input to the incoming machine is reduced, the rotor will fall back from synchronizm by an angle, say δ as shown in Fig. 10.17 (c). The resultant voltage

E_R will circulate a current I_c which will lag E_R by approximately 90°. Now I_c will have a strong in-phase component with V_B which means that the generators connected with the busbar will have to generate more to compensate for the motoring action of the incoming machine.

To sum up, we can say that the *change of excitation of the generator connected to the bus for parallel operation does not affect the sharing of active load. For sharing of active load, the prime-mover input, i.e., for a steam turbine, the steam input has to be increased so that the torque developed is increased.*

10.15 ► Synchronous Motor

10.15.1 Introduction

A Synchronous generator when synchronized with the busbar, floats on the busbar. That is, it neither draws current nor delivers any current. If the prime mover driving the generator is decoupled, the machine will draw current from the busbar and work as a synchronous motor on no load. Now if some mechanical load is connected to the shaft of the motor, its rotor axis will fall back by some more angle from the axis of the rotating magnetic field created by the current of the stator windings drawn from the busbar voltages. As a result, more current will be drawn by the motor. If, however, a synchronous machine has to start as a motor from its standstill condition, three-phase supply has to be given to the stator windings and dc supply has to be given to the field winding. The principle of working of a synchronous motor and the method of starting are discussed as follows.

10.15.2 Principle of Working of a Synchronous Motor

The stator has a three-phase winding which is fed from a 50 Hz three-phase supply as shown in Fig. 10.18. When a three-phase supply is provided to the stator winding, a rotating magnetic field rotating at synchronous speed is produced. The process is same as that described

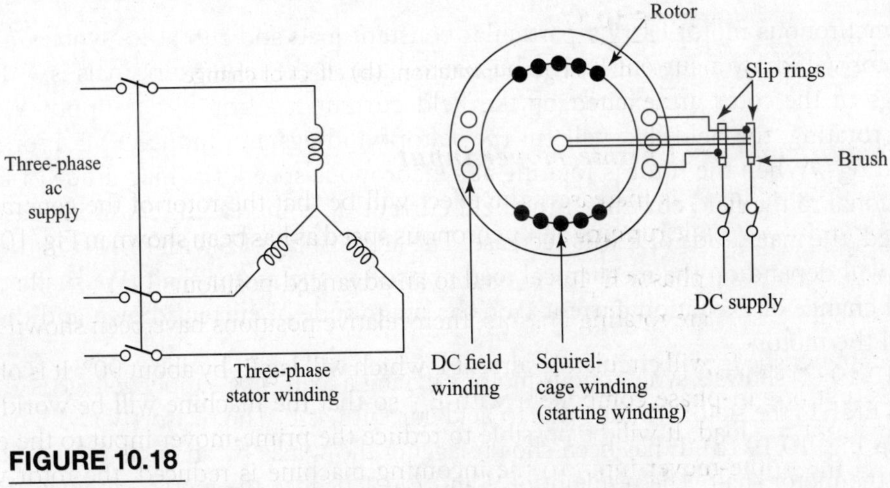

FIGURE 10.18

Synchronous motor with squirrel-cage winding for self-starting in the case of the three-phase induction motor. In a three-phase induction motor the rotor had a closed winding and the rotor got its excitation through electromagnetic induction. In the case of the synchronous motor, the field windings placed on rotor slots are provided with dc excitation through the brush and slip-ring arrangement. Thus, two magnetic fields are produced; one rotating at a synchronous speed, N_s and the other produced by the field winding. The rotor having the field system should start rotating at the same speed as the rotating magnetic field, N_s. The reason is that two magnetic fields will always try to align with each other. However, due to its inertia, the rotor will not pick up speed. That is why a three-phase synchronous motor is not self-starting. To make it self-starting, a squirrel-cage winding is made on the pole faces so that the rotor will start rotating as an induction motor first, without having the field windings excited.

Once the rotor attains a speed near to synchronous speed like a three-phase induction motor, the dc excitation is provided by switching on the field circuit. The rotor immediately attains synchronous speed and gets locked into synchronizm. Thus, the two magnetic fields become stationary with respect to each other and the rotor continues to develop torque. If load is applied on the rotor shaft, the rotor continues to rotate at synchronous speed, but its axis will fall back by angle δ. As a result, more current is drawn from the supply mains. The more is the load applied on the motor, the more will be the angle of lag, δ. The maximum limit of the angle of lag of the rotor field axis from the stator rotating field axis is 90°. The electromagnetic power developed, P, is expressed as

$$P = \frac{VE}{X_S} \sin \delta$$

where, V is the terminal voltage;

E is the induced EMF;

X_S is the synchronous reactance; and

δ is the angle between V and E, also called the power angle or torque angle.

10.15.3 Effect of Change of Excitation of a Synchronous Motor

Let a synchronous motor carry a particular constant load and run at its synchrons speed. The motor will draw a current I_a. The busbar voltage at the motor terminals is V. The field windings in the rotor are excited by the field current fed from a dc supply. When the rotor is rotating, the field flux will cut the stator windings and induce EMF E on the stator windings. When the rotor is rotating at synchronous speed, the magnitude of E will be proportional to the field current. If the field current, I_f, is increased, E will increase; if I_f is decreased, the magnitude of E will decrease. The angle of lag of E with respect to the busbar voltage will depend on the mechanical load applied to the motor shaft. We shall study the effect of change of excitation current I_f on the magnitude of current drawn and the power factor of the motor.

Fig. 10.19 (a) shows a synchronous motor carrying a load. The supply voltage is V and the induced EMF in the stator winding due to field current I_f is E. The phasors V and E have been shown in Fig. 10.19 (b). E has been shown lagging the V axis by an angle δ for a particular load on the motor shaft. The resultant of V and E is E_R. Since the motor windings are highly

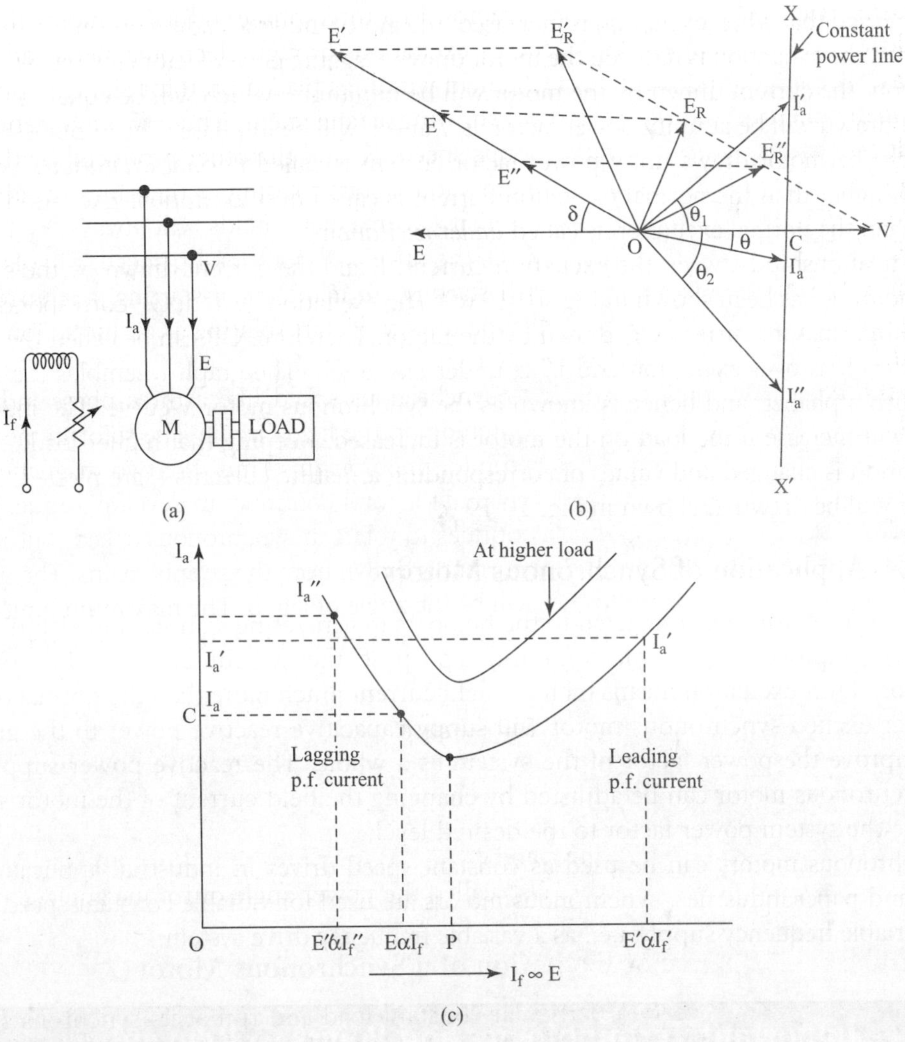

FIGURE 10.19

Effect of change of excitation on the current drawn by a synchronous motor

inductive, I_a drawn by the motor will lag E_R by approximately 90°. The phase angle between V and I_a is the power factor angle θ. The power drawn from the line, i.e., the input power is $VI_a \cos\theta$. As V is constant, $I_a \cos\theta = OC$ will remain constant as long as the mechanical load on the motor remains constant. We can draw a constant power line along XX′ as has been shown in Fig. 10.19 (b). The locus of armature current I_a at a different excitation current I_f will lie on this line. Let excitation current be increased such that E is increased to E′. The resultant of E′ and V is E_R'. Current I_a' lags E_R' by about 90° as shown. The tip of I_a' will lie on line XX′ so that $I_a' \cos\theta_1$ is equal to OC. If excitation current is reduced such that E becomes equal to E″, the resultant of E″ and V is E_R'' and the current which will be lagging E_R'' by about 90° will be I_a''.

It is observed that when excitation is increased, the motor draws a leading power factor current and when the excitation is reduced the motor draws a lagging power factor current. At a certain excitation, the current drawn by the motor will be minimum, which will be equal to OC. The current drawn will be at unity power factor as V and I_a will be in phase. The excitation current at which the motor draws unity power factor current is called ***normal excitation***. Excitation current higher than the normal excitation current is called ***over excitation***. Excitation current lower than the normal excitation is called ***under excitation***.

The relationship between the excitation current, I_f and the current drawn by the synchronous motor, I_a has been shown in Fig. 10.19 (c). The excitation current OP corresponds to the minimum armature current OC drawn by the motor. Therefore, OP can be called the normal excitation. I_f' is over excitation and I_f'' is under excitation. The graph resembles the letter V of English alphabet, and hence is known as the synchronous motor V-curve. The magnitude of OC will increase if the load on the motor is increased. Keeping that higher load constant, if excitation is changed and values of corresponding armature currents I_a are plotted, another V-curve will be drawn as shown in Fig. 10.19 (c).

10.15.4 Application of Synchronous Motors

Synchronous motors are connected to the bus-bars in generating stations and electrical sub-stations to improve the power factor. For this purpose, they are run on no-load over excited condition. Over excitation means increase field current much more than the normal current. The over excited synchronous motor will suppy capacitive reactive power to the grid and help improve the power factor of the system as a whole. The reactive power supplied by the synchronous motor can be adjusted by changing the field current of the motor so as to improve the system power factor to the desired level.

Synchronous motors can be used as constant speed drives in industrial applications. In textile and paper industries, synchronous motors are used for variable constant speed driven with variable frequency supply, i.e., as a variable frequency drive system.

Important Terminologies and Points to Remember

1. Synchronous generators are also called Alternators. They generate electricity for commercial use. Synchronous generators are installed in power houses where a turbine rotates the generator poles so as to produce electricity.
2. Synchronous generators are rotated at synchronous speed given by the formula, $N_S = \frac{120f}{P}$ where N_S is the synchronous speed, f is the frequency of induced emf and P is the number of poles of the generator. To generated voltage of 50 Hertz, a generator having 4 poles, will have to be rotated at a speed of $N_S = \frac{120 \times 50}{4} = 1500$ revolutions per minute (rpm).
3. In synchronous generators, armature winding in which voltage is induced is placed on the stator slots and the magnetic poles are fixed on rotor and are rotated by the prime mover. The rotor may either be projected pole type (salient pole type) or cylindrical type.
4. The rotating field poles are provided with dc supply using brush and slip-rings.
5. The emfs displaced at 120 degrees in time phase are induced in the three-phase stator windings.

6. The winding is distributed throughout the stator and coils used are made short pitched so that a sinusoidal voltage is induced in the windings.
7. A number of generators are used in a power house and they are all connected in parallel. The voltage induced is stepped-up using a step-up transformer and then connected to the high-voltage transmission lines.
8. Voltage regulation of a generator is an important factor. Voltage regulation is the percentage change in terminal voltage when the load on the generator changes from no-load condition to full load.
9. Voltage regulation depends on the load and load power factor. High value of voltage regulation means large variation of terminal voltage when the generator is supplying a load.
10. Voltage drop is due to the resistance and synchronous reactance of the armature windings.
11. Synchronous reactance is the sum of leakage reactance and a fictitions reactance which takes into consideration the effect of armature reaction on voltage induced.
12. Voltage regulation must be calculated so that we will know how the generator will perform when it is supplying an electrical load.
13. Voltage regulation of a synchronous generator can be calculated using synchronous impedance method. Two tests, namely open-circuit test and short-circuit test are to be performed. Synchronous impedance is calculated as,

$$Z_S = \frac{\text{Open-circuit voltage}}{\text{Short-circuit current}} \text{ at a particular } I_f.$$

Per-phase armature resistance, R_a is calculated by using ammeter-voltmeter method. The value of synchronous reactance, X_S is calculated as $X_S = \sqrt{Z_s^2 - R_a^2}$.
14. From the knowledge of R_a and X_S, the voltage regulation is calculated by first calculating the value of induced emf, E as

$$E = \sqrt{(V\cos\phi + I_aR_a)^2 + (V\sin\phi + I_aX_S)^2}$$

Percentage regulation is calculated as $= \frac{E - V}{V} \times 100$.
15. In power stations, a number of alternators are connected in parallel to supply a common load. There are advantages in having a number of alternators instead of having a very high-capacity single alternator.
16. Load sharing of alternators is achieved by changing the field current (excitation current) and/or the prime mover input. Increase in prime mover input of one alternator causes increased active load sharing by it. Change in excitation causes reactive load sharing.
17. A synchronous machine can work as a motor with three-phase supply to the stator (armature) and dc supply to the field windings. In addition, a damper winding placed in slots made on rotor pole faces to make the motor self starting.
18. When three-phase supply is connected across the three-phase stator windings, a rotating magnetic field rotating at synchronous speed, N_S is produced. If we give dc supply to the rotor field windings, no starting torque gets developed. But due to the squirrel cage winding provided on the pole faces, the rotor starts rotating at a speed, N_r which is somewhat less than N_S. At this point, the dc supply to the rotor is provided and the rotor picks up further speed to catch up with the synchronous speed of the rotating magnetic field. The electromagnetic power developed is expressed as $P = \frac{VE}{X_S}\sin\delta$

Where V is the terminal voltage, E is the induced emf, X_S is the synchronous reactance and δ is called the power angle which is the angle between V and E.
19. An over excited (i.e., the field current being more than its normal value) synchronous motor is called as a synchronous condenser and can be used for running a load at constant speed as also to improve the power factor of the existing electrical load.

Important Formulae and Equations

Equation	Remarks
1. Synchronous speed, $N_S = \frac{120f}{P}$ where f is the frequency. P is the number of poles.	1. Expressed in rmp. If expressed in rps, $N_S = \frac{2f}{P}$.
2. 1° mechanical $= \frac{P}{2}$ degree electrical. If P = 2, electrical and mechanical degrees are equal.	2. The distance between two consecutive poles is always taken as 180° electrical.
3. Emf equation is: $E = 4.44\,\phi\, fTK_pK_d$ volts Where ϕ is the flux produced by the field current, f is the frequency, T is the number of turns per phase, K_p is pitch factor and K_d is distribution factor.	3. For full-pitch coil, $K_p = 1$. For concentrated winding $K_d = 1$. Distributed winding and short-pitched coils are used in practice. This helps in producing sinusoidal emf.
4. Pitch factor, $K_p = \cos \beta/2$ Distribution factor, $K_d = \frac{\sin\frac{m\alpha}{2}}{m\sin\frac{\alpha}{2}}$ β is the angle by which the coils are short pitched. α is the slot angle, m is number of slots per pole per phase	4. The values of K_p and K_d are nearly 0.96. Due to K_p and K_d, the magnitude of induced emf gets reduced but its wave shape becomes more sinusoidal.
5. $E = V + I_aR_a + jI_aX_S$ and $X_S = X_\ell + X_a$.	5. X_ℓ is the leakage reactance of the armature winding and X_a is a fictitions (not real) armature reactance.
6. $E = \sqrt{(V\cos\phi + I_aR_a)^2 + (V\sin\phi \pm I_aX_S)^2}$ Voltage regulation $= \frac{E-V}{V} \times 100$. Power developed by a synchronous motor, P is given as $P = \frac{VE}{X_S}\sin\delta$. Where δ is called the power angle whose maximum value is 90°.	6. +sign for lagging power factor load and –sign for leading p.f load. This equation is used to calculate voltage regulation. Voltage regulation of an alternator, i.e., ac generator can be negative if the load power factor is leading. For capacitive load, voltage regulation is negative which means, load voltage becomes more than the no-load voltage. With capacitive load when the load current is increased, the terminal voltage goes on increasing.

REVIEW QUESTIONS

Short Answer Type Questions

1. Explain why we use rotating poles and stationary armature in synchronous machines.
2. Distinguish between salient pole and cylindrical pole rotor construction. Why do we use cylindrical rotors in high-speed turbo generators?
3. What do you mean by pole pitch and coil pitch? What is the relationship between mechanical degrees and electrical degrees?
4. What is meant by synchronous speed. Establish the relation $N_s = 120f/P$, where N_s is the synchronous speed, f is the frequency, and P is the number of poles.

5. For a 50 Hz supply what are the possible synchronous speeds?
6. Draw the cross-sectional view of a salient-pole-type and non-salient pole-type synchronous machine. Why do we use laminated sheets for the construction of the stator and the rotor?
7. Draw a simple two-pole and four-pole stator winding for a synchronous machine showing the flux lines and the position of the poles formed.
8. Distinguish between the following three types of stator windings : (i) lap winding; (ii) wave winding; and (iii) spiral winding.
9. Distinguish between a fractional-pitch winding and a full-pitch winding. Mention the advantages and disadvantages (if any) of using fractional-pitch winding over full-pitch winding.
10. Derive the EMF equation for a three-phase synchronous machine taking into consideration the effect of using distributed winding and short-pitch coils.
11. Explain the constructional details of a synchronous machine. Mention the advantages of the stationary armature and rotating field.
12. Derive the expressions for distribution factor and pitch factor.
13. Distinguish between leakage reactance and synchronous reactance of a synchronous machine.
14. Show how the value of synchronous impedance can be calculated from test results.
15. What is meant by armature reaction? What is the effect of armature reaction on the main field flux at lagging and leading power factor loads?
16. Draw phasor diagrams of a synchronous generator at unity power factor load, lagging power factor load, and leading power factor load, respectively.
17. Derive an expression for voltage regulation of a three-phase synchronous generator. Can the regulation be negative?
18. Explain how you can determine the regulation of a synchronous generator from open-circuit and short-circuit tests.
19. State the conditions for parallel operation of alternators. For parallel operation is it necessary that the alternators be of the same KVA rating?
20. Explain the effect of change of excitation and prime-mover input on the loading of alternators operating in parallel.
21. Explain the construction and working principle of a synchronous motor.
22. A synchronous motor cannot run at any speed other than the synchronous speed. Explain why.
23. Explain how a synchronous motor is made self-starting with the use of squirrel-cage winding on its rotor.
24. Explain the effect of change of excitation on the armature current of a synchronous motor.
25. What are synchronous motor V-curves? Draw and explain V-curves at different loads.
26. Explain why an over-excited synchronous motor is called a synchronous condensor?
27. Explain how a synchronous motor can be used for system power factor correction.
28. State applications of synchronous motors.

Numerical Problems

29. Calculate the EMF induced per phase for a three-phase four-pole synchronous generator having 72 slots on the armature. The number of conductors per slot is 10. The flux per pole is 20 mWb. The alternator is driven at 1,500 rpm. Full-pitch coils have been used for the armature winding.

[**Ans** 510 V]

30. An eight-pole synchronous generator is running at 750 rpm. What is the frequency of induced EMF? At what speed should the generator be run so that the EMF induced will have a frequency of 60 Hz?

[**Ans** 50 Hz, 900 rpm]

31. Calculate the distribution factor for a four-pole, three-phase alternator having 36 slots on the slator.

[**Ans** 0.96]

Multiple Choice Questions

1. In synchronous machines
   (a) Field system is stationary and the armature windings are made rotating
   (b) The armature windings are placed on stator slots and the field system is made rotating
   (c) Both the field system and armature windings are rotating at synchronous speed
   (d) Both the field system and armature windings are rotating at synchronous speed but in opposite directions.

2. In a synchronous machine the speed of rotation of the magnetic field N_s is
   (a) $N_s = \frac{120P}{f}$ (b) $N_s = \frac{120f}{P}$
   (c) $N_s = \frac{Pf}{120}$ (d) $N_s = \frac{120P^2}{f}$.

3. For a synchronous machine with concentrated winding with full-pitch coil, which of the following is true?
   (a) Distribution factor, $K_d < 1$ and pitch factor, $K_p < 1$
   (b) $K_d > 1$ and $K_p > 1$
   (c) $K_d = 1$ and $K_p = 1$
   (d) $K_d = 1$ and $K_p = 0$.

4. With m as the number of slots per pole per phase and α as the slot angle, the distribution factor, K_d is
   (a) $K_d = \frac{m \sin \alpha/2}{\sin m\alpha/2}$ (b) $K_d = \frac{\sin \alpha/2}{\sin m\alpha/2}$
   (c) $K_d = \frac{\sin m\alpha/2}{m \sin \alpha/2}$ (d) $K_d = \frac{\sin^2 m\alpha/2}{m \sin \alpha/2}$.

5. Which of the following statements is not true for a synchronous machine?
   (a) The machine can have a cylindrical rotor
   (b) The machine can have a salient-type rotor
   (c) The machine can have non-salient rotor construction
   (d) The machine can have a squirrel-cage-type rotor construction.

6. In a synchronous machine, armature reaction is
   (a) The effect of leakage flux on the main field flux
   (b) The effect of armature flux on the leakage flux
   (c) The effect of armature flux on the main field flux
   (d) The effect of reduction of air-gap flux due to large air gap between the field system and the armature.

7. In synchronous machine, armature flux aids the main field flux on
   (a) lagging power factor load
   (b) leading power factor load
   (c) resistive load
   (d) resistive-inductive load.

8. When a synchronous generator is loaded, its terminal voltage may increase when the load power factor is
   (a) lagging (b) leading
   (c) unity (d) zero lagging.

9. The voltage regulation of a synchronous generator will always be positive when the load power factor is
   (a) leading (b) lagging
   (c) zero leading (d) above 0.8 leading.

10. The speed regulation of a synchronous machine is
   (a) unity (b) zero
   (c) less than unity (d) none.

11. For lagging power factor load, the relationship between the induced EMF, E and the terminal voltage v, can be expressed as
   (a) $E^2 = (V\cos\phi + I_aR_a)^2 + (V\sin\phi + I_aX_s)^2$
   (b) $E = (V\cos\phi + I_aR_a)^2 + (V\sin\phi + I_aX_s)^2$
   (c) $E = (V\cos\phi + I_aR_a) + (V\sin\phi + I_aX_s)$
   (d) $E = (V\cos\phi + I_aR_a) - (V\sin\phi + I_aX_s)$.

12. For synchronizing an alternator with the busbar which of the following conditions is not applicable?
   (a) The generated voltage of the alternator should be equal to the busbar voltage
   (b) The frequency of the generated voltage should be equal to the busbar frequency
   (c) The phase sequence of the voltage generated should be the same as that of busbar voltage
   (d) The KVA rating of the alternator should be equal to the KVA rating of other alternators already connected to the busbar.

13. Sharing of load by two alternators running in parallel can be achieved by
    (a) change of excitation
    (b) change of speed
    (c) change of prime mover input
    (d) change of excitation.
14. Which of the following statements is not true for a synchronous motor?
    (a) An over-excited synchronous motor draws lagging power factor current
    (b) An over-excited synchronous motor draws leading power factor current
    (c) At normal excitation the current drawn by a synchronous motor is the minimum
    (d) At normal excitation the power factor of the current drawn is unity.
15. In alternators damper windings are used to
    (a) prevent hunting
    (b) reduced eddy current loss
    (c) reduce armature reaction
    (d) reduce both eddy current and hysteresis loss.
16. An infinite busbar should maintain
    (a) infinite frequency but constant voltage
    (b) constant voltage at constant frequency
    (c) constant voltage at variable frequency
    (d) constant voltage but should possess infinite length.

Answers to Multiple Choice Questions

1. (b)	2. (b)	3. (c)	4. (c)	5. (d)	6. (c)
7. (b)	8. (b)	9. (b)	10. (b)	11. (a)	12. (d)
13. (c)	14. (a)	15. (a)	16. (b)		

CHAPTER 11

Measurement and Measuring Instruments

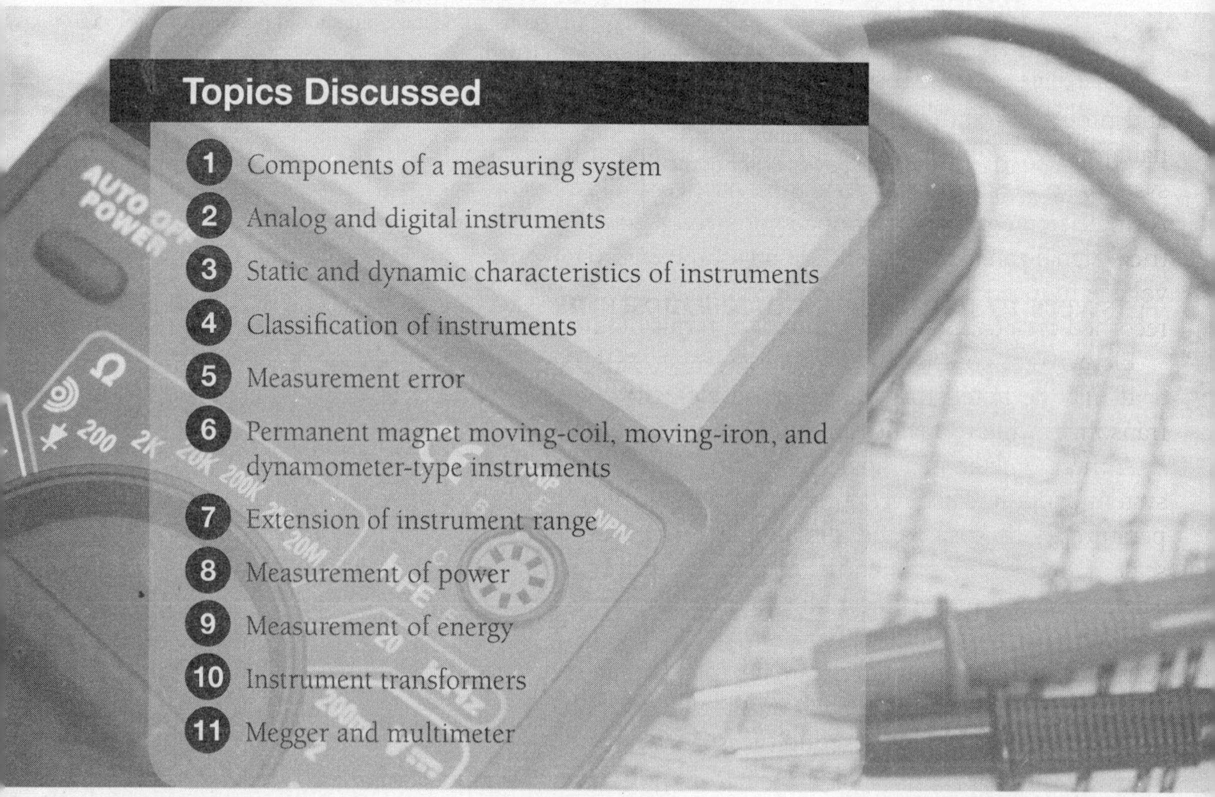

Topics Discussed

1. Components of a measuring system
2. Analog and digital instruments
3. Static and dynamic characteristics of instruments
4. Classification of instruments
5. Measurement error
6. Permanent magnet moving-coil, moving-iron, and dynamometer-type instruments
7. Extension of instrument range
8. Measurement of power
9. Measurement of energy
10. Instrument transformers
11. Megger and multimeter

11.1 ► Introduction

Measurement of any quantity, like length, mass, time, speed, velocity, pressure, temperature, current, voltage, power, etc., is nothing but comparison of the quantity against some standards. As for example, when we measure the length of a piece of cloth with a meter scale, we only compare a particular length against the standard calibrated scale. Using a meter scale to measure the length of a piece of cloth is the direct method of measurement. In engineering applications, for measurement of a variable quantity like pressure, velocity, temperature, etc., indirect methods are used. In an indirect method, a sensing element called transducer converts the quantity to be measured into an analogous electrical signal. This signal is then amplified

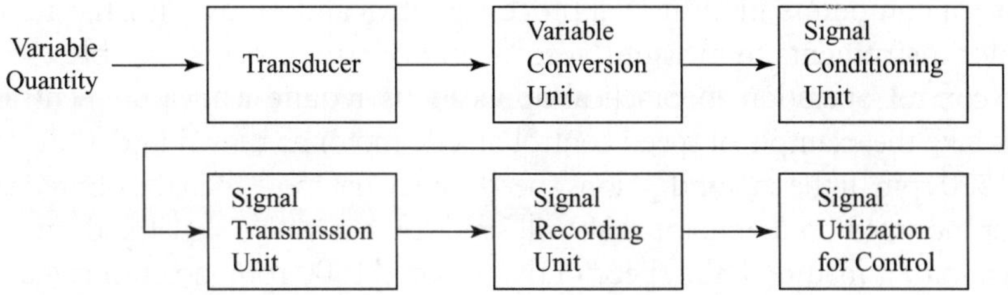

FIGURE 11.1

Basic measuring system elements represented in block diagram form

and processed, and fed to a final recording device. A system comprising a transducer, signal amplifier, converter which changes the signal from analog to digital form, transmission system, and recording or display device, etc., all together, is called instrumentation system. A single element like a thermocouple, which converts an unknown temperature into an electrical signal can also be used as a measurement system for measurement of temperature. The voltage induced across the thermocouple terminals (between hot and cold junctions) can be recorded on a calibrated instrument in terms of temperature of the hot junction.

Elements or components of a measuring system is shown in Fig. 11.1.

It may be noted that the common element of any measuring instrument is the primary transducer which provides an output when input is applied to it.

A thermometer is a single measuring unit but is an example of complete measurement system by itself. The level of mercury in the capillary tube which rises on application of temperature to the base of the tube is used to measure temperature directly against a graduated scale. However, most often the transducer is only a part of the whole measurement system. The output of the transducer often requires conversion to another form to make it suitable for processing or conditioning. For example, a strain gauge used as transducer causes a change in resistance when encountered with a variable quantity. We may have to change resistance into a voltage by some method. Such unit is called variable conversion unit as shown in Fig. 11.1.

Instruments which are used to measure various quantities that normally change with time are called measuring instruments.

The subject instrumentation and measurement is considered most important in any engineering and scientific activity. If we want to control a certain quantity, e.g., the output voltage of a generator, we have to continuously measure the output voltage because it tends to change when the load on the generator changes. If we want to keep the output voltage constant even when the load on the generator changes, we have to measure and send feedback to the generator input devices to make corrections for the desired output. Sometimes measurement has to be done from a distant place. In such cases the measured data have to be acquired, converted to digital form, transmitted, again converted to analog form and recorded. The input signals received through a transducer are required to be processed, i.e., amplified or attenuated, filtered, and converted before transmitting to the desired destination. Signal conditioning, data acquisition, data transmission, etc., are the important components of an instrumentation system.

Application of computers in industrial process control and monitoring has necessitated the requirement for instruments to measure, record, and control process variables.

Automatic control operation in practical applications require a measurement and feedback system. Let us take the example of speed control of a dc motor as shown in Fig. 11.2. The desired speed is, say 1500 rpm. If the motor develops speed somewhat more than the desired speed, a small tachogenerator mounted on the motor shaft will send some negative signal to decrease the voltage applied to the motor armature. If the speed of the motor is 1500 rpm, no error signal is generated.

Input voltage to the motor is changed automatically according o the difference between the measured speed and the required speed. Amplifier, A is used to amplify the signal.

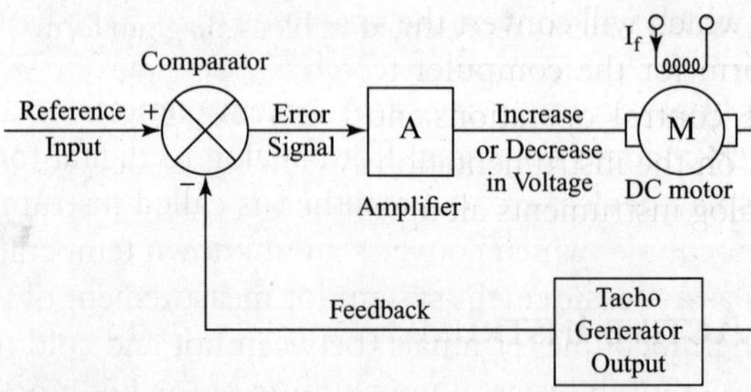

FIGURE 11.2

Speed control scheme for a dc motor

Precision, accuracy, sensitivity, etc., are considered the most important attributes of any measurement system. The more precise the instrument is, the more expensive it becomes.

A measuring instrument provides information about the magnitude of a certain variable quantity. In its simplest form it is a single unit which provides readings on a graduated scale according to the magnitude of the unknown quantity applied to it.

11.2 ► Analog and Digital Instruments

An analog instrument provides output (measured value) continuously as the quantity being measured changes. A deflection-type instrument as shown in Fig. 11.3 is an example of an analog instrument. Here, the pressure of the liquid inside a container is measured by recording the deflection of a pointer on a calibrated scale as shown.

As the pressure of liquid changes due to heating, the pointer moves over a graduated (calibrated) scale slowly and continuously indicating the pressure of the liquid.

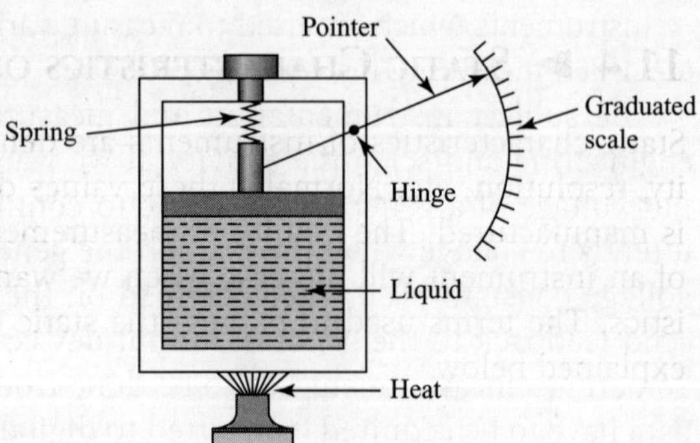

FIGURE 11.3

An analog instrument measuring pressure of liquid

A *digital instrument* provides output readings in discrete steps only. For example a digital instrument used for measurement of speed will read speed in discrete numbers and not as fraction of a revolution. Digital instruments have come into wide use due to the rapid growth of computers in control operations of systems and production processes. Instruments whose output is in digital form is suitable in computer-based control operations as the instrument can be directly interfaced with the computer. However, analog instruments can also be interfaced with computers by an A to D (analog to digital) converter which will convert the analog output signal into digital form for the computer to read and further process. Where control operations must be fast as also the money spent on the instrumentation system has to be kept low, digital instruments are preferred over analog instruments alongwith A to D converters.

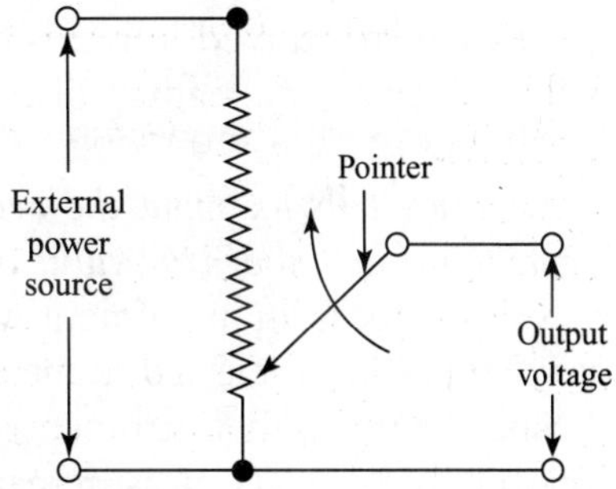

FIGURE 11.4

Passive instrument illustrated

11.3 ► PASSIVE AND ACTIVE INSTRUMENTS

Instruments can either be active or passive. In *active instruments* the output is produced entirely by the quantity being measured.

For example, the instrument shown in Fig. 11.3 is an active instrument because the deflection of the indicator is produced entirely due to the pressure created by the heated liquid.

Passive instruments are provided with some external source of supply. The quantity being measured simply changes the magnitude of the external source of supply as shown in Fig. 11.4. As shown in the figure, the pointer moves due to, say movement of some liquid level in a tank. The output voltage can be calibrated as proportional to the level of liquid in the tank since the pointer will move on the resistance element due to rise of liquid level.

Active instruments do not require any separate source of supply for this operation. Passive instruments require separate power supply.

11.4 ► STATIC CHARACTERISTICS OF INSTRUMENTS

Static characteristics of instruments are defined in terms of accuracy, precision, sensitivity, resolution, etc. Normally, these values do not change with time once the instrument is manufactured. The quality of measurement depends on these characteristics. The cost of an instrument will increase when we want to achieve higher values of these characteristics. The terms used to express the static characteristics of a measuring instrument are explained below.

11.4.1 Accuracy

You must be acquinted with deflecting-type instruments like ammeters and voltmeters used in laboratories. Let us consider a voltmeter which can read a maximum voltage of 100 V. Its accuracy is mentioned in terms of its full-scale deflection. As for example, accuracy of ±1% will mean that for a reading of 100 V, the actual value of voltage could be 100, ±1% of 100, i.e., either 101 V or 99 V.

Accuracy tells us about the nearness of the measured value (indicated value of the instrument) to the actual or true value of the quantity being measured.

Now, suppose an instrument whose measuring range is 0–100 V with an accuracy of say, 1 per cent has been selected to measure a low value of voltage, say 10 V.

Since the error of the instrument is expressed in terms of its full-scale deflection, the maximum error could be 1 V. For a 10 V measurement using the same voltmeter, could give rise to a possible error of 1 V which is 10 per cent of 10 V.

The accuracy of measurement is drastically reduced from 1 per cent to 10 per cent. It is therefore advisable not to use instruments of higher range to measure low values. In your laboratory while doing experiments you must select instruments such that their range is appropriate to the range of values being measured. That is to say, to measure 8 V you should not use a voltmeter of range 0–100 V. Instead you should use a voltmeter of lower range, say 0–10 V.

11.4.2 Precision

Precision is often confused with accuracy. Suppose with a voltmeter of 1 per cent accuracy we take a number of readings of a particular voltage. If the instrument is a high-precision one, the recorded values will not differ much. Thus, for a high-precision instrument the spread of the number of readings taken at a point of time while measuring a particular value will be very narrow. *Precision means* the degree of agreement of several readings taken for the same value. Precision describes the reproducibility to the measurement. High precision does not guarantee anything about the accuracy of measurement. Although several readings taken by a precision instrument may be very close to each other, like 1.1110, 1.1108, 1.1109, 1.1111, etc., the readings may have low accuracy, i.e., they may vary significantly from the true value which could be, say 1.2. Precision and accuracy of instruments are dependent on many design factors like calibration and graduations of scale, the design of sharp-edged pointers, zero adjustment, reduction of error due to parallax, etc. Precision is a measure of reproducibility and repeatability. Reproducibility indicates how close the output readings are for the same input where there are some changes in the method of measurement and the person making the measurement also changes. Repeatability indicates closeness of readings for the same input with no changed conditions of measurement.

The degree of repeatability or reproducibility in measurement is also a way of expressing the precision of an instrument. To further illustrate the difference between accuracy and precision, let us assume that true value of voltage to be measured is 100 V. The voltmeter readings are taken five times. If the readings are, say 98.01 V, 98.05 V, 98.03 V, 98.01 V, 98.02 V, we can say that the instrument is a precision instrument but not an accurate one as there is considerable error in measurement, which is nearly 2 per cent. However, if the consequitive readings taken were closely spread near 100 V, i.e., 100.01, 100.02, 99.09, 100.01, 99.08, we could say that the instrument is precise as well as accurate.

It is an usual practice to record a measurement with all digits which can be read with surity about the true value. For example, a resistor of true value 25 Ω if read as 25 Ω, then we may say that its value is nearer to 25 Ω than to 24 Ω or 26 Ω. If the value of the resistor is expressed as 25.0 Ω, it would mean that the resistor is closer to 25.1 Ω or 24.9 Ω. In 25 Ω there are two significant figures and in 25.0 Ω there are three significant figures. An indication of the precision of measurement is given from the number of significant figures in which the measurement reading is expressed. The more is the number of significant figures used in recording the measurement value, the more is the precision of measurement.

11.4.3 Sensitivity and Resolution

Sensitivity

It is the ratio of the change in the output of the instrument to the change in input variable quantity being measured. *Resolution* is defined as the smallest change in input that can be read or detected by an instrument. For a deflection-type instrument, the torque developed by the instrument should be high and the weight of the moving system of the instrument must be low. If the torque by weight ratio is high, the instrument will have a high resolution.

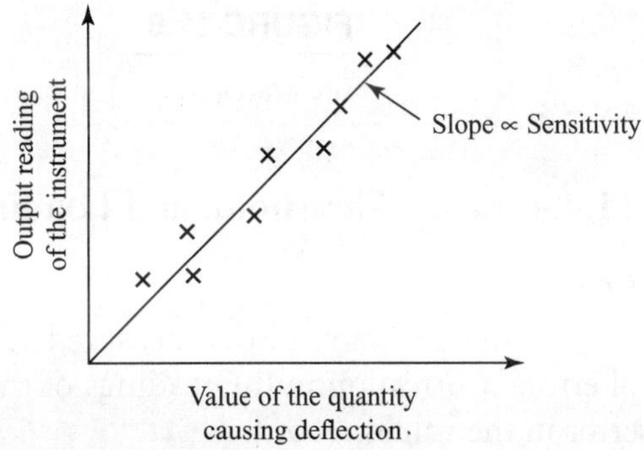

FIGURE 11.5

Instrument output against the value of the quantity being measured

Sensitivity of measurement is a measure of the change in instrument output which will occur when the input quantity, i.e., the quantity being measured changes. Fig. 11.5 shows the output readings of measured quantity of a certain variable, say, current in a circuit. Sensitivity, by definition is the gradient or slope of the straight line drawn as in Fig. 11.5. Higher the slope, higher is the sensitivity. Sensitivity is high when there is large deflection of the instrument pointer for a small value of the quantity being measured. For example, if the deflection of the pointer is by 10° for the input voltage of 1 V, then sensitivity is 10°/Volt.

Resolution

While taking measurement, if the input is slowly increased from a certain value, it may be found that the output does not change until a certain increment is exceeded. This small value of input quantity is expressed as resolution. Hence, we may define resolution as the smallest value of input quantity that can be detected by an instrument with certainty. For example, let us assume that an ammeter has a uniform scale having 10 divisions. The full scale deflection is intended to record 100 V as shown in Fig. 11.6. If each division on the scale is divided into 10 parts, the smallest amount of reading that can be read by the instrument is 1 V. The resolution of the instrument is therefore 1V.

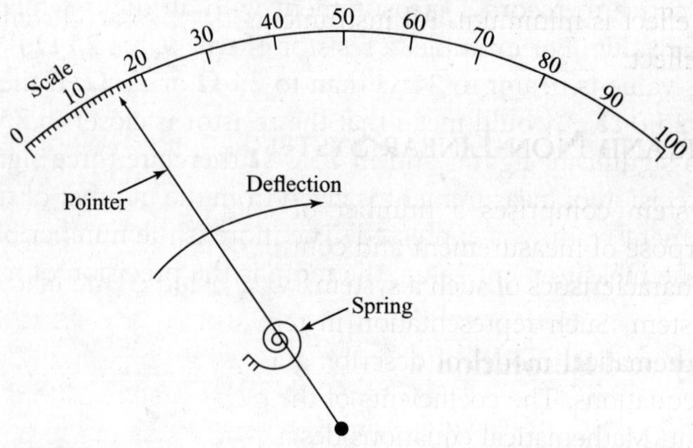

FIGURE 11.6

Illustrates resolution of an instrument

11.4.4 Error, Threshold, and Loading Effect

Error

Error is the deviation of the measured value from the true value. There are different types of errors. Correction in the readings of the instruments is required to be made to eliminate error in the values recorded.

Threshold

If we increase the value of the quantity to be measured slowly from its zero value, there will be some minimum value of the input which provides an output that can be detected by the instrument. This minimum value of input quantity is called threshold. Threshold should be as small as possible.

Loading effect

While making measurement using instruments, the original value of the quantity being measured should not change, otherwise we may get wrong results. For example, when a voltmeter is used to measure voltage by connecting the meter across two terminals, the voltmeter, which although has high resistance, will also draw some current as shown in Fig. 11.7.

If the voltmeter resistance is low as compared to the load resistance, the voltmeter may give misleading readings. However, if the load resistance is low we may get dependable readings. Thus, we may say that the measuring instrument should draw only a very infinitesimal current so that the whole current flows through the load. The effect of measuring instrument on the measuring quantity is called loading effect which causes error in reading. Here, in case of a voltmeter, the resistance of the voltmeter should be very high as compared to the load resistance

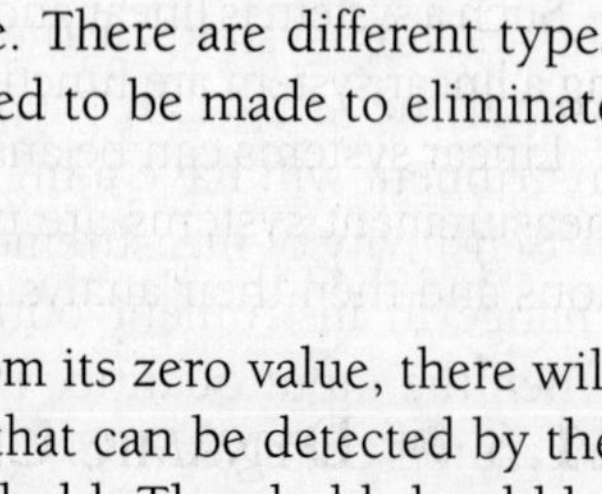

FIGURE 11.7

Loading effect illustrated

so that the loading effect is minimum. All instruments draw some current and hence create undesirable loding effect.

11.5 ► Linear and Non-Linear Systems

A measurement system comprises a number of components or subsystems connected together for the purpose of measurement and control of certain variable quantity.

In studying the characteristics of such a system, we generally write mathematical equations representing the system. Such representation in terms of equations is called mathematical modelling. The mathematical model of describing a system is generally expressed with the help of differential equations. The coefficients of the differential equations can either be constants or time variant. Mathematical equations describing a system can be called linear if the laws of superposition and homogeneity are applicable to the system. Suppose, x_1 (t) and x_2 (t) are the two inputs to the system and the corresponding outputs are y_1 (t) and y_2 (t), respectively, then by applying the principle of superposition we will be able to write

$$a_1 x_1 (t) + a_2 x_2 (t) = a_1 y_1 (t) + a_2 y_2 (t)$$

where a_1, a_2 are constants.

Such a system is linear and time invariant. If the coefficients of the differential equations describing a linear system are functions of time, then the system is called non-linear time-variant system.

Linear systems can be analysed using Laplace transform and Fouriers series. Since almost all measurement systems are non-linear in nature, they are first linearlized through approximations and then their analysis is done to study the dynamic performance of systems.

11.6 ► Dynamic Characteristics of Instruments

Systems are often subjected to inputs which are varying with time. The output of such systems are to be measured. The behaviour of the system under such varying input conditions are called dynamic response of the system. Dynamic inputs can be of two types, viz transient type and steady-state periodic type. Transient inputs die out with time where as periodic type of inputs repeat cyclically. The output of a measurement system may be distorted, and hence may not be a true reflection of the input in terms of magnitude and phase relationship. The difference between the true value (actual value) of the quantity which is varying with time and the value indicated by the measurement system is called the *dynamic error* of measurement. The rapidity of response of a measurement system is its *speed of response*. The delay or lag in response of a measurement system to the changes in the input quantity being measured is called *measuring lag*.

In dynamic analysis, we determine the characteristics like measuring lag, speed of response, dynamic error, etc.

For the purpose of analysis of a system, equations representing the system are written first. System response is studied by applying different kinds of test signals (like step input signal, ramp input signal, etc.) either in time domain or in frequency domain. In time-domain analysis a time-varying test input signal is applied to the system and the output behaviour of the measurement system is studied as a function of time. In frequency response analysis, the behaviour of the system is studied by applying sinusoidal input, as a function of frequency.

11.7 ► Classification of the Instrument System

Measuring instruments can be classified into separate categories on the basis of different criteria. These classifications are useful in knowing the characteristics of instruments and their selection for a particular use. Instruments are classified as primary or absolute instruments, secondary or derived instruments, null-type and deflection-type instruments, indicating instruments, integrating instruments, recording instruments, analog and digital instruments, monitoring and control instruments, electromechanical and electronic instruments, etc. These are discussed in brief as follows.

11.7.1 Active and Passive Instruments

In active instruments, the measurement output is entirely produced by the quantity being measured. In passive instruments, the quantity being measured simply modulates or changes the magnitude of some external source of power. For example, a current flowing through a coil placed in a constant magnetic field produces a torque. The magnitude of the torque can be used as proportional to the current flowing, provided the strength of the magnetic field is constant.

A thermocouple used to measure temperature is a self-generating-type instrument system, and falls under active instrumentation systems or simply active instruments.

The output of a potentiometer having a fixed source of supply can be used to indicate the level of a liquid and is termed as passive instrumentation system. Passive instruments require a source of supply for their operation.

11.7.2 Analog and Digital Instruments

Analog instruments provide output as a function of time, i.e., the output of the instrument varies continuously as the magnitude of the quantity being measured changes. A deflection-type instrument where the output is indicated by a deflecting needle moving over a graduated scale is an analog instrument. As the input changes, the needle or the pointer moves smoothly and continuously till the final deflected position is reached. The range of movement of the pointer can be divided through the scale into very large number of divisions. Thus, in an analog instrument the output is a continuous function of time.

In digital instruments the output varies in discrete steps, and therefore can have only a finite number of values. A digital revolution counter will count the number of revolutions in discrete numbers and not as a fraction of a revolution. Digital instruments and instrumentation system has a number of advantages over the analog system. Digital instruments measuring the magnitude of some signals can be transmitted over long distances without much distortion. The output of a digital instrument can be fed directly into a digital computer system. An instrument whose output is in digital form can be directly interfaced with the computer for monitoring and control operations.

The output of analog instruments has to be converted into digital form using analog to digital converters (A to D converters) before interfacing with computers. However, this conversion costs money and some time is lost which may be critical in a fast-changing control-system operation. That is why digital instruments are preferred where fast measurement and control operations are involved.

11.7.3 Indicating, Recording, and Integrating Instruments

Indicating instruments give the output as a function of time through the movement of a pointer over a graduated scale. These are, therefore, called analog instruments. The deflection-type ammeters, voltmeters and wattmeters in your laboratory are indicating instruments.

Recording instruments create a written record usually on paper of the time-varying quantity. The measurement system carries a pen which is used to record the value of the time-varying quantity on a paper which is driven by a slow-moving motor drive. The curve traced on the paper indicates the actual variation in the value of the quantity being measured. For example, temperature can be measured and recorded continuously using a recording-type instrument. In an ECG machine the status of health of your heart is recorded on a slow-moving paper and can be classified as a recording-type instrument.

Integrating instruments record the total value of the variable quantity over a period of time. For example, the electric meter (kilowatt hour meter) installed in our residences records the total amount of electricity consumed over a period of time. It is a summing or integrating instrument.

11.7.4 Deflection- and Null-Type Instruments

In deflection-type instruments deflection of a movable pointer provides a basis for measurement of the quantity which has created the deflection. This deflection is controlled by an opposing force created by some spring action. At a steady-state deflected position, deflecting torque is equal and opposite to the controlling torque. The deflection is measured on a calibrated scale.

In null-type instruments, a null or zero indication of the pointer is used as criteria for determining the value of an unknown quantity. Generally null-type instruments are uncalibrated instruments with the pointer placed on the middle of the graduated but uncalibrated scale.

DC potentiometer, Wheatstone bridge, and other types of measurement bridges use null-type instruments like a galvanometer as the null indicator representing balanced conditions as a requirement of measurement of an unknown quantity.

Accuracy of a deflection-type instrument depends upon the degree of accuracy of the calibration. The accuracy of null-type instruments are higher than deflection type of instruments because null indication means zero current flowing through the instrument, and hence, there is no effect of error in measurement due to calibration. However, null-type instruments are not suitable for use in the measurement of a quantity which changes with time.

11.8 ► Measurement Error

When we take the measurement of a certain quantity, there may be some error in the measurement due to a number of reasons. It is essential, therefore, to know the causes of such error and find ways to reduce them. Study and analysis of errors will determine the degree of accuracy of our measurements. Measurement error can take place if the reading of the instrument is not recorded correctly; if a large capacity instrument is used to measure a small quantity, ie, because of improper selection of instruments; if the adjustment of the instrument prior to measurement is not done (zero reading adjustment); if there are changes in the environmental conditions like temperature, electromagnetic interference, etc., during the measurement; if there are defects in the instrument itself; if there are random variations in the parameters or the system of measurement etc. Thus, we can see that errors can be due to mistakes made by the person taking the measurement; due to defects in the instrument or an improper use of instrument; due to changes in the environmental conditions or changes in measurement parameters.

Errors are generally classified into three categories, namely, gross error, systematic error, and random error.

Gross errors are human errors; systematic errors are instrumental errors; and random errors are because of random variations in measurement parameters.

Carelessness in taking readings and their recording, incorrect adjustment of the instrument, incorrect choice of instruments, etc., are attributed to gross error. These errors can be avoided by a careful choice of instruments and recording of measured data. Normally, a number of readings are taken instead of depending upon one or two readings.

Systematic errors are due to instrumental error or due to the effect of environment on measurement.

Random errors are due to unknown causes. While measuring some quantity through an instrument we may find that the readings vary for reasons not known to us. Such variations are not attributed to reasons mentioned under gross error and systematic error. While we are particular about the accuracy of measurement, we make corrections in the measured value by using statistical methods of calculations. For this purpose a large number of readings are taken of the quantity and through statistical analysis the best approximation about the true value is arrived at.

EXAMPLE 11.1

It is intended to measure an unknown resistance by the ammeter–voltmeter method. For this, a voltmeter is connected across the unknown resistance and the current flowing through the resistance is measured by an ammeter as shown in Fig. 11.8.

The voltmeter used has full scale range of 0–100 V and has sensitivity of 1000 Ω/V. The milliammeter has negligible resistance. Assume that the voltmeter reads a voltage of 80 V across the terminals P and Q. Let us calculate the value of unknown resistance R and then calculate the error in measurement due to current drawn by the voltmeter.

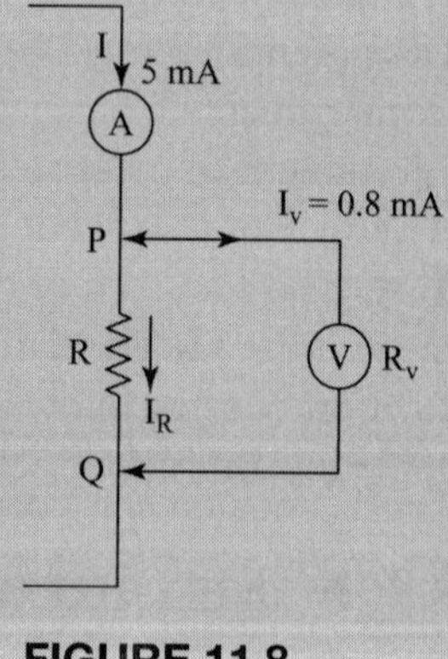

FIGURE 11.8

Loading effect illustrated

SOLUTION

$$R = \frac{V}{I} = \frac{80}{5\times10^{-3}} = 16\times10^{3}\Omega = 16\ k\Omega$$

A voltage of 80 V will also cause a small amount of current flowing through the voltmeter. Voltmeter resistance, R_v = 1000 Ω/V × 100 = 100 kΩ

Current through the voltmeter, $I_v = \frac{80}{100\times10^{3}} = 0.8\ mA$

To calculate the value of R, we had taken current flowing through it as 5 mA. In fact, this has not been the case. Current through R is I_R.

$$I_R = I - R_V = 5\ mA - 0.8\ mA = 4.2\ mA$$

The actual resistance value, $R_a = \frac{80\ V}{4.2\ mA} = 19\ k\Omega$

$$\text{Measurement error in percentage} = \frac{\text{actual value} - \text{measured value}}{\text{actual value}}\times100$$

$$= \frac{19-16}{19}\times100 = 15.8\ \text{per cent}$$

Now, let us examine the effect of using a voltmeter of high input impedance in this measurement. Let the sensitivity of the voltmeter be 5000 Ω/V instead of 1000 Ω/V.

The voltmeter resistance, R_V = 5000 × 100 Ω

= 500 kΩ

Current drawn by the voltmeter, $I_v = \frac{80}{500\times10^{3}} = 0.16\ mA$

Actual value $= \frac{80\ V}{4.84\ mA} = 16.5\ k\Omega$

Percentage error $= \frac{16.5-16}{16.5}\times100 = 3\ \text{per cent}$

It is seen that when we use a voltmeter of higher input impedance, the measurement error gets reduced considerably. Thus, we can conclude that while measuring a high value of resistance by the voltmeter–ammeter method, the resistance of the voltmeter should be very very high. Otherwise there will be considerable measurement error due to the loading effect of the voltmeter. In the ideal case the voltmeter used should not draw any current while measuring the voltage across the terminals.

EXAMPLE 11.2

Calculate the error due to the loading effect of a voltmeter used to measure voltage across the terminals in the circuit as shown in Fig. 11.9. The voltmeter is having an internal resistance of 925 kΩ.

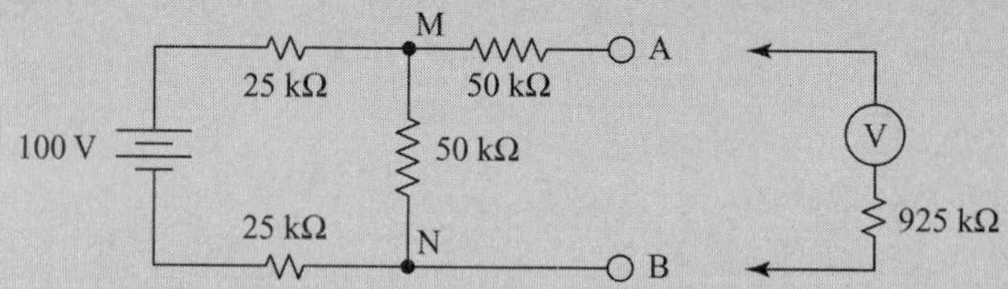

FIGURE 11.9

Loading effect illustrated

SOLUTION

Voltage across terminals A and B, $V_{AB} = \dfrac{100 \times 50}{25 + 25 + 50}$

$= 50\ \text{V}$

Note that voltage across the 50 kΩ resistor is the same as voltage across terminals A and B. That is voltage across terminals MN is the same as across terminals AB.

Thevenin's equivalent resistance across terminals A and B is calculated by replacing the EMF source by its internal resistance which is zero in the case. Hence, we just short the voltage source and calculate the equivalent resistance across terminals AB as shown in Fig. 11.10.

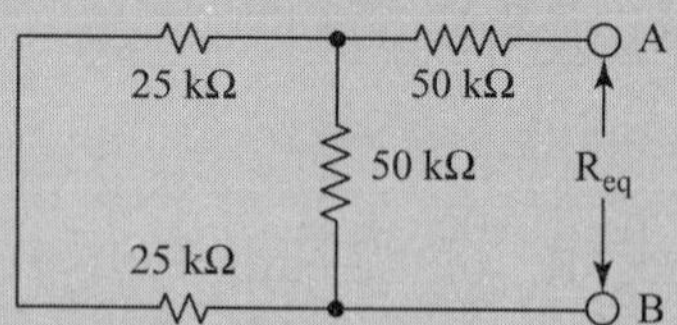

$$Req = \frac{(25+25)50}{(25+25)+50} + 50$$

$$= 75\ \text{k}\Omega$$

FIGURE 11.10

Thevenin's equivalent circuit of Fig. 11.9 is represented as shown in Fig. 11.11.

The voltage measured by the voltmeter as a consequence of its internal resistance is calculated as

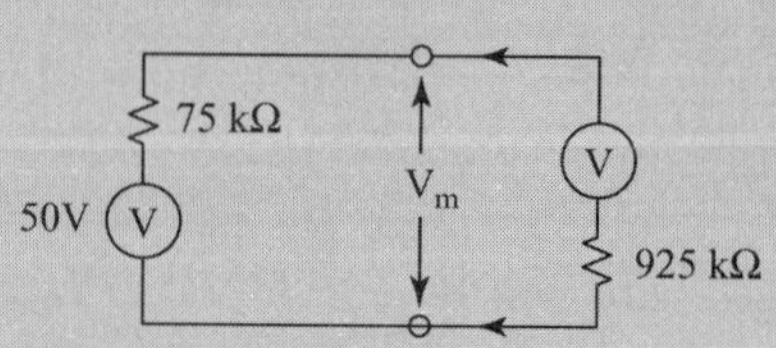

FIGURE 11.11

$$V_m = \frac{50 \times 925}{75 + 925} \text{Volts} = \frac{50 \times 925}{1000} = 46.25 \text{ V}$$

Loading Error $= \dfrac{\text{Actual voltage} - \text{Measured voltage}}{\text{Actual voltage}} \times 100$

$$= \frac{50 - 46.25}{50} \times 100 = 7.5 \text{ per cent}$$

EXAMPLE 11.3

Calculate the error in the measured value of voltage across terminals A and B as shown in Fig. 11.12, using a voltmeter having an internal resistance of 8.5 kΩ. What would be the error if the voltmeter internal resistance is 2 kΩ?

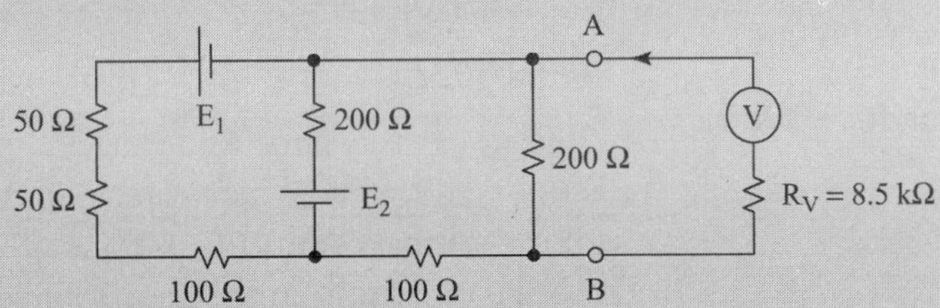

FIGURE 11.12

Loading effect illustrated

SOLUTION

Thevenin's equivalent resistance, R_{AB} of the circuit by assuming zero resistance of the voltage sources is calculated as shown in Fig. 11.13.

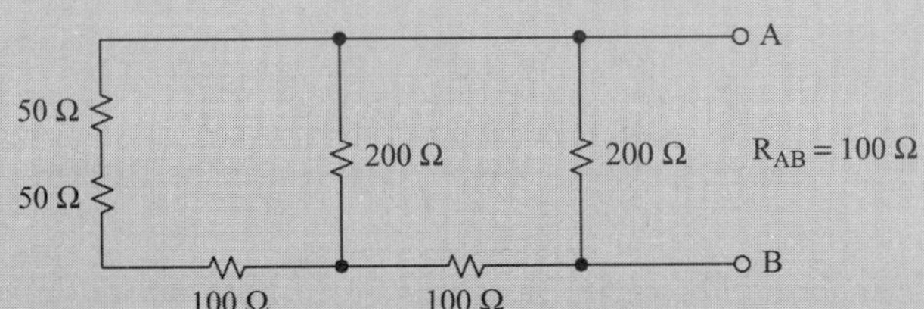

FIGURE 11.13

Let the open-circuit voltage across terminals A and B be E_0

Thevenin's equivalent circuit can be represented as shown in Fig. 11.14.

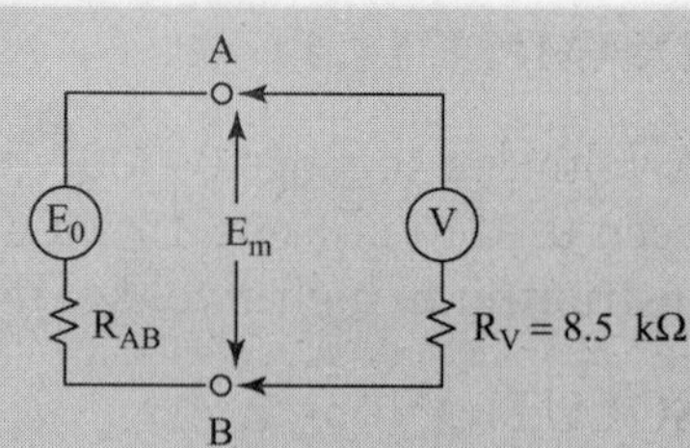

FIGURE 11.14

The measured value by the voltmeter across terminals A and B is E_m.

E_m can be calculated as

$$E_m = \frac{E_0 \times R_V}{R_{AB} + R_V} = \frac{E_0 \times 8500}{100 + 8500}$$

Now, actual value of voltage across AB is E_0, its measured value is E_m.

$$\text{Error in percentage} = \frac{E_0 - E_m}{E_0} \times 100$$

$$= \left[\frac{E_0 - E_0 \times \dfrac{8500}{8600}}{E_0} \right] \times 100$$

$$= 1.16 \text{ per cent}$$

Now, let us assume $R_V = 2000\ \Omega$

$$E_m = \frac{E_0\ R_V}{R_{AB} + R_V} = \frac{E_0 \times 2000}{100 + 2000} = E_0 \frac{2000}{2100}$$

$$\text{Error in percentage} = \frac{E_0 - E_m}{E_0} \times 100$$

$$= \left[\frac{E_0 - E_0 \times \dfrac{20}{21}}{E_0} \right] \times 100$$

$$= 4.76 \text{ per cent}$$

Thus, we see that the voltmeter resistance must be very high as compared to the circuit resistance R_{AB} so as to make the measurement error negligible.

Now, let us consider a case of error in measurement due to wrong selection of instrument range. For example, a voltmeter rated at 0–300 V has its full scale deflection upto 300 V. Its accuracy is specified in terms of its full scale deflection. If the same voltmeter is used to measure a low voltage, the accuracy of measurement will get reduced. That is why, while selecting measuring instruments, we should select instruments of proper range to avoid too much error in measurement. This is explained with the help of one example.

EXAMPLE 11.4

A voltmeter of range 0–300 V, and accuracy of 0.1 per cent on full scale deflection has been used to measure 15 V. Calculate the error in measurement as a result of using an instrument of higher voltage range to measure a comparatively low voltage.

SOLUTION

The magnitude of error on full scale deflection, i.e., 300 V is

$$0.1 \text{ per cent of } 300 \text{ V} = \frac{0.1 \times 300}{100} = 0.3 \text{ V}$$

The maximum error or the limiting error is 0.3 V.
Now on 15 V the limiting error is to be taken as 0.3 V.

$$\text{Therefore, percentage error } = \frac{0.3}{15} \times 100 = 2 \text{ per cent.}$$

It is to be noted that the error is much higher now. That is why one should not use a high-range instrument to measure a low value.

EXAMPLE 11.5

A voltmeter of the range 0–300 V has a guaranteed or limiting error mentioned by the manufacturer as 1.0 per cent. Calculate the limiting error (relative error) when the same instrument is used to measure lower voltages.

SOLUTION

Limiting error for full-scale deflection is 1 per cent magnitude of limiting error for

$$\text{full scale deflection} \quad = \frac{300 \times 1}{100} = 3 \text{ V}$$

If the same instrument is used to measure a voltage, say 150 V, then the relative limiting error would be

$$= \frac{3}{150} \times 100 = 2 \text{ per cent}$$

For measurement of 75 V using the same instrument, the relative limiting error would be equal to

$$\frac{3}{75} \times 100 = 4 \text{ per cent}$$

So, we may conclude that measurement of lower voltages using the same instrument results in less accurate measurement.

11.9 ▶ INDICATING-TYPE INSTRUMENTS

In this type of instruments, measurement is indicated by a pointer moving over a graduated scale. Voltmeters, ammeters, wattmeters, etc., are of indicating type which are extensively used in laboratories and control panels. Different types of indicating instruments are described in the following sections.

11.9.1 Permanent Magnet Moving Coil (PMMC) Instruments

Permanent magnet moving coil (PMMC)-type instrument is the basic dc measuring instrument. In these instruments a permanent magnet, generally of horseshoe type, creates a magnetic field in which a coil of fine wire of number of turns is placed. The coil is wound on a very light aluminium drum and is pivotted on jewel bearings so that the coil is free to move when current flows through it. The current-carrying coil placed in the magnetic field experiences a torque and tries to turn. It's free turning is restricted by spring tension attached to it's shaft. The moving coil produces a *deflecting torque* which is opposed by *control torque* produced by the spring action. A simplified diagram of a PMMC-type instrument has been shown in Fig. 11.15.

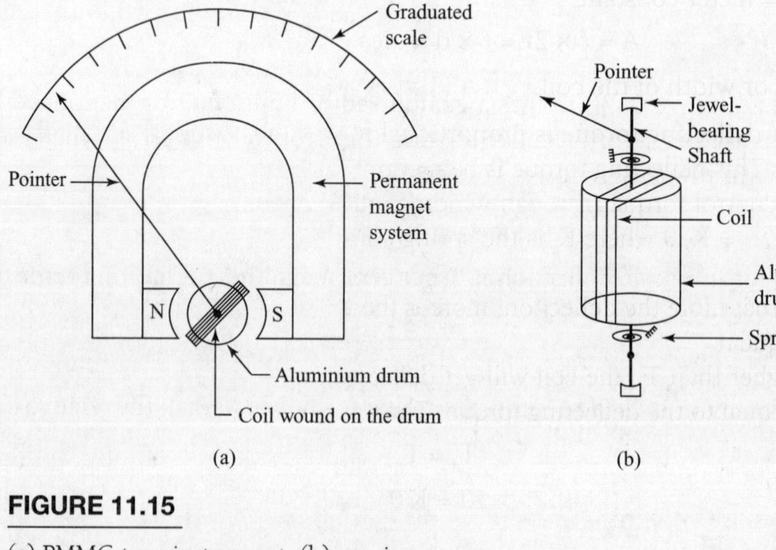

FIGURE 11.15

(a) PMMC-type instrument; (b) moving system

When the coil alongwith its pointer moves because of deflecting torque, it should be free from oscillations, i.e., the pointer should quickly come to its deflected position over a graduated scale so that the reading can be taken. To reduce any oscillation of the pointer, a damping mechanism is provided to produce a *damping torque* which damps or reduces the oscillation.

Thus, there exists three types of torques on the moving system, i.e., ***deflecting torque, control torque,*** and ***damping torque***. Note that if the control torque is not provided for, then for any amount of deflecting torque the pointer will give full-scale deflection.

Equation for deflecting torque

Let I be the current flowing through the moving coil; N be the number of turns of the moving coil; B be the flux density due to the magnetic field created by the magnet in which the moving coil is placed; ℓ be the length of one side of the coil; and r be the radial distance of the coil from its axis of rotation (deflection).

Then, force F on each coil side is BI ℓ

For N number of turns

$$F = NBI\,\ell$$

Torque = Force × Distance

Therefore, deflecting torque, $T_d = NBI\,\ell \times 2r$

$$= NBAI$$

$$= KI$$

Where K = NBA = meter constant.

$$A = \ell \times 2r = \ell \times d = \text{area of the coil}$$

d is the diameter or width of the coil.

Thus, we see that deflecting torque is proportional to the current flowing through the coil.

Control torque: This deflecting torque is being opposed by the control torque. If a spring is used to produce control torque (through winding and unwinding of the spring), the control torque, T_c will be $T_c = K_s\,\theta$ where K_s is the spring constant and θ is the angle of deflection of the moving coil.

It can be seen that more the deflection, more is the winding of the spring, and hence, more is the control torque.

When T_d is higher than T_c, the coil will get deflected more and more, till T_c, the opposing torque becomes equal to the deflecting torque. Thus, for final deflection

$$T_d = T_c$$

or,
$$KI = K_s\theta$$

or,
$$\theta = \frac{k}{k_s} \times I$$

Thus, deflection, θ is directly proportional to the current, I being measured.

Pointer and scale: A pointer is attached to the spindle (shaft) of the moving coil which moves over a graduated scale. The scale is calibrated in terms of amperes or volts depending upon whether the instrument is designed for the measurement of current or voltage, respectively. Since the deflection is directly and linearly related to the current flowing through the coil, the scale of the instrument will be a linear one. That is to say, the calibrations will be equally spaced on the scale as shown in Fig. 11.16.

Torque weight ratio: Fig. 11.16 shows the inside of a PMMC instrument with a U-shaped permanent magnet. For producing high-flux density, magnetic materials such as Alnico (alminium-nickel-cobalt) are used so that magnets of smaller size could be used and the overall size of the instrument could be reduced.

The weight of the moving system is made very low by choosing a light aluminium drum as the armature on which coils are placed. For higher sensitivity, the requirement is that the ratio of torque and weight of the moving system should be high. The requirements are, therefore, for a strong magnetic field, reduced size, and reduced weight of the moving system. The range of deflection of the pointer has to be made high, spread over 120° or more so as to achieve better and correct recording of meter readings. The two hair springs made of phosphor bronze provide the control torque. They also provide path for entry and exit of current to and from the moving coil. Hair springs are attached to the spindle and carry current to the moving coil and also provide the return path.

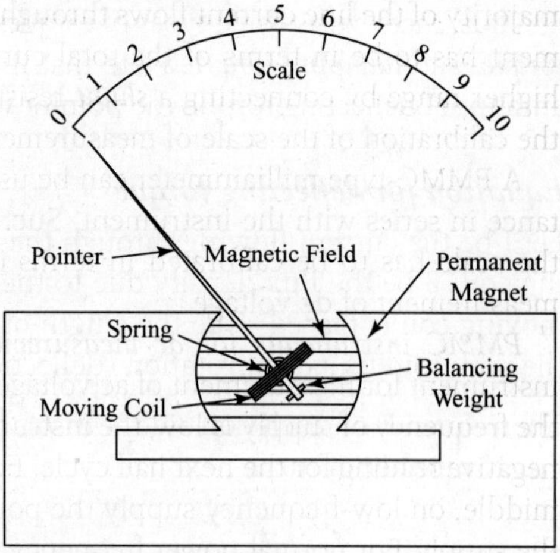

FIGURE 11.16

Constructional details of a PMMC-type measuring instrument

Damping system: Damping of the moving system, (i.e., the reduction of oscillation under the deflected condition), is reduced because of opposing torque developed due to induced eddy currents in the aluminium drum. Interaction between eddy currents and the flux produces opposing torques reducing oscillations of the moving system. This is known as *eddy current damping*. Note that oscillations produce eddy current in the drum, and torque produced by the induced eddy current and magnetic field flux opposes the oscillation. This is an example of the application of Lenz's law.

A properly damped (also called critically damped) moving system will move reasonably fast but without overshoot and oscillations in the deflected position. Damping may be of any of the following types, viz (a) eddy current damping; (b) air friction damping; (c) fluid friction damping.

The pointer fixed on the spindle moves over a graduated scale. To keep the moving system light, the pointer is also made up of a very light-weight material with a fine edge for accurate measurement. The weight of the moving system is balanced by using some counter weight on the pointer on one side. The scale of a PMMC-type instrument is linear, i.e., its divisions are equally spaced. This is because the torque developed is directly proportional to the current flowing through the coil, i.e., the current to be measured.

Use of shunt and multiplier: For low range of currents, say upto 20 mA, the entire current to be measured is allowed to pass through the moving coil. However, for larger currents to be measured, the moving coil is shunted by a parallel resistance of very low value so that the majority of the line current flows through the shunt resistance. However, reading of the instrument has to be in terms of the total current. An instrument of lower range can be used for higher range by connecting a ***shunt*** resistance of appropriate value and accordingly changing the calibration of the scale of measurement.

A PMMC-type milliammeter can be used to measure dc voltage by connecting a high resistance in series with the instrument. Such series resistances are called ***multipliers***. In this case the scale has to be calibrated in terms of voltage. PMMC-type instruments are suitable for measurement of dc voltage.

PMMC instruments for dc measurement only: Now suppose we connect a PMMC-type instrument for measurement of ac voltage or current. Can you imagine what would happen? If the frequency of supply is low, the instrument will show positive reading for one half cycle and negative reading for the next half cycle. For a centre zero instrument where the pointer is at the middle, on low-frequency supply the pointer will be seen oscillating at the same frequency of the supply. For normal power frequency, i.e., 50Hz supply, the oscillations will be very quick. However, due to inertia, the moving system will not be able to respond to the quick positive and negative pulses but will somewhat vibrate in its initial position of rest indicating that the average value of the quantity to be measured is zero. ***PMMC instruments are therefore not suitable for measurement of alternating quantities.***

Temperature compensation: We have seen that torque developed is proportional to magnetic field strength of the permanent magnet used and the current flowing through the coil. The deflecting torque is balanced by control torque produced by spring tension. The temperature variation affects both magnetic field strength and spring tension. The resistance of the coil used also increases with temperature. The reduced spring tension will cause the pointer to read higher than the actual value. The increase in coil resistance (i.e., reduced current) and the reduced magnetic field strength (reduced flux) will cause the pointer to read lower than the actual value. Since these effects are not balancing each other's effect of reduced or enhanced meter reading, some temperature compensating arrangement has to be made to increase the accuracy in meter readings.

If a PMMC instrument is not compensated against temperature effects, it has been observed that for each degree of temperature rise the meter tends to read 0.2 per cent lower than the actual value.

We have known that manganin has negligible temperature coefficient of resistance. If we connect a high-value manganin resistor in series with the moving coil, the total change in resistance of the combination of resistance of manganin resistor and the copper coil will be negligible and the temperature effect due to change in resistance can be avoided. The other two effects, i.e., reduction of spring tension causing increased reading and reduction of flux density causing reduced reading should more or less cancel each other. The manganin resistor which is made 20 to 30 times more than the resistance of the moving coil is called ***swamping resistor***.

Use as a galvanometer: An uncalibrated PMMC instrument is also known as ***galvanometer*** which is generally used as a null detector in a bridge circuit. The galvanometer is used only to detect any flow of current and not the amount of current. When a galvanometer scale is calibrated in terms of current or voltage, it can be used as an ammeter or a voltmeter. So, the basic moving element of a dc ammeter or a dc voltmeter is a PMMC galvanometer. However, since the winding of the moving coil is made up of very thin wire to keep it light, it can carry only a very small amount of current, of the range of few milliamperes. When calibrated in milliamperes, the instrument becomes a milliammeter. When the galvanometer is to be converted into an ammeter of higher ranges, very-low-value resistances are connected in parallel with the moving coil so as to bypass the excess current (i.e., the current which is more than a few milliamperes that the coil can carry safely). These low-value resistances are called ***shunts***.

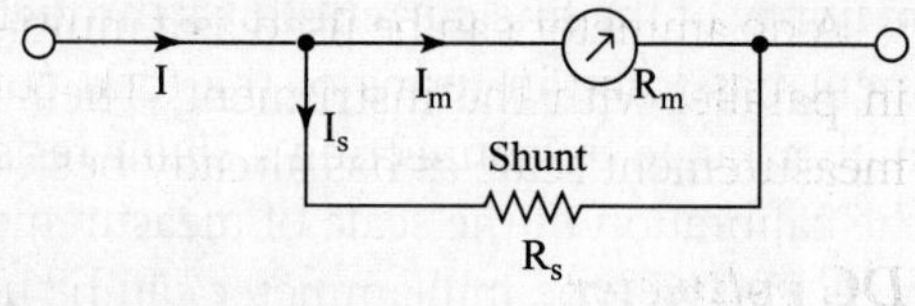

FIGURE 11.17

Basic ammeter circuit with a low-value shunt resistance

The value of a shunt can be calculated by equating the voltage drops across the instrument in two parallel paths as shown in Fig. 11.17.

$$I_m R_m = R_s I_s \quad \text{and} \quad I = I_m + I_s$$

Therefore,
$$R_s = \frac{I_m R_m}{I_s} = \frac{I_m R_m}{I - I_m}$$

The value of shunt resistance, R_s has to be calculated for each range of current measurement. For example let us consider a 0–1 mA instrument having a coil resistance of 80 Ω to be used for measurement of larger current in two ranges, i.e., 0–100 mA and 0–1 A. The values of two shunt resistances are calculated as follows.

For the range of current measurement of 0–100 mA, we have to take, I = 100 mA, i.e., full-scale deflection value of current.

Shunt current, $I_s = I - I_m = 100 - 1 = 99$ mA.

Shunt resistance,
$$R_s = \frac{I_m R_m}{I_s} = \frac{1\text{ mA} \times 80}{99\text{ mA}} = 0.808\ \Omega$$

For the range of current measurement of 0–1 A using the same instrument we need to have a separate shunt whose value is calculated as

$$R_s = \frac{I_m R_m}{I_s} = \frac{1\text{ mA} \times 80}{I - I_m} = \frac{1\text{ mA} \times 80}{1000\text{ mA} - 1\text{ mA}} = 0.080\ \Omega$$

11.9.2 Use of Shunts and Multipliers

Shunts are small resistances connected in parallel to increase the range of an ammeter. For multi-range ammeters, the moving coil remaining same, separate shunts are used to increase the range of measurement of a single instrument.

Multipliers are high resistances connected in series with the moving coil to extend the range of measurement of PMMC-type voltmeters.

A dc ammeter can be used as a multi-range ammeter by using a number of shunt resistances in parallel with the instrument. Their values are to be calculated according to the range of measurement scale as has already been explained with an example.

DC voltmeter

A dc milliammeter can be converted as a dc voltmeter by connecting a high resistance in series with the instrument. Such a high resistance is called ***multiplier***. A dc voltmeter measures the potential difference between two points, and hence is connected across the terminals. The high resistance connected in series with the instrument limits the current flowing through the coil of the instrument. The value of the multiplier is calculated as shown in Fig. 11.18.

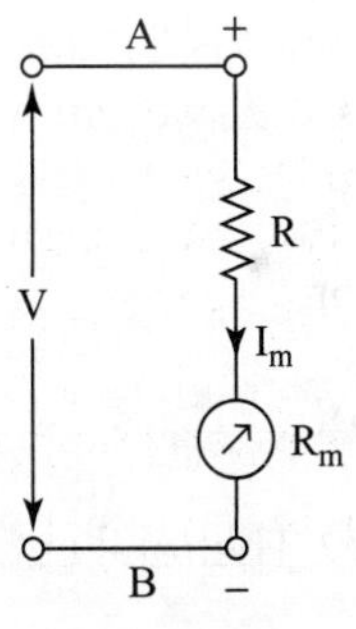

FIGURE 11.18

Let R be the series resistance of the multiplier as shown in Fig. 11.18.

I_m is the full-scale deflection current of the instrument.

R_m is the resistance of the moving coil.

V is the desired full-scale voltage range of the instrument being used as a voltmeter.

From the circuit of Fig. 11.18

$$V = I_m (R_m + R)$$

or,
$$R = \frac{V - I_m R_m}{I_m} = \frac{V}{I_m} - R_m \qquad \text{(i)}$$

For example, let $R_m = 50\ \Omega$, $I_m = 1$ mA and the range of the voltmeter be 0–100 V.

Then the multiplier resistance R is calculated by using relation (i) as

$$R = \frac{100}{1 \times 10^{-3}} - 50 = 100{,}000 - 50$$

$$= 99{,}950\ \Omega$$

This shows that the value of multiplier is very high whereas a shunt resistance used in dc ammeters is very low. The multiplier resistor is placed inside the instrument and only the two terminals are brought out and the range is shown on a calibrated scale. Similar to a multi-range ammeter, voltmeters can be also made for multi-range measurement by using a number of multipliers whose values are calculated for different voltage ranges. While all the multiplier resistances are placed inside the instrument box, a selector switch can be used to choose the particular range of the instrument for the measurement of voltage accurately as shown in Fig. 11.19.

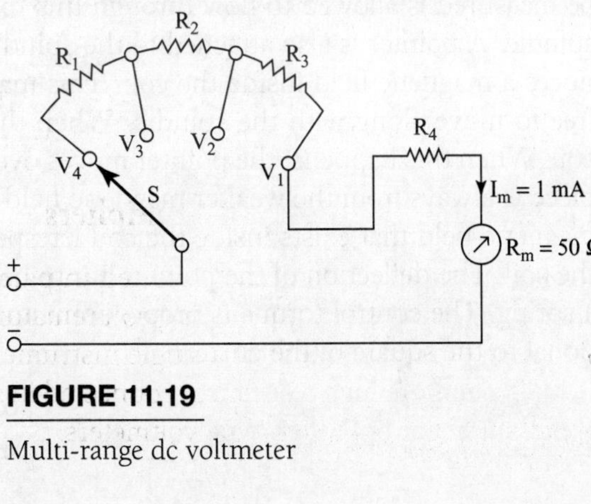

FIGURE 11.19

Multi-range dc voltmeter

V_1 is for the lowest range of measurement of voltage. As the voltage range is higher, higher is the value of multiplier resistance like $R_4 + R_3$ for range V_2, $R_4 + R_3 + R_2$ for range V_3, etc.

Let us assume that the moving coil is to be converted into a voltmeter of different ranges, viz 0–10 V, 0–50 V, 0–50 V, and 0–500 V.

For 0–10 V measurement the selector switch S has to be put at V_1. The series multiplier resistance is R_4. Since, R_4 and R_m are in series across 10 V and for full-scale deflection I_m is equal to 1 mA, we can write

$$V = I_m (R_m + R_4)$$

or,
$$10 = 1 \times 10^{-3} (50 + R_4)$$

or,
$$R_4 = \frac{10}{1 \times 10^{-3}} - 50 = 9{,}950\ \Omega$$

The total circuit resistance is $R_4 + R_m$

$$= 9950 + 50$$

$$= 10{,}000\ \Omega$$

Similarly, the values of R_3, R_2, and R_1 can be catantated for higher voltage ranges.

The *sensitivity* of a voltmeter is expressed as the ratio of the total circuit resistance to the voltmeter rating, i.e., in terms of ohms per volt (Ω/V).

11.9.3 Moving Iron Instruments

Moving iron instruments are used to measure current and voltage. They are made as ammeters and voltmeters and can measure both ac and dc currents and voltages, respectively. Two types of moving iron instruments are manufactured. One type is called attraction type and the other type is called repulsion type.

Attraction-type MI instruments

In this type of instruments there is one fixed coil and one moving iron. The fixed coil is wound on a former. The current to be measured or a current proportional to the voltage to be measured is allowed to flow through this fixed coil. A disc made of iron is attached to the spindle. A pointer is also attached to the spindle. Current passing through the fixed coil produces a magnetic field inside the coil. This magnetic field attracts the piece of iron which is free to move along with the spindle. When the iron disc gets attracted, it moves inside the coil. When this happens, the pointer moves over a graduated scale. The movement of the iron piece is always from the weaker magnetic field that exists outside the coil towards the strong magnetic field that exists inside the coil irrespective of the direction of current flow through the coil. The deflection of the pointer has to be controlled. The control torque is provided by a spring. The control torque is proportional to deflection θ. The deflecting torque is proportional to the square of the current.

$$T_d \propto I^2$$

$$T_c \propto \theta$$

where θ is the angle of deflection of the pointer over the scale.

T_d is the deflecting torque.

T_c is the control torque.

For a steady deflection, $T_d = T_c$

Therefore, $\theta \alpha I^2$

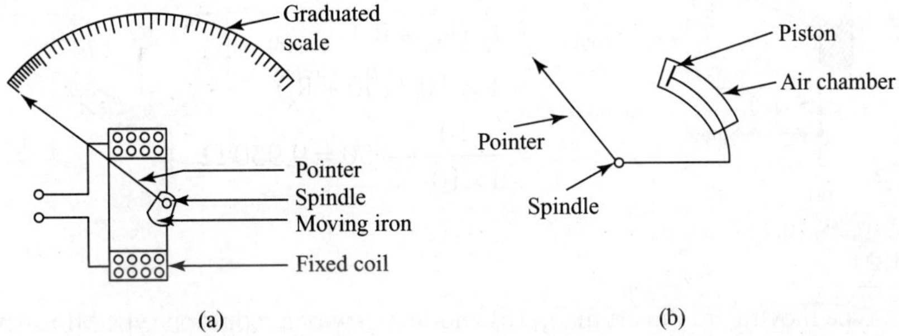

FIGURE 11.20

(a) Attraction-type moving iron instrument; (b) air friction damping mechanism

Deflection of the pointer is proportional to the square of the current flowing through the coil. The scale of the instrument is therefore a cramped scale, i.e., there will be congestion at the beginning of the scale. The instrument will measure both ac and dc and deflection of the pointer will be unidirectional. Damping, i.e., reduction of oscillation of the pointer is achieved by air friction damping mechanism where the base of the pointer is attached to a very light piston placed inside an air chamber as has been shown in Fig. 11.20 (b) where a light aluminium piston attached to the spindle moves in a fixed air chamber which is closed at one end. Due to pressure difference created by the movement of the piston, a damping torque is developed which reduces any oscillation of the pointer on the scale when measurement is taken.

Repulsion-type MI instruments

Like in attraction type, the fixed coil carries the current or current proportional to the voltage to be measured. On the inner surface of the coil, a specially shaped iron piece is attached. See Fig. 11.21 (b). This is the fixed iron piece. On the spindle is attached another piece of iron in the form of a fin. This is the moving iron. The current passing through the coil produces a magnetic field. In this magnetic field are placed both the fixed iron and the moving iron. Both these iron pieces will be similarly magnetized due to the influence of the magnetic field. This will create a force of repulsion among the two pieces of iron. One piece of iron being fixed, the other piece of iron will move which will create a deflecting torque. This deflecting torque has to be opposed by a control torque such that under steady deflection, $T_d = T_c$.

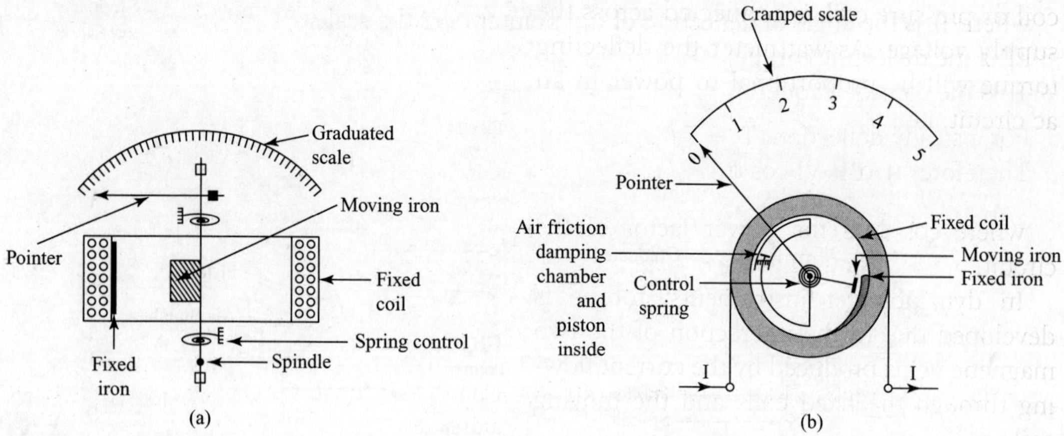

FIGURE 11.21

(a) Repulsion-type moving iron instrument; (b) another view of a repulsion-type MI instrument

If T_d is more than T_c, the pointer will move. Its movement will stop when the deflecting torque, T_d equals the control torque, T_c. Since the magnitude of the deflecting torque depends upon the magnetism of the two pieces of iron, and since this magnetism is produced by the current to be measured, we can write

$$T_d \alpha I^2$$

and $T_c \alpha \theta$, as more the deflection, θ more is the control torque, T_c.

Thus, deflection of the pointer, θ is directly proportional to the square of the current to be measured. Control torque is provided by spring attached to the spindle and damping (reduction of oscillation of the pointer in the deflected position) is provided by the air friction damping mechanism. As $\theta \alpha I^2$, this instrument can be used for both ac and dc measurement.

11.9.4 Dynamometer-Type Moving Coil Instruments

In dynamometer-type instruments, there is one fixed coil and one moving coil. The fixed coil is however, made in two sections, and placed apart as shown in Fig. 11.22. The moving coil is free to move on a spindle. The pointer is attached with it and will move over a graduated scale when the moving coil gets deflected. This instrument can be used as ammeter, voltmeter, and wattmeter for the measurement of current, voltage, and power, respectively. For use as ammeter and voltmeter, both the fixed coils and the moving coil are connected in series as shown in Fig. 11.22. For use as wattmeter, the fixed coils will carry the line current while the moving coil will carry a current proportional to the voltage. That is, in a wattmeter, the fixed coils, which are also called the current coils, are connected in series with the load whose power is being measured and the moving coil, which is also called the voltage coil or pressure coil, is connected across the supply voltage. As wattmeter the deflecting torque will be proportional to power in an ac circuit, i.e.,

$$P = VI \cos \phi$$

where $\cos \phi$ is the power factor of the circuit.

In dynamometer instruments, torque is developed due to the interaction of the two magnetic fields produced by the current flowing through the fixed coils and the moving coil.

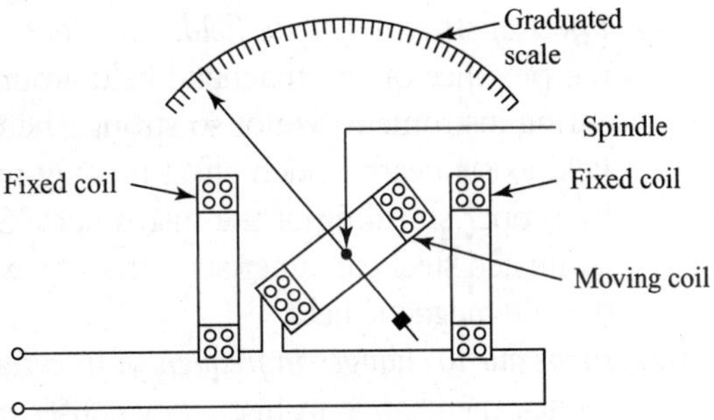

FIGURE 11.22

Connection diagram of a dynamometer-type ammeter or voltmeter

The control torque is provided by a set of helical springs attached to the spindle such that, $T_c = K\theta$. where K is the spring constant.

The deflecting torque T_d is proportional to the product of the currents flowing through the fixed coils and the moving coil. As in ammeter and voltmeter, same current flows through these coils; therefore

$$T_d \propto I^2$$

and

$$T_c \propto \theta$$

Thus,

$$\theta \propto I^2$$

The instrument will follow a square law in its deflection, and hence, the scale will be non-uniform one.

The torque developed will always be positive irrespective of the direction of current. Therefore, the instrument can be used for both ac and dc measurements.

Although dynamometer-type instruments can be used for both ac and dc measurements, they are mostly used for ac measurement.

The weight of the moving system of this type of a instrument is high as compared to other types of instruments. Hence, the torque by weight ratio is low, and therefore the sensitivity is low. Error due to friction is also considerable. Power consumption is considerably high as both the coils will draw currents. Thus, the loading effect will be more.

Dynamometer-type instruments have the advantage of measurement of rms values of ac quantities irrespective of their wave shapes.

Errors that usually occur in such deflecting-type measuring instruments are as follows.

(i) *Frictional error:* This is due to the movement of the deflecting system. If torque developed is high and the weight of the moving system is kept low, frictional error will be minimized.

(ii) *Temperature error:* Current flowing through the coil providing the magnetic field raises the temperature of the copper wires used thereby contributing to error in measurement. Even changes in ambient temperature (operating temperature or room temperature) adversely increases error in measurement.

(iii) *Effect of stray magnetic field:* The deflection of the moving system may get affected due to the presence of any magnetic field around. Since the magnetic field produced by the measuring instruments is not so strong, their operation may get affected if any strong magnetic field exists nearby. Such effect of stray magnetic field (outside magnetic field) is eliminated by proper shielding of the instrument. Shielding involves placing the instruments inside a laminated steel container such that the whole of the instrument is shielded (insulated) from outside magnetic field.

(iv) *Error due to changes in frequency:* In certain types of instruments, like in moving iron instruments, the change in frequency causes error due to the change in magnitude of eddy currents set up in the metal portion of the instruments. Changes in frequency causes changes in reactance of the operating coil ($X = 2\pi fL$). This causes change in the current flowing and field produced as in the case of ac operation of instruments as voltmeters.

11.10 ► MEASUREMENT OF POWER

Power in dc circuit is the product of voltage and current. In ac circuit power is a product of voltage, current, and power factor. Using an ammeter and a voltmeter power cannot be measured in ac circuits as this method would not take care of the power factor. For measuring power in ac circuit we use a wattmeter.

11.10.1 Power in DC and AC Circuits

Power is the product of voltage and current. In dc circuits, $P_{dc} = V \times I$, where V is the voltage across the load and I is the current flowing through the load.

In ac circuits, power is the product of instantaneous value current and instantaneous value voltage. Thus, in an ac circuit with R–L load

AC Power, $p = v \times i$

where $v = V_m \sin \omega t$

and $i = I_m \sin (\omega t - \phi)$ [in R–L circiut current lags the voltage by an angle, ϕ]

ϕ = power factor angle

Substituting

$$P = V_m \sin \omega t \times I_m \sin (\omega t - \phi)$$

$$= \frac{V_m I_m}{2} \times 2 \sin \omega t \sin(\omega t - \phi)$$

$$= \frac{V_m I_m}{2} [\cos\phi - \cos(2\omega t - \phi)]$$ [since 2 sin A sin B = cos (A – B) – cos (A + B)]

Average power is calculated over one cycle.

Thus,

$$P_{av} = \frac{V_m I_m}{2}\left[\frac{1}{2\pi}\int_0^{2\pi}[\cos\phi - \cos(2\omega t - \phi)]d\omega t\right]$$

$$= \frac{V_m}{\sqrt{2}}\frac{I_m}{\sqrt{2}}\cos\phi$$

$$= V I \cos\phi$$

where V and I are the rms values of voltage and current and $\cos\phi$ is the power factor of the circuit.

11.10.2 Measurement of Power in Single-Phase AC Circuit

For measurement of power we generally use an electrodynamometer-type wattmeter. The wattmeter has two coils; one is a stationary or fixed coil and the other is a moving a coil. The moving coil terminals and stationary coil terminals are brought out in the instrument for connection to the circuit for power measurement. The fixed coil is connected in series with the load. The moving coil, in series with a high resistance is connected across the load as shown in Fig. 11.23.

It can be seen from Fig. 11.23 that the coil cc, called the current coil carries the current, i_c which is the same as the load current because this coil is connected in series with the load. The coil pc, called the pressure coil or voltage coil is connected in parallel with the load. This coil carries a current, i_p. Current, i_p flows through the pressure coil and through the high resistance R_s. Neglecting the impedance of the pressure coil, i_p can be taken as equal to $\frac{V}{R_s}$.

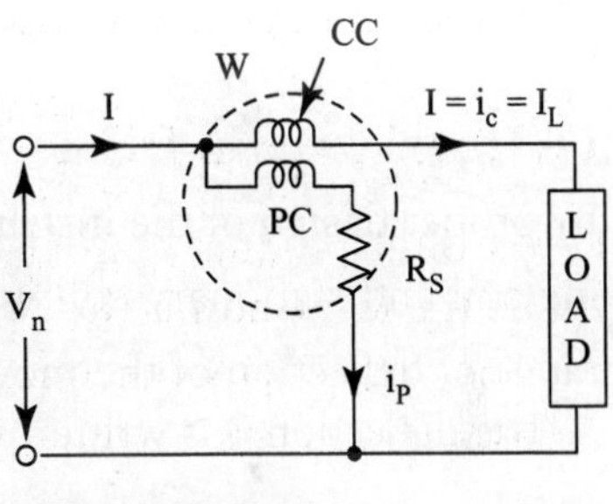

FIGURE 11.23

Measurement of power in an ac circuit using a wattmeter

Instantaneous deflecting torque, T_d will be calculated as,

$$T_d = i_1 i_2 \frac{dM}{d\theta}$$

Here,

$$T_d = i_p i_c \frac{dM}{d\theta}$$

where M is the mutual inductance of the two coils and θ is the angle of deflection.

Let the supply voltage be, $v = V_m \sin\omega t$. Then

$$i_p = \frac{V_m}{Rs}\sin\omega t \quad \text{and} \quad i_c = I_m \sin(\omega t - \phi) \text{ for lagging power factor load}$$

Therefore,

$$T_d = \frac{V_m}{R_s}\sin\omega t\, I_m \sin(\omega t - \phi)\frac{dM}{d\theta}$$

$$= \frac{V_m I_m}{2 R_s}[2\sin\omega t \sin(\omega t - \phi)]\frac{dM}{d\theta}$$

$$= \frac{V_m}{\sqrt{2}} \cdot \frac{I_m}{\sqrt{2}} \frac{1}{R_s} \frac{dM}{d\theta} [2 \sin \omega t \sin(\omega t - \phi)]$$

Average deflecting torque

$$T_{d_{av}} = \frac{VI}{R_S} \frac{dM}{d\theta} \frac{1}{2\pi} \int_0^{2\pi} \cos\phi - \cos(2\omega t - \phi)] d\omega t$$

$$= \frac{VI}{R_s} \cos\phi \frac{dM}{d\theta}$$

The control torque is provided by the spring control mechanism. Therefore,

$$T_c = K_s \theta$$

where θ is the angle of deflection and K_s is the spring constant

Under steady-state deflection, $T_{d_{av}} = T_c$

$$\therefore \quad K_s \theta = \frac{VI \cos\varphi}{R_s} \frac{dM}{d\theta}$$

or,

$$\theta = \frac{VI \cos\varphi}{K_s R_s} \cdot \frac{dM}{d\theta}$$

By proper design of the instrument, the rate of change of mutual inductance, M of the moving coil and stationary coil with respect to deflection, θ, i.e., $\frac{dM}{d\theta}$ is kept constant over the range of deflection of the moving coil.

The deflection θ is written as

$$\theta \alpha VI \cos\varphi$$

i.e., deflection of the moving coil is proportional to the power of the ac circuit.

In ac circuits, the deflection of the moving system will always be positive. This is because in ac as the polarities of supply change, the direction of current through both the current coil and pressure coil will change, and hence, the deflection always remains positive. The wattmeter is suitable for both ac and dc power measurements.

11.10.3 Sources of Error in Measurement Using Dynamometer-Type Wattmeters

A wattmeter has two coils, namely, the current coil which carries the load current and the voltage coil or the pressure coil which, in series with a non-inductive high-value resistance is connected across the load terminals. Torque is produced due to the interaction between the fields created by the current flowing through these two coils. Since the coils are air cored, the fields produced are comparatively weaker than the field produced in a PMMC instrument. For this reason the reading of a wattmeter may get somewhat changed due to any stray magnetic field. Temperature rise due to current flow through the coils is another source of error in the instrument reading.

If the pressure coil inductance is considerable, it will create some error in the instrument reading. There are other reasons for causing error in the reading of a wattmeter unless some remedial measures are taken in the design of the instrument. The various other sources of error and their remedies are discussed in brief as follows.

(a) Error Due to Inductance of the Pressure Coil

The pressure coil of the wattmeter has some inductance as it is made up of a number of turns and the current flowing through it is alternating in nature. While deriving the torque equation for the wattmeter, however, we neglected this inductance and had considered only the resistance of the pressure coil circuit. By assuming the pressure coil circuit as resistive, we had considered the pressure coil circuit current being in phase with the voltage.

In Fig. 11.24, the pressure coil current, i_p has been shown lagging the voltage, V by a small angle, θ. This is by considering the inductive reactance of the pressure coil circuit in addition to the high-value resistance, R_p in the circuit.

The angle of lag of current coil current, I_c which is equal to I is φ, which is the power factor angle of the load.

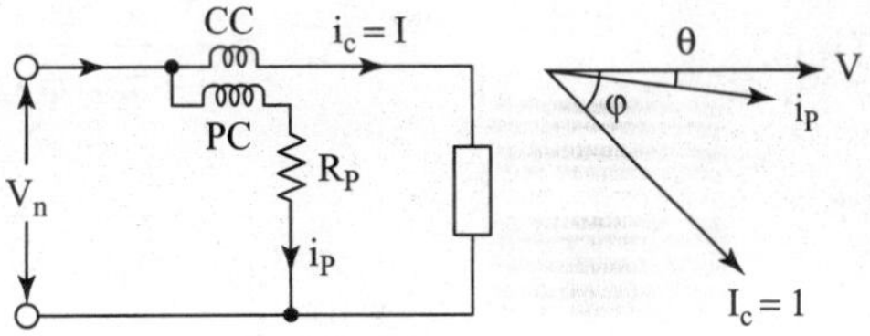

FIGURE 11.24

Actual reading of the wattmeter, W_a considering the inductance of the pressure coil

$$W_a \propto I_p I_c \cos(\varphi - \theta)$$

$$\propto \frac{V}{Z_p} I_c \cos(\varphi - \theta)$$

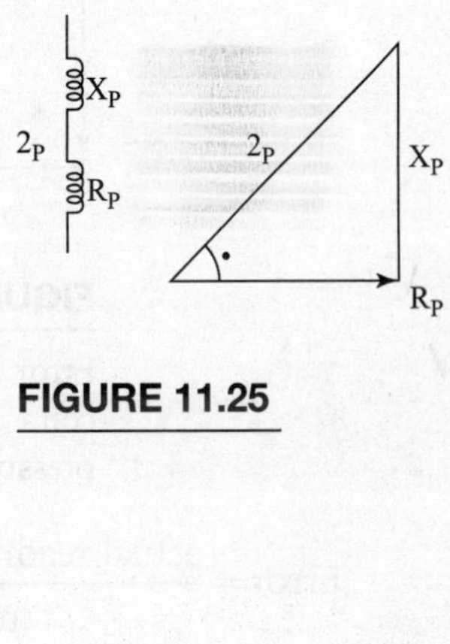

FIGURE 11.25

However, the desirable or true reading, W_d of the wattmeter which will occur when there is no inductive effect of the pressure coil, i.e., when the pressure coil circuit is purely resistive.

Therefore,

$$W_d \propto I_p I_c \cos\varphi$$

$$\propto \frac{V}{R_p} I_c \cos\varphi$$

The ratio of the two wattmeter readings, i.e., W_a and W_d are

$$\frac{W_d}{W_a} = \frac{V}{R_p} I_c \cos\theta \div \frac{V}{Z_p} I_c \cos(\varphi - \theta)$$

or,

$$\frac{W_d}{W_a} = \frac{V I_c \cos\phi}{Z_p \cos\theta} \times \frac{Z_p}{V I_c \cos(\phi - \theta)}$$

$$= \frac{\cos\phi}{\cos\theta \cos(\phi - \theta)}$$

Therefore, the true or desirable reading

$$W_d = \frac{\cos\varphi}{\cos\theta\cos(\varphi-\theta)} \times W_a$$

or,

$$W_d = C.F \times W_a$$

The actual reading of the wattmeter has to be multiplied by a factor called the correction factor, CF, which is

$$CF = \frac{\cos\varphi}{\cos\theta\cos(\varphi-\theta)} = \frac{\sec^2\theta}{\cos(\varphi-\theta)/\cos\varphi\cos\theta} = \frac{\sec^2\theta}{(\cos\varphi\cos\theta+\sin\varphi\sin\theta)/\cos\varphi\cos\theta}$$

$$= \frac{1+\tan^2\theta}{1+\tan\varphi\tan\theta}$$

The angle θ is very small, therefore $\tan^2\theta$ is negligible.

Thus,

$$CF = \frac{1}{1+\tan\varphi\tan\theta}, \text{ Again } W_d = C.F.\ W_a$$

or,

$$\frac{W_a}{W_b} = \frac{1}{C.F.} = 1+\tan\varphi\tan\theta$$

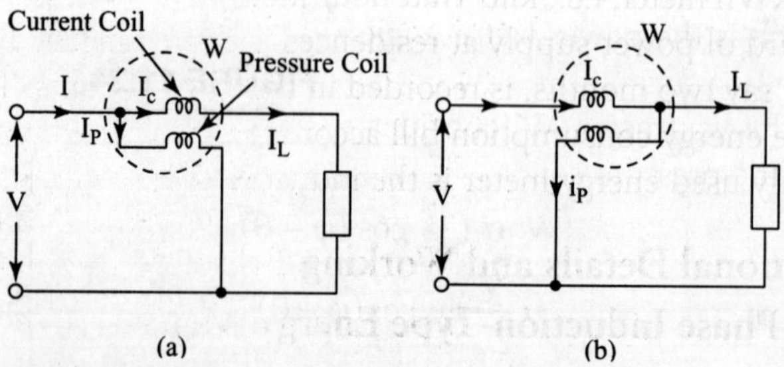

FIGURE 11.26

Error due to wattmeter connections: (a) when pressure coil is connected before the current coil; (b) when the pressure coil is connected after the current coil

$$\text{Error} = \frac{\text{actual reading} - \text{True or desirable reading}}{\text{True or desirable reading}} = \frac{W_a}{W_d} - 1 = 1+\tan\varphi\tan\theta - 1$$

$$= \tan\varphi\tan\theta$$

If the power factor of the load is leading, than the correction factor is calculated as

$$CF = \frac{\cos\phi}{\cos\theta\cos(\phi+\theta)}$$

Instead of applying a correction factor in the reading of the wattmeter, the error due to the inductive effect of the pressure coil can be neutralized by connecting a capacitor in series with the pressure coil so that the inductive reactance is balanced by the capacitive reactance.

(b) Error due to method of connection of the current coil and pressure coil

There will be a certain amount of error due to the two types of connections of the wattmeter current coil and pressure coil as shown in Fig. 11.26 (a) and (b). For correct reading, the current coil should carry the load current and the pressure coil be connected across the load. It may be noted that in Fig. 11.26 (b), the current coil carries a current, i_C which is slightly more than I_L ($i_c = I_L + i_p$). In Fig. 11.26 (a), the current coil carries a current i_c which is equal to I_L but the voltage across the pressure coil is slightly more than voltage across the load. Normally, the current flowing through the pressure coil is kept very low by connecting a high resistance in series with the pressure coil so as to reduce error due to connections as described.

11.11 ▶ MEASUREMENT OF ENERGY

11.11.1 Introduction

We know that electrical energy is the product of electrical power and time. A wattmeter gives the measure of power at a particular instant of time. If electrical power consumed over a period of time is recorded, we can get the measure of electrical energy consumed.

An energy meter is basically a wattmeter fitted with some recording mechanism. An energy meter is also called a kWh meter, i.e., kilo Watt hour meter. The meter which is installed at the main distribution board of power supply at residences is a kWh meter. The energy consumed over a period of time, say two months, is recorded in the form of digits. The electricity supply authority prepares the energy consumption bill according to the consumption made.

The most commonly used energy meter is the induction-type Watt hour meter.

11.11.2 Constructional Details and Working Principle of Single-Phase Induction-Type Energy Meter

The basic components of a kWh meter are as follows.

(i) Pressure coil placed on a magnetic core.

(ii) Current coils placed on a magnetic core; the mmf of these coils produce the driving torque.

(iii) The moving system which is an aluminium disc placed between the pressure coil magnetic core and the current coil magnetic core.

(iv) The braking system which is a U-shaped permanent magnet called the brake magnet which offers a braking effect to the rotating disc.

(v) The registering or counting mechanism which consists of a gear train. The mechanism adds together the revolutions made by the rotating aluminium disc. That is, the counting mechanism integrates the instantaneous values of power consumed by the connected load over a period of time.

The constructional details of an induction-type kWh meter has been shown in Fig. 11.27. The pressure coil is made of a large number of turns and is connected across the supply. The pressure coil is highly inductive as is surround by iron. The core is made of laminated steel sheets. The current coil is shown in two parts wound around a laminated core and is made of thick wire of a few turns. The current coil is connected in series with the load.

Working principle

Let V be the supply voltage and I be the load current. The current flowing through the pressure coil is I_{sh}. Since the pressure coil is highly inductive, I_{sh} will lag voltage V by 90°. Let load power factor be cosϕ. The load current I will lag the supply voltage by an angle ϕ. Current flowing through the series coil is I. The flux produced by the series coil will be in phase with I. The flux produced by the shunt or pressure coil ϕ_{sh} will be in phase with current I_{sh} as shown in Fig. 11.28.

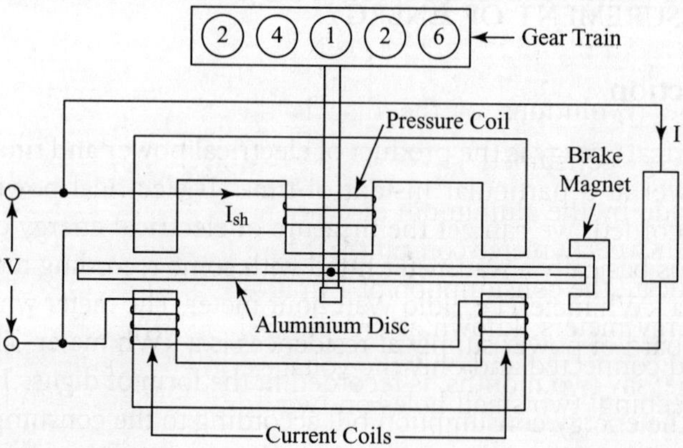

FIGURE 11.27

Induction-type single-phase energy meter

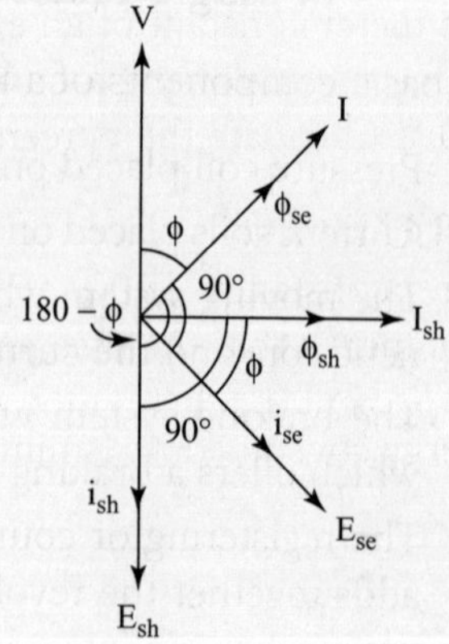

FIGURE 11.28

Phasor diagram showing the relationship between various quantities of an energy meter

The alternating flux ϕ_{se} will induce EMF E_{sc} in the aluminum disc lagging ϕ_{sc} by 90°. Similarly EMF induced by ϕ_{sh} in the disc will be E_{sh}. E_{sh} will lag ϕ_{sh} by 90°. E_{sc} will circulate eddy current i_{se} in the disc.

Similarly, E_{sh} will circulate eddy current i_{sh} in the disc.

Two opposite torques will be developed due to the interaction of ϕ_{sh} with i_{sc} and ϕ_{se} with i_{sh}. From Fig. 11.28, it is noted that the angle between ϕ_{sh} and i_{sc} is ϕ and the angle between ϕ_{sc} and i_{sh} is (180 – ϕ).

Average torque developed, T_d is

$$T_d \propto \left[\phi_{sh}\, i_{se} \cos\phi - \phi_{se}\, i_{sh} \cos(180° - \phi)\right]$$

or,

$$T_d \propto \left[\phi_{sh}\, i_{se} \cos\phi + \phi_{se}\, i_{sh} \cos\phi\right]$$

Now, ϕ_{sh} is proportional to V, i_{sc} is proportional to I, and i_{sh} is proportional to V, and ϕ_{se} is proportional to I.

Therefore, $$T_d \propto [VI\cos\phi + VI\cos\phi]$$

or, $$T_d \propto VI\cos\phi$$

Thus, the deflecting torque is proportional to the power consumed. For a particular period of time, if power consumed is recorded, the recorded value will give the measure of energy consumed.

The brake magnet produces a braking torque, T_b. When the aluminum disc revolves due to T_d, the magnetic flux of the brake magnet induces eddy current in the disc. The more the speed of the disc the more is the magnitude of the induced eddy current. By Lenz's law, EMF a opposite torque T_b is developed which controls the speed of the disc due to T_d. At a steady speed, $T_d = T_b$

and $$T_b \propto N$$

This shows that the revolutions of the disc is the measure of power consumed and $\int_0^t N dt = \int_0^t P dt$ = Energy consumed.

The revolution made by the aluminium disc per hour when the load is 1kW is called the meter constant. When we say meter constant is 1200, it means 1200 revolutions of the aluminium disc will indicate the consumption of 1 unit of electricity, i.e., 1 kWh of electricity.

One defect in energy meters, known as creeping, show that aluminium disc rotates even when there is no load connected and only the voltage coil is excited.

To prevent any creeping, two small holes on two opposite sides of the spindle is made on the aluminium disc.

EXAMPLE 11.6

The meter constant of an energy meter is 1200. The meter disc makes 210 revolutions in one minute when a current of 50 A flows through the load. The supply voltage is 230 V. Calculate the percentage error of the meter. Assume unity power factor load.

SOLUTION

Meter constant = 1200

This means that the disc makes 1200 revolutions per kWh.

$$\text{Energy consumed in 1 minute} = \frac{VI\cos\phi}{1000} \times \text{time}$$

$$= \frac{230 \times 50 \times 1}{1000} \times \frac{1}{60}\,\text{kWh}$$

$$= 0.191\,\text{kWh}$$

As per meter constant, energy registered in

$$\text{One minute} = \frac{210}{1200} = 0.17\,\text{kWh}$$

$$\text{Percentage error} = \frac{\text{energy registered} - \text{actual energy consumed}}{\text{actual energy consumed}}$$

$$\text{Percentage error} = \frac{(0.175 - 0.191)}{0.191} \times 100$$

$$= -8.3 \text{ per cent}$$

The minus sign indicates that the meter is running slow.

11.12 ► Instrument Transformers

We have studied earlier that a low-range ammeter can be used to measure a high value of current by using shunts. Similarly, the measuring range of a voltmeter can be extended by using multipliers. However, there are some limitations to the use of shunts and multipliers for measurement of high current and high voltage, respectively, by using low-range ammeters and voltmeters.

For measurement of high current and high voltage, voltmeters and ammeters of higher ranges are not used. Instead, for such measurement current transformers (CT) and potential transformers (PT) are used along with low-range ammeters and voltmeters. Generally ammeters of range 0–5 A and voltmeter of range 0–110 V are used.

Fig. 11.29 shows the use of CT and PT in the measurement of current and voltage of higher values. High-voltage and high-current measurements in on electrical power generation and a distribution system are done by installing CTs and PTs near the high-voltage and high-current lines

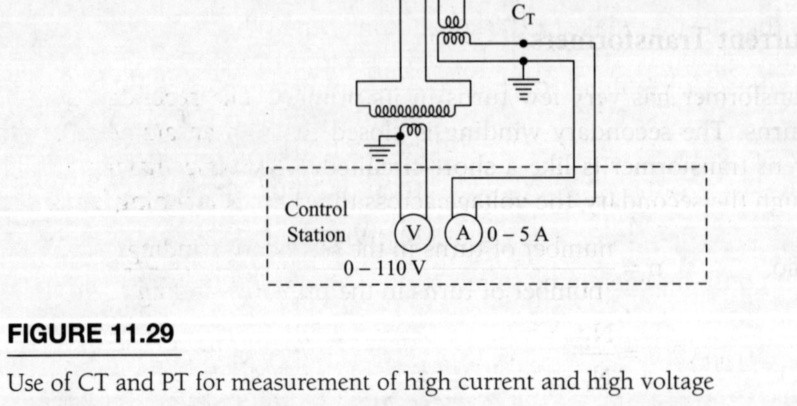

FIGURE 11.29

Use of CT and PT for measurement of high current and high voltage and the output of CTs and PTs are connected to the ammeters and voltmeters of lower rating at the control station. In Fig. 11.29 is shown a power generation and transmission lines. Output of the generator which generates electricity at 11000 V is stepped up at high voltage, say 220 kV by using a step-up transformer and then connected to a long transmission line as shown. Current and potential transformers have been shown connected to the system. Output of the PT is connected to a 0–110 V voltmeter whereas the output of the CT is connected to a 0–5 A ammeter.

Advantages of CT and PT

The advantages of using CT and PT for measurement of high value of current and voltage in preference to directly using ammeters and voltmeters of higher ratings (through shunts and multipliers) are mentioned below.

(i) If we had to use ammeters and voltmeters directly on the high-voltage lines, the instruments would have had to be insulated heavily, and hence, their physical size would be very big. Morever, it would be dangerous to bring high-voltage lines inside the control station. In case of any leakage of currents, the operator using these instruments may get a fatal electrical shock. Thus, the safety of the operator would be at stake.

(ii) With the help of standard instruments of low range, large current and voltage can be measured. Instrument transformers can either be of single range or of multirange.

(iii) From the output of the CT and PT, connections can also be taken to the current coil and pressure coil of a wattmeter for the measurement of power. Similarly, connections can be taken for the frequency meter also. Thus, several instruments can be connected to a single set of instrument transformers.

(iv) Instruments of moderate size and standards can be used for measurement.

It may be noted from Fig. 11.29 that the secondaries of CT and PT have been earthed. This is because, in case of open circuiting of the secondary by the operator, or by other reason the operator may get a severe shock if there is some leakage. That is why it is a warning to the operator as to not to open the secondary winding of a CT or a PT with their primaries connected to the lines.

We will now discuss in some details of current and potential transformers.

11.12.1 Current Transformers

A current transformer has very few turns in its primary. The secondary will have a large number of turns. The secondary winding is closed through an ammeter of range 0–5 A. Thus, a current transformer is like a short-circuited transformer having a limited current flowing through the secondary, the voltage across the primary winding being very small.

The turns ratio,
$$n = \frac{\text{number of turns in the secondary winding}}{\text{number of turns in the primary winding}}$$
$$= \frac{N_2}{N_1}$$

The nominal ratio, $K_n = \dfrac{\text{rated current of primary winding}}{\text{rated current of the secondary winding}}$

$$= \frac{I_1(\text{rated})}{I_2(\text{rated})}$$

The transformation ratio, $K = \dfrac{\text{Primary winding current}}{\text{Secondary winding current}}$

$$= \frac{I_1}{I_2}$$

Since the output current is measured on a 0–5 A ammeter, the current flowing through the primary winding, which is being measured, can be calculated by multiplying the ammeter reading by the turns ratio.

For example, if the primary winding current limit is upto 1000 A, the turns ratio of the CT is 200. For a reading on the ammeter of 2 A, we will calculate the primary current as 2 A × 200 = 400 A. The reading of the ammeter has to be multiplied by the turns ratio factor to get the value of line current.

It may be noted that turns ratio is the ratio of turns of the secondary and primary windings and the transformation ratio is the ratio of current of the primary and secondary windings, If $I_1 N_1 = I_2 N_2$

then $$\frac{N_2}{N_1} = \frac{I_1}{I_2}$$

i.e., Turns ratio = Transformation ratio.

If, however, the turns ratio is not equal to the transformation ratio, it will introduce considerable error in the measurement. The difference between the nominal ratio, i.e., $\dfrac{I_1(\text{rated})}{I_2(\text{rated})}$ and the actual transformation ratio $\dfrac{I_1}{I_2}$ is called ratio error.

Again, if the secondary winding current and primary winding current are not opposing each other at 180°, an error is introduced, which is called phase angle error.

The ratio error and phase angle error in a CT, their causes, and means to reduce these errors are explained now.

For this, let us draw the complete phasor diagram of a CT which is similar to a two-winding transformer.

Calculation of ratio error and phase angle error of a CT

In Fig. 11.30 is shown a CT connected with the circuit where high current, I_1 is flowing through the load. The primary circuit current of the CT is equal to I_1. The secondary circuit current of the CT is I_2 The turn ratio $\dfrac{N_2}{N_1} = \dfrac{I_1}{I_2}$

FIGURE 11.30

Measurement of line current using a CT and a low-range ammeter

or,
$$I_1 = \left(\frac{N_2}{N_1}\right) I_2$$

Thus, by multiplying the reading of the ammeter connected to the secondary circuit of the CT by the turn ratio, i.e., $\frac{N_2}{N_1}$, we can get the value of current flowing through the load circuit. The ammeter is rated at 0–5 A. With this arrangement, i.e., having a CT and low-value ammeter, we are able to measure high-current flowing through a circuit.

However, due to the no-load component of the primary current of a transformer, the actual transformation ratio deviates somewhat from the rated or nominal ratio of currents of the primary and secondary and also of their turn ratio. As we know, a transformer will draw some small amount of current at no load which has two components, viz the magnetizing component of the no-load current, I_m which produces the flux in the core and the loss component, I_c of the primary current, I_0 which corresponds to the losses in the core of the transformer. For the purpose of analysis, we will draw the simplified phasor diagram of a transformer having the core flux, φ as the reference axis as shown in Fig. 11.31.

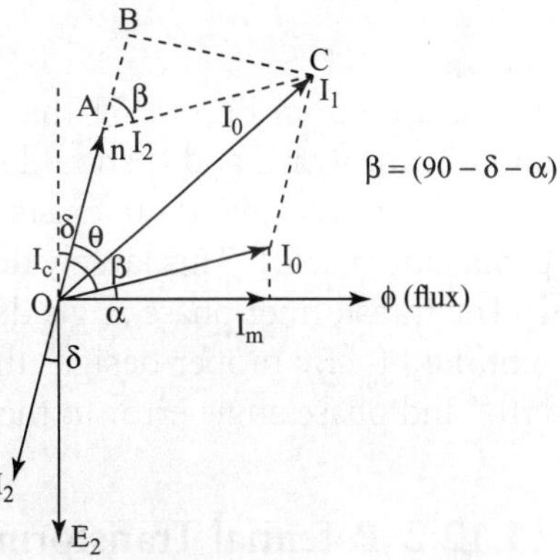

FIGURE 11.31
Simplified phasor diagram of a CT

Actually, nI_2 must be equal to I_1 as in Fig. 11.31. However, due to the presence of the no-load current, I_0, I_1 is somewhat away from nI_2. (Note that in Fig. 11.31, for clarity I_0 has been shown enlarged. Infact, I_0 is only 3 to 5 per cent of I_1.)

The vector sum of nI_2 and I_0 will give the value of I_1 as shown. In the figure, OC is equal to I_1. We extend nI_2 as shown by the dotted line. We draw a perpendicular from C on this extended line to meet at point B. OA is equal to nI_2 and ac is equal to I_0. The vertical line drawn downwards in E_2. E_2 lags the flux, ϕ by 90°.

Current flowing through the secondary winding due to E_2 is I_2. I_2 lags E_2 by an angle δ.

The angle between nI_2 and I_1 is θ.

The angle θ actually is very small, and hence, OC can be considered equal to OB.

Therefore
$$I_1^2 = OC^2 = OB^2 = (OA + AB)^2 = (nI_2 + I_0 \cos\beta)^2$$

or,
$$I_1 = nI_2 + I_0 \cos\beta$$

or,
$$I_1 = nI_2 + I_0 \cos[90 - (\delta + \alpha)]$$
$$= nI_2 + I_0 \sin(\delta + d)$$

Actual transformation ratio,
$$K = \frac{I_1}{I_2} = n + \frac{I_0}{I_2}\sin(\delta + \alpha)$$

$$= n + \frac{I_0}{I_2}[\sin\delta\cos\alpha + \cos\delta\sin\alpha]$$

$$= n + \frac{I_m \sin\delta + I_c \cos\delta}{I_2} \qquad [\therefore I_0 \cos\alpha = I_m \text{ and } I_0 \sin\alpha = I_c]$$

Ratio error of the current transformer
= Nominal ratio – Actual ratio

Since, $\frac{I_1(\text{rated})}{I_2(\text{rated})} = \frac{N_2}{N_1} = n,$

$$\text{Percentage ratio error} = \frac{\text{Nominal ratio} - \text{Actual ratio}}{\text{Actual ratio}} \times 100$$

$$= -\left(\frac{I_m \sin\delta + I_c \cos\delta}{I_2}\right) \times 100$$

It is seen from Fig. 11.31 that the there exists an angle of phase difference of θ between I_1 and nI_2. Idealy nI_2 and I_1 should not have any phase angle between them.

We have seen that there exists a difference between the actual transformation ratio and the turns ratio of a CT. This largely depends upon the loss component of the no-load current, i.e., I_c. The transformer phase angle depends upon the magnetizing component of the no-load current, i.e., I_m. By proper design, the magnitude of I_c and I_m are kept low to maintain the ratio error and phase angle error to the minimum.

11.12.2 Potential Transformers

Potential transformers are similar to two winding transformers except that the secondary volt-ampere loading is very low. The primary winding is the high-voltage winding which is connected across the lines whose voltage is to be measured. The secondary winding is normally rated at 110 V and is connected across a voltmeter or the pressure coil of a wattmeter. For details on potential transformer the students may refer to any advance book on measurement and instrumentation.

11.13 ► Megger and Measurement of Insulation Resistance

The insulation resistance of any cable or electrical system is very high. To measure such high resistance, we use an instrument called megger. The resistance measure is of the order of mega ohms ($\times 10^6\Omega$). The basic constructional details of a megger is explained below.

A megger consists of two coils placed at 90° with each other as shown in Fig. 11.32. One is called the current coil CC and the other is called the pressure coil PC. A magnetic needle is placed in the magnetic field created by the two coils. A pointer is connected to the magnetic needle. The pointer moves over a scale. DC supply is provided across the terminals 1 and 2.

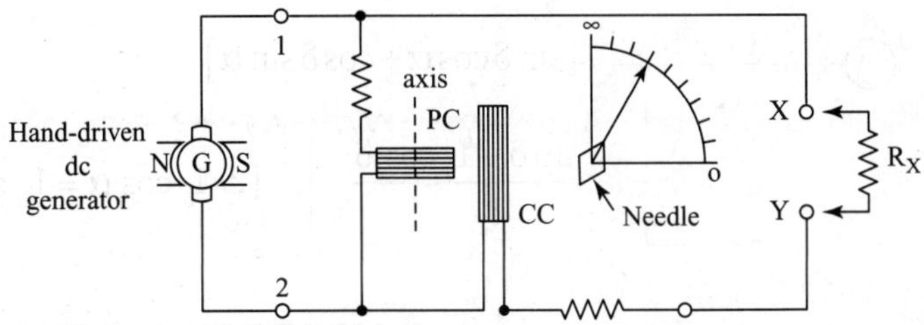

FIGURE 11.32

Principle of operation of an insulation tester illustrated

A megger is a portable instrument, and hence, the dc Supply is made available from a small dc generator which forms a part of the instrument. DC voltage is generated by rotating the generator by hand with the help of a handle provided of the purpose.

R_x is the resistance of the insulation which is to be measured. The hand-driven dc generator provides a constant dc voltage. When dc supply is available, current flows through the pressure coil. The magnetic needle along with the pointer gets aligned with the axis of the pressure coil PC indicating infinite resistance ∞, as shown. On the other hand, when resistance measurement terminals X and Y are closed, current will flow through the current coil CC. The needle and pointer will move in the clockwise direction and align with the magnetic field created by the current coil. The pointer will move to O position on the seak. When R_x is connected for measurement and its value is comparatively low (current high), the torque produced due to coil CC is higher than torque produced by coil PC. The position of the pointer will be towords 0 Ω side on the scale. When the value of R_x is high, very low current will flow through the coil CC, and hence, the needle will align towords the coil PC and indicate resistance towards infinity.

11.14 ► Multimeter and Measurement of Resistance

A single instrument is used to measure three quantities, viz current (i.e., Amperes), voltage (i.e., Volts), and resistance (i.e., Ohms). Multimeters are, therefore, also called AVO meters. Multimeters are available as analog type, electronic type, or digital type.

An analog multimeter is basically a permanent magnet moving coil galvanometer, For measurement of current, a selector switch is operated in the current measurement mode. As shown in Fig. 11.33 (a), a number of low-value resistors, called shunts, are connected in parallel with the instrument through a range selection selector switch. The required range can be selected by moving the selector switch to a particular position as shown in Fig. 11.33 (a).

The use of the multimeter as a multi-range voltmeter has been shown in Fig. 11.33 (b). The rectifier diode D_1 conducts during the positive half cycle of the input waveform and causes the meter to give deflection according to the average value of the half cycle. Diode D_2 prevents

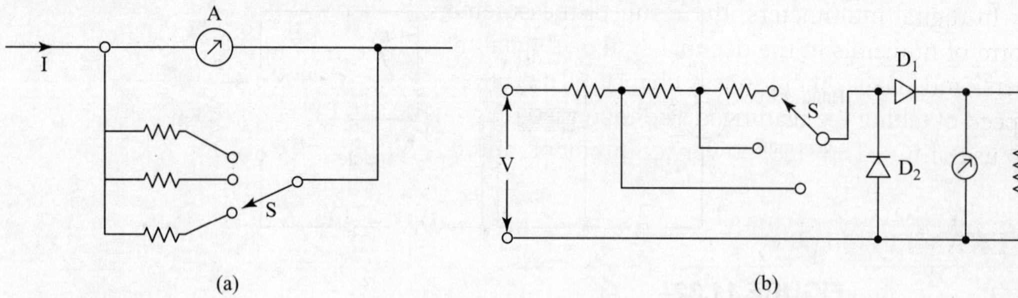

FIGURE 11.33

(a) Multi-range ammeter; (b) multi-range voltmeter

a reverse voltage appearing across the diode D_1 during the negative half cycle of the input voltage. The calibration of the scale is such that the same scale is used for both dc and ac measurements. The rms value of the sine wave is $= \frac{I_m}{\sqrt{2}}$.

The rms value of the half-rectified wave $= \frac{I_m}{\pi}$. The rms value of the half-rectified wave is $\frac{I_m}{\pi} \div \frac{I_m}{\sqrt{2}}$ 0.45 times the rms value of the full sine wave voltage. In order to have the same deflection on dc and corresponding ac voltage range, the value of the multiplier (resistance to be connected in series with the voltmeter) for the ac range must be reduced accordingly.

For multi-range resistance measurement, the basic circuit is similar to an ohm meter and is shown in Fig. 11.34.

When R_X is zero, i.e., the terminals A and B are shorted, meter current is 0. When R_X is open, i.e., when R_X is infinity, current finds path through the meter only. By adjusting R, we can get full deflection. Thus, the scale of the meter will have a zero mark on the left-hand side and on infinity mark on the right-hand side. Electronic multimeters use circuitry using field effect transistors and bipolar junction transistors, fixed and variable resistors, and a deflecting-type permanent-magnet moving-coil instrument.

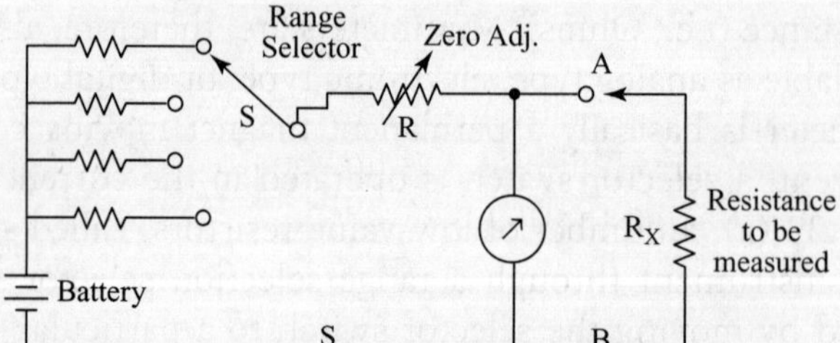

FIGURE 11.34

Measurement of resistance by a multimeter

In digital multimeters, the result of measurement is displayed at discrete intervals in the form of numerals in the decimal system. Digital meters provide readings in the form of numbers, and hence, there is less chance of meter reading error and error due to parallax. The speed of taking the reading is also increased. Digital multimeters have gained popularity due to use of ICs. The size, power requirement, and cost have been reduced drastically.

EXAMPLE 11.7

A pmmc-type instrument has a 4 cm × 3 cm size coil wound on its aluminium drum. The number of turns of the coil is 100. The magnetic field has a flux density fo 0.2 wb/m^2. The control spring provides a control torque of 1×10^{-6} Nm/degree of deflection of the moving coil. Calculate the value of current flowing through the coil when it is deflected by an angle of 48°.

SOLUTION

Area of the coil $= 0.04 \times 0.03$ m^2

$= 12 \times 10^{-4}$ m^2

Deflecting torque, $T_d = \phi NI = BAN\,I$

$= 0.2 \times 12 \times 10^{-4} \times 100 \times I$ Nm

or $Td = 2.4 \times 10^{-2} \times I$ Nm

Control torque produced by the spring when elongated at 48°, $T_c = 1 \times 10^{-6} \times 48$ Nm.

For steady deflection, $T_d = T_c$.

Equating, $2.4 \times 10^{-2} \times I = 48 \times 10^{-6}$

or, $$I = \frac{48 \times 10^{-6}}{2.4 \times 10^{-2}} = 20 \times 10^{-4} \text{ A}$$

$$= 2 \times 10^{-3} \text{ A}$$

or, $$I = 2 \text{ mA}$$

EXAMPLE 11.8

In a moving iron instrument a current of 5 A produces a deflection of 60°. What will be the deflection when a current of 2 A is flowing through the coil of the instrument? Assume that the instrument is spring controlled.

SOLUTION

We know that for a moving iron instrument deflecting torque is proportional to the square of the current flowing through the coil.

Therefore, $T_c \propto I^2$

When spring controlled, the control torque is proportional to the angle of deflection.

Therefore, $T_c \alpha \theta.$

For steady deflection, $T_d = T_c.$

Thus, $T_d \alpha I^2$ and $T_c \alpha \theta$

Equating, $\theta \alpha I^2$

Now $\theta_1 \alpha I_1^2$

and $\theta_2 \alpha I_2^2$

So $$\frac{\theta_2}{\theta_1} = \left(\frac{I_2}{I_1}\right)^2$$

or, $$\theta = \theta_1 \left(\frac{I_2}{I_1}\right)^2$$

Substituting values $$\theta_2 = 60\left(\frac{2}{5}\right)^2 = 60 \times (0.4)^2$$

$$= 9.6°$$

11.15 ► Digital Instruments

Amongst the commonly used digital instruments are the digital Voltmeters and digital Multimeters. In this section these two digital instruments will be discussed.

11.15.1 Digital Voltmeters

We have known that an analog instrument measures the unknown quantity may be a voltage or a current and provides output through the deflection of a pointer over a graduated scale.

A digital voltmeter converts the analog voltage into digital form using an Analog to Digital (A to D) converter and displays the measured value in discrete numericals on a display device. Such a voltmeter can be used to measure both a.c and d.c voltage. This can also be used to measure quantities like temperature, pressure, displacement, stress, etc., using transducers. The transducer converts the non-electrical quantity like the temperature or pressure into a proportional voltage and by measuring that voltage we measure the unknown non-electrical quantity.

A digital voltmeter has a number of advantages over an analog voltmeter. Due to digital display there is no human error in reading the measured value. Accuracy of measurement is as high as 0.005%. The instrument can read a very small amount of voltage of the range of micro volt, i.e., the resolution of digital voltmeter is very high. The digital output can be directly recorded and digital output is suitable for further processing.

Digital voltmeters employ different techniques for analog to digital conversion.

Figure 11.35 shows the block diagram representation of a digital voltmeter. The various blocks and their functions are mentioned below.

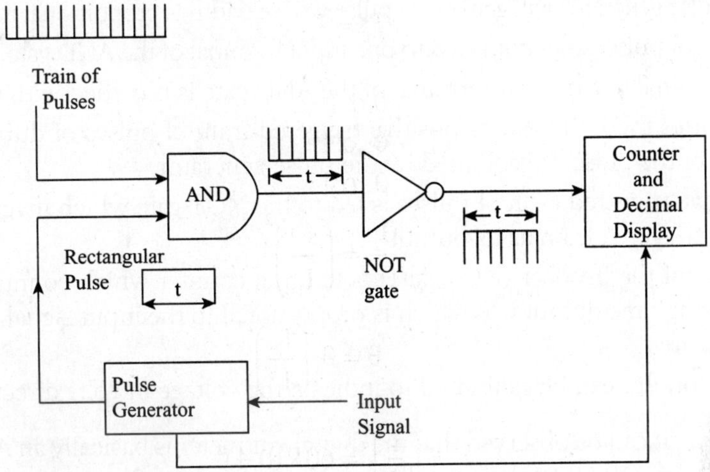

FIGURE 11.35

Block diagram of a simple digital voltmeter

Input signal

It is basically the voltage to be measured. Input signal to the pulse generator may be through an Attenuator which will reduce the voltage an appropriate level.

Pulse generator

The pulse generator generate rectangular pulses using analog, digital or both techniques. The width and frequency of the rectangular pulse is controlled by the digital circuitry inside the pulse generator while the amplitude and rise and fall time are controlled by analog circuitry.

AND gate

An AND gate gives high output only when its both the inputs are high. When a train of pulses along with the rectangular pulse are fed to the AND gate, it provides an output which is a train of pulses having duration '*t*' as same as the rectangular pulse coming from the pulse generator. The number of pulses in the train that passes through the AND gate is therefore proportional to the analog voltage being measured.

NOT gate

It inverts the output of the AND gate.

Counter and digital display

The counter counts the number of pulses during the duration and displays the value on an LED or LCD display after calibrating it.

Working principle of the digital voltmeter.

1. The voltage to be measured after attenuation is fed to the pulse generator in the form of a signal. The pulse generator will generate a pulse whose width is proportional to the input signal.
2. The output of pulse generator is fed to one input terminal of the AND gate.
3. The input signal to the other terminal of the AND gate is provided with a train of pulses.
4. The output of the AND gate is positive triggered train of pulses of duration as same as the width of the pulse as generated by the pulse generator.
5. This positive triggered train of pulses is fed to the NOT gate which inverts its input into a negative triggered train as its output.
6. The output of the inverter (NOT gate) is fed to a counter which counts the number of triggers in the time duration 't' which is proportional to the input signal, i.e., the voltage being measured.
7. Thus, the counter can be calibrated to indicate the voltage in volts directly.

From the above, it can be observed that the digital voltmeter is basically an Analog to Digital converter which converts an analog signal into a train of pulses, the number of which is proportional to the input voltage signal.

A digital voltmeter can be made using any one of the following A to D converter. The digital voltmeters are named according to the type of converters used. They are:

(i) Ramp type digital voltmeter
(ii) Integrating type digital voltmeter
(iii) Potentiometric type digital voltmeter
(iv) Successive approximation type digital voltmeter
(v) Continous balance type digital voltmeter.

The basic block diagram of a digital voltmeter is therefore represented as shown in Figure 11.36.

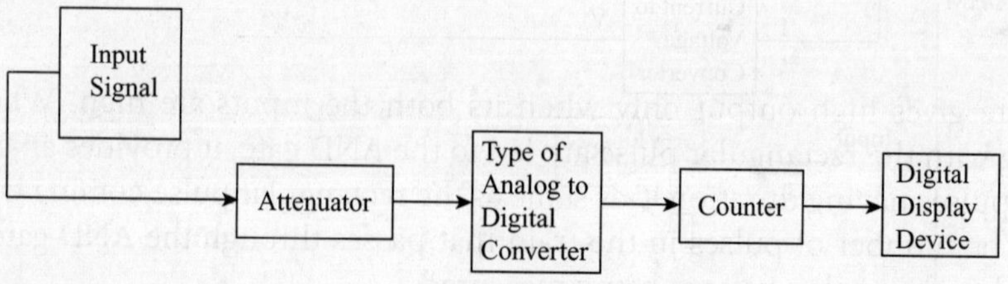

FIGURE 11.36

Basic block diagram of a digital voltmeter

Classification of digital voltmeters is done on the basis of the type of A to D converters used. Depending on the type ADC used, voltmeters are named as Ramp type, Integrating type, Flash type, etc. These days, digital voltmeter is being replaced by digital multimeter (DMM) because a DVM can be used to measure voltage, current and resistance. A separate DVM may not be required when one has a DMM to measure voltage, current and resistance.

11.15.2 Digital Multimeter

A digital multimeter can be used to measure a.c. and d.c. voltage and current and any unknown resistance. Figure 11.37 shows the block diagram representation of a digital multimeter.

The circuit is basically of a d.c. voltmeter where the unknown voltage or current to be measured are converted into low-voltage d.c. using attenuators, rectifiers, and I to V converters as have been shown. This voltage is fed to a A to D converter, a decade counter and ultimately to the digital display unit using three selector switches S_1, S_2 and S_3.

To mesure any unknown resistance, a constant current source has been used. The known constant current is passed through the unknown resistance and the voltage drop is fed to the A to D converter to provide a display of the value of the unknown resistance. It can be seen from the circuit diagram that rectifiers have been used to convert ac to dc for measurement of alternating voltage and current.

There switches S_1, S_2, S_3 have been used in the circuit to select the input quantity to be measured and to use a single A to D converter, a decade counter and a display unit for all measurements one by one.

In the multimeter attenuators, current to voltage converters have been used which are explaned below.

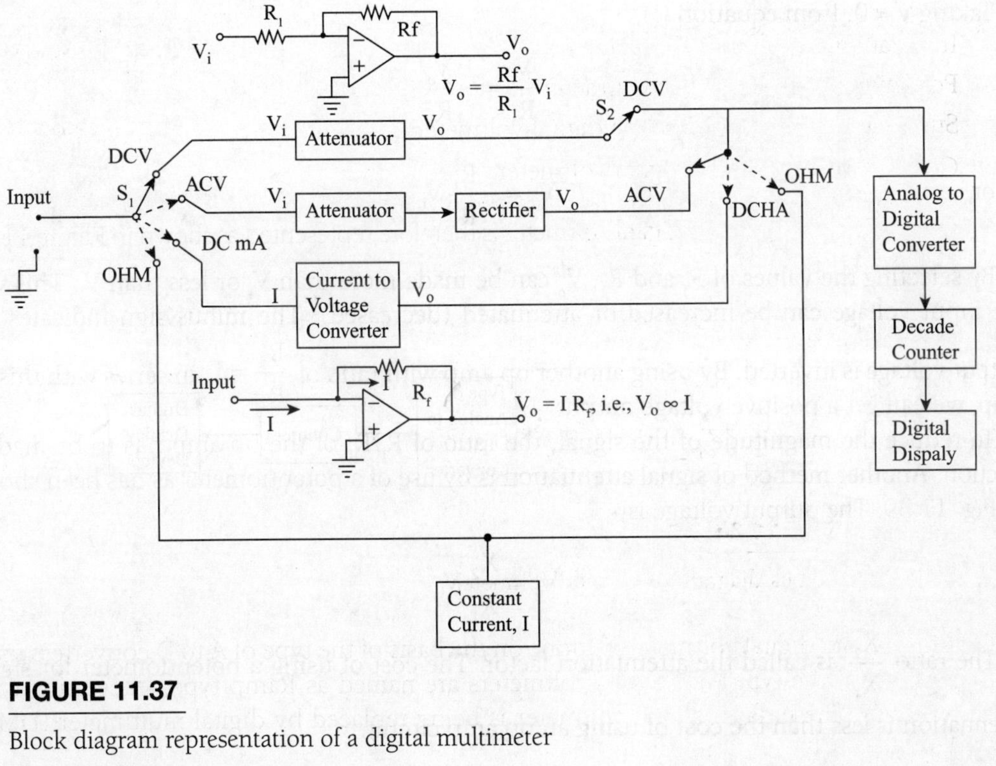

FIGURE 11.37

Block diagram representation of a digital multimeter

Attenuator

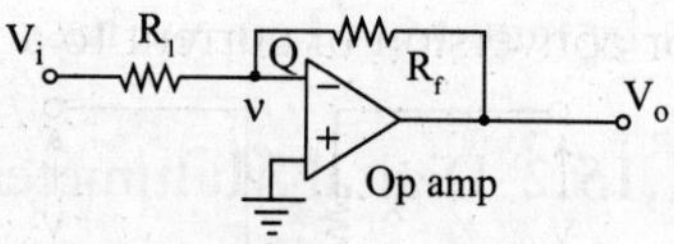

FIGURE 11.38

Op amp based Attenuator

The input impedance of the op amp is very high so that practically negligble current flows through the op amp. Let current through the input resistance R_1 be i. Potential of point Q is v. So, $V_i - v = iR_1$ or, $i = \frac{V_i - v}{R_1}$. As no current flows thing the op amp, the same current i will flow through the feedback resistace R_f. So, $v - V_0 = iR_f$.

So,
$$i = \frac{V_i - v}{R_1} \quad \text{and} \quad i = \frac{v - V_o}{R_f}$$

Equating,
$$\frac{V_i - v}{R_1} = \frac{v - V_o}{R_f} \qquad \text{(i)}$$

The amplifier (op amp) has infinite gain, A. Output voltage V_o is A times the input voltage, v.

∴
$$V_o = Av \text{ or } A = \frac{V_o}{v} = \infty$$

As A tends to ∞, v tends to 0.

Taking v = 0, from equation (i),

$$\frac{V_i}{R_1} = -\frac{V_o}{R_f}$$

or,
$$V_o = -\frac{R_f}{R_1} V_i$$

By selecting the values of R_f and R_1, V_o can be made more than V_i or less than V_i. This way the input voltage can be increased or attenuated (decreased). The minus sign indicates the output voltage is inverted. By using another op amp with ratio of $\frac{R_f}{R_1} = 1$, in series with this op amp, we can get a positive voltage output.

To reduce the magnitude of the signal, the ratio of R_f/R_1 of the op amp has to be made a fraction. Another method of signal attenuation is by use of a potentiometer as has been shown in Fig. 11.39. The output voltage is,

$$V_o = \frac{X_2}{X_1} V_1$$

The ratio $\frac{X_2}{X_1}$ is called the attenuation factor. The cost of using a potentiometer for signal attennation is less than the cost of using an op amp circuit.

Current to voltage converter (I to V conversion)

For conversion of current to a proportional voltage, an op amp can be used as shown.

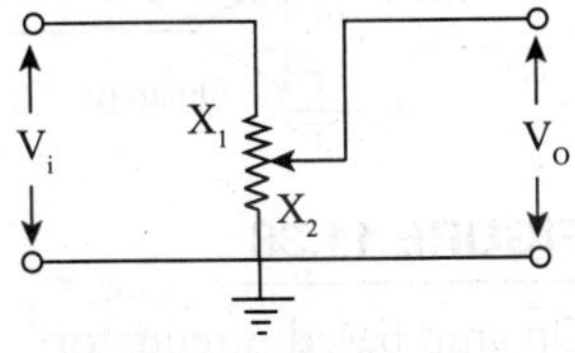

FIGURE 11.39

Potentio meter type attenuator

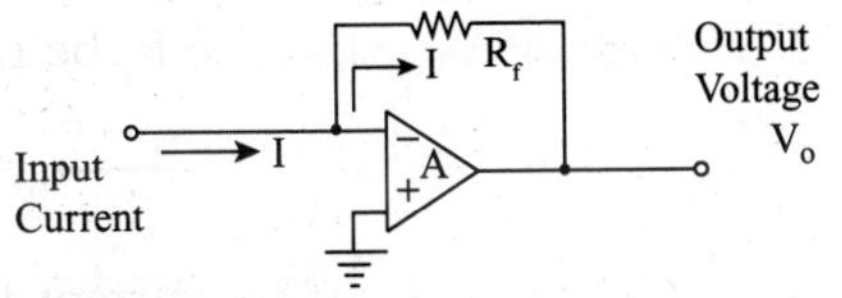

FIGURE 11.40

Op amp based current to voltage converter

Assume that the op amp has infinite gain and infinite input impedance. In that case the amplifier will draw no current. The whole of input current I will flow through the feedback resistor R_f so that $V_o = IR_f$. This way, output voltage V_o is proportional to current I as has been shown in Fig. 11.40.

Multimeter and its main parts on the front panel

The main parts and their functions are described below.

1. *Display Screen:* The LCD screen placed on the upper portion of the multimeter is meant for displaying in four or more digital form the measured value. It may also indicate –ve value, if any. Some of the modern multimeters are provided with illuminated display for better viewing of the measured value in low-light conditions.
2. *Selection Dial:* Using the selection dial, the user can set the multimeter to read different electrical parameters like current (mA),voltage, resistacnce, capacitance, etc.
3. *Ports:* There are two plug-in type ports on the front panel of every multimeter for connections.For making any measurement, two connecting wires, in the form of probes are plugged into these ports.The probes are different in colour, one is red colour and the other is of black colour. In some multimeters four ports are available, the two ports are marked COM and mAVΩ. The terminal or port marked COM means common and is in most cases connected to ground or considered as –ve. Generally, the black colour probe is inserted into it.

   The mAVΩ allows measurement of current in mA, voltage and resistance. This terminal is considred +ve connection of the circuit. The red colour probe is inserted into this port.
4. *The connected leads or probes:* As mentioned earlier there are two leads coloured red and black for connection to the variable to be measured with the help of the multimeter.

Safety precautions in using a digital multimeter

The following are some of the safety precaution points to be remembered while using a digital multimeter:

(i) Do not use damaged meter leads (probes)

(ii) When a test lead is plugged into the ampere or mA input jack, then never touch the probe to a voltage source.

(iii) While making measurement always keep your fingers behind the finger guards on the test probes.
(iv) Replace the battery to avoid false reading, when required.
(v) Be careful when working with voltages above 60 V dc or 30 V ac to avoid any electric shock.

EXAMPLE 11.9

The meter constant of a single-phase energy meter is 1200. When a load of 6 kW is switched on, the disc of the meter made 136 revolutions in 1 minute. Calculate the percentage error.

SOLUTION

Actual energy consumed = 6 kW × 1 minute = $\frac{6\times 1}{60}$ kWh = 0.1 kWh. Meter constant is 1200. This means for 1 kWh of energy consumed, the disc of the meter makes 1200 revolutions.

The reading of the meter corresponding to 136 revolutions = $\frac{136}{60} \times 1$ kWh = 0.113 kWh.

$$\text{Percentage Error} = \frac{(\text{Meter reading} - \text{Actual energy consumed}) \times 100}{\text{Actual energy consumed}}$$

$$= \frac{(0.113 - 0.1)\times 100}{0.1} = \frac{0.013 \times 100}{0.1} = 13 \text{ per cent}$$

EXAMPLE 11.10

A moving-coil-type instrument has an internal resistance of 25 Ohms. The instrument gives full scale deflection when a current of 100 mA flows through it. The same instrument is to be used as an ammeter of range 0–30 A and as a voltmeter of range 0–300 V. Show how this can be done.

SOLUTION

The connection of a multiplier, i.e., a high resistance in series with the instrument is required to use it as a voltmeter. For using the instrument as an ammeter of higher range, a shunt resistance is to be connected in parallel as shown.
From Fig. 11.41 (a),

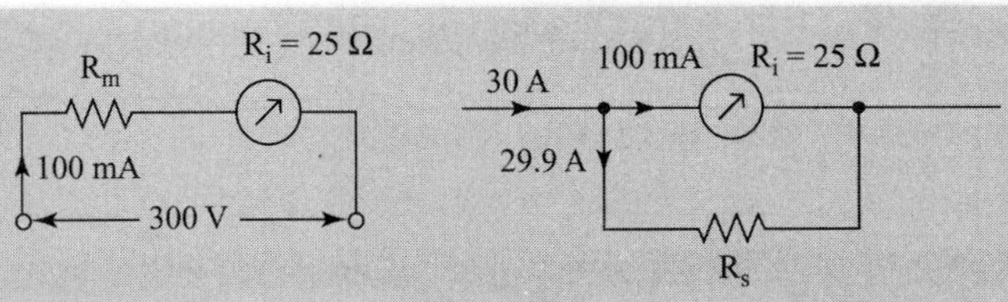

FIGURE 11.41

$$\frac{300}{R_m + R_i} = 100\,\text{mA} = 0.1\text{ A}$$

$$R_m + R_i = \frac{300}{0.1} = 3000$$

$$R_m = 3000 - R_i = 3000 - 25$$

$$= 2975\ \Omega$$

From Fig. 11.41 (b),

$0.1 \times 25 = 29.9 \times R_s$

or, $$= R_s = \frac{2.5}{29.9} = 0.0836\ \Omega$$

11.16 ► Cathode Ray Oscilloscope (C.R.O)

A Cathode Ray Oscilloscope is a very useful measurement and display device that can be used to measure and display signals of a wide range of frequencies.

Deflecting type instruments like an ammeter or a voltmeter cannot be used to measure current or voltage of high frequencies. For such measurements a CRO can be used.

Oscilloscopes are of two different types, namely analog type and digital type. An analog oscilloscope displays a voltage waveform by deflecting an electron beam generated by an electron gun inside a cathode ray tube on a fluorescent screen. Beacause of the use of a cathode ray tube (CRT) an analog oscilloscope is called a cathode ray oscilloscope (CRO).

11.16.1 Main Parts of a CRO

Figure 11.42 shows the main parts of a CRO. The main components parts are mentioned below.

1. *Cathode (K):* The cathode is indirectly heated to produce electrons to form an electron beam inside the CRT.
2. *Anode (P):* It is a plate which is circular with a small hole at the centre. The electron beam from the cathode passes through the hole of the anode plate. The potential of

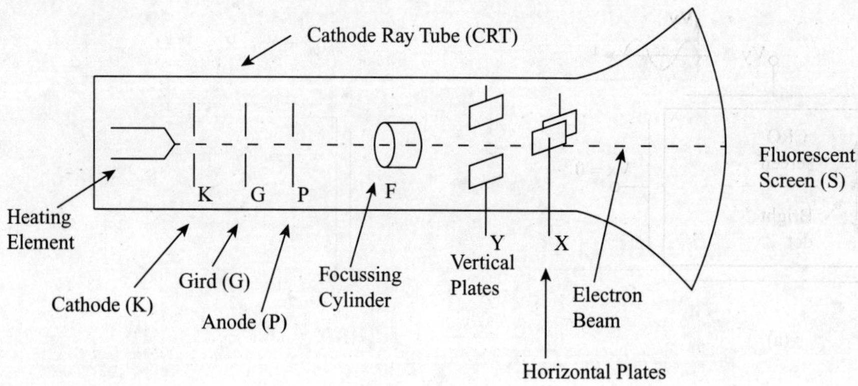

FIGURE 11.42

Main parts of a cathode ray tube

anode (P) creates an electric field which accelerates the electrons to pass through as a fine beam.

3. *Grid (G):* The grid (G) controls the intensity of the electron beam passing through it. This is achieved by controlling the potential of the grid. The grid is also circular type with a central hole placed in between the Cathode and the Anode.
4. *Focussing Cylinder (F):* Through the use of lens the focussing cylinder helps in concentrating the electron beam into a thin and sharp straight line.
5. *Vertical Deflecting Plates (Y):* Using these plates, the electron beam can be deflected vertically on the screen.
6. *Horizontal Deflecting Plates (X):* These plates help deflect the beam more from left to right in the horizontal direction.
7. *Fluorescent Screen (S):* The screen on which the electron beam moves is coated with fluorescent material. When electrons strike the screen, a green light is emitted. The CRO with its cathode ray tube and the associated electronic circuitry has three sections, namely, the electron gun, the deflection system and the display screen.

11.16.2 Displaying a Voltage Waveform on the Screen

Figure 11.43 shows a CRO screen along with a pair of vertical plates and a pair of horizontal plates. As shown in Fig. 11.43 (a), a sinusoidal voltage, V_y has been connected to the vertical plate and horizontal plate is at zero potential. Let us assume that the bright dot created by the electron beam is at the extreme left position on the screen. The bright dot will move up and down in every cycle producing a bright vertical line as shown by arrow lines.

As shown in Fig. 11.43 (b), a voltage V_x which is increasing linearly with time applied to the horizontal plate with vertical plates at zero potential. The bright spot on the screen moves from left to right in time 0 to t_1 and stops. This is because voltage V_x becomes constant at that point

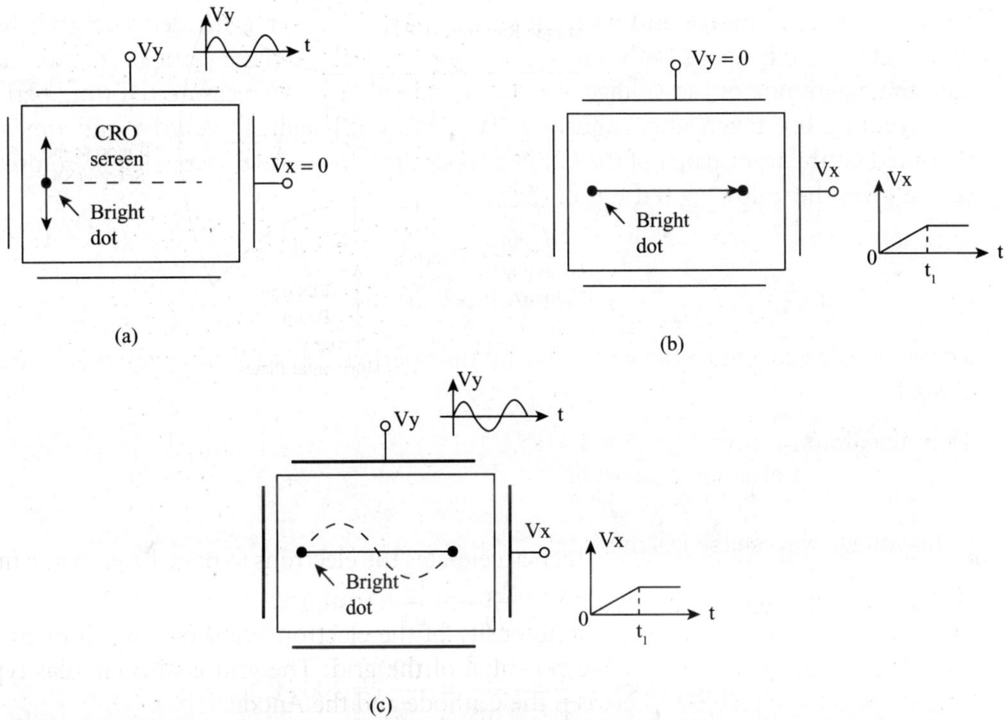

FIGURE 11.43

Display of sinusoidally varying waveform on CRO screen

of time. In Fig. 11.43 (c), it is shown a signal voltage V_y and V_x applied to the plates simultaneously. Voltage V_y is making the spot move up and down while the voltage V_x is pulling the spot horizontally. The resultant of the two forces makes the spot follow a path of the sinusoidal voltage on the screen. Therefore, it is concluded that on the vertical plate we will apply the voltage to be displayed and simultaneously in synchronism we will apply a voltage across the horizontal plate, which will be increasing linearly with time.

A time base circuit is provided in the CRO to produce a voltage in the form of saw-tooth, known as sweep waveform to be applied to the horizontal plate as V_x. The time base voltage is to be synchronized with the input signal such that a steady voltage waveform of the measurement signal is visible on the screen.

11.16.3 Measurement with a CRO

A CRO can be used for the measurement of voltage, frequency, phase difference, etc. In addition, a CRO can be used for testing of diodes and transistors.

(a) **Measurement of voltage and its frequency:** The CRO screen is divided into grids having a number of divisions both horizontally and vertically. On the vertical axis, we measure the magnitude of the voltage and an horizontal axis, we measure the time period. The readings are taken after adjusting the two knobs, namely volt/div and time/div mounted on the front panel of the CRO. Suppose a voltage wave shape displayed on the screen gives the following settings of the knobs:

$$\text{Volt/div} = 5$$

$$\text{Time/div} = 1 \text{ ms/div}$$

The peak value covered 3 div vertically and time period, T for a cycle covered 5 div horizontally.

Then maximum values, $V_m = 5 \times 3 = 15$ V

and time period, $T = 1 \times 5 = 5$ ms

If the voltage waveshape is sinusoidal,

Rms value of voltage, $$V = \frac{V_m}{\sqrt{2}} - \frac{15}{1.414} = 10.6 \text{ V}$$

$$\text{Frequency} = \frac{1}{T} = \frac{1}{5 \text{ ms}} = \frac{1000}{5} = 200 \text{ Hz.}$$

(b) **Measurement of phase difference and frequency using Lissajous figure:**

Phase differcnce

For measuring phase difference between two sinusaidal voltages, they are simultaneously connected to vertical and horizontal plates, i.e., to Y-plates and X-plates. Patterns like the ones shown in Fig. 11.44(a) will appear on the CRO screen for phase difference of 0°, 45°, 90°, 180°, respectively. Suppose, we get a pattern as shown in Fig. 11.44(b) the phase difference has to be calculated by actual measurement as has been shown along with Fig. 11.14(b).

Measurement of frequency

Frequency of unknown voltage wave can be measured by using another voltage wave of known frequency. A signal of known frequency, f_1 is applied to the horizontal plates. This signal can be obtained from a variable frequency oscillator. Signal of unknown frequency, f_2 is connected to the vertical plates. Depending on the ratio of unknown frequency f_2 and known frequency f_1, Lissajous figure are shown on the CRO screen. Lissajous figure becomes tangent to the horizontal axis and vertical axis at number of points from which the unknown frequency is calculated as

$$\frac{f_2(\text{unknown})}{f_1(\text{known})} = \frac{\text{Number of horizontal tangent points}}{\text{Number of vartical tangent points}}$$

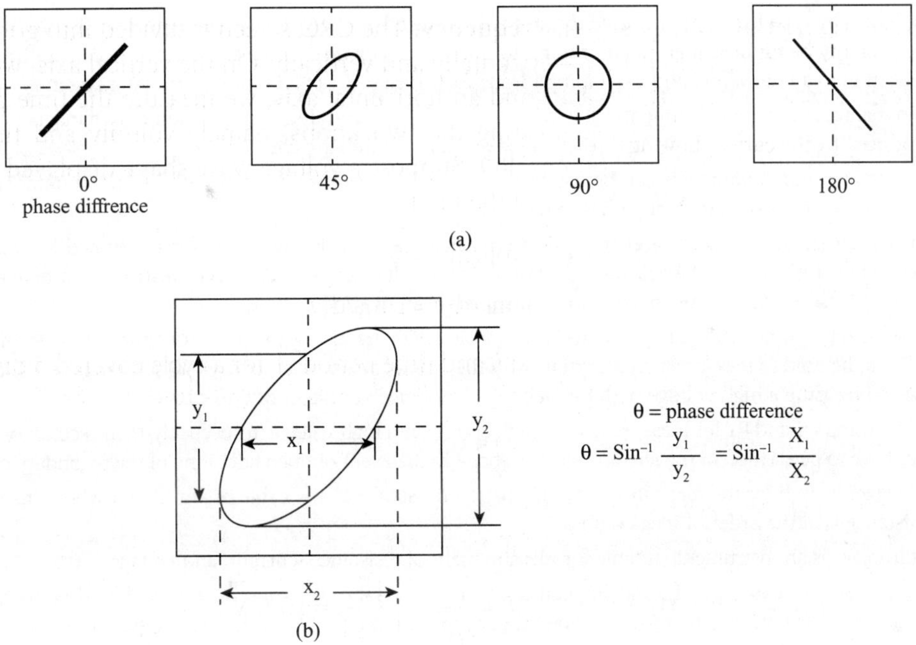

FIGURE 11.44

Calculation/Determination of phase difference from Lissajous figures

Important Terminologies and Points to Remember

1. Measuring instruments like ammeter, voltmeter, multimeter, etc., are used to measure variable electrical quantities.
2. Measuring instruments may be analog type or digital type.
3. A measurement system consists of a basic sensing element, i.e., a transducer which converts the variable quantity to be measured into an equivalent electrical signal. Such a signal is then conditioned, if required, transmitted to a distance place and then recorded or displayed.
4. Static characteristics of instruments are expressed as accuracy, precision, resolution, etc.
5. Instruments may be indicating type (i.e., deflection type), recording type or integrating type.
6. Measurement error may be due to degree of accuracy of the measuring instrument, the mistake made in recording the value, the changes in the environmental conditions, improper use of instruments (using a large capacity instrument to measure a very small quantity), or due to loading effect.
7. Permanent magnet moving coil (pmmc) instruments are used to measure dc quantities like dc voltage, dc current, etc.
8. Shunts and Multipliers are used to extend the range of ammeters and voltmeters, respectively. A single instrument can be used to measure from a low range to a high-range quantities by using shunts and multipliers. Shunts are low-value resistances and multipliers are high-value resistances.
9. A shunt is connected in parallel with an ammeter and a multiplier is used in series with a voltmeter.
10. Moving iron instruments are used for measurement of ac quantities. They can be used for dc measurement also.

11. In deflecting type instruments, three torques are produced, namely the deflecting torque, T_d produced by the quantity being measured; a controlling torque, T_C is provided by spring control system to bring the deflection to a final value; and a damping torque to reduce the oscillation of the moving system.
12. In dynamometer type instruments, deflecting torque is produced due to interaction of two magnetic fields produced by the current flowing through a fixed coil and a moving coil.
13. A voltmeter and an ammeter can be used to measure dc power as power P = VI. For measurement of ac power (P = VI cos ϕ), a wattmeter has to be used.
14. For measurement of electrical energy, an energy meter is used. Energy is power multiplied by time. An energy meter is also called a Kilowatt hour (kWh) meter. It is an integrating type instrument where energy consumed is recorded as a sum of power consumed over a period of time.
15. Current transformer (CT) and Potential transformer (PT) are also known as instrument transformers. A C.T. can be used to measure a high current with the help of an ammeter of say 0 to 5 A range. A P.T. can be used to measure a high voltage with the help of a low range voltmeter of 0 to 110 V.
16. While using CT and PT for measurement of high current and high voltage, respectively, their secondary windings have to be earthed to prevent any electric shock in the event of open circuiting of the secondary circuit.
17. A Megger is used for the measurement of insulation resistance of any cable or any circuit whose insulation resistance is of the order of mega ohms.
18. Multimeter is an instrument suitable for measurement of resistance, current, and voltage.
19. To increase the sensitivity of a moving coil instrument, the ratio of torque by weight should be high. This means that deflecting torque produced should be high while the weight of the moving system should be low.
20. The common type of error in a C.T. is ratio error and phase-angle error.
21. Creeping is an error which may occur in induction-type energy meters. It may happen that the aluminium disc rotates even when there is no load connected. The phenomenon of creeping is prevented by making two holes on the aluminium disc.

Important Formulae and Equations

Equation	Remarks
1. $R_S = \dfrac{I_m R_m}{I - I_m}$ Where R_S is the resistance of the shunt, R_m is the resistance of the measuring instrument, I is the line current to be measured and I_m is the current through the meter.	1. R_S is a shunt of small value resistance connected in parallel with an ammeter to increase its measurement range.
2. $R = \dfrac{V}{I_m} - R_m$ Where R is the resistance of the multiplier, V is the increased voltage range, I_m is the full scale deflection current of the instrument and R_m is the resistance of the meter.	2. R is the resistance of high value connected in series with a voltmeter of low range to increase its measurement range.
3. For permanent magnet moving coil instrument deflecting torque $T_d \propto I$ and for moving iron instrument $T_d \propto I^2$. For dynamometer instruments also $T_d \propto I^2$.	3. Pmmc instruments are suitable for dc measurement only. Others are suitable for both dc and ac measurement.
4. Power = V × I Watts, for dc circuits Power = V I cos ϕ Watts, for ac circuits.	4. Product of voltage and current gives the value of power in dc circuits. For measurement of power in ac circuits, we use a wattmeter.

REVIEW QUESTIONS

Short Answer Type Questions

1. Distinguish between the following three types of measuring instruments:
   (i) deflecting type; (ii) recording type; and (iii) integrating type.
2. Define a transducer, where does it find place in the instrumentation and measurement system? Give one example.
3. Draw a block diagram showing basic instrumentation and measurement system elements.
4. Distinguish between analog and digital instruments.
5. Distinguish between active and passive instruments.
6. How do you define the static characteristics of measuring instruments?
7. Define the following:
   (i) accuracy; (ii) sensitivity; (iii) precision; and (iv) resolution.
8. Distinguish between accuracy and sensitivity of an instrument.
9. What is meant by loading effect of a measuring instrument?
10. Explain the difference between linear and non-linear systems.
11. What are the dynamic characteristics of a measuring instrument?
12. Explain various measurement errors which are likely to occur while using deflecting type instruments.
13. Explain the difference between a galvanometer and a milliammeter.
14. Why should we have high torque by weight ratio of the moving system of an instrument?
15. Explain the constructional details and principle of the working of a permanent-magnet moving-coil instrument.
16. Explain the function of deflecting torque, control torque, and damping torque of a moving coil instrument.
17. Explain eddy current damping and air friction damping.
18. Explain the equation for deflecting torque of a PMMC instrument.
19. Explain why the PMMC-type instruments have a linear scale while moving iron instruments have square scale?
20. What are the sources of error in PMMC-type instruments? Explain them.
21. Where do we use shunts and multipliers?
22. How do we extend the range of ammeters and voltmeters?
23. Distinguish between a PMMC-type ammeter and a voltmeter.
24. Explain the working principle of attraction-type moving iron instrument.
25. Distinguish between an attraction-type and a repulsion-type moving-iron instrument.
26. Draw the constructional details and explain the working principle of a repulsion-type moving-iron instrument.
27. Can we use a PMMC-type instrument for measurement of both dc and ac quantities?
28. Can we use a moving-iron instrument for both ac and dc measurements?
29. Explain the constructional details of a dynamometer-type instrument.
30. What are the sources of error in dynamometer-type instruments?
31. Prove that power in an ac circuit is equal to VI $\cos\varphi$.
32. Explain how a wattmeter reading will be equal to VI $\cos\varphi$.
33. Explain how error in measurement may occur due to the connections of current coil and pressure coil of a wattmeter.
34. Explain the constructional details and working principle of a single-phase induction-type energy meter.
35. What do you mean by creeping of an energy meter? How can creeping be eliminated?

36. Explain the use CTs and PTs.
37. State the advantages of CT and PT.
38. What is meant by ratio error and phase angle error of a current transformer?
39. Explain the basic principle of working of a multimeter.
40. Show how a megger is used for measurement of insulation resistance.

Numerical Problems

41. A PMMC-type instrument gives a full-scale deflection of 1 mA and has an internal resistance of 100 Ω. Calculate the values of shunts required to extend the range of the instrument to 0–100 mA and 0–500 mA.

[**Ans** 1.01 Ω; 0.02 Ω]

42. Calculate the value of multiplier resistance so that an instrument of internal resistance of 100 Ω and full-scale deflection current of 1 mA can be converted into a 0–100 V range voltmeter.

[**Ans** R_m = 99,900 Ω]

43. A PMMC-type instrument gives a full-scale deflection of 100 mA when a potential difference across the terminal is 1 V. How can this instrument be used to measure upto 100 V?

[**Ans** Connecting a series resistance of 990 Ω]

Multiple Choice Questions

1. For measurement of direct current we may use
   (a) a moving-iron-type ammeter
   (b) a galvanometer
   (c) a permanent-magnet moving-coil-type ammeter
   (d) a hot-wire-type ammeter.
2. In permanent-magnet wiring-coil-type instruments expression for deflecting torque can be written as
   (a) $T_d \alpha I^2$ (b) $T_d \alpha I$
   (c) $T_d \alpha \frac{1}{I}$ (d) $T_d \alpha \frac{1}{I^2}$
3. Damping of deflecting-type instruments is done to
   (a) reduce the oscillations of the pointer in the final deflected position
   (b) make the moving system go slow
   (c) make the moving system move fast
   (d) reduce the angle of deflection of the pointer on the graduated acate.
4. For which of the following instruments the calibrated scale is linear?
   (a) Repulsion-type moving-iron instruments
   (b) Attraction-type moving-iron instruments
   (c) Permanent-magnet moving-coil instruments
   (d) Dynamometer-type instruments.
5. The extension of range of an ammeter and a voltmeter can be made, respectively, by
   (a) using multiplier and shunt
   (b) using shunt and multiplier
   (c) using series capacitor and a series inductor
   (d) reducing the spring tension of the deflecting system.
6. To increase the sensitivity of a deflecting-type instrument
   (a) torque–weight ratio of the moving system should be high
   (b) torque–weight ratio of the moving system should be low
   (c) it should be lightly damped
   (d) the control torque must be reduced.
7. A permanent-magnet moving-coil instrument cannot be used for ac measurement because on ac the deflecting system will
   (a) not respond to the quick alternating torques and will stay in its initial position
   (b) respond to ac signal but will not move because of inertia
   (c) get burnt out
   (d) give incorrect reading.
8. In a moving-iron-type instrument the current flowing throgh the coil, I and the torque developed T_d are related as
   (a) $T_d \alpha I$ (b) $T_d \alpha \frac{1}{I}$
   (c) $T_d \alpha I^2$ (d) $T_d \alpha \frac{1}{I_{a^2}}$.

9. Wattmeters are of
   (a) permanent-magnet moving-coil type
   (b) dynamometer type
   (c) moving-iron type
   (d) hot-wire type.

10. Basic difference between an ammeter and a voltmeter is that
    (a) an ammeter is a low-resistance instrument while a voltmeter is a high-resistance instrument
    (b) a voltmeter is a low-resistance instrument while an ammeter is a high-resistance instrument
    (c) a voltmeter is provided with a shunt while an ammeter is provided with a multiplied
    (d) a voltmeter is connected in series with a circuit while an ammeter is connected in parallel.

11. Three-phase power P in terms of voltage and current and power factor can be expressed as
    (a) $P = V_{ph} I_{ph} \cos\phi$ (b) $P = \sqrt{3}\, V_L I_L \cos\phi$
    (c) $P = \sqrt{3}\, V_{ph} I_{ph} \cos\phi$ (d) $P = 3\, V_L I_L \cos\phi$.

12. For measurement of electrical energy, we may use a
    (a) wattmeter
    (b) kWh meter
    (c) multimeter
    (d) voltmeter, ammeter, and a power factor meter.

13. The following could be one of the errors in an energy meter
    (a) clamping (b) clipping
    (c) creeping (d) clogging.

14. The energy meter installed near the main switch in residences and other locations is
    (a) an indicating- or deflecting-type instrument
    (b) an integrating-type instrument
    (c) a recording-type instrument
    (d) an absolute instrument.

15. A voltmeter has a uniform scale with 100 divisions, the full scale reading is 100 V and can be read up to 1/10 of a scale division with a fair degree of certainty. Its resolution is
    (a) 0.1 V (b) 0.2 V
    (c) 0.01 V (d) 0.02 V.

16. The deflection in moving-iron instruments is proportional to
    (a) square of the rms value of the current
    (b) rms value of the current
    (c) square of the maximum value of current
    (d) maximum value of the current.

17. We can distinguish between a moving-iron-type and a moving-coil-type instrument by looking at their
    (a) pointer
    (b) graduated scale
    (c) size
    (d) all the items as in (a), (b), and (c).

18. In a moving iron instrument 8 A current causes a deflection of the needle over the scale by 60°. For a deflection of 15°, the current required will be
    (a) 2 A (b) 4 A
    (c) 16 A (d) 0.5 A.

19. A voltmeter must have very high internal resistance so that
    (a) its accuracy is high
    (b) its resolution is high
    (c) it draws a very small amount of current
    (d) it creates a high-loading effect on the circuit.

20. An instrument with a measurement range of 0–100 V with an accuracy of 1 percent has been used to measure a low voltage of 10 V. The error of measurement will be
    (a) 1 per cent (b) 0.1 per cent
    (c) 0.01 per cent (d) 10 per cent.

Answers to Multiple Choice Questions

1. (c)	2. (b)	3. (a)	4. (c)	5. (b)	6. (a)
7. (a)	8. (c)	9. (b)	10. (a)	11. (b)	12. (b)
13. (c)	14. (b)	15. (a)	16. (a)	17. (b)	18. (b)
19. (c)	20. (d)				

CHAPTER

12 Transducers

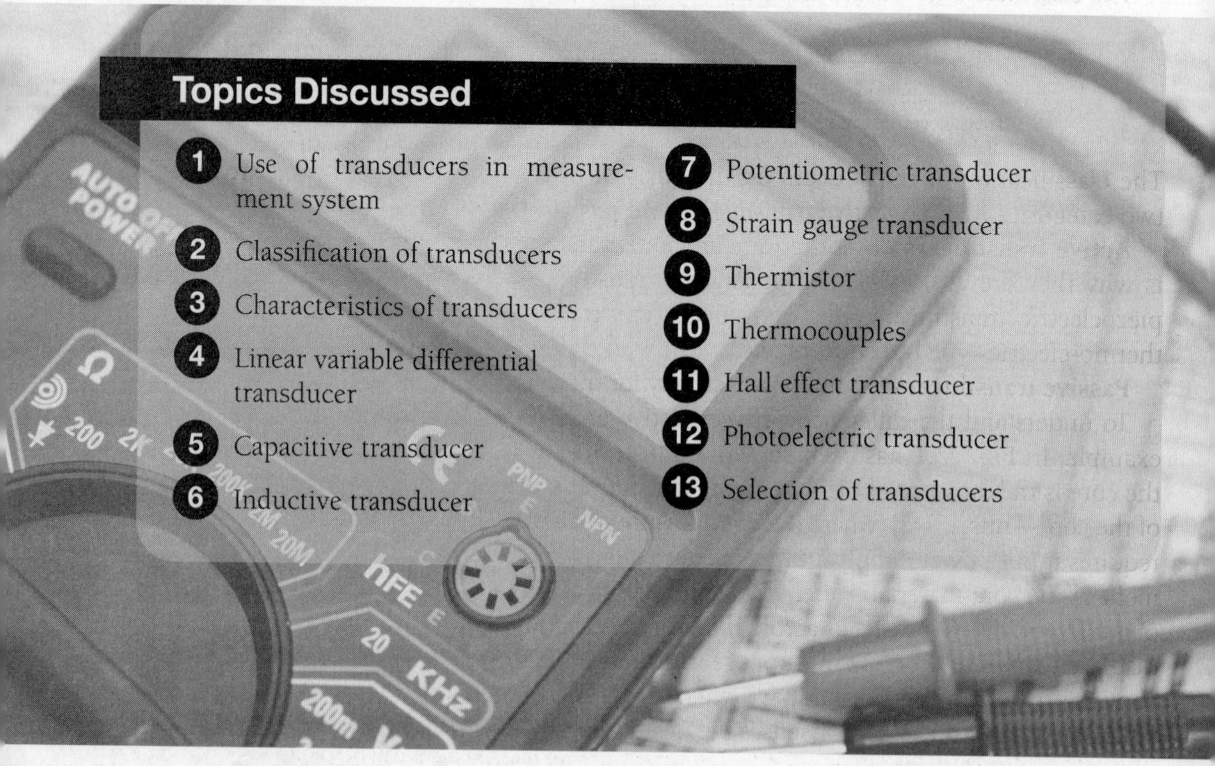

Topics Discussed

1. Use of transducers in measurement system
2. Classification of transducers
3. Characteristics of transducers
4. Linear variable differential transducer
5. Capacitive transducer
6. Inductive transducer
7. Potentiometric transducer
8. Strain gauge transducer
9. Thermistor
10. Thermocouples
11. Hall effect transducer
12. Photoelectric transducer
13. Selection of transducers

12.1 ► INTRODUCTION

When we are to measure any non-electrical quantity like displacement, pressure, temperature, strain, etc., we first convert the measuring quantity into a proportional electrical signal, i.e., voltage or current. This is done with the help of devices called transducers. Therefore, we can state that **a transducer is a device which converts a non-electrical quantity to be measured into a proportional electrical signal.** Such a transducer is also called an electric transducer.

The advantage of creating a proportional electrical signal is that it is easy to process and transmit electrical signals. The electrical signal obtained from the transducer is used in the measurement system to measure, and if required to control the quantity being measured.

12.2 ► Classification of Transducers

Transducers are classified on the basis of quantity to be measured or depending upon whether external power supply is required for their operation or not.

The classification of transducers on the basis of quantity to be measured is shown below:

Transducers

- Temperature Transducers
- Pressure Transducers
- Displacement Transducers
- Vibration Transducers
- Flow Transducers
- Other Transducers

The classification of transducers on the basis of power supply requirement are made into two categories, namely ***active transducers*** and ***passive transducers***.

Active transducers do not require any external source of supply for their operation. That is why they are also called self-generating transducers. Examples of active transducers are piezoelectric transducers, electromagnetic-type transducers, photovoltaic-type transducers, thermo-electric-type transducers, etc.

Passive transducers require power supply for their operation as they are not self generating.

To understand the difference between active and passive transducers, let us take a simple example. In Fig. 12.1 has been shown a transformer with a movable core. The movement of the core is the measure of displacement. The output voltage, e_o will depend upon the position of the core. Thus, $e_o \propto d$, where d is the measure of displacement. Thus, this passive transducer requires input power supply and its output voltage can be made proportional to the displacement of the core material. Here, voltage generated is proportional to displacement. Deflection of a millivoltmeter connected across the output terminal A and B can be calibrated in terms of displacement of the core. Now, suppose we want to use this transducer to measure pressure. To convert pressure to displacement, we will need an input transducer, in this case, a bourdon tube, which we call *sensor*. A sensor senses the desired physical quantity and converts it into another

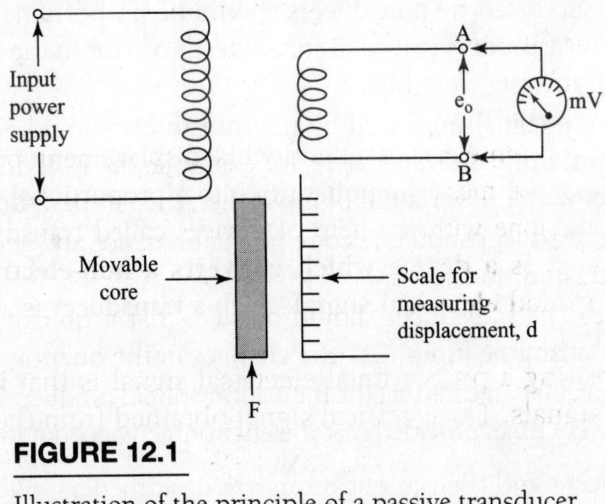

FIGURE 12.1

Illustration of the principle of a passive transducer

energy form. In this case, pressure is converted into displacement by an input transducer, called a sensor, and the displacement is converted to voltage by an output transducer. These are also called *primary transducer* and *secondary transducer*. Here, the primary transducer converts the physical quantity, i.e., the pressure into a mechanical signal, i.e., displacement. The secondary transducer converts the mechanical signal into a proportional electrical signal.

Transducers may also be classified on the basis of the type of their output. If the output is a continuous function of time, such transducers are called *analog transducers*. If the output is in discrete steps, such transducers are called *digital transducers*. The output of transducers like a thermocouple, a tachogenerator, a potentiometer, etc., are voltages which vary with time and is a continuous function of time. These are examples of analog transducers.

Digital transducers convert the input physical phenomenon into an electrical output in the form of a train of digital pulses using a binary system of notation. Since transducers are often required to communicate with computers, transducers with digital output are required. In the case of analog transducers, conversion of analog signals to digital signals will be required for interfacing with a computer system.

Since digital transducers are few, we therefore mostly use analog transducers to produce a voltage signal and use an electronic A to D (Analog to Digital) converter to get the digital data in the binary form, i.e., 0 and 1.

12.3 ▶ Characteristics of a Transducer

Transducers are used in the measurement of system as an input-sensing element. The received signal requires processing. The signal processing involves amplification of the signal, its conversion from analog to digital form we then carry out, its transmission, its display, its recording, etc.

Since transducers generate the basic measurement signal, they must possess certain desirable characteristics which are mentioned below:

(i) **Sensitivity:** It is defined as the output per unit input of the quantity being measured. For example, the output of a thermocouple used as a transducer is expressed as EMF induced per degree centigrade. Sensitivity of a transducer should therefore be as high as possible.

(ii) **Accuracy:** Output produced by transducers should be proportional to the true value of the quantity being measured. Any deviation will lead to error in the measurement. Accuracy of a transducer must be very high.

(iii) **Linearity:** The output signal produced by the transducer should vary linearly with the variation of the input. For a thermocouple, for example, the EMF induced should maintain a linear relation with the variation of input, i.e., the temperature.

(iv) **Ruggedness:** The transducer should require no maintenance and should be able to withstand any over-load conditions.

(v) **Speed of response:** The transducer should be such that it quickly responds, i.e., provides output to the changing input, i.e., the changes in the quantity being measured.

(vi) **Repeatability:** Repeatability means that the transducer should produce the same output signal for the same input every time, provided that the environmental conditions remain the same.

Various types of transducers and their applications are described as follows.

12.4 ► LINEAR VARIABLE DIFFERENTIAL TRANSFORMER (LVDT)

A linear variable differential transformer (LVDT) is an electromagnetic induction-type displacement transducer. It is a reliable and accurate sensing device that can be used to convert linear motion (displacement) to a proportional electrical output. The basic construction of a LVDT is shown in Fig. 12.2. It consists of three elements, viz one primary winding, two identical secondary windings, and a movable magnetic core.

The primary winding is energised with a constant ac voltage source at a frequency of 1 to 10 KHz. This produces an alternating magnetic field in the transducer, which induces a signal into the two secondary windings. The magnitude of induced emf in the two secondary windings will depend upon the position of the iron core inside the coils.

The secondary windings are connected in series opposition (push–pull mode). When the core moves to the centre, i.e., when the core is at null position, the induced emf in the two secondary windings will be equal but oppositely added so that the output signal is zero.

As shown in Fig. 12.2 (a), the position of the core is at null position. In this position the induced EMF in the two secondary windings, E_1 and E_2 are equal. The transducer output voltage is zero. When the core is displaced by a force towards the left as shown in Fig. 12.2 (b), E_1 will be greater than E_2 due to the difference in flux linkage created by the primary winding ampere turns. When the core is moved towords the right as shown in Fig. 12.2 (c), the induced EMF E_2 will be greater than E_1. The magnitude of the differential output voltage E_0 will vary with the change in core position. The output voltage, E_0 at null position is ideally zero and will increase when the core is made to move either towords the left or towords the right as shown in Fig. 12.3. It is observed that the output voltage changes linearly with the displacement of the core. The output voltage is a function of the displacement of the core.

Although the output voltage, E_0 at null position should be zero, it is observed from Fig. 12.3 that a small amount of voltage appears across the output terminals at this position. This is due to either an incomplete electrical or magnetic balance between the output voltages at the null position. This small voltage at null position may also be due to some stray magnetic fields or

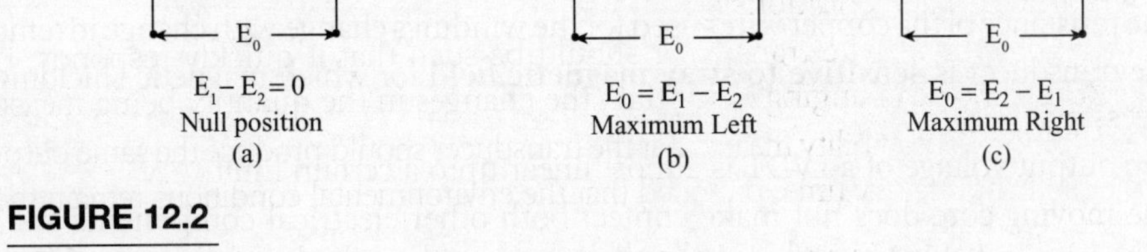

FIGURE 12.2

Basic constructional details of an LVDT transducer

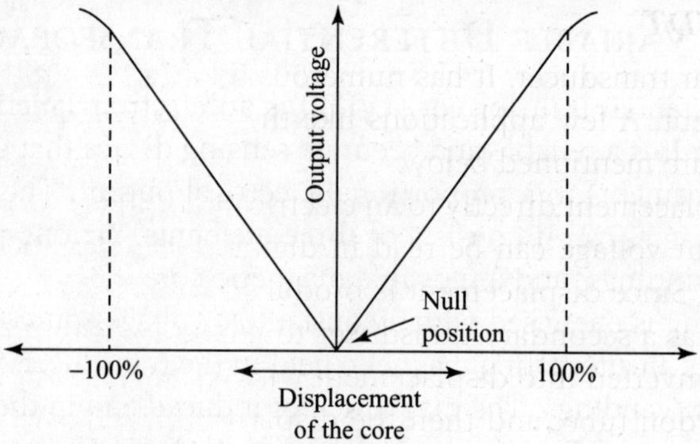

FIGURE 12.3

The output characteristics of an LVDT

temperature effect. The output voltage curve is linear upto a certain range of displacement beyond which the characteristic start deviating from linearity.

Advantages and disadvantages of LVDT transducers

The advantages and disadvantages of a LVDT is expressed in terms of certain performance characteristics like range, sensitivity, resolution, ruggedness, dynamic response, linearity, etc. Some of these are mentioned as follows:

(i) **Range of measurement:** LVDT can be used for the measurement of high range of displacement ranging from 1mm to 250 mm.

(ii) **Frictionless and electrical isolation:** There is no physical contact between the windings and the core. This provides the LVDT a very long life as there is no wear and tear due to friction.

(iii) **Resolution:** A very small movement of the core produces a proportional voltage output. This makes the resolution of LVDT very high.

(iv) **Sensitivity:** It is the amount of voltage induced per mm of displacement of the core. The LVDT gives high-output voltage per mm of displacement of the core making it a high-sensitive transducer. Sensitivity is expressed as Volts/mm.

(v) **Ruggedness:** LVDT transducers are rugged in construction, and therefore can tolerate high degree of vibrations and mechanical shock, especially when the core is loaded with a spring.

(vi) **Power consumption:** LVDT transducers consume very low power, maybe in the range of 1 Watt or so.

(vii) **Temperature effect:** Variation in temperature affects the performance of the transducer. The resistance of the copper wires used for the windings change with change in temperature.

(viii) The transducer is **sensitive to stray magnetic field** for which magnetic shielding may be necessary.

(ix) The output voltage of a LVDT is almost linear upto a certain limit.

(x) The moving core does not make contact both other electrical components and so gives frictionless reliable long life.

Applications of LVDT

LVDT is a very popular transducer. It has numerous applications in the field of instrumentation and measurement. A few applications like the measurement of displacement, force, weight, and pressure are mentioned below.

LVDT converts displacement directly to an electrical output for which it is called primary transducer. The output voltage can be read in digital form through conversion of analog signal to digital signal Since displacement is produced by force, weight, and pressure, etc., an LVDT can be used as a secondary transducer to measure these quantities. For example, the pressure is first converted into displacement with the help of a primary transducer, say a diaphram or a Bourdon tube, and then the displacement is converted into an electrical signal by the LVDT (a Bourdon tube converts pressure into displacement).

Application of LVDT can be seen in weight-measuring machines used in many department stores and shops. The object to be weighed is kept on a tray which is placed on a LVDT shaft. A spring is attached to the shaft which allows the tray to return to its original position once the object from the tray is removed. The object on the tray whose weight is to be measured creates a downward displacement of LVDT shaft (core) and develops a differential voltage. The differential voltage is proportional to displacement, which in turn is proportional to the weight of the object being measured. The generated differential voltage is converted into seven segment digital display to indicate the weight of the object.

There are many other applications of LVDT like in the measurement of thickness of metal sheets being rolled out in industry, in the measurement of tension of a wire in a wire-drawing machine, automatic opening and closing of gates at railway crossings, fluid level measurement, etc.

12.5 ▶ Rotary Variable Differential Transducer (RVDT)

For measurement of linear displacement we use LVDT. For measurement of angular displacement, we can use a rotary variable differential transducer (RVDT). The working principle of RVDT is the same as that of an LVDT except that the core is mounted on a shaft as shown in Fig. 12.4. In LVDT, we have linear movement of the core. In RVDT, the core is mounted on the shaft. The actualing signal makes angular movement of the shaft and hence of the core. The output voltage, e_0 will depend upon the degree of angular movement of the core. A RVDT can be disigned to measure rotational motion from o° to 360°.

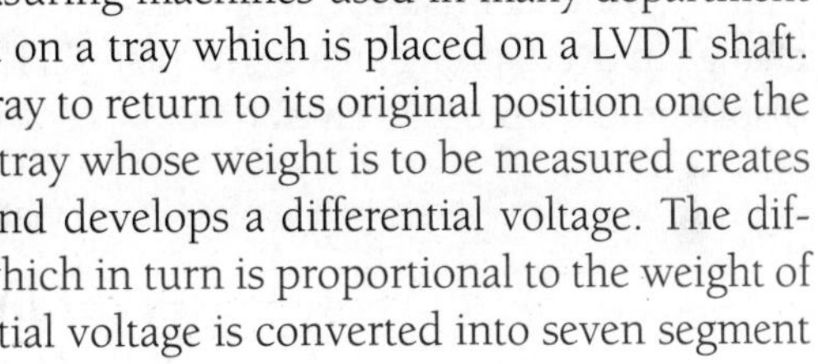

FIGURE 12.4

Rotary variable differential transducer (RVDT)

12.6 ▶ Capacitive Transducers

The capacitance of a parallel plate capacitor is expressed as

$$C = \epsilon\frac{A}{d} \tag{12.1}$$

where, ∈ is the permitivity of the dielectric material placed between the two plates,
A is the overlapping area of the plates, and
d is the distance between the two plates.

Change in the capacitance of a capacitor due to the variation in A, d, or ∈ can be utilized for the measurement of physical variables like displacement, force, pressure, etc. The liquid level in a container can also be measured by measuring the change in capacitance due to the dielectric constant of the liquid poured into the container having two plates inserted inside.

Fig. 12.5 shows a capacitive displacement transducer with variation in (a) dielectric placed between the plates, (b) gap between the plates, and (c) the area of overlap of the capacitor plates.

The change in capacitance is measured using a bridge circuit which is then calibrated in terms of force or pressure to be measured.

In transducers using change in area of the plates, the capacitance changes lenearly with change in overlapping area of the plates. Hence, this type of transducer can be used to measure displacement upto say, 1mm to a few centimeters.

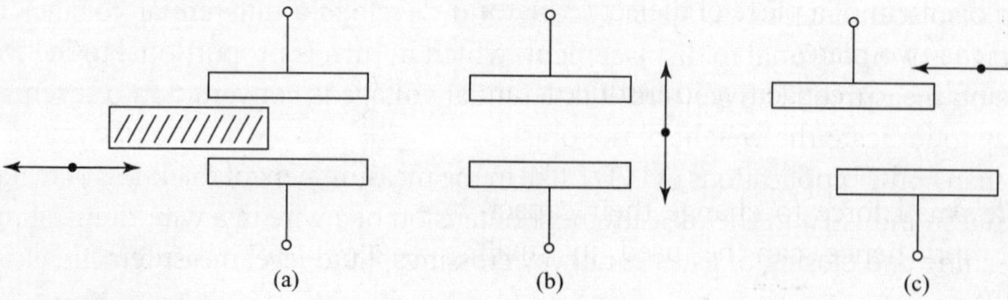

FIGURE 12.5

Capacitive transducer. (a) Variation in dielectric between the plates; (b) variation in gap between the plates; (c) variation in overlapping area

This principle of the change in the capacitance due to the change in area can be used for the measurement of angular displacement using a fixed semicircular plate and a movable semi-circular plate as shown in Fig. 12.6 (a). Fig. 12.6 (b) shows the linear variation of C with the variation of angular displacement, θ of the movable plate.

In a parallel-plate capacitor, the variation of distance between the plates creates a change in capacitance. The value of C is inversely proportional to the distance, d. Therefore, the relationship between displacement and the value of capacitance is hyperbolic. Their relationship can be considered linear only for a very small range of displacement.

By making one of the plates in the form of a flexible diaphragm, a capacitive transducer can be used to measure pressure due to the flow of liquid or gas as in Fig. 12.7. As shown in the figure, the flexible diaphragm, A acts as one of the capacitor plates. The pressure to be measured is applied to the flexible diaphragm, which bends, changing the effective distance, d between the two plates A and B. This will increase the capacitance, C of the capacitor, which can be measured and used to know the value of the pressure applied by the liquid or the gas.

Variation of position of the dielectric material between the plates can be made to create change in capacitance. Linear distance can be measured by measuring the change in capacitance

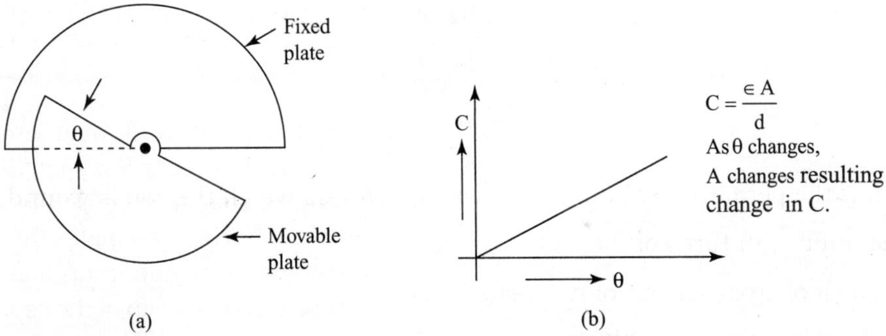

FIGURE 12.6

Measurement of angular displacement by a capacitive transducer.
(a) Fixed and moveable semicircular plates; (b) linear variation of capacitance with angular displacement of the plate

due to the movement of a piece of dielectric material between the two plates.

The advantages of capacitive transducers are that they

(i) require small force to change their capacitance, and hence can be used in small systems;
(ii) are extremely sensitive;
(iii) have high resolution;
(iv) are very little affected by stray magnetic fields;
(v) can be used for measurement of linear and angular displacement, force, pressure, liquid level, etc.

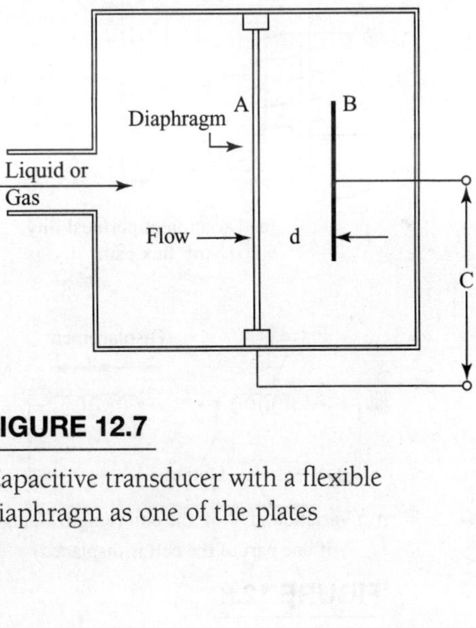

FIGURE 12.7

Capacitive transducer with a flexible diaphragm as one of the plates

12.7 ▶ Inductive Transducers

Inductive transducers operate on the principle of variation of self inductance of a coil or on the principle of variation of mutual inductance.

Inductance of a coil, L is defined as

$$L = N\frac{d\varphi}{di}$$

Assuming linear relationship between flux, φ and the current, i producing the flux, we can write

$$L = N\frac{\varphi}{I} = N\frac{BA}{I} = N\frac{\mu HA}{I} \qquad [\text{as } \phi = B.A \text{ and } B = \mu H]$$

or, $$L = N\frac{\mu NI}{\ell}\cdot\frac{A}{I} \qquad \left[\text{as } H = \frac{NI}{\ell}\right]$$

or, $$L = \frac{\mu N^2 A}{\ell} \tag{12.2}$$

where μ is the permeability of the core material around which the coil is wound,
N is the number of turns of the coil,
A is the area of cross-section of magnetic flux path,
ℓ is the length of the flux path.

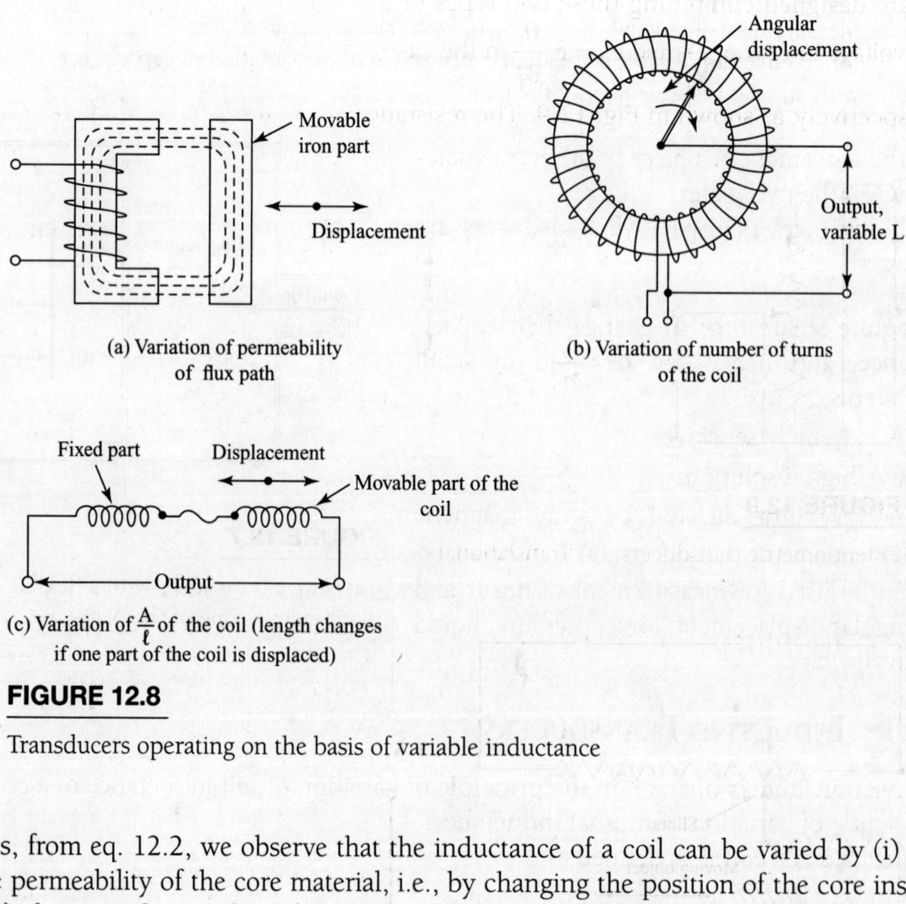

FIGURE 12.8

Transducers operating on the basis of variable inductance

Thus, from eq. 12.2, we observe that the inductance of a coil can be varied by (i) changing the permeability of the core material, i.e., by changing the position of the core inside the coil; (ii) changing the number of turns through which current will flow in the coil, and (iii) changing the ratio of A/ℓ of the coil. These methods are shown in Fig. 12.8. These are passive transducers requiring power supply and the output is analog in nature.

Two coils having inductance L_1 and L_2 will have mutual inductance M as

$$M = K\sqrt{L_1 L_2}\text{, where K is the coefficient of coupling} \tag{12.3}$$

The variation of magnetic coupling between two coils can be utilized to change M. Displacement of the core or a part of the core can cause change in M. The change in M will be measured by a bridge circuit and a proportional signal created for the measurement of displacement.

12.8 ▶ Potentiometric Transducer

A potentiometric transducer is basically an electrical resistive transducer. A potentiometer is an electromechanical device having a resistance element with a sliding facility which enables changes in the resistance value at the output. The sliding contact is known as wiper, whose movement may be translatory or rotary, according to the design. Fig. 12.9 shows potentiometric transducers of translational and rotational design. Some potentiometers, also called 'pots', are designed combining these two types of motions, which is helical in nature. The output voltage is $e_O = e_i \dfrac{x_O}{x_T}$ and $e_O = e_i \dfrac{\theta_i}{\theta_T}$ in the case of translational and rotational potentiometer, respectively, as shown in Fig. 12.9. The resistance element of a potentiometer is excited

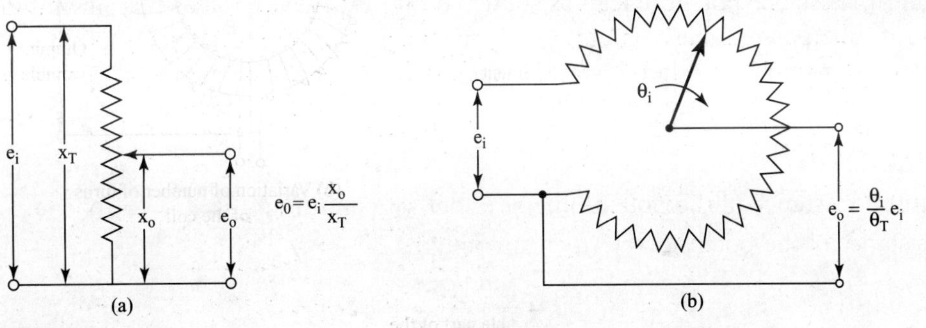

FIGURE 12.9

Potentiometric transducers. (a) Translational design; (b) rotational design

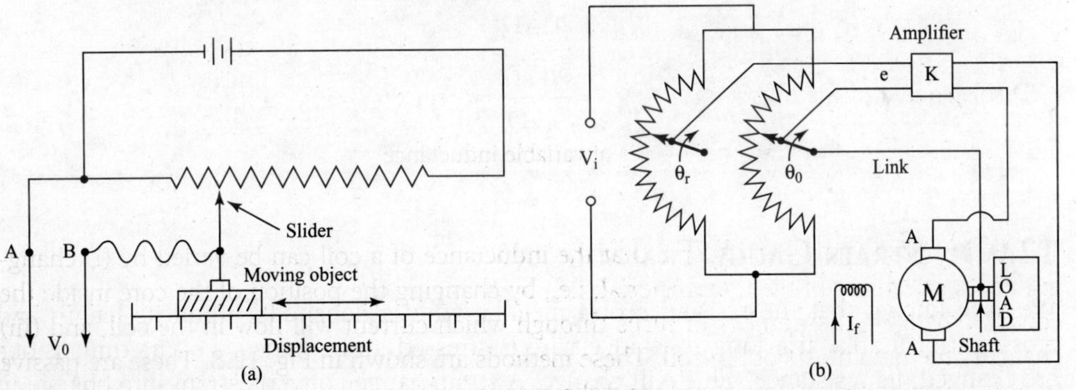

FIGURE 12.10

Applications of potentiometric transducers. (a) Measurement of displacement of a moving object; (b) Error detection in a feedback control system

either by ac or dc voltage. The motion of the wiper or slider makes a resistance change that may be linear, logarithmic, or exponential.

In Fig. 12.10 are shown applications of potentiometric transducers in the measurement of linear and angular displacement. The output voltage, V_0 appearing across the terminals A and B is directly proportional to the displacement of the moving object as has been shown in Fig. 12.10 (a). A typical potentiometric transducer used in a feedback control system as error sensor has been shown in Fig. 12.10 (b). Any derivation of the output, i.e., θ_0 from the input reference angular displacement, θ_r will produce an error signal, e which will be amplified by the amplifier having gain K. The amplifier will supply additional input to the armature terminals AA of the motor, M to produce additional torque to change θ_0, and hence reduce the error.

Potentiometric transducers are widely used in control applications because they provide sufficient output for control operation, in many cases not requiring any amplifier.

EXAMPLE 12.1

A linear resistance potentiometer as shown in Fig. 12.10 (a) has a displacement range of 50 cm. The potentiometer is uniformely wound having a resistance of 200 kΩ. Under normal condition the slider is set at the centre of the potentiometer. Calculate the displacement of the object when the output resistance is measured as 80 kΩ.

SOLUTION

Output resistance of the potentiometer under set condition $= \frac{200\text{ k}\Omega}{2} = 100\text{ k}\Omega$

Resistance of the potentiometer per unit length

$$= \frac{200\text{ k}\Omega}{50\text{ cm}} = 4\text{ k}\Omega/\text{cm}$$

Change in resistance due to displacement

$$= 100\text{ k}\Omega - 80\text{ k}\Omega$$
$$= 20\text{ k}\Omega$$

Displacement of the object $= \frac{20\text{ k}\Omega}{4\text{ k}\Omega} = 5\text{ cm}$

12.9 ► Strain Gauge Transducer

We have known that the resistance of a metallic wire is dependent upon its length and cross-sectional area. If a long resistance wire is stressed and strained, i.e., its dimensions are changed, its resistance value will change. A strain gauge converts strain into change in resistance of the wire under strain.

The basic principle of a strain gauge is that when a resistance wire is stressed within its elastic limit, its dimensions change, and hence, its resistance change. This change in resistance can be used as a measure of the stress and strain of the object on which the strain gauge is attached or embedded. In civil and mechanical engineering fields, strain gauges are used extensively to measure stress on structures.

A strain gauge transducer is made of thin wires of diameter varying from 0.02 to 0.04 mm cemented (permanently fixed) in a zig-zag pattern on a thin flat paper. The zig-zag pattern of winding reduces the inductance and capacitance of the wire to a very low value, which is desirable. The end connecting leads are twisted to minimize any inductive effect.

When a strain gauge transducer is attached to the surface of a structure by means of an elastic cement, any change in the surface dimension of the structure due to stress formation will be reflected as some change in the resistance value of the strain gauge. The sensitivity of a strain gauge is expressed as **gauge factor**. Gauge factor is the change in resistance per unit change in length of the strain wire. To make the transducer more sensitive, the material for the wire should be so chosen that the gauge factor is high, i.e., there is considerable change in resistance due to strain. Flexible silicon strain gauge has a gauge factor much higher than a metallic gauge.

Two types of strain gauges are in use. They are bonded type and semiconductor type. Bonded strain gauges are available in different shapes. These are bonded with elastic cement to the surface whose stress is to be measured. The bonded strain gauge consists of a wire grid whose shape may be square, rectangular, or circular. The wire grid is on a base paper as shown in Fig. 12.11 (a) and (b)

A semiconductor strain gauge is made from a single piece of semiconductor material. There is change in resistivity when the material is strained. Many adhesives have been developed for pasting strain gauges to specimen structures and surfaces. A semiconductor strain gauge has been shown in Fig. 12.11 (c).

Strain gauges are extensively used for the measurement of strain and associated stress in experimental stress analysis. Strain gauges are pasted on the surface of the structures or bodies

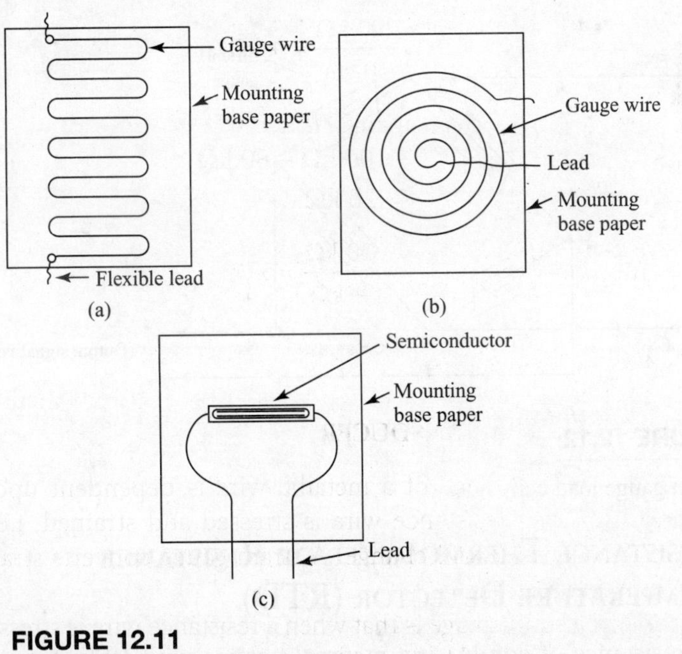

FIGURE 12.11

Strain gauges of different types and shapes

to sense strain under applied load. An electrical output signal proportional to the strain can be obtained from the transducer when the strain gauge forms one of the arms of a bridge circuit and an input voltage is provided. Measurement of stress occurring under varied environmental conditions can be measured using strain gauge. For example, the strain in an aircraft body structure can be measured by using the electrical signal generated by the strain gauges and by transmitting the signal to ground instruments.

12.10 ► Load Cell

A strain gauge is used to measure strain (deformation) on a structure due to loading effect. The electrical resistance of the strain gauge wire changes due to strain of the structure on which the strain gauge is mounted (fixed). Strain gauges are mounted on places in the structure where there is tension or compression. *A load cell usually consists of four strain gauges forming four arms of a wheatstone bridge. When load is applied on to the structure, the resistances of the strain gauges change in proportion to the load.*

The bridge comprising four strain gauges in a wheatstone bridge configuration is excited with stabilized d.c supply of 10 V or so.

When the load is applied, a voltage proportional to the load appears as the output of the bridge as has been shown in Fig. 12.12.

Load cells are also available as other types like Piezoelectric load cell, Hydraulic load cell, Pneumatic load cell, Vibrating wire load cell, etc.

Load cells are used in several types of measuring devices, like laboratory balances, Weighing scales and weighing platforms, in civil engineering structural load measurement and analysis.

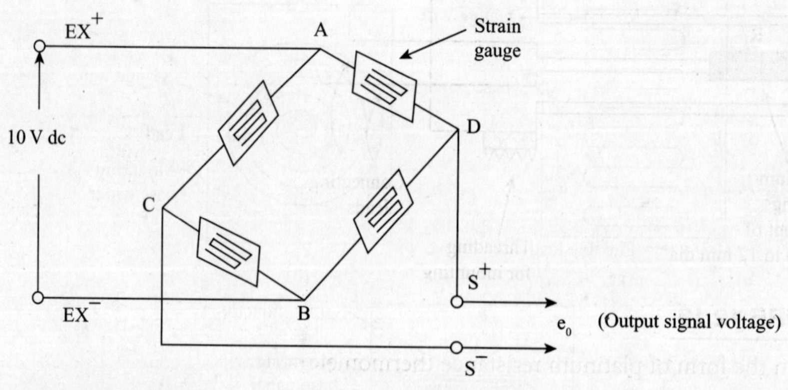

FIGURE 12.12

Strain gauge load cell

12.11 ► Resistance Thermometer or Resistance Temperature Detector (RTD)

We known that resistance of conducting material change with temperature. This property can be utilized to measure unknown temperature. We measure the amount of change in resistance due to change in temperature and equate this change in resistance with change in temperature proportionally.

Materials like platinum, nickle and nickel alloys have high-temperature coefficient of resistance. Their change in resistance with change in temperature is large. These materials therefore, are suitable for making RTDs.

RTD is a primary transducer. It is used as a temperature detector. Its resistance changes with temperature. This change in resistance is converted into a proportional electrical signal by using a secondary transducer.

The relationship between temperature and resistance of RTD is given as

$$R_t = R_o\,(1 + \alpha t)$$

where R_t is the resistance of RTD at t°c

R_o is the resistance of RTD at 0°c

α is the temperature coefficient of resistance of the RTD

t is the rise in temperature.

A high value of α is desirable so that there is considerable change in resistance with temperature.

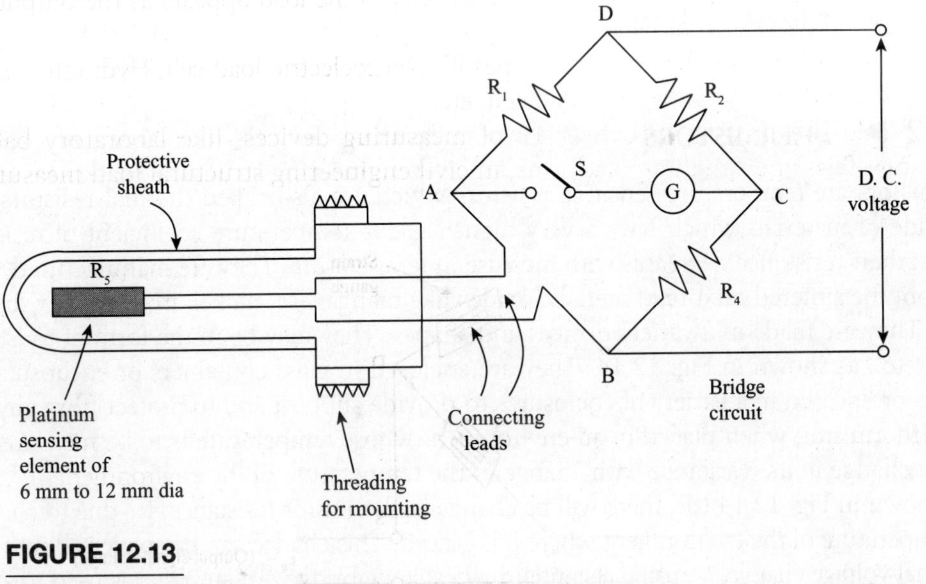

FIGURE 12.13

RTD in the form of platinum resistance thermometer connected to a bridge circuit for temperature measurement

Measurement with a RTD

Figure 12.13 shows a platinum resistance thermometer connected to a bridge circuit. The temperature range for which platinum can be used for temperature measurement is from 250°c to over 1000°c for various industrial applications. As shown in figure R_s is the value of RTD at normal temperature.

The value of R_5 of the RTD will increase due to with increase in temperature of the body in which it is fixed. By properly adjusting the values of R_1, R_2, R_4, the bridge can be balanced for

$$\frac{R_1}{R_2} = \frac{R_5}{R_4}$$

or,

$$R_5 = \frac{R_1}{R_2} \times R_4$$

When the bridge is balanced, the galvanometer will give zero deflection. When the temperature increases, the galvanometer will give some deflection because of the voltage appearing across the bridge terminals A and C. The more temperature rise will create more deflection of the galvanometer.

The deflection of the galvanometer can be calibrated in terms of temperature rise of the body or substance in which the RTD has been fixed. This way RTD can be used to measure temperature of a hot body from a distance.

Measurement of temperature using RTD is quite accurate. RTD can be used to measure temperature over wide range. RTDs can be used for temperature measurement by remaining away from the hot body or hot substance. RTD can be installed easily and measurement can be taken within few seconds only. Their performance is stable over a long period of time. The drawback is their high cost and the requirement of an associated bridge circuit.

12.12 ▶ Thermistors

Thermistors are temperature-sensitive resistors which are also called thermal resistors. They are made of materials which have a very high-negative temperature coefficient of resistance. That is, their resistance decreases with increase in temperature. They are manufactured, generally from the sintered mixure of metallic oxides like manganese, nickel, iron, copper, and uranium. They are made in a variety of sizes and shapes. They may be in the form of rods, discs, beads, etc., as shown in Fig. 12.14. They are enclosed in glass containers or encapsulated in plastics or encased in a variety of enclosures to provide support and to protect from any damage. A thermistor, when placed in an environment whose temperature is to be measured, will show a change in its resistance with change in the temperature of the environment.

A shown in Fig. 12.14 (d), there will be change in thermistor resistance, R_{th} due to change in the temperature of the environment where it is placed. This change can be converted into a proportional voltage change, V_0 using a standard wheatstone bridge. An amplifier is used to amplify the signal generated. Usually the wheatstone bridge is balanced with the thermistor as one of its arms. The wheatstone bridge becomes unbalance a when resistance of the thermistor changes due to variation of temperature being measured.

The voltage generated can be used to make digital output or analog deflection depending upon the arrangement. If calibrated, the output will be a measure of the variation in temperature.

The sensitivity of a thermistor is expressed in terms of variation of resistance per degree change in temperature. High sensitivity of thermistors together with their initial high value of resistance make them suitable for measurement of temperature accurately. They are, therefore,

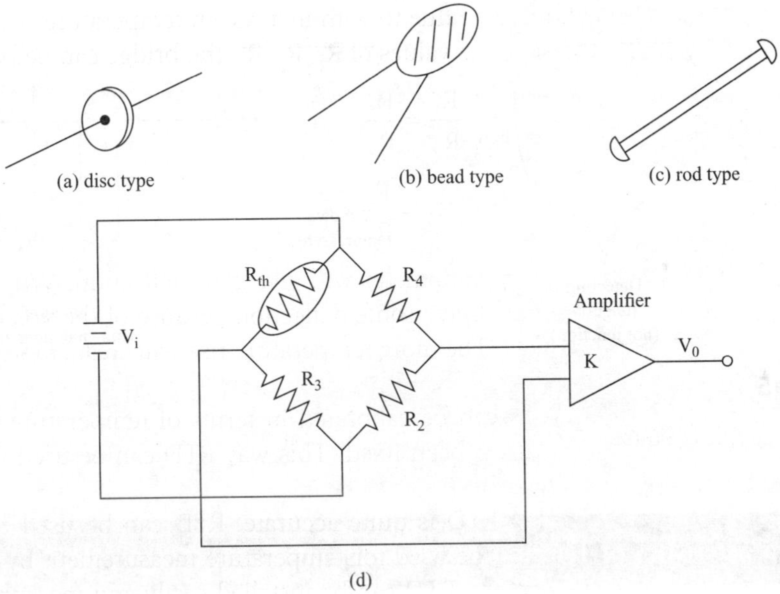

FIGURE 12.14

(a), (b), (c) show three types of thermistors; (d) shows use of a thermistor in a wheatstone bridge for measurement of temperature

suitable for precision temperature measurement and its control. Thermistors can be used to measure minute change of temperature variation, as low as 0.005°C.

Since the resistance versus temperature characteristic of thermistors are non-linear (resistance decreases non-linearly with temperature rise), they can be used for a limited range of temperature measurement. They are passive transducers and require external dc power supply.

Thermistor-based measurement system using digital readouts which read the temperature directly are widely available.

12.13 ► THERMOCOUPLES

Thermocouples are active transducers requiring no power supply. A thermocouple is a junction of two dissimilar materials used for measurement of temperature. It consists of a pair of dissimilar conducting wires joined at two junctions. One junction is maintained at a reference temperature while the other junction is placed at the unknown temperature as shown in Fig. 12.15. The temperature difference between the two junctions produces a thermal EMF which is measured by a milli voltmeter or a digital voltmeter. The use of this thermo EMF as a measure of temperature is known as thermocouple thermometry. The magnitude of induced EMF is generally small and depends on the material used as wires and the temperature difference between the junctions. Sensitivity of a thermocouple transducer is expressed as EMF induced in mV per degree Kelvin. Table 12.1 shows the various combinations of thermocouple materials with their sensitivity and temperature range.

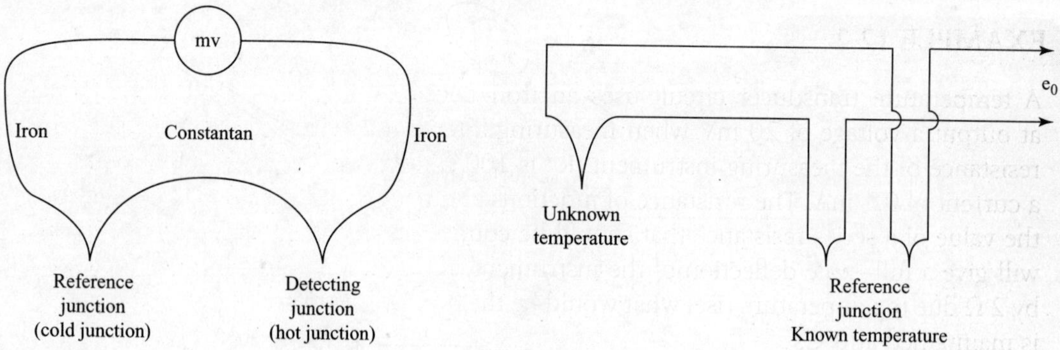

FIGURE 12.15

Simple thermocouple circuits

TABLE 12–1 Characteristics of Thermocouple Materials

Thermocouple material	Sensitivity (EMF in mV per °K)	Temperature range in °K
Copper–constantan	0.05	3–673
Iron–constantan	0.05	63–1473
Chromel–aclumel	0.04	3–1643
Chromel–constantan	0.08	3–1273
Platinum–platinum rhodium	0.01	223–2033

For measurement of temperature at remote places, extension wires are to be used. The connecting wires from the thermocouple lead to the place of measurement are long and are usually not at the same temperature throughout their length. This causes error in the measurement. To avoid such error, the connecting wires are made of the same material as the thermocouple wires. Thermocouple junctions are made by welding or soldering without using any flux. Industrial thermocouples generally have the hot junction placed at the sensing place and the cold junction or the reference junction in the measuring instrument itself. The reference junction is maintained at room temperature. Extension wires are used for connections. For achieving high sensitivity, a number of thermocouples may be connected in series so that the total EMF is the sum of EMF induced in each of the thermocouples. Such arrangement is known as *thermopile*.

Thermocouples are very cheap and handy devices used for temperature measurement in remote and inaccessible places. They are also conveniently used in measuring temperature at one particular point in a piece of equipment. For example, if we want to measure the temperature of the windings of an electrical machine, we can embed a thermocouple there and bring out the connecting leads for connection to the measuring instrument placed in the control room. The accuracy of measurement of temperature using thermocouples is, of course, not very high because of which they are not suitable for precision measurement.

EXAMPLE 12.2

A temperature transducer circuit uses an iron–constantan thermocouple which gives at output a voltage of 20 mV when measuring a temperature difference of 750°C. The resistance of the measuring instrument, R_m is 100 Ω and gives a full-scale deflection for a current of 0.1 mA. The resistance of junctions and connecting wires is 10 Ω, Calculate the value of a series resistance that should be connected so that a temperature of 750°C will give a full–scale deflection of the instrument. If the junction resistance is increased by 2 Ω due to temperature rise, what would be the measurement error? The cold junction is maintained at 0°C.

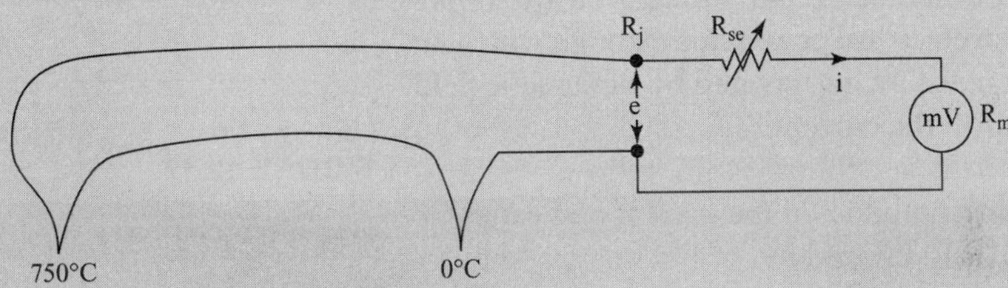

FIGURE 12.16

SOLUTION

Current, i for full-scale deflection $A = 0.1 \times 10^{-3}$

$$i = \frac{e}{R_m + R_j + R_{se}} = \frac{20 \times 10^{-3}}{100 + 10 + R_{se}}$$

or,
$$0.1 \times 10^{-3} = \frac{20 \times 10^{-3}}{110 + R_{se}}$$

or,
$$R_{se} = 90\ \Omega$$

If resistance of the connecting wire junctions rise by 2 Ω, the current flowing will be

$$i = \frac{20 \times 10^{-3}}{100 + 10 + 2 + R_{se}} = \frac{20 \times 10^{-3}}{112 + 90}$$

$$= 0.099 \times 10^{-3} \text{A}$$

$$= 0.099 \text{ mA}$$

When current flowing is 0.1 mA, the temperature read is 750°C.
When current is 0.099 mA, the temperature reading will be

$$= \frac{750 \times 0.099}{0.1} = 742.5°\text{C}$$

Therefore, the error will be equal to 750°C – 742.5°C = 7.5°C

The error is negative as the reading will be less than the previous reading.

12.14 ▶ Hall Effect Transducers

The principle of working of a Hall effect transducer is that when a strip of conducting material carries current in the presence of the transverse magnetic field, as shown in Fig. 12.17, an EMF will be induced between the opposite edges of the conducting strip. The magnitude of the voltage induced will depend upon the material of the strip, the current, and the magnetic field strength. Thus, we can state that **Hall effect refers to the potential difference (Hall voltage) on the opposite sides of an electrical conductor through which an electric current is flowing, created by a magnetic field perpendicular to the current.**

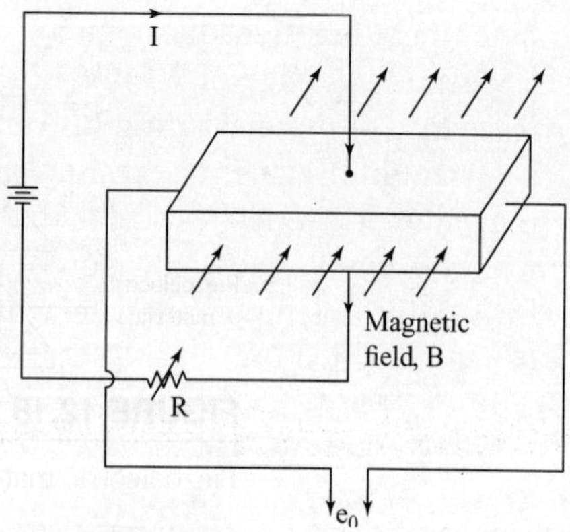

FIGURE 12.17

Hall effect transducer

Hall coefficient is defined as the ratio of the output voltage to the product of the current and magnetic field, i.e., $e_o/(I \times B)$ divided by the thickness of the element, t. Thus,

$$\text{Hall coefficient} = \frac{e_o}{I \times B \times t}$$

Hall coefficient is the characteristic of the material, i.e., it depends on the material from which the conductor is made. To understand how a potential gets developed due to Hall effect, let us consider the following explanation.

When current flows through the conductor, it consists of the movement of charge carriers. In the presence of a perpendicular magnetic field, the moving charges experience a force, called the Lorenz force. This makes the path of the moving charge somewhat curved (and not a straight line flow) so that charges accumulate on one face of the conductor material. Equal and opposite charges start appearing on the other face of the conductor material where there is a shortage of mobile charges. This results in unequal and asymmetric distribution of charges across the conducting element both perpendicular to the direction of a straight line flow of charge (in the absence of a magnetic field) and in the direction of the applied magnetic field. This separation of charge establish an electric field that would oppose the migration of further charge, and therefore an electrical potential gets build up.

Hall effect transducers are non-contact devices, have small size, and high resolution. Such transducers can be used in the measurement of velocity, revolutions per second, magnetic field, charge carrier density, measurement of displacement, etc. Hall effect transducers can be used to measure current in a conductor without actually connecting a meter in the conducting circuit.

12.15 ▶ Piezoelectric Transducer

In certain materials called piezoelectric materials a potential difference appears across their opposite faces as a result of dimensional changes due to application of pressure created by mechanical force. This potential is produced as a result of displacement of charges in

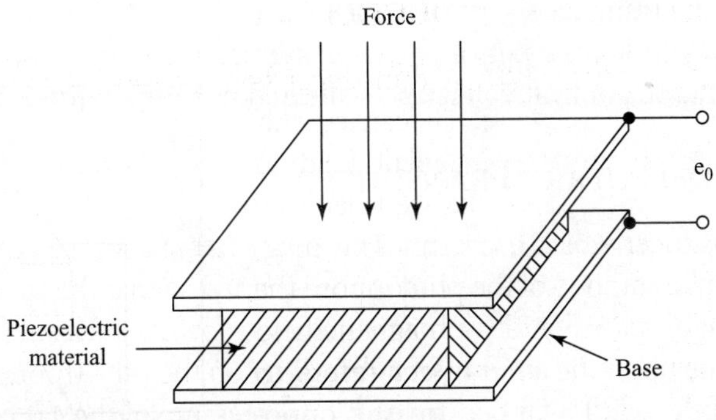

FIGURE 12.18

Piezoelectric material is used to measure force

the body of the material. The effect is reversible, i.e., the reverse happens when a varying potential is applied to the proper axis of the crystal; a change in the dimensions of the crystal occurs. This effect is known as the piezoelectric effect. Commonly used piezoelectric materials are quartz, barium, titanite, lithium sulphate, etc. Fig. 12.18 shows a piezoelectric material used for the measurement of force. Since these transducers are self-generating, i.e., active transducers, the EMF induced due to force applied is directly proportional to the force. As shown in Fig. 12.18, an external force F, which is to be measured, exerts a pressure on the top of the crystal and as a result EMF, e_0 is produced across the crystal.

A piezoelectric material should not be sensitive to temperature and humidity variations. They should lend themselves to forming different shapes. Quatz is the most suitable piezoelectric material on this account. However, the voltage induced is quite small. Rochelle salt provides higher values of induced EMF, but it is affected by temperature variations.

Fig. 12.19 shows a pressure transducer which utilizes the property of piezoelectric crystals. The transducer consists of a diaphragm by which pressure is transmitted to the piezoelectric crystal. The crystal generates an EMF across its two surfaces which is proportional to the magnitude of the applied pressure.

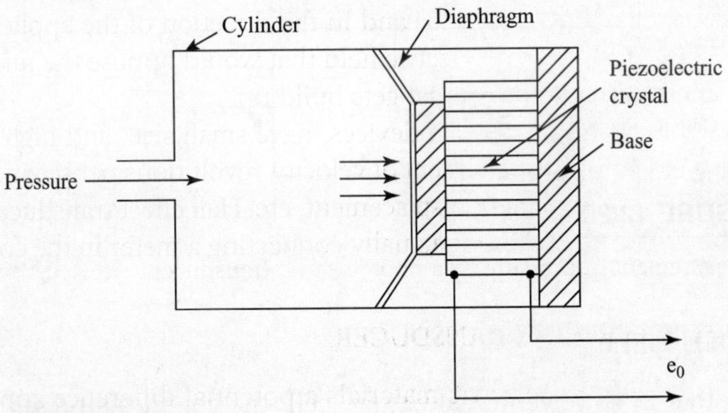

FIGURE 12.19

Piezoelectric pressure transducer

Piezoelectric pressure transducers are used to measure high pressure that changes rapidly like the pressure inside a cylinder of a petrol or diesel engine, or a compressor. The main drawback of this transducer is that the output voltage gets affected by temperature variations of the crystal.

12.16 ▶ Photoelectric Transducer

This is an optical transducer which uses a photo tube and a light source. The amount of light falling on the photosensitive cathode of the photo tube is varied so that the anode current is changed. Fig. 12.20 shows a light source and a photo tube used for the measurement of pressure. The output voltage depends upon the amount of light falling on the tube through the apperture (window). The opening of the window is controlled by the pressure of the gas falling on a membrane.

The output voltage approximately varies linearly with the displacement of the aperture, and hence the pressure.

Photo cells, or solar cells are another type of opto-electric transducer in which light intensity controls the value of electrical potential as output. There are three general types of photo cells, viz photo-emissive, photo-conductive, and photovoltaic type.

A photovoltaic transducer may be considered as a voltage source whose value depends upon the amount of light striking its surface. The materials used are selenium, germanium, and silicon. Photovoltaic transducers are also called solar cells. A large number of solar cells interconnected together form a *solar battery*. Photovoltaic cells, in addition to their application as transducers, are used for illumination in remote areas as a non-conventional source of energy.

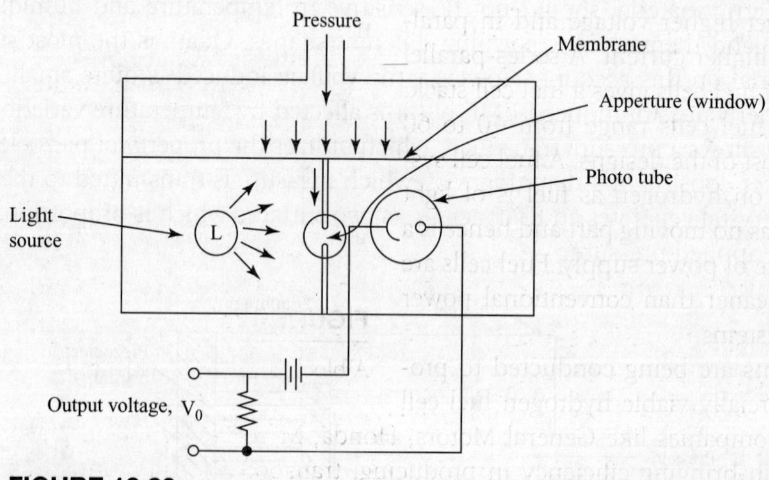

FIGURE 12.20

Measurement of pressure by a photoelectric transducer

12.17 ▶ Fuel Cell

We have known that in a conventional battery cell, the chemicals present in the cell react with each other to generate an electromotive force (emf). These cells do not require any external input to sustain chemical reaction inside.

A fuel cell requires continuous supply of inputs hydrogen ions and an oxidizing agent are required to produce electricity on a continuous basis. Fuel cells are used to generate power for satellites, space capsules, and as backup power for commercial, industrial and residential buildings. They are also used to source of power in buses, boats, motor cycles, submarines etc.

The fuel cell construction is shown in a block diagram from as in Fig. 12.21. As shown at the anode a catalyst oxidizes the hydrogen fuel turning it to positively changed ions and a negatively charged electron. The electrolyte placed in between the anode and cathode is made of such material that ions can pass through it but electrons cannot. So, electrons travel through a conducting wire making connections to the electrical load and hence, current flows through the load. The ions travel to the cathode and get reunited with the electrons entering from the load to the cathode as has been shown. Hydrogen ions reunite with electrons and the combination react with a third chemical, namely O_2 (oxygen) to form water.

- The anode is usually made of fine platinum powder.
- The cathode is usually made of nickel.
- The electrolyte substance defines the type of fuel cell. Acacias alkaline solution, polymer membrane, molten phosphoric acid, molten alkaline carbonate, salt water, etc., are being used.
- The fuel used is hydrogen.

A typical fuel cell produce a voltage of 0.6 V to 0.7 V. Fuel cells can be connected in series to get higher voltage and in parallel to supply higher current. A series-parallel connection of fuel cells gives a fuel cell stack. Efficiency of fuel cells range from 40 to 60 percent in most of the designs. A fuel cell system running on hydrogen as fuel is of light weight and has no moving part and hence is a reliable source of power supply. Fuel cells are also much cleaner than conventional power generation systems.

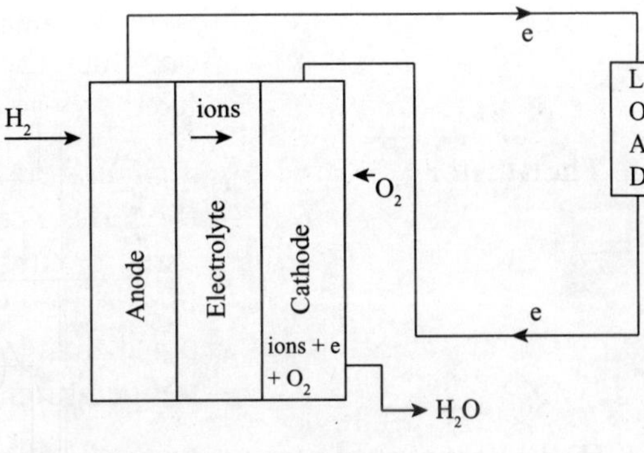

FIGURE 12.21

A block diagram of a fuel cell

Experiments are being conducted to produce commercially viable hydrogen fuel cell vehicles by companies like General Motors, Honda, Mercedes-Benz and Nissan. One of the challenge is in bringing efficiency in producing, transporting and storing of highly inflammable hydrogen gas.

12.18 ► Selection of Transducers

The commonly used transducers, their basic principle of operation, and typical applications are shown in a tabular form in Table 12.2. Passive transducers, i.e., those requiring external power supply and active transducers, which are self-generating have been shown separately in the table.

The factors which decide the selection of a particular transducer for an application are the following:

(i) Sensitivity; (ii) accuracy; (iii) operating range; (iv) ruggedness; (v) environmental effects; (vi) stability and reliability; (vii) linearity, repeatability, and high resolution; (viii) size and shape; and (ix) cost and availability.

It is important that we select the right type of transducer for any application in the area of measurement, instrumentation, and control.

TABLE 12–2 Various Types of Transducers and their Applications

(A) Passive transducers	Conversion principle	Typical applications
1. Potentiometric	Position of the slider is changed by the force or pressure. Resistance, and hence, output voltage gets changed.	Measurement of force, pressure, displacement
2. Strain gauge	Resistance of a wire or a semiconductor gets changed due to stress developed on the surface on which the gauge is pasted/fixed.	Stress, pressure, force, torque, displacement
3. Thermistor	Temperature rise changes the resistance. Resistance of certain materials decrease with increase in temperature. Resistance change is a measure of temperature.	Temperature, thermal conductivity
4. Hall effect transducer or pick-up	Potential difference is attained across a material carrying current in a transverse magnetic field.	Magnetic flux, current, velocity
5. Linear variable differential transformer (LVDT)	When the position of the magnetic core is changed, the differential voltage of two secondary windings of a transformer is changed. Output voltage is proportional to the position of the core, i.e., its displacement.	Measurement of displacement, force, pressure, position

(continued)

TABLE 12–2 Various Types of Transducers and their Applications *(Continued)*

(A) Passive transducers	Conversion principle	Typical applications
6. Capacitive transducer	Variation of distance between the two plates, area of overlap, changing the position of the dielectric material between the plates, changes the capacitance. Change in capacitance is converted into an electrical signal by using a bridge circuit.	Displacement, force, pressure,
7. Inductive transducer	By moving the core material inside the coil, its self-inductance is changed. Mutual inductance between two coils is changed by varying the magnetic coupling of the coils.	Pressure, displacement
(B) Active transducers requiring no power supply	**Conversion principle**	**Typical applications**
1. Thermocouple transducer	EMF is induced when junctions of certain dissimilar materials are heated at different temperatures. One junction is taken as reference junction of known temperature.	Temperature, heat flow
2. *Piezoelectric transducer*	*When force is applied on certain material such as quatz, EMF is induced between the two sides.*	*Pressure, force, vibrations*
3. Photovoltaic transducer	Solar energy or light energy causes voltage to be generated in certain semi-conducting materials.	Lumen, pressure, solar cell

Important Terminologies and Points to Remember

1. Transducer is a device which converts any non-electrical quantity like pressure, temperature, strain, etc., into a proportional electrical signal (voltage).
2. Active transducers do not require any external source of supply for their operation. They are self generating transducers. Examples are piezoelectric type, photovoltaic type, thermoelectric type transducers (thermo couples). Passive transducers require separate power supply for their operation. Linear variable differential transformer (LVDT), used for measurement of displacement, is a passive transducer as it requires a source of ac supply.
3. Transducers can be made utilizing the basic principle of capacitance of parallel plate capacitors where displacement of movable plate caused by actuating force produces a change in capacitance. Similarly, inductive transducers can be made utilizing the variables that determine the value of an inductive coil.
4. A strain gauge transducer which converts strain on a structure into change in resistance of a wire fixed on the body under strain. Sensitivity of a strain gauge is expressed as gauge factor which is equal to the change resistance per unit change in length of the strain gauge wire.
5. Thermistors are temperature sensitive resistors; their resistance change with change in temperature. Thermistors can be used as transducers where change in resistance can be utilized to measure change in temperature.
6. Thermistors are used as one arm of a four arm wheatstone bridge which requires a dc supply as external source of supply. A change in resistance of the thermistor will create an unbalanced voltage output which can be amplified and used for the measurement of unknown temperature which was causing the change in resistance of the thermistor.
7. Thermocouples are used as transducers for the measurement of temperature. Thermocouples are made using a junction of two dissimilar materials like copper-constantan, Iron-constantan, etc. They are active transducers requiring no external power supply. A temperature difference between two junctions produces an emf which is measured. The emf produced can be a measure of the unknown temperature of one of the junctions.
8. Piezoelectric material like a Quartz crystal produces a voltage across their opposite faces when a pressure is applied. This property is utilized to measure pressure by piezoelectric transducers.
9. Photoelectric transducers are optical transducers. A photo cell or a solar cell is a photoelectric transducer where intensity of light controls the voltage produced. Photovoltaic cells, in addition to be used as transducers, are used for illumination where a large number of solar cells are interconnected togather to form a solar battery.

Important Formulae and Equations

Equation	Remarks
1. $C = \in \dfrac{A}{d}$ Where $\in$ is the permittivity of the dielectric material placed in between the two plates of the capacitor, A is the area of the plates and d is the distance between the plates.	1. Capacitive displacement transducers can be made by changing only of the parameters like $\in$, A, and d.
2. $L = \dfrac{\mu N^2 A}{\ell}$ Where μ is the permeability of the core material, N is the number of turns of the inductor coil, A is the area of cross-section of the flux path and ℓ is the length of the flux path.	2. Inductive transducers can be made on the principle of variation of inductance of a coil due to variation of μ, N, A, and ℓ.

REVIEW QUESTIONS

Short Answer Type Questions

1. Define a transducer and give a classification of transducers on the basis of various factors.
2. Name some important characteristics of transducers and explain their significance.
3. Distinguish between a passive transducer and an active transducer giving one example in each case.
4. Explain the working of a LVDT and mention its applications.
5. Explain the working principle of a piezoelectric transducer and show how this can be used for pressure measurement.
6. Explain the basic principle of Hall effect and how this effect can be used to make a transducer.
7. Explain the principle of working of a strain gauge and mention its applications.
8. Explain how an unknown temperature can be measured using a thermo couple transducer.
9. What factors must be considered before selecting a transducer for an instrumentation system?
10. Write short notes on the following:

    (a) photoelectric transducer; (b) thermistor; (c) LVDT; (c) pressure transducer.
11. What kind of transducers would you select for the measurement of the following:

    (i) vibrations; (ii) pressure; (iii) displacement; (iv) temperature; (v) liquid level.
12. Prepare a table mentioning the principle of operation and typical applications of the following transducers:

    (a) LVDT; (b) strain gauge; (c) potentiometer; (d) thermistor; (e) piezoelectric crystal.
13. Explain how displacement can be measured using an inductive transducer.
14. State the difference between a sensor and a transducer. Distinguish between an analog transducer and a digital transducer.
15. Show how a potentiometric transducer can be used in a measurement system. Mention advantages and disadvantages.
16. Explain different types of transducers for the measurement of displacement.

Multiple Choice Questions

1. A transducers is a device which
   (a) transfers a signal from one circuit to the other
   (b) converts a physical quantity to be measured into an equivalent electrical signal
   (c) amplifies a signal for the purpose of measurement
   (d) converts an ac signal into a dc signal.
2. Transducers requiring external power supply for their operation are called
   (a) passive transducers
   (b) active transducers
   (c) separately excited transducers
   (d) self-excited transducers.
3. Which of the following is an active transducer?
   (a) Thermistor
   (b) LVDT
   (c) Photo transistor
   (d) Thermocouple.
4. Sensitivity of a Transducers is
   (a) the quality of output produced by the transducers
   (b) the variation of output produced under any disturbed condition
   (c) the output produced per unit change in the input quantity being measured
   (d) the correctness of the output produced as a proportion to the input variations.
5. A thermistor is a
   (a) temperature-dependent resistor mode of connecting material having negative temperature coefficient of resistance

(b) temperature-dependent resister mode of semiconducting of resistance
(c) temperature-dependent resistor mode of conducting material having positive temperature coefficient of resistance
(d) temperature-dependent resistor mode of semiconducting material having positive temperature coefficient of resistance.

6. Linear variable differential transformer is a
(a) temperature-sensitive transducer
(b) pressure transducer
(c) displacement transducer
(d) vibration measuring transducer.

7. In an electronic weighing machine we can use
(a) LVDT-type transducer
(b) thermistor transducers
(c) thermocouple-type transducer
(d) none of these.

8. Which of the following is not a pressure measurement transducer?
(a) Piezoelectric transducers
(b) Strain gange
(c) LVDT
(d) Thermocouple.

Answers to Multiple Choice Questions

1. (b)	2. (a)	3. (d)	4. (c)	5. (b)	6. (c)
7. (a)	8. (d)				

CHAPTER

13 Power Systems

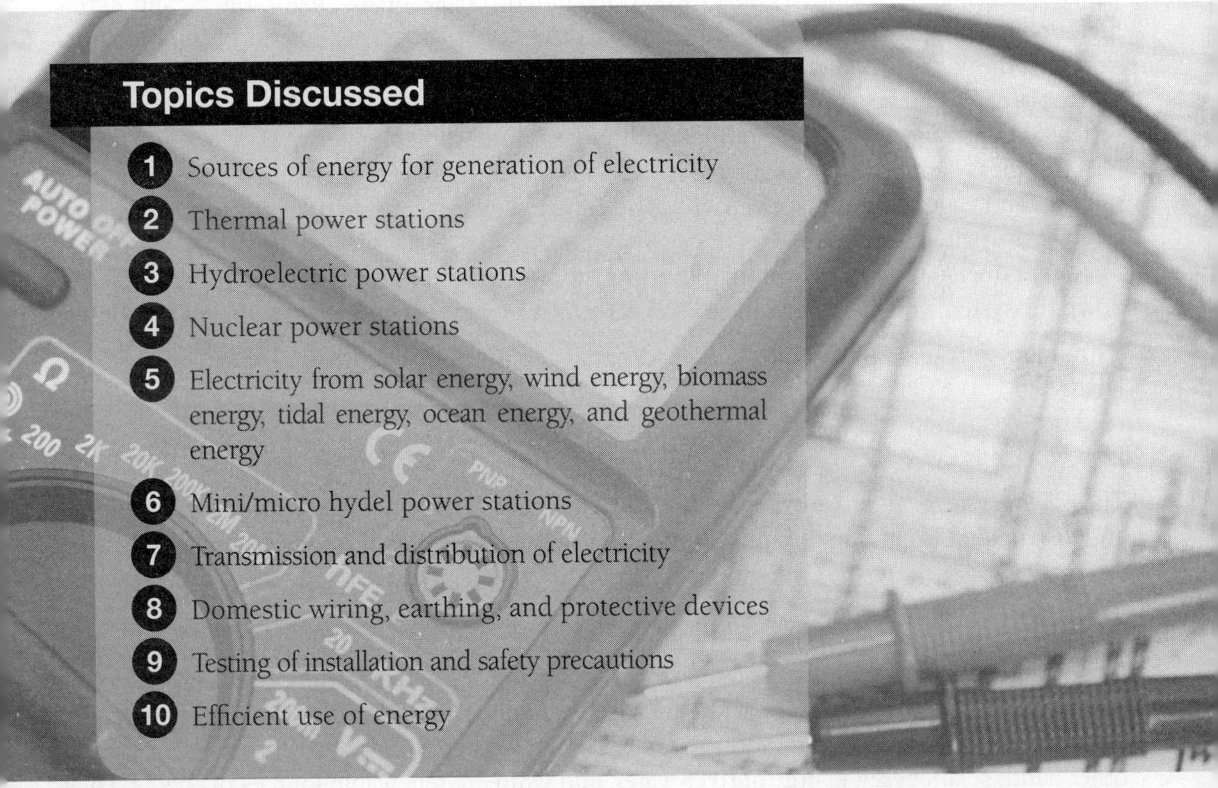

Topics Discussed

1. Sources of energy for generation of electricity
2. Thermal power stations
3. Hydroelectric power stations
4. Nuclear power stations
5. Electricity from solar energy, wind energy, biomass energy, tidal energy, ocean energy, and geothermal energy
6. Mini/micro hydel power stations
7. Transmission and distribution of electricity
8. Domestic wiring, earthing, and protective devices
9. Testing of installation and safety precautions
10. Efficient use of energy

13.1 ► INTRODUCTION

We have known that energy is available in nature in different forms like in fossil fuel such as coal, gas, and oil; high-speed wind; falling water; sun rays, etc. All these forms of energy can be converted into electrical energy which can be transmitted to various places for use. Electrical energy is clean and easily controllable.

Electricity in large quantity is produced in power houses through a turbine–generator set. The turbines are rotated by pressurized steam or the potential energy of water. The generator is rotated by the turbine. The generator produces electricity which is sent for use to various places through transmission lines and distribution networks. An electrical power system consists of generation, transmission, distribution, and utilization of electricity.

The first electric supply system was set up in 1882 in New York City, USA. The system involved electricity generation through a dc generator driven by a steam engine. The electricity generated was distributed for local use by underground cables for lighting purpose only. Gradually, power systems started growing in size. Large size ac generators were installed in power houses to generate electricity in the Mega Watt range and at higher voltage. By using transformers, the generated electricity was stepped up to high voltages such as 132 kV, 220 kV, 400 kV and so on. The magnitude of stepping up of voltage depended upon the distance to which the generated electricity had to be transmitted for use. Thus, high-voltage and extra-high-voltage transmission systems came into existence. At the receiving end of the transmission line, the voltage level is brought down for distribution to different distribution substations where the voltage level is further brought down for supplying to consumers. The distribution system is constituted by overhead lines and underground cables.

A number of control devices are used at various stages of the power system for safe, efficient, reliable, and economic use of electricity. The control devices include different types of relays, switchgear, switch, fuse, etc.

Transformers are used at various places in the power system for stepping-up or stepping-down the system voltage. The voltage level of the power system has been designated as follows.

Extra high voltage, i.e., EHV: above 220 kV and upto 800 kV.
High voltage, i.e., HV: above 66 kV and upto 220 kV.
Medium voltage, i.e., MV: above 1 kV but less than 66 kV.
Low voltage is LV: 1 kV or less than 1 kV.

The power system can broadly be divided into three sub-systems, i.e., ***generation***, ***transmission***, and ***distribution*** excluding the utilization part. These are discussed in some detail as follows.

13.2 ► Generation of Electricity

Electricity for commercial use is generated in power houses by converting primary sources of energy to electricity. The sources of energy are either ***non-renewable*** or ***renewable***. For example coal, oil, gas, etc., are non-renewable sources of energy. They get exhausted once used. On the other hand, solar energy, energy of river water, and wind are renewable sources of energy. They do not get exhausted. The various sources of energy are mentioned in the following section.

13.3 ► Sources of Energy for Electricity Generation

(a) *Coal*: Coal is burnt in a furnace to generate steam. Steam at high pressure and temperature is supplied to a turbo-generator which produces electricity. Use of coal leaves a huge quantity of ash and produces fuel gases. They pollute the environment, and therefore need to be handled carefully.

(b) *Fuel oil*: Fuel oil is a product of oil refinery after the crude oil has been processed. Fuel oil produces no ash. The design of its furnace is simpler as compared to a coal-fired one.

(c) *Gas*: Gas is an important energy source, it contains methane (CH_4). Liquefied petroleum gas (LPG) is one of the petroleum products produced in oil refineries.

(d) *Solar*: Electricity using the sun's rays can be produced through the photovoltaic system or through solar thermal technologies. Solar photovoltaic system converts the sun's energy directly using solar cells (semiconductor device). Solar thermal technologies convert radiant energy of the sun into thermal energy and then into electricity.

(e) *Wind*: Wind energy system converts the kinetic energy of wind into electricity by using wind turbine coupled with a generator. The wind velocity of a particular location depends on the height and the nature of the terrain. Wind results from temperature gradients between the Equator and the Poles, and between land and sea.

(f) *Water*: The potential energy stored in water can be utilized if the water is allowed to fall from a higher to a lower level. This can be achieved by constructing a dam on a river at a suitable place where the gradient is high. Another energy source is from the kinetic energy and the potential energy of ocean waves. A tidal energy of water is extracted by creating a dam at a suitable place, using a turbo-generator and a sluice gate in the dam to allow the tidal flow to enter or leave the tidal basin.

(g) *Biomass*: These are agricultural and forestry wastes, municipal wastes, the waste from the crushing of sugar canes (bagasse) etc. Biomass is burnt to produce steam which will feed a turbo-generator to produce electricity. Alternatively, biomass may be processed to produce gaseous output to run a gas turbine–generator set.

(h) *Geothermal*: Steam and hot water produced inside the earth can be used to generate electricity.

The different technologies used to produce electricity from various energy sources are discussed in brief in the following section.

13.4 ► Thermal Power Generation from Fossil Fuel

In a thermal-power-generating station the heat energy from coal, oil or gas is utilized to produce steam and run steam turbines. The steam turbines are coupled with generators to produce electrical power output. A coal-based thermal plant is described in brief in the following section.

13.4.1 Coal-Fired Thermal Power Stations

Coal is first powdered, i.e., pulverized and burnt in a furnace and the heat energy is used to boil water in the boiler to produce steam. The steam from the boiler is taken to the turbine. The turbine works as a prime mover (drive) to the generator. Thus, the mechanical energy of the turbine is converted into electricity. The generated electricity is transmitted to various places through transmission lines after stepping up the voltage with the help of a transformer. Here, the chemical energy in coal is first converted to heat energy, then to mechanical energy, and finally to electrical energy. The conversion efficiency in a thermal-power-generating station is low due to heat and other types of losses at various stages of conversion. Fig. 13.1 shows a schematic representation of a coal-fired power station. As can be seen from the figure, the steam after passing through the turbine goes to a *Condenser* where it is condensed into water. This water is again fed to the boiler for producing steam. The heat from the used steam is taken out and dissipated in the *Cooling Tower*.

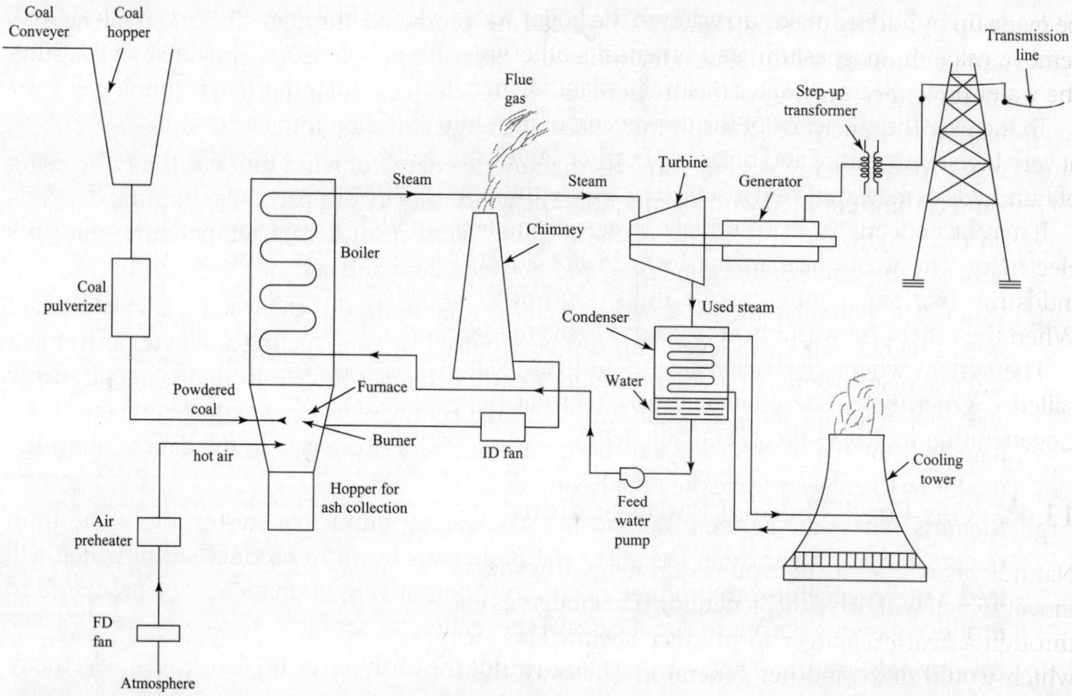

FIGURE 13.1

Schematic diagram of a coal-fired thermal power plant

In coal-fired stations, the coal is brought from the coal storage area through the coal conveyer to the *Coal hopper* placed near the boiler. The coal is then pulverized, i.e., made into powdered form so that complete combustion of coal takes place in the furnace. The *pulverizer* may be a rotating drum grinder or other type of grinders.

Sufficient air is required in the furnace for combustion. Air is drawn from the atmosphere by the use of a *forced draft fan (FD Fan)*. The *FD fan* takes air from the atmosphere and the air is preheated before injecting through air nozzles on the furnace inside wall. The *induced draft fan (ID Fan)* assists the FD fan by drawing out combustible gases from the furnace and maintaining a certain negative pressure in the furnace. Maintaining a certain negative pressure inside the furnace avoids any backfiring through any opening. An *electrostatic precipitator* is used before the ID fan (not shown in the figure) to collect the fine dust particles so that these are not thrown into the atmosphere along with the flue gases and pollute the atmosphere. This is a must requirement provided by the environmental protection law, and the power stations have to abide by this requirement of installing dust collectors from the flue gases.

Steam is taken out continuously from the boiler to the turbine. At the other side of the turbine, used steam is condensed to water and fed back to the boiler through a feed water pump. Thus, there is continuous withdrawal of steam and continuous return of condensed water to the boiler. However, there is some loss of water in the form of steam due to leakage, etc. This has to be made up by adding make-up water to the boiler water system. Boiler water has to be treated to remove calcium, magnesium, and other salts otherwise these may form undesirable deposits in the water flow system. A water treatment plant is, therefore, required in a thermal power plant.

To increase the efficiency of the conversion of heat into mechanical energy, steam is generated at very high temperature and pressure. Further, large capacity turbines and generators of the range of hundreds of megawatts are installed to reduce the capital cost per kilowatt of energy produced.

It may be noticed that in a thermal power station, steam and hot water are produced along with electricity. The waste heat in the form of hot air and steam can be used in certain process industries like paper mills, textile mills, food-processing industry, chemical industries, etc. When the otherwise waste heat is used, the overall plant efficiency is increased.

The system where electricity, steam, and hot water are simultaneously made available, is called *cogeneration*. The efficiency of a thermal power plant is around 40 to 50 per cent. Cogeneration increases the overall efficiency.

13.4.2 Gas-Fired Thermal Power Stations

Natural gas is used as the source of energy to generate electricity. A gas turbine is the prime mover which will drive the generator. The hot gases exhausted from the turbines can be passed through a ***heat exchanger*** to produce steam. This steam can be used to run a steam turbine which would drive another generator. This way, the total thermal efficiency of the plant can be increased. This method is called the ***combined-cycle method of generation of electricity***. If it is decided not to use the hot gases exhausted from the gas turbine for further generation of electricity, the hot gases can be used to produce steam for use in various industrial processes. Gas-fired stations are more environment friendly as compared to coal-fired power stations, as the flue gases emitted contain almost zero sulfur dioxide. As can be imagined, the installation cost of gas-fired stations would be less than the coal-fired ones. However, the operational cost of gas-fired stations will be high due to the higher cost of fuel, i.e., the gas used. In the state of Rajasthan in India, Kota is the place where all types of power stations like nuclear, thermal (coal fired and gas fired), and hydro are in operation.

13.4.3 Oil-and Diesel-Oil-Fired Thermal Power Stations

In oil refineries, the oil left behind after processing the crude oil can be used in an oil-fired steam power station installed near the refinery. The crude oil pumped out of an oil well can also be used directly to generate steam to run the steam turbine which will rotate the generator to produce electricity.

Diesel oil can be used to run an IC engine which will drive the generator. Due to the high cost of diesel, diesel-oil-fired stations are mostly used for standby power supply.

13.5 ► Hydroelectric Power-Generating Stations

Energy from falling water is used to run water turbines. Water turbines are coupled with electrical generators to produce electricity. Hydroelectric power generation depends upon the height of the falling water and the quantity of the falling water. The first use of water power to produce electricity was a water wheel on the Fox River in Wisconsin, USA in the year 1882. Thereafter, many hydroelectric power plants were constructed including one in Niagara falls in USA.

These days hydroelectric power plants are constructed as mega projects with the generating capacity of over 1000 megawatts (MW.). The initial investment on a hydroelectric power plant is huge because of the money required for the construction of a dam. Most often such plants are constructed in hilly areas where there is sufficient gradient available of the river water.

When a dam is constructed to obstruct the flow of water of a river, the upstream areas get flooded with water and creates a water storage area. For example, the dam constructed on the river Satluj at Bhakra–Nangal has created a large water storage area in the form of a lake, called Govind Sagar Lake. This water is used in a controlled way to produce electricity on a continuous basis using water turbine–generator sets installed in the power house at the lower area of the dam. Since a large area gets flooded in the upstream area when a dam is constructed on a river, the people living in such areas are to be shifted permanently from there. Further, there is an environmental impact on the construction of dams in hilly areas due to which the builders often find it difficult to construct new hydroelectric plants in certain areas. However, looking at the large potential of hydroelectric power and its negligibly small running cost, new projects are coming up in many countries.

In India, the installed capacity of hydroelectric power generation was over 25,000 MW by 2001. In a hydroelectric power plant, water at a higher elevation flows downward through large pipes called penstocks. This falling water rotates turbines which drive the generators, thus converting the mechanical energy into electrical energy.

The advantages of hydroelectric power is that it produces no pollution and is continually renewable. Water is continuously available because of rain and melting of snow, and never gets exhausted. Just to get an idea of how much electricity gets generated from falling water, someone calculated that four litres of water falling from a height of 30 meters each second could produce 1 kW of electricity which could run ten 60 W light bulbs and five ceiling fans.

Hydroelectric power plants are of three types, namely high- and medium-head-storage type, run-of-river type, and pumped-storage type.

Storage-type hydroelectric installations use a dam to block and store water in a reservoir. The stored water in the reservoir is released to run the turbines to generate electricity. The flow of water is controlled as per the requirement of output.

In the run-of-river type, use is made of the natural flow of water of a river to run a turbine. This type of installation may not require a dam to be constructed. A low-head diversion structure to direct the water flow to the penstock may be used.

Hydroelectric power plant with pumped storage facility use specially designed turbines. These turbines will drive the generator in a conventional way when water from the reservoir is allowed to flow through the turbines via the penstock, and thereby generate electricity during the day when there is huge demand on electricity from the supply system. However, during night when there is less demand for electricity, some of the turbines can be used as pumps to lift water from the outlet area back to the reservoir for future use. Fig. 13.2 shows the schematic diagram for a

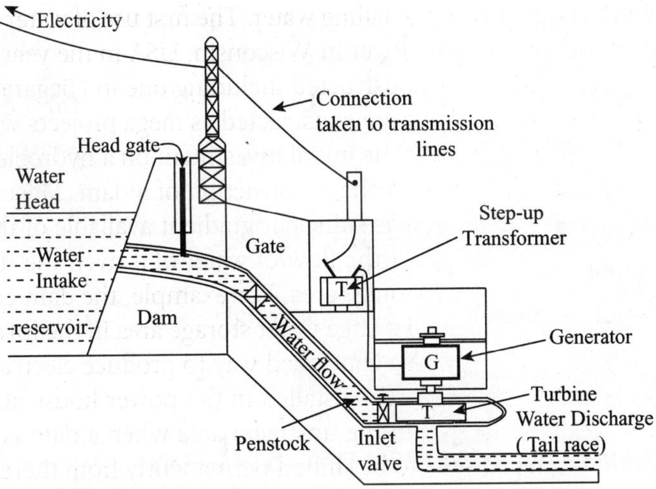

FIGURE 13.2

Layout of Hydroelectric Power Plant

high-head-reservoir-type hydroelectric power plant. The generated power, P in a hydroelectric plant can be expressed as $P = 9.81\ \rho\ Q\ h\ \eta \times 10^{-6}$ MW, where, ρ is the specific weight of water in 1000 Kg/m^3, Q is the quantity of discharge of water through the turbine in m^3/second, h is the head of water in metres, and η is the combined efficiency of the turbine–generator set.

Hydroelectric power generation provides a number of advantages. Some of these are mentioned below.

(i) Hydroelectric plants do not contribute to air pollution, acid rain, ozone depletion or any toxic wastes.
(ii) Many hydroelectric plants provide flood control and water supply for drinking and irrigation.
(iii) The big reservoir in the form of a lake creates recreational facilities like swiming, boating, water-skiing, camping, picknicking, sightseeing, fishing, etc.
(iv) Operation cost of hydroelectric plants is low as there is no fuel cost.
(v) Maintenance cost is low.
(vi) Life of a hydroelectric plant is long.

However, there are certain demerits of hydroelectric plants like heavy investment required for the construction of dam, etc.; time taken to instal a plant is high; opposition from environmentalists and the people living on the land to be used as a reservoir; cannot be constructed anywhere, i.e., near the load centre, etc.

The first hydroelectric power plant was installed in USA in the year 1882 followed by Sweeden and Japan. In India, the first hydroelectric power plant was installed in 1897 in Darjeeling area in West Bengal.

A few hydroelectric power plants in India are mentioned in Table 13.1.

TABLE 13–1 A Few Hydroelectric Plants in India

Location	Units	Total capacity (MW)
• Bhakra, Himachal Pradesh	5 × 108; 5 × 157	1325
• Dehar, Himachal Pradesh	6 × 165	990
• Salal, Jammu & Kashmir	6 × 115	690
• Sardar Soravar, Gujrat	6 × 200; 5 × 140	1450
• Nagarjun Sagar, Andhra Pradesh	1 × 110; 7 × 100.8; 5 × 30	965
• Idukki, Kerala	6 × 130	780

13.6 ▶ Nuclear Power-Generating Stations

We have known that in a thermal power plant, the heat required to produce steam is obtained by burning coal, oil, or natural gas.

In a nuclear power plant the required heat is produced by a process called fission of uranium atoms. Fission or splitting of uranium atoms takes place in a nuclear reactor. Uranium is therefore called nuclear fuel. Nuclear fuel consists of two types of uranium, namely U=238 and U=235. U=235 splits easily, i.e., its nucleus, which is composed of protons and neutrons, break up and release heat as well as neutrons. This splitting of atoms is a chain reaction. Fission in U=235 is started by bombarding it with neutrons. In the fission reaction, neutrons are released and heat is generated. When neutrons hit other uranium atoms, these atoms also split releasing neutrons as well as heat. These neutrons strike other atoms of uranium. Thus, one fission (splitting of uranium atom) leads to more fission, and hence creates a chain reaction. This makes the fission self-sustaining once the process is started.

This chain fission reaction once started will become an uncontrolled one unless some control mechanism is used. In a nuclear reactor, fission reaction of uranium is controlled using control rods which are inserted or withdrawn to slow or accelerate the fission reaction.

In a nuclear power plant, the nuclear reactor replaces the steam boiler used in a thermal power plant using coal or oil as a fuel. Many other components are similar to those used in a coal or oil (fossil fuel) power plant.

Nuclear fuel enrichment. U=235 is more desirable for fuel than U=238 because it is easier to split U=235 atoms than U=238 atoms. In the uranium ore, however, the percentage of U=235 is less (low enrichment). It is therefore necessary to enrich the fuel to be used by U=235 in relation to the number of U=238 atoms. This is called the enrichment process. After enrichment, the uranium is fabricated into pellets and stacked into long metal tubes. The filled rod is called a fuel rod. Schematic diagram of a nuclear power plant has been shown in Fig. 13.3.

At the nuclear power plant the reactor fuel assemblies consisting of fuel rods, spacer grids, the upper-and lower-end fillings, etc., are inserted vertically into the reactor vessel which is a large steel tank filled with water. The fuel is placed in a precise grid pattern and is called the reactor core. The boiling water reactor operates the same way as a boiler in a fossil fuel power plant.

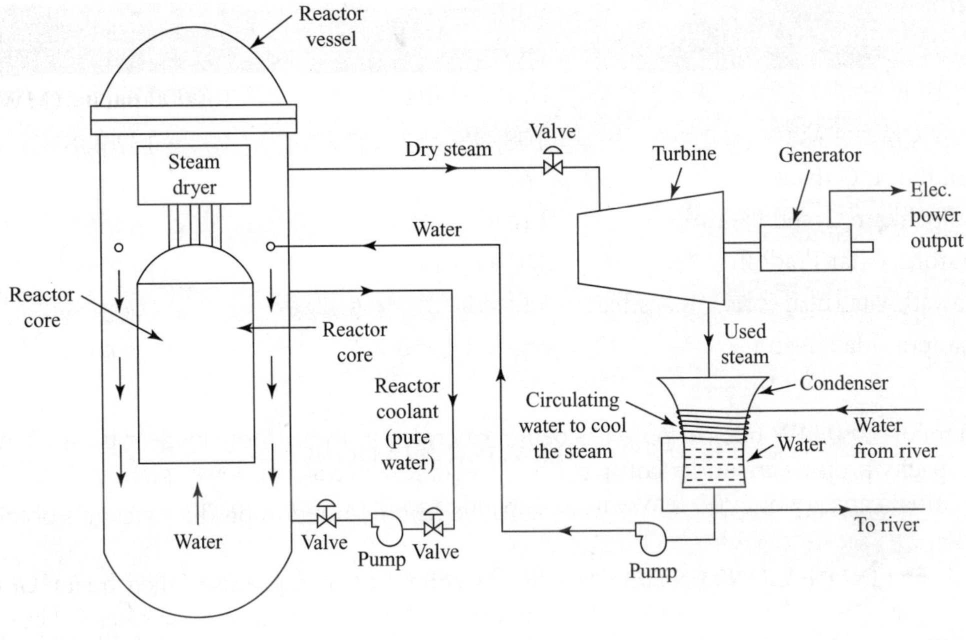

FIGURE 13.3

Schematic diagram of a nuclear power plant

Pure water called reactor coolant moves upwards through the reactor core absorbing the heat generated because of nuclear fission. Due to heat this water becomes steam but contains some water particles. This steam and water mixer moves up and enters the steam dryer and moisture separator. Dry steam, also called saturated steam leaves the reactor vessel through a steam line (insulated pipes) to the steam turbine. The turbine rotates due to the pressure of the steam. Since the generator is coupled with the turbine, the generator also rotates and produces electrical power which is stepped up and transmitted through transmission lines to various places for use. The exhausted steam is collected into the condenser where it is cooled into water. The condensed water is reused for which it is pumped out of the condenser back to the reactor vessel using a number of pumps. The coolant flow through the core can be varied to change the reactor power as and when required. If a nuclear power plant is not situated near a river, the excess heat of the circulated water is removed by using a cooling tower for transferring some of the heat to the air.

As compared to a coal-fired thermal power plant, the quantity of fuel required in a nuclear power plant is much less. The transportation cost of fuel is therefore less as compared to the transportation of coal or oil. Therefore, a nuclear power station can be built near the place where power is to be utilized, i.e., near the load centre. In a nuclear power plant, radioactive fuel waste has to be removed with outmost care and precaution.

Some of the nuclear power plants in India with their installed capacity are shown in Table 13.2.

TABLE 13–2 Some Nuclear Power Plants in Operation in India

Location	Units	Total capacity (MW)
• Kaiga, Karnataka	220 × 3	660
• Kakrapur, Gujarat	220 × 2	440
• Kalpakkam, Tamil Nadu	220 × 2	440
• Narora, Uttar Pradesh	220 × 2	440
• Rawatbhata (near Kota), Rajasthan	100 × 1 200 × 1 220 × 2	740
• Tarapur, Maharastra	160 × 2 540 × 2	1400

In total, 5780 MW nuclear power is being generated in India. More nuclear power generating capacity projects are under construction. Kudan kulam nuclear power plant in Tamil Nadu with initial capacity of 2000 MW will be expanded to 6800 MW capacity.

13.7 ► NON-CONVENTIONAL OR ALTERNATIVE GENERATING STATIONS

Generation of electricity from energy sources like solar energy, wind energy, ocean wave energy, geo-thermal energy, etc., are called non-conventional sources of energy.

As these are renewable sources of energy, efforts are being made to generate more and more electricity from these resources. These are being described in brief in the following sections.

13.7.1 Solar Electricity Generation

Sun's heat energy is the readily available energy that can be used for heating water, cooking food, pumping water, and for generation of electricity for other uses. The energy received from the sun on earth is so much that someone calculated it as equivalent to about 15,000 times the world's annual energy requirement. In India solar energy is available for 300 to 330 days a year, and hence should be utilized to the full through two different ways, i.e., for heating purpose and for generation of electricity. For generation of electricity ***solar photovoltaic cells*** are used. Solar photovoltaic uses sun's heat to produce electricity for lighting, running pump motors, and running of electrical appliances. Solar cell, or photovoltaic cell converts sunlight directly into electricity.

Photovoltaic (PV) is the technical term for Solar electric. PV cells are made of silicon. Silicon releases electrons when exposed to light. The amount of electrons released from silicon depends upon the intensity of light falling on it. One PV cell or Solar cell produces about 1.5 W of electricity when exposed to bright sun. Individual solar cells are connected together to form a panel or module capable of producing upto about 100 W of electric power. A number of such panels are then connected in series and parallel to produce a considerable amount of electricity. Common applications of PV systems are found in street lighting and indoor lighting, particularly in far flung and hilly areas where conventional electric supply may be expensive because of its long route.

Commercial solar cells producing electricity have an efficiency of about 15 per cent only. Low-efficiency system means larger arrays would be required for electricity generation, and hence would cost more. However, research is being conducted to improve the efficiency of solar cells so as to make the system economically viable. Government of India provides subsidy to manufacturers and users of solar appliances to encourage their use.

13.7.2 Electricity from Wind Energy

When solar radiation enters the earth's atmosphere, due to the curvature of the earth, different regions of the atmosphere are heated differently. The equator region gets heated the most and the polar region the least. Since air tends to flow from warmer region to cooler region, air flow takes place. The energy of this air flow is utilized in the wind mills and wind turbines to produce usable power.

Wind energy is converted into electrical energy through wind turbines coupled with a generator. Wind electric generator converts the kinetic energy in wind to electrical energy by using a rotor, a gear box, and a generator is shown schematically in Fig. 13.4.

Due to flow of wind at a considerable speed, the rotor blades rotate which converts wind energy into rotational shaft energy which in turn rotates the generator through the gearbox arrangement. Wind energy generators of rating from nearly 200 kW to 1000 kW have been installed in different parts in India. Wind speed is the most important factor influencing the amount of energy a wind turbine will produce. The rotor is placed on top of towers to take advantage of strong winds available high above the ground.

India is one of the most promising countries for wind power development with an estimated potential of 20,000 MW. Financial and technical assistance is provided by government

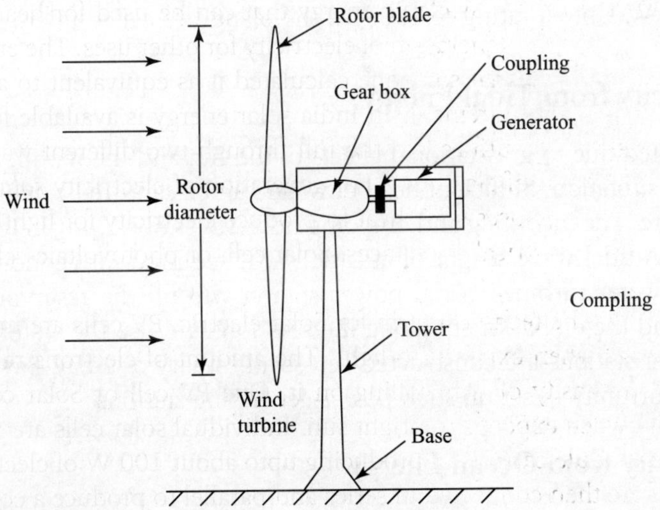

FIGURE 13.4

Wind turbine–generator

of India for wind power development. As on March 2009, the installed capacity of wind power in India was 10,254 MW. Tamil Nadu is the state with most wind-power-generating capacity of 4301.63 MW.

13.7.3 Electricity from Biomass

In most bio power plants, steam is produced by burning of bio-waste material. Bio energy feedstocks are burnt directly to heat the water in the boiler to produce steam. The saturated steam drives the steam turbine which rotates the generator coupled with it. In certain industries the steam produced in the boiler is also used in manufacturing processes or to heat the buildings. These are called combined heat and power facilities. For example, in a paper industry, the waste wood is used to produce both electricity and steam.

The decay of biomass produce methane gas which can be used to produce electricity. By burning methane, steam can be produced in a boiler and this steam will run a turbine coupled with a generator.

13.7.4 Mini/Micro Hydel Power Generation

In hilly areas there is tremendous potential of generation of electricity where there are hydel (water) resources in the form of falling water. The Himalayan belt in India is gifted with such hydel resources. Mini and micro hydel set (turbine–generator set) can be installed on such falling water resources to generate electricity for local use. These small turbine–generator sets are compact and require no maintenance. Instead of using diesel generator sets, small turbine–generator sets can be used in hilly remote areas for the benefit of small cluster of villages to supply electrical power to farms, schools, houses, etc. The mini hydel plants are rated for 10 kW to 1000 kW. Upto a rating of 100 kW, the plants are called micro-generating plants.

13.7.5 Electricity from Tidal Energy

Sea tides are created due to gravitational effects of the sun and the moon, and the centrifugal force of the earth's rotation. Sufficient head of water can be captured by constructing barrages along the sea shore. The incoming and outgoing tidal waves of varying heights can be blocked by the barrages. A tidal wave height of atleast 7 m is required for economical generation of electricity using water turbines. Tidal power station will fill the reservoirs by opening the sluice gates behind the embarkment along the seashore and use the water height to run turbines. Tidal power stations are constructed across an estuary or a bay. Bay of Bengal in India provides the opportunity to set up such power-generating stations.

13.7.6 Electricity from Ocean Energy

Almost two-third of earth's surface is ocean. The surface water gets heated up due to sun's energy. The sun heats up the surface of the water a lot, and hence, the water surface of the ocean works as a solar collector. The warm surface water of the ocean can be used to generate electricity.

The warm water of the ocean can be used to vaporize a working liquid of low boiling point such as amonia. The vapour will expand and can run a gas turbine which will drive a generator.

13.7.7 Electricity from Geothermal Energy

Heat energy is trapped in rocks inside the core of the earth. Water absorbs the heat from these rocks and gets converted into steam. This steam when available on earth's surface at specific places can be used to run steam turbines, and thereby generate electricity. Not much has been achieved in tapping geothermal energy so far. In India, in Ladakh area, feasibility study for construction of a 1 MW power station has been made. Geothermal means heat from the earth's interior.

Geothermal energy is generated in the earth's core about 6400 km below the surface of the earth. Temperatures higher than the temperature of the sun is produced in the core of the earth by slow decay of radioactive particles.

The centre of the earth is of solid iron core and its outer is made of very hot melted rock called magma. The magma is surrounded by rock and magma which is also called mantle. The crust is the outermost layer of the earth. The crust forms the land and ocean floors. See Fig. 13.5.

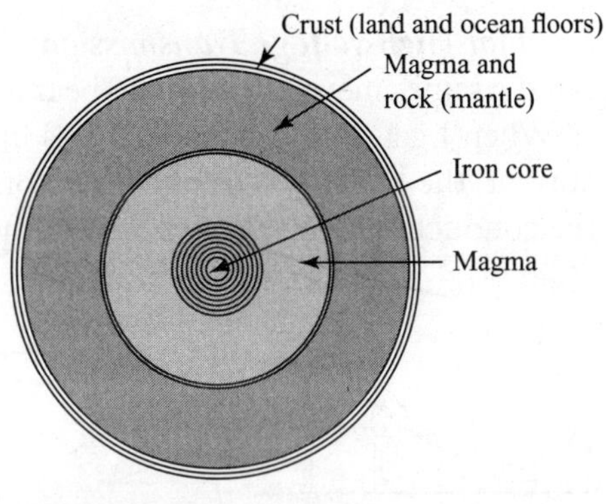

FIGURE 13.5

The interior of the earth showing source of geothermal energy

The crust on land area is about 25 to 55 km thick and on oceans 5 to 8 km thick.

Inside the earth the rocks and water absorb heat from the magma. When magma comes closer to the surface, it heats up the water trapped in porous rocks or the water running along fractured rocky areas. Such naturally formed hydrothermal resource is called a geothermal reservoir.

Electricity generation in power plants requires steam at very high temperature.

Geothermal power plants are built where geothermal reservoirs are available within a depth of a few km from the earth's surface. This requires drilling into the earth to extract hot water and dry steam. California in USA produces the most electricity from geothermal energy. Other uses of geothermal energy is to use hot water from the reservoir or naturally available hot springs for heating buildings, bathing, etc. In India, although electricity generation from geothermal energy is at an initial stage, use is made of hot springs for bathing as many believe that taking bath in mineral-rich natural hot water will have a healing effect.

13.8 ▶ Transmission and Distribution of Electricity

Electricity generated in power stations are brought to the consumer premises through transmission and distribution systems. Power transmission is from the substation near the power plant to the substation near the populated area or industrial area. In substations, the voltage level of the power generated is either stepped up or stepped down using step-up or step-down transformers. Power is transmitted at high voltage using transmission lines. Fig. 13.6 shows the schematic diagram of a power system.

The distribution system includes overhead lines and underground cables drawn from the transformer substation and taken to the consumer premises, i.e., to industries, commercial establishments, and residential areas.

Power transmission lines at a national level or even at an international level are connected through additional paths and lines so that power can be routed from one power plant to any place where there is need. Surplus power from one station can be supplied to deficit areas. Such a network allowing transmission of power nationwide is called a ***national grid.***

Need for High-voltage Transmission: Power is a product of voltage and current ($P = VI \cos \phi$). For the same amount of power to be transmitted, if voltage, V is increased, current I gets reduced.

When I gets reduced, the I^2R loss in the transmission lines gets reduced. Further for lower current, the cross-section of conductors to be used will be less. This will reduce the weight of the conducting material, and hence, the cost of transmission will get reduced.

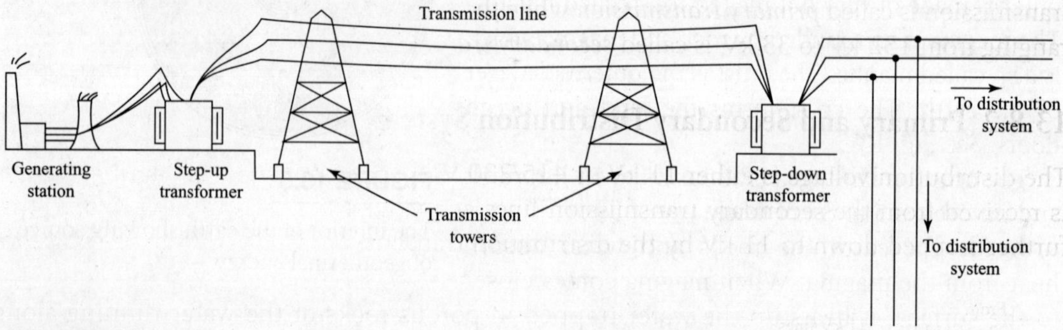

FIGURE 13.6

Shows the schematic diagram of power transmission at high voltage

Again, power transmitted is proportional to the square of the voltage (since $P = VI \cos\phi = V \times \frac{V}{Z} \cos\phi$ or $P = \frac{V^2}{Z} \cos\phi$). With higher transmission voltage it is possible to transmit more power through the transmission lines.

Thus, it is desirable that transmission voltage be made as high as possible to be able to transmit maximum power through a line efficiently. However, the height of the lines above the ground has to be increased to keep them away from the ground level. Thus, there is practical limitation to the level of high-voltage transmission. In India, the highest level of transmission voltage at present is 400 kV. In some countries the highest level is 765 kV or so.

13.8.1 AC and DC Primary and Secondary Transmission

As we have known, $P = VI \cos \Phi$. If the voltage level of transmission is increased, the current level goes down. The cross-sectional area of transmission line conductors get reduced, and hence the cost gets reduced. For economical bulk power transmission, the generated voltage level is stepped up. It has been roughly calculated that most economical transmission voltage is 1 kV per 1.6 km or 1 kV per mile. Thus, the most economical transmission voltage for a distance of around 600 km, works out around 400 kV. In India, the transmission voltage varies from 66 kV to 400 kV. For transmission of power to distances more than 600 km, it has been calculated that dc transmission becomes more economical than ac transmission. Electricity is generated by using an alternator, i.e., an ac generator at 11 kV or more. For ac transmission, the three-phase lines from the generator are connected to the primary side of a step-up transformer. The secondary terminals are connected to the transmission lines.

For dc transmission, the alternating voltage is first converted into dc using an ac to dc converter, and is then connected to the transmission lines. At the receiving end of the transmission lines dc is converted to ac and distributed to the consumers through distribution lines, step-down transformers and underground cables. It may be mentioned here that transmission of power is done in two stages. In the first stage, the voltage level may be 220 kV and in the second stage the voltage level is brought down to 132 kV or less. The level of transmission may be brought down even to 33 kV. The first stage of transmission is called ***primary transmission*** while the second stage of transmission, at voltages ranging from 132 kV to 33 kV, is called ***secondary transmission***.

13.8.2 Primary and Secondary Distribution System

The distribution voltage is either 11 kV or 415/230 V. At the distribution substation, power is received from the secondary transmission lines at a voltage of, say 33 kV. The voltage is further stepped down to 11 kV by the distribution transformer installed at the distribution

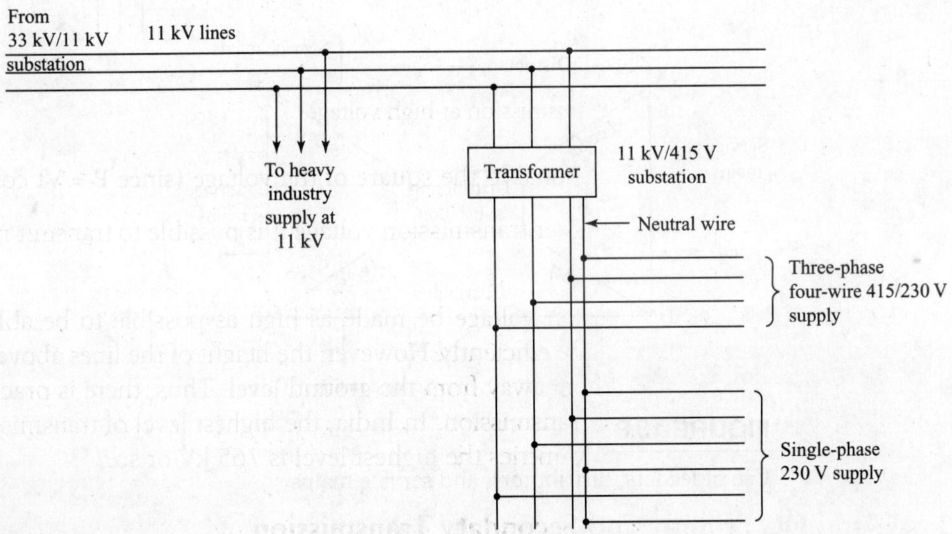

FIGURE 13.7

Distribution of power to industries and residential complexes substation. Heavy industries are supplied with primary distribution voltage of 11 kV, who in turn step down the voltage using their own transformers. For other consumers electricity is supplied at secondary distribution voltage of 415/230 V. Three-phase supply is provided at 415 V and single-phase supply is made at 230 V.

Thus, the part of power system which provides electricity to the customers is called the distribution system as shown in Fig. 13.7.

13.8.3 Overhead Versus Underground Distribution Systems

The distribution system can be installed through overhead lines or through underground cables. Overhead lines are generally mounted on concrete or steel poles. The distribution transformer (11 kV/415 V) is mounted on poles. Supply is taken out from the secondary of the transformer through cables. In an underground system cables are laid below the ground level along the street by the side walk. The choice between overhead lines and underground cables depends upon a number of factors like initial cost, ease of location of any fault, safety consideration, maintenance cost, aesthetics, etc.

13.8.4 Connection Schemes of Distribution System

The distribution network consist of *Feeders, Distributors,* and *Service Mains.*

Feeder is a cable which feeds the distributor. From the distributor, connections are taken to supply the consumers. Such connections are called *Service mains.* Two types of arrangements are made. These are called the *radial system* and the *ring main system*. Fig. 13.8 shows the difference between feeder, distributor, and service mains. Generally, no tappings are taken from the feeder. From the distributor, connections are taken through low-voltage cables to supply the consumers.

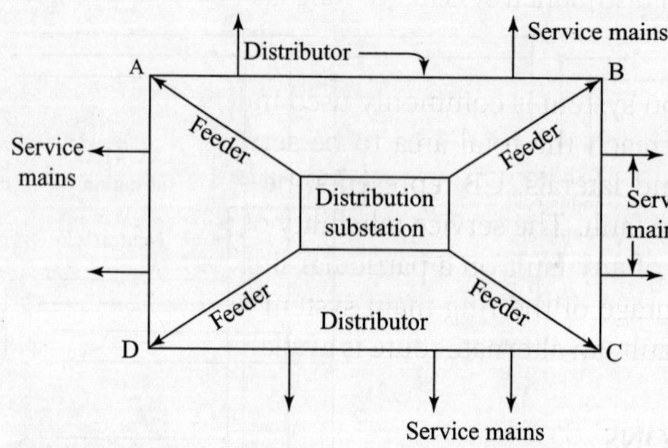

FIGURE 13.8

Use of feeders, distributors, and service mains

The radial system of distribution and the ring main system of distribution are illustrated through schematic diagrams in Fig. 13.9 (a) and (b).

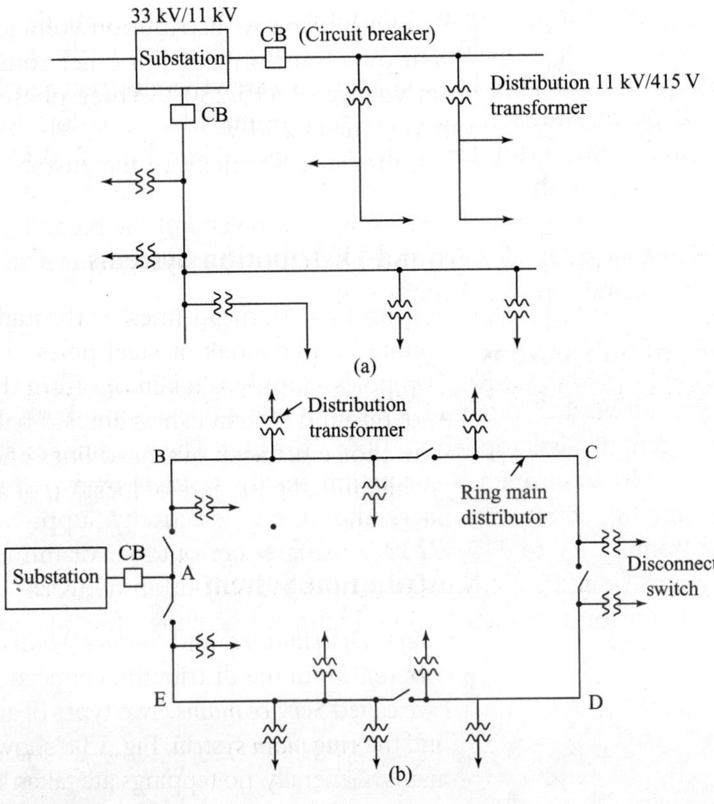

FIGURE 13.9

(a) Radial distribution system; (b) Ring main distribution system

The radial distribution system is commonly used in less populated areas where the primary feeder branches out to reach the total area to be served. Distribution transformers are connected to the feeders and laterals. CB represents the circuit breakers installed to disconnect the system in case of any fault. The service reliability of redial system of distribution of electric power is low as in case of any fault on a particular line, the whole area gets effected, i.e., gets switched off. The advantage of the ring main system over the radial system is that it is more reliable. In case of any fault, an alternate route is available for the maintenance of power supply.

13.9 ▶ Substations

Electricity is generated in power stations. From the generating stations, electricity is supplied to the consumers through transmission and distribution lines. From a number of design considerations, electricity is generated at 11000 V, i.e., 11 kV. For transmission and distribution, the voltage level is raised and then gradually lowered to the level of its used. Electricity is generally used as a voltage level of 415 V/230 V for three-phase and single-phase applications, respectively. From 11 kV the voltage level is raised to 132 kV or 220 kV for high-voltage transmission using a step-up transformer located in an electrical sub-station. The high-voltage transmission lines hanging from tall towers carry the electricity to long distances. The 220 kV or 132 kV transmission lines constitute primary transmission. At the end of the primary transmission distance, the electricity is slipped down from 220 kV/132 kV to 66 kV/33 kV using step-down transformer in a sub-station and transmission is carried forward high secondary transmission lines at 66 kV/33 kV. At the end of the secondary transmission lines, electricity is further stepped down to 11 kV in a sub-station using a step-down transformer from where distribution of electricity starts.

13.9.1 Positions of Substations

Electricity is carried forward through 11 kV distribution lines. Some bulk consumers like heavy industries, electricity is supplied at 11 kV directly. The consumers steps down the electricity to 415 V/230 V through a substation set up at their own premise. For other consumers like domestic, commercial, agricultural, etc., electricity supply authority steps down the voltage from 11 kV to 415 V/230 V using either outdoor or indoor substations. The distribution network at 11 kV is called primary distribution network. Distribution at 415 V/230 V is called secondary distribution. Figure 13.10 shows the positions of different substations in a power system.

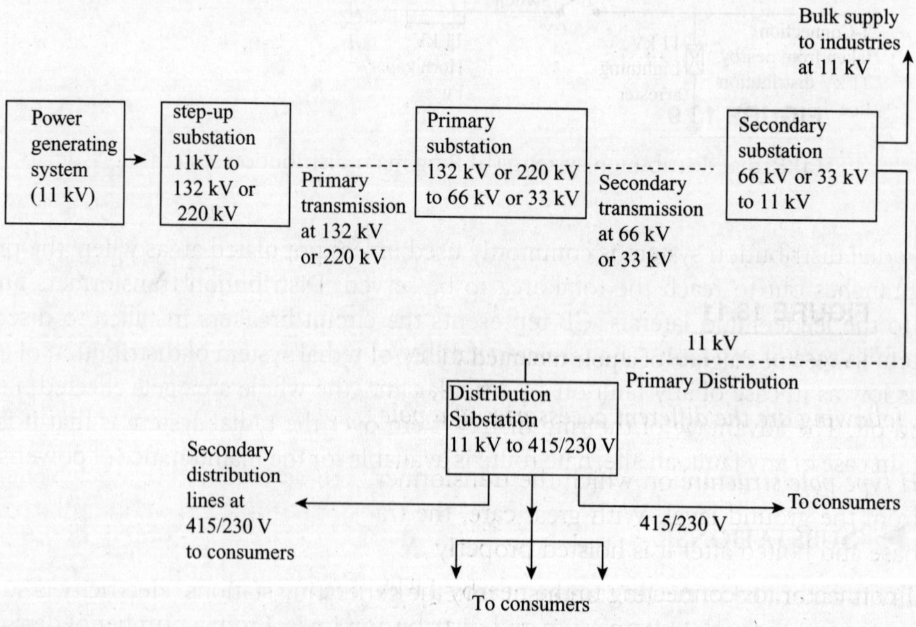

FIGURE 13.10

A line diagram showing primary, secondary transmission and distribution system with the positions of different sub-systems

13.9.2 Distribution Substations

Distribution substations can be located indoor or outdoor. Primary and secondary substations at 220 kV, 132 kV, 66 kV, 33kV are located outdoor. When transformer and associated equipment are installed inside a building, it is called indoor, whereas in an outdoor substation the transformers are either floor mounted or pole mounted. Distribution transformers of rating upto 400 kVA are installed on a foundation or plinth with a fence or wall surrounding it with suffiicient air circulation space. Such substations are often used for factories or large establishments, colleges, and universities where there is enough ground space. Primary distribution lines are taken round the locations where supply is to be provided to different consumers like showing complexes, shops, etc.

13.9.3 Pole Mounted Type Substations

Transformers of low rating, say 100 or 200 kVA are mounted on Ms Channels. Which are rigidly fastened to a two pole structure. This type of substation is quite economical for domestic and commercial loads of small towns, rural electrification, tubewell and agricultural load and medium industry load. Fig. 13.11 shows a line diagram of a pole-mounted distribution substation.

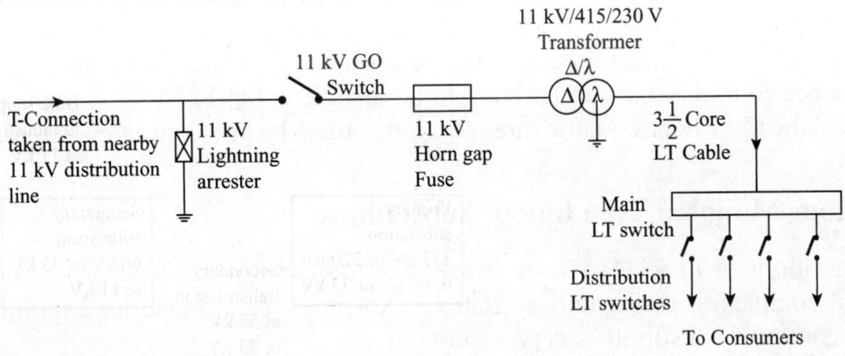

FIGURE 13.11

Single line diagram of a pole mounted 11 kV/415 V distribution substation

The following are the different accessories of a pole mounted type distribution substation:

1. *H type pole structure* on which the transformer is to be mounted at a height of 2.44 m from the ground level. With great care, the transformer is lifted up and placed on the base and bolted after it is hoisted properly.
2. T-connector for connecting to the nearby 11 kV line.
3. 11 kV gang operating (GO) switch (air break switch) to be installed below the 11 kV conductor line before connection to the transformer.
4. Operating handle of the GO switch to be fixed on the pole structure at a height of 1.25 m from the ground.

5. 11 kV lightning arrester (3 nos) to be erected at the top of the H–pole structure for the protection of the transformer against lightning.
6. 11 kV horn gap fuses to be installed below the gang operating switch and 11 kV transformer bushings for protection of transformer on the 11 kV side.
7. 11 kV/415 V three-phase delta-star connected transformer of 400 KVA or lower KVA rating.
8. A three-and-a-half core LT (low tension) cable to be taken from the LT bushings to the LT main switch.
9. Main switch with fuse unit for protection of the transformer against any fault in the feeders.
10. LT distribution switches and LT cables to be taken out for supply to the consumers.
11. Miscellaneous items like connecting jumpers from 11 kV line to the GO switch, from GO switch to the fuse unit, and from fuse unit to the transformer bushings, polythene pipes, insulation tape, barbed wire to be wrapped on poles and stays, danger plate to be provided on the pole structure as a warning measure. All metallic fittings are to be painted with aluminium paint.
12. Earthing electrodes (three nos) to be placed in earth pits on either side of the double pole structure. One of which to be connected to the lightning arresters and the other to be connected to the neutral point of the transformer on LT side. The third electrode is to be connected to the transformer body, to the handle of the 11 kV GO switch and to the body of the LT switches. All the three electrodes are to be interconnected in the earth pits.

13.9.4 Floor Mounted Type Indoor Substations

Indoor substations of 11 kV/415 V are installed in industrial units, food processing plants, commercial complexes, malls, railway stations, hospitals, etc., where load requirement is heavy and continuity of supply is very important.

The substation building should be constructed providing necessary space for the transformer, HT and LT switchgear and cable trenches for the incoming and outgoing cables. Adequate space is to be provided for air circulation and movement of personal for maintenance. Figure 13.12 shows a line diagram showing the equipment and accessories of a 11 kV/415 V, 500 kVA or more of a distribution substation of indoor type. The overhead 11 kV distribution lines are connected to a pole structure outside the indoor substation.

An 11 kV gang operating switch and lightning arresters are installed for isolation and protection, respectively. The supply from the overhead lines is brought inside the substation through an underground cable using an outdoor cable box. The position of the 11 kV circuit breaker is determined by the place where the incoming cable enters the substation building. In the substation, the transformer is the main and most important equipment. The transformer windings are delta-star connected with neutral point of the secondary windings connected to earth. The secondary terminals of the transformer are connected to LT circuit breaker using LT cables. The output terminals of the transformer are connected to the LT busbars. For taking

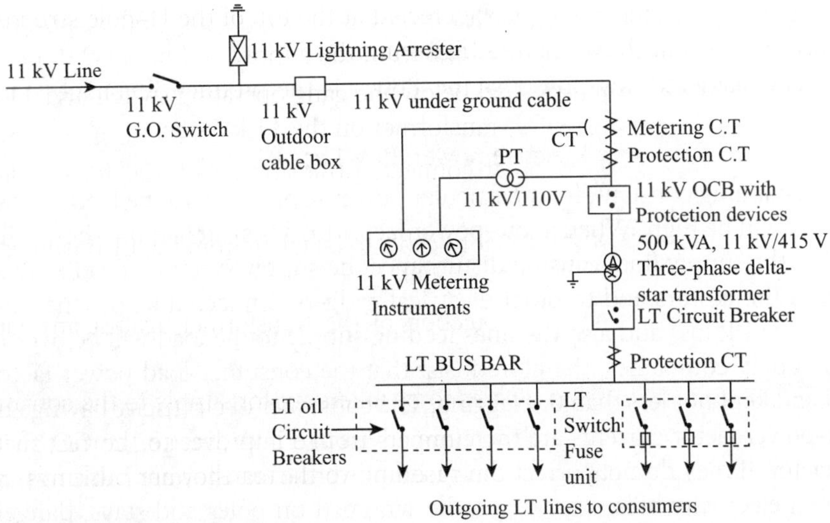

FIGURE 13.12

A line diagram of a 500 kVA 11 kV/415 V floor mounted indoor type Distribution substation

out cables upto current carrying capacity of 300 A, LT switch fuse units are used. For higher currents to be supplied, LT oil circuit breakers are used. The oil circuit breakers and the fuses are used to protect the transformer from any damage in case of excess current is drawn by the consumers. The LT oil circuit breaker, LT bus bars and the LT switch fuse units are installed in a switch board or panel board in convenient location. Adequate fire fighting equipment should be provided in the substation. Fire buckets and fire extinguishers are to be installed in the substation. First aid chart is to be displayed. First aid box containing materials for first aid and danger plates should be visible.

All non-current carrying parts with conducting surface such as switch-boards, tank of the transformer, frame of the oil circuit breakers, CT and PT cases, lightning arresters, etc., should be effectively earthed for protection of operational personnel from electric shock. Depending on the load, a substation may have more than one transformer. For a substation with one transformer, there should be 4 earth electrodes (2 for neutral earthing and 2 for running a common earth wire for connecting to the body of transformer and other equipment). For a two transformer substation, the number of earth electrodes used should be six.

13.10. ► Power Factor Correction

13.10.1. Effect of Low-Power Factor

The average power factor of a load comprising fluorescent lamps (tube lights), fans, small ac motors used in household electrical appliances, three-phase induction motors, arc furnaces, induction furnaces, arc welders, etc., varies from 0.6 to 0.7 lagging. When power factor of the load is low, the current drawn by the load from the supply will be high. A power factor of 0.6 to 0.7 is considered low-power factor. Power is the product of voltage, current, and power factor.

$$\text{Active power, } P = V\, I \cos\theta$$

Voltage, V remaining constant, for a particular power supplied, P to the load, if cos θ is low, current drawn I will be high. When a low-power factor load is switched on, there will be large voltage drop in the supply lines causing disturbance the supply system. Lower voltage will be available across the motors and to other electrical gadgets connected across the lines. There will be higher line I^2R loss and also the lines feeding supply to the load will be over loaded.

Electricity supply companies, therefore insist that the consumer load power factor should be improved and kept not less than 0.9 lagging. The consumer of electricity having equipment drawing low-power factor currents are therefore advised to improve, i.e., correct their overall load power factor. If they do not correct their load power factor, they are liable to pay penalty along with their electricity bills.

13.10.2. Advantages of Power Factor Correction

Low-power factor is due to the inductive component of the load. Raising the system power factor by using some capacitors connected across the load or by some other means, will lead to a number of advantages. *The advantages of power factor correction or improvement are listed below:*

1. If the power factor is improved, there will be less I^2R loss in the lines supplying the load
2. There will be reduction in the magnitude of current drawn thereby not overloading the lines
3. Reduction of voltage drop in the lines thereby maintaining reasonably stable voltage across the load
4. No penalty has to be paid to the electricity supply company

13.10.3. Methods of Power Factor Improvement

The following two methods are most commonly used for power factor improvement:

1. Use of static capacitors
2. Use of synchronous condensers, i.e., over excited synchronous motors.

The above two methods are discussed below.

1. **Use of static capacitors.** Suppose the loads connected draw a lagging current shown in Fig. 13.13. When a capacitor is connected across the load, the capacitor will draw a leading current I_c as shown in Fig. 13.13 (b).

   The original load current is I_L and power factor is $\cos\theta_L$. Current, I_L lags the voltage by an angle θ_L. To improve the power factor a bank of three-phase capacitors are connected across the load terminals. A current of I_C will be drawn by the capacitors. Current I_C will lead the voltage by 90° as has been shown in the phasor diagram. The new load

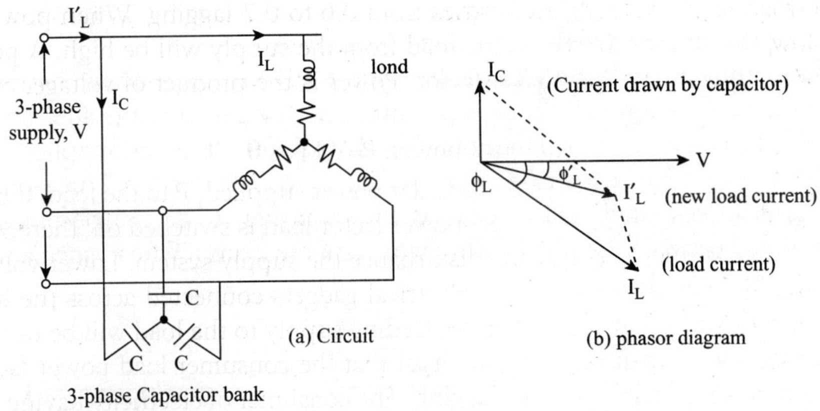

FIGURE 13.13

Power factor improvement using static capacitors

current drawn from the lines will be I_L' which is the vector sum of currents I_L' and I_C. It is observed that the power factor angle between voltage V and resultant current, I_L is reduced and as a result power factor cos θ_L' is increased. This way by proper selection of the rating of the capacitors, power factor can be improved, i.e., corrected.

To improve the load power factor, the consumer, who may be a factory owner, has to invest money in the installation of capacitor banks including their switching arrangement. This will, however, lead to saving in electricity bills for all time to come which may be much more economical.

2. **Use of synchronous condensers.** In an industry, in addition to induction motors, often synchronous motors are also used as constant speed drive. If the field current, also called excitation current, of a synchronous motor is made more than its normal value, the motor is said to be over excited.

   An over excited synchronous motor draws leading power factor current from the lines, and hence, it behaves like a capacitor. Thus, when synchronous motors are used, they can drive a load and at the same time work as power factor improvement device like capacitors. By controlling the excitation current, a smooth control of the system power factor can be obtained. However, use of synchronous condensers over static capacitors for power factor improvement alone will be a costly affair.

13.11 ▶ Electricity Distribution System

Electricity distribution authorities supply power to the consumers at the following two voltages. Single-phase supply: 230 V, 50 Hz, two-wire and three-phase supply: 415 V, 50 HZ, four-wire. In a two-wire single-phase supply there is one phase or live wire and the other is called the neutral wire. Single-phase supply is required for electrical appliances like fan, tube light, lamp, washing machine, refrigerator, electric iron, room heater, room air-conditioner, kitchen electrical appliances like mixer, grinder, microwave oven, etc.

In a four-wire, three-phase supply, power is supplied through three live wires and a neutral wire. The neutral wire is normally at zero potential and is earthed at the substation. Three-phase loads like three-phase induction motors used for water lifting are supplied with three-phase three wire supply. In such a case all the phases get equally loaded. Three-phase supply can also be used to feed single-phase loads as shown in Fig. 13.13. Single-phase electrical loads are connected between the phase or live wires and the neutral wire in such a way that all the phases are equally loaded.

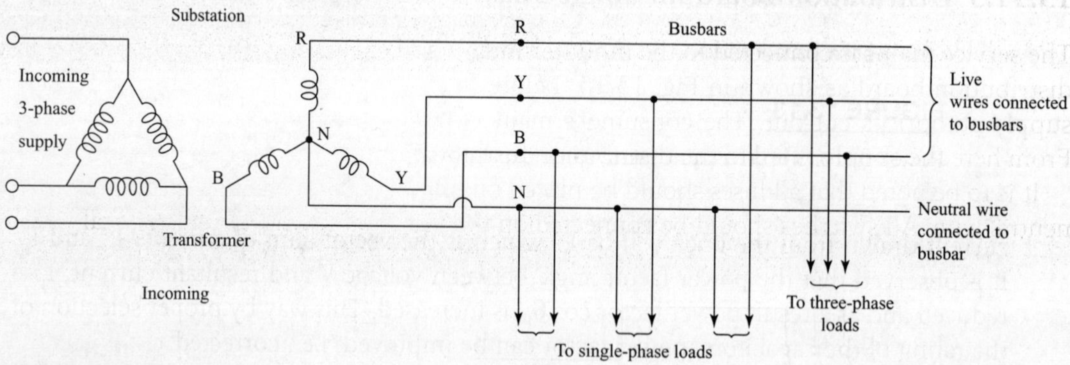

FIGURE 13.14

Three-phase four-wire distribution system

As shown in Fig. 13.14, supply from the secondary of a three-phase transformer is taken through feeder wires to the busbars (busbars are thick copper strips). From the busbars both single-phase supply and three-phase supply are taken out and are connected to the loads.

13.11.1 Service Connection

Electricity supply authorities supply power to the domestic loads, i.e., residential houses through a low-voltage three-phase four-wire distribution system. The distribution system is formed either through overhead lines or through underground cables. In modern cities, only underground cables are used for electric power distribution system. From the electricity supplier's distribution system, power is brought to the premises of the consumer through a cable called the *service line* or service connection. The service line is normally an underground PVC cable. The service cable is brought to the consumer's distribution board.

13.11.2 Service Mains, Meter Board and Distribution Board

In the main distribution board installed in the premises of the consumer where supply comes through the service line, an energy meter is connected so as to measure the amount of electricity consumed by the customer. Thus, the incoming service cable is connected to the energy meter at the distribution board. A cutout, i.e., a fuse wire is connected in the line which will blow off in case the consumer draws more current than the current for which he has been supplied with. If the cutout of proper rating is not provided, the meter will get burnt in case of heavy loads (more than the rated load) drawn by the consumer. For example, if the consumer instals more air-conditioners in his house than permitted, the meter may get burnt. The cutout can only be replaced by the electricity supply authority. The consumer must not tamper with the cutout. The supplier may disconnect supply by removing the cut-out in case the energy bill is not paid by the consumer.

13.11.3 Distribution Board for Single-Phase Installation

The service mains is connected to the input terminals of the single-phase energy meter at the distribution board as shown in Fig. 13.15. Before the energy meter is placed the electricity supply authority's cut-out. The consumer's main switch-fuse is connected after the meter. From here the supply is fed to the distribution fuse board through a busbar and neutral link.

It is to be noted that all fuses should be placed only in the live or phase wire and not in the neutral wire. All switches should be connected on the live lines and not on the neutral line.

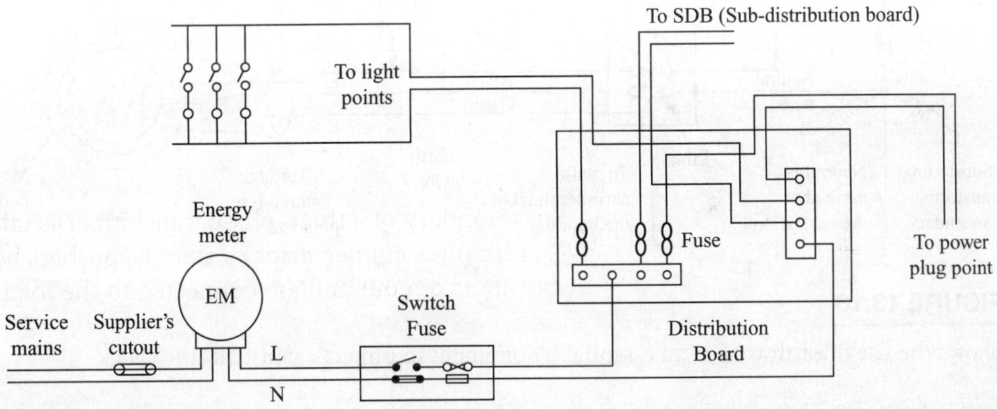

FIGURE 13.15

Layout of distribution board for single-phase installation

13.11.4 Neutral and Earth Wire

The neutral wire is taken out from the star point of the secondary of the distribution transformer. The star point is earthed. The lines are taken out from the terminals R, Y, B of the three-phase windings and the neutral point N. Each phase wire and the neutral wire constitutes single-phase supply.

In addition to the four wires taken out from R, Y, B, and N terminals, a fifth wire, called earth wire is provided. The earth wire is provided for the purpose of protection of persons using electrical appliances. In case of any leakage in the system, the persons using any equipment will not get any electric shock if the metallic body of all electrical appliances are connected to the earth terminal.

The neutral wire is connected to the neutral point of the transformer which is earthed at the substation itself. The earth wire starts at a solid earthing point at the substation and runs along the supply lines. This wire is earthed along its run once in every 1.6 km.

13.11.5 Earthing

All metal parts of electrical appliances are always connected to the earth through an earth wire. If the earth wire is not provided and any part of the current-carrying live wire touches a metal frame, the frame will acquire the same voltage as the live wire. Any person touching the body of the appliance will get a severe electric shock. If a low-resistance earth wire is provided, in case of any fault in the circuit, a large amount of current will flow through the circuit. This will cause the fuse provided in the live wire to blow and protect the circuit as well as the person touching the electrical appliance, gadget or equipment, as the case may be.

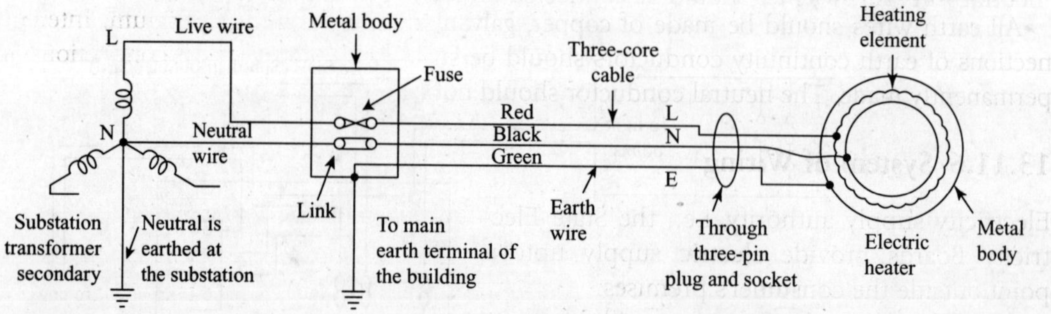

FIGURE 13.16

Shows the use of earth wire as an essential requirement to protect against electric shock

As shown in Fig. 13.16, the live and the neutral wires are used to supply power to an electric device, say a heater. The metal case of the heater is connected to the earth wire of a three-core cable supplying power to the heater. The earth wire from the distribution board is connected to the main earth terminal of the building as has been shown. This earth terminal in turn gets connected to the earth electrode at the substation through the general mass of the earth (earth is the conducting media). Thus, in case of any electric fault, if the live wire touches the metal case, it gets connected to the earth electrode at the substation. At the substation, the neutral point is earthed.

Method of earthing

Earthing means connecting the earth terminal solidly and securely on the ground through an electrode. Conductivity of earth depends on the moisture content of the soil and its chemical composition. The soil near the earth electrode should have low resistivity (i.e., high conductivity). Before placing the earth electrode, the soil around is made highly conducting by adding some agents like common salt, i.e., NaCl, calcium chloride, i.e., CaCl, sodium carbonate, i.e., Na_2Co_3, soft coke, etc.

Earth electrode

Places where there is a network of underground cables, the earth terminal is obtained by making connections with the lead sheath or steel armour of the underground cable. The lead sheath or the steel armour of the cable will serve as the earth electrode. However, where such underground cable is not available, earth electrodes are laid on the ground to get the earth terminal. To avoid corrosion, the material chosen for the earth electrode, either a pipe or a plate, should be made of copper or zinc-coated iron. Typical installation of two types of electrodes, viz the rod or pipe electrode and plate electrode are made. A rod-type electrode has been shown in Fig. 13.17. This method of earthing is called ***pipe earthing***.

In the case of ***plate earthing***, the plate electrode is made of galvanized iron or steel. The size of the plate electrode should be 60 cm × 60 cm and should be burried atleast 1.5 m below the surface of the soil.

The thickness of the plate used should not be less than 6.3 mm. The plate should be set vertically.

All earth wires should be made of copper, galvanized iron, steel or aluminum. Interconnections of earth continuity conductors should be such that good electrical connections are permanently made. The neutral conductor should not be used as earth wire.

13.11.6 System of Wiring

Electricity supply authority, i.e., the State Electricity Boards provide electric supply upto a point outside the consumer's premises.

From this point the consumer will take connection to his main switchboard. Insulated electrical wires will then be taken out to various places in the premises to supply power to different types of electrical loads like lights, fans, refrigerators, room coolers, heaters, etc. There are different types of wirings used. The choice of wiring will depend upon a number of factors like whether temporary or permanent, type of building, aesthetic consideration, cost involved, durability, etc.

The various types of internal wirings are:

(i) Cleat wiring
(ii) Wood Casing wiring
(iii) Batten wiring
(iv) Conduit wiring

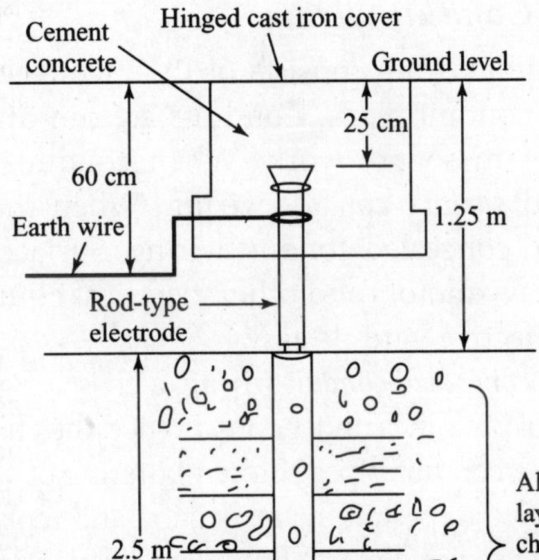

FIGURE 13.17

Earthing terminal from a pipe or rod-type electrode

(i) Cleat wiring

Cleats are made of porcelain and are fixed on walls or ceiling at intervals of 0.6 m. The insulated wires, i.e., the cable is taken through the holes of each cleat. Thus, the cleats support the wire, such a cleat wiring is cheap and is used for temporary installation.

(ii) Wood casing wiring

In wood casing wiring, the cable is run through a wood casing having grooves. The wood casing of a required length is fixed on the walls or ceiling with screws. The cables are placed inside the grooves of the casing. A capping, also made of wood with grooves, is used to cover the cables.

The casing and capping are made from well-seasoned teak wood. The casing should be fixed with flat-headed wooden screws to wooden plugs at an interval of 90 cm. After all the insulated cables are laid inside the grooves of the casing, the capping should be attached to the casing by rust-resistant screws. Care should be taken in fixing the screws on the cappings so that the insulation of the cables inside is not damaged. Wood casing–capping wiring system is used in dry places like in Rajasthan.

(iii) Batten wiring

In batten wiring, insulated wires are run on wooden battens. PVC wires are run on well-seasoned straight teak wood battens. The battens are fixed on the walls or ceilings by plugs and screws. The cables are held on the batten by means of tinned brass link clips. The clips are fixed on the battens with rust-resistant nails. Batten wiring is widely used for indoor installations. Batten wiring is cheap and takes comparatively less time to install.

(iv) Conduit wiring

Conduit wiring consists of PVC wires taken through either steel conduit pipes or through PVC conduit pipes. Conduits are run over the surface of walls and ceiling or are concealed under masonery work. When conduits are run over the surface of walls, the wiring is called *surface conduit wiring*. When the conduits are run inside the walls, the wiring is called concealed conduit wiring. Surface conduit wiring is used in factories for installation of heavy motors and other electrical equipment. The system is water proof and replacement of defective wires is easy.

In *concealed conduit wiring*, a chase or groove is cut on the wall to place the conduit pipes. In case of buildings under construction the chase should be provided on the wall and ceilings for laying the conduit pipes before plastering of walls and ceiling is done. Suitable inspection boxes are provided to permit the inspection and replacement of wires, if necessary. Concealed conduit wiring is used in almost all modern residential, commercial, and public buildings. The appearance of buildings from inside look good with concealed conduit wiring as compared to batten wiring.

13.11.7. Electric Circuits and Sub-Circuits

Once a decision has been taken on the type of wiring to be done, electrical wireman has to plan on the number of circuits and sub-circuits to be formed for carrying out the complete wiring of the building.

Indian Electricity Rules (1956) provide guidelines on the maximum load that should be connected to one sub-circuit consisting of lights, fans, 5A socket, etc. As per rules, the maximum load of one sub-circuit should not be more than 800 Watts and the number of outlet points should be restricted to 10.

Again, light and fan circuits should be separated from power circuits where comparatively heavy loads like air-conditioners, room heaters, etc., are to be connected. That is why, the 5A socket outlets and 15 A socket outlets should not be connected together to a single circuit.

All electrical loads like lamp, fans, etc., should be connected across a 230 V power supply individually. They should not be connected in series. The third point of a socket should be connected to earth terminal for safety reason.

Following the rules, electrical wiring is carried out which may include light and fan circuits, stair-cage lighting circuits, tube light circuits, etc. After the wiring is complete, it is tested for continuity and proper insulation.

Knowledge of energy efficient electrical lamps, chokes, etc., is essential to select the right type of electrical appliances.

13.11.8. Multiway Switching (Two-way and Three-way control)

Multiway switching enables lights, fans, etc., to be controlled from more than one location. A typical '2-way' switch is a Single Pole Single Throw (SPST) switch Single Pole Double Throw (SPDT) and Double Pole Double Throw (DPDT) switches enable devices to be controlled from multiple locations. Externally these switches look similar but their internal connections are different.

Switching on and Switching off an electrical load from two location, e.g., turning a light on or off from either end of a staircase required two SPDT switches as shown in Fig. 13.18 (a). A lamp controlled from three positions using two SPDT switches and one DPDT switch has been shown in Fig. 13.18 (b).

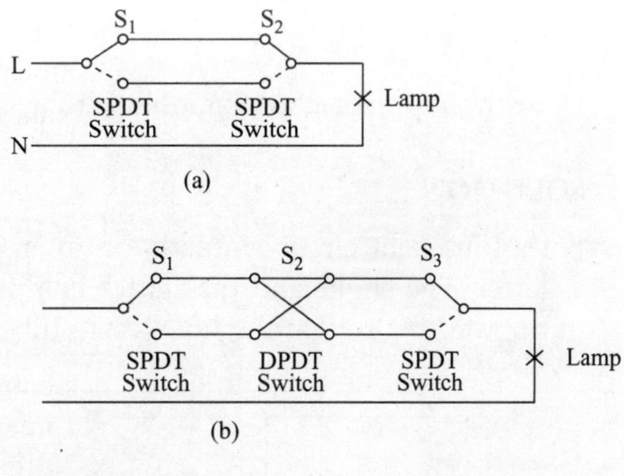

FIGURE 13.18

(a) Circuit diagram of one lamp controlled from two positions. (b) Circuit diagram of one lamp controlled from three positions

A single light or fan or a pump motor can be operated from as many number of points as required. The general formula is the end switches should be SPDT switch and the intermediate switches should be DPDT switches, the number of locations be odd in number, i.e., 3, 5, 7, etc. AS shown in Fig. 13.18 (b) where three switches have been used to control a lamp from any of the three positions. Using three switches, there are eight possible combinations of switch positions, four with the light on and four with the light off, as can be seen by operating the three switches.

13.11.9 System of Connection of Lights, Fans and Other Electrical Loads

All electrical loads are connected in parallel and not in series. There are separate circuits for light and fan loads and for heavy electrical loads like heaters, air conditioners, etc.

Fig. 13.19 shows the connection scheme for two lamps and one fan circuit. Each is operated by an independent switch. The fan is controlled by a regulator to change its speed.

As per Indian Electricity Rules, the number of light and fan points that can be put in one circuit is restricted to eight. This includes the 5 A plug points which are provided to plug-in some small electrical loads like a battery charger, a mosquito repeller, a tape recorder, a TV, etc. For comparatively heavy loads, 15 A plug points are to be provided through separate power circuits. In power circuits the wires used are thicker so that they are able to carry heavy currents.

Some typical house wiring circuits are shown through a few examples.

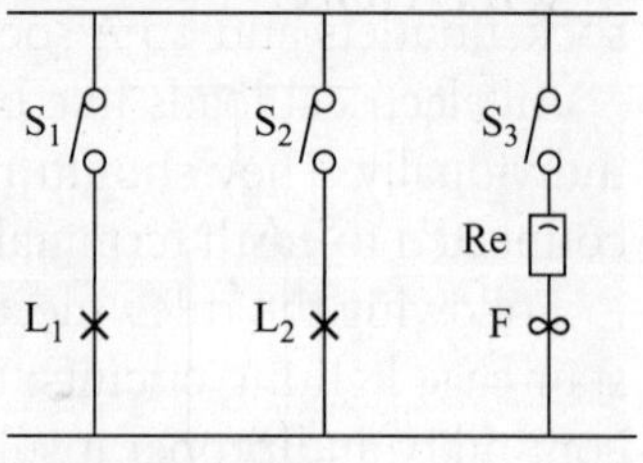

FIGURE 13.19

Light and fan control circuit

EXAMPLE 13.1

Draw the wiring diagram for a single tube light circuit.

SOLUTION

For a tube light circuit, in addition to an On/Off switch we will require a choke and a starter. The choke and the starter help in developing a high voltage across the tube during starting so that the flourescent tube gets illuminated.

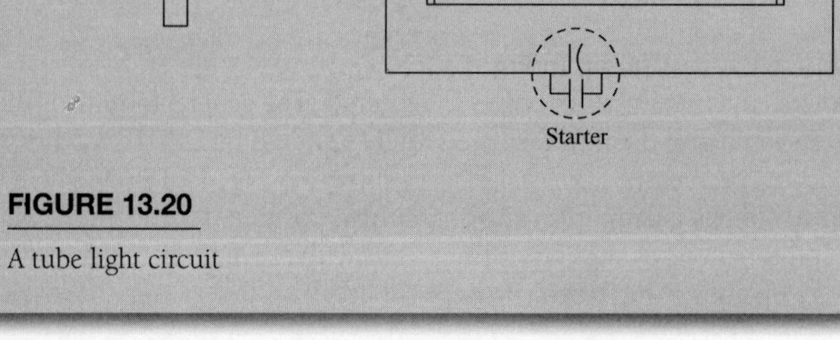

FIGURE 13.20

A tube light circuit

EXAMPLE 13.2

Draw the schematic diagram and wiring diagram for a staircase lamp controlled from two positions. It should be possible to switch-on or switch-off the lamp by any of the two switches, one located upstairs and the other located downstairs.

SOLUTION

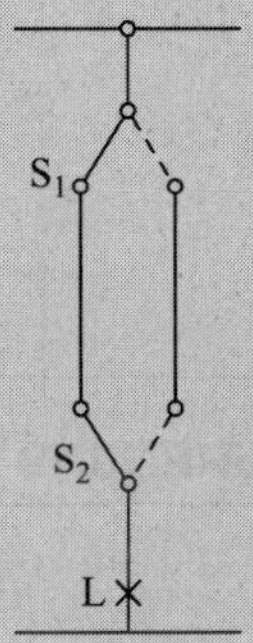

(a) Schematic diagram

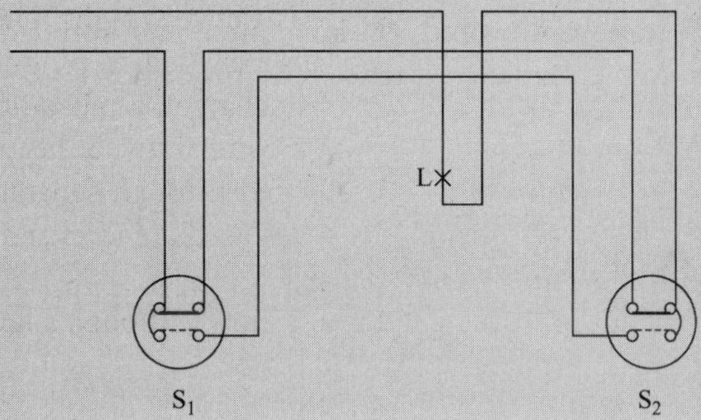

(b) Wiring diagram

FIGURE 13.21

Staircase lighting circuit

Two two-way switches have been used to control the lamp L. Fig. 13.21 (a) shows the schematic diagram whereas Fig. 13.21 (b) shows the actual wiring diagram. Switch S_1 is fixed on the ground floor near the staircase while switch S_2 is fixed on the first floor. While going up S_1 is operated to switch on the light L as shown. After reaching the first floor, switch S_2 is operated. The switch contact moves to position shown by the dotted line. The lamp gits switched off as power supply to the lamp is now cut off.

EXAMPLE 13.3

One lamp, one fan, and one staircase light are to be controlled by independent switches placed on a single switch board. Draw the schematic circuit diagram.

SOLUTION

The schematic circuit is shown in Fig. 13.22.

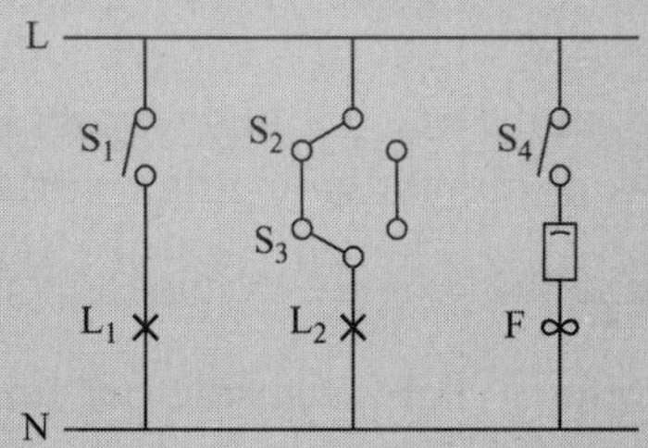

FIGURE 13.22

Schematic diagram of a light and fan circuit

EXAMPLE 13.4

Draw the schematic diagram for one lamp to be switched on and off from any of the three positions.

SOLUTION

In this case we need to use a two two-way switch and an intermediate switch. S_1 and S_3 are the two-way switches and S_2 is an intermediate switch. The working of the single-way, two-way and intermediate switches have been shown in Fig. 13.23 (a), (b), (c), and (d).

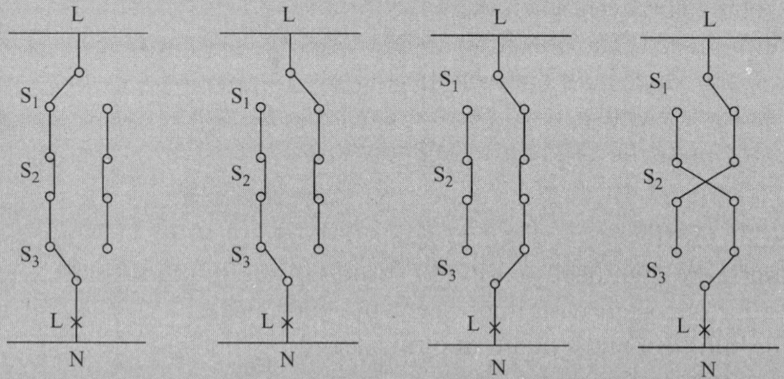

(a) Lamp is ON

(b) Lamp is OFF by S_1

(c) Lamp is ON by S_3

(d) Lamp is OFF by S_2

FIGURE 13.23

(a) Lamp is ON; (b) Lamp is made OFF by switch S_1; (c) Lamp is made ON by switch S_3; (d) Lamp is made OFF by switch S_2.

13.11.10 Electric Lamps

Electric lamps are used for illumination. We know that any hot body radiates energy in all directions. Light is a form of radiant energy. A radiant body which gives out light is termed as a source of light. Light waves emitted by the source fall on a surface to illuminate it. The amount of light falling on a surface is measured in lux. One lux is equal to lumen per unit area. $1 \text{ lux} = \dfrac{\text{lumen}}{\text{area}} = \text{lumen/m}^2$. The efficiency of any light source is determined by the amount of light available per unit of power it is consuming, i.e., by it's "lumen per watt". For example, lumen per watt of a CFL (compact fluorescent lamp) is much higher than an incandescent lamp (filament lamp).

We will now discuss the various types of light sources available in the market. These are incandescent lamps, fluorescent lamps, mercury vapour lamps, sodium vapour lamps, metal helide lamps, and LED lamps.

(i) In candescent lamps

In an incandescent or filament lamp, very thin tungsten filament is enclosed in an evacuated glass bulb. Tungsten can withstand very high temperature. A small quantity of inert gas (nitrogen with argon) is inserted in the bulb. By applying a voltage, the filament is operated at high temperature and it emits light radiations. Coiled coil-type filament is used for getting concentrated light. The lamps are available at a voltage rating of 250 V and power ratings of 15 W, 25 W, 40 W, 60 W,100 W, etc. The light output is 15 lumens per Watt.

Because of low-light output per watt of electricity consumed, incandescent lamps are gradually being replaced by more efficient light sources. The advantage of these lamps, however, is that they give some light, even when there is low voltage. Voltage variation only changes the amount of light produced. The operating life of these lamps is about 1000 hours.

Filament lamps produce both light and heat. Thus, the efficiency of conversion from electricity to light is low. Scientists have been researching on how to convert electricity to light without producing much heat, that is to say how to produce cool bulbs.

(ii) Fluorescent lamps

Fluorescent lamps are manufactured in the form of tube lights. In such lamps, the inside surface of the tube is coated with fluorescent powder which glow when subjected to ultra violet rays. The tube contains small amount of argon gas and a drop of mercury. Oxide coated tungsten filament are used as the two electrodes at the two ends of the tube. A starter and a choke are connected in the tube light circuit.

The wiring diagram of a tube light circuit has been shown in Fig. 13.20. On application of voltage, the starter and choke help to produce a high voltage, three to four times the supply voltage, to appear across the two electrodes and at the same time the tungsten filaments getting heated up. This causes a discharge to occur in the gas inside the tube. The fluorescent tube starts giving light. Once the discharge has occured, a much lower voltage is required to maintain the discharge. This kind of tube lights are available in different lengths and the power rating is usually 40 Watts. The light output of a fluorescent lamp is 70 lumens per Watt and has an average life of 7500 hours. Although the initial cost of a tube light circuit is more than a filament lamp, it is more efficient and lasts longer.

(iii) Mercury vapour lamp

High-pressure mercury vapour lamps consist of an inner discharge tube and an outer evacuated tube. The inner tube is provided with two main electrodes and an auxiliary electrode placed near one main electrode. The tube contains small amount of argon and a little mercury. Supply voltage applied across the main electrodes is not sufficient to cause discharge of mercury vapour between the main electrodes. But this voltage is sufficient enough to cause discharge between one of the main electrodes and the auxiliary electrode. The discharge then spreads rapidly and establishes discharge between the main electrodes. The argon initiates the discharge, and vapourises the mercury and then mercury vapour takes over. The efficiency of mercury vapour lamp is about 40 lumens per Watt.

(iv) Sodium vapour lamps

Sodium vapour lamps are gas-discharge lamps in which sodium vapour in excited state produce yellow light. They are used mainly in street lighting because of their yellow colour. They are available in two varieties as low pressure sodium vapour lamps and high-pressure sodium vapour lamps. High-pressure sodium vapour lamps are quite efficient. They are widely used in outdoor lighting like street lighting. These lamps work on the basis of an electric arc through sodium vapour. Other materials and gases are used to help start the lamp. High-pressure sodium lamps, called HPS lamps produce 80–140 lumens per Watt and has a life of 24,000 hours. Sodium vapour lamps should not be disposed off in the garbage as they may burst under pressure and cause injuries.

(v) Metal halide lamps

The metal halide lamps have an arc tube enclosed inside a larger glass bulb which has a coating to filter out the ultra violet light produced. These lamps produce light by an electric arc through a gaseous mixture of vaporized mercury and metal halides. Metal halides are compounds of metals with bromine and iodine. The most common halide used is sodium iodide. Metal halide lamps are high-intensity gas discharge lamps. Metal halide compounds in the arc tube improves the efficiency of the light source. These lamps operate at a pressure between 4 to 20 times the atmospheric pressure and require special fixures to operate safely. They also require an electrical ballast (choke) for initiating discharge. Their luminous intensity is around 75 to 100 lumens per Watt. Their life is 6000 to 15,000 hours, i.e., they last longer than other types of lamps. Metal halide lamps are used in overhead lighting of commercial, industrial, and public spaces such as parking lots, sports complexes, factories, retail stores, exhibition grounds, head lamps in automobiles, campus lighting, etc.

(vi) LED lamps

Light is a form of electromagnetic wave. In filament lamps, light is produced by conversion of electrical energy into heat energy and then to light energy. In light emitting diodes (LEDs), light energy is produced directly from electrical energy, and hence, LEDs are more efficient than filament lamps and fluorescent lamps. LED lamps also last long and there is no abrupt failure of these lamps. They dim out gradually.

LEDs are basically p–n junction diodes. Electrons release energy when they drop from higher orbital to lower one. Free electrons moving across a diode can fall into empty holes of the p–region. This involves a drop from the conduction band to a lower orbital. Therefore, electrons release energy in the form of photons (called light). This happens in any diode but is visible when the diode is composed of certain material. In an ordinary diode, the semiconductor material absorbs most of the light energy. LEDs are constructed in such away that they release large amount of light outward. Material used in LEDs is basically aluminium–gallium–arsenide. By adding impurity, they are made either p–type or n–type. A p–n junction is made using p–type and n–type material. When a battery is connected to the p–n junction, these is continuous supply of electrons to the n–type material from the battery. As electrons are continuously supplied by the battery, even though electrons and holes unite and release energy in the form of light, they do not become exhausted (depleted) but the process of emission of light from a LED continues. The wavelength of light emitted, and thus, its colours depend on the band gap energy of the materials forming the p–n junction.

LEDs are somewhat expensive. However, in years to come LEDs will replace less efficient light sources like incandescent lamps, and fluorescent lamps. This will eventually lead to energy conservation which is the need of present day situation of energy crisis.

13.12 ▶ Circuit Protective Devices and Safety Precautions

Circuits should be protected against any abnormal condition like overload and short circuit which may be due to any fault conditions or excess load connected to the circuit. Protective devices like fuses and circuit breakers are used, which help protect the circuit from burning out under abnormal conditions.

An important protective device used in all electrical installations is the fuse.

13.12.1 Fuses

A fuse is a device that, by fusion of one of its specially designed and proportional component open the circuit in which it is inserted when the current through it exceeds a given value for a sufficient time.

Fuses are of two types, viz the *rewirable type* and the *cartridge type*.

A rewirable fuse consists of a length of wire made either of tinned copper or some other metal of such size that while it will carry the rated current, it will fuse, i.e., melt, and thus break the circuit if the current rises above the value for which the circuit is rated. Rewirable fuse has the disadvantage that the wire can be replaced by any other wire of any size, which will defeat the purpose for which a fuse is used. Often the user may replace a fuse by a thicker wire which is not desirable at all.

A rewirable fuse also gives an external flash when blown off.

The performance of a fuse is improved by having the fuse wire in a sealed cartridge packed with filler material. Such a fuse is called a cartridge fuse. There are two types of cartridge fuses, viz diazed-type fuse, or simply D-type cartridge fuse and high rupturing capacity, or HRC cartridge fuse. In cartridge fuses, the cartridge, which carries the fuse wire has to be replaced by another cartridge of the same rating. A cartridge of some other rating will not fit into the base. This restricts the user in replacing the fused cartridge by another cartridge of the same rating

For the protection of a power system against any fault condition, e.g., single-phase-to-ground fault or phase-to-phase fault or three-phase fault, protective devices like *relays* and *circuit breakers* are incorporated in the system.

13.12.2 Bimetallic Strip for Overload Protection

We have known that metals expand when heated. The rate of thermal expansion of a metal strip will depend upon the material. Different metals have different coefficient of thermal expansion. For example, steel, copper, brass, etc., have different coefficient of thermal expansion.

A bimetallic strip consists of two different metal strips usually steel and copper joined togather throughout their length by welding or riveting or brazing as shown in Fig. 13.24.

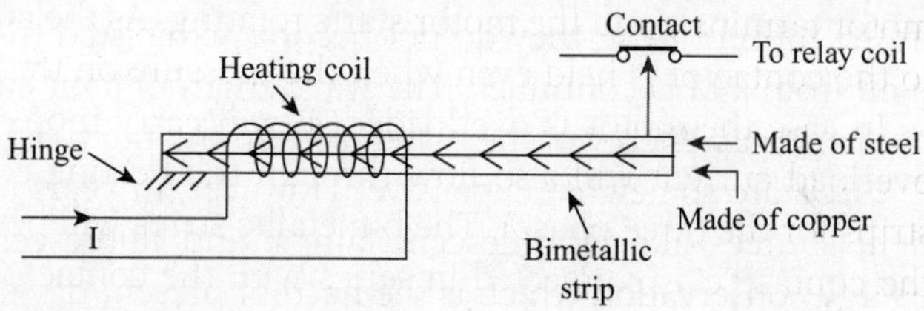

FIGURE 13.24

Bimetallic strip when heated bends upwards to open the contact of a relay circuit

When excessive current flows through the heating coil wound around the bimetallic strip, the strip bends upwords slowly and pushes the contact. The contact is connected in the circuit of a relay coil. The circuit breaks when the relay get deenergized and switches off the main circuit through which excessive current, I was flowing. Note that the heating coil carries the same which to flowing through the device. For understanding of the operation of a bimetallic thermal overload relay, let us consider a motor control circuit shown in Fig. 13.25. Suppose the motor is driving a mechanical load.

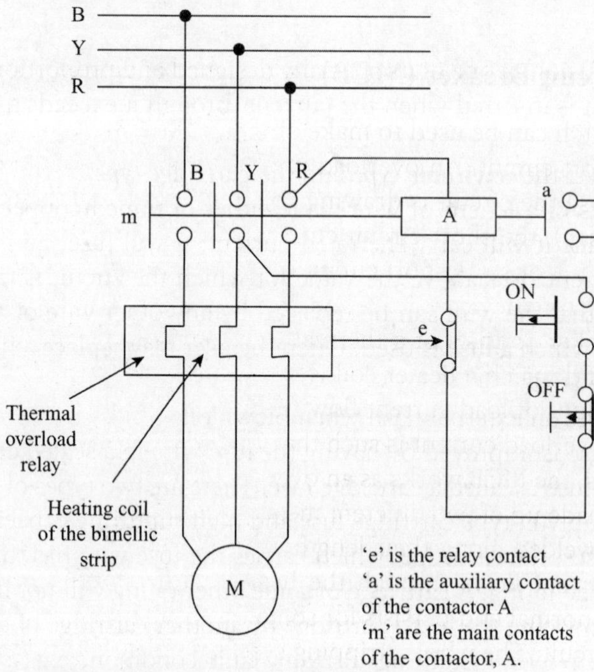

FIGURE 13.25

Motor control circuit being protected against overload using bimetallic thermal overload relay.

As shown in the Fig. 13.25, when the ON push button is pressed, the contactor coil A gets energized as supply becomes available from phase R to coil A to over-load relay contact, e to OFF push button to ON push button (which is now closed) and then to phase Y. The main contacts 'm' of the contactor A will close and three-phase supply becomes available across the motor terminals and the motor starts rotating. As the auxiliary contact 'a' is now closed supply to the contactor is held even when the pressure on the ON push button is released.

In case the motor is overloaded, it will carry more current than the normal current. This overload current will also flow through the heating coils of the bimetallic strips (three such strips for the three phase). The bimetallic strips will bend much and open its contact 'e'. Since the contact 'e' is connected in series with the contactor coil circuit, when e opens the circuit breaks. The contactor coil A gets deenergized and gets switched off to open its main contacts 'm' and this way the motor is stopped.

Bimetallic strips in the form of thermal overload protection device, called thermostats, are provided is may heating devices like in room heaters and hot air blowers, in electric irons, etc.

A circuit breaker will break the circuit automatically under any fault conditions. Circuit breakers can be operated by remote control with the help of relays. Relays detect the fault and initiate tripping of the circuit breaker. Under fault conditions the electrical quantities like current, voltage, and frequency become abnormal and is sensed by the relay. There are various types of relays, like distance or impedance relay, induction-type over-current relay, differential relay, etc.

13.12.3 Miniature Circuit Breaker (MCB)

We have known that a switch can be used to make or break a circuit. A fuse is used to break a circuit in case of any short circuit. An overload protecting device like a thermal overload relay breaks a circuit in case the circuit is drawing more current than its rated capacity. For a rated current of 10 A_F (say), the short-circuit current may be as high as 80 A whereas the overload current may be 15 A or so. A short circuit is a very dangerous situation, and hence, the fuse must be a quick acting device.

A miniature circuit breaker consists of a thermal tripping arrangement. The arrangement has a bimetallic strip around which a heater coil is wound to create heat. The heat produced will depend upon the amount of load current flowing through the coil. The deflection of the bimetallic strip in case of overload current is such that it will activate a tripping mechanism to switch off the supply. Thus, the MCB works as an overload protection device.

The bimetal strips are made up of two different metals, usually brass and steel. These metallic strips are riveted and welded along their length. These are designed such that they will slowly bend when the current flows through the heating coil wound around them. If the current flow is more than normal and persists for some time, the strip bends more to trip the latch and switch off the circuit. The whole stripping system is enclosed in an enclosure made of molded insulting material. The main function of the MCB is to switch on the circuit manually and open the circuit automatically when the current in the circuit is more than rated value. The MCB can also be used to switch on and switch off the circuit manually.

MCBs are time delay tripping devices used in house hold electrical devices like the air-conditioners, refrigerators, pump motors, etc. They automatically switches off the circuit in case overloading condition exists for some time. A fuse is a quick acting protective device and protects the circuit in case of short circuit. An MCB is a slow acting protective device and protects the circuit in case of overload condition.

MCBs are available different current ratings from 6 A to 125 A in modular construction. There is hardly any maintenance required. They are easy to be used by a common man.

13.12.4 Earth Leakage Circuit Breaker (ELCB)

An earth leakage circuit breaker is used as safely device in an electrical installation. It detects any voltage coming in metal enclosures of electrical equipment and disconnects the circuit when the voltage appearing is of high level. It is now somewhat compulsory for all the residential and commercial installations to connect ELCBs for the protection of persons using electrical appliances and gadgets. It detects the leakage current from the body of the device to earth and causes tripping of the circuit breaker. ELCBs is used to protect persons from electric shock and injury. ELCBs are connected between metallic body and earth wire. Without proper earth connection ELCBs will not work.

An ELCB detects fault current from Live to the Earth (ground). If sufficient voltage drop appears across the sensing coil of the ELCB, it will switch off the power. An ELCB has six terminals. Four terminals are for the line and neutral connection (in and out). One E for connection to Earth and one F terminal for connection to the Fault, i.e., the metallic body of the electrical item where there may be fault due to leakage. When an Earth fault occurs, a voltage appears on the body of the equipment and a current will flow to earth from F to E and to the ground. This current will set up a voltage across the trip coil of the relay which is connected between E and F. The relay will get energised and operate a contact to switch off the load from the supply by opening the main circuit breaker.

13.12.5 Residual Current Circuit Breaker (RCCB) or Residual Current Device (RCD)

Residual Current Circuit Breaker (RCCB) is also referred to as a Residual Current Device (RCD). This device breaks an electric circuit when there is any earth fault and prevents persons getting an electric shock.

This device is designed to quickly disconnect a circuit when there is any difference between the incoming current, i.e., the line current and in return current, i.e., the current returning through the neutral wire. The difference between the incoming current and the return current is the residual current. Under balanced condition, incoming current is equal to the return current. Any difference will indicate any abnormality present due to leakage. Leakage current may lead to electric shock to a person and may even cause death. A current nearly 25 to 30 mA is sufficient to cause cardiac arrest (heart attack) or serious harm to a person.

RCCBs are often combined with MCB (Miniature Circuit Breaker) as a single unit as a protective device.

RCCBs are designed to disconnect an equipment or any electrical gadget from the supply if there is any leakage. By disconnecting quickly enough, it saves a person being electrocuted, i.e., getting severe electric shock.

This leakage current protection is in addition to over-load and short circuit protection provided to electrical equipment.

Electricity authorities have provided regulations for use of RCDs as a protective device against leakage current. In most cases RCDs are installed in the Distribution Board. If a single RCD is provided, the whole installation will get disconnected in case of any fault being detected.

13.13 ▶ ILLUMINATION

Light is a form of electromagnetic energy radiated from a source which is capable of being perceived by the human eye. Light source is the cause and illumination is the result of the light falling on a surface. The following are some of the terms used in describing illumination.

Luminous flux is the total quantity of light energy emitted per second from a light source.

Luminous intensity is the luminous flux emitted by the source in a given direction per unit solid angle.

Solid angle, ω is the ratio of the area of the surface to the square of the distance between the area and the point source. Suppose a source of light is subtended by a sphere of radius, r. Area of the sphere is $4\pi r^2$. The solid angle, $\omega = \dfrac{4\pi r^2}{r^2} = 4\pi$ steradians. It is the largest solid angle subtended at a point.

Lumen is the unit of Luminous flux

Candle Power is the number of lumens emitted by a source per unit solid angle in a given direction. It is equal to lumens per solid angle.

Illumination is the luminous flux received by a surface per unit area and is measured in lumens per square metre or flux.

13.13.1 Laws of Illumination

1. ***Illumination:*** Illumination, E of a surface is inversely proportional to the square of the distance, r of the surface from the source.

   This is also called *inverse square law* which is expressed as,

   $$E = \frac{1}{r^2}$$

   Here it is assumed that the distance between the surface and the source is large so that the source can be considered as a point source.

2. ***Lambert's Cosine law:*** According to this law, illumination E is directly proportional to the cosine of the angle made by the normal to the illuminated surface from the source with the direction of the incident flux.

   Using these laws, it is possible to calculate the illumination on any surface or at the working place.

EXAMPLE 13.5

A filament lamp of 100 W is suspended at a height of 4 m over a plane surface. The efficiency of the lamp is 0.8 watt per candle power. Calculate the illumination on the surface normal to the source.

SOLUTION

Candle power of the lamp $= \dfrac{100\text{ W}}{0.8}$

i.e., C.P. = 125

Illumination, $E = \dfrac{\text{C.P.}}{d^2} = \dfrac{120}{4^2} = 7.5$ lux.

EXAMPLE 13.6

A lamp with a reflector is mounted 8 m above the centre of a circular are of 10 m diameter. Candle power of the lamp with the reflector over the circular area is 400. Determine the illumination vertically below the lamp and at the periphery of the circular area.

SOLUTION

Illumination at the centre of the circular area is given as

$$E_1 = \frac{\text{C.P.}}{h^2} = \frac{400}{8^2} = 6.25 \text{ lux (see Fig. 13.26).}$$

Illumination at the periphery of the circular area

$$E_2 = \frac{\text{C.P.}}{d^2}\cos\theta$$

$$\cos\theta = \frac{h}{d}$$

or, $$d = \frac{h}{\cos\theta}$$

$$E_2 = \frac{CP}{\dfrac{h^2}{\cos^2\theta}}\cos\theta = \frac{CP \times \cos^3\theta}{h^2} = \left(\frac{CP}{h^2}\right)\cos^3\theta$$

or, $$E_2 = 6.25 \times \frac{8}{\sqrt{8^2+5^2}} = 6.25 \times \frac{8}{9.42} = 5.3 \text{ lux}$$

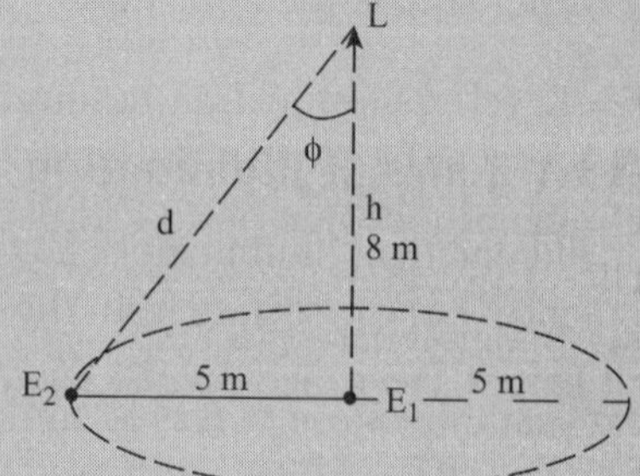

FIGURE 13.26

Refers to example 13.

13.13.2 Requirements of Good Lighting and Types of Lighting

Good lighting should provide visual comfort to your eyes. Sometimes good lighting is confused with high level of illumination. Glare is a very bright light that is difficult to look at. Light sources should be chosen and fixed such that it maintains a comfortable illumination level without glare.

Types of lighting

Different lighting schemes with different types of fixures are used. They are

1. **Direct lighting:** In this a deep reflector directs nearly 90 percent of the light on the working plane.
2. **Semi-direct lighting:** The reflector directs nearby 60 to 90 percent light on the working plane and rest illuminates the walls and ceiling.
3. **Indirect lighting:** The reflector reflects most of the light upwords to the ceiling for diffused reflection. An inverted reflector is used for this purpose. The glare is reduced to a minimum and provides illumination without any shadow.

Different types of work area requires different levels of illumination. Intensity of illumination has to be chosen so that there is no wastage of energy and comfort level is maintained.

13.13.3 Calculation of Illumination

There are several methods of calculation to determine the number of light points and their wattge required for an area.

The following formula can be used to calculate the requirement of fittings required. Let N be the number of fittings required for illuminating an area. Then

$$N = \frac{E \times A}{O \times UF \times MF} \text{ where}$$

E = Level of illumination required, i.e., the lumens required
A = Area to be illuminated in sq.m
O = Luminous flux produced per lamp
UF = Utilization factor
MF = Maintenance factor

An example of use of this formula has been provided as below.

EXAMPLE 13.7

A small workshop 16 m × 10 m and 4 m high is to be illuminated to a comfortable illumination level of 200 lux (lumens per square metre). The utilization factor, i.e., the ratio of total lumens reaching the working plane to the total Illmens giving out by the lamps is 0.75. The maintenance factor, i.e., the ratio of illumination under normal working condition to the illumination when the lamps, reflectors, etc., are clean is 0.85. Calculate the number of 60 W lamps required to illuminate the whole area if the lumens per watt of the lamp chosen in 100.

SOLUTION

Working area, $A = 16 \times 10 = 160\ m^2$

Illumination required, . $E = 200$ lux

Lumen output of one Lamp, $O = 60 \times 100 = 6000$ lumen

$UF = 0.75,\ MF = 0.85$

Number of Lamps required = N

$$N = \frac{EA}{O \times UF \times MF} = \frac{200 \times 160}{6000 \times 0.75 \times 0.85} = 9\ (\text{approx.})$$

13.14 ► Tariff Schemes—Uniform and Differential

Tariff is the rate at which electricity is supplied to the consumer. Tariff refers to the money the consumer has to pay for making electricity available to them. Electricity supplying authority takes into consideration the following factors to calculate the tariff.

1. Recovery of cost of production.
2. Recovery of capital investment over a period of time.
3. Recovery of cost of operation and maintenance of equipment.
4. Earn a suitable profit on capital investment.

Types of tariffs are:

1. Simple tariff,
2. Flat-rate tariff,
3. Block-rate tariff,
4. Two-part tariff,
5. Maximum-demand tariff,
6. Power-factor tariff
   (KVA max. demand, sliding scale, KW and KVAR).

Electrical power system mainly consists of generation, transmission and distribution. For generation, we have many public sector undertakings and private owned generating stations. Electrical transmission system is carried out by the central government body Power Grid Corporation of India (PGCIL). To facilitate this, India is divided into five regions. Northern, Southern, Eastern, Western, and North eastern region. Further, within every state, there are State Load Despatch Centres (SLDCs). The distribution is carried out by many distribution companies (DISCOMS) and state electricity boards/SEBs.

Total cost levied on the consumer on electricity consumption is divided into three parts:
Total cost of electrical energy = Fixed cost + Semi fixed cost + Variable cost

$$= a + b \times kW + c \times kWh.$$

Here, a = fixed cost which is independent of maximum kW demand or actual energy consumed.
b = constant which when multiplied by maximum kW demand gives the semi fixed cost.
c = a constant which when multiplied by actual energy consumed in kWh gives the running cost.

Thus, the total amount to be paid by the consumer depends on the maximum demand, actual energy consumed plus some constant sum of money.

All this is calculated on the assumption that load power factor is not less than 0.8, otherwise a penalty is added.

For residential consumers, the tariff fixed by a particular state electricity supply authority is *uniform for particular level of consumption*. The rate increases for more units of consumption. For example, rates are fixed for 0–100 units, 101–300 units 301–500 units and for above 500 units of electricity consumption.

Differential rates are fixed for other categories of consumers like commercial establishments, industry, railways for low-tension and high-tension supply system. Differential tariffs are fixed in various ways like two-part tariff, Maximum-demand tariff, Power-factor tariff, Block-rate tariff, and off pick-load tariff.

13.15 ▶ Electric Shock, Safety Precautions and Treatment of Electric Shock

Electricity has to be used very carefully. Careless use may lead to severe consequences to the person using electricity. For safety, every electrical circuit should be connected to supply through switch–fuse arrangement. The fuse used should be of correct rating. No connection should taken using naked wires. All metallic parts of electrical equipment should be earthed.

13.15.1 Electric Shock

When a wire carrying current comes in direct contact with our body, there will be a current flow through the body to the earth. The current that flows through the body is the cause of electric shock. The severity of electrical shock will depend upon the system voltage and the resistance of the body through which current will flow.

Our body resistance varies from 10 KΩ to 50 KΩ. This resistance depends upon body mass, skin, moisture, and point of contact of the body with the voltage source. The current that passes through the body determines which tissues or organs will be be affected.

The magnitude of current flow and its effect on the body are shown below:

0.4 mA : Slight sensation.
1.8 mA : Shock but no pain and loss of muscular control.
16 mA : Painful shock.
23 mA : Severe painful shock, muscular contraction, and breathing difficulty.
4 A : Heart paralysis.

If one person of body resistance of 10 KΩ accedently touches a 230 V exposed wire, a current of $230/10 \times 10^3 = 23 \times 10^{-3} = 23$ mA wire flow through his body to earth giving him a severe painful shock which may lead to his death unless given immediate medical help.

13.15.2 Safety Precautions

Electricity is very much required for a number of applications. In our homes, there is presence of electric outlet points including connected lamps, fans, refrigerators, etc. The electrical wiring should be made following electricity rules and good workmanship.

While using electricity, there are safety precautions which should be strictly followed. Some of the do's and dont's are listed below.

1. Switch off main switch first before you work on any electric circuit. Even when you replace a electric bulb, first switch off the supply.
2. Never handle electrical items connected to the supply with wet hands and on bare feet. Always wear shoes and stand on a wooden platform to touch any electrical point for inspection or repair.
3. Never assume that the switch is off, make sure for yourself first. For the negligence of someone else, you should not suffer.
4. Always use three-pin plug for connecting equipment and other electrical gadgets like table lamp, electric iron, mixture grinder, etc.
5. Make sure that the ground pin of all sockets is connected to earth.
6. Avoid playing with terminals of power supply.
7. Report any unsafe electrical wiring to the maintenance department.
8. Capacitors in the circuit store electricity for a long time. They must be properly discharged before you work on them.
9. If any person touches a live electric wire and cannot detach him or her, do not touch her, instead switch off the power immediately. If this is not possible try to separate the body of the person from the conductor using a non-conducting material like a wooden stick, dry piece of cloth, etc.

13.15.3 First Aid for Electric Shock

You need to be very careful while working with electrical appliances, equipment and wiring to avoid getting electric shock.

Electrical shock needs emergency medical attention. But before the patient can be shifted to the hospital some emergency measures, as listed below can be taken.

1. Do not touch the person if he is still in contact with electric wire. If you touch, you will also get an electric shock. Separate the person from current using dry rope or dry wooden stick. Cut off the supply, if possible.
2. Place the person on his back with his feet higher than his head. Keep him warm with adequate covering.
3. If there is visible bleeding, it should be stopped. If there is severe pain, some pain relieving medicine be given.

4. Check for breathing by looking for rise and fall of chest, listening for sound of breathing and putting one hand on the stomach to feel it rising and falling.
5. If the person is not breathing, you have to start CPR (Cardio-Pulmonary Resuscitation). The steps for CPR are:
   5.1. Place the patient on a firm and flat surface.
   5.2. Place your hands one on top of the other on the chest of the patient.
   5.3. Compress the chest up to a depth of 4 to 5 cm rapidly 30 times at the rate of more than one compressions per second. Compression and release should take equal fraction of a second.
   5.4. After 30 compressions, open the airway by lifting the chin and tilting the head back. This will free the tongue from the back of the throat.
   5.5. Seal his nostrils with your fingers and blow air into his mouth by having mouth to mouth contact until you see his chest rise. Remove your mouth to the side and let the chest fall. Inhale some fresh air. Repeat for another time blow of air.
   5.6. If the chest does not rise after giving two rescue breaths as in serial No 5 above, go back to compressions 30 times as in serial no 3 and then try with rescue breaths as in serial no 5.
   5.7. Continue with CPR until the patient shows signs of recovery and emergency medical services can be provided.

13.16 ► Efficient Use of Electricity

Electrical energy is used in almost every place like in running electric trains and metro-rails, heating and cooling of buildings, creating cold storage facilities, illuminating building interiors, street lighting, etc. The demand for electricity is increasing day by day. Non-renewable energy resources like coal and gas are getting depleted. It is therefore important that we save electricity by reducing losses and misuse.

Energy efficient technologies, gadgets, and appliances are to be developed and the less efficient ones be replaced. For example, gradually, electric filament lamps will be replaced by CFLs (compact fluorescent lamps) and LED lamps. Electronic fan regulators have already replaced the traditional variable-resistance-type regulators. For saving energy on heating and cooling, the losses due to leakage have to be reduced. Further, through proper automatic control devices, constant temperature could be maintained irrespective of variation of outside temperature. This will ensure that over cooling and over heating is not done. The persons using electricity need to be made aware of the need for efficient use of electricity.

A subject with the name of Energy Management deals with the reduction of losses in the transmission and distribution system, use of energy efficient technologies, conservation of energy, more and more emphasis on non-conventional sources of energy for generation of electricity, and energy audit to reduce losses, etc.

Important Terminologies and Points to Remember

1. All forms of natural energy like coal, gas, oil, etc, are converted into electrical energy using a turbine and a generator. The turbine rotates the generator in which electricity is produced.
2. Electricity is generated in electric power generating stations at 11000 volts or so. This voltage raised at higher levels to transmit electricity through over head transmission lines. This is done because it is economical to transmit electricity at higher voltages.
3. From the transmission lines, electricity is supplied to consumer after reducing the voltage at lower levels in steps.
4. Transformers are used to step-up and step-down the voltage level at different places in transmission and distribution of electricity.
5. Generation, transmission, distribution and utilization together forms a system which is called the power system.
6. In a power station where electricity is produced with coal as the source is called a thermal power station. The thermal or heat energy of coal is utilized to produce high pressure steam which drives a steam turbine. The turbine is coupled with the alternator. The turbine works as a prime mover which rotates the alternator to produce electricity. Thermal power stations may also be gas or oil and diesel based.
7. The main equipment or devices and raw material in a thermal power plant are—boiler, turbine, A.C. generator, fuel, i.e., coal, gas or oil, furnace, water pumps and fans and heaters, condenser, cooling tower and step-up transformers.
8. Hydroelectric power generating stations produce electricity utilizing the potential energy of water to run a water turbine. We have to construct a dam over a river to hold the water flow of a river and raise the water level. Water from a high head is allowed to flow with a heavy force through the water turbines to rotate. The turbine works as a prime mover to the generator. Hydroelectric power stations are constructed in places where a river is flowing through a gradient like in Bhakra & Dehar in Himachal Pradesh, Salal in Jammu and Kashmir, Sardar Sarovar in Gujrat, Nagarjun Sagar in Andhra Pradesh, Idukki in Kerala, and Damodar Valley Maithon Dam West Bengal.
9. In nuclear power plants, electricity is generated from the heat produced due to fission of uranium atoms. The heat energy is used to produce steam to run a turbine. The turbine drives the generator to produce electricity. Some of the nuclear power plants in India are located in Kaiga in Karnataka, Kalpakkam in Tamil Nadu; Narora in Utter Prasesh; Terapur in Maharastra, Rawatbhata in Rajasthan.
10. Generation of electricity from sources like solar energy, wind energy, geothermal energy, biomass energy, etc., are called non-conventional sources of energy.
11. Solar energy and wind energy are being used to produce electricity in different parts of India on large scales.

    For generation of electricity from sun-rays, solar photovoltaic cells are used. PV cells are made of silicon which releases electrons when exposed to sun.

    The energy of wind is utilized to produce electricity through wind turbines coupled with a generator.
12. Electricity generated in power houses is brought to consumers through transmission and distribution lines. Transmission is done at high voltages because of lower cost involved and of having higher transmission efficiency. In India, highest level of transmission voltage is 400 KV.
13. The distribution voltage is either 11 KV or 415/230 V. Heavy industries are supplied with electricity at 11 KV. They stepdown the voltage using stepdown transformer to the desired voltage level.
14. For all other consumer, electricity is provided at 415 V. For single-phase supply 230 V is the standard supply voltage.
15. Electricity distribution to the consumers can be done either through overhead lines or through underground cables. 11 KV lines are taken from a 33 KV/11 KV substations and the lines are taken around the area where electricity has to be supplied using overhead concrete or steel poles or through underground cables.

16. A distribution network is created consisting of feeders, distributions and service mains to supply electricity at 415/230 V to individual consumers.
17. A three-phase, 4-wire distribution system consists of three-phase wires and a neutral wire. Depending on the requirement, either three-phase supply or single-phase supply can be provided.
18. The consumer is supplied with electricity through a main distribution board where an energy meter is installed to keep a record of the units of electricity consumed for which they are sent monthly or bimonthly electricity bills.
19. Neutral wire is taken out from the secondary of the distribution transformer in addition to the three-phase wires named R, Y, B. A fifth wire called, earth wire is provided for the purpose of earthing of the metallic body parts of all electrical equipment. This is essential from the safety point of view.
20. A consumer has to make the internal wiring of his building. Four types of wiring can be done, namely cleat wiring, wood casing capping wiring, batten wiring, and conduit wiring. In modern residential buildings, concealed conduit wiring is made.
21. Various protective devices like fuses, miniature circuit breakers, earth leakage circuit breakers, etc., are installed in the circuits connected to electrical loads.
22. Efficient use of electricity is the cell of the day. Energy efficient technologies, gadgets and appliances are being used to reduce losses and increase energy utilizing efficiency.

REVIEW QUESTIONS

Short Answer Type Questions

1. What are the various sub-systems of an electrical power system? Explain the significance of each of them.
2. What are the various ways of generation of electricity from natural resources? What are their limitations?
3. Draw a line diagram and explain how electricity is generated in a coal-fired thermal power plant.
4. Explain the basic principle of generation of electricity in a nuclear power plant.
5. What is meant by renewable energy sources? Give two examples.
6. What is the basic principle of solar electricity generation. What are the requirements and limitations?
7. Give brief descriptions of generation of electricity from non-conventional energy sources.
8. Explain how electricity is brought from the generating stations to the consumers of electricity.
9. Compare dc and ac transmission systems. Explain the advantages of high-voltage transmission.
10. Compare overhead versus underground distributions systems.
11. Explain the function of neutral wire and earth wire in an electrical distribution system.
12. Explain the importance of earthing of electrical systems.
13. Explain any one method of earthing.
14. Explain the various types of electrical wiring.
15. Draw the schematic diagram of stair case lighting, i.e., one lamp controlled from two positions. Also draw the wiring diagram.
16. Draw the wiring diagram of a tube light circuit.
17. What are the protective devices used in power systems?
18. What do you mean by an electrical substation? Draw a line diagram of a distribution substation.
19. Prepare a list of equipment and accessories required to set up a 500 kVA, 11 kV/415 V indoor-type distribution substation.
20. Draw a single-line diagram showing all equipment and accessories of a pole-mounted type 100 kVA, 11 kV/415 V distribution transformer.

21. Draw a single-line diagram of a power system starting from generation through transmission and distribution to the consumers. Show also the positions of substations.
22. Write a short note an efficient use of electricity.
23. Write the applications of the following: (i) MCB (ii) Bus-bars (iii) ELCB (iv) Cartridge-type fuse.
24. Name the different types of lamps available in the market. Why are compact fluorescent lamps replacing the incandescent lamps?
25. Briefly explain the uniform tariff and differential tariff systems.

Multiple Choice Questions

1. The voltage range at which power is generated in power stations is
   (a) 440 V to 1 kV (b) 11 kV to 30 kV
   (c) 33 kV to 66 kV (d) 66 kV to 132 kV.
2. The generated voltage is transmitted through transmission lines after stepping up the voltage so that
   (a) the voltage drop in the transmission lines is reduced
   (b) for the same power transmitted the current flowing through the lines is reduced, there by reducing the transmission losses
   (c) the transmission lines are kept far away from the ground level
   (d) large amount of power could be transmitted.
3. In India, transmission voltage level range is
   (a) above 1000 kV
   (b) 765 to 1000 kV
   (c) above 400 kV
   (d) 66 kV 400 kV.
4. Which of the following is used as a prime mover in a nuclear power station?
   (a) Water turbine (b) Gas turbine
   (c) Steam turbine (d) Diesel engine.
5. Which of the following is renewable energy?
   (a) Coal (b) Oil
   (c) Gas (d) Solar.
6. Which one of the following is not used as a source of heat energy in a thermal power station?
   (a) Gas (b) Oil
   (c) Uranium (d) Coal.
7. In which of the following places in India, a nuclear power plant has not yet been constructed?
   (a) Tarapur in Maharastra
   (b) Rawatbhata (near Kota) in Rajasthan
   (c) Mehrauli in Delhi
   (d) Kalpakkam in Tamil Nadu.
8. Cogeneration means
   (a) the system where electricity, steam, and hot water are simultaneously produced
   (b) the system of generating electricity using both coal and diesel oil
   (c) the system of converting heat into mechanical energy at a very high temperature
   (d) the system of generating electricity and other forms of mechanical energy from any primary source of energy.
9. Which of the following is not true for the amount of power generated, P, in a hydroelectric power plant?
   (a) P depends on the quantity of water discharged
   (b) P depends on the head of water being discharged
   (c) P depends on the specific weight of water
   (d) P depends on the capacity of the water reservoir.
10. Which of the following is not the location of a hydroelectric power station in India?
    (a) Salal in Jammu and Kashmir
    (b) Nagarjun Sagar in Andhra Pradesh
    (c) Dehar in Himachal Pradesh
    (d) Faridabad near Delhi on river Yamuna.
11. Commercial solar cells producing electricity has an efficiency of the range.
    (a) 10 to 15 per cent
    (b) 20 to 40 per cent
    (c) 50 to 70 per cent
    (d) above 70 per cent.
12. Which of the following is not a part of the electricity distribution network?
    (a) Feeder (b) Distributor
    (c) Optical fiber (d) Service mains.

13. Which of the following statement is not true?
    (a) The neutral wire is taken out from the star point of the secondary of the distribution transformer
    (b) The neutral wire is taken out from the star point of the secondary of a step-up transformer
    (c) Each phase wire and the neutral wire constitute single-phase supply
    (d) The neutral wire is taken out from the star point of the secondary of the distribution transformer and is earthed of the substation.

14. In a distribution transformer
    (a) the primary windings are star connected and the secondary windings are delta connected
    (b) both the primary and secondary windings are delta connected
    (c) the primary windings are delta connected and the secondary windings are star connected
    (d) both the primary and secondary windings are star connected.

15. Which of the following types of wiring is used for purely temporary installations?
    (a) Concealed conduit wiring
    (b) Surface conduit wiring
    (c) Wood casing–capping wiring
    (d) Cleat wiring.

16. The function of choke and starter in a tube light circuit is to
    (a) create a high voltage across the tube during starting
    (b) improve the power factor of the tube light circuit
    (c) reduce the power consumed by the tube light circuit
    (d) help draw very high current during starting.

Answers to Multiple Choice Questions

1. (b)	2. (b)	3. (d)	4. (c)	5. (d)	6. (c)
7. (c)	8. (a)	9. (d)	10. (d)	11. (a)	12. (c)
13. (b)	14. (c)	15. (d)	16. (a)		

CHAPTER 14

Semiconductor Devices

Topics Discussed

1. Atomic theory
2. P-type and n-type semiconductor material
3. P–n junction
4. Forward and reverse-biased p–n junction
5. Semiconductor diodes
6. Zener diode
7. Bipolar junction transistors
8. Transistor configurations
9. Transistor characteristics
10. Transistor as an amplifier
11. Field effect transistors
12. MOSFET
13. SCR
14. SCR applications
15. TRIAC
16. LDR
17. LED
18. Seven segment displays
19. Photodiodes
20. Photovoltaic cells
21. Phototransistor
22. Optocoupler

14.1 ► Introduction

The transistor, most commonly used in electronic circuits, was invented in 1948. The first integrated circuit (IC) came in the market in the mid 1960s. Through continuous research and development, very large scale integrated (VLSI) circuits have been brought in use, resulting in gradual miniaturization of electronic circuits and devices. The sizes of calculators, computers, communication satellite, and all such electronic gadgets and systems have become smaller but powerful. It is predicted that nano-technology will further revolutionize the industry in general in the years to come.

Electronics deals with the flow of electrons through vacuum, gas or semiconductors. Electronic devices like diodes, transistors, field effect transistors (FETs), silicon controlled rectifiers (SCRs), opto-electronic devices, and resistors, inductors, capacitors, etc., form circuits of electronic gadgets, equipment, and control systems.

An electronic device consists of integrated circuits which have several diodes, transistors, resistors, capacitors, etc., mounted on a single chip. Electronic components like diodes, transistors, SCRs, etc., are made of semiconductor materials.

As we know, all materials are classified into three categories, namely conductors, semiconductors, and insulators. Gold, silver, copper, aluminium, etc., are conducting materials. Conducting materials have a large number of free electrons in their atomic structure which allow flow of current.

Rubber, ceramic, glass, wood, paper, bakelite, mica, etc., are insulating materials. In these materials no free electrons are available and as such no current should flow through them.

Substances like germanium, silicon, carbon, etc., are called semiconducting materials. Atoms of these materials binds themselves through sharing of electrons in their outermost shell or orbit. Such bonds are called covalent bonds. At absolute zero degree temperature, semiconducting materials behave like insulators as no free electrons are available for conduction. However, with increase of temperature or on application of voltage, some electrons become free electrons by breaking away from their covalent bonds and create a current flow. That is why these materials are called semiconductors. To have a complete understanding of behaviour of semiconducting materials we will first have a quick review of atomic theory.

14.2 ► Review of Atomic Theory

An atom of any material consists of a central nucleus around which electrons are orbiting in different shells or orbits. Electrons are held in orbits due to the electrostatic force between them and the nucleus. Electrons are negatively charged while the protons in the nucleus are positively charged. Because there is an equal number of protons with positive charge and orbiting electrons of equal but opposite charge, an atom is electrically neutral. If, however, an atom loses an electron it will lose some negative charge and will become a positively charged ion. Similarly, if an atom gains an electron it will become a negatively charged ion. Different materials are made up of different types of atoms. However, their electrons and protons are identical. An electron from one atom can replace an electron of any other atom of different material.

Electrons orbit in different shells around the nucleus. Each shell contains a definite number of electrons. For example, the total number of electrons in first, second, and third shells are 2, 8, and 18, respectively. The number of electrons in the outermost shell of the atom determines the electrical property of the material. The number of electrons present in the outermost shell are called *valence electrons*. The outermost shell of an atom may be completely filled or partially filled. The outermost orbit must have eight electrons. If the outermost orbit has less than eight electrons, we call them vacancies or holes (i.e., empty spaces).

Let us consider atoms of two semiconducting materials, silicon and germanium. Silicon has 14 electrons orbiting in three orbits and they are distributed as 2, 8, 4. The nucleus has

14 protons which are positively charged. Thus, the atom is electrically neutral. Germanium has 32 electrons orbiting in four shells as 2, 8, 18, 4. The nucleus has 32 protons, and hence, the atom is electrically neutral. Two-dimensional representations of silicon and germanium atom has been shown in Fig. 14.1. Both silicon and germanium have four valence electrons and four holes in their outermost orbits. The closer an electron is to the nucleus, the stronger is the binding force between the electrons and the protons in the nucleus. Electrons in the outermost orbit are comparatively loosely bound with the nucleus. This means, a small amount of energy will be required to take out an electron from the outermost orbit. When an electron leaves its orbit, it becomes a free electron.

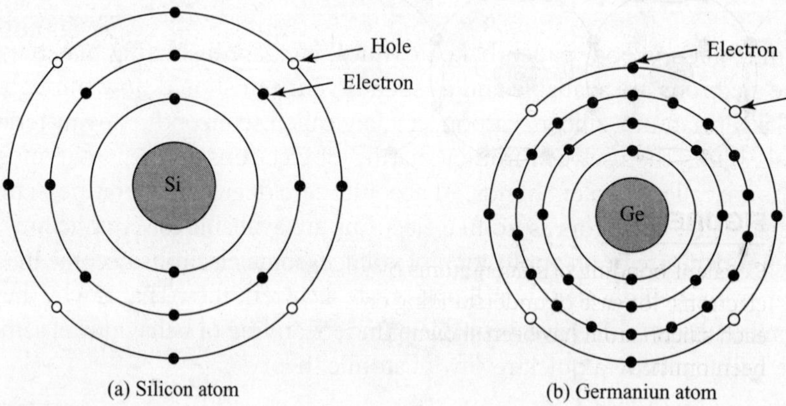

FIGURE 14.1

Atomic structure of silicon and germanium atom (outermost orbit in both the cases has four electrons leaving four unfilled spaces or vacancies, called holes)

14.3 ► BINDING FORCES BETWEEN ATOMS IN SEMICONDUCTOR MATERIALS

A semiconductor atom having four valence electrons and four holes require four more electrons so as to make the outermost orbit completely filled (total number must be eight). The atoms in a crystal are arranged so closely that electrons orbit in valence shells of two atoms. Each valence shell electron fills the hole of the neighbouring atom as shown in Fig. 14.2. In the figure, atoms of silicon material have been shown. For ease of understanding, only the outermost orbits of atoms have been shown. Sharing of electrons of the neighbouring atoms to satisfy the need to have eight electrons on the valence shell in an atom is called *covalent bonding*. Because of covalent bonding, i.e., bonding through sharing of electrons, it is seen that the valence shells of all the electrons are full, i.e., all of them have eight electrons in their outermost orbit. At absolute zero temperature there will be no free electrons in the crystal. However, although the electrons are bound to their atoms due to covalent bonding, a rise in temperature breaks some of the covalent bonding

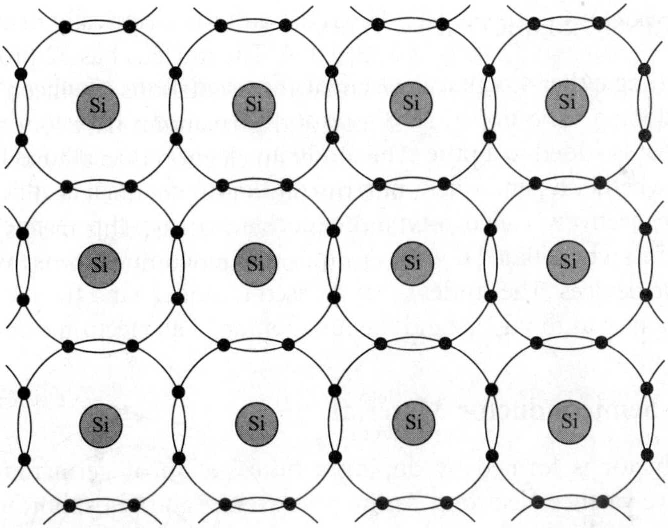

FIGURE 14.2

Covalent bonding of silicon atoms by sharing of valence electrons, for ease of understanding only the outermost orbit of each silicon atom has been shown. The inner two orbits have been omitted.

and make some electrons free. A semiconductor material, silicon or germanium where the electrons are bound to their respective atoms and are not free to conduct electric current, are called ***intrinsic***, or ***pure semiconductors***. When the temperature of the crystal is raised, external energy in the form of heat gets applied to the semiconductor material. This heat energy enables the valence electrons to acquire sufficient energy to break away from the atoms and become free electrons. When an electron leaves to become free, it leaves a vacant space called a hole. For every free electron there will be a corresponding hole produced, which is called an electron–hole pair. A large number of such electron–hole pairs are formed due to rise in temperature of the semiconductor. When electrons become free, they get attracted and fall into a hole created by another electron. This merging of free electrons and holes is called recombination. The time of creation of a free electron and its falling into a hole is very quick and the time taken is of the order of nanoseconds. Thus, pure silicon or germanium is not of much use in electronics except for the manufacturing of heat- or light-sensitive resistance.

The conductivity of semiconductor materials can be increased by adding some amount of another material having either three or five valence electrons. Adding such materials with the pure semiconductor material is called ***doping***. The material formed by doping is called extrinsic semiconductors.

14.4 ► EXTRINSIC SEMICONDUCTORS

The process of adding either a pentavalent element or a trivalent element to a pure semiconductor is called doping. The doped semiconductor is called extrinsic semiconductor. The doping material that is added to a pure semiconductor is also referred to as impurity material. Depending upon whether a pentavalent or trivalent doping material is added, extrinsic semiconductors are, respectively, called n-type semiconductor or p-type semiconductors. These two types of materials when joined together form, a p–n junction which is the basis of working of all the electronic devices. The students are advised to understand the mechanism of working of a p–n junction so as to understand the functioning of all electronic devices and circuits.

14.4.1 N-Type Semiconductor Material

N-Type semiconductor is formed by doping a pure silicon or germanium crystal with a material having five valence electrons. Antimony, arsenic, and phosphorous are pentavalent materials as can be read from the periodic table. If arsenic in very small quantity is added to a silicon crystal, four out of five valence electrons will form covalent bonds with silicon atoms with one electron left free. Thus, for each arsenic atom there will be one free electron. Although the percentage of arsenic added is very small, the number of atoms being very large, a huge amount of free electrons will be available in the n-type semiconductor. These electrons being free (not taking part in any covalent bonding) are loosely bound to their parent atom and are free to conduct electricity. Fig. 14.3 shows impurity atoms of antimony

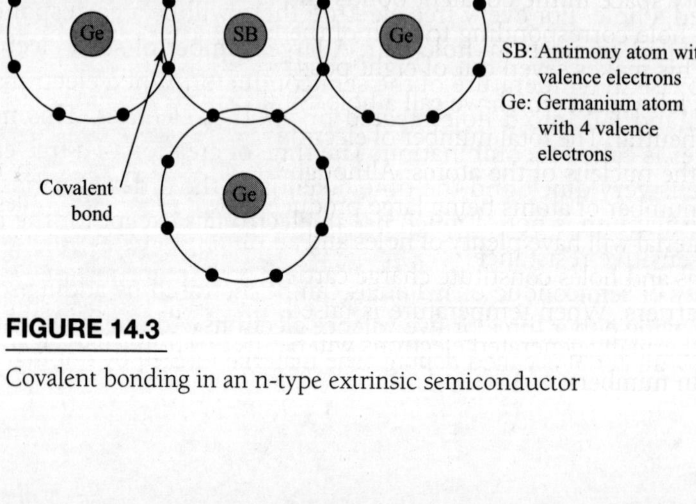

FIGURE 14.3

Covalent bonding in an n-type extrinsic semiconductor

having five valence electrons forming covalent bonds with germanium having four valence electrons. There is one extra electron for each impurity atom added. This extra electron is not a part of any covalent bond and is called free electron. This free electron has been shown out of the orbit.

It is noticed that each antimony atom is making covalent bonds with four neighbouring germanium atoms. Each bond has one electron belonging to germanium atom and one electron belonging to antimony atom. Each antimony atom will make four covalent bonds with four germanium atoms. By sharing of electrons in the covalent bonds, all the atoms will satisfy their need to have all the eight positions filled in their outermost orbit, i.e., their valence shells.

It may be noted that the n-type semiconductor, thus formed remains electrically neutral, i.e., neither positively charged nor negatively charged. This is because the total number of electrons including the free electrons is equal to the total number of protons in the nuclei of the atoms.

The added impurity material has infact donated one free electron per atom to the extrinsic semiconductor, and hence are called *donor atoms*. Donor atoms create free electrons which form the majority charge carrier (responsible for current flow) in an n-type material.

Temperature rise above absolute zero also creates free electrons and holes due to breaking of covalent bonds, thus increasing the total number of free electrons. However, a certain amount of holes are also formed.

When electrons leave their positions creating holes, the movement of electrons gets associated with the movement of holes. The holes therefore form charge carriers, and since they are in minority, they are called minority charge carriers in the n-type semiconductor.

Thus, in an n-type semiconductor the majority charge carriers are the electrons and the minority charge carriers are thermally generated holes.

14.4.2 P-Type Semiconductor Material

P-Type material is formed when silicon or germanium crystal is doped with (added with) a small percentage of trivalent impurity material like boron, gallium or indium. When covalent bonds are formed between boron having three valence electrons with silicon having four valence electrons, there will be shortage of one electron in the covalent bonds. This is represented by an empty space in the covalent bonds and is called a hole as shown in Fig. 14.4. There will be one hole corresponding to each of the impurity atoms taking part in forming covalent bonds. This makes seven out of eight positions filled.

One position is left vacant which we call a hole. Similar to n-type material, p-type material is also electrically neutral. The total number of electrons in the orbits is equal to the total number of protons in the nucleus of the atoms. Although the amount of impurity material added is small, the total number of atoms being large produce a large number of holes in the crystal. Thus, a p-type material will have plenty of holes and an n-type material will have plenty of free electrons. Electrons and holes constitute charge carriers. In a p-type material the holes are the majority charge carriers. When temperature is raised, there will be creation of free electrons and holes. These thermally generated electrons will be the minority charge carriers because of their being small in numbers.

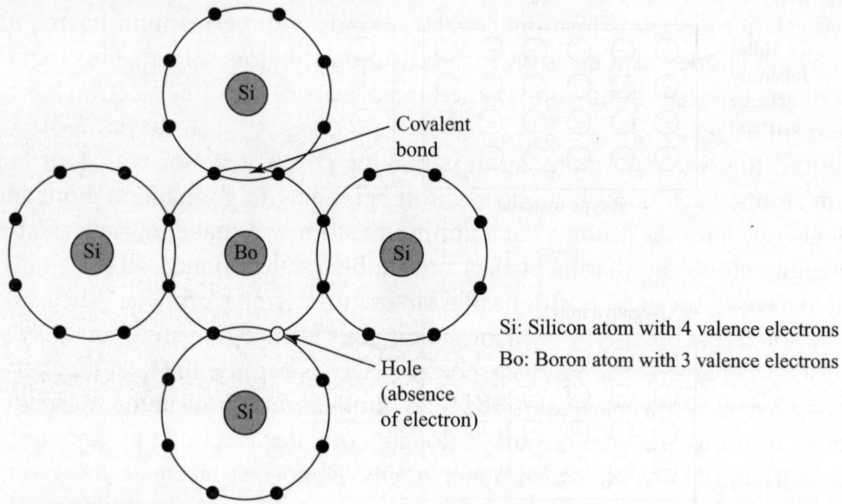

FIGURE 14.4

P-type semiconductor material

14.4.3 The p–n Junction

A p-type semiconductor can be represented by holes as the majority charge carriers. The trivalent impurities that produce a p-type semiconductor are called acceptor impurities because the holes are ready to accept any free electrons. A free electron coming from elsewhere occupying a hole in a p-type material will create negative ions on the p-side because the atom will gain one more electron. This way, the number of electrons with negative charge will be one more than the number of protons with positive charge on the nucleus. Thus, a p-type semiconductor will have negative acceptor ions and holes as majority carrier.

An n-type semiconductor can be represented by donor impurities because they give one free electrons to the semiconductor crystal and become positive ions. p-type and n-type semiconductor materials have been shown side by side in Fig. 14.5 (a). In the p-type material small circles represent the holes as the majority carriers. In the n-type material the black dots represent the free electrons as the majority charge carriers. Electrons are negative charge carriers while holes are positive charge carriers.

It may be noted that when an electron moves out of an atom, the atom becomes a positively charged ion which is immobile, i.e., unable to move. Similarly, addition of an electron in a hole makes an atom a negatively charged immobile ion. The minority carriers produced due to thermal effect have not been shown in Fig. 14.5. When a p-type semiconductor is joined with an n-type semiconductor, as shown in Fig. 14.5 (b), through a special technique, a junction called p–n junction is formed. We will examine what happens to the electrons and holes at a p–n junction.

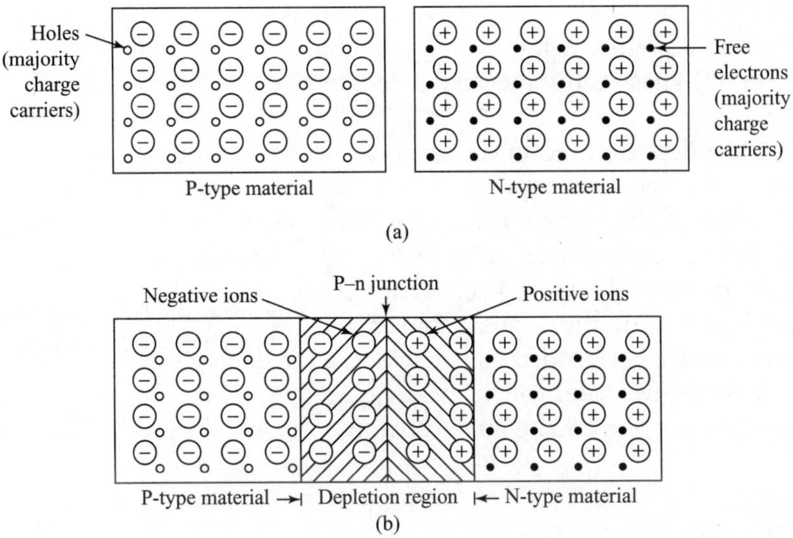

FIGURE 14.5

(a) P-type and n-type materials represented side by side; (b) formation of p–n junction

At the p–n junction there will be a tendency of the free electrons from the n-type material to *deffuse* (move from a high-concertration area to a low-concertion area) into the p-side and combine with a hole nearest to the junction. The free electron crossing over from the n-side to the p-side will leave behind positive immobile ions on the n-side of the junction. The electrons crossing over the junction will occupy the holes in the p-type material making the atoms negatively charged immobile ions. The atoms accepting the negative charge become negatively charged ions which were earlier neutral atoms. Thus, looking at Fig. 14.5 (b) we see that on one side of the junction there is an accumulation of negative ions and on the other side there is accumulation of positive ions. Negative ions created on the p-side close to the junction will acquire a negative voltage and the positive ions created on the n-side close to the junction will acquire a positive voltage. The negative voltage on the p-side will repel further diffusion of electrons from the n-side. The positive voltage on the n-side will repel diffusion of holes from the p-side. *When a p–n junction is made there is an initial diffusion of electrons and holes which creates a barrier voltage at the junction, which stops any further diffusion of charge carriers. This initial diffusion of charge carriers at the junction, and the development resultant barrier voltage take place when a p–n junction is formed during the manufacturing process. The barrier voltage depends upon the amount of doping, charge carriers, and the junction temperature. For germanium the barrier voltage is 0.3 V and for silicon the barrier voltage is 0.7 V at room temperature (25°C)*. The shaded portion on both sides of the p–n junction is having only immobile ions of opposite polarities which creates a potential difference, i.e., barrier voltage. This portion is devoid of any electron or hole, i.e., any charge carriers. This region is depleted of any charge carrier, and hence is called the ***depletion region***. It may be noted that the thickness of the depletion region in Fig. 14.5 (b) has been shown expanded. In fact, this layer is very thin, of the order of micrometer.

The p-side of the depletion region consists of acceptor impurity atoms that have lost their holes associated with them by accepting electrons and have become negatively charged ions. Similarly the n-side of the depletion region consists of donor atoms (pentavalent atoms) that have lost their associated free electrons and have become positive ions. As shown in Fig. 14.5 (b), the depletion layer is equally divided on both sides of the p–n junction. This is because both the p-type and the n-type materials have been equally doped. Equal percentage of doping materials have been added on both sides. If the doping is different, the width of the depletion region on the two sides will be different. Application of some voltage across the p–n junction is called biasing. Depending on the polarity of biasing the width of the depletion layer will change.

14.4.4 Biasing of p–n Junction

Biasing of a p–n junction means application of some external voltage across the two sides of the p–n junction. When the p-side is connected to the positive terminal of a battery and the n-side is connected to the negative terminal, the p–n junction is said to be a ***forward-biased*** junction. If the positive terminal of the battery is connected to the n-side and the negative terminal on the p-side, the p–n junction is said to be a ***reverse-biased*** junction.

(a) A forward-biased p–n junction

In Fig. 14.6 (a) is shown the p-side connected to the positive terminal of a battery and the n-side connected to the negative terminal. The holes on the p-side are positively charged and the electrons on the n-side are negatively charged. When forward biased, the positive terminal of the battery will repel the holes from the terminal. Similarly, electrons on the n-side will be repelled from the negative terminal of the battery. As a result, the width of the depletion layer will be reduced. The potential barrier will also get reduced. If the applied voltage is gradually increased, the depletion region and barrier potential will disappear as shown in Fig. 14.6 (b).

The resistance R in the circuit is connected to limit the current flowing in the circuit. Fig. 14.6 (a) shows the p–n junction before the forward voltage is applied. As in Fig. 14.6 (b), when the switch S is closed, the forward voltage gets applied. When the voltage is gradually increased from zero voltage to 0.3 V, for the germanium semiconductor the barrier voltage is overcome. When the barrier voltage is overcome, the depletion layer disappears. Electrons from the n-side are attracted by the positive terminal A of the p-side and the holes from the p-side get attracted by the negative terminal B of the n-side.

Electrons are the negatively charged particles and the holes are assumed to be positively charged particles. As soon as the potential barrier is overcome at 0.3 V for germanium and at 0.7 V for silicon, the majority charge carriers start moving across the p–n junction establishing

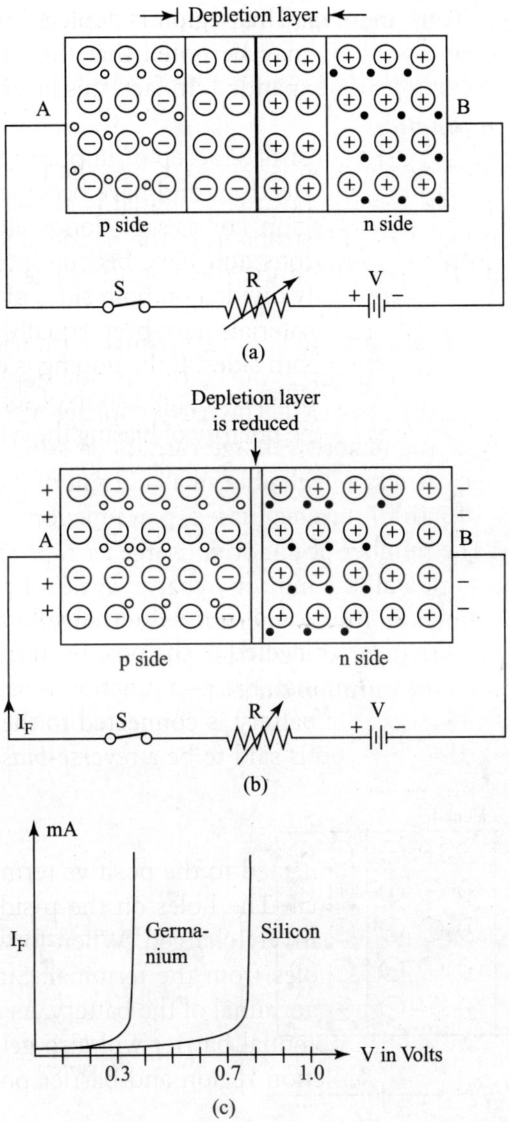

FIGURE 14.6

Forward biasing of a p–n junction. (a) Before switch S is; (b) switch S is closed and forward voltage gradually increased; (c) forward characteristics of the p–n junction

a forward current, I_F to flow as shown in Fig. 14.6 (c). The forward voltage, V versus forward current, I_F characteristics for germanium and silicon have been shown. With increase of forward voltage beyond 0.3 V or 0.7 V for germanium and silicon, respectively, the forward current increases as shown. Thus, in forward-biased p–n junction, the potential barrier is neutralized allowing current flow.

(b) A reverse-biased p–n junction

In reverse biasing, the battery connection is reversed, i.e., the negative terminal of the battery is connected to the p-side and the positive terminal is connected to the n-side of the p–n junction as shown in Fig. 14.7. Electrons from the n-side are attracted to the positive terminal of the battery and the holes from the p-side get attracted to the negative terminal of the battery.

Since the holes of the p-side are attracted by the negative terminal and the electrons of the n-side are attracted by the positive terminal of the battery, the depletion layer gets widened as the applied voltage is gradually increased. The barrier voltage also gets gradually increased and as such the possibility of the majority charge carriers crossing the barrier is reduced to zero. Due to minority charge carriers a negligibly small current of the order of micro amperes will flow as shown in Fig. 14.7 (b). A reverse-biased, p–n junction, therefore, offers very high resistance to current flow. The number of minority charge carriers are small and a very small reverse voltage is required to pull all the minority charge carriers by the two terminals of the applied voltage across the junction. Any further increase in reverse voltage does not increase the negligible small reverse current produced. Hence, this very small amount of reverse current is also referred to as *reverse saturation current*.

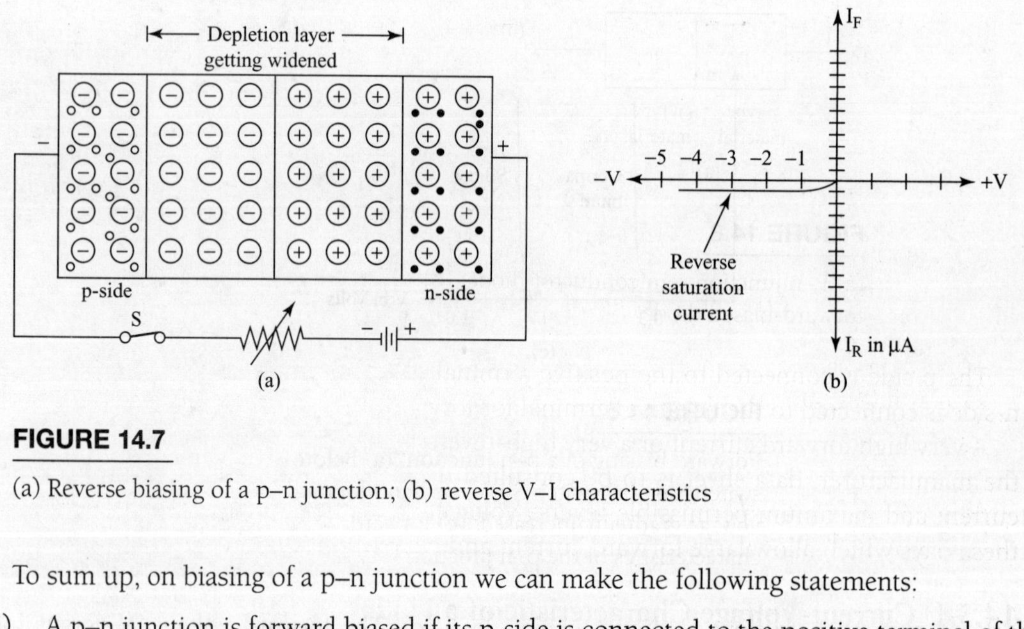

FIGURE 14.7

(a) Reverse biasing of a p–n junction; (b) reverse V–I characteristics

To sum up, on biasing of a p–n junction we can make the following statements:

(i) A p–n junction is forward biased if its p-side is connected to the positive terminal of the supply and n-side is connected to the negative terminal of the supply.

(ii) The depletion layer width gets narrowed down on application of forward voltage.

(iii) The majority charge carriers current is established in a forward-biased p–n junction.

(iv) Barrier potential for germanium is 0.3 V and for silicon is 0.7 V at room temperature. These are the voltage drops across the p–n junction when current flows.

(v) When p-side is connected to the negative terminal and n-side is connected to the positive terminal of the supply, the p–n junction is said to be reverse biased.

(vi) The width of the depletion layer, and hence, the barrier potential increases when the junction is reverse biased.

(vii) A minutely small current flows through a reverse-biased p–n junction due to the minority charge carriers.

(viii) A forward-biased p–n junction offers very small resistance to current flow while a reverse-biased p–n junction offers very high resistance to current flow.

14.5 ▶ Semiconductor Diodes

A semiconductor diode is simply a p–n junction which offers very low resistance when forward biased and very high resistance when reverse biased. Diodes are available in different current ratings. Low-current-rated diodes are used in switching circuits as the diode works like a switch allowing current to flow in one direction. A p–n junction with connecting leads on both sides form a p–n junction diode as shown in Fig. 14.8 (a). The symbolic representation of a p–n junction diode has been shown in Fig. 14.8 (b).

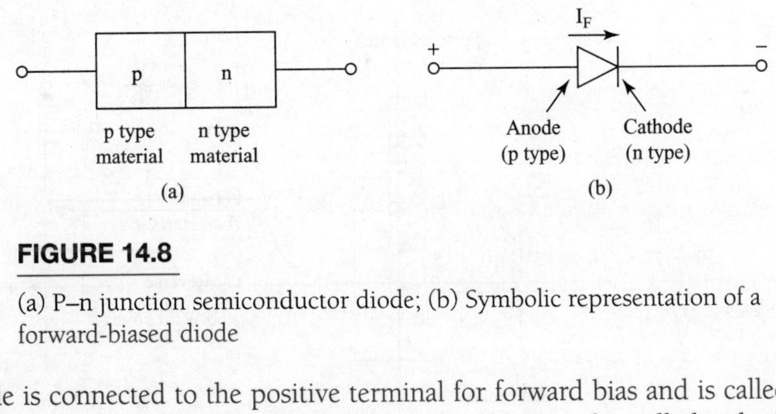

FIGURE 14.8

(a) P–n junction semiconductor diode; (b) Symbolic representation of a forward-biased diode

The p-side is connected to the positive terminal for forward bias and is called anode. The n-side is connected to the negative terminal for forward bias and is called cathode.

A very high forward current or a very high-reverse voltage can destroy a diode. That is why the manufacturer, data sheet is to be consulted to note the maximum permissible forward current and maximum permissible reverse voltage. High-current power diodes are available these days which allow large forward current and considerable amount of reverse voltage.

14.5.1 Current-Voltage Characteristic of a Diode

When a p–n junction diode is connected to a source of supply in such a way that it is forward biased, the relationship between the voltage applied and current flowing will give us a forward

I–V characteristic. The connection diagram for finding the I–V characteristic has been shown in Fig. 14.9. When the applied voltage is gradually increased, at a small value of forward voltage the forward current is negligible small. At a voltage near 0.3 V, the current suddenly increases. This voltage at which the forward current starts increasing is called the cut-in voltage of the diode. The voltage drop across the diode while it is conducting remains almost constant. For the germanium semiconductor diode, the forward voltage drop is 0.3 V and for the silicon diode, the forward voltage drop is 0.7 V.

For determining the reverse characteristic, the supply connection has to be reversed. Under the reverse-biased condition, the junction resistance is very high and ideally no current

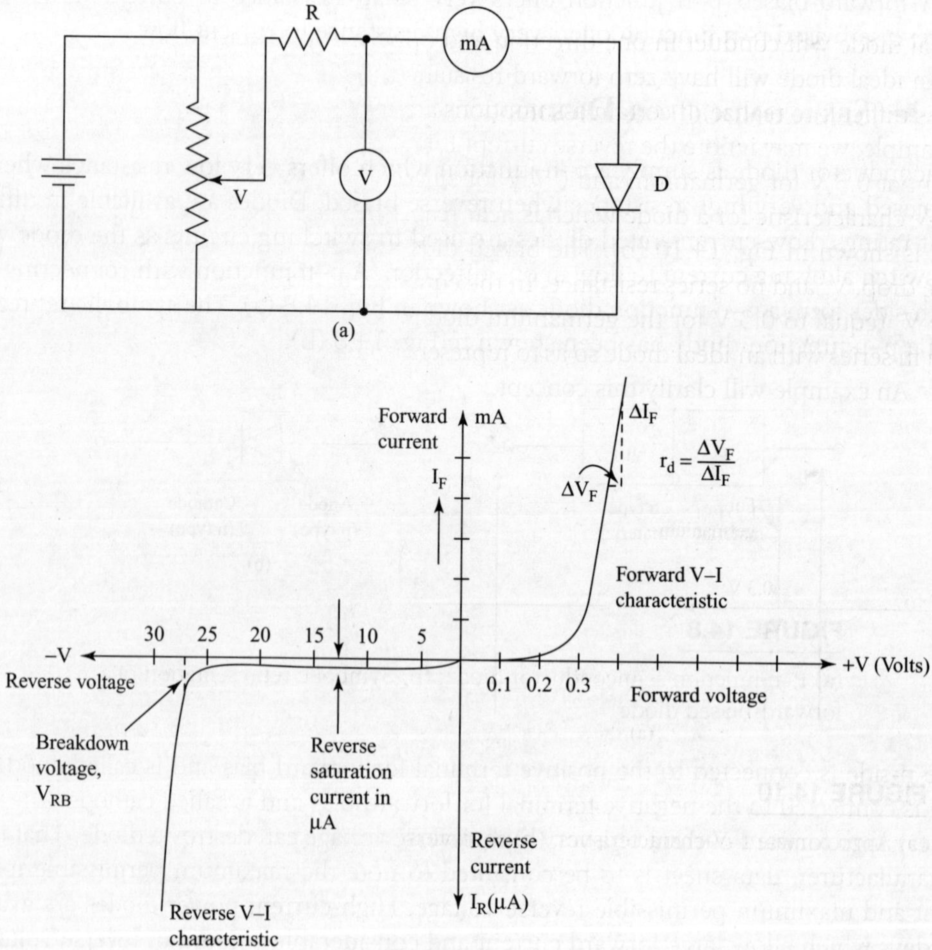

FIGURE 14.9

(a) Circuit diagram; (b) I–V characteristic

should flow. But due to minority charge carriers, a negligibly small current of the order of microamperes will flow. This current is also called leakage current of the diode. It gets saturated to its initial value of a few microamperes or even less than that. Increase of negative biasing, i.e., increase of negative voltage across the diode does not increase this reverse current. However, if the reverse voltage is increased to a large value, at one stage, the p–n junction will break down with a sudden rise in reverse current. The reverse voltage at which the diode breaks down and a large reverse current starts flowing is called the breakdown voltage. At this reverse breakdown voltage, current continues to increase.

14.5.2 An Ideal Diode

An ideal diode will conduct in one direction and oppose any current flow in the other direction. An ideal diode will have zero forward resistance and infinite reverse resistance. An ideal diode is difficult to realize. If certain assumptions are made, we may realize a near ideal diode. For example, we may ignore the reverse current I_R and assume that forward voltage drop, V_F as constant at 0.3 V for germanium and 0.7 V for silicon (for an ideal diode, $I_R = 0$ and $V_F = 0$). The I–V characteristic for a diode which is near real is shown in Fig. 14.10 (a). The equivalent circuit is shown in Fig. 14.10 (b). The biased diode is assumed to have a constant forward voltage drop, V_F and no series resistance. In the equivalent circuit of a practical diode a voltage source V_F (equal to 0.3 V for the germanium diode and 0.7 V for the silicon diode) has been shown in series with an ideal diode so as to represent voltage drop across an ideal divode equal to zero. An example will clarify this concept.

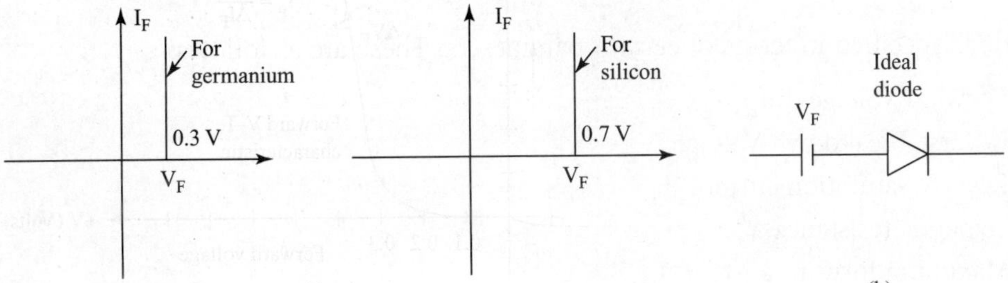

(a) (b)

FIGURE 14.10

(a) Approximate I–V characteristics; (b) equivalent circuit

EXAMPLE 14.1

A silicon diode is connected across a 3 V supply with a series resistance of 20 Ω. Neglecting diode resistance, calculate the diode current.

SOLUTION

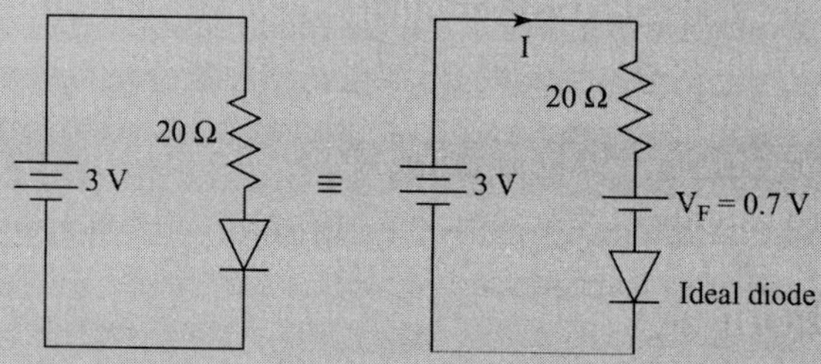

FIGURE 14.11

A silicon diode has $V_F = 0.7$ V. The diode equivalent circuit has been shown. Applying Kirchhoff's voltage law,

$$+3\text{ V} - 20\text{ I} - 0.7\text{ V} = 0$$

or,

$$20\text{ I} = 3 - 0.7 = 2.3\text{ V}$$

$$\text{I} = \frac{2.3}{20} = 0.115\text{ AI}$$

14.5.3 Diode Parameters and Diode Ratings

A diode is specified in terms of certain parameters. These are as follows:

(i) Forward Voltage drop, V_F
(ii) Reverse Breakdown Voltage, V_{RB}
(iii) Reverse saturation current, I_R
(iv) Dynamic resistance, r_d
(v) Maximum forward current, I_{FM}

The dynamic resistance, r_d is calculated from the slope of the forward I–V characteristic as shown in Fig. 14.9 (b)

$$r_d = \frac{\Delta V_F}{\Delta I_F}\,\Omega$$

The values of these parameters are normally provided by the manufacturers in their specification sheet.

Diodes are available in low-, medium-, and high-current ratings. Diodes of low-current ratings are used in electronic switching circuits, i.e., they work as switches. Their forward current ranges from a few mA to a maximum of 100 mA. The safe reverse bias that can be applied is around 75 V. The reverse saturation current is very small, usually less than a micro-ampere.

Medium current diodes have a maximum current rating of 400 mA and reverse voltage of about 200 V.

High-current diodes are also called ***power diodes***. They are rated for high current and high reverse voltage ratings. Metal heat sinks are used for dissipation of heat produced in a diode when it is conducting.

In addition to their use in switching circuits, diodes are used in rectifier circuits for half-wave and full-wave rectification.

14.6 ▶ Zener Diode

Like as ordinary diode, zener diode allows current flow in the forward direction and also in the reverse direction at certain reverse voltage, called zener breakdown voltage.

We have known that when a diode is reverse based, only a minutely small current called saturature current flows (ideally no current should flow). If the reverse voltage is increased continuously, the junction breaks down and suddenly a large reverse current flows. This reverse current is controlled or limited by connecting a suitable series resistance so that excessive heat produced due to heavy current flow may not burn the diode. If the reverse breakdown current is limited to the current-carrying capacity of the diode, it can be operated under reverse breakdown condition. The I–V characteristic of the diode under the reverse-biased condition can be made dropping down almost vertically by proper doping of the semiconductor material. A diode with a very sharp breakdown voltage is called a ***zener diode***. Diodes designed to operate under the reverse breakdown condition, maintain a fairly constant voltage over a wide range of current levels. When the reverse voltage is reduced below the breakdown voltage, the current level returns to the very low-saturation current level.

There are two ways that breakdown of a zener diode may occur. One is called ***zener breakdown*** and the other is called ***avalanche breakdown***. If the depletion layer of a diode is narrow and we apply a reverse voltage, the voltage per unit of width of the depletion layer becomes high. This establishes a strong electric field intensity which causes electrons to break away from their parent atoms. Thus, a depletion layer which was insulating in nature, becomes a conducting path. This kind of breakdown due to the creation of a strong electric field intensity, i.e., V/μm is called zener breakdown.

If the width of the depletion layer is wide for a zener breakdown, a sufficient reverse voltage may provide the free electrons (minority carriers causing saturation current) to gain sufficient energy to knockout electrons from the atoms of the semiconductor in the depletion region. This is called ionization by collision. The breakdown occurring this way is called avalanche breakdown.

Zener breakdown occurs at a voltage less than 5 V and avalanche breakdown voltage is higher than 5 V. The symbol, the circuit, and the I–V characteristic of a zener diode are shown in Fig. 14.12. The forward I–V characteristic of a zener diode is the same as an ordinary diode.

As shown in Fig. 14.12 (b), a zener diode is operating under the reverse-biased condition. A resistance, R is connected to limit the current beyond the normal current-carrying capacity. A constant voltage, V_R will be available across the zener diode, even if the input voltage changes. Manufacturers specify the zener breakdown voltage and the maximum zener current. Type of zener diodes are numbered by manufactures as IN 746, IN 747, IN 755, IN 759, etc.

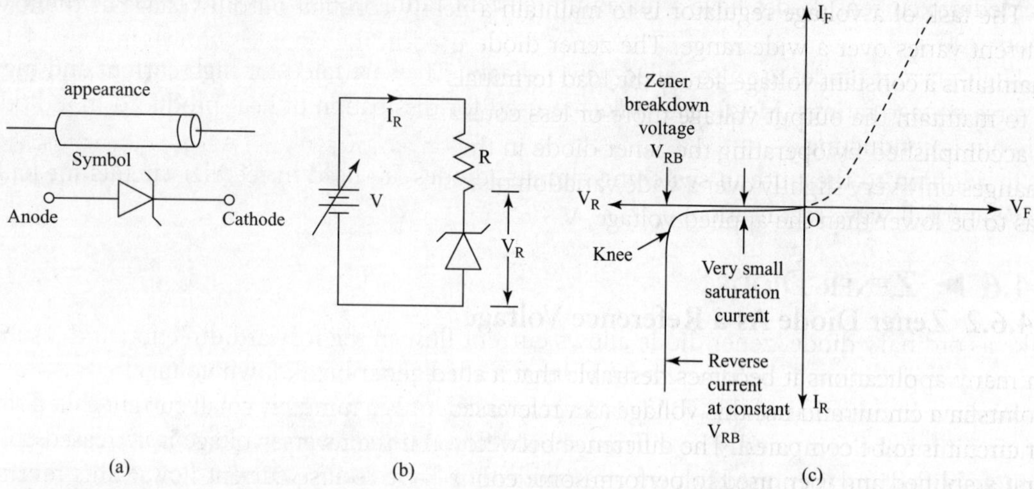

FIGURE 14.12

(a) Appearance and (b) Symbol of a zener diode; (b) zener diode circuit; (c) I–V characteristic

Zener resistance

It is the dynamic resistance of the zener diode somewhat below the knee point on the reverse V–I characteristic. It is similar to the dynamic resistance of a forward-biased diode. Zener resistance, R_Z is given by

$$R_z = \frac{\Delta V_z}{\Delta I_z}$$

The value of RZ is to be small so that it will indicate a steep curve where due to a small change in voltage, DVZ a large change in current, DIZ takes place. The value of zener resistance varies with the current. It decreases with increase in current.

An application of zener diode as a voltage regulator or stabilizer, and often used as reference voltage in electronic circuits has been shown in Fig. 14.13. How the zener diode helps in maintaining a constant voltage when the load current changes and when the input voltage changes can be understood from the following two applications.

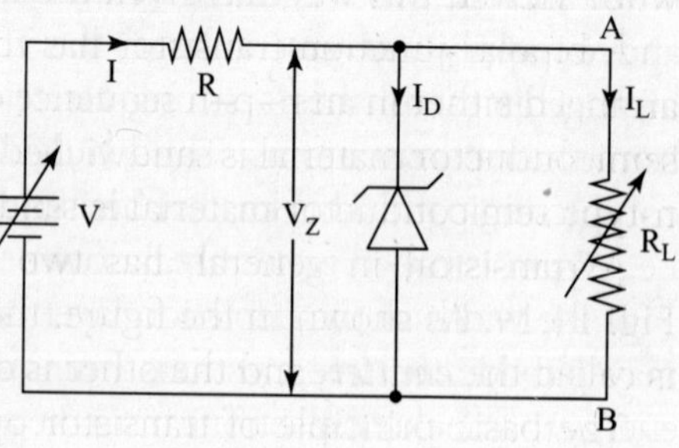

FIGURE 14.13

Zener diode voltage regulator circuit

14.6.1 Zener Diode As Voltage Regulator

A voltage regulator maintains nearly constant voltage output across the load over a wide range of variation of load current. A zener diode voltage regulator circuit has been shown in Fig. 14.13. The zener diode has been reverse biased.

The task of a voltage regulator is to maintain a nearly constant output voltage as the load current varies over a wide range. The zener diode used in the circuit is shown in Fig. 14.13 maintains a constant voltage across the load terminals A and B. The function of the zener diode is to maintain the output voltage more or less constant even if the load current changes. This is accomplished by operating the zener diode in the breakdown region when voltage across it changes only very slightly over a wide variation of zener current. The zener breakdown voltage has to be lower than the applied voltage, V.

14.6.2 Zener Diode As a Reference Voltage

In many applications it becomes desirable that a constant voltage is maintained between two points in a circuit and use this voltage as a reference voltage to which voltage of another point or circuit is to be compared. The difference between reference voltage and compared voltage is first amplified and then used to perform some control operations. A zener diode will maintain a constant voltage across its terminals even if there is a change in the supply voltage, V in the circuit as shown in Fig. 14.13.

14.7 ▶ Bipolar Junction Transistors

Transistors are used in almost all electronic circuits. The ability to amplify or switch electrical signals accounts for their wide use. The word transistor is the short form of the word "transfer resistor". The signal amplification in a transistor is achieved by transfering the signal from a region of low resistance to a region of high resistance. The concept of transfer of resistance, when viewed this way, has given the name transistor.

A bipolar junction transistor has three layers of semiconductor material. These layers are arranged either in an n–p–n sequence or in a p–n–p sequence. In an n–p–n transistor, a p-type semiconductor material is sandwiched between two n-type materials. In a p–n–p transistor, an n-type semiconductor material is sandwiched between two p-type materials.

A transistor, in general, has two p–n junctions connected back to back as shown in Fig. 14.14. As shown in the figure, the central layer is called the ***base***, one of the outer layers is called the *emitter*, and the other is called the *collector*.

The basic principle of transistor operation is that a small current in the base region can control a much larger current flow through the transistor, i.e., from the emitter to the collector. A transistor can be used as current amplification or a voltage amplification device. Since a transistor combines two junction diodes, it works on the basis of p–n junction theory as has already been explained.

The symbolic representation of a transistor has also been shown. The symbol for an n–p–n or a p–n–p transistor is the same except for the direction of the arrow head. The arrow head has to be shown from p terminal to n terminal between the emitter and the base. The emitter of a transistor is heavily doped. The base is lightly doped while the collector is less heavily doped than the emitter.

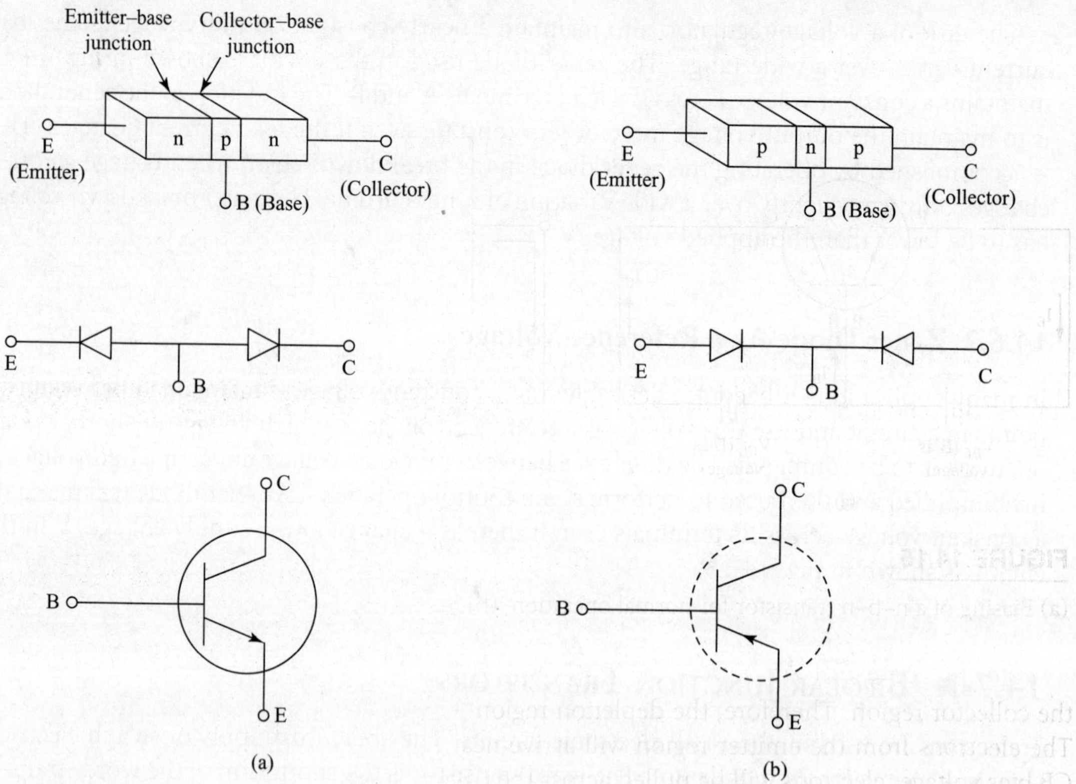

FIGURE 14.14

(a) Block diagram representation, two-diode transistor analogy, and symbol of a n–p–n transistor; (b) block diagram representation, two-diode transistor analogy, and symbol of a p–n–p transistor

14.7.1 Working of n–p–n Transistor

For a transistor to work, it has to be biased by applying external voltage supply with proper polarity. For operation in the active region, a transistor's emitter–base junction must be forward biased while the collector base region reverse biased as shown in Fig. 14.15 (a) and (b).

The forward bias of the EB junction reduces the width of the depletion region and the barrier voltage gets reduced.

The reverse bias of the CB junction increases the width of the depletion region and the barrier voltage gets increased.

Since the EB junction is forward biased, electrons form the majority charge carriers and would flow from the emitter to the base region. Since the base is lightly doped, there will be a smaller number of holes present there. Only a small percentage of electrons from the emitter will recombine with holes in the base region. Only around two per cent of the electrons from the emitter recombine with the holes that are present in base region. The reverse bias of the CB junction causes expansion of the depletion layer. The width of the base region is thinner than

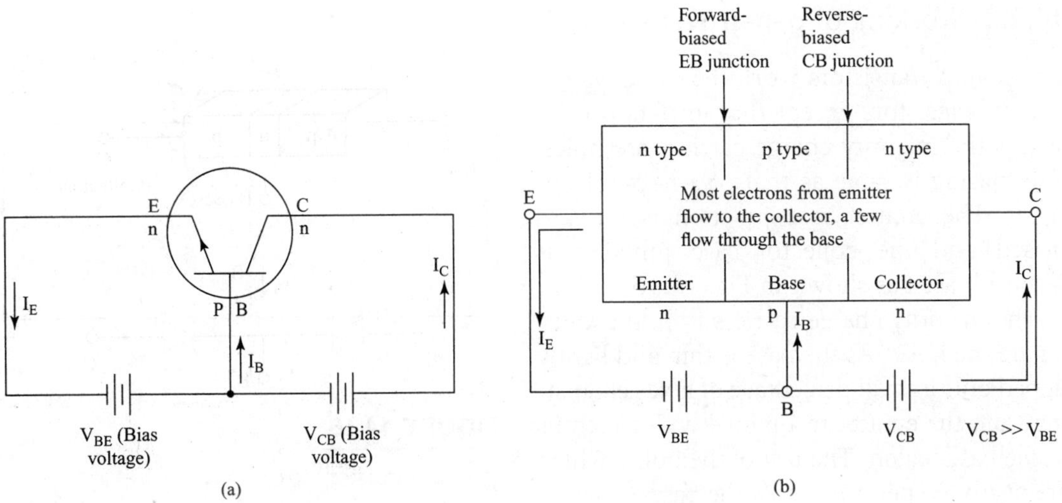

FIGURE 14.15

(a) Biasing of a n–p–n transistor for normal operation; (b) operation of a n–p–n transistor

the collector region. Therefore, the depletion region will penetrate deep into the base region. The electrons from the emitter region will arrive near the CB depletion region. Due to large CB bias voltage, electrons will be pulled across the CB junction by the positive terminal of the collector. The collector, thus collects the 98 per cent of the electrons emitted by the emitter.

The quantity of charge carriers crossing the emitter to the base in controlled by the base–emitter bias voltage. Thus, it can be said that emitter and collector current levels can be controlled by the base–emitter bias voltage. Note that the conventional direction of current flow, i.e., the directions of emitter current, I_E, base current, I_B and collector current, I_C have been shown opposite to the direction of flow of charge carriers.

Note that for a silicon transistor, substantial current will start flowing only when the bias voltage V_{BE} is about 0.7 V and for the germanium transistor V_{BE} is 0.3 V. Beyond this voltage, a small variation of V_{BE} will control I_E and I_C, and V_{BE} has to supply only a small I_B.

To sum up, the following statements can be made for the operation of a n–p–n transistor:

(i) The outer layers of n–p–n transistors are called emitter and collector, respectively, the central layer is called the base.
(ii) The base–emitter junction, EB is forward biased and the collector–base junction is reverse biased, and V_{CB} is higher than V_{BE}.
(iii) The base section is made very thin and is lightly doped so that majority of the charge carriers (electrons in n–p–n transistor) can move from the emitter to the collector.
(iv) Most charge carriers flow from emitter to collector and only a small percentage flow through the base material.
(v) Variation of base–emitter voltage changes base, emitter, and collector currents.

14.7.2 Working of p–n–p Transistor

The p–n–p transistors work the same way as n–p–n transistors except that in p–n–p transistors the majority charge carriers are holes. The biasing is same as that of a n–p–n transistor. The emitter-base junction is forward biased and the collector–base junction is reverse biased as shown in Fig. 14.16.

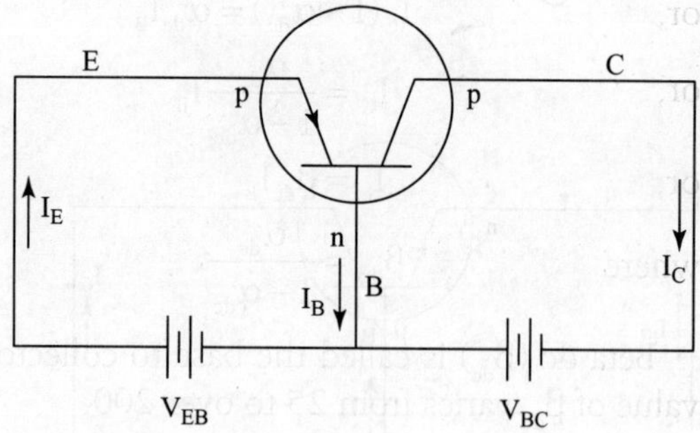

FIGURE 14.16

P–n–p transistor

The majority charge carriers from the emitter are the holes. As the base is thin and lightly doped only a small percentage of holes emanating from the emitter recombine with electrons in the base region. The rest of the holes, which are nearly 98 per cent, cross the base–collector barrier potential, because they are attracted by the negative terminal of the base–collector bias voltage.

Thus, holes are emitted from the p-type emitter across the forward-biased emitter–base junction and only a few of them find electrons in the base region to recombine with. Most of the holes get attracted to the collector side by the reverse-biased collector–base junction.

By varying the forward bias voltage at the base–emitter junction, we can control the large emitter and collector current through small variations of base current.

Both n–p–n and p–n–p type of transistors are called bipolar junction transistors, or simply BJTs because the charge carriers for both types of transistors are electrons, and holes, although for n–p–n transistors the majority charge carriers are electrons and for p–n–p transistors the majority charge carriers are the holes.

When both junctions of a transistor are reverse biased, the transistor is said to be cut off and operating in the cut-off region.

The ratio of the collector current to the emitter current of a transistor is called α_{dc} which is greater than 0.95. We can express

$$I_E = I_C + I_B \quad \text{(i)}$$

and $$IC = \alpha_{dc} I_E \text{ neglecting the reverse saturation current} \quad \text{(ii)}$$

Therefore, $$I_B = (1 - \alpha_{dc}) I_E \quad \text{(iii)}$$

where α_{dc} is the emitter to collector current gain.

The value of alpha dc (α_{dc}) is normally 0.95 to 0.99. As the collector–base junction is reverse biased, a small reverse saturation current flows across the junction due to minority charge carriers which are very small in number, and can be neglected.

From eqs. (i) and (ii)

$$I_C = \alpha_{dc} I_E = \alpha_{dc}(I_C + I_B)$$

or, $$I_C(1-\alpha_{dc}) = \alpha_{dc} I_B$$

or, $$I_C = \frac{\alpha_{dc}}{1-\alpha_{dc}} I_B \tag{iv}$$

or, $$I_c = \beta_{dc} I_B \tag{v}$$

where $$\beta_{dc} = \frac{\alpha_{dc}}{1-\alpha_{dc}}$$

Beta dc (β_{dc}) is called the base to collector current gain. This is the ratio of I_C and I_B. The value of β_{dc} varies from 25 to over 200.

EXAMPLE 14.2

An n–p–n transistor shown in Fig. 14.17 has been provided with biasing voltage V_{BE} and V_{CB}. Calculate the values of I_C and I_E if α_{dc} is 0.96 and I_B is 80 μA. Also calculate the value of β_{dc}.

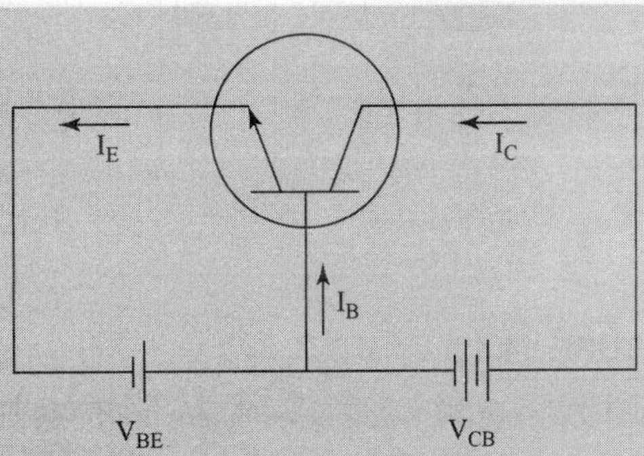

FIGURE 14.17

Relates to example 14.2

SOLUTION

Given $\alpha_{dc} = 0.96, I_B = 80\times10^{-6}$ A

$$\beta_{dc} = \frac{\alpha_{dc}}{1-\alpha_{dc}} = \frac{0.96}{1-0.96} = \frac{0.96}{.04} = 24$$

$$I_C = \frac{\alpha_{dc}}{1-\alpha_{dc}} I_B = \frac{0.96}{1-0.96}\times 80\times10^{-6}$$

$$= 1.92 \text{ mA}$$

Again, $$\frac{I_C}{I_E} = \alpha_{dc}$$

or, $$I_E = \frac{I_C}{\alpha_{dc}} = \frac{1.92}{0.96} \text{ mA} = 2 \text{ mA}$$

14.8 ► Transistor Configurations

A transistor has three terminals, namely emitter, base and collector. For the connection of input and output, one of the transistor terminals is made common.

For example, the base terminal can be made common to both input and output terminal connections. This will be called common base configuration. Similarly, the emitter or the collector can be connected in common configurations.

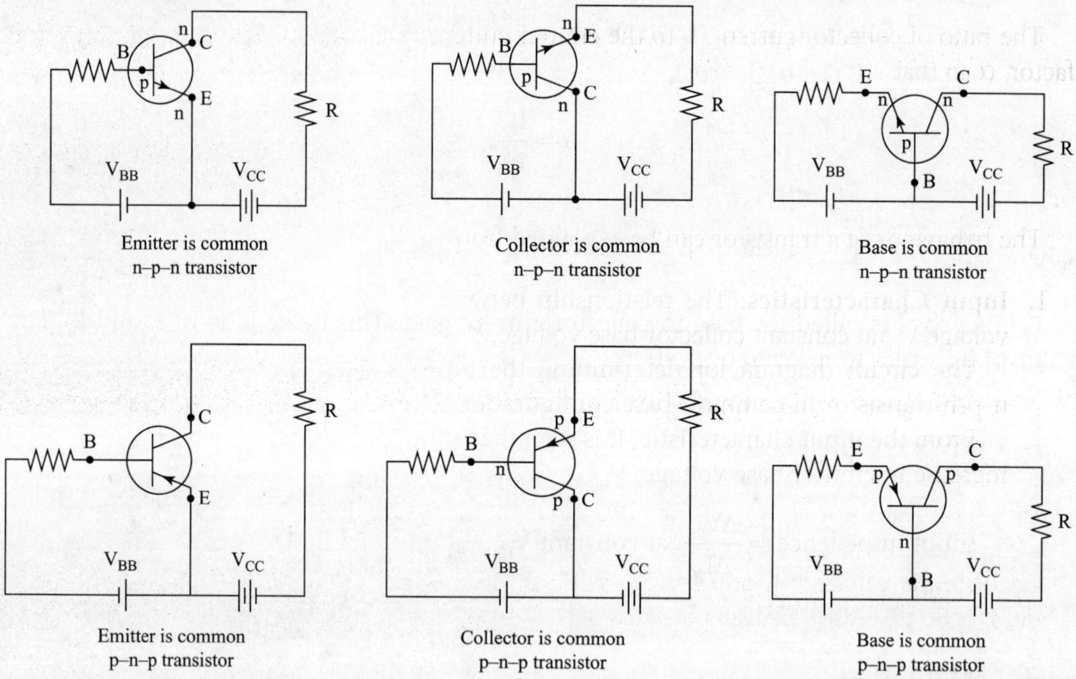

FIGURE 14.18

Common-emitter, common-collector, and common-base configurations of n-p-n and p-n-p transistors

These three types of connections or configurations can be made for both n-p-n and p-n-p transistors as shown in Fig. 14.18.

Each type of configuration has its advantages and disadvantages. Out of these three types of configurations, the common emitter configuration is widely used because of their very high voltage and power gain.

The characteristics of transistor for the three types of configurations are explain as follows.

14.8.1 Common-Base Transistor Characteristics

In common-base (CB) configuration, the input is connected between emitter, E and base, B. The output is taken from collector, C and base, B as has been shown. Note that the emitter-base junction is to be forward biased and collector-base junction is to be reverse biased. The emitter current, I_E flows in the input circuit and collector current, I_c flows in the output circuit. A change in emitter current produces similar change in the collector current. The impedance of the input and output circuits are different. The emitter-base circuit has low impedance while the collector-base circuit has high impedance. A high voltage and power gain can be achieved in a transistor because of transfer of voltage or power from input to output circuit.

The ratio of collector current, I_c to the emitter cuttent, I_E is called the current amplification factor, α so that

$$\alpha = \frac{I_C}{I_E}$$

The behaviour of a transistor can be explained from its input and output characteristics.

1. **Input Characteristics:** The relationship between emitter current, I_E and emitter-base voltage V_{EB} at constant collector-base voltage, V_{CB} is called the input characteristic.

   The circuit diagram for determining the input and output characteristics for an n-p-n transistor in common-base configuration is shown Fig. 14.19.

   From the input characteristic, it is seen that emitter current I_E increases rapidly with increase in emitter base voltage, V_{EB}

   $$\text{Input impedence} = \frac{\Delta V_{EB}}{\Delta I_E} \text{ at constant } V_{CB}.$$

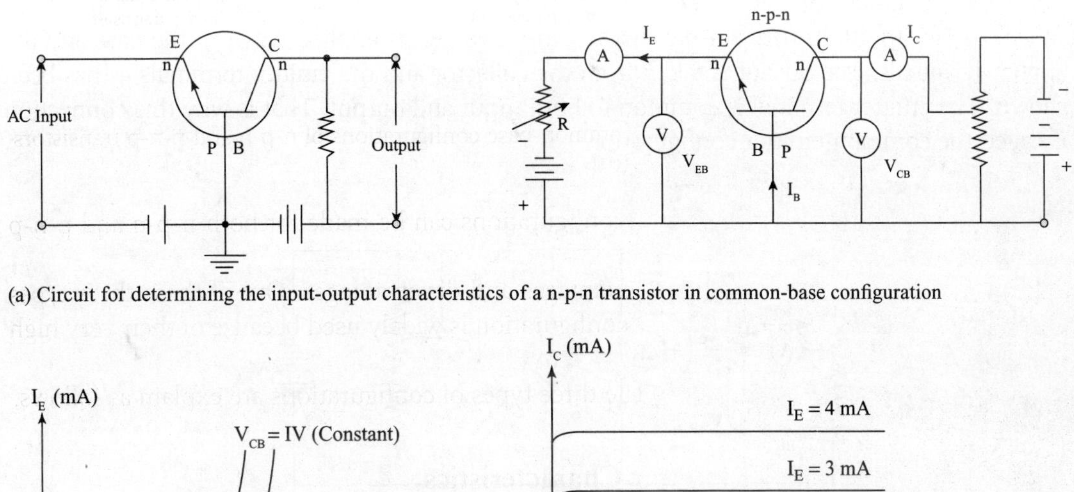

(a) Circuit for determining the input-output characteristics of a n-p-n transistor in common-base configuration

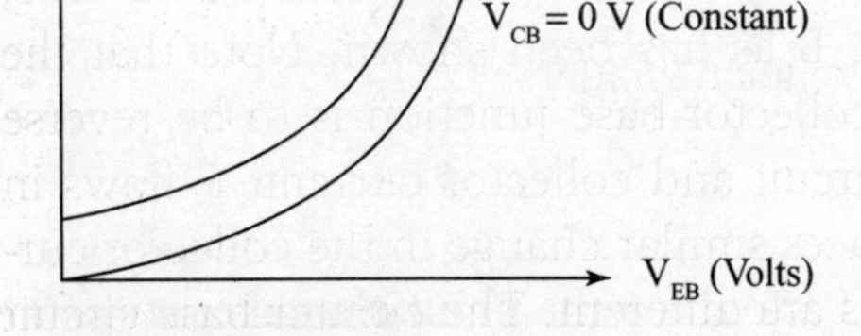

(b) Input characteristics

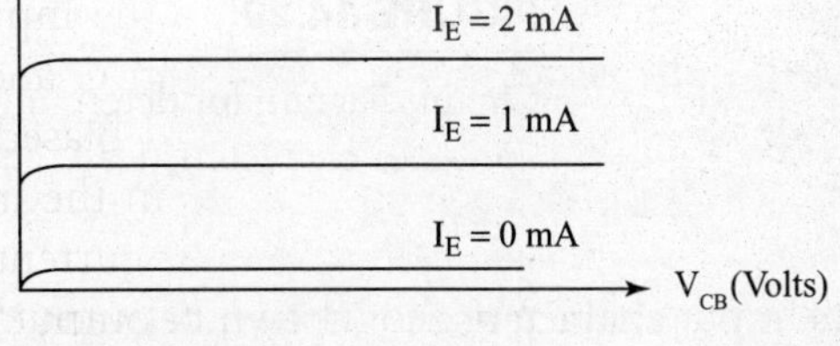

(c) Output characteristics

FIGURE 14.19

Determination of input and output characteristics in common-base configuration of a n-p-n transistor

This input impedance is very small because a very small V_{EB} is sufficient to produce a large flow of emitter current I_E.

2. **Output characteristics:** The relationship between collector current, I_C and collector-base voltage, V_{CB} at constant emitter current, I_E is called the output characteristic. Figure 14.19(c) shows the output characteristics at different values of emitter currents.

It can be observed from the characteristics that when the V_{CB} is increased by a small amount, the collector current becomes constant as shown by almost horizontal curves. This shows that I_C is independent of change in V_{CB} and depends on I_E only. Almost the whole of I_E flows to the collector circuit as I_C. A large change in V_{CB} creates only a minor change in I_C. The output impedance is given as

$$\text{Output impedance} = \frac{\Delta V_{CB}}{\Delta I_C} \text{ at constant } I_L.$$

The output impedance of the circuit is very high.

14.8.2 Common-Emitter Transistor Characteristics

The circuit diagram for determining the common-emitter transistor characteristics has been shown in Fig. 14.20. In this p-n-p transistor, input voltage is applied between the base and the emitter terminals. The output is taken from the collector and the emitter terminals as has been shown. The emitter terminal is common to both input and output. That is why this connection is called the common-emitter configuration.

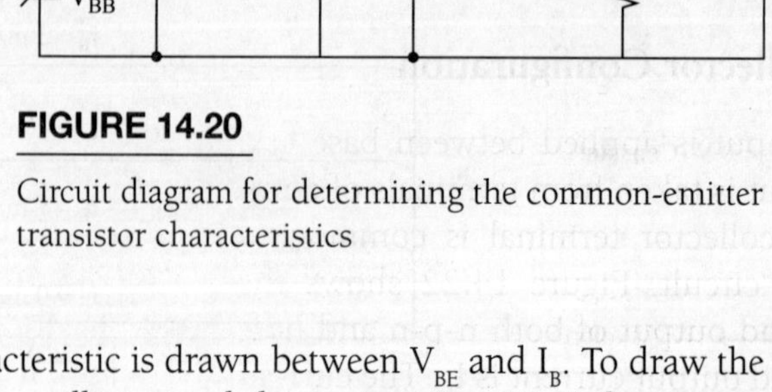

FIGURE 14.20

Circuit diagram for determining the common-emitter transistor characteristics

The input characteristic is drawn between V_{BE} and I_B. To draw the input characteristics, the voltage between the collector and the emitter, i.e., V_{CE} is kept constant at a value. By changing V_{BE} through V_{BB}, current I_B is recorded for atleast, say five readings. A plot of I_B against V_{BE} when made shows that the characteristic is similar to characteristic of a forward-biased p-n junction. The value of I_B is very small, is of the order of several microamperes only.

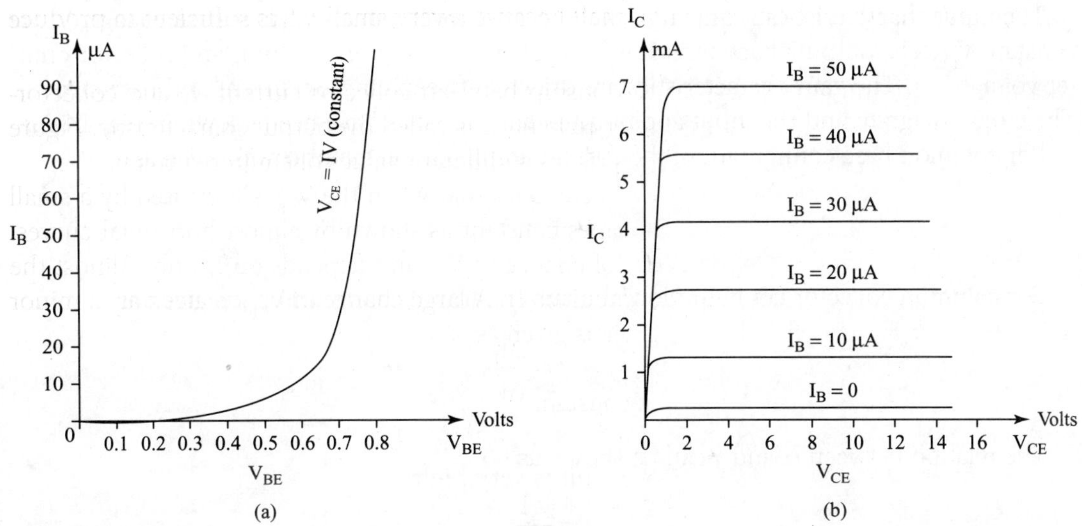

FIGURE 14.21

(a) Common-emitter input characteristics; (b) common-emitter output characteristics

The output characteristics are drawn between I_C and V_{CE} keeping I_B constant at different values. For each value of I_B, V_{CE} is adjusted in steps and the values of I_C are recorded. The values of I_C are plotted against V_{CE} for each value of I_B as shown in Fig. 14.21, Fig. 14.21(a) shows the input characteristic and Fig. 14.21(b) shows the output characteristics.

From the characteristics, it is seen that for a small change in base current (in μA) there is a large change (in mA) in collector current and emitter current. The *current gain* from the base to the collector can be stated as

$$\beta_{dc} = \frac{\Delta I_C}{\Delta I_B}$$

Thus, the transistor can be used as a current-amplication device.

The *voltage gain*, A_V of transistor is defined as the ratio of output voltage to input voltage.

14.8.3 Common-Collector Configuration

In this configuration, input is applied between base and collector while the output is taken from terminals of emitter and collector. The collector terminal is common to both input and output circuit. Figure 14.22 shows the connections for input and output of both n-p-n and here the input current is I_B and output current is I_E. The current amplification is given as

$$\gamma = \frac{\Delta I_E}{\Delta I_B}$$

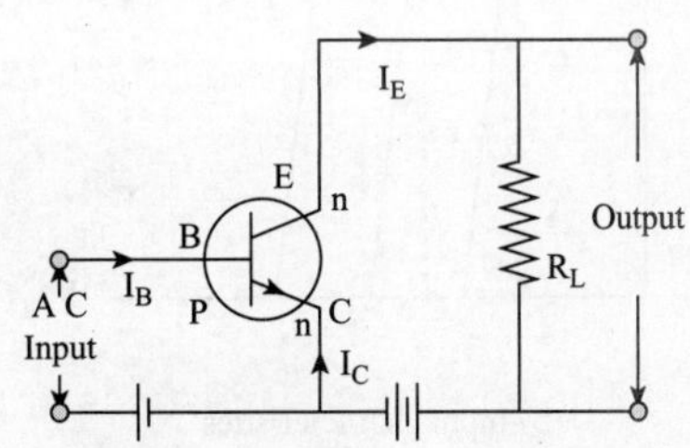

FIGURE 14.22

Common-collector configuration of n-p-n transistor

The input characteristics is drawn between base current I_B and collector-base voltage, V_{CB} at constant V_{CE}. The output characteristic is drawn between emitter current I_E and collector-emitter voltage V_{CE}. The gain characteristics are drawn between emitter current and base current. The circuit diagram and the input and output characteristics have been shown in Fig. 14.23.

For common base configuration, the current amplification factor α was given as

$$\alpha = \frac{\Delta I_C}{\Delta I_L}$$

For common collector configuration, the current amplification factor γ is given as

$$\gamma = \frac{\Delta I_E}{\Delta I_B}$$

The relation between α and γ can be shown as

$$\gamma = \frac{1}{1-\alpha}$$

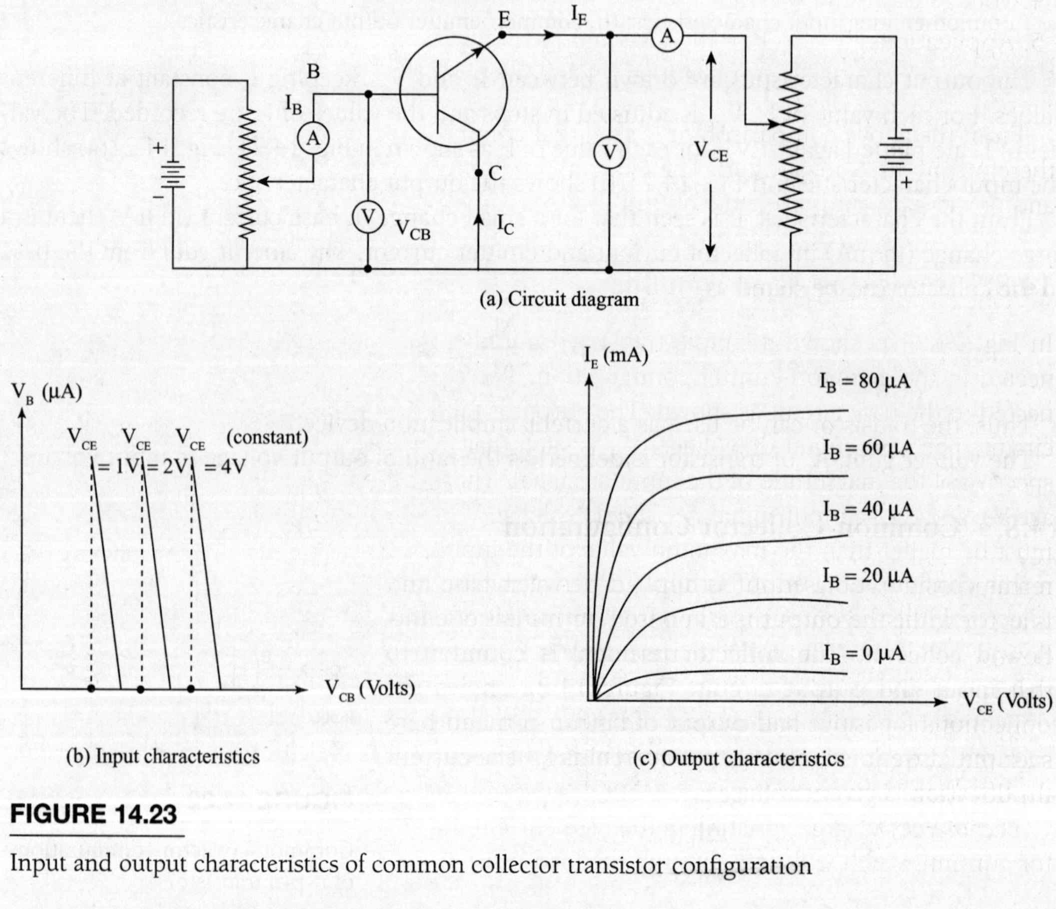

(a) Circuit diagram

(b) Input characteristics

(c) Output characteristics

FIGURE 14.23

Input and output characteristics of common collector transistor configuration

The common-collector (CC) circuit has very high input impedance and very low-output impedance. So the voltage gain will be low. Therefore CC circuit is rarely used for voltage amplification.

This circuit, however, is used for impedance matching because of its high-input impedance and low-output impedance.

14.8.4 Comparison of Transistor Characteristics of Different Configurations

The basic characteristic parameters for comparison are input impedance, output impedance, current gain, voltage gain, and application areas. These are shown in a tabular from.

Characteristics	Common base	Common emitter	Common collector
1. Input Impedence	Low in ohms	Low in ohms	High in Kilo-ohms
2. Output Impedence	Very high in Kilo-ohms	High in Kilo-ohms	Low in ohms
3. Current Gain	Less than unity	High	High
4. Voltage Gain	High	High	Less than unity
5. Applications	For High frequency applications	For Audio frequency application	For impedance matching

From the above comparison, it is seen that common emitter circuits are most efficient and therefore are used most often. The CE configuration provides high current gain, high voltage and power gain, and moderate output to input impedance ratio.

14.8.5 Transistor As an Amplifier

In Fig. 14.24 is shown a simple transistor amplifier circuit using an n–p–n transistor connected in the common-emitter configuration. The ac signal which is to be amplified is connected to the base circuit as shown. The output is taken across a resistance, R_L in the collector circuit. The base circuit dc voltage, V_{BB} is such that the base will always remain positive irrespective of the magnitude of the input ac signal. The voltage, V_{BE} is the summation of dc voltage V_{BB} and the ac input signal, V_i. The dc voltage, V_{BB} is the bias voltage. The magnitude of V_{BB} must be higher than the maximum value of the input signal. Then, only the base will always remain positively biased in both half cycles of the input ac signal voltage. When the ac signal is applied, this becomes superimposed on the battery voltage, V_{BB} and the base current I_B will flow, which is the sum of the dc base current $I_{B\alpha}$ and the ac signal current. It can be observed that I_B is always positive.

During the positive half cycle of the input signal, the dc and ac voltages are added up and as such the base current is highly positive. During the negative half cycle, ac voltage is subtracted from the dc voltage. The net voltage is low but positive. The base current now will be positive but low.

Because of the large variation in the base current there will be a large variation in the collector current, which will flow through the load resistance. An amplified output voltage is, thus available across the load. The amplified collector ac current is superimposed on the dc current, I_{CQ} which will flow through the collector when the ac input signal is not applied. It is the current when the base current is I_{BQ}.

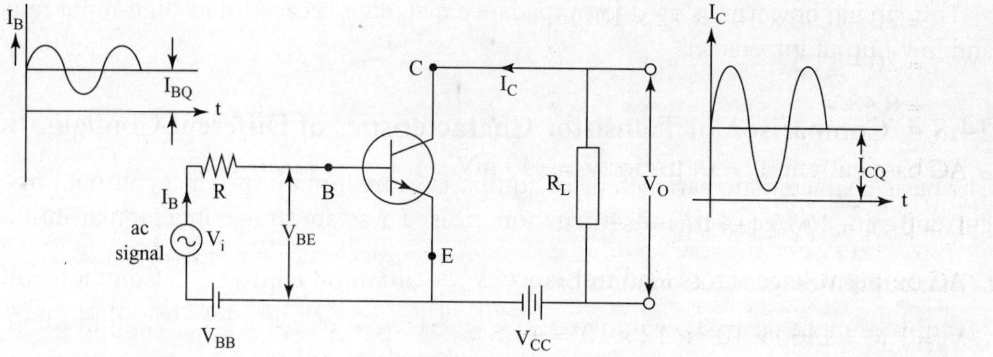

FIGURE 14.24

Transistor as an amplifier

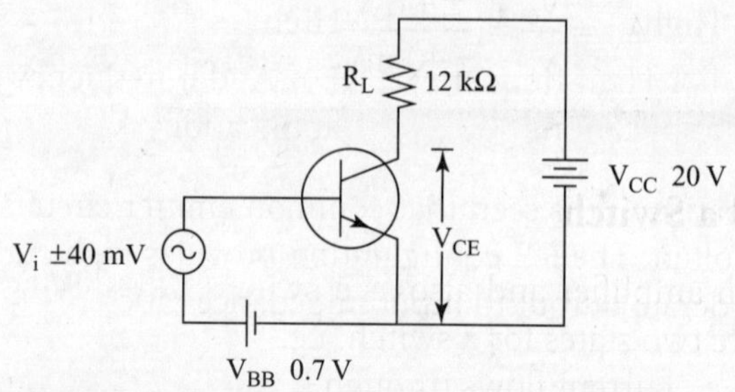

FIGURE 14.25

Transistor amplifier circuit

EXAMPLE 14.3

In an n–p–n transistor in the common emitter configuration, an ac input signal of ±40 mV is applied as shown in Fig. 14.25. The dc current gain, β_{dc} and ac current gain β_{ac} are given as 80 and 100, respectively. Calculate the voltage amplification, A_V of the amplifier. The I_B versus V_{BE} characteristic is such that for $V_B = 0.7$ V, $I_B = 12$ μA and for $V_i = \pm 40$ mV, $I_b = \pm 4$ μA. Also calculate the dc collector voltage.

SOLUTION

DC base current, $I_B = 12$ μA for dc voltage, $V_{BB} = 0.7$ V and $\beta_{dc} = 80$

$I_C = \beta_{dc}\, I_B = 80 \times 12\ \mu\text{A} = 0.96$ mA

The collector voltage V_{CE} is calculated as

$$V_{CE} = V_{CC} - I_C R_L$$
$$= 20 - 0.96 \times 10^{-3} \times 12 \times 10^3$$
$$= 20 - 11.52$$
$$= 8.48 \text{ V}$$

AC base current, $I_b = \pm 4\ \mu A$ for $V_i = \pm 40$ mV

$I_C = \beta_{ac} I_b = 100 \times (\pm 4\ \mu A) = \pm 400\ \mu A$

AC output voltage across load resistance, V_0 is calculated as

$V_0 = I_C R_L = \pm 400 \times 10^{-6} \times 12 \times 10^3 = \pm 4.8$ V

AC voltage amplification factor, A_V is calculated as

$$A_v = \frac{v_0}{v_i} = \frac{\pm 4.8 \text{ V}}{\pm 40 \text{ mV}} = \frac{\pm 4.8}{\pm 40 \times 10^{-3}} = 60$$

14.8.6 Transistor As a Switch

A BJT can be used as an amplifier and also as a switch. A switch either closes a circuit or opens a circuit. There are two states for a switch, i.e., either there is no current flow (cut off) or the switch is closed, i.e., current flows through it with the minimum of resistance offered.

These two conditions can be created by applying a pulse wave input to the base of the BJT. In the case of an amplifier we had applied a bias voltage plus an ac signal that had to be amplified to the base circuit. In the case of switching operation, a pulse voltage of appropriate level has to be applied as shown in Fig. 14.26 (a) and (b).

The base voltage level is either at zero level or at an appropriate positive level. When the input voltage, V_i is at zero level, the base current is zero and there is no collector current, i.e., $I_C = 0$ as shown is Fig. 14.26 (a).

The transistor is cut off and works like an open switch.

From Fig. 14.26 (b),

V_{CE} = Supply voltage – voltage drop across R_2

$V_{CE} = V_{CC} - I_C R_2$

When the base-emitter voltage is at zero level, the transistor is not working, and hence, $I_C = 0$. Therefore,

$$V_{CE} = V_{CC}$$

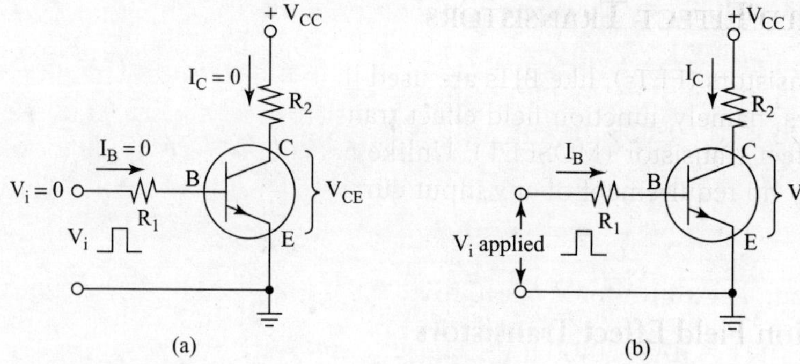

FIGURE 14.26

(a) Off state of a BJT; (b) on state of a BJT

When V_i is at positive level, base current, I_B will flow. If the BJT is to be used as a switching device, the level of I_B is made high enough so that the transistor is saturated. At saturated state the level of I_C will be such that $I_C R_2$ will be equal to V_{CC} for which $V_{CE} = 0$. The transistor will operate as a closed switch between the collector and emitter.

EXAMPLE 14.4

What minimum input voltage level is required to switch a BJT into saturation (on state) when $V_{CC} = 10$ V, R1 = 16 kW, $R_2 = 6.2$ kW and $\beta_{dc} = 20$ in an n–p–n CE configuration BJT, as has been shown in Fig. 14.27.

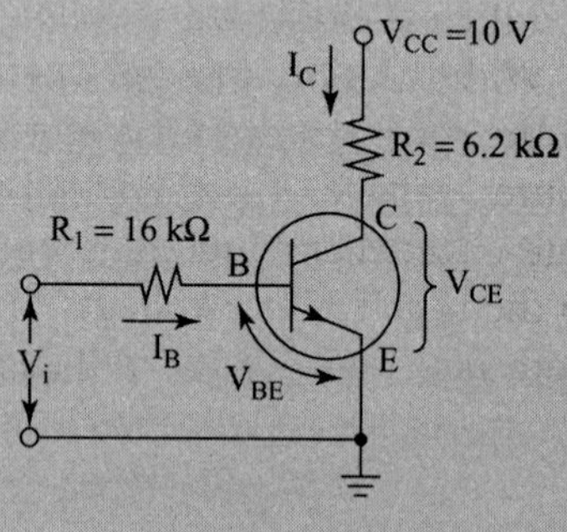

FIGURE 14.27

SOLUTION

$$V_{CC} = I_C R_2 + V_{CE}$$

for $V_{CE} = 0$

$$I_C = \frac{V_{CC}}{R_2} = \frac{10}{6.2\times10^3} = 1.612 \text{ mA}$$

$$= 1612\ \mu\text{A}$$

$$I_C = \beta_{dc} I_B$$

or,

$$I_B = \frac{I_C}{\beta_{dc}} = \frac{1612}{20} = 80.6\ \mu\text{A}$$

Taking $V_{BE} = 0.7$ V,

$$V_i - I_B R_1 = V_{BE}$$

or,

$$V_i = V_{BE} + I_B R_1$$

$$= 0.7\text{V} + 80.6\times10^{-6}\times16\times10^3\text{ V}$$

$$= (0.7+1.29)\text{V} = 1.99\text{ V}$$

14.9 ▶ FIELD EFFECT TRANSISTORS

Field effect transistors (FETs), like BJTs are used in amplifier and switching circuits. FETs are of two types, namely, junction field effect transistor (JFET) and metal oxide semiconductor field effect transistor (MOSFET). Unlike a BJT, FET is a voltage operated device having virtually no requirement of any input current. The two types of FET are described below.

14.9.1 Junction Field Effect Transistors

A JFET has been shown in the form of a block diagram in Fig. 14.28 along with its symbol. A JFET is made of an n-type semiconductor material called channel, sandwiched between two p-type materials. The two p-type materials are connected togather to form a gate.

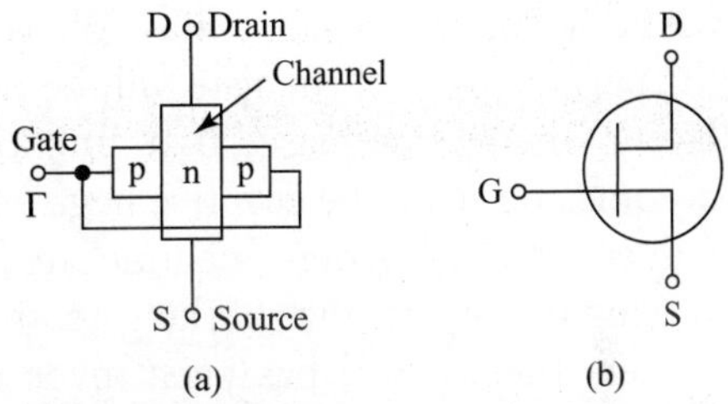

FIGURE 14.28

(a) Construction of a JFET; (b) symbol of a JFET

The two ends of the central n-type material, i.e., the channel has two end terminals. One terminal of the channel is called ***drain*** and the other terminal is called *source*. The gate material is highly doped as compared to the channel. The principle of operation of a JFET is explained as follows.

With the gate terminal open, when a positive voltage is applied to the drain with respect to the source, a drain current, I_D will flow as has been shown in Fig. 14.29. Now, if a gate-source voltage, V_{GS} is applied with the gate having connected to the negative terminal, the gate-channel p–n junctions will be reverse biased. Since the gate material is heavily doped, due to the negative bias voltage, the depletion region will expand and penetrate into the channel from both sides. If the voltage, V_{GS} is increased, the penetration of the depletion region will

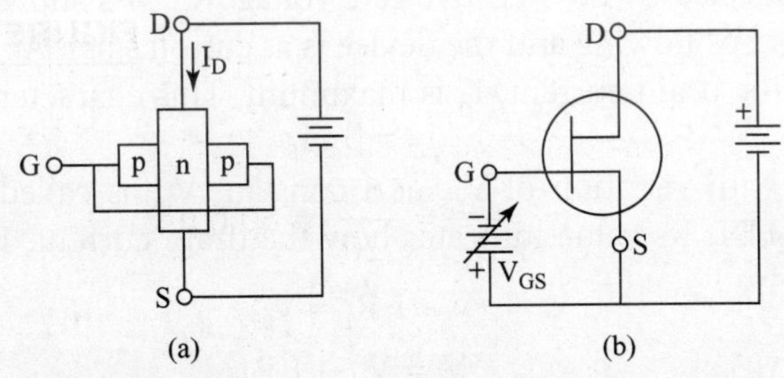

FIGURE 14.29

(a) A JFET with no gate-source bias voltage;
(b) variable gate-source voltage applied to a JFET

be so high that it will stop the flow of current, I_D through the channel. Thus, by varying the voltage, V_{GS}, the depletion region at the gate-channel region can be varied. This will result in the passage for current I_D at the gate region narrowed down, causing high resistance to the current flow. This way the drain to source current can be varied. If the voltage V_{GS} is increased more, the depletion region from both sides will expand and close the passage of current, I_D, and hence, the device will come to non-conducting or cut-off state. The two gate-channel p–n junctions are kept reverse biased, and the gate current is normally very low.

Now, assume that an ac signal is applied to the gate circuit. The signal will be superimposed on the negative dc bias voltage. Since the gate is negatively biased, in the negative half cycle of the ac signal, the negative bias voltage of the gate will increase. During the positive half cycle the negative bias of the gate will be reduced. When the signal goes negative of the depletion layer of the reverse-biased p–n Junction will widen, causing a reduction of I_D. When the signal becomes positive, the effective negative bias of the gate will decrease, the depletion region will reduce, channel widens, channel resistance decreases, and hence, I_D increases. This way, by varying the reverse bias of the gate, the drain current is controlled.

Here the channel has been made of an n-type semiconductor material. That is why the device is called n-channel JFET. If the channel is made of p-type material, the device will be called p-channel JFET.

The reverse bias voltage produces an electric field which changes the depletion region at the gate-channel junction. The effect of the electric field produces the transistor action, and hence, the device has been named field effect transistor. An FET is also often referred to as unipolar transistor because here electrons are the only charge carriers when the channel is made of an n-type material. For a p-type channel, the charge carriers will be only the holes.

For a p-channel JFET, the channel is made of a p-type material sandwiched between two n-type materials. The battery connection is reversed, i.e., the drain is connected to the negative terminal of the battery and the source is connected to the positive terminal. The drain current flows from the source to the drain. The gate channel junctions are reverse biased. The relationship between drain current, I_D and voltage across the device, V_{DS} when external gate-source bias, V_{GS} is applied is shown in Fig. 14.30.

In Fig. 14.30 (b) is shown the I_D versus V_{GS} characteristic. This characteristic shows how the drain current, I_D is controlled by the negative gate voltage V_{GS}. As shown in Fig. 14.30 (b), at –4V the drain current stops flowing and the device is at cut-off state. At no voltage at the gate, i.e., when V_{GS} is zero, the drain current, I_D is maximum. This characteristic is also called the ***transfer characteristic.***

The variation of I_D with variation of V_{GS} at a constant V_{DS} is called the forward transfer admittance, Y_{fs}, of the FET. Its value indicates how the drain current, I_D is controlled by the gate-source voltage, V_{GS}.

$$Y_{fs} = \frac{\Delta I_D}{\Delta V_{GS}} \text{ at constant } V_{DS}$$

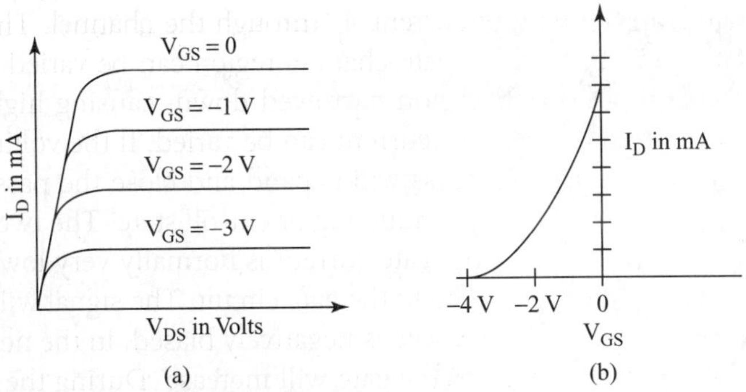

FIGURE 14.30

(a) I_D versus V_{DS} characteristics for variable V_{GS} of a JFET; (b) I_D versus V_{GS} characteristics (also called transfer characteristic) for a JFET

14.9.2 FET Applications

FETs, like BJTs are used for voltage amplification and in switching circuits. Fig. 14.31 shows an FET voltage amplifier circuit. The gate–source junctions are reverse biased. The drain-source terminals are provided with dc voltage supply with the positive terminal connected to the drain. An ac signal voltage is connected in series with the gate circuit. R_1 is a resistance connected in the drain circuit of the FET. An input ac voltage, vi will produce an output voltage change across the device, i.e., VD. The ratio of the output voltage to the input voltage is called voltage amplification of the circuit, A_v.

EXAMPLE 14.5

In an FET voltage amplifier circuit as shown in Fig. 14.31, an input voltage of ±50 mV is applied. Calculate the voltage gain of the amplifier. The following are given:

VDD = 18 V, $R_1 = 5\ k\Omega$, V_G is so adjusted that I_D is 2.5 mA.

Forward transfer admittance of the FET, $Y_{FS} = 5000\ \mu S$

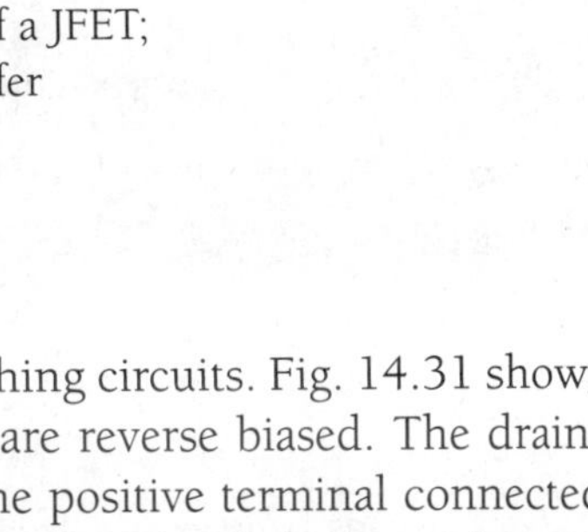

FIGURE 14.31

A voltage amplifier circuit using an FET

SOLUTION

We have to calculate the output voltage change for an input of ±50 mV.

The dc level of the drain voltage is

$$V_D = V_{DD} - I_D \times R_1 = 18 - 2.5 \times 10^{-3} \times 5 \times 10^3$$
$$= 7.5\ V$$

When the positive half cycle of the ac input is flowing, V_i = +50 mV. For this current,

$$\Delta I_D = Y_{fs} \times v_i = 5000 \times 10^{-6} \times 50 \times 10^{-3}$$
$$= 0.25 \text{ mA}$$

The new drain voltage is

$$V_D = V_{DD} - (I_D + \Delta I_D)\, R_1$$
$$= 18 - (2.5 + 0.25) \times 10^{-3} \times 5 \times 10^{-3}$$
$$= 6.25 \text{ V}$$

Thus, V_D has changed from V_D = 7.5 V to V_D = 6.25 V. The change of V_D is equal to −1.25 V For the negative half cycle of input cycle, the change of V_D can be calculated as +1.25 V. Thus, we see that for an input voltage variation of ±50 mV with VG adjusted for ID = 2.5 mA, the output voltage change obtained is ±1.25 V. The voltage amplification of the circuit, A_v is

$$A_V = \frac{1.25}{50 \times 10^{-3}} = 25$$

Thus, the input voltage is amplified 25 times.

14.10 ► MOSFET

MOSFET is one of the most widely used devices incorporated in integrated circuits (IC). It is a three-terminal device, the three-terminals are drain, source, and the gate. A MOSFET is also known as insulated gate field effect transistor (IGFET) as the gate of the MOSFET is insulated from the channel between the drain and the source. A MOSFET is commercially more important than JFET because the MOS devices are suitable for very large-scale integration. MOSFET can be *n-channel*-type or *p-channel*-type channel formed between the source and the Drain. MOSFET, either n-channel or p-channel type can further be categorized into enhancement MOSFET or depletion MOSFET.

14.10.1 The Enhancement MOSFET (EMOSFET)

The constructional details of an EMOSFET has been shown in Fig. 14.32. In an n-channel EMOSFET two blocks of heavily doped n-type material have been diffused into a p-type substrate or a base (a substrate is a surface or material on which a device is made or developed/grown). The terminals for external connection have been brought out. The whole of the surface has been coated with a layer of silicon dioxide. A thin metal plate has been placed on the surface of the silicon dioxide layer wherefrom the gate terminal connection has been taken. In fact, the metal plate itself will function as the gate of the MOSFET. The substrate is a high-resistive material.

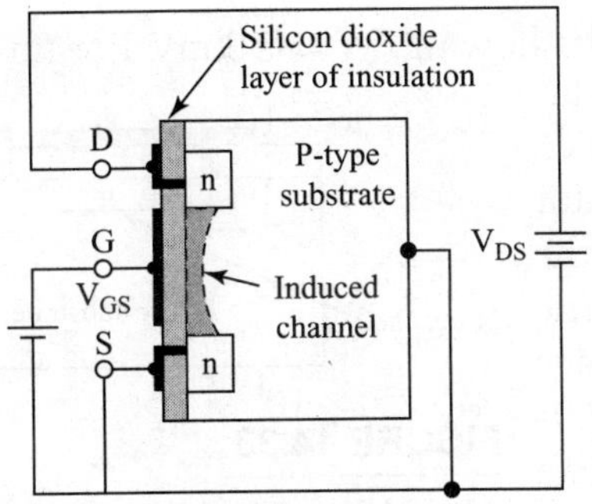

(a) N-channel EMOSFET

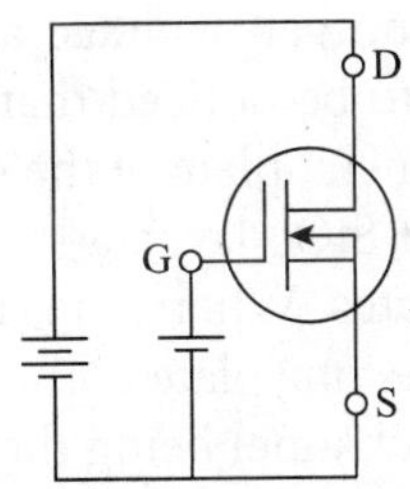

(b) N-channel EMOSFET with positive gate bias symbolic representation

FIGURE 14.32

EMOSFET constructional details

It may be noted that no continuous channel exists between the source and the drain. The two n-blocks forming the drain and the source, makes two p–n junctions with the p-type substrate. When a positive voltage, V_{DS} is applied between the drain and the source terminals, the p–n junction close to the drain is reverse biased while the other junction close to the source is forward biased. Since one of the p–n junctions is reverse biased between terminals D and S, only a negligible small current will flow between the drain and the source due to the minority charge carriers. Virtually no continuous channel exists between the drain and the source with the result that virtually no current flows between the drain and the source without any gate voltage applied. In the symbolic representation, the broken line represents this condition. When a positive voltage, V_{GS} is applied on the gate, the gate voltage will induce a channel by pulling the minority charge carriers of the p-type substrate towards the gate. If the gate voltage is increased more and more, negatively charged minority carriers, i.e., the electrons in this case, from the p-type substrate will get collected in the region between the drain and the source as shown in Fig. 14.32 (a). These electrons cannot cross the SiO_2 layer, and hence, constitute an n-type-induced channel between the drain and the source. A minimum or threshold gate voltage, V_{GST} is required before drain current will flow. The magnitude of the current between the drain and the source will depend on the gate potential. Thus, the magnitude of drain current can be controlled by changing the gate voltage. The conductivity of the channel is enhanced due to the gate voltage. With no gate voltage, the device is non-conducting, i.e., off. When the device is to be switched on, a gate voltage higher than the threshold voltage has to be applied. This type of MOSFET is therefore suitable for switching operations.

14.10.2 The Depletion MOSFET

As shown in Fig. 14.33, the region between the drain and the source on the top surface of the p-type substrate is called the channel. The channel is produced by diffusing an n-type impurity material between the drain and source terminals before the insulating SiO_2 layer is deposited. When a positive voltage V_{DS} is applied between the drain and the source, a drain current I_D flows with no gate voltage applied, i.e., when $V_{GS} = 0$. It is to be noticed that a capacitor exists between the metal plate of the gate and the channel having the SiO_2 layer as the dielectric between the two plates. When a negative voltage is applied to the gate, one plate of the capacitor is made negative, the channel being the other plate will have positive charges induced in it. This will deplete the channel majority charge carriers, i.e., the electrons, and hence, the channel conductivity will decrease. The depletion of the majority charge carriers justifies its name as depletion MOSFET. The drain current will get reduced if V_{GS} is made more negative.

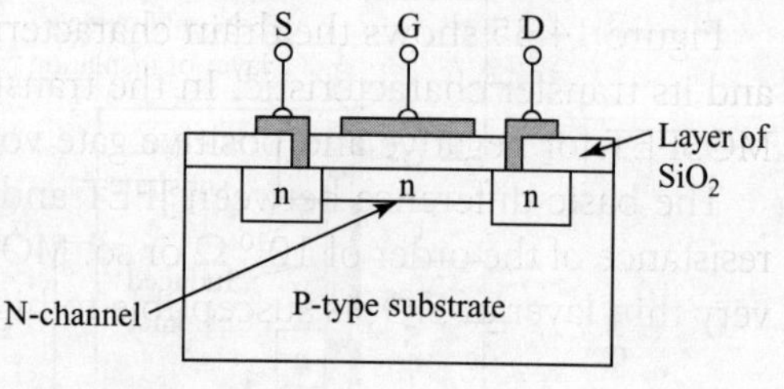

FIGURE 14.33

An n-channel depletion MOSFET

A depletion MOSFET can also be made to operate in enhancement mode by applying a positive gate voltage. When a positive voltage is applied to the gate due to capacitor action, negative charges will get induced in the channel thereby increasing its conductivity. As a result drain current will increase.

14.10.3 Static Characteristics of MOSFET

Figure 14.34 shows the typical drain characteristics of an n-channel enhancement type of MOSFET. Each characteristic curve shows the variation of drain current, I_D with drain voltage V_{DS} for a fixed gate to source voltage, i.e., V_{GS}. The forward transfer characteristic which is the relation between I_D and V_{GS} has also been shown.

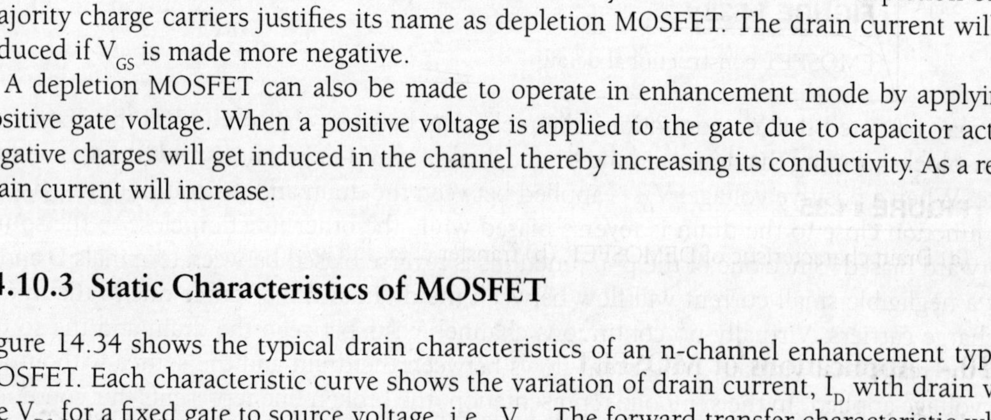

FIGURE 14.34

(a) Typical drain characteristic of an EMOSFET; (b) transfer characteristic

Figure 14.35 shows the drain characteristics for a depletion-type MOSFET, or DEMOSFET and its transfer characteristic. In the transfer characteristic the two modes of operation of the MOSFET for negative and positive gate voltage, respectively, have been shown.

The basic difference between JFET and a MOSFET is that a MOSFET has very high input resistance of the order of $10^{10}\ \Omega$ or so. MOSFETs should be handled very carefully because the very thin layer of SiO_2 is susceptible to high voltages and can be easily punctured.

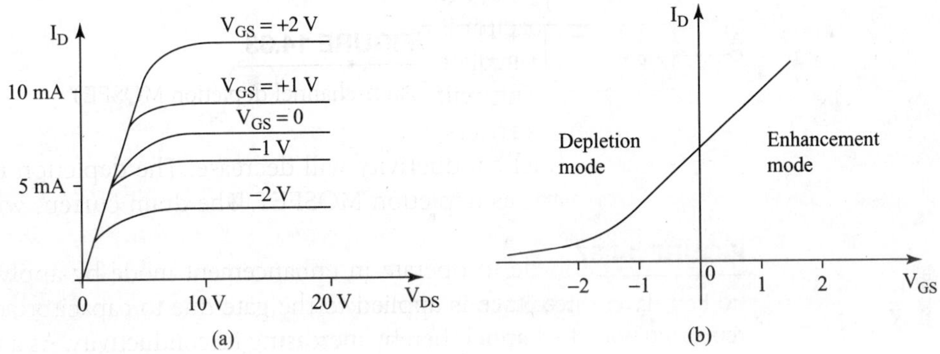

FIGURE 14.35

(a) Drain characteristic of DEMOSFET; (b) transfer characteristics

14.10.4 Applications of MOSFET

Like JFETs, MOSFETs are also used in a wide range of applications. In fact we may use either JFET or MOSFET in most of the applications interchangeably. But the MOSFETs require very careful handling. MOSFETs are used in switching circuits, as phase shift oscillator, as an amplifier, etc. Fig. 14.36 shows a capacitor-coupled MOSFET switching circuit. As shown in Fig. 14.36 (a), no gate voltage is applied to the device, i.e., $V_{GS} = 0$. The device is off. A positive gate voltage higher than the threshold voltage is required to turn the device on.

Through the voltage divider R_1 and R_2 as in Fig. 14.36 (b) a positive voltage when applied, the circuit is switched on. The device can be set to be on for a desired level of drain current, by determining the V_{GS} from the transfer characteristic provided by the manufacturer. To make sure that the device gets switched off, the V_{GS} must be brought below the minimum threshold voltage, i.e., V_{GST}.

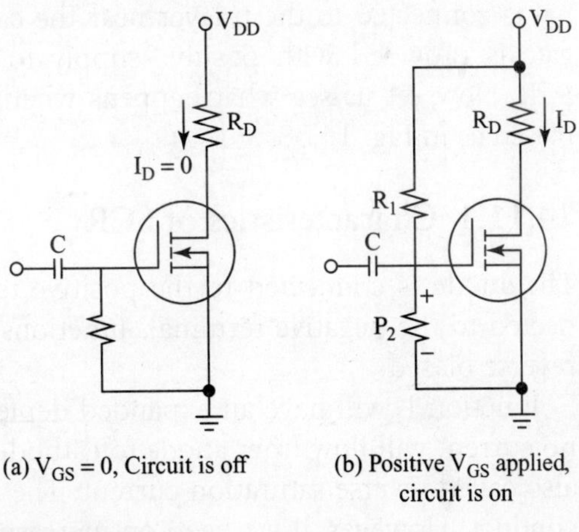

FIGURE 14.36

MOSFET switching circuit

14.11 ► Silicon-Controlled Rectifier

A silicon controlled rectifier (SCR) is a four-layer three-junction device. Thyristor is the general name given to a family of such semiconductor devices having four layers and three junctions. After semiconductor diodes and transistors, thyristors were introduced in a big way in almost all control applications in industry.

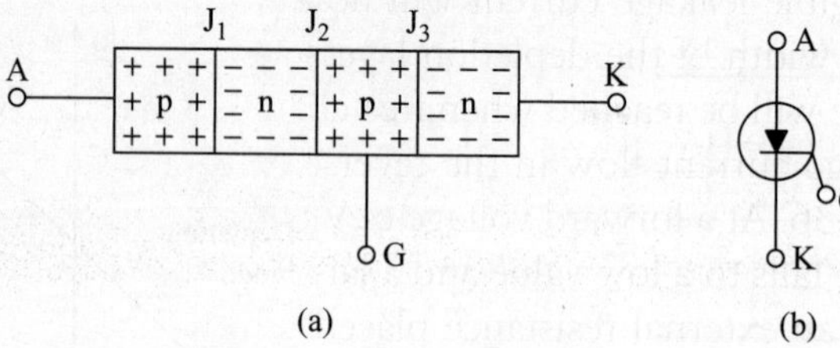

FIGURE 14.37

(a) Four-layer three-junction diagramatic representation of a silicon controlled rectifier; (b) symbolic representation

A SCR shown in Fig. 14.37 has four alternate layers of p, n, p, and n. It has three terminals, viz A (anode), K (cathode), and G (gate). The three p–n junctions are J_1, J_2, and J_3.

The majority charge carriers in the p-type material are the holes and in the n-type material are the electrons, respectively. Electrons are negatively charged while the holes are positively charged. The gate terminal is connected to the p-layer near the cathode. The gate is provided with positive supply to trigger the SCR. Now, let us see what happens when the SCR is biased as in Fig. 14.38.

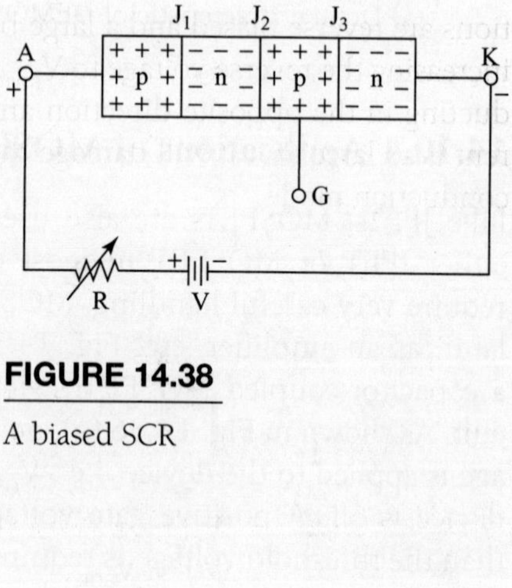

FIGURE 14.38

A biased SCR

14.11.1 Characteristics of SCR

The anode is connected to the positive terminal of the battery while the cathode is connected to the negative terminal. Junctions J_1 and J_3 are forward biased while junction J_3 is reverse biased.

Junction J_3 will have an expanded depletion layer because of its reverse biasing. As such no current will flow from anode to cathode except a very minute amount of leakage current also called reverse saturation current. The SCR under the forward-biased condition will not conduct. However, if we keep on increasing the voltage between the terminals A and K, a stage comes when the depletion layer at junction, J_2 will breakdown due to a large voltage gradient across the depletion layer. This phenomenon is called *avalanche breakdown*, the other two junctions, J_1 and J_3 being forward biased, there will be free movement of charge carriers resulting in a large current flow through the device. The SCR is said to be in conducting state.

For the reverse biasing, the cathode is made positive with respect to the anode by connecting the negative terminal of the dc source to the anode and the positive terminal to the cathode. Junctions J_1 and J_3 are reverse biased and junction J_2 is forward biased. No current, except some negligible leakage current will flow. However, when the reverse voltage is gradually increased, the width of the depletion layers at junctions J_1 and J_3 will become thin and a breakdown voltage will be reached when free charge carriers will be flowing through the device resulting in a large current flow in the reverse direction. The V–I characteristic of an SCR is shown in Fig. 14.36. At a forward voltage of $V_{AK} = V_{BOF}$, the SCR starts conducting. The voltage across the device falls to a low value and a large amount of current starts flowing. This current is controlled by an external resistance placed in the circuit. The voltage at which the SCR starts conducting is called its triggering voltage. By gate control, it is possible to reduce this triggering voltage.

In Fig. 14.39 it is seen that the application of gate current I_{g1} reduces the voltage at which the device starts conducting. A further increase in gate current to I_{g2}, V_{AK} required to make the device conducting is again reduced. This process is called gate control process.

Under the reverse-biased condition, the device is in blocking state because two of its junctions are reverse biased and a large barrier potential is built up across the junctions. But upon increasing the reverse voltage to V_{BDR}, there is a breakdown of junctions. The device starts conducting in the opposite direction and a large reverse current starts flowing. The reverse current is so large that it will damage the SCR. That is why an SCR is not operated in the reverse conduction mode.

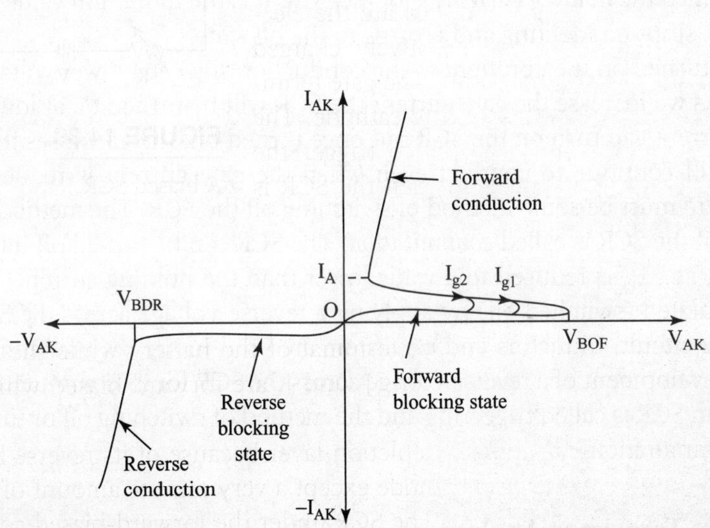

FIGURE 14.39

V–I characteristic of an SCR

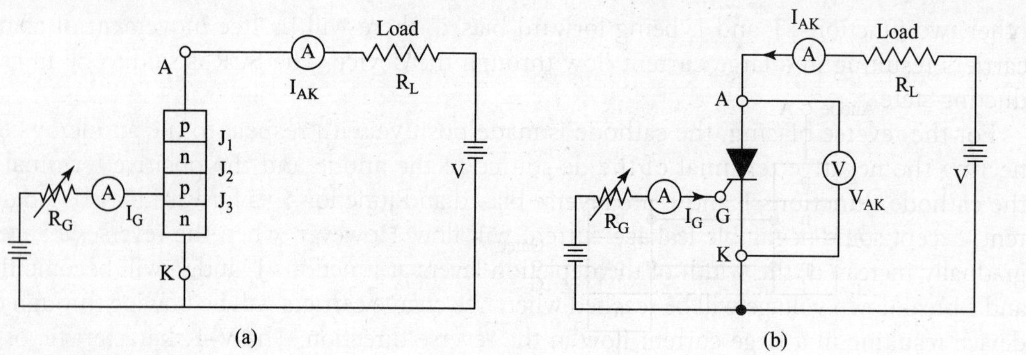

FIGURE 14.40

(a) Operation of an SCR with gate control; (b) connection diagram for plotting the V–I characteristic

An SCR can be operated with and without gate control as explained above. With gate control the forward bias voltage required gets reduced. Operation of an SCR with gate control has been shown in Fig. 14.40. The gate is biased with a positive voltage. The gate current, I_G is controlled by adjusting the resistance R_G. At certain voltage V_{AK} and gate current I_G the junction J_2 will break down and the SCR will start conducting. Current will flow through the load. The conduction will continue on any value of load. However, the conduction current I_{AK} should not fall below a certain value when the depletion layer will start building up due to reduced number of charge carriers. This minimum value of current required to hold the conduction of the device is called the ***holding current***, I_H of the SCR. It is the minimum value of current below which the SCR stops conducting and returns to the off state.

The SCR is turned on, i.e., brought to the conduction state at a lower voltage by applying a gate current. As we increase the gate current the SCR will be turned on at lower values of V_{AK}.

The gate current can turn on the SCR but once turned on, the gate loses its control.

The SCR will continue to conduct even when the gate current is reduced to zero. This means that there must be some method of switching off the SCR. The method of switching off (turning off) of the SCR is called *commutation*. The SCR can be turned off if the current flowing through it, i.e., I_{AK} is reduced to a value lower than the holding current, I_H. The SCR can also be commutated (switched off) by applying a reverse voltage across the SCR by means of a commutating circuit. Inductors and capacitors are used in the commutating circuits which facilitate the development of a reverse voltage across the SCR for its commutation. The method of turning on an SCR is called triggering and the method of switching off or turning off an SCR is called its commutation.

14.11.2 Two-Transistor Analogy of an SCR

An SCR can be considered made up of one p–n–p and one n–p–n transistor sandwiched together as shown in Fig. 14.41 (a). Two-transistor-equivalent circuit has been shown in

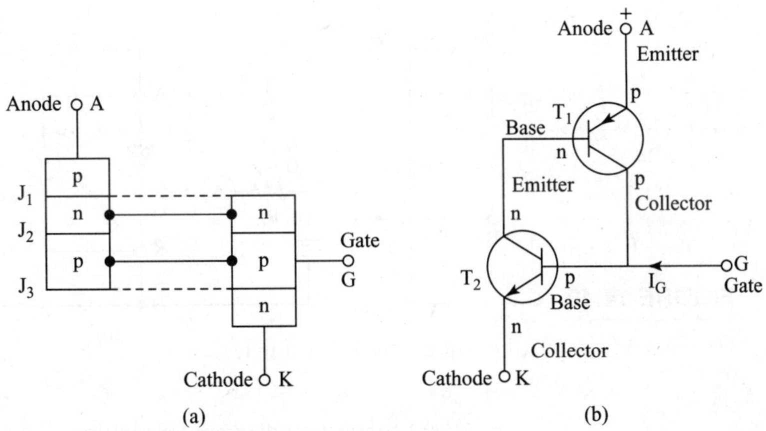

FIGURE 14.41

Two-transistor analogy of an SCR. (a) Block diagram; (b) equivalent circuit

Fig. 14.41 (b). T_1 and T_2 are the two equivalent transistors of the SCR. The gate current, I_G acts as the base current for transistor, T_2.

As shown in Fig. 14.41 (b) the anode is made positive with respect to the cathode and a gate current, I_G is supplied. This gate current acts as the base current of transistor, T_2. This transistor gets turned on. The emitter current of T_2 now becomes the base current of transistor, T_1. Transistor T_2 now gets turned on. The collector current of T_1 adds to the base current of transistor T_2. This way current multiplication takes place. Even if the gate current is removed, the device will not get turned off as long as the current flow does not come down to the level below the holding current.

14.11.3 Applications of SCR

SCR is a members of the thyristor family. DIAC (bidirectional diode thyristor), TRIAC (bidirectional triode thyristor), etc., are some of the other members. An SCR is often referred to as a thyristor. Thyristors find wide applications in the field of industrial electronics and control. They are extensively used as controlled rectifiers, as inverters, converters, cycloconverters, in ac and dc motor control circuits, dc circuit breakers, in electronic control of heating and welding circuits, etc. We will discuss the use of SCR as a controlled rectifier.

Single-phase half-wave uncontrolled and controlled rectifier

We have known earlier that a diode can be used as a half-wave rectifier. The circuit for a half-wave rectifier is shown again in Fig. 14.42 with the input and output voltage wave. When the input terminal A is positive, current flows through the diode and the load. When terminal B is negative, the diode is reverse biased and no current flows through it and the

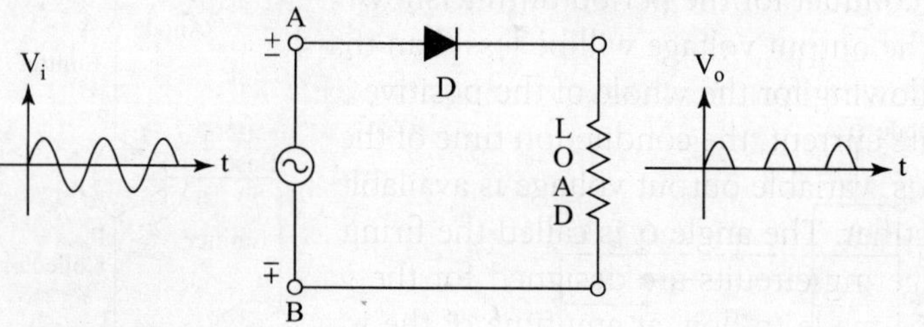

FIGURE 14.42

Half-wave diode rectifier (uncontrolled rectifier)

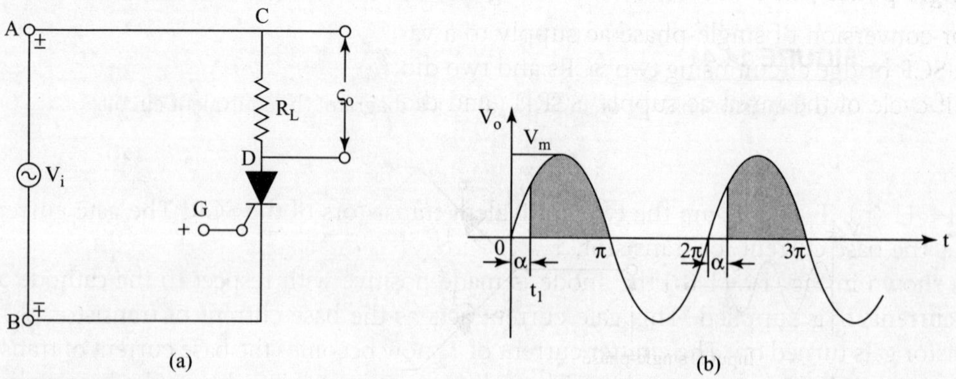

FIGURE 14.43

(a) Half-wave-controlled rectifier circuit; (b) output voltage wave shape

load. The output voltage wave shape is half-wave rectified voltage wave whose average value is $\frac{V_m}{\pi}$, where Vm is the maximum value of the input voltage wave.

In a controlled rectifier, instead of a diode we will use an SCR as shown in Fig. 14.43 (a). We have just seen that a diode allows current flow for the whole of the positive half cycles. In case of an SCR control, if the SCR gate is provided with sufficient bias voltage, the SCR will be on all the time, provided the anode to cathode voltage is positive. We will get a half-wave rectified output across the load as we get in the case of a diode circuit. If a gate current can be provided at an instant of time, say t_1 as shown in Fig. 14.44 (b), the SCR will be turned on at that instant and current will flow. Current will flow for the period starting from t_1, at an angle α to π. This is called the conduction time as has been shown by the shaded portion of voltage wave in Fig. 14.43 (b). In the negative half cycle, the SCR will be reverse biased and no current will flow as is in the case of a diode circuit also. In the second positive half cycle again, the gate will be provided with a gate current so as to turn on the SCR with a delay angle of α;

the SCR will conduct for the period of time shown by shading on the voltage wave. The average value of the output voltage will be less than that the average value of the output voltage if current was flowing for the whole of the positive cycles. By adjusting the time of the application of the gate current, the conduction time of the SCR, and hence, the output voltage can be changed. Thus, variable output voltage is available across the load. This is called a controlled half-wave rectifier. The angle α is called the firing angle of the SCR. The conduction angle is $(\pi - \alpha)$. Triggering circuits are designed for the gate circuit to switch the SCR on. The gate current can be made to flow at any time of the input voltage wave, and hence, we can get a variable output voltage.

Similar to a full-wave rectifier using diodes, we can have a full-wave-controlled rectifier using more number of SCRs.

Single-phase full-wave-controlled bridge rectifier

For conversion of single-phase ac supply to a variable dc output across a load, we can use an SCR bridge circuit using two SCRs and two diodes as shown in Fig. 14.44. In the positive half cycle of the input ac supply, SCR T_1 and diode D_1 will be conducting. In the negative

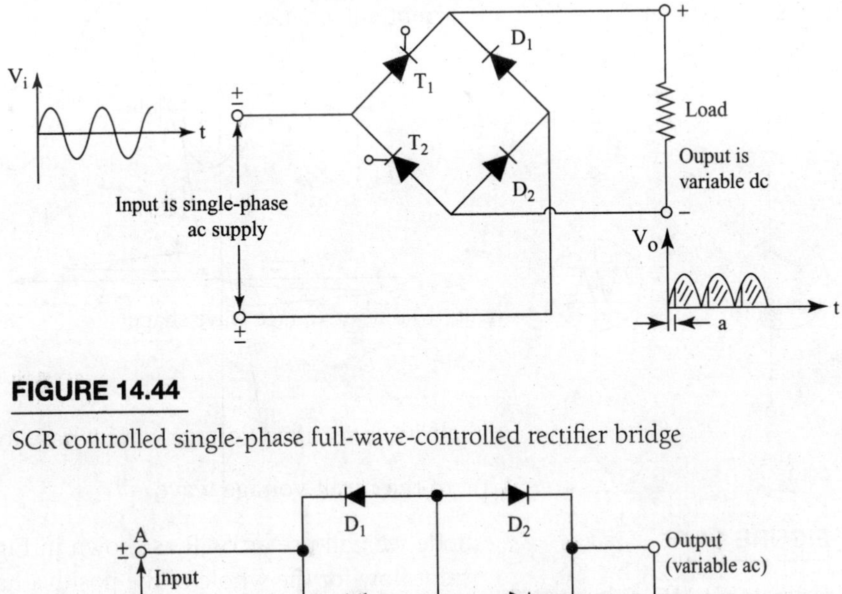

FIGURE 14.44

SCR controlled single-phase full-wave-controlled rectifier bridge

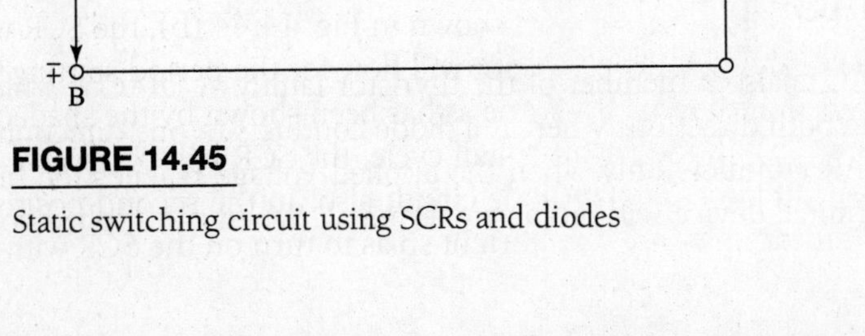

FIGURE 14.45

Static switching circuit using SCRs and diodes

half cycle of the input voltage, diode D_1 and SCR T_2 will be conducting. The firing angle of both the SCRs are controlled so that the average output voltage is variable dc.

If we used four diodes in the bridge circuit, the output voltage would have been a full-wave-rectified dc. By using two SCRs T_1 and T_2, we are able to vary the magnitude of the average output voltage by changing the firing angle α of the two SCRs. The output voltage is variable dc because the angle α can be changed by changing the time at which the gate current will be made available to the two SCRs. With this variable dc we can control the speed of a dc motor, for example. In Fig. 14.45, the gate control circuit for the SCRs have not been shown.

Static switching circuit using SCR-diode combinations

From a fixed ac voltage, a variable ac output voltage can be made available using SCR–diode combinations as shown in Fig. 14.46.

In the positive half cycle of the input voltage when terminal A is positive, SCR T_1 and diode D_2 will conduct. The conduction angle is decided by the time at which the SCR, T_1 will be triggered. In the negative half cycle, terminal B is positive, and hence, SCR T_2 and diode D_1 will conduct. The time of triggering of the gate of SCR, T_2 will decide the conduction angle in the negative half cycle. By changing the firing angle of the SCRs, a variable ac voltage will be available across the load, and hence, load current will change.

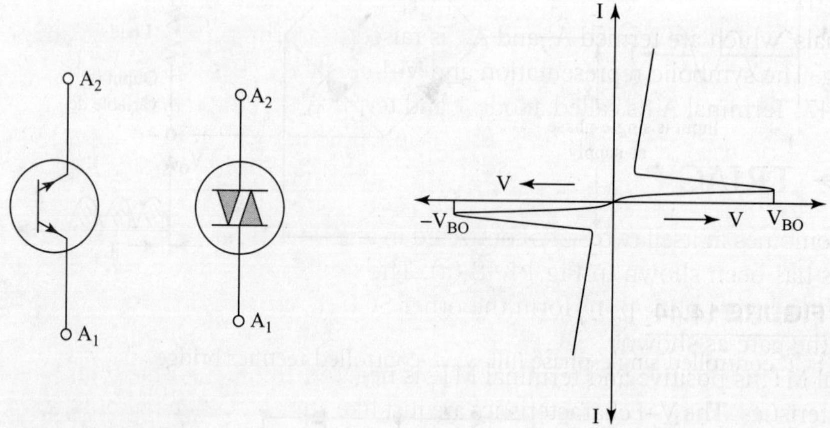

FIGURE 14.46

Symbols of DIAC and its V–I characteristics

14.12 ► DIAC

Like an SCR, DIAC is also a member of the thyristor family. A DIAC is a bidirectional diode which conducts in both directions whereas a diode conducts in one direction. Unlike an ordinary diode, a DIAC conducts only when the applied voltage reaches the breakover voltage. DIAC is a two-terminal device without a gate terminal. When the applied voltage between its

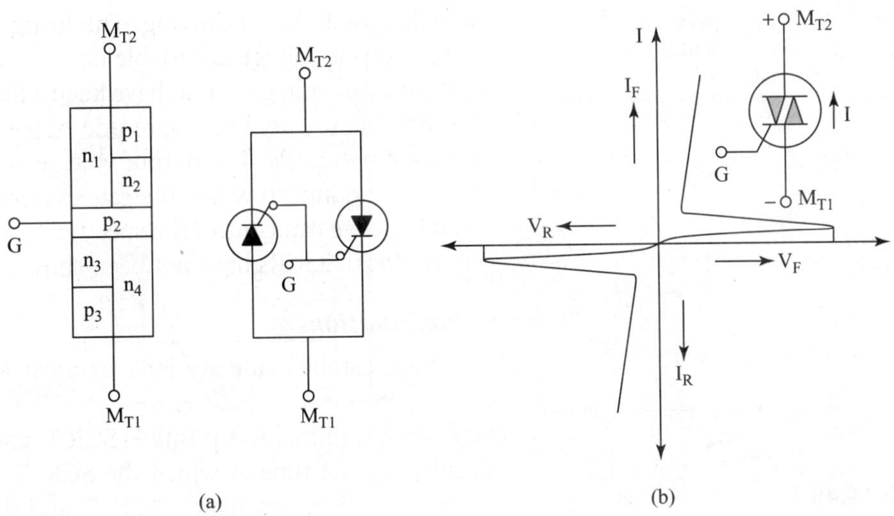

FIGURE 14.47

(a) Construction and symbol of TRIAC; (b) V–I characteristics of a Triac

two terminals, which are termed A_1 and A_2, is raised to the breakover voltage, the DIAC starts conducting. The symbolic representation and V–I characteristics of a DIAC have been shown in Fig. 14.47. Terminal A_1 is called anode 1 and terminal A_2 is called anode 2.

14.13 ► TRIAC

A TRIAC combines in itself two SCRs connected in anti-parallel or inverse-parallel with one gate terminal as has been shown in Fig.14.48 (a). The four layers n_1 p_2 n_3 p_3 form one SCR and the other four layers p_1 n_2 p_2 n_4 form the other SCR. Layer p_2 is common to both SCRs and is used as the gate as shown.

Terminal MT_2 is positive and terminal MT_1 is negative in the first quadrant as shown in the V–I characteristics. The V–I characteristics are just like an SCR. When the polarities of its terminals are changed, the TRIAC conducts in the opposite direction and the V–I characteristics are shown in the third quadrant. The TRIAC when starts conducting in any direction, a heavy current will flow which has to be controlled by some external resistance. The TRIAC's time of triggering, and hence, its conduction time is controlled by a common gate. By applying a proper signal, the firing angle of the TRIAC can be changed, and hence, a variable voltage can be made available across the load. The TRIAC is the most widely used thyristors in control circuits. TRIACS have replaced SCRs in many applications because of their bidirectional characteristics.

A TRIAC can be triggered by applying a voltage to its gate terminal. When terminal MT_2 is positive with respect to MT_1, a positive gate voltage is to be applied to make the TRIAC conducting. When terminal MT_1 is positive with respect to MT_2, a negative gate voltage will make the triac start conducting.

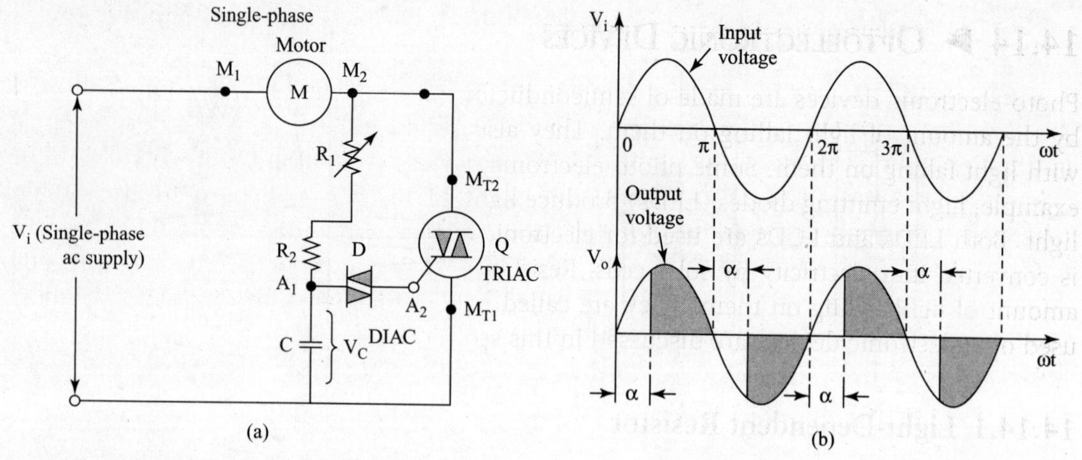

FIGURE 14.48

(a) Phase control circuit using a TRIAC; (b) input and output voltage waveforms

TRIAC phase-control circuit

The circuit for control of ac voltage using a TRIAC has been shown in Fig. 14.48 (a). From a fixed input ac voltage, a variable ac voltage can be made available across the load. The triac will control the voltage in either direction, and as such, control is achieved on both positive and negative half cycles of the input voltage as has been shown in Fig. 14.48(b).

A TRIAC control circuit for the speed control of a single-phase induction motor has been shown in Fig. 14.48(a). A DIAC has been used to provide the gate voltage for triggering the TRIAC. These days a DIAC–TRIAC combination is available in the market. The input and variable output voltage wave forms have been shown in Fig. 14.49 (b). Let us consider the time when the TRIAC is off and the input voltage has been applied. During the positive half cycle of the input voltage, the capacitor is charged as current will flow through R_1, R_2, and C. When the voltage across the capacitor, VC becomes equal to the sum of the DIAC switching voltage and the gate triggering voltage of the TRIAC, the DIAC, D conducts and passes a gate current to trigger the TRIAC, Q. Now the capacitor will start discharging. The capacitor will continue to discharge until the capacitor current falls below the holding current level of D. The TRIAC will get switched off (commutated) at the end of the positive half cycle of the input voltage. The same process of the capacitor being charged in the negative half cycle of the supply will take place and help make the TRIAC getting switched on through the DIAC supplying gate voltage. The firing angle, α of the TRIAC can be controlled by adjusting the value of the variable resistance R_1. The time constant of the R–C circuit (R_1 + R_2 and C) get altered when R_1 is changed, and hence, the rate of charge of C changes.

14.14 ► Optoelectronic Devices

Photo-electronic devices are made of semiconductor material whose conduction is affected by the amount of light falling on them. They also produce light; their resistance change with light falling on them. Some photo-electronic devices emit light and modify light. For example, Light emitting diodes (LEDs) produce light. Liquid crystal displays (LCDs) modify light. Both LEDs and LCDs are used for electronic display of numerals and alphabets. Light is converted into electricity by solar cells. Resistance of certain material is affected by the amount of light falling on them. They are called light dependent resistors (LDRs). Widely used optoelectronic devices are discussed in this section.

14.14.1 Light-Dependent Resistor

An LDR is a semiconductor device whose resistance varies with the amount of light falling on it. Light energy falling on its surface provides sufficient energy to the valence electrons to break away from their atoms. Thus, these charge carriers, i.e., free electrons and holes make the material more conductive. A material becoming more conductive means its resistance is getting reduced. That is why these resistors are also called photoresistors, i.e., light-dependent resistors.

Cadmium sulfide and cadmium selenide are used in making LDRs. The material in the form of a long strip is arranged in a zigzag form on a base. A glass or plastic cover is provided for protection. The symbol and the constructional arrangement of an LDR have been shown in Fig. 14.49.

The response time of these photo-sensitive material is quite small, of the order of few milliseconds. The variation of resistance with falling light on these materials is quite high. For example, in the absence of light, i.e., in darkness, the resistance could be 100 kΩ, while in bright sunlight while the resistance may change to 10 to 20 kΩ.

The principle of operation of a street light circuit using an LDR has been illustrated in Fig. 14.50. In sunlight, the streat lights should be off. Under dark conditions, the lights should automatically get switched on. A 12 V battery supply is connected to the relay coil which has

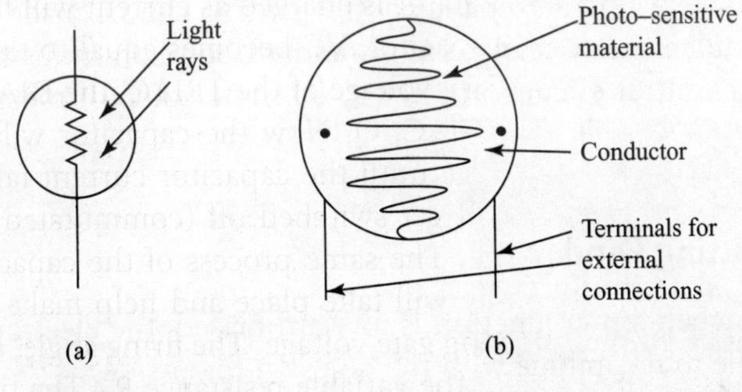

FIGURE 14.49

(a) Symbol of an LDR; (b) constructional details of an LDR

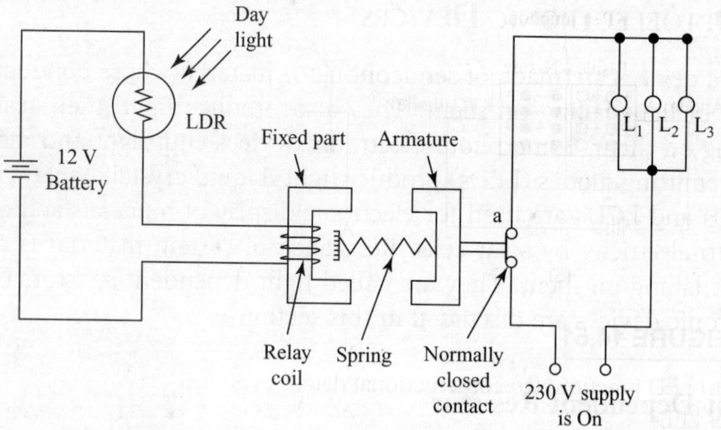

FIGURE 14.50

Lights controlled by relay-operated LDR circuit

an LDR resistance connected in series as shown in Fig. 14.50. During night darkness the LDR resistance is very high, and hence, a very small amount of current will flow through the relay coil. During night, or during darkness, and even during cloudy days, the lights L_1, L_2, L_3 should be on as they get 230 V supply through the normally closed contact of the relay. Current through the relay coil is very low because the LDR resistance during darkness is very high. The armature (the movable part of the relay) of the relay will stay at the position shown. When during day time light falls on the LDR which may be placed in a convenient position, its resistance will decrease, and hence, current through the relay will increase. At a particular value of current, the armature will be attracted and pulled towards the fixed part of the relay against the spring pressure. Then the relay contact, 'a' will open. The lights will not get any supply, and hence will be off. When darkness comes, the LDR resistance will increase, current through the relay will decrease, the attractive force of the relay coil will decrease, the armature will return to its original position due to the spring action. The relay contact, 'a' will close, and hence, the lights will get full supply and will be switched on.

A number of such interesting automatic control circuits can be made using LDR, DIAC, TRIAC, and SCRs. Students may refer to any standard book on industrial electronics and control to study such control circuits and try to fabricate some of them as a part of their project work.

14.14.2 Light-Emitting Diodes

We have known that when a p–n junction is forward biased, the electrons from the n-side cross over to the p-side to recombine with the holes on the p-side.

The free electrons have higher energy than the holes. When a negatively charged electron from the n-side enters the p-side and recombines with a positively charged hole, some amount of energy is emitted in the form of heat and light. Similarly, a hole from the p-side has a

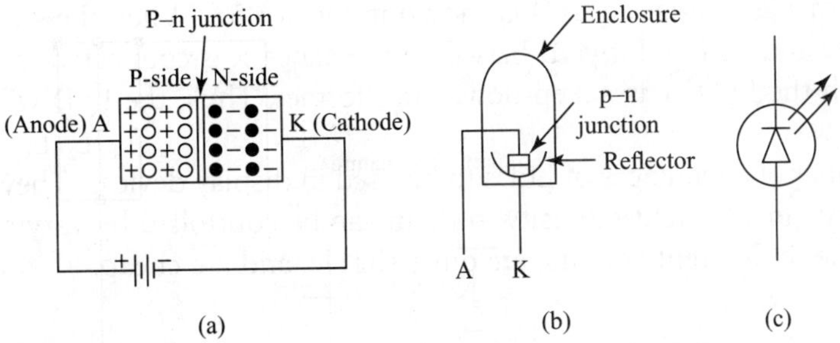

FIGURE 14.51

(a) LED junction; (b) constructional details of an LED; (c) symbol of LED

tendency to cross the junction and recombine with an electron on the n-side. Each recombination causes radiation of energy in the form of heat and light. If the semiconductor material is translucent, it will emit light. LEDs are made from special semiconductor materials such as arsenide phosphide or gallium phosphide. The LED junction, constructional details, and symbol have been shown in Fig. 14.51.

The intensity of light energy emitted from an LED will depend upon the forward voltage applied. The applied voltage is low, of the order of 1 or 2 V. An LED is a compact device which will emit a point source of light. A group of LEDs can be used to illuminate a display of numbers and alphabets. Segmented illuminated display of circuits, diagrams, photographs, etc., can be arranged with the help of an array of LEDs.

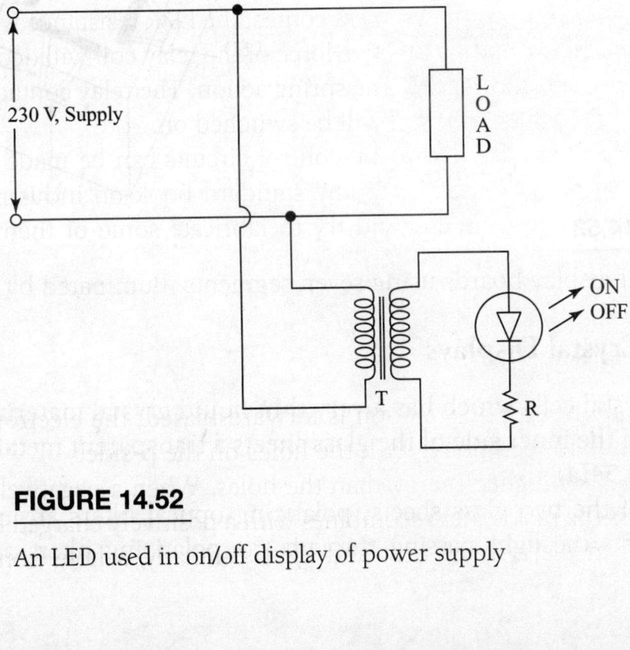

FIGURE 14.52

An LED used in on/off display of power supply

A simple on/off display using an LED is shown in Fig. 14.52. When the supply is on, the LED will get supply and will be forward biased. A resistance R is connected in series to limit the current through the LED. T is a step-down transformer. Thus, the LED will show on/off status of supply.

There are a number of advantages of LEDs to be used in display devices. They are compact, are illuminated very quickly, light intensity in them can be controlled by varying the applied voltage, are available in different colours, are quite simple, and are cheap.

14.14.3 Seven Segment Displays

A seven segment display board using seven LEDs is most widely used in display devices in the field of electronics and instrumentation.

Some LED displays, you might have noticed in control panels, calculators, mobile phones, electronic digital instruments, digital watch, etc., the LEDs are very minute structures. Often plastic light pipes are used to enlarge the lighted area as shown in Fig. 14.53. Display of digits ranging from 1000 to 1999 can be made on the display board by illuminating the segments in a particular order.

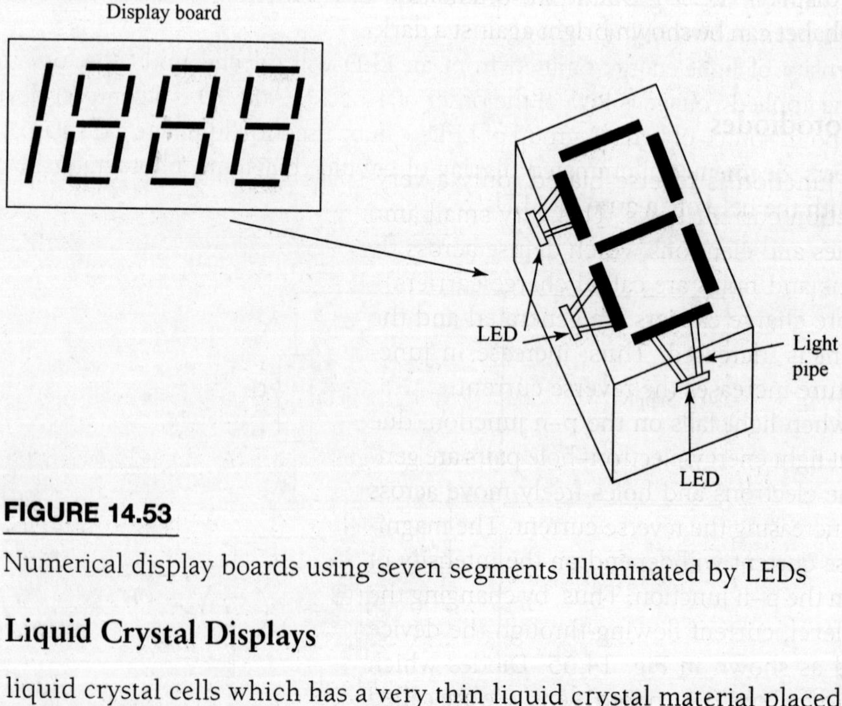

FIGURE 14.53

Numerical display boards using seven segments illuminated by LEDs

14.14.4 Liquid Crystal Displays

LCDs use liquid crystal cells which has a very thin liquid crystal material placed between two glass sheets. On the inner side of the glass sheets a transparent metal film is deposited as shown in Fig. 14.54(a).

On the surface of the two glass sheets, polarizing optical filters are placed. The liquid crystal material twists the light passing through the polarizing filters and the cell looks

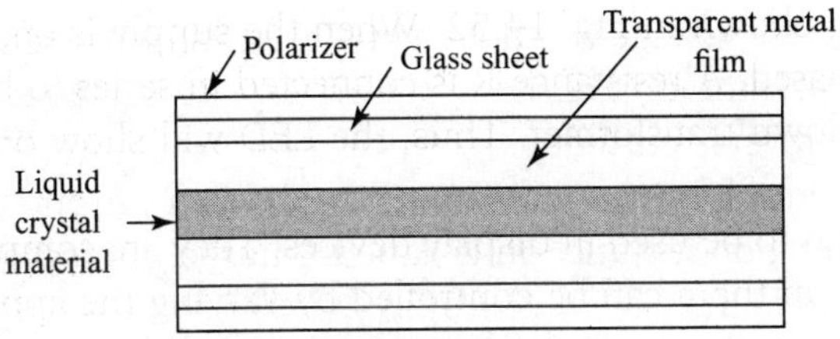

(a)

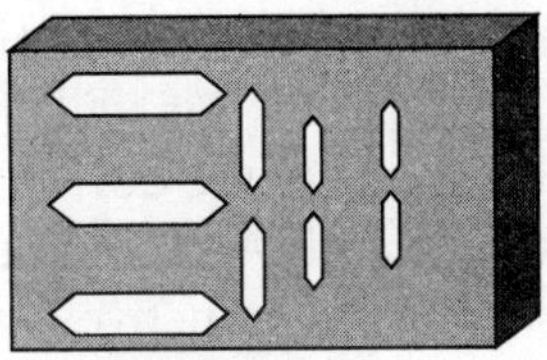

(b)

FIGURE 14.54

(a) Sectional view of a liquid crystal cell; (b) LCD

semi-transparent. When the cell is energized with electrical signal, the liquid crystal molecules orient in a definite crystal pattern and no twisting of liquid occurs and the cell appears dark. Liquid crystal cell can also be made to appear bright when the background is dark. Seven segment displays of numerical numbers, alphabets, and others are often made from liquid crystal displays (LCDs). Due to the orientation of the molecules of the liquid crystal, a digit or an alphabet can be shown bright against a dark background as shown in Fig. 14.54(b).

14.14.5 Photodiodes

When a p–n junction is reverse biased, only a very small current called the leakage current or saturation current flows. This very small amount of current is due to the thermally generated holes and electrons which defuse across the p–n junction as minority charge carriers (electrons and holes are called charge carriers). If the junction temperature increases, more and more charge carriers are generated and the reverse current is increased. Thus, increase in junction temperature increases the reverse current.

Similarly, when light falls on the p–n junction, due to the incident light energy, electron–hole pairs are generated and the electrons and holes freely move across the junction increasing the reverse current. The magnitude of reverse current will depend on the intensity of light falling on the p–n junction. Thus, by changing the illumination level, current flowing through the device can be varied as shown in Fig. 14.55. Diodes which are designed to be light (i.e., photo) sensitive are called photodiodes. Photodiodes operate under the reverse-biased condition and are designed to be light sensitive. There is minority charge carrier current flow when the p–n junction is illuminated.

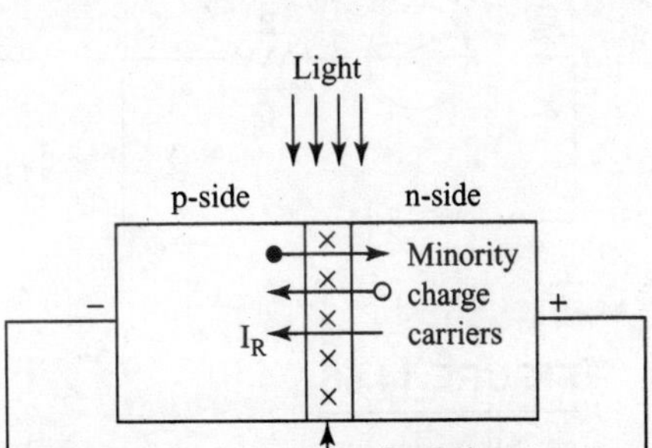

FIGURE 14.55

Photodiode with reverse-bias (only a very small reverse current flows across the junction due to minority carriers)

It is to be noted that increasing the reverse voltage will not change the reverse current. That is why the reverse current is called reverse saturation current. But increase in the illumination level of the p–n junction will change the reverse current.

The resistance of the device at a different illumination level will be different. A photodiode can work like an LDR when the light falling on its junction is varied. Thus, a photodiode can work as a photo-conductive device like an LDR.

14.14.6 Photovoltaic Cells or Solar Cells

A photovoltaic cell or a solar cell is a photodiode with no reverse bias voltage applied across it. Even when the reverse bias voltage across a photodiode is removed, the minority charge carriers continue to cross the junction when light continues to fall on the p–n junction. Reverse current will continue to flow through the circuit with the bias voltage removed but the circuit kept closed as shown in Fig. 14.56(a) and (b).

Electrons from the p-side (minority charge carriers) will flow to the n-side and return to the p-terminal through the n-terminal of the device. This way the device will work as a voltage source, i.e., a small voltage cell as shown in Fig. 14.56(c). p-Side is the positive terminal and n-side is the negative terminal as has been shown. Thus, a junction-illuminated photodiode is a photovoltaic cell

A solar cell is essentially a photodiode. A large number of photodiodes are connected in series and parallel fixed on a panel so as to get considerable voltage and current output. This way sun's energy (solar energy) is converted into electrical energy. Much research is being conducted to increase the energy conversion efficiency and to reduce the cost of manufacturing of solar cells and solar panels.

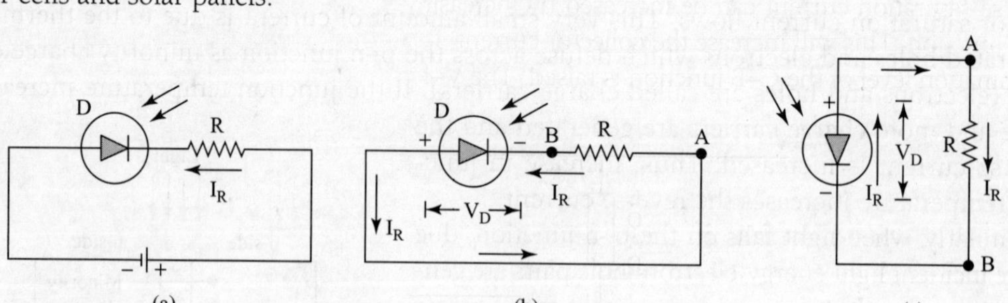

FIGURE 14.56

(a) Photodiode with reverse bias; (b) an illuminated photodiode without any bias voltage; (c) photovoltaic cell

14.14.7 Phototransistors

A phototransistor is similar to a bipolar junction transistor, generally an n–p–n type with a bias voltage applied between the collector and the emitter with the base terminal kept open. No supply is given to the base. The transistor action is achieved by the light energy falling

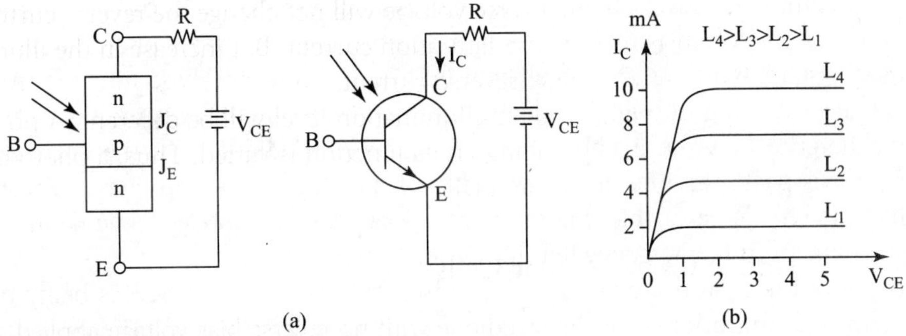

FIGURE 14.57

(a) Transistor action achieved by light energy falling on base-emitter junction
(b) V-I characteristic of a phototransistor

on its base–emitter junction. Thus, instead of base current, light energy provides the input to the base.

An n–p–n type BJT with the base terminal open has been shown in Fig. 14.57. The collector base junction, J_C is reverse biased while the base–emitter junction, J_E is forward biased. The collector–base junction is constructed like a photodiode. In the absence of light falling on the junction, only a small amount of reverse saturation current, also called leakage current, will flow through the device. This current of very low value is due to the minority charge carriers flowing at room temperature. When light energy falls on the C–B junction, additional charge carriers are created to add to the reverse saturation current. Thus, light energy increases the base current. The reverse saturation current can be increased by increasing the illumination level of the collector–base junction. This will increase the collector current, I_C. Collector current will increase when the illumination level of the C–B junction is raised. The V–I characteristic is drawn between V_{CE} and

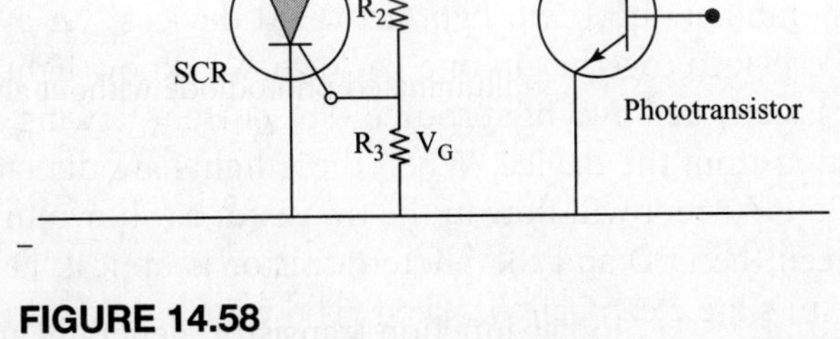

FIGURE 14.58

A phototransistor triggering an SCR on low illumination level of the surrounding area

I_C at different illumination levels, L_1, L_2, L_3, L_4, etc., as shown in Fig. 14.57(b). For an equal level of illumination, a phototransistor provides more output current than a photodiode. A biased phototransistor starts working as soon as light falls on the collector–base junction. An amplified current starts flowing through the device, and consequently the output current increases. The main difference between a phototransistor and a photodiode is in the current gain. A phototransistor is packaged in a metal case with a lens filled on the top. Plastic-encapsulated phototransistors are also available. Phototransistors can be used in electronic control circuits like illumination control, emergency lighting system, relays, etc.

A simple phototransistor circuit has been shown in Fig. 14.58. An SCR is being triggered with the help of a phototransistor. When the illumination level is high, the phototransistor will allow sufficient current to flow through it and resistance R_1. Resistance R_2 and R_3 are chosen so that when the phototransistor is on, very little current will flow through them and the voltage V_G across the gate will not trigger the SCR. When the illumination level of the surrounding falls, the phototransistor will stop conducting and current will flow through resistors R_1, R_2, and R_3. The value of these resistors are so chosen that, voltage across the gate, V_G, will trigger the SCR, and the SCR will start conducting. When the SCR is conducting, current will flow through the load, which could be a lighting load.

14.14.8 Photo-Darlington

A phototransistor can be connected with another transistor to produce higher output. The arrangement as shown in Fig. 14.59 is called a photo-darlington. Photo-darlington is a phototransistor connected in darlington with another bipolar junction transistor. With this arrangement, the output current obtained will be higher than what could be obtained in a phototransistor or a photodiode. The sensitivity of the device to illumination is high but the switching time is somewhat more here as two transistors are getting involved.

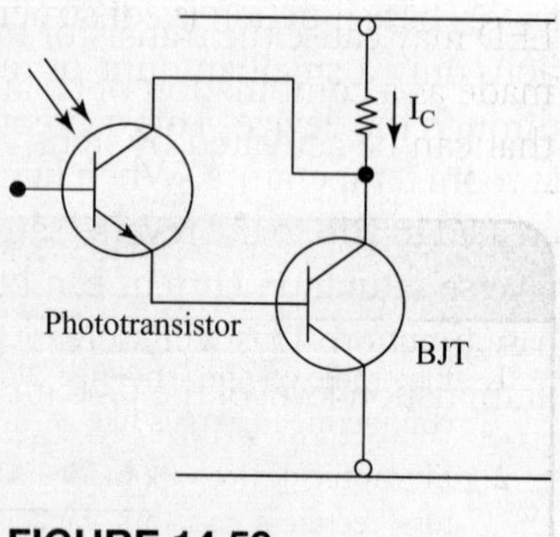

FIGURE 14.59

Photo-darlington transistor

14.14.9 Optocouplers

An optocoupler is a combination of two light-activated devices, viz a phototransistor and an LED. These two are placed together in one package. We know light emits from an LED and a phototransistor is activated by a light source. The current flowing through an LED will cause light to be emitted from the device. When these lights are directed to illuminate the phototransistor junction, current will flow in the transistor as shown in Fig. 14.60.

The coupling between the LED and the phototransistor is optical, i.e., through the emitted light. The two circuits are electrically isolated. The voltage level of the two circuits may be made different. Thus, from a low voltage side, control can be made of the high-voltage side. Fig. 14.60(b) shows an optocoupler contained in a transparent insulating base material.

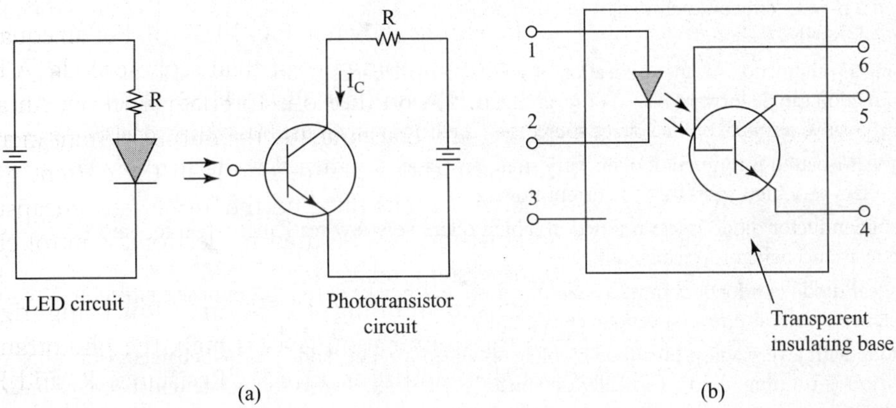

FIGURE 14.60

(a) Optocoupler circuit, an LED is driving a phototransistor; (b) optocoupler package with terminals

This base material will allow light emitted by the LED to be transmitted to the base of the phototransistor. However, the two, i.e., the LED and the phototransistor are electrically isolated.

An optocoupler can be operated as an electronic switch. A pulse of current through the LED may cause the transistor to be switched on which is otherwise off. An optocoupler can be made as a combination of (i) an LED and a photo-darlington; (ii) a LED and any other device that can be activated by light.

Important Terminologies and Points to Remember

1. Electronic components like diodes, transistors, silicon controlled rectificrs (SCRs), etc., are made of semiconducting materials like germanium, silicon, carbon, etc.
2. Electronic devices are made using integrated circuits (ICs) which contain a large number of diodes, transistors, rectifiers, resistors, capacitors, etc., all fabricated on a single IC chip. This IC technology has resulted in miniaturization of electronic devices and also in increasing their reliability.
3. Atoms in a semiconductor material bond themselves through covalent bonds by sharing of their four valence electrons. At absolute zero degree temperature, they behave like insulators. With increase in temperature or with the application of a voltage, these materials behave like conductors due to availability of free electrons. That is why materials like germanium and silicon are called semiconducting materials.
4. N-type material is formed by doping (adding) a pure silicon or germanium crystal with a material with five valence electrons like antimony, phosphorous, arsenic, etc., making one free electron for each bonding.
5. P-type materials are formed when silicon or germanium crystal is doped with a small percentage of trivalent material like boron, gallium or indium, thereby creating a shortage of electrons in the covalent bonds formed. This shortage of one electron in each bond is called a hole.
6. A p-type material is represented by holes as the majority carrier while an n-type material is represented a free electron as the majority carrier.
7. When a p-n junction is made, there is an initial diffusion of electrons and holes from opposite sides of a region or layer, called depletion layer is formed at the junction. The layer has no electrons or holes to further diffuse. A barrier potential of 0.3 V for germanium and 0.7 V for silicon is developed at the junction at room temperature.

8. When a p-type semiconductor is joined with an n-type semiconductor by a special technique, a p-n junction is formed.
9. Biasing a p-n junction means application of some external voltage across the two sides of the p-n junction. A p-n junction can be forward biased or reverse biased. In forward bias, p-side of the p-n junction is connected to the positive terminal of the voltage source, i.e., the battery. In reverse bias battery connection is reversed.
10. A forward-biased p-n junction offers very small resistance to current flow while a reverse-biased p-n junction offers very high resistance to current flow.
11. A semiconductor diode is a p-n junction which offers very low resistance when forward biased and very high resistance when reverse biased.
12. An ideal diode conducts current in one direction and blocks current in opposite direction. The forward resistance is zero and reverse resistance is infinite.
13. A diode with a very sharp breakdown voltage is called a zener diode. The V-I characteristic in the reverse direction is running almost vertically downwards. Diodes designed to operate under reverse breakdown condition will maintain a fairly constant voltage over a wide range of load current.
14. A zener diode voltage regulator will maintain a nearly constant voltage across the load over a wide range of load current variation.
15. Transistor is a three layer p-n-p or n-p-n semiconductor material. The ability of a transistor to amplify a signal accounts for its wide use in electronic circuits.
16. The word transistor is a short form of "transfer resistor" which means signal amplification in a transistor is achieved by transferring the signal from a region of low resistance to a region of high resistance.
17. A transistor has two p-n junction connected back to back to have three terminals, viz., emitter, base, and collector. There is one emitter-base (EB) junction and one collector-base (CB) junction in a transistor.
18. To operate a transistor, we need to bias it by applying external voltage supply. In a p-n-p or n-p-n transistors, when p-terminal is made positive with respect to n-terminal, it is called forward biasing and the reverse is called reverse biasing.
19. Normally the emitter-base junction is forward biased and collector-base junction is reverse biased. In an n-p-n or p-n-p transistor, the central terminal is the base.
20. To bias an n-p-n transistor, we have to connect positive terminal of the battery to the base terminal, p and the negative terminal to the emitter terminal, i.e., left-hand side n-terminal. This way emitter-base terminal gets forward biased. The collection-base terminal is reverse biased by connecting the other n-terminal, i.e., collector terminal to the positive terminal of a battery while the negative terminal is connected to the base terminal. When a junction is forward biased, it reduces the width of the depletion region and barrier voltage gets reduced.
21. Due to large CB bias, voltage electrons from the emitter region will be pulled towards the positive terminal of the battery. The collector will collect about 98 percent of the electrons emitted by the emitter. In an n-p-n transistor, majority change carriers are electrons.
22. In a p-n-p transistor, the majority change carriers are the holes. Biasing is done in the same way as n-p-n transistor. By varying the forward bias voltage at the emitter-base junction, we can control a large emitter-collector current through small variation of base current.
23. When both the junctions of a transistor are reverse biased, the transistor is paid to be operating in the cut-off region.
24. There are three types of connections that can be made for both n-p-n and p-n-p transistors, they are common-emitter, common-collector, and common-base connections or configurations. In common-emitter connection, the emitter terminal is common to both input and output circuits. Similar is the case with the word, 'common' in other two cases.
25. Both n-p-n and p-n-p transistor are called bipolar junction transistor (BJT) because the change carriers (both majority and minority) for both types are electrons and holes although for n-p-n the majority change carriers are electrons and for p-n-p transistors majority change carriers are the holes.

26. After biasing the transistor, when an ac signal is connected to the emitter-base circuit, the signal gets amplied in the collector circuit. This way the transistor works as a signal amplifier. A transistor can also be used as a switch.
27. There are two types of field effect transistors (FETs), namely junction field effect transistor (JFET) and metal oxide semiconductor field effect transistor (MOSFET).
28. A JFET is made using an n-type material called channel sandwiched between two p-type material. One terminal of the n-type material is called drain and the other terminal is called source. The two p-type materials are connected together to form the gate. The gate material is highly doped as compared to the channel. The two gate-channel p-n junctions are kept reverse biased.

    The reverse bias voltage produces an electric field which changes the depletion region at the gate-channel junction. The effect of the electric field produces the transistor action. Life BJTs, FETs are used for signal amplification and in switching circuits.
29. MOSFET also has three-terminals, viz., drain, source and gate. MOSFETs can be n-channel type or p-channel type and are widely used and are incorporated in integrated circuits (ICs).
30. A Silicon Controlled Rectifier (SCR) is a four layer three-junction device. It has three terminals, viz., a (anode), K (cathode) and G (gate). The three p-n junctions are J_1, J_2 and J_3. The gate is biased with a positive voltage and the gate current I_G is controlled by adjusting a resistor in the circuit. A dc voltage V_{AK} is connected is required to make the SCR conducting. The gate voltage at which the SCR starts conducting and current flows through the load, is called the trigging voltage. Triggering current causes the thyristor ON but then the gate loses its control on the SCR. The method of switching off of the SCR is called commutation.
31. The minimum value of current at which the SCR remains conducting is called th holding current. Current level has to be brought below that level to switch off the SCR.
32. The method of turning ON an SCR is called triggering and the method of switching it off is called commutation.
33. A two terminal diode is a rectifier and a three terminal SCR with gate control mechanism is called a controlled rectifier.
34. By triggering the SCR at a particular time, the angle α, called the firing angle, can be changed. By changing the firing angle, the output voltage can be varied. When SCR is used instead of diode, the output dc voltage can be made a controlled one.
35. SCRs, also called thyristors, are used in many applications like in controlled bridge rectifiers, in making inverters, choppers, and cycloconverters. These circuits are very much in used in industrial electronic circuits for control operations.
36. LDR is a light-dependent resistor whose resistance varies with the amount of light falling on it. LED is a light emitting diode which produce light. Light is converted into electricity by solar cells. LCD is a liquid crystal display which modifies light. All these are widely used in various control and display devices. Seven segment illuminated display boards are made using LEDs.

REVIEW QUESTIONS

Short Answer Type Questions

1. Distinguish between an intrinsic and extrinsic semiconductor material. What is a p-type material and an n-type material?
2. Explain how a p–n junction can be used to work as a diode.
3. What is the significance of 'barrier potential' in a p–n junction?
4. A reverse-biased p–n junction has a wide depletion region. Explain.

5. What do you mean by forward biasing and reverse biasing of a p–n junction?
6. Sketch the typical voltage/current characteristics for a forward-biased and reverse-biased p–n junction. Briefly mention the salient points.
7. What is an ideal diode? Draw its characteristic.
8. How is a zener diode different from a conventional diode?
9. Show how a zener diode can be used as a voltage regulator.
10. What are the diode parameters?
11. Show two applications of zener diodes.
12. Explain the operation of a zener diode under forward- and reverse-biased conditions.
13. Explain clearly the difference between acceptor-type and donor-type impurities and state what types of charge carriers are contributed by them.
14. Distinguish between intrinsic and extrinsic semiconductors. Give examples.
15. Explain the difference between p-type and n-type semiconductor materials. Draw the atomic structure of silicon and insert an impurity of arsenic in silicon.
16. What are p-type and n-type semiconductors? Draw and explain the V–I characteristic of a p–n junction diode.
17. Explain for a semiconductor diode the following terms: peak inverse voltage; forward and reverse biasing; potential barrier; depletion layer and reverse saturation current.
18. Explain covalent bonding of semiconductor materials.
19. Distinguish between intrinsic and extrinsic semiconductor. What is doping? Give one example.
20. Show how n-type and p-type semiconductor materials can be formed.
21. Explain the formation of depletion region in a p–n junction. What is barrier voltage?
22. Explain biasing of a p–n junction and draw forward characteristic of the p–n junction.
23. Draw the reverse V–I characteristic of a p–n junction. What is reverse saturation current?
24. Draw and explain the complete volt-ampere characteristi c of a p–n junction diode. What is an ideal diode?
25. Mention the specification parameters of a diode.
26. What is a zener diode? How is it different from a p–n junction diode?
27. Explain the construction and working principle of a bipolar junction transistor.
28. Explain the working principle of an n–p–n transistor.
29. Explain the working principle of a p–n–p transistor.
30. Show the three types of transistor configurations.
31. Draw and explain the common–emitter transistor characteristics.
32. Explain with the help of a circuit the working of a transistor as an amplifier.
33. Show how a transistor can be used as a static switch.
34. What is a field effect transistor? How is an FET different from a BJT?
35. Explain the construction and working of a junction field effect transistor.
36. Draw and explain the characteristic of a JFET.
37. What do you mean by transfer characteristic of a JFET?
38. Draw and explain a single voltage amplification circuit using a field effect transistor.
39. Explain the constructional details and the working principle of a MOSFET.
40. Draw and explain a MOSFET switching circuit.
41. Explain the working principle of a silicon control rectifier.
42. Draw the forward and reverse characteristics of an SCR.
43. Explain the operation of an SCR with gate control.
44. Show two transistor analogy of an SCR.
45. Draw and explain a single-phase half-wave-controlled rectifier.
46. Draw and explain a single-phase full-wave-controlled bridge rectifier.
47. Draw and explain a static switching circuit using two diodes and two SCRs.
48. What is a Diac? Draw its V–I characteristics.
49. Explain how a TRIAC is different than a DIAC.
50. Explain the construction and V–I characteristic of a Triac.
51. Draw and explain a phase control circuit using a Triac.
52. What are optoelectronic devices. Name two of them.
53. What is an LDR? Show one LDR-operated circuit.
54. Explain the construction and principle of working of a light-emitting diode. Give one practical LED circuit.
55. Write a brief explanation of the following devices: (i) seven segment displays; (ii) light emitting

diodes; (iii) light-dependent resistor, and (iv) liquid crystal displays.

56. What is a photovoltaic cell. How is a solar battery constructed?

57. Explain the difference between a phototransistor and a BJT. Explain its working.

58. Give one practical application of a phototransistor.

59. Write short explanations of the following:

(i) phototransistor; (ii) photo-darlington; (iii) optocoupler, and (iv) solar cell.

60. Explain the formation of the depletion region in an unbiased p–n junction.

61. What are the barrier potentials of silicon diodes and germanium diodes?

62. Explain how a depletion-type MOSFET can work either in an enhancement mode or in a depletion mode.

63. Name two major differences between a JFET and a MOSFET.

Multiple Choice Questions

1. Which of the following is a semiconductor material?
   (a) Ceramic (b) Silicon
   (c) Cadmium (d) Rhodium.

2. Which of the following statements is correct?
   (a) Germanium valence electrons will need a smaller amount of additional energy to escape from the atom than silicon valence electrons
   (b) Valence electrons of silicon are in the fourth shell while those of germanium are in the third shell
   (c) Silicon is more widely used semiconductor material than germanium
   (d) Germanium produces less number of electron hole pairs than silicon.

3. Which of the following is not shown in the energy band diagram?
   (a) Permeable energy band (b) Conduction band
   (c) Valence band (d) Forbidden gap.

4. Which of the following statements is not true?
   (a) Semiconductors have negative temperature coefficient of resistance
   (b) Insulators have positive temperature coefficient of resistance
   (c) Insulators have negative temperature coefficient of resistance
   (d) Conductors have positive temperature coefficient of resistance.

5. Which of the following statements is not true?
   (a) In both silicon and germanium atoms there are four valence electrons
   (b) Intrinsic semiconductors behave like conductors at low temperatures
   (c) Intrinsic semiconductors behave like insulators at low temperatures
   (d) When an electron breaks a covalent bond and becomes free, a vacancy called hole is created.

6. Which of the following statements is not true?
   (a) An *n*-type material is formed by adding a small amount of pentavalent impurity to the pure silicon
   (b) Examples of pentavalent materials are antimony, arsenic, and phosphorous
   (c) An *n*-type material, inspite of the presence of a large number of free electrons is still electrically neutral
   (d) In an *n*-type material holes are the majority carriers and electrons are the minority carriers.

7. The width of the depletion layer of a *p–n* junction will increase when
   (a) the *p–n* junction is forward biased
   (b) the *p–n* junction is reverse biased
   (c) the *p–n* junction is forward biased with a voltage which is higher than the normal rating
   (d) the *p–n* junction is kept unbiased for a long time.

8. In the reverse-biased state of a diode current flowing, which is called reverse saturation current is due to
   (a) majority carriers
   (b) minority carriers
   (c) increase in reverse voltage applied externally
   (d) both majority and minority carriers.

9. Silicon diode is more popular than germanium diode because
   (a) reverse saturation current of silicon is higher than that of germanium
   (b) reverse saturation current of silicon is much lower than that of germanium

(c) temperature effect on reverse saturation current of silicon diode is very high
(d) break down voltage of germanium is higher than that of silicon.

10. Which of the following statements is not true for a zener diode?
(a) A forward-biased zener diode behaves identical to a forward-biased diode
(b) Zener diode under the reverse-biased condition is used as a voltage regulator
(c) Zener diode under the forward-biased condition is used as a voltage regulator
(d) The operation of a reverse-biased zener diode is different than that of a reverse-biased diode.

11. Which of the following statements is not true about the rectified sinusoidal wave?
(a) The ripple factor of a full-wave rectifier is 0.48
(b) The ripple factor of a half-wave rectifier is 1.21
(c) The ripple factor of an ideal rectified wave should be zero
(d) The ripple factor of an ideal rectified wave should be unity.

12. Which of the following is not true for a bridge rectifier?
(a) A centre-tapped input side transformer is required
(b) Bridge rectifiers offer full-wave rectification
(c) At a given time only one pair of diodes are conducting
(d) Ripple factor of a bridge rectifier is 0.48.

13. Conduction in a bipolar transistor takes place due to
(a) electrons only
(b) holes only
(c) both electrons and holes
(d) majority carriers only.

14. Which one of the following is not a region of operation of a transistor?
(a) Cut-off region (b) Active region
(c) Saturation region (d) Cut-in region.

15. Which one of the following is not the configuration of a transistor?
(a) Common–collector configuration
(b) Common–emitter configuration
(c) Common–base configuration
(d) Maximum current gain configuration.

16. Transistor can be used as an amplifier when it is operated
(a) in the saturation region
(b) in the cut-off region
(c) in the active region
(d) in both saturation and cut-off regions.

17. Which of the following is not true for a properly biased transistor?
(a) Can work as a current amplifier
(b) Can work as a voltage amplifier
(c) Can work as a switch
(d) Can work as a rectifier.

18. MOSFET stands for
(a) metal oxide silicon field effect transistor
(b) metal oxide semiconductor field excited transistor
(c) metal oxide semiconductor field effect transistor
(d) metal oxide silicon field excited transistor.

19. Which of the following does not belong to the thyristor family?
(a) DIAC (b) TRIAC
(c) SCR (d) SCS.

20. We turn off a conducting SCR by
(a) reducing the anode current below the holding current
(b) decreasing the voltage applied across the SCR
(c) removing the gate terminal connection
(d) applying a negative gate voltage.

21. Which of the following is not an opto-electronic device?
(a) Light-emitting diode
(b) Liquid crystal display
(c) Photovoltaic cell
(d) Silicon-controlled rectifier.

Answers to Multiple Choice Questions

1. (b)	2. (a)	3. (a)	4. (b)	5. (b)	6. (d)
7. (b)	8. (b)	9. (b)	10. (c)	11. (d)	12. (a)
13. (c)	14. (d)	15. (d)	16. (c)	17. (d)	18. (c)
19. (a)	20. (a)	21. (d)			

CHAPTER 15

Rectifiers and Other Diode Circuits

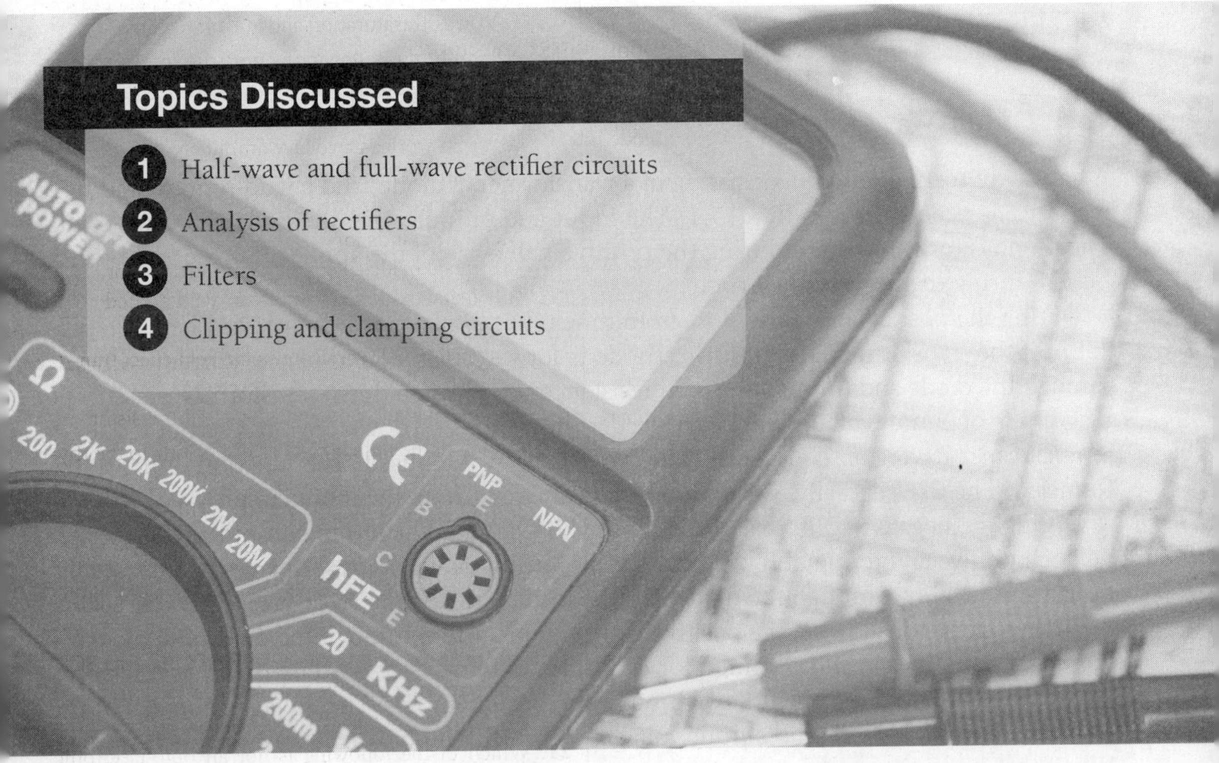

Topics Discussed

1. Half-wave and full-wave rectifier circuits
2. Analysis of rectifiers
3. Filters
4. Clipping and clamping circuits

15.1 ► RECTIFIERS

15.1.1 Introduction

A rectifier is a device that converts ac supply into dc using diodes. Rectification can be done by half-wave or full-wave rectification circuits. All electronic circuits need a low dc voltage for their operation. The supply voltage available is from the ac mains which is 230 V, 50 Hz supply. A rectifier will first step down the ac supply voltage to the required level by using a step-down transformer. A single diode can be used to rectify the ac voltage into half-wave rectified dc voltage. Since the rectified voltage is a unidirectionally changing dc, filter circuits are used to get steady dc output.

In this section we will discuss half-wave and full-wave rectifier circuits.

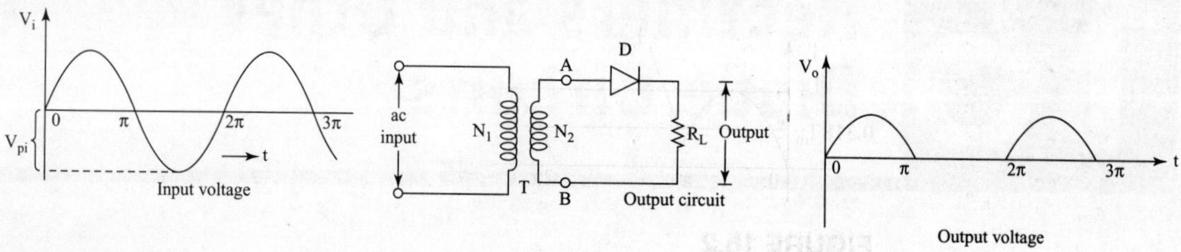

FIGURE 15.1

Half-wave rectifier circuit with input and output voltage waveforms

15.1.2 Half-Wave Rectifier

A half-wave rectifier circuit consisting of a transformer, a diode, and a load represented by a resistor has been shown in Fig. 15.1. The input and output wave shapes have been shown. The diode is forward biased during the positive half cycle of the applied voltage and reverse biased during the negative half cycle.

During the positive half cycle, i.e., from time 0 to π, voltage is positive, and hence, terminal A of the transformer, T is positive. Diode is forward biased, and hence, current will flow through the diode and through the load resistance, R_L. If a step down transformer is used, the magnitude of output voltage will be reduced. If however, step down of voltage is not required, a transformer having an equal turn ratio, i.e., N_2/N_1 will be used. The function of the transformer will be to electrically isolate the dc output circuit from the input circuit.

During the negative half cycle of the input voltage wave, the terminal A of the transformer will be negative and terminal B will be positive. The diode will be reverse biased and virtually no current will flow through the diode and the load.

This sequence of allowing the positive half cycle of current through the diode and blocking the negative half cycle will continue for each and every cycle of power supply. As a result, a half-wave rectified current will flow through the output circuit. The voltage drop across the load resistance, R_L will be the output voltage, V_O which will be a half-wave rectified voltage. When the diode is reverse biased, the maximum or peak voltage of the negative half cycle of the input will be appearing across the diode terminals. This is the peak of the reverse voltage V_{pi}(see Fig. 15.1) or peak-inverse voltage (PIV) which is getting applied to the diode in every alternate half cycles.

It is seen that a diode works as a closed switch during the positive half cycle of the input voltage and works as an open switch during the negative half cycle. The output voltage appears across the load during the positive half cycle only. The load voltage and load current although positive all the time (i.e., unidirectional) are fluctuating dc as its magnitudes changing. Our aim will be to obtain steady dc at the output.

The average value of this fluctuating dc as also its RMS values can be calculated as I_m/π and $I_m/2$, respectively.

The performance parameters of a half-wave rectifier output are calculated in terms of the output dc current (i.e., the average value of the rectified wave), the RMS value of the output current, the output voltage, ripple factor, peak inverse voltage, etc.

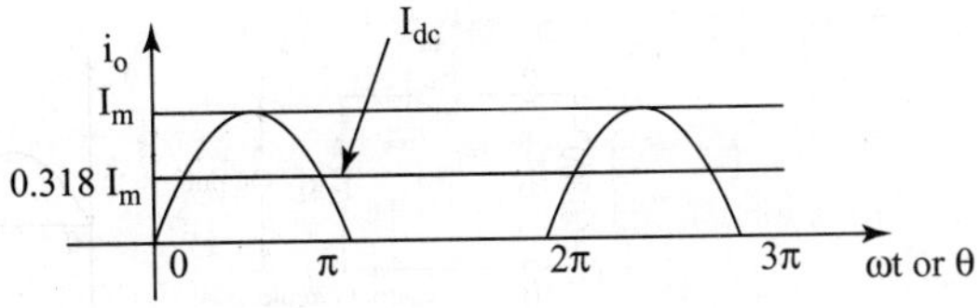

FIGURE 15.2

Half-wave rectified current

15.1.3 Analysis of Half-Wave Rectifier

The equation for a sinusoidal voltage is written as $v = V_m \sin\theta$ and for a sinusoidal current,

$$i = I_m \sin\theta,$$

where i is the instantaneous value, I_m is the maximum value, and $\theta = \omega t$.

The average value is calculated by integrating the current for a period $\theta = 0$ to $\theta = \pi$ and averaging it for the entire cycle, i.e., $\theta = 0$ to $q = 2\pi$.

I_{dc} is the average value of the rectified current.

$$I_{dc} = I_{av} = \frac{1}{2\pi}\int_0^{\pi} i\,d\theta = \frac{1}{2\pi}\int_0^{\pi} I_m \sin\theta\,d\theta$$

$$= \frac{I_m}{2\pi}[-\cos\theta]_0^{\pi}$$

$$= \frac{I_m}{2\pi}[-(-1)-(-1)]$$

$$= \frac{I_m}{\pi} = 0.318\ I_m$$

The RMS value, I is calculated by first squaring the current, then taking its mean, and then taking its root as

$$I = \sqrt{\frac{1}{2\pi}\int_0^{\pi} i^2 d\theta} \quad \text{or,} \quad I^2 = \frac{1}{2\pi}\int_0^{\pi} i^2 d\theta$$

$$I^2 = \frac{1}{2\pi}\int_0^{\pi} I_m^2 \sin^2\theta\,d\theta$$

or,

$$I^2 = \frac{I_m^2}{2\pi}\int_0^{\pi}\left[\frac{1-\cos 2\theta}{2}\right]d\theta = \frac{I_m^2}{4\pi}\left[\theta - \frac{\sin 2\theta}{2}\right]_0^{\pi}$$

or,

$$= \frac{I_m^2}{4\pi}\left[\pi - \frac{\sin 2\pi}{2} - \left(0 - \frac{\sin 2\times 0}{2}\right)\right]$$

$$= \frac{I_m^2}{4\pi}\times\pi$$

$$= \frac{I_m^2}{4}$$

or,

$$I_{rms} = \frac{I_m}{2}$$

Ripple Factor: The output of a half-wave rectifier is a pulsating dc. If we analyse, we will see that it has a steady dc component and an ac component. The ac component is called the ripple. Ripple factor is defined as the ratio of the RMS value of ac component to the value of dc component. The ripple factor indicates the level of fluctuation of the output voltage from its steady value. Ripple is an undesired effect and should be minimized. The ripple factor, r for a half-wave rectifier is calculated as

$$\text{Ripple Factor, } r = \frac{\text{RMS value of ac component}}{\text{value of dc component}}$$

$$= \frac{I_{ac}}{I_{dc}}$$

Again, $I_{rms}^2 = I_{dc}^2 + I_1^2 + I_2^2 + I_{44}^{22} + \cdots = I_{dc}^2 + I_{ac}^2$

where I_1, I_2, I_4, etc., are the fundamental and harmonics of the ac component

or,
$$I_{ems} = \sqrt{I_{dc}^2 + I_{ac}^2}$$

or,
$$I_{ac} = \sqrt{I_{rms}^2 - I_{dc}^2}$$

where I_{ac} is the RMS value of the ac component of the output current.

I_{dc} is the dc component of the output current and I_{rms} is the RMS value of the output current.

Substituting

$$r = \frac{\sqrt{I_{rms}^2 - I_{dc}^2}}{I_{dc}} = \sqrt{\left(\frac{I_{rms}}{I_{dc}}\right)^2 - 1}$$

Considering $I_{rms} = \frac{I_m}{2}$ and $I_{dc} = \frac{I_m}{\pi}$ a half-wave rectifier,

$$r = \sqrt{\left[\frac{I_m/2}{I_m/\pi}\right]^2 - 1}$$

$$= \sqrt{\frac{\pi^2}{4} - 1}$$

ripple factor, $= 1.21$

Output voltage, V_{dc} across the load is

$$V_{dc} = I_{dc} R_L = \frac{I_m}{\pi} R_L = \frac{V_m}{\pi} = 0.318\ V_m$$

Rectifier efficiency: It is calculated as the ratio of output power to input power.

DC output power, $P_{dc} = I_{dc}^2 R_L = \left(\frac{I_m}{\pi}\right)^2 R_L$

AC input power, P_{ac} = Power dissipated in diode junction + Power dissipated in the load $= I_{rms}^2 R_f + I_{rms}^2 R_L$

The forward resistance, R_f of the diode is very small, and hence, $I_{rms}^2 R_f$ can be neglected in comparison with $I_{rms}^2 R_L$.

$$\text{Therefore, rectifier efficiency} = \frac{P_{dc}}{P_{ac}} = \frac{I_m^2}{\pi^2} RL \div I_{rms}^2 R_L$$

$$= \frac{I_m^2 \, RL}{\pi^2 I_{rms}^2 RL} = \frac{I_m^2}{\pi^2 \cdot \left(\frac{I_m}{2}\right)^2}$$

$$= \frac{4}{\pi^2} = 0.406 = 40.6 \text{ per cent}$$

This value of efficiency is considered low and ripple factor is considered very high. A filter circuit has to be used to minimize the ripples.

Peak inverse voltage (PIV): As mentioned earlier, PIV is the maximum value of reverse voltage that appears across the diode when it gets reverse biased. Here

$$PIV = V_m$$

Rectifier diodes are specified for their average forward current-carrying capacity and their reverse voltage capacity, i.e., their PIV capacity. For example, low-power rectifier diode series, IN 4000 to IN 4007 are rated for forward current of 1000 mA and maximum reverse voltage of value varying from 50 V to 1000 V.

Voltage Regulation of the rectifier is calculated using the relation,

$$\text{Voltage regulation} = \frac{V_{dc} \text{ at no load} - V_{dc} \text{ on full load}}{V_{dc} \text{ on full load}}$$

The difference between no-load voltage and full-load voltage is the voltage drop in the transformer winding and across the diode. The value of voltage regulation, which is generally expressed in percentage should be low.

EXAMPLE 15.1

A half-wave diode rectifier has a forward voltage drop, i.e., voltage drop across the diode when conducting is 0.7 V. The load resistance is 600 Ω. The RMS value of the ac input is 28.87 V. Calculate I_{dc}, I_{rms}, PIV, and form factor.

SOLUTION

The half-wave rectifier circuit with the given data is shown in Fig. 15.3.

$V_i(\text{RMS}) = 28.27$ V

$V_i(\max) = \sqrt{2}\, V_i(\text{RMS})$

$= 1.414 \times 28.27$

i.e., $V_m = 40$ V

$PIV = V_m = 40$ V

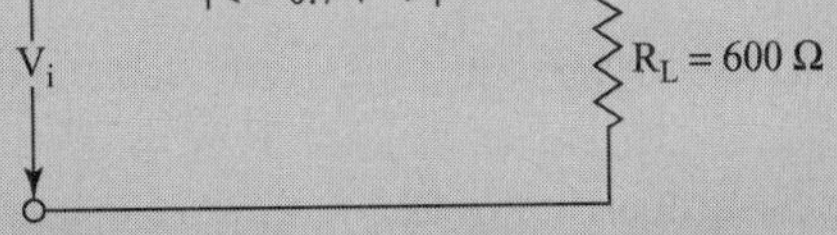

FIGURE 15.3

Half-wave rectifier circuit

$$I_{dc} = \frac{I_m}{\pi}; \qquad I_m = \frac{V_m - 0.7}{R_L} = \frac{40 - 0.7}{600} = \frac{39.3}{600} A$$

$$I_{dc} = \frac{39.3}{600 \times \pi} = 0.0208\ A = 20.8\ mA$$

$$I_{rms} = \frac{I_m}{2} \frac{39.3}{600 \times 2} = 0.0327\ A = 32.7\ mA$$

$$\text{Form factor} = \frac{\text{RMS value}}{\text{Average value}} = \frac{I_{rms}}{I_{dc}} = \frac{32.7}{20.8}$$

$$= 1.57$$

EXAMPLE 15.2

A half-wave rectifier produces a maximum load current (peak value) of 50 mA through a 1200 Ω resistor. Calculate the PIV of the diode. The diode is of silicon material.

SOLUTION

Assuming a voltage drop of 0.7 V across the silicon diode, the peak value of current, I_m is

$$I_m = \frac{V_m - 0.7}{R_L} = \frac{V_m - 0.7}{1200}$$

I_m is given as 40 mA.
Therefore,

$$40 \times 10^{-3} = \frac{V_m - 0.7}{1200}$$

or,

$$V_m - 0.7 = 1200 \times 40 \times 10^{-3} = 48\ V$$

or,

$$V_m = 48 + 0.7 = 48.7\ V$$

$PIV = V_m = 48.7\ V$

EXAMPLE 15.3

A half-wave rectifier circuit has been made using a step-down transformer of turn ratio 10:1. The input voltage is $v = 325 \sin \omega t$ the diode forward resistance is 25 Ω. A load resistance of 1.2 kΩ has been connected in the circuit. Assuming a secondary winding resistance of the transformer as 1Ω, calculate the following: (a) RMS value of load current (b) rectification efficiency, and (c) ripple factor.

SOLUTION

Input voltage, $v = 325 \sin \omega t$

Input, $V_m = 325$

Transformer shown in Fig. 15.4, has a turn ratio of 10:1

The output $V_m = \frac{325}{10} = 32.5$ V

$$I_m = \frac{V_m}{R_2 + R_F + R_L}$$

where R_2 is the secondary winding resistance, R_F is the forward resistance of the diode and R_L is the load resistance.

$$I_m = \frac{32.5}{1+25+1200}$$

$$= \frac{32.5}{1226} \text{ A}$$

$$= 26.5 \text{ mA}$$

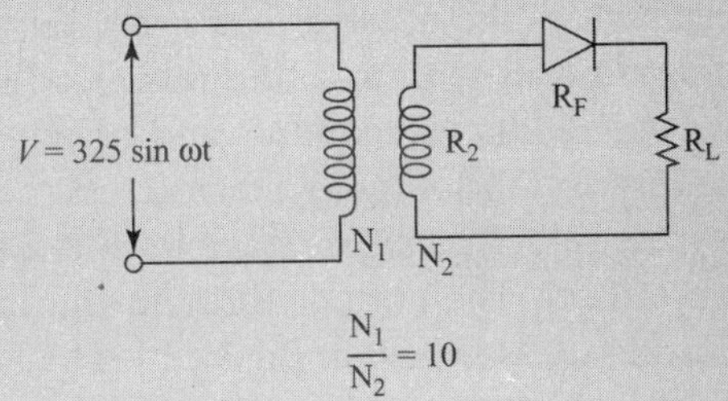

FIGURE 15.4

Half-wave rectifier

Since it is a half-wave rectifier circuit,

$$I_{rms} = \frac{I_m}{2} = \frac{26.5}{2} = 13.25 \text{ mA}$$

$$I_{dc} = \frac{I_m}{\pi} = \frac{26.5}{3.14} = 8.44 \text{ mA}$$

$$\text{output dc power} = I_{dc}^2 R_L = (8.44 \times 10^{-3})^2 \times 1200$$

$$= 85.48 \text{ mW}$$

$$\text{AC input power} = (I_{rms})^2 [R_2 + R_F + R_L]$$

$$= (13.25 \times 10^{-3})^2 \times 1226$$

$$= 0.215 \text{ W} = 215 \text{ mW}$$

$$\text{Rectifier efficiency} = \frac{\text{Output dc power}}{\text{Input ac power}} \times 100$$

$$= \frac{85.48 \times 100}{215} = 39.75 \text{ per cent}$$

$$\text{Ripple factor, } r = \sqrt{\left(\frac{I_{rms}}{I_{dc}}\right)^2 - 1} = \sqrt{\left(\frac{13.25}{8.44}\right)^2 - 1} = 1.21$$

15.1.4 Full-Wave Rectifier

Full-wave rectifiers can be made using two diodes and a centre-tapped transformer. Full-wave rectifiers are also made using a two-winding transformer and four diodes. Such rectifiers are called bridge rectifiers. These are discussed as follows.

Two-diode full-wave rectifier: here ac input voltage is supplied from the secondary of a centre-tapped transformer. The circuit consists of the transformer, two diodes and the load resistance. The circuit is essentially the summation of two half-wave rectifiers as shown in Fig. 15.5.

For the positive half cycle of the input voltage, diode D_1 will conduct. This is because terminal A is positive and the diode D_1 is forward biased. As terminal B is negative, diode D_2 will not conduct. In the negative half cycle, terminal B is positive and terminal A is negative. Hence, diode D_2 will conduct and diode D_1 will be reverse biased. This way in each half cycle one of the two diodes will conduct and current will flow through the load resistance, R_L. The output current and the output voltage across the load will be a full-wave rectified current and voltage, respectively. The output wave form is a series of consecutive positive half cycles of sinusoidal wave form. The current through the load resistance is unidirectional but its magnitude is fluctuating as shown in Fig. 15.5. The PIV is the maximum voltage that would appear across a diode when it is reverse biased. Here, when D_1 is conducting, D_2 is reverse biased and vice versa. When D_1 is conducting, the voltage that would appear across diode D_2 is the sum of the voltage across the lower half of the transformer secondary winding and the voltage appearing across the load. PIV of the diode is equal to 2 V_m.

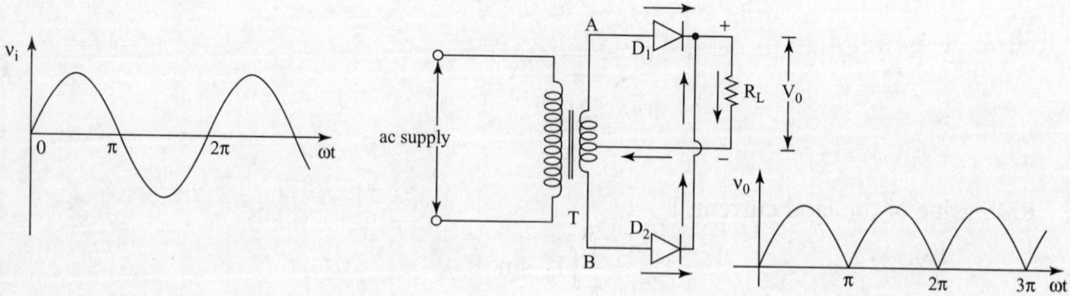

FIGURE 15.5

Full-wave rectifier using two diodes and a centre-tapped transformer

15.1.5 Full-Wave Bridge Rectifier

A bridge rectifier circuit uses four diodes connected in the form of a bridge. The various ways the four diodes in the bridge circuit are drawn have been shown in Fig. 15.6 (a) and (b). As shown in Fig. 15.6 (a), the arrow head symbols of all the diodes are pointing towards the positive terminal of the output, i.e., the load.

During the positive half cycle of the input voltage, terminal A of the transformer is positive. Current will flow from the positive terminal through diode D_1, load R_L and diode D_4, and back to the negative terminal, B of the transformer. The direction of current through the load has been shown to be from C to D, i.e., from top to bottom. The polarities of the load terminals have been shown. During this period diodes D_2 and D_3 are reverse biased.

During the negative half cycle of the input voltage diodes D_3 and D_2 are forward biased while diodes D_1 and D_4 are reverse biased. Current through the load will flow in the same direction, i.e., from terminal C to D. During both positive and negative half cycles of the input voltage, current will pass through the load in the same direction. The output voltage wave shape is a series of positive half cycles of the sinusoidal voltage. This is dc output but having a varying magnitude. The transformer provides isolation of the dc output from the supply ac input.

The important parameters of a full-wave bridge rectifier are determined as follows.

15.1.6 Analysis of Full-Wave Rectifiers

Average value or dc value of load current, I_{dc}: The average value or dc value will be the same if calculated for a period 0 to π or 0 to 2 π

$$I_{dc} = \frac{1}{\pi}\int_0^{\pi} I_m \sin\theta \, d\theta$$

$$= \frac{I_m}{\pi}\int_0^{\pi} \sin\theta \, d\theta$$

$$= \frac{I_m}{\pi}[-\cos\theta]_0^{\pi}$$

$$= \frac{I_m}{\pi}[-(-1)-(-1)]$$

$$= \frac{2I_m}{\pi}$$

RMS value of the laod current, I_{rms}

$$I_{rms}^2 = \frac{1}{\pi}\int_0^{\pi} I_m^2 \sin^2\theta \, d\theta$$

$$= \frac{I_m^2}{2\pi}\int_0^{\pi}(1-\cos 2\theta)\, d\theta$$

$$= \frac{I_m^2}{2\pi}\left[\theta - \frac{\sin 2\theta}{2}\right]_0^{\pi}$$

$$= \frac{I_m^2}{2\pi} \times \pi = \frac{I_m^2}{2}$$

$$I_{rms} = \frac{I_m}{\sqrt{2}}$$

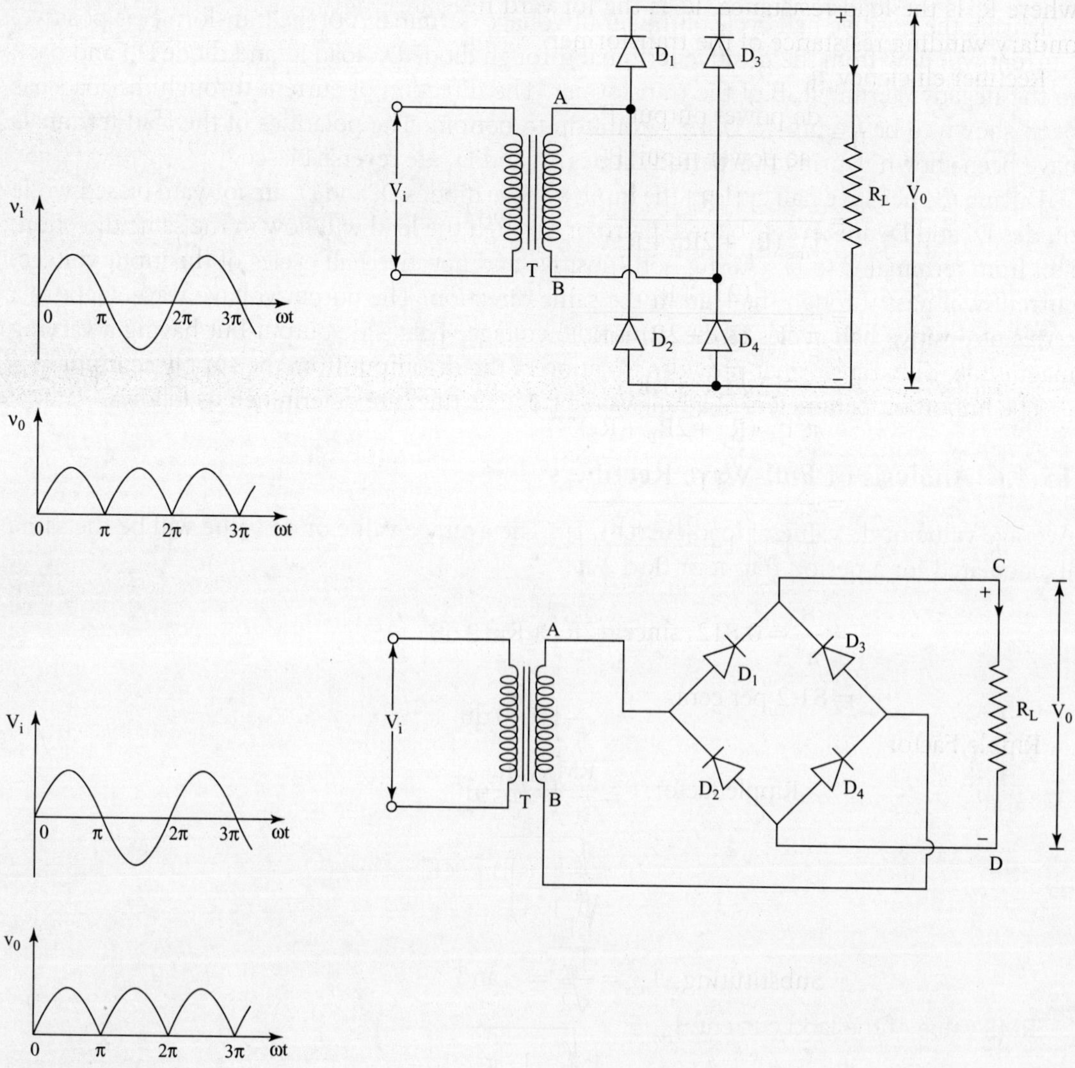

FIGURE 15.6

A bridge rectifier circuit for full-wave rectification using four diodes and a transformer

Output voltage, V_{dc}

$$\begin{aligned} V_{dc} &= I_{dc} R_L \\ &= \frac{2\ I_m}{\pi} R_L \\ &= \frac{2\ R_L}{\pi} \frac{V_m}{(R_L + 2R_F + R_2)} \end{aligned}$$

where R_L is the load resistance, R_F is the forward resistance of the diode, and R_2 is the secondary winding resistance of the transformer.

Rectifier efficiency, η

$$\eta = \frac{\text{dc power output, } P_{dc}}{\text{ac power input, } P_{ac}}$$

$$= \frac{I_{dc}^2 R_L}{I_{rms}^2 (R_L + 2R_F + R_2)} \qquad \text{(Two diodes being in series)}$$

$$= \frac{(2I_m)^2 R_L}{\pi^{22} I_{rms}^2 (R_L + 2R_F + R_2)}$$

$$= \frac{(2\sqrt{2}\, I_{rms})^2 R_L}{\pi^2 I_{rms}^2 (R_L + 2R_F + R_2)}$$

$$= \frac{8\, I_{rms}^2}{\pi^2 I_{rms}^2 \left(1 + \dfrac{2R_F + R_2}{R_L}\right)}$$

$$= \frac{8}{\pi^2} = 0.812 \ \text{ since } \ (2R_F + R_2) << R_L$$

$$= 81.2 \text{ per cent}$$

Ripple Factor

$$\text{Ripple factor, } r = \frac{\text{RMS value of ac component}}{\text{dc component}}$$

$$= \sqrt{\left(\frac{I_{rms}}{I_{dc}}\right)^2 - 1}$$

$$\text{Substituting, } I_{rms} = \frac{I_m}{\sqrt{2}} = \ \text{ and } \ I_{dc} = \frac{2I_m}{\pi}$$

$$r = \sqrt{\left(\frac{I_m \, \pi}{\sqrt{2}\, 2\, I_m}\right)^2 - 1}$$

$$= \sqrt{\frac{\pi^2}{8} - 1}$$

$$= 0.48$$

We had earlier calculated the ripple factor for a half-wave rectifier as 1.21. For a full-wave rectifier, the ripple factor is reduced to 0.48. This indicates that the fluctuation of dc output is reduced. PIV for a bridge rectifier = V_m.

It is now possible to compare the performance parameters of a half-wave rectifier with a bridge rectifier. This has been shown in Table 15.1. Although the performance parameters of a bridge rectifier are superior than a half-wave rectifie, the quality of output voltage is still not acceptable and has ripple content which must be further minimized. This is achieved by using filters. Filters are discussed in the section that follows.

15.1.7 Comparison of Half-Wave and Full-Wave Rectifiers

The performance of half-wave and full-wave rectifiers with respect to certain parameters has been compared as in Table 15.1.

TABLE 15–1 Comparison of Half-wave and Full-wave Rectifiers Against Their Salient Parameters

Parameters	Half-wave rectifier	Full-wave rectifier with centre-tapped transformer	Bridge rectifier
Number of diodes reqd.	1	2	4
DC load current, I_{dc}	$\frac{I_m}{\pi}$	$\frac{2I_m}{\pi}$	$\frac{2I_m}{\pi}$
RMS value or load current I_{rms}	$\frac{I_m}{2}$	$\frac{I_m}{\sqrt{2}}$	$\frac{I_m}{\sqrt{2}}$
DC output, P_{dc}	$I_{dc}^2 R_L = \frac{I_m^2}{\pi^2} R_L$	$\frac{4I_m^2}{\pi^2} R_L$	$\frac{4I_m^2}{\pi^2} R_L$
AC input, P_{ac}	$I_{rms}^2 (R_L + R_F + R_2) = \frac{I_m^2}{4}(R_L + R_F + R_2)$	$I_{rms}^2 (R_L + 2R_F + R_2) = \frac{I_m^2}{2}(R_L + 2R_F + R_2)$	$I_{rms}^2 (R_L + 2R_F + R_2) = \frac{I_m^2}{2}(R_L + 2R_F + R_2)$
Maximum rectification efficiency	40%	81.2%	81.2%
Ripple factor	1.21	0.48	0.48
PIV	V_m	$2V_m$	V_m

(continued)

TABLE 15–1 Comparison of Half-wave and Full-wave Rectifiers Against Their Salient Parameters *(continued)*

Parameters	Half-wave rectifier	Full-wave rectifier with centre-tapped transformer	Bridge rectifier
Ripple frequency	Fr = f	Fr = 2f	Fr = 2f
Centre tap transformer	Not required	Required	Not required
Transformer utilization factor	28.7%	69.2%	81.2%

EXAMPLE 15.4

The input to a bridge rectifier is through a step-down transformer of turn ratio 10:1. The supply voltage is 230 V at 50 Hz. The load resistance is 1.2 kΩ secondary winding resistance of the transformer is 4 Ω diode forward resistance is 2 Ω. Calculate the efficiency of the bridge rectifier.

SOLUTION

Given V_i (RMS) = 230 V

$R_F = 2\ \Omega, R_2 = 4\ \Omega, R_L = 1200\ \Omega$

The RMS value of the emf in transformer secondary

$$V_s\,(\text{RMS}) = 230\left(\frac{N_2}{N_1}\right)$$

$$= 230\left(\frac{1}{10}\right)$$

$$= 23\ \text{V}$$

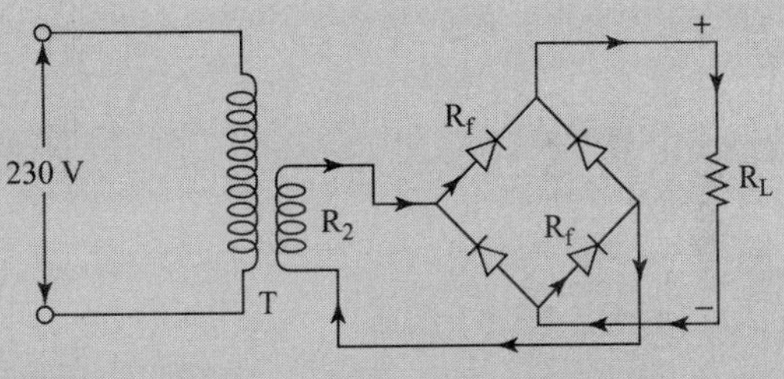

FIGURE 15.7

Bridge-rectifier circuit

Peak secondary voltage, V_m is

$$V_m = \sqrt{2}\,V_s(\text{RMS})$$

$$= \sqrt{2} \times 23$$

$$= 32.5\ \text{V}$$

Current through the load will flow from the transformer secondary via two diodes. Therefore, following the current path during the positive half cycle

$$I_m = \frac{V_m}{R_L + 2\,R_F + R_2} = \frac{32.5}{1200+4+4} = 26.8 \text{ mA}$$

For a bridge rectifier,

$$I_{dc} = \frac{2I_m}{\pi} = \frac{2\times 26.8}{3.14} = 17 \text{ mA}$$

DC power output, $P_{dc} = I_{dc}^2 R_L = (17\times 10^{-3})^2 \times 1200$

$$= 346.8 \text{ mW}$$

AC power input, $P_{ac} = (I_{rms})^2 (R_L + 2R_F + R_S)$

$$= \left(\frac{I_m}{\sqrt{2}}\right)^2 (R_L + 2R_F + R_S)$$

$$= \left(\frac{26.8\times 10^{-3}}{2}\right)^2 \times (1200 + 2\times 2 + 4)$$

$$= 432 \text{ mW}$$

Rectifier efficiency, $\eta = \frac{P_{dc}}{P_{ac}} \times 100 = \frac{346.8\times 100}{432} = 80$ per cent.

EXAMPLE 15.5

Determine for the bridge circuit shown in Fig. 15.8, the peak value of load current when $V_i = 15$ V, $R_l = 600\ \Omega$ and the forward voltage drop of the diode is 0.7 V. Also calculate the average value of the output current

SOLUTION

$$V_i = 15 \text{ V}$$

$$V_i(\text{max}) = \sqrt{2}\times 15 = 21.21 \text{ V}$$

Maximum value of voltage appearing across the load will be $V_i\,(\text{max}) - 2\,V_F$.
This is because two diodes are involved in the current flow through the load at any point of time

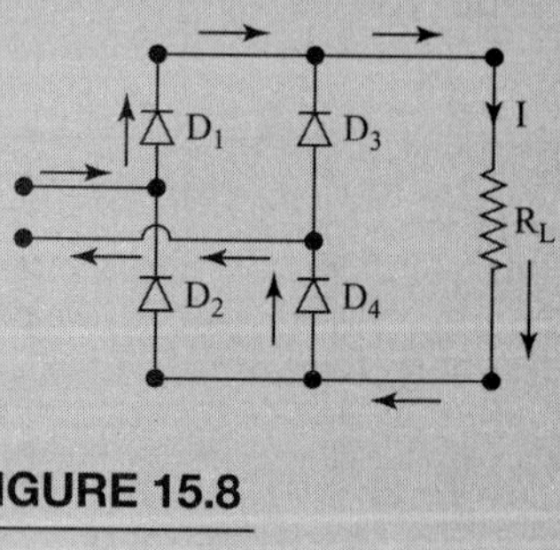

FIGURE 15.8

Bridge rectifier circuit

$$V_0(\max) = V_i(\max) - 2\,V_F$$
$$= 21.21 - 2 \times 0.7$$
$$= 19.81\text{ V}$$
$$I_0(\max) = \frac{V_0(\max)}{R_L} = \frac{19.81}{600} = 36\text{ mA}$$
$$I_0(\text{average}) = I_{dc} = \frac{2\,I_m}{\pi} = \frac{2 \times 36}{3.14} = 22.93\text{ mA}$$

15.2 ► FILTERS

We have seen that the wave form of the rectified voltage is a series of positive half cycles of the input voltage wave form either of equal or of reduced magnitude. For a half-wave rectifier we get a series of positive half cycles with one missing in between. Our objective is to get a steady-value dc output. To convert the fluctuating output voltage into a steady dc, smoothing circuits called filters must be used. The simplest filter is a capacitor which is connected across the load. Fig. 15.9 shows a capacitor C connected across the load resistance R_L in a half-wave rectifier. The effect of the use of a capacitor on the output voltage wave has been shown.

During the positive half cycle of the input voltage the diode D_1 is forward biased. Current flows through the diode and the load resistor, R_L. At the same time the capacitor, C gets charged upto the peak value, V_m of the input voltage.

After attaining the pick value, the input voltage starts reducing, its value becoming less and less than V_m. But the capacitor has been charged to a voltage V_m. Thus, the potential of terminal B becomes higher than the potential of terminal A. As a result, diode D_1 gets reverse biased but the capacitor voltage remains close to V_m. With the diode D_1 reverse biased, the changing of the capacitor stops. The capacitor now starts getting discharged through the load resistor R_L. The voltage across the capacitor, V_C starts falling, as has been shown in Fig. 15.9, through a thick horizontally inclined line. The diode, D_1 remains reverse biased throughout the rest of the positive half cycle and also during the negative half cycle, and further to the next positive half cycle until at θ_2 when the input voltage starts becoming higher than the capacitor voltage, V_C once again. At this point the

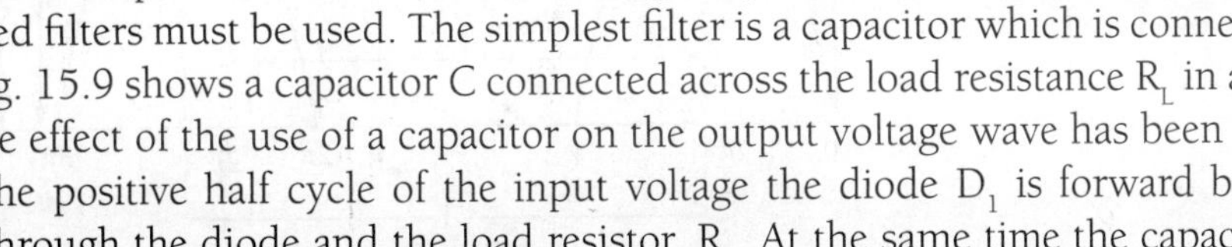

FIGURE 15.9

Half-wave rectifier circuit with a capacitor filter

diode becomes conducting supplying current to the load as also charging the capacitor once again. This process of charging and discharging of the capacitor continues in every cycle and an output voltage waveform, as shown by a thick line, is achieved. This wave shape of the output voltage is superior to the wave shape of the output voltage obtained when no capacitor was connected. The ripple of the output voltage is now reduced and a near-steady dc output voltage obtained.

Amplitude of ripple voltage and selection of capacitor

The half-wave rectified voltage with a capacitor filter has been shown again in Fig. 15.10. V_r represents the peak to peak ripple voltage. The time of discharge of the capacitor is represented by t_1 as shown in Fig. 15.10. The output voltage fluctuates between V_0 (min) to V_0 (max).

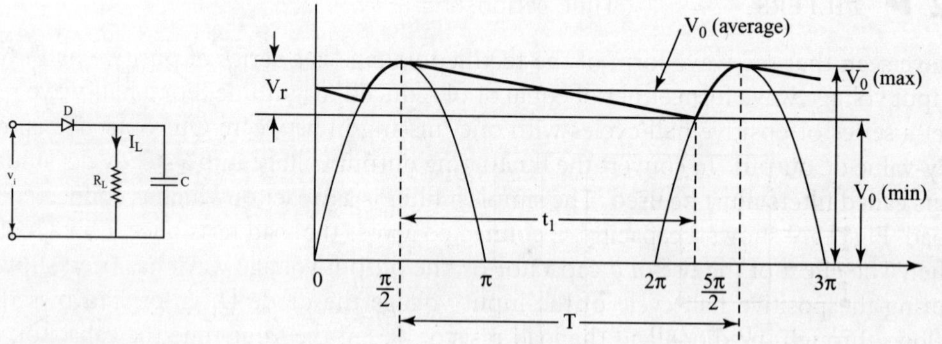

FIGURE 15.10

Half-wave rectified voltage with a capacitor filter

Peak to peak, $V_r = V_0\,(\text{max}) - V_0\,(\text{min})$.

Average value of output voltage, $V_0(\text{average}) = \dfrac{V_0(\text{max}) + V_0(\text{min})}{2}$

The capacitor C gets discharged during the time t_1 when the voltage across it drops by V_r causing a load current, I_L to flow through the resistance R_L. So we can write

$$Q = CV_r = I_L \times t$$

or,
$$C = \frac{I_L \times t}{V_r} \qquad \text{(i)}$$

Calculation of the value of the capacitor to be used depends on the allowable ripple voltage, the average output voltage, the load resistance, and the supply frequency. The approximate value is calculated using the procedure illustrated through an example. The standard manufacture's list is then consulted to select the next higher value of the capacitor available in the market.

EXAMPLE 15.6

A half-wave capacitor filter rectifier has maintained an average output voltage of 15 V with a peak to peak ripple of not more than 3 V. The load resistance is 100 Ω Calculate the value of the capacitor filter. The ac supply frequency is 50 Hz.

SOLUTION

$$\text{The load current, } I_L = \frac{V_0(\text{average})}{R_L}$$

$$= \frac{15}{100} A = 150 \text{ mA}$$

$$\text{Time period, } T = \frac{1}{f} = \frac{1}{50} \text{ seconds}$$

$$= \frac{1000}{50} \text{ ms} = 20 \text{ ms}$$

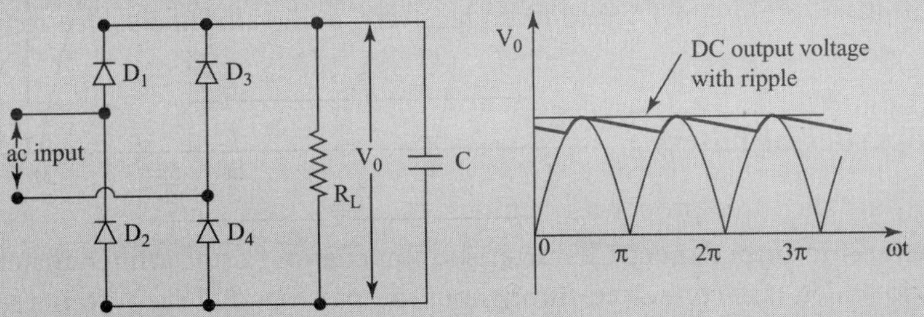

FIGURE 15.11

Bridge rectifier with a smoothing capacitor

The time of discharge of the capacitor t_1 can approximately by considered equal to T. Therefore,

$$t_1 = T = 20 \text{ ms}$$

using the relation,

$$CV_r = I_L t_1$$

$$C = \frac{I_L t_1}{V_r} = \frac{150 \times 10^{-3} \times 20 \times 10^{-3}}{3} = 1000 \ \mu F$$

Capacitors of 100 μF are available as can be checked from the manufacturer's list.

Similar to the capacitor filter used in half-wave rectifiers, capacitor filters are used in full-wave rectifier also as shown in Fig. 15.11. The circuit works exactly the same way as has been explained in the case of half-wave rectifier with a capacitor filter. From the dc output voltage wave shape it is observed that the ripple is minimized to a very small level.

The ripple voltage that appears across the capacitor can further be reduced by use of another resistor and a capacitor. The resistor is connected in series while the capacitors are in parallel making a π formation. Such filters are called R–C filters or simply π filters. Similarly, we can use L–C filters also to further smoothen the output voltage. An inductor in series with the load also works as a filter as the inductor allows dc current to flow and opposes ac current flow.

15.3. ▶ Introduction to Design of L and C Filters

15.3.1 Purpose of Filters

A filter is an electrical network that can transmit signals within a specified frequency range. This frequency range is called pass band and other frequency band where signals are suppressed is called attenuation band or stop band. The frequency that separates the pass band and stop band (attenuation band) is denoted by f_c, called cut-off frequency.

In ideal filters, the cut off frequency, f_c is sharp, i.e., the filter cuts off the signal totally at that frequency. Such a filter is called ideal filter. In a practical filter, the cut off is not so sharp. Filters are used in applications like in T.V. and radio receivers, audio amplifiers, electronic measuring instruments, etc.

15.3.2 Properties of Filters

The following are the three properties of filters.

1. **Characteristic impedance:** Characteristic impedance, Z_0 of a filter matches with the circuit to which it is connected throughout the pass band. This prevents reflection loss in the combination.

   Care must taken in selecting a filter so that it can be inserted into a given line or in between two pieces of equipment.
2. **Pass band characteristic:** A filter used should have minimum attenuation in its pass-band range and high attenuation in the stop-band range. The degree of attenuation is expressed by attenuation constant α, and its unit is Neper or Decibel.
3. **Cut-off frequency characteristic:** The filter should possess frequency distinguishing property in the pass band and stop band. It should be capable of identifying lower as well as higher cut off frequency for transmitting signals through it. The transition frequency region between the pass band and stop band must be negligible.

15.3.3 Types of Filters

Filters and their pass band and stop band are shown below.

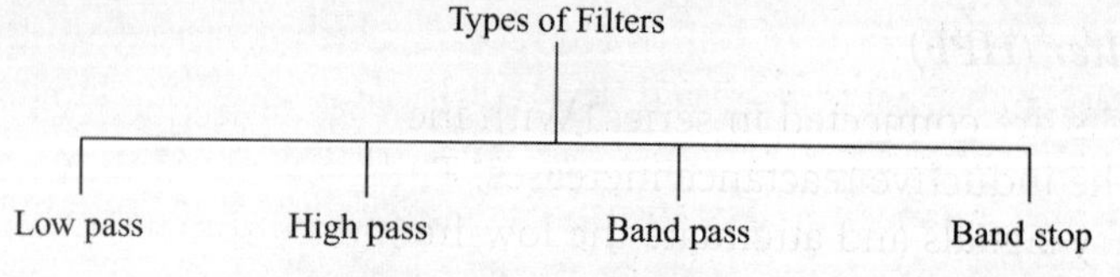

Types of Filters	Pass band	Attenuation or stop band
Low-pass filter	$0 \rightarrow f_c$	$f_c \rightarrow \infty$
High-pass filter	$f_c \rightarrow \infty$	$0 \rightarrow f_c$
Band-pass filter	$f_2 \rightarrow f_1$	$0 \rightarrow f_2, f_1 \rightarrow \infty$
Band-stop filter	$0 \rightarrow f_1, f_2 \rightarrow \infty$	$f_1 \rightarrow f_1$

Filters are also classified into *Passive filters* and *Active filters*.

Passive filters are made using passive circuit components like resistor, inductor and capacitor. Passive filters are generally used above 1 MHz frequency, have no power gain, and are relatively difficult to tune.

Active Components like transistors, op-amps are used to make active filters. Active filters are useful below 1 MHZ frequency, have power gain and are relatively easy to tune. Active filters are also light in weight as there is no inductors used which are bulky.

15.3.4 Design of Filters Using L and C

Inductors and capacitors are used to make filters. Inductive and capacitive reactances are related to frequency as

$$X_L = \omega L = 2\pi f L$$

and

$$X_C = \frac{1}{\omega C} = \frac{1}{2\pi f C}$$

Inductors and capacitors are energy storage devices.

(a) Low-pass filter (LPF)

These filters can be made connecting inductors and capacitors in either L-section or π–section as shown in Fig. 15.12.

Desired signals should pass through the filter and be available as output. Inductor can pass low-frequency signals. So, here low-frequency signals will pass through the filter while high-frequency signals will be blocked. At high frequencies, inductive reactance of the inductor will be high whereas the capacitive reactance of the capacitors will be low. High-frequency signals will return to the input through the capacitors. However, the cut off frequency is not sharp as the pass band is extended beyond the cut off frequency f_C creating a transition band as has been shown. If we make a combination of π–section and L-section in series, then the cut off may be sharper and the Transition band is reduced.

(b) High-pass filter (HPF)

Here the capacitors are connected in series. With increase in frequency capacitive reactance decreases while the inductive reactance increases. Thus, the filters shown in figure will pass the high-frequency signals and attenuate the low-frequency signals.

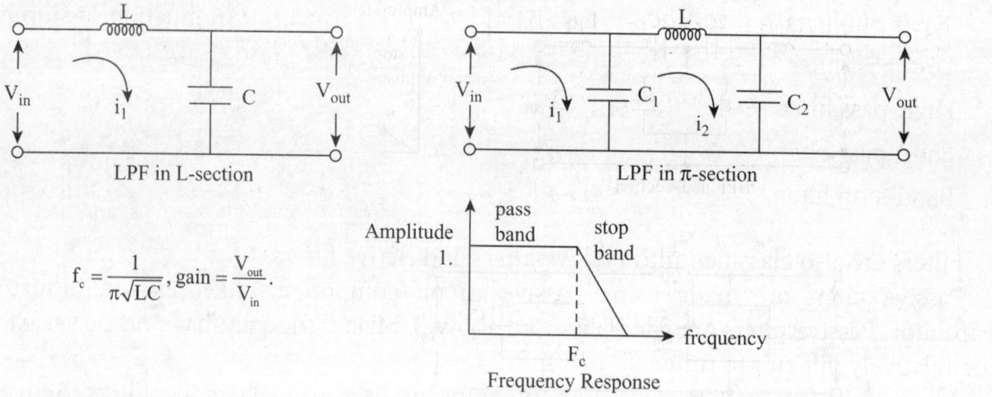

FIGURE 15.12

Low-pass filter and its frequency response

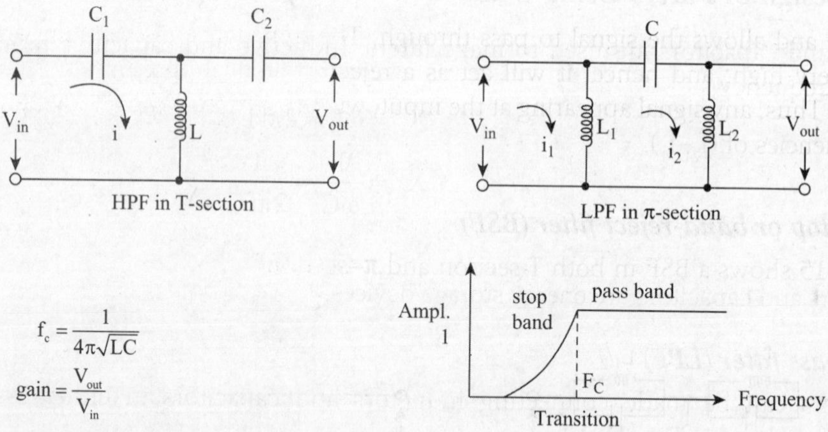

FIGURE 15.13

High-pass filter and its frequency response

Again from the frequency response graph it is seen that f_C is not sharp. To get a sharper cut off, we can add T-section and π–section in series to make a better HPF.

(c) Band-pass filter (BPF)

A band-pass filter in T-section and π–section is shown in Fig. 15.14. Its frequency response has also been shown.

In this filter, signal for a band of frequency f_2–f_1 has to pass through it and be available as output. The filter has series arm and shunt (parallel) arm. Within the pass band both the series arm and the shunt arm are in resonance. At resonance the series arm offers very low

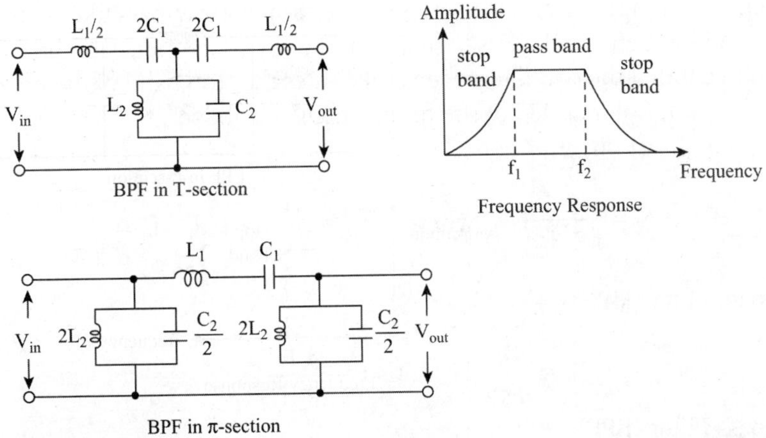

FIGURE 15.14

Band Pass filter and its frequency response

impedance and allows the signal to pass through. The impedance of the shunt arm at resonance is very high, and hence, it will act as a rejector circuit opposing any signal to pass through it. Thus, any signal appearing at the input, will pass through the filter within the passband frequencies of (f_2–f_1).

4. Band-stop or band-reject filter (BSF)

Figure 15.15 shows a BSF in both T-section and π–section

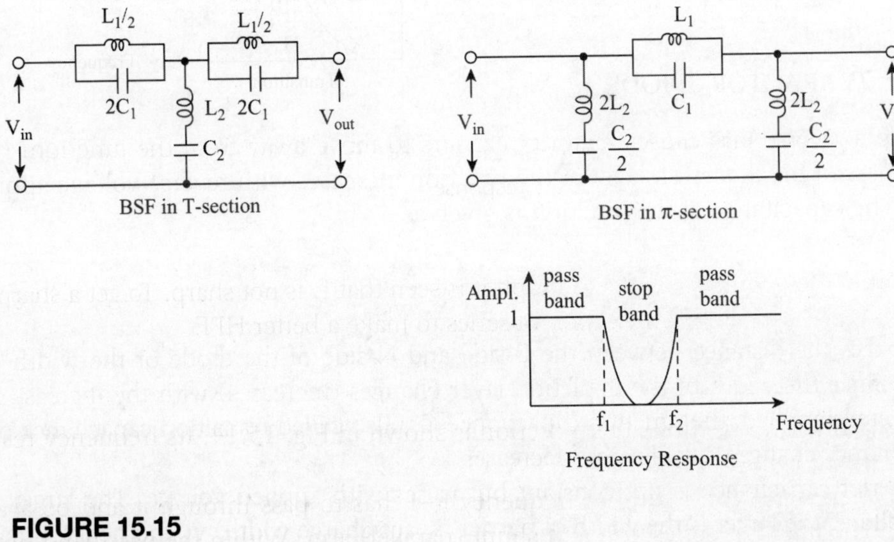

FIGURE 15.15

BSF in T-section and π-section

A band-stop filter, in its simplest from, is a series connection of a low-pass filter with a high-pass filter. Here the cut off frequency of HPF is higher than the cut off frequency of the LPF. The overlapping attenuation band of the two filters then constitutes the stop band of the band stop filter.

Determining the values of L and C of the filters described above

(i) For Low-pass Filter (LPF)

$$L = \frac{R_O}{\pi f_C} H; \quad C = \frac{1}{\pi R_O f_C} F \quad \text{where} \quad R_O = \sqrt{\frac{L}{C}} \quad \text{and} \quad f_C = \sqrt{\frac{1}{\pi\sqrt{L}}}$$

(ii) For High-pass Filter (HPF)

$$L = \frac{R_O}{4\pi f_C} H; \quad C = \frac{1}{4\pi R_O f_C} F$$

(iii) For Band-pass Filter (BPF)

$$C_1 = \frac{(f_2 - f_1)}{4\pi f_1 f_2 R_0} F; \quad L_1 = \frac{R_0}{\pi(f_2 - f_1)} H$$

$$C_2 = \frac{1}{\pi R_0 (f_2 - f_1)} F; \quad L_2 = \frac{R_0(f_2 - f_1)}{4\pi f_1 f_2} H$$

(iv) For Band-stop Filter (BSF)

$$C_1 = \frac{1}{4\pi R_0 (f_2 - f_1)} F; \quad L_1 = \frac{R_0(f_2 - f_1)}{\pi f_1 f_2} H$$

$$C_2 = \frac{f_2 - f_1}{\pi f_1 f_2 R_0} F; \quad L_2 = \frac{R_0}{4\pi^2 (f_2 - f_1)} H$$

15.4 ► Varactor Diode

In a diode a reverse bias causes majority carriers to move away from the junction. Hence, the thickness of the depletion layer at the junction increases with reverse voltage applied to a diode. The capacitance of the junction is given as

$$C_T = \frac{\varepsilon A}{d} = \frac{\varepsilon A}{w}$$

Where d is the distance between the P-side and N-side of the diode or the width of the junction. Since the width of the depletion layer changes (increases) with the increase in the reverse voltage applied, the junction capacitance C_T, also called transition capacitance or barrier capacitance changes, in this case, decreases.

The barrier capacitance is not constant but varies with applied voltage. The larger is the reverse voltage, the larger is the width of barrier (space charge width) w, and hence, smaller is the capacitance C_T.

This voltage-variable capacitance of a p-n junction diode biased in the reverse direction is called varactor diode. Varactor diodes are useful in a number of electronic circuits like in voltage tuning of an LC resonant circuit, self-balancing bridge circuits, special types of amplifiers, etc.

It should be noted that if the capacitance C_T due to reverse voltage is large, the capacitive reactance will be low as $X_C = \frac{1}{\omega c} = \frac{1}{2\pi fc}$. The diode has to restrict flow of current through it when it is reverse biased. But if capacitive reactance is low, current will flow through the capacitor, C_T.

For high frequency applications, it is required that the transition capacitance, C_T of a diode should be as small as possible. Varactors, also called varicaps or voltacaps are designed to have low-junction capacitance under reverse voltage applications.

15.5 ▶ Applications of Diodes in Clipping and Clamping Circuits

A clipping circuit removes, i.e., clips off a certain portion of the input voltage waveform. Clipping circuits are used when it becomes necessary to protect a circuit or a device that might get destroyed due to large amplitude signal. In fact, the half-wave rectifier described earlier is a clipper circuit. It clips off the negative half cycle and allows only the positive half cycle.

A clipping circuit, or also called a clipper is often used to clip off certain unwanted portions of the wave form which may lead to noise, and deteriote the performance of the device through which such currents pass. Let us see one noise clipper circuit as shown in Fig. 15.16. Here the noise level is lower than the forward voltage drop V_F of the diodes, i.e., $V_F = 0.7$ V. Two diodes, D_1 and D_2 clear both the halves of the main signal but do not allow the noise signal to pass.

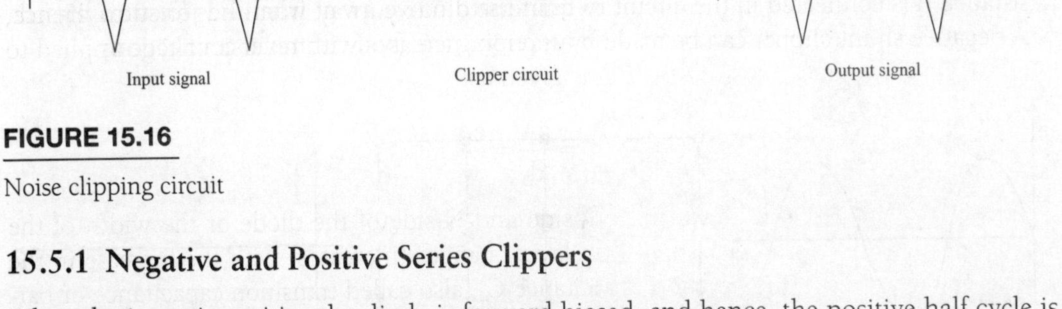

FIGURE 15.16

Noise clipping circuit

15.5.1 Negative and Positive Series Clippers

When the input is positive the diode is forward biased, and hence, the positive half cycle is passed to the output as shown in Fig. 15.17 (a). During the negative half cycle the diode is reverse biased, and hence, the output will remain zero. This way we can say that the circuit cuts off the negative half cycles of the input voltage.

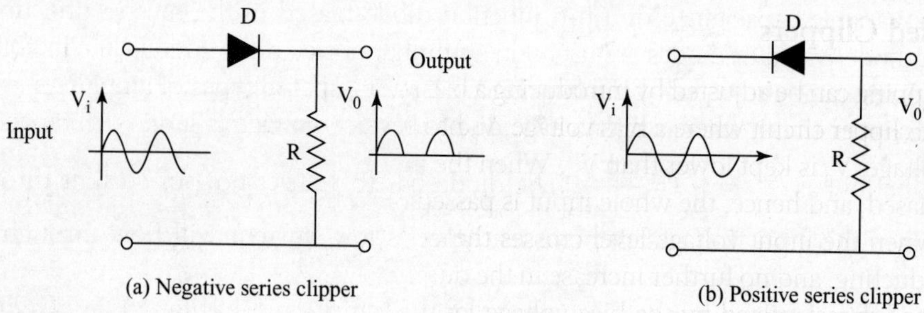

FIGURE 15.17

Negative and positive series clippers

In Fig. 15.17 (b), the positive half cycles are clipped off as the diode is reverse biased during the positive half cycles of the input. The input voltage wave in these two clippers are sinusoidal. However, the input waveform could be of any other shapes like square, rectangular, triangular, etc.

15.5.2 Shunt Clippers

Like series clippers, shunt clippers can also clip off the positive half cycle or the negative half cycle of the input.

Figure 15.18 shows a positive shunt clipper circuit where the diode is connected in parallel, i.e., in shunt with the load resistance R_L.

When the input voltage is positive, the diode is forward biased and is in conducting mode. The voltage drop across the diode V_F, which is only 0.7 V for a silicon diode will appear across the load. Thus, effectively the positive half cycle of the input voltage is clipped off and the output voltage will be nearly zero. When the input voltage is negative, the diode is reverse biased. Current flows through the load resistance R_L and the series resistance R. The voltage drop across R is I_L R which is small as compared to the voltage drop across the load resistance R_L. The output voltage V_0 is nearly equal to the negative input voltage as has been shown. Resistance R is connected in the circuit to limit the diode current when it is forward biased.

A negative shunt clipper can be made by reversing the diode terminal connections.

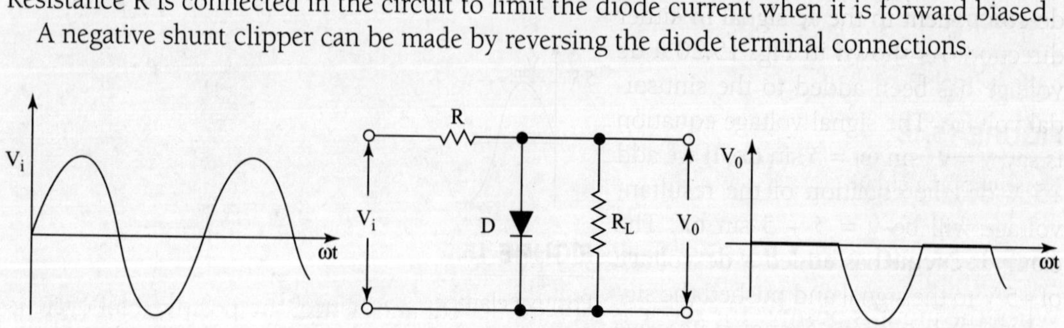

FIGURE 15.18

Positive shunt clipper

15.5.3 Biased Clippers

The level of clipping can be adjusted by introducing a bias voltage in the circuit. Fig. 15.19 shows a biased positive clipper circuit where a bias voltage, V_B has been connected in series with the diode.

The bias voltage, V_B is kept lower than V_m. When the input voltage is less than V_B, the diode is not forward biased, and hence, the whole input is passed on to the load and a load current flows through R_L. When the input voltage level crosses the voltage V_B, the diode gets forward biased and starts conducting, and no further increase in the output current is possible. Thus, the current through the load is resistricted by the bias voltage level. A combination biased clipping circuit can be made by using another diode together with a bias voltage. The bias voltage for the two diodes can be made different to achieve the clipping of the positive and negative half cycles at different voltage levels. Biased clipping circuits are used to protect circuits and devices from over voltages. We have observed that clipper circuits can change the output voltage wave shape also.

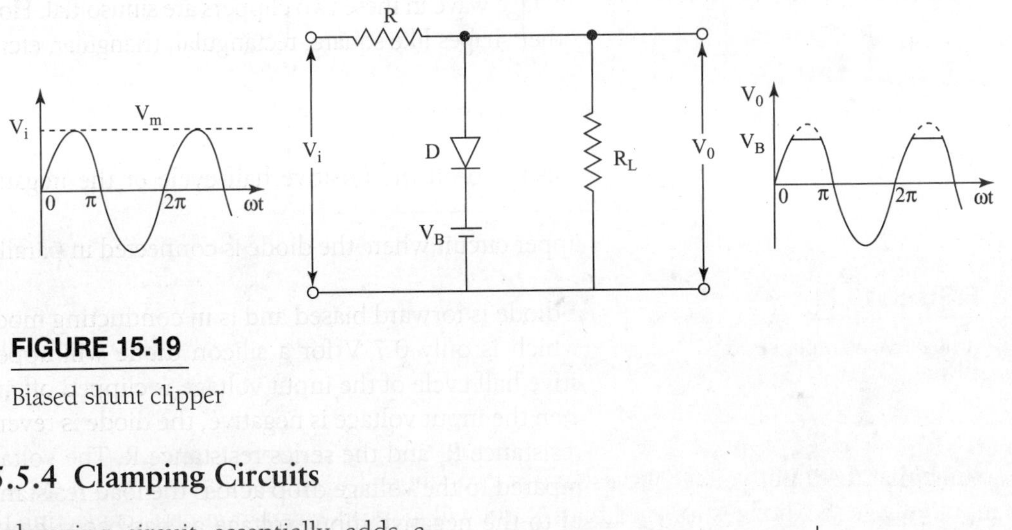

FIGURE 15.19

Biased shunt clipper

15.5.4 Clamping Circuits

A clamping circuit essentially adds a dc component to the ac signal in either direction. As shown in Fig. 15.20 a dc voltage has been added to the sinusoidal voltage. The signal voltage equation is say, $v = v_m \sin \omega t = 5 \sin \omega t$. If we add +5 V dc, the equation of the resultant voltage will be $v = 5 + 5 \sin \omega t$. The clamper circuit has added a dc voltage of +5 V to the signal and pushed the signal upwards without changing its wave shape. This is called a positive clamper. If a negative dc voltage is added, the signal will be brought downwards, and such a clamper will be called a negative clamper. A negative clamper circuit has been explained in Fig. 15.21.

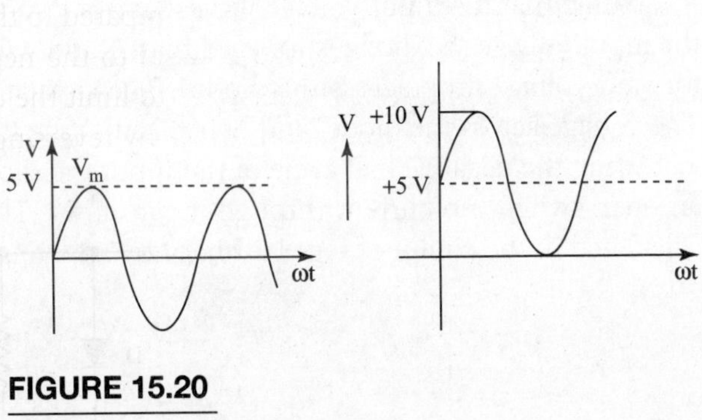

FIGURE 15.20

Positive clamper's wave shape

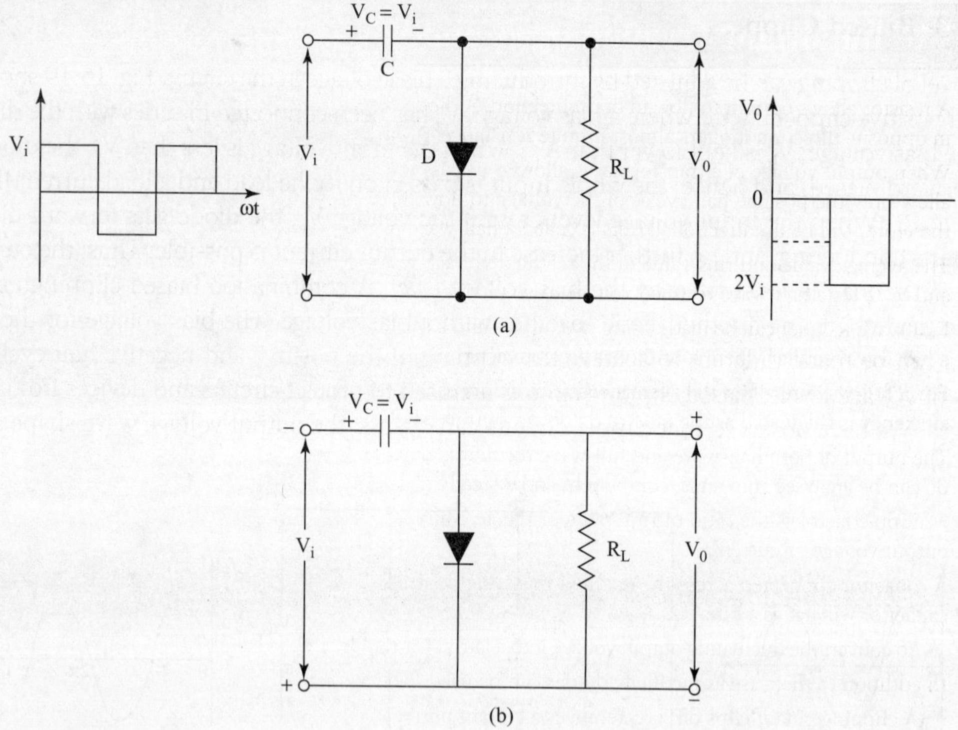

FIGURE 15.21

A negative clamper circuit

Assume that the input voltage waveshape is a square one. During the positive half cycle of the input voltage the diode is forward biased and will work like a short circuit (neglecting the forward voltage drop across the diode). The capacitor will be charged to input voltage level. The voltage across the diode, and hence across the load resistor R_L will be zero.

During the negative half cycle of the input the diode will be reverse biased and will be like an open switch (no current through it can flow). The voltage across the load, i.e., V_0 can be calculated in the circuit of Fig. 15.21 (b) by applying Kirchhoff's voltage law as

$$-V_i - V_c - V_0 = 0$$

or, $$V_0 = -V_i - V_i \text{ (as } V_c = V_i)$$

or, $$V_0 = -2V_i$$

This shows that the input signal is negatively clamped by a dc voltage equal to the magnitude of the input voltage.

Important Terminologies and Points to Remember

1. A resistor allows current to flow in both direction. A diode conducts current in one direction only because in opposite direction it offers almost infinite resistance. A diode can therefore be used as rectifier.
2. When output voltage of a transformer is allowed to pass through the load through a diode, the diode will allow only the positive half cycles of the voltage to flow through the load. The current through the load, therefore, will be unidirectional. This is called a half-wave rectifier circuit. A rectifier converts ac into dc.
3. The average value and rms value of the rectified voltage wave are calculated which are, respectively, V_m/π and $V_m/2$ for a half-wave rectified voltage. Also rectifier efficiency and ripple factor are calculated.
4. Full-wave rectifier can be made using two diodes and a centre tapped transformer. A bridge rectifier is also a full-wave rectifier where four diodes are connected in the form of a bridge.
5. For a full-wave rectifier, the average value of dc output voltage is $2V_m/\pi$, the rms value is $I_m/\sqrt{2}$. Rectifier efficiency is calculated as the ratio of dc power output to ac power input.
6. The output of both half-wave and full-wave rectifier is not a steady dc but are pulsating dc. The pulsating dc can be analysed into an ac component and a steady dc component.

   Ripple factor is the ratio of rms value of the ac component and the value of the dc component of the output voltage or current.
7. A capacitor connected across the rectified pulsating dc will reduce pulsating nature of the output, i.e., the capacitor will act as a filter.

   To convert the fluctuating output voltage into a fairly steady, smoothing circuits, called filters must be used.
8. In addition to their use as rectifiers, diodes can be used in clipping and clamping circuits.

   A clipping circuit clips off, i.e., removes a certain portion of the unwanted signal waveform. For example, an audio signal may, in addition to the desired signal, a noise signal of higher frequency. To obtain a clear audio signal, the high frequency noise signal has to be clipped off.
9. Using clamping circuits, a dc component can be added to any sinusoidal voltage in either direction.

Important Formulae and Equations

For half-wave rectifier:

1. $I_{dc} = I_{av} = \dfrac{1}{2\pi}\int_0^{\pi} I_m \sin\theta d\theta = \dfrac{I_m}{\pi}$	1. Average value of half-wave rectified current.
2. $I_{rms} = \sqrt{\dfrac{1}{2\pi}\int_0^{\pi} I_m^2 \sin^2\theta d\theta} = \dfrac{I_m}{2}$	2. R_{ms} value for a half-wave rectifier current.
3. Ripple factor, $r = \dfrac{\text{rms value of ac component}}{\text{value of dc component}} = \sqrt{\left(\dfrac{I_{rms}}{I_{dc}}\right)^2 - 1} = 1.21$	3. Ripple factor of the output voltage or current of a half-wave rectifier. Ripple is high.

(continued)

Important Formulae and Equations

4. Rectifier efficiency, $\eta = \frac{P_{dc}}{P_{ac}}$ $= \frac{4}{\pi^2} = 0.406 = 40.6\%$	4. Efficiency of rectification is low.
5. For Full-wave Rectifier: $I_{av} = \frac{1}{2\pi}\int_{0}^{2\pi} I_m \sin\theta d\theta = \frac{2I_m}{\pi}$ $I_{rms} = \sqrt{\frac{1}{2\pi}\int_{0}^{2\pi} I_m^2 \sin^2\theta d\theta} = \frac{I_m}{\sqrt{2}}$ Efficiency, $\eta = \frac{8}{\pi^2} = 0.812 = 81.2\%$ Ripple factor, $r = \sqrt{\left(\frac{I_{rms}}{I_{dc}}\right)^2 - 1} = 0.48$	5. As compared to half-wave rectifier, I_{av}, I_{rms}, η are more and ripple is less.

REVIEW QUESTIONS

Short Answer Type Questions

1. Draw a half-wave rectified circuit and show the input and output voltage waveforms.
2. List the performance parameters of a half-wave rectifier and explain this significance.
3. Draw the circuit diagram for a centre-tap full-wave rectifier and explain its operation with the help of input and output voltage wave forms.
4. Draw and explain the circuit for a bridge rectifier and draw the input and output voltage waveforms.
5. Derive the expression for the following in the case of a half-wave rectifier:

   (a) rectifier efficiency (b) ripple factor (c) average value of output dc current (d) output voltage, V_{dc}.
6. Compare the performance parameters of a half-wave rectifier and a full-wave bridge rectifier.
7. What is meant by ripple factor and what is its significance. Calculate its value for a half-wave rectifier and a full-wave rectifier.
8. What are the disadvantages of a half-wave rectifier?
9. What is meant by rectifier efficiency? What is its value for a full-wave and a half-wave rectifier?
10. What is the purpose of using a filter circuit in a rectifier and how does it work?
11. Draw the circuit for a capacitor filter and explain how does it half reduce the ripple factor?
12. Sketch the nature of ripple in the output of a capacitor filter used in a full-wave bridge rectifier.
13. Draw and explain the half-wave rectifier circuit with a capacitor filter. Also draw the input and output voltage wave form.

14. Draw and explain a bridge rectifier with a smoothing capacitor. Draw the output voltage wave shape.
15. What is a clipping and a clamping circuit? Where are they used?
16. Draw and explain negative and positive series clipper circuits with their input and output voltage waveforms, respectively.
17. What is a shunt clipper circuit? Draw the input and output voltage waveforms.
18. Draw the circuit for a biased clipper. Also draw the input and output voltage waveforms.
19. What are clamping circuits and what are their applications?
20. Draw and explain positive and negative clamper circuits.

Multiple Choice Questions

1. The average or dc value of output current for a half-wave rectifier is
   (a) $\frac{2I_m}{\pi}$ (b) $\frac{I_m}{\pi}$
   (c) $\frac{I_m}{2\pi}$ (d) $\frac{I_m}{\sqrt{2}}$.
2. The RMS value of load current for a half-wave rectifier is
   (a) $\frac{I_m}{\sqrt{2}}$ (b) $\frac{I_m}{\pi}$
   (c) $\frac{I_m}{2}$ (d) $\frac{I_m}{2\pi}$.
3. The output voltage V_{dc} for a half-wave rectifier is
   (a) $\frac{V_m}{\pi}$ (b) $\frac{V_m}{2\pi}$
   (c) $\frac{V_m}{\sqrt{2}}$ (d) $\frac{V_m}{2}$.
4. Rectifier efficiency of a half-wave and full-wave rectifier circuit, respectively, are
   (a) 40.6% and 81.2%
   (b) 20.3% and 81.2%
   (c) 40.6% and 91.2%
   (d) 20.3% and 40.6%.
5. Ripple factor for a half-wave and full-wave rectifier circuit, respectively, are
   (a) 0.48 and 1.21 (b) 0.48 and 0.121
   (c) 4.8 and 1.21 (d) 8.21 and 0.48.
6. Peak inverse voltage is defined as the
   (a) Minimum voltage that appears across the diode when it is forward biased
   (b) Minimum voltage that appears across the diode when it is reverse biased
   (c) Maximum voltage that appears across the diode when it is reverse biased
   (d) Minimum voltage that appears across the diode when it is reverse biased.
7. In a capacitor filter circuit, a capacitor is connected
   (a) In series with the load and it allows dc current to flow through it
   (b) In parallel with the load and it allows ac but blocks dc
   (c) In parallel with the load and it allows dc but blocks ac
   (d) In series with the load and it allows ac and blocks dc.
8. A clamping circuit
   (a) adds a dc component to an ac signal in either direction
   (b) adds an ac component to a dc signal in either direction
   (c) adds a dc component to an ac signal in positive direction only
   (d) adds a dc component to an ac signal in negative direction only.

Answers to Multiple Choice Questions

1. (b)	2. (c)	3. (a)	4. (a)
5. (d)	6. (c)	7. (b)	8. (a)

CHAPTER

16 Digital Electronics

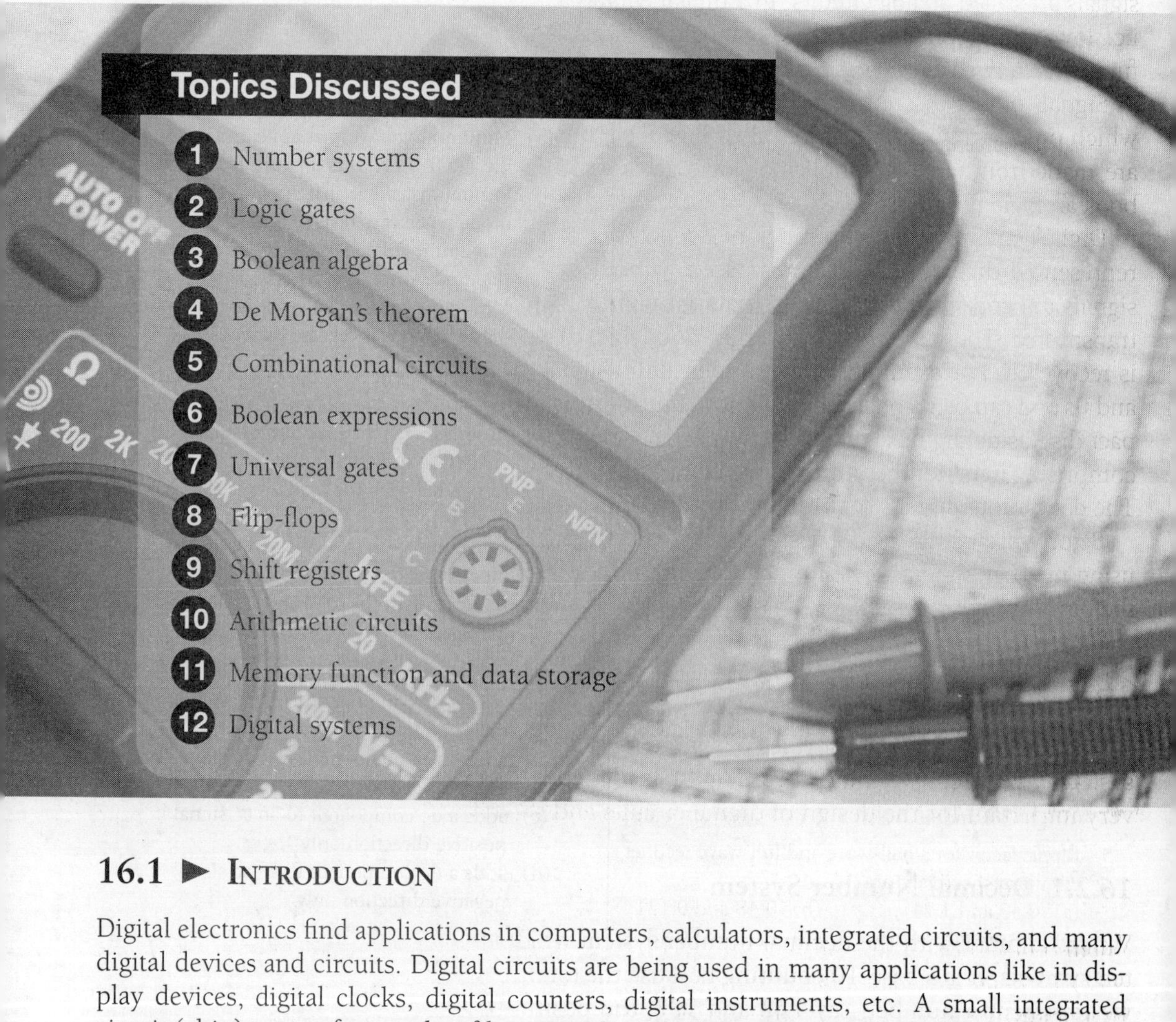

Topics Discussed

1. Number systems
2. Logic gates
3. Boolean algebra
4. De Morgan's theorem
5. Combinational circuits
6. Boolean expressions
7. Universal gates
8. Flip-flops
9. Shift registers
10. Arithmetic circuits
11. Memory function and data storage
12. Digital systems

16.1 ▶ Introduction

Digital electronics find applications in computers, calculators, integrated circuits, and many digital devices and circuits. Digital circuits are being used in many applications like in display devices, digital clocks, digital counters, digital instruments, etc. A small integrated circuit (chip) can perform tasks of large number of transistors, diodes, resistors, etc. Digital electronics is an interesting subject of study.

Digital electronics includes electronic systems that use digital signals. Digital electronics or any digital circuit are usually made from large assemblies of logic gates. Digital electronics, digital circuits, and digital systems are often used interchangeably. Digital circuits and systems used in computers and electronic controls are ***logic gates*** and ***multivibrators***. Logic gates are combinational (i.e., with no memory) and sequential (i.e., with memory). Combinational gates are AND, NOT, OR, NAND and NOR gates. Sequential gates include flip-flops (bistable multivibrators) which are available as registers and counters.

In this chapter brief descriptions of the above mentioned digital circuits and systems will be provided. Electrical signals that are continuous and can have any value over a given range are called analog signals. Electronic circuits that are used to process (i.e., amplify or modify) analog signals are called analog circuits. In a digital circuit, a signal is represented in one of two states, i.e., low or high (on or off). These two voltage levels are called logic levels. Thus, a digital circuit functions in two states, i.e., in a binary manner.

Signals of two discrete values or levels, high and low, are called digital signals. The circuits which process digital signals are called digital circuits. Digital electronics, or any digital circuit, are made from large assemblies of logic gates which are simple electronic representation of boolean logic functions.

Digital circuits have a number of advantages over analog circuits. For example, signals represented digitally can be transmitted without degradation of quality. That is why analog signals are converted to digital signals through A to D converters and the digital signals are transmitted. These are again converted through D to A converters and the actual analog signal is recovered. For example, a continuous audio signal can be transmitted as a sequence of 1s and 0s and can be reconstructed without much error. An hour of music can be stored in a compact disc using digital signals which may be a few billion binary digits (1s and 0s). All systems comprising light, temperature, conductivity, pressure, magnetic fields, etc., are analog systems. The continuous signals of these are converted into discrete digital signals.

Before discussing logic gates and some digital circuits we will discuss data representation using number systems. The number system used to represent data in a computer is known as the binary number system.

16.2 ► Number Systems

A number system is a set of rules of symbols used to represent numbers. There are number systems like decimal, binary, octal, and hexadecimal. The knowledge of number systems is very important for the design of digital circuits and systems.

16.2.1 Decimal Number System

We are familiar with the decimal number system which uses 0, 1, 2, 3, 4, 5, 6, 7, 8, 9, i.e., ten symbols. It is a base-10 counting because the total number of symbols used is ten. When we represent a number, say 752 and 907, the positional values of the numbers and their decimal equivalent would be represented.

A decimal number of, say, 692 is represented as

$$\text{Decimal } 692 = 6 \times 10^2 + 9 \times 10^1 + 2 \times 10^0$$

Number	Value	Position	Positional value	Decimal equivalent
7 5 2				
	2	0	10°	$2 \times 10^\circ = 2$
	5	1	10^1	$5 \times 10^1 = 50$
	7	2	10^2	$7 \times 10^2 = 700$
				752
9 0 7				
	7	0	10°	$7 \times 10^\circ = 7$
	0	1	10^1	$0 \times 10^1 = 0$
	9	2	10^2	$9 \times 10^2 = 900$
				907

A decimal number of say, 824.48 is represented as

$$\text{Decimal } 824.48 = 8\times10^2 + 2\times10^1 + 4\times10^0 \cdot 4\times10^{-1} + 8\times10^{-2}$$

The multiples of 10 used are called weighted values. The decimal number 824.48 is represented as $(824.24)_{10}$ as 10 is the base.

16.2.2 Binary Number System

In the binary system a number is expressed by two symbols or digits, 0 and 1. Since the total number of symbols or digits used is two, it has a base 2. The binary symbols 0 and 1 are called binary digits or bits. A group of eight bits is called a byte. In this system the positional value is used in the same way as the decimal system. For example, two binary numbers, 11010 and 110011 are represented as their decimal equivalent as

Number	Value	Position	Positional value	Decimal equivalent
1 1 0 1 0				
	0	0	2^0	$0 \times 2^0 = 0$
	1	1	2^1	$1 \times 2^1 = 2$
	0	2	2^2	$0 \times 2^2 = 0$
	1	3	2^3	$1 \times 2^3 = 8$
	1	4	2^4	$1 \times 2^4 = 16$
				26
1 1 0 0 1 1				
	1	0	2^0	$1 \times 2^0 = 1$
	1	1	2^1	$1 \times 2^1 = 2$
	0	2	2^2	$0 \times 2^2 = 0$
	0	3	2^3	$1 \times 2^3 = 0$
	1	4	2^4	$1 \times 2^4 = 16$
	1	5	2^5	$1 \times 2^5 = 32$
				51

16.2.3 Conversion of Binary to Decimal

A binary number can be converted into its decimal equivalent number by noting that successive digits from the extreme right of a binary number are the coefficients of ascending power of 2 beginning with zeroeth power of 2. A few examples will illustrate this conversion.

EXAMPLE 16.1

Binary number is 1100101. Convert this number into its decimal equivalent.

SOLUTION

1	1	0	0	1	0	1
1×2^6	1×2^5	0×2^4	0×2^3	1×2^2	0×2^1	1×2^0

$64 + 32 + 0 + 0 + 4 + 0 + 1 = 101$

Decimal number is 101. Therefore, $(1100101)_2 = (101)_{10}$

EXAMPLE 16.2

Binary number is 11010. Convert this number into its decimal equivalent.

SOLUTION

1	1	0	1	0
1×2^4	1×2^3	0×2^2	1×2^1	0×2^0

$16 + 8 + 0 + 2 + 0 = 26$

Decimal number is 26. Therefore, $(11010)_2 = (26)_{10}$

EXAMPLE 16.3

Binary number is 101101. Convert this number into its decimal equivalent.

SOLUTION

1	0	1	1	0	1
1×2^5	0×2^4	1×2^3	1×2^2	0×2^1	1×2^0

$32 + 0 + 8 + 4 + 0 + 1 = 45$

Therefore, $(101101)_2 = (45)_{10}$

Fractional binary number to fractional decimal number

The decimal equivalent of a binary fraction is determined by multiplying each digit in the fraction by 2^{-1}, 2^{-2} 2^{-3}, 2^{-4}, . . . etc., beginning with the first digit after the binary point. The products are then added to get the decimal fraction. A few examples will illustrate this procedure.

EXAMPLE 16.4

Convert the fractional binary number 0.1011 into decimal fraction.

SOLUTION

The number 0.1011 is converted into decimal number as

$1 \times 2^{-1} + 0 \times 2^{-2} + 1 \times 2^{-3} + 1 \times 2^{-4}$

$= 0.5 + 0 + 0.125 + 0.0625$

$= 0.6875$

EXAMPLE 16.5

Convert the fractional binary number 0.1101 into decimal fraction

SOLUTION

The number 0.1101 is converted into decimal number as

$1 \times 2^{-1} + 1 \times 2^{-2} + 0 \times 2^{-3} + 1 \times 2^{-4}$

$= 0.5 + 0.25 + 0 + 0.625$

$= 0.\ 8125$

16.2.4 Conversion of Decimal to Binary

To convert a decimal number into a binary equivalent number, the decimal number is expressed as a sum of ascending power of 2. The positional values of the binary number system are

…	2^7	2^6	2^5	2^4	2^3	2^2	2^1	2^0
	128	64	32	16	8	4	2	1

Suppose we are to find the binary equivalent of the decimal number 45. The number 45 is arrived at by adding 32 + 8 + 4 + 1. The 1S and 0S are placed as shown

2^5		2^4		2^3		2^2		2^1		2^0
32	+	0	+	8	+	4	+	0	+	1 = 45

The binary number is 101101

The binary equivalent of 19 is arrived as

16 + 2 + 1

2^4		2^3		2^2		2^1		2^0
16	+	0 +		0	+	2	+	1 = 19

The binary number is 10011

The binary equivalent of decimal numbers 7, 9, 11, respectively, are

Example 1

$$7 = 4 + 2 + 1$$
$$2^2 + 2^1 + 2^0$$

The binary number of 7 is 111

Example 2

$$9 = 8 + 0 + 0 + 1$$
$$2^3 + 0 + 0 + 2^0$$

The binary number of 9 is 1001

Example 3

$$11 = 8 + 0 + 2 + 1$$
$$2^3 + 0 + 2^1 + 2^0$$

The binary number of 11 is 1011.

We can use an alternative method of converting a decimal number into a binary equivalent by dividing the decimal number progressively by 2 till the quotient is zero. The remainders of the successive divisions, expressed in the reverse order, gives the binary equivalent number. A few examples will illustrate the procedure.

EXAMPLE 16.6

Convert decimal 23 into its binary equivalent number.

SOLUTION

			Remainder
Divide	23 by 2	$2\lfloor 23$ → 11	1
Divide	11 by 2	$2\lfloor 11$ → 5	1
Divide	5 by 2	$2\lfloor 5$ → 2	1
Divide	2 by 2	$2\lfloor 2$ → 1	0
Divide	1 by 2	$2\lfloor 1$ → 0	1

Now we will arrange the remainders in the reverse order, i.e., from bottom upwards. The binary equivalent number is 10111.

EXAMPLE 16.7

Determine the binary equivalent of decimal number 17.

SOLUTION

			Remainder
Divide	17 by 2	2⌊17 → 8	1
Divide	8 by 2	2⌊8 → 4	0
Divide	4 by 2	2⌊4 → 2	0
Divide	2 by 2	2⌊2 → 1	0
Divide	1 by 2	2⌊1 → 0	1

The remainder when put in reverse order give the binary equivalent a 10001.

EXAMPLE 16.8

Determine the binary equivalent of decimal number 10.

SOLUTION

			Remainder
Divide	10 by 2	2⌊10 → 5	0
Divide	5 by 2	2⌊5 → 2	1
Divide	2 by 2	2⌊2 → 1	0
Divide	1 by 2	2⌊1 → 0	1

The remainders are impressed in reverse order to get the binary equivalent as 1010.

Conversion of fractional decimal number to fractional binary number

When the decimal number is a fraction, the conversion can be carried out through the following steps.

First, multiply the decimal fraction by 2. The result contains an integer part and a fractional part. Separate the integer part and write it in a column. The fractional part becomes the new fraction. Repeat these steps till the fractional part becomes zero or the desired number of binary places are obtained. The 1s and 0s separated placed together gives the required fractional binary number. A few examples will illustrate the procedure.

EXAMPLE 16.9

Convert decimal 0.8125 into its equivalent binary number.

SOLUTION

		Fractional part	Integer part
0.8125 × 2	= 1.6250	0.6250	1
0.6250 × 2	= 1.2500	0.2500	1
0.2500 × 2	= 0.5000	0.5000	0
0.5000 × 2	= 1.0000	0	1

The binary decimal number is 0.1101

EXAMPLE 16.10

Convert 0.65 into its binary fraction.

SOLUTION

		Fractional pat	Integer part
0.65 × 2	= 1.3	0.3	1
0.3 × 2	= 0.6	0.6	0
0.6 × 2	= 1.2	0.2	1
0.2 × 2	= 0.4	0.4	0
0.4 × 2	= 0.8	0.8	0
0.8 × 2	= 1.6	0.6	1
0.6 × 2	= 1.2	0.2	1

In the last step we get back 0.2 as the fractional part. To get the approximate result, we can terminate the process at this stage. Thus, the approximate binary fraction is 0.1010011.

16.2.5 Binary Addition

The binary addition is performed by using the following rules.

(i) 0 + 0 = 0

(ii) 1 + 0 = 1

(iii) 0 + 1 = 1
(iv) 1 + 1 = 10 (∵ decimal equivalent of 10 is 2)

Examples

(i) Add

$$\begin{array}{r} 10 \\ +11 \\ \hline 101(=5) \end{array} \qquad \begin{array}{c} 10+11=101 \\ 2+3=5 \end{array}$$

(ii) Add

$$\begin{array}{r} 111(=7) \\ +111(=7) \\ \hline 1110(=14) \end{array}$$

Here the first column gives 1 + 1 = 10. We put 0 below this column and add the carry 1 to the second column. In the second column 1 + 1 = 10. We add the carry 1 with 0 to get 1 and carry 1 to the third column. In the third column 1 + 1 = 10. We add the carry 1 to 0 to get 1 and the carry 1 is placed in the fourth column.

(iii) Add

$$\begin{array}{r} 10001(=17) \\ +111(=7) \\ \hline 11000(=24) \end{array}$$

First column 1 + 1 = 10, place 0 and carry 1
Second column, 1 + 0 = 1, 1 + 1 = 10, place 0 and carry 1
Third column, 1 + 0 = 1, 1 + 1 = 10, place 0 and carry 1
Fourth column, 1 + 0 = 1, place 1 and carry 0
Fifth column, 1 + 0 = 1, place 1

16.2.6 Binary Subtraction

The following rules are used for binary subtraction

(i) 0 – 0 = 0
(ii) 1 – 0 = 1
(iii) 1 – 1 = 0
(iv) 10 – 1 = 1 (decimal equivalent of 10 is binary 2)

Example

(i) Subtract

$$\begin{array}{r} 1011(=11) \\ -1000(=8) \\ \hline 0011(=3) \end{array}$$

(ii) Subtract

$$\begin{array}{r} 110(=6) \\ -11(=3) \\ \hline 011(=3) \end{array}$$

In the second example, the first column is 10 – 1 = 1 (after borrowing 1 from the second column). The borrow is carried and added to the lower number of second column to get 1 + 1 = 10. Therefore, the second column gives 1 – 0 = 1 and the third column gives 1 – 1 = 0.

16.2.7 Binary Multiplication

Binary multiplication is carried out using the following rules.

(i) 0 × 0 = 0

(ii) 1 × 0 = 0

(iii) 0 × 1 = 0

(iv) 1 × 1 = 1

Example

			1	1	1	(= 7)
			1	0	1	(= 5)
			1	1	1	
		0	0	0	×	
	1	1	1	×	×	
1	0	0	0	1	1	(= 35)
1 + 1 = 10		1 + 1 = 10 carry 1	1 + 1 = 10 carry 1	1 + 0 = 1	1 + 0 = 1	

16.3 ► Octal Number System

Octal system has a base of 8 counting from 0 to 7. To convert a decimal number into octal equivalent of decimal number, we divide the integer repeatedly by 8 till the quotient is zero. The remainders written in reverse order gives the octal equivalent number.

For example let us find the octal equivalent of decimal number 756.

756 ÷ 8 = 94	Remainder = 4
94 ÷ 8 = 11	Remainder = 6
11 ÷ 8 = 1	Remainder = 3
1 ÷ 8 = 0	Remainder = 1

The octal equivalent of 756 is 1364.

To convert an octal to binary, each octal symbol is replaced by a 3-bit binary equivalent as shown

Octal	Binary
0	000
1	001
2	010
3	011
4	100
5	101
6	110
7	111

Conversion of binary number into an equivalent octal number can be done in the following way.

Split the binary digits into groups of three digits, each beginning from the right end for integers and from the left end for fractions. Replace each group by its equivalent octal number. A few examples are given to illustrate the procedure.

Example (i)

Convert binary 11001 into octal equivalent
The binary number is grouped as 011 001
Replacing each group by its octal equivalent we get the number 31

Example (ii)

Convert binary 0.111 001110 into its equivalent octal
The binary number is grouped into 3-bit form as
0.111 001 110
The equivalent is 0.716

To convert an octal number into its decimal equivalent, the following steps be followed.

Consider the octal number from the extreme right as the coefficient of the ascending power of 8, the starting power of 8 at the extreme right has to be taken as zero. If the octal number has a fraction, the fractional numbers have to be considered as the coefficient of ascending power of 8^{-1}.

Examples

(i) $415.2 = 4 \times 8^2 + 1 \times 8^1 + 5 \times 8^0 + 2 \times 8^{-1}$
$= 256 + 8 + 5 + 0.25$
$= 269.25$

(ii) $732.14 = 7 \times 8^2 + 3 \times 8^1 + 2 \times 8^0 + 1 \times 8^{-1} + 4 \times 8^{-2}$
$= 448 + 24 + 2 + 0.125 + 0.0625$
$= 474.1875$

16.4 ► Hexadecimal Number System

In hexadecimal system we use 16 symbols, viz 0 to 9 and the capital letters A B C D E F for 10 to 15. Hence, the hexadecimal number system uses a base of 16.

To convert a binary number into a hexadecimal number (Hex), the binary digits are grouped into 4 binary bits, counting from the right end for integral numbers and from the left end for fractions. Each group is replaced by its equivalent.

Binary	Hex	Binary	Hex	Binary	Hex	Binary	Hex
0000	0	0100	4	1000	8	1100	C
0001	1	0101	5	1001	9	1101	D
0010	2	0110	6	1010	A	1110	E
0011	3	0111	7	1011	B	1111	F

Examples

(i) Convert 1111 1110 0011 into hexadecimal equivalent
Group the binary as (from right to left) 1111 1110 0011
Replace the group by their equivalent as FE3

(ii) Convert 1111 10001.100 1100 1101 into hexadecimal equivalent
Group the binary as
0001 1111 0001 . 1001 1001 1010
Replace the groups by their equivalent to get
1F1.99A

(iii) Convert Hex number 6D5 into its binary equivalent.
Starting from right we replace binary equivalent of 5, D, and 4 as

(6)	(D = 13)	(5)
0110	1101	0101

It is seen from the above examples that conversion between Hex and binary is very quick.

For conversion of decimal number into binary equivalent, or useful system, a binary coded decimal system, or BCD system is often used.

It is a code in which individual decimal digits are replaced by a group of 4 binary bits in a row. The weights of the 4 bits from left to right of the binary string are 8,4,2, and 1, respectively. The BCD code is also called eight four two one code.

Example Express 2534 in BCD form.

Decimal form	2	5	3	4
BCD form	0010	0101	0011	0100

Thus, $(2534)_{10} = (0010010100110100)_{BCD}$.

It is to be noted that this system is totally different from the method of conversion of decimal numbers into binary numbers.

The binary equivalent of 2534 is calculated as

2^{11}	2^{10}	2^9	2^8	2^7	2^6	2^5	2^4	2^3	2^2	2^1	2^0
2048	1024	512	256	128	64	32	16	8	4	2	1
1	0	0	1	1	1	1	0	0	1	1	0

$2048 + 0 + 0 + 256 + 128 + 64 + 32 + 0 + 0 + 4 + 2 + 0 = 2534$

Therefore, $(2534)_{10} = (100111100110)_2$

16.4.1 Application of Binary Numbers in Computers

The computer works with binary numbers, and hence all numbers, letters, symbols, etc., are to be converted into binary equivalents. Computers are made using very-large-scale integrated digital circuits. The main circuit element being transistors which work in two modes, i.e., on or off, i.e., conducting or non-conducting. These transistors, in fact, work like a switch. The term binary or bistable means two states of operation either conducting or non-conducting giving rise to voltage level high or voltage level low, i.e., either 1 or 0. The computer works on the basis of binary logic circuits made of logic gates. Various logic gates are discussed in the following section.

16.5 ► Logic Gates

A logic gate is a simple device used to make digital integrated circuits. Diodes and transistors are used to perform switching functions in logic gates. Logic gates basically have one or more inputs and only one output. The output of the gate depends upon the way the inputs are applied and they are named as AND gate, OR gate, NOT gate, etc. Logic gates may be classified into two categories; combinational type (with no memory) and sequential type (with memory).

In digital circuits or systems there are two possible states or conditions. A switch may be closed or open, i.e., on or off. These two conditions may be represented by 1 and 0, respectively. The two logic levels, i.e., 1 and 0 represent, respectively, voltage high or voltage low. Logic systems can be positive or negative. In positive logic voltage high is represented by 1 and voltage low is represented by 0. In negative logic, voltage high is 0 and voltage low is 1. Binary number system may be used to represent the states. In a digital system, analog quantities are first converted to digital forms using the binary code. After the signal is processed, it is again converted into analog form for display.

The possible states or conditions in digital systems are expressed as 1 and 0. The closed switch or the existence of a signal may be called 1 and open switch or non-existence of a signal may be called 0. 1 and 0 are referred to as binary numbers. As mentioned earlier, in decimal system we have 0 to 9 as ten numbers. Using these numbers, we can represent any number, e.g., year 2009 is a representation of

$$2 \times 10^3 + 0 \times 10^2 + 0 \times 10^1 + 9 \times 10^0$$

Here, the base is 10 as we use 10 digits. In the binary system we only use 1 and 0 with 2 as the base. The number 82 is represented as

2^7	2^6	2^5	2^4	2^3	2^2	2^1	2^0
128	$\underline{64}$	32	$\underline{16}$	8	$\underline{4}$	2	$\underline{1}$

$$1 \times 2^6 + 0 \times 2^5 + 1 \times 2^4 + 0 \times 2^3 + 1 \times 2^2 + 0 \times 2^1 + 1 \times 2^0$$

The number 85 is represented as 1010101.

Logic circuits or logic gates provide an output of logical 1 or logical 0 with a certain logical input. The logic gates are classified into two categories viz three logic gates and two universal gates. There are also two special types of gates. These are shown below.

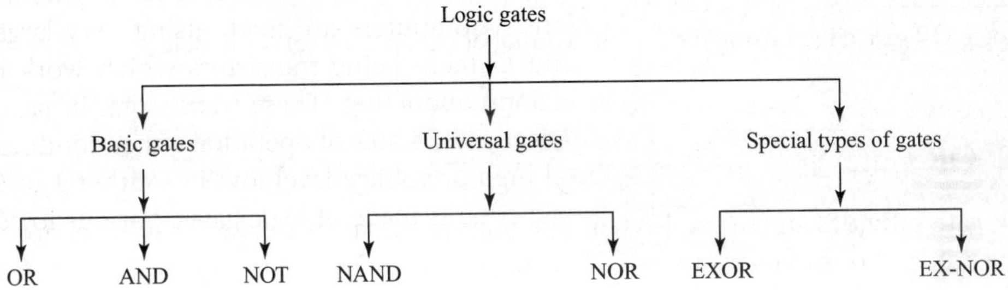

16.5.1 NOT Gate

A NOT gate works as an inverter. It has a single input and a single output. The output signal is the negative of the input signal. The output is not the same as input, the output negates the input. If input is A, the output is $\bar{A}$ (A-bar).

Figure 16.1 shows a transistor circuit. When input A is 0 V, the transistor is off because V_{BE} = 0. The output voltage is the same as V_{CC}, i.e., 5 V which is referred to as 1. If input A is 1, i.e., 5 V, the transistor goes into saturation and the output w ill be low which is referred to as 0. The logic symbol and truth table for a NOT gate have also been shown in Fig. 16.1 (b) and (c), respectively.

Thus, the NOT gate is a logic gate that gives an output that is opposite to the state of its input.

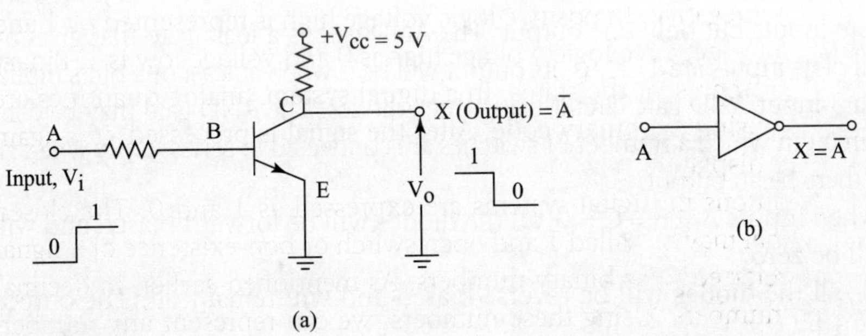

Input, A	Output, X
0	1
1	0

(c)

FIGURE 16.1

(a) Circuit for a positive logic NOT gate; (b) symbol of a NOT gate; (c) truth table of a NOT circuit

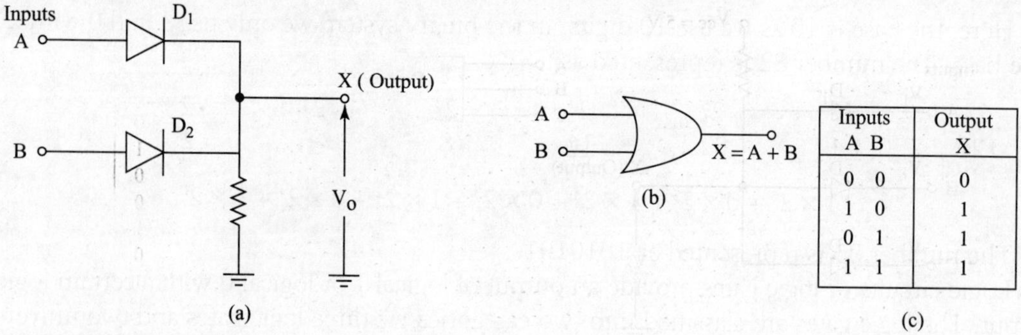

FIGURE 16.2

(a) Logic OR gate circuit using diodes; (b) symbol of an OR gate; (c) truth table

16.5.2 OR Gate

The OR gate provides an output 1 when one of its inputs is in 1 state. That is, its output is 1 whenever at least one of its inputs is 1. The output of an OR gate is 0 only when all its inputs are 0. The circuit using diodes for such a logic has been shown in Fig. 16.2 (a). The standard symbol for an OR gate and its truth table have also been shown along with the figure.

Output will occur when there is input in any one of the inputs A or B or at both. If A = B = 0, then the two diodes D_1 and D_2 will be reverse biased, and hence X output, X will be zero. If A = 1 and B = 0 or if A = 0 and B = 1, i.e., when any one of the inputs is high, the output, X will be 1. If A = 1 and B = 1, i.e., when both the diodes output X will be 1. If A = 1 and B = i.e., when will be in forward biased Condition, the output X will be 1.

Although two diodes have been used to show an OR gate, in actual practice OR function is obtained using a transistor–transistor logic (TTL). The 2-input OR gates is available in the TTL 74 xx series. 7432 is a 2-input TTL OR gate integrated circuit. The IC chip has four OR gates.

16.5.3 AND Gate

This gate has two or more inputs but only one output. The AND gate is a logic gate that gives an output of 1 only when all of its inputs are 1. Thus, its output will be 0 when at least one of its inputs is 0. Figure 16.3 shows a 3-input AND gate circuit using three diodes. The AND gate may be considered equivalent to a circuit in which a number of switches are connected in series. Only when all the switches are on, will there be an output.

As in Fig. 16.3 (a), when inputs A = B = C = 0, all the diodes will be forward biased and will conduct, the output will be zero.

When A = B = C = 1, all the diodes will be reverse biased and will remain off. The output will be high, i.e., equal to 1.

When any of the inputs is high, the corresponding diode will conduct, and thus, output will be zero.

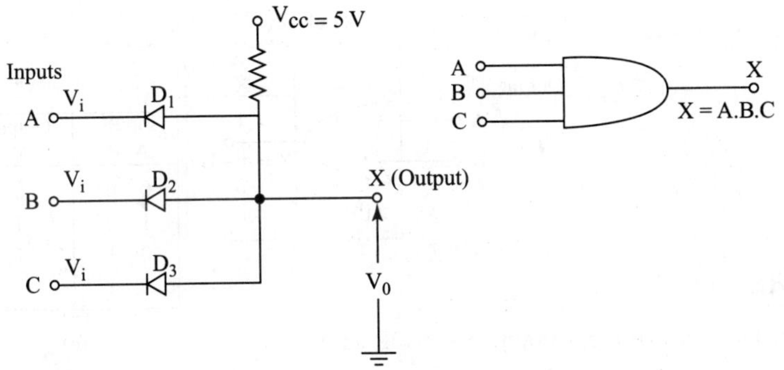

A	B	C	X
0	0	0	0
0	0	1	0
0	1	0	0
1	0	0	0
1	0	1	0
1	1	0	0
1	1	1	1

FIGURE 16.3

(a) Logic AND gate circuit; (b) symbol; (c) truth table

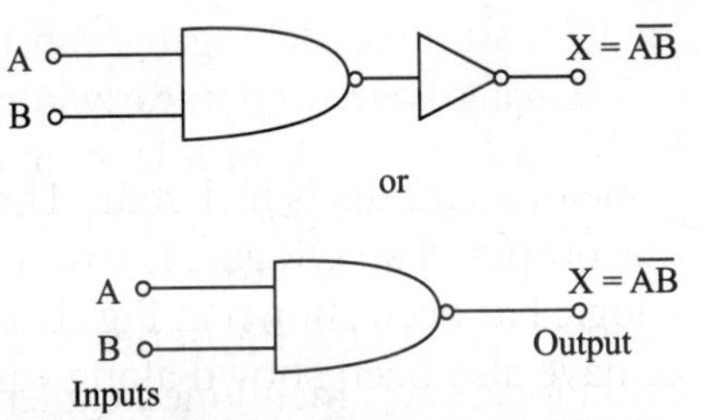

Truth table

Inputs		Output
A	B	X
0	0	1
0	1	1
1	0	1
1	1	0

FIGURE 16.4

Standard symbol and truth table of a NAND gate

As shown in figure, X = 1 only when A = B = C = 1. The 2-input AND gate is available in IC 7408.

There are such four AND gates in the IC chip.

16.5.4 NAND Gate

The NAND gate is an AND gate with a NOT gate at its end. For the same combination of inputs, the output of a NAND gate will be opposite to that of an AND gate. The combination of AND and NOT gate can be made using a diode transistor logic circuit with minor modification. The symbol of a 2-input NAND gate along with its truth table have been shown in Fig. 16.4.

16.5.5 NOR Gate

The NOR is an OR gate with a NOT gate at its end. For the same combination of inputs, the output of a NOR gate will be opposite to that of the output of a NOR gate. The symbolic representation and truth table of a NOR gate have been shown in Fig. 16.5.

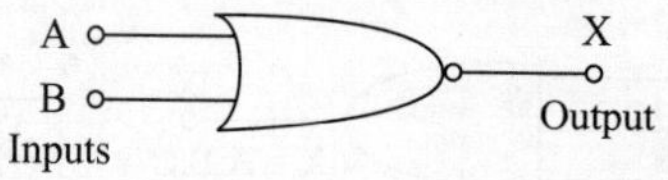

Truth table

Inputs A B	Output X
0 0	1
0 1	0
1 0	0
1 1	0

FIGURE 16.5

Standard symbol and truth table of a NOR gate

A truth table describes the behaviour of a logic gate. It shows the value of the output for every possible combination of inputs. The simplest form of electronic logic for AND and OR gates can be built using diodes. But a NOT gate cannot be made using diodes. To build a complete logic system a resistor–transistor logic, or RTL are used. RTL gates can be cascaded indefinitely to produce complex logic function. RTL gates were used in early integrated circuits. For achieving higher speed of functioning, diode–transistor logic or DTL were used. To reduce the space required, transistor–transistor logic (TTL) was created in the fabrication of ICs. Complementary metal oxide semiconductor (CMOS) logic has been replacing TTL logic to reduce size and power consumption. For small-scale logic, TTL 7400 series and CMOS 4000 series of ICs are used. Gradually, the fixed function logic gates mentioned earlier are being replaced by programmable logic devices, which permit a large number of fixed function logic gates packed into an integrated circuit.

16.6 ► BOOLEAN ALGEBRA

In 1854, George Boole developed a new form of algebra which has come to be known as Boolean algebra. Boolean algebra is being applied for the solution of logical problems.

The operation of logic gates in relation to one another may be represented and analysed using a branch of mathematics called Boolean algebra. Expressions used in Boolean algebra are called Boolean expressions. In Boolean algebra only capital letters like A, B, C, D, etc., are used.

In Boolean algebra, the AND operation is represented by multiplication, since the result of multiplication of a combination of binary numbers 1 and 0 will be equal to 1 only if all its inputs are equal to 1. The Boolean expression for A and B is A.B.

The OR operation is represented by addition because the only way to obtain a result of an addition operation equal to 0 is to make all inputs equal to 0, which basically describes an OR operation. The Boolean expression for A or B is A + B.

NOT operation means inversion. The Boolean expression for NOT operation is A = NOT(A) = $\bar{A}$. NAND operation is an AND operation followed by a NOT operation. The NOR operation is an OR operation followed by a NOT operation. Boolean expressions for the three basic gates are given below.

16.6.1 Boolean Expressions

Boolean expressions are equivalent expressions of logic states of gates. For example, the Boolean expressions for the basic gates are

a NOT gate with input A and output C, C = NOT A, or $C = \bar{A}$

The NOT gate

The output of a NOT gate is the inverse of its input. If the input is low, the output is high, and if the input is high, the output is low. NOT gates can have only one input and one output. A NOT gate is also called an inverter. The symbolic representation along with its Boolean expression and truth table are shown below.

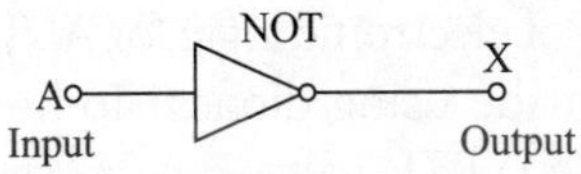

Input, A	Output, X
0	1
1	0

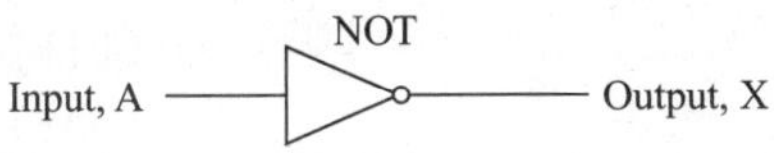

X NOT A

or X = A 0 1

1 0

The OR gate

The output of an OR gate is high if any of its inputs is high. An OR gate has two or more inputs and one output. The Boolean expression for an OR gate is written as

$$X = A \text{ or } B$$

or,

$$X = A + B$$

The symbol and truth tables are shown below.

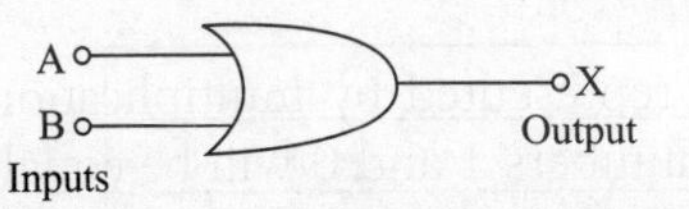

Input A	Input B	Output X
0	0	0
0	1	1
1	0	1
1	1	1

Truth table

Input A	Input B	Output X
0	0	0
0	1	1
1	0	1
1	1	1

The AND gate

The output of an AND gate is high if all inputs are high, otherwise the output is low. The AND gate has two or more inputs and one output. The Boolean expression, truth table, and symbolic representation of an AND gate are shown below.

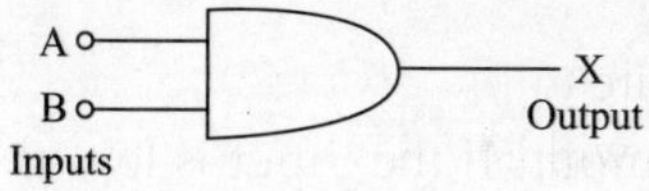

Truth table

Input A	Input B	Output X
0	0	0
0	1	0
1	0	0
1	1	1

X = A and B
or, X = A.B

Truth table

Input A	Input B	Output X
0	0	0
0	1	0
1	0	0
1	1	1

The EXOR gate

The EXOR gate stands for Exclusive OR gate. The EXOR gate is a logic gate that gives an output 1 when only one of its inputs is 1. The symbol and truth table of an EXOR gate have been shown in Fig. 16.6 .

Symbol

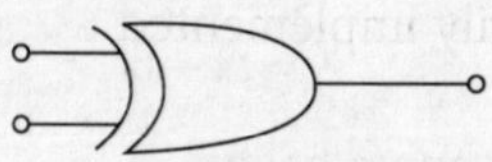

Truth table

Inputs		Output
A	B	X
0	0	0
0	1	1
1	0	1
1	1	0

FIGURE 16.6

Standard symbol and truth table of an EXOR gate

Note the difference between an OR gate and EXOR gate. In an OR gate. The output is 1 when any one of the inputs is 1 or both the inputs are 1. In an EXOR gate the output is 1 when any one of its inputs is 1. If both the inputs are 1 the output is 0. That is why an OR gate as such is also called an inclusive OR gate whereas EXOR gate is an exclusive OR gate.

EX-NOR gate

The operation of an EX-NOR gate is just opposite to the operation of an EX OR gate. Inputs are two or more and output is one. The output will be high for similar types of inputs only. The symbol and truth table have been shown.

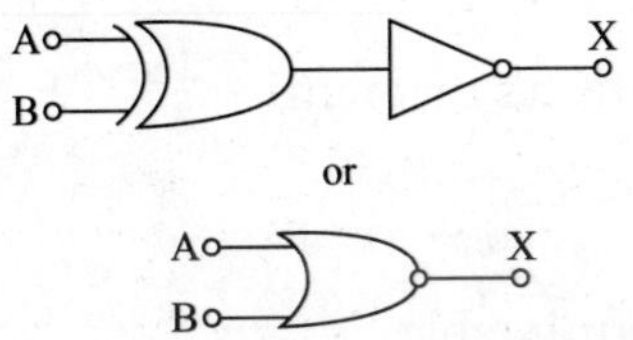

Inputs		Output
A	B	X
0	0	1
0	1	0
1	0	0
1	1	1

Inputs		Output
A	B	X
0	0	1
0	1	0
1	0	0
1	1	1

Thus, we have seen seven types of gates. Three basic gates are NOT, OR, and AND gates and the combinational gates are NAND, NOR, EXOR and EX-NOR gates.

NAND and NOR gates are called the universal gates as all other gates can be created from a suitable network of just NAND or just NOR gates.

A logic gate performs a logical operation on one or more logic inputs and produces a single logic output. Logic gates are primarily implemented using resistors, diodes, and transistors.

16.7 ► De Morgan's Theorem

The theorem states how an AND operation can be converted into an OR operation, as long as a NOT operation is available. The theorem is usually expressed in two equations as

$$\overline{A+B} = \overline{A}\cdot\overline{B}$$

$$\overline{A\cdot B} = \overline{A}+\overline{B}$$

and

De Morgan's theorem has a practical use in digital electronics. Any logic circuit may be implemented by the basic gates described earlier. A designer may eliminate the use of more ICs by substituting gates with the equivalent combination of other gates wherever possible. De Morgan's theorem basically implies that any Boolean operation may be simulated with NAND or NOR gates.

16.8 ► Combinational Circuits

Combinational logic circuits are constructed by interconnecting different logic gates. A combinational circuit with its truth table has been shown in Fig. 16.7. The same circuit can also be implemented using NAND gates only as shown in Fig. 16.7 (b).

The logic gates and related Boolean equations are shown in tabular form in Table 16.1.

Output $S = P \cdot \overline{\overline{Q \cdot R}}$

$= \overline{P} + Q \cdot P$ (according to De Morgan's theorem)

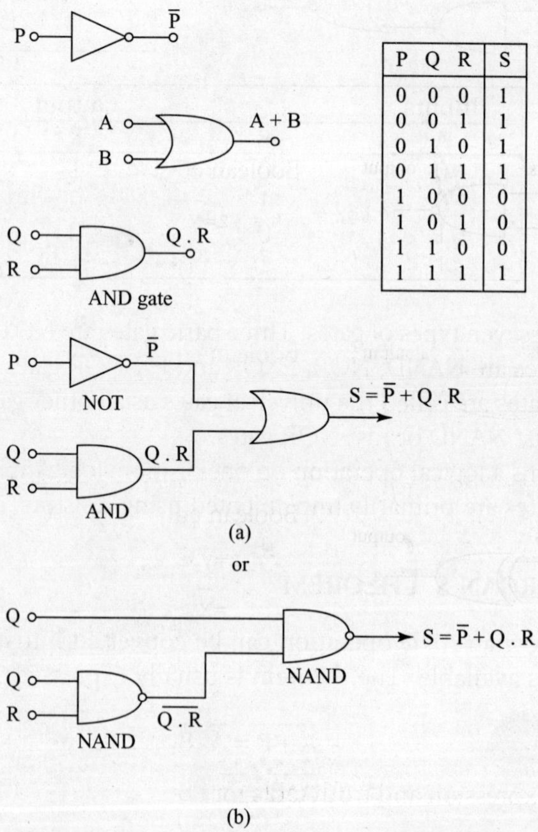

P	Q	R	S
0	0	0	1
0	0	1	1
0	1	0	1
0	1	1	1
1	0	0	0
1	0	1	0
1	1	0	0
1	1	1	1

FIGURE 16.7

(a) Truth table for a combinational circuit;
(b) NAND gate implementation of the same logic

TABLE 16–1 Logic Gates and Their Truth Tables

Gate name	Gate symbol	Boolean equation	Truth table
NOT	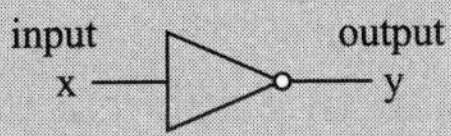	Boolean equation: $y=\bar{x}$	input x | output y 0 | 1 1 | 0
OR	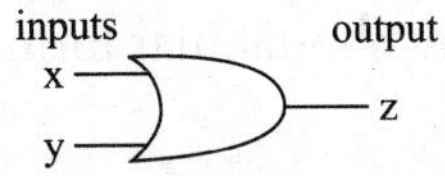	Boolean equation: $z=x+y$	input x y | output z 0 0 | 0 0 1 | 1 1 0 | 1 1 1 | 1
AND		Boolean equation: $z=x.y$	input x y | output z 0 0 | 0 0 1 | 0 1 0 | 0 1 1 | 1
NAND	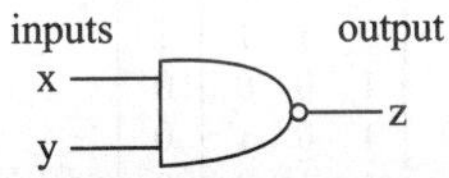	Boolean equation: $z=\overline{x.y}$	input x y | output z 0 0 | 1 0 1 | 1 1 0 | 1 1 1 | 0
NOR	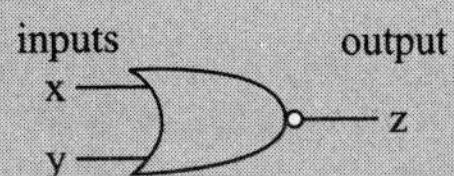	Boolean equation: $z=\overline{x+y}$	input x y | output z 0 0 | 1 0 1 | 0 1 0 | 0 1 1 | 0
EXOR	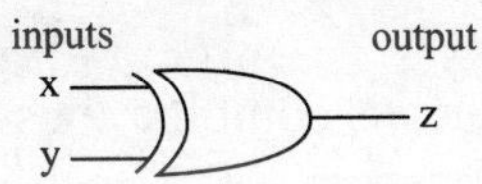	Boolean equation: $z=y \oplus y$ $=\bar{x}y+x\bar{y}$	input x y | output z 0 0 | 0 0 1 | 1 1 0 | 1 1 1 | 0

EXAMPLE 16.11

Write the Boolean expression and truth table for the logic gate circuit shown in Fig. 16.8.

SOLUTION

The Boolean expression is developed as shown in Fig. 16.9.

The Boolean expression is

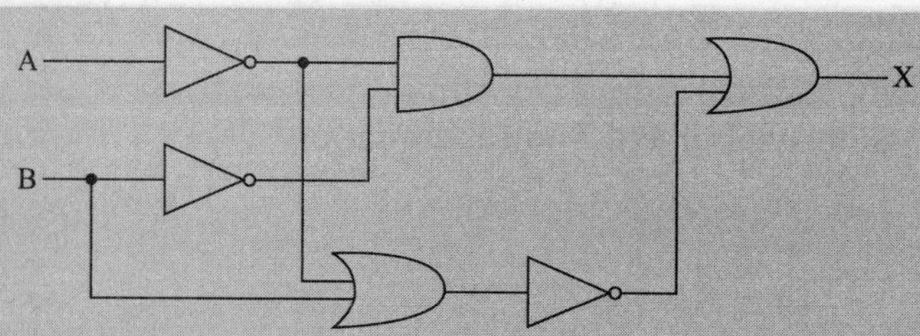

FIGURE 16.8

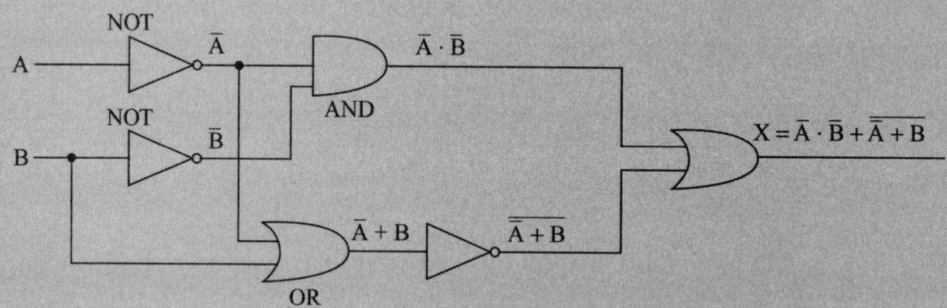

FIGURE 16.9

$$X = \overline{A}.\overline{B} + \overline{\overline{A}+B}$$

$$= Y + Z$$

where $$Y = \overline{A}.\overline{B}$$

and $$Z = \overline{\overline{A}+B}$$

Now let us write the truth table. We first write the various combinations of 0 and 1 for the two inputs, A and B. There are $2^N = 2^2 = 4$ combinations. For each row of inputs, we will calculate Y and Z, respectively. Then we would add Y and Z, to get X.

Truth table

A	B	Y	Z	X
0	0	1	0	1
0	1	0	0	0
1	0	0	1	1
1	1	0	0	0

EXAMPLE 16.12

For the logic circuit shown in Fig. 16.10, write the Boolean expression and construct the truth table.

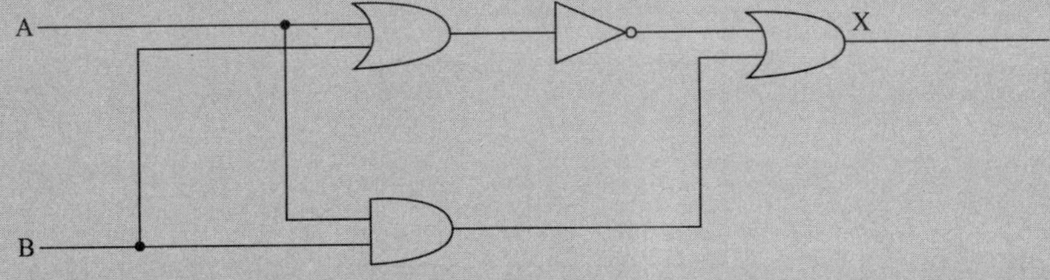

FIGURE 16.10

SOLUTION

The Boolean expression is developed from Fig. 16.11 as

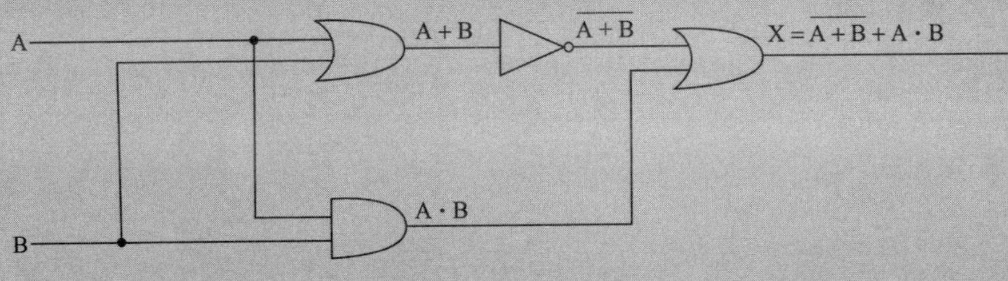

FIGURE 16.11

The Boolean expression is

$$X = \overline{A + B} + A.B$$
$$= Y + Z$$

where $Y = \overline{A + B}$ and $Z = A.B$

For constructing the truth table, we will first write the various combinations of the inputs at A and B in the form of binary numbers (the total number of possible combinations being Z^N where N is the number of inputs). We then calculate Y and Z, and then X is calculated.

Truth table

A	B	Y	Z	X = Y + Z
0	0	1	0	1
0	1	0	0	0
1	0	0	0	0
1	1	0	1	1

Boolean expression for any logic function can be obtained from the truth table. This is done by taking the sum of all terms which correspond to all those combinations of inputs for which the output attains a high value. Let us consider the truth table of the above example.

A	B	X
0	0	1
0	1	0
1	0	0
1	1	1

The first row and fourth row have to be considered where output is high, i.e., 1. Inputs A and B are to be considered 1. The complement $\bar{A}$ and $\bar{B}$ therefore 0.
For row 1

$$Y = \overline{A + B}$$

For row 4

$$Z = A.B$$

The Boolean expression for the output is

$$X = Y + Z = \overline{A + B} + A.B$$

EXAMPLE 16.13

The truth table for three variable inputs and single output logic circuit is shown below. Note that for three inputs the combinations are $2^3 = 8$. Construct the logic circuit using AND and OR gates.

A	B	C	X
0	0	0	1
0	0	1	0
0	1	0	1
0	1	1	0
1	0	0	0
1	0	1	1
1	1	0	0
1	1	1	0

SOLUTION

From observing the truth table we find that output is high, i.e., 1 for the inputs at rows 1, 3, and 6. Inputs at A, B, and C is 1 and complement $\bar{A}$, $\bar{B}$ $\bar{C}$ is 0.

By writing the sum of all terms for which the output attains a high

$$X = \overline{ABC} + \bar{A}B\bar{C} + A\bar{B}C$$

Using the above Boolean expression, the logic circuit is drawn as in Fig. 16.12.

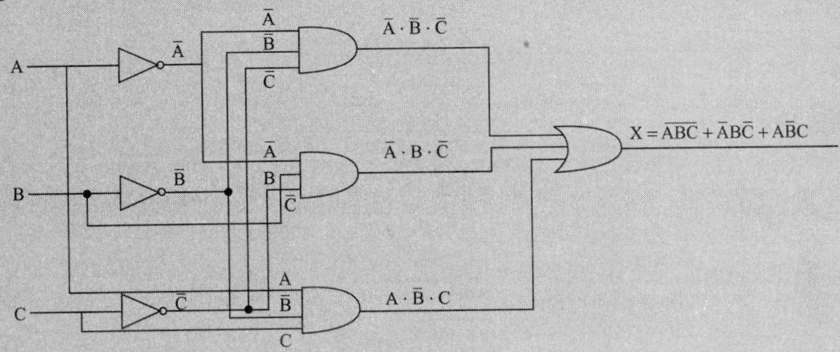

FIGURE 16.12

EXAMPLE 16.14

Draw the logic circuit for the Boolean expression expressed in two forms. Also write the truth table

$$Y = \bar{A}\,\bar{C} + B\,\bar{C} \quad \text{(i)}$$

and

$$Y = \bar{C}\,(\bar{A} + B) \quad \text{(ii)}$$

SOLUTION

for form (i), the logic circuit is drawn as shown in Fig. 16.13.

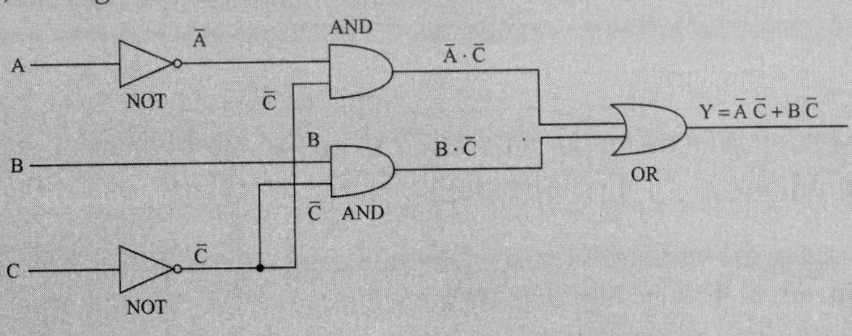

FIGURE 16.13

for form (ii), the logic circiut is drawn as shown in Fig. 16.14.

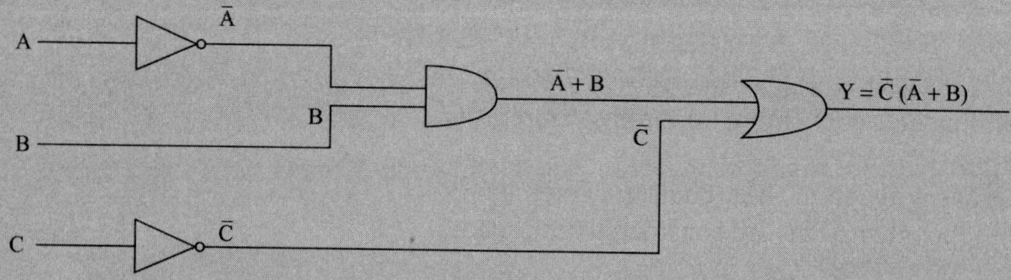

FIGURE 16.14

The truth table for

$$Y = \overline{A}\,\overline{C} + B\overline{C}$$

$$= P + Q$$

is shown as

A	B	C	P	Q	Y
0	0	0	1	0	1
0	0	1	1	0	1
0	1	0	0	1	1
1	0	0	0	0	0
1	0	1	0	0	0
1	1	0	0	1	1
0	1	1	0	0	0
1	1	1	0	0	0

16.9 ► Simplification of Boolean Expressions Using De Morgan's Theorem

De Morgan's theorem are very useful in simplifying expressions in which a product or a sum of variables is inverted. The De Morgan's theorem is reproduced as two forms

(i) Complement of a sum is equal to the product of complements, e.g.,

$$\overline{A + B + C} = \overline{A}.\overline{B}.\overline{C}$$

(ii) Complement of a product is equal to sum of complements, e.g.,

$$\overline{A.B.C} = \overline{A} + \overline{B} + \overline{C}$$

We will include some more important theorems which help in the simplification of Boolean logic expression.

Redundancy theorem: In a Boolean expression, if the same factor is present in more than one term, the other terms are considered redundant, e.g.,

$$A + AB = A$$

(here term A is present in both the terms, and hence, the other term is neglected)

$$AB + ABC + A\overline{B} = AB + A\overline{B}$$
$$= A$$

AC + ABC + ACD = AC, (here the factor AC is present in all the terms, and hence, the other terms are neglected)

Consensus theorem: If in a Boolean expression, one term contains a variable and the second term contains the complement of this variable, then in the expression a third term can be added without effecting the value of the expression, e.g., if $X = AC + B\overline{C}$ where $\overline{C}$ is the complement of C. We can add another term in the expression as

$$X = AC + B\overline{C} + AB$$

To prove, we write

$$X = AC + B\overline{C} + AB(C + \overline{C})$$
$$= AC + B\overline{C} + ABC + AB\overline{C}$$
$$= AC(B + 1) + B\overline{C}(A + 1)$$
$$= AC + B\overline{C}$$

16.10 ► Universal Gates

As mentioned earlier NAND and NOR gates are known as universal gates because they can be used to build any logic circuit. The logic function of NOT, OR, and AND can be performed using either NAND or NOR gates. First we shall see how a NAND gate, functions as NOT, AND, and OR gates, respectively. A NAND gate with its two inputs tied together is a single input NAND gate.

16.10.1 Use of NAND Gate to Form the Three Basic Gates

Using NAND gate, the three basic gates, i.e., NOT, AND, and OR gates are formed as shown in Fig. 16.15.

(i) NAND gate as NOT gate

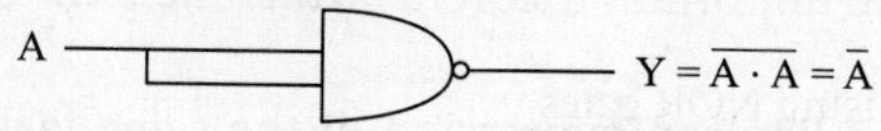

(ii) NAND gate as AND gate

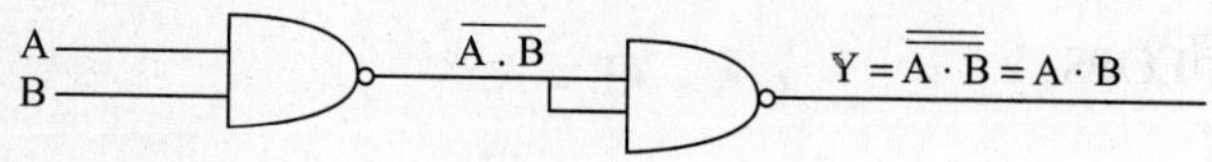

(iii) NAND gate as OR gate

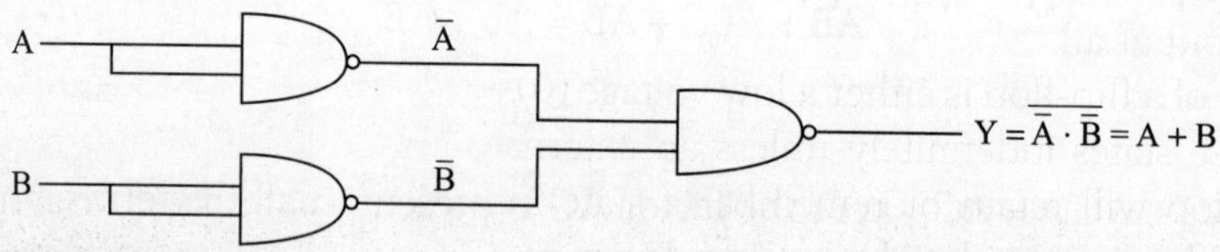

FIGURE 16.15

Three basic gates using NAND gates

16.10.2 Use of NOR Gate to Form the Three Basic Gates

Now we shall see the use of a NOR gate to form the three basic gates as shown in Fig. 16.16.

(i) NOR gate as NOT gate

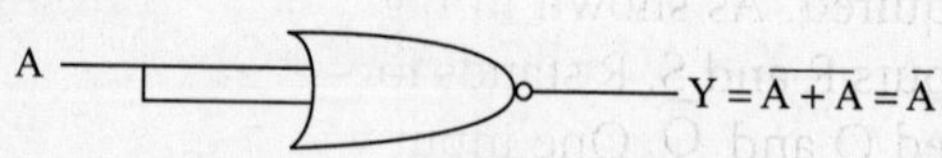

If all the inputs of the NOR gate are connected together, we obtain a NOT gate as shown above.

(ii) NOR gate as OR gate

The OR operation is obtained by a NOR gate followed by a single input NOR gate as shown.

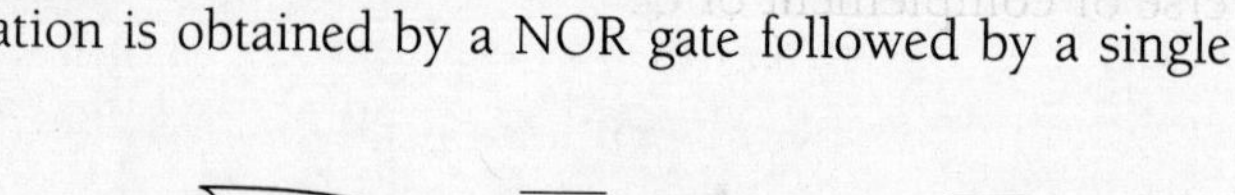

(iii) NOR gate as AND gate

The AND operation by NOR gates is achieved as shown. The inputs at A and B are inverted by 1-input NOR gates. The inverted inputs are given to a NOR gate.

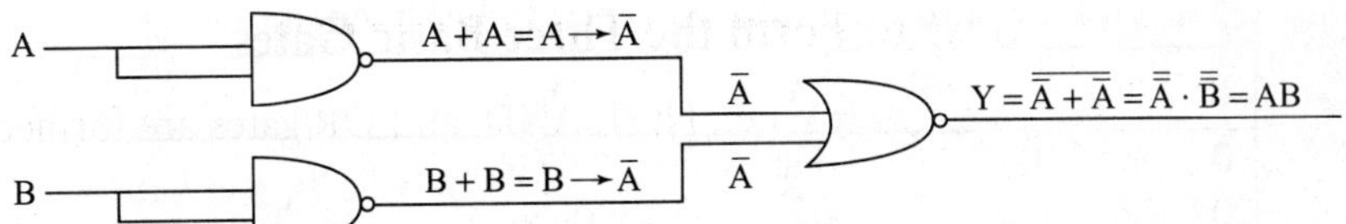

FIGURE 16.16

Three basic gates using NOR gates

16.11 ► Flip-Flops

In the previous section we have seen logic gates of combinational type where the output at a given time depends upon the inputs at that time. The previous records of inputs does not effect the output at all.

The output of a flip-flop is either a low voltage (0) or a high voltage (1). The output remains in one of these states indefinitely unless an external trigger is applied to change that state. Thus, a flip-flop will retain or remember a state indefinitely after the removal of the input trigger. A flip-flop is considered a storing device of 1-bit memory. Since a flip-flop will have two stable states (0 or 1), it is also referred to as ***bistable multivibrator***.

In most flip-flop circuits the change of state of the flip-flop is of a definite rate, and hence, they are of clocked type.

16.11.1 RS Flip-Flop

The RS flip-flop is used to temporarily hold or store information until it is required. A single RS circuit will store one binary digit, i.e., either 1 or 0. For storing a six digit binary number, six RS flip-flops will be required. As shown in Fig. 16.17, a RS flip flop which is a one-bit memory device has two inputs R and S. R stands for 'RESET' and S stands for 'SET'. The two output terminals are marked Q and $\bar{Q}$ One input will set the device and another will reset the device back to its original state. The output will either be at logic level 1 or 0 depending upon the set/reset condition. The simplest way to make a 1-bit set/reset RS flip-flop is to connect together a pair of NAND gates which are cross-coupled as shown in Fig. 16.17. The output $\bar{Q}$ is the inverse or complement of Q.

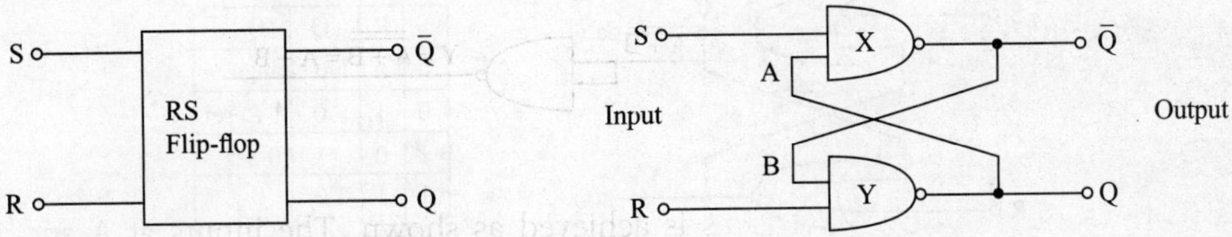

FIGURE 16.17

RS flip-flop using NAND gates

The truth table of a RS flip-flop has been shown

R	S	Q	$\overline{Q}$
0	0	Last value * * (No change)	
0	1	1	0 (Set)
1	0	0	1 (Reset)
1	1	Illegal	

Application of a positive trigger at the input S will be setting the flip-flop. The output at Q will be 1 and at $\overline{Q}$ will be 0. The application of positive trigger at R will be resetting the flip-flop. The output as $\overline{Q}$ will be 1 and at Q will be 0.

For the set state, the input at R is at logic level 0 and input at S is at logic level 1, the output of NAND gate Y has atleast one of its inputs at logic level 0, and therefore the output Q must be at logic level 1. The output Q is fed back to input A so that both the inputs to the NAND gate X are at logic level 1. The output $\overline{Q}$ must be at 0 level.

For reset state, R = 1 and S = 0. The NAND gate X will have one of its inputs at level 0, and hence, output at $\overline{Q}$ isQ 1. The flip-flop will function as follows.

(a) For R = 0, S = 0, i.e., with no input at the input terminals, the output will retain its previous state, i.e., output at Q may be 0 or 1.

(b) For R = 0, S = 1, i.e., trigger input is provided in the set input terminal, the output at Q will be 1.

(c) For R = 1, S = 0, Q = 0

(d) For R = 1, S = 1 the output is forced to become 0 and 1 simultaneously. Such input condition is called illegal or invalid forbidden input condition.

An RS flip-flop can also be created by two single input NAND gates and two 2-input NAND gates connected in a fashion as shown in Fig. 16.18.

Flip-flop can also be formed by using 2-input NOR gates as shown in Fig. 16.19.

The circuit works in a similar way as in the case of NAND gate circuit. When both the inputs are at logic level 1, an invalid condition is created.

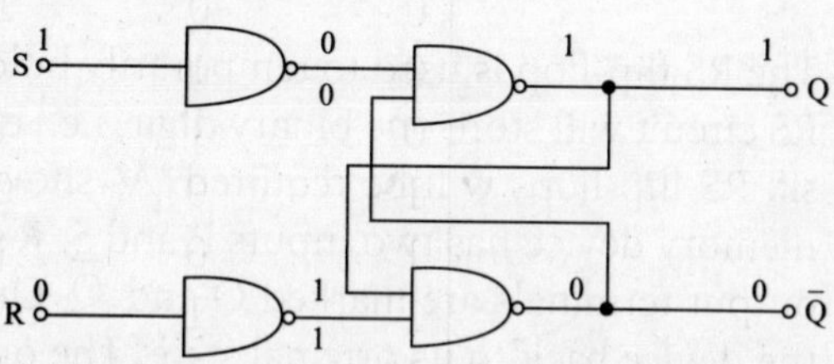

FIGURE 16.18

RS flip-flop using combination of 1-input and 2-input NAND gates

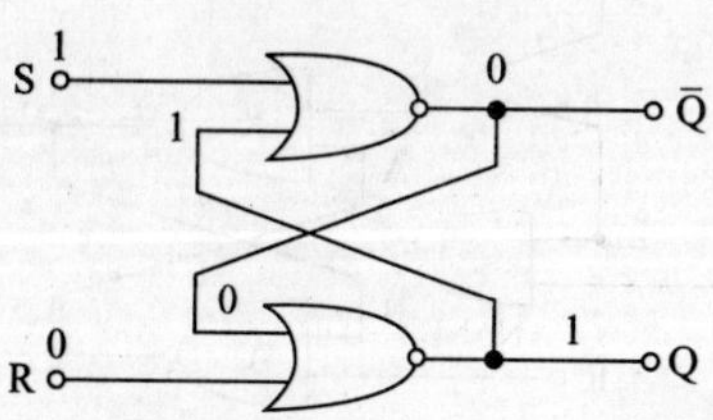

S	R	Q	$\overline{Q}$
0	0	No change	
0	1	0	1 Reset
1	0	1	0 Set
1	1	0	0 Invalid

FIGURE 16.19

Flip-flop using NOR gates

16.11.2 Gated or Clocked RS Flip-Flop

An RS flip-flop that has a clock input as shown in Fig. 16.20 is called a clocked RS flip-flop. The clock input is a square wave. When the clock is at low, the output at Q and $\overline{Q}$ will remain unchanged.

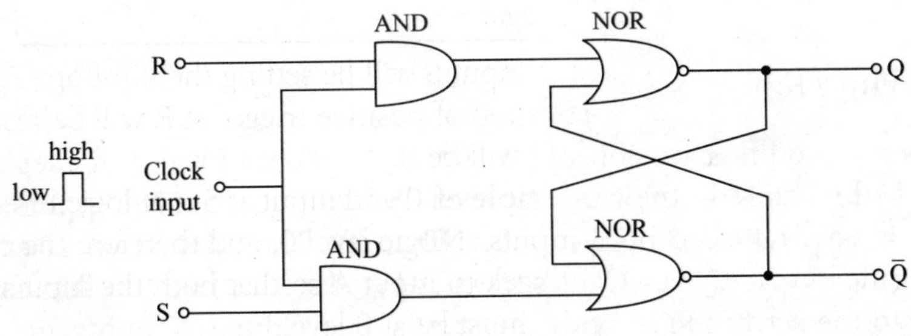

FIGURE 16.20

Clocked RS flip-flop

When the clock input is high (i.e., 1), the flop-flip will be set for R = 0 and S = 1 and will be reset for R = 1 and S = 0 as has been shown in the truth table.

Truth table for a clocked RS flip-flop

CLK	R	S	Q
0	0	0	No change
0	0	1	No change
0	1	0	No change
0	1	1	No change
1	0	0	No change
1	0	1	1 Set
1	1	0	0 Reset
1	1	1	Invalid*

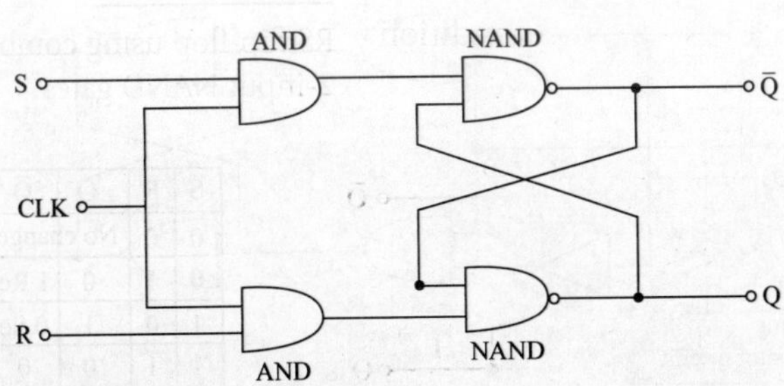

FIGURE 16.21

Clocked SR flip-flop

Clocked RS flip-flop can also be made using 2-input gates as shown in Fig. 16.21.

When the CLK input is at 0 level, the outputs of the two AND gates are also at logic level 0. The output of the Flip-flop at Q and $\overline{Q}$ will be uneffected and will be latching at the last known state. With a high-clock input (i.e., 1) the flip-flop is set and reset as has been shown in the truth table. Synchronization of the clock timing signal to the flip-flop creates what is sometimes called a clocked SR flip-flop.

16.11.3 JK Flip-Flop

JK flip-flop is the modified version of RS flip-flop with no invalid or illegal output state (see state marked invalid* in the truth table of RS flip-flop). The two inputs of RS flip-flop of Fig. 16.17 is now replaced by 3-input AND gates. The third input of each gate is the feedback from the output Q and $\overline{Q}$ as shown in Fig. 16.22. In RS flip-flop the invalid state occurred due to the activation of both the inputs. Here, the two inputs are interlocked so that they cannot be activated simultaneously.

16.11.4 D Flip-Flops

The output Q and $\overline{Q}$ of a D flip-flop change only on the positive going edge of the incoming clock pulse as shown in Fig. 16.23. If D = 1 and the positive going clock edge appears, then output Q = 1 and $\overline{Q}$ =0. When D = 0 and positive going pulse appears, then output Q = 0 and $\overline{Q}$=1 For negative going edge of the incoming clock pulse, the flip-flop is inactive don't care.

For edge-triggered flip-flops, the clock signal is applied in the form of sharp positive and negative spikes instead of in the form of square or rectangular pulse train. Such sharp

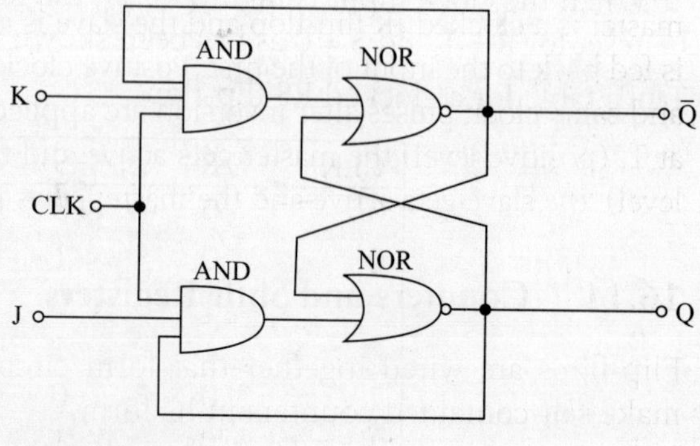

FIGURE 16.22

JK flip-flop

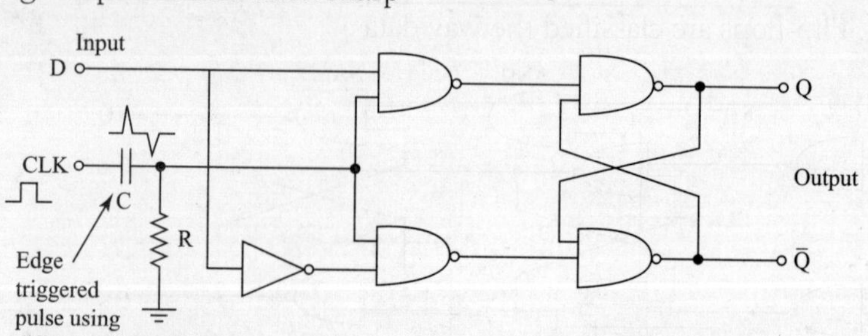

FIGURE 16.23

D flip-flop using edge-triggered NAND gates

spikes are obtained from the rectangular pulse with the help of a passive differentiator circuit as has been shown in Fig. 16.23. Edge-trigger flip-flops can be of two types, viz positive edge triggered and negative edge triggered.

16.11.5 T Flip-Flops (Toggle Flip-Flop)

These are basically JK flip-flops with both J and K terminals connected together permanently. Thus, such a flip-flop will have only one input terminal, T as has been shown in Fig. 16.24.

If T = 0, J = 0 and K = 0, the output Q and $\overline{Q}$ will remain unchanged. When T = 1, J = 1 and K = 1, the output will toggle corresponding to each and every edge of the clock signal.

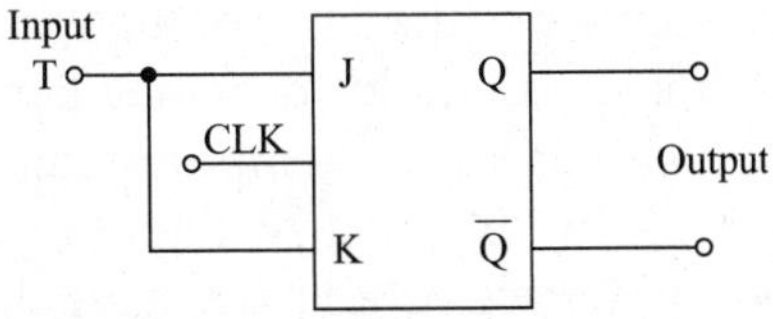

FIGURE 16.24

A JK flip-flop converted into T flip-flop

16.11.6 Master–Slave JK Flip-Flop

Master-slave JK flip-flop as shown in Fig. 16.25 is a cascade connection of two flip-flops. The master is a clocked JK flip-flop and the slave is a RS flip-flop. The output of the second flip-flop is fed back to the input of the first. Positive clock pulses are applied to the first, called the master and same clock pulses after inversion are applied to the slave Flip-flop. When the clock pulse is at 1, (positive level) the master gets active and the slave stays inactive. When clock is at 0 (low level), the slave gets active and the master stays inactive

16.11.7 Counters and Shift Registers

Flip-flops are wired together that form circuits to do counting functions. Manufacturers make self-contained counters in IC form.

A register is a group of flip-flops used to store binary numbers. The output on each flip-flop is connected to the input of the adjacent input to form a register. The connection of the flip-flops are made in such a way that each data bits get transferred or shifts one flip-flop to the left or to the right. Flip-flops are classified the way data are entered and retrieved. For example, in

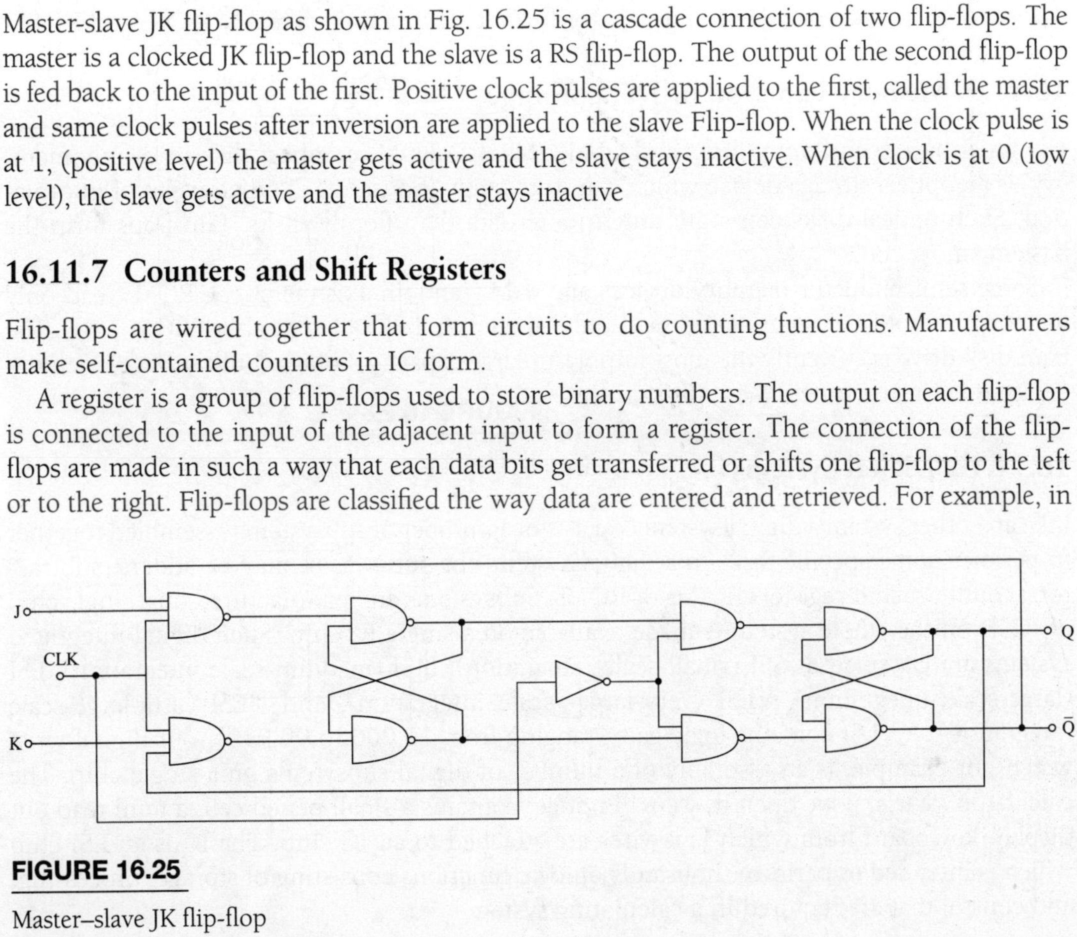

FIGURE 16.25

Master–slave JK flip-flop

serial-in serial-out registers, input data are applied one bit at a time to the first flip-flop in the form of a chain and is retrieved or read out from the last flip-flop in the form of a chain, one bit at a time. You must have seen the use of shift registers in your calculators. As we enter any digit on the key board, the numbers shift to the left on the display.

Taken, for example, what happens when we enter 65. We press 6 on the key board. 6 appears on the extreme right on the display. Next we press 5 on the key board. The number 6 gets shifted to the left creating space for number 5. This way the shift register carries out two functions, viz it holds or remembers the number 6 even after the pressure on the key board is released as it has a temporary memory, then it shifts the number to the left on the display each time we press a new digit or a character on the key board. The short time memory and shifting characteristics have made shift registers extremely useful in digital electronic systems.

16.11.8 Arithmetic Circuits

Combinational logic circuits are used to make adders and subtractors. In the central processing unit (CPU) of a computer, arithmetic functions are performed in a section called arithmetic logic unit (ALU). This section can perform functions like add, subtract, multiply, dvide, compare, complement, shift, and perform logic functions like AND, OR, and XOR.

16.11.9 Memory Function or Data Storage

You must have seen floppy disks and CD-ROMs. CD-ROM (compact disk read-only memory) is an optical storage device which can store many thousands of typed pages of information. Such optical disks can store much more data than floppy disks. Flip-flops form the basic memory cell.

Some semiconductor memory devices are RAM (random-axis memory), ROM (read-only memory), PROM (programmable read-only memory), EPROM (erasable PROM), etc. The hard disk drive is currently the most important large capacity storage (bulk storage) memory device used in computers these days.

16.11.10 Digital Systems

Like any other system a digital system consists of a number of subsystems assembled together to perform some specific task. In a digital system, the sub-system may be adders, subtractors, counters, shift registers, RAMs or ROMs. Subsystems are manufactured on a single chip of IC. Even the whole system is made available on a single IC chip. Such digital integrated systems are classified as SSI (small-scale integration), MSI (medium-scale integration), LSI (large-scale integration), VLSI (very-large- scale integration), and ULSI (utra-large-scale integration). A VLSI contains logic gates ranging from 10,000 to 99,999. Your digital wrist watch, for example, is an assembly of a number of digital subsytems on a single chip. The calculator we use, if we open it, we will notice, contains a small pencil cell, a mini read out display, key board from which few wires are attached to an IC chip. The IC is an LSI chip which is intended to perform thousands of logic functions consisting of storage, processing, and control that are required in a calculating system.

Important Terminologies and Points to Remember

1. Digital circuits used in computers and other electronic devices contain logic gates and multi vibrators. Logic gates work on number systems. The number system rules allows conversion of decimal number into binary equivalent number.
2. Computer works with binary numbers, and hence, all numbers, letters, symbols, etc., are to be converted into binary equivalents.
3. Logic gates use diodes, transistors to perform switching fractions. A switch may be close or open representing 0 and 1, respectively. The output of a gate depends upon the way the inputs are applied to a gate. For example, a NOT gate is a single input single output gate. The output signal is negative of the input signal, i.e., if input is 1 (one), the output will be 0 (zero).

   Gates are named AND gate, OR gate, NOT gate, etc.
4. Using these basic gates of AND, OR, and NOT, NAND gate and NOR gate can be built.
5. An AND gate and a NOT gate can make a NAND gate. Similarly as OR gate and a NOT gate can make a NOR gate.
6. The operation of logic gates in relation to one another may be represented and analyzed using Boolean algebra. In Boolean algebra, capital letters like A, B, C, D, etc., are used.
7. De Morgan's theorem shows how an AND operation can be converted into an OR operation, as long as a NOT operation is available. Boolean expressions can be simplified using DE Morgans theorem.
8. Logic gates are represented by their input–output symbols, the Boolean expression, and their truth tables.
9. NAND and NOR gates are known as universal gates because they can be used to build any logic circuit.
10. A flip-flop is considered a storing device of 1-bit memory. The output of a flip-flop is either low voltage meaning 0, or high voltage meaning 1. The output remains in a state indefinitely after the removal of the input trigger. A flip-flop is referred to as bistable multivibrator.
11. In most flip-flop circuits the change of stable of the flip-flop is of definite rate, and hence, they are of clocked type.
12. Using gates, different types of flip-flops can be made. For example, R S flip-flop can be formed using NAND gates.
13. Flip-flops can be connected together to form circuits to do counting functions. Counters are available in IC form.
14. In the central processing unit (CPU) of a computer, arithmetic functions are performed by arithmetic logic unit (ALU). The functions include add, subtract, multiply, divide, compare, complement, shift, and perform logic functions.

REVIEW QUESTIONS

Short Answer Type Questions

1. State the advantages of digital signals over analog signals.
2. Differentiate between analog and digital signals. What are the disadvantages of analog systems?
3. What are the different types of number systems? Explain each of them.
4. What is a binary number system. Mention its applications
5. Explain with example how binary numbers can be converted into decimal numbers.
6. Illustrate how decimal numbers can be converted into binary numbers.

7. Convert the following decimal numbers into binary numbers:
(a) 49; (b) 756; (c) 0.578; (d) 110.58;
(e) 109; (f) 18.33; (g) 212.
8. Convert the following fractional decimal numbers into binary numbers:
(a) 0.782; (b) 0.359; (c) 0.568; (d) 0.075.
9. Convert the following binary numbers into decimal numbers:
(a) 1101; (b) 1111; (c) 1001; (d) 1100101.
10. Convert the following fractional binary numbers to fractional decimal numbers:
(a) 0.1111; (b) 0.1001; (c) 0.0101.
11. Add the following binary numbers:
(a) 1011+1110; (b) 101010+111011.
12. Explain the difference between hexadecimal and octal systems.
13. Perform the following binary multiplication:
(i) $(111)_2 \times (1.2)_2$; (ii) $(10101)_2 \times (1.1)_2$;
(iii) $(10100)_2 \times (0.1)_2$.
14. State De Morgan's theorem
15. Write down truth tables for the following gates:
(a) NAND gate; (b) AND gate; (c) NOR gate.
16. Explain a logic OR gate using diodes. Also write the truth table.
17. What are universal gates? Explain with an example.
18. What ate flip-flops? Explain one of them.
19. Explain RS flip-flop. What is D flip-flop?
20. Explain JK flip-flop. What is T flip flop?
21. Draw and explain the circuit for master–slave JK flip-flop.
22. What is a digital system? Give an example of a digital system and name its subsystems.

Numerical Problems

23. Convert the binary number 1001 into decimal number, and decimal number 19 into binary number.
[**Ans** 9, 10011]
24. Convert decimal numbers 7547 and 100.625 into hexadecimal numbers.
[**Ans** $(7547)_{10} = (107\text{ B})_{16}$; $(100.625)_{10} = (64.\text{A})_{16}$]
25. Convert the following hexadecimal numbers to binary.
(a) AEO; (b) A75 F; (c) E 25.
[**Ans** (a) 101011100000; (b) 011101101111(c) 111000100101]
26. Convert the following octal into binary.
(a) 71; (b) 560; (c) 173.
[**Ans** (a) 111001; (b) 101110000; (c) 001111011]
27. Perform the following binary operations.
(a) $(1011)_2 \times (1001)_2$;
(b) $(111)_2 \times (190)_2$.
[**Ans** (a) 1100011; (b) 101010]
28. Express the decimal number 9 into (a) binary; (b) BCD; (c) octal; (d) hexadecimal.
[**Ans** (a) 1001; (b) 1001; (c) 11; (d) 9]
29. Add hexadecimal numbers 3E91 and 2F93.
[**Ans** 6E 24]
30. Solve the following conversion.
(a) $(110001)_2 = (?)_{10}$; (b) $(23.6)_{10} = (?)_2$;
(c) $(\text{BCA 3.AD})_{16} = (?)_2$.
[**Ans** (a) 49; (b) 10111.1001100; (c) 1011110010100011.10101101]

Multiple Choice Questions

1. A NOR gate is equivalent to
(a) inverters connected to the output of an AND gate
(b) an inverter connected to the inputs of a NAND gate
(c) inverters connected to the inputs of an OR gate
(d) an inverter connected to the output of an OR gate.

2. The sum of binary 10011 and 0111 is
   (a) 01011 (b) 11010
   (c) 11101 (d) 10111.
3. Which of the following statements is true according to De Morgan's theorems?
   (a) $\overline{A+B}=\overline{A}\cdot\overline{B}$ (b) $\overline{A}+\overline{B}=\overline{A}\cdot\overline{B}$
   (c) $A+B=\overline{A}\cdot\overline{B}$ (d) $\overline{A\cdot B}=\overline{A}\cdot\overline{B}$.
4. Connecting of inverters at all the inputs of an AND gate produces a
   (a) NAND gate (b) OR gate
   (c) NOR gate (d) XOR gate.
5. Which of the following gates is represented by the Boolean expression: A + B + C + D = Y
   (a) 4-input AND gate
   (b) 4-input OR gate
   (c) 4-input NAND gate
   (d) 4-input NOR gate.
6. Which of the following is the Boolean expression for a 3-input NAND gate?
   (a) $\overline{A\cdot B\cdot C}=Y$ (b) $\overline{A+B+C}=Y$
   (c) $\overline{A}\cdot\overline{B}\cdot\overline{C}=Y$ (d) $A\cdot B\cdot C=Y$.
7. Which flip-flop has only one data input?
   (a) RS flip-flop
   (b) D flip-flop
   (c) JK flip-flop
   (d) Triggering flip-flop.
8. Which of the following can be called an universal flip flop?
   (a) JK flip-flop (b) D flip-flop
   (c) RS flip-flop (d) Clocked RS flip-flop.
9. In a JK flip-flop, repeated clock pulses causes the output to turn on-off-on-off—when
   (a) both inputs J and K are at 0
   (b) J is at 0 and K is at 1
   (c) K is at 0 and J at 1
   (d) both inputs J and K are at 1.
10. The master–slave JK flip-flop is an example of
    (a) level-triggered device
    (b) positive edge-triggered device
    (c) negative edge-triggered device
    (d) pulse-triggered device.
11. An astable multivibrator
    (a) generates a continuous flow of pulses
    (b) is a flip-flop
    (c) generates a single short pulse
    (d) always is in one of the two stable states.
12. A monstable multivibrator is also called a
    (a) flip-flop
    (b) clock
    (c) one shot multivibrator
    (d) free running multivibrator.

Answers to Multiple Choice Questions

1. (d)	2. (b)	3. (a)	4. (c)	5. (b)	6. (a)
7. (b)	8. (a)	9. (d)	10. (d)	11. (a)	12. (c)

CHAPTER 17

Integrated Circuits

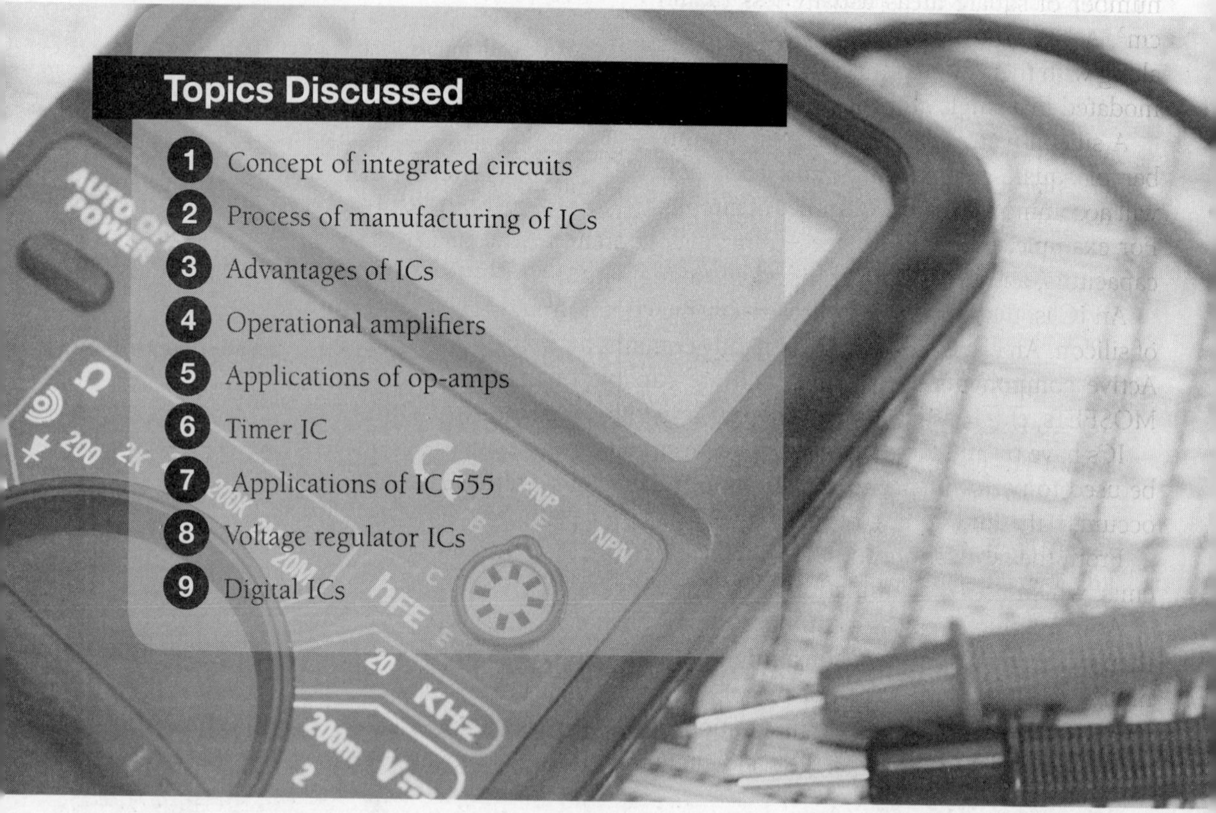

Topics Discussed

1. Concept of integrated circuits
2. Process of manufacturing of ICs
3. Advantages of ICs
4. Operational amplifiers
5. Applications of op-amps
6. Timer IC
7. Applications of IC 555
8. Voltage regulator ICs
9. Digital ICs

17.1 ► INTRODUCTION

Electronic components and circuits have undergone tremendous changes in terms of reduction of their size, i.e., miniaturization, reduction of cost, improvement of reliability, reduction in power consumption, ease of replacement, etc. Going by history, the initial effort was to reduce the size of discrete components. This was followed by the development of printed circuit boards (PCBs) which eliminated connections of components by wires. The components were placed on PCBs to develop the circuits. In early 1960s micro electronics was introduced, which drastically reduced the size of electronic equipment and gadgets. This was possible due to the introduction of integrated circuits called ICs. An integrated circuit is a combination of components like resistors, capacitors, diodes, transistors, etc., and their interconnections, fabricated into an extremely tiny single chip of silicon. A chip is an extremely small part of silicon wafer on which an IC is fabricated or grown. Fig. 17.1 shows a silicon wafer of thickness less than 1mm. The wafer contains a large number of square areas usually less than 1 cm^2. A large number of circuit components along with their interconnections are accommodated in a single chip.

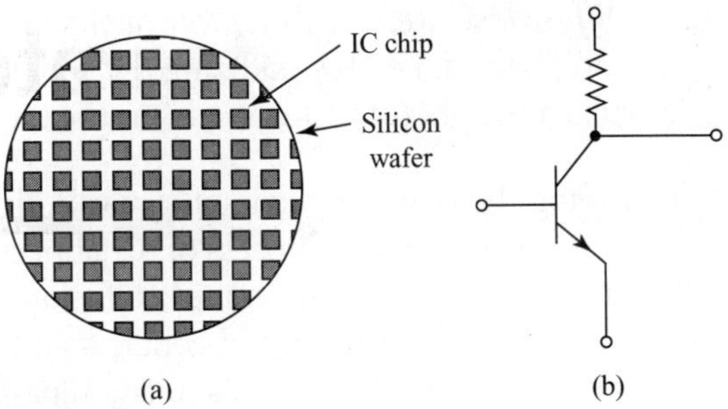

FIGURE 17.1

(a) A silicon wafer having a large number of square-shaped chips made on it; (b) a sample of circuits to be constructed on a single chip

A silicon wafer will contain a large number of square-shaped chips and each chip will accommodate a large number of circuits. For example, one chip of size 4 mm^2 may contain 40 transistors, 35 resistors, a number of capacitors, and other such components along with their interconnections.

An IC is, therefore, a complete electronic circuit in miniature form grown into a single chip of silicon. An electronic circuit generally contains a number of active and passive components. Active components are those which are capable of producing gains like transistors, FETs, MOSFETs, etc. Passive components are resistors, inductors, capacitors, etc.

ICs have by and large replaced discrete circuit design using individual components. ICs can be used to perform specific functions and they can be replaced as a whole in case any defect occurs in the functioning of the IC.

From the fabrication point of view ICs are classified into monolithic circuits or hybrid circuits, which uses a combination of different processes.

A monolithic circuit is a complete circuit having active and passive circuit components and their interconnections made on a single chip (the word monolithic is derived from a Greek word mono = single, lithic = stone or piece). Integrated circuit technology has developed small scale integration to very large scale integration. Small scale integration (SSI) has less than 12 gates on a single chip while very large scale integration (VLSI) has over 1000 gates in a single chip.

17.2 ► Fabrication of Monolithic ICs

The fabrication of ICs involves a very special and sophisticated technology. Here we will briefly mention only the steps involved. The steps are as follows:

(i) *Wafer preparation:* Wafer is a very thin surface of p-type silicon which provides the base or substrate on which the circuit elements are grown or developed. Wafers are made by slicing a p-type round silicon material of certain diameter. These wafers are polished to a mirror finish surface.

(ii) *Epitaxial growth:* On the p-type substrate an n-type layer is grown by placing the wafer in a furnace in an atmosphere of phosphorous gas at 1200°C. Thus, an epitaxial layer of n-type material is formed on the p-type wafer.

(iii) *Diffused isolation:* A thin layer of silicon dioxide is formed on the n-type layer by oxidation method, i.e., by oxidizing the wafer in dry oxygen. Next, a thin coating of a chemical, called photoresist is made on the SiO_2 layer.

Following these, use is made of a mask, ultraviolet exposure, etching, scrubbing and diffusion. Thus, the monolithic process of IC fabrication involves wafer preparation, epitaxial growth, diffused isolation, base and emitter diffusion, etching, metallization (making of contacts and interconnections), checking of circuitry, separating the wafer into individual chips, mounting, packaging, sealing, and testing. Fig. 17.2 shows some of the IC packages available. Integrated circuit components like resistors, capacitors, diodes, transistors, etc., are made from the monolithic structure. For example, resistors are made using the resistivity of the diffused areas, diodes are made utilizing the p–n junctions available in the diffused structure, both n–p–n and p–n–p transistors are produced on the substrate by the diffusion process.

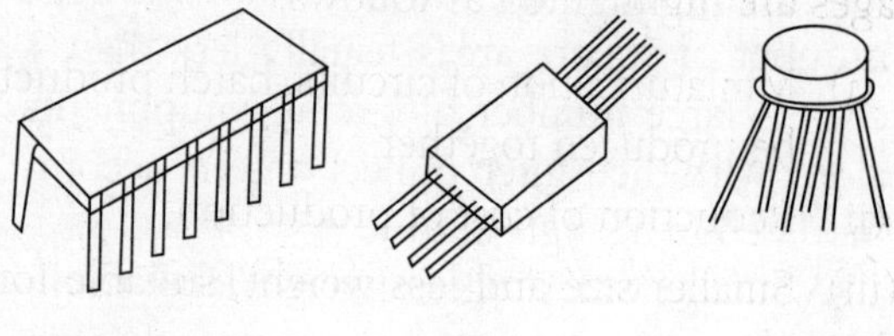

FIGURE 17.2

IC packages of various types

17.3 ► Hybrid Integrated Circuits

Hybrid or multichip integrated circuits are made by interconnecting a number of monolithic ICs. Hybrid ICs are also made by using a combination of monolithic technique and their filming technique. Here, the active components of the desired circuit is first made using the monolithic technique which has a layer of SiO_2 as its cover. Thin film technique is then employed to form the passive components on the SiO_2 surface. Connections are then made from the film to the monolithic structure.

17.4 ► Linear and Digital ICs

Integrated circuits are classified into linear and digital circuits according to their mode of operation. In linear ICs, the input–output relation is linear. The output is directly proportional to the input. Linear ICs are made for amplifiers, oscillators, filters, multipliers, modulators, voltage regulators, etc. In linear ICs the electrical signals are analogous to the physical quantities, and hence, they are often referred to as analog circuits.

The major application of ICs is in the field of computers and logic circuits. Digital circuits are concerned with two levels of voltage, i.e., high or low which in turn can make a switch closed or open. The two states are referred to as 1 and 0. A closed switch is referred to as a binary 1, and an open switch or the absence of a signal is referred to as a binary 0.

Most commonly used linear IC is found in operational amplifiers (op-amps). With some suitable external components like resistors and capacitors, op-amps are used in amplifiers as integrators and differentiators, in filters and other such applications. A general purpose op-amp is numbered IC 741.

Digital ICs find applications in switching circuits, flip-flops, counters, registers, microprocessors, clock chips, calculator chips, memory chips, multivibrators, etc.

Advantages of IC technology over discrete circuits

Integrated circuit technology has brought in a number of advantages over the fabrication of electronic circuits using individual circuit components, i.e., discrete circuits. The advantages are highlighted as follows:

(i) Miniaturization of circuits, batch production, i.e., a large number of identical circuits can be produced together.

(ii) Reduction of cost of production.

(iii) Smaller size and less weight, suitable for space applications like in aircraft, space vehicles, etc.

(iv) Maintenance is easy, the whole of the IC is replaced in case of any fault. There is no need for repair of components.

(v) Less likely to be faulty. Since ICs are manufactured in monolithic structure, there is very little scope for the circuit to go faulty except for misuse.

(vi) Performance of ICs are better than discrete circuits in high-frequency applications.

Digital ICs are also classified according to the number of components or gates placed on one chip.

Large-scale integration (LSI) will have more than 100 components and very large scale integration (VLSI) will have more than 1000 components in a single chip.

Integrated circuits are manufactured by various manufactures like Fairchild, Motorola, Texas instruments, National semiconductor, Signetics, etc. They put their brand name along with a particular number on each type of ICs they produce. For example, operational amplifier, manufactured by various manufacturers name them as

Motorola	MCI 741
National Semiconductor	LM 741
Texas Instruments	SN52 741

The last three digits in each case is 741. Irrespective of who manufactures an op-amp, the specification for an IC 741 is the same. We shall discuss in brief IC 741 which is an op-amp, IC555 which is a timer, and IC78XX series which are voltage regulators.

17.5 ▶ Operational Amplifiers

An operational amplifier is abbreviated as op-amp. An op-amp is the best-known linear-integrated circuit. Originally op-amps were meant for performing operations like integration, subtraction, differentiation, etc., and hence, the name was given.

Such operations are useful in analog computers. However, op-amps can also be used in signal amplification, filters, oscillators, voltage regulators, analog to digital (A to D) and digital to

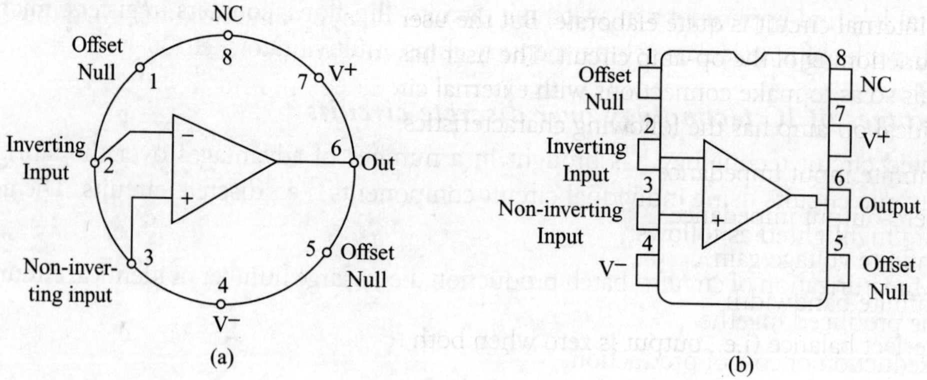

FIGURE 17.3

Pin diagrams of IC op-amp 741. (a) 8-pin mini dip packaging; (b) 8-pin metal can packaging

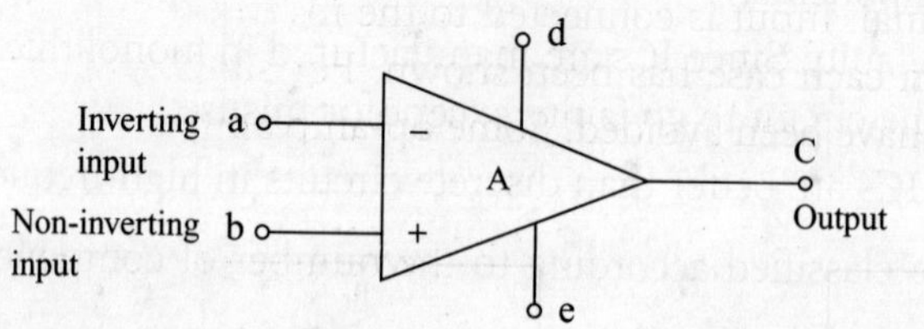

FIGURE 17.4

Symbolic representation of a basic op-amp

analog (D to A) converters, etc. Manufacturers provide a number of informations of the IC they supply including their pin diagrams. The informations provided include intended applications, maximum ratings, electrical characteristics, performance limitations, equivalent circuit, etc. Maximum ratings may include supply voltage, internal power dissipation, temperature range, etc. The pin diagrams of an IC 741 in eight-pin metal can package and eight-pin mini dip package are shown in Fig. 17.3.

An op-amp is a high-gain directly coupled linear differential amplifier. The performance of an op-amp is controlled by negative feedback from the output to the input. It is called a differential amplifier since it amplifies the difference between the two input signals. An op-amp amplifies both ac and dc input signals.

The circuit symbol of an op-amp has been shown in Fig. 17.4. The input terminals are marked a and b. Terminal a is marked negative while terminal b is marked positive. Terminal a is the inverting terminal. The input provided at this terminal appears inverted at the output with amplification. Terminal b is non-inverting. The output for input at b appears non-inverted but amplified . When signals are provided at both a and b terminals, the output at terminal c is proportional to the difference of the two signals.

The internal circuit is quite elaborate. But the user of an op-amp need not go to the details of the functioning of the op-amp circuit. The user has to know the basic specifications and the terminals so as to make connections with external circuit components for a purpose.

An ideal op-amp has the following characteristics:

(i) Infinite input impedance
(ii) Zero-output impedance
(iii) Infinite voltage gain
(iv) Infinite bandwidth
(v) Perfect balance (i.e., output is zero when both the inputs are equal)

A practical op-amp will have finite but very high-input impedance, very high-voltage gain, finite bandwidth and low-output impedance. A practical op-amp may not have perfect balance. The output is fed back to the inverting input terminal to provide negative feedback for the op-amp. Fig. 17.5 shows an op-amp with negative feedback. The output is connected back to the –ve input terminal. Input is connected to the inverting or non-inverting terminals and the output obtained in each case has been shown. For simplicity, analysis of the circuit and derivation of formula have been avoided. Some op-amp applications are explained as follows.

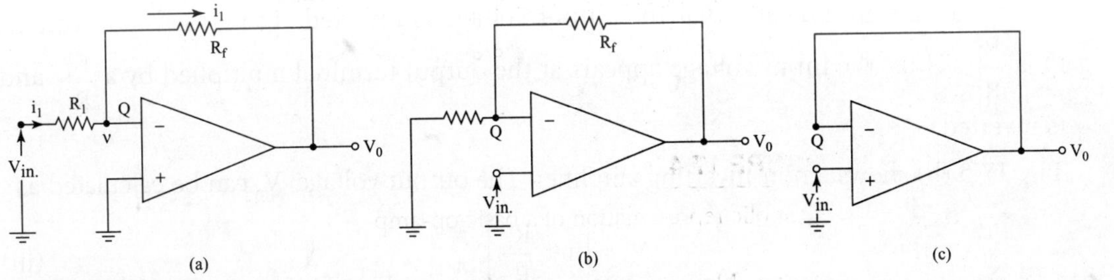

FIGURE 17.5

Op-amp with negative feedback working as (a) inverting amplifier; (b) non-inverting amplifier; (c) voltage follower

17.6 ► OP-AMP APPLICATIONS

In all the cases shown in Fig. 17.5, the output has been fed back to the inverting input terminal so as to provide a negative feedback. As shown in Fig. 17.5 (a), the output terminal is connected to the inverting input terminal through a feedback resistance R_f. The non-inverting terminal has been connected to the ground. The input voltage is V_{in} and the output voltage is V_0. The voltage at point Q is υ.

Current i_1 through the resistance R_1 is

$$i_1 = \frac{V_{in} - \upsilon}{R_1}$$

As the input impedance of the op-amp is infinite, the whole of i_1 will pass through R_f. The op-amp having infinite input impedance will not allow any current to flow through it.

Therefore,

$$i_1 = \frac{V_{in} - \upsilon}{R_1} = \frac{\upsilon - V_0}{R_f} \qquad \text{(i)}$$

As the open-loop gain, A of the amplifier is infinite, and the output voltage, V_0 is finite, we have

$$V_0 = A\upsilon \qquad \text{or, } \upsilon = \frac{V_0}{A}$$

If A tends to infinity, v will tend to zero.

Thus, the potential of point Q can be considered equal to zero. This is also referred to as a virtual ground potential. With v = 0, from equation (i),

$$\frac{V_{in}}{R_1} = \frac{-V_0}{R_f}$$

or,

$$V_0 = -\left(\frac{R_f}{R_1}\right)V_{in} \qquad \text{(ii)}$$

The negative sign indicates that the output voltage is inverted. The closed loop voltage gain is $\frac{R_f}{R_1}$. Thus, the input voltage appears at the output terminal multiplied by a $\frac{R_f}{R_1}$ and gets inverted.

Fig. 17.5 (b) shows a non-inverting amplifier. The output voltage, V_0 can be calculated as

$$V_0 = \left(1 + \frac{R_f}{R_1}\right)V_{in} \qquad \text{(iii)}$$

The output is positive, i.e., non-inverted and is multiplied by a factor, $\left(1 + \frac{R_f}{R_1}\right)$.

In Fig. 17.5 (c), R_f has been made zero, i.e., output terminal is directly connected to point Q so that using equation (iii), we can write

$$V_0 = (1+0)V_{in} = V_{in}$$

This circuit is called a unity gain voltage follower. Such a circuit is used for impedance matching in electronic devices and circuits.

17.6.1 Op-Amp As a Summing Amplifier

Op-amps can be used for adding and subtracting two input signals. They can also be used as differentiators and integrators. Analog computers use extensive op-amp circuits for these functions. These are discussed in brief as follows.

Fig. 17.6 shows an adder, i.e., a summing amplifier and a subtractor, i.e., a differential amplifier circuit.

Assuming point Q at potential υ, and assuming no current flowing through the amplifier

$$i_1 + i_2 + i_3 = i$$

$$\frac{V_1 - \upsilon}{R_1} + \frac{V_2 - \upsilon}{R_2} + \frac{V_3 - \upsilon}{R_3} = \frac{\upsilon - V_0}{R_f}$$

and

$$V_0 = A\upsilon$$

or,

$$\upsilon = \frac{V_0}{A}$$

If $A \rightarrow a$, $v \rightarrow 0$ (virtual ground)

we can write

$$\frac{V_1}{R_1} + \frac{V_2}{R_2} + \frac{V_3}{R_3} = -\frac{V_0}{R_f}$$

If

$$R_1 = R_2 = R_3,$$

$$V_0 = -\frac{R_f}{R}(V_1 + V_2 + V_3)$$

Now if we make

$$R_f = R,$$

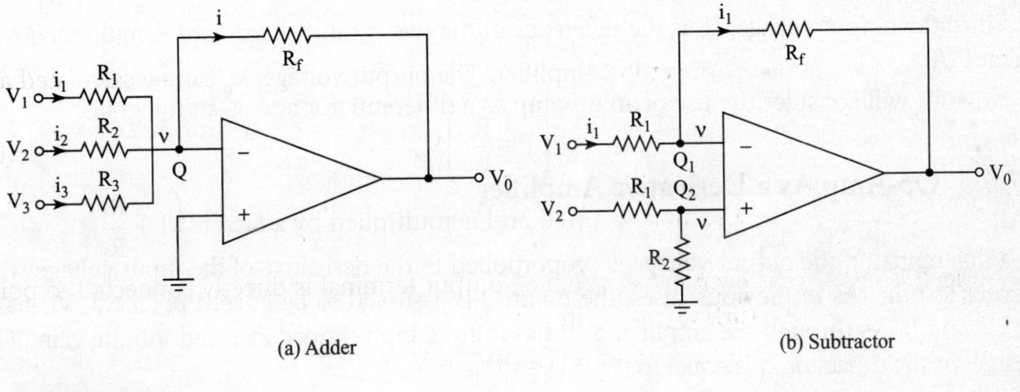

FIGURE 17.6

(a) Op-amp used as an adder or summer; (b) op-amp used as a differential amplifier

Then,

$$V_0 = -(V_1 + V_2 + V_3)$$

The output voltage, V_0 is equal to the algebraic sum of the input voltages. By choosing a suitable ratio of R_f/R, the output voltage can be made equal to the sum of the desired ratio of the input voltages. The circuit is called a summing amplifier circuit.

17.6.2 Op-Amp As a Differential Amplifier (Subtractor)

Fig. 17.6 (b) shows a differential amplifier, or a difference amplifier, or simply a subtractor. Here, the difference between two voltages, viz V_2 and V_1 can be amplified using this circuit. Again, assume that the amplifier has infinite gain and infinite input impedance, point Q_1 and Q_2 at virtual ground, i.e., zero potential. Amplifier allows no current flow through it. Looking at the inverting input terminal and feedback path

$$i_1 = \frac{V_1 - \upsilon}{R_1} = \frac{\upsilon - V_0}{R_2} \tag{iv}$$

From the non-inverting input terminal

$$\frac{V_2 - \upsilon}{R_1} = \frac{\upsilon}{R_2} \tag{v}$$

Subtracting equation (iv) from equation (v), we have

$$\frac{(V_2 - V_1)}{R_1} = \frac{V_0}{R_2}$$

or,

$$V_0 = \frac{R_2}{R_1}(V_2 - V_1)$$

Thus, the output voltage, V_0 is the difference of the two input voltages and is multiplied by a factor R_2/R_1.

Now we will consider the use of an op-amp as a differentiator and as an integrator.

17.6.3 Op-Amp As a Derivative Amplifier

In a differentiator, the output voltage is proportioned to the derivative of the input voltage with respect to time. As in previous cases, the point Q is assumed to be at zero potential. Virtually no current flows through the amplifier as it has infinite input impedance and infinite gain. The charge on the capacitor, q is equal to

$$q = CV_1$$

$$\int i dt = CV_1$$

or,

$$i = C\frac{dV_1}{dt} \tag{vi}$$

Again

$$\frac{\upsilon - V_0}{R} = i$$

Assuming v = 0,

$$i = -\frac{V_0}{R} \qquad \text{(vii)}$$

using equations (vi) and (vii),

$$-\frac{V_0}{R} = C\frac{dV_1}{dt}$$

or,

$$V_0 = (-CR)\frac{dV_1}{dt}$$

Thus, the output voltage, V_0 is the derivative of the input voltage V_1 and the multiplying factor is (–CR).

17.6.4 Op-Amp As an Integrator

In an integrator, the output voltage is proportional to the integration of the input voltage with respect to time

From Fig. 17.7 (b),

$$\frac{V_1 - \upsilon}{R} = i$$

and

$$\upsilon - V_0 = \frac{Q}{C}$$

Again $V_0 = A\upsilon$

or,

$$\upsilon = \frac{V_0}{A}$$

If $A \rightarrow \alpha$, $v \rightarrow 0$.

The point Q of which potential is v tends to zero.

Assuming $\upsilon = 0$

$$\frac{V_1}{R} = i$$

and

$$-V_0 = \frac{q}{C} = \frac{\int idt}{C}$$

or,

$$V_0 = -\frac{\int idt}{C} = \int\frac{V_1\,dt}{CR}$$

or,

$$V_0 = \left(-\frac{1}{CR}\right)\int V_1 dt$$

The output voltage V_0 is an integral of input voltage V_1 multiplied by proportionality constant $\left(-\frac{1}{CR}\right)$. This circuit is therefore called an integrator.

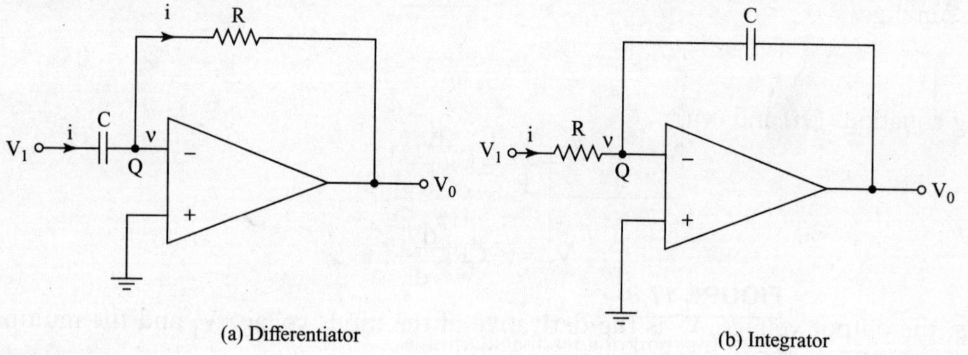

FIGURE 17.7

(a) Op-amp used as a differentiator; (b) op-amp used as an integrator

17.7 ► Feedback Amplifier

17.7.1 Introduction

We require an amplifier to amplify the signal without any distortion. While using an amplifier, one has to consider the variables introduced like voltage gain, input and output impedances, bandwidth, linearity, noise level, etc.

Feedback is a technique by which the above-mentioned performance parameters can be changed to get the desired output of the amplifier.

By feedback, we mean that some amount of output is returned back or fed back to the input either negatively or positively.

In positive feedback, the feedback voltage taken from the output is in phase with the input signal, and thus aids it.

In negative feedback, the feedback voltage from the output is 180° out of phase with the input signal, and thus opposes it.

Positive feedback increases the gain but at the same time increases the noise, brings distortion, and leads to poor stability.

Negative feedback decreases the gain but also decreases the noise level.

17.7.2 Negative Feedback Amplifier

Fig. 17.8 shows the block diagram representation of an ideal feedback amplifier with an input signal V_S and output signal V_O.

For negative feedback

We have, $V_i = V_S - V_f$; $V_O = AV_i$; $V_f = \beta V_O$

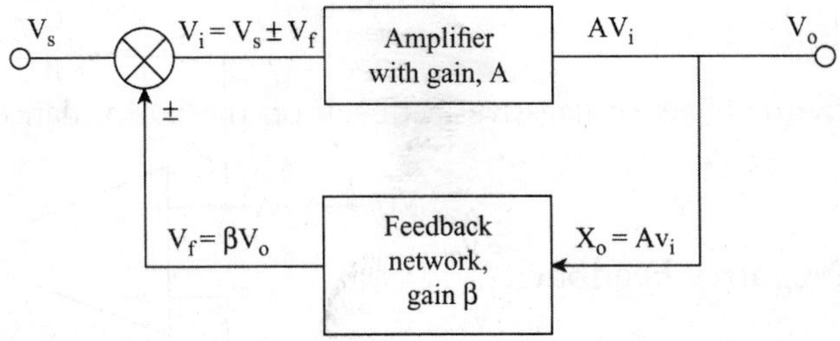

FIGURE 17.8

Block diagram of a feedback amplifier

So,
$$V_O = AV_i = A(V_S - V_f)$$
$$= AV_S - AV_f$$
or,
$$V_O = AV_S - A\beta V_O$$
or,
$$V_O(1 + A\beta) = AV_S$$
or,
$$\frac{V_O}{V_S} = \frac{A}{1 + A\beta}$$

Amplifier gain with negative feedback, $A_f = \dfrac{V_O}{V_S} = \dfrac{A}{1 + A\beta}$

If we consider positive feedback, i.e., if $V_i = V_S + V_f$,

Then
$$A_f = \frac{V_O}{V_S} = \frac{A}{1 - A\beta}.$$

Thus, the general expression of gain of a feedback amplifier is, $A_f = \dfrac{A}{1 \pm A\beta}$ where +ve sign is for negative feedback an –ve sign is for positive feedback. Gain A of the amplifier is it's open-loop gain. Whereas gain A_f is called the closed loop gain.

A is the gain of the basic amplifier and β is the gain of the feedback network. The product (Aβ) is called the feedback factor.

From the expression, $A_f = \dfrac{A}{1 + A\beta}$, it is observed that

(i) If the magnitude $|1 + A\beta| > 1$ then $|A_f| < |A_f|$

(ii) If the magnitude $|1 + A\beta| > 1$ then $|A_f| > |A|$

(iii) If $|(1 + A\beta)| = 1$ then $|A_f| = |A|$

(iv) If $|1+A\beta|=0, |A_f|=\infty$

We will now study the effect of negative feedback on the performance parameters of the amplifier.

17.7.3 Effect of Negative Feedback

Let the gain of the Amplifier is, A=10,000. The value of gain of the feedback circuit, β =0.01. Then the gain of the feedback amplifier, A_f is calculated as

$$A_f = \frac{A}{1+A\beta} = \frac{10,000}{1+10,000\times 0.01} = \frac{10,000}{101} = 99$$

This shows that the gain of the feedback amplifier with negative feedback has been reduced. *Therefore, we can state that negative feedback reduces the overall gain of the amplifier. The effect of negative feedback on all performance parameters are summarized below.*

SL. no	Performance parameters	Effect of negative feedback
1.	Overall gain	Overall gain is reduced
2.	Gain stability	Gain of the feedback amplifier gets stabilized, i.e., the gain does not change with variations in circuit due to temperature, ageing, change in components, etc. The condition for gain stability is that $|1+A\beta| >> 1$.
3.	Input impedance	High-input impedance is desirable for an amplifier. Due to negative feedback, the input impedance increases by a factor of $|1+A\beta|$.
4.	Output impedance	Output impedance gets reduced by the factor $|1+A\beta|$. This is desirable.
5.	Bandwidth	Bandwidth is the range of frequencies between which the gain remains within 70.7 percent of its value. Negative feedback increases the bandwidth which is desirable.
6.	Noise level	Negative feedback will cause decrease in the noise level at the output which is desirable.
7.	Harmonic distortion	
8.	Sensitivity	

17.7.4 Effect of Positive Feedback

The amplifier overall gain with positive feedback is given as

$$A_f = \frac{A}{1 - A\beta}$$

The gain of the amplifier is increased but makes the amplifier performance unstable. It therefore does not make any sense in increasing the gain of the amplifier and make its performance unstable. If we continue to increase the overall gain of the amplifier by increasing the value of $|A\beta|$ to almost equal to unity so that $1 - A\beta = 0$, there will be a spontaneous oscillation of the output voltage. When $|1 - A\beta| = 0$, $A_f \to \infty$. That is the gain becomes infinity. Gain infinity means there is output without any input. Hence, a positive feedback amplifier with unity loop gain acts as an oscillator. We had $A_f = \dfrac{A}{1 - A\beta}$

If the loop gain $|A\beta| = 1$, then

$$A_f = \frac{V_O}{V_S} = \frac{1}{1 - A\beta} = \infty$$

If $\dfrac{V_O}{V_S} = \infty$, then V_S must be zero.

Thus, has been shown in Fig. 17.9.

$$V_i = V_f = \beta V_O.$$

The amplifier out is $AV_i = A\beta V_O$

If this signal is equal to V_0, then $A\beta V_O = V_O$ or $|A\beta| = 1$,

Where $A\beta$ is the loop gain with unity loop gain, output voltage is available with $V_S = 0$, i.e., with no signal at the input. There will be a regeneration of output voltage and the amplifier will provide an oscillatory output, i.e., the amplifier circuit will act as an oscillator.

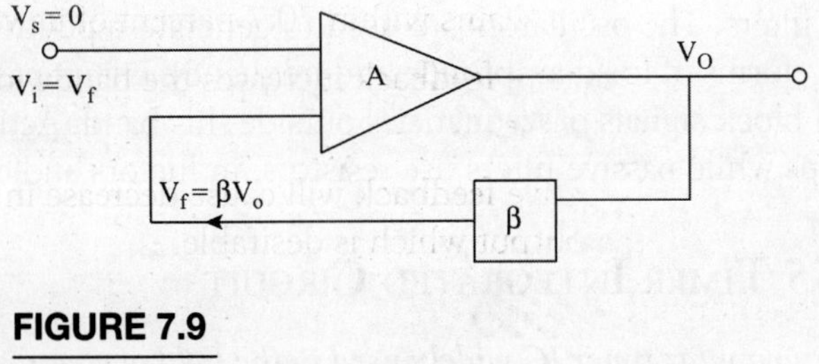

FIGURE 7.9

Condition of oscillation illustrated

17.7.5 Conditions for Oscillation (Barkhausen Criterion)

The conditions for oscillation are

(i) The loop gain must be unity, i.e., $|A\beta| = 1$

(ii) The phase angle of $\angle A\beta$ must be zero or an integrnal multiple of 360°. This implies that the feedback must be regenerative.

The first condition states that the feedback signal to the input must be the same as the input signal. The second condition states that the feedback should be positive with zero phase shift.

The above conditions or criteria for oscillations is known as *Barkhausen criterion*. Barkhausen criterion is stated as:

It states the conditions for oscillation of a feedback amplifier. The conditions are:

(i) Positive feedback

(ii) The loop gain $|A\beta|$ must be equal to 1 and phase shift of $A\beta$ is zero.

The amplifier will continue to work as an oscillator. If $|A\beta|$ becomes less than unity, the oscillations of the output will die out soon.

If $|A\beta|$ is greater than unity, the amplitude of oscillation will continue to increase to be limited however by saturation effect. The amplifier is unstable if the response increases with time. The amplifier is critically stable if the response continues to persist indefinitely.

17.7.6 Other Applications of Op-Amps

There are many applications of op-amps in the field of electronics. Some of them are mentioned below.

(i) Current to voltage converter where the output voltage is proportional to the input current.

(ii) Digital to analog converter.

(iii) Analog computer which is constructed by using op-amp integrators and adders to solve differential equations.

(iv) Waveform generators or function generators. Function generators are also incorporated in analog computers to solve non-linear differential equations.

(v) Oscillators and filters. The oscillators are used to generate repetitive alternating current and voltage waveforms of fixed amplitude and frequency. Filters pass a specified band of frequencies and block signals of frequencies outside this band. Active filters use transistors and op-amps while passive filters use resistors, inductors and capacitors.

17.8 ► The 555 Timer Integrated Circuit

The 555 timer is a very popular timer IC widely used in the field of electronics and control engineering. Signetics Corporation of USA first developed this chip in the 1970s. Nowadays this IC is being manufactured by many semiconductor companies. However, regardless of the manufacturer, all the IC 555 chips provide the same basic functions and are packaged the same way.

The 555 timer contains 23 transistors, 2 diodes, and 16 resistors on a single silicon chip.

17.8.1 Three Operating Modes of IC 555

This timer has three operating modes:

(i) *Monostable mode:* in this mode the timer functions as a 'one shot' multivibrator. A multivibrator generates non-sinusoidal waveforms. It is an oscillator operating in two modes or states. At stable state the output is low. When a trigger pulse is applied, the output becomes high but automatically returns to stable state (i.e., low output) after a time interval determined by externally connected RC network. In a monostable multivibrator a trigger input will cause the timer to get switched on for a time determined by the external components, R and C connected to it. A monostable circuit has only one stable state and one quasi-stable state. A triggering signal is required to induce a transition from stable to quasi-stable state The circuit will remain in quasi-stable state for some time. However, eventually it will return to its stable state with no external signal required to change that state. Since, the circuit is ***stable in one state*** it is called monostable. The stable state will mean low or no output voltage and quasi-stable state will mean high-output voltage (also called cut-off state and saturation state).

(ii) *Astable mode:* the astable mode of operation has two states, both of which are quasi-stable. The astable mode of operation will, therefore, make successive transitions from one quasi-stable state to other without the aid of an external triggering signal. An astable multivibrator will alternate automatically and continuously between two states at a rate determined by the circuit components.

(iii) *Bistable mode:* this is also referred to as Schmitt trigger. The 555 IC can operate as a flip-flop. A flip-flop multivibrator operates in two states but requires an external trigger pulse to change from one state of operation to the other.

Monostable multivibrator generates a single cycle, and hence can be used to provide gate pulse to other circuits. Astable circuit is used to generate square waves. Both these circuits are used extensively in pulse circuitry.

17.8.2 Pin Configuration of IC 555

The 555 timer is an extensively used IC. The 8-pin diagram of a mini dual-in-line package (DIP-8) has been shown in Fig. 17.10. The IC 556 is a 14-pin DIP that combines two 555s in a single chip.

The functions of each pin of IC 555 are mentioned below:

Pin 1 ground: this is the ground pin connected to 0 V rail. This is also called negative, or 0 V or earth rail. All the voltages are measured with respect to this terminal.

Pin 2 trigger: This pin connects to a comparator and is used to set the control flip-flop. When it is taken low, it causes the output to go high. Triggering is accomplished by taking the pin below a certain voltage, called low.

Pin 3 output: Output is taken from pin 3. The load can be connected between pin 3 and pin 1 or between pin 3 and pin 8. To make the output high, the trigger pin is momentarily taken from high to low. The output can be turned to a low by making the threshold pin (pin 6) go from a low to a high. The output can also be made to go low by taking the reset pin (pin 4) to a low state.

Pin 4 reset: The reset pin has an over riding function. It will force the output pin to go low regardless of the state of the trigger pin (pin 2). It can be used to terminate an output pulse prematurely when not in use. It is recommended that the reset pin be tied to the positive rail, i.e., to $+V_{CC}$ to avoid the possibility of false resetting.

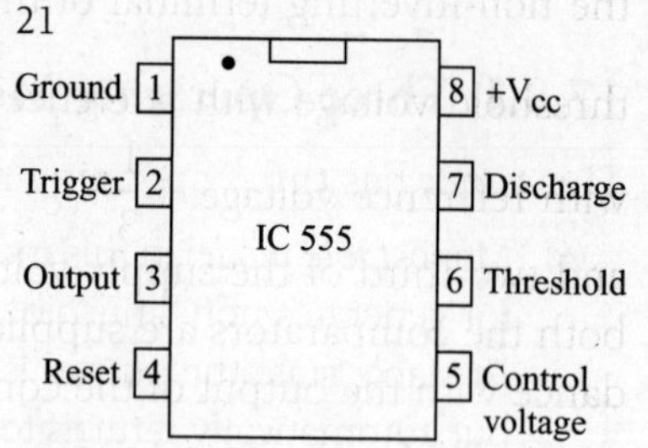

FIGURE 17.10

Block diagram representation of an IC 555 timer

Pin 5 control voltage: By applying a voltage to this pin, it is possible to vary the timing of the timer chip independent of the RC network. The control voltage may be varied from 45 to 90 per cent of the V_{CC} in the monostable mode to control the width of the output pulse. When in astable mode, the control voltage can be varied from 1.7 V to the full V_{CC}. Frequency-modulated output is produced by varying the control voltage. When not in use the control voltage pin should be connected to ground with a 10 nano farad capacitor so as to avoid any noise entering the circuit.

Pin 6 threshold: Pin 6 is connected to one of the non-inverting input terminals of comparator 1 or upper comparator inside the IC 555. The comparator compares the voltage applied to this terminal with a reference voltage of $+\frac{2}{3}V_{CC}$. To make the output to go low, the threshold pin is taken from a low to a level above $\frac{2}{3}V_{CC}$.

Pin 7 discharge: This pin is connected to the open collector of an n–p–n transistor inside the chip. A capacitor is connected between pin 7 and ground which gets discharged when the transistor is turned on. That is to say, when the transistor is turned on, pin 7 is effectively shorted.

Pin 8 supply terminal: This pin is the positive supply terminal for the 555, also referred to as $+V_{CC}$. The supply voltage operating range is from +5V to +18V.

17.8.3 Functional Block Diagram of IC 555

Fig. 17.11 shows the functional block diagram of 555 timer IC. The timer consists of two comparators (these are op-amps), an R–S flip-flop, two transistors and three equal-value (5 kΩ) resistors in series forming a voltage divider. Because of the use of three 5 kΩ resistors, the timer was given the name 555.

The three resistors of equal value, i.e., R divides the V_{CC} into $\frac{1}{3}V_{CC}$.

The voltage across the inverting terminal of the comparator 1 is $+\frac{2}{3}V_{CC}$. Voltage across the non-inverting terminal of the comparator 2 is $+\frac{1}{3}V_{CC}$. The comparator 1 compares the threshold voltage with reference voltage $+\frac{2}{3}V_{CC}$. Comparator 2 compares the trigger voltage with reference voltage $+\frac{1}{3}V_{CC}$. Thus, the reference voltage of the comparators are one-third and two-third of the supply voltage which may be between + 5 V and + 18 V. The output of both the comparators are supplied to the RS flip-flop. The flip-flop changes its state in accordance with the output of the comparators. The flip-flop changes states when the trigger input at pin 2 is brought down below $+\frac{1}{3}V_{CC}$. When this occurs the output (at pin 3) changes state to $+V_{CC}$. However, if the threshold input (at pin 6) is now raised above $+\frac{2}{3}V_{CC}$, the output will return to ground. The n–p–n transistor, T_1 is the discharge transistor. Pin 7, i.e., the discharge pin is connected to the collector of this transistor. The emitter is connected to the ground. When this transistor is on, this pin, i.e., pin 7 gets connected to ground through a timing capacitor (not shown in figure) connected externally between pin 7 and ground when the transistor is on.

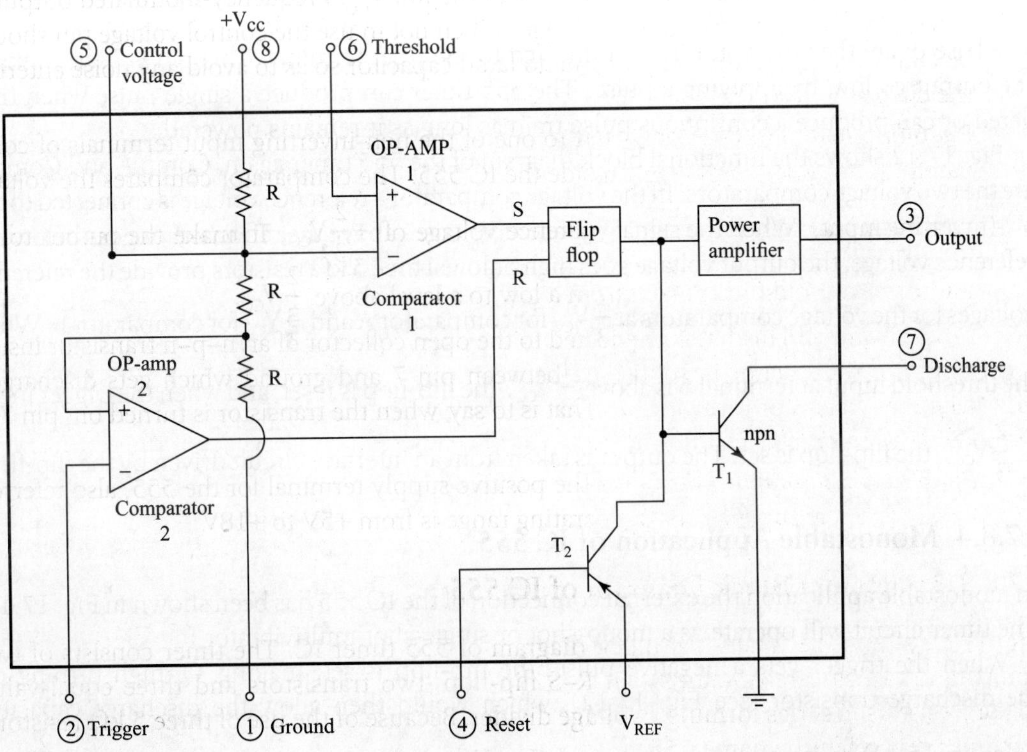

FIGURE 17.11

Block diagram representation of an IC 555 timer

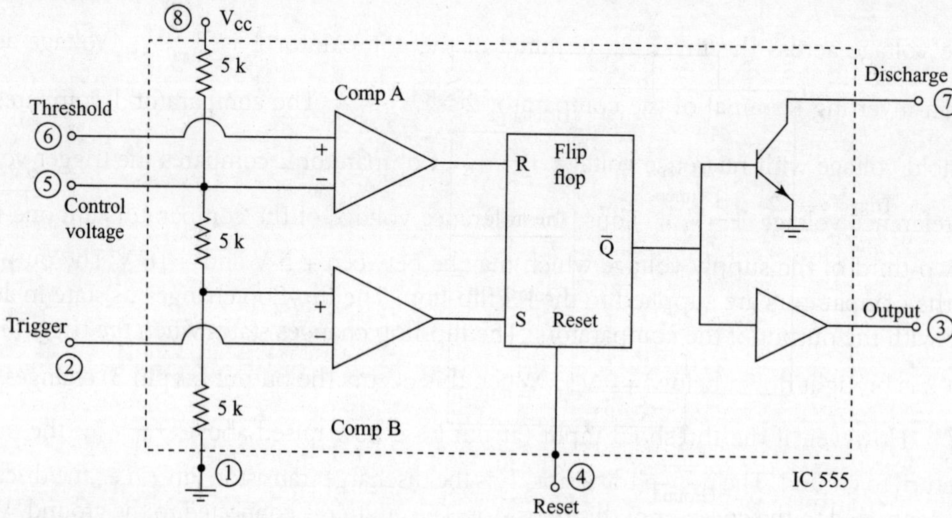

FIGURE 17.12

Functional block diagram of IC 555 timer (same diagram as in Fig. 17.11 redrawn)

The base of another transistor T_2 is connected to the reset terminal. This pin is used to make the output go low, by applying a pulse. The 555 timer can produce a single pulse when triggered or can produce a continuous pulse train as long as it remains powered.

Fig. 17.12 shows the functional block diagram of the 555 timer again. Comp A and Comp B are the two voltage comparators. In the voltage comparator a reference voltage is connected to the V_{in}-(inverting input). When the signal voltage at V_{in+} rises above the reference, or goes below the reference voltage, the output voltage goes high or low. Three 5 kΩ resistors provide the reference voltages for the voltage comparators at $\frac{2}{3}V_{CC}$ for comparator A and $\frac{1}{3}V_{CC}$ for comparator B. When the threshold input at terminal 6 is above $\frac{1}{3}V_{CC}$, the flip-flop is reset, and when the trigger input is $\frac{2}{3}V_{CC}$, the flip-flop is set. The output is taken from an interface circuit driven by the flip-flop.

17.8.4 Monostable Application of IC 555

In monostable application the external connection of the IC 555 has been shown in Fig. 17.13. The timer circuit will operate as a mono-shot or single-shot multivibrator.

When the trigger gets a negative pulse, the flip-flop is set, making $\overline{Q}$ high turning off the discharge transistor (see Fig. 17.12), which would then allow the discharge capacitor C (as in 15.12) to be charged up towards V_{CC}.When the capacitor voltage reaches $\frac{2}{3}V_{CC}$, the threshold signal causes the flip-flop to be reset, dischargingthe capacitor again. The voltage waveforms of monostable operation have been shown in Fig. 17.13 (b).

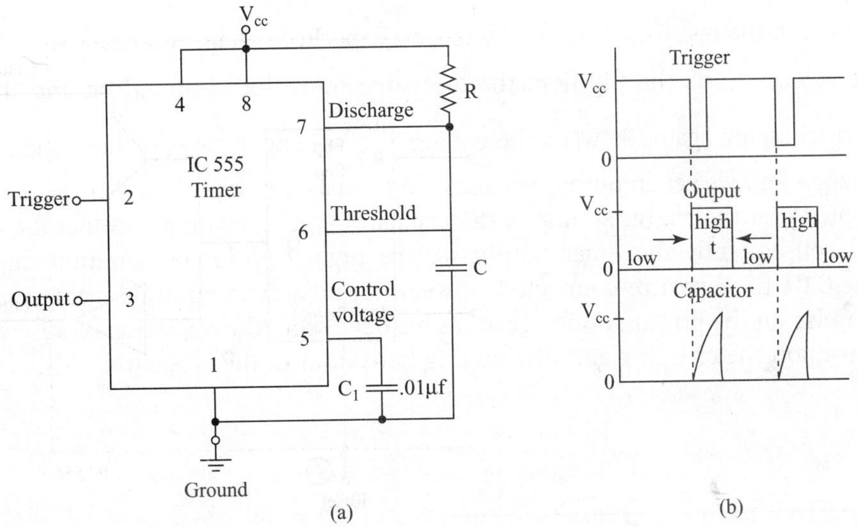

FIGURE 17.13

(a) IC 555 timer in monostable operating mode; (b) waveforms of voltages in monostable operation

It can be seen from the waveforms that the output remains low until the trigger signal is applied. With the application of negative trigger voltage, the output goes high while the capacitor gets charged and then goes back to low, and it remains low until another trigger pulse is received. The name single shot is appropriately given because multiple trigger shots during charging of the capacitor has no effect on the output. However, the trigger voltage must return to the high level again before the flip-flop can be reset by the threshold voltage signal. The time t of the output wave depends on the time taken by the capacitor to get charged up from a discharged state from near zero voltage to $\frac{2}{3}V_{CC}$. The capacitor has been shown getting discharged instantaneously; however, the rate of discharge will depend upon the discharge transistor.

17.8.5 Astable Application of IC 555

For astable application, the external connection of the timer has been shown in Fig. 17.13. In astable application, the output will alternate between two states, i.e., high and low automatically and continuously without any trigger pulse.

Here, both the trigger input and the threshold input are connected to the capacitor. An additional resistance R_1 has been connected between the capacitor and the discharge transistor to slow down the discharge of the capacitor. When the capacitor discharges to $\frac{1}{3}V_{CC}$, the comparator B (i.e., the trigger comparator as shown in Fig. 17.11) sets the flip-flop which in turn switches the discharge transistor off, allowing the capacitor getting charged through the resistors R and R_1. When the capacitor voltage reaches $\frac{2}{3}V_{CC}$, the threshold input voltage resets the flip-flop, the discharge transistor is turned on and the capacitor starts discharging again. Between the voltage $+\frac{1}{3}V_{CC}$ and $+\frac{2}{3}V_{CC}$, the capacitor charges and discharges and we get an output voltage waveform as has been shown in Fig. 17.14 (b). It is to be noted that the charging time of the capacitor can be made more than the discharge time, and consequently the timer output will be high for a longer duration than its low value. The 0.01 μF capacitor connected between terminal 5 and ground, acts to reduce the transient noise on the power supply. Thus, as we have seen, the operation of 555 timer IC is dependent upon the charging and discharging behaviour of the capacitor.

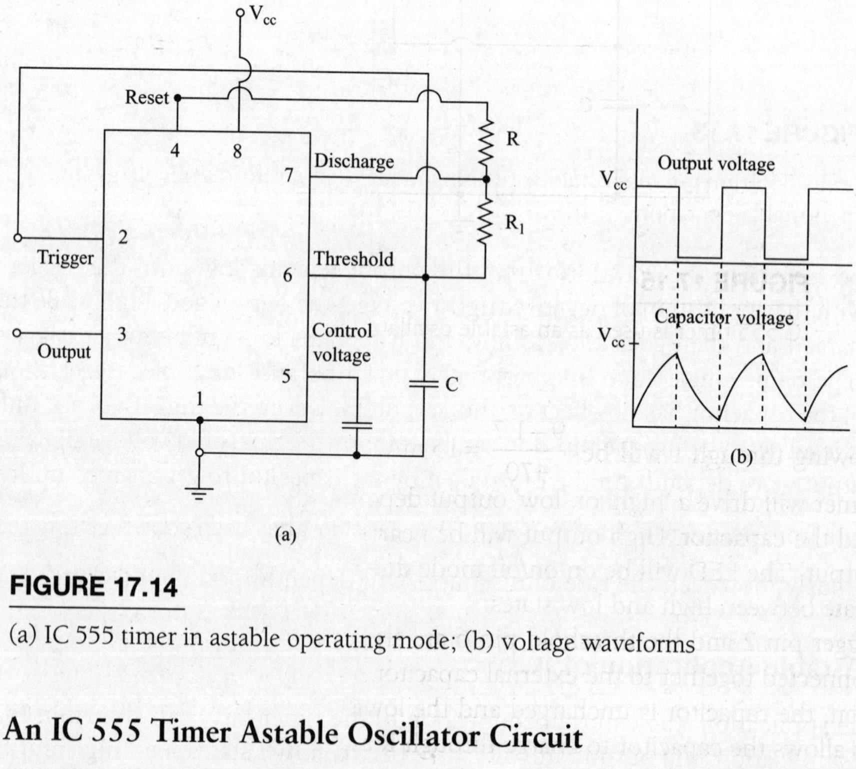

FIGURE 17.14

(a) IC 555 timer in astable operating mode; (b) voltage waveforms

17.8.6 An IC 555 Timer Astable Oscillator Circuit

A practical circuit of an IC 555 timer astable oscillator circuit has been shown in Fig. 17.15 drawn in a slightly different way. The students may try out this circuit for its working. We may choose $R = R_1 = 10$ kΩ, $C = 10$ μF, $C_1 = 0.01$μF, supply voltage to the IC, i.e., $V_{CC} = +9$ V, a red LED, a series resistance of 470 Ω. The LED red light has been used to indicate clearly that the circuit is working. A LED will require about 5 mA to 20 mA of current to emit light.

The current flowing through the LED circuit will be equal to the output voltage divided by the LED series resistance of 470 Ω. Assuming a voltage drop across the LED of 1.7 V, the

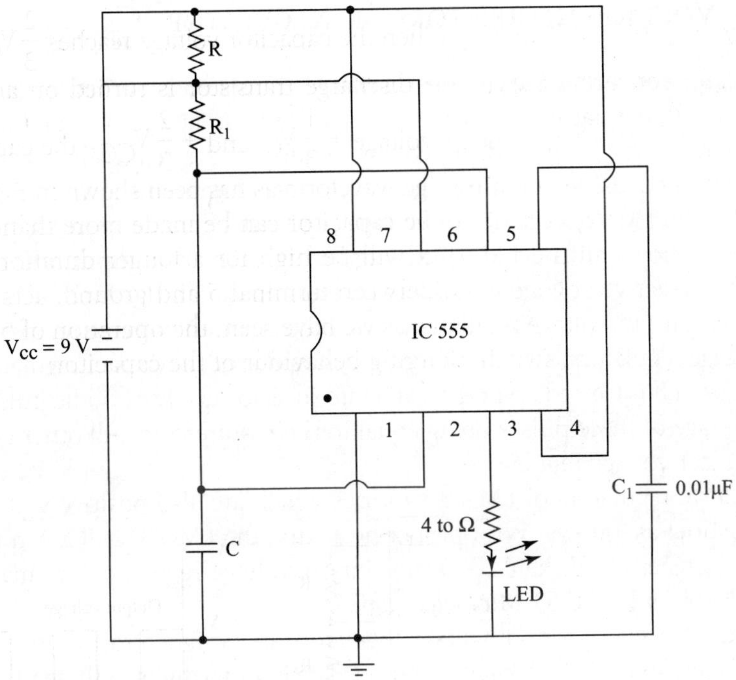

FIGURE 17.15

IC 555 timer is used as an astable oscillator

current flowing through t will be $\dfrac{9-1.7}{470} = 15$ mA, which is sufficient for the LED to light up. The timer will drive a 'high' or 'low' output depending on the charging cycle of the series resistor and the capacitor. High output will be nearly V_{CC} and low output will be nearly zero voltage output. The LED will be on on/off mode due to the periodic output waveform, which will alternate between high and low states.

The trigger pin 2 and the threshold pin 6 are attached to the two comparators (op-amps) and are connected together to the external capacitor. When power is first supplied, i.e., V_{CC} is switched on, the capacitor is uncharged and the lower comparator sets the output voltage to high. This allows the capacitor to charge through R and R_1. When the charge of the capacitor reaches $\frac{2}{3}V_{CC}$, the upper comparator is triggered, causing the output to go low. When the voltage across the capacitor is $\frac{1}{3}V_{CC}$, the lower comparator is triggered, setting the output to high.

This type of oscillators find many applications like in quartz watch to keep track of time in AM radio to create a carrier wave for the station, in computers where a specialized oscillator called a clock serves as a sort of pacemaker for the microprocessor, in wireless receivers and transmitters, in cell phones, pagers, in music synthesizers, etc.

17.9 ▶ IC Voltage Regulators or Regulator ICs

A voltage regulator converts a varying input voltage into a constant 'regulated' output voltage. Thus, a voltage regulator supplies a constant voltage at the output regardless of the magnitude of load current supplied. IC voltage regulators are relatively cheap as compared to regulators designed using op-amps. IC voltage regulators are available in a variety of output voltages. The LM 78 xx series of voltage regulators are designed for positive voltage input. Table 17.1 shows all the LM 78 xx regulators with their input and output voltage ranges. There are negative voltage regulators that are marked LM 79 xx.

These are fixed output voltage regulators. The other types of voltage regulators are adjustable output voltage regulators, switching regulators, and special types of regulators. In switching regulators a switch is turned on and off at a rate so as to supply periodic pulse of current to the load, the average of these pulses being equal to the requirement. All other types of regulators are called linear voltage regulators.

The fixed voltage regulators of LM 78 xx series which are also positive voltage regulators, are described in brief as follows. With proper heat sink, the LM 78 xx ICs can handle output current even somewhat more than 1000 mA. Linear voltage regulators are manufactured by companies like *Fair Child* and *ST Microelectronics.*

As mentioned, LM 78 xx series of voltage regulators are designed for positive voltage input. For negative voltage input LM 79 xx series is used. The last two digits in these numbers indicate the output voltage. For example, the LM 7805 regulator gives a regulated output voltage of +5 V and LM 7905 gives an output voltage of –5 V. These ICs, although internally complex are inexpensive and easy to use. In Fig. 17.16 (a) has been shown a three terminal voltage regulator IC.

Two capacitors C_1 and C_2 are connected on the input and output sides. The output capacitor, C_2 helps in isolating the effect of transients that may appear on the regulated supply line. C_2 is a high-quality tantalum capacitor with capacitance of around 1.0 μF connected close to the

TABLE 17–1 IC 78 xx Voltage Regulators

Device type	Output voltage	Input voltage range
LM 7805	5 V	7–25 V
LM 7806	6 V	8–25 V
LM 7808	8 V	10.5–25 V
LM 7809	9 V	11.5–25 V
LM 7812	12 V	14.5–30 V
LM 7815	15 V	17.5–30 V
LM 7818	18 V	21–33 V
LM 7824	24 V	27–38 V

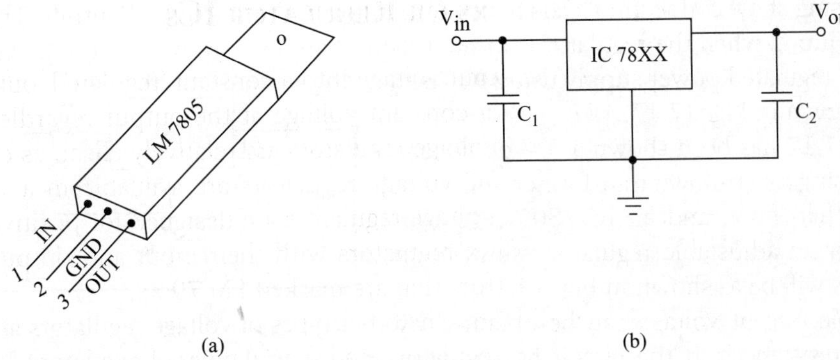

FIGURE 17.16

(a) The physical configuration of a regulator IC; (b) typical circuit connection

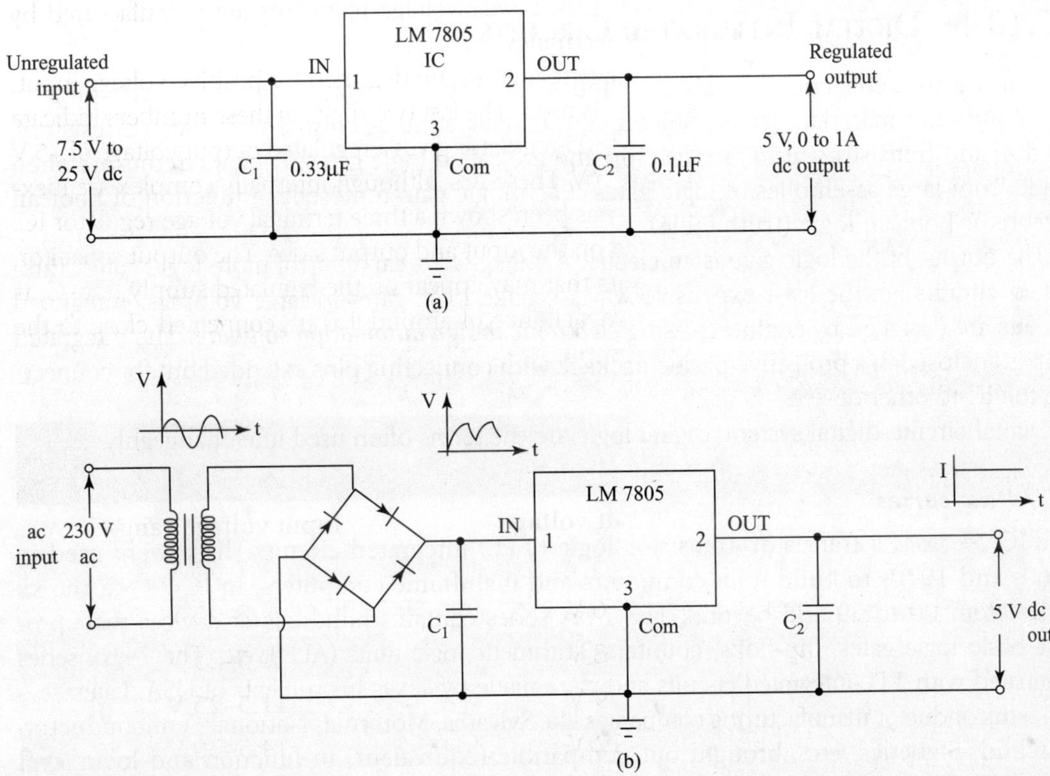

FIGURE 17.17

(a) Regulated power supply with dc input and dc output; (b) regulated 5 V dc. Power supply with 230 V, 50 Hz ac supply as input.

regulator using short connecting leads in order to improve the stability of output. The capacitor C_1 is required when the regulator is located more than 5 cm away from the power supply filter. A 5 V regulated power supply using IC 7805 has been shown in Fig. 17.17.

In Fig. 17.17 has been shown a 5 V dc voltage regulator using a step-down transformer, full-wave bridge rectifier, filter, and an IC 7805 regulator. A circuit for an adjustable regulator using 78 xx series of ICs will be as shown in Fig. 17.18.

Adjustable output voltage can be obtained by using two resistances in the circuit as has been shown. We may use the input and output side capacitors in the circuit. Voltage regulators are available either in plastic package or in metal can package.

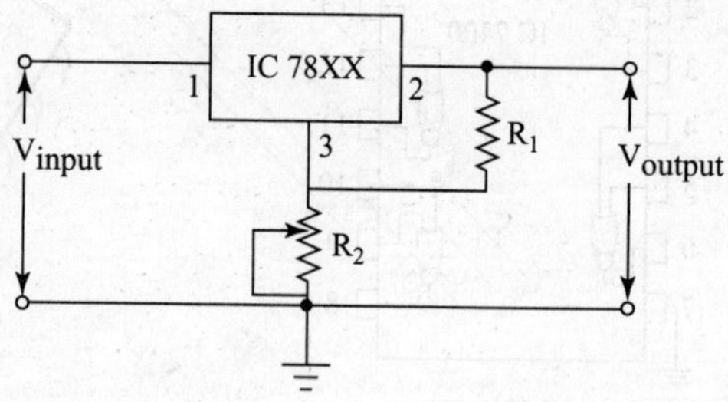

FIGURE 17.18

Voltage regulator with adjustable output

17.10 ► DIGITAL INTEGRATED CIRCUITS

Digital ICs are a collection of resistors, diodes, and transistors fabricated on a single piece of semiconductor material, usually silicon, and is referred to as chip. The fabricated resistors, diodes, and transistors reside in the chip and are called logic gates. Digital circuits are often made from large assemblies of logic gates. Each logic gate represents a function of Boolean algebra or Boolean logic (truth table).

The output of the logic gate is an electrical voltage that can control more logic gates. Integrated circuits are the least expensive way to make logic gates in large volumes. Integrated circuits are designed by engineers using ***electronic design automation software***. The integrated chip is enclosed in a protective plastic package with connecting pins extended out for connecting the IC to other devices.

Digital circuit, digital system, digital logic are the terms often used interchangeably.

IC 74 xx series

The IC 74 xx is a transistor–transistor logic (TTL) integrated circuit. These were used in 1960s and 1970s to build mini computers and mainframe computers. In IC 74 xx, the xx varies from 00 to 99 and beyond. The 74xx series contains hundreds of devices that provide basic logic gates, flip-flops, counters, arithmetic logic units (ALU), etc. The 74 xx series originated with TTL-integrated circuits and were made by Texas Instruments of USA. Later, several semiconductor manufacturing companies like Sylvania, Motorola, National Semiconductor, Fairchild, Signetics, etc., brought out compatible (equivalent) in function and logic level newer sub-series of ICs. Digital integrated circuits are constructed using bipolar transistors. Nowadays the sub-series of 74 xx ICs are made using CMOS (complementary metal oxide semiconductor) technology. Digital ICs produced for various logic functions are shown in Table 17.1. Fig. 17.19 shows an IC 7400 which is one of the 74 xx digital logic devices.

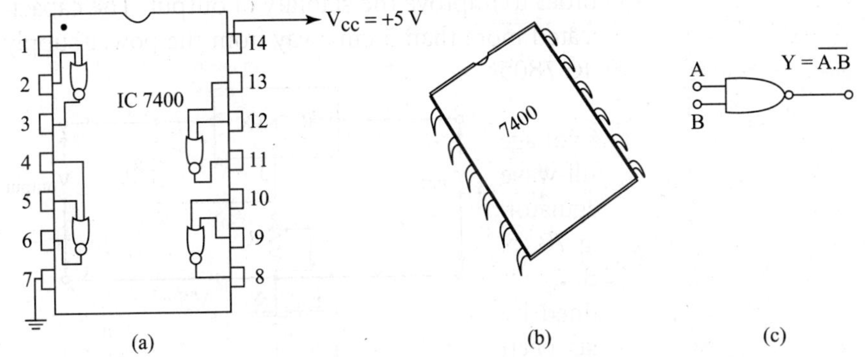

Truth table

A	B	Y
0	0	I
0	I	I
I	0	I
I	I	0

(d)

FIGURE 17.19

(a) IC 7400 with four 2-input NAND gates; (b) sketch of the IC chip;
(c) symbolic representation of the NAND gate; (d) the truth table for the gate

TABLE 17–2 Integrated Circuits with Their Level of Complexity

Complexity	Approximate number of gates per chip	Typical products
Small-scale integration (SSI)	Less than 12	Logic gates, flip-flops
Medium-scale integration (MSI)	12 to 99	Counters, multiplexers, adders
Large-scale integration (LSI)	100 to 9999	8 bit microprocessors, ROM, RAM
Very Large-scale integration (VLSI)	10,000 to 99,999	16 bit and 32 bit microprocessors, sophisticated computer peripherals
Ultra large-scale integration (ULSI)	100,000 or more	64 bit microprocessors, real-time image processing

This IC chip contains four numbers of two input NAND gates. We have known that NOT gates can be built using transistors and AND gates can be built using diodes or transistors. The logic circuit in which a NOT gate follows an AND gate is called a NOT AND or NAND gate. NAND gates can be made with diode–transistors logic (DTL) or transistor–transistor logic (TTL).

The IC 7400 chip shown in the figure has 14 pins marked 1 to 14 in the anticlockwise direction with the notch placed vertically up. The pin 14 is marked for power supply, V_{CC} which is +5 V and

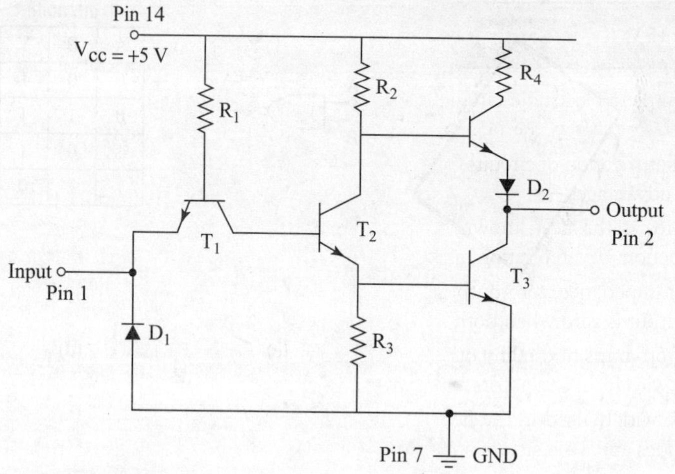

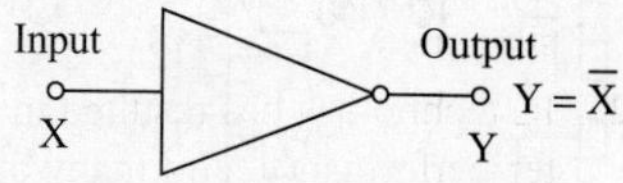

Input, X	Output, Y
0	1
1	0

FIGURE 17.20

A TTL inverter circuit

pin 7 is the ground pin. All the other pins are the input and output terminals of the four NAND gates. Each gate uses two pins for input and one pin for its output. The truth table for the two input NAND gate has been shown in Fig. 17.19 (d).

From the list of 7400 series of integrated circuits it is seen that IC 7402 contains two-input NOR gates while IC 7410 contains three-input NAND gates.

Digital IC chips are commonly available in dual-in-line pakage (DIP). We had seen a 14-pin IC 7400. ICs with 16-, 20-, 24-, 28-, 40-, and 64-pin packages are also available.

Digital ICs are often categorized according to their circuit complexity. The complexity is measured by the number of equivalent logic gates in an IC. There are at present five standard levels of complexity of integrated circuits as indicated in Table 17.2.

With ICs, electronic circuits and components are becoming smaller and less expensive. Modern electronics using ICs find applications in mobile communication, satellite and aerospace communication, high-speed computers, home appliances and in automatic control systems. Electronics using ICs are becoming the brains and nerves of our complex society.

TTL 74 series is the most widely used family of digital ICs in the SSI and MSI categories. Fig. 17.20 shows a standard TTL inverter circuit. It contains several bipolar transistors, and hence, the name TTL (transistor–transistor logic) is given.

As shown in Fig. 17.20, dc power supply voltage of +5 V has been connected to pin 14, and pin 7 has been connected to ground. Pin 1 is for the input. The inverted output is obtained at pin 2. A number of inverter circuits are made available through IC 7405.

While using integrated circuits, one has to refer to the manufacturer's data book. The data book provides all the specifications and functions of a particular chip. From the data book one can also know about all the ICs that are available and the functions they would perform.

Important Terminologies and Points to Remember

1. An integrated circuit (IC) is a complete electronic circuit in miniature form grown into a single chip of silicon.
2. IC technology has resulted in miniaturization of circuits, reduction in cost of production, smaller size, better performance, and many other advantages.
3. An operational amplifier (Op-amp) is the best known linear-integrated circuit. Op-amps can perform operations like integration, subtraction, differentiation, addition, etc.
4. An ideal op-amp has infinite input impedance, zero-output impedance, infinite voltage gain, infinite bandwidth and perfect balance, i.e., output is zero when both the inputs are equal.
5. In addition to perform operation, op-amps find many other applications like in current to voltage (I to V) conversion, in analog computers, waveform generators, oscillators, filters, etc.
6. A 555 is a timer integrated circuit widely used in the field of electronics and control engineering. A 555 timer contains 23 transistor, 2 diodes and 16 resistors on a single silicon chip.
7. Voltage regulators converts a varying input voltage into a nearby constant 'regulated' output voltage. Voltage regulators are designed using op-amps. IC voltage regulators are available series and are cheaper than op-amp-based voltage regulators.

REVIEW QUESTIONS

Short Answer Type Questions

1. State the different levels of integration in integration circuits.
2. Briefly describe the manufacturing processes of monolithic integrated circuits.
3. Mention the advantages of ICs over discrete components.
4. Describe various types of IC packages.
5. Distinguish between linear and digital ICs.
6. What is an operational amplifier? Mention few applications of op-amps.
7. Show Application of an op-amp as an interting amplifier.
8. Explain how an op-amp can be used as a non-inverting amplifier.
9. Show the pin diagram of an IC op-amp, 741.
10. Show the use of an op-amp as a differentiator and as an integrator.
11. Mention any five applications of operational Amplifiers.
12. Draw the PIN diagram of an IC 555 timer, Explain the function of each pin.
13. Draw and explain the functional block diagram of an IC 555.
14. Explain the working of an IC 555 timer in monostable operating mode. Draw the voltage waveforms.
15. Explain the working of an IC 555 timer in astable operating mode. Draw the voltage waveforms.
16. Explain the working of an IC 555 timer as an astable multivibrator.
17. Show the use of a voltage regulator IC for converting 230 V ac into a low-voltage regulated dc.
18. What are the different types of IC voltage regulators?
19. What are digital ICs? Name any two digital IC, and the function they perform.
20. Draw the PIN diagram of an IC 7400 with four two-input NAND gates. Draw the truth table for the gate.
21. Show the classification of integrated circuits with their leave of complexity or miniaturization.
22. Draw and explain or TTL inverter circuit.

Multiple Choice Questions

1. Which of the following statements is not true for an integrated circuit?
   (a) Integrated circuits are basically microelectronic circuits
   (b) Integrated circuit is basically a complete circuit
   (c) Integrated circuits are very small in size
   (d) Integrated circuits consists only of active circuit components.
2. Microprocessors are
   (a) small-scale integrated circuits (SSI)
   (b) medium-scale integrated circuits (MSI)
   (c) large-scale integrated circuits (LSI)
   (d) very-large scale integrated circuits (VLSI).
3. Logic gate is an example of
   (a) SSI (b) MSI
   (c) LSI (d) VLSI.
4. Which of the following is a timer?
   (a) IC 741 (b) IC 555
   (c) IC 78XX (d) IC 79XX.
5. For which of the following functions op-amps are not used
   (a) integrator (b) differentiator
   (c) summing amplifier (d) voltage regulator.
6. An RS flip-flop is an integral part of
   (a) IC 555 (b) IC 741
   (c) MC 1741 (d) IC 78XX.
7. A IC 555 will have
   (a) 2 comparators, 1 RS flip-flop, 1 discharge transistor
   (b) 2 comparators, 2 RS flip-flop, 1 discharge transistor
   (c) 1 comparator, 2 RS flip-flop, 1 discharge transistor
   (d) 1 comparator, 1 RS flip-flop, 1 discharge transistor.
8. Which of the following is not a voltage regulator IC?
   (a) IC 7805
   (b) IC 7808
   (c) IC 7812
   (d) IC 741.
9. An operational amplifier (op-amp) is basically
   (a) a negative feedback amplifier
   (b) a positive feedback amplifier
   (c) a low-gain amplifier
   (d) none of these.
10. Which of the following statements is not true?
   (a) Logic gates, flip-flops, counters, shift registers, etc., form a digital integrated circuit
   (b) The NAND or NOR gates are the universal building block of digital systems
   (c) TTL stands for transistor–transistor logic
   (d) In an integrated circuit only active components incorporated in a single chip.

Answers to Multiple Choice Questions

1. (d) 2. (a) 3. (b) 4. (d) 5. (a) 6. (a)
7. (a) 8. (d) 9. (a) 10. (d)

CHAPTER 18

Communication Systems

Topics Discussed

1. Elements of a communication system
3. Baseband and broadband communication
3. Modulation techniques
4. Bandwidth requirement
5. Demodulation
6. Multiplexing
7. Radio communication receivers
8. Microwave and satellite communication
9. Fibre-optic communication
10. Television, mobile and fax communication

18.1 ► INTRODUCTION

The function of a communication system is to transfer information from one place to the other through some communication link or medium. Technological development has provided us various means of communication. Modern communication systems in use include telephony, radio broad-casting, mobile communication, Fax communiication, satellite communication, and so on.

If we look back, the main barriers of communication were the language and the distance. Long- distance communication was restricted to sending message letters through ships, trains, and aeroplanes. With the discovery of electricity in the late nineteenth century, human communication took a dramatic turn. Telephone was invented in 1876. Speech was converted into electrical signals which were sent using wires to long distances and were received through a receiver set where the electrical signal was converted to sound. Radio was introduced in 1895.

Communication systems like telephone, radio, television, e-mail and communication through personal computers have increased the speed of communication. These communication systems, infact, have changed our lives.

Access to information has changed the way we do things today be it in office or at home. Internet facility, for example, is so essential that without it we cannot communicate quickly and achieve much success at our workplace.

18.2 ► Elements of a Communication System

The basic components of a communication system are a transmitter, a communication channel or medium, and a receiver. In most cases, human beings generate the information to be transmitted. The information is called the intelligence signal or simply intelligence. The signal is given as input to the transmitter which then transmits the information through the communication channel or the medium. The medium of communication may be wires or it may be wireless. The message sent is received by the receiver at the receiving end and is sent to places for use. The basic elements of a communication system are shown in Fig. 18.1. When a message is sent through long distances, any sort of interference reduces the quality of the message being communicated. The interference may be due to some undesired electrical voltage or current appearing at the output.

Transmitter is an electronic device that converts the message into signals which are easy to transmit through the communication channel or the medium. A simple form of a transmitter is the microphone we often use to communicate. The speech is converted into electrical signal and is transmitted through the medium, that is, the wire connecting the microphone with the loudspeaker.

The communication channel is used to send the message from one place to another. Communication channel could be a pair of wires, a fibre-optic cable, or it could be wireless. Wireless communication is referred to as radio. Signals are communicated from one place to another (usually long distances) using electromagnetic field waves. Signals are converted into electric and magnetic fields that propagate easily over long distances in a minute of a second.

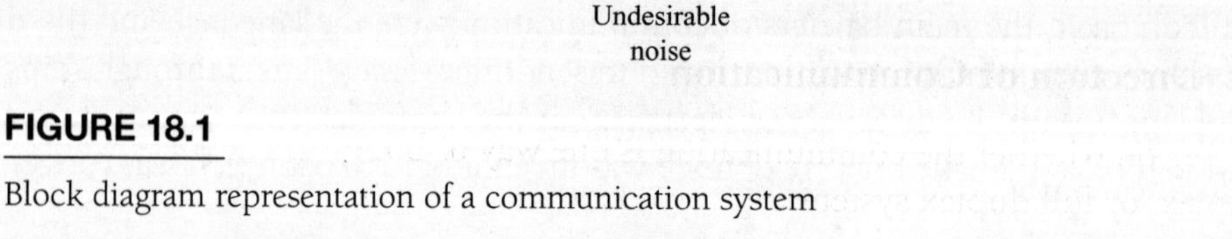

FIGURE 18.1

Block diagram representation of a communication system

When a signal is transmitted, it gets reduced in magnitude considerably. That is why it becomes necessary to amplify the message signal at the transmitter and at the receiver using electronic amplifiers. Radio communication is wireless and requires no electrical connection between the transmitter and the receiver through wires or cables. At the transmitter end, one antenna transmits the signal through free space in the form of electromagnetic waves using a high-frequency carrier wave (called modulation). The receiver picks up the signal through another antenna (receiving aerial) and then separates the signal from the carrier (demodulation).

Receiver converts the signal received back to its original form, that is, it separates the message signal from the carrier (called demodulation). The receiver is a complex electronic circuit.

Noise is undesirable electric energy which enters the communication channel because of a number of reasons. The noise signal, when amplified, creates disturbances in the signal received. Noise comes from the atmosphere in the form of radiation energy, from electrical interference created by the electronic equipment used, etc.

Receiver receives the signal and converts it back to the original information. For example, when we use an earphone, it is a receiver which converts the transmitted signal into the original message which may be music or some other form of message.

18.3 ► Basics of Electronic Communication

In this section, we shall discuss types of signals, types of communication, and the need for modulation.

18.3.1 Types of Signals Used in Communication

Transducers are used to convert information into a proportional signal. The signals are of two types. They are called analog signals and digital signals.

1. *Analog signals*: Analog signal is a continuously varying voltage or current. When we speak using a microphone, we generate an equivalent analog electrical voltage signal which continuously varies with time.
2. *Digital signals*: Data used in computers is digital where binary codes are used representing numbers, letters, symbols, etc. In earlier times, telegraph system used Morse code and radio telegraphy used code of dots and dashes for communication.

Analog signals are often converted into digital form using analog to digital converters for communication. There are a number of advantages of digital communication over analog communication.

18.3.2 Direction of Communication

Depending on whether the communication is one way or both ways, it is known as simplex, half duplex, or full duplex system.

For example, radio and TV communicate information in one direction only. They simply transmit and do not receive any information. Such systems are called *simplex systems*. When communication is both ways, the system is called *duplex system*. Telephone communication is a two-way communication and is a duplex system. There are communication systems which are bidirectional but not simultaneous, that is, communication is possible in one direction at a time. Such systems are called *half duplex systems*. Walky-talky set used by security forces is an example of a half duplex system.

18.3.3 Baseband and Broadband Communication

The original information signals, whether analog or digital, are called *baseband signals*.

When baseband signals are directly communicated through the channel or the medium, it is called *baseband transmission system*. In telephone communication system, the human voice gets converted into analog electrical signal and then the signal travels through the wires over some distance. In local area computer network (LAN), co-axial cables are used to directly transmit data from one computer to the other. Thus, in baseband transmission, the original information signals are directly transmitted. However, there are limitations to baseband transmission.

Communication using a carrier frequency wave is called ***broadband communication***. The process of using a high-frequency carrier wave to transmit a baseband signal containing the information is called *modulation*.

18.3.4 Modulation

Modulation is a process of using a high-frequency carrier wave for sending information which is comparatively of low frequency to distant places. The baseband signal containing the information modulates (alters) the carrier frequency in three ways, that is, by modulating the amplitude or the frequency or the phase. The information signal is said to be impressed upon the carrier wave. In other words, we may say that the information signal is made to ride on the carrier wave and get transmitted. The carrier wave is a high frequency sine wave. Its frequency is higher than the information signal frequency. The information signal is called the modulating signal or wave. The characteristics of the carrier wave in terms of its amplitude, frequency, and phase can be changed by the modulating signal. Accordingly, the modulating methods are called *amplitude modulation, frequency modulation*, and *phase modulation*, respectively. Fig. 18.2 shows the wave shape of an amplitude modulated carrier signal. The information signal, that is, the baseband, is a sine wave of low frequency. The carrier is a high-frequency sine wave. The carrier frequency wave is generated by the transmitter electronic circuitry. The message signal, that is, the information signal, is made to ride over the carrier signal through a process called modulation. Thus, ***modulation is the process of placing or fixing the information signal over a high-frequency transmitter generated carrier wave for long-distance transmission***.

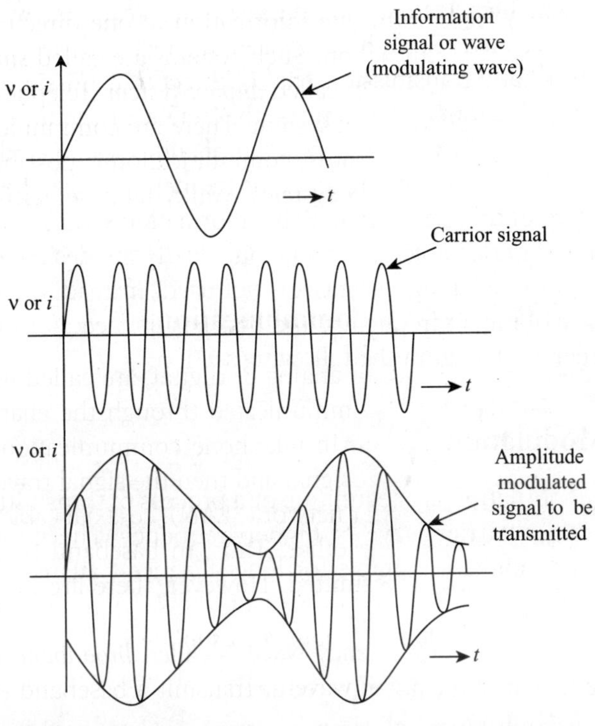

FIGURE 18.2

Process of amplitude modulation illustrated

The question that may be asked at this stage is to why should we require a carrier wave for long- distance communication of an information signal? What is the need for modulation of a carrier wave by the information signal? This is explained as follows.

The information signals are generally low-frequency signals. For example, the frequency of voice signal ranges from 300 to 3000 Hz. They are used to modulate a carrier of higher frequency, of the range of megahertz. The modulated electromagnetic wave travels through the free space at a high speed. A receiver picks up this modulated electromagnetic wave. This is then converted into electrical signal. The carrier and the information signals are separated out through a process called demodulation, and thus the original message is recovered at the user end. However, if we directly send the information signal, let us examine the problem that we would face. This is explained with the help of an example. Let us consider that the speech delivered by someone has to be transmitted to a distant place. The frequency of voice signal created through the microphone ranges from 300 to 3000 Hz. A sine voice signal could be amplified through an electronic amplifier and then connected to an antenna through a wire connection (as you are aware, for short-distance communication in large gatherings, we use microphones, amplifiers, and speakers).

Antenna is a device which can radiate and receive electromagnetic waves. The electromagnetic waves will get propagated through the space and received by another antenna at the receiving station. For baseband transmission, that is, direct transmission of the low-frequency information signal using electromagnetic waves, the height of the antenna required will be very large. Again, if everyone transmits large number of audio, video, and data simultaneously using baseband communication occupying the same frequency range, the frequency range will be nothing but mixing up of different signals. The information signals will interfere with each other and the receiver will not be able to separate out clearly the desired signal. For these reasons, baseband transmission for long distances is not used. Instead, the information signal is sent using a carrier wave of higher frequency, and hence is the need for modulation. The need for modulation is further explained in the following section.

18.3.5 Need for Modulation

As mentioned earlier, modulation is a technique or a process of transmitting a low-frequency information signal using a carrier wave of higher frequency. The message signal is super imposed on the carrier to make it suitable for long-distance transmission through the free space as electromagnetic waves.

The advantages of using modulation technique are as follows:

1. Reduction in the height of the antenna required
2. Increase in the range (distance) of communication
3. Avoidance of mixing up of information signals
4. Improvement of quality of reception of the information sent
5. Allowing multiplexing of signals, that is, making possible transmission of more signals simultaneously using the same channel
6. Allowing adjustment of bandwidth.

Here, let us, for the sake of understanding of requirement of height of antenna, take an example.

EXAMPLE 1.1

Calculate the height of the antenna required for baseband transmission of a signal of frequency 3 kHz. Also calculate the height of antenna required when a carrier frequency of 2 MHz is used for transmitting a modulated signal.

SOLUTION

The height of the antenna is taken as fraction of the wavelength of the frequency of transmission signal. The minimum height of antenna required is generally taken as one fourth of the wavelength. The wavelength is calculated using the relation.

$$\lambda = \frac{c}{f} \tag{18.1}$$

Where λ is the wavelength, c is the velocity of light, and f is the frequency. From equation (18.1) it is observed that if frequency, f, is low wavelength, λ is high.

The height of antenna required when $f = 3000$ Hz is

$$h = \frac{\lambda}{4} = \frac{c}{f \times 4} = \frac{3\times10^8}{3\times1000\times4} = 25000 \text{ m} = 25 \text{ km}$$

An antenna of 25 km height is practically impossible to build, and hence, this concept is rejected. Now, let us calculate the height of antenna required if a modulated frequency wave of $f = 4$ MHz is used. The height of antenna required is

$$h = \frac{\lambda}{4} = \frac{c}{f \times 4} = \frac{3\times10^8}{4\times10^6\times4} = 18.75 \text{ m}$$

An antenna of height 18.75 m is suitable for installation. This justifies the use of modulation technique in long-distance communication.

Modulation helps to avoid mixing of signals sent from different sources. A communication system using modulation technique is represented in block diagram form as in Fig. 18.3.

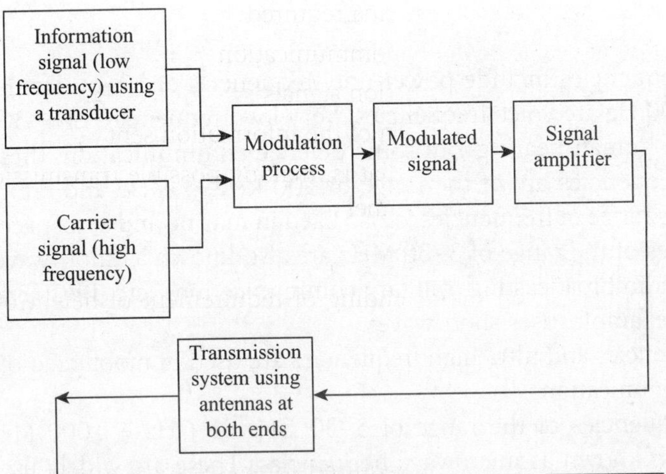

FIGURE 18.3

Block diagram representation of a long-distance communication system using modulation technique

Before we discuss the different types of modulation, let us have a look at the various frequency ranges of the electromagnetic spectrum. Table 18.1 depicts the various segments of radio frequency spectrum.

TABLE 18–1 Radio Frequency Spectrum

Frequency range	Name or Designation	Wave length
30–300 Hz	Extremely low frequencies (ELF)	10^7–10^6 m
300–3000 Hz	Voice frequencies (VF)	10^6–10^5 m
3–30 KHz	Very low frequencies (VLF)	10^5–10^4 m
30–300 KHz	Low frequencies (LF)	10^4–10^3 m
300 KHz–3 MHz	Medium frequencies (MF)	10^3–10^2 m
3–30 MHz	High frequencies (HF)	10^2–10^1 m
30–300 MHz	Very high frequencies (VHF)	10^1–1 m
300 MHz–3 GHz	Ultra high frequencies (UHF)	1–10^{-1} m
3–30 GHz	Super high frequencies (SHF)	10^{-1}–10^{-2} m
30–300 GHz	Extra high frequencies (EHF)	10^{-2}–10^{-3} m

Extremely low frequencies include power line frequencies of 50–60 Hz. Frequencies of the range of 300–3000 Hz are voice frequencies. Very low frequencies of 3–30 kHz range are at the higher end of human hearings. In some defence communication, these frequencies are used. Medium frequencies are of the range of 300 kHz–3 MHz and are used in AM radio broadcasting. This range of frequencies is also used in marine and aerospace communication.

High frequencies of the range of 3–30 MHz are also known as short waves (SW). Applications are in SW radio broadcasting, military communications, etc. BBC (British Broadcasting Corporation), for example, uses short waves.

Very high frequencies and ultra high frequencies are used in mobile radio, FM radio, television channel communications.

Super high frequencies of the range of 3–30 GHz (1 GHz = 1000 MHz and 1 MHz = 1000 kHz) are also known as microwave frequencies. These are widely used in satellite and radar communications.

Extra or extremely high frequencies of the range of 30–300 GHz are yet to be used extensively. Equipments required to generate and use these frequencies are complex and very costly.

Electromagnetic waves higher than 300 GHz are also called *microwaves*. Their wavelengths are extremely small, of the order of millimeters and nanometers. For example, infrared signals are produced by radiation from heat source. Infrared signals are used in some special applications like guided missiles, astronomical field to detect position of stars, etc. TV remote control set uses infrared signals to change TV channels, volume control, etc. Spectrum of waves above the infrared region is the visible spectrum referred to as light. Light is electromagnetic radiation of extremely minute wavelengths.

Light waves are used in communication in many ways. Optical fibre communication uses light waves and is becoming very popular these days. Light waves can be modulated by the information signals and sent through glass fibres very efficiently. Lasers can generate light beams which can be modulated by information signals like voice, video, etc.

Above the visible spectrum are the x-rays, gamma rays, and the cosmic rays which are also electromagnetic radiations but are not used in communication systems.

18.4 ▶ Bandwidth and Its Significance

Bandwidth is the difference between the upper and lower frequency limits of a signal. Fig. 18.4 shows the bandwidth of a signal having frequency range of f_1 and f_2.

The bandwidth BW is

$$BW = f_2 - f_1 = 3000 - 300 = 2700 \text{ Hz}.$$

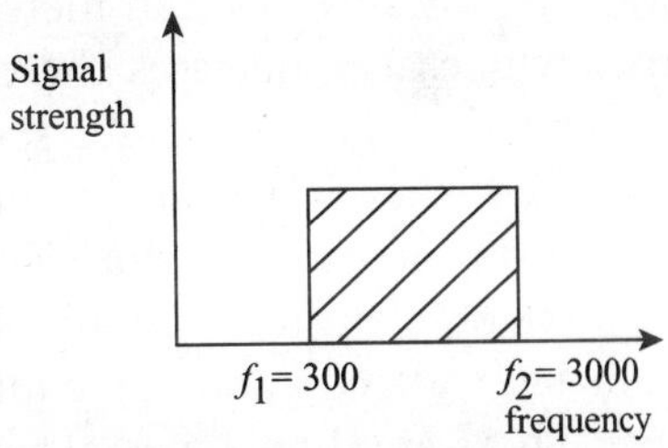

FIGURE 18.4

Bandwidth of a signal occupying a portion of the frequency spectrum

Bandwidth may be defined as the range of frequencies which is occupied by a signal.

The same signal when modulated on to the carrier will occupy only a small portion of the available electromagnetic spectrum. Suppose an audio signal of 5 kHz is modulating a carrier of 1000 kHz, then the range of frequencies that would contain the signal will be 1000 ±5 kHz, that is, 995 kHz to 1005 kHz. The channel bandwidth is 1005 – 995 = 10 kHz.

Because of the use of carrier of higher frequency, the spectrum space occupied is less now. The signal with a bandwidth of 10 kHz will occupy only a small portion of the electromagnetic spectrum at 1000 kHz. It is just 1% of the spectrum. If a carrier of higher than 1000 kHz is used, a smaller percentage of spectrum space will be used. For low-frequency information signals there is more space available when higher value of carrier frequency is used. There is more spectrum space for the information signal at higher frequencies. Signal with high bandwidth must be transmitted using carrier of very high frequencies (VHF) or ultra high frequencies (UHF). A TV signal of bandwidth of say 5 MHz should use a carrier of a few gigahertz because in that range of frequencies there will be enough space for such high-bandwidth signals.

The VHF and UHF portion of spectrum have plenty of space. Modulation technique can be used to send information signal having considerable bandwidth. However, there is lot of competition between the companies for using these frequencies for communication. Different countries are utilizing these spectrum space for their communication needs. Thus, there is small space left unused in the high-frequency spectrum also. Research is now being carried out on the methods of reduction of bandwidth requirement, that is, to find out means as to how to send maximum information utilizing minimum of bandwidth.

As we have seen earlier, bandwidth is the range of frequencies that contain an information. When modulation is done, other signals called ***sidebands*** are created. Sideband frequencies are above and below the carrier frequency. If the carrier frequency is 2000 kHz and the modulating frequency is 10 kHz, then sidebands will be produced at both lower end and upper end of the carrier frequency. In this case, the sidebands will be at 2000 – 10 = 1990 kHz and 2000 + 10 = 2010 kHz. Here, the bandwidth is 2010 – 1990 = 20 kHz. This 20 kHz can also be called the ***channel bandwidth*** as it occupies 20 kHz of the frequency spectrum at 2000 kHz range, that is, from 1990 kHz to 2010 kHz.

18.5 ▶ Types of Modulation

We have realized that for long-distance transmission of information signals having low frequencies like voice signal, video signal, or digital data, modulation technique has to be used. Information signals like sound signals are low frequency and due to less energy they cannot travel long distance. We can increase the energy by modulating them using high-frequency carrier which also increases the efficiency of transmission. A high-frequency carrier wave is generated at the transmitter station and the information signal is used to modulate the carrier frequency wave. In this process, the low-frequency information signal gets superimposed on the carrier signal. The process of modulation creates a higher frequency modulated wave containing the information signal. There are three types of modulation used in communication.

1. Amplitude Modulation (AM)
2. Frequency Modulation (FM)
3. Phase Modulation (PM)

18.5.1 Amplitude Modulation

In amplitude modulation, the amplitude of the carrier is varied by the modulating signal. As shown in Fig. 18.5, the information signal is called the ***modulating signal***. A high-frequency sinusoidal wave is used as the carrier wave. The carrier wave or signal gets modulated by the modulating signal.

As can be seen from Fig. 18.5(c), a higher frequency signal containing the original information signal is created for long-distance transmission. The amplitude of the modulated signal varies in accordance with the information signal. The magnitude of

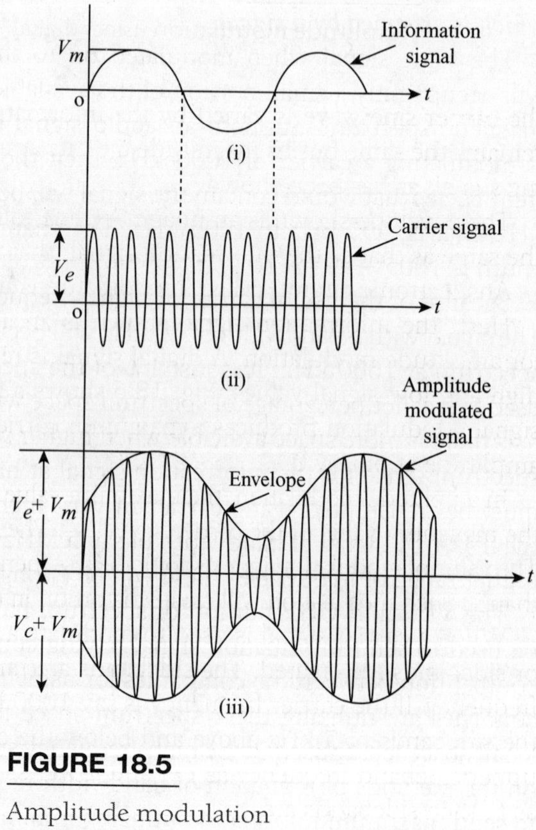

FIGURE 18.5

Amplitude modulation

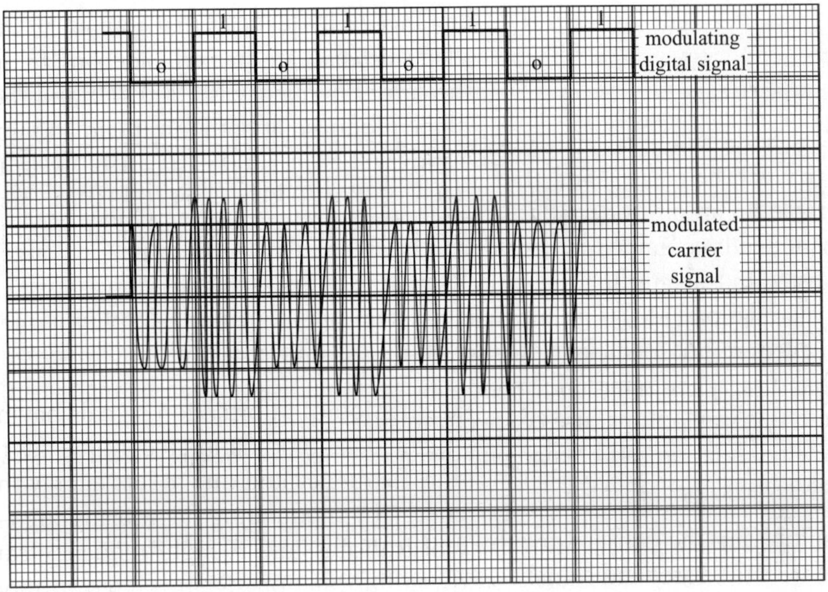

FIGURE 18.6

Amplitude modulation using digital information signal

the carrier sine wave is varied by the information signal. The frequency of the carrier signal remains the same but its magnitude, i.e., its amplitude varies in accordance with the modulating signal, as has been shown.

The envelope shown is an imaginary line joining the peaks of the modulated wave. This is the same as that of the information signal, that is, the modulating signal.

An electronic circuit, called a modulator, is used for creating amplitude modulation.

Here, the information signal shown is an analog signal. Digital signals can also be used for amplitude modulation. A digital signal is represented by binary numbers 1 and 0, that is, high and low-signal values. Fig. 18.6 shows a binary modulating signal modulating a carrier signal. Modulation produces a maximum carrier amplitude of binary 1 and minimum carrier amplitude at binary 0.

In amplitude modulation, the maximum value of the modulating signal, V_m, should be less than the maximum value of the carrier signal, V_c. The ratio of V_m and V_c is called modulation index, *m*. The value of modulation index should be less then 1. If V_m is higher than V_c, then m becomes greater than 1. Such modulation will create distortion in the modulated signal which is not desirable.

When a carrier signal is used to modulate a high-frequency carrier signal, side frequencies or sidebands are created. The sidebands are range of frequencies below and above the carrier frequency. If the carrier frequency is say 1000 kHz, and the modulating frequency is 10 kHz, the sidebands of 5 kHz above and below the carrier frequency will be created. If we call the upper sideband frequency as f_{USB} and lower sideband frequency as f_{LSB}, and carrier frequency as f_c, then sidebands are

$$f_{\text{USB}} = f_{\text{c}} + f_{\text{m}}$$

$$f_{\text{LSB}} = f_{\text{c}} - f_{\text{m}}$$

When f_{m} is not a single frequency wave, a range of frequencies above and below the carrier frequency will be created and they will occupy a large chunk of spectrum space which is not desirable. In ampli- tude modulation there is waste of power. Only a small part of the power transmitted is contained in the sidebands which carry the information. Large amount of power is wasted in transmitting the carrier signal together with the information signal. Thus, amplitude modulation (AM) is an inefficient system of modulation. Amplitude modulation is simple and effective but inefficient. It is interesting to note that both the upper sideband and lower sideband carry the same information and communication can be made using only one sideband and this way optimal use of spectrum space can be made.

Amplitude modulation can be made by using modulation circuits. This can be achieved by any of the two ways, viz

1. by multiplying the carrier wave by a gain or attenuation factor that changes in accordance with the baseband signal or the intelligence signal or
2. by combining the information signal and the carrier signal and then the output applied to a non-linear circuit.

Modulation index is defined as the ratio of the amplitude of the modulating wave to that of the carrier wave.

18.5.2 Frequency Modulation

In frequency modulation (FM), the information signal modulates the frequency of the carrier signal. As the amplitude of the information signal (i.e., the modulating signal) increases, the frequency of the carrier signal increases. Similarly, as the amplitude of the information signal decreases, the frequency of the carrier signal decreases, as has been shown in Fig. 18.7. Maximum frequency of the modulated signal occurs when the amplitude of the modulating signal reaches its maximum value. When the modulating signal becomes negative, the frequency of the carrier will decrease and will continue to decrease till the modulating signal reaches its negative maximum value.

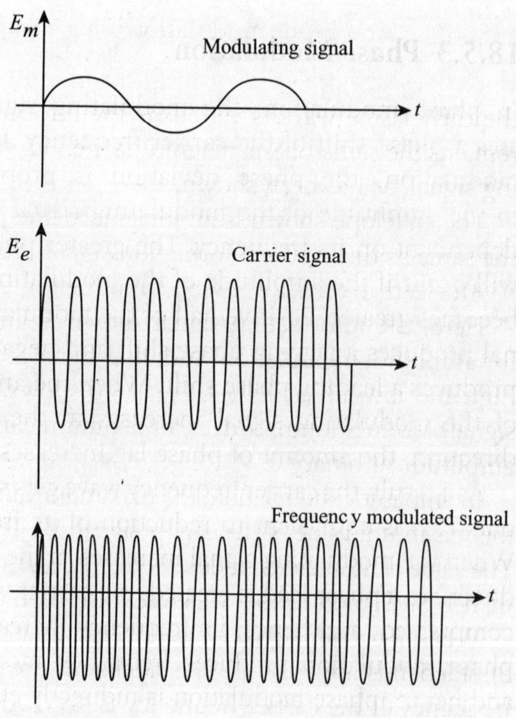

FIGURE 18.7

Frequency modulated signal

Thus, frequency modulation creates compression and expansion of the modulated signal. The information signal is transmitted by the transmitter in the form of variation of frequency of the modulated signal.

The primary advantage of frequency modulation over amplitude modulation is that a frequency modulated signal does not get affected by noise. Noise is short-duration amplitude variations. Noise is caused by factors like lightening, power transients, fluorescent tube light operations, etc. Further, the advantage of frequency modulation over amplitude modulation is its higher transmitter efficiency.

In frequency modulation, the modulating frequency creates a deviation in the frequency of the carrier signal. The carrier frequency increases and decreases in every half cycle of the modulating signal. The maximum deviation occurs at the maximum values of the modulating signal. The number of times this deviation of the frequency of the carrier will make above and below its normal frequency will depend upon the frequency of the modulating signal. The frequency deviation rate is decided by the frequency of the modulating signal, and the amount of deviation is decided by the amplitude of the modulating signal. Thus, in frequency modulation the message or the intelligence is transmitted by varying the frequency of the modulating signal. Modulation index, here, is the frequency deviation to the modulating frequency.

18.5.3 Phase Modulation

In phase modulation, the modulating signal creates a phase shift of the carrier frequency. In phase modulation, the phase deviation is proportional to the amplitude of the modulating signal and not dependent on its frequency. The greater phase shift will occur if the amplitude of the modulating signal becomes greater. Positive half of the modulating signal produces a lagging phase shift and negative half produces a leading phase shift. When the amplitude of the modulating signal increases in the positive direction, the amount of phase lag increases.

As a result, the carrier frequency wave gets stretched out. This is equivalent to reduction of its frequency. When the modulating signal increases in the negative direction, a phase lead is created, the carrier wave gets compressed, and hence, its frequency increases. The phase modulation produces frequency modulation, and hence, phase modulation is indirectly equivalent to frequency modulation. A phase modulated signal has been shown in Fig. 18.8.

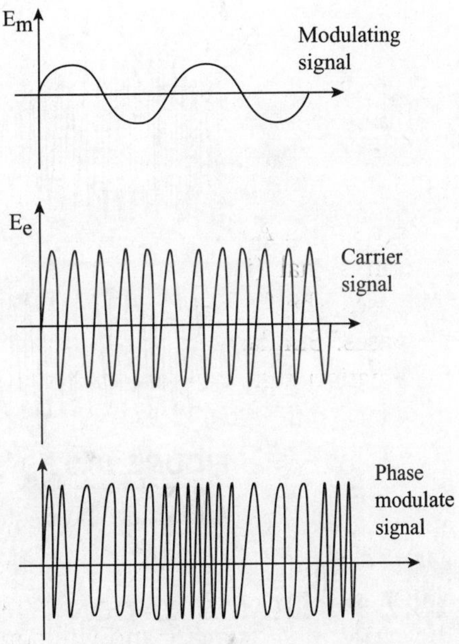

FIGURE 18.8

Phase modulated signal

18.6 ► Comparison of Amplitude Modulation and Frequency Modulation

Both AM and FM are suitable for long-distance communication. However, FM is considered superior method than AM. The advantages of FM over AM are as follows:

1. Transmitting efficiency is higher
2. Rejects interfering signals
3. Immune to external noise

FM however occupies more spectrum space because bandwidth of an FM signal is wider than that of an AM signal. Further, the circuits necessary to produce and demodulate an FM signal are more complicated than those required for an AM signal.

To sum up, we can say that modulation and demodulation are the essential processes of a communication system. For modulation, a high-frequency carrier wave is needed. The three parameters of this carrier wave, that is, its amplitude or frequency or phase are varied by the message signal. Demodulation is the inverse of modulation. In demodulation, we get back the message signal from the modulated signal. Fig. 18.9 shows how the message signal modulates the amplitude and frequency of the carrier wave.

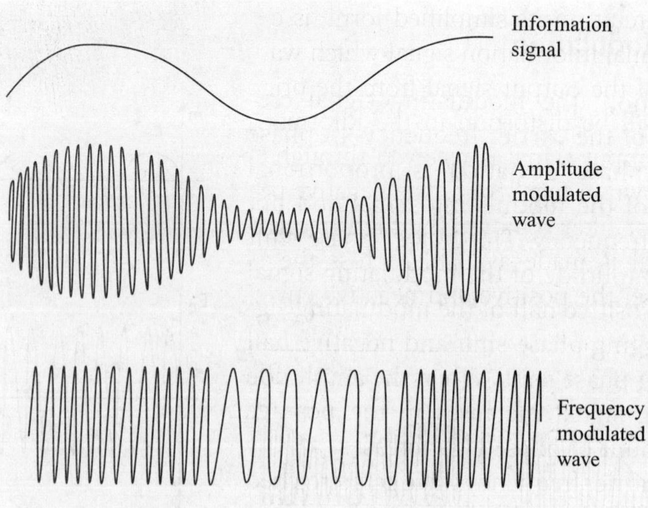

FIGURE 18.9

Amplitude and frequency modulation

18.7 ► Demodulation

After modulation, the modulated signal is transmitted in the form of electromagnetic waves using an antenna. At the receiving end, the message communicated is detected. The detection or demodulation is the process by which the intelligent signal, that is, the message, is recovered from the modulated signal. *Demodulation is the process of recovering the original message signal from the modulated carrier signal or wave. In other words, we can state that demodulation is the process of extracting the original information containing signal from the modulated carrier wave.*

A demodulator is an electronic circuit that is used to recover the information contained in the modulated signal received by the receiver. Different kinds of circuits are used as demodulator depending upon whether the carrier signal is amplitude modulated or frequency modulated or phase modulated.

It is necessary to separate out the information signal from the carrier. This is because the modulated carrier frequency as such cannot be used directly at the output. For example, in case of sound signal, the modulated frequency cannot be given to the loudspeaker because the loudspeaker diaphragm cannot vibrate at this high frequency. Audio frequency signal has to be separated out from the radio frequency carrier wave. Radio frequency wave is also not audible by human ear. Hence, we need to recover the original signal from the modulated carrier signal or wave.

Amplitude demodulation process has been illustrated graphically in Fig. 18.10. The modulated signal containing the message signal is received through the antenna at the receiving end of the trans- mission—reception system. The signal is first rectified and then filtered in the detector.

For the purpose of demodulation, the modulated wave is fed to a circuit as shown in Fig. 18.11. The circuit, in its simplified form, is called detector. The output of the detector should be the original information signal which was transmitted after modulation process.

Any variation of the output signal from the original information signal must be avoided, otherwise there will be distortion in the information received, be it a voice signal or any data. The desired input signal is received through the tuning circuit. The diode rectifies the input signal as shown in Fig. 18.11. The negative portion of the modulated signal is removed. A suitable capacitor filter removes the high-frequency carrier wave from the rectified wave. The original signal is made available across the resistor. Rectification is necessary before filtering. Otherwise, the positive and negative envelopes of the modulated signal will cancel

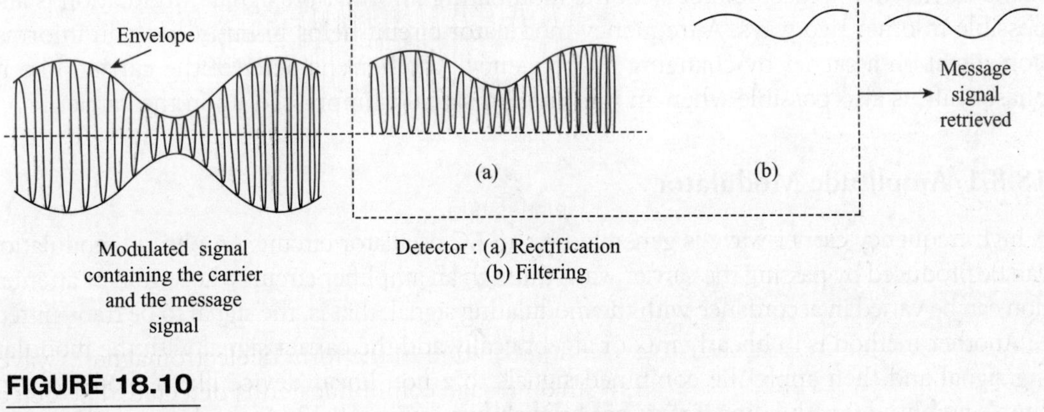

FIGURE 18.10

Process of demodulation

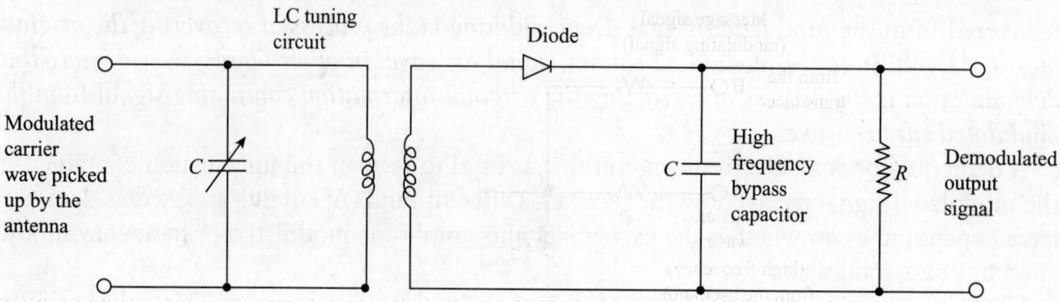

FIGURE 18.11

Basic circuit of a diode AM detector (demodulator)

each other after filtering and the original signal will get lost. That is why rectification is done before filtering. In case the modulated signal is not symmetrical, that is, upper side is not the same as the lower side, one half cannot be removed.

The quality of a detector is expressed in terms of linearity, sensitivity, and signal handling capacity. Linearity is measured as to how accurately the output of the detector follows the input. That is to say that the amplitude of the output of the detector needs to be linearly proportional to the input amplitude. Sensitivity is a measure of how much of the input signal is delivered as the output. Signal handling capacity is measured in terms of the ability of the detector in accepting a signal without any distortion of the original signal. So far, we have discussed demodulation of an amplitude modulated wave.

Now let us discuss in brief modulators used to modulate a carrier wave.

18.8 ► MODULATORS

Modulators are electronic circuits used to modify a carrier wave in accordance with the information signal to be transmitted. An amplitude modulation circuit will cause amplitude of the carrier vary in accordance with the modulating information signal. Modulation is also possible in other two ways. A frequency modulator circuit helps in impressing an information signal on a carrier by changing its frequency. Another character of the carrier, like its phase shift, is also possible when an information signal is impressed upon the carrier.

18.8.1 Amplitude Modulator

A high-frequency carrier wave is generated by an LC oscillator circuit. Amplitude modulation can be produced by passing the carrier wave through an amplifier circuit whose gain or attenuation can be varied in accordance with the modulating signal, that is, the signal to be transmitted.

Another method is to linearly mix or algebraically add the carrier signal with the modulating signal and then apply the combined signals to a non-linear device like a diode. A very simple amplitude modulation circuit has been shown in Fig. 18.12. As can be seen, the circuit

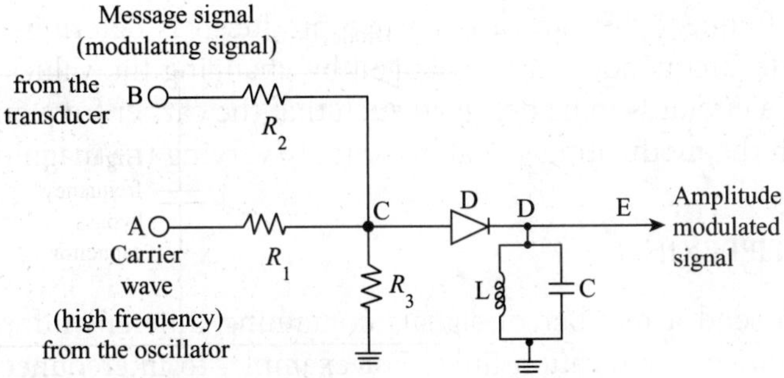

FIGURE 18.12

Amplitude modulation circuit in its simplest form

contains a mixing network of resistors (the voltage drops are added), a diode rectifier, and LC tuning circuit. The carrier wave generated by the oscillator is connected to resistor R_1 and the modulating message signal from the transducer is connected to resistor, R_2. A transducer converts the information signal into an equivalent—electrical signal, for example, a microphone converts voice into an equivalent electrical signal. Therefore microphone is called a transducer.

The parallel sensitive network containing R_1 and R_2 adds the two signals linearly. Assuming that the two signals are sine waves of different frequencies, their sum will be as shown in Fig. 18.13. Diode D in Fig. 18.12 cuts off the negative portion of the input signals. The positive pulses are applied to the tuned circuit made up of L and C. Fig. 18.13 is to be examined with reference to the circuit shown in Fig. 18.12.

High-amplitude positive pulses create a corresponding negative pulse for every positive pulse. This is because every time diode conducts, a pulse curve flows through the tuned circuit. The capacitor and the inductor interchange energy creating an oscillation. The output of the oscillator is therefore an amplitude modulated signal.

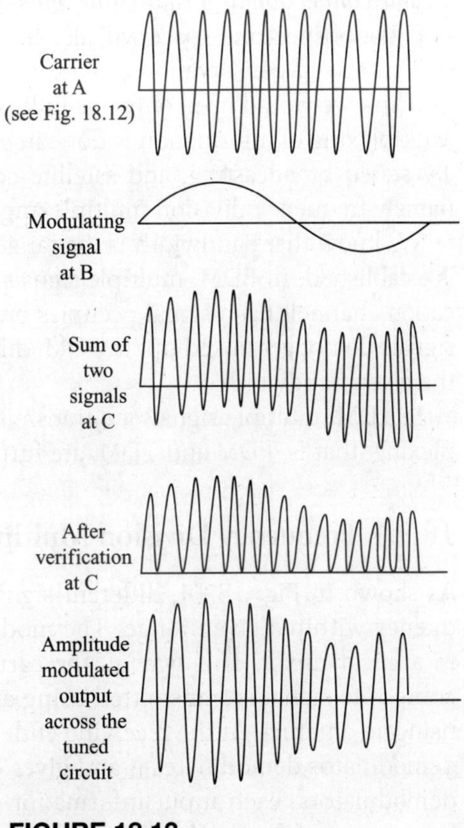

FIGURE 18.13

Amplitude modulator wave forms

18.8.2 Frequency Modulator

Frequency modulator is an electronic circuit where a carrier frequency is generated using either an LC circuit oscillator or a crystal oscillator. In an

LC oscillator, the frequency of the carrier wave produced is determined by the values of L and C. The carrier frequency can be changed by changing the values of inductance and capacitance. Now, a circuit is to be designed such that the carrier frequency can be changed in accordance with the modulating signal by suitably varying the magnitudes of L and C.

18.9 ► MULTIPLEXING

When we have to send a number of signals containing different information, we would require a number of communication links. For example, to interconnect all the telephone lines with each other we would require a large number of wires to be connected making it a complicated network. When large number of data signals are to be communicated, multiple cable connections and radio links will have to be established. This would be practically a complex task and also very expensive. It has been possible to use a single cable connection or a single radio link to transmit a large number of information.

Multiplexing is a technique or a process by which a large number of information signal can be communicated through a single communication channel. The communication channel could be a cable connection or a radio link. This way communication will be less complex and inexpensive. Lot of spectrum space will also be conserved. Thus, the advantage of multiplexing is cost reduction and conservation of spectrum space. Instead of establishing multiple communication link we would need only one link and of course the equipment required for multiplexing. Multiplexing of information is done in case of telephone communication, telemetry, radio and TV stereo broadcasting, and satellite communication. There are two types of multiplexing, namely frequency division multiplexing (FDM) and time division multiplexing (TDM).

We know that bandwidth is the range of frequencies over which a communication channel is established. In FDM, multiple signals are adjusted within the bandwidth of one communication channel. Each signal occupies only a small portion of the available bandwidth and the signals do not get mixed up. In FDM, different signals are assigned different frequencies within the given bandwidth.

In TDM, multiple signals are transmitted using different time slots. The two types of multiplexing, that is, FDM and TDM, are further discussed in the following sections.

18.9.1 Frequency Division Multiplexing (FDM)

As shown in Fig. 18.14, different signals modulate carriers. Each carrier has different frequency within a given range. The modulated carriers are then mixed, that is, summed up, in a linear mixer. This way all the carriers are combined into a composite signal and this composite signal is transmitted using a transmitter over the single communication channel using an antenna. At the receiving end, an antenna receives the electromagnetic signal and a demodulator demodulates it and gives out the composite signal. Using bandpass filters and demodulators, each input information signal is recovered as has been shown. The carrier frequencies of f_{c1}, f_{c2}, f_{c3}, etc., are the carrier frequencies used for each information signal.

The bandpass filter used at the receiving end passes only its channel frequency and rejects the others. As mentioned earlier, FDM is used in telemetry, telephone systems, stereo FM, etc.

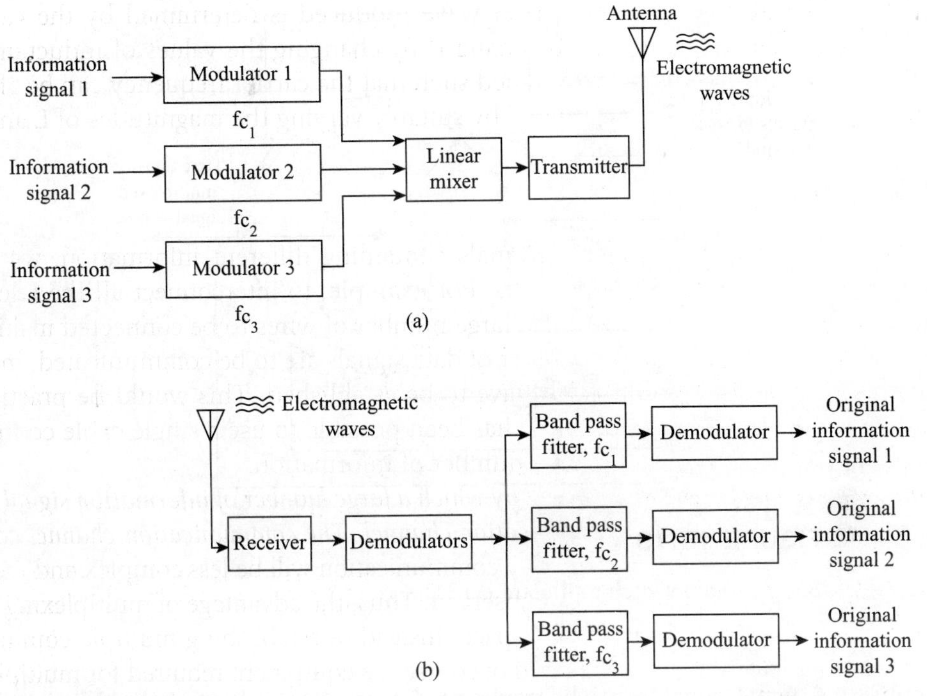

FIGURE 18.14

Schematic representation of frequency division multiplexing: (a) transmitting end; (b) receiving end

To sum up, we can state that in FDM large number of signals are transmitted using a single channel sharing the channel bandwidth range. This is achieved by assigning each information signal a small portion of spectrum space within the channel bandwidth. This way lot of spectrum space is preserved and cost and complexity of transmission of multiple signals get reduced.

18.9.2 Time Division Multiplexing (TDM)

Here also by using TDM multiple signals can be transmitted over a single channel by time sharing mechanism. The whole of the bandwidth available can be utilized by each signal but on time sharing basis. Of the multiple signals to be transmitted, each signal is allowed to use the channel for a short period of time turn by turn. Some mechanism of switching off each of the signals in a sequence for a very brief period of time is made, as shown in Fig. 18.15. Here for simplicity of understanding a mechanical switching arrangement has been shown. This, however, is done using electronic switching circuits. Here transmission of all the signals is done continuously by time sharing. Time division multiplexing is possible for both analog and digital signals.

For transmission of digital signals using TDM, data to be transmitted take their turn for transmission. Equal time slot is allocated for each signal and a particular sequence is maintained.

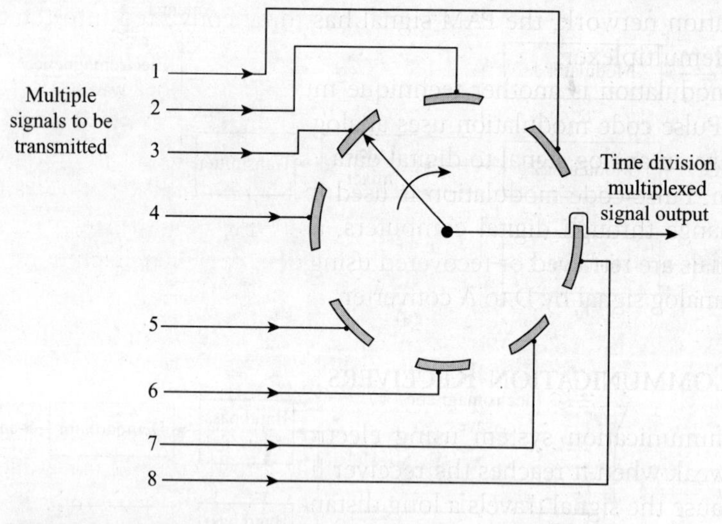

FIGURE 18.15

Simple rotary switch to illustrate a TDM

For example, if 4 digital signals are to be transmitted over a single channel, the signals will take their turn as 1 2 3 4 1 2 3 4 1 2 . . . . This is the basic concept of TDM.

Analog signals, like voice signals, can also be transmitted using TDM system. The analog signal is sampled at a very high rate using electronic switching technique. Samples are the values of the analog signal after each short interval of time. By taking samples of the analog signal at a very high rate, pulses whose magnitude vary in accordance with the analog signal can be produced. The pulses produced can be considered as equivalent to the magnitude of the analog signal with very short intervals of time. This process of sampling an analog signal is called pulse amplitude modulation (PAM). Using PAM technique, an analog signal is converted into series of pulses proportional to the magnitude of the analog signal as represented in Fig. 18.16.

The electronic circuit required for conversion of a continuously varying analog signal into equivalent pulses is essentially a switching circuit comprising mono-shot multivibrator which generates narrow fixed width pulses. These pulses open and close a gate for very small durations continuously allowing the signal to pass through when the gate opens. The gate circuit is designed using field effect transistors (FETs). Using PAM, the analog signal is converted into a series of pulses. These pulses now can be transmitted using TDM. The multiplexer is basically a rotary switch as has been explained earlier. At the receiving end of

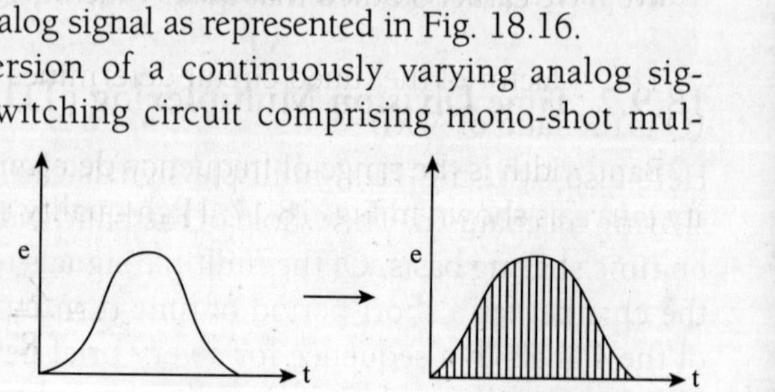

FIGURE 18.16

Pulse amplitude modulation

the communication network, the PAM signal has to be converted into the original analog signal using a demultiplexer.

Pulse code modulation is another technique much in use these days for transmission of analog signals. Pulse code modulation uses analog to digital conversion (A to D conversion) process to change an analog signal to digital data. The digital data is multiplexed using TDM for transmission. Pulse code modulation is used in telephone systems to transmit voice signals, data exchange through digital computers, and video data through TDM. Pulse code modulation signals are retrieved or recovered using demultiplexing technique and conversion from digital to analog signal by D to A converter.

18.10 ▶ Communication Receivers

In wireless communication system using electromagnetic waves, the transmitted signal becomes very weak when it reaches the receiver like a radio receiver or a television receiver set. This is because the signal travels a long distance. Further, the signal gets somewhat distorted due to interference from a large number of other radio signals on its way. These interfering signals are called noise signals or unwanted signals. The function of a radio receiver is to have high level of sensitivity and selectivity so as to reproduce the original modulating signal. *Thus, the function of a receiver is to receive only the desired signal, that is, the modulating signal, and reject all other signals.*

A simple receiver is a tuned circuit. There are LC circuits tuned to resonate at a desired signal frequency. If the resonant frequency matches with the frequency of the signal to be received, then the circuit will pass that useful signal and reject other unwanted signals. The electrical signal received can be converted into sound using a microphone.

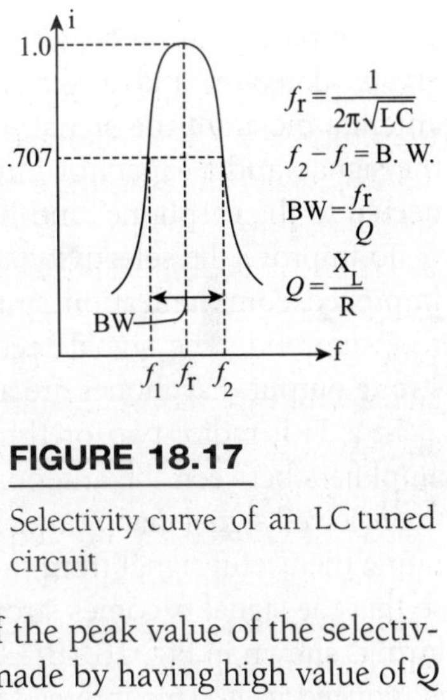

FIGURE 18.17

Selectivity curve of an LC tuned circuit

We have earlier studied that quality factor, $Q = \frac{X_L}{R}$.

The Q-factor of the tuned circuit determines the selectivity (Q is the ratio of *X/R*).

Bandwidth is the range of frequency determined by 0.707 of the peak value of the selectivity curve as shown in Fig. 18.17. High quality tuned circuit is made by having high value of Q

$$Q = \frac{X_L}{R} = \frac{WL}{R}.$$

where L is the inductance of the coil and R is its resistance.

If Q is high, bandwidth will be narrow. This means that the selectivity curve will be more selective.

We can make a simple radio receiver using an antenna, an LC tuned circuit called the detector, and a headphone. The headphone converts the received signal into sound.

The *selectivity* of a receiver is the ability to select a desired frequency signal while rejecting the unwanted noise generating signals.

Sensitivity of a communication receiver is the ability to pick up weak signals. The signal so received must be amplified using an amplifier and then demodulated.

The functions of a radio receiver are as follows:

1. Select the desired signal from all other unwanted signals using LC tuned circuits
2. Amplify the desired signal
3. Demodulate the amplified signal, that is, separate the message signal from the carrier
4. After demodulation, to amplify the original modulating signal
5. Apply the amplified demodulated signal to the transducer, for example, a microphone

Two types of radio receivers are discussed in brief here.

1. Tuned radio frequency (TRF) receiver
2. Superheterodyne receiver.

18.10.1 Tuned Radio Frequency Receiver (TRF Radio)

The simplest radio receiver is a crystal radio consisting of a tuned circuit, a diode called crystal detector, and a head phone (also called an earphone). In this radio receiver, an antenna picks up the signal, the tuned circuit selects the desired signal, the diode rectifies the signal, and a capacitor filters out the carrier frequency wave. The original signal is connected to the earphone, and hence, the message can be heard.

To improve the selectivity and sensitivity, a number of amplifiers are used in a TRF radio. Improved communication system should clearly separate out the information signal at the receiving end. The signal received should also be strong enough to produce a reasonably strong output. Earphones are also replaced by speakers for convenient use.

In a TRF radio, two or three stages of radio frequency amplification are done using RF amplifiers between the antenna and the detector. The RF amplifiers amplify the amplitude of the received signal many times before it is taken to the detector or the demodulator. After separating the useful signal from the carrier, the signal is amplified by one or two audio amplifiers so that the signal becomes strong enough to drive a speaker. A TRF receiver in block diagram form is shown in Fig. 18.18. Electromagnetic waves are received by the antenna and a voltage is induced in it. This induced voltage is due to the various modulated signals from different transmitters sending different kinds of messages.

Radio frequency amplifiers are tuned to select and amplify the desired signal. Tuning of RF amplifiers is done simultaneously. Tuning means making the resonant frequency of the LC tuning circuit equal to the frequency of the desired signal. The amplified signal is then detected by a diode detector and demodulated using a bypass capacitor. Due to demodulation, the carrier signal is bypassed and the original information signal is recovered. This signal is amplified to the necessary power level using audio and power amplifiers before connecting to the speaker. Although TRF receivers were in use for a long time, they had some problems with respect of

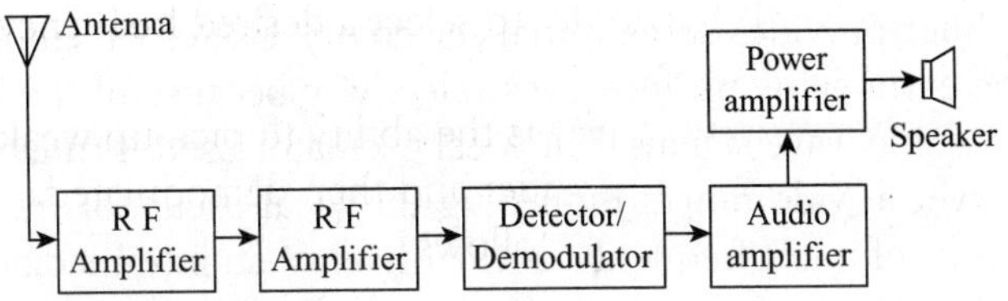

FIGURE 18.18

Block diagrammatic representation of a tuned radio frequency (TRF) radio receiver

selectivity and variation of bandwidth over the tuning range. These problems were overcome in an improved type of receiver called *superheterodyne receiver*.

18.10.2 Superheterodyne Receivers

A superheterodyne receiver uses frequency mixing technique to convert all incoming signals into a fixed lower intermediate frequency which can be more conveniently processed than the radio carrier frequency. Fixed tuned amplifiers can be used to provide fixed level of sensitivity and selectivity using this technique. For gain and selectivity, intermediate frequency (IF) amplifiers are used. Block diagram representation of a superheterodyne receiver has been shown in Fig. 18.19.

To receive a radio signal, a suitable antenna is used. The antenna picks up the radio signal and feeds to the RF amplifier. Amplification of the signal at the input stage is required as the signal received is very weak, of the order of a few microvolts. A tuned circuit used along

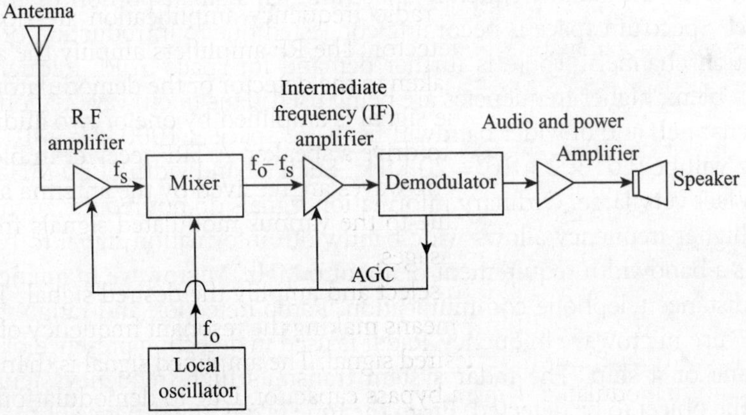

FIGURE 18.19

Block diagram representation of a superheterodyne receiver

with the RF amplifier provides some initial selectivity. Use of RF amplifier improves sensitivity because of the extra gain and improves selectivity because of the tuned circuit. The local oscillator produces sine wave signals. The local oscillator used is tunable, that is, its frequency can be adjusted over a wide range. By tuning the local oscillator frequency, it is possible to convert a wide range of input frequency at the mixer to an intermediate lower frequency. This is called ***super-heterodyning***. If f_s is the frequency of the signal received and f_o is the frequency of the signal generated by the local oscillator, the output of the mixer will select a frequency of $f_o - f_s$. This is called intermediate frequency or IF. This original audio signal containing IF is amplified by the IF amplifier. The original audio signal is recovered using a demodulator. The recovered signal is further amplified and applied to the loudspeaker.

An automatic gain control circuit (AGC) controls the gain of RF and IF amplifiers to maintain constant output if the strength of the signal received at the input stage is fluctuating.

Thus, we have seen that in a superheterodyne receiver a mixer is used to translate or change the incoming signal to a lower frequency, called the intermediate frequency, so as to achieve fixed gain and selectivity. The mixer performs the analog multiplication of the input signal, that is, the signal received, and the signal generated by the local oscillator.

18.11 ► Microwave and Satellite Communication

18.11.1 Microwave Communication

The frequency spectrum used for radio communication has become overcrowded due to large number of users. Increased number of users has crowded the airspace. Advancement of technology has enabled us to extend radio communication into higher frequencies. Microwaves are signals with frequencies higher than 1 GHz (1 GHz = 10^9 Hz).

In electronic communication system, the modulated signal along with its sidebands occupies certain space in the radio frequency spectrum. This small portion of the spectrum is called a *channel*. Spectrum space is becoming congested due to introduction of large number of communication channels. There is further demand for space in RF frequency spectrum. To solve this problem, higher frequencies are being used. This microwave region has sufficient space for new channels and of wider bandwidths. If 1% space at 1000 kHz is 10 kHz, then 1% space at 2 GHz will be $0.01 \times 2 \times 10^9 = 20$ MHz. Such a bandwidth of 20 MHz as compared to 10 kHz is in fact very large. Ordinary information signals do not require such a wide bandwidth. Use of higher frequency allows wide bandwidth information signal to be transmitted. A TV signal has a bandwidth requirement of about 6 MHz. Microwave frequency spectrum is used for long-distance telephone communication. Radio detection and ranging (radar) communication also use microwave frequency. Radar is used in detecting presence of distant object like an aeroplane or a ship. The radar system transmits high-frequency signal which gets reflected from the object to be detected. From the time taken, the distance of the object can be determined by the radar system. Television stations also use microwave to transmit TV signals. In space communication using satellites, microwaves are used.

It is interesting to note that microwave has other applications like in heating, for example, microwave ovens used in many kitchens, in hospitals for heating of injured muscles (diathermy machines), etc.

18.11.2 Satellite Communication

We know that the earth moves round the sun. Thus, the earth is the satellite of the sun. The moon is the satellite of the earth. These are examples of natural satellites. Artificial satellites are physical objects that are made to revolve around a celestial body like the earth. Artificial satellites are launched to go into orbit around the earth. While launching a satellite, it is given a vertical as well as some forward motion. The combined effect of centripetal acceleration and the earth's gravitational pull keeps the satellite orbiting round the earth. The distance between the earth and the communication satellite is about 36,000 km. At this distance the satellite needs to travel at a speed of nearly 11,000 km/h to stay in the orbit. With this speed, the satellite will revolve round the earth at an orbital height of 36,000 km in 24 h. Fig. 18.20 shows an artificial communication satellite revolving round the earth. As mentioned earlier, satellite is launched with both vertical and forward motion. The gravitational force tends to pull the satellite towards earth. The inertia force and gravitational pull balance each other and keeps the satellite in motion. The lowest orbit of a satellite can be as low as 160 km or so. At this height the earth's pull is high and at this level the satellite must move at higher speed to create sufficient inertia to stay in orbit. At 160 km height, speed of the satellite would be 28,000 km/h. We know that the earth in rotates or spins around its own axis once in 24 h. If the satellite also revolves around the earth in 24 h, the satellite will appear to be stationary with respect to any point on the earth. The satellites orbit the earth in either circular or elliptical path.

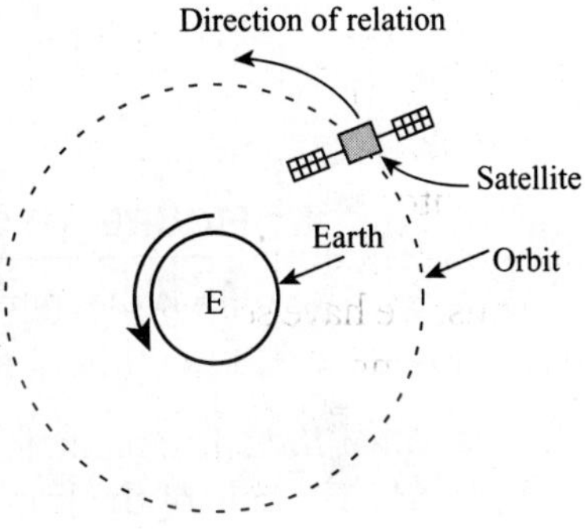

FIGURE 18.20

An artificial satellite orbiting round the earth

Geostationary Orbit: A satellite that orbits directly over the equator at a height of 36,000 km and travels at a speed of 11,000 km/h is said to be in a *geostationary orbit* and the satellite is said to be *geosynchronous*. However, there are other satellites whose height varies from 160 km to less than 36,000 km and their speed varies. The nearer the satellite from the earth, higher is its speed. Solar panels attached to the satellite supply power to the satellite. A communication satellite works as a relay station. The transmitting station sends information to the satellite which in turn sends the signal to the receiving station. An earth station transmits information to the satellite. A receiver is placed on the satellite which receives the signal, amplifies it, and translates to another frequency. The information is transmitted back to the earth to the receiving stations. The action of transmission of signal to the satellite from ground station is called ***uplink***. Sending the received signals by the satellite back to the ground station

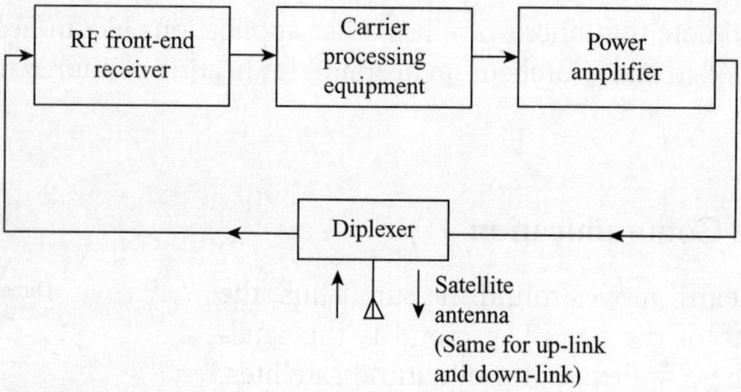

FIGURE 18.21

Block diagram representation of a satellite transponder

is called *down-link*. The name for the receiver–transmitter combination in a satellite is called *transponder*. The role of the transponder is to amplify the signal before it is transmitted back to the earth stations. Application of satellites is mainly in television and mobile communication. Satellites are also used in surveillance or observation. Satellites are also used in geological information system (GIS) where information about the earth's resources can be acquired.

The advantages of satellite communication are mentioned below: Satellite communication is meant for long-distance communication. Almost one-third of the earth except the polar region is visible from a geostationary satellite. It is possible to have a very wide transmission–reception area irrespective of the geographical irregularities using a satellite. Communication link can be established to very distant remote areas. The reliability and quality of communication is very high. Cost of satellite communication is independent of the distance and the quality of transmission remains the same. Mobile communication can be provided by mounting an earth station on a vehicle within a short period of time. Fig. 18.21 shows a block diagram of a satellite transponder which works as a relay station.

The satellite transponder receives uplink transmission information and retransmits the signal through its downlink mechanism. The uplink information is received by the antenna of the satellite and through the diplexer it is given to RF front-end receiver. The power received by the antenna is very weak. The RF receiver increases the signal to noise ratio of the signal received and provides amplification. Carrier processing includes demodulation and remodulation. Uplink carrier frequency is demodulated and remodulation of the information signal is done at downlink carrier frequency. This is done so that the uplink and downlink frequencies do not interfere with each other.

This way, antenna used for uplink and downlink operation can be the same. Since the two frequencies are different, simultaneous reception and transmission is possible. The diplexer facilitates simultaneous reception and transmission through the single antenna. The power amplifier with sufficient gain is used to increase the power level of the transmission signal so that information sent reaches the earth station. A block diagram of a simplified earth station has been shown in Figure 18.22.

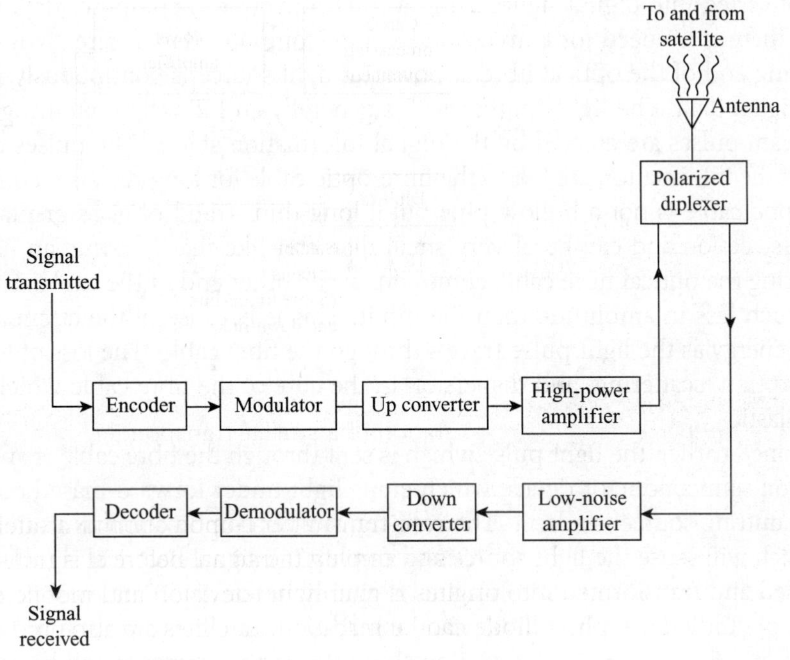

FIGURE 18.22
Block diagram of an earth station

The encoder converts the baseband information signal into a format ready for modulation. The carrier is modulated by the encoded information signal. The modulated frequency is then converted to the uplink frequency of the satellite. The amplifier amplifies the signal which then passes through polarized diplexer of the antenna. The signal received from the satellite through the same antenna is of different frequency which goes through the demodulation process and through the decoder and the information signal is received in its original form.

18.12 ► Fibre-Optic Communication

In radio communication or microwave communication system, the information signal modulates a carrier frequency wave. Communication can also be made using light wave. Information can be put on a light beam and transmitted through space or through a special cable called fibre-optic cable. Communication by light beam in free space over a long distance is impractical because of attenuation of light due to atmospheric disturbances like fog, smog, rain, snow, cyclonic conditions, etc. Instead of using free space, special cables called fibre-optic cables are used these days extensively. All communication systems using cables, these days, are being done using fibre-optic cable. Fibre-optic cable is essentially a narrow light pipe that is used to carry light beam from one end to the other. The information signal, which may be a voice signal, is converted into digital signal using A to D converter. If computer data is to be communicated, there is no need for conversion because computer data is already in digital form. At the sending end of the optical fibre, a powerful light source is continuously made on and off by the digital data. The light source used is generally an LED (light emitting diode). This way light beam pulses are created by the digital information signal. The pulses of light beam representing the information are fed to the fibre-optic cable for long-distance communication. The fibre-optic cable is not a hollow pipe but a long thin strand of glass or plastic having a circular cross-section and can be of very small diameter like that of a human hair. The light source entering the optical fibre cable comes out at the other end of the cable. The light pulse output is much less in amplitude than the input. This is because of the attenuation, that is, loss of light energy as the light pulse travels through the fibre cable. The loss of light energy is due to absorption, scattering, and dispersion in the core of the fibre cable which is made up of glass or plastic.

As mentioned earlier, the light pulse which is sent through the fibre cable is an LED. LED is a p-n junction semiconductor device which emits light under forward biased condition. The other light emitting source is a laser. The receiver of a fibre-optic communication system is a detector which will sense the light source and convert it into an electrical signal. The signal is then amplified and transformed into original digital signal or data. The most commonly used receiver is a photodiode. A photodiode can be made to work in converting light into voltage pulses.

Figure 18.23 shows the basic elements of a fibre-optic communication system.

It is to be understood that for the communication to be very fast the speed at which the optical pulses are created must be very high. This is called the pulse rate. For high speed long-distance communication, the light source used is an injection laser diode rather than an LED. At the receiving end of the fibre cable, avalanche photodiodes are used instead of ordinary photodiodes because they are very fast and highly sensitive.

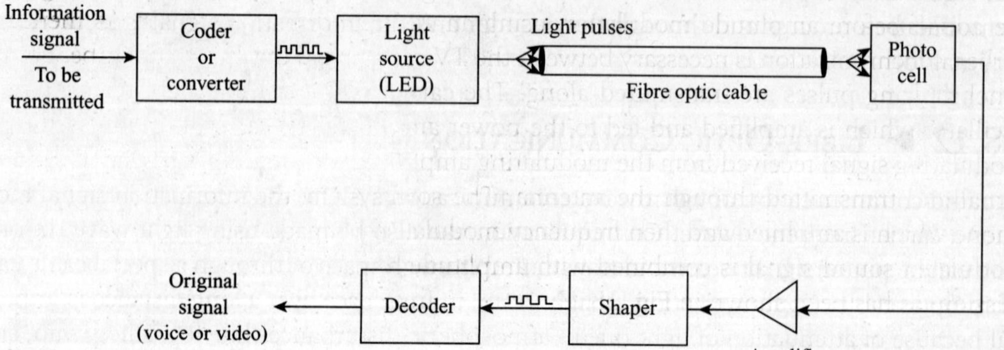

FIGURE 18.23

Fibre-optic communication system represented by a block diagram

18.13 ▶ Television

18.13.1 Introduction

A television system transmits a picture or a scene along with sound to a distant place. The scene or the picture signal is generated by a TV camera and the sound signal by a microphone. The sound may be stereo and the picture may be coloured. This way a TV signal is a complex one. The picture signal is amplitude modulated. The sound signal is frequency modulated. The two carrier frequencies are separated by a wide margin to avoid mixing of the two. The two modulated frequencies are transmitted through a common antenna. For stereo sound, signal multiplexing technique is used. For colour transmission of coloured pictures, FDM is used.

18.13.2 TV Transmitter

The picture information is converted into electrical signal by considering the light intensity and colour details of a picture or a scene. The scene to be transmitted is focused by the lens of the TV camera upon a light-sensitive imaging device. Using a method of scanning, the optical information is converted into an electrical signal in a sequential manner line by line so as to cover the entire scene. Scanning is done at a very fast rate and is repeated a very large number of times. Use of synchronization technique of scanning by the TV camera and its reproduction at the TV receiver is necessary. This creates an illusion, that is, an impression on the eye of simultaneously viewing the whole scene at a time.

A colour TV camera is used to develop signal voltages proportional to the intensity of light of three primary or basic colours, namely red, green, and blue. By mixing these colours in right proportion, any colour can be created. This forms the basis of colour television. By mixing any two or three of the basic colours, other colours can be created. The colour TV camera develops voltages proportional to the intensity of the primary colours.

A simple diagram of a black and white (monochrome) TV transmitter is shown in Fig. 18.24. The signal from the video camera is amplified using a video amplifier. Synchronizing pulses are added before amplitude modulation using an AM modulating amplifier. As mentioned earlier, synchronization is necessary between the TV camera and the TV picture tube. For this, synchronizing pulses are transmitted along. The carrier frequency is generated from a local oscillator which is amplified and fed to the power amplifier. The carrier is modulated by the modulating signal received from the modulating amplifier. This signal is mixed with the audio signal and transmitted through the antenna. The sound signal is picked up using a microphone which is amplified and then frequency modulated using a carrier frequency. Frequency modulated sound signal is combined with amplitude modulated picture signal before transmission, as has been shown in Fig. 18.24.

18.13.3 TV Receiver

Television transmitter antenna radiates electromagnetic waves from its antenna. At the receiving end, a receiving antenna intercepts the radiated RF signals. An RF tuner circuit selects

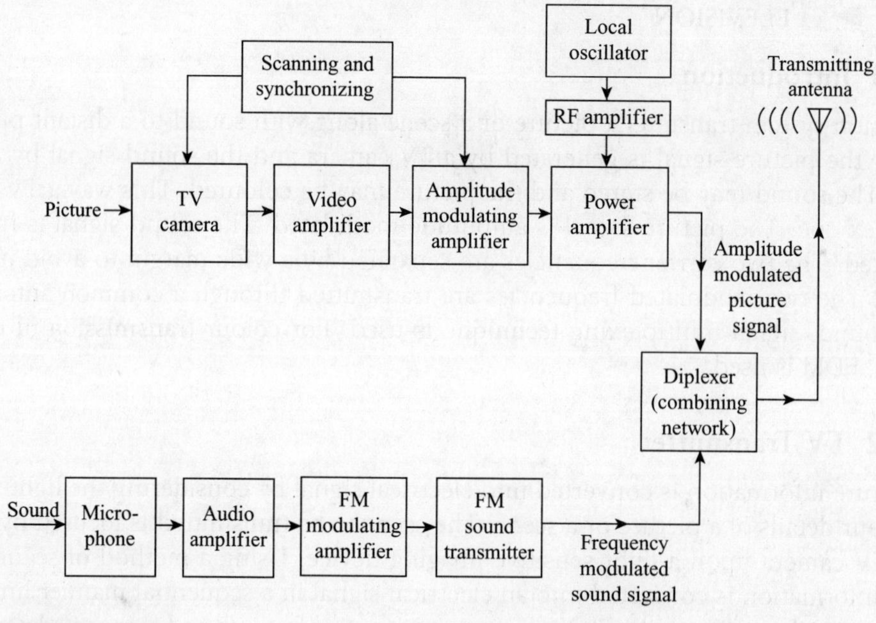

FIGURE 18.24

Block diagram of a television transmitter

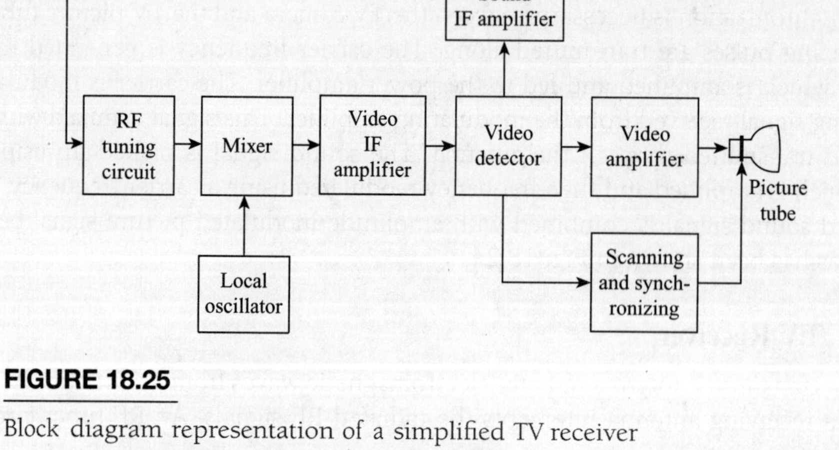

FIGURE 18.25

Block diagram representation of a simplified TV receiver

the desired channel signals of a particular frequency band and converts IF band of frequencies using a local oscillator. Fig. 18.25 shows a simple block diagram of a TV receiver. The video detector isolates the video and audio signals. The signal that carrier picture is amplified using a video amplifier and then connected to TV picture tube. In the picture tube the electrical signal is converted to picture on its illuminated screen.

The sound signal along with the video signal travels together from the antenna up to the video detector. At the video detector, the two signals are separated. The frequency modulated sound signal is demodulated after amplification. The audio output is connected to a loudspeaker where electrical signal is converted to sound. It may be noted that the block diagram represents a black and white TV receiver. A colour receiver is similar to a black and white receiver. However, a colour receiver circuit will have a colour subsystem to accept and process colour signals.

A colour TV camera will have three colour pick up tubes and a colour picture tube will have three colour guns: red, green, and blue. The screen of the picture tube has red, green, and blue phosphors arranged as alternate strips. The electron guns produce electron beam to illuminate corresponding colour phosphor on the screen. The TV viewer's eye integrates coloured information and their luminance to visualize the actual picture.

18.13.4 Need for Synchronization

When a TV camera starts scanning line 1, the receiver, that is, the picture tube should also simultaneously start scanning line 1. If this does not happen, then top line of camera will appear as a middle line at the receiver which is not desirable. The speed of scanning of the video camera should be exactly the same as the speed of scanning process of the receiver. Thus, the transmission reception synchronization is essential to avoid any distortion of the picture received on the TV screen.

18.13.5 Satellite TV

Communication satellites orbit round the earth in synchronism with the earth, and hence, these satellites appear to be stationary when viewed from any point of the earth. TV signal can be transmitted via communication satellite using a microwave carrier. The signal to be transmitted is used to modulate a microwave carrier and transmitted to the satellite. This is called uplink. The transponder in the satellite after changing the signal to another frequency relays back the signal to an earth station. This is called downlink. At the receiving side it may be a cable company or an individual viewer. Dish antennas of less than a meter diameter can be used to receive the TV signal.

18.14 ► Mobile Communication

18.14.1 Introduction

We know that a telephone set converts sound energy into electrical signals and vice versa. A simple telephone transmission system is a pair of wires connecting two telephone sets

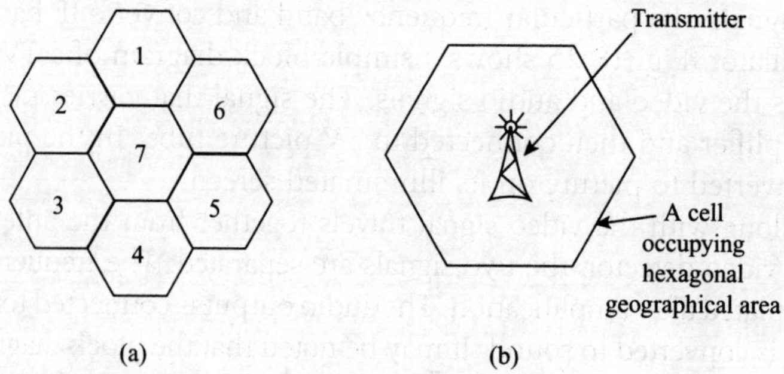

FIGURE 18.26

(a) Honeycomb cell pattern formed by subdividing an area into hexagonal cells; (b) A transmitter located at the centre of a cell

together. A subscriber is the operator or the user of a telephone set. To enable a large number of subscribers communicate with each other from different locations, telephone network, switching, and signalling systems have been developed over the years. The telephone network equipment includes instruments, local loops, trunk circuits, and telephone exchanges. Computer controlled automated switching system has been introduced. Early telephone systems were based on analog signals and analog signal transmission. Digital signalling systems modernized the telephone industry.

Mobile telephone systems involve complex communicating networks and computer controlled switching centers. Mobile telephone stations are small handsets which can be carried in a pocket. A mobile telephone is a wireless telephone capable of operating while moving, small in size which can be carried by a person. Mobile telephone communication problems were solved by cellular telephone systems which comparatively offer high-user capacity within a limited frequency spectrum. Cellular telephone is sometimes called cellular radio. Cellular telephone systems allow large number of users to share the limited number of common usage radio channels available in that area. The basic concept of cellular telephone is that each radio coverage area is divided into hexagon-shaped cells that fit together to form a honeycomb pattern as shown in Fig. 18.26. The concept of frequency reuse is used to increase the capacity of a mobile telephone channel. A geographical area is divided into small cells and using a low-power transmitter with low antenna at base station in each cell, a cellular communication network is developed.

The design process involves allocation of channel groups for all the cellular base stations within a system and is called *frequency reuse planning*. The size of a cell area varies depending upon user density and calling pattern. A typical cell may have an area of 10 sq km or so. Microcells are used in large cities. Fig. 18.26(b) shows a cell with its base station transmitter located at the centre with omnidirectional antenna. Omnidirectional antennas radiate and receive signals equally in all directions.

Instead of having a high-powered fixed base station located at the centre of a city at very high altitude or on a hill top, a number of low-powered duplicates distributed over the covered area using a cellular telephone concept are used with advantage.

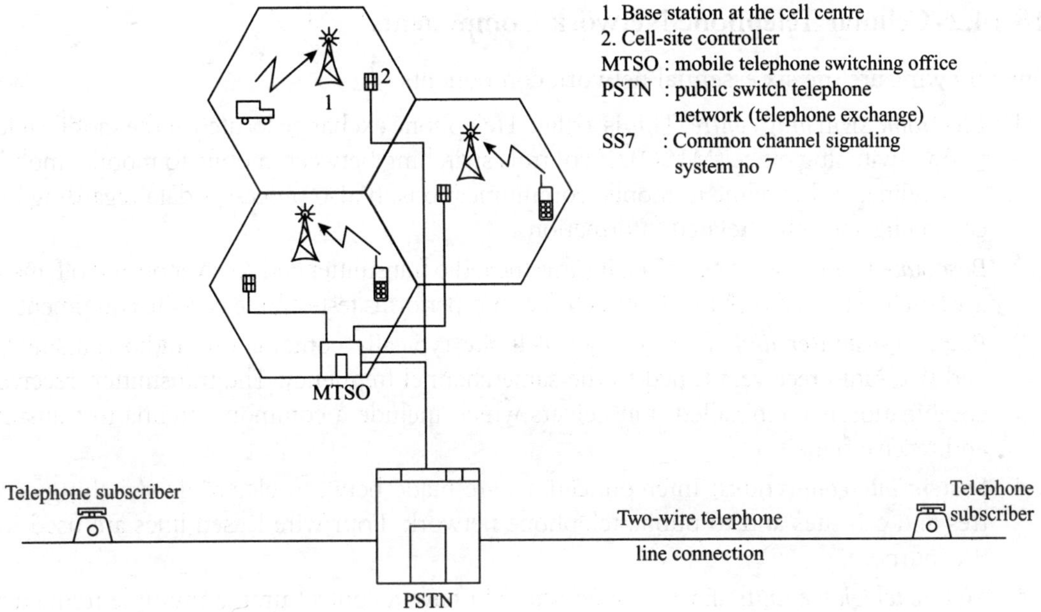

FIGURE 18.27

A simplified cellular telephone system

Frequency reuse concept allows the use of same set of frequencies (called channels) to be allocated to more than one cell. The cells, however, must be separated by sufficient distance. If we reuse same channels at different cells located at a distance, there will be significant increase of number of channels available in a given cellular system as compared to a non-cellular system. Fig. 18.27 shows a simplified cellular telephone system covering a set of geographical areas called cells inside which people can communicate using mobile phones. A base station located at each cell serves as a control station for all users within that cell. Mobile units can communicate directly with the base stations. The base stations can communicate directly with a mobile telephone switching office (MTSO). Function of MTSO is to control assignment of channels, cell, setup, and termination. These include signalling, switching, supervision, allocation of radio frequency channels, etc. MTSO establishes link with public telephone network. MTSOs are connected to the common channel signaling system (SS7) which allows cell phone to operate outside their service area.

The base stations provide an interface between mobile telephone sets and the MTSOs. MTSO provides centralized administration to the entire network. It also interfaces with the public switch telephone network (PSTN) so that contact from a mobile phone can be made to a wireline telephone subscriber.

Roaming facility to a mobile phone is provided when one user moves from one cell area to another.

18.14.2 Cellular Telephone Network Components

The following are the six essential network components.

1. *Electronic switching centre*: This is a digital telephone exchange located at the mobile telephone switching office (MTSO). It controls switching between mobile to mobile, mobile to wireline, and wireline to mobile communications. It also processes data regarding bill compiling and other related information.
2. *Base station or cell site controllers*: It turns the radio transmitter and receiver on and off, manages each of the radio channels at each site, and performs tests on the cell-site equipment.
3. *Radio transmitter and receiver*: Each cell site typically contains one radio transmitter and two radio receivers tuned to the same channel frequency. The transmitter–receiver combination is often called transceivers which include a common antenna to transmit and receive signals.
4. *System interconnections*: Interconnections are made between electronic switching centres and cell sites and to public telephone network. Four wire leased lines are used for this purpose.
5. *Mobile telephone units*: Each mobile unit will have a control unit, a multiple frequency radio transceiver, a logic unit, and an antenna. The transceiver tunes into any cellular system channel using a frequency synthesizer.
6. *Communications Protocol*: This governs the way cells are established and disconnected.

Within mobile (cellular) telephone system calls can be made from mobile to wireline, mobile to mobile, and from wireline to mobile. A mobile phone can also support additional services such as SMS messaging, MMS, e-mail, internet access, infrared, bluetooth, etc.

18.14.3 Concept of Frequency Reuse

If we plan for a single base station to serve an area having a large number of users, a high-power transmitter will be required. The maximum number of simultaneous users will be limited by the radio frequency spectrum allocated.

In a cellular system, each service area is divided into number of small hexagonal areas called cells. Each cell is allocated a subset of allocated frequencies or frequency channels. Group of cells can use same set of frequencies without increasing the interference. This way within the available spectrum of frequencies a large number of customers can be served. Power transmitters of lower rating with lower height antennas can be used in each cell base station. Using the concept of frequency reuse, efficiency of spectrum utilization is increased and service area capability in terms of numbers of customers served is increased.

Mobile users communicate only via the base station located at each cell. Each cell is allocated a fixed number of radio frequency (RF) channels. This will of course depend upon the number of simultaneous users within a cell. A cell which is located somewhat away from another cell can reuse the same set of frequencies without much interference between these cells.

18.14.4 Cellular Antenna

An antenna is a device meant to convert electrical signals travelling along the RF cable from the transmitter to electromagnetic waves to be radiated in space. Various types of antennas differ in the amount of radiation emitted in various directions. Omnidirectional antennas radiate radio signals of equal strength in all directions.

Antennas used in cellular telephone system will not only radiate electromagnetic signal to provide adequate radio coverage but also reduce the interference arising out of frequency reuse by the cells.

18.14.5 Multiple Access Techniques

In wireless communication, the challenge is to find out ways and means to maximize the use of the available radio spectrum. Multiple access techniques are used to share the available channels by a large number of subscribers (users). There are four basic forms or types of multiple access techniques.

1. *Frequency-division multiple access (FDMA)*: The available frequency spectrum is shared by assigning specific frequency channels to users on a permanent or temporary basis. FDMA has been widely used in all first generation (1G) analog cellular systems.
2. *Time-division multiple access (TDMA)*: The available channel bandwidth is shared between a number of subscribers on time sharing basis. In a TDMA system, the channel bandwidth is divided into time slots, and in each time slot only one subscriber is allowed to communicate.
3. *Spread-spectrum multiple access (SSMA)*: This technique allows simultaneous use of the same spectrum by many subscribers in adjacent cells. Code-division multiple access (CDMA) is a multiple access technique in which an individual subscriber occupies the complete spectrum to communicate.

   CDMA is a form of spread spectrum technique in which subscribers are allowed to use the available spectrum with their signal with a specific code to distinguish from one another. In CDMA technique, one unique code is assigned to each subscriber. This code word is used to encode and decode information signal. Some of the second generation (2G) digital cellular systems like IS–95 and most of the third generation (3G) cellular systems use CDMA technique.
4. *Space division multiple access (SDMA)*: We have mentioned that in FDMA, TDMA, and SSMA the spectral efficiency is increased by increasing reuse of frequency, time sharing, and use of codes, respectively. SDMA technique employs control of radiated energy from the antennas for each subscriber by using directional antennas or spot beam antennas at each cell site. Smart antennas like sector antennas, switched-beam antennas, and adaptive antennas are used at the cell site to make possible greater reuse of the radio spectrum.

The four multiple access techniques mentioned above are intended to increase the efficiency of use of the available radio spectrum for communication. The techniques involve dividing the spectrum into different frequency channels (FDMA), time sharing of the spectrum by the subscribers (TDMA), using the same spectrum simultaneously by all the subscribers by use of coding (CDMA), and sharing of the spectrum in angular directions by using smart antennas (SDMA).

18.14.6 Generations of Cellular Telephone System

First generation analog cellular telephone (1G)

In 1971, Bell Telephone laboratories proposed the cellular telephone concept as advanced mobile telephone system (AMPS). AMPS is the first generation analog cellular radio communication technology. Telephone subscribers can access the AMPS system using a technique called FDMA. Each channel is allocated a carrier frequency and channel bandwidth. Subscribers are assigned a voice channel for the duration of their call. For large number of users, a considerable bandwidth is required. First generation telephone systems were designed for a limited number of users like business houses and wealthy individual customers. When demand for cellular phone increased, the manufacturers looked for a newer technology. The drawbacks with 1G systems, like poor battery performance and unavailability of channels, were considered. AMPS, the first generation analog cellular telephone system, was not designed to support a large customer base like in large cities.

In North America, two 25 MHz bands were allocated to AMPS standard. One for transmission from the base station to the mobile subscriber (869 MHz to 894 MHz) and other for transmission from mobile to the base station (825 MHz to 849 MHz). Mobile users were assigned a pair of voice channels for forward and reverse communication.

Narrow-band AMPS permitted use of three mobiles simultaneously in a 30 kHz channel, thereby increasing the user capacity of the system.

Expanding demand for cellular telephones has thrown challenge to the manufacturers for newer technology for increasing the system capacity and enhancing the transmission reception quality within the allocated frequency spectrum.

Second generation analog cellular telephone (2G)

The second generation (2G) cellular systems have been developed to provide greater system capacity, higher data communication rate, and improved signal quality using digital technique. 1G system is purely analog while 2G system is purely digital. In a 1G system, each cell supports a number of channels. At any given time a channel is allocated to one subscriber. In a 2G system, each channel in a cell is dynamically shared by a number of subscriber using TDMA or CDMA technique.

Global System for Mobile (GSM)

The GSM system has 50 MHz allocation in the 900 MHz frequency band. It uses two 25 MHz frequency bands. 890 MHz to 915 MHz band is for mobile to base transmission and 935 MHz to 960 MHz band is for base to mobile communication. It utilizes combination of TDMA and FDMA techniques to increase the system capacity. Using FDMA the band is divided into 124 channels, each channel having a bandwidth of 200 kHz. Each of these channels is further divided into 8 time slots using TDMA. Thus, a maximum of 124 × 8 = 992 channels can be made for communication.

A smart card called subscriber identity module (SIM) is issued to a customer at the subscription time. SIM card is unique to the GSM system. It allows a subscriber to use any GSM mobile phone anywhere in the world where GSM service is available. GSM provides a number of value added services like telephone services, data communication services, and short messaging services (SMS).

CDMA digital cellular standards

Code division multiple access system is a fast growing digital cellular communication technology. It offers a number of advantages over FDMA and TDMA where frequency division or time sharing is used. Users are differentiated by a code in CDMA and subscribers with different codes do not interfere with each other. CDMA digital cellular standard is the basis for third generation (3G) CDMA 2000 cellular technology. CDMA permits use of all frequencies in all cells, and hence, the system capacity increases many times.

CDMA is a 2G digital cellular technology based on the principle of spread spectrum technique. GSM is also a 2G digital cellular technology which uses TDMA and FDMA techniques.

CDMA digital cellular technology

In order to provide enhanced mobile services like high-speed data transmission, web browsing, e-mailing, and e-commerce, superior technologies, named 2.5G, 3G, 4G have been developed. For example, General packet radio service (GPRS) allows very efficient packet data transfer compared to GSM data transfer services. A subscriber using electronic data transfer facilities is assigned a channel on temporary basis and the channel is released after the completion of data transfer. Enhanced data rates for GSM (EDGE) evolution is an example of 2.5G+ standard. By employing EDGE on an existing GSM, a mobile user can be provided with landline quality voice or data transfer facilities. The complete range of communication services like voice, data, video, multimedia, and internet browsing facilities on mobile phone is possible by 3G digital cellular networks.

18.15 ► Fax Communication

Fax or facsimile is used to send printed pages through telephone lines or through radio communication. Letters, drawings, photographs, documents, etc., can be communicated using fax. Here no sound or voice is communicated. The information contained in a paper is first converted into electrical signal and then communicated. Scanning technique is used that is someway similar to the process involved in TV transmission. The printed material is divided into a large number of horizontal scan lines which are scanned at a very high rate and transmitted and received serially and simultaneously.

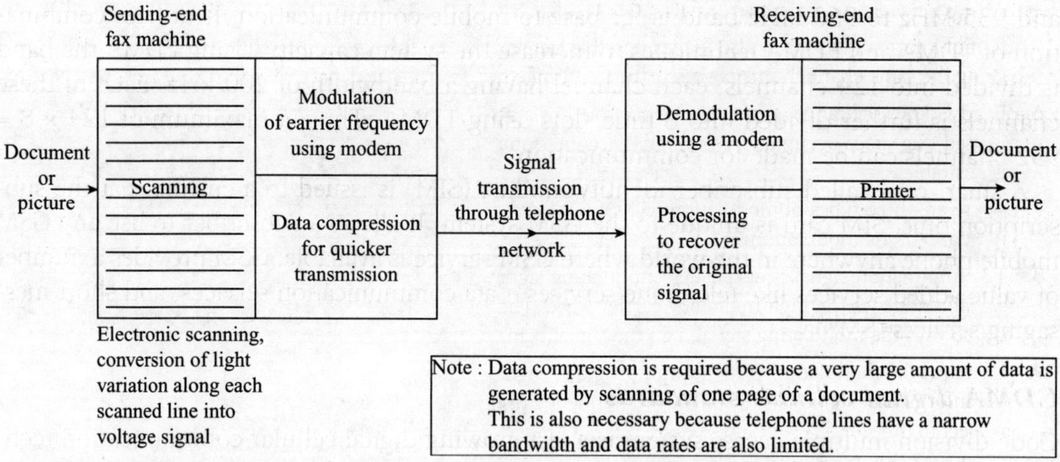

FIGURE 18.28

Representation of fax communication

Modern fax machines are sophisticated electro-optical devices. Scanning is carried out electronically at a very fast speed and the scanned signal is converted into binary signals. An image scanner converts the document into hundreds of horizontal scan lines. A photosensitive device converts the variations of light along a horizontal scan line into electrical voltage signal. The signal is then processed and data compression technique is used to compress the data. Modulation is done using a *modem* where the information signal modulates a carrier. The signal is then transmitted to another fax machine at the receiving end through the telephone lines.

The fax machine at the receiving end reproduces the document and provides a printout through its printer. The fax machine first demodulates the signal and then processes to recover the original data. The data is decompressed and the printer provides a printout. At both ends, the fax machine can transmit as well as receive printed information. Fax machines are provided with built-in telephone and printer. The process of a fax communication is illustrated in the form of a block diagram in Fig. 18.28.

Important Terminologies and Points to Remember

1. The basic components of a communication system are the signal, transmitter, medium, i.e., the channel, and the receiver. Two types of signals are used, viz., analog signals and digital signals.
2. Communication may be one way or both ways. TV communication is in one way only. Such a system is called simplex system. Telephone communication is two-way and the system is called duplex system.
3. The original information signal is called baseband signal. Communication using a carrier frequency is called broadband communication.
4. Modulation is a process of placing or fixing the information signal over a high-frequency transmitter generated carrier wave for long-distance transmission.

5. Modulation of signals increases the range (distance) of communication with improved quality of reception of information sent. It also makes possible transmission of number of different signals simultaneously using the same channel, this process called multiplexing.
6. Bandwidth is the range of frequencies occupied by a signal. For modulation when a high-carrier frequency is used, the space occupied by the signal becomes only few percentage of available spectrum space which is desirable.
7. There are three types of modulation techniques used, namely amplitude modulation, frequency modulation, and phase modulation.
8. The information signal is also called modulating signal. In Amplitude modulation (AM), the modulating signal modulates the high frequency carrier wave. The amplitude of the modulated signal varies in accordance with the modulating signal, i.e., the signal being sent.
9. In frequency modulation (FM), the information signal modulates the frequency of the carrier signal. If the information signal is sinusoidal, the frequency of the modulated signal also change sinusoidally with respect to time.
10. In phase modulation, the modulating signal creates a phase shift between the modulating signal and modulated signal.
11. Both AM and FM are suitable for long distance information communication.
12. Demodulation the process of recovering or separating the original information signal from the modulated carrier wave at the receiving end.
13. Modulators and Demodulators are electronic circuits used at the sending-end and receiving-end of the information.
14. Multiplexing is a technique or a process by which a large number of information signals can be communicated through a single-communication channel, the channel may be a cable connection or a radio link using electromagnetic waves.
15. There are two types of multiplexing, namely frequency division multiplexing (FDM) and time division multiplexing (TDM).
16. In FDM, a large number of signals are transmitted through a single channel by sharing, i.e., by assigning a small portion of the spectrum space to each signal within the channel bandwidth (range of frequencies).
17. In TDM, the whole of the spectrum space within the channel bandwidth can be utilized by the signals to be transmitted by time sharing mechanism, i.e., using the channel for short period of time turn by turn.
18. The function of the signal receiver is to receive only the desired signal and reject all other signals. A simple receiver is a tuned L-C circuit. The L-C circuit is tuned to resonate at the desired signal frequency to be received.
19. Radio frequency spectrum space has become conjusted due to the introduction of large number of communication channels, each channel occupying some space. To solve this problem, higher frequencies are being used.
20. Microwave region has sufficient space for new channels with requirement of wider bandwidth. Microwave frequencies are higher than 1 Giga Hertz (1 GHz = 10^9 Hz).

    Television signals has a bandwidth requirement of about 6 Mega Hertz (1 MHz = 10^6 Hz). Television stations use microwave for transmission of TV signals.
21. Microwave has other applications like in heating (microwave ovens), in diathermy machines (for heating of injured body muscles), etc.
22. Earth is a natural satellite of Sun, because the earth revolves round the Sun. Moon is the satellite of earth.

    Artificial satellites are launched to revolve round the earth for the purpose of quality TV and Mobile communications.

    While launching a satellite, it is given a vertical as well as some forward motion.

The distance between Earth and the satellite is 36,000 Km. At this height, the satellite needs to have a speed of 11000 Km/hour to stay in orbit. As the earth spins around it once in 24 hours, the satellite at a height of 36000 Km has to have a speed on 11000 Km/h to remain geostationary and the satellite is said to be geosynchronous, i.e., in synchronism with the earth.

23. Satellite can have orbits at a height lower than 36000 Km. The lowest orbit can be as low as 160 Km or so. At this height the satellite must have higher speed of 28,000 Km/h to remain geostationary. Nearer the satellite, higher should be the speed. Solar panels attached to the satellite supply electrical power to the satellite.
24. The name of the signal receiver-transmitter combination in a satellite is called transponder. The satellite transponder receiver uplink transmission information and retransmits the signal through the downlink mechanism. This way a satellite works as relay station.
25. In addition to radio communication and microwave communication, information can be put on a light beam and transmitted through special cable called fibre-optic cable.

    Fibre-optic cable is essentially a narrow light pipe that is used to carry light beam from one end to the other.
26. Any analog signal to be transmitted is first converted into digital form using an analog to digital converter. The digital data makes a strong light source at the sending end into continuous on and off light beam and feed to the optical fibre for transmission. The strong light source is from an LED or a LASER.
27. At the receiving end of the optical fibre, the light beam pulses are converted into voltage pulses by photodiode. This signal is then amplified and transformed into original form.
28. Mobile communication started with utilizing techniques involving technologies ranging from FDMA, TDMA, CDMA, and SDMA.

    In 2G system, each channel in a cell is dynamically shared by a number of subscriber using TDMA or CDMA technique.

    Global system for mobile (GSM) provides number of value-added services like telephone services, data communication services and short message services (SMS).

    In order to provide high-speed data transmission superior techniques named, 2.5G, 3G, 4G have been developed.
29. In FAX communication, the information contained in a printed paper is first converted into electrical signal and then communicated. Scanning technique is used for this process. At both end, the fax machine can transmit as well as receive printed information.

REVIEW QUESTIONS

Short Answer Type Questions

1. Represent the basic elements of a communication system and briefly explain them.
2. What is meant by electromagnetic spectrum?
3. Why do we use a carrier frequency in communication?
4. Define bandwidth. Explain its significance in communication.
5. Define modulation. Explain any one method of modulation.
6. Distinguish between amplitude modulation and frequency modulation.
7. Distinguish between frequency modulation and phase modulation.
8. What is multiplexing? Explain the two types of multiplexing.
9. Distinguish between frequency division multiplexing and time division multiplexing.

10. What is the advantage of microwave transmission over radio transmission? What are the requirements?
11. What is a satellite? What is geostationary orbit? What are the advantages of satellite communication?
12. Draw a basic block diagram of a fibre-optic communication system and write the function of each part.
13. State advantages of fibre-optic communication over using electrical cables for communication.
14. Explain the process of television communication system.
15. Draw and explain simplified block diagram of (a) TV transmitter and (b) TV receiver.
16. Explain the principle of fax communication.
17. Explain the basic principle of mobile communication. What are the latest advancements in mobile technology?
18. Explain the concept of frequency reuse in a communication system.
19. Distinguish between FDMA and TDMA techniques in communication.

Multiple Choice Questions

1. Which of the following is not a communication medium?
   (a) free space (b) vacuum
   (c) twisted cables (d) fibre-optic cable
2. The original information signal to be transmitted is called
   (a) carrier (b) baseband signal
   (c) modulating signal (d) modulated signal
3. Modifying a high-frequency carrier with the information signal to be transmitted is called
   (a) modulation (b) demodulation
   (c) multiplexing (d) detection
4. Recovering information from a modulated signal is called
   (a) modulation (b) multiplexing
   (c) demodulation (d) carrier recovery
5. Voice frequency is of the range
   (a) 300–3000 Hz
   (b) 30–300 kHz
   (c) 3–30 MHz
   (d) 30–300 MHz
6. Which of the following statements is not correct?
   (a) bandwidth is that portion of the electromagnetic spectrum occupied by a signal
   (b) bandwidth is the frequency range over which an information signal is transmitted
   (c) bandwidth is the frequency range over which a receiver operates
   (d) bandwidth is the sum of upper and lower frequency limits of the signal
7. Which of the following is not true for amplitude modulation?
   (a) the information signal varies the amplitude of carrier sine wave
   (b) frequency of the carrier remains constant
   (c) frequency of carrier changes
   (d) instantaneous value of carrier amplitude changes in accordance with amplitude and frequency variations of the modulating signal
8. In frequency modulation
   (a) carrier amplitude is changed by the modulating signal
   (b) carrier frequency is changed by the modulating signal
   (c) if amplitude of modulating signal increases the carrier frequency remains constant
   (d) frequency of the carrier must be lower than the modulating frequency
9. Superheterodyne receivers convert all incoming signals to
   (a) a higher frequency signal
   (b) a lower intermediate frequency signal
   (c) a stronger signal
   (d) a resonant signal
10. Multiplexing is a process of
   (a) transmitting multiple signals over a single channel
   (b) transmitting multiple signals over multiple channels

(c) transmitting same signal over multiple channels
(d) transmitting multiple signals simultaneously to a receiver on a common frequency

11. Microwaves are signals with frequency of
(a) greater than 1 MHz
(b) greater than 1 GHz
(c) greater than 100 MHz
(d) lower than 100 Hz

12. Which of the following is not a common microwave application?
(a) mobile telephone
(b) satellite communication
(c) radar
(d) fax communication

13. Main advantage of using microwave communication is
(a) low cost
(b) simple equipment
(c) greater transmission distances
(d) availability of more spectrum space of signals

14. As the height of the orbit of a satellite from the earth increases, the speed of the satellite
(a) increases
(b) decreases
(c) remains constant
(d) increases proportionately

15. The main electronic equipment in a satellite is
(a) transponder (b) reflector
(c) solar cells (d) modulator

16. Communication satellites are approximately
(a) 36,000 km away from earth
(b) 22,300 km away from earth
(c) 6800 km away from earth
(d) 11,000 km away from earth

17. Speed of a communication satellite orbiting at a height of 36,000 Kms is approximately
(a) 36,000 km/h
(b) 22,300 km/h
(c) 6800 km/h
(d) 11,000 km/h

18. A geostationary satellite orbits the earth around the equator at a distance of approximately
(a) 36,000 km (b) 22,300 km
(c) 6800 km (d) 11,000 km

19. A satellite stays in its orbit because of the balance of the following two factors
(a) centripetal force and speed
(b) gravitational pull and centripetal force
(c) weight of the satellite and speed
(d) weight of the satellite and the gravitational pull

20. State which one of the alternatives is not true: The technique of frequency reuse in electronic communication is applied because
(a) efficiency of spectrum utilization is increased
(b) number of customers served is increased
(c) same set of channels can be allocated to more than one cell
(d) the rate of data transfer through a channel will increase

Answers to Multiple Choice Questions

1. (b)	2. (c)	3. (a)	4. (c)	5. (a)	6. (d)
7. (e)	8. (b)	9. (b)	10. (a)	11. (b)	12. (d)
13. (d)	14. (b)	15. (a)	16. (a)	17. (d)	18. (a)
19. (b)	20. (d)				

Index

C

D

Q

R